# HANDBUCH DER PHYSIK

HERAUSGEGEBEN VON

## S. FLÜGGE

BAND XLVII

## GEOPHYSIK I

REDAKTION

## J. BARTELS

MIT 289 FIGUREN

SPRINGER-VERLAG

BERLIN · GÖTTINGEN · HEIDELBERG

1956

# ENCYCLOPEDIA OF PHYSICS

EDITED BY

## S. FLÜGGE

VOLUME XLVII

## GEOPHYSICS I

GROUP EDITOR

## J. BARTELS

WITH 289 FIGURES

SPRINGER-VERLAG
BERLIN · GÖTTINGEN · HEIDELBERG
1956

ISBN-13: 978-3-642-45857-6     e-ISBN-13: 978-3-642-45855-2
DOI: 10.1007/978-3-642-45855-2

# Inhaltsverzeichnis.

# Inhaltsverzeichnis.

# The Rotation of the Earth.

By

Sir HAROLD SPENCER-JONES.

With 3 Figures.

## I. The Unit of Time.

**1.** The three fundamental units in physics, which form the basis of the c.g.s. system, are the centimetre, the unit of length; the gram, the unit of mass; and the second, the unit of time. Material standards of length and mass are provided by the International Prototype Metre and Kilogram, preserved at the International Bureau of Weights and Measures at Sèvres, near Paris. ·Any length or mass can be compared either directly or indirectly with sub-standards, which have themselves been compared with the actual prototype standards.

But with the unit of time it is otherwise. There is no material standard of time which can be used as an invariable standard of reference. When an atomic clock has been successfully developed, it may provide an absolute standard of time; but as no clock will function perpetually without stopping, it would be necessary to employ several atomic clocks, regularly intercompared with one another, so as to carry on the measure of time in the event of a stoppage of any one clock.

Lacking any material absolute standard of time, the rotation of the Earth has been adopted as a standard. The second, the unit of time in both the metric and the imperial (f.p.s.) system of measurement, is the mean solar second, which is the 1/86,400th part of the mean solar day. The mean solar day is a day of a length equal to the average of the true or apparent solar days throughout the year. Astronomers use the sidereal day, the time interval required for the Earth to make one complete rotation through 360°. It is the interval provided by two consecutive transits of a star across any given meridian. Sidereal time is determined by relating the instants at which stars of accurately known right ascension transit across the meridian with the corresponding time by the standard clock of the observatory. By a sequence of observations, the error and rate of the clock are controlled, so that the true sidereal time corresponding to any clock time is known.

**2.** The use of the Earth as a clock tacitly assumes that its rotation is uniform. If this assumption is not correct, so that the length of the day is variable, discordances will be produced between the observed and ephemeris positions of bodies in the solar system. The ephemeris positions of the Sun, Moon and planets which are tabulated in the various national ephemerides (*Nautical Almanac, Astronomisches Jahrbuch, Connaissance des Temps* etc.) are based on gravitational theory with the assumption that the length of the day is constant. Differences between observation and theory would be produced by errors of observation, by errors or incompleteness of the theory, or by variations in the length of the day. The theories of the motions of the Moon, Sun, Mercury and Venus, which are the four bodies of main interest in the investigation of variations in the length of the day, are sufficiently complete and exact for this purpose.

The discordances between the observed and theoretical positions caused by variations in the length of the day will be greater the more rapid the geocentric motion of the body, for they are attributable to a time error on the part of the Earth; the discordances in position represent the motions in longitude during that time. The Moon, which has the most rapid motion in longitude (0.″55 in one second of time) is consequently the most favourable object for detecting changes in the Earth's rotation; Mercury is the next most favourable; the Sun and Venus can also be used. Each of these four bodies can serve as a clock; their ephemerides are based on a uniform time which we shall term *ephemeris time*. To each observed position, an ephemeris time can be found, by interpolation in the ephemeris, at which the ephemeris position agrees with the observed position. This Ephemeris Time can be compared with the Universal Time (or G.M.T.) of the observation. The discordances between the two times must be due to a lack of uniformity in Universal Time, assuming that there are no defects in the theory of the motion of the body. Thus each of the four bodies, the Moon, Sun, Mercury and Venus, can be used as a clock, against which the rotation of the Earth can be checked. If the four bodies agree with each other but not with the Earth, it can safely be concluded that the Earth is at fault and that its rotation is not uniform or that, in other words, our adopted standard of time is variable.

The errors inherent in the astronomical observations of position of the four bodies make it impossible to detect in this way small discordances in position which are of short period or which fluctuate rapidly. The astronomical observations of position are suitable, however, for detecting *slow secular changes* in the Earth's rotation or the cumulative effect over a long time interval of small changes. Short period or *rapid changes* occurring within a year or so can best be investigated by comparing the time provided by the Earth's rotation with the time given by modern precision quartz-crystal oscillators. These oscillators are capable of very high precision over short time intervals, but because of ageing effects, and the possibility of involuntary stoppages, they are not suitable for the control of slow secular changes or of long period changes in the Earth's rotation.

**3.** It has been established within recent years that the rotation of the Earth is not strictly uniform and that the departures from uniformity are of three different types, which are due to different causes. They are:

a) A slow secular increase in the length of the day.

b) Irregular fluctuations, the length of the day sometimes increasing and sometimes decreasing.

c) A seasonal variation in length.

Of these, a) and b) have been detected by observations of discordances between observations and theory in the positions of the Moon, Mercury, Venus and the Sun; c) has been detected through the high precision of modern clocks. The first two of these effects are somewhat inter-related and must be considered together.

## II. The Secular Acceleration of the Motions of the Sun and Moon.

**4.** Halley[1], in 1695, by comparing the positions of the Moon derived from early observations of eclipses with observations in his time, concluded that the motion of the Moon was being accelerated. This was confirmed by Dunthorne in 1749, by Mayer in 1753, and by Lalande in 1757. An acceleration of the mean

---

[1] E. Halley: Phil. Trans. **19**, 174, (1695).

motion of the Moon implies that its mean longitude (which differs from the true longitude by the omission of all periodic terms) can be represented by an expression of the form

$$L = a + b\,T + c\,T^2 \tag{4.1}$$

in which $T$ denotes the time (which is usually expressed in Julian centuries of 36525 days). The longitude at epoch $T = 0$ is $a$; the mean motion at that epoch $(dL/dT)$ is $b$. The mean motion at any epoch $T$ is represented by $(b + 2c\,T)$. The acceleration of the mean motion in unit time is thus $2c$. But from long usage it has become customary to designate the coefficient $c$ of the term in $T^2$ in the mean longitude as *the secular acceleration of the mean motion*. The investigations of DUNTHORNE, MAYER and LALANDE indicated that the secular acceleration of the Moon's mean motion was about $10''$ a century.

In the latter half of the 18th century much attention was being given by mathematicians to the detailed explanation of the motions of the bodies in the solar system under the action of NEWTON's law of gravitation. EULER and LAGRANGE both attempted, without success, to account for the secular acceleration of the Moon's motion. At length in 1787 LAPLACE[1] announced that he had discovered the explanation, which may be briefly stated as follows. The mean action of the Sun upon the Moon tends to diminish the Moon's gravity towards the Earth, and thereby to cause her angular velocity to decrease. The diminution being once supposed to have occurred, the angular velocity will thereafter remain constant, if the mean solar action remains constant. But the mean action of the Sun depends to a certain extent upon the eccentricity of the orbit of the Earth, and this eccentricity is decreasing secularly as a result of the action of the planets on the Earth. This gradual decrease of the eccentricity will cause a gradual decrease in the mean action of the Sun, which will result in a slow increase in the Moon's mean motion.

On the basis of his mathematical investigation, LAPLACE computed the amount of the acceleration as $10''.18$ per century, in close agreement with the observed value. But in 1853 J. C. ADAMS[2] found that the calculations of LAPLACE were not complete. LAPLACE had assumed that the areal velocity of the Moon remained unaltered, so that the tangential disturbing forces produced no permanent effect. ADAMS pointed out that this was correct if the eccentricity were constant but that, as a result of the gradual change of shape of the orbit, there was an uncompensated effect. Allowing for this, the theoretical value of the secular acceleration of the Moon's motion was reduced to $5''.70$, little more than half the observed value.

**5.** As there seemed to be no way in which the residual acceleration could be accounted for by gravitational action, it was concluded that it must be an effect arising from a secular retardation of the rotation of the Earth. In 1754 KANT had published an essay entitled "Untersuchung der Frage, ob die Erde in ihrer Umdrehung um die Achse, wodurch sie die Abwechselung des Tages und der Nacht hervorbringt, eine Veränderung seit den ersten Zeiten ihres Ursprunges erlitten habe, und woraus man sich ihrer versichern könne". In this essay he pointed out that the action of the Moon in raising tides in the oceans must have a secondary effect in a slight retardation of the Earth's motion by tidal friction, and explained the fact the Moon always turns the same face to the Earth as a

---

[1] P. S. DE LAPLACE: Mém. Acad. Sci. **235**, 1788 (1786). See also Mécanique Celeste, Book 7, Chapter I, § 16 and Chapter IV, § 23.

[2] J. C. ADAMS: Phil. Trans. A **143**, 397 (1853).

consequence of the retardation of the Moon's rotation in an early period of existence by bodily tides raised by the Earth on the Moon. LAPLACE had considered
and rejected this suggestion on the ground that, if such an effect existed, accelerations of the mean motions of the planets, as well as of the Moon, should be observed,
whereas no such accelerations had been detected by observation. DELAUNAY in
1865 revived the hypothesis of tidal friction to account for the acceleration of
the mean motion of the Moon. It was not until 1905, however, that COWELL[1]
found that there was a small secular acceleration of the orbital motion of the
Earth or, otherwise expressed, that the mean motion of the Sun was being
accelerated. Secular accelerations of the mean motions of Mercury and Venus
have since been established. Tidal friction appears qualitatively to be a *vera
causa* for the secular retardation of the rotation of the Earth.

**6.** The rate of dissipation of energy required to account for the observed
secular accelerations of the Moon and Sun can be estimated. The following
discussion is based on the investigation by JEFFREYS [1].

$M, m, m'$  = denote the masses of the Earth, Moon and Sun;

$n, n'$   = are the mean angular velocities of the Moon and Sun about the
Earth;

$c, c'$   = are the distances of the Moon and Sun from the Earth;

$-N, -N'$ = are the couples acting on the Earth due to lunar and solar tides;

$\omega$    = is the angular velocity of the Earth;

$C$    = is the moment of inertia of the Earth about its axis of rotation;

$f$    = denotes the constant of gravitation.

Then

$$n^2 c^3 = f(M + m), \qquad n'^2 c'^3 = f(M + m'). \tag{6.1}$$

Put

$$c = c_0 \xi^2, \qquad n = n_0 \xi^{-3}, \qquad c' = c_0' \xi'^2, \qquad n' = n_0' \xi'^{-3} \tag{6.2}$$

the suffix 0 denoting the present value.

The angular momentum of the orbital motion of the Moon and Earth about
their centre of mass is

$$\frac{M\,m\,c^2\,n}{M + m} = \frac{M\,m\,c_0^2\,n_0}{M + m}\,\xi. \tag{6.3}$$

To the couple $-N$ acting on the Earth's rotation must correspond an equal
and opposite couple $+N$ tending to increase the orbital angular momentum.
Because of the difference in periods of the lunar and solar tides, the solar tides
will have no secular effect on the Moon and the lunar tides will have no secular
effect on the Sun. Thus we must have

$$\frac{M\,m\,c_0^2\,n_0}{M + m}\,\frac{d\xi}{dt} = N, \tag{6.4}$$

$$\frac{M\,m'\,c_0'^2\,n_0'}{M + m'}\,\frac{d\xi'}{dt} = N', \tag{6.5}$$

$$C\frac{d\omega}{dt} = -N - N'. \tag{6.6}$$

---

[1] P. H. COWELL: M. N. **66**, 13 (1905). — M. N. = Monthly Notices Roy. Astron. Soc.

If $E$ is the total mechanical energy in the system, its rate of decrease must be equal to the rate of performance of work by the angular motions in overcoming the couples. Consequently

$$- \frac{dE}{dt} = (N + N') \omega - N n - N' n'. \tag{6.7}$$

The part of the right-hand side of this equation due to the lunar tides is $N(\omega - n)$, and must be positive, since the couples arise from dissipation of energy. As $(\omega - n)$ is positive, $N$ must be positive. Similarly $N'$ must be positive. Hence $d\xi/dt, d\xi'/dt$ must both be positive, from which it follows that $c, c'$ must be increasing and accordingly the mean motions of the Sun and Moon must be decreasing. The rate of rotation of the Earth must be decreasing, since $d\omega/dt$ must be negative.

7. We must now consider the effect of the changes in the angular velocities on observation. If $\omega, n, n'$ are assumed to be derived from observations at time zero, then the effect of the variation in the Earth's rotation is to put the Earth ahead in time $T$ by an angular amount $\frac{1}{2} T^2 \frac{d\omega}{dt}$. The time of transit of a fixed star is earlier by $\frac{T^2}{2\omega} \frac{d\omega}{dt}$. As the mean angular velocity of the Moon relative to the stars is $n$, the alteration of the time of observation results in the Moon being behind its calculated position when the star transits by an angle $\frac{n T^2}{2\omega} \frac{d\omega}{dt}$. But the change in the Moon's angular velocity puts it ahead by an angle $\frac{1}{2} T^2 \frac{dn}{dt}$. The combined effect is for the Moon to appear to have gained on the stars by $\frac{1}{2} T^2 \left( \frac{dn}{dt} - \frac{n}{\omega} \frac{d\omega}{dt} \right)$. If $\nu, \nu'$ denote the secular accelerations of the Moon and Sun, we therefore have

$$\nu = \frac{1}{2} \left( \frac{dn}{dt} - \frac{n}{\omega} \frac{d\omega}{dt} \right) \tag{7.1}$$

and similarly

$$\nu' = \frac{1}{2} \left( \frac{dn'}{dt} - \frac{n'}{\omega} \frac{d\omega}{dt} \right). \tag{7.2}$$

But from (6.2)

$$\frac{dn}{dt} - - 3 n_0 \xi^{-4} \frac{d\xi}{dt}, \quad \frac{dn'}{dt} - - 3 n_0' \xi'^{-4} \frac{d\xi'}{dt} \tag{7.3}$$

whence, using (6.4) to (6.6), and assuming $c$ is constant

$$\nu = - \frac{3}{2} \frac{M+m}{M m} \frac{N \xi^{-4}}{c_0^2} + \frac{N + N'}{2 C \omega} n_0 \xi^{-3}, \tag{7.4}$$

$$\nu' = - \frac{3}{2} \frac{M+m'}{M m'} \frac{N' \xi'^{-4}}{c_0'^2} + \frac{N + N'}{2 C \omega} n_0' \xi'^{-3}. \tag{7.5}$$

For intervals of time of a few thousand years, since the earliest observations of eclipses, it is sufficient to take $\xi$ and $\xi'$ both equal to 1.

If we denote by $\varkappa$ the present ratio of the orbital angular momentum to the angular momentum of the Earth's rotation

$$\varkappa = \frac{M m}{M + m} \frac{c_0^2 n_0}{C \omega_0}, \tag{7.6}$$

we obtain

$$\nu = \frac{M + m}{2 M m c^2} \{ \varkappa (N + N') - 3 N \}. \tag{7.7}$$

The ratio of the first term in $v'$ to the second term is of the order of $10^{-6}$, and the first term can consequently be neglected, whence

$$v' = \frac{M+m}{2Mmc^2} \varkappa (N+N') \frac{n'}{n}. \tag{7.8}$$

The present value of $\varkappa$ is 4.82 and of $n/n'$ is 13.4.

**8.** The precise ratio of the retarding couples due to the lunar and solar tides, $N/N'$, can not be calculated, because a detailed knowledge of the forces that are acting is not available. On the assumption that the equations of motion are linear and that the system is far from resonance, Jeffreys [1] obtained a value of 5.1 for the ratio. But the equations are much more likely to be non-linear; on the assumption that the friction is proportional to the square of the velocity, he obtained a value of 3.4. Using this estimate and the observed value of the secular acceleration of the Moon, he computed the rate of dissipation of energy to be about $1.4 \times 10^{19}$ ergs/sec.

Several discussions of the tides in mid-ocean have been made, by G. H. Darwin[1], Jeffreys [1] and others[2]. It appears that neither viscosity nor turbulence in the oceans can cause sufficient friction to account for the required rate of dissipation of energy. Jeffreys, for instance, estimates the dissipation of energy in the open oceans to be of the order of $10^{16}$ ergs/sec. The reason for this low value is that the rate of dissipation is proportional to the third power of the velocity of the water, while the velocities in the open oceans are small, of the order of 1 cm./sec.

In the shallow seas, however, the tidal currents can have much greater velocities and the rate of dissipation of energy can be considerable. But the areas of the shallow seas are relatively small, and detailed computation is needed to decide whether the total rate of dissipation of energy in these seas is adequate to account for the Moon's secular acceleration.

The rate of dissipation by the tides in the Irish Sea was determined by G. I. Taylor[3] in 1919 by two different methods. The mean rate of dissipation in this sea was found to be about 30 times the total rate of dissipation for the whole of the open oceans and about 2% of the total rate needed to account for the lunar secular acceleration.

Taylor's method was extended by Jeffreys [1] to include most of the shallow seas of the globe, using the tide ranges and currents given in the *Admiralty Pilots*. The Bering Sea was found to be much the most important, this one sea accounting for about 70% of the total dissipation for the whole of the shallow seas. The total dissipation computed by Jeffreys amounted to $2.2 \times 10^{19}$ ergs/sec. at spring tides, the average rate being $1.1 \times 10^{19}$ ergs/sec., which is about 80% of the amount needed to account for the secular acceleration of the Moon as determined by Fotheringham.

**9.** There is, however, another effect which has to be considered, to which attention has been called by Holmberg[4]. It is well known that there is a large semi-diurnal variation in the height of the barometer; the phase of this variation is remarkably constant over the globe. In 1882 Lord Kelvin[5] suggested that the close coincidence between the period of the Earth's rotation and the natural

---

[1] G. H. Darwin: Collected Papers. Cambridge 1907.1916. (See Vol. 2.) Also Phil. Trans. A **170**, 447 (1879).

[2] See Defant's article on Ocean Tides in Vol. XLVIII of this Encyclopedia.

[3] G. I. Taylor: Phil. Trans. A **220**, 1 (1919).

[4] E. R. R. Holmberg: M. N. Geophys. Suppl. **6**, 325 (1952).

[5] W. Thomson (Lord Kelvin): Proc. Roy. Soc. Edinburgh **11**, 396 (1882).

period of resonance of the atmosphere might provide the explanation of the large amplitude of the semi-diurnal barometric variation. He pointed out that the phase of the variation (with maxima occurring in the second and fourth quadrants after midnight) is such that the gravitational attraction of the Sun on the atmospheric tides exerts an accelerating couple on the Earth, whose amount he was able to estimate. This paper by KELVIN has been generally overlooked and the effect of the accelerating couple has been neglected in the discussions of the lunar secular acceleration.

More recently the resonance between the natural period of the atmosphere and the period of the rotation of the Earth has been called upon to account for the magnitude of the diurnal variation of the Earth's magnetic field. This requires the resonance to be very sharp, the periods not differing by more than a few minutes[1].

The acceleration of the rotation of the Earth by the atmospheric tidal couple requires a flow of angular momentum from the Earth's heliocentric orbit, which increases the mechanical energy of the system. HOLMBERG suggests that this energy is extracted from the solar energy falling on the Earth's surface by a heat-engine effect. His investigation of this effect shows that the maxima of the diurnal variations of temperature and pressure—the temperature maximum tending to occur in the afternoon and the maximum of the 24-hour pressure component in the forenoon—are such that work is done by the atmosphere. The mechanical energy produced in the atmosphere maintains a back pressure, which keeps the semi-diurnal maxima in the barometric height from slipping round into the first and third quadrants under the action of frictional drag from the Earth's surface.

A very large number of determinations of the semi-diurnal pressure variation have been made. Using SIMPSON'S[2] data for this variation

$$\Delta P = 1.25 \times 10^3 \sin^3 \vartheta \cos (2t + 64°) \ \text{dyn./cm.}^2 \tag{9.1}$$

in which $\vartheta$ denotes the colatitude and $t$ the local time, and following Lord KELVIN's argument, the total couple is found by HOLMBERG to be $3.7 \times 10^{22}$ dyn./cm., working at a rate of $2.7 \times 10^{18}$ erg./sec., which can be compared with the rate of working calculated by JEFFREYS for the oceanic tidal couple of $1.1 \times 10^{19}$ erg./sec.

**10.** The value of the atmospheric couple is unlikely to be in error by more than a few per cent, whereas JEFFREYS considered that his computation of the tidal couple might be in error by half its amount. HOLMBERG has raised the question whether the two couples might not actually be equal in their rate of working. This possibility is related to the question whether the present close agreement between the rate of rotation of the Earth and the natural period of resonance of the atmosphere is fortuitous or not. He suggests that the period of rotation of the Earth was progressively slowed down by oceanic friction until it approached the resonance period of the semi-diurnal atmospheric tide. This approach was accompanied by a progressive increase in the amplitude of the atmospheric tide, which resulted in a progressive increase in the atmospheric accelerating couple. The retardation of the Earth's rotation continued until the two couples were brought into quasi-equilibrium, with the rotational and vibrational periods closely coincident.

---

[1] The sharpness of this resonance has recently been questioned by M. SIEBERT: Naturwiss. **41**, 446 (1954). See the article by W. KERTZ on Atmospheric Tides in Vol. XLVIII of this Encyclopedia.

[2] G. C. SIMPSON: Quart J. Roy. Met. Soc. **44**, 1 (1918).

This possibility will be discussed further when the determinations of the secular accelerations of the Sun and Moon have been considered. But it is necessary first to discuss another aspect of the Earth's rotation.

## III. Fluctuations of the Motion of the Moon.

**11.** In 1870 Newcomb called attention to the existence of fluctuations in the motion of the Moon which were either of long period or of an irregular nature, and which did not seem to be accounted for by current theories. The reason that such fluctuations had not been detected earlier was that the theories of the motion of the Moon were not sufficiently complete. The theory of the motion of the Moon is of extreme complexity and has been gradually developed and refined by the researches of many eminent investigators. In 1857 the British Admiralty had published new tables of the Moon, based upon the investigations of the Danish astronomer P. A. Hansen, to whom the British Government made a grant of £ 1,000 in recognition of his work. The Astronomer Royal, Airy, said of Hansen's Tables that "probably in no recorded instance has practical science ever advanced so far by a single stride". Hansen's Tables were adopted as the basis for the computation of the ephemerides of the Moon published in the *Nautical Almanac*. They were believed to represent closely all observations of the Moon since 1750 and it was expected that they would continue to represent the motion of the Moon for many years to come. By the year 1870 this hope had been destroyed; the Moon had deviated from the positions computed on the basis of Hansen's Tables by an amount that could not be attributed to errors of observation. The departure between the tabular and observed positions went on increasing year by year.

In 1878 Newcomb [2] published a large memoir in which he discussed all available observations of the Moon, particularly those before 1750, and compared them with Hansen's Tables. He found that there were many observations of occultations of stars by the Moon recorded in the observation books of the Paris Observatory, which had not hitherto been published. He was able to extend the history of the Moon's motion back from 1750 to 1675 and, with a lesser degree of accuracy, thirty years farther still. He found that these earlier observations also showed an increasingly large deviation from the tabular positions. He concluded that either a) Hansen's theory was inadequate and that some important terms of long period had been omitted, which a careful revision of the theory should reveal; or b) the rotation of the Earth on its axis is subject to fluctuations of irregular character. He proposed the following criterion by which the two possibilities could be separated: if other celestial phenomena present fluctuations of the same general type, we must suspect the rotation of the Earth; if they do not, there must be some error in Hansen's theory or in his calculations.

Newcomb found that the fluctuations in the motion of the Moon from about 1650 onwards could be approximately represented by an empirical periodic term, with a period of 260 years, for which there was no gravitational explanation, and that superposed on it were irregular fluctuations of minor extent. His separation of the fluctuations into what became known as the *Great Empirical Term*, and the minor fluctuations, was followed by other investigators.

**12.** Though Newcomb spent many years in the attempt to account for the fluctuations, their origin was still undecided at the time of his death in 1909. The first essential question for their further elucidation was a decision whether the existing theories of the Moon (Hansen's and Delaunay's) were inadequate. A complete determination of the perturbations in the motion of the Moon was needed. This work was undertaken by E. W. BROWN at the suggestion of Sir GEORGE DARWIN and the new theory of the motion of the Moon was published by him in five important memoirs[1]. On the completion of the theoretical investigations, the construction of new lunar tables, based on the theory, was undertaken. The tables were published in three large volumes in 1919 and have been used for predicting the place of the Moon in the *Nautical Almanac* from 1923 onwards. The accuracy of the tables has recently been checked by numerical integration with an electronic computing machine; they have been found to agree satisfactorily with the numerical integration, except for a few minor errors which are immaterial for the purposes of this discussion.

In his Tables BROWN included that portion of the total secular acceleration which is attributable to the gravitational attraction of the planets as computed by ADAMS, but not the portion which is attributable to the retardation of the rotation of the Earth by tidal friction. He incorporated NEWCOMB's great empirical term, though it remained without any theoretical justification, in order to obtain an approximate representation of the observations from about 1650 onwards. Observations since 1923 have shown a changing difference between the tables and observation, whether the great empirical term is included or excluded. As it is now certain that there is no defect in theory to which the fluctuations can be attributed, the presumption is that they are caused by fluctuations in the speed of rotation of the Earth, and therefore in our measure of time.

The decision whether this is the correct explanation depends upon whether similar fluctuations are present in the observed motions of Mercury, Venus and the Sun. Several investigations were devoted to this problem, by BROWN[2], DE SITTER[3] and SPENCER JONES [3], the question being finally settled by the latter.

**13.** It is necessary to adopt a value for the secular acceleration of the Moon due to tidal action, from the discussion of early observations of solar and lunar eclipses and of occultations. The ancient observations were very thoroughly discussed by FOTHERINGHAM in a series of papers and by SCHOCH. These discussions were co-ordinated by DE SITTER[3] and from a least squares solution, the following values of the secular accelerations were derived:

$$\left.\begin{array}{l} \text{for the Sun} \quad +1\overset{''}{.}80 \pm 0\overset{''}{.}16 \\ \text{for the Moon} \ +5\overset{''}{.}22 \pm 0\overset{''}{.}30. \end{array}\right\} \tag{13.1}$$

In the investigation by SPENCER JONES [3] the fluctuation in the position of the Moon is denoted by $B$, defined as

$$B = \text{Observed Longitude} - C \tag{13.2}$$

where $C = $ BROWN's Tables $- 10\overset{''}{.}71 \sin (140\overset{\circ}{.}0\,T + 240\overset{\circ}{.}7)$

$$+ 5\overset{''}{.}22\,T^2 + 12\overset{''}{.}96\,T + 4\overset{''}{.}65 \tag{13.3}$$

in which $T$ denotes centuries from 1900.0. The sine term removes the great empirical term, the term in $T^2$ incorporates the effect of the secular retardation of the Earth's rotation, and the other terms represent the consequential

---

[1] E. W. BROWN: Mem. Roy. Astronom. Soc. **53**, 39, 163 (1899); **54**, 1 (1904); **57**, 51 (1905); **59**, 1 (1908).

[2] E. W. BROWN: Trans. Yale U. Obs. **3**, pt. 6 (1926).

[3] W. DE SITTER: Bull. Astronom. Inst. Netherlands **4**, 21 (1927).

corrections to the mean motion and longitude at epoch in order to secure close agreement with modern observations.

It is assumed provisionally that the fluctuations in the Moon's position are due to some cause or causes affecting the Earth only, such as changes in its moment of inertia, or changes in its angular momentum which are compensated by changes in the angular momentum of the atmosphere or the oceans. The effects on the mean longitudes of other bodies will then be similar to the effects on the mean longitude of the Moon, but reduced in proportion to the ratio of their mean motions in longitude.

When we come to consider the effects of tidal friction, there is an important difference. Tidal friction does not concern the Earth alone; it has an influence on the motion of the Moon. But in order to calculate the exact quantitative effects produced on the motion of the Moon, it would be necessary to have a detailed knowledge of the forces that are acting, which is lacking. The apparent discordances in the positions of the Sun and planets, consequent upon the retardation of the Earth's rotation by tidal friction, will have the same value when expressed in time and will therefore be proportional, when expressed as differences of longitude, to the mean motions. In this respect tidal friction behaves in a similar way to change of moment of inertia of the Earth. The discordance in position of the Moon, due to tidal friction, is not proportional to its mean motion, however, and in this respect tidal friction behaves differently from change of moment of inertia.

**14.** Accordingly it is to be expected that the observed minus tabular differences in longitude of the Moon, Sun, Mercury and Venus should be capable of representation by the following expressions:

$$\text{Moon} \qquad \varDelta L = a \quad + b\,T \quad + 5\overset{''}{.}22\,T^2 \quad + B, \tag{14.1}$$

$$\text{Sun} \qquad \varDelta l' = a' \quad + b'\,T \quad + c'\,T^2 \qquad + \frac{n'}{n}B, \tag{14.2}$$

$$\text{Mercury} \quad \varDelta l'' = a'' + b''\,T + \frac{n''}{n'}c'\,T^2 + \frac{n''}{n}B, \tag{14.3}$$

$$\text{Venus} \qquad \varDelta l''' = a''' + b'''\,T + \frac{n'''}{n'}c'\,T^2 + \frac{n'''}{n}B, \tag{14.4}$$

where $n$, $n'$, $n''$, $n'''$ are the mean motions of the Moon, Sun, Mercury and Venus respectively; $5\overset{''}{.}22$ is adopted as the secular acceleration of the Moon; $c'$ denotes the secular acceleration of the Sun; and $B$ denotes the fluctuation in the motion of the Moon, defined as above.

The observations of the transits of Mercury across the Sun, which are available from the transit of November 1677, provide the most complete data. Observations of the Sun's declination are more reliable for the present purpose than observations of its right ascension, being less affected by personal errors of observation and by changes in methods. Observations of the Sun's declination from 1760 can be used, and of its right ascension from 1835. Observations of the right ascension of Venus from 1835 can be used.

The data were analysed by Spencer Jones [3] and the constants in the above formulae were determined. If the observational data are satisfactorily represented by the formulae, the values of $B$ derived for each body on substituting the values of $a$, $a'$, $a''$, $a'''$, $b$, $b'$, $b''$, $b'''$ and $c'$ should be in satisfactory agreement. That this is in fact the case is shown by Figs. 1 and 2. The first

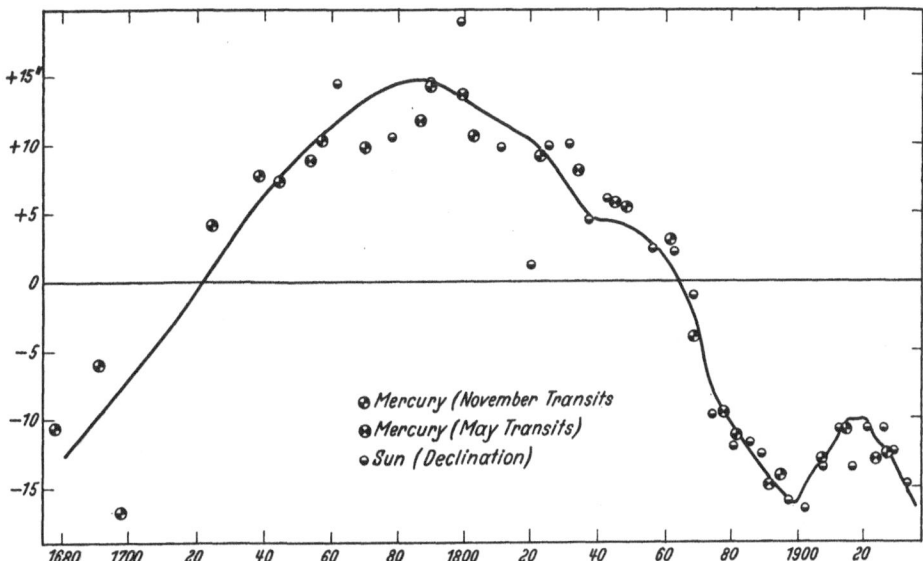

Fig. 1. Fluctuations in the Moon's mean longitude (continuous curve) and as derived from observations of transits of Mercury and of the declination of the Sun (1680 to 1940).

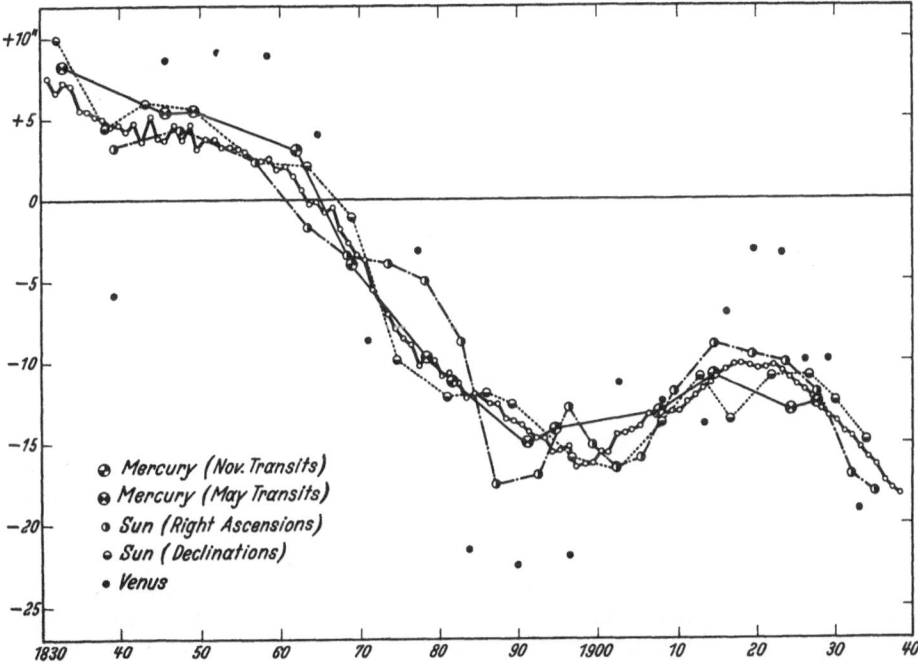

Fig. 2. Fluctuations in the Moon's mean longitude derived from observations of occultations of stars by the Moon, of transits of Mercury, of Venus, and of the Sun (1830 to 1940).

of these shows the fluctuation in the Moon's motion by a continuous line and the data from Mercury and the Sun's declination for the period from 1675 onwards. The second shows all the data available from 1830 onwards.

The numerical values of the solutions were as follows:

$$\text{Sun} \qquad \Delta l' \; = + 1''\!.00 + 2''\!.97\, T \; + 1''\!.23\, T^2 + 0.0748\, B, \qquad (14.5)$$

$$\text{Mercury} \quad \Delta l'' = + 4''\!.96 + 13''\!.08\, T + 5''\!.10\, T^2 + 0.310\, B, \qquad (14.6)$$

$$\text{Venus} \quad \Delta l''' = + 2''\!.26 + 5''\!.39\, T \; + 2''\!.00\, T^2 + 0.112\, B. \qquad (14.7)$$

These corrections to the tabular longitudes thus provide a satisfactory representation of the available observational data, when $B$ is defined as above, and the assumptions on which the discussion was based are thereby confirmed.

**15.** The value derived for the secular acceleration of the Sun is $+1''\!.23 \pm 0''\!.04$, the secular acceleration of the Moon having been assumed to be $+5''\!.22$. On comparing with the values derived by DE SITTER, given in (13.1), it will be seen that the discordance between DE SITTER's value of $1''\!.80$, based on ancient eclipse data in conjunction with current observations, and the value of $1''\!.23$, derived from observations since 1675, is much greater than would be expected from the assigned probable errors.

The explanation of this discordance may lie in the fact that it is not possible to make a unique separation between the secular acceleration and the fluctuation terms. The quantity $B$ has been defined in such a way that its extreme positive and negative values between 1675 and 1925 are approximately equal. If the true value of the secular acceleration of the Moon were known, the fluctuation might be found to have values considerably different from our $B$ values. Furthermore, we do not know what the fluctuation may have amounted to at the time of the early eclipse observations; we have tacitly assumed that it can be neglected.

If we suppose that the true secular acceleration of the Moon is not $+5''\!.22$, as has been assumed, but $+5''\!.22 + s$, then we must replace $B$ by a quantity $B'$, which we may define by

$$B' = B - s\, S \qquad (15.1)$$

where $S = (T - \alpha)\,(T - \beta)$, so that $S$ can be made zero at any two epochs ($T = \alpha$ and $T = \beta$) that may be chosen.

The terms in $T^2$ and $B'$ for the Sun and planets then become

$$\text{Sun} \qquad \left(c' + \frac{n'}{n}\, s\right) T^2 + \frac{n'}{n}\, B', \qquad (15.2)$$

$$\text{Mercury} \quad \frac{n''}{n'} \left(c' + \frac{n'}{n}\, s\right) T^2 + \frac{n''}{n}\, B', \qquad (15.3)$$

$$\text{Venus} \quad \frac{n'''}{n'} \left(c' + \frac{n'}{n}\, s\right) T^2 + \frac{n'''}{n}\, B', \qquad (15.4)$$

which are of precisely the same form as before but with $B$ replaced by $B'$ and $c'$ by $c' + n's/n$, which is equal to $+1''\!.23 + 0.0748\, s$. The representation of the observations by these changes will be unaltered, but the corrections to the longitudes at epoch and the mean motions for each body will depend upon the values assigned to $\alpha$ and $\beta$ in the expression for $S$.

Now it is conceivable that the dissipation of energy by tidal friction which, as we have seen, occurs almost entirely in the shallow seas of the globe, may have changed during the past 20 centuries as a result of changes in sea level. Though the amount of tidal friction may be variable, it seems reasonable to suppose that the secular accelerations of the Moon and Sun will remain in a constant ratio. With this assumption we have the relationship

$$\frac{1.80 \pm 0.16}{5.22 \pm 0.30} = \frac{(1.23 \pm 0.04) + 0.0747 s}{5.22 + s}$$

which gives

$$s = -2.11 \pm 0.57.$$

The average values of the secular accelerations of the Sun and Moon during the past 250 years then become

$$\left.\begin{array}{ll} \text{for the Moon} & +3''.11 \pm 0''.57, \\ \text{for the Sun} & +1''.07 \pm 0''.06. \end{array}\right\} \tag{15.5}$$

**16.** The curve representing the fluctuations in the Moon's longitude (Fig. 1) has many changes of slope, some of which appear to be rather sudden. The observations are not sufficiently accurate, however, to decide whether these major changes in slope occurred instantaneously or whether they were spread over several months or even a couple of years. The major changes of slope, such as those around 1900 and 1918, correspond to changes in the length of the day of about 4 or 5 milliseconds. It would be difficult to account for changes of this magnitude if they occurred instantaneously. The general appearance of the curve is very similar to the error curve of a good quality free-pendulum clock. After quartz crystal clocks had been introduced into the time service of the Greenwich Observatory, their performance for short-term predication was found to be so superior to that of the free-pendulum clocks, which had previously been used as standards of time, that the latter clocks were soon discarded. An analysis by GREAVES and SYMMS[1] of the comparisons between the quartz crystal clocks and the free-pendulum clocks established that the pendulum clocks were subject to frequent random small changes of rate, whose integrated effect appeared as an irregular wandering of the clock error. It seemed reasonable to suppose by analogy that the fluctuations in the motions of the Moon, Sun and planets are the consequence of the integrated effect of random small changes in the rate of rotation of the Earth.

On this hypothesis, BROUWER[2] showed that the mean error of the fluctuation in the Earth's rotation should increase with time proportionally to $T^{\frac{3}{2}}$ measured from the present epoch. The precision with which the secular accelerations of the Moon and Sun can be determined from ancient eclipse observations is consequently much reduced, because the integrated effect of the random changes in the course of some 20 centuries may be very large. VAN WOERKOM[3] has discussed the general statistical properties of cumulative series of random numbers and has applied the method of variate analysis to the annual values of the observed fluctuations in the Moon's mean longitude. He has confirmed that the observed changes in the Earth's rotation are due primarily to cumulative random changes. On this hypothesis the values of the secular accelerations, from the discussions of BROUWER[4] and VAN WOERKOM, become

$$\left.\begin{array}{ll} \text{for the Moon} & +2''.2 \pm 9''.5 \text{ (m.e.)}, \\ \text{for the Sun} & +1''.01 \pm 0''.70 \text{ (m.e.)}. \end{array}\right\} \tag{16.1}$$

These values, considered in relation to their mean errors, leave it uncertain whether there is, in fact, any secular acceleration of the mean motions of the

[1] W. M. H. GREAVES and L. S. T. SYMMS: M. N. **103**, 196 (1943).
[2] D. BROUWER: Proc. Nat. Acad. Sci. Wash. **38**, 1 (1952).
[3] A. J. J. VAN WOERKOM: Astronom. J. **58**, 10 (1953).
[4] D. BROUWER: Astronom. J. **57**, 125 (1952).

Sun and Moon, and tend to support Holmberg's conjecture that there may be a balance between the atmospheric accelerating couple and the oceanic retarding couple. The secular accelerations inferred from the ancient observations of eclipses would in that case not be real but should be attributed to the accumulated effect of the random changes in the rotation of the Earth.

**17.** We return, therefore, to the theoretical expressions derived by Jeffreys [1] for the secular accelerations of the Moon and Sun, given in formulae (7.7) and (7.8). These formulae did not include the solar atmospheric accelerating couple and it is necessary to replace $N'$ by $N' - N''$, where $N''$ denotes the latter couple. The formulae (7.7) and (7.8) for the secular accelerations $v$ and $v'$ now become

$$v = \frac{M + m}{2 M m c^2} \{ (\varkappa - 3) N + \varkappa (N' - N'') \}, \tag{17.1}$$

$$v' = \frac{M + m}{2 M m c^2} \varkappa (N + N' - N'') \frac{n'}{n} \tag{17.2}$$

giving the ratio

$$\frac{v}{v'} = \frac{\dfrac{\varkappa - 3}{\varkappa} N + N' - N''}{N + N' - N''} \frac{n}{n'}. \tag{17.3}$$

When the accelerating couple $N''$ is omitted, the value of the ratio $v/v'$ becomes 6.3, with Jeffreys' estimated value of 5.1 for $N/N'$, on the assumption that the equations of motion are linear; and becomes 7.2, with the value of 3.4 for $N/N'$, on the assumption that friction is proportional to the square of the velocity. The minimum value is 5.0 when $N' = 0$, so that all the friction is in the lunar tides. None of the estimates of the secular accelerations given above lead to a ratio of $v/v'$ as high as this minimum value. This discordance between theory and observation has for long been puzzling. It is removed by the inclusion of the accelerating couple $N''$.

We put

$$N'' = \lambda (N + N')$$

so that $\lambda$ denotes the ratio between the atmospheric accelerating couple and the total oceanic retarding couple.

With $\varkappa = 4.82$, $n/n' = 13.4$, and $N/N' = 3.4$

$$\frac{v}{v'} = \frac{30.69 - 59.0\lambda}{4.40 - 4.4\lambda}. \tag{17.4}$$

If $\lambda$ has the value 0.4, $v/v'$ has the value 2.7, which is in satisfactorily close agreement with the various values deduced from the observations. Corresponding to the accelerating atmospheric couple working at a rate of $2.7 \times 10^{18}$ erg./sec., this value of $\lambda$ would require a retarding oceanic couple working at the rate of $0.7 \times 10^{19}$ erg./sec., which is well within the limit of uncertainty of the value computed by Jeffreys. If $\lambda$ has the value 0.52, the secular acceleration of the Moon becomes zero, while that of the Sun remains positive, in satisfactory agreement with the results of Brouwer and van Woerkom. What is needed now is a more accurate determination of the total dissipation of energy by tidal friction in narrow seas. All that we can say at present is that there is no longer any contradiction between theory and observation.

Holmberg[1] has suggested that there may be a statistical balance between the accelerating and retarding couples, but that, as changes in sea level, which

---

[1] E. R. R. Holmberg: M. N., Geophys. Suppl. **6**, 325 (1952).

might occur within times that are short geologically and astronomically, would have their greatest proportionate effect on the shallow seas, the magnitude of the oceanic couple may fluctuate appreciably. The balance of the evidence suggests that at the present time it may be somewhat larger than the atmospheric couple and that in consequence there has been some retardation in the rate of the Earth's rotation within historical times.

If a secular acceleration of the Moon's mean motion of about 5" per century can be attributed to a slowing down of the Earth's rotation, the increase in the length of the day in the course of a century is between one and two milliseconds. The causes of the random changes in the length of the day can most conveniently be considered along with those of the seasonal variation.

## IV. Seasonal Changes in the Length of the Day.

**18.** The first discussions to indicate with some confidence that the rate of rotation of the Earth has a seasonal variation were made by STOYKO[1] at the Bureau de l'Heure, Paris, based on the performance of clocks at the observatories of Paris, Washington and Berlin during the three years 1934–1937. The evidence was not convincing, as it depended mainly on data from pendulum clocks; all pendulum clocks are subject to erratic changes of rate which, by integration, produce an irregular wandering; the quartz clocks of that period, moreover, did not have a high standard of performance. PAVEL and UHINK[2] in 1935, for instance, had concluded from the performance of two quartz clocks at the Geodetic Institute, Potsdam, that there was a range of about 9 milliseconds in the length of the day during the year, a result that is certainly spurious.

Subsequent discussions, based on the performance of the clocks at the Physikalisch-Technischen Reichsanstalt from 1934 to 1944 by SCHEIBE and ADELSBERGER[3]; of the clocks at the Geodetic Institute, Potsdam, from 1938 to 1944 by UHINK[4]; and of the clocks at the Royal Greenwich Observatory from 1943 to 1951 by FINCH[5], have nevertheless closely confirmed the results obtained by STOYKO.

The great advantage of quartz crystal clocks over pendulum clocks is that they are almost entirely free from erratic changes of rate. A seasonal variation in the rate of rotation of the Earth will be reflected as an apparent seasonal variation in the error of the clock: comparisons between different clocks prove that it is the rate of rotation of the Earth and not the rate of the clock that is variable.

It is necessary, however, in analysing the time determinations, first to correct them for the effects of the polar motion. The geographical poles have an irregular motion within a maximum distance of about 40 feet from their mean positions; this motion contains a component with an annual period and another component with a period of about 14 months. A motion of the pole along the meridian of any place causes a variation in its latitude; a motion eastwards or westwards displaces the meridian and therefore has an effect on time determinations. The motion of the pole is derived from observations of latitude at five observatories on the same parallel of latitude 39° 8' N., which are analysed at the Central Bureau of the International Latitude Service, now in Turin. The

[1] N. STOYKO: C. R. Acad. Sci. Paris **203**, 39 (1936); **205**, 79 (1937).
[2] F. PAVEL and W. UHINK: Astronom. Nachr. **257**, 365 (1935).
[3] A. SCHEIBE and U. ADELSBERGER: Z. Physik **127**, 416 (1950).
[4] W. UHINK: Astron. Nachr. **278**, 97 (1949).
[5] H. F. FINCH: M. N. **110**, 3 (1950).

results are not available, however, until some months later. At Greenwich the time determinations are corrected by using data of the latitude variation determined by the U.S. Naval Observatory, Washington, which are supplied week by week; the longitude of this Observatory is $77°$ W., which enables a substantially correct allowance for the effect of the motion of the pole on the determinations of time to be made.

**19.** Every quartz crystal clock is subject to a secular change of rate or frequency drift, for whose accurate determination the error of the clock must be known over a time interval of a year or longer. A seasonal variation in the rate of rotation of the Earth complicates the determination of this frequency drift. The method used by Finch was to assume that the observed clock corrections could be fitted to an ephemeris of the form

$$E = a + b\,t + c\,t^2 + f \tag{19.1}$$

where the term $f$ is a function with a period of one year. Mean monthly values of $E$ were formed for each clock, and their third differences derived. The terms in $a, b, c$ are thereby eliminated and only the third differences of the function $f$ remain, which should be the same for all clocks, apart from accidental errors. It was found that the third differences from the different clocks were negative at one period of the year and positive at another. The mean values from all the clocks were formed in order to reduce accidental errors; then, by successive integration, the second and first differences and the function itself were obtained.

From the mean of the investigations of the performance of clocks at the observatories of Greenwich, Paris, Washington; the Geodetic Institute, Potsdam; the P.T.R., Berlin; and the Deutsche Seewarte, Hamburg, for various periods between 1934 and 1949, Smith[1] found that the amount $f$ by which the Earth is slow, expressed in milliseconds, can be represented by

$$f = 55.1 \sin n\,(d - 58) + 5.8 \sin 2n\,(d - 117) \tag{19.2}$$

where $n$ is the diurnal motion of the Sun in degrees (0.986) and $d$ is the day of the year reckoned from Jan. 0.

The Earth gets slow in the spring and fast in the autumn. The total variation in the length of the day in the course of a year from these results is about $2\frac{1}{2}$ milliseconds. The annual fluctuation in the Earth's rotation persists from year to year, but is variable both in amplitude and phase.

For 1950–1951 the Greenwich clocks gave an appreciably smaller annual fluctuation, represented by

$$f = 33.0 \sin n\,(d - 14) + 5.8 \sin 2n\,(d - 76) \tag{19.3}$$

and this decrease was confirmed by the Potsdam clocks. The decrease may have been real, though it is not impossible that the larger value obtained from the earlier period may be to some extent attributable to the clocks on which it was based having been less accurate. The annual fluctuation has remained of about the same amplitude during 1952–1954 as during 1950–1951[2].

**20.** The system of star places on which the time determinations at all observatories are based is, by resolution of the International Astronomical Union, that of the fundamental catalogue prepared at the Astronomisches Rechen-Institut and known as the FK 3 Catalogue. If the right ascensions of this catalogue are affected by an error which varies with right ascension, such an error will

---

[1] H. M. Smith: Institution Electr. Eng. Monograph. **1952**, No. 39.
[2] See H. M. Smith and R. H. Tucker: M. N. **113**, 251 (1953).

enter into the derived annual fluctuation of the Earth's rotation, since at most observatories the observations for the determination of clock error are made in the early evening after sunset. The new "Catalogue of 5268 Standard Stars, 1950.0, based on the Normal System N 30", prepared by H. R. Morgan[1], shows that there are periodic errors in the right ascensions of the FK 3 system and that these errors depend also on declination.

When the right ascensions of the N 30 system are used, the annual fluctuation derived from the Greenwich time determinations is somewhat reduced, and a closer agreement with the annual fluctuations derived from the time determinations at other observatories is obtained.

The following table gives the values of the constants in the expression

$$f = a_1 \sin n (d - d_1) + a_2 \sin 2n (d - d_2) \qquad (20.1)$$

derived by Smith and Tucker[2] for Greenwich, Mount Stromlo, Washington and Richmond (Florida), when the N 30 system is used, based on the time determinations at those observatories, corrected for the polar motion, referred to the Greenwich quartz crystal clocks through the link provided by radio time signals.

| | Period | $a_1$ | $d_1$ | $a_2$ | $d_2$ |
|---|---|---|---|---|---|
| Greenwich . . . . . | Jan. 1951–Sept. 1952 | 18.5 | 26 | 10.5 | 87 |
| Mount Stromlo . . . | Jan. 1951–June 1952 | 23.9 | 364 | 8.5 | 84 |
| Washington . . . . . | Jan. 1951–June 1952 | 22.3 | 35 | 11.0 | 119 |
| Richmond . . . . . | Jan. 1951–June 1952 | 23.0 | 36 | 10.5 | 111 |

The agreement between these results is satisfactory.

# V. The Causes of the Irregular and Seasonal Variations in the Rotation.

**21.** The origin of the irregularities in the rate of rotation of the Earth has been discussed by Munk and Revelle [4], who considered possible causes in the atmosphere, oceans, mantle and core of the Earth. Two different types of effect have to be considered: those arising from displacements of matter, which would change the moment of inertia; and those arising from motion, which would change the angular velocity. Any effect which is not symmetrical about the axis of rotation would produce a tilt of the axis, and therefore a displacement of the poles of rotation. The observations of the International Latitude Service have shown that in the past 60 years the mean position of the pole has not changed by more than 15 feet, while during this period there have been changes in the length of the day of four or five milliseconds. The possible effects giving rise to the irregularities in the rotation must therefore be nearly symmetrical about the axis of rotation.

There is a seasonal variation in the distribution of air mass over the globe, which is responsible for the annual term in the variation of latitude, but which contributes only a small amount to the seasonal change in the rotation. A substantial change in the circulation of the atmosphere would have only a small effect on the length of the day, and it is concluded that a long-term variation in the zonal circulation could not account at the most for more than 5 to 10% of the observed fluctuations.

---

[1] H. R. Morgan: Ast. Papers of the American Ephemeris, vol. XIII, pt. III. 1953.
[2] H. M. Smith and R. H. Tucker: M. N. **113**, 250 (1953).

**22.** Variations in oceanic circulation have a negligible effect on the length of the day. But changes in sea level caused by growth or melting of the ice-caps over Greenland and Antarctica have to be considered. Munk and Revelle collected the data for mean annual sea level at 132 tide-gauge stations all over the world. Combining the results from nearby stations, the average sea levels at 50 localities for three consecutive decades, starting in 1901, were obtained. Changes in mean level by as much as 5 cm. per decade are not uncommon, but they differ in amount and sign from one location to the next and are variable with respect to time. From the analysis of the data, and on three alternative assumptions (melting in Antarctica only, in Greenland only, and in both regions), they conclude that the maximum possible eustatic change in sea level from 1910–1930 cannot have changed the length of the day by more than 0.6 ms., and this is on the assumption that there is no isostatic adjustment. The observed change in the length of the day in this period would require a lowering in sea level by 63 cm., if the whole effect were attributed to this cause.

Furthermore, if the observed change in the length of day between 1901–1930 were to be attributed to changes in mean sea level, the asymmetry in the distribution of land and sea would entail a displacement of the pole of rotation far greater than indicated by the latitude observations and in the opposite direction.

Observed changes in mean sea level might be due to a heating or cooling of the oceans. A heating by $1°$ C of the entire oceans would raise the sea level by 60 cms. Munk and Revelle[1] find that this part of the changes in sea level can be neglected as far as the effect on the Earth's rotation is concerned, because an addition of mass at the surface requires a corresponding removal at depth. They conclude that recorded changes in sea level can account for not more than 20% of the changes in rotation in recent years.

Young[2] has discussed the question whether the secular acceleration of the Moon might be attributed in whole or in part to a secular change in sea level. He finds that a progressive rise in sea level at the rate of 1 cm. per century, due to melting of ice in glaciated regions, would increase the moment of inertia of the Earth at a rate sufficient to cause a secular acceleration of the Moon of $2''$ per century, provided there was no isostatic compensation. The changes would be negligible, however, if isostatic compensation were immediate and complete. Appreciable changes in mean sea level may have occurred in the course of thousands of years and their effects may not be negligible.

If there is a secular change in the moment of inertia, $C$, the equation (6.6) must be modified and becomes

$$C \frac{d\omega}{dt} = - N - N' - \omega \frac{dC}{dt}.$$

If we can assume that over a period of centuries $\omega \, dC/dt$ has a constant value, $\delta$, the value of the ratio of the secular accelerations of the Moon and Sun becomes

$$\frac{v}{v'} = \frac{\dfrac{\varkappa - 3}{\varkappa} N + N' + \delta}{N + N' + \delta} \frac{n}{n'}.$$

This equation is of the same form as (17.3), and $\delta$ must have a negative value for agreement with the observational ratio to be possible, whereas Young's hypothesis, which gives it a positive value, would increase the discordance. Urey[3], in fact,

---

[1] W. Munk and R. Revelle: Amer. J. Sci. **250**, 829 (1952).
[2] A. Young: M. N., Geophys. Suppl. **6**, 482 (1953).
[3] H. C. Urey: Geochim. et Cosmochim. Acta **1**, 209 (1951). See Sect. VIII.

has suggested that the moment of inertia of the Earth is decreasing secularly, as a consequence of the continuous formation of an iron-nickel core, which sinks towards the centre as it is formed. This is an *ad hoc* hypothesis for which independent support is lacking. The inclusion of the atmospheric accelerating couple has removed the discordance between the observed and theoretical ratio, as has already been shown.

**23.** The effect of possible random fluctuations in the elevation of continental blocks was discussed by MUNK and REVELLE. Quite large displacements in elevation have only a small effect on the Earth's rotation but may cause a considerable shift of the positions of the poles. DE SITTER estimated that if the whole of the Central Asian high plateau, including the whole of the Himalayas, were brought down to sea level, the change in the length of the say would not exceed 1 ms. The observed irregularities in the length of the day during the past 250 years are much too large to be accounted for by any possible localised crustal displacements. It seems more probable that an explanation is to be found in some phenomena affecting the Earth as a whole. If, for instance, the radius of the Earth increased or decreased by about 6 inches, the largest observed changes in the length of the day could be accounted for. Such a change would not displace the poles and would be much too small to be detected by observation. It is not obvious, however, what phenomena within the Earth could cause its radius to change in the way suggested, which entails a considerable change in the volume of the Earth.

MUNK and REVELLE considered the effect of a symmetrical oscillation of the crust, with a rise or fall in high latitudes, accompanied by a movement in the opposite sense in low latitudes, the volume of the Earth remaining constant. If elastic yield is neglected, an apparent rise in sea level of a few centimetres in median latitudes, with a larger rise near the equator and an even larger lowering at high latitudes, would be quantitatively adequate to account for observed changes in the length of the day. When elastic yielding is taken into account, the required changes in sea level are increased almost three-fold and appear to be too large for acceptance. Further, there have been occassions when an appreciable increase and an appreciable decrease in the length of the day have both occurred within a few years of each other. Not only does it seem *a priori* impossible for such rapid reversal of a large-scale phenomenon, but there is no evidence that at such times there has been a universal change in the trend of tide records.

The final phenomenon discussed by MUNK and REVELLE is electromagnetic coupling of the Earth's mantle to a turbulent core. There is a general westward drift of the Earth's magnetic field: VESTINE has shown that this westward drift since 1890 has not been at a uniform rate and that the variations in the rate of drift are correlated with the observed variations in the length of the day. BULLARD[1] explains the drift on the hypothesis that thermal convection, due to radioactive heating, causes a flux of matter between the inner and outer portions of the core: the associated flux of angular momentum will tend to increase the rate of rotation of the inner portion and to decrease that of the outer portion of the core, so that features in the outer core drift westward relative to the mantle. Electromagnetic coupling between the core and the mantle changes the angular velocity of the mantle. The quantitative investigation of the relationship proves to be in reasonably good agreement with observation.

---

[1] E. C. BULLARD and others: Phil. Trans. A **243**, 67 (1950). See RUNCORN's article on Magnetism of the Earth's Body in this volume.

**24.** Thus, amongst a wide range of phenomena that have been discussed in an attempt to account for the irregularities in the rotation of the Earth, there is only one that appears quantitatively adequate, viz. the electromagnetic coupling of the mantle to a turbulent core. Further observations of the changes in the westward drift of the Earth's magnetic field and the investigation of their correlation with the changes in the rate of rotation are needed, however, before the reality of the relationship between the two phenomena can be regarded as definitely established.

**25.** The explanation of the seasonal fluctuation of the Earth's rotation is more certain. A variety of seasonal phenomena, such as shifts in air masses, melting of polar ice-caps, variations in the angular momentum of the atmosphere, and variations in oceanic circulation, may be expected to have a seasonal effect on the rate of rotation. The question is whether they can account for the amplitude and phase of the observed fluctuations.

The most detailed and thorough investigation of possible causes of the seasonal fluctuation has been made by Munk and Miller[1], supplemented by a re-examination of certain details and of some other effects by Mintz and Munk[2].

Let $I_e, I_a, I_0$ denote the moments of inertia of the solid earth, atmosphere, and ocean respectively; $\Omega_e$ the angular velocity of the Earth, $\omega_a, \omega_0$ the components along the Earth's axis of the angular velocities of atmosphere and ocean relative to the solid earth. For seasonal effects we can neglect secular changes caused by tidal friction. Hence

$$I_e \Omega_e + I_a (\Omega_e + \omega_a) + I_0 (\Omega_e + \omega_0) = \text{constant}. \tag{25.1}$$

By differentiation we have

$$-\frac{\Delta \Omega_e}{\Omega_e} = \frac{\Delta (I_a \omega_a)}{I \Omega_e} + \frac{\Delta (I_0 \omega_0)}{I \Omega_e} + \frac{\Delta I}{I} \tag{25.2}$$

where $I = I_e + I_a + I_0$ is the total moment of inertia.

Munk and Miller computed the relative angular momentum of the atmosphere for the months of January and July, these two months being assumed to represent the extreme conditions of winter and summer. For the northern hemisphere the computations were made for longitudes 100° E. and 80° W., by numerical integration from the surface to 17 km. It was assumed that these sections were representative of conditions throughout 90° of longitude on either side. For the southern hemisphere data were available only for a section in longitude 150° E., which is not altogether satisfactory. The angular momentum of the oceanic circulation was computed for the antarctic circumpolar current and the equatorial easterly current system, these being the two largest current systems. Various effects that would lead to changes of inertia of atmosphere, ocean, and earth were computed or estimated. The magnitudes of the estimated differences between January and July were as follows:

$$\Delta (I_a \omega_a)/I \Omega_e \sim + 1.6 \times 10^{-8},$$

$$\Delta (I_0 \omega_0)/I \Omega_e \sim 0.15 \times 10^{-8} \quad \text{(plus or minus)},$$

$$\Delta I/I \qquad \sim 0.12 \times 10^{-8} \quad \text{(plus or minus)}.$$

[1] W. H. Munk and R. L. Miller: Tellus 2, 93 (1950).
[2] Y. Mintz and W. H. Munk: Tellus 3, 117 (1951).

The value of $-\varDelta\Omega_e/\Omega_e$ (January minus July) derived by FINCH from the Greenwich clocks was $2.0 \times 10^{-8}$ for 1943–1947 and $2.3 \times 10^{-8}$ for 1948–1949.

The seasonal variation in the angular momentum of the atmosphere is thus the primary cause of the annual fluctuation in the rate of rotation, and from this investigation it appears to be quantitatively adequate.

**26.** The estimate of the effect of the seasonal winds was recomputed by MINTZ and MUNK[1] using new data on zonal winds: for the northern hemisphere data for the latitude belt 20–90° N. averaged over all longitudes were used; for the southern hemisphere, data for the longitudes of New Zealand and the east coast of Australia for the latitude belt 20–75° S. were used. The result of the computation was to reduce the value of $\varDelta(I_a\omega_a)/I\Omega_e$ from $1.6 \times 10^{-8}$ to $0.6 \times 10^{-8}$, little more than one-third of the previous value. MINTZ and MUNK considered also the effect of seasonal changes in the semidiurnal earth tides. As the Earth is not perfectly rigid, it is tidally distorted and the equatorial protuberance increases the moment of inertia beyond what it would be if the Sun were absent. This tidal effect varies through the year because of the seasonal changes in the distance and latitude of the Sun. The tidal

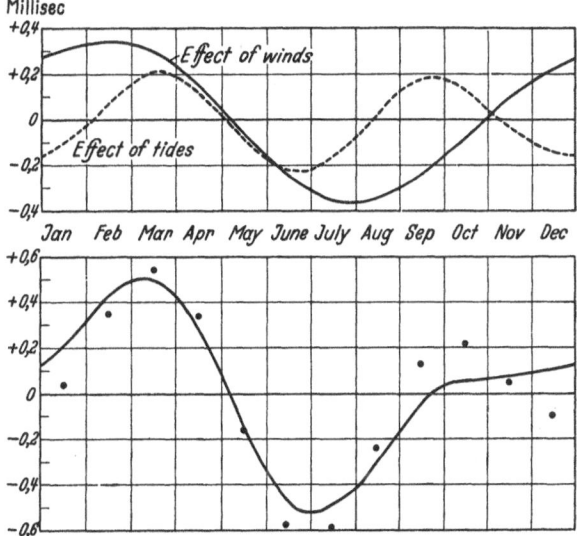

Fig. 3. Annual fluctuation in the length of the day. Computed effects of winds and of earth tides (above); combined computed effect compared with observed effect from Greenwich time determinations (below).

effects produce a small, though not negligible, variation in the rate of rotation with a semi-annual period.

The annual fluctuation as derived at Greenwich does not have its extreme range between January and July. MINTZ and MUNK[2] have extended their computation to other months and have made a revision to allow for the fact that the fluid core of the Earth hardly participates in the annual change of rotation. Fig. 3 represents their results: the effects of winds alone and of tides alone are shown separately. Their combined effect is compared with the Greenwich values for the annual fluctuation, in the form of monthly mean values based on the time determination referred to the N30 Catalogue and for the period since 1950. The agreement is better than is to be expected from the uncertainties in the meteorological data. It thus appears that the seasonal variation in the rate of rotation of the Earth is satisfactorily accounted for, as regards both amplitude and phase, by the combination of the seasonal variation in the angular momentum of the winds and of the earth tide effect. Though other factors may contribute, their combined effect appears to be small.

---

[1] Y. MINTZ and W. H. MUNK: Tellus **3**, 117 (1951).

[2] Y. MINTZ and W. H. MUNK: Private communication 1953.

# VI. Derivation of a constant Unit of Time.

**27.** The preceding discussion has shown that the length of the day is not constant: it has a slow secular increase, is subject to irregular fluctuations in length, and has a fairly regular seasonal variation. It is unsatisfactory that one of the fundamental units in physics should be variable. The question arises how a constant unit of time is to be defined and how it is to be related to the variable unit provided by current observations.

This question has been considered by the International Astronomical Union, which has recommended that:

"In all cases where the mean solar second is unsatisfactory as a unit of time by reason of its variability, the unit adopted should be the sidereal year at 1900.0; that the time reckoned in this unit be designated *Ephemeris Time*; that the change of mean solar time to ephemeris time be accomplished by the following correction:

$$\Delta T = + 24^s\!.349 + 72^s\!.3165\,T + 29^s\!.949\,T^2 + 1\cdot821\,B \tag{27.1}$$

where $T$ is reckoned in Julian centuries from 1900.0 January 0 Greenwich Mean Noon and $B$ has the meaning given by Spencer Jones in M.N., R.A.S. **99**, 541 (1939), and that the above formula define also the second."

It has since been agreed that it is preferable to adopt the *tropical year* at 1900.0 as the fundamental unit, in preference to the *siderel year*. The tropical year is the time required for the mean longitude of the Sun, referred to the mean true or moving equinox, to increase by 360°, whereas the siderel year is the time required for the mean longitude of the Sun, referred to a fixed equinox, to increase by 360°. The length of the tropical year is obtained by direct observation, while the length of the siderel year can be derived from it when the constant of precession is known. Any change in the adopted constant of precession would therefore change slightly the length of the siderel year but would not affect the length of the tropical year.

In terms of the length of the tropical year, the second can be derived as the fraction 1/31 556 925.975 of the tropical year for 1900.0.

This definition of the second, as the fundamental unit of time, has been given a legal status by its adoption by the Comité Permanent des Poids et Mesures.

The correction to the Sun's longitude, on the basis of Newcomb's Tables, was derived by Spencer Jones, and is given in formula (14.5) above. The time required for the mean longitude of the Sun to increase by $1''$ is $24^s\!.349$. Multiplying $\Delta L$ by $24^s\!.349$ gives the above expression (27.1) for $\Delta T$.

If the observed position of the Sun at *astronomical time* (G.M.T.) $T_1$ agrees with the position at time $T$ *(ephemeris time)* interpolated from the ephemeris, the difference between them is $\Delta T$; the ephemeris or uniform time corresponding to a particular instant of the non-uniform astronomical time can therefore be derived. But the slow motion of the Sun in longitude makes this method an insensitive one. It is much more convenient to use the Moon, whose motion in longitude is 13.4 times that of the Sun. The observations of the Moon, compared with its position on gravitational theory, provide the value of $B$, the fluctuation in the Moon's longitude, which enters into formula (27.1). The formula then gives the value of $\Delta T$.

From the rate of change of $B$, the length of the mean solar day at time $T$ can be related to its length at 1900.0, and thus the length of the second at any time $T$ can be related to its value at 1900.0. The value of the Moon's fluctuation $B$ is derived from recent observations and is therefore always known somewhat in arrear, but any necessary correction required in physical or astronomical investigations can always be applied retrospectively.

## Bibliography.

[1] JEFFREYS, H.: The Earth: its Origin, History and Physical Constitution. Third Edition. Cambridge: University Press 1952. Chapter VIII is concerned with tidal friction. Theoretical expressions are derived for the secular accelerations of the Sun and Moon. The dissipation of energy by tidal friction is discussed and quantitative estimates given.

[2] NEWCOMB, S.: Researches on the Motion of the Moon, pt. I. Reduction and Discussion of Observations of the Moon before 1750. Washington. 1878. Pt. II. The Mean Motion of the Moon and Astronomical Elements, based on Observations extending from the Era of the Babylonians until A. D. 1908. Washington. 1912.
SPENCER JONES, H.: Ann. of Cape Observatory 13, 3 (1932) contains a revision of Pt. II, with a reduction of the observations to the basis of BROWN's Theory of the Moon.

[3] SPENCER JONES, H.: M. N. 99, 541 (1939). This paper contains the analysis of observations of the Moon, Mercury, Venus and the Sun for the derivation of the secular accelerations and of the irregularities in the rotation of the Earth. The results of this paper have been adopted by the International Astronomical Union for reduction from astronomical time to ephemeris time.

[4] MUNK, W., and REVELLE, R.: M. N., Geophys. Suppl. 6, 331 (1952). This paper contains the most exhaustive discussion of the various causes which might affect the rotation of the Earth and give rise to its irregular variations.

# Séismométrie.

Par

JEAN COULOMB.

Avec 33 Figures.

## Introduction.

**1.** La séismométrie mesure, ou plutôt enregistre en fonction du temps, les ébranlements du sol causés par les tremblements de terre ou séismes. Les vibrations correspondantes sont extrêmement irrégulières; on a coutume de parler néanmoins de leur amplitude et de leur période en entendant par là, assez vaguement il faut l'avouer, l'amplitude et la période du mouvement sinusoïdal qui s'en approcherait le plus, en moyenne, pendant un certain intervalle de temps. Amplitudes et périodes varient beaucoup suivant l'importance du séisme, la distance de l'épicentre et la phase envisagée. On peut vouloir mettre en évidence des périodes d'une fraction de seconde dans les séismes proches ou s'intéresser aux ondes longues de quelques minutes ayant fait plusieurs fois le tour de la terre. C'est dire qu'en séismologie aucun type d'appareil ne suffit à tous les usages. Pour étudier les ébranlements produits par des explosions artificielles, par exemple ceux qui servent à la prospection, on doit enregistrer des périodes encore beaucoup plus courtes, comptées en centièmes ou en millièmes de seconde, tandis que les marées terrestres mettent en jeu des périodes de douze heures et au delà. Nous ne parlerons pas systématiquement de ces deux phénomènes mais nous y ferons parfois allusion et certains des résultats obtenus leur seraient applicables.

Les appareils que nous étudierons, et qu'on appelle des *séismographes* se divisent en deux grandes classes: les séismographes pendulaires, qui comportent un système oscillant, et les enregistreurs de déformation séismique, qui mesurent des déplacements relatifs. Les premiers sont de beaucoup les plus nombreux et nous nous en occuperons bien plus longuement. Nous insisterons sur les questions théoriques, renvoyant d'avance à BERLAGE [3] pour la description des modèles classiques. Mais nous signalerons les particularités essentielles des réalisations modernes.

## A. Théorie linéarisée des Séismographes pendulaires.

**2. L'équation générale des séismographes.** Le sol est un milieu déformable; mais si l'on installe à sa surface un pilier de béton massif, fondé si possible sur le roc, en tous cas tel que les vibrations propres du système sol-pilier aient des périodes beaucoup plus courtes que les périodes à mesurer, le problème est ramené à celui des mouvements de la surface du pilier, considérée comme un solide indéformable. En prospection séismique au contraire, il peut être nécessaire de tenir compte des résonances entre le sol et le séismographe[1].

---

[1] H. WASHBURN and H. WILEY: Geophysics 6, 116 (1941). — A. WOLF: Geophysic 9, 29 (1944). — I. P. PASSETCHNIK: Izv. Akad. Nauk. SSSR. Sér. Géophys. 1, 21; 3, 34; 5, 25 (1952).

Sur le pilier est fixé le bâti du séismographe, lui-même massif et pratiquement indéformable. Soit alors $S_2$ un système de coordonnées $O_2\, x_2\, y_2\, z_2$ lié au bâti, $O_2$ étant un point situé vers le centre du pilier. Le problème est d'étudier les mouvements du système $S_2$ par rapport à un système $S_1$ qui ne participe pas au mouvement, et dont les axes $O_1\, x_1\, y_1\, z_1$ coïncident avec la position de repos des axes $O_2\, x_2\, y_2\, z_2$. Les rotations du sol au cours des séismes sont extrêmement faibles, une fraction de seconde d'arc (ANGENHEISTER l'a vérifié en observant une étoile pendant un tremblement de terre). L'enregistrement des inclinaisons séismiques, tenté notamment par SCHLÜTER et GALITZINE, n'a jamais été réalisé, celui des rotations verticales pas davantage. Ils ne présenteraient à vrai dire qu'un intérêt restreint, car les équations de l'élasticité permettent de calculer les rotations à partier des déplacements; les discordances éventuelles mettraient seulement en évidence des écarts très locaux du sol à la loi de HOOKE. On ne cherche plus guère aujourd'hui que les composantes du déplacement $O_1 O_2$ suivant les trois axes fixes $O_1\, x_1\, y_1\, z_1$. Vu la faiblesse des rotations, ce déplacement est pratiquement le même en tous les points du pilier. C'est lui que nous devrons déterminer, tout en conservant provisoirement dans les équations les termes correspondant aux rotations, afin de pouvoir nous rendre compte dans quelle mesure ils sont négligeables.

La plupart des séismographes modernes déterminent seulement l'une des trois composantes NS, EW ou verticale de la translation. Dans les deux premiers cas on dit qu'on a un séismographe horizontal, dans le troisième un séismographe vertical.

En exposant la théorie des séismographes pendulaires, nous nous inspirerons de WIECHERT [1] et GASSMANN[1]. Un cas particulier a été étudié en détail par BYERLY[2]. Considérons de façon générale un système de masses, mobiles par rapport au bâti, mais que des forces rappellent vers une configuration où il serait en équilibre stable si $S_2$ était en repos par rapport à $S_1$. L'exemple le plus simple d'un tel système de masses inertes, celui qui constituait les premiers séismographes en service, est un pendule supendu à l'intérieur du bâti; nous donnerons le nom de *pendule*, de façon générale, à la partie mobile des séismographes de ce type.

Une fois écarté de sa position d'équilibre le pendule va tendre à osciller pour son compte. Il faudra amortir ces vibrations propres, et pour cela introduire à l'intérieur du système $S_2$ de nouvelles forces qu'on tâchera de rendre proportionnelles à la vitesse pour leur effet soit aisément calculable. A vrai dire il serait nécessaire de tenir compte d'autres forces, dues par exemple à l'amplification ou à l'enregistrement; nous examinerons ces complications éventuelles sur des cas particuliers. Dans le même esprit nous supposerons que le pendule n'a qu'un degré de liberté. Or la plupart des appareils en service comprennent des parties peu déformables assemblées entre elles par des liaisons plus ou moins élastiques (ressorts); il faut qu'aucune vibration interne, notamment aucune oscillation propre des ressorts ne puisse être excitée; on s'arrange en général pour qu'elles aient une période très courte. La position, dans le système matériel $S_2$, d'un point matériel quelconque $P$ du pendule défini par le vecteur $\overrightarrow{O_2 P} = \boldsymbol{p}$ est alors fonction d'un seul paramètre, par exemple l'ordonnée du point inscripteur c'est à dire la distance $q$, entre la position instantanée et la position au repos, de la pointe du style ou du point lumineux qui trace l'enregistrement. Nous admettrons que $q$ est toujours petit et qu'on peut écrire:

$$\boldsymbol{p} = \boldsymbol{p}_0 + q\, \boldsymbol{p}'_0 + \tfrac{1}{2} q^2\, \boldsymbol{p}''_0,$$

[1] F. GASSMANN: Arch. Meteorol., Geophys. u. Bioklim., Ser. A **4**, 408 (1951).
[2] P. BYERLY: Bull. Seism. Soc. Amer. **42**, 251 (1952).

en notant d'un accent la dérivation par rapport à $q$; nous noterons d'un point la dérivation par rapport au temps $t$. *Dans ce qui suit nous ne garderons en principe que les termes d'ordre zéro et un par rapport à $q$ et ses dérivées.*

Nous utiliserons pour la mise en équation des mouvements du système le principe des travaux virtuels [1], en nous plaçant dans le système $S_2$. La masse du point $P$ est soumise aux forces suivantes:

$\alpha$) *La force d'inertie relative* $-m\ddot{p} = -m\ddot{q}\,p_0'$ due au mouvement de $m$ à l'intérieur de $S_2$. Son travail virtuel dans le déplacement:

$$\delta p = (p_0' + q\,p_0'')\,\delta q \quad \text{est} \quad -m\,p_0'^2\,\ddot{q}\,\delta q.$$

Le travail virtuel total de ces forces est donc:

$$-\mathfrak{M}\ddot{q}\,\delta q,$$

en posant:

$$\mathfrak{M} = \Sigma\,m\,p_0'^2.$$

La somme représentée par $\Sigma$ est étendue à tous les points matériels du pendule. Si $q$ était le temps, $\mathfrak{M}$ serait l'énergie cinétique. On l'appelle *masse résultante* ou masse réduite au point inscripteur [1], parce que la masse $m\,p_0'^2$ placée au point inscripteur joue le même rôle au point de vue de l'inertie que la masse $m$ placée en $P$. Comme conséquence immédiate, si l'amplification est réalisée au moyen de leviers, pour lesquels $p_0'^2$ augmente en général lorsqu'on s'approche du style, la masse des derniers leviers jouera un rôle très important.

Fixons les idées concernant la masse résultante par deux exemples extrêmement simples, dans lesquels les masses mobiles constituent un solide indéformable dont le mouvement dans $S_2$ est restreint à une translation (premier exemple) ou à une rotation (deuxième exemple) par des liaisons sans frottement.

Dans le *pendule à translation*, $p_0'$ est indépendant de $P$ et tel que le déplacement d'un point quelconque soit multiplié par $1/p_0'$. Si $M$ est la masse du pendule:

$$\mathfrak{M} = M\,p_0'^2.$$

Pour une *masse tournante*, $\vartheta$ étant l'angle de rotation, l'analogie avec l'énergie cinétique permet d'écrire immédiatement: $\mathfrak{M} = K\vartheta_0'^2$, où $K$ est le moment d'inertie du pendule autour de l'axe de rotation.

$\beta$) *Les forces d'inertie d'entraînement* dues à la *translation* $\overrightarrow{O_1O_2} = u$ et à la rotation $\Omega$ autour d'un axe variable passant par $O_2$, qui amèneraient $S_1$ sur $S_2$. Les rotations du sol, nous l'avons dit, sont très petites; et de même la vitesse de rotation $\dot{\Omega}$ autour de l'axe instantané. Nous pourrons donc nous contenter de remplacer $p$ par $p_0$ et $\delta p$ par $p_0'\delta q$ dans le travail virtuel des forces d'inertie de rotation. L'accélération d'entraînement est $\gamma(p) = \ddot{u} + \ddot{\Omega} \times p + \dot{\Omega} \times (\dot{\Omega} \times p)$. Le travail virtuel total des forces d'inertie d'entraînement sera donc:

$$\left. \begin{aligned} -\Sigma m\,\gamma \cdot \delta p = -\{ &\ddot{u} \cdot \Sigma m\,p_0' + q\,\ddot{u} \cdot \Sigma m\,p_0'' + \ddot{\Omega} \cdot \Sigma m\,p_0 \times p_0' - \\ &- \Sigma m(\dot{\Omega} \times p_0) \cdot (\dot{\Omega} \times p_0')\}\,\delta q. \end{aligned} \right\} \quad (2.1)$$

Le premier terme dépend de l'accélération de translation. En introduisant le centre de gravité $G$ et la masse totale $M$ on peut l'écrire:

$$\ddot{u}\,\Sigma m\,p_0' = M\,\ddot{u} \cdot L_0' \tag{2.2}$$

où

$$L = \overrightarrow{O_2 G} = \frac{1}{M}\,\Sigma m\,p. \tag{2.3}$$

C'est le terme important dans les *séismomètres* proprement dits. On voit que $\ddot{u}$ y entre par sa composante suivant la direction initiale de déplacement du centre de gravité dans $S_1$. *Cette direction $L_0'$ définit la composante de $u$ qui sera mesurée par l'appareil.*

Le second terme:

$$q\,\ddot{u} \cdot \Sigma m\,p_0'' = q\,M\,\ddot{u} \cdot L_0'', \qquad (2.4)$$

est nul pour $q=0$, donc peu important tant qu'on reste au voisinage de la position de repos. Mais comme il dépend de la composante de $u$ suivant $L_0''$, il n'est pas toujours possible de le négliger devant le premier terme, qui dépend d'une autre composante. Supposons par exemple que le déplacement du point inscripteur soit proportionnel à celui du centre de gravité: $L'^2$ est constant, $L' \cdot L'' = 0$; la nouvelle composante, si elle n'est pas nulle, est perpendiculaire à la précédente. On a une «accélération en bout». Il est difficile d'en apprécier l'importance car il s'agit de termes non linéaires. Nous y reviendrons (Sect. 9).

Le troisième terme de (2.1) est proportionnel à l'accélération de rotation; il devra donc être prépondérant dans les *gyro-accéléromètres* qui la mesurent. Enfin le dernier terme dépend de la vitesse de rotation, abstraction faite de son sens. Il définirait les *gyromètres*.

Reprenons les deux exemples du pendule à translation et de la masse tournante. Explicitons les composantes de $u$ et de $\dot{\Omega}$:

$$u = [\xi, \eta, \zeta], \qquad \dot{\Omega} = [\alpha, \beta, \gamma].$$

Dans le premier exemple, supposons $O_2 x_2$ parallèle à la translation. Soit $L = [x, y, z]$. $L_0''$ est parallèle à $L_0'$ et le second terme peut être négligé devant le premier. On obtient:

$$\ddot{u}\,\Sigma m\,p_0' = M\,x_0'\,\ddot{\xi},$$

$$\dot{\Omega}\,\Sigma m\,p_0 \times p_0' = M\,x_0'\,(\dot{\beta}\,z_0 - \dot{\gamma}\,y_0),$$

$$\Sigma m\,(\dot{\Omega} \times p_0)\,(\dot{\Omega} \times p_0') = M\,x_0'\,[(\beta^2 + \gamma^2)\,x_0 - \alpha\beta\,y_0 - \alpha\gamma\,z_0].$$

L'appareil est un pur séismomètre si $x_0 = y_0 = z_0 = 0$, c'est à dire si le centre de gravité coïncide avec le point dont on mesure le déplacement. Bien entendu, vu la petitesse des rotations, l'appareil reste correct si $x_0, y_0, z_0$ ne sont pas trop grands. Mais si $x_0$ est important, avec $y_0 = z_0 = 0$, la vitesse de rotation $\sqrt{\beta^2 + \gamma^2}$ perpendiculaire à $O_1 x_1$ intervient. On pourrait même la mesurer au moyen de deux appareils de ce type, correspondant à des valeurs opposées de $x_0$, dont les indications se retrancheraient de façon à faire disparaître l'effet du terme en $\ddot{\xi}$. Ce gyromètre serait analogue au régulateur à boules des machines à vapeur.

De même supposons $x_0 = y_0 = 0$, mais $z_0$ important, l'accélération de rotation $\dot{\beta}$ autour de $O_1 y_1$ intervient; elle pourrait être mesurée au moyen de deux appareils de ce type, correspondant par exemple à une valeur nulle et à une valeur importante de $z_0$, dont les indications se retrancheraient de façon à faire disparaître l'effet du terme en $\ddot{\xi}$. Cette idée de mesurer les accélérations d'inclinaison par différence entre les indications de deux séismographes placés au pied et au sommet d'un pylône remonte à GALITZINE.

Dans le second exemple, supposons que la rotation se fasse autour de $O_2 z_2$, que le centre de gravité au repos $G_0$ soit sur $O_2 x_2$ et que $O_2 x_2$, $O_2 y_2$ et $O_2 z_2$ soient axes principaux d'inertie avec $I, J, K$ pour moments d'inertie. Si $q$ est proportionnel à l'angle de rotation $\vartheta$, on obtient:

$$\ddot{\boldsymbol{u}} \cdot \Sigma m \, \boldsymbol{p}'_0 = M L \vartheta'_0 \ddot{\eta},$$

$$q \, \ddot{\boldsymbol{u}} \cdot \Sigma m \, \boldsymbol{p}''_0 = - M L \vartheta'^2_0 \, q \, \ddot{\xi},$$

$$\ddot{\boldsymbol{\Omega}} \cdot \Sigma m \, \boldsymbol{p}_0 \times \boldsymbol{p}'_0 = K \vartheta'_0 \dot{\gamma},$$

$$\Sigma m (\dot{\boldsymbol{\Omega}} \times \boldsymbol{p}_0) \cdot (\dot{\boldsymbol{\Omega}} \times \boldsymbol{p}'_0) = (I - J) \vartheta'_0 \, \alpha \cdot \beta.$$

Il n'existe aucun moyen simple d'éviter la présence des termes de rotation pour obtenir rigoureusement un séismographe. Mais on peut construire un pur gyro-accéléromètre en faisant $L = 0$, $J = I$.

Pour examiner quantitativement l'importance des deux termes de rotation relativement au terme de translation, on se place pour simplifier [1] dans le cas schématique où un mouvement sinusoïdal du sol se propage horizontalement à vitesse constante, comme une onde superficielle:

$$\boldsymbol{u} = [\xi, \eta, \zeta]$$
$$= [U \sin(a x + b y - p t + \varphi), V \sin(a x + b y - p t + \chi), W \sin(a x + b y - p t + \psi)];$$

$$\boldsymbol{\Omega} = \left[ \frac{1}{2} \left( \frac{\partial \zeta}{\partial y} - \frac{\partial \eta}{\partial z} \right), \frac{1}{2} \left( \frac{\partial \xi}{\partial z} - \frac{\partial \eta}{\partial x} \right), \frac{1}{2} \left( \frac{\partial \eta}{\partial x} - \frac{\partial \zeta}{\partial y} \right) \right]$$

$$= \frac{1}{2} [b W \cos(a x + b y - p t + \psi), - a W \cos(a x + b y - p t + \psi),$$
$$a V \cos(a x + b y - p t + \chi) - b U \cos(a x + b y - p t + \varphi)].$$

Pour le séismographe horizontal formé par un pendule à translation on doit comparer entre eux les termes:

$$\ddot{\xi}; \quad \dot{\beta} z_0 - \dot{\gamma} y_0; \quad (\beta^2 + \gamma^2) x_0 - \alpha \beta y_0 - \alpha \gamma z_0.$$

En supposant que $U$, $V$ et $W$ soient de même ordre et qu'il en soit de même pour $x_0$, $y_0$ et $z_0$, $a$ et $b$, les ordres de grandeur relatifs pour les amplitudes des trois termes sont:

$$1, \quad x_0 a, \quad U a.$$

Or la longueur d'onde séismique $2\pi/a$ se compte au moins en kilomètres; tandis que $x_0$ se compte au plus en mètres, et $U$ en millimètres. Le second terme est déjà très petit, le troisième est insignifiant.

De même, pour le séismographe horizontal à masse tournante, on doit comparer entre eux les termes: $\ddot{\eta}$, $\frac{K}{ML} \dot{\gamma}$, $\frac{I - J}{ML} \cdot \alpha \beta$. Leurs ordres de grandeur relatifs sont $1$, $\frac{K}{ML} a$, $\frac{I - J}{ML} a^2 U$. Or $\frac{K}{ML}$ et $\frac{(I - J)}{ML}$ se comptent au plus en mètres; la conclusion est analogue à la précédente.

Nous laissons le lecteur faire les calculs pour les séismographes verticaux.

$\gamma$) *La force d'inertie complémentaire* ou force de Coriolis, dont le travail virtuel est nul puisque dans un système à un degré de liberté les déplacements virtuels compatibles avec les liaisons se font suivant la vitesse relative.

$\delta$) *La force de pesanteur mg*: sa direction est fixe dans $S_1$; il serait facile de passer de $S_1$ à $S_2$ en introduisant des angles analogues aux angles d'Euler (on sait que l'emploi de ceux-ci est incommode lorsque l'angle des deux verticales est voisin de zéro). Bornons-nous à supposer petite la rotation $\Omega$ correspondante. On peut alors écrire:

$$\boldsymbol{g} = \boldsymbol{g}_0 - \boldsymbol{\Omega} \times \boldsymbol{g}_0$$

où $g_0$ a des composantes fixes $0, 0, -g$ dans $S_2$. Le travail des forces de pesanteur est alors:

$$\sum m\,\boldsymbol{g}\,\delta p = [M\boldsymbol{g}_0 \cdot \boldsymbol{L}_0' + q\,M\boldsymbol{g}_0 \cdot \boldsymbol{L}_0'' - M\,(\boldsymbol{\Omega} \times \boldsymbol{g}_0) \cdot \boldsymbol{L}_0']\,\delta q.$$

$\boldsymbol{L}_0'$ définissant comme nous l'avons vu (2.2) la composante mesurée, le premier terme est nul pour un séismographe horizontal, dans tous les cas indépendant de $q$; s'il s'agit d'un séismographe vertical, on devra trouver une contrepartie à ce terme pour que le pendule ait une position d'équilibre.

Le second terme, qui est le terme fondamental dans les *gravimètres*, peut représenter suivant son signe une force de rappel (c'est le cas dans le pendule ordinaire) ou une force astatisante (c'est le cas dans le pendule inversé). On pourrait croire que sa présence rende négligeable le terme $q\,M\ddot{u} \cdot \boldsymbol{L}_0''$ qui correspond aux accélération en bout; car on sait que l'accélération du sol n'est jamais comparable à celle de la pesanteur, sauf peut-être au voisinage de l'épicentre des plus grands séismes. Mais leurs composantes peuvent devenir comparables si $\boldsymbol{L}_0''$ est voisin de l'horizontale; c'est le cas pour un pendule horizontal (Sect. 9 et 10).

Le dernier terme permet de mesurer la composante horizontale de la rotation; il caractérise les indicateurs de pente ou *clinomètres* (niveau à pendule, clinographe de SCHLÜTER). On se placera ici encore dans le cas d'un mouvement sinusoïdal du sol pour comparer ce terme au terme principal $M\ddot{u}\boldsymbol{L}_0'$, ou encore la «rotation vraie» $\frac{1}{2}\left(\frac{\partial \zeta}{\partial y} - \frac{\partial \eta}{\partial z}\right)$ autour de $O\,x$ par exemple à la «rotation apparente» $\ddot{\eta}/g$. Le rapport de leurs amplitudes est de l'ordre de $ga/2p^2$. Si on fait $g \sim 10$ m./sec.$^2$, $p/a \sim 5$ km./sec., la période $2\pi/a$ des ondes devrait être d'environ 100 minutes (contre un petit nombre de minutes dans les ondes les plus longues) pour que ce rapport soit voisin de l'unité.

ε) *Les forces exercées sur la masse m par les masses voisines.* Dans l'ensemble le travail $A\,\delta q$ de ces forces n'est pas nul bien que certaines d'entre elles (cohésion) soient opposées à des réactions sur les masses voisines. Pour que le pendule ait une position d'équilibre, s'il s'agit d'un séismographe horizontal, $A$ ne doit pas comprendre de terme indépendant de $q$ et de ses dérivées, mais s'il s'agit d'un séismographe vertical il doit comporter un terme $-M\boldsymbol{g}_0 \cdot \boldsymbol{L}_0'$, annulant l'effet de la pesanteur considérée précédemment; ce terme est produit par des ressorts avec parfois un appoint magnétique plus ou moins volontaire. (Il n'est peut-être pas inutile de signaler l'impossibilité d'employer une poussée hydrostatique qui réintroduirait les accélérations d'entraînement.) $A$ peut encore comprendre un terme de déformation, de nouveau dû à des ressorts et qu'on supposera en première approximation linéaire par rapport à $q$; nous représenterons par $-Cq$ l'ensemble de ce terme et du terme $q\,M\boldsymbol{g}_0 \cdot \boldsymbol{L}_0''$ provenant de la pesanteur, $C$ étant positif pour assurer la stabilité de l'équilibre. Enfin $A$ comprendra un terme d'amortissement que nous supposerons linéaire par rapport à $\dot{q}$, soit $-f\dot{q}$ avec $f > 0$.

Nous pouvons maintenant écrire que le travail virtuel est nul. Gardons seulement les termes principaux. Appelons $u$ la composante de $\boldsymbol{u}$ suivant $-\boldsymbol{L}_0'$[†]. Il vient finalement:

$$\mathfrak{M}\ddot{q} + f\dot{q} + Cq = -M L_0'\ddot{u}. \tag{2.5}$$

---

[†] On aurait pu choisir comme sens celui de $+\boldsymbol{L}_0'$, mais en pratique on compte suivant les mêmes axes le mouvement du sol et celui du centre de gravité, qui se déplacent en sens inverse par suite de l'inertie; et d'autre part on compte positivement le déplacement de l'indicateur qui correspond à un déplacement positif du centre de gravité.

C'est l'équation générale linéarisée des séismographes pendulaires; nous allons l'étudier complètement, avant de considérer les écarts possibles à la linéarité.

**3. Constantes caractéristiques.** Nous écrirons l'équation précédente sous la forme:

$$\ddot{q} + 2\alpha\omega\dot{q} + \omega^2 q = -V\ddot{u}, \tag{3.1}$$

$$V = \frac{ML_0'}{\mathfrak{M}} \quad \text{est le } \textit{grandissement} \tag{3.2}$$

Il est intéressant d'écrire son expression dans les deux exemples déjà utilisés:

Pour le *pendule de translation*, le grandissement est égal à $1/L_0'$, c'est à dire au facteur par lequel est multiplié le déplacement du centre de gravité. Ceci suggère de considérer d'une façon générale le quotient:

$$Q_0 = \frac{ML_0'^2}{\mathfrak{M}} \tag{3.3}$$

qu'on peut appeler avec Grenet[1] *coefficient d'utilisation de la masse*. En vertu du théorème de Koenig sur le moment d'inertie autour du centre de gravité, ce coefficient est inférieur à 1 dans tous les cas autres que celui du pendule de translation. C'est un facteur de qualité, important pour les séismographes à grande masse avec amplification par leviers (dans le cas des séismographes à faible masse il est facile d'adopter une forme ramassée conduisant à un coefficient de l'ordre de 0,8, et d'ailleurs les méthodes d'amplification plus puissantes qu'on utilise alors rendent sa considération presque superflue).

Pour un *séismographe à masse tournante* on a $Q_0 = ML^2/K$, d'autre part le grandissement prend la forme:

$$V = \frac{ML}{K\vartheta_0'},$$

$K$ étant toujours le moment d'inertie par rapport à l'axe de rotation, $L$ la distance de $G$ à l'axe. On appelle *longueur réduite* du pendule la longueur:

$$l = \frac{K}{ML} = L + \frac{\varrho^2}{L} \tag{3.4}$$

($\varrho$ rayon de giration). La longueur réduite permet d'exprimer le grandissement et le coefficient d'utilisation de la masse sous la forme simple:

$$V = 1/l\vartheta_0', \tag{3.5}$$

$$Q_0 = L/l. \tag{3.6}$$

Tout se passe comme si la masse était concentrée au « centre d'oscillation » c'est à dire au point situé à la distance $l$ de l'axe sur la perpendiculaire issue de $G$.

On utilise souvent la notion de longueur réduite dans la cas où le séismographe contient, outre la masse principale qui tourne autour d'un axe, des leviers ou des ressorts. On doit alors considérer (3.5) comme définissant $l$; ni (3.4), ni (3.6) ne sont plus strictement valables.

Pour avoir une bonne utilisation de la masse, on fera les leviers d'amplification très légers, surtout les derniers. Au contraire, la masse principale sera en plomb. Si elle tourne autour d'un axe, on lui donnera la forme d'un cylindre

---

[1] G. Grenet: C. R. Acad. Sci. Paris **214**, 916 (1942).

allongé parallèlement à cet axe (MAINKA). MILNE laissait la masse pivoter autour de son centre de gravité; tout se passe alors comme si elle y était concentrée; le montage est encore en service dans les séismographes MILNE-SHAW (fig. 7).

A titre d'exemple numérique prenons $V = 800$, $l = 40$ cm. L'équation (3.5) donne $1/\vartheta_0' = 320$ m., ce qui montre que l'on ne peut pas se contenter d'un simple levier optique.

Revenons à (2.5) et (3.1)

$$\omega = \sqrt{\frac{C}{\mathfrak{M}}} \qquad \text{est la pulsation propre;} \qquad (3.7)$$

$$\alpha = \frac{f}{2\sqrt{C\mathfrak{M}}} \qquad \text{est l'amortissement.} \qquad (3.8)$$

On peut introduire au lieu de $\omega$, la *période propre* $T_0 = 2\pi/\omega$ ou la longueur du pendule simple synchrone, en abrégé *longueur synchrone*:

$$\Lambda = \frac{g}{\omega^2} = \frac{\mathfrak{M} g}{C} .$$

Dans le pendule ordinaire $\Lambda = l$. Les périodes utiles en séismologie, 10 secondes par exemple, correspondent à $\Lambda$ de l'ordre de 25 mètres. Pour avoir une bonne utilisation de la masse, $L$ devrait approcher de $l$ donc ici de $\Lambda$, ce qui conduirait à des dimensions prohibitives. Aussi les pendules ordinaires sont-ils abandonnés depuis longtemps pour l'enregistrement des séismes lointains. Pour résoudre le problème de *l'astatisation* (obtenir de grandes périodes) en adoptant des modes de construction appropriés il faut diminuer $\omega$ sans beaucoup changer $Q$, donc diminuer les forces de rappel. Nous verrons plus loin comment. Si l'on voulait au contraire obtenir des périodes très courtes (séismographes de prospection), on augmenterait $C$; pour maintenir un amortissement suffisant, il faudrait alors augmenter $f$ ce qui pose la question de puissance des amortisseurs.

**4. Du séismographe au mouvement du sol. Méthode de la double intégration.** Reprenons l'équation fondamentale:

$$\ddot{q} + 2\alpha\omega\dot{q} + \omega^2 q = -V\ddot{x} . \qquad (4.1)$$

$q$ est donné par l'enregistrement. Il semble donc possible d'obtenir l'accélération $\ddot{x}$ en calculant les trois termes du premier membre. La dérivation d'une fonction empirique est sujette à erreur, mais les deux dérivations pourraient être faites électriquement, par induction. LIPPMANN a proposé dès 1909 un accéléromètre fondé sur ce principe. Sauf pour les séismographes de prospection, le courant proportionnel à $\ddot{q}$ serait faible et la réalisation difficile. On la faciliterait en enregistrant $q$ sur un ruban magnétique qu'on déroulerait beaucoup plus vite.

Si l'on veut calculer le déplacement lui-même, il faut tenir compte des conditions initiales. Nous supposerons d'abord, ce qui n'est jamais parfaitement exact, qu'au début du séisme le sol et le pendule partent tous deux de leur position d'équilibre sans vitesse initiale. Intégrons alors à deux reprises les deux membres de l'équation fondamentale (4.1).

On obtient d'abord:

$$\dot{q} + 2\alpha\omega q + \omega^2 \int_0^t q\, dt = -V\dot{x} \qquad (4.2)$$

puis

$$q + 2\alpha\omega \int\limits_0^t q\,dt + \omega^2 \int\limits_0^t dt \int\limits_0^t q\,dt = -Vx. \tag{4.3}$$

L'intégration graphique ou mécanique d'une fonction empirique n'offre pas de difficultés de principe. Malheureusement certaines erreurs de mesure s'accumulent. Ainsi une erreur constante de $\varepsilon$ sur la position de la ligne $q=0$, qui n'est jamais très bien connue, entraînerait une erreur $\varepsilon(1 + 2\alpha\omega t + \frac12\omega^2 t^2)$ sur

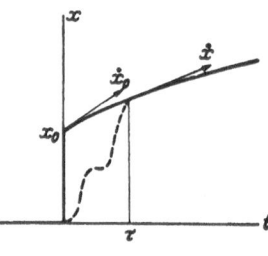

$Vx$. La courbe $x(t)$ obtenue subit toujours une dérive progressive. En examinant cette courbe [ainsi que la courbe intermédiaire $x'(t)$] on peut lui apporter une correction empirique raisonnable, en général parabolique, et parfois dépister des erreurs instrumentales. Le procédé est employé par le *U.S. Coast and Geodetic Survey* pour interpréter les données des séismographes à courte période et à déclenchement automatique destinés aux études des bâtiments de Californie[1].

A un autre point de vue, (4.2) et (4.3) permettent d'étudier mathématiquement l'effet sur le pendule d'un mouvement brusque du sol.

Supposons en effet que le sol et le pendule soient en repos au temps $t=0$, pour lequel on a $x=0$, $\dot x = 0$, $q = 0$, $\dot q = 0$, mais que les valeurs $x$ et $\dot x$ prises par le sol au temps $t>0$ tendent vers $x_0$ et $\dot x_0$ lorsque $t$ tend vers $O$. Admettons qu'à ces discontinuités sur $x, \dot x$ correspondent des discontinuités analogues $q_0, \dot q_0$ sur $q, \dot q$. Il s'agit de déterminer $q_0, \dot q_0$. On suppose pour cela le mouvement $q(t)$ remplacé pendant un temps

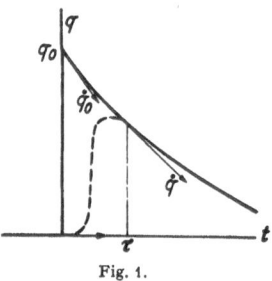

Fig. 1.

infiniment petit $\tau$ par un mouvement physiquement plus vraisemblable qui raccorde la courbe $q(t)$ pour $t > \tau$ avec la droite $q(t) = 0$ pour $t < 0$ (fig. 1). On peut choisir ce mouvement de raccordement tel que $q(t)$ reste borné entre 0 et $\tau$, lorsque $\tau \to 0$; peu importe si les fonctions $\dot q(t)$ et $\ddot x(t)$ qui correspondent à cette fonction $q(t)$ ne satisfont pas à la même condition.

Dans ces conditions $\int\limits_0^\tau q\,dt$ est infiniment petit d'ordre $\tau$, $\int\limits_0^\tau dt \int\limits_0^t q\,dt$ infiniment petit d'ordre $\tau^2$. Il suffit alors d'écrire les équations (4.2) et (4.3) pour $t=\tau$, et de faire tendre $\tau$ vers zéro pour obtenir les relations cherchées:

$$\left.\begin{aligned} \dot q + 2\alpha\omega q_0 &= -V\dot x_0, \\ q_0 &= -Vx_0. \end{aligned}\right\} \tag{4.4}$$

La seconde montre que $V$ est l'amplification des déplacements brusques. La première, en y faisant $q_0 = 0$, montre que $V$ est aussi l'amplification des percussions.

**5. Du mouvement du sol au séismogramme. Les oscillations propres.** Pour éviter la double intégration et surtout pour apprécier les propriétés du séismographe, on part d'une expression analytique simple pour $x(t)$, qui est supposée représenter au moins le début d'une phase, et on calcule la «réponse» $q(t)$ correspondante, qu'on cherche ensuite à identifier avec certains aspects de l'enregistrement. Le nombre des expressions proposées pour $x(t)$ est considérable. Citons

---

[1] F. Neumann: Bull. Seism. Soc. Amer. **33**, 1 (1943).

celle de BERLAGE [3]:

$$x = A t e^{-Kt} \sin \omega t,$$

qui évite les discontinuités de vitesse. Nous nous contenterons d'approfondir deux cas très simples: les déplacements brusques et les ondes sinusoïdales, parce qu'on peut décomposer un mouvement quelconque en mouvements élémentaires de l'un ou l'autre type[1].

Il faut étudier d'abord les oscillations propres; la solution générale de l'équation (4.1) privée de second membre est:

$$q = A_1 e^{\lambda_1 t} + A_2 e^{\lambda_2 t}, \qquad (5.1)$$

$\lambda_1$ et $\lambda_2$ étant les racines, supposées distinctes, de l'équation caractéristique: $\lambda^2 + 2\alpha \omega \lambda + \omega^2 = 0$; $A_1$ et $A_2$ deux constantes arbitraires. Le cas de la racine double peut être traité comme un cas limite. On désire voir disparaître rapidement les phénomènes transitoires correspondant à ces oscillations propres, pour qu'ils ne troublent pas les phases ultérieures. Aussi n'emploie-t-on jamais les amortissements (3.8) supérieurs à 1, pour lesquels $\lambda_1$ et $\lambda_2$ sont réelles, car l'une des deux racines est supérieure à $-\alpha\omega$ et le mouvement correspondant s'éteint moins vite que pour $\alpha = 1$.

L'amortissement $\alpha = 1$, $q = A t e^{-\omega t}$, dit *amortissement critique*, est généralement employé dans les appareils récents. Mais un grand nombre de séismographes en service correspondent à $0 < \alpha < 1$. Plaçons-nous dans ce cas. Si on pose:

$$\mu = \sqrt{1 - \alpha^2}, \qquad (5.2)$$

on pourra écrire:

$$q = A e^{-\alpha \omega t} \sin \mu \omega (t - \tau).$$

Le mouvement est oscillatoire amorti. Il serait représenté par une courbe tangente alternativement aux courbes exponentielles $q = \pm A e^{-\alpha \omega t}$. Il y a un intervalle de temps constant $\pi/\mu\omega$ entre deux points de tangence consécutifs, entre un maximum et un minimum consécutifs (ou l'inverse), entre deux points d'inflexion consécutifs, etc. ...

$$T = \frac{2\pi}{\mu \omega} = \frac{T_0}{\mu} \qquad \text{est la } \textit{pseudo-période}.$$

Deux amplitudes successives sont toujours dans un rapport constant:

$$R = e^{\frac{\pi \alpha}{\mu}} \qquad \text{appelé } \textit{rapport d'amortissement}. \qquad (5.3)$$

Pour obtenir pratiquement la *période propre* $T_0$ d'un séismographe, on supprime l'amortissement imposé. Il subsiste un faible amortissement naturel pour lequel on peut commodément déterminer le nouveau rapport d'amortissement $R$. Soit par exemple $-q_1$ un minimum, $+q_2$ le maximum suivant, $-q_3$ le minimum qui suit ce maximum:

$$R = \frac{q_1}{q_2} = \frac{q_2}{q_3} = \frac{q_1 + q_2}{q_2 + q_3}.$$

Le dernière expression dispense de tracer la ligne des zéros. Si l'on opère par enregistrement (5.2) et (5.3) permettent, connaissant $R$, d'en déduire $\mu$ à l'aide

---

[1] Il est parfois commode de prendre pour variable $\omega t$, qu'on peut appeler le «temps réduit». Nous ferons un changement de variable analogue pour les séismographes électromagnétiques où son emploi est plus justifié.

de tables [4], [6], [7]. D'autre part on détermine la *pseudo-période* $T$ soit en pointant deux passages au zéro, soit en mesurant l'intervalle entre deux maxima sur un enregistrement. On a enfin $T_0 = \mu T$.

Parmi ces oscillations propres, celle qui suit un déplacement brusque du sol offre un intérêt particulier. Appelons « *déplacement unité* » le mouvement du sol défini par $x = 0$ pour $t < 0$, $x = 1$ pour $t \geq 0$, d'où $\dot{x} = 0$. Soit $q = D(t)$ la réponse au déplacement unité. On a (4.4)

$$D(0) = -V, \qquad \dot{D}(0) = 2\alpha\omega V.$$

On peut considérer un mouvement quelconque $x(t)$ du sol comme superposition d'un déplacement brusque initial et de petits déplacements brusques ultérieurs successifs. La réponse au déplacement $dx$ subi au temps $\tau$ est:

$$dq = D(t - \tau)\, dx = D(t - \tau)\, \dot{x}(\tau)\, d\tau,$$

d'où

$$q = x(0)\, D(t) + \int_0^\tau D(t - \tau)\, \dot{x}(\tau)\, d\tau,$$

ou encore:

$$q = x(0)\, D(t) + \int_0^t D(v)\, \dot{x}(t - v)\, dv. \tag{5.4}$$

Appliquons ceci à un mouvement du sol défini par $x = 0$ pour $t < 0$, $x = t$ pour $t \geq 0$, que nous appellerons « *choc unité* » et qui bien entendu, moins encore que le déplacement unité, n'est pas physiquement susceptible de se maintenir. Soit $C(t)$ la réponse au choc unité; on aura:

$$C(t) = \int_0^t D(v)\, dv,$$

ou

$$D(t) = \dot{C}(t).$$

Indiquons deux variantes de l'équation (5.4) qu'on obtient aisément en intégrant par parties. La première s'écrit, en remplaçant $D(0)$ par $-V$:

$$q = -V x(t) + \int_0^t \ddot{C}(v)\, x(t - v)\, dv; \tag{5.5}$$

elle met en évidence l'écart à un grandissement proportionnel. La seconde:

$$q = x(0)\, \dot{C}(t) + \dot{x}(0)\, C(t) + \int_0^t C(v)\, \ddot{x}(t - v)\, dv, \tag{5.6}$$

sole l'effet du déplacement et du choc à l'origine.

Explicitons $C(t)$. C'est l'oscillation propre correspondant aux conditions initiales $C(0) = 0$, $\dot{C}(0) = -V$, donc:

$$\left. \begin{aligned} C(t) &= -V \frac{e^{\lambda_1 t} - e^{\lambda_2 t}}{\lambda_1 - \lambda_2} \\ &= -\frac{V}{\mu\omega} e^{-\alpha\omega t} \sin\mu\omega t. \end{aligned} \right\} \tag{5.7}$$

La forme même des réponses $C(t)$ et $D(t)$ fournit quelques indications sur les propriétés d'un séismographe. Mais il faut se méfier de certains paradoxes:

Si l'appareil était parfait on aurait: $C(t) = -Vt$; or au début, d'après (5.7):

$$C(t) = -Vt(1 - \alpha \omega t + \cdots).$$

A égalité de grandissement, il semblerait donc préférable, contrairement à ce qu'on attendait, de diminuer $\alpha$ et $\omega$ pour améliorer l'enregistrement des départs. Mais il ne suffit pas de considérer ce qui se passe pour un temps infiniment petit: Pour que le mouvement $C(t)$ soit perçu il faut que son premier maximum dépasse le seuil de l'agitation microséismique; ce maximum a lieu pour:

$$\operatorname{tg} \mu \omega t = \mu/\alpha, \quad \text{ou} \quad \omega t = \varphi/\sin \varphi \quad \text{en posant}$$

$\alpha = \cos \varphi$; son amplitude est $C_M = \dfrac{V}{\omega} e^{-\varphi \cot \varphi}$. Comparons alors deux enregistrements correspondant à la même valeur de $C_M$:

$$\frac{|C(t)|}{C_M} = \omega \cdot e^{\varphi \cot \varphi} t (1 - \alpha \omega t + \cdots).$$

La pente au départ est multipliée par $e$ lorsque $\alpha$ passe de 0 à 1, et elle est directement proportionelle à $\omega$, ce qui supprime le paradoxe.

**6. Les oscillations forcées. La résonance.** Étudions maintenant l'effet d'un mouvement du sol sinusoïdal amorti:

$$
\begin{aligned}
x &= 0 && \text{pour} \quad t < 0 \\
x &= P e^{-kt} \sin nt && \text{pour} \quad t \geq 0.
\end{aligned}
$$

Il sera commode de considérer le mouvement complexe

$$X = P e^{\lambda t} \quad \text{où} \quad \lambda = -k + jn.$$

(5.5) donne, pour $\lambda \neq \lambda_1 \neq \lambda_2$,

$$-\frac{1}{VP} Q = \frac{\lambda^2 e^{\lambda t}}{(\lambda - \lambda_1)(\lambda - \lambda_2)} + \frac{\lambda_1^2 e^{\lambda_1'}}{(\lambda_1 - \lambda_2)(\lambda_1 - \lambda)} + \frac{\lambda_2^2 e^{\lambda_2 t}}{(\lambda_2 - \lambda)(\lambda_2 - \lambda_1)}. \qquad (6.1)$$

La réponse comprend un premier terme dont la pulsation et l'amortissement sont ceux du mouvement du sol (c'est l'oscillation forcée), puis deux termes dont la pulsation et l'amortissement sont ceux du pendule et qui constituent l'oscillation propre. Amplitude et phase dépendent de $\lambda$. En revenant au mouvement réel, *ceci implique que l'oscillation forcée reproduit parfaitement à une certaine échelle ou avec un certain retard un mouvement sinusoïdal amorti du sol, mais déforme les mouvements plus généraux*, par exemple le mouvement obtenu en superposant plusieurs mouvements sinusoïdaux amortis ou non.

Si $\lambda$ tend vers une des racines de l'équation caractéristique, $\lambda_1$ par exemple, l'amplitude de l'oscillation forcée devient infiniment grande. C'est ce qu'on appelle la « *résonance* ». Mais le premier terme d'oscillation propre devient également infini. Il est facile de voir que la somme de ce terme et de l'oscillation forcée reste finie, et que $Q$ a une limite donnée par:

$$-\frac{1}{VP} Q = \frac{\lambda_1}{\lambda_1 - \lambda_2} \left( \frac{\lambda_1 - 2\lambda_2}{\lambda_1 - \lambda_2} + \lambda_1 t \right) e^{\lambda_1 t} + \frac{\lambda_2^2}{(\lambda_1 - \lambda_2)^2} e^{\lambda_2 t}.$$

Le cas est analogue à celui d'une racine double de l'équation caractéristique. Dans le cas où l'amortissement est nul ($\alpha = k = 0$, $\lambda_1$ imaginaire pur), le terme en $t$ exp. $(\lambda_1 t)$ conduirait à vrai dire à des amplitudes infinies, mais au bout d'un temps infini et, comme on le verrait, en dépensant une énergie infinie (si nos équations, qui ont été établies en supposant les amplitudes infiniment petites, conservent leur validité). Autrement dit la résonance n'offre aucun des paradoxes qu'on peut y voir lorsqu'on se contente d'étudier un état de régime.

Bornons-nous désormais au cas *sinusoïdal* $k = 0$. L'oscillation forcée $Q_1$, non amortie, finit alors par prédominer sur les oscillations propres amorties. On a:

$$-\frac{1}{VP} Q_1 = \frac{n^2}{n^2 - \omega^2 - 2\alpha\omega j n} e^{j n t}.$$

On en déduit aisément, pour l'oscillation réelle $q_1 = a \sin(nt - \varphi)$ l'amplitude $a$ et la phase $\varphi$:

$$a = \varepsilon \frac{n^2 VP}{N}, \qquad \frac{\cos\varphi}{n^2 - \omega^2} = \frac{\sin\varphi}{2\alpha\omega n} = -\frac{1}{\varepsilon N}.$$

On a posé:

$$N = \sqrt{(n^2 - \omega^2)^2 + 4\alpha^2\omega^2 n^2}, \qquad \varepsilon = \pm 1.$$

Le double signe correspond à l'indétermination de $\pi$ dans la définition de la phase. Si nous convenons que $V > 0$ et que $a$ et $P$ sont de même signe, nous devons prendre $\varepsilon = +1$, $0 < \varphi < \pi$. C'est une pure convention, dont on ne peut conclure à un retard de phase.

Appelons *amplification* (on dit parfois *grandissement dynamique*) le rapport $W = a/P$. Pour $\alpha$ et $\omega$ donnés l'amplification d'une période quelconque est proportionnelle au grandissement $V$ et l'on a:

$$\frac{W}{V} = \frac{n^2}{N}.$$

Si $n/\omega$ est petit,

$$N \sim \omega^2, \qquad W \sim \frac{n^2}{\omega^2} V,$$

si $n/\omega$ est voisin de 1,

$$N \sim 2\alpha\omega n, \qquad W \sim \frac{n}{2\alpha\omega} V,$$

si $n/\omega$ est grand,

$$N \sim n^2, \qquad W \sim V;$$

$q_1$ est donc proportionnel

à *l'accélération* $\ddot{x}$ si la période du mouvement est grande par rapport à la période propre,

à *la vitesse* $\dot{x}$ si les périodes sont voisines,

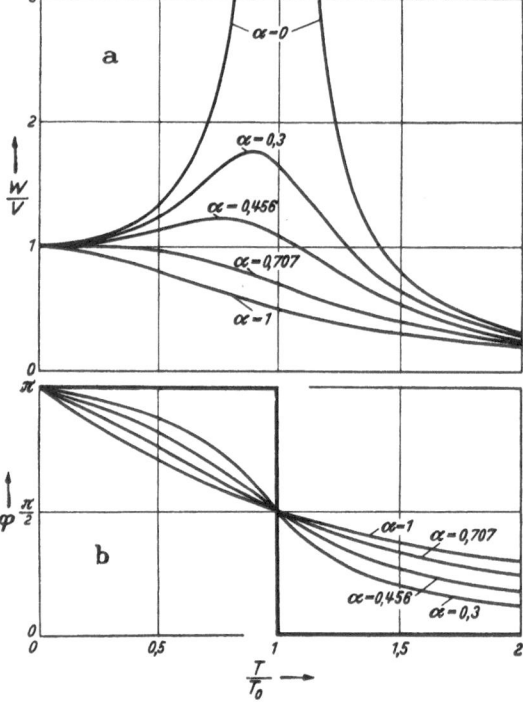

Fig. 2. Amplification et déphasage de l'oscillation forcée dans un séismographe à inscription directe. $T$ période du mouvement du sol, $T_0$ période du pendule.

au *déplacement* $x$ si la période du mouvement est courte par rapport à la période propre.

Ces résultats s'étendent par superposition à des mouvements plus complexes. La fig. 2 représente $W/V$ et $\varphi$ en fonction de $\omega/n = T/T_0$ pour les valeurs 0; 0,3; 0,456; $1/\sqrt{2}$; 1 de $\alpha$.

Si $\alpha$ est nul ($R=1$) on a résonance pour $\omega/n=1$; l'amplitude devient infinie, la phase saute de $\pi$ à 0. Pour $\alpha$ un peu plus grand l'amplification passe par un maximum pour $\omega/n=\sqrt{1-2\alpha^2}$. Pour $\alpha=0,46$ ($R=5$) elle reste la même à 25% près de $T=0$ à $T=1,3\,T_0$. Ce rapport d'amortissement 5, préconisé par WIE-CHERT, est généralement adopté pour les séismographes mécaniques. Pour $\alpha=\frac{1}{2}$, $W/V=1$ lorsque $T=T_0$. Pour $\alpha=1/\sqrt{2}$ ($R=23,1$) la courbe $W/V$ a, au départ, un contact du 4ème ordre et non plus seulement du second avec sa tangente horizontale; l'amplification décroît de 10% seulement entre $T=0$ et $T=0,7\,T_0$. Un amortissement de cet ordre est utilisé dans le séismographe MILNE-SHAW. On a alors $N^2=n^4+\omega^4$. Mais l'amortissement critique $\alpha=1$ ($R=\infty$) conduit à des équations plus simples encore, puisque $N=n^2+\omega^2$, $\tan\dfrac{\varphi}{2}=\dfrac{\omega}{n}$.

A l'oscillation forcée $q_1$ il faut maintenant ajouter l'oscillation propre $q_2$ en déterminant ses constantes de façon que $q=q_1+q_2$ satisfasse aux conditions

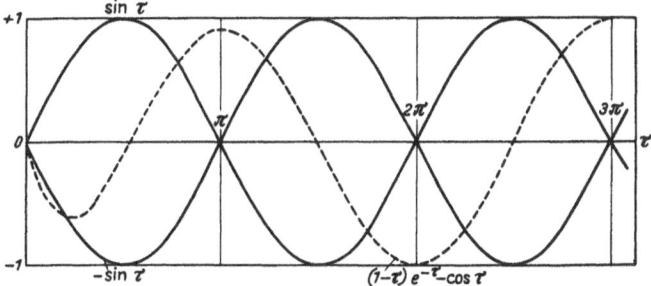

Fig. 3. Effet d'un mouvement sinusoïdal $x$ du sol, débutant brusquement, sur un pendule de même période à l'amortissement critique. La figure est faite avec le grandissement 2, en portant $x$ vers le haut (retard apparent du pendule) et vers le bas (avance apparente).

initiales $q_0=0$, $\dot{q}_0=-V\,n\,P$. La fig. 3 montre le résultat obtenu pour $V=\pm1$, $\alpha=1$, $n=\omega$ par exemple. Suivant que l'on oriente dans le même sens ($V=-1$) ou dans le sens contraire ($V=+1$) les axes $x$ et $q$, l'enregistrement paraît présenter un retard ou une avance de phase par rapport au sol.

## B. Effets non linéaires.

**7. Frottement dans l'inscription mécanique.** Pour diminuer les frottements, les séismographes à inscription mécanique, qui ne font appel à aucune énergie auxiliaire, utilisent un style (c'est à dire une tige légère à l'extrémité de laquelle est fixée perpendiculairement une pointe fine) inscrivant sur papier enfumé. L'action du papier sur le style est complexe: la pointe frotte sur une surface glacée en écartant le noir de fumée qui y est déposé et dont l'épaisseur est loin d'être uniforme. Admettons cependant que cette action se réduise à une résistance constante $F$ suivant la direction opposée à la vitesse relative de la pointe par rapport au papier et que le style soit assez long pour qu'on puisse considérer le déplacement $q(t)$ de l'inscripteur comme rectiligne. Si $W$ est la vitesse de déroulement du papier, la composante de $F$ suivant l'axe des $q$ est:

$$F_q=-F\,\frac{\dot{q}}{\sqrt{\dot{q}^2+W^2}}\,. \qquad (7.1)$$

Supposons $\dot{q}$ assez grand pour qu'on puisse négliger $W$:

$$F_q=-F \quad \text{si} \quad \dot{q}>0,$$
$$F_q=+F \quad \text{si} \quad \dot{q}<0.$$

(Le cas échéant, on peut inclure dans $F_q$ les frottements des pivots mais ils sont en général faibles devant celui du style.) (2.5) et (3.1) deviennent:

$$\ddot{q} + 2\alpha\omega\dot{q} + \omega^2 q = -V\ddot{x} \pm \frac{F}{\mathfrak{M}}$$

ou

$$\ddot{\varrho} + 2\alpha\omega\dot{\varrho} + \omega^2\varrho = -V\ddot{x}$$

en posant

$$r = \frac{F}{\mathfrak{M}\omega^2} \quad \text{et} \quad \varrho = q \mp r.$$

L'enregistrement est décalé de $r$ dans le sens opposé au déplacement.

En particulier, pour les oscillations propres, la durée d'une demi oscillation n'est pas changée, donc on peut toujours déterminer la pseudo-période $T$ en mesurant sur un enregistrement l'intervalle de deux extremums. Cherchons le rapport d'amortissement $R$. De $q = -q_1 < 0$ à $q = +q_2 > 0$ l'oscillation se fait autour de l'axe $q = -r$ donc:

$$R = \frac{q_1 - r}{q_2 + r}$$

et de même:

$$R = \frac{q_2 - r}{q_3 + r} = \frac{q_1 + q_2 - 2r}{q_2 + q_3 + 2r} = \frac{q_2 + q_3 - 2r}{q_3 + q_4 + 2r} = \frac{(q_1 + q_2) - (q_2 + q_3)}{(q_2 + q_3) - (q_3 + q_4)}.$$

Des trois amplitudes $q_1 + q_2$, $q_2 + q_3$, $q_3 + q_4$ on déduit donc $R$ et $r$. En fait le résultat dépend un peu de l'amplitude moyenne utilisée. Néanmoins $R$ (ou $\alpha$), $r$ (ou $\omega^2 r$), $V$ et $T_0$ sont des constantes d'étalonnage couramment utilisées dans les séismographes mécaniques (la période $T_0$ se déduit des valeurs de $T$ et $R$ correspondant à un amortissement réduit, ainsi que nous l'avons expliqué dans le cas sans frottement, Sect. 5).

Dans la théorie ainsi simplifiée, le point inscripteur a une plage d'équilibre $-r < q < r$ (un peu plus étendue si l'on tient compte de la différence du frottement au repos et du frottement pendant le mouvement). En réalité, au voisinage des maximums et des minimums de $q$ et surtout à la fin du séisme, $\dot{q}$ devient faible devant $W$. On déduit alors de (7.1):

$$F_q = -F\frac{\dot{q}}{W}.$$

Le frottement est proportionnel à la vitesse; il joue le rôle d'un amortissement et ne peut s'opposer au retour à zéro.

L'étude complète de la formule (7.1) n'a pas été abordée. Reid[1] suppose le mouvement sinusoïdal:

$$q = Q\cos nt.$$

Dans le développement de $F_q$ en série de Fourier:

$$F_q = +F\frac{\sin nt}{\sqrt{\sin^2 nt + W^2/Q^2 n^2}} = +F(A_1\sin nt + A_3\sin 3nt + \cdots)$$

bornons-nous au premier terme (Reid a montré que cela n'entraînait pratiquement pas d'erreur supérieure à 3%).

$$F_q = +FA_1\sin nt = -\frac{F}{Q\omega}A_1\dot{q}.$$

---

[1] H. F. Reid: Bull. Seism. Soc. Amer. **25**, 222 (1925).

On en déduit pour le supplément d'amortissement dû au frottement:

$$\alpha_q = \frac{1}{2\mathfrak{M}\omega} \frac{F}{Qn} A_1 = \frac{r}{2Q} \frac{\omega}{n} A_1 = \frac{r}{2WT_0} B_1$$

en posant

$$B_1 = \frac{WT}{Q} A_1.$$

$\dfrac{r}{2WT_0}$ est une constante de l'instrument. On aura $\alpha_q$ si l'on sait calculer $B_1$. Or $A_1$ et par conséquent $B_1$ sont des fonctions de $Qn/W$ ou de $Q/WT$ que l'on peut mesurer sur l'enregistrement: $WT$ est la longueur d'onde des oscillations, $Q/WT$ leur hauteur relative. Ayant $Q/WT$, on tire $B_1$ de la table suivante:

| $Q/WT$ | 2,00 | 1,00 | 0,50 | 0,29 | 0,20 | 0,14 | 0,10 | 0,07 | 0,05 | 0,03 | 0,01 | 0,00 |
|---|---|---|---|---|---|---|---|---|---|---|---|---|
| $B_1$ | 0,625 | 1,23 | 2,29 | 3,50 | 4,32 | 4,99 | 5,53 | 5,88 | 6,06 | 6,2 | 6,2 | 6,3 |

L'amortissement $\alpha_q$ s'ajoute à l'amortissement principal $\alpha$, qu'on peut déterminer en déséquilibrant légèrement le style de façon qu'il n'appuie plus sur le papier. $\alpha$ est en général de l'ordre de 0,45; deux exemples numériques montreront que $\alpha_q$ est parfois notable:

Pour $r = 0,4$ mm., $W = 1$ mm./s., $T_0 = 2$ s. (pendule à grande masse), $Q = 5$mm, $T = 5$ s. (ondes transversales), on trouve $\alpha_q = 0,12$.

Pour $r = 0,4$ mm., $W = 0,5$ mm./s., $T_0 = 8$ s. (pendule ordinaire pour les séismes lointains), $Q = 5$ mm., $T = 50$ s. (ondes longues), on trouve $\alpha_q = 0,22$.

Enfin le frottement peut créer un couple de rappel si, au cours du mouvement, l'angle que fait le style avec la vitesse de déroulement du papier devient notable (cela est d'ailleurs assez rare, à moins que le style n'ait été tordu). Négligeons de nouveau $\dot q$; le couple exercé par la force de frottement sur l'axe du style est $Fq$. Le frottement influerait ainsi sur la période.

Les conditions du frottement sont trop mal définies pour que la théorie de REID puisse inspirer confiance. Le mieux serait de rendre tous les frottements négligeables. Pour voir si c'est possible, bornons-nous avec SOHON [4] à la théorie simplifiée. Le frottement sera négligeable si le décalage $r = F/\mathfrak{M}\omega^2$ imposé à l'enregistrement, tombe au dessous de la limite appréciable, soit par exemple 0,025 cm. On ne peut guère faire descendre $F$ au dessous de 1 dyne sauf en recourant à des artifices spéciaux comme des vibrations du style. On devra donc avoir:

$$\mathfrak{M}\omega^2 > 40$$

soit environ

$$\mathfrak{M} > T_0^2.$$

Si on suppose égal à 1 le coefficient d'utilisation de la masse (3.3) la condition s'écrit $M > V^2 T_0^2$ grammes. La limite inférieure $0,001\ V^2 T_0^2$ pour la masse en kilogs est:

| pour | $T_0 =$ 0,1 s. | 1 s. | 10 s. |
|---|---|---|---|
| si $V =$ 100, | $M =$ 0,1 kg. | 10 kg. | 1000 kg. |
| si $V =$ 1000, | $M =$ 10 kg. | 1000 kg. | 100000 kg. |

Des périodes propres de l'ordre de 2 secondes peuvent suffire pour l'enregistrement des séismes rapprochés; il leur correspond à la rigueur des appareils raisonnables. Mais l'inscription mécanique correcte des séismes lointains exigerait qu'on suspendît des masses gigantesques, auxquelles on pourrait difficilement donner les périodes nécessaires. Pratiquement les appareils en service approchent seulement des conditions théoriques.

**8. Forces de rappel non linéaires.** Nous avons admis dans la Sect. 2 que le travail virtuel des forces de rappel était linéaire. Cela revient, si nous l'écrivons $C(q) \delta q$, à négliger les termes supérieurs dans le développement de la fonction $C(q)$:

$$C(q) = C\, q + C_2\, q^2 + C_3\, q^3 + \cdots.$$

Si l'on cherche à astatiser, c'est à dire à diminuer $\omega^2 = C/\mathfrak{M}$, le terme en $C_2 q^2$ peut devenir gênant. Les oscillations deviennent alors dissymétriques (le pendule boîte). A l'extrême il deviendrait instable et se calerait dans un sens ou dans l'autre suivant le signe de $C_2$.

En pratique, la dissymétrie des forces de rappel, et par conséquent des oscillations, est peu importante, sauf pour les séismographes verticaux antérieurs à l'adoption du ressort La Coste (Sect. 11). Par contre le terme $C_3 q^3$, déjà présent dans le pendule ordinaire, peut devenir sensible avant le terme en $C_2 q^2$, en sorte qu'il conviendrait de substituer à (2.1) une équation comportant simultanément ces deux termes.

La difficulté fondamentale des problème non linéaires, c'est qu'il est impossible d'obtenir leur solution générale par superposition. Dans ces conditions on se contente généralement d'étudier, soit les oscillations propres pour essayer d'en tirer $C(q)$ et même éventuellement la fonction analogue $f(\dot q)$ qui caractériserait un amortissement non linéaire[1], soit comme à la Sect. 7 l'oscillation forcée engendrée par une cause sinusoïdale. L'intérêt séismologique de ces derniers travaux n'est pas très grand; la plupart ont en vue la théorie des oscillateurs et supposent l'amortissement faible. Les complications qui interviennent alors et qui ont été étudiées par Duffing [8], [9] pour le terme en $q^3$ doivent disparaître quand l'amortissement devient suffisant; mais l'étude complète n'a pas été faite à notre connaissance. C'est d'autant plus regrettable que, des résultats relatifs aux oscillations sinusoïdales, on a parfois conclu que la présence de termes en $q^3$ favoriserait la raideur des impetus[2] et qu'on en a tiré des propositions de construction. Or les deux premiers termes de (2.1) interviennent seuls dans la détermination des conditions initiales et (4.4) reste valable. On ne voit donc pas qu'on puisse, indépendamment des amplitudes, améliorer le début des phases en introduisant un terme en $q^3$. L'idée mériterait cependant d'être approfondie: des forces de rappel mollissantes ($C_3 < 0$) sembleraient favoriser les grandes amplitudes relativement à l'agitation microséismique; mais on côtoierait l'instabilité dès que l'effet deviendrait intéressant. Au contraire des forces de rappel durcissantes ($C_3 > 0$) s'opposeraient aux grandes amplitudes; cela revient à favoriser les petites si on est maître de l'amplification, et peut-être à transformer par là emersio en impetus.

**9. Les accélérations en bout.** Un des termes négligés dans (2.1) demande encore à être examiné; c'est celui qui provient des accélérations en bout. L'équation d'un séismographe tournant autour d'un axe s'écrit compte tenu de ce terme:

$$\ddot\vartheta + 2\alpha\,\omega\,\dot\vartheta + \left(\omega^2 + \frac{\ddot x}{l}\right)\vartheta = -\frac{\ddot y}{l}. \tag{9.1}$$

---

[1] M. Weber: Geofisica Pura e Applicata **23**, 1 (1952). — Eidg. Techn. Hochschule Zürich, Mitt. aus dem Inst. für Geophysik Nr. 23.

[2] La dissymétrie rend la présence des termes en $q^2$ toujours fâcheuse.

Les seuls résultats qui aient été obtenus concernent le cas où $x$ et $y$ sont des fonctions sinusoïdales du temps, auxquelles on peut se contenter de supposer la même période, correspondant à une phase déterminée du séisme:

$$\frac{x}{l} = \xi \cos \Omega\, t, \tag{9.2}$$

$$\frac{y}{l} = A \cos \Omega\, t + B \sin \Omega\, t. \tag{9.3}$$

Admettons d'abord que $y$ soit constamment nul, ce qui ne correspond pas à une véritable oscillation propre puisqu'il y a «excitation paramétrique» ou «modulation de fréquence» par le terme $\ddot{x}/l$. En posant:

$$\vartheta = z\, e^{-\alpha \omega t} \quad \text{et} \quad \Omega\, t = \tau,$$

on obtient pour $z$ l'équation classique de MATHIEU [8], [9], [10]:

$$\frac{d^2 z}{d\tau^2} + \left(\mu^2 \frac{\omega^2}{\Omega^2} - \cos \tau\right) z = 0 \quad \text{où} \quad \mu^2 = 1 - \alpha^2. \tag{9.4}$$

Les valeurs de $\xi$ à considérer sont très petites, de l'ordre de 0,01 au grand maximum, au cours des séismes et même sur une table à secousses. On sait que l'équation de MATHIEU présente pour certaines valeurs de $\mu\frac{\omega}{\Omega}$ et $\xi$ des solutions instables comportant un facteur exp. $(K\tau)$ où l'exposant caractéristique $K$ a une partie réelle positive. Les domaines d'instabilité s'étendent en forme de coins à partir des points $\xi = 0$, $\mu\frac{\omega}{\Omega} = \frac{n}{2}$, au voisinage desquels la solution est encore peu différente d'une solution périodique (de période $2\pi$ en $\tau$ si $n$ est un entier pair, $4\pi$ si $n$ est un entier impair). Mais ce qui importe ce sont les instabilités en $\vartheta$ donc celles des instabilités en $z$ pour lesquelles $K$ dépasse $\alpha\omega/\Omega$. La question a été étudiée quantitativement par G. KOTOWSKI[1] et résumée par elle en un graphique (sa figure 6, reproduite partiellement dans [10] fig. 11 et dans [9] fig. 38). Un exposant d'amortissement $\alpha\omega/\Omega$ égal à 0,01 seulement suffit à déplacer les régions d'instabilité jusqu'à des valeurs de $\xi$ supérieures à 0,02 pour $n = 1$, à 0,28 pour $n = 2$, vraisemblablement beaucoup plus grandes pour les valeurs suivantes de $n$ pour lesquelles (9.4) est voisine d'une équation à coefficients constants[2]. Même si les amplitudes correspondantes étaient atteintes et si le pendule était faiblement amorti, à $\alpha = \frac{1}{5}$ par exemple, il faudrait qu'il soit encore sensible au cinquantième de la période propre pour que l'effet soit appréciable.

En disposant des pendules non amortis sur une table à secousses animée d'un mouvement sinusoïdal parallèle au bras, KANAI et SEZAWA[3] ont observé les instabilités correspondant à $n = 1$ et à $n = 2, 4, 6$ mais non pas celles qui correspondent à $n = 3, 5, 7\ldots$; ils ont reconnu que l'amortissement rétablissait la stabilité dans tous les cas pratiques.

Revenons à l'équation (9.1), G. KOTOWSKI étudie les effets d'un second membre périodique et notamment les résonances possibles avec les solutions périodiques de l'équation sans second membre. Des solutions de cette équation

---

[1] G. KOTOWSKI: Z. angew. Math. Mech. **23**, 213 (1943).

[2] L'équation de MATHIEU présente également un domaine d'instabilité pour $\xi$ petit et $\mu^2 \omega^2/\Omega^2$ voisin de zéro par valeurs négatives, qu'on pourrait paradoxalement vouloir assigner à un pendule suramorti. Mais l'exposant caractéristique est alors de la forme $\frac{\omega}{\Omega}\sqrt{\alpha^2 - 1} + O(\xi^2)$ donc supérieur à l'exposant d'amortissement.

[3] K. KANAI et K. SEZAWA: Bull. Earthq. Res. Inst. **18**, 483 (1940); **19**, 9, 177 (1940/41).

existent sur les frontières des régions d'instabilité (compte tenu de l'amortissement), mais seulement pour les régions correspondant à $n$ pair. Ceci permet de
penser que KANAI et SEZAWA ont vu apparaître les instabilités $n = 2, 4, 6$ par
suite d'un faible écart involontaire dans l'orientation de leurs pendules, entraînant
la présence d'un terme en $y$ proportionnel à leur terme principal en $x$. Quant à
l'instabilité en $n = 1$, relativement facile à faire apparaître (expérience de MELDE,
synchronisation sous-harmonique) elle a pu provenir dans leurs recherches d'une
excitation imparfaitement sinusoïdale.

La résonance véritable, étudiée par G. KOTOWSKI, se traduit par l'apparition d'un facteur dépendant linéairement du temps. Elle suppose que $\xi$ est
suffisamment grand pour atteindre la
frontière ce qui, nous l'avons vu, ne
peut se produire que très exceptionnellement. Sans aller jusque là on pourrait atteindre des points où la vibration soit bornée dans le temps mais
néanmoins considérable; ce sont les
points qui correspondraient à une résonance vraie si l'amortissement était
plus faible. Supposons $\xi$ assez petit
pour qu'on puisse en s'inspirant de
GRENET[1] ou de PENDSE[2] résoudre
l'équation (9.1) par approximations

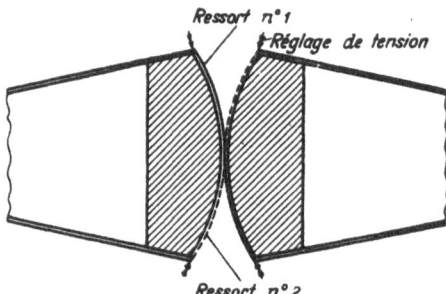

Fig. 4. Pendule duplex de MATUZAWA.

successives: on néglige d'abord le terme en $\vartheta \ddot{x}/l$, puis on remplace dans ce terme $\vartheta$
par l'oscillation forcée $\vartheta_1$ correspondant à l'équation modifiée, et ainsi de suite.
On obtient, outre des termes harmoniques, une déviation fixe par rapport au
temps dont l'expression se réduit pour $\omega/\Omega$ petit à:

$$\vartheta_0 = \frac{\xi}{2} \frac{\Omega^2}{\omega^2} \left( A + 2\alpha \frac{\omega}{\Omega} B \right).$$

On voit que cet effet de redressement est dû surtout à la composante $A \cos \Omega t$
de $y$ qui est en phase avec $x$. En supposant $B = 0$, le rapport de $\vartheta_0$ à l'amplitude
de $\vartheta_1$ est $(\xi/2) \cdot \Omega^2/\omega^2$; il faut avoir $\Omega/\omega = 20$ pour obtenir $\xi = 0,005$ par exemple.

Cherchant une période propre de l'ordre de la minute pour l'enregistrement
de forts tremblements de terre locaux, MATUZAWA a adopté une construction
«duplex» qui permet d'éviter les accélérations en bout: deux masses tournantes
sont reliées par des lames ressorts qui fournissent le couple de rappel (fig. 4).
Les accélérations principales s'ajoutent, mais les accélérations en bout se dé-
truisent.

## C. Réalisation des Séismographes pendulaires.

**10. Séismographes horizontaux.** Les pendules ordinaires ne sont plus employés
comme séismographes horizontaux, sauf s'ils sont munis de dispositifs d'astatisation analogues à ceux que nous décrirons pour les séismographes verticaux,
ou encore comme parties mécaniques d'un séismographe électromagnétique, la
période du galvanomètre intervenant dans la détermination de la période propre
(Sect. 13). BENIOFF emploie pour son séismographe électromagnétique horizontal
(Sect. 16) un pendule à translation (fig. 5) constitué par une masse cylindrique
suspendue, guidée, et rappelée par des rubans d'acier. Mais la théorie générale des

[1] G. GRENET: Rev. sci. **80**, 89 (1942).
[2] C. G. PENDSE: Phil. Mag. **35**, 706 (1944); **39**, 985 (1948).

séismographes a montré la nécessité pour obtenir de grandes périodes de diminuer les forces de rappel; dans la grande majorité des séismographes horizontaux en service on tente de supprimer le couple de rappel du pendule ordinaire en rendant

l'axe de rotation quasi vertical; chaque point de la masse se déplace alors dans un plan horizontal, d'où le nom de *pendule horizontal*. En pratique il faut maintenir un faible couple de rappel; c'est la résultante des couples produit par les articulations et par le petit angle résiduel entre l'axe de rotation et la verticale.

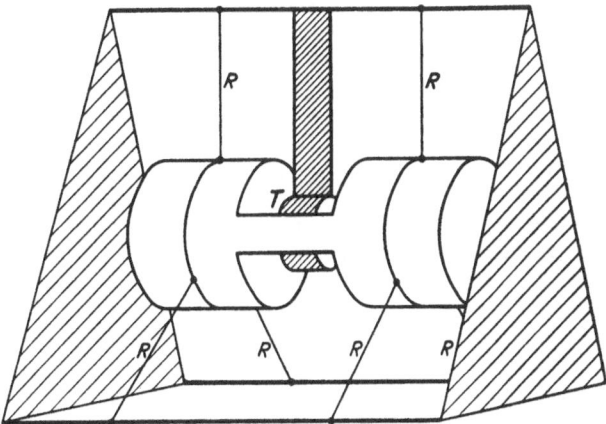

Notons immédiatement que les pendules horizontaux sont très sensibles aux inclinaisons de leur bâti: une petite

Fig. 5. Schéma de pendule à translation: BENIOFF, masse 100 kg., période 0,5 ou 1 s., séismographe électromagnétique. *T* traducteur électromagnétique; *R ... R* rubans d'acier tendus.

rotation autour d'un axe horizontal perpendiculaire à la position d'équilibre $OG_0$ du «bras» $OG$ (fig. 6) change seulement leur période. Mais une rotation $\Omega$ autour d'un axe parallèle à $OG_0$ (roulis, «*tilt*» des anglais) déplace la position d'équilibre d'un angle $\vartheta$ que les formules de la Sect. 2 permettent d'obtenir immédiatement:

On a en effet:

$$- Cq - M(\boldsymbol{\Omega} \times \boldsymbol{g_0}) \cdot \boldsymbol{L_0'} = 0;$$

ou en prenant $\vartheta$ pour variable au lieu de $q$:
$$C\vartheta = - MLg\Omega.$$

Si on introduit la longueur synchrone $\Lambda = \mathfrak{M}g/C = Kg/C$ et la longueur réduite $l = K/ML$, on a simplement:

$$\vartheta = - \Omega \frac{\Lambda}{l}. \qquad (10.1)$$

D'autre part, à un déplacement rapide $x$ correspond un angle:

$$\vartheta = - \frac{x}{l}.$$

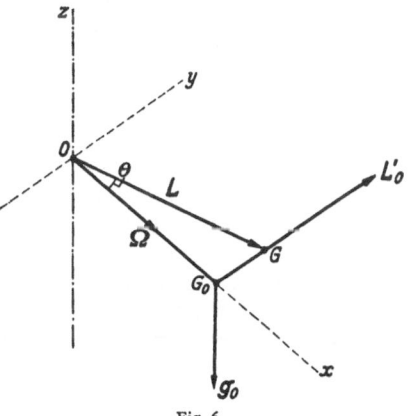

Fig. 6.

Si la longueur synchrone est $\Lambda = 25$ m. (période de l'ordre de 10 secondes), on a la même déviation pour $\Omega = 0''2 = 10^{-6}$ radian qui est une valeur malheureusement assez courante des inclinaisons du pilier produites par l'échauffement diurne du sol et des bâtiments et pour $x = 25\ \mu$ qui correspond à un séisme notable.

Pour pallier à cette dérive du zéro, on peut, au moins pour les appareils à grande masse, insérer dans la transmission un joint spécial imaginé par ROMBERG: une coupe remplie d'huile épaisse où trempe une palette reliée au miroir ou au style inscripteur (les deux parties de l'appareil comportent des couples de rappel, mais la première seule est sensible aux mouvements du sol). L'huile transmet les variations rapides comme si elle était solide; elle cède au contraire aux déplacements lents.

Pour construire les pendules horizontaux, et de façon générale dans toute espèce de séismographe, trois sortes de liaisons sont employées parce qu'elles donnent peu de frottements sans être trop fragiles: les pointes d'acier en forme de cône très ouvert (90° par exemple) pour éviter qu'elles ne s'émoussent, appuyant sur une cupule en acier ou en pierre dure; les fils métalliques; enfin, depuis WIECHERT, les lamelles d'acier comme celles qui supportent les balanciers d'horloge, peu épaisses (0,01 à 0,1 mm) et qu'on fait travailler à la traction pour ne pas les fausser. Les fig. 7 représentent schématiquement quelques dispositions en service. On ne construit plus guère d'appareils nouveaux utilisant les deux premières — a) fil en haut, pointe en bas — et — b) fil en haut, lamelle en bas —. La troisième — c) fils en haut et en bas —, connue sous le nom de suspension ZÖLLNER, possède un degré de liberté supplémentaire (balancement dans le plan de symétrie) gênant pour l'enregistrement des courtes périodes. Mais cette suspension, qui permet d'obtenir aisément des frottements très faibles, est bien adaptée à la construction de clinomètres pour l'étude des faibles déformations du sol (marées terrestres). La période cherchée est alors grande, le couple de torsion des fils n'est plus négligeable; or il varie avec la température. Pour éviter tous les effets thermiques, ISHIMOTO a construit ses clinomètres entièrement en silice: bâti, fils, et masse mobile.

Si les deux points d'attache sur le bras se rapprochent, la suspension ZÖLLNER a pour limite la suspension unifilaire du séismographe ANDERSON-WOOD, dans lequel une masse inférieure au gramme est portée excentriquement par un fil vertical. En général le fil est tendu et le couple de rappel est entièrement dû à la torsion. Les balancements de la suspension ZÖLLNER sont remplacés par les vibrations transversales du fil, qu'on amortit en lui faisant traverser des gouttes d'huile épaisse placées aux ventres de vibration. L'orientation de la petite masse est un peu incertaine, la période (0,8 seconde ou 6 secondes à Pasadena) limitée par les instabilités du zéro.

La dernière combinaison — d) lamelles en haut et en bas — est utilisée par la plupart des constructeurs actuels. On peut, comme l'a fait WENNER, remplacer une lamelle par un couple de lamelles en croix pour mieux définir l'axe de rotation, ou se contenter de lamelles minces dont la longueur libre soit très faible.

La détermination de la longueur réduite est une partie essentielle de l'étalonnage d'un séismographe. On peut la calculer si la distribution géométrique des masses est assez simple, ce qui est rare. Une méthode très simple consiste à rendre l'axe de rotation horizontal, obtenant ainsi un pendule ordinaire; la longueur réduite est égale à sa longueur synchrone. Cette méthode est applicable au séismographe WILIP, dont la lamelle inférieure est assez courte (une fraction de millimètre) pour pouvoir travailler à la compression.

Dans le séismographe MILNE SHAW on utilise la formule (10.1) en donnant au pendule un roulis $\Omega$ extrêmement petit au moyen d'une vis micrométrique actionnée à distance, dont on lit la rotation sur une échelle graduée. Il est bon de réduire auparavant la période propre.

On peut aussi partir de $l = K/MC$, déterminer la masse totale, la place du centre de gravité (par équilibre sur un couteau), enfin le moment d'inertie en appliquant au pendule un couple $\mathfrak{L}$ connu. Car si $\vartheta_0$ est la déviation correspondante on a $\mathfrak{L} = C \vartheta_0$ donc $C$ puis $K = C/\omega^2$. Dans le séismographe MAINKA (fig. 7b) le couple est produit par un poids connu $mg$ qui est suspendu à un fil et qui, au moyen d'une poulie de renvoi, tire horizontalement sur le centre de gravité; la mesure de $L$ n'a plus à être faite, car $\mathfrak{L} = mgL$, $l = mg/M\omega^2\vartheta_0$.

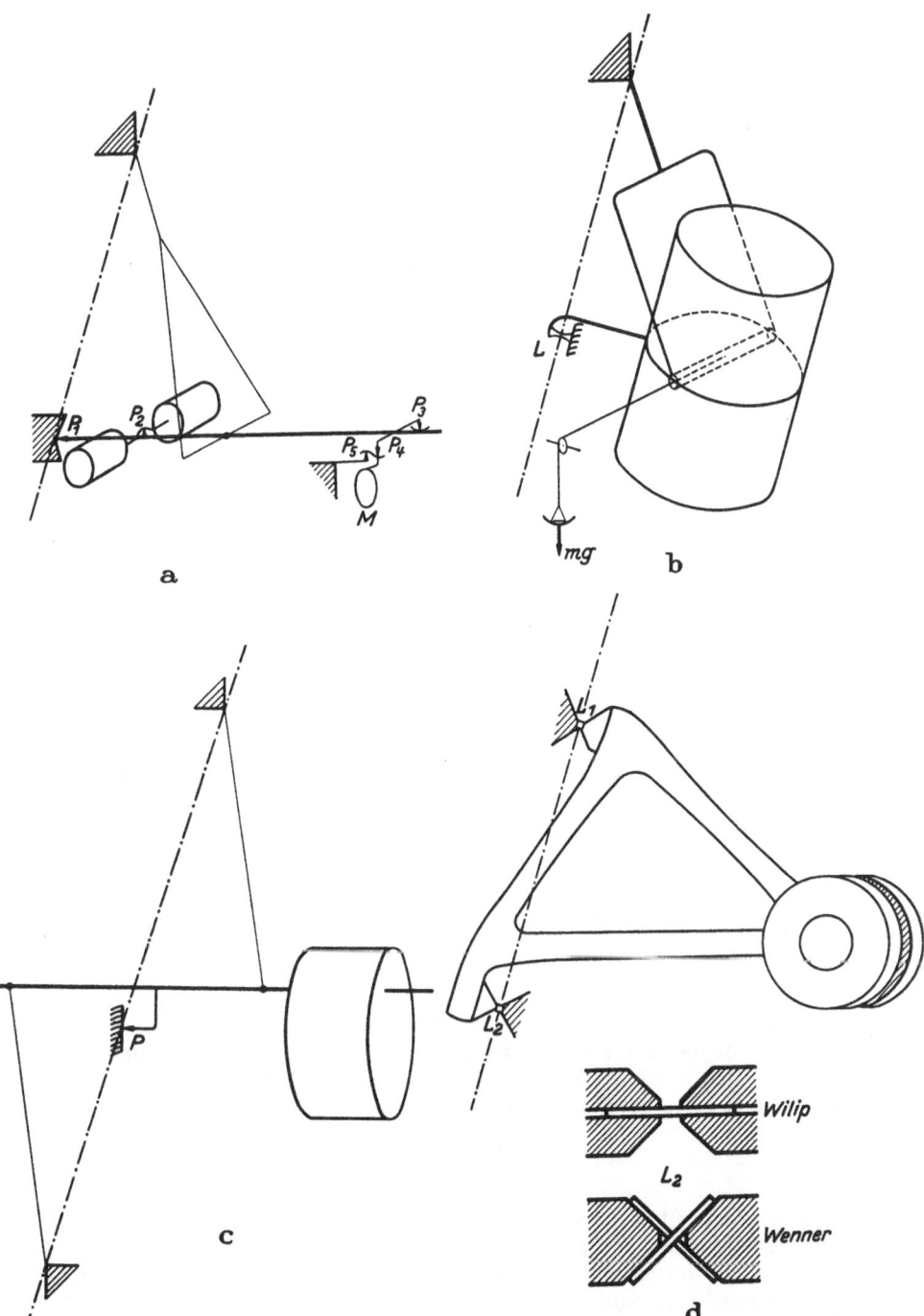

Fig. 7a—d. Schéma de pendules horizontaux: a MILNE-SHAW, masse 450 gr., période 10 ou 12 s., enregistrement optique $P_1 \ldots P_5$ pointes, $M$ miroir. $P_1$ est doublée d'une pointe plus forte qui sert pendant les réglages. La masse tourne librement autour de $P_2$. b MAINKA, Masse 450 kg., période 8 s., enregistrement mécanique. Le ressort $L$ est évidé pour mieux définir l'axe. Le poids mg sert à l'étalonnage. c GALITZINE, Masse 7 kg., période 24 s., séismographe électromagnétique. La pointe $P$ a pour but d'éviter les balancements de la suspension, mais son réglage est délicat. d WILIP, Masse 7 kg., WENNER, Masse 500 g.; périodes 12 s. séismographes électromagnétiques.

Enfin le procédé suivant est applicable lorsqu'une partie importante du couple de rappel est due au défaut de verticalité de l'axe de rotation. On a alors (Sect. 2)

$$C = C_0 - M\boldsymbol{g}_0 \cdot \boldsymbol{L}_0'' = C_0 + M g L \sin i$$

où $C_0$ est le couple de rappel dû à la suspension[1], $i$ l'angle de l'axe de rotation avec la verticale (fig. 8). Si $C_0$ n'est pas trop grand, le second terme est déjà

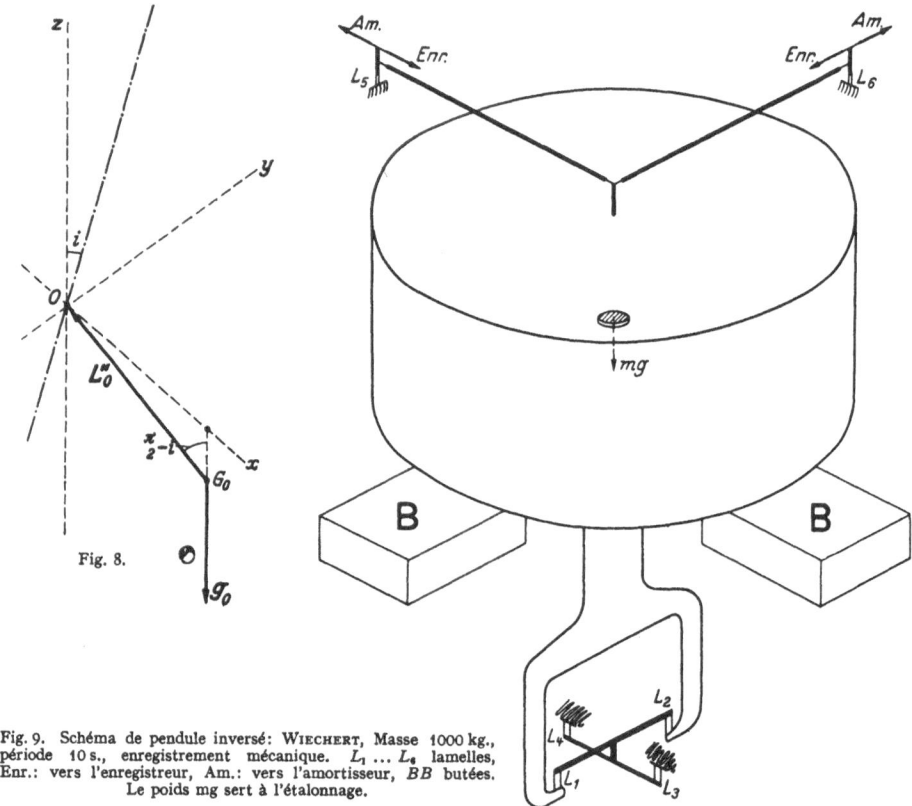

Fig. 8.

Fig. 9. Schéma de pendule inversé: Wiechert, Masse 1000 kg., période 10 s., enregistrement mécanique. $L_1 \dots L_6$ lamelles, Enr.: vers l'enregistreur, Am.: vers l'amortisseur, $BB$ butées. Le poids $mg$ sert à l'étalonnage.

sensible même si l'inclinaison est assez faible pour qu'on puisse confondre $\sin i$ avec $i$. On peut mesurer avec précision ses variations $\varDelta i$ à partir d'une valeur inconnue en utilisant un miroir lié au bâti, et déterminer en même temps la pulsation propre $\omega$ correspondant à chaque inclinaison. On a:

$$\varDelta i = \frac{1}{M g L} \varDelta C = \frac{l}{g} \varDelta \omega^2.$$

En dehors du pendule horizontal on n'emploie guère pour faire un séismographe horizontal que le principe du *pendule inversé*, dans lequel, au repos, le centre de gravité de la masse tournante est dans le plan vertical de l'axe de rotation, au dessus de cet axe. La pesanteur tend à écarter le pendule de sa position d'équilibre, vers laquelle il est rappelé par des ressorts. Ce type d'appareil se prête à l'adoption

---

[1] On suppose implicitement que la position d'équilibre $\vartheta = 0$ correspond à un couple de pesanteur et à un couple de suspension tous deux nuls. S'il y avait simplement compensation des deux couples, l'appareil serait sensible à la composante verticale du mouvement.

d'une masse d'inertie commune aux deux composantes horizontales, posée sur un joint de cardan à lamelles. C'est le cas dans les séismographes WIECHERT, en service depuis 1904 (fig. 9). La théorie de la Sect. 2 s'étend immédiatement à des séismographes de ce type, pourvu qu'il n'existe aucune interaction entre les deux degrés de liberté. Comme on s'efforce de donner aux deux compo-santes même période et même amortissement pour rendre leurs indications comparables, chacune se trouve dans les conditions de la résonance pour agir sur l'autre. Mais l'amortissement élevé atténue les effets de couplage.

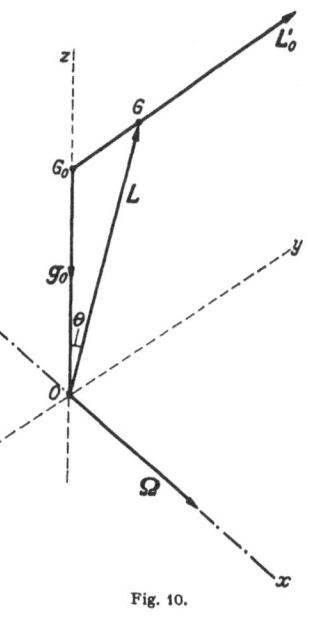

Fig. 10.

Comme les pendules horizontaux, les pendules inversés sont très sensibles aux inclinaisons de leur bâti. Malgré la différence de disposition (fig. 10), (10.1) et les conclusions que nous en avons tirées restent valables. C'est ainsi que le pendule inversé en quartz supporté par une lame élastique d'élinvar qui constitue les gravimètres HOLWECK-LEJAY a pu être employé au nivelle-ment des instruments astronomiques[1].

La détermination de la longueur réduite d'un pendule inversé pourrait se faire, au moins théori-quement, par le calcul, par transformation en pendule ordinaire, par inclinaison connue, ou par observation de la période à différentes inclinaisons. Dans le WIECHERT, on dépose un faible poids au bout de la masse, symétriquement par rapport aux deux directions d'enregistrement (fig. 9) et on enregistre les deux déviations correspondantes.

En accouplant par une tige horizontale deux séismographes inversés on peut obtenir un séismographe inversé à translation, qui peut être commode pour certains dispositifs d'amplification[2].

**11. Séismographes verticaux.** Dans un séismographe vertical la masse $M$ est suspendue à des ressorts, qui jouent deux rôles distincts: compenser le poids de $M$ à l'équilibre et contribuer aux forces de rappel vers cet équilibre. Si on prend des ressorts faibles pour diminuer l'effet de rappel, ils s'allongent lentement sous la charge, et même si la vitesse de déformation décroît avec le temps, on est conduit à utiliser le séismographe sans attendre la stabilisation complète qui prendrait des mois.

Dans le cas le plus simple la masse $M$ est suspendue à un ressort à boudin. Cette disposition n'est plus guère employée que comme partie mécanique d'un séismographe électromagnétique tel que celui de BENIOFF (fig. 11). Soit $C$ la «constante du ressort», rapport de la tension à l'allongement au voisinage de l'équilibre. Si le ressort est fait d'un fil de longueur $L$, de diamètre $d$, de rigidité $\mu$, et si $R$ est le rayon moyen des spires, la théorie de l'élasticité permet d'écrire $C = \frac{\pi \mu d^4}{32 L R^2}$. La longueur synchrone est $\Lambda = Mg/C$. Ce serait l'allongement dû au poids $Mg$ si les lois de l'élasticité s'appliquaient pour une variation de

---

[1] P. M. BURGAUD: Le pendule élastique inversé. Son application au nivellement des instruments astronomiques. Observatoire de Zi-Ka-Wei. Shanghai 1939.

[2] L. NÉEL: J. de Phys. sér. VII, **4**, 118 (1933).

charge aussi importante. L'effort tangentiel maximum subi par les éléments superficiels du ressort serait alors: $T = \dfrac{16\,M g\,R}{\pi\,d^3}$.

On peut mesurer, même en dehors du domaine élastique, la fatigue du ressort par cette quantité $\dfrac{16\,M g\,R}{\pi\,d^3}$ pour laquelle on s'imposera une valeur $T$ supportable par le métal employé. Soit alors $\varrho$ la densité du ressort, $m = \pi d^2 L \varrho$ sa masse: on peut écrire avec GRENET:

$$\varLambda = \frac{T^2}{2\,\mu\,\varrho\,g}\,\frac{m}{M}.$$

Ainsi, pour un métal donné, la longueur synchrone dépend du rapport de la masse du ressort à la masse

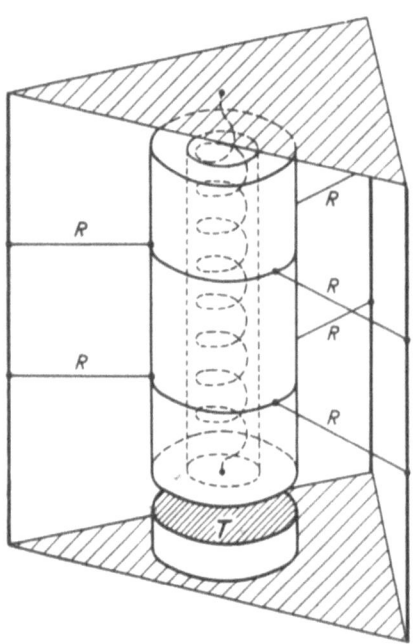

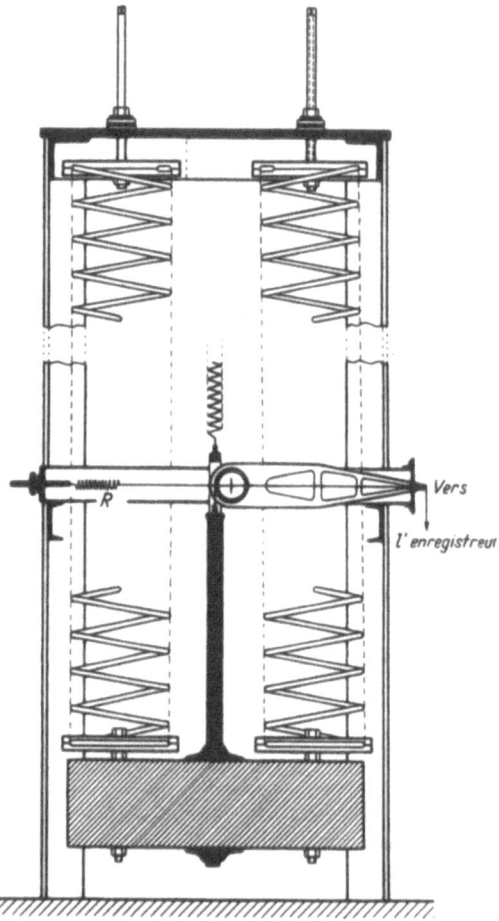

Fig. 11. Schéma de séismographe vertical simple: BENIOFF, Masse 100 kg., période 0,5 ou 1 s., séismographe électromagnétique. $T$: Traducteur électromagnétique; $R \ldots R$ rubans d'acier guidant la masse; celle-ci est creuse pour loger le ressort et diminuer l'encombrement.

Fig. 12. Astatisation d'un séismographe vertical: KREIS-WANNER, Masse 1000 kg., enregistrement mécanique. La période passe de 3 s. sans astatisation à 25 s. grâce au dispositif figuré; elle est réduite à 9 s. par les leviers suivants.

suspendue. En dehors des questions d'encombrement, ce rapport doit rester faible si l'on veut éviter que les oscillations propres du ressort n'interviennent. L'appareil ne se prête pas à l'augmentation des périodes.

Pour astatiser un séismographe vertical, on oppose aux forces de rappel dues aux ressorts de suspension des forces qui soient sans effet dans la position d'équilibre, mais qui accentuent les déviations; le plus simple est de les faire agir «en bout» sur un levier basculant, comme la pesanteur dans le pendule inversé.

Dans le séismographe KREIS-WANNER[1] par example, un fil attaché à un ressort $R$ (fig. 12) indépendant des 4 ressorts de suspension tire sur l'extrémité du levier d'astatisation. Dans les séismographes de faible masse c'est le ressort de suspension lui-même qui crée le couple d'astatisation: Le pendule, symétrique par rapport

au plan $xoz$, attaché à un ressort à boudin $PQ$, tourne autour de l'axe horizontal $y'y$ (fig. 13) défini par deux paires (horizontale et verticale) de lamelles croisées. À l'équilibre le centre de gravité $G$ de la masse doit se trouver en $G_0$ dans le plan horizontal passant par $y'y$, sans quoi les accélérations horizontales prendraient de l'influence. Soit $T$ la tension du ressort, $h$ son bras de levier, $\vartheta$ l'angle d'écart. Le couple de rappel est $C(\vartheta)=MgL\cos\vartheta-Th$. Soit encore $\overline{OP}=p$, $\overline{OQ}=q$, $\overline{PQ}=\zeta, \widehat{G_0OP}=\alpha, \widehat{QOG}=\beta$. $T$ est une fonction de $\zeta$, linéaire si l'élasticité est parfaite. $\zeta$ est lié à $\vartheta$ par l'équation: $\zeta^2 = p^2 + q^2 - 2pq\cos(\alpha+\beta-\vartheta)$. Enfin $h$ est lié à $\zeta$ et $\vartheta$ par

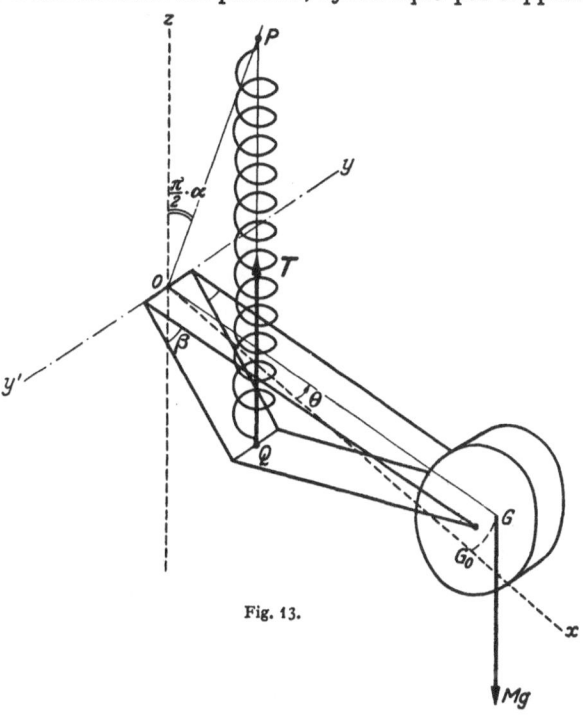

Fig. 13.

$h\zeta = pq\sin(\alpha+\beta-\vartheta)$. On peut donc calculer $C(\vartheta)$ et le développer en série de puissances. On constate que si l'on veut de grandes périodes, le terme en $\vartheta^2$ est en général important. On peut corriger plus ou moins cette dissymétrie

en utilisant plusieurs ressorts, mais LA COSTE a découvert en 1935 la méthode simple suivante[2]: $C(\vartheta)$ est identiquement nul si on a simultan-ément $\alpha+\beta=\pi/2$ et $T=\dfrac{MgL}{pq}\zeta$. La première condition exprime simple-ment qu'à l'équilibre ($\vartheta=0$) le ressort est vu du point 0 sous un angle droit. La seconde exige que dans certaines limites la tension du ressort soit proportionnelle à son allongement. On sait effectivement construire des

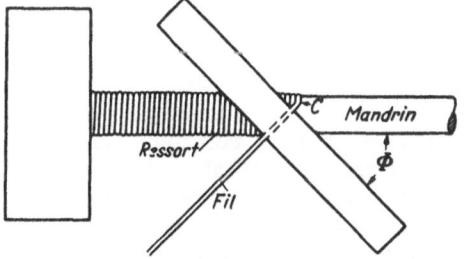

Fig. 14. Procédé LA COSTE pour faire un ressort de longueur nulle. Le fil se tord en $C$ pour s'appliquer sur les spires précédentes.

ressorts dont les spires soient, au repos, appuyées les unes sur les autres en sorte que les premiers efforts servent seulement à les décoller: LA COSTE enroulait le fil sur le mandrin en le faisant arriver sous un angle fixe $\Phi$ à travers un trou percé dans une barre plate (fig. 14); on peut aussi, comme l'a proposé

[1] A. KREIS et E. WANNER: Ann. schweiz. meteorol. Zentral-Anstalt, Anhang 6, 1 (1937).
[2] L. J. B. LA COSTE: Physics 5, 178 (1934). Bull. Seism. Soc. Amer. 25, 176 (1935).

Grenet, placer un ressort ordinaire dans un tube et le retourner sur le tube, spire par spire, comme une manche de veste. Dans les deux cas, on s'arrangera pour que le ressort obtenu ait une «longueur au repos» négative (toujours faible pour éviter l'instabilité) qu'on déterminera expérimentalement, soit $-l_0$; on ajoutera alors une tige de longueur $l_0$, et l'ensemble aura une longueur au repos exactement nulle.

L'équilibre indifférent n'est d'ailleurs ni possible ni désirable. On s'attachera à réaliser avec précision la condition de longueur nulle, soit $T = K\zeta$, $K$ étant une constante. On s'approchera seulement de la condition de l'angle droit, soit $\alpha + \beta = \pi/2 - \gamma$, $\gamma$ étant un petit angle. En tenant compte de la condition d'équilibre $C(o) = 0$ le couple se réduit à $C(\vartheta) = K p q \sin \gamma \sin \vartheta$, proportionnel au sinus de l'angle d'écart, comme dans les séismographes horizontaux classiques.

Les difficultés dites géométriques sont donc supprimées par le ressort LaCoste. Il reste celles qui sont dues à l'hystérésis élastique et qui sont considérables. La plupart des appareils verticaux en service ne dépassent guère 10 secondes et

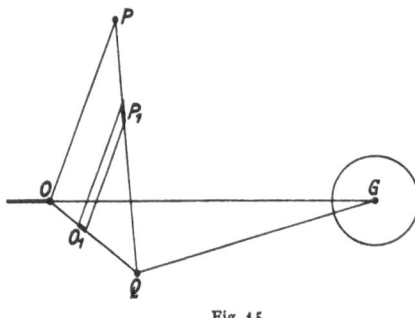

Fig. 15.

pourtant leur amortissement par friction interne est tel qu'ils ne peuvent exécuter plus d'un petit nombre d'oscillations propres. Ceci conduit parfois à employer pour les ressorts un acier plus dur que l'élinvar. La variation de rigidité avec la température devient alors gênante et pose, même pour les appareils de période moyenne, les questions d'isolement et de compensation thermique. L'isolant sera avantageusement placé entre deux enceintes métalliques pour uniformiser le flux de chaleur et supprimer les courants de convection dans la cage, qui créent de faux microséismes[1]. Les systèmes de compensation les plus simples déplacent le point d'attache du ressort ou déforment la masse par un effet de dilatation (grille du Wiechert vertical, bilame du Wilip).

Comme l'ont montré Press et Ewing il y a grand intérêt à compenser en outre les appareils à grande période des variations de la pression barométrique en disposant un corps creux équilibrant la poussée sur la masse[2].

La mise en service d'un séismographe vertical comporte, outre la détermination de la longueur réduite, le réglage de l'horizontalité du bras $O G_0$. Or il ne suffit pas de déterminer la longueur réduite et la position du centre de gravité de la masse tournante; il faut tenir compte des masses constituant le ressort; la correction peut être notable: Galitzine [2] trouve qu'elle modifie seulement de $1/400^e$ la longueur réduite de son pendule; mais dans un cas étudié par Grenet l'effet sur l'horizontalité atteignait $4°$. Si on opère par le calcul, on peut supposer la masse du ressort répartie à chaque instant sur la droite qui joint les points d'attache (nous avons négligé les oscillations propres des ressorts en établissant les équations générales). On peut même, comme l'a remarqué Grenet, considérer cette masse comme transportée sur la droite $O Q$ (fig. 15) parallèlement à $PO$ et la lier rigidement au pendule, réduit ainsi à une simple masse tournante. En effet deux éléments correspondants entourant $P_1$ sur $PQ$ et $O_1$ sur $OQ$ ont même déplacement virtuel, savoir celui de $Q$ réduit dans le rapport $\overline{PP_1}/\overline{PQ} = \overline{OO_1}/\overline{OQ}$.

[1] P. Bernard: Ann. Géophys. 3, 96 (1947).
[2] M. Ewing et F. Press: Trans. Amer. Geophys. Un. 34, 95 (1953).

Une méthode de détermination de la longueur réduite plus pratique que le simple calcul consiste à accrocher effectivement le ressort suivant $OQ$ (GRENET) et à faire basculer l'appareil de 90° autour de son axe pour le transformer en un pendule ordinaire. Il est alors suspendu à l'ancienne paire horizontale de lamelles, du moins si le ressort à boudin était vertical (GALITZINE). Si l'inclinaison du ressort obligeait primitivement à saisir en arrière les lamelles horizontales, elles travaillent à la compression dans la position basculée; GRENET leur adjoint donc des couteaux spéciaux.

Une dernière méthode, due à GRENET et à Mme. DUCLAUX [6], consiste à étudier les déplacements du pendule par rapport à sa plateforme lorsqu'on incline le bâti d'un petit angle autour d'un axe parallèle à l'axe de rotation. Soit $i$ l'angle de la plateforme avec le plan horizontal, $\varphi$ l'angle du bras $OG$ avec la plateforme, en sorte que $\vartheta = i + \varphi$. Le couple de rappel est:

$$C(\varphi, i) = MgL\cos\vartheta - f(\varphi),$$

car si le ressort a peu d'hystérésis élastique, le couple du ressort, qui ne dépend pas de $i$, peut être considéré comme une fonction de $\varphi$. Si le pendule est suffisamment astatisé, cette fonction $f(\varphi)$ décroît pour compenser la décroissance du couple de pesanteur dans la position normale d'équilibre $i = 0$. Pour chaque angle $i$ il existe une position d'équilibre $\varphi_0$ définie par $f(\varphi_0) = MgL\cos\vartheta_0$; $f(\varphi_0)$ passe par un maximum pour $\vartheta_0 = 0$, et par conséquent $\varphi_0$ passe par un minimum. L'observation optique de ce minimum permet aisément le réglage de l'horizontalité à 5' près.

Le couple de rappel au voisinage d'une position quelconque d'équilibre est $(\varphi - \varphi_0)\frac{\partial}{\partial\varphi_0}C(\varphi_0, i)$. Si $\omega$ et $l$ sont la longueur réduite et la pulsation propre dans cette position, on en déduit que:

$$\omega^2 l = \frac{1}{ML}\frac{\partial}{\partial\varphi_0}C(\varphi_0, i) = -g\sin\vartheta - \frac{1}{ML}f'(\varphi_0).$$

En toute rigueur, $l$ est une fonction de $i$; mais l'effet du ressort est assez petit pour que la variation puisse être négligée. On cherche alors, de part et d'autre du minimum de $\varphi_0$, deux positions de la plateforme pour lesquelles $\varphi_0$ ait la même valeur. On mesure les pulsations $\omega_1$, $\omega_2$ correspondantes. Les valeurs $\vartheta_1$, $\vartheta_2$ de $\vartheta$ sont petites et on peut les confondre avec leur sinus. Finalement:

$$l = -\frac{g(i_1 - i_2)}{\omega_1^2 - \omega_2^2}.$$

Avec des précautions on obtient $l$ au centième près.

**12. Amortissement. Amplification.** Pour la simplicité des théories, la force d'amortissement doit être proportionnelle à la vitesse. En outre, on doit pouvoir, pour l'étalonnage, la réduire beaucoup sans toucher aux masses mobiles. Les amortisseurs à liquide étant trop variables, la première condition limite le choix aux amortisseurs à air ou aux amortisseurs magnétiques.

Les amortisseurs à air, dont le prototype est celui de WIECHERT [3] dérivent de l'amortisseur des balances de CURIE (1889). Pour que l'amortissement soit proportionnel à la vitesse, l'air doit s'écouler en régime laminaire sous l'influence d'une très faible différence de pression. La distance entre parois d'écoulement doit être réduite à une fraction de millimètre, ce qui nécessite un réglage soigné, et le volume d'air doit être assez grand. La suppression de l'amortissement s'obtient en ouvrant à l'air un large chemin.

4*

L'amortissement magnétique est en général obtenu par courants de Foucault, et il est bien proportionnel à la vitesse. Il s'est peu à peu étendu aux séismographes de toutes masses, de l'Anderson-Wood (0,7 g.) au de Quervain-Piccard (21 tonnes). Malheureusement l'amortissement s'accompagne ordinairement d'attractions ou de répulsions qui modifient les forces de rappel. D'après Somville le diamagnétisme de la plaque de cuivre servant à l'amortissement suffit à expliquer la variation de période propre des pendules Galitzine (Sect. 14) lorsqu'on éloigne les aimants amortisseurs. L'effet a été corrigé par Wilip en employant un bronze d'aluminium mais il vaut mieux remplacer la plaque par un circuit bobiné; on ouvre ce circuit pour supprimer l'amortissement.

Nos descriptions de séismographes se sont bornées au mouvement de la masse principale. Or, dans l'enregistrement mécanique, toute l'amplification se fait au moyen de leviers. Le problème consiste alors à éviter les frottements aux articulations. Celles-ci sont en général constituées par des lamelles qui peuvent subir des tractions importantes, des poussées ou des flexions modérées; en cas de besoin on fait supporter les efforts perpendiculaires au plan d'une lamelle par une deuxième lamelle perpendiculaire à la première. La flexion des lamelles produit des couple de rappel qui ne sont pas toujours négligeables. On emploie également des pointes maintenues en contact avec leur cupule par la pesanteur ou par un faible ressort; exceptionnellement enfin des liaisons par attraction de pièces aimantées avec ou sans contact.

Dans l'enregistrement optique l'amplification peut commencer par des leviers (fig. 7a). Dans le séismographe Anderson-Wood, pour obtenir un grand trajet optique le rayon lumineux revient sur le miroir mobile après avoir rencontré un miroir fixe. Dans un mode d'enregistrement qui remonte à Galitzine [2] le mouvement de la masse produit par induction un courant électrique qu'on fait agir sur un galvanomètre. Ce soi-disant enregistrement galvanométrique modifie profondément la nature des enregistrements; nous allons étudier l'ensemble de l'appareil sous le nom de séismographe électromagnétique. Enfin nous réservons les amplificateurs électroniques pour une étude spéciale (Sect. 17 et 18).

## D. Séismographes électromagnétiques.

**13. Généralités sur les séismographes électromagnétiques.** Dans un séismographe électromagnétique le pendule produit un courant d'induction dans une bobine qui se déplace dans l'entrefer d'un aimant (Galitzine) ou qui embrasse un circuit magnétique dont la réluctance varie (Benioff). Les équations sont analogues dans les deux cas; pour les établir nous supposerons qu'il s'agit du premier dispositif. De toute façon le courant est envoyé dans un galvanomètre à cadre dont les rotations sont enregistrées photographiquement. On peut ainsi séparer la cave profonde où doit se trouver le pendule, de la pièce obscure d'enregistrement.

Le galvanomètre peut réagir sur le pendule lui-même; Wenner a le premier tiré parti de cette possibilité[1]. L'idée s'étendrait à des séismographes qui comporteraient plusieurs galvanomètres, dont certains mêmes ne seraient pas observés (Grenet).

Soit $R$ la résistance intérieure du pendule, $r$ celle du galvanomètre. Ils peuvent être shuntés par une résistance $S$ qui rend leurs amortissements indépendants (Wenner). Soient $I, i, I+i$, les courants respectifs dans les trois résistances; $E, e$, zéro, les forces électromotrices d'induction, le tout compté positivement dans le sens des flèches (fig. 16). Les phénomènes sont à variation assez lente

---

[1] F. Wenner: Research Papers 66. Bur. Stand. J. Res. 2, 963 (1929).

pour que la self-induction soit négligeable. La différence de potentiel aux bornes du shunt est:

$$V_B - V_A = E - IR = (I + i)\,S = e - i\,r.$$

D'où

$$Q^2 I = (r + s)\,E - Se$$
$$Q^2 i = -SE + (R + S)\,e$$

où

$$Q^2 = Rr + (R + r)\,S.$$

Les flux traversant la bobine du pendule et le cadre du galvanomètre seront positifs s'ils ont le sens des flux produits par des courants positifs. Supposons que le pendule se réduise à une masse tournante et soit $\Theta$ son angle de rotation, $\vartheta$ étant celui du cadre; le sens positif des angles est défini par la convention suivante: les variations dues au flux à partir de la position d'équilibre seront données par:

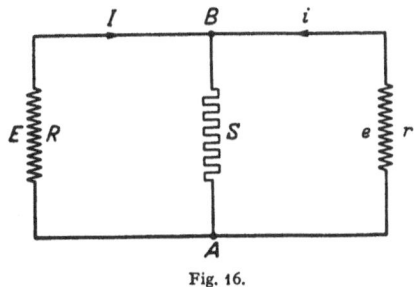

$$\Phi = -G\Theta, \qquad \varphi = -g\vartheta$$

où $G$ et $g$, constantes électrodynamiques du pendule et du galvanomètre, sont essentiellement positives. Avec cette convention:

$$E = -\frac{d\Phi}{dt} = G\dot{\Theta}$$

$$e = -\frac{d\varphi}{dt} = g\dot{\vartheta}.$$

Fig. 16.

L'équation du mouvement du pendule comprend, en sus des termes habituels, le couple produit par les forces électromagnétiques. Le travail de ce couple est $I\Phi = IG\Theta$, le couple est donc $-GI$, et l'équation s'écrit:

$$K\ddot{\Theta} + F\dot{\Theta} + U\Theta = -ML\ddot{x} - \frac{G^2(r+s)}{Q^2}\dot{\Theta} + \frac{GgS}{Q^2}\dot{\vartheta}.$$

$K$ est le moment d'inertie du pendule, $M$ sa masse, $L$ la distance du centre de gravité à l'axe de rotation, $-U\Theta$ et $-F\dot{\Theta}$ les couples de rappel et d'amortissement.

L'équation du galvanomètre est analogue, sauf que le centre de gravité de l'équipage est sur l'axe par construction:

$$k\ddot{\vartheta} + f\dot{\vartheta} + u\vartheta = -g\,i = +\frac{GgS}{Q^2}\dot{\Theta} - \frac{g^2(R+S)}{Q^2}\dot{\vartheta}.$$

Nous mettrons les deux équations sous la forme fondamentale:

$$\ddot{\Theta} + 2\beta\Omega\,(\dot{\Theta} - \Gamma\dot{\vartheta}) + \Omega^2\Theta = -\frac{1}{l}\ddot{x}, \qquad (13.1)$$

$$\ddot{\vartheta} + 2\alpha\omega\,(\dot{\vartheta} - \gamma\dot{\Theta}) + \omega^2\vartheta = 0 \qquad (13.2)$$

en posant:

$$l = \frac{K}{ML} \quad \text{(longueur réduite)}$$

$$\Omega^2 = \frac{U}{K}; \qquad\qquad \omega^2 = \frac{u}{k},$$

$$2\beta\Omega = \frac{1}{K}\left(F + G^2\frac{r+s}{Q^2}\right); \qquad 2\alpha\omega = \frac{1}{k}\left(f + g^2\frac{R+S}{Q^2}\right),$$

$$2\beta\Omega\Gamma = \frac{GgS}{KQ^2}; \qquad\qquad 2\alpha\omega\gamma = \frac{GgS}{kQ^2} = \chi.$$

$$\left.\begin{array}{l}\\ \\ \\ \\ \\ \end{array}\right\} \quad (13.3)$$

$\gamma$ et $\Gamma$ sont les «réactions» (pendule sur galvanomètre et galvanomètre sur pendule). Au lieu de $\gamma$ il est souvent commode d'utiliser $\chi$ qui est dit *«coefficient de transfert de* Galitzine*»* [2], [6]. Les amortissements $\beta$ et $\alpha$ comprennent non seulement les amortissements en circuit ouvert $F/2K\Omega$, $f/2k\omega$, mais les amortissements électromagnétiques produits par le circuit d'utilisation.

(13.1) et (13.2) montrent que le pendule et le galvanomètre constituent deux systèmes oscillants couplés par les termes d'amortissement. On appellera coefficient de couplage, ou simplement «couplage», la quantité

$$\sigma = \sqrt{\Gamma\gamma}.$$

On a toujours $0 < \sigma < 1$. En effet:

$$\sigma^2 < \frac{S^2}{(R+S)(r+S)},$$

la limite étant atteinte si tout l'amortissement du pendule et du galvanomètre est produit par le circuit lui-même. Cette limite est 1 si en outre $S$ est infini (absent). A l'opposé, si $K$ est très grand, $\Gamma$ et $\sigma$ sont très petits. Si on les néglige, les équations (13.1) et (13.2) peuvent être résolues successivement, ce qui est une simplification considérable.

Eliminons $\Theta$ entre (13.1) et (13.2). Il suffit de dériver la première et d'y substituer $\dot{\Theta}$ tiré de la seconde. Il est commode d'utiliser la mise en facteur symbolique des symboles de dérivation:

$$\left\{\left[\frac{d^2}{dt^2} + 2\beta\Omega\frac{d}{dt} + \Omega^2\right]\left[\frac{d^2}{dt^2} + 2\alpha\omega\frac{d}{dt} + \omega^2\right] - 4\beta\alpha\Omega\omega\sigma^2\frac{d^2}{dt^2}\right\}\vartheta = -\frac{\chi}{l}\dddot{x}. \quad (13.4)$$

Le mouvement dépend d'une équation différentielle linéaire à coefficients constants comme pour un séismographe simple, mais du quatrième ordre et non plus du second.

Les déviations du galvanomètre sont enregistrées optiquement; soit $q = D\vartheta$, $D$ étant la longueur du levier optique. Le grandissement est $V = \chi D/l$. Pour des mouvements très rapides $V$ représente le rapport de la vitesse du point inscripteur au déplacement du sol. Pour des mouvements très lents $q\omega^2\Omega^2 = -V\ddot{x}$. On voit s'introduire la *«pulsation moyenne»* $\overline{\omega} = \sqrt{\Omega\omega}$ dont le choix est déterminé par le domaine de fréquences à étudier. Prenons le «temps réduit» $\tau = \overline{\omega}T$ pour variable dans (13.4)

$$\frac{d^4\vartheta}{d\tau^4} + A\frac{d^3\vartheta}{d\tau^3} + B\frac{d^2\vartheta}{d\tau^2} + C\frac{d\vartheta}{d\tau} + \vartheta = -\frac{\chi}{\overline{\omega}l}\frac{d^3x}{d\tau^3}, \quad (13.5)$$

$$\left.\begin{aligned}
A &= 2\left(\beta\varrho + \alpha\frac{1}{\varrho}\right), \\
B &= \varrho^2 + \frac{1}{\varrho^2} + 4(1-\sigma^2)\beta\alpha, \\
C &= 2\left(\beta\frac{1}{\varrho} + \alpha\varrho\right), \\
\varrho &= \sqrt{\frac{\Omega}{\omega}} = \frac{\Omega}{\overline{\omega}} = \frac{\overline{\omega}}{\omega}.
\end{aligned}\right\} \quad (13.6)$$

Les questions de grandissement et de période moyenne étant mises à part, la façon dont le séismographe traduit les mouvements du sol exprimés en temps réduit dépend uniquement des coefficients $A, B, C$. Si on permute les caractéristiques $\beta, \Omega$ du pendule avec celles $\alpha, \omega$ du galvanomètre, ces coefficients ne changent pas. L'existence de galvanomètres à grande période permet de mettre à profit cette remarque. Mais ceci amène à se poser une question générale: si l'on choisit des caractéristiques intéressantes pour un séismographe pourra-t-on le construire?

Si on se donne pour $A, B, C$ trois valeurs positives, les oscillations propres $\vartheta = \exp.(\lambda t)$ du système s'obtiendront en cherchant les racines de l'équation caractéristique:

$$\lambda^4 + A\,\lambda^3 + B\,\lambda^2 + C\,\lambda + 1 = 0.$$

Il faut que ces oscillations propres soient amorties; trois cas sont admissibles, si on excepte les cas limites: Ou bien toutes les racines sont réelles et négatives; l'appareil est «apériodique». Ou bien deux racines sont réelles et négatives, deux racines sont complexes, conjuguées, à partie réelle négative; l'appareil est «simplement oscillatoire». Ou bien enfin les quatre racines sont complexes, deux à deux conjuguées, à partie réelles négatives; l'appareil est «doublement oscillatoire».

Pour que l'on se trouve dans l'un de ces trois cas, c'est à dire que les parties réelles des racines soient négatives, il faut et il suffit que l'on ait la condition de ROUTH[1]

$$A\,BC - A^2 - C^2 > 0.$$

On peut montrer [5] que si la condition est vérifiée et si on choisit arbitrairement une valeur $\sigma$ du couplage $(0 < \sigma < 1)$, les équations (13.6) permettent d'obtenir au moins un système de valeurs $\beta, \alpha, \varrho$. Il peut même en exister trois. Il y a donc une infinité d'appareils équivalents à un appareil donné[2].

En particulier il existe un appareil au moins tel que la réaction du galvanomètre soit négligeable ($\sigma = 0$, donc $\Gamma = 0$). Sa construction effective pourrait d'ailleurs exiger des masses prohibitives. On voit aisément à quoi correspond dans ce cas l'existence de trois solutions distinctes: Le premier membre de (13.5), soit:

$$\left[\frac{d^2}{d\tau^2} + 2\beta\,\varrho\,\frac{d}{d\tau} + \varrho^2\right]\left[\frac{d^2}{d\tau^2} + 2\alpha\,\frac{1}{\varrho}\,\frac{d}{d\tau} + \frac{1}{\varrho^2}\right]\vartheta,$$

doit être identique à:

$$\left[\frac{d}{d\tau} - \lambda_1\right]\left[\frac{d}{d\tau} - \lambda_2\right]\left[\frac{d}{d\tau} - \lambda_3\right]\left[\frac{d}{d\tau} - \lambda_4\right]\vartheta,$$

où $\lambda_1, \lambda_2, \lambda_3, \lambda_4$, sont les racines de l'équation caractéristique. Pour cela, chaque facteur de la première expression doit être identique au produit de deux facteurs pris dans la seconde. Lorsque les racines sont toutes réelles il y a six façons de faire l'identification. Mais ce montage apériodique est rarement utilisé.

Le premier soin lorsqu'on veut étudier un séismographe électromagnétique est de déterminer un séismographe sans réaction équivalent, dans lequel on puisse considérer successivement le pendule et le galvanomètre. On obtient explicitement ses constantes $\beta_0, \alpha_0, \varrho_0$, si on a résolu l'équation caractéristique. On doit prendre en outre;

$$\gamma_0 = \frac{\alpha\,\varrho_0}{\alpha_0\,\varrho}\,\gamma$$

si on veut que les deux appareils aient le même grandissement.

---

[1] ROUTH: Advanced rigid dynamics, 6ème éd., § 287. London 1905.
[2] La détermination de constantes de construction se poursuivrait aisément, puisque (13.3) fournit seulement 3 équations entre $F, f, G, g, K, k, R, r$ et $S$.

Les méthodes qui nous ont servi à étudier la réponse du pendule aux mouvements simples du sol s'étendent aisément au séismographe électromagnétique tout entier. On démontrerait les propriétés que nous allons énoncer soit directement, soit en s'adressant à l'appareil sans réaction équivalent et en utilisant deux fois les résultats établis pour un pendule seul.

Si le sol, et par conséquent le pendule, d'abord au repos subissent un déplacement brusque, le galvanomètre part du repos avec une vitesse finie. Si le sol, et par conséquent le pendule, partent du repos avec une vitesse finie, le galvanomètre part du repos avec une vitesse nulle et une accélération finie. Pour cette raison on a reproché aux séismographes électromagnétiques de fournir des débuts moins nets que les séismographes à inscription directe. Ce reproche serait fondé si l'amplification était la même dans les deux cas. Mais les séismographes électromagnétiques permettent une amplification bien plus grande.

Un mouvement sinusoïdal du sol débutant brusquement entraîne une réponse du galvanomètre composée de deux parties: l'oscillation propre qui s'amortit et l'oscillation forcée qui persiste. L'amplification de l'oscillation forcée est nulle pour les mouvements de période courte, ce qui peut paraître un défaut grave, le même au fond que le précédent. Afin d'y porter remède on adoptera une période courte pour le séismographe lui-même lorsqu'on s'intéressera aux mouvements rapides. Ainsi des périodes de l'ordre de la seconde sont indispensables pour étudier les séismes proches et certaines ondes longitudinales des séismes lointains.

Lorsque la période du sol augmente, le grandissement dynamique présente un maximum (théoriquement deux peuvent se présenter), puis tend vers zéro pour les mouvements lents: il est alors proportionnel à $n^3$, $n$ étant la pulsation imposée, tandis que celui des séismographes à inscription directe était seulement proportionnel à $n^2$. Les séismographes électromagnétiques sont donc peu sensibles aux mouvements lents, aux inclinaisons du pilier dans les appareils horizontaux ou aux variations de longueur dans les appareils verticaux. Si leur galvanomètre est bon, *les séismographes électromagnétiques n'ont pas de déplacement du zéro* et permettent de réaliser une économie de papier photographique en réduisant l'intervalle des lignes successives.

L'oscillation forcée présente un retard de phase $\pi/2$ pour les mouvements rapides; ce retard diminue lorsque la période augmente et se transforme en une avance de phase qui croît jusqu'à $3\pi/2$ pour les mouvements sinusoïdaux les plus lents.

Revenons au grandissement $V = \chi D/l$ et voyons comment lui donner des valeurs importantes. On dépasse rarement 2 mètres pour $D$. D'autre part on transfèrera plus d'énergie du sol au galvanomètre si l'on réduit les amortissements improductifs $F$ et $f$. Supposons les négligeables; on tire de (13.3):

$$\frac{\chi^2}{\varrho^2} = 4\alpha\beta\,\overline{\omega}^2\,Q_0^2\,\sigma\,\frac{1}{L^2}\,\frac{K}{k}$$

où $Q_0$ est le coefficient d'utilisation de la masse (Sect. 3).

$\alpha, \beta, \overline{\omega}, Q_0$ étant déjà choisis, on augmente $\chi/l$ en augmentant $\sigma$, donc le shunt; mais pour maintenir $\alpha$ et $\beta$ on doit augmenter la puissance des aimants. Par contre on peut diminuer $L$ sans toucher à $G$ en disposant une masse importante près de l'axe et une bobine légère au bout d'un bras rigide; c'est la solution de Galitzine (Sect. 14); mais on est alors limité par la nécessité de renforcer la suspension, ce qui augmenterait $U$ et $\Omega$. Reste le facteur $K/k$, qui est le plus important. Si on augmente $K$, il n'y a plus d'inconvénient à augmenter $U$, mais il faut aussi augmenter $G$. L'augmentation proportionnelle de $G^2$ et $K$ peut se faire simultanément en augmentant le cuivre de la bobine pourvu que sa masse constitue l'essentiel de la masse inerte; mais il faut augmenter l'entrefer en améliorant le circuit magnétique; c'est la solution de WENNER (Sect. 15). Enfin la diminution de $k$ est limitée théoriquement par le poids de l'isolant et celui du miroir, pratiquement par la fragilité et le coût du galvanomètre.

En fait les conditions sont assez larges pour qu'on puisse atteindre sans dimensions exagérées du pendule et sans prix de revient excessif du galvanomètre les amplifications permises par l'agitation microséismique, qui sont de l'ordre de 1000 pour les appareils de période moyenne. Pour les séismographes portatifs à courte période, on se reportera à la discussion approfondie qu'en a fait WILLMORE[1].

Arrêtant ici les indications générales, renvoyant pour les questions de réglage à un article de GRENET [6], nous allons donner quelques détails sur les trois types les plus importants de séismographes électromagnétiques.

**14. Séismographes électromagnétiques de GALITZINE.** GALITZINE composait ses appareils d'un pendule à l'amortissement critique et d'un galvanomètre de même période (24 secondes pour les horizontaux, 12 pour les verticaux), également à l'amortissement critique. Son successeur WILIP et aujourd'hui SPRENGNETHER ont utilisé des périodes plus courtes qu'on puisse maintenir sans peine sur les trois composantes.

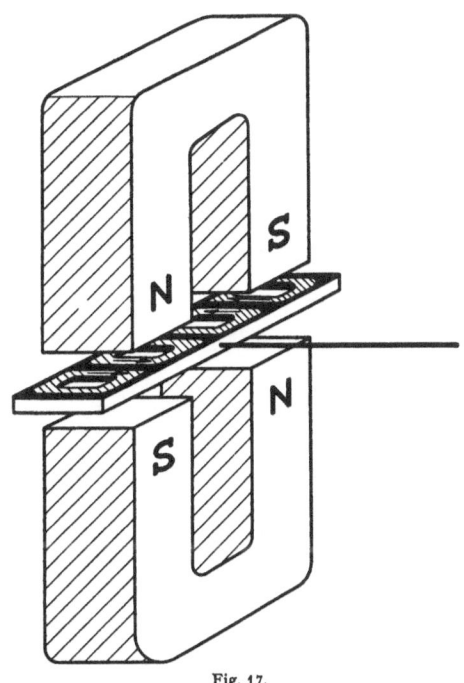

Fig. 17.

La fig. 17 schématise le dispositif d'induction: quatre bobines rectangulaires juxtaposées ont chacune un côté plongé dans l'entrefer créé par deux aimants en fer à cheval. Elles sont mises en série de façon que les forces électromotrices s'ajoutent. Il n'y a pas de shunt; le pendule est amorti par courants de FOUCAULT (Sect. 12) dans une plaque portée par le même bras que les bobines, et placée également entre deux aimants en fer à cheval.

Dans les séismographes originaux de GALITZINE, la réaction du galvanomètre sur le pendule est négligeable. Avec les notations de la Sect. 13

$$\alpha = \beta = 1, \quad \varrho = 1, \quad \sigma = 0.$$

L'équation caractéristique admet alors $-1$ comme racine quadruple. Cette propriété, qui simplifie beaucoup les équations, est imparfaitement réalisée dans les modèles de WILIP, dont la masse est moindre et où le couplage atteint quelques dixièmes. La théorie générale nous a d'ailleurs appris qu'on pouvait obtenir les caractéristiques que l'on désire, ici:

$$A = C = 4, \quad B = 6,$$

---

[1] P. L. WILLMORE: Mon. Not. Roy. Astronom. Soc., Geophys. Suppl. **6**, 129 (1950).

avec un couplage quelconque. On en tire:

$$\beta = \alpha, \qquad \varrho + \frac{1}{\varrho} = \frac{2}{\alpha}, \qquad \sigma = \frac{1}{\alpha^2} - 1,$$

pourvu que:

$$\frac{1}{2} \sqrt{2} < \alpha < 1,$$

$$3 - 2\sqrt{2} = 0{,}172 \cdots < \frac{\Omega}{\omega} < 3 + 2\sqrt{2} = 5{,}83 \ldots.$$

On peut par exemple constituer un Galitzine de 12 s. à réaction avec un galvano-mètre original de Galitzine ayant 24 s. de période et un séismographe vertical de 6 s.

On trouve aisément à partir de l'équation du 4ème ordre l'oscillation forcée qui correspond au mouvement du sol $x = P \sin \overline{\omega} t = P \sin n \tau$ soit

$$\vartheta = \frac{4\gamma}{\varrho^2 + 1} \frac{P}{l} \frac{n^3}{(1+n^2)^4} \left[ 4n(1-n^2) \sin n\tau + (1 - 6n^2 + n^4) \cos n\tau \right].$$

Si on écrit

$$\vartheta = a \sin (n\tau - \psi)$$

on doit prendre

$$\frac{-\sin \psi}{1 - 6n^2 - n^4} = \frac{\cos \psi}{4n(1-n^2)} = \frac{1}{\varepsilon (1+n^2)^2}$$

$$\frac{a}{P} = \frac{4\gamma}{\varrho^2 + 1} \frac{1}{l} \frac{\varepsilon n^3}{(1+n^2)^2}$$

où $\varepsilon = \pm 1$. Si nous convenons que $a$ est de signe contraire à $P$ nous devons prendre $\varepsilon = -1$.

Appelons encore amplification $W$ le rapport de l'amplitude enregistrée à l'amplitude du mouvement du sol. $V$ étant le grandissement,

$$\frac{W}{V} = \frac{n^3}{(1+n^2)^2}.$$

D'autre part, si on pose $\tan \varphi = \frac{2n}{1-n^2}$ avec $-\pi < \varphi < 0$, on a

$$\psi = 2\varphi + \frac{\pi}{2}.$$

Ces expressions de $W/V$ et $\psi$ se retrouvent immédiatement sur l'appareil sans réaction: on applique les formules de la Sect. 6 au pendule, puis au galvanomètre, en intercalant une intégration puisque le second membre de l'équation du galvanomètre comprend seulement la dérivée première de $\Theta$.

Les courbes de la fig. 18 représentent $W/V$ et $\psi$ en fonction de $1/n = T/T_0$. $W/V$ présente pour $n = \sqrt{3} = 1{,}732$ un maximum égal à $3\sqrt{3}/16 = 0{,}3248$. Si on compare à la fig. 2 on voit que les séismographes de Galitzine sont plus sélectifs que les séismographes à inscription directe amortis critiquement.

On obtiendrait de même la réponse de l'appareil, partant du repos, au choc unité

$$x = 0 \quad \text{pour} \quad \tau < 0$$

$$x = \tau \quad \text{pour} \quad \tau > 0$$

soit:

$$\vartheta = \frac{\gamma_0}{l}\left(\tau^2 - \frac{\tau^3}{3}\right)\exp.(-\tau). \tag{14.1}$$

$\dfrac{l\vartheta}{\gamma_0}$ passe par un maximum 0,261 pour $\tau = 3 - \sqrt{3} = 1{,}27$; par zéro pour $\tau = 3$; par un minimum $-0{,}114$ pour $\tau = 3 + \sqrt{3} = 4{,}73$ (fig. 19). Le rapport du maximum

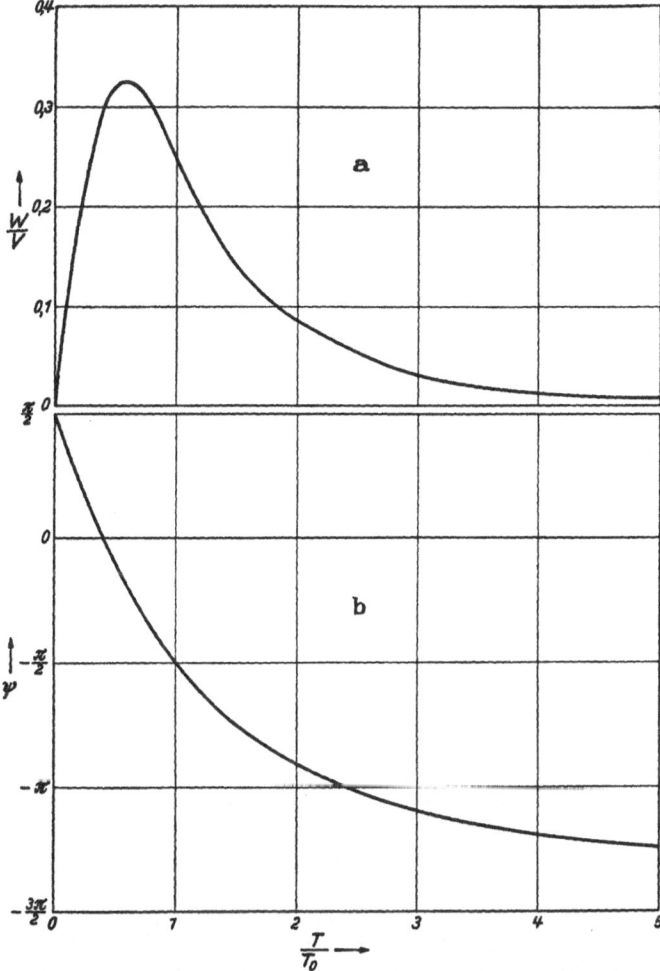

Fig. 18. Amplification et déphasage de l'oscillation forcée dans un séismographe GALITZINE. $T$ période du mouvement du sol, $T_0$ période commune du pendule et du galvanomètre.

au minimum, seul intéressant lorsqu'on ignore l'intensité du choc est:

$$e^{2\sqrt{3}}\left(7 - 4\sqrt{3}\right) = 2{,}29.$$

Connaissant le mouvement du galvanomètre on remonte sans difficulté au mouvement du pendule dans le séismographe avec réaction.

$$-l\Theta = \left[\tau - \frac{\varrho^2 - 1}{\varrho^2 + 1}\tau^2 + \frac{(\varrho^2 - 1)^2}{2\varrho^2(\varrho^2 + 1)}\frac{\tau^3}{3}\right]\exp.(-\tau). \tag{14.2}$$

En appliquant cette formule au séismographe sans réaction ($\varrho = 1$) et à l'appareil décrit plus haut ($\varrho = 2$) on verrait combien la réaction modifie le mouvement du pendule.

Ecrivons enfin les mouvements du galvanomètre et du pendule, lorsque le pendule écarté de $\Theta_0$ est abandonné sans initiale tandis que le galvanomètre part du repos:

$$\vartheta = -\gamma_0 \Theta_0 \frac{\tau^3}{3} \exp. (-\tau), \tag{14.3}$$

$$\Theta = \Theta_0 \left[ 1 + \tau + \frac{\varrho^2 - 1}{\varrho^2 + 1} \frac{\tau^3}{3} \right] \exp. (-\tau). \tag{14.4}$$

Le galvanomètre s'écarte très lentement de sa position d'équilibre. $\left| \dfrac{\vartheta}{\gamma_0 \Theta_0} \right|$ passe par un maximum 0,448 pour $\tau = 3$, et revient lentement au zéro. La réaction

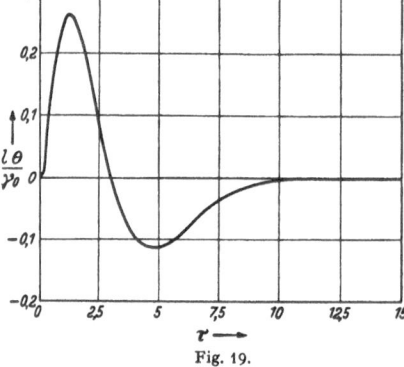

Fig. 19.

gêne le retour du pendule à sa position d'équilibre, mais modifie moins son mouvement que dans le cas précédent.

Les séismographes du modèle original de GALITZINE sont encore les plus répandus des séismographes électromagnétiques. A ce titre insistons un peu sur leur réglage, très étudié par GALITZINE, SOMVILLE, RYBNER. Les méthodes qui suivent ont été étendues par ROUAUD[1] au cas où $\varrho$ est quelconque. On mesure d'abord la période propre du galvanomètre et son amortissement lorsqu'il est fermé sur des résistances connues. On en déduit la valeur de la résistance critique et on donne cette valeur au circuit du pendule dans lequel on a préalablement introduit une résistance réglable. On écarte du pendule les aimants amortisseurs, on ouvre son circuit d'utilisation et on modifie sa période (en inclinant l'axe dans les horizontaux ou en déplaçant le point d'attache du ressort dans les verticaux) jusqu'à ce qu'elle devienne égale à celle du galvanomètre. Enfin on rapproche les aimants amortisseurs jusqu'à ce que le pendule n'oscille plus. Théoriquement le séismographe est réglé. En réalité le couple dû au magnétisme de la plaque de cuivre dans les anciens GALITZINE, la réaction du galvanomètre dans les GALITZINE-WILIP, obligent à vérifier ce réglage.

Une première vérification s'obtient en donnant un «choc au pendule» au moyen d'un petit marteau commandé. Si le séismographe était parfaitement ajusté, les mouvements du pendule et du galvanomètre seraient donnés par (14.1) et (14.2) (avec $\varrho = 1$). De toute façon ils en diffèrent peu. On mesure les amplitudes maximum $\vartheta_1$ et minimum $-\vartheta_2$ du galvanomètre, ce qui est facile, et le temps $t_0$ du passage au zéro, ce qui l'est moins. On regarde si $\vartheta_1/\vartheta_2$ et $\omega t_0$ sont proches de leur valeur théorique. Si on a déterminé l'amplitude $\Theta_m$ du mouvement du pendule avec une précision suffisante, ce qui nécessite un grand levier optique, on obtiendra en outre le grandissement.

On complète la vérification par un «abandon de pendule»: on l'écarte d'un angle connu $\Theta_0$, on attend que le galvanomètre soit immobile et on laisse revenir le pendule au zéro sans vitesse initiale. Si l'appareil est réglé, les mouvements

[1] A. ROUAUD: Ann. Géophys. 4, 124 (1948).

du pendule et du galvanomètre sont donnés par (14.3) et (14.4) (avec $\varrho = 1$). Les mouvements réels en diffèrent peu. On mesure l'amplitude maxima $\vartheta_m$ du galvanomètre. On regarde si $\vartheta_m/\gamma_0\Theta_0$ est proche de sa valeur théorique.

L'ensemble de ces mesures permet de préciser l'écart à l'appareil théorique. Supposons le galvanomètre au critique et posons:

$$\xi = 1 - \frac{\Omega}{\omega}, \qquad \mu^2 = 1 - \beta^2.$$

Admettons que $\xi, \mu^2$, et $\sigma^2$ soient petits du même ordre, quelques centièmes par exemple; RYBNER obtient en conservant seulement les termes du premier ordre:

$$\mu^2 = 2{,}222\,P - 2{,}050\,Q,$$
$$\sigma^2 = 0{,}778\,P - 1{,}728\,Q,$$

où:

$$P = 4{,}754\,\frac{\Theta_m}{\vartheta_1}\,\frac{\vartheta_m}{\Theta_0} + \omega_0 t_0 - 6,$$

$$Q = 1 - 0{,}436\,\frac{\vartheta_1}{\vartheta_2},$$

puis $\xi = 0{,}667\,\omega_0 t_0 - 2 + 0{,}1\,\mu^2 + 0{,}17\,\sigma^2$. Pour le calcul de l'amplification on peut prendre simplement $\gamma = 1{,}408\,\frac{\vartheta_1}{\Theta_m}$ et contrôler au moyen de la formule: $\gamma = 3{,}231\,\frac{\vartheta_2}{\Theta_m}\,(1 - 0{,}34\,\mu^2 - 0{,}99\,\sigma^2)$. Ces valeurs de $\xi, \mu^2, \sigma^2$, permettent de calculer les réponses de l'appareil réel ou de remonter d'un enregistrement supposé sinusoïdal au mouvement original du sol. Mais les calculs sont trop pénibles pour être d'un usage fréquent. Par exemple le grandissement de l'oscillation forcée est:

$$W = \frac{2\alpha\gamma D}{l}\,\frac{n^3}{(1+n^2)^2}\left\{1 + \frac{2}{1+n^2}\,\xi + \frac{2n^2}{(1+n^2)^2}\,\mu^2 - \frac{4n^2(1-6n^2+n^4)}{(1+n^2)^4}\,\sigma^2\right\}.$$

Pour $\xi = \mu^2 = 0$ et $n = 1$ par exemple, l'erreur relative sur $W$ est $\sigma^2$. Il semble que pour les appareils originaux de GALITZINE $\sigma^2$ soit inférieur à 0,04 donc négligeable. Au contraire WENNER et McCOMB ont trouvé (implicitement) $\sigma^2 = 0{,}3$ pour un modèle de WILIP. Ce cas sort du champ des formules précédentes. RYBNER les a complétées[1] en supposant que $\sigma^2$ soit seulement de l'ordre de $\xi^{\frac{1}{2}}$ (ou de $\mu$) et que l'on conserve toujours les termes ayant l'ordre de $\xi$.

Les méthodes précédentes n'indiquent pas les manoeuvres qui réduiraient $\xi, \mu, \sigma$ aux valeurs théoriques. Lorsque $\sigma$ est négligeable, il suffit de régler le pendule seul. Dans ce cas, GRENET a proposé de suivre les variations de la déviation permanente due à un courant constant et de la déviation maximum due à la décharge d'un condensateur, en fonction de la distance des aimants amortisseurs.

**15. Séismographes électromagnétiques de WENNER.** Revenons à l'équation (12.2) du galvanomètre:

$$\ddot{\vartheta} + 2\alpha\omega\,(\dot{\vartheta} - \gamma\,\dot{\Theta}) + \omega^2\vartheta = 0. \tag{15.1}$$

Si $\alpha$ est grand, par exemple $\alpha = 10$ et si la période du mouvement n'est ni très grande ni très petite par rapport à la période propre, le terme central est prépondérant et l'on a approximativement $\dot{\vartheta} = \gamma\,\dot{\Theta}$ d'où $\vartheta = \gamma\Theta$ si le pendule et le

---

[1] J. RYBNER: Gerlands Beitr. Geophys. **55**, 303 (1939). — Geodaetisk Institut København Meddelelse Nr. 11.

galvanomètre partent du repos. Ainsi le galvanomètre suramorti reproduit le mouvement du pendule. Substituons dans (14.1):

$$\ddot{\vartheta} + 2\beta\Omega\,(1 - \sigma^2)\,\dot{\vartheta} + \Omega^2\vartheta = -\,\gamma\frac{\ddot{x}}{l}. \tag{15.2}$$

L'enregistrement est analogue à celui d'un séismographe à inscription directe ayant même pulsation que le pendule, un amortissement réduit par la réaction (mais qu'on peut amener au critique en disposant de $\beta$), et un grandissement proportionnel à $\gamma$. C'est le principe du séismographe original de Wenner[1]. Il permet de concilier la grande amplification des séismographes électromagnétiques avec les bons débuts des séismographes directs.

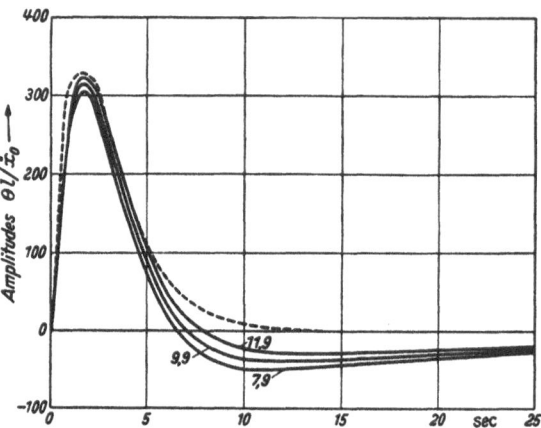

Fig. 20. Réponse, à une impulsion, d'un faux Wenner de période 9,9 s. (courbe centrale) et de l'appareil à inscription directe équivalent (en pointillé). Variation de la réponse lorsque la période du pendule passe de 7,9 à 11,9 s.

Le galvanomètre de Wenner était un fluxmètre du commerce, à faible moment d'inertie, permettant une amplification de 1000 avec uns masse inerte de 500 g., constituée surtout par le cuivre de la bobine. Le courant induit dans celle-ci par un aimant à champ radial, servait en outre à l'amortissement grâce à la présence d'un shunt. Wenner supprimait ainsi les difficultés de réglage: En faisant osciller le galvanomètre et le pendule séparément sur des résistances connues, on peut calculer la résistance extérieure nécessaire pour donner à chacun l'amortissement désiré puis les valeurs de $R$, $r$ et $S$, qu'on réalise en s'aidant de résistances d'appoint.

L'analogie avec un séismographe direct (ou encore la présence dans l'équation caractéristique (12.6) d'une racine petite et d'une grande) définit de façon générale les séismographes du type de Wenner. On en obtient un à partir du

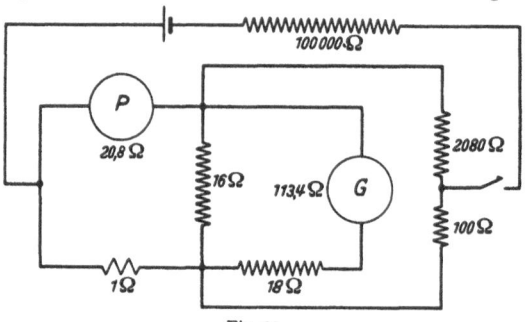

Fig. 21.

Wenner original par échange entre les caractéristiques du pendule et celles du galvanomètre. C'est le «faux Wenner» [5] qui exige un aimant puissant mais permet d'obtenir une période propre élevée dans les appareils verticaux et rend les caractéristiques du séismographe indépendantes des petites variations de la période du pendule. C'est ce que montre la fig. 20 pour un appareil réel[2] dans lequel on avait seulement $\beta = 6{,}95$. La variation, avec la période, de la réponse à une impulsion a été calculée par différentiation[3]. Ce faux Wenner de Grenet est inséré dans un pont (fig. 21) permettant d'envoyer périodiquement un courant

[1] F. Wenner: Research Paper Nr. 66. Bur. Stand. J. Res. **2**, 963 (1929).
[2] G. Grenet et Mme F. Bayard Duclaux: Ann. Géophys. **2**, 104 (1946).
[3] Ph. Pluvinage: Ann. Géophys. **2**, 179 (1946).

constant dans le pendule sans affecter le galvanomètre; si l'appareil est exempt de défauts mécaniques, il enregistre la réponse à une accélération constante du sol. L'examen de la feuille avertit d'un déréglage éventuel. Dans un but analogue McComb et Nelson laissent revenir au zéro le pendule artificiellement dévié[1].

**16. Séismographes électromagnétiques de Benioff.** Les séismographes électromagnétiques de Benioff[2] sont caractérisés par un dispositif instrumental plutôt que par des propriétés théoriques: les variations de flux qui produisent le courant sont dues à des variations d'entrefer et non au déplacement d'une bobine.

Le plus récent *«traducteur électromagnétique»* de Benioff[3], schématisé sur la fig. 22, comprend d'une part un aimant faisant partie de la masse mobile avec ses pièces polaires N et S, d'autre part deux armatures fixes en tôles au silicium orientées $A_1$ et $A_2$, identiques, portant des bobines symétriques.

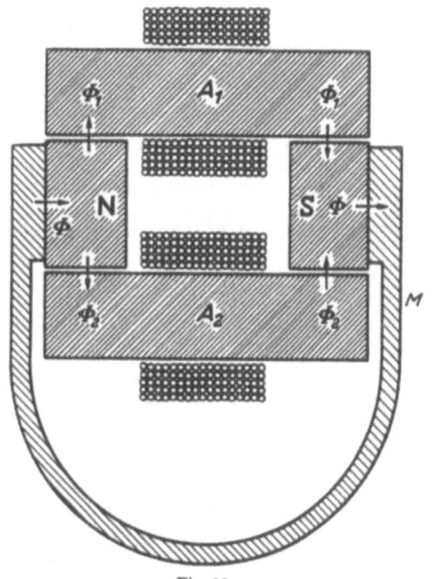

Négligeons les pertes, la réluctance des armatures, les courants de Foucault. Le flux $\Phi$ de l'aimant se partage entre $\Phi_1$ et $\Phi_2$ dans les armatures:

$$\Phi = \Phi_1 + \Phi_2.$$

Soit $s$ la section des entrefers, $\varepsilon$ leur largeur à l'équilibre, $x$ le déplacement vers $A_1$. En u.e.m.,

Fig. 22.

$$2\frac{\varepsilon - x}{s}\Phi_1 = 2\frac{\varepsilon + x}{s}\Phi_2 = F - R\Phi,$$

où $F$ est la force magnétomotrice de l'aimant, $R$ sa réluctance. On en tire:

$$\frac{\Phi}{\varepsilon} = \frac{\Phi_1 - \Phi_2}{x} = \frac{Fs}{\varepsilon^2 + Rs\varepsilon}\left(1 + \frac{x^2}{\varepsilon^2 + Rs\varepsilon} + \cdots\right).$$

Si on se limite aux déplacements de l'ordre de 0,1 mm. sur un entrefer de 3 mm. par exemple, le flux dans l'aimant est constant à mieux que 1/900 près, ce qui évite les effets d'hystérésis. A la même approximation, la force exercée sur l'aimant, comptée vers $A_1$, est proportionnelle à $\Phi_1 - \Phi_2$ donc à $x$, avec un coefficient positif. Elle accroît la période propre.

Enfin, si on oppose les forces électromotrices induites dans les deux bobines, le courant produit est proportionnel à:

$$\frac{d}{dt}(\Phi_1 - \Phi_2) = \frac{Fs\dot{x}}{\varepsilon^2 + Rs\varepsilon}\left(1 + \frac{3x^2}{\varepsilon^2 + Rs\varepsilon} + \cdots\right)$$

donc, à 1/300 près, proportionnel à $\dot{x}$ comme dans les séismographes à bobines mobiles.

Chaque pendule est muni de deux couples de bobines permettant l'emploi de deux galvanomètres. La masse de 100 kg. rend leurs réactions négligeables. La

[1] H. E. McComb et J. H. Nelson: Trans. Amer. Geophys. Un. **1940**, 236.
[2] H. Benioff: Bull. Seism. Soc. Amer. **22**, 155 (1932).
[3] H. Benioff: Bull. Seism. Soc. Amer. **25**, 283 (1935).

fig. 23 montre en coordonnées doublement logarithmiques les courbes d'amplification correspondant aux galvanomètres de période 0,23 et 90 secondes utilisés
à Pasadena. La grande masse de l'appareil est précieuse pour de pareilles asso-

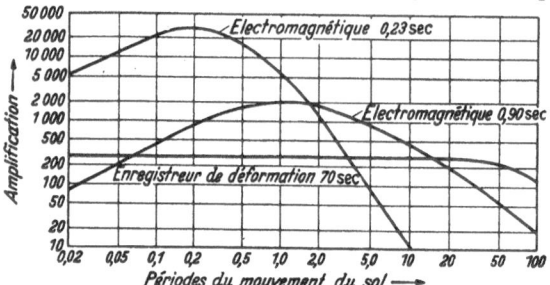

ciations; aucun séismographe
courant ne donne de résultats
comparables. Par contre l'emploi des pendules BENIOFF
avec des galvanomètres de
période moyenne n'offre pas
d'avantages marqués sur les
séismographes à réaction beaucoup plus légers à périodes
de l'ordre de la seconde, qu'ils
soient à bobine mobile (GRE-
NET et Mme. BAYARD DU-
CLAUX, BENIOFF lui-même) ou

Fig. 23. Courbes d'amplification des séismographes BENIOFF de Pasadena (d'après GUTENBERG et RICHTER). (Explication de la troisième
courbe à la Sect. 17.)

à réluctance variable (HILLER, PETERSCHMITT). WILLMORE a montré[1] que la
bobine mobile était préférable pour un appareil portatif.

# E. Enregistreurs de Déformation séismique
## (Strain Seismographs).

**17.** A l'aide de son traducteur électromagnétique BENIOFF a réalisé à Pasadena
les premiers séismographes non pendulaires[2]: deux piliers $A$ et $B$ sont enfoncés
dans le roc à 20 mètres l'un de l'autre. Un tube de fer fixé en $A$ s'avance

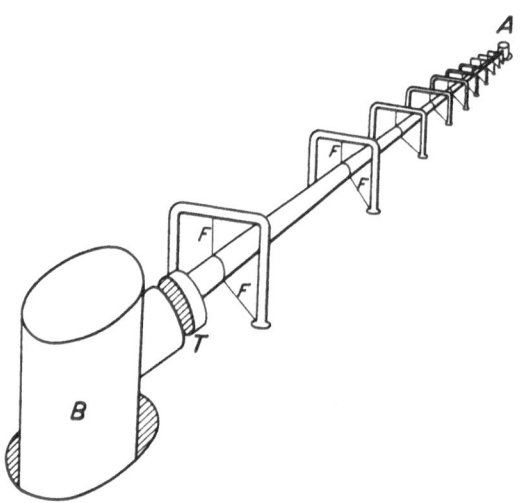

horizontalement jusqu'auprès
de $B$. Il porte l'aimant du
traducteur; $B$ porte les armatures bobinées. On mesure
ainsi les variations de distance
entre $A$ et $B$ produites par la
déformation du sol.

Le tube traverse 12 ar-
ceaux; chacun d'eux supporte
et maintient latéralement le
tube au moyen de 3 fils d'acier
tendus ne gênant pas les
déplacements longitudinaux
(fig. 24). Enfin le tube est
recouvert de 2 cm. d'amiante
pour éviter les sautes de température (la longueur du modèle
récent installé à Palomar ayant

Fig. 24. Schéma des enregistreurs de déformation de BENIOFF.

été portée à 50 m., son tube
est en quartz).

Les ondes séismiques attaquant le pilier $A$ ne sont pas susceptibles de provoquer
de vibrations appréciables du tube dont la période propre est de l'ordre du centième
de seconde. On peut donc admettre qu'il se comporte comme s'il était parfaitement rigide. Soit $L$ sa longueur, petite par rapport à la longueur d'onde

---

[1] P. L. WILLMORE: Mon. Not. Roy. Astronom. Soc., Geophys. Suppl. **6**, 129 (1950).
[2] H. BENIOFF: Bull. Seism. Soc. Amer. **25**, 283 (1935).

séismique; $u$ la composante du déplacement du sol suivant la direction $AB$, prise pour axe des $x$. Le déplacement relatif $\xi$ de $B$ par rapport à $A$ est donné par:

$$\xi = u_B - u_A = L\,\frac{\partial u}{\partial x}.\qquad(17.1)$$

Si, au cours d'une phase particulière d'un séisme, le déplacement longitudinal $u$ se propage le long de $AB$ avec une vitesse constante $c$, $u$ est une fonction de $x - ct$ et on peut remplacer (17.1) par:

$$\xi = \frac{L}{c}\,\frac{\partial u}{\partial t},$$

tant que la phase durera.

La force électromotrice engendrée par le traducteur est proportionnelle à $\partial\xi/\partial t$ donc à $\partial^2 u/\partial t^2$, et l'équation du galvanomètre sera de la forme:

$$\vartheta'' + 2\alpha\,\omega\,\vartheta' + \omega^2\vartheta = \frac{P}{c}\,\frac{\partial^2 u}{\partial t^2},$$

$P$ étant une constante. Elle a même forme que l'équation d'un séismographe pendulaire à inscription directe ayant même période et même amortissement. La comparaison des deux appareils pourrait fournir la vitesse apparente $c$.

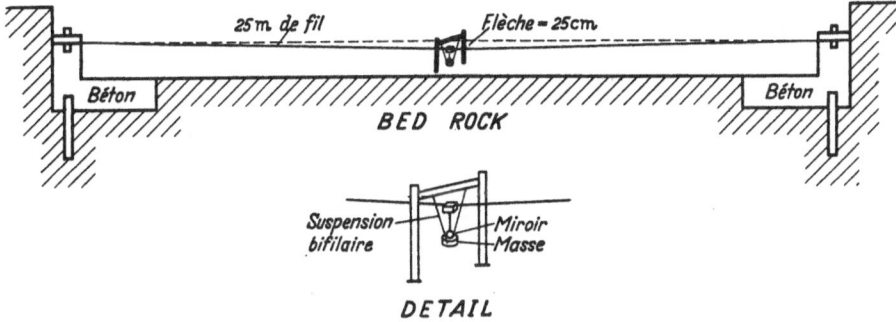

Fig. 25. Enregistreur de déformation de Sassa.

L'enregistreur de déformation est surtout intéressant pour les longues périodes car il n'est pas gêné par les inclinaisons lentes des piliers. La fig. 23 contient une courbe d'amplification avec galvanomètre de 70 secondes. La sensibilité de cette combinaison a été augmentée en 1952, tandis qu'on en ajoutait une autre de sensibilité réduite, comportant un galvanomètre de 3 minutes de période.

Sassa a remplacé le tube par un fil d'invar tendu par un ressort à boudin et déplaçant une bobine dans le champ d'un aimant fixe. Dans un autre appareil (fig. 25) destiné aux mouvements lents, Sassa a tenté de mesurer directement la variation de distance de deux ancrages, donc $\xi$ et non $\partial\xi/\partial t$, au moyen de la variation de flèche d'un fil de superinvar chargé en son milieu. Un bifilaire transforme en rotations les déplacements de la charge.

Un séismographe fournissant le *gradient vertical du déplacement vertical* $\partial w/\partial z$, dont le principe avait également été indiqué par Benioff, a été employé pour l'étude des ondes engendrées par des explosions[1]; sa longueur était seulement 2 mètres et son déplacement mesuré par un pick-up de phonographe (fig. 26).

---

[1] Lynn G. Howell, E. F. Neuenschwander et A. L. Pierson III: Geophysics **18**, 41 (1953).

BENIOFF a également proposé de mesurer la dilatation cubique $\frac{\partial u}{\partial x} + \frac{\partial v}{\partial y} + \frac{\partial w}{\partial z}$ au moyen d'un liquide remplissant une cavité du sol (dont le séisme ferait varier le volume) fermée par un diaphragme actionnant le traducteur. Une fuite capillaire éliminerait les variations lentes.

Reste à voir l'effet des diverses espèces d'ondes séis·miques sur les enregistreurs de déformation. Nous nous bornerons à l'appareil classique mesurant $\xi = L\,\partial u/\partial x$ (17.1). Dans les ondes $P$ ou $S$, ce qui compte (comme d'ailleurs pour un séismographe horizontal pendulaire) c'est la projection horizontale du mouvement complet produit à la surface par l'onde incidente et par ses ondes réfléchies. La vitesse apparente de propagation $c$ de ce mouvement horizontal est supérieure à la vitesse réelle $c \sin i$ ($i$ angle d'incidence); pour $i = 0$ l'appareil n'enregistre rien. Pour les ondes $R$ de RAYLEIGH ou $Q$ de LOVE, $i = \pi/2$, le mouvement apparent se confond avec le mouvement réel.

Fig. 26. Enregistreur de déformation verticale.

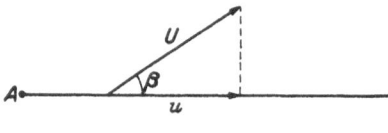

Soit $U$ le déplacement horizontal, $r$ la coordonnée suivant la propagation. On a (fig. 27): $u = U \cos \beta$ et $x = r \sec \alpha$ d'où
$$\xi = L \cos \beta \cos \alpha \frac{\partial U}{\partial r} = -\frac{L}{c} \cos \beta \cos \alpha \frac{\partial U}{\partial t}.$$

Pour les ondes $P$, $SV$ ou $R$, $\beta = \alpha$; le diagramme polaire de la quantité mesurée dépend de $\cos^2 \alpha$ au lieu de $\cos \alpha$ pour un pendule. En particulier la réponse est la même pour $\alpha$ et $\alpha + \pi$. La direction de l'épicentre n'étant connue qu'à $\pi$ près par l'impetus des $P$ sur des pendules horizontaux, des séismographes à déformations indiqueront s'il s'agit d'une condensation ou d'une dilatation, levant ainsi l'ambiguïté d'azimut.

Pour les ondes $SH$ ou $Q$, $\beta = \alpha + \pi/2$; le diagramme polaire dépend de $\sin \alpha \cos \alpha$ au lieu de $\sin \alpha$ pour un pendule. L'appareil n'enregistre ni pour $\alpha = 0$, ni pour $\alpha = \pi/2$.

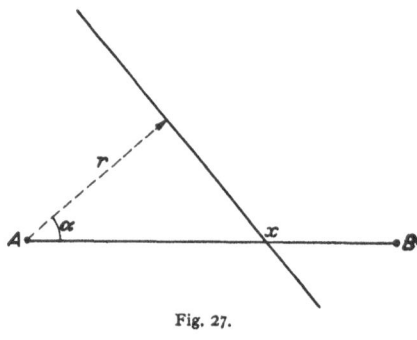

Fig. 27.

Si l'on ajoute les courants provenant de deux appareils à angle droit, la réponse aux ondes de RAYLEIGH est indépendant de l'azimut, la réponse aux ondes de LOVE est nulle. Si on possède en outre les données de séismographes pendulaires, on séparera $Q$ et $R$ et on précisera les directions et les sens de propagation[1].

---

[1] H. BENIOFF et B. GUTENBERG: Bull. Seism. Soc. Amer. **43**, 229 (1952).

# F. Séismographes électroniques.

**18. Amplification électronique.** En séismologie, l'électronique a trouvé sa première application dans les séismographes de prospection. Nous n'en parlerons guère, car les problèmes à résoudre sont nettement différents des nôtres (périodes très courtes, amplification très grande mais inconnue, filtrage des phases utiles, mixage, emploi à des dates prévues mais en des lieux variables, etc. ...). C'est assez récemment qu'on a cherché à faire profiter les séismographes véritables des gains élevés obtenus par l'amplification électronique, pour pouvoir utiliser des galvanomètres rustiques, pour centraliser des indications (stations tripartites pour l'étude des microséismes), ou enfin pour éviter les inconvénients de l'enregistrement photographique (prix, durée du développement).

La variété des principes qui peuvent être appliqués est très grande. Nous tâcherons d'abord de classer les exemples les plus suggestifs[1].

*α) Amplification directe des tensions de sortie données par le pendule d'un séismographe électromagnétique.* Les tensions sont proportionnelles à la vitesse de déplacement de la masse; d'autre part l'impédance de sortie est forcément assez faible. Bien qu'on réalise aujourd'hui des amplificateurs transmettant les mouvements lents avec une amplification suffisante (ceux qui sont employés en électro-encéphalographie sont sensibles au microvolt en courant continu), la solution doit être délicate et coûteuse sauf pour les courtes périodes, où elle semble avoir été essayée.

*β) Amplification photo-électrique des mouvements d'un pendule.* Un dispositif de ce genre a déjà été utilisé par NÉEL[2]: un écran rectiligne porté par un pendule inversé à translation coupait l'image réelle d'une fente éclairée et faisait ainsi varier le flux reçu sur une cellule. Pour certaines mesures (magnétiques), NÉEL opposait une seconde cellule éclairée directement; ce dispositif s'adapterait aisément à l'enregistrement.

HUGHES[3] projette un spot sur deux cellules dont les tensions sont amplifiées symétriquement (la plupart des dispositifs que nous mentionnerons sont symétriques, pour éliminer les variations d'alimentation). Le courant de sortie commande d'une part le galvanomètre à enregistrement photographique, d'autre part l'alimentation de la lampe d'enregistrement; l'intensité lumineuse augmente avec la vitesse des mouvements et la densité de la trace varie peu.

*γ) Amplification des mouvements du galvanomètre d'un séismographe électromagnétique.* Cette amplification est en général photo-électrique. Dans le dispositif le plus simple, employé déjà par WOLFE en 1934, le pinceau lumineux éclaire une surface variable de la cellule. Le séismographe plus moderne de KELLER[4] comporte une amplificateur à grande constante de temps de liaison (5 à 7 sec.) se terminant par un push-pull attaquant les deux enroulements en opposition d'un grand galvanomètre à inscription mécanique. SULKOWSKI[5] utilise dans le même but un multiplicateur d'électrons.

Dans le dispositif de VOLK[6], le pinceau lumineux éclaire un point variable d'un coin optique d'opacité croissante placé devant la cellule. Le courant de celle-ci est modulé en la polarisant par un courant pulsé (200 à 1000 fois par

[1] L'auteur a été considérablement aidé par P. MOLARD, directeur de l'Observatoire de la Martinique, dans la rédaction de cette section et de la suivante.
[2] L. NÉEL: J. de Phys., sér. VII **4**, 118 (1939).
[3] D. S. HUGHES: Trans. Amer. Geophys. Un. **28**, 691 (1947).
[4] F. KELLER: Trans. Amer. Geophys. Un. **27**, 636 (1946).
[5] E. L. SULKOWSKI: Bull. Seism. Soc. Amer. **40**, 165 (1950).
[6] J. A. VOLK: Bull. Seism. Soc. Amer. **40**, 169 (1950).

seconde) et peut ainsi être amplifié par un amplificateur normal pour courant alternatif. En changeant le coin ou les impulsions commandant la cellule on obtient une grande variété de courbes de réponse en fonction de *l'amplitude* du mouvement.

Wilson et Burgess[1] emploient deux cellules symétriques dont les tensions de sortie attaquent un amplificateur symétrique à montage cathodyne. Celui-ci joue le rôle de transformateur d'impédance: le gain en tension est légèrement inférieur à 1, mais comme les résistances de charge des cellules sont de $10^7\omega$ le gain en puissance, de l'ordre de 5700, permet l'emploi du galvanomètre de Keller. L'ensemble est très stable, suffisamment indépendant des variations résiduelles des tensions d'alimentation, mais présente une résonance à 1,5 sec., période du galvanomètre.

Donnons enfin un exemple d'amplification non photo-électrique. Dans un séismographe exposé au Palais de la Découverte, à Paris, le courant alternatif du secteur est envoyé (sans grand inconvénient) dans un enroulement placé sur l'aimant· d'un galvanomètre. Le courant induit dans le cadre est repris par un transformateur, amplifié, et envoyé dans un wattmètre enregistreur (pour obtenir une déviation dans les deux sens).

δ) *Variation de fréquence d'un courant de haute fréquence, produite par le déplacement d'un pendule.* Dans les dispositifs passant par l'intermédiaire de courants de haute fréquence, le bras du pendule commande en général deux capacités symétriques (lorsque le pendule dévie, l'une augmente, l'autre diminue de la même quantité) par déplacement de la lame centrale d'un condensateur à 3 armatures, parallèlement aus lames fixes (Molard), ou perpendiculairement. Dans ce dernier cas, la variation est moins linéaire et la sensibilité n'est pas nécessairement plus grande, car elle est limitée par l'amplitude des mouvements lents qu'on dcit accepter sans déréglage de l'amplificateur (Sect. 19).

Un premier système, employé déjà par Haeno (1931), retrouvé par Molard et probablement par Blumberg (1948) fonctionne sans faillir depuis 1942 à l'observatoire de la Martinique[2]: Deux oscillatrices dont les circuits d'accord contiennent les capacités variables ont leurs fréquences décalées (autour de 3500 et 3950 Kc respectivement). Les battements des circuits sont détectés, amplifiés et reçus sur un discriminateur qui transforme les variations de la fréquence autour de 450 Kc en variations d'intensité. Les courants de sortie attaquent deux amplificateurs basse fréquence commandant deux galvanomètres; l'une des combinaisons a une réponse voisine de celle d'un séismographe mécanique de 10 secondes (contre 12 s. au pendule) mais avec un grandissement 2000; l'autre a une réponse voisine de celle d'un séismographe mécanique de 1 s. mais avec un grandissement 20000. Dans un montage plus récent les capacités variables sont remplacées par des selfs (ferrites liées au bras plongeant dans des bobines fixes).

Un principe voisin est appliqué dans le séismographe de Leet et Linehan[3] et dans un appareil de Benioff, non encore décrit en détail. Ils comportent un oscillateur unique, à fréquence fixe (Chez Benioff un oscillateur à quartz dont on utilise le 2ème harmonique pour éviter la réaction des circuits). Les capacités variables font partie de deux circuits résonants à faible inductance mutuelle, couplés par induction à l'oscillateur. Quand le bras se déplace, un des circuits s'approche de la résonance, l'autre s'en éloigne. On amplifie, on redresse et on

[1] R. M. Wilson et L. R. Burgess: Bull. Seism. Soc. Amer. **42**, 341 (1952).
[2] P. Molard: Ann. Géophys. **3**, 24 (1947).
[3] L. D. Leet and D. Linehan: Earthquake Notes **17**, 7 (1947).

oppose les tensions produites. Le zéro est indifférent aux variations de fréquence de l'oscillateur dans la mesure où celui-ci reste sur la tangente d'inflexion à la courbe de résonance. LEET et LINEHAN commandent un potentiomètre enregistreur par un étage cathodyne. BENIOFF termine par un pont de résistances: le courant de sortie du pont traverse une forte capacité dont la constante de temps avec les résistances du pont peut atteindre 1000 secondes; nous en verrons le but dans la section suivante.

Citons enfin le séismographe de VOLK et ROBERTSON[1] où la variation de fréquence n'apparaît pas explicitement: les condensateurs du pendule font partie d'un pont de capacités dont une diagonale est alimentée en H.F. Le pont est équilibré pour une des déviations extrêmes du pendule. Aucune tension n'apparaît alors dans la seconde diagonale dont les extrémités sont reliées à la cathode et à la grille d'une lampe penthode. Si le bras se déplace, la tension H.F. sur la grille est proportionnelle à la valeur absolue du déplacement. La position du zéro dépend évidemment de la stabilité du générateur H.F. Cet appareillage centralise les enregistrements des stations tripartites pour l'étude des microséismes de période 0,5 sec. L'amplification atteint $10^6$.

ε) *Amplification électrostatique.* Les séismographes GANE[2] (période 0,23 sec., suspension à torsion même pour les verticaux) sont seulement destinés à l'enregistrement des coups de toit dans les mines du Witwatersrand.

L'armature isolée est un tube d'aluminium auquel le fil de torsion est fixé suivant une génératrice

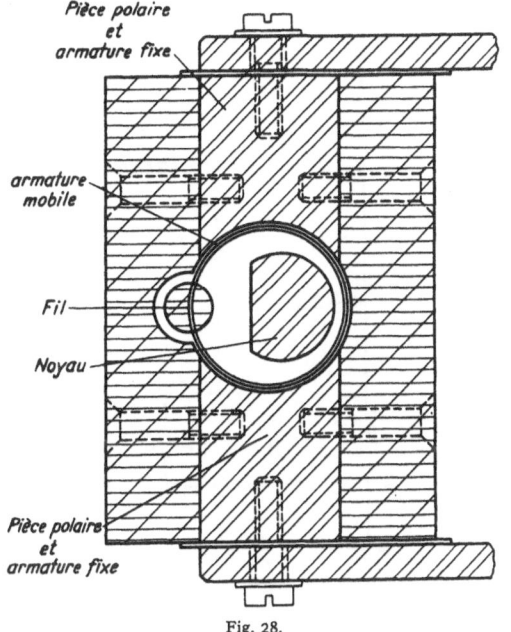

Fig. 28.

(fig. 28). Les armatures fixes, portées au potentiel du sol et à 150 V., servent de pièces polaires pour l'amortissement magnétique. Les tensions induites sont appliquées sur la grille d'une penthode montée en cathodyne avec grille isolée (résistance équivalente $10^9\omega$). Les tensions recueillies sur la cathode sont amplifiées et enregistrées photographiquement par un galvanomètre à courte période. Dans une installation ultérieure, centralisée, le courant de sortie module en fréquence des ondes porteuses de fréquence audible transmises elles-mêmes par modulation d'amplitude d'un émetteur à ondes courtes[3].

Un séismographe construit par MOLARD pour la Guadeloupe utilise un pendule analogue à celui de la Martinique décrit en δ). Les armatures fixes sont liées chacune à la grille d'un tube électromètre; l'armature mobile est portée à + 200 V. La constante de temps des circuits de grille est de l'ordre de 1 sec. Les courants de plaque sont appliqués à travers une liaison par capacités à un ampli basse

[1] J. A. VOLK et FL. ROBERTSON: Bull. Seism. Soc. Amer. **40**, 81 (1950).

[2] P. G. GANE: Bull. Seism. Soc. Amer. **38**, 95 (1948).

[3] P. G. GANE, H. J. LOGIE et J. H. STEPHEN: Bull. Seism. Soc. Amer. **39**, 117 (1949).

fréquence à deux chemins actionnant chacun un galvanomètre à inscription mécanique (réponses équivalentes à des pendules de 1 et 5 sec., de grandissement 15000 et 2000).

ζ) *Variation de phase d'un courant de haute fréquence par le déplacement d'un pendule.* Ce montage, en essai à l'Institut de Physique du Globe de Paris, utilise d'une part le fait que la tension aux bornes d'un circuit oscillant à forte surtension varie très rapidement au voisinage de la résonance et, d'autre part, l'existence de tubes radio (EQ 40) fournissant un courant proportionnel au temps pendant lequel deux de leurs grilles sont positives, d'une façon presque indépendante

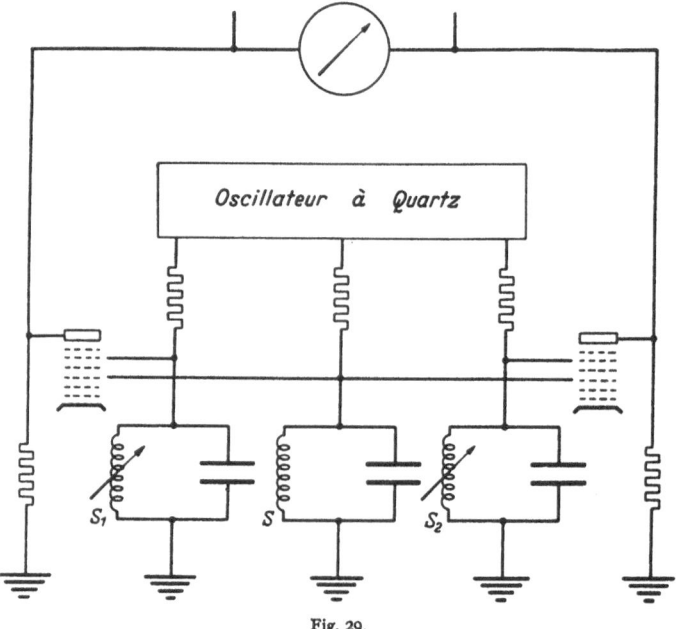

Fig. 29.

de la valeur des tensions appliquées. Si ces tensions sont sinusoïdales et décalées en phase d'un angle $\varphi$ (qui ne soit pas trop grand) le tube fournit un courant proportionnel à $\pi/2 - \varphi$.

Un oscillateur à quartz alimente trois circuits résonants à travers des résistances égales à l'impédance de ces circuits pour la résonance. L'un des circuits, dit de référence, a une self fixe $S$ (fig. 29). Les selfs $S_1$ et $S_2$ des deux autres circuits varient symétriquement lorsque le pendule se déplace, comme dans le séismographe MOLARD cité en δ). A l'équilibre, ces circuits sont sur la résonance, le circuit de référence étant décalé en phase. L'angle de phase entre le circuit de référence et chacun des deux autres commande deux tubes discriminateurs dont on oppose les courants de sortie.

**19. Possibilités des séismographes électroniques.** Bornons-nous aux séismographes électroniques non électromagnétiques dans lesquels le courant à la sortie de l'amplificateur principal est proportionnel au déplacement du pendule. Si la période du galvanomètre est courte, la courbe donnant l'amplification en fonction de la pulsation $n$ est celle d'un séismographe direct. Si la période du galvanomètre est notable, le facteur d'amplification est de la forme:

$$\frac{n^2}{\sqrt{(n^2 - \omega^2)^2 + 4\alpha^2 \omega^2 n^2}\,\sqrt{(n^2 - \Omega^2)^2 + 4\beta^2 \Omega^2 n^2}}$$

avec des notations analogues à celles de la Sect. 13. On a $n^2$ au numérateur au lieu de $n^3$ dans les séismographes électromagnétiques, où l'induction crée une dérivation supplémentaire. Cette substitution déplace le maximum d'amplification vers les grandes périodes. Ainsi pour $\omega = \Omega$, $\alpha = \beta$, le maximum a lieu pour la période propre $T_0$ contre $0{,}58\,T_0$ dans le GALITZINE. Mais, pas plus que pour les séismographes électromagnétiques, cette égalité des périodes et des amortissements ne représente forcément la meilleure combinaison. Par exemple, la puissance de l'amplification électronique permet d'envisager la réalisation d'appareils où $\alpha$ et $\beta$ seraient simultanément très grands, $\omega$ et $\Omega$ très écartées. L'amplification serait alors constante dans une gamme étendue.

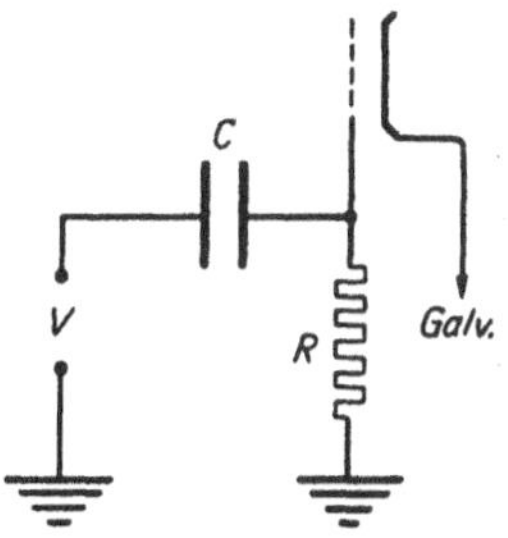

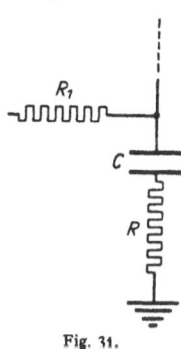

Fig. 30.

On peut éviter d'inscrire les mouvements de dérive lente du pendule ou de l'amplificateur en introduisant entre l'amplificateur et le galvanomètre une ou plusieurs liaisons basse fréquence par capacité. Le circuit est schématisé dans la fig. 30. $V$ est la tension appliquée. Si on suppose le fonctionnement linéaire et le galvanomètre de période très courte l'équation donnant la charge du condensateur est du 3ème ordre. Sa discussion[1] montre que les mouvements dont la période est faible devant la constante de temps $RC$ ou du même ordre qu'elle sont transmis fidèlement; les autres sont atténués et déphasés. Dans le cas d'un mouvement sinusoïdal permanent par exemple, le couplage introduit dans l'amplification un facteur $RCn(1 + R^2C^2n^2)^{-\frac{1}{2}}$ qui la rend analogue à celle d'un pendule de période plus courte.

Le même circuit, dans lequel la tension appliquée $V$ est fixe tandis que la capacité $C$ est variable avec la position du pendule, peut schématiser les séismographes électrostatiques; il faut seulement que la variation $\Delta C$ de capacité soit assez faible (quelques %) pour qu'elle soit équivalente à une variation $V\dfrac{\Delta C}{C}$ de tension sur la grille. Les séismographes électrostatiques ne transmettent pas la dérive lente du pendule. Mais il n'y a pas de difficultés pour la transmission des périodes utiles, les lampes électromètres permettant de donner au circuit de grille des constantes de temps énormes.

Fig. 31.

Des circuits plus complexes permettraient de modifier les caractéristiques du séismographe de façon presque arbitraire. Donnons quelques exemples: plusieurs couplages successifs par condensateurs donneront une coupure brusque à partir d'une période déterminée. Un circuit comprenant une résistance en série et une capacité en dérivation supprimera les mouvements de courte période. Le circuit de la fig. 31 introduit un facteur $[1 + R^2C^2n^2]^{\frac{1}{2}}\,[1 + (R + R_1)^2C^2n^2]^{-\frac{1}{2}}$ qui permet, tout en conservant une réponse aux mouvements rapides, de compenser pour une certaine gamme de périodes la diminution d'amplification due au pendule.

Théoriquement des possibilités analogues existent avec les dispositifs électromagnétiques, mais la faible impédance des circuits conduit à des capacités énormes

---

[1] P. MOLARD: Ann. Géophys. 3, 24 (1947).

pour avoir des constantes de temps de l'ordre des périodes séismiques, tandis que les impédances de grille sont normalement de l'ordre du mégohm.

Enfin on dispose, à la sortie d'un amplificateur électronique, d'une puissance suffisante pour l'utiliser en partie à modifier le mouvement du pendule en l'envoyant dans les bobines d'amortissement. Molard[1] a envisagé d'abord une réaction positive à traverse une liaison par capacité pour allonger la période du pendule sans diminuer sa stabilité, puis une réaction négative appliquée au pendule par l'intermédiaire d'un circuit comprenant un condensateur en dérivation; ce dernier montage peut compenser à 5% près par exemple, les variations lentes du pendule ou de l'amplificateur. Il est en essai, sous forme symétrique

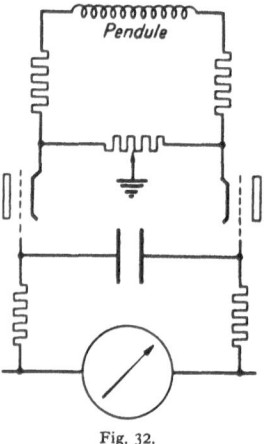

Fig. 32.

(fig. 32) avec l'amplificateur mentionné en $\zeta$ (Sect. 18.) Dans un montage non décrit en détail, Benioff utilise même un circuit intégrateur de Miller avec une constante de temps de 20 minutes pour supprimer seulement les mouvements les plus lents.

Augmenter l'amplification est inutile si les fluctuations dues à l'amplificateur sont du même ordre que l'agitation microséismique. Dans les dispositifs photoélectriques l'instabilité provient surtout des variations d'éclat de la lampe. Dans les dispositifs utilisant une discrimination de fréquence l'instabilité est surtout attribuable aux variations de fréquence des oscillatrices. On a donc intérêt à avoir une modulation aussi profonde que possible pour un déplacement donné. Mais on est vite limité par la dérive du pendule: en un jour, la variation de fréquence correspondante ne doit pas faire sortir de la bande passante du discriminateur. L'emploi d'une réaction permet d'éluder cette condition. Il semble néanmoins que le maximum soit une amplification de $5\cdot10^4$ pour les périodes longues (contre $10^6$ pour les périodes courtes) qui serait d'ailleurs inutile à cause de l'agitation microséismique si elle ne permettait d'imposer à l'appareil des conditions supplémentaires.

Enfin dans les séismographes électrostatiques on peut admettre une limite $V = 10^{-4}$ V. pour la tension détectable par la lampe électromètre. Si l'armature centrale est à 100 V. et si la capacité varie de 0 à son maximum pour 1 cm. de déviation du pendule on peut théoriquement mesurer $10^{-6}$ cm. donc atteindre une amplification de l'ordre de $10^4$.

## G. Dispositifs accessoires.

**20.** Nous groupons dans la présente section quelques indications brèves et non techniques sur les dispositifs accessoires de la séismologie. D'abord les enregistreurs: bien qu'on ait cherché récemment, de divers côtés, à utiliser des enregistrements «vectoriels» qui cinématographient la projection horizontale du mouvement, rappelant ainsi les débuts de la séismologie, la quasi totalité des inscriptions se fait par composantes séparées. Comme les séismes sont rares, la ligne des zéros (non tracée) est formée de parallèles successives aussi rapprochées que le permettent la dérive et l'agitation microséismique.

Dans l'enregistrement mécanique ordinaire, un style équilibré par un contrepoids inscrit un trait fin sur une bande de papier dont les deux extrémités sont collées. Cette bande passe sur deux rouleaux horizontaux; celui du haut est

---

[1] P. Molard: Ann. Géophys. **3**, 24 (1947).

moteur, l'autre tend le papier, leurs axes font un petit angle (fig. 33), et le glissement du papier décale progressivement les lignes (MAINKA). A l'inscription classique sur noir de fumée, qui donne le minimum de frottement, BENIOFF substitue une inscription par style chauffant sur papier spécial[1]. Dans les séismographes électroniques, c'est en général l'encre qui est préférée.

Dans l'enregistrement photographique le trait est moins uniforme car il dépend de la vitesse du spot. Le papier est appliqué sur un tambour de grand diamètre animé d'un mouvement hélicoïdal. BENIOFF a aussi employé le film de 35 mm., qui est économique mais qui doit être lu au microscope ou en projection[2].

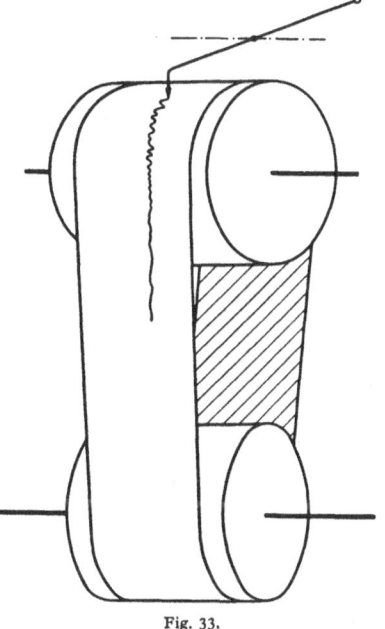

Pour l'enregistrement à grande vitesse des séismes proches un intermédiaire intéressant est un enregistrement sur ruban magnétique, soit direct, soit par modulation d'une onde porteuse. On efface le ruban après dépouillement visuel[3] ou automatique[4], et reproduction des séismes enregistrés; on peut également faire une lecture accélérée pour analyser l'énergie des ondes ou le spectre de l'agitation microséismique par les moyens employés en acoustique[5].

Les enregistreurs sont entraînés par des mouvements d'horlogerie à régulateur centrifuge ou des moteurs synchrones. Leur régularité et celle de l'horloge devraient permettre l'interpolation à 0,1 seconde près entre les marques de temps faites en général chaque minute; c'est un idéal rarement atteint. Dans l'enregistrement mécanique, les marques de temps s'obtiennent en soulevant le style ou parfois en le déplaçant latéralement; ce dernier procédé, peu lisible au cours des séismes, est nécessaire dans

Fig. 33.

l'inscription à l'encre, qui reprend mal après une perte de contact. Il convient surtout à l'enregistrement photographique, où le décalage du spot amène deux interruptions nettes. Enfin la marche de l'horloge elle-même pourra être suivie si l'on inscrit sur la feuille d'enregistrement les signaux horaires radio-électriques au moyen de relais à retard négligeable.

Pour éviter les microséismes dûs au vent, on installe en général les séismographes dans des caves, sous un terrain plat. Sauf si l'humidité du sous-sol est à craindre, les piliers sont isolés du plancher pour éviter la transmission des pas.

En été, l'air provenant du dehors gagne à être refroidi comme à Iéna ou condensé sur une paroi froide comme à Clermont-Ferrand. Les pendules à faible masse sont sensibles aux mouvements de l'air et leurs cages doivent être étanches. Pour éviter l'effet du courant d'air sur la cage elle-même, WILIP supprime les vis calantes et boulonne la cage sur le pilier.

[1] H. BENIOFF, B. GUTENBERG et C. F. RICHTER: Trans. Amer. Geophys. Un. **34**, 785 (1953).
[2] D. S. CARDER: Bull. Seism. Soc. Amer. **35**, 175 (1945).
[3] H. DÖRMANN: Ann. Géophys. **8**, 286 (1952).
[4] P. GANE, H. J. LOGIE et J. H. STEPHEN: Bull. Seism. Soc. Amer. **39**, 117 (1949).
[5] Engineering and Science Monthly, Cal. Inst. of Technology, Nov. 1951.

Signalons pour finir l'intérêt que présente, pour mettre en évidence les défauts mécaniques des séismographes légers (résonances dans la suspension) les plateformes auxquelles on peut communiquer des mouvements connus. Il y en a de nombreux modèles, certains très perfectionnés pour les mouvements horizontaux des séismographes ordinaires[1] ou pour les déplacements quelconques des séismographes de prospection[2].

## Bibliographie Générale.

[1] WIECHERT, E.: Theorie der automatischen Seismographen, Abh. Ges. Wiss. Göttingen, Math.-Phys. Kl., N.F. **2**, 1 (1903).

[2] GALITZINE, B.: Vorlesungen über Seismometrie, Deutsche Bearbeitung. Leipzig u. Berlin: Teubner 1914.

[3] BERLAGE, H. P. jr.: Seismometer. In Handbuch der Geophysik, Bd. IV, Abschnitt IV, S. 299. 1932.

[4] SOHON, F.W.: Seismometry, Introduction to theoretical Seismology, Part II, New-York: Wiley 1932.

[5] COULOMB, J. et G. GRENET: Nouveaux principes de construction des séismographes électromagnétiques. Ann. Phys., Paris, Sér. XI, **3**, 321 (1935).

[6] GRENET, G.: L'étalonnage des séismographes électromagnétiques modernes. Ann. Géophys. **2**, 329 (1946).

[7] GALITZINE, B.: Seismometrische Tabellen. St. Pétersbourg 1911.

[8] STOKER, J. J.: Nonlinear vibrations. New York: Interscience Publishers 1950.

[9] Mc. LACHLAN, N.W.: Ordinary non linear differential equations in Engineering and Physical Sciences. Oxford: Clarendon 1950.

[10] Mc. LACHLAN, N.W.: Theory and Application of Mathieu functions. Oxford: Clarendon 1947.

---

[1] A. C. RUGE: Bull. Seism. Soc. Amer. **26**, 201 (1936).

[2] M. WEBER: Helv. phys. Acta **22**, 425 (1949). — Eidg. Techn. Hochschule Zürich, Mitt. aus dem Inst. für Geophysik Nr. 14. P. M. HONNELL: Geophysics **18**, 160 (1953). Ces articles contiennent une bibliographie suffisante.

# Seismic Wave Transmission.

By

## K. E. BULLEN.

With 9 Figures.

## Introduction.

When an earthquake occurs, energy, which may reach the order of $10^{25}$ ergs in the greatest earthquakes is released, within a few seconds or less, from a focal region inside the Earth whose linear dimensions may be of the order of several kilometres. The (somewhat indefinite) centre of the focal region is called the *focus*.

Seismic waves are transmitted outwards from the focal region. They include bodily waves which may penetrate to all parts of the Earth's interior, and also seismic surface waves which are transmitted along surfaces of discontinuity, especially over the Earth's outside surface. There are at present some six hundred seismological observatories distributed over the Earth, each of which has one or more *seismographs* which respond to arriving earthquake waves from foci that may be as far away as the antipodes in large earthquakes. A seismograph records a trace or *seismogram* on paper on a revolving drum, and the seismograms provide the primary observational data on seismic waves. They bear evidence on the origin-time of each earthquake, the location of the focus, the mechanism of energy release at the focus, the size of the earthquake, and the mechanical properties of the regions traversed by the ensuing waves. A summary of the routine readings of seismograms at all observatories, together with determinations of origin-times and focal positions, is published in the *International Seismological Summary*[1]. The data in the *I. S. S.*, supplemented as required from studies of the original seismograms, constitute the principal raw material which has to be fitted by the theory of seismic wave transmission.

The seismic evidence shows that the outermost 30 to 40 km. of the Earth, often referred to as the crustal layers, are much more heterogeneous than the region below, at least in continental regions. (There is evidence that the crustal layers are appreciably thinner below the oceans.) The present chapter will be concerned largely with *bodily waves* that have penetrated some distance below the crustal layers, and with their bearing on problems of the physical properties of the Earth's deeper interior. Problems of crustal structure, which involve "near-earthquake" theory and surface-wave theory, and wider problems of the physics of the Earth, are being treated in chapters by other authors.

## A. Deformation theory.

**1. General remarks.** The Earth transmits seismic waves because it consists of deformable matter, so that the investigation of the properties of seismic waves must rest on elasticity theory. (For details, complementary to those in the present Section, of the elasticity theory needed in seismology, see Reference [1]. Alternative accounts may be found in References [8], [11] and [16].) On the small

---

[1] At present, the *I.S.S.* is prepared at Kew Observatory under the direction of Sir HAROLD JEFFREYS of Cambridge. The volume for 1946 has appeared in 1955.

scale, the material of the Earth obviously deviates from simple conditions such as perfect elasticity and isotropic behaviour. The early problems in seismology are, however, large-scale ones in which the chief concern is with the transmission of waves over distances that are large compared with the dimensions of the focal region. Comparison with the observational data shows that in these problems it is permissible to disregard the deviations from a simple isotropic perfect elasticity theory. In considering strains caused by seismic waves, in excess of the large strains already present in the equilibrium state in the Earth's deep interior, it is further permissible to use infinitesimal strain theory, in which only the lowest powers of the components of strain are retained in equations. Also, although the Earth contains innumerable small-scale discontinuities, it is customary in general to treat functions of position as differentiable except at certain notable surfaces of discontinuity. As occasion warrants, comments will be made on the influence of certain departures from these simple ideal conditions.

The observed periods of bodily seismic waves do not greatly exceed the order of a few seconds. It is therefore sufficiently accurate for seismological purposes to use cartesian frames of reference assumed fixed in the Earth, ignoring such longer-period effects as tidal deformations.

**2. Strain, dilatation and rotation.** Referred to the cartesian frame $O\,x_1\,x_2\,x_3$, let $x_i\,(i=1, 2, 3)$ give the position of a particle $Q$ when the Earth is undisturbed. At time $t$ during the passage of a seismic disturbance, let $\boldsymbol{u}$ be the (small) displacement of $Q$, and let $u_i$ be the components of $\boldsymbol{u}$. The displacement of a neighbouring particle, initially at $x_i+d\,x_i$, is then, to sufficient accuracy, given by

$$u_i + \sum_{j=1}^{3} \frac{\partial u_i}{\partial x_j}\, d\,x_j. \tag{2.1}$$

Now $\partial u_i/\partial x_j$ is a second-order cartesian tensor, expressible as the sum of symmetrical and antisymmetrical tensors in the form

$$\frac{\partial u_i}{\partial x_j} = e_{ij} - \xi_{ij}, \tag{2.2}$$

where

$$e_{ij} = \frac{1}{2}\left(\frac{\partial u_j}{\partial x_i} + \frac{\partial u_i}{\partial x_j}\right) \tag{2.3}$$

and

$$\xi_{ij} = \frac{1}{2}\left(\frac{\partial u_j}{\partial x_i} - \frac{\partial u_i}{\partial x_j}\right). \tag{2.4}$$

The symmetrical tensor $e_{ij}$ is the *strain tensor* at $Q$, and represents the deformation in the vicinity of $Q$. Consider, for example, a small block of material containing $Q$ which, before the deformation, was rectangular with edges parallel to the axes. The strain components $e_{11}, e_{22}$ and $e_{33}$ are the proportionate *extensions* of the block in the directions of the axes, while $2e_{23}, 2e_{31}$ and $2e_{12}$ are equal to the changes in the angles at the corners of the block and give the *angles of shear*. It is always possible to select the axes so that $e_{23}, e_{31}$ and $e_{12}$ are all zero; $e_{11}, e_{22}$ and $e_{33}$ are then called the *principal strains*, and the axes are the *principal axes of strain*.

Contracting the tensor $e_{ij}$ gives the (scalar) *dilatation* $\vartheta$ at $Q$. Thus

$$\vartheta = \sum_i e_{ii} = \sum_i \frac{\partial u_i}{\partial x_i} = \operatorname{div} \boldsymbol{u}, \tag{2.5}$$

A *compression* is a negative dilatation. The compression at a point may be measured by the proportionate increase in density in the vicinity of the point.

For certain seismological purposes, it is convenient to introduce the *strain deviation tensor* $E_{ij}$, defined by

$$e_{ij} = \tfrac{1}{3}\vartheta\,\delta_{ij} + E_{ij}, \tag{2.6}$$

where $\delta_{ij}$ denotes the KRONECKER delta; ($\delta_{ij}=1$ if $i=j$; $\delta_{ij}=0$ if $i \neq j$). Thus $E_{ij}$ gives the deviation of the strain from a purely symmetrical dilatation. When $i \neq j$, $E_{ij}=e_{ij}$.

The antisymmetrical tensor $\xi_{ij}$ is the *rotation tensor* at $Q$. Its three independent components $\xi_{23}, \xi_{31}, \xi_{12}$, are equal to the components of the vector $\tfrac{1}{2}\boldsymbol{\xi}$, where

$$\boldsymbol{\xi} = \operatorname{curl} \boldsymbol{u}. \tag{2.7}$$

It is easy to show that

$$\sum \frac{\partial E_{ij}}{\partial x_j} = \frac{2}{3}\frac{\partial \vartheta}{\partial x_i} - \sum_j \frac{\partial \xi_{ij}}{\partial x_j} \tag{2.8}$$

$$= \frac{2}{3}\frac{\partial \vartheta}{\partial x_i} - \frac{1}{2}(\operatorname{curl} \boldsymbol{\xi})_i, \tag{2.9}$$

where the last term in (2.9) denotes the $i$th component of curl $\boldsymbol{\xi}$.

By (2.1) and (2.2), the displacement in the neighbourhood of $Q$ is given as the resultant of three types of movement: the translation $u_i$, the deformation $\sum_j e_{ij}\, dx_j$, and the rotation $-\sum_j \xi_{ij}\, dx_j$. The kinematic effect of an earthquake on a given element of matter at any given time is compounded of these three movements. The usual seismograph is designed to measure only the changes in the translation components $u_i$ at given points, although BENIOFF[1] has made important progress in measuring the ground deformation caused by earthquakes. As yet, little progress has been made with instruments designed to measure ground rotations.

**3. Stress and pressure.** The stress at the point $Q$ is defined in terms of the forces or *tractions*, per unit area, across small plane interfaces containing the point. In general the stress varies with the orientation of the interface and is inclined to the normal to the plane. The stress across any small plane interface through $Q$ can be determined from knowledge of the nine components $p_{ij}$ $(i,j = 1, 2, 3)$ referred to any set of cartesian axes $O x_1 x_2 x_3$, where $p_{ij}$ denotes the component parallel to $O x_j$ of the stress at $Q$ across an interface normal to the axis $O x_i$, acting on the matter on that side of the interface for which $O x_i$ is the outward normal. The set of components $p_{ij}$ constitutes the *stress tensor* at $Q$. In ordinary elasticity theory, and in particular in seismology, couples across vanishingly small plane interfaces are disregarded; it then follows that the tensor $p_{ij}$ is symmetrical in $i, j$, and that only six of the nine components can be independent. *Principal* stresses and axes of stress can then be defined analogously to the case of strain.

Let $P$ be the mean of the three principal stresses. Then

$$P = \tfrac{1}{3}\sum_i p_{ii}, \tag{3.1}$$

the right-hand side of (3.1) being invariant for changes of the axes, so that $P$ is a scalar. In the particular case in which the three principal stresses are equal (which is always the case in a perfect fluid), the stress tensor takes the form

---

[1] H. BENIOFF: Bull. Seism. Soc. Amer. **25**, 283—309 (1935). — Progr. Rept., Seism. Laboratory, Calif. Inst. of Technology. Trans. Amer. Geophys. Un. **35**, 979—987 (1954). See also the chapter Séismométrie, Sect. 17, in this volume, p. 64.

$P\delta_{ij}$ and corresponds to a hydrostatic pressure $-P$. (The minus sign is involved because a positive pressure acts *inwardly* across an interface.)

The *stress deviator* $P_{ij}$ is defined, analogously to $E_{ij}$, by

$$p_{ij} = P\,\delta_{ij} + P_{ij}, \tag{3.2}$$

and represents the deviation of the stress $p_{ij}$ from a fully symmetrical stress. For $i \neq j$, $P_{ij} = p_{ij}$, and $p_{ij}$ is called a *shear component of stress*.

The scalar $(3 \sum_i \sum_j P_{ij}^2)^{\frac{1}{2}}$ is important in seismology since the value which it assumes when a given solid material is on the point of fracture is, on the theory of Mises[1], taken as the *breaking strength* of the solid. By (3.1) and (3.2),

$$3 \sum_i \sum_j P_{ij}^2 = (p_2 - p_3)^2 + (p_3 - p_1)^2 + (p_1 - p_2)^2, \tag{3.3}$$

where $p_1, p_2$ and $p_3$ are the principal stresses.

**4. Stress-strain relations; elastic parameters.** According to Hooke's law, generalised, the stress and strain components at the point $Q$ are connected by linear relations of the form

$$p_{kl} = \sum_i \sum_j A_{ijkl}\, e_{ij} \tag{4.1}$$

where the $A_{ijkl}$ constitute a set of 81 parameters, of which at most 36 are independent because $p_{ij}$ and $e_{ij}$ are both symmetrical in $i, j$. The number of independent parameters is reducible to 21 if the deformation takes place adiabatically or isothermally, for then the right-hand side of (5.2) (see Sect. 5) is a perfect differential, so that $\partial p_{kl}/\partial e_{ij} = \partial p_{ij}/\partial e_{kl}$ and $A_{ijkl} = A_{klij}$. The number is further reducible if there is some measure of symmetry in the elastic behaviour of the material, and in the case of completely isotropic behaviour reduces to two. In this last case, the relations (4.1) can be written in the form

$$p_{ij} = \lambda\,\vartheta\,\delta_{ij} + 2\mu\,e_{ij}, \tag{4.2}$$

where $\lambda$ and $\mu$ are the *Lamé elastic parameters*. The parameters in equations such as (4.1) and (4.2) depend on the material and on the thermodynamical conditions under which deformation takes place.

The relations (4.2) are characteristic of a perfectly elastic isotropic body on the infinitesimal strain theory. In this case, since $p_{ij} = 2\mu e_{ij}$ when $i \neq j$, it follows that the principal axes of stress and strain coincide. In seismology, it is convenient to replace the parameter $\lambda$ by $k$, where

$$k = \lambda + \tfrac{2}{3}\mu. \tag{4.3}$$

Then (4.2) become

$$p_{ij} = k\,\vartheta\,\delta_{ij} + 2\mu\,(e_{ij} - \tfrac{1}{3}\vartheta\,\delta_{ij}). \tag{4.4}$$

On contracting (4.4), we derive, using (2.5) and (3.1),

$$P = k\,\vartheta. \tag{4.5}$$

By (3.2), (4.4) then yield

$$P_{ij} = 2\mu\,(e_{ij} - \tfrac{1}{3}\vartheta\,\delta_{ij}),$$

i.e., by (2.6),

$$P_i = 2\mu\,E_{ij}. \tag{4.6}$$

The stress-strain relations (4.4) are equivalent to (4.5) and (4.6) together, only five of the equations (4.6) being independent since $\sum_i P_{ii} = \sum_i E_{ii} = 0$.

---

[1] R. v. Mises: Z. angew. Math. Mech. **8**, 161—185 (1928).

One advantage of using (4.5) and (4.6) is that the significances of the para-meters $k$ and $\mu$ are made clear. Thus $k$, being equal to the ratio of a hydrostatic pressure to the corresponding compression, measures the *incompressibility* of the material; the reciprocal of $k$ is the *compressibility*. And $\mu$, which gives the ratio of a shear-component of stress to the corresponding angle of shear, measures the *rigidity*. The value of $\mu$ discriminates between ideal fluids, for which $\mu = 0$, and solids; for ordinary solids $\mu$ is of the order of $10^{11}$ to $10^{12}$ dyn./cm.² The value of $k$ discriminates between liquids, for which $k$ is high (of order $10^{10}$ dyn./cm.²), and gases, for which $k$ is moderate (of order $10^6$ dyn./cm.²).

The incompressibility (or *bulk-modulus*) $k$ is commonly defined alternatively by

$$k = -\varrho \frac{dP}{d\varrho}, \tag{4.7}$$

where $\varrho$ denotes the density, a relation which is readily derivable from (4.5).

Poisson's *ratio* $\sigma$ is given in terms of $k$ and $\mu$ by

$$\sigma = \frac{3k - 2\mu}{2(3k + \mu)}; \tag{4.8}$$

In theory, $k$ and $\mu$ may range from zero to infinity, so that $\sigma$ may range from $-1.0$ to $+0.5$. In practice, however, $\sigma$ is found to be positive for all ordinary materials. For many solids, the value of $\sigma$ is found to be moderately close to 0.25, and the relation

$$\sigma = 0.25, \quad \text{or} \quad k = 1.67\mu, \tag{4.9}$$

referred to as Poisson's *relation*, has been frequently used in obtaining specimen numerical results in seismology when the algebra would otherwise be formidable. For a perfect fluid, $\sigma$ takes its greatest value of 0.5.

**5. Strain energy.** Consider a deformation in which the strain at the point $Q$ increases by the differential amount $de_{ij}$; let $dW'$ be the work per unit volume done by the external forces on a small block containing $Q$; and let $dW$ be the work done per unit volume against the internal forces. Let $dT$ be the increase in the kinetic energy of the block, $dq$ the external heat it receives, and $dU$ the increase in its internal energy, all measured per unit volume. By the principle of mechanical energy, and by the first law of thermodynamics, we then have

$$dW' - dW = dT,$$

and

$$dW' + dq = dT + dU,$$

so that

$$dW = dU - dq. \tag{5.1}$$

On interpreting $e_{11}, e_{23}$, etc., as for the strained block in Sect. 2, and total-ling the work (per unit volume) done by the separate components of $p_{ij}$, or of $P$ and $P_{ij}$, it is found that

$$dW = \sum_i \sum_j p_{ij} de_{ij}, \tag{5.2}$$

or

$$dW = P \, d\vartheta + \sum_i \sum_j P_{ij} \, dE_{ij}. \tag{5.3}$$

The formal equivalence of (5.2) and (5.3) is checked on applying (2.6) and (3.2). [For a more complete demonstration of (5.2), see Reference (1), Sect. 2.35.]

We now limit consideration to the case in which the deformation takes place adiabatically or isothermally. Then, by the first two laws of thermodynamics, $dU - dq$, and therefore $dW$, is a perfect differential in the $e_{ij}$, or in $\vartheta$ and the $E_{ij}$, and $W$ is called the *strain-energy function*. In the adiabatic case, (5.1) shows that $W$ is the internal energy per unit volume.

Now (on the infinitesimal strain theory) $W$ is, by (4.5), (4.6) and (5.3), a homogeneous quadratic function of $\vartheta$ and the $E_{ij}$ (neglecting an additive constant), so that by EULER's theorem

$$\left.\begin{aligned}
W &= \frac{1}{2}\vartheta\frac{\partial W}{\partial \vartheta} + \frac{1}{2}\sum_i\sum_j E_{ij}\frac{\partial W}{\partial E_{ij}} \\
&= \frac{1}{2}P\vartheta + \frac{1}{2}\sum_i\sum_j P_{ij}E_{ij},
\end{aligned}\right\} \tag{5.4}$$

by (5.3).

The relations (5.1) to (5.4) all apply to aeolotropic as well as isotropic materials. The right-hand side of (5.4) is equal to $\frac{1}{2}\sum\sum p_{ij}e_{ij}$, so that, by (4.1),

$$W = \frac{1}{2}\sum_i\sum_j\sum_k\sum_l A_{ijkl}e_{ij}e_{kl} \tag{5.5}$$

in the general aeolotropic case.

When, in particular, the stress-strain relations conform to the isotropic relations (4.5) and (4.6), $W$ takes the simple form

$$W = \frac{1}{2}k\vartheta^2 + \mu\sum_i\sum_j E_{ij}^2. \tag{5.6}$$

The formula (5.6) is important in connection with the accumulation of strain energy which leads to the emission of seismic waves from an earthquake focal region.

**6. The fundamental equation of seismic wave transmission.** At time $t$ during the transmission of a seismic (or other) disturbance, let $p_{ij}$ denote the excess of the stress at the point $Q$ above that in the steady equilibrium state. Let $u_i$, $\vartheta$ and $E_{ij}$ be the corresponding increases in the displacement, dilatation and deviatoric strain, and let $X_i$ be the change in the body force per unit mass. Application of the principle of linear momentum to an element of matter of density $\varrho$ containing $Q$ then yields

$$\varrho\frac{d^2 u_i}{dt^2} = \sum_j\frac{\partial p_{ij}}{\partial x_j} + \varrho X_i. \tag{6.1}$$

In (6.1) it is sufficiently accurate, on the elementary theory, to replace $d^2 u_i/dt^2$ by $\partial^2 u_i/\partial t^2$, and we shall use (3.2) to replace $p_{ij}$ by $P_{ij} + P\delta_{ij}$.

Substituting the stress-strain relations (4.5) and (4.6), we then derive, if we treat $k$ and $\mu$ as constants,

$$\varrho\frac{\partial^2 u_i}{\partial t^2} - \varrho X_i = k\frac{\partial\vartheta}{\partial x_i} + 2\mu\sum_j\frac{\partial E_{ij}}{\partial x_j} = \left(k + \frac{4}{3}\mu\right)\frac{\partial\vartheta}{\partial x_i} - 2\mu\sum_j\frac{\partial\xi_{ij}}{\partial x_j}, \tag{6.2}$$

by (2.8); or, in vector form, using (2.9),

$$\varrho\frac{\partial^2 u}{\partial t^2} + \varrho\,\mathrm{grad}\,V = \left(k + \frac{4}{3}\mu\right)\mathrm{grad}\,\vartheta - \mu\,\mathrm{curl}\,\xi, \tag{6.3}$$

taking $X_i$ to be derivable from a potential function $V$.

In the Earth, $k$ and $\mu$ are variable functions of position, so that, strictly, additional terms involving grad $k$ and grad $\mu$ should appear on the right-hand side of (6.3). The influence of such terms would be to introduce dispersion effects into the transmitted disturbances. Calculation shows that these effects can be important if the proportionate changes in $k$ and $\mu$ are not small over distances comparable with the predominating wave-lengths of the disturbance. This is likely to be the case in the crustal layers of the Earth, where, however, there are many other complications in the transmission of seismic waves; the seismology of this region is being treated by other authors. Below a depth of about 33 km. (the most probable value of the thickness of the crustal layers in continental regions), it is not normally necessary to correct (6.2) or (6.3) for variation in $k$ and $\mu$ except near a few levels of discontinuity. The variation will, however, cause the direction of the normal to wave-fronts to be changed as the waves progress—see Sect. 17.

The equation (6.3) is of central importance in the routine theory of seismic wave transmission. As already indicated, the equation leads to theory that accords well with observed large-scale behaviour. In special problems, it may be necessary to use more elaborate equations.

# B. The initial conditions.

**7. Causes of earthquakes.** Work of H. F. Reid[1] showed that generally, and this is invariably the case in larger shallow earthquakes, the immediate cause of an earthquake is the sudden release of elastic strain energy that has been accumulating for some time beforehand in the focal region. Earthquakes arising in this way are *tectonic earthquakes*. Some smaller earthquakes are caused by volcanic action, for example when moving magma below a volcano is suddenly stopped or when accumulations of high gas pressure lead to explosions. Small earthquakes may also be caused by rockfalls and landslides.

**8. Extent of accumulated strain before an earthquake.** By measuring the amplitudes of ground movements as shown on seismograms, it is possible to estimate the energy released in an earthquake. Various methods have been devised for this purpose, notably by Galitzin and Jeffreys[2]. In 1935, Richter[3] defined a "magnitude" scale, based on observed amplitudes in Californian earthquakes, and subsequent empirical studies by Gutenberg and Richter have enabled magnitudes to be assigned to earthquakes in all parts of the world. According to their latest results[4], the magnitude, $M$ say, is roughly connected with the energy $E$ in larger earthquakes by the formula

$$\text{Log } E = 11 + 1.6 M, \qquad (8.1)$$

where $E$ is measured in ergs. The values of $M$ range from zero in earthquakes that are just perceptible to 8.6 in the greatest earthquakes of the present century. (The two earthquakes known to have reached a magnitude of 8.6 are the Columbia-Ecuador earthquake of 1906 January 31 and the Assam earthquake of 1950 August 15; Gutenberg considers that the magnitude of the great Lisbon earthquake of 1755 November 1 may have been a little greater than 8.6.) The greatest energy released in any one earthquake is thus of the order of $10^{25}$ ergs, although Gutenberg and Richter point out that the empirical relation (8.1) is as yet tentative and that the indicated value of $E$ is uncertain by a factor of order ten.

---

[1] H. F. Reid: Bull. Dep. Geol. Univ. Calif. **6**, 413—444 (1911).
[2] See Reference [*I*], pp. 228—233.
[3] C. F. Richter: Bull. Seism. Soc. Amer. **25**, 1—32 (1935).
[4] Reference [7], 2nd. edn., (in course of publication) 1954. (Details have been kindly made dvailable to the writer in advance of publication.) Dr. M. Báth has suggested the exeression $8.9 + 1.8 M$ for the right-hand side of (8.1).

It may be remarked, incidentally, that the amplitudes of $P$ waves from the Bikini atom bomb explosion of 1946 would, on the data of Gutenberg and Richter, correspond to those in earthquakes of magnitude 5.5.

From the energy results, supplemented by certain experimental data, it is possible by (5.6) to estimate the extent of the strained region[1] just prior to a great earthquake. Observations of seismograms indicate that, in the focal region, the compressional strain and the deviatoric strain are of comparable magnitudes, so that the values of the two terms on the right-hand side of (5.6) will be comparable in the zone of strain. Data on aftershocks indicate that the energy $E$ in seismic waves in a great earthquake is of the order of one-half, or a little less, of the accumulated strain energy. Hence, neglecting variations in $\mu$, we have

$$q E = \mu \iiint \sum_i \sum_j E_{ij}^2 \, d\tau,$$

where $q$ is of the order of unity, or a little greater, and the integral is taken throughout the volume $V$ of the strained region. If $S$ denotes the breaking-strength as defined in Sect. 3, and if the simplifying assumption is made that $S$ is reached throughout the whole of $V$, we have, by (4.6),

$$12 q \mu E = V S^2. \tag{8.2}$$

Taking broad representative values of $\mu$ and $S$ as $4 \times 10^{11}$ and $10^9$ dyn./cm.$^2$, respectively, as indicated from experimental data on rocks near the Earth's outer surface, and taking $q = 2$, (8.2) then yields, for the greatest shallow-focus earthquakes, $V = 6 \times 10^{19}$ cm.$^3$. It follows that, just prior to an extreme earthquake, the material around the focus is near breaking-point throughout a volume equal to that of a sphere of radius of order 25 km.

One or two of the three dimensions of the strained region may substantially exceed 25 km. For example, surveys in the region of the Californian earthquake of 1906 April 18 revealed slipping along a fault-face over a distance of 430 km.

**9. The focal region.** In a tectonic earthquake, the accumulated strain is relieved by fractures which in most cases involve relative movement across interfaces which may or may not be visible at the Earth's outer surface. The movements from which the main seismic energy emanates may take place inside a volume somewhat less than the volume $V$ of Sect. 8. Jeffreys has calculated[2] that an earthquake of energy of order $10^{22}$ ergs would be generated by a movement of 4 cm. taking place along a plane fault-face of linear dimensions 16 km. The figure of 4 cm. is his estimate of the greatest permissible throw in a single movement along a fresh fault in which no pseudotachylyte is formed, a feature that applies to the majority of moderately great earthquakes. (Absence of pseudotachylyte implies that the temperature during the slipping is not much greater than 1000° C.)

The calculation gives an estimate, for a moderately large earthquake, of the size of the focal region, defined as the region from which the main seismic energy emanates. The *epicentre* of an earthquake is defined as the point of the Earth's surface vertically above the focus. It is evident from this discussion that the points defined as the epicentre and focus are likely to be indefinite at least to the extent of several km. in many earthquakes; (in practice, the uncertainty is further increased by uncertainties in the recording of earthquakes).

---

[1] In the present context, "strained region" means the region in which there is removable strain in excess of that caused by the equilibrium hydrostatic pressure.

[2] Reference [11], pp. 340—341.

**10. Characteristics of release of energy at the focus.** Tectonic earthquakes which have nearly the same focus and magnitude sometimes produce nearly identical seismograms. This has led BENIOFF to suggest that the wave-forms near the foci of tectonic earthquakes are in general fairly simple, and such as would conform fairly well to movements of the type assumed by JEFFREYS, as referred to in Sect. 9. This simplicity is not directly evident from seismograms, the appearance of which is generally exceedingly complicated. Work of JEFFREYS[1] has, however, shown that the predominating cause of the complicated appearance of seismograms lies in the heterogeneity of the Earth's crustal layers rather than in the mode of energy release at the focus.

While the initial wave-forms may thus be relatively simple, they are of course not symmetrical about the focus in earthquakes caused by shearing along a fault-face. This is illustrated in studies of BYERLY, KAWASUMI[2] and others who have shown that with many tectonic earthquakes there is a recognisable pattern of the directions of the first ground movement at observatories whose distances and azimuths from the epicentre are widely distributed; for any of these earthquakes, the surface of the Earth can be divided into zones inside each of which the direction (upward or downward) of the first movement is the same. BYERLY, HODGSON[3] and others have devised means of using this pattern to determine features of the faulting in the focal region. Mention may also be made of an interesting "dislocation theory" of earthquakes put forward by HOUSNER[4]; this theory sets up a detailed mechanism of the slipping along a fault-face which fits observations of strong ground accelerations close to an earthquake epicentre.

On account of the asymmetry near the focus, it is to be expected that diffraction effects will be important in seismic wave transmission. The effects are, however, of most consequence at distances from the focal region not more than a few times the linear dimensions of the focal region, and therefore normally in the heterogeneous crustal region which is not part of the subject-matter of the present chapter. (Diffraction needs, however, to be also considered at certain deeper-seated surfaces of discontinuity in the Earth—see Sect. 32.) At greater distances from the focus, it then becomes permissible to use ray theory for a great many purposes. It will in fact be seen that seismic ray theory has to date been perhaps the most fruitful source of knowledge of important physical properties of the Earth's deep interior. As in optics, FERMAT's principle of stationary time holds along seismic rays.

Investigations made by GUTENBERG and BENIOFF[5] of the Kern County earthquake of 1952 July 21 showed that, possibly as a result of the asymmetry at the focus, the amplitudes of surface waves recorded at New Zealand and Australian observatories were only one-tenth of those at European stations at a similar distance but different azimuth. They found, however, that there were only relatively small differences in the amplitudes of the bodily waves recorded at stations at similar distances and different azimuths, and this observation applies also to many other earthquakes. It is therefore justifiable for many purposes, in investigating the order of the effects of bodily waves at distant stations, to take symmetrical initial conditions, in spite of the actual asymmetry at the focus.

[1] H. JEFFREYS: Mon. Not. Roy. Astronom. Soc., Geophys. Suppl. **2**, 407—416 (1931).
[2] P. BYERLY: Bull. Seism. Soc. Amer. **28**, 1—13 (1938). — H. KAWASUMI: Bur. Centr. Séism. Internat. A **15** (ii), 258—330 (1937).
[3] J. H. HODGSON and W. G. MILNE: Bull. Seism. Soc. Amer. **41**, 221—242 (1951).
[4] G. W. HOUSNER: Calif. Inst. of Technology, Office of Naval Research, N6 onr—244, 1—34, 1953.
[5] Progress Report, Seismological Laboratory, Calif. Inst. of Technology 1953. Trans. Amer. Geophys. Un. **35**, 979—987 (1954).

Examples of both symmetrical and simple asymmetrical initial conditions will be referred to in Sect. 14 and 16.

**11. Foreshocks and aftershocks.** In major earthquakes, i.e. earthquakes of magnitude about $7\frac{1}{2}$ or more, the accumulated strain energy is not in general all released with the one earthquake. Usually there are *aftershocks*, whose total energy may be comparable with, or even exceed, that of the original shock; sometimes the aftershocks are many hundreds in number and last over many months. Aftershocks are also common with many earthquakes of magnitude less than $7\frac{1}{2}$. BENIOFF[1] has drawn an interesting analogy between the occurrence of aftershocks and creep phenomena as observed in laboratory tests on rocks. Occasionally, but comparatively rarely, major earthquakes are preceded by *foreshocks*; a notable example is the Japanese Idu earthquake of 1930 November 25, the recorded foreshocks of which extended over the preceding eighteen days, reaching a total number of nearly 700 on November 24. Foreshocks and aftershocks, as well as the main shocks of a series, are utilised to provide important information on seismic wave transmission.

**12. Focal depth.** The great majority of earthquakes originate within the crustal layers. In the decade between 1920 and 1930, WADATI and TURNER[2] found evidence from travel-time data pointing to the existence of foci at appreciably greater depths. JEFFREYS suggested a dynamical test based on the theorem that if an impulse is applied near a node for one normal mode of vibration of a dynamical system, then that mode will be very subdued among the ensuing vibrations. As applied to seismology, this theorem would entail that if deep-focus earthquakes were a reality, they should reveal themselves through abnormally small surface waves. STONELEY[3] in 1931 showed that, in certain earthquakes for which WADATI and TURNER had found evidence of deep focus, the ratios of the amplitudes of the surface waves to those of bodily waves were indeed significantly less than in the case of normal (shallow-focus) earthquakes, and the existence of deep-focus earthquakes became well confirmed. Soon afterwards, SCRASE[4] detected in certain earthquakes seismic phases that were associated with rays reflected at the outside surface after rising from the focus through several hundred kilometres.

Subsequently, there have been notable improvements in the seismic travel-time tables (see Sect. 20 and 21) with the result that focal depths can be estimated to within 20 km. or less in well-recorded earthquakes. Estimates of the greatest focal depth of any earthquake range from 650 to 720 km. It is an interesting fact that no earthquake has ever been recorded with a focal depth greater than this value.

In 1930, the first map indicating the probable distribution of deep foci was published by TURNER[5]. This was followed by work of WADATI, VISSER, BERLAGE, GUTENBERG and RICHTER and the distribution is now well known. GUTENBERG and RICHTER[6] estimate that 85% of the total energy released in earthquakes comes from foci at depths not exceeding 70 km., 12% from "intermediate" foci at depths between 70 and 300 km., and 3% from "deep" foci at depths between 300 and 700 km. The "intermediate" foci are largely, and the "deep" foci are

---

[1] H. BENIOFF: Bull. Seism. Soc. Amer. **41**, 31—62 (1951). — Bull. Geol. Soc. Amer. **65**, 385—400 (1954).

[2] H. H. TURNER: Mon. Not. Roy. Astronom. Soc., Geophys. Suppl. **1**, 1—13 (1922). — K. WADATI: Geophys. Mag., Tokyo **1**, 162—202 (1928); **2**, 1—36 (1929); **4**, 231—283 (1931).

[3] R. STONELEY: Gerlands Beitr. Geophys. **29**, 417—435 (1931).

[4] F. J. SCRASE: Phil. Trans. Roy. Soc. Lond., Ser. A **231**, 207—234 (1932).

[5] H. H. TURNER: International Seismological Summary for 1927, p. 108, 1930.

[6] Reference [7].

almost wholly, confined to a circum-Pacific earthquake-belt[1]. About 80% of all seismic energy comes from this same belt, and about 15% from a belt which passes through the Mediterranean region eastward through Asia to join the circum-Pacific belt in the East Indies.

# C. $P$ and $S$ waves and rays.

**13. Introduction.** From the account given in Sect. 8—12 of the conditions in which an earthquake is generated, we can proceed to discuss the ensuing wave motion in the Earth at a distance from the focus. It will be shown in Sect. 14 that the solid material of the Earth will transmit dilatational and rotational waves with different speeds. These are the primary or $P$, and the secondary or $S$, waves, respectively, of seismology. At distances from the focus where diffraction effects have become relatively unimportant, the $P$ and $S$ waves will be sufficiently separated for us to be able to treat them independently wherever the elastic parameters $k$ and $\mu$ and the density $\varrho$ are sufficiently slowly varying functions of position. As will be seen in Sect. 30, discontinuities in these parameters can result in a change of type between $P$ and $S$ waves; and there are complexities wherever $k$, $\mu$ and $\varrho$ change appreciably over a distance comparable with a wave-length.

**14. The existence of $P$ and $S$ waves[2].** Apply the operation curl to (6.3), and treat $k$, $\mu$ and $\varrho$ as constants. Then since curl grad $\Phi = 0$ if $\Phi$ is any scalar, and curl curl $\boldsymbol{\xi} = -\nabla^2 \boldsymbol{\xi}$ (using the fact that $\boldsymbol{\xi} = $ curl $\boldsymbol{u}$), we obtain

$$\varrho \frac{\partial^2 \boldsymbol{\xi}}{\partial t^2} = \mu \nabla^2 \boldsymbol{\xi} . \tag{14.1}$$

By (2.7) and (14.1), a rotation given by curl $\boldsymbol{u}$ can be transmitted through the medium with speed $\beta$, where

$$\varrho \beta^2 = \mu . \tag{14.2}$$

The corresponding waves are the $S$ waves of seismology, and, since (14.1) has the form of the classical wave equation, the waves are, on the assumptions made, transmitted without dispersion. It is evident from (2.9) that $S$ waves will be generated in general whenever the released strain includes deviatoric components.

Next, apply the divergence operation to (6.3). We obtain

$$\varrho \frac{\partial^2 \vartheta}{\partial t^2} + \varrho \nabla^2 V = \left(k + \frac{4}{3}\mu\right) \nabla^2 \vartheta. \tag{14.3}$$

In the seismological case, $V$ is the gravitational potential associated with the change, $\delta \varrho$ say, in $\varrho$ from the equilibrium value, so that, if $G$ is the gravitation constant,

$$\nabla^2 V = 4\pi G \delta \varrho = -4\pi G \varrho \vartheta.$$

Thus (14.3) becomes

$$\varrho \frac{\partial^2 \vartheta}{\partial t^2} = \left\{\left(k + \frac{4}{3}\mu\right) \nabla^2 + 4\pi G \varrho^2\right\} \vartheta. \tag{14.4}$$

The presence of the term in $G$ makes (14.4) more complicated in form than (14.1). Equations of the type (14.4) have been treated in a number of contexts[3].

---

[1] A preliminary calculation by the United States Coast and Geodetic Survey indicates that on 1954 March 29 there occurred in the region of Spain an earthquake of focal depth near 650 km. If, when more complete data become available, this calculation is confirmed, this earthquake will have been shown to be unique over the half-century or so since reliable recording of earthquakes began.

[2] See References [18], [19], [21].

[3] See e.g. R. YOSIYAMA: Bull. Earthquake Res. Inst., Tokyo, **19**, 185—205 (1941).

A model solution by Jeffreys[1] is adequate for the main purposes of seismology. Jeffreys takes the simple case in which the medium is initially in equilibrium and a symmetrical stress of the form $P\delta_{ij}$ (in the notation of Sect. 2), where $P$ is proportional to Heaviside's unit function $H(t)$, acts inside the medium on the surface of a sphere whose radius is of the order of the dimensions of strained regions in earthquakes; it is also assumed that $k = 1.67\mu$ [Poisson's relation (4.9)]. The solution shows that, in the conditions holding in the Earth the influence of the term in $G$ in (14.4) on the displacement at a distant point $Q$ is negligible until a time of the order of several hundred seconds after the first disturbance has reached $Q$.

It follows that, except in special contexts, it is permissible to ignore the term in $G$, in which case (14.4) reduces to the classical wave equation again. By this equation, a dilatation $\vartheta$ is transmitted through the Earth with speed $\alpha$, where

$$\varrho\,\alpha^2 = k + 4\mu/3. \tag{14.5}$$

The corresponding waves are the $P$ waves, their speed $\alpha$ being necessarily greater than $\beta$.

The equations (14.1) and (14.4) would need to be modified if there were significant aeolotropy. Stoneley[2] has examined the influence on seismic wave transmission of the condition referred to by Love[3] as transverse isotropy. For this case it is necessary to introduce three additional elastic parameters into the equations. When the axis $Ox_3$ is taken to be vertical, the corresponding strain-energy function takes the form

$$W = A\,\vartheta^2 + B\sum E_{ij}^2 + C\,E_{33}^2 + D\vartheta\,E_{33} + E\,(E_{12}^2 - E_{11}\,E_{22}), \tag{14.6}$$

where $A$, $B$, $C$, $D$ and $E$ are the five parameters required. The work of Stoneley shows that the influence of the terms in $C$, $D$ and $E$ is of possible significance only at short distances from the focus.

The equations also need modification if there is significant departure from perfect elasticity. According to laboratory investigations such departures are much more prominent under deviatoric rather than under purely symmetrical stress. Hence, in modifying the equations, the stress-strain relation (4.5) is preserved, while additional terms are added to (4.6). In seismology, where knowledge is required only of the broad effect of imperfections, it is sufficient to take model representations involving the introduction of just one or two additional parameters (see e.g. Reference [1], pp. 31–38). A sufficient indication for present purposes is obtained by taking, in place of (4.6), the relations

$$P_{ij} = 2\mu\,E_{ij} + 2\eta\,\frac{dE_{ij}}{dt}, \tag{14.7}$$

where the additional parameter is $\eta$. Substituting (14.7) instead of (4.6) into (6.1) s equivalent to replacing $\mu$ by $\mu + \eta\,d/dt$ in (6.3).

For the case of transmission of waves through a solid, for which $\mu \neq 0$, Jeffreys has shown[4] that the use of (14.7) would involve the introduction of a damping factor, of order $\exp\,(-\eta\gamma^2 s/2\mu v)$ over a distance $s$, into the amplitudes of seismic waves of velocity $v$ and period $2\pi/\gamma$. Comparison with amplitudes measured on seismograms shows that, for bodily waves penetrating the solid part of the

[1] H. Jeffreys: Mon. Not. Roy. Astronom. Soc., Geophys. Suppl., 2, 407—416 (1931).
[2] R. Stoneley: Mon. Not. Roy. Astronom. Soc., Geophys. Suppl. 5, 343—353 (1949).
[3] A. E. H. Love: Mathematical Theory of Elasticity, 3rd. edn., p. 158. Cambridge University Press, 1920.
[4] Reference [11], p. 107.

Earth's interior (down to a depth of 2900 km.—see Sect. 20), $\eta/\mu$ is not more than 0.003 sec. The smallness of this value shows that the equations of perfect elasticity will lead to highly accurate results in many problems. JEFFREYS has shown that the principal effect of imperfect elasticity is to introduce a slight blunting into what would be otherwise a sharp onset of a seismic wave at a recording station. He has also shown that scattering rather than the usual laboratory type of imperfection would be closely associated with the form (14.7), and has put forward evidence to show that the principal source of damping in seismic waves is scattering due to the presence of small-scale irregularities in the solid Earth, chiefly near the outer surface. According to GUTENBERG[1], the extent of damping is such as would reduce the amplitudes of $P$ waves by a factor of $e^{-1}$ over a distance of 8000 km.

When $\mu$ is put equal to zero in (6.3), the theory becomes the ordinary theory of sound waves in a fluid. In this case the $P$ velocity $\alpha$ is given by

$$\varrho\, \alpha^2 = k,\tag{14.8}$$

and there are no $S$ waves.

Imperfect elasticity in a fluid is represented by taking

$$P_{ij} = 2\eta\, \frac{dE_{ij}}{dt},\tag{14.9}$$

where $\eta$ is the usual fluid viscosity. [It will be noticed that (14.7) is a generalisation of (4.6) and (14.9).] In the fluid zone of the Earth (below a depth of 2900 km.), GUTENBERG finds the damping to be of the same order as in the solid part. JEFFREYS[2] has pointed out that, since the Earth's fluid zone passes $P$ waves of periods as small as 1 sec., a firm upper limit of $5 \times 10^8$ poises can be set for the viscosity $\eta$ of the core, although he considers that the actual value is probably appreciably less than this figure. Under stresses of period $\tau$, (14.9) yields

$$P_{ij} = 4\pi\,\eta\,\tau^{-1}E_{ij},$$

(apart from a phase difference between $P_{ij}$ and $E_{ij}$), and this by (4.6) and the numerical result for $\eta$ corresponds to an effective rigidity of order not greater than $3 \times 10^9$ dyn./cm.$^2$ for waves of periods of one second or more. It will be seen (Sect. 35) that the rigidity of the Earth's rocks is of the order of 100 or more times this value, from which it follows that there will be no effective transmission of $S$ waves in the Earth's fluid zone. On the direct observational side, there has been no authentic observation of $S$ waves in this zone.

It may therefore be concluded that (4.5) and (4.6), and the $P$ and $S$ wave equations (14.1) and (14.4) (with the term in $G$ omitted), can be reliably used in seismic bodily wave transmission; and (14.2) and (14.5) satisfactorily give the $P$ and $S$ velocities. Strictly, $k$ and $\mu$ are to be interpreted as the adiabatic incompressibility and rigidity, but they may be interpreted as isothermal parameters with little error[3]. We shall work in terms of these equations in the following sections.

**15. Plane $P$ and $S$ waves.** In the majority of problems on bodily seismic waves, the distance from the focus is so large compared with the dimensions of the focal region that the curvature of wave fronts may be neglected. Thus it is commonly sufficient to treat the waves as plane waves.

---

[1] B. GUTENBERG: Bull. Seism. Soc. Amer. **35**, 57—69 (1945).
[2] Reference [*11*], p. 244.
[3] H. JEFFREYS: Proc. Cambridge Phil. Soc. **26**, 101—106 (1930). — See also F. BIRCH: J. Geophys. Res. **57**, 254—255 (1952).

We shall take the normal to the plane in the direction of the $x_1$-axis, so that the displacements in the wave motion may be written in the form

$$u_i = F_i(x_1 - v t). \tag{15.1}$$

By (2.5) and (2.7), (15.1) yield

$$\vartheta = F_1'; \quad \xi = (0, -F_3', F_2'); \quad \operatorname{curl} \xi = (0, -F_2'', -F_3''). \tag{15.2}$$

Substituting from (15.1) and (15.2) into (6.3) (neglecting the term in $V$), we then have

$$\varrho v^2 (F_1'', F_2'', F_3'') = (k + \tfrac{4}{3} \mu)(F_1'', 0, 0) + \mu(0, F_2'', F_3''). \tag{15.3}$$

From (15.3), it is seen that the only possible plane wave motions compatible with (6.3) (neglecting the term in $V$) are: (a) a disturbance $(u_1, 0, 0)$ travelling with speed $\{(k + 4\mu/3)/\varrho\}^{\frac{1}{2}}$, i.e. $\alpha$; (b) a disturbance $(0, u_2, 0)$ travelling with speed $(\mu/\varrho)^{\frac{1}{2}}$, i.e. $\beta$; (c) a disturbance $(0, 0, u_3)$ travelling with speed $\beta$. The waves (a) are *longitudinal* and are the one-dimensional case of dilatational waves. Each of (b) and (c) gives a *transverse* wave corresponding to the rotational waves of the preceding Section.

The solution shows that the $P$ and $S$ are the only possible types of plane bodily waves in a medium to which (6.3) applies.

Further, the separate existence of the solutions (b) and (c) shows that the $S$ waves can be polarised. In seismology, the notation $SV$ is used for polarised $S$ waves in which the particles of the medium move in vertical planes, and $SH$ when the particles move horizontally.

**16. Seismic phases.** Taking the same initial conditions as in Sect. 14, JEFFREYS[1] has shown that, at appreciable distances from an earthquake focus, the displacement in the ensuing $P$ waves in a simple homogeneous medium would be essentially a single swing from the equilibrium position followed by a rapid return to equilibrium. JEFFREYS shows also that a similar result holds in the case of $S$ waves, taking the waves to be generated by the action of a tangential impulse over a sphere of focal dimensions, the impulse being symmetrical about a diameter.

Slightly more complicated initial conditions have been considered by RICKER[2] who takes a case in which the initial wave-form consists of sudden up and down swings and then return to zero. RICKER likewise finds a rapid return to the equilibrium position of a disturbed distant particle of the medium after one or two swings, and uses the term "wavelet" to describe this motion. Other initial conditions have been considered by SEZAWA and KANAI[3].

In contrast to the theoretical results, $P$ and $S$ bodily waves as shown on seismograms are generally in the form of long irregular trains with no quiescence becoming apparent after one or two swings. An exhaustive study by JEFFREYS[4] of the possible causes of these oscillatory movements has shown that they must arise from the heterogeneity of the crustal layers. This is observationally confirmed in work of BENIOFF and others who have, moreover, shown that, in favorable circumstances, for example when waves travel from a deep focus nearly vertically through the crustal layers to an observatory, motions approximating to the wavelet type frequently stand out sharply from background oscillatory motions on the seismograms.

[1] H. JEFFREYS: Mon. Not. Roy. Astronom. Soc., Geophys. Suppl. 2, 318—323, 407—416 (1931).
[2] N. RICKER: Geophysics 5, 348—366 (1940).
[3] K. SEZAWA and K. KANAI: Bull. Earthquake Res. Inst., Tokyo 19, 162—175, 443—457 (1941).
[4] H. JEFFREYS: Mon. Not. Roy. Astronom. Soc., Geophys. Suppl. 2, 407—416 (1931).

Apart from the crustal layers, there are surfaces of discontinuity in the Earth's deep interior which cause incident $P$ and $S$ waves to be reflected and refracted and therefore introduce additional onsets on seismograms. In the present chapter we shall be interested in the direct $P$ and $S$ onsets, and in those onsets arising from reflection at the outer surface and from reflection or refraction at levels below the crustal layers.

Such is the complication caused by the crustal layers that, to date, it is from the times of these various onsets rather than their form that the most reliable information on the Earth's interior has emerged. In the following Sections (to Sect. 26) we shall be considering in some detail the question of the travel-times along rays corresponding to the onsets, and of inferences that can be made from these times.

The onsets as recorded on a seismogram mark the beginnings of the various *phases* of the seismogram, and the motions that give rise to the phases are called *pulses*. When the onset is sharp, the pulse is called an *impetus*, and when gradual, an *emersio*.

It will be seen later that the Earth consists of a solid *mantle* 2900 km. thick, surrounding an essentially fluid zone about 2200 km. thick, which in turn surrounds an *inner core* of radius about 1250 km. The whole region below the mantle, of radius 3470 km., has been referred to as the *central core* (the *inner* core having been a comparatively recent discovery—see Sect. 32).

Phases from $P$ and $S$ rays which lie entirely in the mantle are denoted as $P$ and $S$ respectively. The symbol $K$ is used to correspond to a path along a $P$ ray in the fluid zone. (We have seen [Sect. 14] that there would be no significant $S$ waves in the fluid zone; in the use of the symbol $K$, the initial letter of *Kernwelle*, tribute is paid to the early work of the Göttingen school of geophysicists.) The symbol $I$ is used to correspond to a path along a $P$ ray in the inner core; ($J$ has been proposed for $S$ rays, not as yet observed—but see Sect. 32—in the inner core). A separate letter, $P$, $S$, $K$, $I$ or $J$ is used for each segment of an entire ray between two surfaces of discontinuity. Thus, for example, the ray for the phase $PP$ has two segments, each of $P$ type, with one reflection at the outer surface; for $PS$, the first segment is of $P$ type, but the second segment corresponds to a reflection in $S$ type; with rays such as $SPP$ there are two reflections at the outer surface; $PKP$, $SKP$, etc., have two segments in the mantle separated by a segment which lies inside the fluid zone; $PKKP$, $SKKP$, etc. have the further complication of a reflection inside the fluid zone at its outer boundary; $PKIKP$ is the simplest ray that has entered the inner core, and has five segments all of $P$ type. The notation for rays reflected upward at the base of the mantle, e.g. $PcP$, $PcS$, etc., includes the symbol $c$ in order to avoid confusion with $PP$, $PS$, etc. The notation $P'$ is commonly used for $PKP$ or $PKIKP$, and $P'_2$ for these same phases in cases where the onsets correspond to a second branch in the travel-time-distance curves (see Sect. 24); thus, for example, a ray in which the path is from the focus through the centre to the anticentre (antipodean point) and back to the epicentre may be denoted as $P'P'$, in place of the more cumbrous $PKIKPPKIKP$. A few typical paths are illustrated in Fig. 1.

In deep-focus earthquakes, the symbols $p$ and $s$ are used to denote relatively short upward paths from the focus prior to reflection at the outer surface, this notation having been introduced by SCRASE in the work referred to in Sect. 12. Thus deep-focus earthquake phases can include $pP$, $sP$, $pS$, $sS$, $pPP$, $sScS$, etc. The entire family of $PP$ and $pP$ rays has a cusp[1] in the travel-time-distance

---

[1] V. C. STECHSCHULTE: Bull. Seism. Soc. Amer. **22**, 81—137 (1932). — K. E. BULLEN: Mon. Not. Roy. Astronom. Soc., Geophys. Suppl. (in course of publication) 1955.

curve which corresponds to a ray which leaves the focus at a finite angle above the horizontal. One convention has been to use the notation $PP$ for rays corresponding to the upper of the two branches which start from the cusp, and $pP$ for the lower. There are, however, certain advantages in using $PP$ and $pP$ to denote rays which leave the focus in directions inclined below and above the horizontal through the focus, respectively. Similar remarks apply to $sS$ and $SS$.

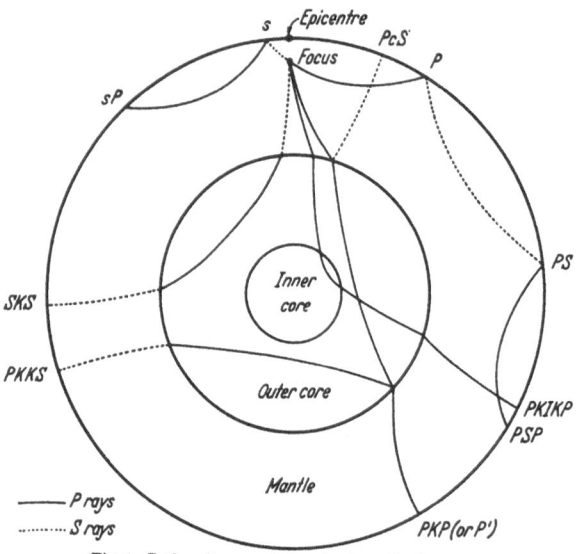

Fig. 1. Paths of some of the simpler seismic rays.

## 17. Reflection and refraction of seismic rays.

The elastic parameters $k$ and $\mu$ and the density $\varrho$ vary with depth in the Earth. In general they vary so that the velocities $\alpha$ and $\beta$ increase with increase of depth; (there may be some exceptions — see Sect. 24). Hence in general seismic rays will be steadily refracted so as to be concave upward. Using Fermat's principle, and taking a spherically symmetrical model Earth in which the velocity $v$ (which may be either $\alpha$ or $\beta$) is a function only of the distance $r$ from the centre $O$ we now derive laws of reflection and refraction analogous to Snell's laws in optics.

Let $T$ be the travel-time along a ray from a focus $F$ to a point $Q$ of the Earth's outer surface, and let $(r, \vartheta)$ be the polar coordinates of any point $P$ of the ray as shown in Fig. 2. Let $P_1$ be the lowest point of the ray, distant $r_1$ from $O$, and let $s$ be the arc-length $P_1P$. Denote $ds/d\vartheta$, expressed in terms of $r$ and $\vartheta$, by $\zeta$. Thus

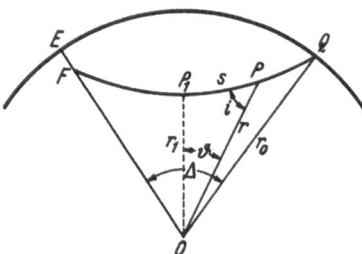

$$\zeta = \left\{r^2 + \left(\frac{dr}{d\vartheta}\right)^2\right\}^{\frac{1}{2}}. \qquad (17.1)$$

Since $T = \int ds/v$, taken along the ray, we have

$$T = \int v^{-1}\zeta\,d\vartheta. \qquad (17.2)$$

Fig. 2. Notation used in Sect. 17 for a simple seismic ray $FQ$.

By Fermat's principle, the integral in (17.2) is to be stationary for variations in which the end points of the ray are kept unvaried. The Euler differential equation for this is

$$\frac{\partial(v^{-1}\zeta)}{\partial r} - \frac{d}{d\vartheta}\frac{\partial(v^{-1}\zeta)}{\partial r'} = 0,$$

where $r' \equiv dr/d\vartheta$, and yields the first integral

$$v^{-1}\zeta = r'\frac{\partial(v^{-1}\zeta)}{\partial r'} + \eta^0, \qquad (17.3)$$

where the integration constant $\eta^0$ is a parameter, constant for the ray $FQ$, but variable if the angular distance $\Delta$ between the epicentre $E$ and $Q$ is varied. Thus $\eta^0$, as well as $T$ and $r_1$, are functions of $\Delta$.

By (17.1) and (17.3),
$$v^{-1}\zeta = (v\,\zeta)^{-1} r'^2 + \eta^0,$$

so that
$$v\,\eta^0\frac{ds}{d\vartheta} = \zeta^2 - r'^2 = r^2. \tag{17.4}$$

Now let $i$ be the angle between the tangent to the ray at $P$ and the radius vector $OP$. Thus $i$ is the angle of incidence of the ray against the surface of constant velocity through $P$; its complement, denoted as $e$ in the case of $P$ rays (or $f$ in the case of $S$ rays), is called the *angle of emergence*. Then, taking as representative the case of a $P$ ray, we have $\sin i = \cos e = r\,d\vartheta/ds$, so that by (17.4)

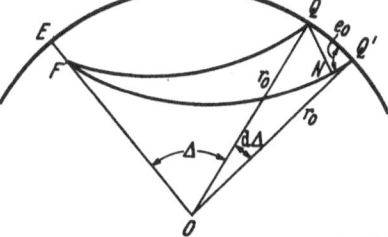

$$r\,v^{-1}\cos e = \eta^0, \quad \text{or} \quad \eta\cos e = \eta^0, \tag{17.5}$$

where $\eta = r\,v^{-1}$.

Since $\eta^0$ is constant along the ray, (17.5) shows that the ray satisfies SNELL's law of refraction.

Fig. 3. Diagram for neighbouring rays $FQ$, $FQ'$ (Sect. 17).

If there is a discontinuity in $v$ at a certain depth, the integral in (17.2) may need to be broken up into more than one part, but the result (17.5) still holds. At a discontinuity surface, however, an incident $P$ or $S$ ray may be reflected as well as refracted, and in general the reflected and refracted rays include both $P$ and $S$ types. In such cases, the left-hand sides in (17.5) have the same value for every one of the rays concerned ($e$ being replaced by $f$ in the case of $S$ rays)[1].

By (17.5), the parameter $\eta^0$ is the value of $\eta$ or $r/v$ at the lowest point of the ray. There is an important simple connection between $\eta^0$, $T$ and $\Delta$. Let $Q'$ be a neighbouring point to $Q$ on the outside surface in the plane of the ray $FQ$, let $FQ'$ be the corresponding neighbouring ray, and let $QN$ be drawn normally to $FQ'$, as in Fig. 3. Then if $d\Delta$ is the angle $QOQ'$ and $dT$ the corresponding increment to the travel-time, and if the suffix zero denotes values at the outside surface, we have by (17.5)

$$\begin{aligned}\eta^0 &= \eta_0\cos e_0 \\ &= \frac{v_0}{r_0}\frac{NQ'}{QQ'} \\ &= \frac{dT}{d\Delta}.\end{aligned} \tag{17.6}$$

# D. Seismic travel-times.

**18. Introduction.** One of the principal achievements of twentieth-century seismology has been the evolution of tables giving the travel-times $T$ in terms of the epicentral distances $\Delta$ for a large number of the phases referred to in Sect. 16. The procedure has involved long series of approximations starting from crude

---

[1] For other methods of deriving this result, see Reference [1], Sect. 6.1, and Reference [2], pp. 82—83.

beginnings. The empirical data used come from seismograms and routine observatory reports, and the theory of the preceding sections is brought to bear in conjunction with statistical theory. In addition, a good part of the theory of Sect. 23, 24 and 25 on $P$ and $S$ velocity distributions is brought to bear.

**19. Seismograms.** The essential principle of the seismograph is that, in response to a component, $u(t)$ say (where $t$ denotes the time), of the ground motion [the component is usually north-south, east-west or vertical $(Z)$], the seismograph records the changes in a variable $\vartheta(t)$ which satisfies[1] an equation of the form

$$\ddot{\vartheta} + 2\lambda\omega\dot{\vartheta} + \omega^2\vartheta = -Ku''(t), \tag{19.1}$$

where $\lambda$, $\omega$ and $K$ are known instrumental constants. Through being attached to the moving ground, the seismograph cannot record $u(t)$ directly, so that there is the problem of inferring $u(t)$ from the variation of $\vartheta$ as read from the seismogram.

In well-recorded earthquakes, this problem is not acute in respect of the arrival-time of the main $P$ phase, since the ground is usually near a state of rest when this pulse arrives; (sometimes there are difficulties caused by the presence of "microseismic" ground movements). For this reason, the $P$ arrival-times as recorded at the world's observatories are the firmest seismological data available. For the same reason, a cardinal requirement in recording earthquakes is accurate timing; a mechanism marks the seismogram at intervals of a minute, and the absolute time is carefully checked in reliable observatories. The best observatories aim at accuracy to 0.1 sec., although local geological influences and emersio onsets frequently introduce errors well outside this uncertainty.

For most other phases, the seismograph is already in motion when they arrive, so that their arrival-times are generally much less accurately determined than the $P$. The uncertainty is minimised by arranging the damping coefficient $\lambda$ so that earlier movements do not mask later ones too seriously; ($\lambda$ commonly has a value between 0.7 and unity). As with $P$, the arrival-times of these phases are read directly from the seismogram without analysis of (19.1).

More detailed examination of (19.1) is required to obtain knowledge of the amplitudes of the ground motion (amplitudes will be further discussed in Sect. 27 to 32), since the amplitude of $\vartheta$ depends not only on that of $u$, and on $\lambda$, $\omega$ and $K$, but also on the predominating period of the movement $u$. For this reason, particular values of the instrumental period are selected in order to show prominently particular periods of the spectrum of the ground movement and minimise other periods. A good observatory has several instruments of different periods. For a given instrument, tables are usually prepared to facilitate the estimation of the ground movement.

Modern seismographs of the Galitzin, Benioff and related types have galvanometers coupled to pendulums, and the equations needed to connect the recorded motion $\vartheta$ with the ground motion $u$ are more complicated than (19.1). For details of instrumental seismology, reference may be made to the chapter in this series by Coulomb.

**20. Construction of travel-time curves.** The starting-point in the construction of travel-time curves is knowledge of $P$ arrival-times from a suitable selection of the well-recorded earthquakes to date. Given this data, and, to begin with, treating the Earth as spherically symmetrical, a least squares procedure would, in theory, determine at the one operation the most probable values of the origin-times, epicentres and focal depths of the earthquakes used, as well as the best

---

[1] See Reference [1], Chapter IX, or eqn. (3.1) in the chapter Séismométrie.

values of the $P$ travel-times. In practice, the direct process would involve a prohibitive quantity of labour, and there are as well a number of sources of systematic error. Thus, starting from the dawn of the present century[1], there have been many successive approximations, each resulting usually in improved tables, and in improved origin-times, etc., for the earthquakes used.

During the period 1900 to 1906, travel-times of $P$ and $S$ waves for distances up to 90° were published by OLDHAM, MILNE, BENNDORF and ZÖPPRITZ. ZÖPPRITZ[2] also gave times for additional phases, including $PP$, $SS$, etc. In 1914, GUTENBERG[3] published the first travel-times for phases influenced by the central core; this paper was noted, among other things, for its estimate (near 2900 km.) of the depth of the outer boundary of the central core. The tables of ZÖPPRITZ in the form used by TURNER, the ZÖPPRITZ-TURNER tables, were used in compiling the International Seismological Summary for earthquakes up to 1929. Many seismologists have contributed to improvements to travel-times, including LEHMANN, BYERLY, HODGSON, MACELWANE and BRUNNER[4]. The most comprehensive investigations since 1930 have been made by JEFFREYS and BULLEN[5] and by GUTENBERG and RICHTER[6].

The problem of constructing travel-time tables has involved special statistical considerations[7]; for example, residuals often show marked deviations from normality. In addition, departures from steady variation of velocity with depth (see Sect. 24) introduce such complications as triplication of times over certain ranges of values of $\Delta$, cusps, abnormal curvature of the $T-\Delta$ curves, and shadow zones.

By 1933, the extant $P$ tables had reached the point where it had become necessary to allow for the Earth's oblateness. GUTENBERG, RICHTER[8] and COMRIE[9] showed that the residuals arising from this oblateness would be much reduced if geocentric latitudes were used instead of geographic latitudes in calculating $\Delta$. Subsequently, JEFFREYS defined a standard sphere in which a surface of given constant $P$ or $S$ velocity encloses the same volume as the corresponding level surface of the Earth. If $\delta r$ is the excess of the radius vector in the Earth over that of the corresponding point of the standard sphere, JEFFREYS[10] showed that the corresponding excess $\delta T$ in the travel-time along a ray of epicentral distance $\Delta$ and parameter $\eta^0$ is given by

$$\delta T = \eta^0 \left[ r^{-2} \frac{d r}{d \vartheta} \delta r \right]_0^{\Delta} + \eta^0 \int_0^{\Delta} \left( \frac{d^2 r^{-1}}{d \vartheta^2} + r^{-1} \right) \delta r \, d\vartheta, \qquad (20.1)$$

the notation being as in Sect. 17.

---

[1] For a brief history of earlier twentieth century seismology, see GUTENBERG, Seismology, Fiftieth Anniversary Volume of the Geol. Soc. of Amer., p. 439—470, 1941.

[2] Reference [23].

[3] B. GUTENBERG: Nachr. Ges. Wiss. Göttingen, Math.-phys. Kl. 1914, 1—52, 125—176.

[4] See Reference [16]. See also I. LEHMANN: Gerlands Beitr. Geophys. 26, 402—412 (1930); Bur. Centr. Séism. Internat. A 14, 109—136 (1935) and A 14, 87—115 (1936).

[5] H. JEFFREYS and K. E. BULLEN: Bur. Centr. Séism. Internat. A 11, 1—202 (1935). Seismological Tables, British Assoc. publication, p. 1—50, 1940. The last-mentioned tables, referred to as the JEFFREYS-BULLEN tables, incorporate the results of a large series of papers over the years 1933—39; the JEFFREYS-BULLEN tables are used at present in compiling the I.S.S.

[6] B. GUTENBERG and C. F. RICHTER: Gerlands Beitr. Geophys. 43, 56—133 (1934); 45, 280—360 (1935); 47, 73—131 (1936); 54, 94—136 (1939).

[7] See, for example, references to seismological problems in JEFFREYS, Theory of Probability, 2nd. edn. Oxford: Clarendon Press 1948. See also H. JEFFREYS, Bur. Centr. Séism. Internat., A 14, 1—86 (1936).

[8] B. GUTENBERG and C. F. RICHTER: Gerlands Beitr. Geophys. 40, 380—389 (1933).

[9] Reference [11], p. 57.

[10] H. JEFFREYS: Mon. Not. Roy. Astronom. Soc., Geophys. Suppl. 3, 271—274 (1935).

For the sake of a definite model, the standard sphere assumes that the crustal layers are uniformly spread over the whole Earth, the total thickness being taken as 33 km. This is at variance with the evidence of marked differences between the crustal layers over continents and oceans (see the chapter by EWING and PRESS), but is a desirable simplification in studying the deeper interior since details of the crustal layering are still very uncertain.

Apart from certain difficulties associated with abnormal velocity variation in the Earth, the $P$ travel-time tables have now reached the point where regional crustal-structure influences and possible (though fairly small) deeper-seated differences below continents and oceans are the main sources of uncertainty.

The problem of determining travel-times for other phases is in general treated separately from that of origin-time and epicentral determinations. In some cases, especially in the first $PKP$ or $PKIKP$ arrivals (see Sect. 21), which, like $P$, are not confused by earlier ground motion, the precision of the tables is comparable with that for $P$, i.e. mostly of the order a second. But with other phases there are uncertainties up to several seconds or more. In the case of phases such as $PP$, $SS$, etc., there is a further difficulty in that the theoretical travel-time $T$, though stationary, is not a minimum for variations of the point of reflection; in addition there is the complicating effect of the crustal layers on the mechanism of reflection; for these phases, the dispersion of the observed travel-times for given $\Delta$ is considerable.

**21. Numerical results.** From a knowledge of the travel-times for the phases $P$, $S$, $PcP$ and $ScS$, and of the times for paths corresponding to $K$ and $I$ in the outer and inner core, it is possible to deduce the travel-times for all other observed phases, since for a compound phase (for example, $PcSPKP$) the value of $\eta^0$, or $dT/d\Delta$, is, by (17.6), the same for every segment of the whole ray. The JEFFREYS-BULLEN tables include these travel-times and also the times for many other phases for a range of focal depths extending from zero down to 700 km. The JEFFREYS-BULLEN tables and those of GUTENBERG and RICHTER are in good agreement for the principal phases; JEFFREYS[1] has given an important summary of the degree of reliability of the tables and of various problems connected with their construction.

Table 1 below gives the JEFFREYS-BULLEN travel-times for the $P$, $S$, $PcP$ and $ScS$ phases, and also the values of the parameter $\eta^0$, for hypothetical surface foci. The table applies to the standard sphere defined in Sect. 20; the tables terminate at $\Delta = 105°$ since, near this distance, the rays meet the boundary of the central core at grazing incidence. Table 2 gives the JEFFREYS-BULLEN travel-times and $\eta^0$ for $K$ and $I$ rays. In addition, theoretical times and values of $\eta^0$ for $J$, which corresponds to $S$ waves in the inner core, have been included; these times have been derived[2] assuming values of the density and rigidity indicated in Sect. 36; no observations of phases including a segment of $J$ type have as yet been certainly recorded, the expected amplitude in these phases being extremely small (see Sect. 32).

It needs to be remarked that the $P$ times in Table 1 are associated with a $P$ velocity of 7.8 km./sec. just below the crustal layers, this having been the most probable value at the time the JEFFREYS-BULLEN tables were constructed. The most probable value of the $P$ velocity just below the crustal layers is now near 8.1 km./sec. Thus the table needs some small amendment for the shorter distances.

---

[1] Reference [11], pp. 85—100.
[2] K. E. BULLEN: Mon. Not. Roy. Astronom. Soc., Geophys. Suppl. **6**, 125—128 (1950).

Table 1. *Travel-times T and parameters $\eta^0$ for P, S, PcP and ScS, for a surface focus.*
[$\eta^0$ is defined in (17.6).]

| $\Delta$ | P | | S | | PcP | | ScS | |
|---|---|---|---|---|---|---|---|---|
| | T | $\eta^0$ | T | $\eta^0$ | T | $\eta^0$ | T | $\eta'$ |
| | m. s. | | m. s. | | m. s. | | m. s. | |
| 0° | | 14.3 | | 25.4 | 8 34 | 0.0 | 15 36 | 0.0 |
| 5 | 1 18 | 14.1 | 2 17 | 25.2 | 8 35 | 0.5 | 15 38 | 0.9 |
| 10 | 2 28 | 13.7 | 4 22 | 24.5 | 8 39 | 1.0 | 15 45 | 1.8 |
| 15 | 3 35 | 12.0 | 6 23 | 23.6 | 8 45 | 1.4 | 15 55 | 2.6 |
| 20 | 4 37 | 11.0 | 8 17 | 21.2 | 8 53 | 1.8 | 16 10 | 3,4 |
| 25 | 5 27 | 9.5 | 9 49 | 16.8 | 9 3 | 2.1 | 16 29 | 4.1 |
| 30 | 6 12 | 8.9 | 11 10 | 15.8 | 9 15 | 2.5 | 16 51 | 4.7 |
| 35 | 6 56 | 8.5 | 12 28 | 15.5 | 9 29 | 2.9 | 17 16 | 5.3 |
| 40 | 7 38 | 8.2 | 13 44 | 15.0 | 9 44 | 3.2 | 17 45 | 5.8 |
| 45 | 8 19 | 8.0 | 14 58 | 14.4 | 10 0 | 3.4 | 18 15 | 6.3 |
| 50 | 8 58 | 7.6 | 16 9 | 13.8 | 10 18 | 3.6 | 18 48 | 6.7 |
| 55 | 9 35 | 7.3 | 17 17 | 13.4 | 10 37 | 3.8 | 19 22 | 7.1 |
| 60 | 10 11 | 6.8 | 18 23 | 12.8 | 10 57 | 4.0 | 19 59 | 7.4 |
| 65 | 10 44 | 6.4 | 19 25 | 12.3 | 11 17 | 4.1 | 20 36 | 7.7 |
| 70 | 11 15 | 6.1 | 20 26 | 11.7 | 11 38 | 4.2 | 21 15 | 7.9 |
| 75 | 11 45 | 5.7 | 21 23 | 11.1 | 11 59 | 4.3 | 21 55 | 8.0 |
| 80 | 12 13 | 5.3 | 22 16 | 10.4 | 12 21 | 4.3 | 22 35 | 8.1 |
| 85 | 12 38 | 5.0 | 23 7 | 9.8 | 12 42 | 4.4 | 23 16 | 8.2 |
| 90 | 13 3 | 4.6 | 23 54 | 9.0 | 13 4 | 4.4 | 23 58 | 8.3 |
| 95 | 13 26 | 4.5 | 24 38 | 8.5 | 13 26 | 4.4 | 24 39 | 8.3 |
| 100 | 13 48 | 4.5 | 25 20 | 8.4 | 13 48 | 4.4 | 25 21 | 8.3 |
| 105 | 14 11 | 4.4 | 26 2 | 8.3 | 14 11 | 4.4 | 26 2 | 8.3 |

Table 2. *Travel-times and parameters for K, I and J.*

| $\Delta$ | K | | I | | J | |
|---|---|---|---|---|---|---|
| | T | $\eta^0$ | T | $\eta^0$ | T | $\eta^0$ |
| | m. s. | | m. s. | | m. s. | |
| 0° | 0 0 | 7.5 | 0 0 | 2.0 | 0 0 | 4.5 |
| 10 | 1 15 | 7.4 | 0 20 | 1.9 | 0 45 | 4.5 |
| 20 | 2 28 | 7.2 | 0 39 | 1.9 | 1 30 | 4.4 |
| 30 | 3 39 | 6.9 | 0 58 | 1.9 | 2 14 | 4.4 |
| 40 | 4 46 | 6.5 | 1 17 | 1.8 | 2 57 | 4.2 |
| 50 | 5 48 | 5.9 | 1 35 | 1.8 | 3 38 | 4.1 |
| 60 | 6 45 | 5.4 | 1 52 | 1.7 | 4 18 | 3.9 |
| 70 | 7 36 | 4.9 | 2 8 | 1.6 | 4 56 | 3.7 |
| 80 | 8 22 | 4.4 | 2 23 | 1.5 | 5 31 | 3.4 |
| 90 | 9 3 | 3.9 | 2 38 | 1.4 | 6 4 | 3.2 |
| 100 | 9 40 | 3.3 | 2 51 | 1.2 | 6 34 | 2.9 |
| 110 | 10 11 | 2.9 | 3 2 | 1.1 | 7 1 | 2.5 |
| 120 | 10 37 | 2.3 | 3 13 | 1.0 | 7 25 | 2.2 |
| 130 | | | 3 22 | 0.8 | 7 46 | 1.9 |
| 140 | | | 3 29 | 0.6 | 8 2 | 1.5 |
| 150 | | | 3 35 | 0.5 | 8 16 | 1.1 |
| 160 | | | 3 39 | 0.3 | 8 25 | 0.8 |
| 170 | | | 3 41 | 0.2 | 8 31 | 0.4 |
| 180 | | | 3 42 | 0.0 | 8 33 | 0.0 |

The necessary correction has not yet been carried out because of other complications, not yet finally resolved, in the outermost several hundred kilometres of the Earth—see Sect. 24 (i).

The values of $\eta^0$ have been included in the tables since they enable the travel-times for compound phases to be readily computed from the data given. For example, the travel-time for $PcS$ at $\varDelta = \frac{1}{2}(65° + 25°) = 45°$ is $T = \frac{1}{2}(11 \text{ m. } 17\text{s.}$ $+ 16 \text{ m. } 29 \text{ s.}) = 13 \text{ m. } 53 \text{ s.}$; this follows since $\eta^0$ has the same value, namely 4.1, for $PcP$ at 65° as for $ScS$ at 25°. Again, the travel-time for $SKSP$ at $\varDelta = 45°$ $+ 42° + 67° = 154°$ is seen, on combining the results for $ScS$, $K$ and $P$ at $\eta^0 = 6.3$ to be approximately $18 \text{ m. } 15 \text{ s.} + 5 \text{ m. } 11 \text{ s.} + 10 \text{ m. } 56 \text{ s.} = 34 \text{ m. } 22 \text{ s.}$

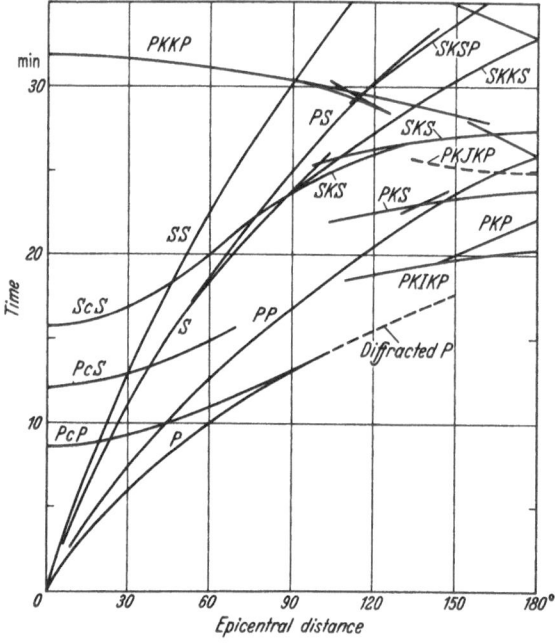

Fig. 4. Travel-time curves (Jeffreys-Bullen) for some of the principal seismic phases.

In Fig. 4 the travel-time curves are shown for a number of the main phases.

As already stated, the times in Tables 1 and 2 apply to a standard sphere. The formula (20.1) has been adapted[1] to a form enabling the allowances for oblateness to be computed. It is found that,

Table 3. *Values of $f(\varDelta)$ for $P$ and $S$ ellipticity corrections.*

| $\varDelta$ | $P$ phase sec./km. | $S$ phase sec./km. |
|---|---|---|
| 0° | 0.000 | 0.000 |
| 10 | 0.010 | 0.018 |
| 20 | 0.028 | 0.050 |
| 30 | 0.035 | 0.063 |
| 40 | 0.042 | 0.077 |
| 50 | 0.047 | 0.085 |
| 60 | 0.050 | 0.090 |
| 70 | 0.060 | 0.105 |
| 80 | 0.066 | 0.115 |
| 90 | 0.066 | 0.115 |
| 100 | 0.066 | 0.115 |

for an epicentre and observatory distant $\varDelta$ apart, the allowances in the case of the $P$ and $S$ phases are sufficiently accurately expressible in the form

$$\delta T = f(\varDelta)\,(h_0 + h_1), \tag{21.1}$$

where $h_0$ and $h_1$ are the heights of the epicentre and station above the standard sphere. The values of $f(\varDelta)$ are as shown in Table 3.

Tables are available[2] which give the values of $h_1$ for the world's seismological observatories; these tables also give the observatories' direction-cosines which are needed in epicentral distance calculations. Values of $h_0$ for various latitudes are shown in Table 4.

Table 4. *$h_0$ in terms of latitude.*

| Latitude | 0° | 10° | 20° | 30° | 40° | 50° | 60° | 70° | 80° | 90° |
|---|---|---|---|---|---|---|---|---|---|---|
| $h_0$ (km.) | 7 | 6 | 5 | 2 | −2 | −6 | −9 | −12 | −14 | −14 |

[1] K. E. Bullen: Mon. Not. Roy. Astronom. Soc., Geophys. Suppl. 4, 143—157 (1937) and later papers on Ellipticity in the same journal over the period 1937—39.

[2] H. Jeffreys (Editor): The Geocentric Direction Cosines of Seismological Observatories, 3rd. edn. Kew Observatory 1951.

Formulae of the type (21.1) can also be applied in the case of several other phases, including $PcP$, $ScS$, $PKP$ and $SKS$; but with certain other phases some elaboration is needed.

It will be noticed from Fig. 4 that the travel-time curve denoted as $PKP$ branches from that denoted as $PKIKP$ near $\Delta = 142°$. For $105° < \Delta < 142°$, the amplitudes of the first movements of $P$ type are abnormally small, and this range of distance is a partial, though not complete, shadow zone. The $PKP$ travel-times are somewhat more complicated than appears from Fig. 4 in that there is a cusp near 142° which is associated with large amplitudes. The cusp arises from the sudden decrease in $P$ velocity across the boundary of the outer core (see Sect. 24), as a result of which there are two branches of the $T-\Delta$ curve, the lower extending from 142 to 147°, and the upper from 142 to 180°. The notation $PKP_2$ is commonly given to the upper branch which is essentially that shown in Fig. 4. The $PKIKP$ phase, however, partly overlaps these branches at distances between 142 and 147°, and makes the interpretation of the readings somewhat complicated. (The notation $PKP$ has also been commonly used for the curve labelled $PKIKP$ in Fig. 4; this notation, and also $PKP_2$, was introduced before the recognition of the existence of the inner core.) The curve for the diffracted $P$ wave, also shown in Fig. 4, applies to an extension, for $\Delta > 105°$, of the $P$ curve corresponding to waves diffracted round the boundary of the central core.

# E. $P$ and $S$ velocity distributions.

**22. Introduction.** The travel-time tables are the key to the determination of the $P$ and $S$ velocity distributions in the Earth. Early this century, the determinations were attempted by trial and error procedures. Then HERGLOTZ in 1907 and BATEMAN in 1910 independently devised a direct mathematical method of obtaining the velocity $v$ as a function of the distance $r$ from the centre $O$ of the Earth, and WIECHERT and GEIGER in 1910 adapted the method to provide a more convenient means of computing $v$. In the following sub-sections, the necessary theory will be outlined, including consideration of normal and abnormal types of velocity variation. It will be shown that the results lead to an important division of the Earth into regions according to the character of the velocity variation with depth. Throughout Sect. 23 to 26, a spherically symmetrical Earth will be assumed, and the relevant travel-time tables will be those relating to the standard sphere of Sect. 20.

**23. Theory of determination of the velocities from travel-time data.** In the notation of Sect. 17, we have for the angular length $\Delta$ of a seismic ray

$$\Delta = \int \frac{d\vartheta}{dr} \, dr = \int \frac{\cot e}{r} \, dr,$$

where $e$ is the angle of emergence, and the integral is taken along the ray. Using (17.5), we can write this in the form

$$\Delta = 2\eta^0 \int_{\eta^0}^{\eta_0} r^{-1} (\eta^2 - \eta^{0\,2})^{-\frac{1}{2}} \frac{dr}{d\eta} \, d\eta, \qquad (23.1)$$

where $\eta = r/v$, the subscript zero applies to values at the outside surface, and we are taking the case in which the ray terminates at the outside surface at both ends.

The integrand in (23.1) becomes infinite (a) when $\eta^0 = \eta$, and (b) when $d\eta/dr$ is zero. The condition (a) occurs at the lowest point of the ray, which is why the right-hand side of (23.1) has been expressed as twice the integral taken from the lowest to the highest point of the ray.

Now consider a family of rays to which (23.1) applies. The value of the parameter $\eta^0$ decreases continuously from $\eta_0$ in the highest ray, for which $\Delta = 0$, to the value $\eta_1$ say, in the lowest ray, for which $\Delta = \Delta_1$, say.

In order that the condition (b) shall not be realised, we require that

$$\frac{dv}{dr} < \frac{v}{r}, \qquad (23.2)$$

at all points reached by the family of rays. Now the downward radius of curvature $\varrho$ at a point $P$ of a ray is given by $r\, dr/dp$, where $p$ is the length of the perpendicular from the Earth's centre $O$ to the tangent to the ray at $P$; thus $p$ is therefore equal to $r \cos e$, i.e. to $\eta^0 v$. Hence

$$\varrho = \frac{r}{\eta^0} \frac{dr}{dv}. \qquad (23.3)$$

By (17.5) and (23.2), we require $dr/dv$ at $P$ to be greater than the value of $\eta^0$ for a ray whose tangent is horizontal at $P$. Thus, by (23.3), the avoidance of the condition (b) is equivalent to requiring the downward curvature of such a ray to be less than that of the sphere of centre $O$ through $P$. If this were not the case, it is obvious that there would be no ray with its lowest point at $P$ and that $\Delta$ could not then be a continuous function of $\eta^0$.

Now multiply both sides of (23.1) by $(\eta^{02} - \eta_1^2)^{-\frac{1}{2}}$ and integrate over the range of values of $\eta^0$, i.e. over the whole family of rays. This operation can be validly carried out if the condition (23.2) is satisfied, and also, as pointed out by SLICHTER[1], if at some level in the range of values of $r$ involved $\eta$ decreases discontinuously with increase of depth. (In the latter case, appropriate subdivision of the range of integration with respect to $\eta$ needs to be understood.)

We then obtain

$$
\begin{aligned}
&\int_{\eta_1}^{\eta_0} \Delta\, (\eta^{02} - \eta_1^2)^{-\frac{1}{2}}\, d\eta^0 \\
&= \int_{\eta_1}^{\eta_0} 2\eta^0\, d\eta^0 \int_{\eta^0}^{\eta_0} r^{-1}\, (dr/d\eta)\, \{(\eta^{02} - \eta_1^2)(\eta^2 - \eta^{02})\}^{-\frac{1}{2}}\, d\eta \\
&= \int_{\eta_1}^{\eta_0} r^{-1}\, (dr/d\eta)\, d\eta \int_{\eta_1}^{\eta} 2\eta^0\, \{(\eta^{02} - \eta_1^2)(\eta^2 - \eta^{02})\}^{-\frac{1}{2}}\, d\eta^0 \\
&= \pi \log\left(\frac{r_0}{r_1}\right),
\end{aligned}
\qquad (23.4)
$$

where $r_1$ is the distance from $O$ of the deepest point of the deepest ray of the family.

Now the travel-time tables give an empirical relation between $\Delta$ and $\eta^0$, so that $r_1$ is, by (23.4), determined as a function of $\eta_1$ for any given family of rays. In this way, $v$ can be found as a function of $r$ for the range of depth involved. The formula (23.4) is the one found by HERGLOTZ and BATEMAN, and the above simple formulation of the solution is due to RASCH (as published by JEFFREYS)[2].

[1] L. B. SLICHTER: Physics 3, 273—295 (1932).
[2] G. HERGLOTZ: Phys. Z. 8, 145—147 (1907). — H. BATEMAN: Phil. Mag. (6) 19, 576—587 (1910). — H. JEFFREYS: Mon. Not. Roy. Astronom. Soc., Geophys. Suppl. 4, 508—509 (1939).

Integrating by parts the left-hand side of (23.4), we obtain

$$\left.\begin{array}{l} \pi \log\left(\dfrac{r_0}{r_1}\right) = \left[\varDelta \operatorname{Ar} \operatorname{Cos}\left(\dfrac{\eta^0}{\eta_1}\right)\right]_{\varDelta=\varDelta_1}^{\varDelta=0} - \int\limits_{\varDelta_1}^{0} \operatorname{Ar} \operatorname{Cos}\left(\dfrac{\eta^0}{\eta_1}\right) d\varDelta \\[2mm] = \int\limits_{0}^{\varDelta_1} \operatorname{Ar} \operatorname{Cos}\left(\dfrac{\eta^0}{\eta_1}\right) d\varDelta , \end{array}\right\} \qquad (23.5)$$

since the term in square brackets vanishes at both terminals; (when $\varDelta = \varDelta_1$, $\eta^0$ takes the value $\eta_1$). The form (23.5) is the one first derived by WIECHERT and GEIGER, and is simpler to use than (23.4).

**24. Abnormal velocity variation.** In theory, the procedure of Sect. 23 is sufficient to determine the Earth's $P$ and $S$ seismic velocity variation except where complexities arise through violation of the condition (23.2) or through a sudden increase of $\eta$ with increase of depth. In practice, both in constructing the travel-time tables and in deriving the velocity values, it is necessary to give special attention to places where the rate of change of $v$, and therefore of $\eta$, with $r$ is greater or less than normal.

From the base of the crustal layers to the centre of the Earth, it transpires that the ratio of $dv/dr$ to $v/r$ for $P$ waves increases fairly steadily from about $-2.5$ to zero, except for a limited number of fairly small ranges of depth. For $S$ waves the ratio increases from about $-2.0$ to $-0.2$ through the mantle, and again the variation is fairly steady except in the same limited ranges of depth. The $P$ or $S$ variation will be called abnormal wherever there is significant deviation from this fairly steady behaviour.

An analytical discussion of the various abnormal velocity variations is contained in Reference [1], Sect. 7.3 to 7.39[1]. The following is a brief discussion of locations where there is some evidence of abnormality. The evidence is based mainly on independent work of JEFFREYS and GUTENBERG, who have used much more extensive data than other authors.

(i) 33 to 150 km. Investigations of a number of near earthquakes have indicated that the $P$ velocity just below the crustal layers is 8.1 km./sec. within 1%, with the possible exception of the Japanese region. Below 33 km., the JEFFREYS-BULLEN tables imply fairly steady $P$ and $S$ velocity gradients. GUTENBERG[2] considers that the $P$ velocity diminishes steadily with depth below the crustal layers and reaches a minimum at a depth between 100 and 150 km. and that the $S$ velocity behaves similarly; and considers that the condition (23.2) is violated in this region. In this case the theory of Sect. 24 formally fails, and the practical effect is that some uncertainty attaches to estimates of velocities at greater depths. Evidence from earthquakes with foci in the region where (23.2) may be violated helps to improve the velocity determinations. On the latest work of JEFFREYS[3], the $P$ travel-time curve for European earthquakes is nearly a straight line up to $\varDelta = 12°$; in this case it is just possible to apply the HERGLOTZ-BATEMAN method. Miss LEHMANN considers that abnormal behaviour of the velocities in this region arises from abnormality in the behaviour of $\mu$.

---

[1] See also L. B. SLICHTER: Physics 3, 273—295 (1932). — I. LEHMANN: Mon. Not. Roy. Astronom. Soc., Geophys. Suppl. 4, 250—271 (1937). — K. E. BULLEN: Mon. Not. Roy. Astronom. Soc., Geophys. Suppl. 5, 91—98 (1945). — C. Y. FU: Bull. Seism. Soc. Amer. 37, 331—346 (1947).
[2] B. GUTENBERG: Bull. Seism. Soc. Amer., 43, 223—232 (1953).
[3] H. JEFFREYS: Mon. Not. Roy. Astronom. Soc., Geophys. Suppl. 6, 557—565 (1954).

(ii) 400 to 1000 km. Changes of gradient in the Jeffreys-Bullen travel-times near $\Delta = 20°$ imply that either (a) there are sharp increases in the velocity gradients setting in near a depth of 410 km., the gradients then steadily diminishing until normality is resumed near 1000 km., or (b) the $P$ and $S$ velocities jump discontinuously at a depth of the order of 400 km. Jeffreys states that on the whole the evidence favours (a), though the discrimination is not clear-cut. The changes taking place near the depth of 410 km. have been referred to as the "20° discontinuity", the first evidence for which appeared in Byerly's study[1] of the Montana earthquake of 1925. Gutenberg[2], on the other hand, considers that (c) the velocity gradients (with respect to increase of depth) are

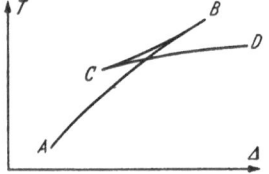

Fig. 5. Triplication of travel-time curve, corresponding to a rapid rise in velocity with depth.

steady but greater than normal between depths of 200 km. through 400 km. to 1000 km., and that (d) there is a sharp reversion to the normal variation at 1000 km. There have been still other suggestions made, but a consideration of (a), (b), (c) and (d) covers the main theoretical points.

In the condition (a), there is a range of depth in which $dr/dv$ is negative and numerically much less than normal, so that by (23.3), the upward curvature of rays in this region is much greater than normal. Consequently, rays which penetrate a little below 410 km. would emerge with smaller values of $\Delta$ than rays just above; for rays which penetrate further below this level a value of $\eta^0$ would in due course be reached at which $\Delta$ would start to increase again (provided there is no further source of abnormality). Thus the effect of

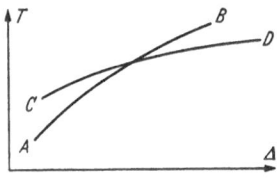

Fig. 6.
Form of travel-time curve when velocity increases discontinuously with depth.

(a) is that the travel-time curve would show triplication over a range of values of $\Delta$, as illustrated in Fig. 5.

If (b) is the case, the first rays refracted into the region below the discontinuity would emerge at a smaller value of $\Delta$ than rays which do not quite reach the discontinuity, and the $T$—$\Delta$ relation would be as illustrated in Fig. 6. In this case there is duplication of the travel-time curve for a range of values of $\Delta$, and not triplication, except in so far as there may be upward reflection at the discontinuity surface. In Fig. 5 and 6 (and also in Fig. 7), the order of the letters $A$, $B$, $C$, $D$ is that of decreasing $\eta^0$.

The condition (c) would, by (23.3) imply that the upward curvature of rays at points between depths of 200 and 1000 km. is greater than normal, though on Gutenberg's data the curvature would not be great enough to produce the "reversed segment" of Fig. 5. Thus $d\Delta/d\eta^0$ would not change sign, but would be numerically smaller than normal, so that $|d^2 T/d\Delta^2|$, being equal to $|d\eta^0/d\Delta|$ by (17.6), would be greater than normal, and the curvature of the $T$—$\Delta$ curve would be greater than normal.

The circumstances (d) would be opposite in effect to (c) and result in a reduced curvature of the $T$—$\Delta$ curve over a range of distance.

(iii) 2700 to 2900 km. The evidence of both Jeffreys and Gutenberg indicates that the $P$ and $S$ velocity gradients are practically zero in this region. As with (ii, d), this corresponds to an abnormally low curvature of the $T$—$\Delta$

[1] P. Byerly: Bull. Seism. Soc. Amer. 16, 209—265 (1926).
[2] Reference [6], p. 409.

curves. Certain other writers, notably DAHM[1], have suggested that there are discontinuous velocity changes near 2700 km.; in this case Fig. 6 would be relevant.

(iv) 2900 km. This is the boundary between the mantle and central core, where all investigators agree that the $P$ velocity sharply diminishes. According to JEFFREYS, the change is from 13.6 to 8.1 km./sec., and GUTENBERG's values agree with this within 1 %. By (17.5), $P$ rays just entering the core will therefore be refracted downward so that there will be a discontinuous jump in $\Delta$, and therefore, by (17.6), also in $T$, as $\eta^0$ continuously diminishes.

For rays which enter the core, (23.1) has to be replaced by

$$\Delta = 2\eta^0 \int_{\eta_a}^{\eta_0} r^{-1}(\eta^2 - \eta^{0\,2})^{-\frac{1}{2}} \frac{dr}{d\eta}\, d\eta + 2\eta^0 \int_{\eta^0}^{\eta_b} r^{-1}(\eta^2 - \eta^{0\,2})^{-\frac{1}{2}} \frac{dr}{d\eta}\, d\eta$$

$$= \Delta_1 + \Delta_2$$

say, where $\eta_a$ and $\eta_b$ are the values of $\eta$ at the base of the mantle and top of the core, respectively. Now

$$\frac{d\Delta_1}{d\eta^0} = \frac{\Delta_1}{\eta^0} + 2\eta^0 \int_{\eta_a}^{\eta_0} r^{-1}\eta^0\,(\eta^2 - \eta^{0\,2})^{-\frac{3}{2}}\frac{dr}{d\eta}\, d\eta. \tag{24.1}$$

For rays which are just steep enough to enter the central core, $\eta^0$ is nearly equal to $\eta_a$, in which case the second term on the right-hand side of (24.1) is large. On the other hand, $d\Delta_2/d\eta^0$ is moderate in value so long as the rays in question do not encounter further abnormal regions in the central core; and of course the term $\Delta_1/\eta^0$ in (24.1) is also moderate in value. Hence, for the first rays to enter the core, $d\Delta/d\eta^0$ is large and positive, so that $\Delta$ decreases with increasing steepness of these rays until in due course a minimum value of $\Delta$ is reached, analogous to the minimum deviation of light through a prism or lens. After the minimum is reached, $\Delta$ increases again [until new complications are met for deeper rays which encounter the abnormality (v) below]. By (17.6) it is then easy to deduce the corresponding form of the $T$—$\Delta$ relation, which is as in Fig. 7.

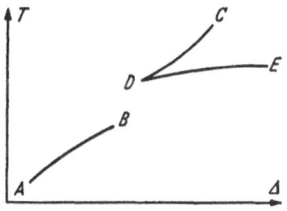

Fig. 7.
Form of travel-time curve when velocity decreases discontinuously with depth.

(v) 4980 to 5120 km. JEFFREYS finds that the travel-time data cannot be fitted unless the $P$ velocity decreases with depth in a range of depth near 5000 km. The theory of Sect. 23 may again fail, and JEFFREYS assumes (arbitrarily) that $v$ is proportional to $r$ over a range of depth which he calculates as 4980 to 5120 km., these values being uncertain to the extent of some 50 km. GUTENBERG does not find evidence of this particular abnormality although his data do not permit the rejection of it.

(vi) 5120 km. At this depth, JEFFREYS finds a sudden jump in the $P$ velocity from the somewhat arbitrary value of 9.5 km./sec. on the upper side to 11.2 km./sec. on the lower side. On GUTENBERG's data, the velocity gradient but not the velocity itself increases sharply near this level, and the gradient is steep for the next 150 km.

(vii) 5120 to 6370 km. In this region, which is the *inner core*, JEFFREYS finds the $P$ velocity gradient to be abnormally small. The same applies on GUTENBERG's data apart from the uppermost 150 km. where the gradient is steep.

[1] C. G. DAHM: Bull. Seism. Soc. Amer. 26, 159—171 (1936).

**25. Further travel-time and velocity relations.** The following two relations are sometimes found useful:

$$T = \int_{\eta^0}^{\eta_0} 2\eta^2 r^{-1}(\eta^2 - \eta^{0\,2})^{-\frac{1}{2}} \frac{dr}{d\eta}\, d\eta, \tag{25.1}$$

$$T = \eta^0 \varDelta + \int_{\eta^0}^{\eta_0} 2 r^{-1}(\eta^2 - \eta^{0\,2})^{\frac{1}{2}} \frac{dr}{d\eta}\, d\eta. \tag{25.2}$$

If the variation of velocity is normal near the outside surface, then, for values of $\varDelta$ not too great,

$$T \approx \eta_0 \varDelta - a\varDelta^3, \tag{25.3}$$

where $a$ is a positive constant. For rays of the type $PKIKP$,

$$T \approx a - b\,(\pi - \varDelta)^2, \tag{25.4}$$

where $a$ and $b$ are positive constants, provided $\pi - \varDelta$ is not too great. Near the cusp in Fig. 7, the form is approximately

$$T - T_1 = a\,(\varDelta - \varDelta_1) \pm b(\varDelta - \varDelta_1)^{\frac{3}{2}}, \tag{25.5}$$

where $T_1$ and $\varDelta_1$ apply at the cusp.

Proofs of the results (25.1) to (25.5) will be found in Reference [1] pp. 111–117.

A number of travel-time relations have been worked out for simple assumed velocity distributions, including the following:

(i) $v$ *constant.* In this case the $T$—$\varDelta$ relation is obviously

$$T = 2\eta_0 \sin \tfrac{1}{2}\varDelta. \tag{25.6}$$

(ii) $v = A r^B$ $(A, B$ *constants;* $B < 1)$: We have in this case $\eta = r/v = a r^b$, where $a$ and $b$ are positive constants. Then

$$\eta^{-1} d\eta = b r^{-1} dr,$$

so that by (23.1)

$$\varDelta = 2\eta^0 \int_{\eta^0}^{\eta_0} (b\eta)^{-1}\,(\eta^2 - \eta^{0\,2})^{-\frac{1}{2}}\, d\eta$$

$$= 2 b^{-1} \operatorname{ar} \cos\,(\eta^0/\eta_0).$$

Hence, by (17.6),

$$\frac{dT}{d\varDelta} = \eta^0 = \eta_0 \cos\left(\frac{1}{2} b\varDelta\right),$$

and

$$T = 2\eta_0 b^{-1} \sin\left(\frac{1}{2} b\varDelta\right). \tag{25.7}$$

(iii) $v = a - b r^2$ $(a, b$ *positive constants*). By (23.3), this velocity relation gives circular rays, concave upward, of radius $(2\eta^0 b)^{-1}$. In this case, it can be shown (see Reference [1], Sect. 7.53) that

$$T = 2\eta_0(\lambda^2 - 1)^{-\frac{1}{2}} \operatorname{Ar\,Sin}\{(\lambda^2 - 1)^{\frac{1}{2}} \sin \tfrac{1}{2}\varDelta\}, \tag{25.8}$$

where

$$\lambda = \frac{2a}{v_0} - 1.$$

**26. Numerical results.** From the numerical travel-time data described in Sect. 21, and the theory of Sect. 23, supplemented as required by recourse to detail of the type referred to in the two preceding Sections, numerical values of the $P$ velocities throughout the Earth, and of $S$ velocities throughout the mantle, have been derived. The results obtained by JEFFREYS are given in Tables 5 and 6, and corre-spond to the JEFFREYS-BUL-LEN tables — see Sect. 21. In Fig. 8, the velocities of both JEFFREYS and GUTEN-BERG are shown graphically; the results of other authors are mostly close to the curves shown. An indication of the uncertainties has been given by JEFFREYS[1].

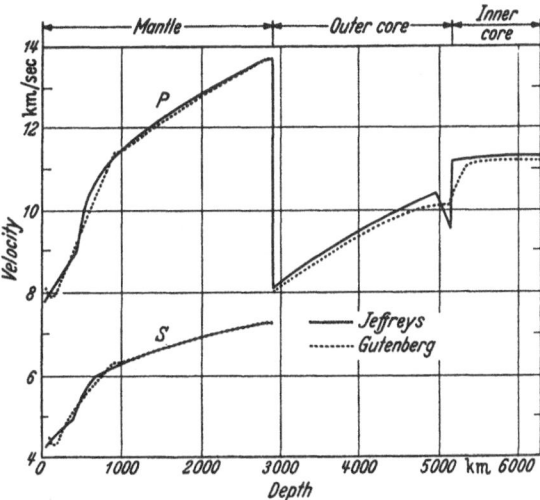

Fig. 8. Variation of the $P$ and $S$ velocities with depth in the Earth.

It follows from Sect. 24 (iv), that, because of the decrease in the $P$ velocity across the boundary of the central core, it is not possible from the $P$ and $PKP$ travel-time data to infer the $P$ velocities in the outer part of the core. The velocities, however, can be determined using $S$ rays in the mantle and $SKS$ rays, since the $P$ velocities at all points of the core exceed the $S$ velocities at all points of the mantle.

On the basis of the velocity results shown in Tables 5 and 6, the Earth has been divided into regions $A$, $B$, $C$, $D$, $E$, $F$ and $G$, the region $D$ having been later subdivided into $D'$ and $D''$. The division, which is shown in Table 7, is pro-visional, since there are some uncertainties, as indicated in Sect. 24, especially in respect of details in the regions $B$, $C$ and $F$. The steady variations in $D'$ and

Table 5. *P velocities below the crustal layers.*

| Depth km. | P velocity in mantle km./sec. | Depth km. | P velocity in core km./sec. |
|---|---|---|---|
| 100 | 7.95 | 2900 | 8.10 |
| 200 | 8.26 | 3000 | 8.22 |
| 410 | 8.97 | 3200 | 8.47 |
| 600 | 10.25 | 3400 | 8.76 |
| 800 | 11.00 | 3600 | 9.04 |
| 1000 | 11.42 | 3800 | 9.28 |
| 1200 | 11.71 | 4000 | 9.51 |
| 1400 | 11.99 | 4200 | 9.70 |
| 1600 | 12.26 | 4400 | 9.88 |
| 1800 | 12.53 | 4600 | 10.06 |
| 2000 | 12.79 | 4800 | 10.25 |
| 2200 | 13.03 | 4980 | 10.44 |
| 2400 | 13,27 | } 5120 { | 9.5 |
| 2600 | 13.50 | | 11.16 |
| 2700 | 13.63 | 5700 | 11.26 |
| 2900 | 13.64 | 6370 | 11.31 |

Table 6. *S velocities in the mantle.*

| Depth km. | S velocity km./sec. | Depth km. | S velocity km./sec. |
|---|---|---|---|
| 100 | 4.45 | 1600 | 6.73 |
| 200 | 4.60 | 1800 | 6.83 |
| 410 | 4.96 | 2000 | 6.93 |
| 600 | 5.66 | 2200 | 7.02 |
| 800 | 6.13 | 2400 | 7.12 |
| 1000 | 6.36 | 2600 | 7.22 |
| 1200 | 6.50 | 2700 | 7.29 |
| 1400 | 6.62 | 2900 | 7.30 |

[1] Reference [*11*], pp. 115—116.

Table 7. *Regions of the Earth's interior.*

| Region | Range of depth km. | Description | P velocities km./sec. | S velocities km./sec. | Character of gradients |
|--------|--------|--------|--------|--------|--------|
| A | 0–33 | Crustal Layers | Very variable | Very variable | Unsteady |
| B | 33–410 | Upper mantle | (8)–8.97 | 4.4–4.96 | Steady |
| C | 410–1000 | Transition region | 8.97–11.42 | 4.96–6.36 | Greater than normal |
| D' | 1000–2700 | Lower Mantle | 11.42–13.63 | 6.36–7.29 | Steady |
| D'' | 2700–2900 | | 13.64 | 7.30 | Nearly zero |
| E | 2900–4980 | Outer core | 8.10–10.44 | | Steady |
| F | 4980–5120 | Transition region | 10.44–9.5 | (Not observed) | Negative (with respect to depth) |
| G | 5120–6370 | Inner core | 11.16–11.31 | | Smaller than normal |

$E$, the sudden reduction in the $P$ velocity between $D''$ and $E$, the sudden jump in either the $P$ velocity or its gradient between $E$ and $G$, and the smallness of the velocity gradient in $G$, are, however, all well established.

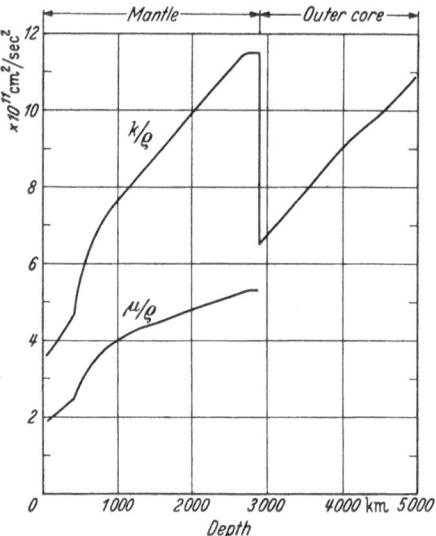

Fig. 9. Values of $k/\varrho$ and $\mu/\varrho$ down to a depth of 5000 km. in the Earth. ($\varrho$ = density; $k$ = incompressibility; $\mu$ = rigidity.)

Using equations (14.2) and (14.5), values of $k/\varrho$ and $\mu/\varrho$ can be computed throughout the mantle from the data in Tables 5 and 6. In the outer core, the effective value of $\mu/\varrho$ is less than one-hundredth of the value of $\mu/\varrho$ in the upper mantle, so that $k/\varrho$ is well determined by the $P$ velocity distribution in the outer core. Values of $\mu/\varrho$ in the mantle and of $k/\varrho$ down to the base of the region $E$, as deduced from Tables 5 and 6, are shown in Table 8, Table 9 and Fig. 9. (The question of values of $k/\varrho$ and $\mu/\varrho$ in the inner core is discussed in Sect. 35 and 36.) These results are of special physical significance in that they constitute the firmest quantitative information which seismology has to offer on specific mechanical properties of the Earth's interior. The data and curves in Tables 8 and 9 and Fig. 9 are derived by fairly direct calculation from the basic observational data of seismology, and the errors in them come largely from uncertainties in the observational data itself. In contrast to many other sources of evidence on properties of the Earth's interior, they are very largely uncluttered by questions of hypothesis and interpretation.

Table 8. *Values of $k/\varrho$ in mantle and outer core.*     Table 9.  *Values of $\mu/\varrho$ in mantle.*

| Depth km. | $k/\varrho$ in mantle cm.²/sec.² | Depth km. | $k/\varrho$ in outer core cm.²/sec.² | Depth km. | $\mu/\varrho$ cm.²/sec.² | Depth km. | $\mu/\varrho$ cm.²/sec.² |
|---|---|---|---|---|---|---|---|
| 100 | $3.68 \times 10^{11}$ | 2900 | $6.56 \times 10^{11}$ | 100 | $1.98 \times 10^{11}$ | 1600 | $4.53 \times 10^{11}$ |
| 200 | 4.00 | 3000 | 6.76 | 200 | 2.12 | 1800 | 4.66 |
| 410 | 4.77 | 3200 | 7.17 | 410 | 2.46 | 2000 | 4.80 |
| 600 | 6.23 | 3400 | 7.67 | 600 | 3.20 | 2200 | 4.93 |
| 800 | 7.09 | 3600 | 8.17 | 800 | 3.76 | 2400 | 5.07 |
| 1000 | 7.65 | 3800 | 8.61 | 1000 | 4.04 | 2600 | 5.20 |
| 1200 | 8.08 | 4000 | 9.04 | 1200 | 4.22 | 2700 | 5.32 |
| 1400 | 8.53 | 4200 | 9.41 | 1400 | 4.38 | 2900 | 5.33 |
| 1600 | 8.99 | 4400 | 9.76 | | | | |
| 1800 | 9.48 | 4600 | 10.12 | | | | |
| 2000 | 9.95 | 4800 | 10.51 | | | | |
| 2200 | 10.41 | 4980 | 10.90 | | | | |
| 2400 | 10.85 | | | | | | |
| 2600 | 11.30 | | | | | | |
| 2700 | 11.50 | | | | | | |
| 2900 | 11.50 | | | | | | |

# F. Amplitude theory.

**27. Introduction.** In the simple conditions of spherical symmetry about a focus $F$, the classical wave equation of form $\partial^2\Phi/\partial t^2 = v^2 \nabla^2 \Phi$ reduces to

$$\frac{\partial^2 (r\,\Phi)}{\partial t^2} = v^2 \frac{\partial^2 (r\,\Phi)}{\partial r^2},$$

where $r$ is the distance from $F$, and the velocity $v$ is assumed constant. For waves advancing outward from $F$, this equation has the simple solution

$$\Phi = r^{-1} f(r - vt), \tag{27.1}$$

where the form of $f$ depends on the initial conditions. By (27.1), the amplitude varies inversely as the distance from the focus in these simple conditions.

This law of amplitude variation is, by (14.1) and (14.4) (omitting the term in $G$), relevant to seismic waves at distances not too close to the focus, except for the effects of damping and of variation in the $P$ and $S$ velocities. The effects of damping are relatively slight, and have already been discussed in Sect. 14. The effects of variation in the $P$ and $S$ velocities are specially great over epicentral distances up to about 1000 km. because of the dominating influence of the heterogeneous crustal layers. At greater distances, the law holds as a first approximation, the distances being measured along rays. But, in addition, the various types of abnormal velocity variation have considerable influence, including cases of abnormal velocity gradients as well as discontinuous velocity changes.

It may be remarked that the amplitudes of surface waves (apart from dispersion effects) essentially diminish as $r^{-\frac{1}{2}}$, so that surfaces waves become relatively more prominent on seismograms than bodily waves as the focal distance increases. In earthquakes of normal focal depth, the surface waves are usually greater than bodily waves at distances beyond some 10 degrees.

**28. Connection between amplitudes and $dT^2/d\Delta^2$.** For simplicity of exposition, consider seismic waves emanating from a focus at the Earth's outer surface, carrying energy $I$ per unit solid angle, and neglect any dependence of $I$ on direction. (The case in which the focus is below the surface is treated in Reference

[1], pp. 123—125.) For a ray which emerges at angular distance $\Delta$, let the angle of emergence at the outer surface be $e$. Then, for the whole set of rays emerging at angles between $e$ and $e+de$, the energy carried is $2\pi I \cos e\, de$. The area of the Earth's surface over which these rays emerge is $2\pi r_0^2 \sin \Delta d\Delta$, where $r_0$ is the Earth's radius; and the area of the corresponding portion of wave-front near the surface is $2\pi r_0^2 \sin \Delta \sin e d\Delta$. The energy $E(\Delta)$ per unit area of wave-front emerging at angle $e$ is therefore given by

$$E(\Delta) = \frac{I}{r_0^2} \frac{\cot e}{\sin \Delta} \left| \frac{de}{d\Delta} \right|.$$

If $T$ is the travel-time corresponding to $\Delta$, we have, by (17.5) and (17.6)

$$\left. \begin{aligned} \left| \frac{d^2 T}{d\Delta^2} \right| &= \left| \frac{d}{d\Delta} (\eta_0 \cos e) \right| \\ &= \eta_0 \sin e \left| \frac{de}{d\Delta} \right| \\ &= E\, I^{-1} \eta_0 r_0^2 \sin \Delta \sin e \tan e. \end{aligned} \right\} \tag{28.1}$$

It follows from (28.1) that, when an abnormal variation results in a strong curvature of the $T$—$\Delta$ curve, and so in a large value of $|d^2 T/d\Delta^2|$, then $E$ is in general abnormally large. Since $E$ is proportional to the square of the amplitude, it follows that the amplitudes are in this case abnormally large. Similarly, at distances for which the $T$—$\Delta$ curvature is abnormally small, the amplitudes are likewise generally small. Thus amplitude data, as well as travel-time data, are capable of indicating the presence of abnormal velocity gradients in the Earth. In practice, the use of amplitudes leads to less precise results, however, because of the distorting influence of the crustal layers.

GUTENBERG and RICHTER have made extensive use of amplitude data in constructing the magnitude scale referred to in Sect. 8. In addition, GUTENBERG has brought amplitude data to bear in support of his theory of the existence of low velocity layers in the outer part of the mantle.

**29. Energy in plane sinoidal waves.** Consider a plane sinoidal wave represented by the form

$$a \cos \left\{ 2\pi \left( \frac{x}{\lambda} - \frac{t}{\tau} \right) \right\}, \tag{29.1}$$

where $a$, $\lambda$ and $\tau$ denote the amplitude, wave-length and period. The energy in one wave-length, per unit area of wave-front, being double the kinetic energy, is then equal to

$$\int_0^{2\pi} \varrho \, (2\pi a \tau^{-1})^2 \left( \frac{\lambda}{2\pi} \right) \sin^2 \xi \, d\xi,$$

i.e. $2\pi^2 a^2 \tau^{-2} \lambda \varrho$, where $\varrho$ is the density. If the waves emerge at an angle $e$ against a plane, the energy reaching unit area of the plane in one period (during which time the waves advance through one wave-length) is then

$$2\pi^2 a^2 \tau^{-2} \lambda \varrho \sin e. \tag{29.2}$$

Wave-forms of the type

$$A \exp \{i q (x_2 + x_3 \tan e - vt \sec e)\} \tag{29.3}$$

will shortly be considered, corresponding to sinoidal waves of amplitude $A$, velocity $v$, wave-length $2\pi/(q \sec e)$ and period $2\pi/(vq \sec e)$. In the case (29.3), normals to the wave-front are parallel to the plane $O x_2 x_3$, and the rays emerge at an angle $e$ against any plane normal to the $x_3$-axis. By (29.2), the energy reaching unit area of a plane normal to $O x_3$ during one period is

$$2 \pi^3 A^2 \tau^{-2} q^{-1} \varrho \sin 2e. \tag{29.4}$$

**30. Reflection and refraction of seismic waves.** For many seismological purposes it is sufficient to consider the case of plane sinoidal waves incident against a plane boundary separating two homogeneous media of indefinite extent, and this will apply in the following treatment. Reference may be made to CAGNIARD[1] for an important direct investigation of the case of incident waves of more general types than plane sinoidal waves.

In the present treatment, the boundary will be taken to be the plane $O x_1 x_2$. For the purpose of using the symbolism $SV$ and $SH$ for polarised $S$ waves, the boundary will be conventionally taken as horizontal. Symbols such as $\alpha, \beta, \varrho, k, \mu$, defined in Sect. 13 and 14, will apply to the lower medium $M$, and primes will be used with these symbols for the upper medium $M'$.

Corresponding to (29.3), we take an $SH$ wave to be represented in $M$ by the form

$$u_1 = A \exp\{i q (x_2 + x_3 \tan f - \beta t \sec f)\}, \tag{30.1}$$

where $f$ is the angle of emergence in $M$ against the boundary.

While the displacement component $u_1$ is always associated purely with $SH$ waves, each of the components $u_2$ and $u_3$, however, contributes to both $P$ and $SV$ waves in general. In order to represent the $P$ and $SV$ waves separately, it is customary to introduce functions $\Phi$ and $\psi$, where

$$u_2 = \frac{\partial \Phi}{\partial x_2} + \frac{\partial \psi}{\partial x_3}, \qquad u_3 = \frac{\partial \Phi}{\partial x_3} - \frac{\partial \psi}{\partial x_2}. \tag{30.2}$$

Thus $\Phi$ is a displacement potential analogous to the velocity potential in hydrodynamics, while $\psi$ is analogous to a hydrodynamical stream-function. Then in the wave-form

$$\Phi = (i q)^{-1} B \cos e \exp\{i q (x_2 + x_3 \tan e - \alpha t \sec e)\}, \tag{30.3}$$

the components $\partial \Phi / \partial x_2$ and $\partial \Phi / \partial x_3$ give a displacement in the direction of the normal to the wave-front, as in $P$ waves, and $B$ is the amplitude of the $P$ waves. And in the wave-form

$$\psi = (i q)^{-1} C \cos f \exp\{i q (x_2 + x_3 \tan f - \beta t \sec f)\}, \tag{30.4}$$

the components $\partial \psi / \partial x_3$ and $- \partial \psi / \partial x_2$ give transverse motion of type $SV$, $C$ being the amplitude of the $SV$ waves. From (30.2), it is, moreover, readily checked that, for $i = 2, 3$, the equations (6.3) (taking grad $V = 0$) are satisfied if

$$\frac{\partial^2 \Phi}{\partial t^2} = \alpha^2 \nabla^2 \Phi, \qquad \frac{\partial^2 \psi}{\partial t^2} = \beta^2 \nabla^2 \psi,$$

showing again the connection of $\Phi$ and $\psi$ with $P$ and $S$ waves, respectively.

In treating reflection and refraction at the boundary, we take an incident wave represented by (30.1), (30.3) or (30.4) according as it is of $SH$, $P$, or $SV$ type, and contemplate the possibility of the generation of $SH$, $P$ and $SV$ reflected waves in $M$ and of $SH$, $P$ and $SV$ refracted waves in $M'$. A little study of the boundary conditions shows that when the incident wave is $SH$, there can be only $SH$ reflected and refracted waves. When the incident wave is $P$ or $SV$, there are

---

[1] Reference [4].

no reflected or refracted *SH* waves. Thus there are polarisation effects, as in optics, but more general in that there may be longitudinal as well as transverse waves involved.

For incident *SH* waves, we then assume in $M$ the form

$$u_1 = A_0 \exp\{i\,q\,(x_2 + x_3 \tan f - \beta\,t \sec f)\} + \\ + A \exp\{i\,q\,(x_2 - x_3 \tan f - \beta\,t \sec f)\},$$

the suffix zero applying to the incident wave; and in $M'$

$$u_1 = A' \exp\{i\,q\,(x_2 + x_3 \tan f' - \beta'\,t \sec f')\}.$$

For incident *P* waves, we assume in $M$

$$i\,q\,\Phi = B_0 \cos e \exp\{i\,q\,(x_2 + x_3 \tan e - \alpha\,t \sec e)\} + \\ + B \cos e \exp\{i\,q\,(x_2 - x_3 \tan e - \alpha\,t \sec e)\},$$

$$i\,q\,\psi = C \cos f \exp\{i\,q\,(x_2 - x_3 \tan f - \beta\,t \sec f)\};$$

and in $M'$

$$i\,q\,\Phi = B' \cos e' \exp\{i\,q\,(x_2 + x_3 \tan e' - \alpha'\,t \sec e')\},$$

$$i\,q\,\psi = C' \cos f' \exp\{i\,q\,(x_2 + x_3 \tan f' - \beta'\,t \sec f')\}.$$

For incident *SV* waves, we assume in $M$

$$i\,q\,\Phi = B \cos e \exp\{i\,q\,(x_2 - x_3 \tan e - \alpha\,t \sec e)\},$$

$$i\,q\,\psi = C_0 \cos f \exp\{i\,q\,(x_2 + x_3 \tan f - \beta\,t \sec f)\} + \\ + C \cos f \exp\{i\,q\,(x_2 - x_3 \tan f - \beta\,t \sec f)\};$$

and in $M'$

$$i\,q\,\Phi = B' \cos e' \exp\{i\,q\,(x_2 + x_3 \tan e' - \alpha'\,t \sec e')\},$$

$$i\,q\,\psi = C' \cos f' \exp\{i\,q\,(x_2 + x_3 \tan f' - \beta'\,t \sec f')\}.$$

In each group of the above equations, $q$ is taken to be the same since the periods $2\pi/(\alpha q \sec e)$, $2\pi/(\beta q \sec f)$, $2\pi/(\alpha' q \sec e')$ and $2\pi/(\beta' q \sec f')$ must be the same, and since

$$\alpha \sec e = \beta \sec f = \alpha' \sec e' = \beta' \sec f',$$

by (17.5).

The conditions that have to be satisfied at the boundary when the media are solid and non-slipping are that $u_1$, $u_2$, $u_3$, $p_{31}$, $p_{32}$ and $p_{33}$ are continuous at the boundary. If one of the media is a fluid, so that (neglecting viscosity) there will be slipping, then $u_1$ and $u_2$ do not have to be continuous at the boundary, and $p_{31}$ and $p_{32}$ will be zero at the boundary. If there is only one semi-infinite medium, bounded by a free plane surface, the boundary conditions are that $p_{31}$, $p_{32}$ and $p_{33}$ are zero at this surface.

The stress-components $p_{31}$, $p_{32}$ and $p_{33}$ may, by the stress-strain relations (4.4), and the strain-component definitions (2.3), be expressed in terms of $u_1$, $u_2$ and $u_3$, and then, by (30.2), in terms of $u_1$, $\Phi$ and $\psi$. From the boundary conditions, there then emerge sufficient equations to determine the ratios of the amplitudes in the reflected and refracted waves to that in the incident waves. By means of (17.5), the ratios are determined in terms of the angle of emergence of the incident wave and the elastic properties of the two media. In general, the algebra is heavy, but a number of cases relevant to conditions at discontinuities in the Earth have been worked out in some detail. In practice, the algebra is often simplified by assuming that Poisson's relation (4.9) holds.

It is not possible, in the space here available, to give details of these calculations. The reader who is interested will find cases worked out in the references given below[1]. But it is of interest to comment on a few further properties of reflected and refracted waves.

For particular angles of emergence it can happen that the amplitude of one of the reflected waves is zero. For example, when POISSON's relation applies, JEFFREYS has shown that from $P$ waves incident at 60° or 77°.2 there are refracted $P$ and $SV$ waves, and reflected $SV$ waves, but the reflected $P$ waves have zero amplitude. From $SV$ waves incident at 30° or 34°.3, there are no reflected $SV$ waves. The amplitudes of one type of reflected wave may, moreover, be quite small over an appreciable range of angles of incidence.

It can be shown that, with incident $SH$ waves through $M$ against the boundary between $M$ and $M'$, the amplitudes of the reflected $SH$ waves are zero if the angle of emergence is such as to satisfy the relations $\mu \tan f = \mu' \tan f'$ and $\beta \sec f = \beta' \sec f'$. In this case, the whole of the energy is transferred into refracted waves.

As in optics, there may also be total reflection. In the case of $SH$ waves incident through $M$ at angle $\frac{1}{2}\pi - f$, there will be no refracted waves if $\beta < \beta'$ and $f < \text{ar} \cos (\beta/\beta')$, since this inequality would, by (17.5), imply an unreal value of $f'$. In place of refracted waves, there would be a surface motion in $M'$ advancing with velocity $\beta \sec f$ along the boundary, the amplitude of this movement diminishing exponentially in $M'$ with increasing distance from the boundary. There is also in general a change of phase between the incident and reflected waves. Incident $SV$ waves are also totally reflected if $\beta < \beta'$ and $f < \text{ar} \cos (\beta/\beta')$; and incident $P$ waves, if $\alpha < \beta'$ and $e < \text{ar} \cos (\alpha/\beta')$.

If $v$ and $w$ are the $x_2$- and $x_3$-components of the displacement at a point of the Earth's surface, due to seismic waves emerging from below, it is necessary to discriminate between the "apparent angle of emergence" $\bar{e}$, as indicated from the ratio of $v$ to $w$, and the actual angle of emergence $e$. For the case of $P$ waves, taking $\tan \bar{e} = w/v$, it can be shown that $e$ and $\bar{e}$ are connected by

$$2 \cos^2 e = (\alpha^2/\beta^2)(1 - \sin \bar{e}). \qquad (30.5)$$

**31. Partitioning of energy.** When the amplitude ratios of the incident, reflected and refracted waves have been determined, simple use of (29.4) gives the partitioning of energy between the various waves. For example, in the case of incident $P$ waves, it follows from (29.4) that, since $q$ and $\tau$ are the same for all the waves involved, the ratios of the energies in the reflected $P$, reflected $SV$, refracted $P$ and refracted $SV$ waves to that in the incident $P$ waves will be

$$\frac{B^2}{B_0^2}, \quad \frac{C^2 \sin 2f}{B_0^2 \sin 2e}, \quad \frac{\varrho' B'^2 \sin 2e'}{\varrho B_0^2 \sin 2e}, \quad \frac{\varrho' C'^2 \sin 2f'}{\varrho B_0^2 \sin 2e}, \qquad (31.1)$$

respectively.

It needs to be remarked that, in the preceding discussions, only sinoidal waves have been considered. Because of the orthogonal property of terms of Fourier series, the results for the energy partitioning at plane boundaries by the processes that have been indicated will, however, remain valid for more general plane wave-forms.

In the preceding sections, we have considered amplitude effects where the velocity gradient is abnormally high and where the velocity changes discontinuously.

[1] Reference [13]. — H. JEFFREYS: Mon. Not. Roy. Astronom. Soc., Geophys. Suppl. 1, 321—334 (1926). — L. B. SLICHTER and V. G. GABRIEL: Gerlands Beitr. Geophys. 38, 228—256 (1933). — Reference [1], Chapter VI. — Reference [2], pp. 29—34.

In the case of normal velocity variation, for example inside the regions $D'$ and $E$, it is desirable to check that the energy is transmitted along rays. JEFFREYS has shown that the energy is transmitted with negligible loss, provided the wave-length is not comparable with the distance through which the velocity changes by a significant fraction. There can also be significant departures from ray theory where waves are close to grazing incidence against level surfaces of velocity. The proof of these results comes from considering a region in which the velocity changes continuously as the limiting case of one in which there are an indefinitely large number of thin homogeneous layers across each of which the velocity changes slightly, and applying the theory of this and the preceding Section.

It is evident from (31.1) that in making inferences from observed amplitudes of bodily seismic waves, it is necessary to allow for conversion of energy into other wave types at each discontinuity surface encountered, in addition to the effects discussed in Sect. 28. The use of amplitude theory is therefore seen to become specially difficult because of the marked heterogeneity of the crustal layers through which all bodily waves must pass before being observed.

**32. Amplitude effects and the Earth's core.** The primary evidence for the existence of the Earth's central core was the relatively low amplitudes of the first waves of $P$ type arriving from earthquakes at distances between 105 and 142°. This "shadow zone" could be explained only by a large and sharp fall in the $P$ velocity at a depth which GUTENBERG computed as near 2900 km.

Such observations as there were of $P$ waves in the range $105° < \Delta < 142°$ (apart from compound types such as $PP$, $SP$, etc.) were for a considerable time attributed to diffraction effects connected with the outer boundary of the central core. Observations for distances a few degrees greater than 105° could be attributed to a simple extension of the $T—\Delta$ relation for the ordinary $P$ phase, while those at distances extending backwards from 142° were linked with $PKP$. In 1936, LEHMANN[1], followed by GUTENBERG and RICHTER[2], proposed the existence of an *inner* core, across the boundary of which the $P$ velocity increased sharply with increase of depth, in order to account for the observed waves of $P$ type at distances short of 142°.

JEFFREYS[3] applied amplitude theory to test whether the inner core theory or the diffraction theory was the better hypothesis to explain the readings before 142°. Adapting AIRY's theory (for the optical case) of diffraction near a caustic, JEFFREYS showed that a wave train having the original form $A \sin (2\pi t/\tau)$ would, due to diffraction, emerge at the surface at an epicentral distance $\Delta$ degrees with amplitudes proportional to those given by the AIRY integral

$$\int_0^\infty \cos \left\{ \frac{2\pi}{\tau} [0.04(142 - \Delta)\, x + 0.0004\, x^3] \right\} dx, \tag{32.1}$$

the numerical values in this expression being derived from seismic travel-time data near $\Delta = 142°$. The expression (32.1) is equivalent to

$$\int_0^\infty \cos \left\{ \tfrac{1}{3} \pi (w^3 + m\, w) \right\} dx, \tag{32.2}$$

[1] I. LEHMANN: Bur. Centr. Séism. Internat. A. **14**, 3—31 (1936).
[2] B. GUTENBERG and C. F. RICHTER: Mon. Not. Roy. Astronom. Soc., Geophys. Suppl. **4**, 363—372 (1938).
[3] H. JEFFREYS: Mon. Not. Roy. Astronom. Soc., Geophys. Suppl. **4**, 548—561 (1939).

where $m = 1.4 (142 - \varDelta)\, \tau^{-\frac{1}{3}}$, and $x = 8.5\, w\, \tau^{\frac{1}{3}}$. By the properties of AIRY integrals, JEFFREYS showed that (32.2) has its first maximum for $\varDelta < 142°$ at $m = -1.1$, and that, as $\varDelta$ decreases, (32.2) falls to 0.01 of this maximum at $m = 3.3$; and he thence took a range of 4.4 in $m$ as corresponding to the range of appreciable amplitude of the diffracted $PKP$ waves near $142°$. For waves of period of the order of a second, the range of values of $\varDelta$ in which the diffracted waves would be significant would then be about 3 degrees. Observations of GUTENBERG and RICHTER made with BENIOFF short-period seismographs had shown appreciable amplitudes with periods of the order of a second at distances as far as 19 degrees below $142°$, so that diffraction was indicated to be an inadequate explanation of the observed $P$ waves between 123 and $139°$. This argument of JEFFREYS then indicates that the inner core must be responsible for these observations.

A further question on amplitudes arises from the existence of an inner core. It will be seen (Sect. 36) that evidence points to the inner core being solid in the sense that its elastic behaviour conforms fairly closely to the stress-strain relations (4.5) and (4.6), with significant $\mu$. Thus the inner core is probably capable of transmitting $S$ waves. With a view to testing the theory of a solid inner core, theoretical travel-time tables have been worked out for the phase $PKJKP$, where $J$ refers to motion in $S$ type in the inner core; (relevant numerical details are given in Sect. 21). But in order that the test should be useful, it is necessary that $P$ waves incident through the outer core should be capable of exciting $S$ waves in the inner core, and that these $S$ waves on emerging from the inner core should excite $P$ waves again in the outer core having sufficient energy to be observable on seismograms. By the method outlined in Sect. 30, 31, it is possible to compare the energy in $PKJKP$ waves with that in $PKIKP$ waves emerging at the same epicentral distances. If the JEFFREYS $P$ velocity data are used, it transpires that the phase $PKJKP$ is most likely to be detectable in the range $130° < \varDelta < 140°$, approximately. The energy[1], however, would be only 0.04 of that in $PKIKP$ waves, and in practice that would mean that the phase $PKJKP$ would be only on the border of observability with present seismograph resolving power. If GUTENBERG's velocity data are correct, the phase would be unobservable since the discontinuity at the inner core boundary, being in the gradient instead of the actual velocity, would result in less energy being excited in the form of $S$ waves in the inner core. On the observational side, there have been several suggestions of the existence of the phase $PKJKP$, but, in order to establish the solidity of the inner core by this method, it will be necessary to collate a large number of readings.

# G. Applications of seismic data to the determination of properties of the Earth's deep interior.

**33. Introduction.** From Sect. 26, we have seen that seismology provides a good knowledge of the $S$ velocities in the Earth's mantle and of the $P$ velocities throughout the whole Earth. By (14.2) and (14.5), and the evidence on the fluidity of the outer core, it follows that $k/\varrho$ and $\mu/\varrho$ are well determined down to a depth of nearly 5000 km. If there were data which could give numerical values of a further independent function of $\varrho$, $k$ and $\mu$ in the Earth, it would be a simple matter to deduce the values of all three of these quantities. No such simple data are as yet available, but by proceeding indirectly it is nevertheless possible to make good progress with the problem of the Earth's density variation.

[1] K. E. BULLEN: Mon. Not. Roy. Astronom. Soc., Geophys. Suppl. 6, 125—128, 163—167 (1950/51).

**34. The Earth's density variation.** The main rocks which occur in the earth's outer mantle are found to yield when the stress-difference, i.e. the greatest difference between any two of the three principal stresses, $p_1$, $p_2$, $p_3$, say, reaches the order of $10^9$ dyn./cm².

At the base of the crustal layers, the greatest principal stress is already nearly $10^{10}$ dyn./cm.², and increases steadily throughout the Earth to values exceeding $10^{12}$ dyn./cm.² (see Sect. 35). Hence, throughout most of the Earth, $p_1 \approx p_2 \approx p_3$, and for present purposes it is therefore sufficiently accurate to treat the Earth as being in a hydrostatic state. Hence we can argue in terms of the pressure, $p$ say, and ignore the deviatoric components $P_{ij}$. By the principles of hydrostatics and of gravitational attraction, we then have

$$\frac{dp}{dz} = g\varrho, \quad \text{where} \quad g = \frac{Gm}{r^2};\qquad(34.1)$$

in (34.1), $z$ denotes the depth at distance $r$ from the Earth's centre $O$, $G$ the constant of gravitation, and $m$ the mass inside a sphere of centre $O$ and radius $r$ (the Earth being here treated as spherically symmetrical).

We shall use the term *homogeneous* to apply to a region of the Earth in which there are no changes of chemical composition, and no polymorphic transitions; in a homogeneous region as thus defined, there will, however, be variations in density arising from changes in pressure and temperature. In considering a homogeneous region, we shall take the independent variables as the pressure $p$ and the entropy $S$.

Let $\tau$ denote the excess of the gradient, with respect to $z$, of the actual temperature $T$ in the Earth over the adiabatic gradient. Then

$$\tau = \frac{dT}{dz} - \left(\frac{\partial T}{\partial p}\right)_S \frac{dp}{dz}$$

$$= \left(\frac{\partial T}{\partial S}\right)_p \frac{dS}{dz}.$$

Since

$$\alpha_p = -\frac{1}{\varrho}\left(\frac{\partial \varrho}{\partial T}\right)_p,$$

where $\alpha_p$ is the coefficient of thermal expansion at constant pressure, we then have

$$\tau = -\frac{1}{\varrho \alpha_p}\left(\frac{\partial \varrho}{\partial S}\right)_p \frac{dS}{dz}.\qquad(34.2)$$

The adiabatic incompressibility $k$, which is involved in the transmission of $P$ seismic waves, is connected with $p$ and $\varrho$ by the appropriate form of (4.7), namely

$$\left(\frac{\partial \varrho}{\partial p}\right)_S = \frac{\varrho}{k}.\qquad(34.3)$$

Using (34.1), (34.2) and (34.3), we then obtain

$$\left.\begin{aligned}
\frac{d\varrho}{dz} &= \left(\frac{\partial \varrho}{\partial p}\right)_S \frac{dp}{dz} + \left(\frac{\partial \varrho}{\partial S}\right)_p \frac{dS}{dz} \\
&= \frac{Gm\,\varrho^2}{k\,r^2} - \varrho\,\tau\,\alpha_p \\
&= \frac{Gm\,\varrho}{\Phi\,r^2}(1 - \delta),
\end{aligned}\right\}\qquad(34.4)$$

where

$$\Phi = \frac{k}{\varrho} \tag{34.5}$$

and

$$\delta = \frac{\alpha_p k \tau}{g \varrho}. \tag{34.6}$$

BIRCH[1] has presented evidence to show that $\delta$ is of the order of $0.1\,\tau$. There is no direct evidence on the magnitude of $\tau$. Such evidence as there is indicates that $\tau$ may be great enough for the term in $\delta$ in (34.4) to be of some significance in the Earth's outermost 1000 km. Inhomogeneity in this region would, however, affect $d\varrho/dr$ in the opposite direction. Below 1000 km., it is unlikely that the $\delta$ term would be large. In these circumstances, it is possible to make useful progress taking the simple equation

$$\frac{d\varrho}{dz} = \frac{G m \varrho}{\Phi r^2}. \tag{34.7}$$

As a provisional step towards determining the Earth's density variation, (34.7) was first applied to the whole of the mantle below the crustal layers, a starting value of $m$ being obtained by subtracting from the mass of the Earth a conventional allowance for the crustal layers. In (34.7), $\Phi$ is given from the seismic data (see Table 8 and Fig. 9 of Sect. 26), but a starting value of $\varrho$ is required. This was taken as 3.32 g./cm.³, on evidence from petrological correlations with seismic velocities, from geophysical experiments and from the mean density of the Moon. Writing

$$I = x M a^2,$$

where $I$, $M$ and $a$ are the moment of inertia, mass and radius of the central core, and using the known value of $8.10 \times 10^{44}$ g.-cm.² for the moment of inertia of the whole Earth, it transpires that the density variation in the mantle if given by (34.7) would lead to the impossible result $x = 0.57$. The essential assumptions involved in deriving this result are: (a) homogeneity in the mantle (below the crust); (b) the assumed starting value of $\varrho$; (c) the accuracy of the JEFFREYS velocity data; (d) neglect of the temperature term $\delta$. The assumption (d) cannot be the cause of the error since allowance for $\delta$ would be in the direction of diminishing the density gradient and this would lead to an increase, not decrease, in $x$. While there is some uncertainty in the seismic data in the regions $B$ and $C$, it is practically certain that this could not account for more than a small portion of the error. The starting value of $\varrho$ would need to be least 3.7 g./cm.³ to reduce $x$ to 0.40, the value for a homogeneous sphere. In view of the strength of the evidence for a figure less than 3.7 g./cm.³ immediately below the crustal layers, it then becomes strongly probable that (a) is the main source of the impossible value of $x$, i.e. that there is significant inhomogeneity in the Earth's mantle below the crust.

The JEFFREYS velocity data suggest that such inhomogeneity would occur in the region $C$, and a model Earth has been constructed in which (34.7) is used in the regions $B$ and $D$, but not in $C$; in $C$ the form of the density variation was chosen to match the general features of the velocity behaviour. This procedure introduces some small indeterminacy into the solution, but it was found that the density at a depth of 1000 km. was nevertheless formally determined within 0.1 g./cm.³ in order to obtain an acceptable value for the coefficient $x$.

[1] F. BIRCH: J. Geophys. Res. 57, 227—286 (1952).

On applying (34.7) further to the region $E$, it was possible to obtain values for $\varrho$ in $E$, subject to a formal uncertainty of order 0.5 g./cm.³. It may be remarked that the use of (34.7) in the regions $D'$ and $E$ receives posterior justification from work of Birch[1] on the homogeneity of these regions, and also from the available evidence on the relatively small temperature gradient in these regions. Since the regions $D'$ and $E$ together occupy the greater part of the Earth, the density variation obtained in this way is likely to be a good approximation to the actual figures.

By the procedure just outlined, it is not possible to infer the density in the inner core, beyond noting that the density at the centre cannot be less than 12.3 g./cm.³.

An Earth model, Model $A$, has been set up[2] in which the density at the centre is arbitrarily chosen as 17.3 g./cm.³. In Model $A$, the ranges of density in the regions $B$, $C$, $D$ and $E$ are 3.32–3.64, 3.64–4.68, 4.68–5.69 and 9.4–11.5 g./cm.³, respectively. A cardinal feature of the density variation is the jump in density in the ratio 1.65 at the boundary of the central core. The calculations show that the mass of the central core is nearly one-third of that of the whole Earth.

**35. Pressure, gravity, compressibility and rigidity.** From (34.1) the pressure variation is immediately deducible from the density solution. For Model $A$, the pressures at the tops of the regions $B$, $C$, $D$, $E$, $F$ and $G$ are 0.009, 0.14, 0.39, 1.37, 3.17 and $3.27 \times 10^{12}$ dyn./cm.², and the pressure at the centre is $3.64 \times 10^{12}$ dyn./cm.².

The incompressibility $k$, the rigidity $\mu$, and Poisson's ratio $\sigma$, as deduced using the seismic data on $k/\varrho$ and $\mu/\varrho$, are given in Table 10.

Table 10. *Incompressibility, rigidity and* Poisson's *ratio in the Earth's mantle and outer core.*

| Region | $k$ dyn./cm.³ | $\mu$ dyn./cm.³ | $\sigma$ |
|---|---|---|---|
| $B$ | $1.2–1.7 \times 10^{12}$ | $0.6–0.9 \times 10^{12}$ | 0.27–0.28 |
| $C$ | 1.7–3.6 | 0.9–1.9 | 0.29 |
| $D$ | 3.6–6.5 | 1.9–3.0 | 0.28–0.30 |
| $E$ | 6.2–12.6 | 0.0 | 0.50 |

From the values for $\mu$ in the mantle, and from data on the Earth's bodily tides and on movements of the poles which lead to an estimate of the overall rigidity of the Earth, it is possible to estimate the rigidity of the core. The latest and most detailed calculation is that of Takeuchi[3] who finds that, on the assumption of a homogeneous core, the rigidity of the core cannot exceed $10^{10}$ dyn./cm.². This implies that most of the core, and in particular the outer core $E$, must be in an essentially fluid state, as previously pointed out in Sect. 16. The failure of seismologists to detect $S$ waves in the core is confirmatory evidence.

The variation in $g$ as given using the second equation in (34.1) is interesting: it transpires that, as a result of the high density of the central core, $g$ remains within 1 % of 990 cm./sec.² between the Earth's surface and a depth of nearly 2500 km. This result appears to be firmly established since various other Earth models give a similar result. In Model $A$, the value of $g$ formally reaches a maximum of 1037 cm./sec.² at the outer boundary of the central core, and then diminishes monotonely to its value zero at the Earth's centre.

[1] F. Birch: J. Geophys. Res. **57**, 227—286 (1952).
[2] See Reference [*1*], pp. 217—218.
[3] H. Takeuchi: Trans. Amer. Geophys. Un. **31**, 651—689 (1950).

**36. Compressibility-pressure hypothesis.** In a homogeneous region, the term $\delta$ being neglected, we have, by (34.3) and (34.5),

$$\frac{dp}{d\varrho} = \Phi.$$

Hence

$$\left.\begin{aligned}
\frac{dk}{dp} &= \frac{d(\Phi\varrho)}{\Phi\, d\varrho} \\
&= 1 + \frac{\varrho}{\Phi}\frac{d\Phi}{d\varrho} \\
&= 1 + g^{-1}\frac{d\Phi}{dz},
\end{aligned}\right\} \tag{34.8}$$

by (34.1) and (34.7). On applying (34.8) between depths of 2500 and 3100 km., using Model $A$ values of $g$ and the seismic data, we obtain the following results:

| Region | Depth km. | $d\Phi/dz$ c.g.s. units | $dk/dp$ | Region | Depth km. | $d\Phi/dz$ c.g.s. units | $dk/dp$ | Region | Depth km. | $d\Phi/dz$ c.g.s. units | $dk/dp$ |
|---|---|---|---|---|---|---|---|---|---|---|---|
| $D'$ | 2500 | $2.2 \times 10^3$ | 3.2 | $D''$ | 2700 | $0.0 \times 10^3$ | 1.0 | $E$ | 2900 | $2.0 \times 10^3$ | 2.9 |
| | 2600 | 2.2 | 3.2 | | 2800 | 0.0 | 1.0 | | 3000 | 2.0 | 3.1 |
| | 2700 | 2.2 | 3.2 | | 2900 | 0.0 | 1.0 | | 3100 | 2.1 | 3.2 |

It will be noticed first that the indicated values of $dk/dp$ in the region $D''$ are markedly different from those near the bottom of $D'$. This implies that $D''$ is in some important physical respect different from $D'$; and, since the $P$ and $S$ velocities vary continuously from $D'$ to $D''$, it must follow that the region $D''$ is inhomogeneous. It further follows that (34.8) is inapplicable to $D''$ and that the values of $dk/dp$ indicated in the table have no validity for $D''$.

Secondly, it will be noticed that the indicated values of $dk/dp$ near the bottom of $D'$ and near the top of $E$ are not markedly different; these differences are in fact within the uncertainties of the seismic data and the postulates underlying Model $A$. Thus there is no established discontinuity in $dk/dp$ at the boundary between the mantle and the inner core.

Next, it is seen from Sect. 35 that, according to the formal calculations underlying Model $A$, $k$ changes only by 5% across this boundary. Further, the indicated change is a reduction from the upper to the lower side, which is contrary to the general trend of increasing $k$ with increase of atomic number at high pressures.

This set of results has led to the setting up of a compressibility-pressure hypothesis, to the effect that, at pressures of the order of a million atmospheres (or $10^{12}$ dyn./cm.²), the compressibility $k^{-1}$ changes little with chemical composition, at least for the range of atomic numbers relevant to the Earth's deep interior. On this hypothesis, $k$ and its gradient would be fairly smoothly varying functions of pressure throughout the Earth below the region $C$.

A second Earth model, Model $B$, has been set up[1], assuming this hypothesis in conjunction with the seismic data, applying the equation (34.7) in the regions $D'$ and $E$ but not in $C$, $D''$, $F$ or $G$, (and with little reference to the region $B$), and taking the usual values of the mass and moment of inertia of the Earth. This procedure formally determines the Earth's density variation from the

---

[1] K. E. Bullen: Mon. Not. Roy. Astronom. Soc., Geophys. Suppl. 6, 50—59 (1950).

centre of the Earth to within about 100 km. of the outside surface, apart from an uncertainty of 0.5 g./cm.[3] in the inner core, and does not require the assumption of any starting density value in this whole region. The stated postulates also lead to the following inferences:

(a) The density at the centre of the Earth is between 17 and 18 g./cm.[3]. (This result depends on the use of the Jeffreys velocity values. If, as in Gutenberg's results, no provision is made for the region $F$, the density at the centre could be as low as $14\frac{1}{2}$ g./cm.[3].)

(b) The density gradient in $D''$ is about three times that in $D'$, implying the accumulation of a quantity of denser matter at the base of the mantle. This is seen as follows. The compressibility-pressure hypothesis would imply that $dk/dp$ is approximately 3 units in $D''$. Since $k = \varrho \Phi$, and, by the seismic data, $\Phi$ is nearly constant in $D''$, we have $dk/d\varrho = \Phi$. Hence, in $D''$,

$$\frac{d\varrho}{dz} = \frac{d\varrho}{dk} \frac{dk}{dp} \frac{dp}{dz}$$

$$\approx \frac{1}{\Phi} \times 3 \times g\varrho,$$

as against the value $g\varrho/\Phi$ relevant to a homogeneous layer.

(c) By a similar argument to that in (b), it follows from the abnormally low velocity gradient in the inner core $G$ that the density gradient in $G$ would be somewhat greater than normal, again implying some degree of progressive change of composition in the inner core.

(d) The inner core would be significantly rigid. To show this, we have, on the Jeffreys velocity data, a jump of 18% in the $P$ velocity $\alpha$ between $F$ and $G$, and therefore a 39% jump in $\alpha^2$. If the inner core were fluid, the relation

$$k = \varrho \alpha^2$$

would apply; and, since $\varrho$ could not at this depth decrease with increasing depth, there would be a jump in $k$ of at least 39%, in contradiction to the compressibility-pressure hypothesis. If, on the other hand, the inner core were solid, we should have

$$k + \tfrac{4}{3}\mu = \varrho \alpha^2,$$

and $k$ could be continuous if the value of $\mu$ for the inner core were about $3.9 \times 10^{12}$ dyn./cm.[2]. The Gutenberg velocity data would lead to a similar result, except that the implied rigidity would be smaller, namely $2.1 \times 10^{12}$ dyn./cm.[2], but still about $2\frac{1}{4}$ times that of steel at zero pressure.

It may be remarked that additional support for the solidity of the inner core comes from a comparison between data on the density of Thomas-Fermi-Dirac matter at pressures above $10^7$ atmospheres, geophysical data, and high-pressure experimental results. Adaptation[1] of an extrapolation made by W. M. Elsasser shows that it is probable that any jump in $k$ between the regions $E$ and $G$ will be less than or equal to 5%, which is appreciably less than the amount that would be required if the inner core were fluid.

(e) In Model $\boldsymbol{B}$, the density jump between the base of the mantle and the top of the core is from 5.5 to 9.7 g./cm.[3].

(f) In Model $\boldsymbol{B}$, the density gradient between depths of about 100 to 1000 km. is fairly close to that in the region $D'$, so that Model $\boldsymbol{B}$ is significantly different

[1] K. E. Bullen: Mon. Not. Roy. Astronom. Soc., Geophys. Suppl. 6, 383—401 (1952). — Ann. Geofisica, Roma 6, 1—10 (1953).

from Model $A$ in giving no direct evidence of the existence of the transition region $C$. If the representation in Model $B$ is closer than that in Model $A$ to conditions in the actual Earth, the suggestion then arises that such peculiarities as there are in the $P$ and $S$ seismic velocity gradients in the outermost 100 to 1000 km. arise from abnormal behaviour of the elasticity parameters rather than of the density. [Cf. the suggestion of LEHMANN in Sect. 24(i).] On all the available evidence, it is, however, not yet possible to discriminate sharply between the two models.

The two models, $A$ and $B$, have been taken as the basis for many inferences on the Earth's interior extending well beyond seismology. These inferences, which will be discussed in the chapter by JACOBS, concern questions of temperature in the Earth, the composition of the mantle and the core[1], the internal constitution of the terrestrial planets, and the ELSASSER-BULLARD theory of the Earth's magnetism. On present evidence, there is a likelihood that the actual conditions in the Earth at most levels are somewhere intermediate between the indications of the two models.

### Acknowledgment.

The work on this chapter was carried out while the author was on leave of absence from the University of Sydney. A large part of the work was carried out while the author was at the California Institute of Technology and the Dominion Observatory, Ottawa, and he wishes to record his gratitude for the use of the facilities of both these Institutions. The author is also indebted to Miss I. LEHMANN for a valuable set of criticisms.

## General References.

[1] BULLEN, K. E.: An Introduction to the Theory of Seismology, 2nd edn. Cambridge: University Press 1953. — Contains, especially in Chapters II–IV, VI–VIII, X and XIII, much of the detail of the mathematical argument outlined by the author in the present article. The second edition also contains an extensive bibliography.

[2] BULLEN, K. E.: Seismology. Methuen 1954. — A short summary of the main features of seismic wave propagation designed to give the general student of physics an indication of the present-day problems of seismology.

[3] BYERLY, P.: Seismology. New York: Prentice-Hall 1942. — A fairly elementary introduction to seismological theory, with a long and useful chapter on instrumental seismology.

[4] CAGNIARD, L.: Réflexion et Réfraction des Ondes Séismiques Progressives. Paris: Gauthier Villars 1939. — Contains important mathematical theory relevant to seismic ray propagation. In the problem of reflexion and refraction, account is taken of initial conditions and curvature of wave fronts.

[5] DAVISON, C.: The Founders of Seismology. Cambridge: University Press 1927. — Noted for its account of the earlier history of seismology.

[6] GUTENBERG, B. (Editor): Internal Constitution of the Earth, Physics of the Earth VII, 2nd edn. New York: Dover 1951. — An important composite work, with useful chapters on the application of seismological theory to determining mechanical properties of the Earth's interior.

[7] GUTENBERG, B., and C. F. RICHTER: Seismicity of the Earth and Related Phenomena. Princeton University Press, 1st edn. 1949; 2nd edn. in course of publication 1954. — A thorough account of the magnitudes, focal depths and geographical distribution of the more important earthquakes from 1904 to the present day, and an important source of general statistical information on earthquakes.

[8] GUTENBERG, B.: Theorie der Erdbebenwellen, Beobachtungen von Erdbebenwellen. Die seismische Bodenunruhe. In Handbuch der Geophysik, Bd. 4, S. 1—298. Berlin 1929. — An important account of theoretical seismology, with special reference to the progress over the first quarter of the present century.

[9] HODGSON, E. A., W. G. MILNE and W. E. T. SMITH: Bibliography of Seismology. Ottawa: Dominion Observatory (1929 to date). — Contains a most complete list, with occasional short reviews, of papers relevant to seismology and border subjects.

---

[1] For a recent summary of the evidence on the composition of the core, see K. E. BULLEN, Physical properties of the Earth's core, Annales de Géophysique, **11**, 53—64 (1955).

[10] Imamura, A.: Theoretical and Applied Seismology. Tokyo: Maruzen 1937. — Valuable because of its discussion of Japanese contributions to seismology up to the date of publication.

[11] Jeffreys, H.: The Earth, 3rd edn. Cambridge: University Press 1952. — The early chapters contain a most important account of seismic wave propagation by one of the most noted contributors to the subject. The later chapters are concerned with other aspects of the Earth's interior, but the information in these chapters is important to the sound appreciation of seismology.

[12] Jeffreys, H.: Earthquakes and Mountains, 2nd edn. Methuen 1950. — Contains an exposition, in non-mathematical terms, of salient features of seismology.

[13] Knott, C. G.: Reflexion and Refraction of Elastic Waves, with Seismological Applications. Phil. Mag. (5) **48**, 64—97, 567—569 (1899). — Historically important as one of the first papers concerned with the mathematics of reflected and refracted seismic waves.

[14] Lamb, H.: On the Propagation of Tremors over the Surface of an Elastic Solid. Phil. Trans. Roy. Soc. Lond., Ser. A **203**, 1—42 (1904). — Concerned primarily with a surface wave problem admitting of an exact solution, and noted as an important contribution to the mathematics of seismic wave propagation.

[15] Love, A. E. H.: Some Problems of Geodynamics. Cambridge: University Press 1911. — This book, which is entirely composed of the author's original work in theoretical geophysics, contains a long chapter on the theory of seismic wave propagation, and includes Love's famous contribution to surface wave theory.

[16] Macelwane, J. B., and F. W. Sohon: Theoretical Seismology. New York: Wiley, Part I 1936; Part II 1932. — Part I contains a very detailed exposition of the theory of seismic wave propagation using relatively elementary mathematical methods. (Part II is concerned with instrumental seismology.)

[17] Milne, J.: Earthquakes and other Earth Movements, rev. by A. W. Lee, London: Routledge, Kegan Paul 1939. — Although not primarily theoretical, this work derives importance through the great contributions made by Milne to seismology towards the close of the last century. In his revision, Dr. Lee has brought Milne's earlier work into line with later developments.

[18] Oldham, R. D.: On the Propagation of Earthquake Motion to Great Distances. Phil. Trans. Roy. Soc. Lond., Ser. A **194**, 135—174 (1900). — Oldham's work is historically important in its demonstration of the presence on seismograms of phases corresponding to P, S and surface waves.

[19] Poisson, S. D.: Note sur les Vibrations des Corps Sonores. Ann. Chim. Phys. **36**, 86—93 (1827). — Historically important as showing the existence of two types of bodily waves in a perfectly elastic medium.

[20] Lord Rayleigh (J. W. Strutt): On Waves Propagated along the Plane Surface of an Elastic Solid. Proc. Lond. Math. Soc. **17**, 4—11 (1885). — Historically important for its theoretical demonstration of the existence of "Rayleigh waves".

[21] Stokes, G. G.: Propagation of an Arbitrary Disturbance in an Elastic Medium. Cambridge Math. Papers **2**, 257—280 (1880). — This contains Stokes's historically important demonstration of the irrotational and equivoluminal character of P and S waves, respectively.

[22] Wiechert, E.: Über Erdbebenwellen. I. Nachr. Ges. Wiss. Göttingen, Math.-phys. Kl. **1907**, 415—529. — The work of Wiechert, as the founder of the famous Göttingen school of seismology, is of much historical importance.

[23] Zöppritz, K.: Über Erdbebenwellen. II. Nachr. Ges. Wiss. Göttingen, Math.-phys. Kl. **1907**, 529—549. — This paper contains the travel-time tables of Zöppritz which were extensively used from 1907 to 1930.

# Surface waves and guided waves.

By

MAURICE EWING and FRANK PRESS.

With 19 Figures.

**1. Introduction.** Although surface waves and guided waves may exist under a wide variety of conditions our main concern will be those generated in the earth by earthquakes or explosions. These waves are of interest to seismologists not only because they form the most prominent features of seismograms but also for their usefulness in determining the structure of the earth's crust.

We shall consider propagation of elastic waves in a medium in which the elastic constants are functions of a single variable, $x_3$, the vertical coordinate. When a disturbance occurs in such a medium a part of the radiated energy travels outward as body waves[1] similar to the compressional and shear waves which would propagate in an infinite homogeneous medium with velocities $\alpha$ and $\beta$ respectively. Part propagates in a horizontal direction, being confined to one or more *channels* associated with particular variations in elastic properties. In the most important practical cases the free surface—of a halfspace or of a sphere—forms the upper boundary of the channel, and an elastic discontinuity such as that at the bottom of the crust forms the lower boundary. In all cases the motion decreases rapidly with distance from the channel, usually as an exponential function, and the most efficient excitation occurs when the source is within the channel. Unlike body waves, whose velocity in the vicinity of any point is simply $\alpha$ or $\beta$, the surface waves propagate at a velocity which depends upon the elastic constants within and adjacent to the channel. In most cases velocity depends on wave length and it is necessary to consider both a *phase velocity* and a *group velocity*.

*Two broad categories of surface waves* may be defined, according as the phase velocity is *greater* than or *less* than the lowest velocity of body waves in or near the channel. In the *first category* we may consider the surface waves to be composed of body waves travelling in the channel, confined there by total reflections at its boundaries or by refractions if the channel is formed by velocity gradients. We may draw ray paths for the body waves and assign a phase velocity $\alpha/\cos\vartheta$ or $\beta/\cos\vartheta$ to each path, where $\vartheta$ is the inclination of the path to the horizontal at any point and $\alpha$ or $\beta$ is the corresponding body wave velocity at the point. The period associated with a given ray path, hence with a given phase velocity, is determined by the requirement for *constructive interference* between successive orders of reflection. An example of this type of propagation is LOVE waves, composed of horizontally polarized shear waves multiply reflected between the earth's surface and the bottom of the crust.

In the *second category* the phase velocity is less than the lowest velocity for body waves in or near the channel, and it is not possible to represent the phenomenon by body waves propagated over ray paths. The simplest case of this

---

[1] See the preceding chapter by BULLEN.

type of propagation is the RAYLEIGH wave on the free surface of a *homogeneous* solid half-space. This wave is *non-dispersive*, it involves particle motion in vertical and longitudinal directions which is roughly similar to that of gravity waves in deep water. In actual cases of RAYLEIGH wave propagation in the earth, variations in elastic properties with depth introduce *dispersion*.

All surface waves or guided waves may be considered as modifications of these two basic types, in which the effect of additional layers or gradients must be considered. Surface waves form the most prominent features on seismograms of distant shallow-focus earthquakes. The usual explanation has been that they are confined to channels and therefore their decrease with distance is less rapid than that of body waves. However, this is an over-simplification because the effect of dispersion is to add an additional factor $r^{-1/2}$ in the amplitude of surface waves.

The principal layers in the earth affecting surface waves are ocean, crust, mantle, and core. The details of these layers are described in other chapters[1]. For first approximation calculations on surface waves we may usually consider each layer homogeneous and restrict our attention to two layers at a time, the lower one having infinite thickness, and the layers above the upper one being ignored or included with it. In a few cases it is necessary to consider the effects of velocity gradients within the layers, and less frequently the effect of sphericity of the earth.

There are numerous channels within the earth which transmit surface waves and guided waves. These may be listed as follows:

(a) LOVE waves in the continental and oceanic crust.

(b) *G*-waves in the mantle.

(c) RAYLEIGH waves in the continental and oceanic crust.

(d) RAYLEIGH waves in the mantle.

(e) Crustal velocity gradient: *Lg*.

(f) Low velocity layer in the mantle.

(g) Curvature of the MOHOROVIČIĆ discontinuity.

(h) Sofar waves and *T* phases in the ocean.

(i) Explosion-generated surface waves.

(j) Fundamental modes of vibration, and low order overtones of the earth.

In view of space limitations we can only give derivations of the dispersion formulae for the most commonly observed waves. Only *steady state, plane wave solutions* will be considered. More elaborate solutions for impulsive sources of limited extent are necessary if one wishes to determine amplitudes, and quantitative effect of depth of focus. Thus far, only the dispersion formulas have been used in actual earthquake studies.

In the derivations which follow the symbols and nomenclature of the preceding Chapter are utilized.

**2. Love waves in the crust.** The presence of large transverse waves in the "main tremor" of earthquake seismograms was one of the first established facts of seismology. It was not until 1911 when LOVE showed that these waves could be explained as the trapping of horizontally polarized shear (SH) waves by repeated total reflection from the boundaries of a superficial layer.

*α) LOVE wave dispersion.* To derive the period equation it is only necessary to consider simple harmonic plane waves. Take the origin of coordinates in the

---

[1] See chapters on Structure of the Earth's Crust, and on Seismic Wave Propagation.

lower interface with the $x_2$-axis in the direction of propagation, and the $x_3$-axis positive downward. The plane $x_3 = -H$ represents the free surface. Take the transverse displacement $u_1$ as independent of the coordinate $x_1$. The displacements $u_1$ satisfy the wave equations

$$\nabla^2 u_1 = \frac{1}{\beta^2} \frac{\partial^2 u_1}{\partial t^2} \qquad \text{for the layer,} \tag{2.1}$$

$$\nabla^2 u_1' = \frac{1}{\beta'^2} \frac{\partial^2 u_1'}{\partial t^2} \qquad \text{for the substratum} \tag{2.2}$$

where $\beta$ and $\beta'$ are the velocities of shear waves in the layer and substratum respectively. Using well known solutions of the wave equation we can write

$$u_1 = [A \exp(i\,q\,\sigma\,x_3) + B \exp(-i\,q\,\sigma\,x_3)] \exp[i\,q\,(x_2 - ct)] \tag{2.3}$$

and

$$u_1' = C \exp[i\,q\,(\sigma'\,x_3 + x_2 - ct)] \tag{2.4}$$

where $\sigma = (c^2/\beta^2 - 1)^{\frac{1}{2}}$, $\sigma' = (c^2/\beta'^2 - 1)^{\frac{1}{2}}$ and $c$ is the phase velocity, $2\pi/q =$ wavelength.

In order that the energy be confined to the superficial layer it is required that $c < \beta'$ and that $\sigma'$ be positive imaginary. The three constants $A$, $B$, $C$ may be determined by the boundary conditions

$$\left.\begin{array}{lll} p_{31} = 0 & \text{at} & x_3 = -H, \\ p_{31} = p_{31}' & \text{at} & x_3 = 0, \\ u_1 = u_1' & \text{at} & x_3 = 0 \end{array}\right\} \tag{2.5}$$

where from p. 76, (2.3) and p. 78 (4.2), $p_{31} = \mu\,\partial u_1/\partial x_3$. Upon substitution of equations (2.3) and (2.4) in (2.5), three homogeneous linear equations in $A$, $B$, $C$ are obtained with solutions different from zero if

$$\tan\left[q\,H\left(\frac{c^2}{\beta^2} - 1\right)^{\frac{1}{2}}\right] = \frac{\mu'}{\mu}\left(1 - \frac{c^2}{\beta'^2}\right)^{\frac{1}{2}}\left(\frac{c^2}{\beta^2} - 1\right)^{-\frac{1}{2}}. \tag{2.6}$$

This equation relates implicitly the period $T = 2\pi/qc$ to the phase velocity $c$. Real roots correspond to the case $\beta' \geq c \geq \beta$. For a given value of phase velocity, the period is multiple valued, each value corresponding to a different mode of propagation. The existence of an infinite number of modes follows from the periodicity of the tangent function and it is readily shown on writing $B$ in terms of $A$ in (2.3) that the different *modes* correspond to 0, 1, 2, ... nodal planes within the layer. It can also be shown that the period equation expresses the conditions of constructive interference between plane waves undergoing multiple reflection in the layer at angles of incidence beyond the critical angle $\sin^{-1}\beta/\beta'$.

Waves corresponding to the *lowest mode* only have been observed by seismologists. From equation (2.6) we find that $0 < qH\sigma < \pi/2$ in the first mode. Also $T \to \infty$ as $c \to \beta'$ and $qH \to \infty$, $T \to 0$ as $c \to \beta$.

To study the propagation of an impulsive disturbance the FOURIER integral is usually used to represent the initial disturbance as the summation of a complete spectrum of simple harmonic waves with proper initial phases and amplitudes. If the phase velocity is a function of period, as is the case for LOVE waves, distortion of the initial pulse occurs, since the component waves travel outward with different velocities. At a distant point the initial pulse is transformed into a train of sinusoidal waves in which period and amplitude vary gradually along the train. The sequence of arrivals is governed by the *group velocity* $U = qH\, dc/d(qH)$ *.

β) *Continental dispersion.* A theoretical group velocity curve is plotted as a heavy line in Fig. 1[1] for the case $\beta = 3.51$ km./sec., $\beta' = 4.70$ km./sec., $H = 35$ km., $\mu'/\mu = 2.22$. These are representative conditions for the earth's continental crust[2]. According to the group velocity curve, the first LOVE waves to arrive consist of long period oscillations traveling with the velocity of $\beta'$. The wave periods thereafter decrease gradually, in accordance with the portion of the curve to the right of the minimum. At a time corresponding to propagation

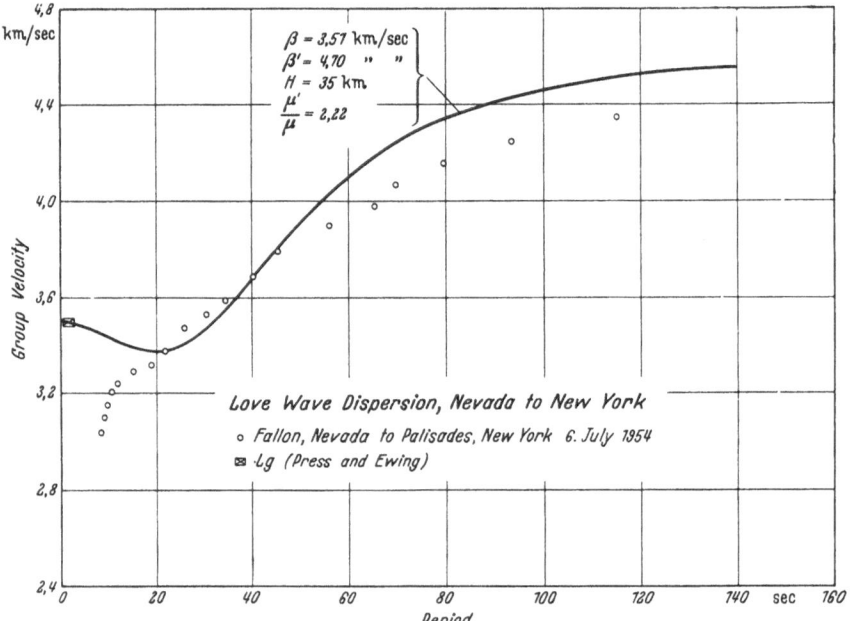

Fig. 1. Theoretical and observed dispersion of LOVE waves in continental crust.

with the velocity $\beta$ very short period waves arrive, superposed on the continuing longer waves and showing inverse dispersion. These short waves are governed by the portion of the curve to the left of the minimum. Both trains of waves merge finally in a phase of large amplitude at a time corresponding to propagation at the minimum value of group velocity.

In practice only the long period branch of the group velocity curve is observed on earthquake seismograms.

STONELEY[3] has derived the appropriate period equation and has computed dispersion curves for a layered crust. In view of the recent indications that the crust occurs as a single layer with a slight velocity gradient, these calculations serve only as approximations to the actual conditions.

---

* For details of the generalization of plane, steady state surface waves to an impulsive source of limited extent see K. SEZAWA, Love Waves Generated from a Source at a Certain Depth. Bull. Earth. Res. Inst. Tokyo **13**, 245—250 (1935). — C. L. PEKERIS: Theory of Propagation of Explosive Sounds in Shallow Water, Mem. 27. Geol. Soc. Amer. **1948**, 1—117.

[1] After F. PRESS and M. EWING, Earthquake Surface Waves and Crustal Structure. Crust of the Earth Symposium. Geol. Soc. Amer., paper 62, 1955.

[2] See chapter on the Earth's Crust in this volume.

[3] R. STONELEY: Love Waves in a Triple Surface Layer. Mon. Not. Roy. Astronom. Soc., Geophys. Suppl. **4**, 43—50 (1937).

The Palisades seismogram of the earthquake of 6 July 1954 at Fallon Nevada provided the first data covering a large range of periods (8 to 140 sec.) on continental LOVE wave dispersion. It is shown in Fig. 2. Period as a function of arrival time is obtained from the seismogram by graphical means, and the values of group velocity are shown in Fig. 1 as a function of period. Before comparison between the theoretical curve and the observations is made, it is important to note that the effect of a gradient of only 3 parts in 10,000 in the mantle[1] is to lower the calculated group velocity curve by about 0.1 km./sec. for periods greater than about 30 sec. Allowing for this effect, the theoretical curve would fit the observed data to within 0.1 km./sec. for most of the period range 20 to 140 sec. The agreement is excellent. For periods less than 20 sec. the experimental points fall increasingly below the curve. This phenomenon is observed in almost all studies of surface waves. It corresponds to a prolongation of the wave train beyond the limit imposed by a minimum value of group velocity. Two explanations for this behavior of the short period waves have been advanced. One involves reflection, scattering and refraction of the short period waves along paths considerably longer than the great circle path. The other suggests that it is an effect of low velocity sedimentary layers.

LOVE waves with periods less than 20 sec., showing

[1] PRESS and EWING: loc. cit., in press.

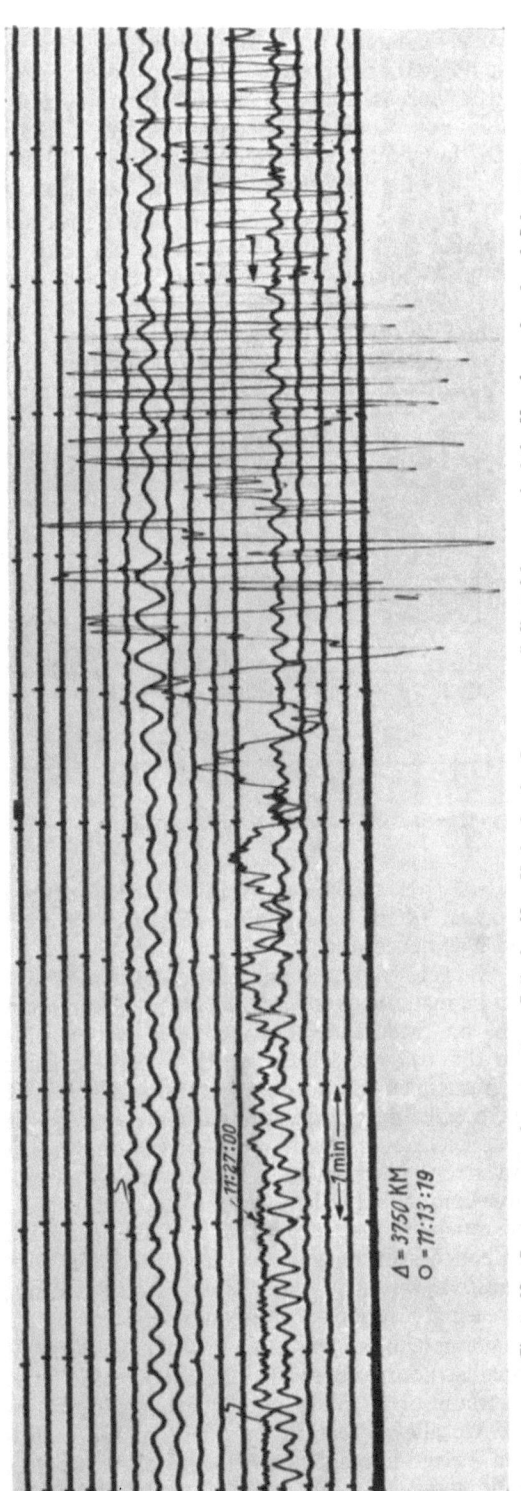

Fig. 2. N.—S. component seismogram from Palisades, New York, showing LOVE waves (indicated by arrows) of the Nevada earthquake, 6 July 1954.

inverse dispersion according to the theoretical curve, have not been observed. The discrepancies for periods less than 20 sec. may be due to a gradient in crust or to the increased effects of crustal heterogeneity on the shorter period waves, or both. For periods greater than 20 sec., Love wave dispersion is found to agree with the results of seismic refractions in finding a nonlayered, silicic continental crust having an average thickness of 35 km. (see the preceding Chapter).

*γ) Oceanic dispersion.* Until seismic refraction measurements at sea became available it was not realized that the crustal thickness under oceans is only about 1/6 that under continents. Since the water layer cannot *influence* Love waves, it may be seen from Equation (2.6) that the principal change from continental to oceanic dispersion will be that for a given group velocity the period will be decreased in the same proportion as the layer thickness.

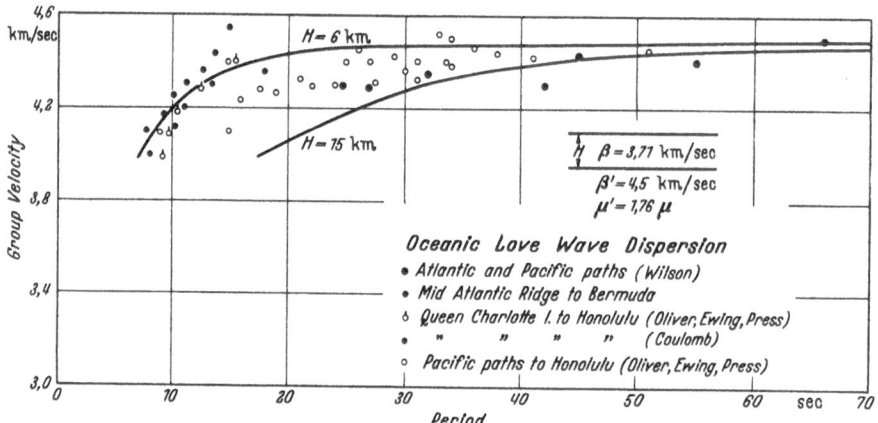

Fig. 3. Theoretical dispersion of Love waves for crustal thickness $H = 6$ km. and $H = 15$ km. compared with observed dispersion.

In Fig. 3 are shown group velocity curves calculated for crustal thicknesses of 6 and 15 km. with points observed by many investigators for the Atlantic and Pacific oceans.

We note that only with data for periods less than about 25 sec. can a distinction be made between a 6 km. and a 15 km. crust, and until very recently observations on these shorter waves were not considered trustworthy. One objection was that they are infrequently recorded, occurring almost exclusively on seismograms from island observatories and only from a small fraction of the shocks which have purely oceanic paths. Secondly, at a period of about 8 seconds the Love waves merge with a prolonged train of waves having almost constant period and strong vertical and longitudinal components. With the use of matched three-component seismographs the two types of oscillations are separable and the crucial Love wave data become available. The vertical and longitudinal components of motion are accounted for by scattering and transformation of the Love waves in the vicinity of the seismograph station, an effect which is increasingly serious for the shorter waves.

We conclude that oceanic Love wave results may be explained by a theoretical structure consisting of a 6 km. mafic crust with shear velocity 3.71 km./sec. underlain by a thick homogeneous stratum with shear velocity 4.50 km./sec. Had we allowed for the gradient in the mantle, as in the previous section, an even better agreement with the observations for periods greater than 30 sec. would have been obtained.

*δ) G-waves.* A large amplitude, long period, transverse oscillation is often found in seismograms of severe earthquakes. The wave is pulse-like, shows no evidence of dispersion, and is best recorded by the BENIOFF linear strain seismograph (Fig. 4). The propagation velocity is about 4.5 km./sec. and it is not unusual to record the wave repeatedly as it circles the earth several times. The channel which guides the G wave has not been identified. It has been suggested however that the phase may correspond to a LOVE wave in the mantle, controlled mostly by the gradient in shear velocity.

**3. RAYLEIGH waves in the crust.** *α) General theory.* The RAYLEIGH wave is well known as a surface wave associated with a homogeneous half space, with phase velocity $c$ independent of period. When $\lambda = \mu$ (POISSON's constant having the value 1/4) $c = 0.9194\beta$ and the particle displacements $u_2$ and $u_3$ are given by[1]

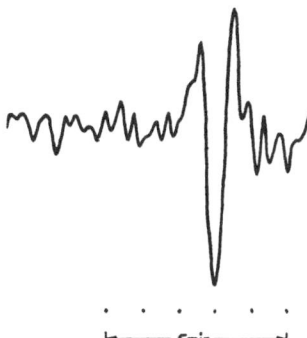

Fig. 4. Pasadena N.—S. linear strain seismogram showing G wave from Kamchatka earthquake of November 4, 1952 (courtesy Prof. H. BENIOFF).

$$u_2 = a \{\exp (0.85\, q\, x_3) - 0.58 \exp (0.39\, q\, x_3)\} \sin [q (x_2 - ct)] ,$$
$$u_3 = a \{-0.85 \exp (0.85\, q\, x_3) + 1.47 \exp (0.39\, q\, x_3)\} \cos [q (x_2 - ct)] \qquad (3.1)$$

where $a$ is a constant depending on $q$.

From (3.1) one finds that during passage of a RAYLEIGH wave the motion of a surface particle is elliptic and retrograde, the major axis occurring in the vertical direction with about 3/2 the amplitude of the displacement in the direction of propagation.

At the interface between two elastic half spaces in welded contact an analagous boundary wave known as the STONELEY wave can exist under certain restrictive conditions on the values of the elastic parameters of the media.

In most cases of importance layering occurs and the simple theory of RAYLEIGH and STONELEY waves fails except in the limit of very short wave length. Dispersion and the occurrence of an infinite number of modes of propagation are the principal effects of layering. To derive the dispersion it is convenient to use the displacement potentials $\Phi$ and $\Psi$ which satisfy respectively wave equations for compressional and shear waves (see Sect. 30, p. 107). For a single surface layer over an infinite substratum solutions of the wave equations may be written in the form

$$\Phi = A \exp [-\nu x_3 + iq(ct - x_2)] + B \exp [\nu x_3 + iq(ct - x_2)],$$
$$\Psi = C \exp [-\zeta x_3 + iq(ct - x_2)] + D \exp [\zeta x_3 + iq(ct - x_2)] \qquad (3.2)$$

for the layer, and

$$\Phi' = E \exp [-\nu' x_3 + iq(ct - x_2)],$$
$$\Psi' = F \exp [-\zeta' x_3 + iq(ct - x_2)] \qquad (3.3)$$

for the substratum, where

$$\nu = q \left(1 - \frac{c^2}{\alpha^2}\right)^{\frac{1}{2}}, \qquad \zeta = q \left(1 - \frac{c^2}{\beta^2}\right)^{\frac{1}{2}},$$
$$\nu' = q \left(1 - \frac{c^2}{\alpha'^2}\right)^{\frac{1}{2}}, \qquad \zeta' = q \left(1 - \frac{c^2}{\beta'^2}\right)^{\frac{1}{2}}.$$

---

[1] See, for example, K. E. BULLEN: An Introduction to the Theory of Seismology, 2nd Ed. Cambridge, England: University Press 1953. 296 p.

At the free surface $x_3 = -H$ the stresses $p_{32}$ and $p_{33}$ vanish. The displacements $u_2$, $u_3$ and the stresses $p_{32}$ and $p_{33}$ are continuous across the interface $x_3 = 0$. Using (2.3), (4.4) and (30.2) of the preceding Chapter, the six boundary conditions may be expressed in terms of the potentials in equations (3.2) and (3.3) and six homogeneous linear equations emerge. Since the coefficients $A \ldots F$ must have values different from zero the determinant of the system of equations must vanish. This last condition yields the period equation. A form useful for computation was given by Lee[1]. If $X, Y, Z, W$ are defined by

$$X = \frac{\mu'}{\mu} \frac{c^2}{\beta^2} - 2\left(\frac{\mu'}{\mu} - 1\right), \qquad Y = \frac{c^2}{\beta^2} + 2\left(\frac{\mu'}{\mu} - 1\right),$$

$$Z = \frac{\mu'}{\mu} \frac{c^2}{\beta'^2} - \frac{c^2}{\beta^2} - 2\left(\frac{\mu'}{\mu} - 1\right), \qquad W = 2\left(\frac{\mu'}{\mu} - 1\right),$$

and

$$r^2 = (i\,\nu)^2, \qquad r'^2 = \nu'^2,$$
$$s^2 = (i\,\zeta)^2, \qquad s'^2 = \zeta'^2,$$

the period equation may be written as

$$\xi\,\eta' - \xi'\,\eta = 0 \tag{3.4}$$

where

$$\xi = \left(2 - \frac{c^2}{\beta^2}\right)\left[X\cos r\,H + \frac{r'}{r}\,Y\sin r\,H\right] + 2\frac{s}{q}\left[\frac{r'}{q}\,W\sin s\,H - \frac{q}{s}\,Z\cos s\,H\right],$$

$$\xi' = \left(2 - \frac{c^2}{\beta^2}\right)\left[\frac{s'}{q}\,W\cos r\,H + \frac{q}{r}\,Z\sin r\,H\right] + 2\frac{s}{q}\left[X\sin s\,H - \frac{s'}{s}\,Y\cos s\,H\right],$$

$$\eta = \left(2 - \frac{c^2}{\beta^2}\right)\left[\frac{r'}{q}\,W\cos s\,H + \frac{q}{s}\,Z\sin s\,H\right] + 2\frac{r}{q}\left[X\sin r\,H - \frac{r'}{r}\,Y\cos r\,H\right],$$

$$\eta' = \left(2 - \frac{c^2}{\beta^2}\right)\left[X\cos s\,H + \frac{s'}{s}\,Y\sin s\,H\right] + 2\frac{r}{q}\left[\frac{s'}{q}\,W\sin r\,H - \frac{q}{r}\,Z\cos r\,H\right].$$

An implicit relation between phase velocity $c$ and wavelength $2\pi/q$ is provided by equation (3.4). Alternately $c$ may be obtained as a function of period from the relation $T = 2\pi/cq$. Two branches of this function occur, corresponding to the symmetrical ($M_1$) and antisymmetrical ($M_2$) vibrations of a free plate. These branches were first found by Sezawa and Kanai and later studied by others[2].

In the $M_1$ branch the retrograde elliptic orbital motion normal for Rayleigh waves occurs, whereas in the $M_2$ branch the motion is of the opposite type. Each branch includes an infinite number of modes $M_{1n}$, $M_{2n}$, $n = 1 \cdots \infty$. In the lowest mode of the symmetric branch, $M_{11}$ the long wave limit of phase velocity approaches the speed of Rayleigh waves in the bottom. In all other modes there is a greatest value of wavelength $2\pi/q$ at which the phase velocity is that of shear waves in the substratum, the cutoff wavelength decreasing with increasing mode numbers. In the limit of infinitely short wavelength the phase velocity in the mode $M_{11}$ is the speed of Rayleigh waves in the layer. Under the very stringent conditions necessary for the existence of Stoneley waves[3]

[1] A. W. Lee: The effect of geological structure upon microseismic disturbance. Mon. Not. Roy. Astronom. Soc., Geophys. Suppl. 3, 83—105 (1932).
[2] See for example I. Tolstoy and E. Usdin: Dispersion properties of stratified elastic and liquid media; a ray theory. Geophysics 18, 844—870 (1953).
[3] J. G. Scholte: The range of existence of Stoneley and Rayleigh waves. Mon. Not. Roy. Astronom. Soc., Geoph. Suppl. 5, 120—126 (1947).

at the interface between two solid half spaces, there would be an additional mode in the $M_2$ branch for which the short wave limit of phase velocity is the speed of STONELEY waves at the interface. For all other modes the short wave limit of phase velocity is the speed of shear waves in the layer.

Discussions of methods of numerical computation of phase velocity from equation (3.4) have been given by several investigators[1]. A discussion of amplitudes as well as dispersion, involving a solution for an impulsive source of limited extent, was given by NEWLANDS[2].

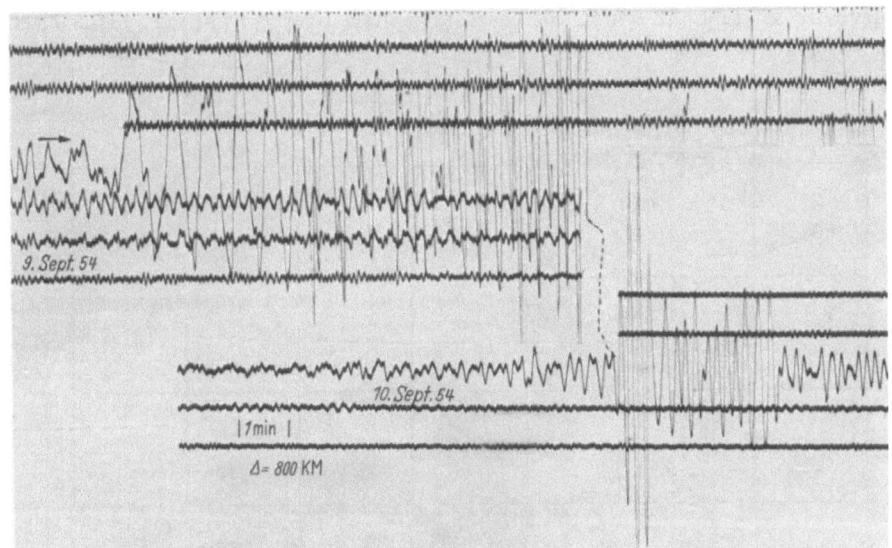

Fig. 5. RAYLEIGH waves on Pietermaritzburg, S. Africa seismograms of the Algerian earthquake of 9 September 1954 and aftershock 10 September 1954. Points of corresponding travel time are connected by dashed line.

$\beta$) *Propagation of* RAYLEIGH *waves across continents.* Only the mode $M_{11}$ appears to be relevant to the propagation of earthquake-generated RAYLEIGH waves. The problem of finding a continental structure for which the calculated RAYLEIGH wave dispersion agrees with the observed dispersion has been studied by many investigators[3], but no conclusive solution was possible until a precise dispersion curve covering a wide range of periods became available

For many reasons observations on RAYLEIGH waves for pure continental paths have been meager. A long path is required because the dispersion is far less than for RAYLEIGH waves across oceans. In addition, the need for long period vertical seismographs to separate RAYLEIGH waves from LOVE waves is rarely met. Recently a Columbia University long period seismograph was installed at Pietermaritzburg, S. Africa. Seismograms of the shocks of 9 and 10 September, 1954, gave excellent data for periods from 10 to 70 sec. at a distance of 8000 km. Records for the main shock and the principal aftershock are shown in Fig. 5.

---

[1] See for example N. A. HASKELL: The dispersion of surface waves on multilayered media. Bull. Seism. Soc. Amer. **43**, 17—34 (1953). For use of models as analog computers see J. OLIVER, F. PRESS and M. EWING: Two dimensional model seismology. Geophysics **19**, 202—219 (1954).
[2] M. NEWLANDS: The disturbance due to a line source in a semi-infinite elastic medium with a single surface layer. Phil. Trans. Roy. Soc., Ser. A **245**, 213—308 (1952).
[3] For references see N. A. HASKELL: loc. cit.

Dispersion data from these seismograms are plotted in Fig. 6 as circles. Over the entire range of periods the agreement is quite good between the observed points and the theoretical curve for a single homogeneous layer, with $\beta = 3.51$ km./sec., $\beta' = 4.68$ km./sec., $H = 35$ km., $\varrho_2 = 1.25\,\varrho_1$, $\sigma = 1/4$. For periods from 18 to 30 sec. the observed points lie above the theoretical curve by no more than 0.2 km./sec., and the minimum of group velocity occurs at about 17 sec. instead of the theoretical value of 22 sec. Both of these discrepancies are in the proper direction to be explained by an increase of velocity with depth in the crust.

For periods greater than 38 sec. the observed points fall below the theoretical curve by an amount which increases to a little over 0.1 km./sec. This effect

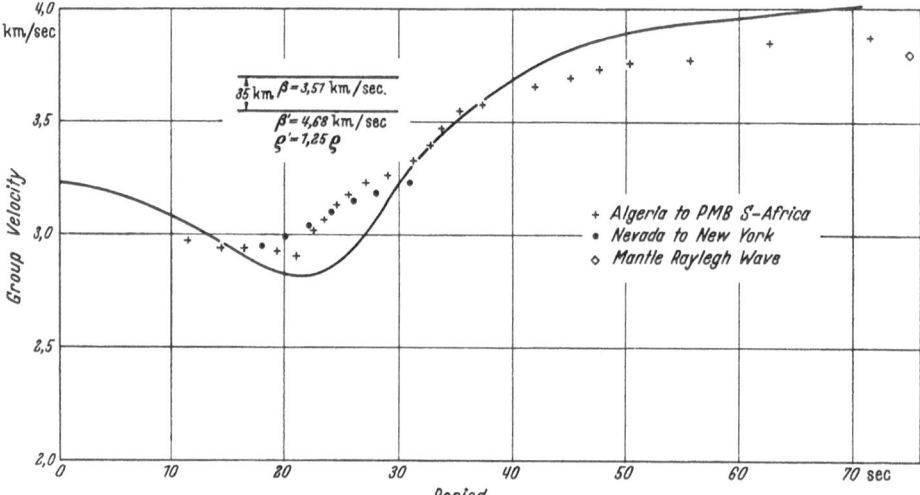

Fig. 6. Theoretical and observed dispersion for continental RAYLEIGH waves.

is of proper magnitude and direction to be accounted for by a velocity gradient in the mantle, as is shown by the dispersion data for mantle RAYLEIGH waves (Sect. 3 $\delta$).

Also plotted in the figure are the data for RAYLEIGH wave dispersion along a path crossing North America. The agreement for Africa and North America is remarkable, and is indicative of great similarity of the crust of these continents.

$\gamma$) *Propagation of* RAYLEIGH *waves across oceans.* The principal features of RAYLEIGH wave dispersion for oceanic paths are determined by the water layer. STONELEY[1] considered the effect of a superficial liquid layer and derived the following equation:

$$\tan\left[q\,H\left(\frac{c^2}{\alpha^2}-1\right)^{\frac{1}{2}}\right]=\frac{\varrho'}{\varrho}\,\frac{\beta'^4}{c^4}\,\frac{\left(\frac{c^2}{\alpha^2}-1\right)^{\frac{1}{2}}}{\left(1-\frac{c^2}{\alpha'^2}\right)^{\frac{1}{2}}}\left[4\left(1-\frac{c^2}{\alpha'^2}\right)^{\frac{1}{2}}\left(1-\frac{c^2}{\beta'^2}\right)^{\frac{1}{2}}-\left(2-\frac{c^2}{\beta'^2}\right)^2\right] \quad (3.5)$$

EWING and PRESS[2] identified the long train of regular surface waves (Fig. 7) characteristic of most seismograms for long oceanic paths, as RAYLEIGH waves

[1] R. STONELEY: The effect of the ocean on RAYLEIGH waves. Mon. Not. Roy. Astronom. Soc., Geophys. Suppl. 1, 349—356 (1926).
[2] M. EWING and F. PRESS: Crustal structure and surface wave dispersion, part II. Bull. Seism. Soc. Amer. 42, 315—325 (1950).

in which the group velocity was far smaller than any values previously admitted. Using equation (3.5) they computed dispersion curves and demonstrated that the observed dispersion could be explained by the effect of a water layer underlain by a thick ultramafic substratum. As a further refinement, the thin mafic layer known from seismic refraction measurements was included in the calculations, and it was finally established that RAYLEIGH wave determination of crustal structure under ocean basins was completely consistent with seismic refraction measurements. The depth for the liquid layer exceeds the water depth by an amount equal to the sedimentary layer thickness as found by seismic refraction.

In Fig. 8 theoretical and observed dispersion are compared for paths across the Pacific Ocean. The agreement over the large range of periods (15—40 sec.) and group velocities (1.5 — 4.0 km./sec.) is satisfactory. Even more striking than the gap in the RAYLEIGH dispersion curve for continents is that between periods 0 and 15 seconds in the present curve.

Waves corresponding to second modes or higher modes have been sought on seismograms, but are definitely absent, except for the $T$ phase (Sect. 6). Short period surface waves ($T < 15$ sec.) often observed on seismograms for oceanic paths probably have travelled as LOVE waves. The apparent inability of short period RAYLEIGH

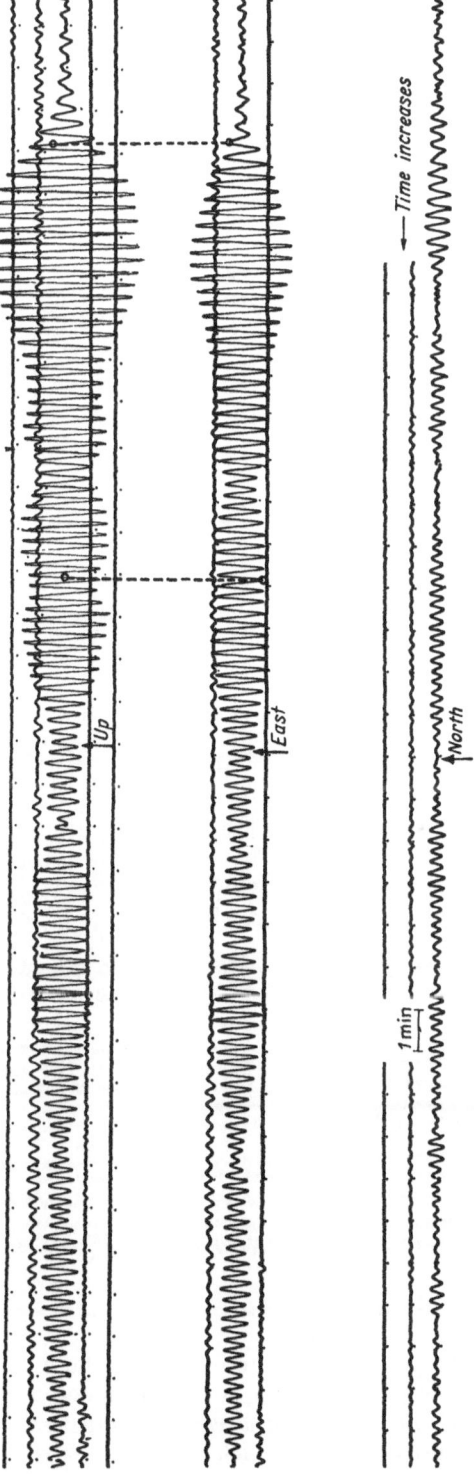

Fig. 7. Palisades 3-component seismograms showing RAYLEIGH waves for long oceanic path. Tonga Islands (22° S., 175° W.) to Palisades, New York. $\Delta = 12.450$ km. 12 August 1953. Magnitude $M = 6\frac{1}{4}$. Retrograde orbital motion corresponding to arrival from west is indicated.

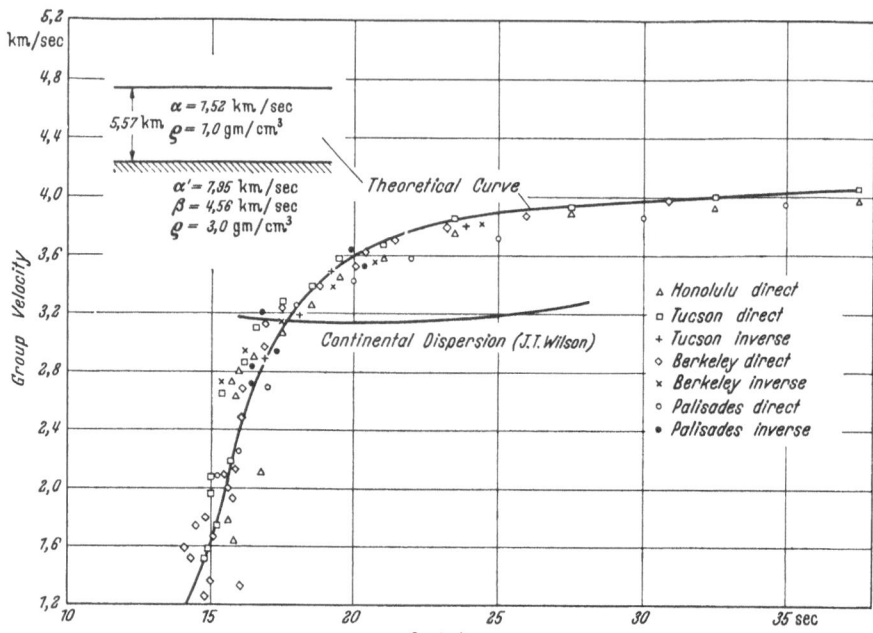

Fig. 8. Theoretical and observed RAYLEIGH wave dispersion for oceanic paths. Where correction for continental travel was necessary continental dispersion curve was used.

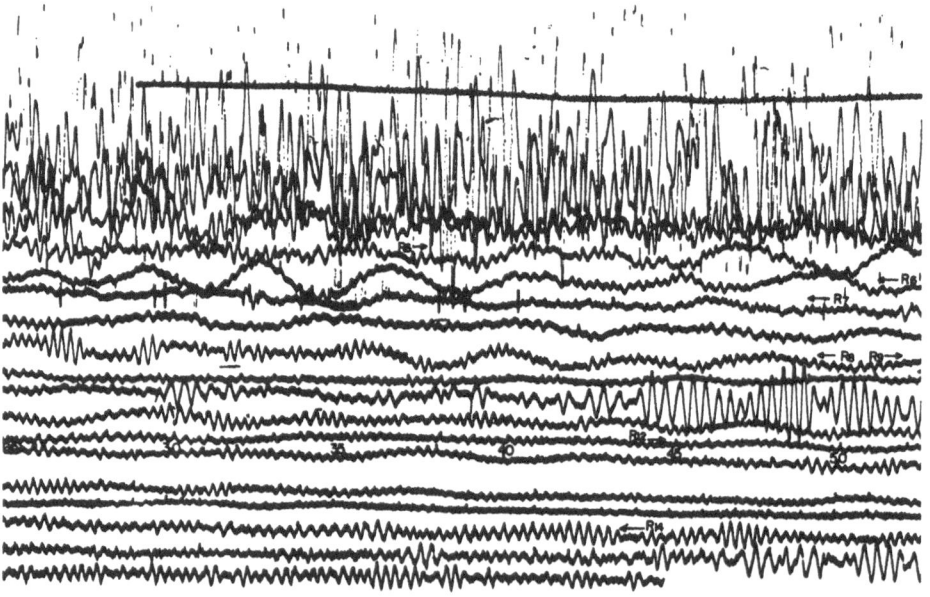

Fig. 9. Palisades vertical seismogram showing mantle RAYLEIGH waves of orders $R_4 - R_{15}$ from Kamchatka earthquake

waves to propagate in the deep ocean poses a serious difficulty for all theories of microseism generation in the open ocean (see p. 146).

δ) *Mantle* RAYLEIGH *waves.* The greatest earthquakes usually generate long period surface waves, which circle the earth several times. These waves have been known for a long time but it is only with the aid of modern long period seismographs, particularly vertical component instruments, that the RAYLEIGH waves could be isolated and studied in sufficient detail to extend the dispersion curve. Remarkable seismograms were obtained from the Kamchatka earthquake of November 4, 1952 on the BENIOFF linear strain seismograph at Pasadena and on the long period vertical pendulum seismograph ($T_0 = 15$ sec., $T_g = 90$ sec.) at Palisades, Fig. 9. On the Palisades seismogram RAYLEIGH waves are observed which have circled the earth as many as seven times. Period and arrival time were read from the seismogram by the usual method and group velocity was calculated using the epicentral distance appropriate for the number of circuits of the earth. Wave trains of orders $R_6$ to $R_{15}$ are indicated on the seismogram, corresponding to epicentral distances (in degrees) $\Delta_n = (n-1) \cdot 180 + \Delta$ for $n$ odd and $\Delta_n = n \cdot 180 - \Delta$ for $n$ even where $\Delta$ is the least distance between station and epicenter. $R_1$ arrives on the direct path, $R_2$ via the antipodes, $R_3$ is $R_1$ after one complete circuit, etc. Comparison of the Pasadena strain and pendulum seismograms demonstrated that the orbital motion is retrograde elliptical. Observed group velocity is plotted as a function of period in Fig. 10 for various orders. Striking features are the occurrence of a minimum value of group velocity (3.54 km./sec.) at a period of 225 sec., a short period limit at 3.8 km./sec. and 70 sec., and a tendency toward flattening of the curve as though a maximum of group velocity were being approached for periods greater than 400 sec. There can be no doubt that the dispersion is the result of the known increase of shear wave velocity with depth in the mantle. As a first approximation the known

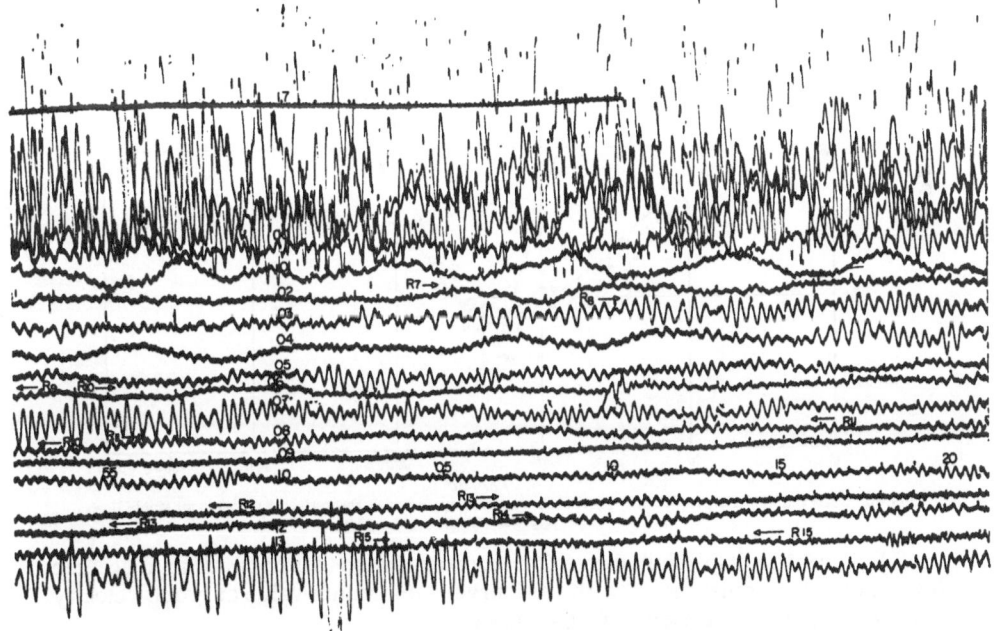

of November 4, 1952.—Each trace from left to right gives one hour; $R_6$ begins about 00 h 37 min., $R_{15}$ ends 13 h 03 min.

velocity gradient in the mantle may be replaced by two homogeneous layers and the theory of Sect. 3 used to compute the theoretical curve shown in the figure. A theoretical curve of HASKELL was used and the constants $\beta' = 6.15$ km./sec., $\beta = 4.48$ km./sec., $H = 516$ km. were chosen to fit the observed minimum group velocity. Better agreement with observation could be obtained by using the existing theory for dispersion of RAYLEIGH waves in the presence of a velocity gradient in the substratum[1], but these calculations are very lengthy.

Several important results emerge from the study of mantle RAYLEIGH wave dispersion.

(a) The short period limit $T \sim 75$ sec. corresponds to the least wave length for which the continent-to-ocean transition is a negligible barrier.

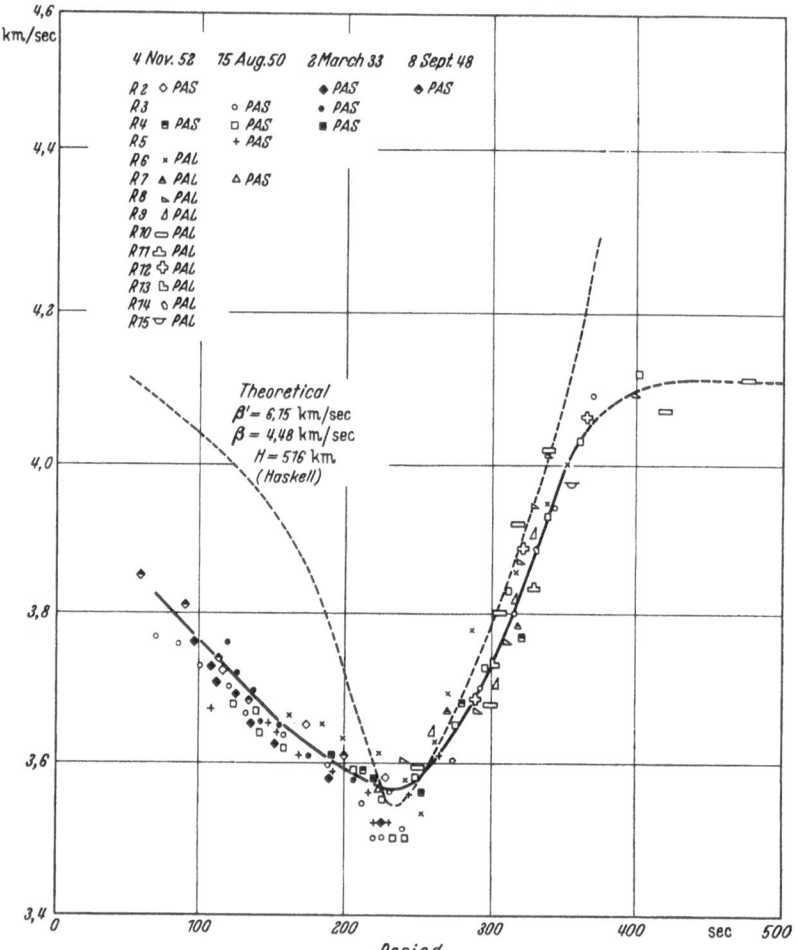

Fig. 10. Observed disperion for mantle RAYLEIGH waves compared with theoretical dispersion for single layer approximation.

(b) Since the dispersion curves for shorter period continental and oceanic RAYLEIGH waves must merge with this curve, a maximum value of group velocity must exist between periods of $50-70$ sec.

(c) The flattening of the dispersion curve for $T > 400$ sec. is interpreted as an effect of the vanishing rigidity of the earth's core.

(d) Since mantle RAYLEIGH waves are sufficiently long to be unaffected by crustal irregularities, amplitude measurements of suitable precision for studying

[1] M. NEWLANDS: RAYLEIGH waves in a two-layer heterogeneous medium, Mon. Not Roy. Astronom. Soc., Geophys. Suppl. **6**, 109—128 (1950).

decrement can be made. For amplitude decrement $\exp(-\gamma \Delta)$ where $\gamma = \pi/QcT$, it was found that $1/Q = 665 \times 10^{-5}$ at $T = 215$ sec., after allowance for effects of geometric spreading and dispersion.

(e) It is seen from Fig. 10 that for the highest orders only periods associated with the minimum group velocity can be read, verifying the theoretical result that at stationary value of group velocity amplitudes diminish less rapidly with distance.

**4. *Lg* waves.** The *Lg* phase[1] is a short period, large amplitude arrival in which the motion is predominantly transverse, but accompanied by appreciable longitudinal

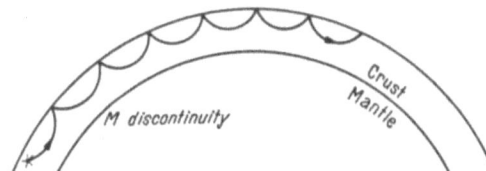

Fig. 12. Schematic illustration of mechanism for *Lg* propagation.

and vertical components (Fig. 11). The phase occurs only when the path between epicenter and station is continental, less than 2 degrees of oceanic path being required to eliminate it entirely. Its velocity of propagation is 3.51 km./sec., a value essentially identical with the velocity of propagation of shear waves in the upper part of the continental crust.

Although the precise mechanism of propagation is not understood, it is certain that transmission of a shear wave through a very efficient waveguide is involved. The thickness of the waveguide must be at least 10 km., and it must lie very near the surface. It is probable that the free surface behaving as an ideal reflector is the upper boundary of the waveguide, and that a velocity gradient in the crust serves as a lower boundary by refracting energy to the surface. Fig. 12 illustrates the propagation paths postulated.

GUTENBERG[2] offered the alternative hypothesis that the waveguide is a low velocity layer in the continental crust.

The *Lg* phase may be used to determine whether the crust beneath a given area is continental or oceanic. Such determinations have been made for Euroasia[3]

[1] F. PRESS and M. EWING: Two slow surface waves across North America. Bull. Seism. Soc. Amer. **43**, 219—228 (1952).
[2] B. GUTENBERG: Channel waves in the Earth's crust. Geophysics **20**, 283—294 (1955).
[3] M. BÅTH: The elastic waves *Lg* and *Rg* along Euroasiatic paths. Arkiv f. Geofysik Stockholm **2**, 295—342 (1954).

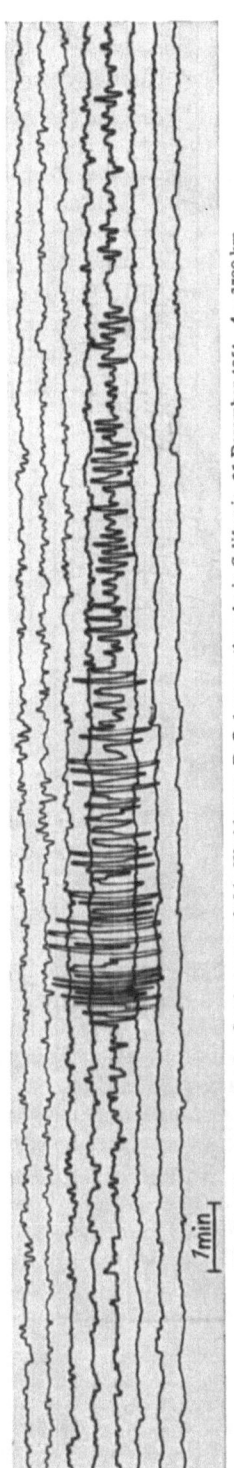

Fig. 11. Seismogram showing *Lg* phase recorded in Washington, D. C. from earthquake in California, 25 Dezember 1951, $\Delta = 3790$ km.

and the Arctic regions[1]. A principal problem for the $Lg$ phase is the great duration, which is probably caused by reflections from continental boundaries rather than by very low values of group velocity.

**5. Waveguides in the upper mantle.** Two guides are theoretically possible in the upper mantle, as illustrated in Fig. 13. The Mohorovičić discontinuity can act as a reflector for waves incident from below at angles near grazing, in a manner analogous to the *"whispering gallery"* effect described by Rayleigh.

A low velocity layer beneath the Mohorovičić can trap waves by repeated refraction like the Sofar channel in the ocean (Sect. 6).

Recently Caloi[2] reported observations on compressional and shear waves, called Pa and Sa respectively. The velocities reported are 7.9–8.0 km./sec. and

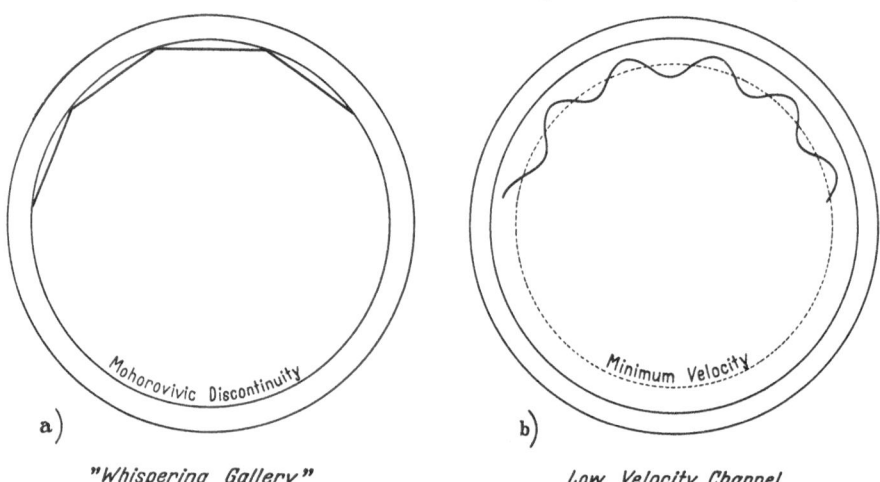

a)    "Whispering Gallery"          b)    Low Velocity Channel

Fig. 13 a and b. Schematic illustration of two possible waveguides in the mantle.

4.4 km./sec. respectively. He finds that these waves are best excited by earthquakes having foci in the neighborhood of the low velocity layer postulated by Gutenberg (Fig. 13 b).

Press and Ewing[3] independently discovered phases which almost certainly are the same as Caloi's Pa and Sa, the main difference being (a) no obvious correlation with focal depth could be established; (b) the velocities of 7.98–8.24 km./sec. and 4.58 km./sec. are larger than Caloi's, being nearly those of Pn and Sn. For these reasons Press and Ewing postulate the propagation of these waves may be represented by Fig. 13 a.

In Fig. 14 a seismogram showing both phases is presented. The compressional wave occurs on the vertical component as a train of sinusoidal oscillations with periods 8–12 sec. The shear wave occurs as a large pulselike oscillation with period about 20–30 sec. on the vertical component. The longitudinal component is incoherent with respect to the vertical and of longer duration. The transverse component does not record this phase.

[1] J. Oliver, M. Ewing and F. Press: Crustal structure of the Arctic regions from the $Lg$ phase. Bull. Geol. Soc. Amer. **66**, 1063—1074 (1955).

[2] P. Caloi: Onde longitudinali e traversali dall astenasfera. Rend Accad. naz. Lincei, Ser. VIII **15**, 352—357 (1953).

[3] F. Press and M. Ewing: Waves with Pn and Sn velocity at great distances. Proc. Nat. Acad. Sci. USA, **41** (1955).

The compressional phase has been observed in the range 69–79°. The shear phase has been observed in the range 50–125°.

**6. *T* phase and Sofar waves in the Ocean.** Two effects combine to produce a low velocity channel in the ocean. From the surface to the channel axis the velocity of sound decreases due to the temperature drop in the main thermocline. Below the axis, the temperature gradient is small, and the sound velocity increases, due to the pressure effect, until the bottom is reached. The depth of the channel axes varies gradually with latitude and certain oceanographic factors, its maximum value being about one mile. A typical velocity-depth curve for the Atlantic ocean is shown in Fig. 15 (at the right). EWING and WORZEL[1] established that extremely efficient propagation of sound to long distances was possible through this channel. Sound energy released by explosion of a bomb in this channel is confined to the channel by refractions in the regions of velocity gradient above and below the channel axis (Fig. 15). Due to the low attenuation of sound in water it is probable that explosion of a 4 lb charge can be heard to at least 10,000 miles. EWING and WORZEL recommended that this *Sofar-channel* be utilized in an air-sea rescue system in which three or more stations receiving a signal could locate the source within a mile.

Sofar signals may also be utilized to locate submarine mountains by topographic echoes and by acoustical shadows.

[1] M. EWING and J. L. WORZEL: Long range sound transmission, Propagation of Sound in the Ocean, Memoir 27. Geol. Soc. Amer. **1948**, p. 1—35.

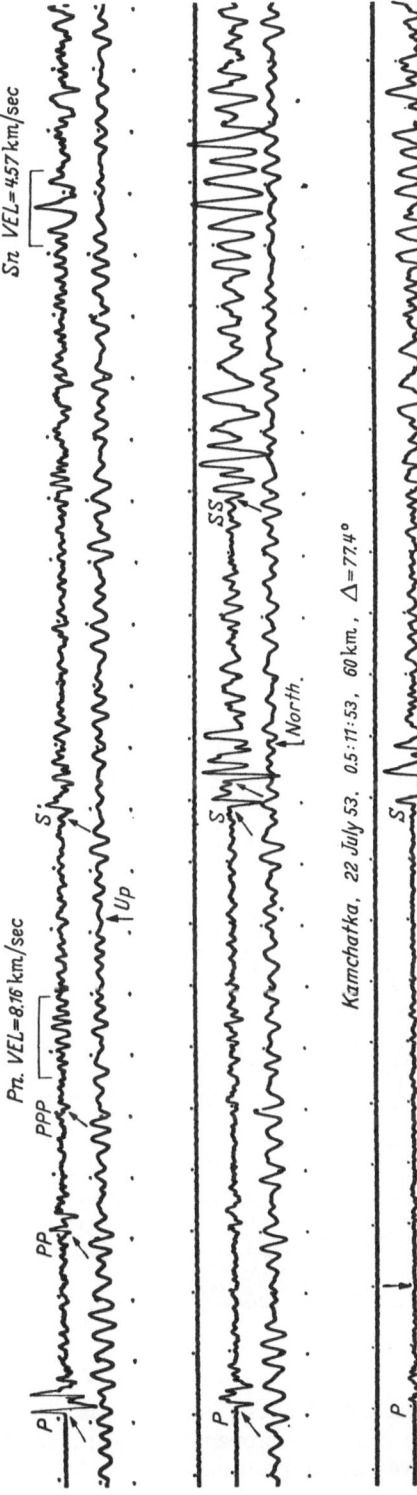

Fig. 14. Three component Palisades seismograms showing phases with Pn and Sn velocity (from Kamchatka earthquake, 22 July 1953, $\Delta = 8590$ km.). Time marks are dots at 1 minute intervals.

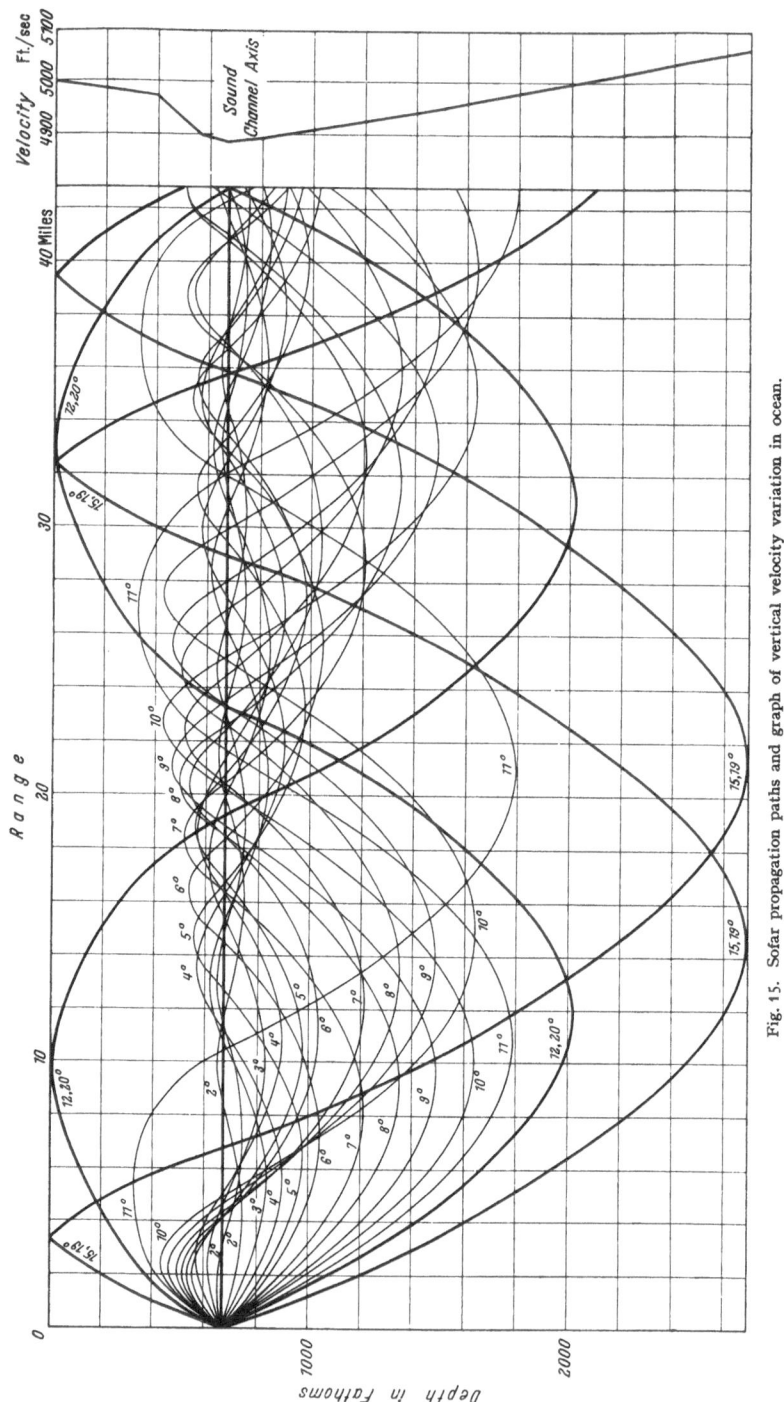

Fig. 15. Sofar propagation paths and graph of vertical velocity variation in ocean.

A short period phase, $T < 1$ sec., is often observed on seismograms of earthquakes in which the propagation path is mostly oceanic. It has been established that the velocity of propagation over the oceanic part of the path is 1.49 km./sec.,

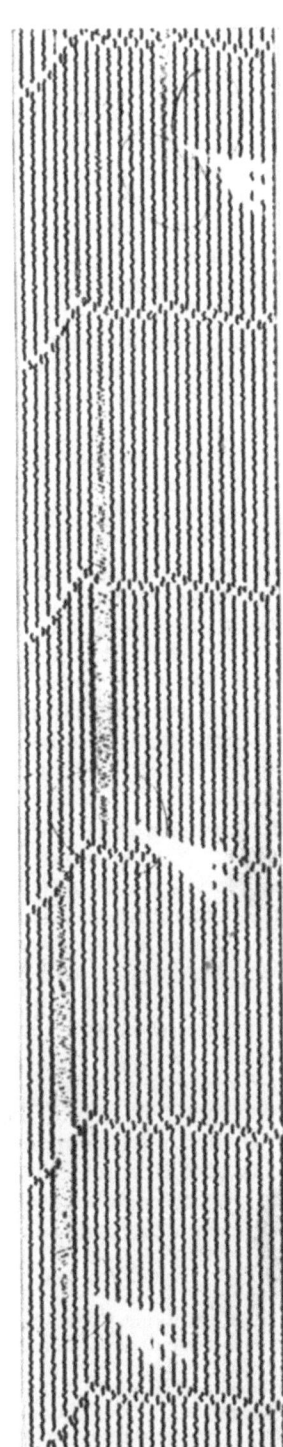

Fig. 16. Bermuda seismogram of three *T* phases (indicated by the white arrows) from West Indies tremors.

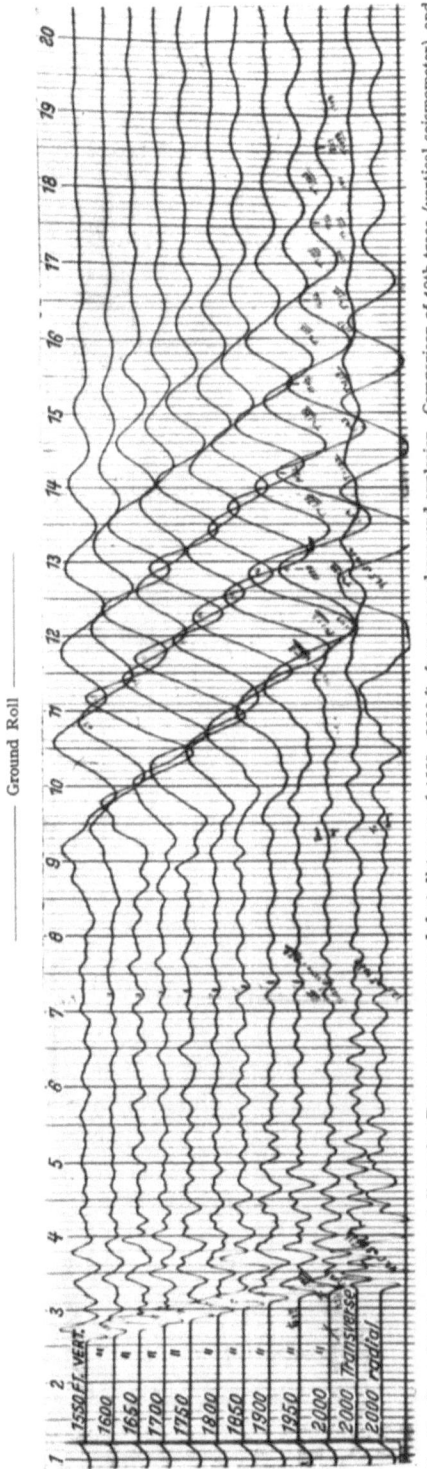

Fig. 17. Ground roll consisting of dispersive RAYLEIGH waves recorded at distances of 1550—2000 ft. from an underground explosion. Comparison of 10th trace (vertical seismometer) and 12th trace (radial horizontal seismometer) establishes RAYLEIGH wave orbital motion.

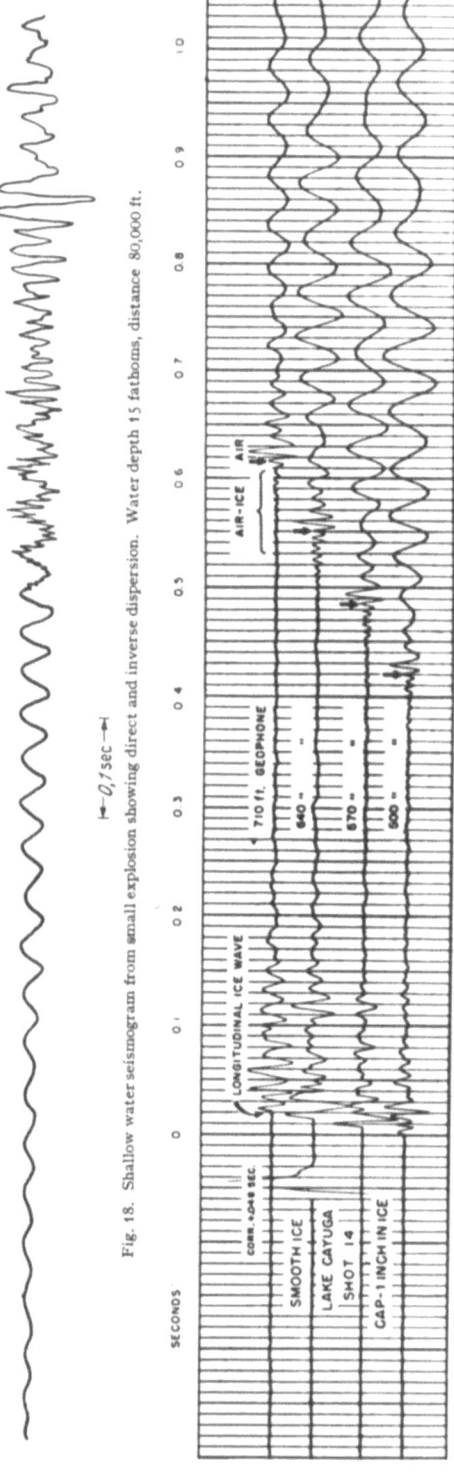

Fig. 18. Shallow water seismogram from small explosion showing direct and inverse dispersion. Water depth 15 fathoms, distance 80,000 ft.

Fig. 19. Seismogram from small explosion obtained on floating lake ice. Dispersed flexural waves follow arrows.

the speed of sound in sea water[1]. There is little doubt that the energy crosses the ocean as sound waves. But compressional, shear, and in some cases surface waves are involved in propagation across land. For the ocean part of the path the upper boundary of the channel consists of the free surface, the lower boundary being provided by the longest refraction path not intercepted by the bottom (see Fig. 15). A seismogram with a typical $T$ phase is shown in Fig. 16.

Shurbet[2] reported observations of $T$ phases recorded at Bermuda from South American shocks in which the energy was transmitted over as much as 51° as $P$, before entering the ocean at the scarp north of Puerto Rico. The most efficient generators of these $T$ phases are shocks with greater than normal focal depth, which produce short period $P$ phases at San Juan, Puerto Rico.

**7. Explosion-generated surface waves.** For the short period surface waves generated by explosions of moderate size, layering in the sedimentary rocks, depth to the ground water, and thickness of the weathered layer become important. Surface waves encountered in geophysical prospecting, commonly called *ground roll*, have received the most attention[3]. In their most common form they consist of a train of Rayleigh waves strongly influenced by

[1] M. Ewing, F. Press, and J. L. Worzel: Further study of the $T$ phase. Bull. Seism. Soc. Amer. **42**, 37—51 (1952).

[2] D. H. Shurbet: Bermuda $T$ phases with large oceanic paths. Bull. Seism. Soc. Amer. **43**, 23—35 (1955).

[3] See, for example, M. B. Dobrin, R. F. Simon and P. L. Lawrence: Rayleigh waves from small explosions. Trans. Amer. Geophys. Un. **32**, 822—832 (1951).

dispersion, and controlled primarily by the variation of shear wave velocity with depth in the weathered layer and the substratum. In Fig. 17 a seismogram illustrating these waves is presented. In most cases, ground roll is more complicated, usually due to simultaneous excitation of several modes and to deviations from ideal layering and perfect elasticity.

Attempts to analyze simple cases of ground roll, using equation (3.4) have met with some success[1].

A very efficient waveguide, which has been studied experimentally[2] and theoretically[3], occurs in *shallow water*, usually on the continental shelf. A seismogram depicting the two superposed dispersive trains of waves, corresponding to two branches of a dispersion curve (like that in Fig. 6) and gradually merging to a common frequency as the minimum group velocity is approached, is shown in Fig. 18. Despite PEKERIS' result that the underlying sediments may usually be treated as liquids in the calculations, experiments made with this shallow water sound channel serve as excellent scale model experiments for RAYLEIGH wave propagation in the oceanic crust. As was the case for earthquake waves, analysis of the dispersion can provide information about the elastic properties of the ocean bottom material.

A *sheet of ice floating on water* is an efficient waveguide for the *flexural mode*[4] *of vibration*, in which the velocity of propagation is less than the speed of sound in water. In Fig. 19, a train of flexural waves showing the typical inverse dispersion may be seen on a seismogram of waves from a small explosion in lake ice. The thickness of the ice sheet may be readily computed from observations of dispersion of flexural waves if the quality and elastic constants of the ice are known, otherwise the flexural rigidity of the plate is determined.

An interesting special case is when the explosion occurs in the air above the ice. Only those flexural waves are excited whose phase velocity equals the speed of sound in air[5]. Instead of a dispersed train of waves, a constant-frequency train is observed, the frequency depending on the ice thickness. Ice thickness measurement by the use of flexural waves is of some importance in high latitudes.

Similar interaction with the air is observed in studies of ground roll or in any case where the phase velocity of the surface waves becomes equal to the speed of sound in air.

---

[1] M. B. DOBRIN et al.: loc. cit.

[2] J. L. WORZEL and M. EWING: Explosion sounds in shallow water, Memoir 27. Geol. Soc. Amer. **1948**, 1—53.

[3] C. L. PEKERIS: Theory of propagation of explosive sound in shallow water. Geol. Soc. Amer. **1948**, 1—117.

[4] F. PRESS, A. P. CRARY, J. OLIVER and S. KATZ: Air-coupled flexural waves in floating ice. Trans. Amer. Geophys. Un. **32**, 166—172 (1951). Observations of flexural waves on lake ice made in 1943 are described by O. FOERTSCH: Gerlands Beitr. Geophysik **61**, 272—290 (1950); see also G. ANGENHEISTER, Fortschreitende elastische Wellen in planparallelen Platten. Gerlands Beitr. Geophysik **61**, 296—308 (1950).

[5] F. PRESS et al.: loc. cit., 1951.

# L'agitation microséismique.

Par

J. COULOMB.

Avec 7 figures.

**1. Les différentes formes de microséismes.** α) *Introduction.* En dehors des tremblements de terre ou séismes le sol n'est jamais en repos complet. Ces mouvements, qu'on groupe sous le nom d'agitation microséismique ou de microséismes, sont d'origine très variée. Comme il arrive pour les séismes proprement dits, l'aspect des enregistrements dépend beaucoup du séismographe utilisé; leur amplitude est particulièrement grande dans les stations où le sous-sol est mal consolidé. Depuis BERTELLI (1869) les microséismes ont été étudiés par tant d'auteurs professant les opinions les plus contradictoires (468 références dans [7]), qu'il est difficile de voir clairement leurs causes. Décrivons d'abord les phénomènes observés.

β) *Microséismes industriels.* Au voisinage des centres d'activité humaine, les séismographes à courte période enregistrent les vibrations dues au trafic ferroviaire ou routier, aux machines, aux explosions, etc... Les périodes observées ne dépassent guère la seconde. Des causes naturelles comme les cascades et surtout les volcans fournissent des inscriptions analogues.

γ) *Effet du vent.* Les rafales de vent, transmises au sol par les collines, les bâtiments, les arbres, voisins de la station séismique, y produisent des oscillations irrégulières. La «période», au sens peu précis des séismologues, est beaucoup plus grande que dans le cas précédent (jusqu'à une vingtaine de secondes). L'amplitude peut atteindre plusieurs microns.

δ) *Effet des vagues.* Il est classique d'attribuer aux vagues qui, par mauvais temps, se brisent sur une côte rocheuse, l'agitation désordonnée, «en dents de scie» (GHERZI), qui se manifeste aux stations voisines. Mais cette interprétation ne va pas sans difficultés: On doit s'attendre tout près du rivage à une agitation à courte période, non enregistrée par les séismographes usuels. Plus loin devrait apparaître le rythme d'arrivée des grosses vagues; il semble qu'on ait observé à Sylt [7] une période de 7 secondes ayant cette origine, disparaissant d'ailleurs vers 10 km. Mais en général on observe surtout des périodes intermédiaires: 2 s. à Héligoland, 4 s. à Apia ou à Alger; à Casablanca[1], sur fond sableux, une analyse harmonique montre un spectre étendu de périodes avec un maximum quasi permanent vers 2—4 secondes et, lorsque la houle est forte, un second maximum de période environ moitié de celle de la houle. Ce second genre d'agitation est toujours lié à l'état de la mer, mais son mécanisme est différent (Sect. 2, α et 3, β).

ε) *Agitation par temps froid.* On observe en hiver, aux latitudes moyennes, une agitation à longue période (parfois plus de 30 s.), très irrégulière. GUTENBERG l'a attribuée au gel des couches superficielles du sol, dont l'effet se propagerait

---

[1] J. DEBRACH: Soc. Sci. Nat. du Maroc, Comptes rendus des séances mensuelles, 35, 1953.

à des distances de l'ordre de la longueur d'onde. Mais la période atteint 2 heures sur les enregistrements de déviation de la verticale avec le double pendule de LETTAU! Une partie, sinon la totalité, de cette agitation est due à des courants de convection dans la cage du séismographe, refroidie à sa partie supérieure par l'air froid qui pénètre dans la cave[1].

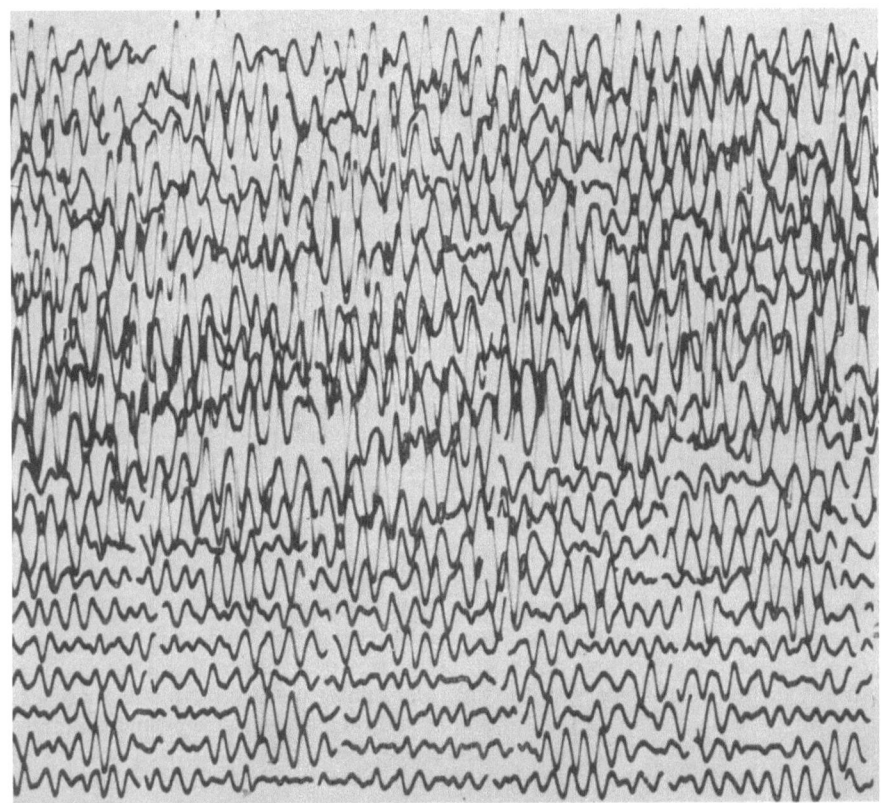

Fig. 1. Exemple d'agitation générale. Tempête microséismique se renforçant rapidement. Séismographe vertical GALIT-ZINE, St.Maur, 15 au 16 Janvier 1954. — Chaque trace montre cinq minutes (à peu près); les lacunes sont produites chaque minute. Les 24 traces successives sont des parties de chaque intervalle horaire d'un jour complet.

ζ) *Poussée de l'air.* Sur les séismographes verticaux à grande période non compensés pour les variations barométriques apparaissent des oscillations beaucoup plus régulières que les précédentes, allant de 30 secondes à plusieurs minutes; GHERZI les appelait ondes Z.

η) *Agitation générale.* Tous ces phénomènes locaux une fois écartés (mais c'est souvent là la difficulté!) il reste un phénomène d'aspect beaucoup plus régulier, à très grande extension (il intéresse des pays entiers sans considération de leur activité séismique). C'est l'agitation générale, dont il sera exclusivement question dans la suite. Son type le plus parfait, tel qu'il apparaît sur les séismographes exempts de frottements solides, comporte une succession de groupes (composés chacun d'une demi douzaine d'ondes dont l'amplitude croît puis décroît) séparés par des intervalles durant à peu près le même temps, pendant lesquels l'amplitude reste faible (fig. 1). Ce caractère est artificiellement renforcé

---

[1] P. BERNARD: Ann. Géophys. **3**, 96 (1947).

par la sélectivité des séismographes. Mais même avec un séismographe peu sélectif la période des ondes composant les groupes est assez bien définie. Sa valeur est généralement comprise entre 2 et 10 secondes. L'amplitude est plus forte, nous l'avons dit, si le sol est formé de sédiments meubles. Inférieure au demi-micron dans les stations favorisées elle atteint quelques microns dans la plupart des autres, très exceptionnellement quelques dizaines de microns.

**2. Théorie de l'agitation générale.** α) *Circonstances d'apparition.* On a trouvé à l'agitation générale observée sur les séismographes mécaniques une période diurne dont Whipple a montré l'origine artificielle: les microséismes industriels (Sect. 1, β) font vibrer le style, surmontent le frottement et augmentent l'amplitude. Par contre l'agitation générale a une période annuelle incontestable, faible dans les régions équatoriales (Apia, la Martinique) mais présentant un maximum net en Janvier pour l'Europe, en Juin pour l'hémisphère austral (La Plata, Terre Adélie); elle est donc liée aux saisons météorologiques [1].

La variation d'amplitude est d'ailleurs loin d'être progressive de l'été à l'hiver: l'agitation, qui n'est jamais nulle, comporte des crises plus fréquentes en hiver, que l'on appelle des «*tempêtes microséismiques*» au cours desquelles l'amplitude augmente parfois jusqu'au décuple ou même davantage, puis revient au seuil en un petit nombre de jours. Les heures du début et de la fin peuvent d'ailleurs différer notablement suivant les périodes auxquelles sont sensibles les séismographes utilisés [3]. Les tempêtes microséismiques fournissent des enregistrements non seulement beaucoup plus amples que le «bruit de fond» mais aussi beaucoup plus réguliers; la plupart des efforts ont porté sur leur interprétation.

Sous les tropiques la situation météorologique par tempête microséismique correspond fréquemment à la présence d'un typhon sur une mer voisine, à une distance pouvant atteindre 3000 km. Dans les régions tempérées, il apparaît souvent des relations analogues avec les perturbations météorologiques sur la mer (ainsi Klotz a reconnu dès 1908 l'effet des dépressions du Golfe du Saint-Laurent sur les microséismes à Ottawa). Par contre on ne trouve aucune relation de détail entre l'agitation et la situation météorologique sur le continent intéressé, pourvu que l'effet du vent sur la station soit bien éliminé. Bien plus, quelque temps après qu'un typhon ou une dépression sont passés de la mer sur la terre, les microséismes s'atténuent puis disparaissent; les bancs très peu profonds comme ceux de Terre Neuve [3] ou des îles Bahama (Gilmore 1947) se comportent comme la terre ferme à ce point de vue.

Les systèmes dépressionnaires des latitudes moyennes sont en général étendus et multiples; d'autre part les cartes météorologiques des Océans sont très incertaines. Il est alors difficile, ou trop facile, d'attribuer la tempête microséismique à telle dépression et plus particulièrement à tel phénomène qui l'accompagne. Par exemple une dépression peut donner des vagues qui vont se briser sur les côtes. Wiechert suggérait en 1903 que par diffusion des composantes de courte période les microséismes correspondants (Sect. 1, δ) pouvaient se régulariser en s'éloignant de la côte; l'agitation générale se créerait ainsi progressivement. Depuis, on a très souvent obervé des micrnséismes avant de que fortes vagues ou que de la houle organisée soient présentes sur les côtes voisines; bien plus, au Groënland (Mlle Lehmann) et en Terre Adélie (Imbert), la houle à la côte est pratiquement supprimée par les champs de glace pendant la saison des plus forts microséismes [6]. Ceci n'exclut pas la possibilité d'expliquer quelques cas par le mécanisme de Wiechert, comme le soutient toujours Gutenberg [2a]; mais nous avons déjà fait des réserves sur les périodes observées (Sect. 1, δ).

Nous admettrons donc, au moins pour l'instant, que les microséismes proviennent de la mer elle-même, ou exceptionnellement de surfaces d'eau moins étendues, comme ceux de 2 s. de période attribués par Lynch aux Grands lacs américains.

β) *Existence d'une propagation.* Il a été longtemps affirmé, au Japon notamment, que l'agitation était constituée par des vibrations stationnaires du sous-sol de la station. RAMIREZ[1] et BERNARD [1] ont montré indépendamment en 1940 qu'il s'agissait d'ondes progressives.

BERNARD cherche d'où provient une augmentation rapide d'amplitude observée sur les séismographes verticaux Galitzine de Scoresby-Sund (Groënland), de Ksara (Liban) et de diverses stations d'Europe. Les différences de temps entre couples de stations lui fournissent comme origine une dépression située au SSW de l'Islande et comme vitesse de propagation $1,9 \pm 0,4$ km./s.

RAMIREZ utilise trois stations rapprochées, avec un repérage des temps plus précis que dans les essais antérieurs et peut identifier sur les trois courbes les accidents produits par une même onde. Des différences de temps il déduit la vitesse (2,7 km./s.) et surtout la direction, qui est celle d'une dépression atlantique. A partir de 1944 cette méthode des «stations tripartites» [2 b] a été appliquée par la Marine américaine à la détection des typhons. Mais sa faveur a diminué: son principal protagoniste conseille aujourd'hui [9] d'utiliser pour cette détection le rapport des amplitudes moyennes en deux stations distantes

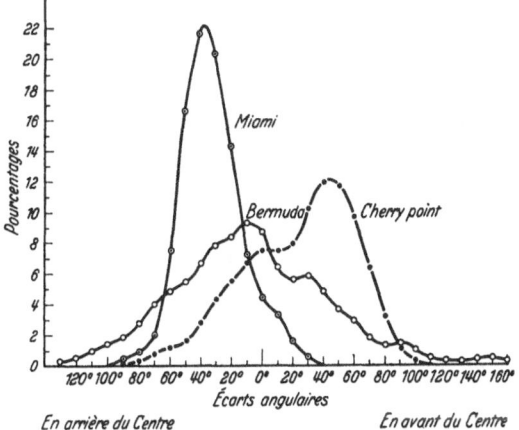

Fig. 2. Pourcentage d'erreurs dans la détermination angulaire du centre de la tempête au moyen des stations tripartites, d'après DONN et BLAIK.

de quelques centaines de km., rapport dont on détermine d'abord la valeur pour un grand nombre de situations connues (c'est là une méthode grossière, car si les amplitudes dépendent de la position du typhon et de son intensité, elles dépendent aussi de son histoire).

DONN et BLAIK[2] ont examiné avec soin les erreurs des stations tripartites et leurs causes. La fig. 2 montre, en trois de ces stations, la distribution des différences angulaires entre les azimuts obtenus et l'azimut du centre déduit des cartes météorologiques. Les différences systématiques que présentent deux des stations peuvent être dues à une déviation horizontale des ondes en terrain non homogène, ou au décalage de l'aire active par rapport au centre de la dépression (Sect. 3, α). Les écarts autour des différences moyennes peuvent dépendre aussi, pour partie, de ce décalage, et plus encore de l'inexactitude des cartes météorologiques; une autre part est expérimentale et pourrait être un peu améliorée. Mais l'erreur la plus importante provient certainement du fait que *des ondes arrivent en même temps de directions différentes.*

Ouvrons ici une parenthèse, pour examiner l'origine des «groupes» caractéristiques de l'agitation générale. On a voulu les expliquer précisément par des interférences entre des ondes provenant de directions différentes et on a même

[1] E. RAMIREZ: Bull. Seism. Soc. Amer. **30**, 35, 139 (1940).
[2] W. L. DONN et M. BLAIK: Bull. Seism. Soc. Amer. **43**, 311 (1953).

proposé des méthodes pour séparer ces directions[1]. Or les phénomènes physiques qui présentent des groupes nets sont ceux dont le spectre est étroit; cette condition générale a été retrouvée par Donn [3] pour les microséismes. Si la coexistence de plusieurs sources est fréquente, elles ne sont pas synchrones et la superposition des mouvements qu'elles engendrent n'est guère susceptible de régularité. L'interférence est possible, et même probable, entre des ondes provenant d'une même source par des trajets continentaux différents, mais on ne voit pas qu'elle puisse réduire l'étendue du spectre. Au contraire, l'interférence (plus parfaite par suite de l'homogénéité de l'eau) des ondes provenant d'une même source par des trajets maritimes différents peut jouer ce rôle. Dans le schéma le plus simple (Sect. 2, $\varepsilon$) leur superposition crée à une distance suffisante de la source une onde superficielle douée de dispersion; la dispersion crée ensuite les groupes en favorisant certaines périodes.

La disposition des maximums et minimums constituant le groupe change au cours de la propagation. Un maximum avance avec la vitesse de phase, le groupe avec une vitesse inférieure. Si l'on veut qu'une crête puisse être identifiée sur les trois enregistrements d'une station tripartite, la distance entre les séismographes utilisés ne doit pas dépasser un petit nombre de kilomètres pour que la progression de la crête à l'intérieur du groupe reste négligeable par rapport à la longueur de celui-ci[2].

Numériquement la vitesse 1,9 km./s. trouvée par Bernard est une vitesse moyenne de groupe. Voici selon Gilmore des vitesses locales de phase, sous-produit des stations tripartites: 2,6 à Cuba sur des sédiments marins; 3,3 en Floride sur un rocher corallien partiellement décomposé; 4 à Porto-Rico sur du roc solide. Malheureusement les vitesses fournies par les stations tripartites fluctuent comme les directions, et pour les mêmes raisons sur lesquelles nous allons maintenant revenir: Nous avons repoussé l'idée que la superposition des ondes arrivant de directions différentes puisse former des groupes plus réguliers que les ondes composantes. Au contraire il est clair que cette superposition compliquera l'enregistrement des groupes déjà formés. Les battements entre deux ondes sinusoïdales pures sont précisément un exemple d'une telle complication. Sur cet exemple Kammer et Dinger[3] [9] ont montré que la vitesse apparente était toujours plus grande que la vitesse vraie. En fait on améliore beaucoup les directions fournies par les stations tripartites si on conserve seulement les observations correspondant aux vitesses les plus faibles.

La longueur des ondes microséismiques étant de l'ordre de 20 km., elles intéressent des profondeurs comparables et la structure superficielle de la croûte influe sur leur propagation. Quoique, en séismologie, les comparaisons d'amplitude soient toujours délicates, il semble bien que les continents et les océans se comportent différemment dans leur ensemble: dans le premier cas, les amplitudes décroîtraient beaucoup moins vite avec la distance[4] [9], ce qui est assez paradoxal et a fait penser (Båth, Gutenberg) à la présence d'un guide d'ondes analogue à celui qui expliquerait la propagation des ondes $Lg$, $Rg$ de Ewing sous le continent américain. La transition devrait correspondre au bord de la plateforme continentale. Or, tandis que Båth[5] ne trouve à celle de

[1] R. Bungers: Z. Geophys. 17, 114 (1941/42).
[2] J. Coulomb: Ann. Géophys. 4, 163 (1948).
[3] E. W. Kammer et J. E. Dinger: J. Meteorology 8, 347 (1951).
[4] D. S. Carder: Earthq. Notes 22, 26 (1941); 24, 21 (1953). — Trans. Amer. Geophys. Un. 33, 315 (1952).
[5] M. Båth: Geol. Fören. Stockholm Förh. 1952, 427. — Met. Inst. Uppsala Medd. 1953, Nr. 28.

Norvège aucune influence, Mlle Charpentier[1] observe que les dépressions sont actives à Paris lorsqu'elles gagnent des fonds inférieurs à 1000 mètres, et Donn [3] fait une observation plus précise sur un cyclone tropical au large de la Nouvelle-Angleterre.

Sur mer, les grandes frontières géologiques, comme celle qui borne les régions à andésite du Pacifique Ouest, s'opposeraient à la transmission de l'énergie [2a]. Beaucoup d'effets de ce genre ont été invoqués, par exemple celui des grandes failles caraïbes par Gilmore, Murphy et Gutenberg, et à l'opposé un effet conducteur de grands alignements basaltiques par Bernard.

*γ) Nature des ondes enregistrées.* Sur terre, l'agitation paraît surtout composée d'ondes de Rayleigh[2]. Lee [8] en a donné en 1932 une preuve indirecte: à la surface d'un milieu homogène, le rapport de l'amplitude verticale à l'amplitude horizontale dans la vibration sinusoïdale de Rayleigh est 1,47. Dans tous les autres cas le rapport dépend de la période. Lee l'a calculé pour des couches sédimentaires variées surmontant une croûte granitique. En tenant compte des effets locaux (piliers des séismographes, etc....), il a trouvé un bon accord avec les observations européennes. Mais il opérait sur des moyennes, en prenant pour définir l'amplitude horizontale celle des deux composantes horizontales qui fournissait l'amplitude la plus grande. Le contraire lui eût été difficile, car il est rare de trouver, sur les enregistrements, des ondes de Rayleigh assez pures pour que les élongations maximums soient simultanées sur les deux composantes. Les reconstitutions du mouvement horizontal montrent généralement des phénomènes complexes, parfois des rotations d'azimut curieuses (fig. 3): on invoque pour expliquer ces phénomènes la présence d'ondes de Love, bien prouvée dans certains cas[3]; elles peuvent être engendrées par les ondes de Rayleigh rencontrant des accidents du sous-sol. On invoque aussi, plus rarement, la multiplicité des sources ou encore les réfractions et réflexions latérales[4].

---

[1] J. Charpentier: Ann. Géophys. **4**, 1 (1948).

[2] Il faudrait donc, contrairement à l'habitude, employer pour les stations tripartites des séismographes verticaux.

[3] L. don Leet: Geophysics **12**, 639 (1947).

[4] Ajouté sur épreuves: M. Blaik et W. L. Donn: Bull. Seism. Soc. Amer. **44**, 597 (1954) confirment les résultats de Lee, la rareté des ondes de Love, et la présence de réfractions latérales.

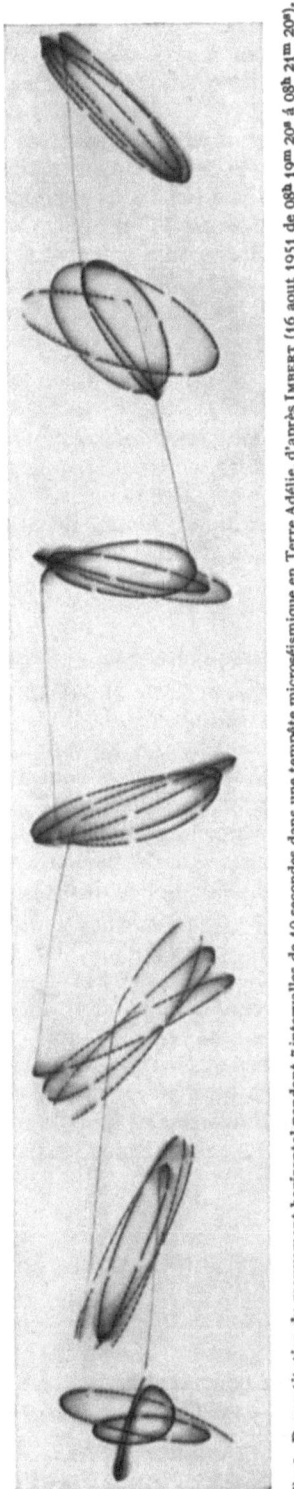

Fig. 3. Reconstitution du mouvement horizontal pendant 7 intervalles de 10 secondes dans une tempête microséismique en Terre Adélie, d'après Imbert (16 aout 1951 de 08ʰ 19ᵐ 20ˢ à 08ʰ 21ᵐ 20ˢ).

Darbyshire[1] montre comment obtenir l'azimut de la source, supposé bien défini, à partir des enregistrements d'une seule station, malgré la présence d'ondes de Love: pour définir les moyennes des amplitudes horizontales $x$ (EW), $y$ (NS) et $z$ (verticale) il fait l'analyse spectrale des enregistrements, forme les carrés des amplitudes relatives aux diverses périodes, les ajoute et prend la racine carrée, le tout électriquement. Sur une quarantaine d'heures pendant lesquelles la direction d'une dépression certainement responsable de l'agitation avait tourné de 90° il ne trouve aucune variation systématique des rapports $\bar{x}/\bar{z}$, $\bar{y}/\bar{z}$ (la barre supérieure désigne les moyennes quadratiques). Or, si le déplacement horizontal se compose d'une onde de Rayleigh $R$ et d'une onde de Love $Q$ et si $\Theta$ est l'azimut de la source:

$$\bar{x}^2 = \bar{R}^2 \sin^2 \Theta + \bar{Q}^2 \cos^2 \Theta,$$
$$\bar{y}^2 = \bar{R}^2 \cos^2 \Theta + \bar{Q}^2 \sin^2 \Theta.$$

Darbyshire en conclut que $\bar{R}$ est approximativement égal à $\bar{Q}$ ce qui est un résultat surprenant. Les spectres de fréquence de $Q$ et $R$ devant être analogues[2], on peut alors déduire $\Theta$ des coefficients de corrélation $r_{xy}$, $r_{xz}$, $r_{yz}$, $z$ étant décalé pour tenir compte de la différence de phase entre les composantes de Raýleigh. Les meilleurs résultats sont obtenus en admettant qu'on a rigoureusement $\bar{R} = \bar{Q}$, auquel cas:

$$\tan \Theta = r_{xz}/r_{yz}.$$

Il serait intéressant d'essayer cette méthode en divers pays.

Selon Gutenberg [2a] la période des microséismes augmente, sous le continent européen, de l'Atlantique vers l'Asie (1 sec. par 1000 km d'après Båth). Carder trouve au contraire la même période sur toute l'Amérique du Nord. Si la variation de période existe il est difficile de l'attribuer à une diffusion des composantes de courte période, aidée ou non par la dispersion: diffusion et dispersion agiraient peu sur un spectre déjà étroit et elles seraient accompagnées d'une régularisation progressive des ondes que l'on n'observe guère. La friction interne du milieu de propagation [2a] fournit un schéma plutôt qu'une explication.

$\delta$) *Ondes au fond de la mer.* Par analogie avec la propagation continentale on peut penser qu'au fond des océans les microséismes se propagent surtout sous forme d'ondes de Rayleigh, du moins à grande distance de leur région d'origine. Le cas théorique d'un fond homogène surmonté d'une couche liquide d'épaisseur constante, schéma applicable à quelque distance des côtes, a été étudié par Press et Ewing [4]. Soit $h$ la profondeur de la mer, $\varrho_0$ sa densité, $\varrho$ la densité du fond, $v$ la vitesse du son dans la mer, $\alpha$ et $\beta$ les vitesses des ondes longitudinales et transversales dans le fond. Le nombre d'ondes $\xi$ et la pulsation $\omega$ des ondes de Rayleigh sont liés par l'équation de dispersion

$$G(\omega, \xi) = 0 \tag{2.1}$$

où:

$$G(\omega, \xi) = \left(\frac{\beta}{\omega}\right)^4 \left[\left(2\xi^2 - \frac{\omega^2}{\beta^2}\right)^2 \left(\xi^2 - \frac{\omega^2}{\alpha^2}\right)^{-\frac{1}{2}} - 4\xi^2 \left(\xi^2 - \frac{\omega^2}{\beta^2}\right)^{\frac{1}{2}}\right] \cos\left(\frac{\omega^2}{v^2} - \xi^2\right)^{\frac{1}{2}} h +$$
$$+ \frac{\varrho_0}{\varrho}\left(\frac{\omega^2}{v^2} - \xi^2\right)^{-\frac{1}{2}} \sin\left(\frac{\omega^2}{v^2} - \xi^2\right)^{\frac{1}{2}} h.$$

(On pourrait bien entendu introduire dans cette équation la période $T = 2\pi/\omega$, la longueur d'onde $\Lambda = 2\pi/\xi$ ou la vitesse de phase $c = \omega/\xi$.)

---

[1] J. Darbyshire: Proc. Roy. Soc. Lond., Ser. A **223**, 96 (1954).
[2] Voir cependant J. T. Wilson: Trans. Amer. Geophys. Un., Part II **1942**, 228.

Si la propagation se fait sans amortissement ($\omega$ réel) on a toujours $c < \beta$ mais $c$ peut devenir inférieur à $v$; l'équation reste réelle si on introduit des fonctions hyperboliques. Pour $\omega$ croissant à partir de zéro, l'équation (2.1) n'a d'abord qu'une racine en $\xi$, puis 2, 3, ..., $m$, ... Il y a une infinité de modes de vibration possibles, comportant un nombre croissant de plans nodaux dans la mer. La fig. 4, qui est celle de PRESS et EWING légèrement rectifiée, fournit le rapport $u/v$, où $u$ est la vitesse de groupe $d\omega/d\xi$, en fonction de $\gamma = \omega h/2\pi v$ pour les deux premiers modes, lorsque $\alpha = \beta \sqrt{3}$ (coefficient de POISSON 1/4); $\varrho = 2{,}5\varrho_0$; $\beta = 2v$. Les modes supérieurs correspondent à des périodes plus courtes, plus diffusibles par les accidents du fond.

Comme dans toutes les propagations avec dispersion les périodes correspondant à un maximum ou à un minimum de la vitesse de groupe tendent à prédominer: dans une propagation rectiligne de parcours $x$ l'amplitude des groupes correspondants décroît comme $x^{-\frac{1}{3}}$ contre $x^{-\frac{1}{2}}$ pour un point ordinaire[1]. De plus, dans la fig. 4, le premier minimum et le premier maximum du deuxième mode sont voisins et se rapprochent encore si l'on suppose la densité ou la rigidité du fond un peu plus faibles, comme il conviendrait pour un fond sédimentaire. S'ils

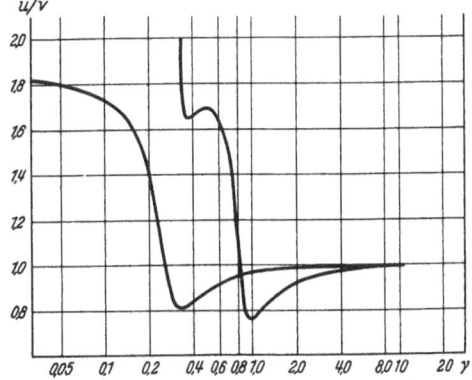

Fig. 4. Dispersion des ondes de RAYLEIGH au fond de la mer, d'après PRESS et EWING.

se confondaient en un palier l'amplitude du groupe décroîtrait seulement comme $x^{-\frac{1}{2}}$; leur simple proximité peut suffire à assurer une certaine prépondérance des périodes correspondantes[2]. Pour $\varrho = 2{,}5\varrho_0$; $\beta = v\sqrt{3}$ par exemple, le palier correspond à $u/v = 1{,}45$; $\gamma = 0{,}45$. Si on prend $v = 1{,}5$ km./s., on a $u = 2{,}2$ km./s., du même ordre que la valeur de BERNARD. Quant à la période $T = h/\gamma v = 1{,}5$ h$^{km}$, on peut seulement la comparer à la relation empirique $T = 1{,}2$ h$^{km}$, établie par IMBERT [6] sur une dizaine de cas, entre la période microséismique moyenne dans un observatoire côtier et la profondeur moyenne des bassins océaniques dans un rayon de 10° (majorée de 1,3 km. pour tenir compte des sédiments non consolidés). De telles vérifications ne sont guère susceptibles d'une plus grande précision.

$\varepsilon$) *Transmission de l'énergie entre la surface et le fond.* Comme l'a reconnu WHIPPLE dès 1933, la *compressibilité de l'eau* joue un rôle essentiel dans la transmission du mouvement entre la couche superficielle de la mer, perturbée par les actions météorologiques, et le fond: Si, dans les conditions du paragraphe précédent, on écrit qu'une onde plane de compression ayant subi deux réflexions totales[3] au fond et à la surface se retrouve en phase avec l'onde initiale (ce que Pekeris appelle une «*interférence constructive*») on est conduit, pour l'angle d'incidence arc sin $(v/c)$ de ces ondes à une condition qui équivaut à (2.1); de telles interférences, nous l'avons vu (Sect. 2, $\beta$), peuvent expliquer les groupes et DONN [3] trouve que la régularité des microséismes dépend de l'uniformité de profondeur dans la région où ils se forment.

---

[1] H. et B. S. JEFFREYS: Methods of Mathematical Physics, Chapter 17. Cambridge 1946.

[2] J. COULOMB: C. R. Acad. Sci. Paris **227**, 1163 (1948).

[3] Si la réflexion au fond n'est pas totale, on a des ondes amorties.

Scholte[1] cherche l'effet sur le fond des ondes de compression produites par une force verticale $Q e^{i\omega t}$ concentrée en un point de la surface de la mer (pression sur une petite aire). Le déplacement vertical $w$ du fond[2], si la distance horizontale $r$ au point d'application de la force est grande par rapport à la profondeur $h$, est donné par la formule de Longuet-Higgins [5]:

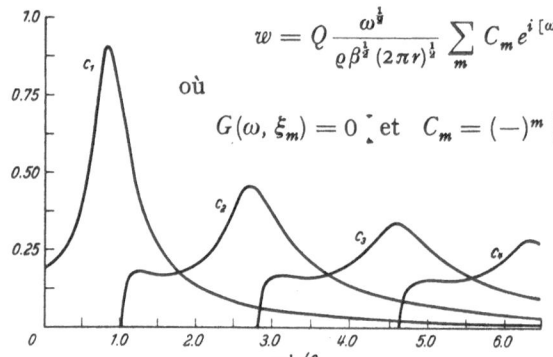

$$w = Q \frac{\omega^{\frac{5}{2}}}{\varrho \beta^{\frac{3}{2}} (2\pi r)^{\frac{1}{2}}} \sum_m C_m e^{i[\omega t - \xi_m r + (m+\frac{1}{4})\pi]} \tag{2.2}$$

où

$$G(\omega, \xi_m) = 0 \quad \text{et} \quad C_m = (-)^m \left(\frac{\beta}{\omega}\right)^{5/2} \xi_m^{\frac{1}{2}} \left(\frac{\partial G}{\partial \xi_m}\right)^{-1}.$$

Fig. 5. Coefficients d'amplitude des divers modes, pour une pression sinusoïdale à la surface de l'eau, d'après Longuet-Higgins.

La fig. 5 empruntée à Longuet-Higgins, fournit les coefficients $C_m$ en fonction de $\omega h/\beta$, avec les constantes de Press et Ewing ($\alpha = \beta \sqrt{3}$; $\beta = 2v$; $\varrho = 2,5 \varrho_0$). $C_1$ a un maximum pointu; Scholte (1953) déduit de (2.2) que, pour ce maximum, l' amplitude $Q$ qui produirait une agitation de $5\mu$ à 3000 km. est $3 \times 10^{15}$ dynes. $C_2$, $C_3$, $C_4$ ont un maximum, puis un minimum, et enfin le maximum correspondant à celui de $C_1$.

Longuet-Higgins rapproche les valeurs de $\omega h$ qui correspondent respectivement aux extremums de $C_m$ (fig. 5) et de $u$. (fig. 4); mais les uns correspondent à $d\left[\frac{\partial G}{\partial \omega} \middle/ \frac{\partial G}{\partial \xi}\right] = 0$, les autres à $d\left[\xi^{\frac{1}{2}} \middle/ \frac{\partial G}{\partial \xi}\right] = 0$. En fait, on doit choisir entre deux points de vue: Longuet-Higgins raisonne comme si les effets relatifs aux diverses fréquences avaient leurs phases distribuées au hasard, et il se borne à étudier l'énergie reçue en fonction de l'énergie moyenne du phénomène origine en superposant simplement les *énergies* des composantes spectrales. A priori ce schéma semble exprimer correctement la turbulence météorologique. Pourtant si l'on veut expliquer la présence, dans les enregistrements, des groupes caractéristiques, on ne peut guère se dispenser d'admettre une certaine solidarité entre fréquences voisines. Par exemple Press, Ewing et Tolstoy[3] considèrent l'effet d'une force instantanée, et non plus sinusoïdale, qu'ils supposent d'ailleurs appliquée au fond, et ils retrouvent bien entendu les groupes stationnaires; il en serait de même pour toute perturbation à spectre bien défini, donc de durée limitée, et certainement aussi pour des liaisons spectrales de caractère aléatoire convenable[4].

$\zeta$) *Pression et houle à la surface de la mer.* Nous abordons la question la plus controversée: savoir quel phénomène superficiel produit l'ébranlement de l'eau. Deux opinons s'affrontent. La première admet *l'effet direct des variations de pression atmosphérique*; c'est ainsi que Gherzi affirme depuis 1923 l'existence et

[1] J. G. Scholte: I en II, Kon. Versl. Ned. Akad. Wetensch. **52**, 669 (1943); Koninklijk Nederlandsch Meteorologisch Instituut 1854—1954, p. 419, 1953. Le second article comporte de menues erreurs.

[2] Scholte donne des formules pour les deux composantes.

[3] F. Press, M. Ewing et I. Tolstoy: Bull. Seism. Soc. Amer. **40**, III (1950). Les calculs sont faits pour un cas ($\alpha = \beta\sqrt{3}$; $\beta = 3 V$; $\varrho = 2,5\varrho_0$) où l'effet de palier (Sect. 2, $\delta$) est peu important.

[4] A. Blanc Lapierre et P. Lapostolle: Rev. sci. **84**, 579 (1946).

l'efficacité d'oscillations de pression («*pumping*») au centre des cyclones; d'autres auteurs supposent une action analogue le long des fronts froids, action que PRESS et EWING [9] ou DONN considèrent comme une résonance entre l'air et l'eau. Mais HASKELL [9] montre l'insignifiance du phénomène. D'ailleurs, de son exemple déjà cité (Sect. 2, ε) SCHOLTE conclut à la nécessité d'admettre des variations de l'ordre du millibar sur 10 km. de rayon; une concordance des variations barométriques sur une aire aussi étendue n'est pas vraisemblable. D'autre part les oscillations irrégulières de la pression enregistrées au cours des cyclones

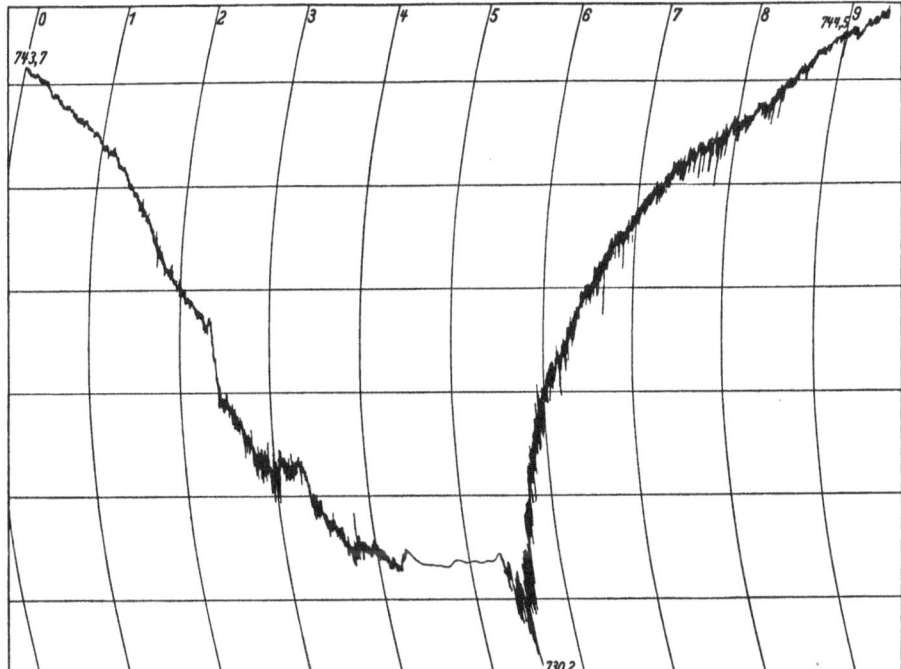

Fig. 6. Barogramme lors du passage d'un cyclone sur l'Observatoire Central de l'Indochine à Phu-Lién le 16 Juillet 1928.

correspondent simplement aux rafales de vent et cessent dans le calme central (fig. 6). A distance enfin, les quelques microbarographes en service n'ont guère montré de corrélation entre les ondes de pression et les microséismes[1]. Pour toutes ces raisons, faire appel à un pumping hypothétique ne serait légitime qu'en l'absence de toute autre cause vraisemblable.

La seconde opinion, que les microséismes sont *l'effet des vagues produites par le vent*, clairement exprimée par WALKER en 1931 [1], s'est heurtée à l'objection que la pression décroît exponentiellement avec la profondeur sous les houles progressives, telles que:

$$\zeta = a \cos (kx - \omega t) + O (a^2) \tag{2.3}$$

où $\zeta$ est le soulèvement au point d'abscisse $x$ et où:

$$\omega^2 = g\,k\,\mathrm{Tan}\,k\,h.$$

Cette objection est levée grâce aux travaux publiés depuis 1947 par l'Amirauté Anglaise. Indépendamment BERNARD[2] avait émis l'idée qu'un clapotis était

---

[1] M. EWING et F. PRESS: Trans. Amer. Geophys. Un. **34**, 95 (1953). — B. GUTENBERG: Trans. Amer. Geophys. Un. **34**, 161 (1953).

[2] P. BERNARD: Bull. Inst. océanogr. Monaco **38**, Nr. 800 (1941).

nécessaire, et Miche[1] avait montré qu'une houle stationnaire:

$$\zeta = a \cos kx \cos \omega t + O(a^2) \qquad (2.4)$$

entraînait sur un fond situé à grande profondeur (pour laquelle la variation de premier ordre en $a$ est devenue négligeable) une variation de pression:

$$\Delta p = -\tfrac{1}{2} a^2 \omega^2 \varrho_0 \cos 2\omega t + O(a^3) \qquad (2.5)$$

en phase pour tous les points du fond, mais de période $\pi/\omega$. Longuet-Higgins [5] en donne l'interprétation suivante: Partons de l'instant où le soulèvement de la surface est approximativement nul. Pendant un quart de période, l'eau monte de l'emplacement des creux à celui des crêtes; le centre de gravité d'une tranche liquide correspondant à une longueur d'onde s'élève, donc la force que le fond exerce sur cette tranche augmente, pour reprendre sa valeur au bout de la demi-période.

Il serait tentant d'appliquer directement ce résultat en attribuant les micro-séismes aux variations de pression sur le fond; numériquement l'exemple de Scholte (Sect. 2, $\varepsilon$) demande une aire de 1 km.$^2$ pour un clapotis de $10^m$ d'amplitude et $10^s$ de période, ce qui n'est pas invraisemblable. Mais on a négligé la compressibilité de l'eau. Or Longuet-Higgins montre qu'elle se fait sentir dès que la profondeur est du même ordre que la longueur d'onde du clapotis, soit à quelques dizaines de mètres. A ces profondeurs, négligeables devant celle de la mer, la pression moyenne sur une aire horizontale suit la loi de Miche; tout se passe comme si le mouvement de faible amplitude qui se propage plus bas était produit par les variations de pression correspondantes. On peut donc appliquer la théorie de Scholte (Sect. 2, $\varepsilon$) avec ou sans cohérence entre les composantes des diverses périodes.

Il n'est d'ailleurs pas nécessaire que le clapotis soit parfait. Si on représente avec Longuet-Higgins le soulèvement par une intégrale de Fourier à deux dimensions, l'essentiel est qu'il existe dans la région perturbée des couples de composantes spectrales, éléments de l'intégrale double, dont la période soit la même, l'amplitude comparable, mais le sens de propagation opposé. Cela peut a priori se produire «dans la zone centrale des cyclones et, d'une façon moins accentuée, aux endroits où se produisent des réflexions de la houle»[2].

**3. Applications.** $\alpha$) *Dépressions au large.* Au centre des cyclones tropicaux et même des dépressions extratropicales apparaît un clapotis gigantesque sous forme de vagues coniques ou pyramidales se dressant verticalement (voir les descriptions d'Everett[3] et de Charcot [1]). L'existence de composantes de houle de directions opposées est due à la disposition giratoire des vents (notamment à leur rotation au passage des fronts si ceux-ci existent) jointe à la persistance de certaines houles si la dépression évolue. Dans un cas cité par Darbyshire[4] les vents sur le trajet d'une dépression tournaient de 180° en 3 ou 4 heures. La régularité des microséismes est d'autant plus grande que l'aire génératrice est plus petite [3].

Il faudrait préciser. On sait que dans l'hémisphère Nord par exemple, les vagues sont plus hautes dans la moitié du cyclone située à droite de son axe de déplacement, car elles y restent soumises plus longtemps au vent soufflant dans le sens de la marche (à condition que le déplacement ne soit pas assez rapide pour les dépasser). En outre, s'il s'agit d'une dépression extra-tropicale, les seules dont on ait tenté d'analyser en détail les effets microséismiques, les vagues sont aussi plus hautes dans les secteurs froids que dans le secteur chaud (20% de plus d'après Roll). Finalement, le maximum d'activité est à attendre dans le secteur froid, à l'arrière de la dépression [6].

[1] M. Miche: Ann. Ponts Chaussées **114**, 42 (1944).
[2] P. Bernard: Bull. Inst. océanogr. Monaco **38**, Nr. 800 (1941).
[3] E. Gherzi: Notes Séismol. Zi-Ka-Wei **8** (1926).
[4] J. Darbyshire: Proc. Roy. Soc. Lond., Ser. A **223**, 96 (1954).

Ce sont là des résultats moyens, donc adoucis. La répartition réelle est mal connue, mais les vents forts, soufflant par rafales, qui accompagnent le front froid principal, ou les fronts froids secondaires du secteur postérieur, jouent certainement un rôle extrêmement important. Le maximum de houle est en général en arrière du front froid (cela dépend de la durée d'action du vent et de la vitesse de la dépression). Le décalage a été étudié par SALLARD, et trouvé par IMBERT [6] en accord raisonnable avec les microséismes de l'Antarctique. La houle produite dans le secteur froid postérieur interférait pour engendrer les microséismes avec la houle du secteur chaud, ou du secteur froid antérieur; IMBERT a pu préciser numériquement la façon dont l'amplitude des microséismes à l'île Macquarie augmentait avec l'angle des houles, plus exactement avec l'angle des vents moyens.

β) *Houle à la côte.* Peut-il y avoir génération de microséismes au voisinage immédiat de la côte? BERNARD invoquait, nous l'avons vu, la réflexion de la

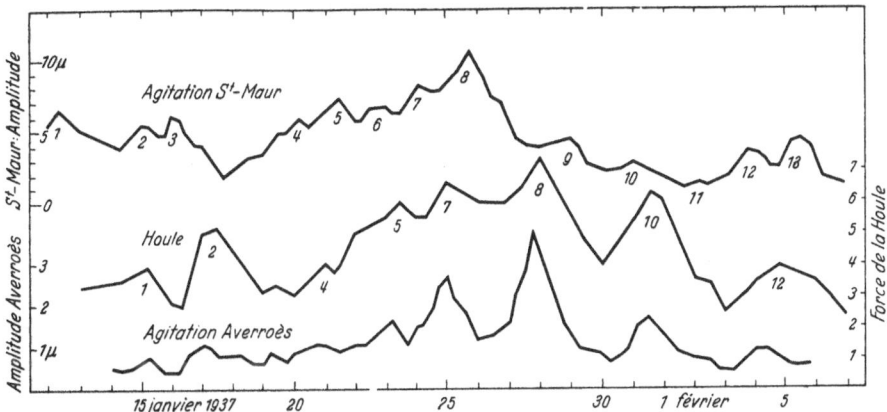

Fig. 7. Comparaison de l'agitation microséismique à Paris et au Maroc avec la houle au Maroc, d'après BERNARD.

houle et donnait l'exemple de la fig. 7, dans lequel l'agitation à l'observatoire Averroès, proche de Casablanca, accompagne la houle sur la côte marocaine ou la précède de peu, contrairement à l'agitation à St. Maur que BERNARD attribue à la dépression elle-même. BERNARD observait enfin que la période de l'agitation à Averroès suivait la période de la houle, mais en restait approximativement la moitié; le fait a été retrouvé par DEACON (1947) et rapproché par lui de la formule de MICHE (2.5). Il infirme les idées de WIECHERT (Sect. 2, α) qui impliqueraient l'égalité entre la période des vagues et celle des microséismes.

DARBYSHIRE[1] a comparé des analyses spectrales de la houle en Cornouaille et des microséismes à Kew près de Londres. Il place parfois au large l'origine de ceux-ci. Mais il les attribue plus souvent à des réflexions de houle sur la côte anglaise; la loi des périodes se vérifie bien, sauf pour des houles très longues. Dans deux cas, DARBYSHIRE fait appel à des réflexions sur la côte du Groënland. Ces résultats, les derniers en particulier, paraissent trop beaux quand on pense à la complexité des réfractions et réflexions de la houle arrivant sur une côte accidentée. On peut donc se demander s'il n'existait vraiment aucune cause d'interférences proprement météorologique, comme cela apparaît plus clairement dans les cas suivants.

A l'est des Etats Unis, il arrive fréquemment [3] qu'un front froid sensiblement parallèle à la côte passe de la terre sur la mer et que l'on observe peu après des microséismes à groupes nets et de grande amplitude, qui semblent provenir

[1] J. DARBYSHIRE: Proc. Roy. Soc. Lond., Ser. A **202**, 439 (1950); **223**, 96 (1954).

d'abord du centre de la dépression s'il est observable, puis progressivement de tout le front[1]. La période, d'abord courte, augmente et tend vers une limite lorsque le front gagne le large [3]. L'interférence de la houle engendrée derrière le front froid avec la houle résiduelle du secteur chaud paraît expliquer convenablement ces phénomènes qu'on ne peut en tout cas attribuer à une réflexion. Donn repousse cependant cette interprétation qu'il juge incompatible avec la rapidité du début (l'agitation s'installe parfois en un petit nombre d'heures).

Les circonstances ne semblent pas très différentes à l'Ouest des Etats-Unis[2] ou dans le cas de la Scandinavie étudié par Bâth[3], bien que les fronts traversent les côtes de la mer vers la terre. Aux latitudes plus basses les fronts sont moins nets mais on soupçonne néanmoins un mécanisme analogue, au Maroc ou en Italie par exemple.

Ainsi les phénomènes importants près des côtes seraient les mêmes qu'au large, et leur étude serait plus facile puisque les périodes ne seraient pas altérées par la propagation.

Il est cependant des microséismes que Donn[4] a pu attribuer à un clapotis par réflexion: de fortes vagues du secteur chaud atteignant obliquement la côte de Nouvelle Angleterre, y produisent une très faible agitation, satisfaisant d'ailleurs assez mal à la loi de Miche (périodes 2,6 s. et 7 s. au maximum d'amplitude dans l'un des cas). Mais au total il ne semble pas qu'il faille attendre de la réflexion des effets comparables à ceux de la rotation des vents. Un calcul numérique conduit Longuet-Higgins au même résultat.

γ) *Conclusion.* Le problème des microséismes ne peut être posé correctement que si l'installation des séismographes élimine les effets locaux, si l'on a étudié au préalable l'influence du sous-sol de la station (qui ne s'exprime pas seulement par un facteur d'amplification ou par un schéma de stratification), enfin si l'on possède la carte météorologique détaillée des mers voisines, avec des enregistrements de houle sur les côtes les plus proches. Ces conditions n'étant jamais réunies, la plupart des auteurs comparent simplement la période ou l'amplitude, dont ils forment la moyenne par divers procédés [1] [3][3], à l'élément dont ils disposent: (vent local, vagues à la côte, minimum barométrique, front froid, masse d'air polaire, noyau isallobarique, etc.). L'ensemble de ces travaux est difficile à coordonner; on gagnerait à confronter chaque observation avec une interprétation mécanique précise.

## Bibliographie générale.

[1] Bernard, P.: Thèse Paris 1940. — Ann. Inst. Phys. Globe Paris **19**, 1 (1941).
[2] *Compendium of Meteorology:* Amer. Met. Soc., Boston **1951**: a) B. Gutenberg, 1303; b) J. B. Macelwane, 1312.
[3] Donn, W. L.: J. Meteorology **8**, 406 (1951); **9**, 61 (1952).
[4] Press, F., and M. Ewing: Trans Amer. Geophys. Un. **29**, 163 (1948).
[5] Longuet-Higgins, M. S.: Phil. Trans. Roy. Soc., Ser. A **243**, 1 (1950).
[6] Imbert, B.: Ann. Géophys. **10**, 175 (1954).
[7] Gutenberg, B., and F. Andrews: Bibliography on Microseims, Second edit. Calif. Inst. of Techn., Division of Earth Sciences, Contrib. No 602. 1952.
[8] Milne, J., and A. W. Lee: Earthquakes and other movements, Chapter XIV. London: Kegan 1939.
[9] *Symposium on Microseisms:* Nat. Acad. of Sci., Nat. Res. Council, Publ. 306, 1953.

---

[1] L. don Leet: Geophysics **12**, 639 (1947).

[2] B. Gutenberg: Trans. Amer. Geophys. Un. **34**, 161 (1953). Le même article décrit un cas de cyclone tropical longeant la côte mexicaine.

[3] M. Bâth: Tellus **5**, 109 (1953). — Met. Inst. Uppsala Medd. **190**, Nr. 32, citant ses travaux antérieurs.

[4] W. L. Donn: Trans. Amer. Geophys. Un. **34**, 471 (1953). Une situation météorologique analogue donne la plupart des tempêtes locales sur la côte Est des Etats-Unis d'après, D. K. Todd et R. L. Wiegel: Trans. Amer. Geophys. Un. **33**, 217 (1952).

# Seismic Prospecting.

By

MAURICE EWING and FRANK PRESS.

With 19 Figures.

**1. Introduction.** Seismic prospecting has its basis in the classical physical principles of transmission, reflection, refraction, and scattering of elastic waves in a layered solid half-space. The great increase in application of the methods and the continual effort to improve them since 1925 have resulted in elaboration and refinement of instruments, field methods, and interpretation techniques. It is the most expensive method of geophysical prospecting, but also the most powerful. In many areas it can map beds many thousands of feet deep and detect depth variations of the order of a few feet.

The basic procedure is to generate elastic waves by a near-surface explosion, to record the resulting waves reaching the surface at various distances, and to deduce the positions of reflecting and refracting interfaces by analysis of the travel times and characteristics of identifiable wave groups. The techniques using refracted waves differ completely from those based on reflected waves. *"Refraction shooting"* is useful only for mapping a bed in which the velocity is greater than in those above it. The shot to detector distance must be several times greater than the depth of the bed, since the refracted waves must travel a considerable horizontal distance through the bed on a minimum time path.

*"Reflection shooting"* uses near vertical reflections of compressional waves, hence the shot to detector distance is small compared to the depth of the reflecting bed. The principal problem is to isolate the reflection from scattered waves and low velocity surface waves by filtering, mixing signals from large arrays of detectors, automatic control of gain, and advantageous arrangement of the explosive charges.

Complete references to original publications may be found in several textbooks[1]. Several journals publish articles in this field[2]. Compilations of geophysical case histories are available[3].

**2. Reflection Shooting.** α) *Physical Principles.* In Chapters 3 and 4 it is demonstrated that *compressional* and *shear* waves are propagated in an elastic solid body and that *surface* waves of several types occur if the body has a layered structure. Small departures from perfect elasticity are shown to produce *attenuation*, which for earth materials usually *increases with frequency*. Formulae for

---

[1] See, for example, M. B. DOBRIN: Introduction to Geophysical Prospecting, New York: McGraw-Hill 1952. 435 p. — C. H. DIX: Seismic Prospecting for Oil. New York: Harper a. Bros. 1952. 414 p. — E. ROTHÉ and J.-P. ROTHÉ: Prospection Géophysique, Tome I. Paris: Gauthier-Villars 1950. 438 p. — A historical account (100 years of earthquake physics and explosion seismology) was given by L. MINTROP. Naturwiss. **34**, 257 and 289 (1947); also Z. Geophys. **1953**, 101.

[2] *Geophysics,* a Quarterly Journal published by Society of Exploration Geophysicists, Tulsa, Oklahoma; *Geophysical Prospecting,* a Quarterly Journal published by the European Assoc. of Exploration Geophysicists.

[3] Geophysical Case Histories, Vol. I. Tulsa: Soc. of Explor. Geophys., 1948, 671 pp. Vol. II. in preparation.

the amplitudes and phases of the waves reflected, transmitted and transformed at an interface are given and discussed. For almost all of these phenomena it is adequate to consider simple harmonic plane waves and the laws of geometrical optics. Only the generation of the "ground roll" and similar waves, and questions such as energy flux near the source and distortion of the initial pulse require special solutions of the wave equations. The *useful signals* in reflection shooting consist only of *compressional waves*; the shear wave and surface waves contribute only to the noise.

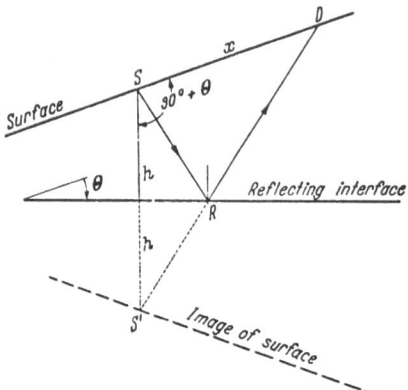

Fig. 1. Ray path from shot-point $S$ to detector $D$ via reflexion point $R$ on interface, with slope $\Theta$. Formula (2.1) is obtained with $S'$=image of $S$, and path length $SRD = S'RD = VT$.

To a first approximation the velocity above the reflecting surface may be taken as constant $V$. The travel time, $T$ of a reflection to a surface point at distance $x$ is given by (Fig. 1)

$$V^2 T^2 = 4h^2 + x^2 + 4h\,x \sin \Theta \quad (2.1)$$

where $h$ is the perpendicular distance from the shot point to the interface, and $\Theta$ is the component of interface slope in the vertical section through shot and detector positions. If the shot point is down dip from the detectors, $\Theta$ is negative. In all cases a series of detectors along the section is used to facilitate identification of the reflection, hence the change in travel time, $\Delta T$, for an increment in distance, $\Delta x$ is also available for each shot. By differentiation, we obtain

$$\sin \Theta = (V^2 T \Delta T/2h\Delta x) - x/2h. \quad (2.2)$$

By shooting in reverse directions to eliminate the effect of slope, $V$ and $h$ may be determined from the first equation and $\Theta$ from the second.

In practice the velocity is determined by *well shooting*, in which travel times between a surface explosion and detectors in a bore hole are used. Depths and dips are found graphically using $T$ and $\Delta T$ from seismograms and charts computed for the observed velocity-depth function. When either the change in velocity with depth or the shot to recorder distance is large, it is necessary to include the effect of ray curvature in preparing these charts.

The *frequency* of most reflections lies in the range 20–100 cycles/sec. For lower frequencies, wavelengths are large compared to the bed thickness, the reflection coefficient is lower, and little energy is returned to the surface. Higher frequency waves are scattered and selectively attenuated.

*β) Field Operations.* The principal field problems of reflection shooting are to obtain the maximum signal to noise ratio over a suitable range of reflection times and to survey a given area as rapidly and efficiently as possible. Since the terrain and climate encountered range over wide extremes, various systems of equipment are required, their characteristics being determined largely by the available transportation.

The *recording apparatus* consists of geophones, amplifiers, mixing circuits, and oscillographs. Most geophones (Fig. 2) contain a mass constrained to move in the vertical direction, while supported by a spring or membrane chosen to place the natural frequency in the range 5–30 cycles per sec. Motion is usually detected by a *transducer* consisting of a coil of wire attached to the mass, situated in the field of a magnet attached to the case (Fig. 2) which also supplies damping for the system. A *calibration curve* for a typical geophone subject to simple harmonic ground motion is shown in Fig. 3.

*Amplifiers* are designed to provide the following characteristics.

1. Shock proof construction, small size, freedom from microphonics and other noise, and identical response characteristics.

2. Incorporation of non-resonant *band-pass filters* with adjustable band width frequency, and sharpness.

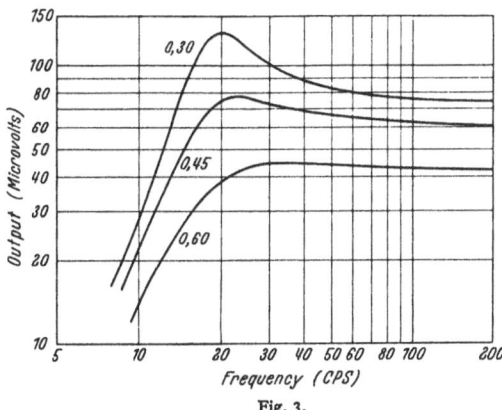

Fig. 2.                                     Fig. 3.

Fig. 2. Section through a moving coil type geophone. The mass *6*, suspended by an elastic membrane *9*, oscillates when the ground moves vertically, and the coil *5*, attached to the mass, moves in the field of a strong permanent magnet *3* with pole caps *4* so that a voltage is induced in it which, through the attachment *7* and the cable is fed to the galvanometer in the truck. The case *2*, which also closes the magnetic circuit, ends below in the spur *1* with which the geophone is fastened on the ground. (After W. ZETTEL, Prakla-Hannover.)

Fig. 3. Calibration curve for a geophone for damping constants 0.30, 0.45, and 0.60 of critical. (Southwestern Industrial Electronics Co.).

Fig. 4. Response curves of seismic reflection amplifiers for various filter combinations. (Southwestern Industrial Electronics Co.).

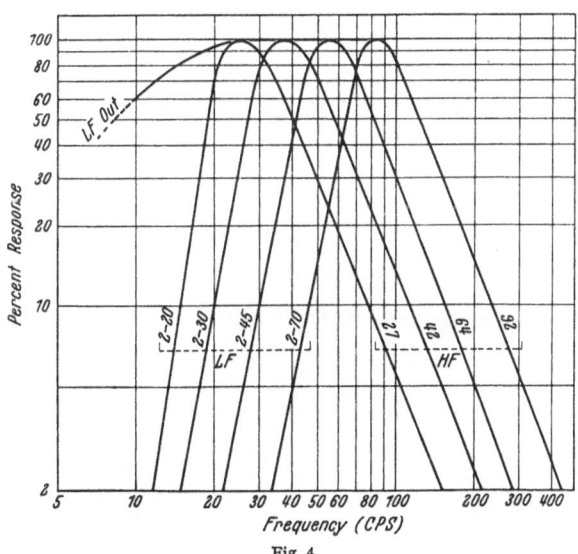

Fig. 4.

3. *Automatic gain control* to provide a readable record from the arrival of the explosion shock wave until signal strength drops below the level of background noise.

In Fig. 4 typical response curves for reflection amplifiers are shown. The recording of signals from the amplifiers is usually made with an oscillograph

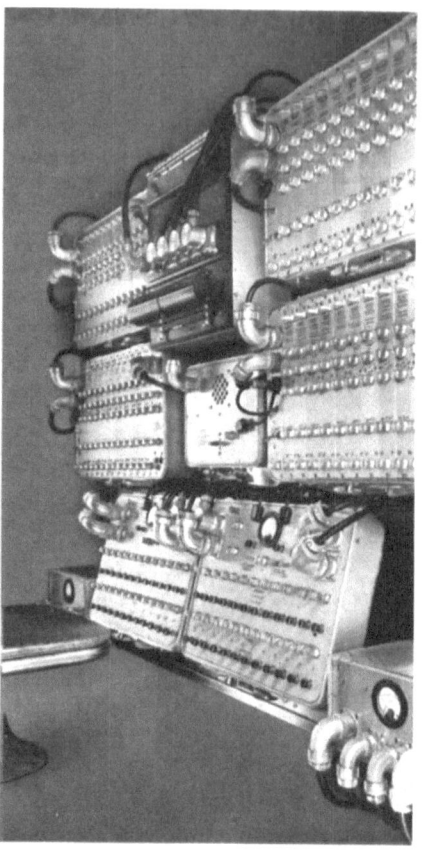

Fig. 5a. 48 trace seismic reflection apparatus installed in a truck, showing amplifiers, oscillograph, mixing circuits. telephone and power supply. (Courtesy Southwestern Industrial Electronics Company.)

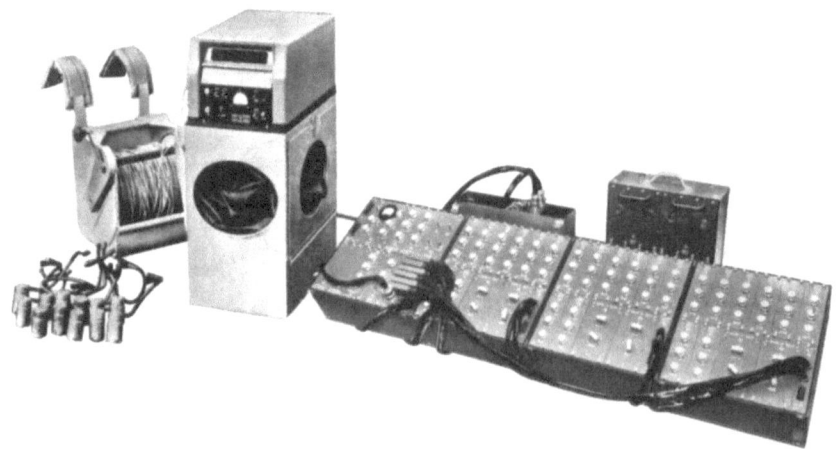

Fig. 5b. 24 trace portable seismic reflection apparatus showing, from left to right, geophones, cable, reel, oscillograph and developing tank, amplifiers and filters, power supply. (Courtesy Texas Instrument Co.)

containing a bank of galvanometers, a device for transporting photographic paper and a mechanism for placing time marks and the instant of the explosion on the paper. Provision is usually made for reading travel times with an accuracy of about 1 millisecond. A typical oscillograph uses paper 8 inches wide, records at a rate of 15 inches/sec., contains as many as 50 "pencil" galvanometers, with natural frequency in the range 100–200 cycles/sec., and electromagnetic critical damping. Mixing circuits are commonly used to compound the outputs of several geophones into each galvanometer. This reinforces the vertical reflection and tends to cancel horizontal traveling ground roll. In some difficult

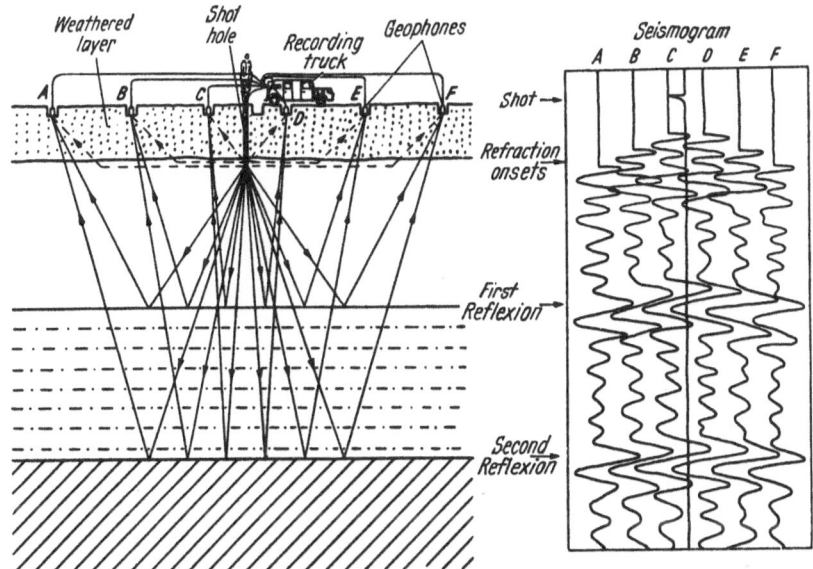

Fig. 6. Schematic diagram showing the production of a reflexion seismogram. Under the weathered layer, two layers are separated by reflecting interfaces. A number of geophones (24 or 32) are placed at distances of 20 to 50 m. along a line (profile), and an explosive charge is fired at the center of the profile, buried below the weathered layer, preferably in the ground water. The responses of all geophones are conducted by cables to the recording truck and collected on a film. The seismogram is given schematically at the right. The center trace records the moment of the shot. The geophone traces begin with waves refraction onsets which have run along the surface below the weathered layer. Their travel times are proportional to the distance. Later, reflected waves from the two interfaces arrive; their onsets at the successive geophones differ less than the first onsets, because the ray paths increase less than the horizontal distance. (After W. ZETTEL-Prakla.)

areas as many as several hundred geophones have been compounded for each trace on the seismogram[1]. This apparatus is usually installed in a truck or boat, together with facilities for rapid processing of the photographic record. A typical installation is shown in Fig. 5, and a schematic diagram (Fig. 6) shows the principle of the reflection method.

Use of *magnetic tape* and drum recording in seismic exploration is increasing rapidly. This instrumentation allows greater flexibility because of the wide dynamic range and the playback feature. Standard seismograms for a variety of filter settings and different geophone combinations can be made from a magnetic record of a single shot[2].

In addition to the recording truck, vehicles equipped for surveying, drilling shot holes, and shooting the explosive charges are used. In a typical operation the holes extend below the "weathered" layer, usually from 30 to 300 ft., and

[1] A. E. McKay: Review of pattern shooting. Geophysics 19, 420—437 (1954).

[2] G. B. Loper and R. R. Pittman: Seismic recording on magnetic tape. Geophysics 19, 104—115 (1954).

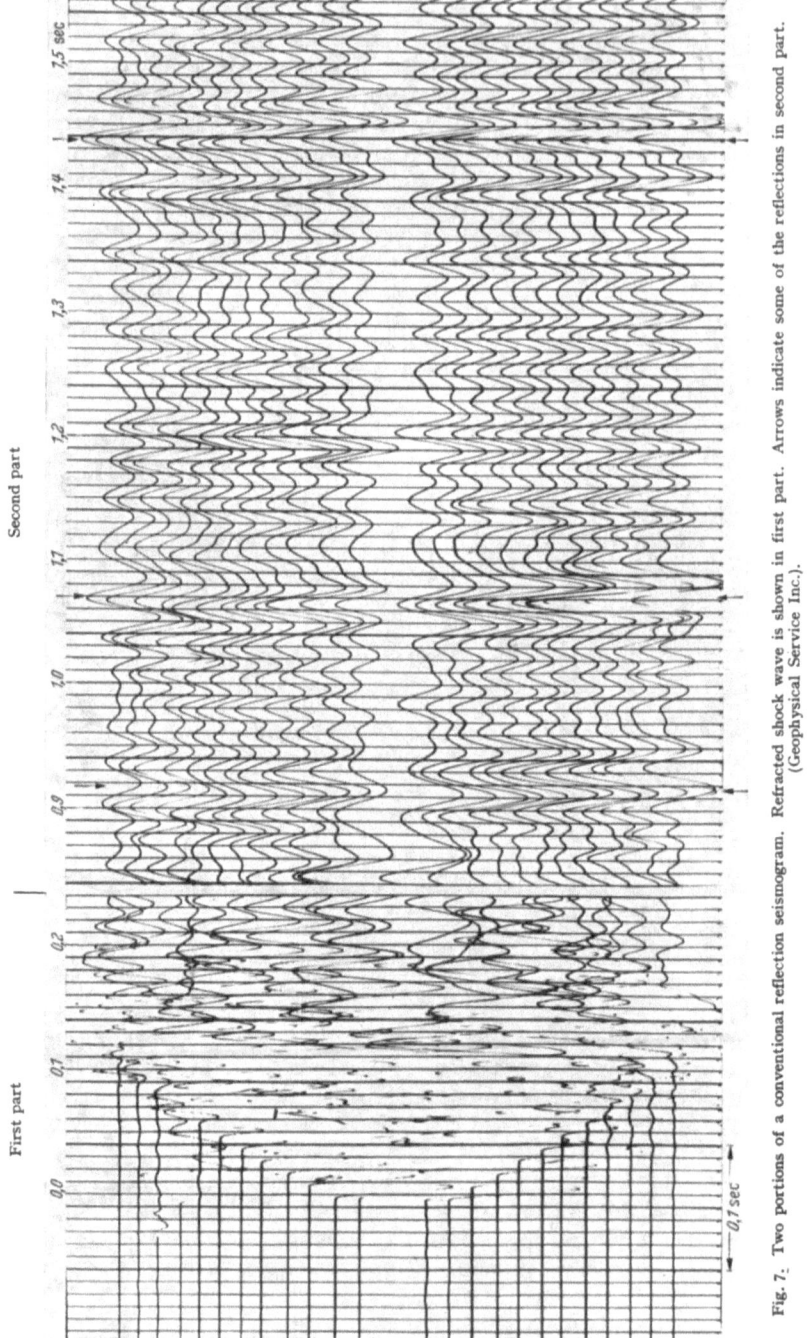

Fig. 7. Two portions of a conventional reflection seismogram. Refracted shock wave is shown in first part. Arrows indicate some of the reflections in second part. (Geophysical Service Inc.).

the explosive charge is usually less than 50 lbs. Air shooting techniques in which an array of shots is detonated a few feet above the ground have met with some success. A conventional reflection seismogram is shown in Fig. 7. Reflections can be improved by arranging the charge in patterns (Fig. 8).

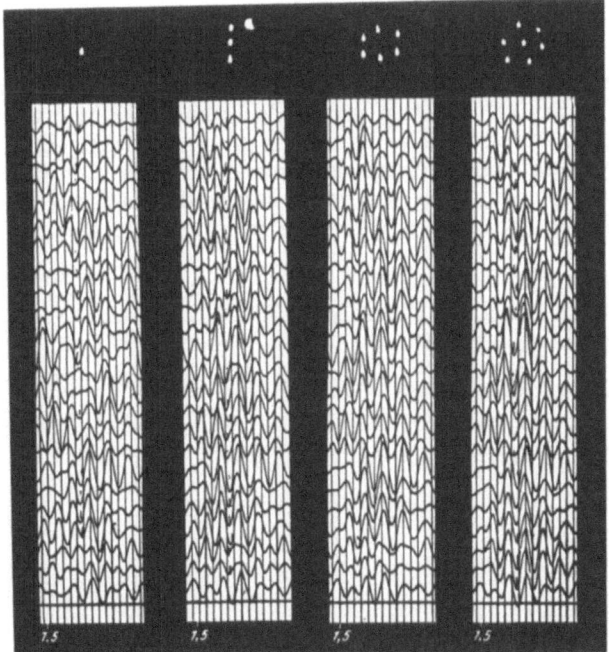

Fig. 8. Improvement of a reflexion by arrangement of the charge in groups: The four sections from seismograms shown demonstrate how the reflexion from one and the same reflecting horizon changes with the distribution of the charge according to the four patterns indicated in the upper part of the figure.

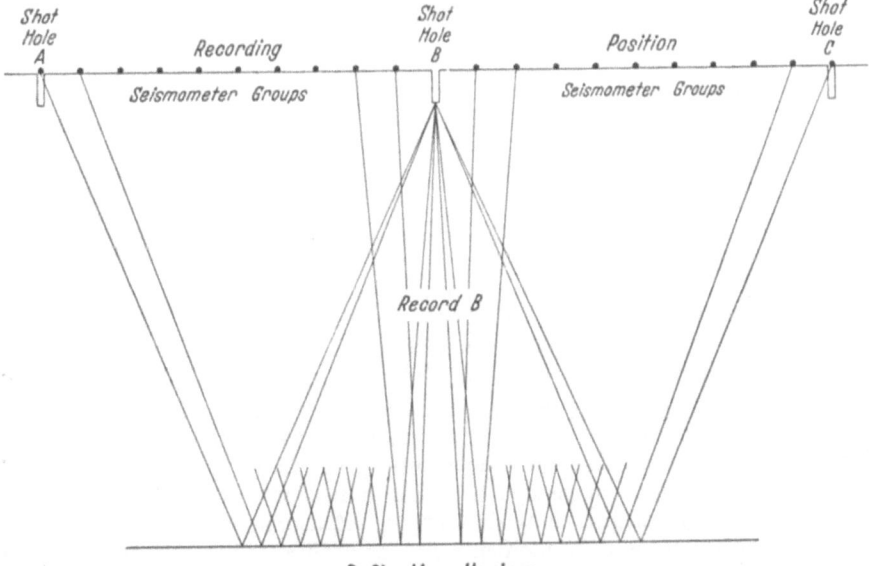

Fig. 9. Arrangement of geophones and shot holes for continuous correlation shooting: Split spread. 20 trace equipment. (Geophysical Service Inc.).

The arrangement of shots and detectors is often modified to meet local conditions. If continuous coverage is necessary, a common arrangement ("*split spread*") is that shown in Fig. 9. When the reflections maintain identifiable characteristics over wide areas it may be allowable to use *"jump correlations"* across gaps. Such an arrangement is shown in Fig. 10.

For reconnaissance surveys or where the character of the reflections changes rapidly, *"dip shooting"* is often used. Using an array of detectors along the line of traverse, and sometimes an additional one at right angles to it, depths and dips of several reflectors are determined for each shot point. The "split spread" pattern shown in Fig. 9 is commonly used to facilitate dip measurement by providing data in opposite directions from a single shot.

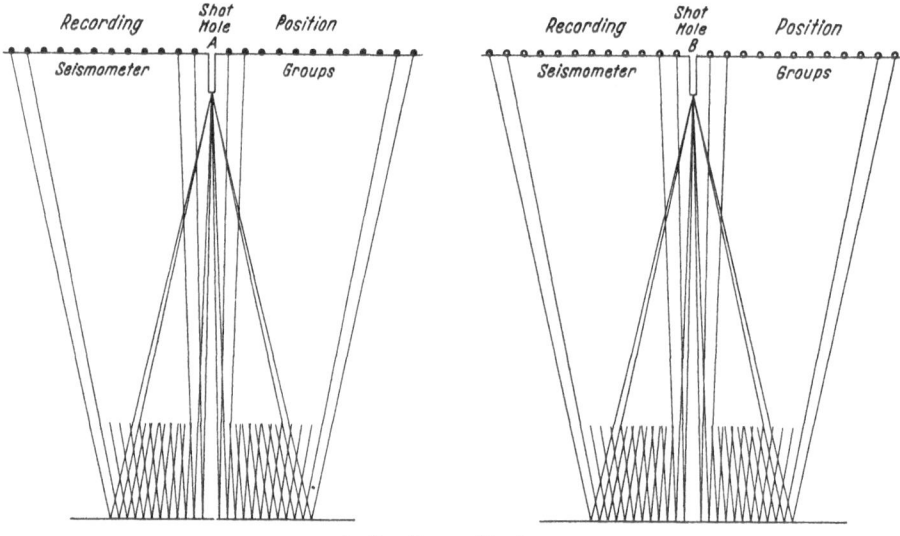

Fig. 10. Arrangement of geophones and shot holes for jump correlation. 24 trace equipment. (Geophysical Service Inc.)

The thickness of the weathered layer is variable, and the velocity of seismic waves in it is very low. Correction must be made at each shot point for the delay in the weathered layer, usually equal to the travel time to a surface detector at the shot point. In extreme cases allowance must be made for variation over the detector spread, by using the travel time to each detector of the waves received by refraction at the base of the weathered layer. Except at the shortest distances these are the first waves received.

Reflection shooting is widely used in *water covered areas*, principally in depths up to 200 ft. (Fig. 11). A single ship serves for shooting and recording. Charges are suspended from floats at such a depth that the gas bubbles break the water surface. Detectors are either electromagnetic geophones or piezoelectric transducers. They are either suspended and maintained at a small constant depth by surface floats or dragged along the bottom. Measurements are rapidly made as the ship either stops only momentarily for each shot, or has provision for recording while under way. Locations are determined by the use of various electronic navigating devices.

*γ) Interpretation Procedures.* To convert reflection times, after correction for weathering and shot and detector elevations, it is necessary to know the relation between velocity and depth. This is best done by measurement of the time of travel of direct waves from a surface explosion to detectors in a *bore*

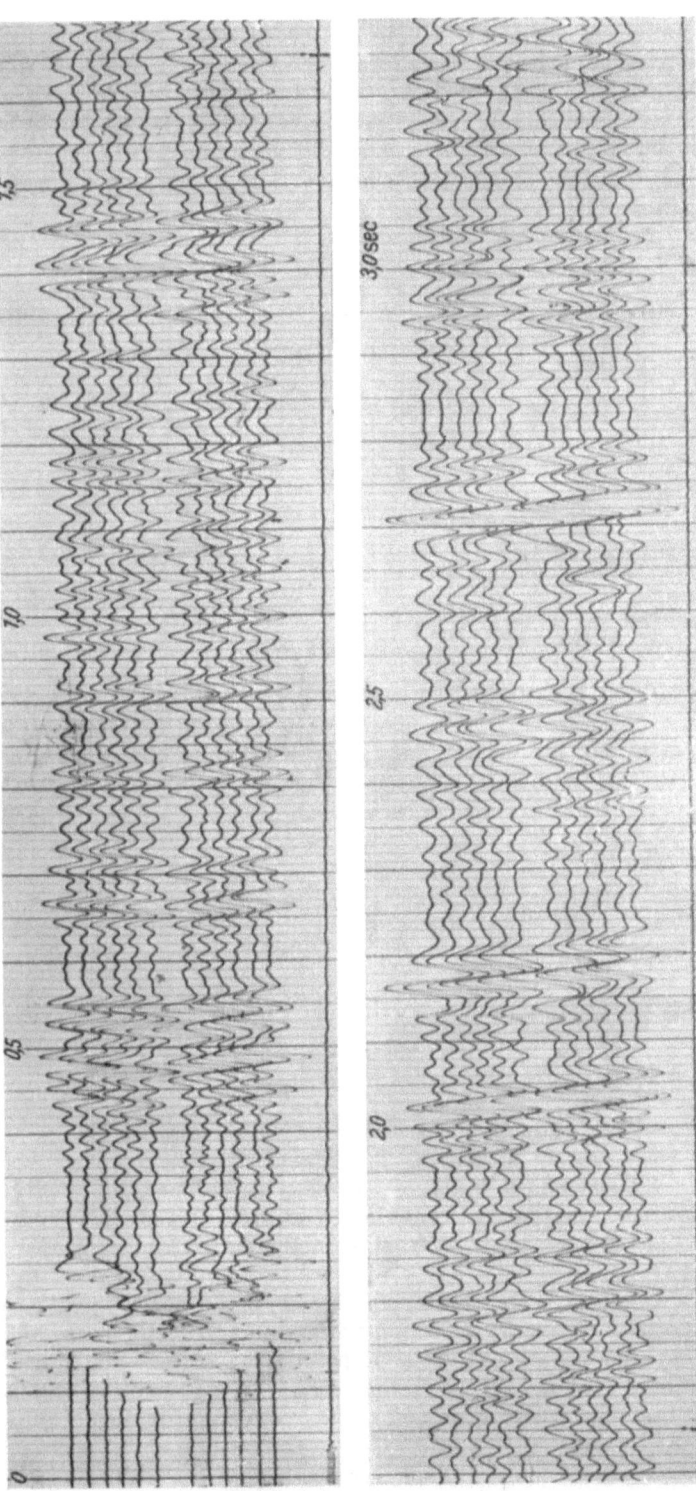

Fig. 11. The first 3.3 sec. of a reflection seismogram from the North Sea obtained with a 12 trace arrangement, showing reflections from horizons below the sea bottom down to 4500 m. (Courtesy Prakla-Hannover.)

*hole.* A typical arrangement is shown in Fig. 12, together with graphs of interval velocity and average velocity. An alternative method which is used when bore holes are not available is the plot of $T^2$ as a function of $x^2$. In the absence of dip it may be seen from equation (2.1) that the plot should give a straight line with slope $1/V^2$ if the velocity is constant. Usually the points depart very little from a straight line, the slope of which is taken to give the average velocity $\overline{V}$.

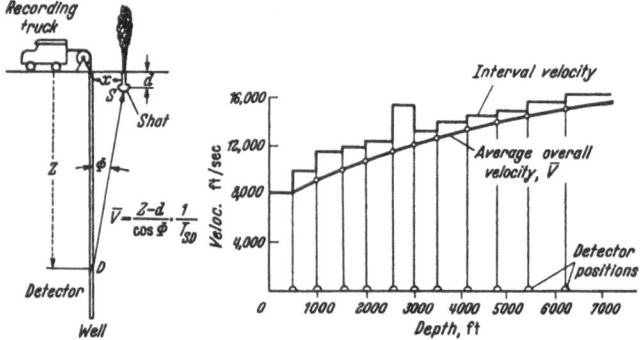

Fig. 12. Typical arrangement for hole shooting. Velocity-depth graph gives interval velocity and average overall velocity $V$. (M. B. Dobrin, Copyright 1952, McGraw-Hill Book Co.)

A more commonly used method is the "$T\,\varDelta T$" method, which makes use of equation (2.2) in the form

$$V^2 = L^2/\varDelta T\,(T_0 + T_1) \qquad (2.3)$$

where $T_0$ and $T_1$ are the reflection times at the two ends of the detector spread, and $\varDelta T = T_1 - T_0$. Here $L$ is the length of the spread, one end of which is at the shot point. When there are many $T\,\varDelta T$ velocity determinations in a small area, the average of these is usually found to have errors of only a few percent when compared with the results of well shooting[1].

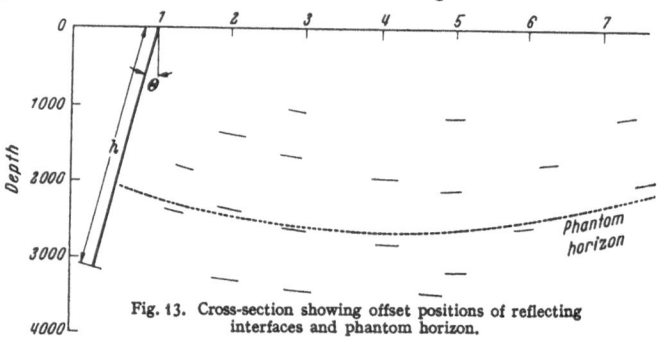

Fig. 13. Cross-section showing offset positions of reflecting interfaces and phantom horizon.

When the velocity-depth function has been determined for a given area, corrected reflection times can be converted to depths. This is often done graphically, with the aid of a chart. An alternative method is to plot corrected reflection times directly.

When the reflecting beds dip steeply the computed depth must be plotted in an offset position, since the reflection path is not vertical. In an area where the record character changes as the shot point is moved, so that no reflection may be correlated over a large area, the method of dip shooting and *phantom horizon* is used, as shown in Fig. 13; on each seismogram the dips of all suitable

[1] B. G. Swan and A. Becker: Comparison of velocities obtained by Delta-Time Analysis and well velocity surveys. Geophysics 17, 575—597 (1952).

reflections are read and plotted on a cross section, offset if necessary. Phantom horizons are constructed by drawing a series of curves at chosen depths, each curve being everywhere tangent to the local dip segments.

It is not uncommon to find *multiple reflections* on seismograms[1]. These waves have undergone one or more additional reflections either from the surface of the earth or from some lower interface. When recognized and identified, multiple reflections can aid the interpretation. Otherwise they complicate the interpretation by introducing spurious reflecting horizons.

**3. Seismic Refraction Measurements.** In the earliest days of seismic prospecting, the refraction method was the only one available. In the form of *"fan shooting"*

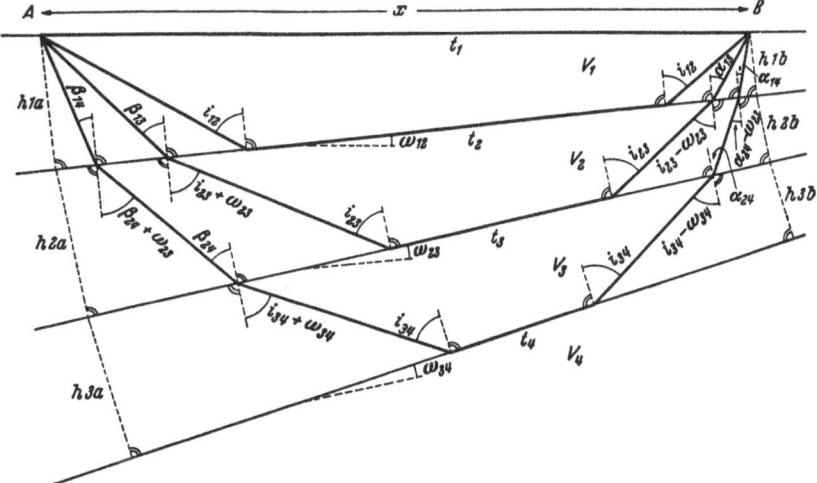

Fig. 14. Refraction paths for the case of four layers with sloping interfaces.

it provided an efficient and successful method for locating shallow salt domes. In later years it was almost entirely replaced by the reflection method, but it is now used frequently in difficult areas.

*α) Physical Principles.* The equations for reflection and transmission of elastic waves at an interface (p. 107) apply only to plane waves, but are adequate for almost all problems in reflection seismology. The propagation paths of the principal waves used in refraction seismology are shown in Fig. 14. It is necessary to consider the curvature of the wave front to account for the transmission of energy along such a path. It can be shown[2] that the amplitude of the refracted wave generated by an impulsive point source decreases with distance as $r^{-2}$ in contrast to the $r^{-1}$ decrease of dispersed surface waves and of body waves in an infinite medium.

For all calculations except those of energy and dependence upon frequency, we may use the concept of energy propagation along *rays* for problems of refraction seismology. By plotting arrival time as a function of distance a *travel time curve* like that shown in Fig. 15 is obtained where a four-layered structure occurs. The velocities of propagation in the layers may be obtained from the slopes of the travel-time curve, and the layer thickness from these and the intercepts $T_{2a}, T_{2b}$ etc.

---

[1] Symposium on multiple reflections. Geophysics **13**, 1—91 (1948).

[2] H. HONDA and H. NAKAMURA: On the Reflection and Refraction of the Explosive Sounds at the Ocean Bottom, II, Sci. Reports, Tohoku Univ. Sendai, Jap. **16**, 70—84 (1954).

In general, many layers are present and the interfaces between them are inclined. The slopes of the travel time curve then give only apparent velocities, and it is necessary to observe travel times in opposite directions (reversed profiles, as in Fig. 15) along the line of survey in order to obtain a solution. If there are several layers in the earth, each with higher sound velocity than those above it, the travel time curve will include a line for each layer. If one uses only the first arriving wave for each shot (heavy lines in Fig. 15), each layer is represented

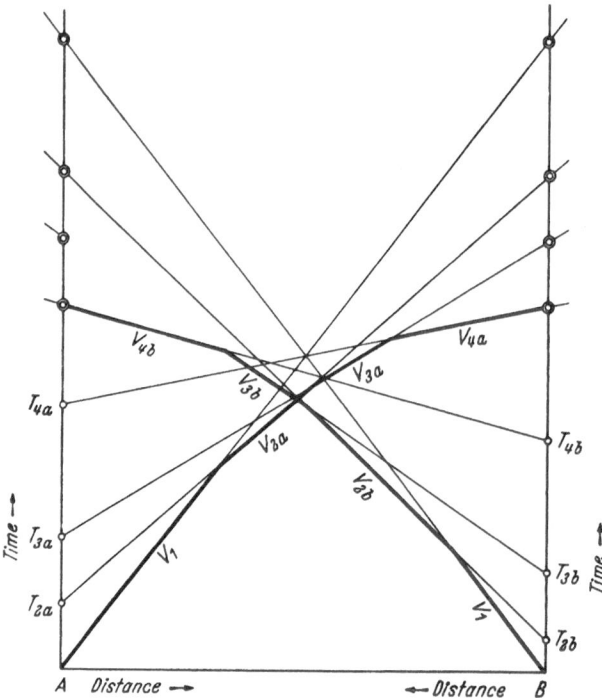

Fig. 15. Travel time curves for a reversed refraction profile. Heavy line segments indicate first arrivals.

at best by only a short segment of the travel time curve, and some layers may be entirely "masked" so that the waves through them can never arrive ahead of all others. This occurs when the layer is very thin or when the one beneath it has a far greater sound velocity. By identifying waves arriving later in the seismogram, one may often extend the range of observation of each segment of the travel time curve and investigate masked layers. If a layer has a sound velocity *lower* than that above it, it cannot be investigated or detected by standard refraction methods and will introduce an error into the depth determination for all deeper beds.

Formulae[1] for obtaining velocities, layer thicknesses, and slopes may be derived on the basis of the following statements:

1. Each layer is bounded by planes and transmits elastic waves at a constant velocity.

2. At the interface between two layers the path of the seismic waves is bent according to Snell's law.

[1] M. Ewing, G. P. Woolard and A. C. Vine: Geophysical investigations in the emerged and submerged Atlantic coastal plain, Part III. Bull. Geol. Soc. Amer. 50, 257—296 (1939).

3. A wave travelling in any layer with velocity $V$ and incident on the surface of the layer at an angle $\alpha$ with the normal, has an apparent velocity $V/\sin \alpha$ along the surface.

4. A travel time curve is unchanged if shot point and recording point are interchanged.

If the angle $\omega_{ij}$ represents the dip of the $ij$ interface with respect to the one above and $\alpha, \beta, i$ are defined in Fig. 14, then all angles and true velocities may be determined from equations derived on the basis of statements (2.2) and (2.3):

$$\frac{V_1}{V_{2a}} = \sin (i_{12} - \omega_{12}), \qquad \frac{V_1}{V_{2b}} = \sin (i_{12} + \omega_{12}), \qquad \frac{V_1}{V_2} = \sin i_{12},$$

$$\frac{V_1}{V_{3a}} = \sin (\alpha_{13} - \omega_{12}), \qquad \frac{V_1}{V_{3b}} = \sin (\beta_{13} + \omega_{12}),$$

$$\frac{V_1}{V_2} = \frac{\sin \alpha_{13}}{\sin (i_{23} - \omega_{23})} = \frac{\sin \beta_{13}}{\sin (i_{23} + \omega_{23})}, \qquad \frac{V_2}{V_3} = \sin i_{23} \quad \text{etc.}$$

Equations relating the time intercepts to layer thicknesses are readily derived:

$$T_{2a} = 2h_1 \cos i_{12}/V_1$$

$$T_{3a} = 2h_2 \cos i_{23}/V_2 + h_1 (\cos \alpha_{13} + \cos \beta_{13})/V_1$$

$$T_{4a} = 2h_3 \cos i_{34}/V_3 + h_2 (\cos \alpha_{24} + \cos \beta_{24})/V_2 + h_1 (\cos \alpha_{14} + \cos \beta_{14})/V_1 \quad \text{etc.}$$

When the velocity varies continuously with depth the slope of the travel time curve varies continuously with distance, and the methods of Chapter 3 must be used to derive the velocity-depth function.

$\beta)$ *Field Procedures:* The equipment used for refraction measurements differs from that used for reflections in several respects. A single detector is usually used to supply signal to a single galvanometer trace. The detectors and amplifiers emphasize lower frequencies, usually about 2 to 40 cycles per sec. The maximum shot to detector distances are usually 10 to 20 miles, hence radio is used for transmission of shot instant and general communication and charges up to 600 lbs. are required. Spacing between geophones in the spread is often about 200 ft., and as many as 24 are used. Profiles are usually reversed by having the detector spread unchanged while shooting first beyond one end, then beyond the other end of it.

It is customary when undertaking refraction measurements in a new area to make observations for a reversed travel time curve to a distance of at least three times as great as the depth of the deepest bed which is of interest. Once the arrivals from this bed have been identified, and the range of distances for which it is well observed has been established, subsequent profiles in the area include data for this range of distances only.

A refraction seismogram recorded 5 miles from a 5 lb. shot is shown in Fig. 16. In the most widely used arrangement shots and detectors are located along a straight line or "profile". Both shot points and detectors are shifted progressively along the line in such a way as to keep within the range of detectable arrivals from the bed being mapped. Either first or second arrivals may be used. As in any form of surveying it is highly desirable to combine lines of survey into closed polygons and check for closure.

A technique used in the earliest seismic surveys for locating shallow salt domes is illustrated in Fig. 17. A normal travel time curve for the area is first obtained from a standard profile. A shot point is chosen as the center of a roughly circular array of detector points, the radius of the circle determining the effective

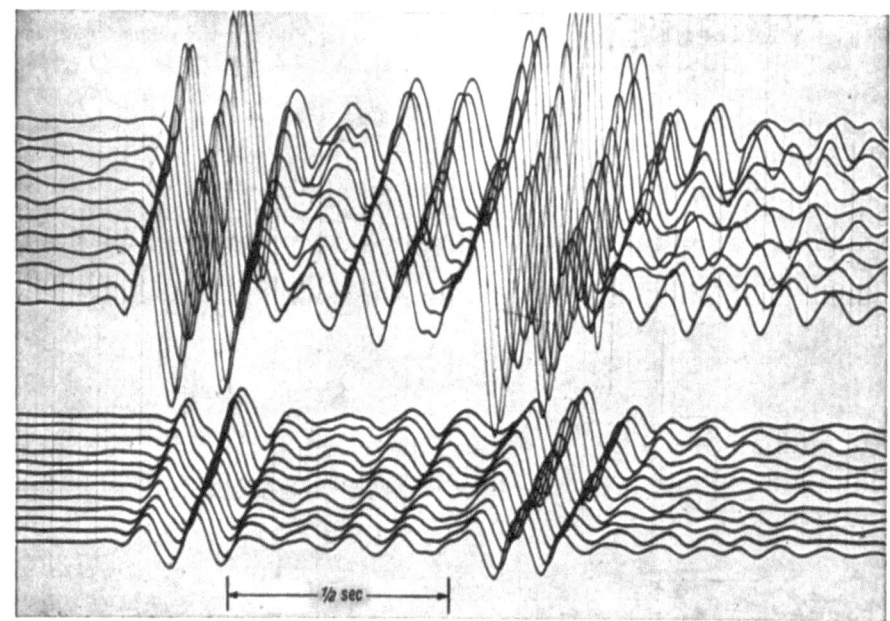

Fig. 16. Refraction seismogram recorded 5 miles from 5 lb. shot. Seismometer spacing 100 feet, frequency pass band 2—24 cycles/sec. (Houston Technical Laboratories.)

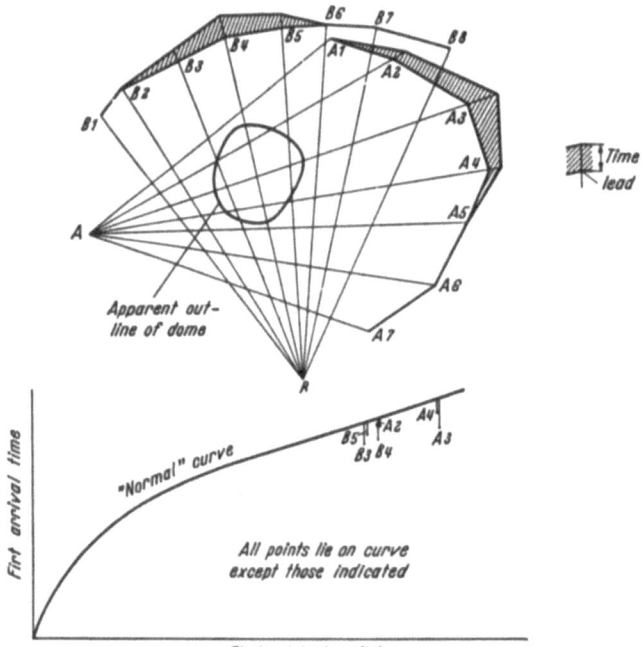

Fig. 17. Fan shooting method of locating salt dome. Time leads from shots *A* and *B* are plotted on map to indicate dome location. (M. B. Dobrin, Copyright 1952, McGraw-Hill Book Co.)

depth searched. The travel time anomaly for each detector position is found by comparison of the observed value with the normal curve, and the anomalies are plotted on a map as shown in Fig. 17. Due to the high velocity of sound in salt, a very obvious pattern of anomalies (travel times several tenths of a second less than normal) was produced by a shallow salt dome[1].

Methods for seismic refraction measurements in oceanic areas will be discussed in the Chapter on Structure of the Earth's Crust.

*γ) Methods of Interpretation.* The first step in reducing refraction data is to apply elevation and weathering corrections as discussed in Sect. 2c. In refraction shooting it is desirable to make auxiliary measurements with small auxiliary shots at short distances from each geophone spread in order to determine the local thickness of the weathered layer. For an isolated profile, the method of

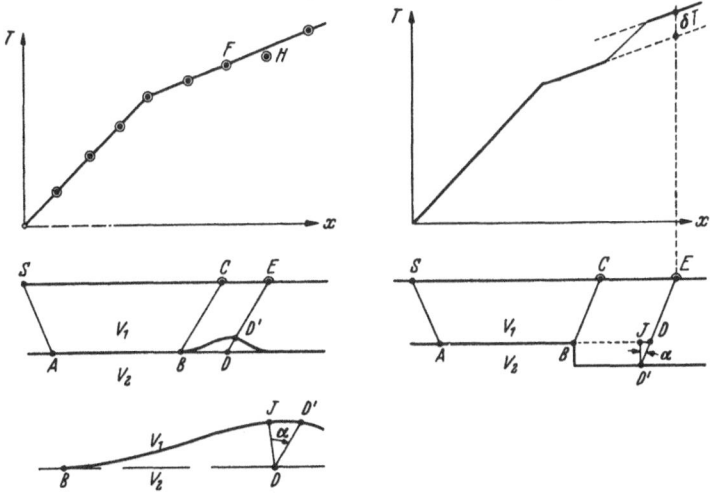

Fig. 18. Effect of small elevation (left) or fault (right) on refraction travel time curve. Deviation of travel time $\partial T$ from normal refraction curve gives magnitude of elevation or throw of fault. (C. H. Dix, Courtesy Harper & Bros.)

interpretation follows that outlined in Sect. 3a, but when a selected bed is to be "detailed" by continuous or closely spaced profiles the deviation of each travel-time from a standard curve is the quantity from which structure is deduced. Fig. 18 shows the effect of a small elevation in the marked bed, and of a fault. It may readily be shown that if $\partial T$ is the travel time anomaly, the deviation $\overline{DD'}$ in the bed is given by

$$\overline{DD'} = -V_1\,\partial T/(1 - V_1^2/V_2^2)$$

which must be offset by $h\,(V_2^2/V_1^2 - 1)^{-\frac{1}{2}}$.

An elaboration of this method uses quantities known as *"delay times"*[2]. The intercept time $T - x/V_1$ is the sum of the delay times $D_1$ and $D_2$ at the shot and detector respectively:

$$D_1 = h_1\,(V_2^2 - V_1^2)^{\frac{1}{2}}/V_2 V_1$$
$$D_2 = h_2\,(V_2^2 - V_1^2)^{\frac{1}{2}}/V_2 V_1.$$

After determining the delay time at one point, the delay times at all other points can be obtained from the intercept times. Variations in delay time are sensitive

---

[1] A great number of salt-domes were detected in North-West Germany by this method. See the map of travel-times for standard distance 4 km. shown in H. REICH, Geophysikalische Lagerstättenforschung. In Naturforschung und Medizin in Deutschland 1939—1946 (= FIAT Review of German Science), Vol. 17, Geophysik, Teil I. Wiesbaden 1948.

[2] L. W. GARDNER: An areal plan of mapping subsurface structure by refraction shooting. Geophysics **4**, 247—259 (1939).

indications of variations in depth of the marked bed provided no unsuspected changes in mean velocity of overlying beds is encountered.

**4. Engineering and Mining Applications.** Under engineering applications are included search for water table and aquifers and determination of bedrock depth for foundation design or roadbed planning[1]. The objective is to provide

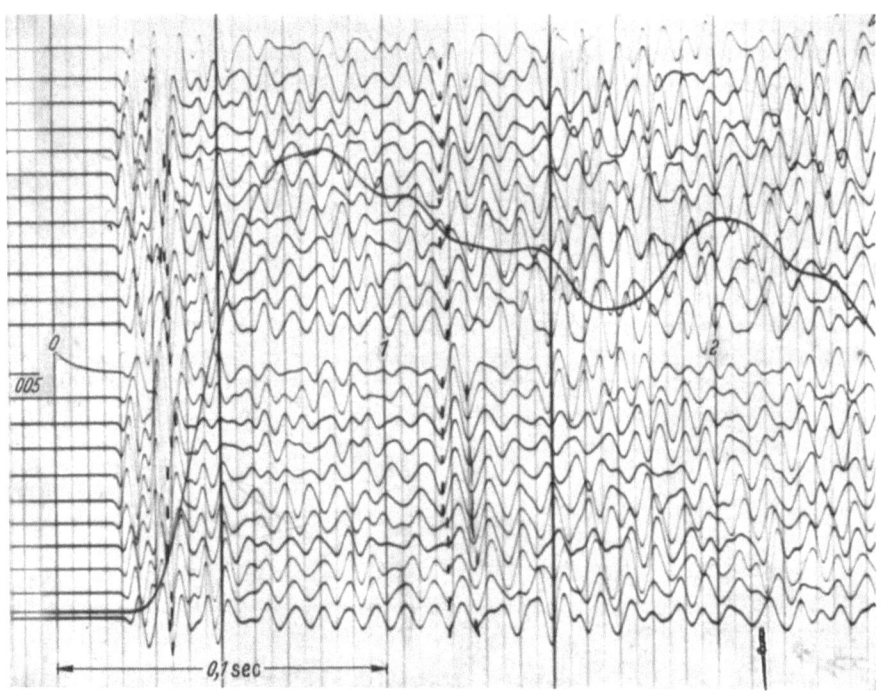

Fig. 19. A high resolution seismogram made in connection with ground water survey. Shallow reflection occurs at 125 milliseconds. Charge 4 lbs. at 16 feet depth. (Houston Technical Laboratories.)

rapid reconnaissance and to minimize the need for drilling. Applications to mining occasionally involve direct detection of an ore body, but usually reduce to mapping bedrock configuration. The modification of the seismic technique is principally one of scale, for in almost all of these problems the depths and lateral dimensions range from a few feet to a few hundred feet.

The modifications in apparatus include faster paper speeds and higher frequency response to permit precise measurement of short travel times. Refraction measurements are principally used, but in some instances reflection methods have met with success. Portability of the apparatus is essential. Explosive charges are generally small, occasionally being replaced by falling weights or hammer blows.

A high resolution recording reflection seismogram from a ground water survey is shown in Fig. 19.

---

[1] C. F. ALLEN, L. V. LOMBARDI and W. M. WELLS: The application of the reflection seismograph to near-surface exploration. Geophysics **17**, 859—866 (1952).

# Messung elastischer Eigenschaften von Gesteinen.

Von

HEINRICH BAULE und ERICH MÜLLER.

Mit 26 Figuren.

**1. Einleitung.** Die Ausbreitungs-Geschwindigkeiten elastischer Wellen und die elastischen Konstanten von Gesteinen hängen von mehreren Parametern ab, unter anderem von der Mineral-Zusammensetzung, der Druckbeanspruchung, dem Feuchtigkeitsgehalt und dem Zersetzungsgrad der Proben. Grundsätzlich sind Messungen *in situ*, also am Gesteinskörper unter seinen natürlichen Bedingungen in der Natur möglich und vorzuziehen. Demgegenüber besitzen Untersuchungen an Proben *im Laboratorium* vor allem meßtechnische Vorteile, wobei jedoch auf die oben genannten Einflüsse geachtet werden muß, wenn man die an kleinen Proben gewonnenen Ergebnisse auf das umgebende Gestein übertragen will.

Unsere Ausführungen befassen sich im wesentlichen mit den Untersuchungen im Laboratorium und schließen nur neuartige Messungen in situ ein. Die Laboratoriumsmethoden gleichen in vieler Hinsicht denjenigen, die zur Untersuchung der Schallausbreitung in Metallen oder Kunststoffen und zur Bestimmung ihrer elastischen Konstanten angewendet werden. Im Gegensatz zu den meisten Metallen muß bei Messungen an Gesteinen berücksichtigt werden, daß das HOOKEsche Gesetz nicht streng befolgt wird und die elastischen Eigenschaften sich oft auf kleinstem Raum ändern können. Die Dehnungs-Spannungs-Abhängigkeit wird hier durch eine Kurve dargestellt, die der magnetischen Hysteresis ähnlich ist. Die Elastizitätsgrößen können bei räumlich wenig auseinanderliegenden Proben selbst dann streuen, wenn keinerlei mineralogisch-petrographische Veränderungen wahrzunehmen sind, so daß die Meßwerte oft nur auf die unmittelbare Umgebung der untersuchten Probe übertragen werden können.

Der Elastizitäts-, der Scherungs- und der Kompressionsmodul, sowie die POISSONsche Zahl, können mittels statischer, dynamischer oder Impulslaufzeit-Methoden bestimmt werden, bei den letztgenannten Verfahren auf dem Umweg über die Messung der Geschwindigkeit von longitudinalen und transversalen oder Biege-Wellen. Der vorliegende Beitrag befaßt sich im ersten Abschnitt (A) mit statischen Untersuchungen, besonders mit den wichtigen Kompressibilitäts-Messungen. In den folgenden Abschnitten werden dann Methoden zur dynamischen Bestimmung der Elastizitätskonstanten und zur Messung der Ausbreitungs-Geschwindigkeit elastischer Wellen beschrieben. Bei den dynamischen Verfahren (Abschnitt B) wird die gesamte stabförmige Probe in freie oder erzwungene Schwingungen versetzt, um die Eigenfrequenz des Prüflings zu bestimmen. Im Abschnitt C werden dann moderne Ultraschall-Methoden angeführt, mit denen die Gruppengeschwindigkeit hochfrequenter Wellen, die sich im Untersuchungskörper ausbreiten, gemessen wird. Diese Gliederung entspricht auch der zeitlichen Entwicklung der Methodik zur Messung elastischer Eigenschaften von Gesteinen.

## Bezeichnungen.

| | |
|---|---|
| $E$ = Youngscher Elastizitätsmodul | $p$ = Druck |
| $G$ = Scherungsmodul | $\varepsilon$ = relative Längenänderung $\Delta l/l$ |
| $\sigma$ = Poissonsche Zahl | $\varepsilon_q$ = Querkontraktion, relative Änderung |
| $\varkappa = K^{-1}$ = .Kompressibilität | des Stabdurchmessers |
| $K$ = Kompressionsmodul | $\Delta V/V$ = relative Volumenänderung |
| $V_l$ = Longitudinalgeschwindigkeit im unend- | $\tau_N$ = Normalspannung |
| lich ausgedehnten Medium | $\tau_s$ = Schubspannung |
| $V_t$ = Transversalgeschwindigkeit im unendlich | $\alpha$ = Schubwinkel |
| ausgedehnten Medium | $f$ = Frequenz |
| $V_d$ = Geschwindigkeit der Stabdehnwelle | $\lambda$ = Wellenlänge |
| $\varrho$ = Dichte | |

Krafteinheit kp (= Kilopond = 0,980665 · 10⁶ dyn) oder p (= pond).

# A. Statische Methoden.

**2. Einfluß der elastischen Hysteresis.** Die elastischen Konstanten fester Körper können alle oder einzeln statisch gemessen werden, indem der Untersuchungskörper einseitigem oder allseitigem Druck oder Zug, sowie Biegung und Torsion ausgesetzt wird. Hierbei mißt man die *Spannung*, die das Volumen oder die Gestalt der Probe ändert, und die *Deformation* des Körpers. Wird völlig elastisches und isotropes Verhalten vorausgesetzt, so berechnen sich die elastischen Konstanten je nach der Art des Versuches aus den verschiedenen Formen des Hookeschen Gesetzes.

Für eine zylinderförmige Probe folgt, bei einseitiger Druck- oder Zugbeanspruchung, $E$ aus

$$E = \tau_N/\varepsilon, \tag{2.1}$$

bei reiner Torsionsbeanspruchung $G$ aus

$$G = \tau_s/\alpha \tag{2.2}$$

und bei allseitigem Druck $\varkappa$ aus

$$\varkappa = K^{-1} = - (\Delta V/V)/p. \tag{2.3}$$

Schließlich kann $\sigma$ aus dem Verhältnis

$$\sigma = - \varepsilon_q/\varepsilon \tag{2.4}$$

bestimmt werden.

Bei unvollkommen elastischen Festkörpern, zu denen Gesteine zählen, hängen die Meßergebnisse mehr oder minder stark vom Betrag der benutzten Spannung ab. Wird diese als Abszisse und die Deformation als Ordinate in einem rechtwinkligen Koordinatensystem aufgetragen, so stimmt die Kurve für *abnehmende* Belastung nicht mit derjenigen für *wachsende* überein. Geht man von positiven Spannungen kontinuierlich zu negativen über und wiederholt den Meßvorgang dann in umgekehrter Richtung, so liefert das Dehnungs-Spannungsdiagramm eine als „*elastische Hysteresis*" bezeichnete geschlossene Schleife. Hierin drückt sich die Abweichung des Untersuchungskörpers vom ideal-elastischen Zustand aus.

Bei Gesteinen sind die Verhältnisse nach Zisman[1] noch komplizierter, da die Hysteresis ihre Form ändert, wenn die Kurve mehrfach durchlaufen wird. Insbesondere nimmt die Breite der Hysteresis bei den ersten beiden Umläufen stark ab, um dann ungefähr konstant zu bleiben. Zisman diskutiert auch die Frage,

---

[1] W. A. Zisman: Young's Modulus and Poisson's Ratio with Reference to Geophysical Applications. Proc. Nat. Acad. Sci. **19**, 653 (1933).

wie man den Elastizitätsmodul und die POISSON-Zahl aus einer solchen Kurve berechnet.

Ein Beispiel für eine elastische Hysteresis veranschaulicht Fig. 1. Rechts im Bilde ist das Schema der zugehörigen Versuchsanordnung wiedergegeben. Die auf der Abszisse aufgetragene Kraft wurde hierbei sinusförmig mit einer Periode von 400 sec (Dauer eines Umlaufs) verändert und dazu der jeweilige Ausschlag $x$ gemessen[1,2].

Auch die Porosität und die Klüftung wirken sich erheblich auf die Elastizität des Gesteins aus. So wird bei niedrigen Drucken eine höhere Kompressibilität

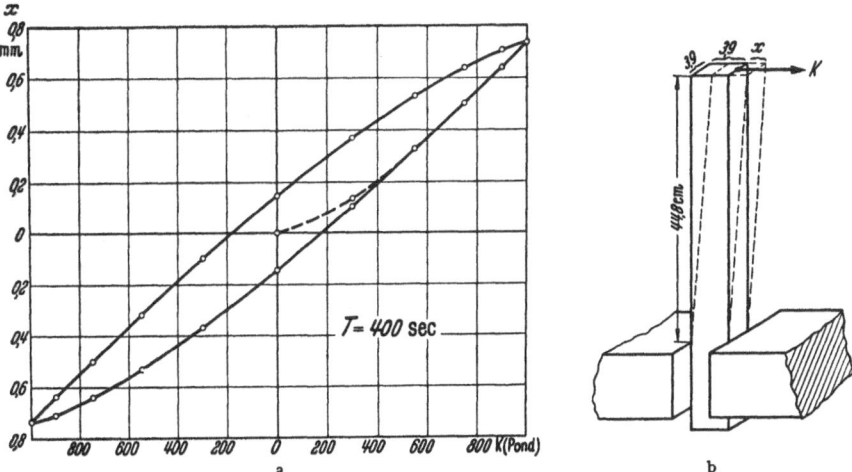

a                    b

Fig. 1 a u. b. a Die elastische Hyteresis eines porösen trockenen Sandsteins aus Ostermundigen bei Bern (obere Meeres-molasse, $\varrho = 2,2$ g/cm³). b Die zugehörige Versuchsanordnung. $K = $ Kraft in Pond ($=$ Gramm-Gewicht), $x = $ Ausschlag in mm. (Nach GASSMANN, WEBER und VÖGTLI.)

vorgetäuscht, da die Hohlräume zusammengepreßt werden und sich das Gesamt-volumen stärker ändert, als bei einem gleichen Material in kompakter Form. Aus diesen Gründen hat man schon frühzeitig erkannt, daß den statischen Gesteins-untersuchungen manche prinzipielle Mängel anhaften. Obwohl ausdrücklich an-erkannt wird, daß wir den statischen Messungen wertvolle Erkenntnisse über das elastische Verhalten unserer Gesteine und den Aufbau unserer Erdkruste verdanken, müssen heute die meisten dieser Untersuchungen als überholt be-zeichnet werden. Einige Bedeutung haben nach wie vor nur die Kompressibili-täts-Messungen an Gesteinen und statische Spezialuntersuchungen, zu denen Messungen am Gesteinskörper in situ und gesteinsmechanische Bestimmungen der Fließ- und Bruchgrenze an Proben im Laboratorium gehören.

**3. Messung der Kompressibilität.** Die statische Kompressibilität von Gesteinen und Mineralien wurde unter anderen von ADAMS[3], ADAMS-WILLIAMSON[4], ZISMAN[5],

[1] F. GASSMANN, M. WEBER u. K. VÖGTLI: Beitrag zur Ermittlung der inneren Dämpfung (Werkstoffdämpfung) von Gesteinsstäben. Mitt. Inst. Geophys. E.T.H. Zürich **1952**, Nr. 22.

[2] Über dynamischen Messungen der elastischen Hysteresis siehe Ziff. 14.

[3] L. H. ADAMS: Elastic Properties of Materials of the Earth's Crust. Internal Constitution of the Earth, ed. by Gutenberg, 2nd edit., Chapt. IV. 1949. — A Simplified Apparatus for High Hydrostatic Pressures. Rev. Sci. Instrum. **7**, 174 (1936).

[4] L. H. ADAMS u. E. D. WILLIAMSON: The Compressibility of Minerals and Rocks at High Pressures. J. Franklin Inst. **195**, 475 (1923).

[5] W. A. ZISMAN: Compressibility and Anisotropy of Rocks at and near the Earth's Surface. Proc. Nat. Acad. Sci. **19**, 666 (1933).

Griggs[1], Birch[2], Bancroft[3] und Bridgman[4] gemessen. Dazu wird die zylindrische Probe in eine Stahlkammer gebracht, die mit Flüssigkeit gefüllt ist, und in die man einerseits einen abgedichteten Kolben, andererseits ein Widerlager einführt. Der auf die Flüssigkeit ausgeübte Druck deformiert nicht nur den Prüfling, sondern auch die Flüssigkeit und die Wände der Kammer. Um diese Fehlerquellen auszuschalten, pflegt man Vergleichsmessungen an einer Stahlprobe mit genau bekannter Kompressibilität durchzuführen. Wenn diese Korrektionen beachtet werden, so liefert der gemessene Kolbenhub direkt die relative Volumenabnahme der Probe und man berechnet die *kubische Kompressibilität* aus Gl. (2.3).

Bridgman baute eine Apparatur zur genauen Messung der Volumenänderung mit Hilfe von Kapillaren, die mit der Kammer verbunden sind, und in denen Flüssigkeitshöhen abgelesen werden. Ein mit dem Druckstempel starr verbundener Ausgleichskolben eliminiert den Einfluß der bei Belastung verdrängten Flüssigkeit.

An Stelle der Volumenabnahme mißt man häufig die *lineare Kompression* der Proben. Diese Methode ist empfindlicher und von Bridgman zu höchster Vollkommenheit entwickelt worden. Hierbei wird die Längenänderung der Probe mit Hilfe von Dehnungs-Meßstreifen umgewandelt in eine geeichte elektrische Widerstandsänderung, die in Brückenschaltung gemessen wird.

Wenn der Prüfling isotrop ist, kann die kubische Kompressibilität aus der linearen berechnet werden. Gilt für die lineare Längenänderung als Funktion des Druckes $p$ die Beziehung

$$- \Delta l/l = a \, p - b \, p^2,$$

wo $a$ und $b$ Konstante sind, so folgt für die relative Volumenverkleinerung

$$- \Delta V/V = 3 \, a \, p - 3 \, (a^2 + b) \cdot p^2.$$

Für die heterogen aufgebauten Gesteine ist die „Volumenmethode" in der Regel nicht nur ausreichend genau, sondern auch besser geeignet als die „lineare Methode", die höhere Anforderungen an die Homogenität der Proben und an die Isotropie ihres elastischen Verhaltens stellt.

In der Stahlkammer lassen sich Drucke bis etwa 15000 Atm erzeugen. Höhere Drucke bis 50000 Atm können angewendet werden, wenn man die Druckkammer außen konisch formt und sie in einen konischen Stützmantel aus Stahl preßt, während der Druck innen zunimmt. Dabei reicht die Festigkeit von Stahl nicht mehr aus und man verwendet Carboloy, einen gesinterten Stoff aus Wolframkarbid und Kobalt.

---

[1] D. T. Griggs: Deformation of Rocks under High Confining Pressures. J. Geology **44**, 541 (1936). — Deformation of Single Calcite Crystals under High Confining Pressures. Amer. Mineral. **23**, 28 (1938).

[2] F. Birch: The Effect of Pressure upon the Elastic Parameters of Isotropic Solids. J. Appl. Phys. **9**, 279 (1938). — F. Birch u. R. R. Law: Measurements of Compressibility at High Pressures and High Temperatures. Bull. Geol. Soc. Amer. **46**, 1219 (1935). — F. Birch u. R. B. Dow: Compressibility of Rocks and Glasses at High Temperatures and Pressures. Bull. Geol. Soc. Amer. **47**, 1235 (1936).

[3] F. Birch u. D. Bancroft: The Effect of Pressure on the Rigidity of Rocks. I. J. Geology **46**, 59 (1938). — II. J. Geology **46**, 113 (1938).

[4] P. W. Bridgman: The Measurement of Hydrostatic Pressure to 30000 kg/cm². Proc. Amer. Acad. Arts Sci. **74**, 1 (1940). — Recent Work in the Field of High Pressures. Sci. Progr. **3**, 108 (1945). — The Compression of 39 Substances to 100000 kg/cm². Proc. Amer. Acad. Arts Sci. **76**, 55 (1948). — Linear Compression to 30000 kg/cm², including relatively incompressible Substances. Proc. Amer. Acad. Arts Sci. **77**, 187 (1949). — Viele weitere Arbeiten ebenda in früheren Jahrgängen, zusammenfassender Überblick: Endeavour **10**, 63 (1951).

BRIDGMAN erzeugt auch Drucke bis $10^5$ Atm, indem er die Druckkammer mit Flüssigkeit umgibt, die selber hohem hydrostatischen Druck ausgesetzt wird. Kammer und Kolben sind hierbei aus Carboloy gefertigt. Bei diesen extrem hohen Drucken erstarrt die Flüssigkeit in der Druckkammer zu einem Festkörper, so daß man die Flüssigkeit von vornherein durch weiches Zinn oder Indium ersetzt, um den — nun nicht mehr rein hydrostatischen — Druck auf die Probe zu übertragen.

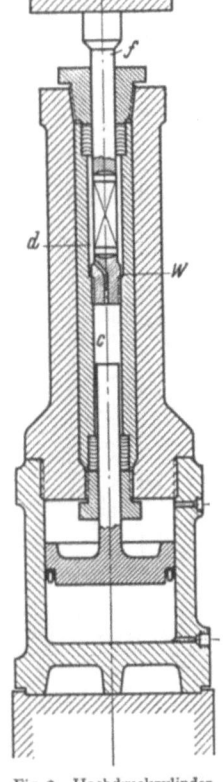

Eine alte Versuchsanordnung von KÁRMÁN[1] gewinnt erneut Bedeutung und sei nachfolgend beschrieben. Der Prüfling wird hierbei nicht nur einem allseitigen (Manteldruck), sondern gleichzeitig auch einem einseitigen (axialen) Druck ausgesetzt. Beide Drucke können unabhängig voneinander geändert und gemessen werden. Die Möglichkeit, genau definierte dreiachsige Spannungszustände im Laboratorium zu erzeugen, ist gerade für die Gebirgsdruckforschung von erheblicher Bedeutung.

Aus Fig. 2 ist zu ersehen, daß die Stahlkammer $d$ auf der unteren Seite von einem durchbohrten Widerlager $W$ begrenzt ist, und durch diese Bohrung pflanzt sich der in $c$ erzeugte Manteldruck in den Raum $d$ fort und wirkt allseitig auf die Probe ein, die sich in $d$ befindet. Die Probe wird außerdem an ihren Enden von dem Widerlager $W$ und dem Kolben $f$ einer gewöhnlichen Druckpresse gefaßt und axial gedrückt.

Der KÁRMÁNsche Hochdruckzylinder von 50 mm Druckraum-Durchmesser bestand aus zwei warm aufeinander gezogenen Stahlrohren und war für einen Innendruck von 6000 Atm berechnet. Als Kraftmesser dienten zwei Manometer, die Längenänderungen wurden mit Mikrometerschrauben auf 0,01 mm genau gemessen. Die Proben hatten die Abmessungen 40 mm × 100 (110) mm und waren von einer Messingfolie umgeben. Untersucht wurden die Formänderungen und das elastische Verhalten von Marmor und Sandstein in Abhängigkeit vom Spannungszustand.

## 4. Messung des Elastizitätsmoduls und der POISSONschen Zahl.

Die ersten einachsigen Messungen zur Bestimmung des Elastizitätsmoduls von Gesteinen führte BAUSCHINGER[2] 1874 aus, nachdem man sich bis dahin nur auf die Prüfung der Druckfestigkeit, Wasseraufnahme usw. der bautechnisch wichtigen Gesteine beschränkt hatte. In den Jahren 1900 bis 1905 folgten NAGAOKA[3] und KUSAKABE[4] mit zahlreichen statischen und dynamischen Gesteinsuntersuchungen. Hierbei wurde bereits der Zersetzungsgrad oder, was damit meist zusammenhängt, die Durchfeuchtung des Gesteins mit berücksichtigt. Über statische Elastizitätsmessungen an Gesteinsstäben berichteten ADAMS und COKER[5] und GRAF[6]. Untersuchungen an Karbongesteinen zur Klärung von

Fig. 2. Hochdruckzylinder nach KÁRMÁN. Mantel- und axialer Druck können unabhängig voneinander erzeugt und geändert werden.

[1] TH. V. KÁRMÁN: Festigkeitsversuche unter allseitigem Druck. Z. VDI **55**, 1749 (1911). — Mitt. über Forschungsarbeiten a. d. Gebiete des Ing.-Wesens, Heft 118, S. 37, 1912.

[2] BAUSCHINGER: Untersuchungen über die Elastizität und Festigkeit der wichtigsten natürlichen Bausteine in Bayern. Mitt. a. d. mech. tech. Labor der T. H. München, Heft 4 und 10, 1874 und 1884.

[3] NAGAOKA: Elastic Constants of Rocks and the Velocity of Seismic Waves. Publ. Earthquake Invest. No. 4, Tokyo 1900.

[4] KUSAKABE: On the Modulus of Rigidity of Rocks. Publ. Earthquake Invest. No. 4, Tokyo 1900. — Modul of Elasticity of Rocks. Publ. Earthquake Invest. No. 17, 1904.

[5] F. D. ADAMS u. E. G. COKER: An Investigation into the Elastic Constants of Rocks. Carnegie Instn. Wash. Publ. **46** (1906).

[6] GRAF: Versuche über Druckelastizität. Beton u. Eisen, 399, 1926.

Gebirgsdruckfragen wurden 1930 von O. Müller[1] veröffentlicht. Ähnliche Messungen wurden 1929 von J. K. F. Breyer[2] und 1933 von W. A. Zisman[3] durchgeführt.

In sämtlichen Fällen untersuchte man Gesteinsstäbe mit kreisförmigem oder rechteckigem Querschnitt und planparallelen Endflächen, wobei die Proben einseitig unter Druck gesetzt wurden. Bestimmt wurde der Betrag der jeweiligen Spannung und die relative Längenänderung. Der Elastizitätsmodul berechnet sich dann unter Annahme idealer Elastizität aus Gl. (2.1). Der Einfluß der Hysteresis wurde erst in den Arbeiten von Zisman berücksichtigt und diskutiert.

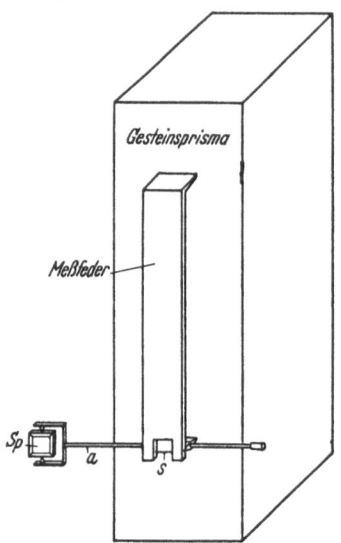

Gesteinsprisma

Meßfeder

Sp

a

s

Fig. 3. Schema des Gauss-Martensschen Spiegelapparates. Längenänderungen des Prismas werden durch den starr mit der drehbaren Schneide s verbundenen Planspiegel Sp angezeigt.

Fig. 3 veranschaulicht das Schema des von Breyer zur Messung der Längenänderung benutzten Gauss-Martensschen Spiegelapparates[4]. An das Gesteinsprisma wird eine Meßfeder gedrückt, die an der einen Seite von einer starren, an der anderen von einer drehbaren Schneide $s$ begrenzt ist, welche in ihrer Verlängerung $a$ einen kleinen Planspiegel $Sp$ trägt. Wenn die Probe unter Druck ihre Länge um einen kleinen Betrag $x$ verringert, so dreht sich die Schneide um den Winkel $\varphi = x/r$ ($r =$ Breite der Schneide). Der Spiegel dreht sich dann ebenfalls um $\varphi$ und man beobachtet bei subjektiver Betrachtung im Fernrohr das Skalenstück $A = 2\varphi L$ ($L =$ Abstand Skala—Spiegel). Für $r = 4$ mm und $L = 1$ m erhielt Breyer eine Übersetzung $A/x = 2L/r = 500 : 1$. Da man 0,1 mm auf der Skala schätzen konnte, wurden Längenänderungen von $2 \cdot 10^{-4}$ mm erfaßt.

Zisman verbesserte die Empfindlichkeit der Apparatur und konnte $\varepsilon$ auf $10^{-7}$ genau messen. Breyer arbeitete mit zwei Spiegelapparaten an den beiden gegenüberliegenden Seiten der Gesteinsprismen und bildete dann den Mittelwert aus seinen Ablesungen. Seine Gesteinsstäbe hatten die Abmessungen $6 \times 6 \times 30$ cm; die zylinderförmigen Proben von Zisman waren etwa 5 cm stark und 25 cm lang.

Mißt man außer der Länge auch die gleichzeitige Änderung des Querschnitts der Probe mit Hilfe eines zweiten, sinngemäß angewandten Anzeigegerätes der oben beschriebenen Art, so kann die Poissonsche Konstante $\sigma$ aus dem Verhältnis Querkontraktion zu Längsdilatation statisch bestimmt werden[5].

Solche Messungen sind jedoch sehr problematisch, da vorausgesetzt wird, daß der Untersuchungskörper völlig elastisch ist und keinerlei Poren oder Klüfte aufweist. Gerade die letzte Bedingung wird von den Gesteinen nicht erfüllt und hat zur Folge, daß eine Längenverkürzung nicht von einer entsprechenden Querschnittsvergrößerung begleitet wird, die Poisson-Zahlen daher druckabhängig sind und in der Regel zu klein ausfallen[6].

Die unvollkommene Elastizität des Gesteins wurde von Zisman berücksichtigt, indem er die Hysteresisschleife der Probe aufnahm. Er vermeidet es, ein einziges $E$ oder $\sigma$ für das Gestein anzugeben und nennt stets mehrere, meist recht unterschiedliche Werte mit den zugehörigen Drucken als Parameter.

[1] O. Müller: Untersuchungen an Karbongesteinen zur Klärung von Gebirgsdruckfragen. Glückauf **1930**, 1601.

[2] J. K. F. Breyer: Über die Elastizität von Gesteinen. Diss. T. H. Berlin 1929. — Z. Geophys. **6**, 98 (1930).

[3] Zisman a. a. O.

[4] Siehe auch F. Kohlrausch: Praktische Physik, 19. Aufl., Bd. 1, S. 76. 1944.

[5] Eine solche Apparatur beschrieb Zisman in Rev. Sci. Instrum. **4**, 342 (1933).

[6] Siehe Ziff. 5.

Der Elastizitätsmodul kann auch aus der statischen Biegung von frei aufliegenden oder einseitig eingeklemmten Stäben berechnet werden. Messungen dieser Art wurden von FÖRTSCH und REGULA ausgeführt (s. Ziff. 9 und Tabelle 6). REGULA bestimmte auch den Scherungsmodul von Gesteinsstäben mit Hilfe statischer Torsion nach Gl. (2.2).

**5. Abhängigkeit der elastischen Konstanten von Druck und Porosität.** Die an Einzelproben erzielten Meßergebnisse können nur dann exakt auf das Gestein in situ übertragen werden, wenn sich dieses physikalisch genau so verhält wie die untersuchte Probe im Laboratorium. Diese Voraussetzung ist stets nur näherungsweise erfüllt. Insbesondere verändert der Druck die physikalischen Eigenschaften der Gesteine. Bei Druckentlastung lockert sich das Gesteinsgefüge und es öffnen sich Poren und Hohlräume aller Art. Daher werden im Laboratorium unter niedrigen Drucken meist zu kleine $E$-Moduln und zu große Kompressibilitäten gemessen[1]. Auch die unterschiedliche Feuchtigkeit und Temperatur beeinflußt die Elastizität des Prüflings.

Messungen der kubischen Kompressibilität an Gesteinsproben wurden 1923 von ADAMS und WILLIAMSON[2], lineare Messungen 1933 von ZISMAN[3] veröffentlicht. Beide Autoren untersuchten die Gesteine unter zwei verschiedenen Bedingungen: Einmal wurde die Probe „unbedeckt" in die Flüssigkeit gebettet, die somit in die Poren und Hohlräume eindringen konnte; das andere Mal wurde die luftgetrocknete Probe von einer dünnen Metallfolie „bedeckt", die für Flüssigkeit undurchlässig war. Bei Drucken unter 1000 Atm erhielt man für die Kompressibilität im bedeckten Zustand einen erheblich größeren Wert, als bei Messungen an unbedeckten Proben. Das Volumen nahm somit bei Belastung im bedeckten Zustand stärker ab, da die mit Luft gefüllten Poren leichter deformiert werden konnten als die mit Flüssigkeit gefüllten. Mit zunehmenden Drucken wurde die Kompression der bedeckten Probe kleiner, und nach und nach verringerte sich die Differenz zwischen beiden $\varkappa$-Werten. Beispiele veranschaulichen die Fig. 4 bei Drucken bis 700 bar und die Fig. 5 bei Drucken bis 12000 bar.

Vergleicht man statische Ergebnisse mit dynamischen Kompressibilitäts-Messungen, so stimmen die letzten recht gut mit den $\varkappa$-Werten der unbedeckten Proben überein. Jedoch zeigten BIRCH[4], BANCROFT[5] und GORANSON[6], daß die lineare Methode bei Gesteinen nur dann sinnvoll angewendet werden kann, wenn man die Proben bedeckt und sehr hohen Drucken von etwa 10000 Atm aussetzt. Bei unbedeckten Proben ist nämlich stets nur ein unkontrollierbarer und von Fall zu Fall unterschiedlicher Prozentsatz der Poren mit Flüssigkeit, der Rest mit Luft gefüllt. Und bei niedrigen Drucken sind die Ergebnisse nicht reproduzierbar und können nicht streng nach den Gleichungen der Elastizitätstheorie behandelt werden.

BIRCH und BANCROFT untersuchten einen Diabas bei Drucken von 4000 bar und erhielten an der bedeckten Probe mit der kubischen und der linearen Methode die gut übereinstimmenden Kompressibilitäten 1,19 bzw. 1,16·$10^{-12}$ cm²/dyn.

Bei Gesteinsuntersuchungen unter extrem hohen Drucken fand BRIDGMAN[7] nicht nur die erwartete Volumenabnahme, sondern auch einen entgegengesetzten

---

[1] Siehe F. BIRCH u. D. BANCROFT: J. Geology **48**, 752 (1940). — Bull. Geol. Soc. Amer **54**, 263 (1943).

[2] ADAMS u. Williamson: a. a. O.

[3] ZISMAN: a. a. O.

[4] BIRCH: a. a. O.

[5] BANCROFT: a. a. O.

[6] R. W. GORANSON: A Note on the Elastic Properties of Rocks. J. Acad. Sci. Washington **24**, 419 (1934).

[7] P. W. BRIDGMAN: Volume Changes in the Plastic Stages of Simple Compression. J. Appl. Phys. **20**, 1241 (1949).

Effekt, nämlich eine Volumenzunahme bei wachsendem Druck. Die Ursache hierfür dürften Zwischenräume sein, die sich öffnen, um einen Bruch der Probe vorzubereiten. Der Effekt ist reversibel, denn das Volumen des Prüflings nimmt bei anschließender Drucksenkung zunächst wieder ab.

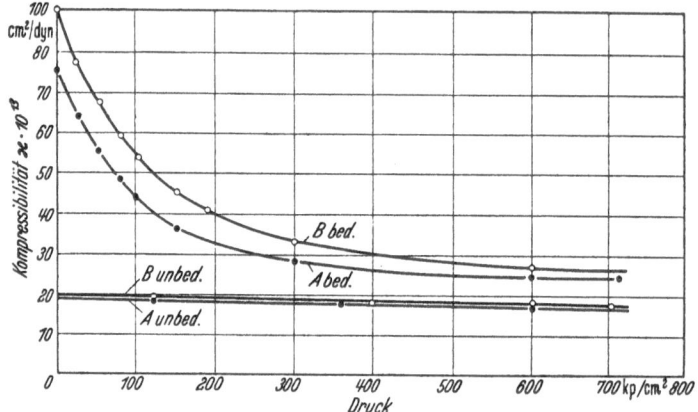

Fig. 4. Die Kompressibilität $\varkappa$ von Quincy-Granit ($A$) und Rockport-Granit ($B$) als Funktion des Druckes. Bei den niedrigen Drucken waren die Proben einmal von einer Metallfolie „bedeckt", das andere mal „unbedeckt". (Nach W. A. ZISMAN.)

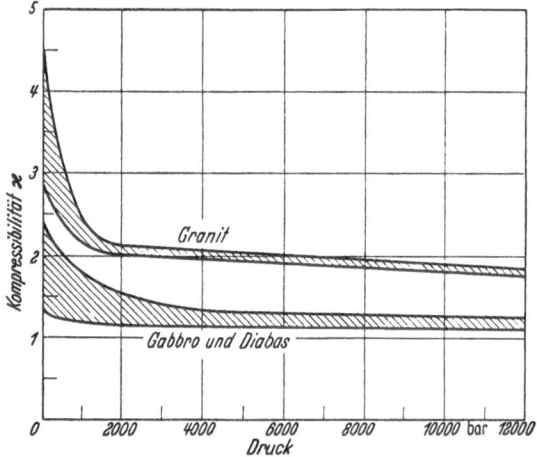

Fig. 5. Die Kompressibilität $\varkappa$ von Granit, Gabbro und Diabas als Funktion des Druckes. $\varkappa$ liegt für alle Gesteine des jeweiligen Typs in dem gestrichelt gezeichneten Bereich. Die Messungen erfolgten im „bedeckten" Zustand. (Nach L. H. ADAMS.) — Einheit der Ordinaten ist $10^{-12}\,\text{cm}^2/\text{dyn}$.

Im Bereich niedriger Drucke sind auch der Elastizitätsmodul und die POISSON-Zahl druckabhängig. Fig. 6 zeigt den $E$-Modul verschiedener Gesteine als Funktion des Druckes nach Untersuchungen von BREYER[1]. In Tabelle 1 sind $E$ und $\sigma$ nach Messungen von ZISMAN[2] wiedergegeben. Man sieht, daß die statisch bestimmten Elastizitätsmoduln stark vom Druck abhängen und findet darin eine Erklärung für die mangelnde Übereinstimmung der vielen unter verschiedenen Versuchsbedingungen ausgeführten Messungen.

In der Regel reagieren $E$ und $\sigma$ eines Gesteins um so stärker auf den Druck, je poröser das Gestein ist. Für Proben eines bestimmten Gesteinstyps pflegen

---

[1] BREYER: a. a. O.
[2] ZISMAN: a. a. O.

Elastizitätsmodul und POISSON-Zahl mit wachsender Porosität abzunehmen. Beide Werte ändern sich in gleichem Sinne, und außerdem sind kleine $E$-Moduln meist mit kleinen $\sigma$-Werten verknüpft und umgekehrt. Im allgemeinen ändert sich somit die Elastizität eines Gesteins um so stärker mit dem Druck, je kleiner $E$ und $\sigma$ sind. Die Druckabhängigkeit statisch gemessener Elastizitätskonstanten ist daher in erheblichem Maße auf Hohlräume und Klüfte im Gestein zurückzuführen. Einwandfreie Messungen erfordern hohe Drucke, bei denen diese Hohlräume geschlossen sind, doch läßt sich diese Bedingung praktisch nur bei Kompressibilitäts-Messungen realisieren.

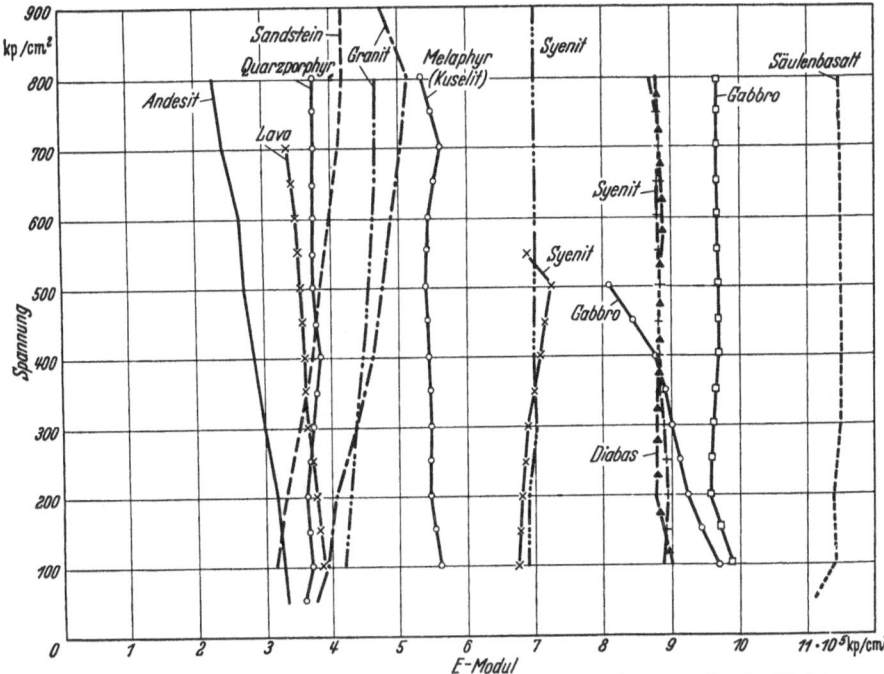

Fig. 6. Statisch ermittelte Elastizitätsmoduln verschiedener Gesteine in Abhängigkeit vom Druck. (Nach BREYER.)

Tabelle 1. *Beispiel für die Änderung des statisch bestimmten Elastizitätsmoduls und der* POISSON*zahl im Bereich niedriger Drucke.* Nach ZISMAN.

| Quincy Oberflächengranit | | Rockport-Granit | | | |
|---|---|---|---|---|---|
| mittlerer Druck kp/cm² | $E$ dyn/cm² | mittlerer Druck kp/cm² | $E$ dyn/cm² | mittlere Belastung in Pfund | $\sigma$ |
| 11,2 | $2,58 \cdot 10^{11}$ | 11,2 | $3,95 \cdot 10^{11}$ | 1000 | 0,093 |
| 22,4 | 3,07 | 22,4 | 4,10 | 3000 | 0,111 |
| 33,6 | 3,64 | 89,6 | 4,85 | 5000 | 0,129 |
| 56,0 | 4,05 | 157 | 5,60 | 7000 | 0,149 |
| 78,4 | 4,73 | 202 | 6,10 | 9000 | 0,172 |

**6. Über einige Ergebnisse statischer Messungen.** Schon frühzeitig wurde erkannt, daß sich der Chemismus eines Gesteins in seinem elastischen Verhalten bemerkbar macht. Für Quarz fand man z.B. einen niedrigeren $E$-Modul als für Glimmer, und dementsprechend besitzen saure Gesteine auch ein kleineres $E$ als die basischen (vgl. Fig. 5 und 6). Quarz beeinflußt mit seinem niedrigen $E$-Modul zwei so verschiedene Gesteine wie Granit und Sandstein dahingehend, daß beide annähernd denselben Elastizitätsmodul besitzen (vgl. Fig. 11). Dieser steigt mit zunehmender Basizität der Proben an.

Adams und Williamson[1] haben gezeigt, daß man die Kompressibilität eines frischen Gesteins mit ausreichender Genauigkeit berechnen kann, wenn die $x$-Werte seiner Mineralien und deren Anteil am Mineralbestand des Gesteins bekannt sind. Nach ihren Angaben stimmen die berechneten mit den gemessenen Werten bei Drucken von 10000 Atm auf etwa 2%, bei 2000 Atm auf etwa 5% überein; bei niedrigen Spannungen differieren sie aber stark. Dies hat einige praktische Bedeutung, da $x$ bei den genannten hohen Drucken nur mit erheblichem Aufwand gemessen werden kann. Oberhalb von 2000 Atm unterscheiden sich die Kompressibilitäten verschiedener Gesteine eines bestimmten Typs nur wenig voneinander, was ebenfalls von Bedeutung ist und in Fig. 5 veranschaulicht wird.

Auch das Gefüge eines Gesteins macht sich bei der Messung seiner elastischen Eigenschaften bemerkbar. Nach Breyer ist ferner der $E$-Modul bei wassersatten Proben geringer, als bei trockenen[2].

Den statischen Untersuchungen verdanken wir wertvolle Erkenntnisse über den Aufbau unserer Erdkruste. So bestimmten Adams und Gibson[3] die Kompressibilität von Dunit und berechneten daraus die Geschwindigkeit der longitudinalen Welle, bezogen auf 60 km Tiefe, zu 8,4 km/sec. Um mit der seismischen Forschung übereinzustimmen, mußte angenommen werden, daß das Gestein in 60 km Tiefe basischer ist als Gabbro und sich in seiner Zusammensetzung dem Dunit nähert. Für basische Gläser ergaben sich geringere $V_l$-Werte als für das kristalline Gestein gleicher Zusammensetzung, so daß auf einen kristallinen Aufbau der tiefen Erdschichten geschlossen werden konnte. Vorher schon hatte Bridgman vulkanische Gläser bei Drucken bis 12000 Atm untersucht und — wie erwartet — gefunden, daß sauere Gesteine kompressibler sind als basische. Aus Bridgmans Arbeiten ist zu ersehen, daß wir die Frage nach dem Zustand des Erdinnern nicht ohne weiteres mit Hilfe normaler Laboratoriumsmessungen an Gesteinsproben klären können. Infolge hoher Drucke und Temperaturen nimmt die Materie in der Tiefe sicher Formen an, deren physikalische Eigenschaften uns nicht vertraut sind.

1931 untersuchte Adams[4] die Frage, wie sich der Peridotit in 50 km Tiefe zusammensetzen muß, um eine Geschwindigkeit von 8.2 km/sec. für die Kompressionswelle zu erhalten. Er fand, daß der molekulare Anteil von MgO und FeO im Peridotit sich dann verhalten muß wie 4:1, denn $V_l$ beträgt beim reinen $Mg_2SiO_4$ 8,6 km/sec, beim reinen $Fe_2SiO_4$ 7,1 km/sec, bei einem Druck von 15000 bar.

**7. Statische Messungen in situ.** Bei Baugrunduntersuchungen sollten die Laboratoriumsmessungen durch Feldversuche an Ort und Stelle ergänzt werden, da die Elastizität und Plastizität eines Gesteins im Verbande nicht immer mit derjenigen eines Handstückes übereinzustimmen pflegt. Die Ursache ist in der Klüftung und Schichtung des Gesteinskörpers zu suchen. Dadurch findet man die elastische Deformation im Verbande meist kleiner, die plastische Verformung größer als bei einer Einzelprobe.

---

[1] Adams u. Williamson: a. a. O.

[2] Vgl. dazu Ziff. 13, Tabellen 8 und 9.

[3] L. H. Adams u. R. E. Gibson: The Compressibilities of Dunite and of Basalt Glass und their Bearing on the Composition of the Earth. Proc. Nat. Acad. Sci. **12**, 275 (1926); deutsch in Gerlands Beitr. Geophys. **15**, 241 (1926).

[4] L. H. Adams: The Compressibility of Fayalite, and the Velocity of Elastic Waves in Peridotite with Different Iron-Magnesium Ratios. Gerlands Beitr. Geophys. **31**, 315 (1931).

Bei Lockergesteinen können statische Belastungsversuche ausgeführt werden, wobei man stufenweise belastet und entlastet und die Deformation durch präzises Nivellieren mißt. Bei festen Gesteinen wurden von verschiedenen Autoren[1] hydraulische Pressen oder mechanische Schraubenwinden eingesetzt. Meist führt man solche Messungen im Stollen aus, wobei die einzelnen Laststufen stunden- oder tagelang konstant gehalten werden.

Nach der Theorie hat die Verteilung des Druckes auf der belasteten Fläche keinen nennens- werten Einfluß auf den Stauchungsbetrag, der im wesentlichen von der Gesamtlast $Q$ ab- hängt. Verwendet man einen kreisförmigen Druckstempel vom Radius $a$, so gehorcht die Deformation $w$ der Beziehung

$$w = c\,Q\cdot\frac{1-\sigma^2}{a\,E}.$$

Der Faktor $c$ beträgt bei gleichförmiger Druckverteilung 0,54. Für eine starre Kreisfläche mit der Druckverteilung

$$p(r) = \frac{Q}{2\pi a\,\sqrt{a^2-r^2}}$$

ist $c = 0,50$. — Man sieht ferner, daß die gemessene Deformation von $E$ und $\sigma$ abhängt. Es müssen somit spezielle Annahmen für $\sigma$ gemacht werden, um $E$ näherungsweise zu bestimmen.

Bei einer anderen Methode werden Teile des Stollens durch Spezialtüren ab- geschlossen und mit Wasser gefüllt, welches unter Druck gesetzt wird. Die be- nötigten Kräfte liegen in der Größenordnung von einigen hundert Tonnen. Mit Hilfe geeigneter Komparatoren, deren Haltevorrichtungen außerhalb des Ver- suchsgebietes liegen, wird dann die Distanzänderung diametral gelegener Be- lastungsflächen als Funktion des Druckes gemessen.

Aus dem Spannungs-Dehnungsdiagramm kann der elastische und plastische Verformungsanteil entnommen werden. Sofern eine plastische Komponente vor- handen ist, wird der Elastizitätsmodul $E$ häufig durch den „*Zusammendrückungs- Modul*" $M$ ersetzt, der die elastischen *und* plastischen Verformungen einschließt. Letztere betrugen z.B. bei einem Serizitschiefer 28% der Gesamtdeformation, während bei einem Biotitgneis keine meßbare Plastizität gefunden wurde.

Im einzelnen schwanken die Ergebnisse stark von Ort zu Ort. Der so bestimmte Elastizitätsmodul ist kleiner, als der im Laboratorium gemessene. In Tabelle 2 sind einige Werte für den „Zusammendrückungs-Modul $M$" wiedergegeben.

Tabelle 2. *Beispiele für den „Zusammendrückungs-Modul" M bei statischen Versuchen in situ. M tritt an die Stelle des Elastizitätsmoduls E und umfaßt den elastischen und plastischen Verformungsanteil (nach* Moos *und* Quervain, *Technische Gesteinskunde, 1948).*

| Gestein | Druck (kp/cm²) | Schieferung | $M$ (kp/cm²) |
|---|---|---|---|
| Granit | 60 | ‖ | etwa $4{,}0\cdot10^5$ |
|  | 60 | ⊥ | etwa $2{,}5\cdot10^5$ |
| Biotitschiefer | 60 | ‖ | etwa $2{,}8\cdot10^5$ |
|  | 60 | ⊥ | etwa $0{,}8\cdot10^5$ |

---

[1] A. Schrafl: Kurzer Bericht über die Druckstollenversuche der schweizerischen Bundes- bahnen. Schweiz. Bauztg. **83** (1924). — Urserenkraftwerke. Techn. Rdsch. **1945**. — O. Frey- Baer: Die Berechnung der Betonauskleidung von Druckstollen. Schweiz. Bauztg. **124** (1944). — E. Meyer-Peter u. Th. Frey: Das Projekt 1943/44 der Urserenkraftwerke. Schweiz. Bauztg. **126** (1945). — O. Frey-Baer: Die Dehnungsmessungen im Druckstollen des Kraft- werkes Lucendo. Schweiz. Bauztg. **65** (1947). — P. Habib: Détermination du Module d'Éla- sticité des Roches en Place. Inst. Techn. du Bâtiment et des Travaux Publics, Sols et Fon- dations, No. 3, 27, 1950. — A. v. Moos u. F. de Quervain: Technische Gesteinskunde. Basel 1948.

Die plastische Verformung wurde auch bei Einzelproben berücksichtigt, und wir finden in Tabelle 3 einige Ergebnisse für $E$ und $M$ gegenübergestellt, außerdem die reziproken Werte $m = 1/\sigma$ der POISSON-Zahlen, und zwar für den rein elastischen ($m_e$) und den totalen Deformationsanteil ($m_t$).

Tabelle 3. *Der Elastizitätsmodul E, der Zusammendrückungsmodul M (Einheiten $10^5$ kp/cm² für E und M) und die reziproken Werte $m = 1/\sigma$ der POISSON-Zahlen ($m_e$ für den rein elastischen, $m_t$ für den totalen Verformungszustand) für einige Gesteinsproben (nach MOOS und QUERVAIN, Technische Gesteinskunde 1948).*

| Gestein | $E$ | $M$ | $m_e$ | $m_t$ |
|---|---|---|---|---|
| Granit Handeck, Bern 1. Probe . . . . | 2,16 | 1,94 | 10,0 | 11,5 |
| 2. Probe . . . . | 4,03 | 2,92 | 7,0 | 7,6 |
| Gneis, Tessin 1. Probe . . . . . . . | 3,22 | 2,80 | 3,3 | 3,4 |
| 2. Probe . . . . . . . | 2,61 | 2,10 | 6,6 | 6,0 |
| Diabas, Württemberg . . . . . . . | 5,60 | 5,60 | 3,1 | 3,1 |
| Marmor, Carrara . . . . . . . . . | 8,65 | 8,65 | 4,3 | 4,3 |
| Kalkstein, Arvel (Waadt) . . . . . | 8,60 | 8,53 | 2,9 | 2,9 |
| Grauer Sandstein, St. Margarethen . . | 0,88 | 0,47 | 7,5 | 2,8 |

## B. Dynamische Methoden.

**8. Übersicht. Viscosität und innere Reibung.** Man kann dieElastizität von Gesteinen mit Hilfe verschiedener dynamischer Methoden untersuchen, von denen nachfolgend die Rede sein soll. Im vorliegenden Abschnitt werden solche Verfahren genannt, bei denen der gesamte stabförmige Probekörper in freie oder erzwungene Schwingungen versetzt wird. Die Untersuchungen können hierbei sowohl mit longitudinalen, als auch mit Biege- und Torsionsschwingungen im Frequenzbereich von einigen Hertz bis zu mehreren Megahertz ausgeführt werden. Häufig wendet man verschiedene der hier genannten Methoden nacheinander auf dieselbe Probe an, um deren Konstanten mit größerer Sicherheit zu bestimmen.

Während in der Kontinuumsmechanik üblicherweise mit idealer Elastizität der festen Körper bei kleinen Deformationen gerechnet wird, findet man die Gesteine nur unvollständig elastisch. So zeigen sie bei langandauernden Belastungen irreversible viscose Eigenschaften, d.h. die Gesteine „fließen" ein wenig und verformen sich *plastisch*. Allerdings tritt dieser Effekt bei kristallinen Festkörpern erst oberhalb einer gewissen Mindestspannung („*strength*" oder „*Fließwiderstand*") auf. Sind die Spannungen kleiner, so gehorcht die Deformation dem HOOKEschen Gesetz.

Andererseits erklärt die Elastizitätstheorie nicht die Dämpfung mechanischer Schwingungen durch die innere Reibung, durch die die kinetische Energie der Schwingung in Wärme umgewandelt wird. Dadurch klingt die Amplitude einer Schwingung nicht nur mit der Zeit (Dämpfung), sondern auch mit der Entfernung (Absorption) ab[1].

Diese Abweichungen vom HOOKEschen Gesetz pflegt man für jeden Festkörper durch zwei neue, vermutlich voneinander unabhängige Größen zu erfassen, nämlich durch den „*Viscositätskoeffizienten $\nu$*" und die „Konstante der *inneren Reibung $\eta$*"[2]. An Stelle von $\eta$ wird meist die *Dämpfung*[3] gemessen, welche jedoch

---

[1] Eine einfache Beziehung zwischen Dämpfung und Absorption wurde von O. FÖRTSCH [Z. Geophys. **16**, 57 (1940)] hergeleitet.

[2] Näheres in H. JEFFREYS: The Earth, 3nd edit., p. 5. 1952. — B. GUTENBERG: Internal Constitution of the Earth, 2nd edit., p. 382, 1951; enthält ausführliches Literaturverzeichnis. — B. GUTENBERG: Handbuch der Geophysik, Bd. II, S. 529, Berlin 1933; hier werden auch Meßmethoden beschrieben. — F. GASSMANN: Über kleine Bewegungen in nicht vollkommen elastischen Körpern. Schweiz. Bauztg. **4** (1949) und Mitt. Inst. Geophys. E. T. H. Zürich **1949**, Nr. 11.

[3] Einige Werte für die Dämpfung sind in Tabelle 6 zusammengestellt.

außer der „*inneren Dämpfung*" (die wohl im wesentlichen, aber nicht unbedingt allein durch die innere Reibung hervorgerufen wird) auch Anteile einer „*äußeren Dämpfung*" (z.B. durch den Luftwiderstand) und einer „*Abstrahlungsdämpfung*" (Energieabstrahlung, insbesondere in die vom schwingenden Körper deformierten Haltevorrichtungen) umfaßt[1].

**9. Freie und erzwungene Schwingungen.** Bei den nachfolgend beschriebenen Methoden werden stabförmige Proben in freie oder erzwungene longitudinale, Biege- oder Torsionsschwingungen versetzt und die Eigenfrequenz und Dämpfung des Prüflings bestimmt. Wenn die Frequenz der anregenden Kraft variiert wird, so kann eine vollständige Resonanzkurve des Probekörpers aufgenommen werden *(Resonanzmethode).* Je nach der Versuchsanordnung kann dann der Elastizitäts- und Scherungsmodul, sowie die Geschwindigkeit $V_d = \sqrt{E/\varrho}$ der Stab-Dehn-Welle und die Transversalgeschwindigkeit $V_t = \sqrt{G/\varrho}$ berechnet werden.

Die ersten dynamischen Messungen an Gesteinsproben wurden 1905 von KUSAKABE[2] veröffentlicht. Der Verfasser schlug mit einem Hammer auf eine einseitig eingeklemmte Probe und beobachtete am freien Ende die mechanisch vergrößerten Schwingungen. Mit diesem unausgereiften Verfahren wurden Messungen an 158 Proben ausgeführt und zum Teil mit den statisch gewonnenen Ergebnissen von NAGAOKA[3] verglichen.

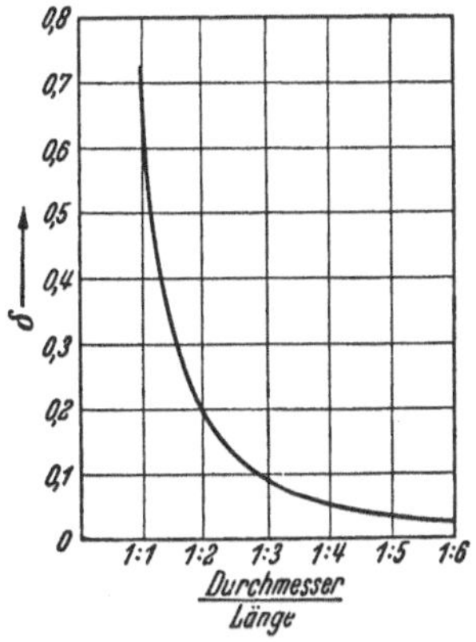

Die Methodik dieser Untersuchungen soll am Beispiel einer von THYSSEN und RÜLKE[4] entwickelten Meßapparatur veranschaulicht werden. Hierbei wird eine freistehende stabförmige Probe in erzwungene Schwingungen versetzt und diejenige Schwingungszahl ermittelt, bei der die Probe ihre longitudinale Grundschwingung ausführt, also mit der aufgedrückten Schwingung in Resonanz ist. Ist $f_0$ die Eigenfrequenz des zylindrischen Stabes vom Radius $r$ und von der Länge $l$, so folgt für die Dehnwellengeschwindigkeit $V_d$ und den Elastizitätsmodul $E$

$$V_d = 2l f_0, \qquad E = (2l f_0)^2 \cdot \varrho.$$

Fig. 7. RAYLEIGHsche Korrektionsgröße $\delta = (\pi \sigma r/2l)^2$ für $\sigma = 0{,}27$ als Funktion von Stabdurchmesser zu Stablänge. Bei kurzen und dicken Proben muß $V_d$ mit $(1 + \delta)$ multipliziert werden. (Nach VON THYSSEN und RÜLKE.)

Dabei wird vorausgesetzt, daß $l$ groß ist im Vergleich zu $r$. Anderenfalls muß $V_d$ mit dem RAYLEIGHschen Korrektionsfaktor $[1 + (\pi \sigma r/2l)^2]$ multipliziert werden, der die Querkontraktion des Stabes berücksichtigt[5]. In Fig. 7 ist die Korrektions-

---

[1] Siehe F. GASSMANN: Über Dämpfung durch Abstrahlung elastischer Wellen und über gedämpfte Schwingungen von Stäben. Z. angew. Math. Phys. **2** (1951) und Mitt. Inst. Geophys. E. T. H. Zürich. **1951**, Nr. 20. — W. KELLENBERGER: Über Biegeschwingungen stabförmiger Bauelemente mit Abstrahlungsdämpfung. Mitt. Inst. Geophys. E. T. H. Zürich **1954**, Nr. 24.

[2] KUSAKABE: Kinetic Measurements of the Modulus of Elasticity for 158 Spec. of Rocks. Comm. Coll. Sci. Imp. Univ. Tokyo **20**, Art 9. (1905).

[3] NAGAOKA: a. a. O.

[4] ST. V. THYSSEN u. O. RÜLKE: Beschreibung des neuen Gerätes zur Bestimmung der Fortpflanzungsgeschwindigkeit elastischer Wellen in Gesteinsproben und einige Meßergebnisse. Z. Geophys. **15**, 130 (1939). — ST. V. THYSSEN: Ein neues Gerät zur Schnellbestimmung von Laufzeiten elastischer Wellen in Bohrkernen. Erdöl u. Kohle **1938**, H. 46.

[5] KUMIZI IIDA berichtet in „Determining Young's Modulus and the Solid Viscosity Coefficient of Rocks by the Vibration Method" (Bull. Earthqu. Res. Inst. Tokyo **17**, 79 1939), daß der nach $E = (2l f_0)^2 \cdot \varrho$ bestimmte Elastizitätsmodul von der Länge $l$ der Probe abhängt und führt dies auf die Wirkung der „solid viscosity" zurück. Für jede Probe wird neben $E$ ein Beiwert $\gamma$ (solid viscosity coefficient) berechnet, der durch $E/\varrho = (\gamma/2\varrho\, l)^2 + (2l f_0)^2$ definiert ist und für Marmor 7,18, Granit 3,13, Sandstein 2,07, Aluminium 1,78 und Stahl 7,32 ($\times 10^6$ g cm$^{-1}$ sec$^{-1}$) beträgt.

größe $\delta = (\pi\sigma r/2l)^2$ für $\sigma = 0{,}27$ als Funktion des Verhältnisses Stabdurchmesser zu Stablänge dargestellt.

Die Probe, an deren Basis eine Metallfolie befestigt ist, wird auf ein Dielektrikum gestellt, welches sich auf einer leitenden Grundplatte befindet. Zwischen Grundplatte und Folie wird eine Wechselspannung angelegt, die in der Frequenz regelbar ist und von einem Schwebungssummer erzeugt wird. Die beschriebene Anordnung stellt einen kapazitiven Geber dar, der die Probe in Schwingungen versetzt. Schaltungstechnisch ist erreicht, daß die Frequenz der Schwingung mit der Frequenz der angelegten Wechselspannung übereinstimmt. Ein Piezoquarzempfänger am anderen Ende der Probe zeigt die Resonanzstelle an. Zwei Resonanzkurven für Materialien mit verschieden großer Dämpfung sind in Fig. 8 wiedergegeben, einige Meßergebnisse in Tabelle 4 zusammengestellt.

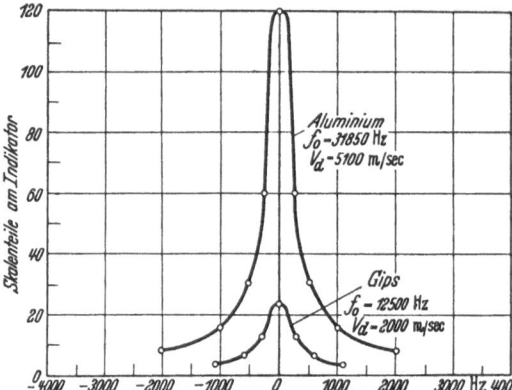

Fig. 8. Resonanzkurven für Aluminium und Gips (longitudinale Schwingungen). (Nach von Thyssen und Rülke.) Abszisse: Frequenzen in Abweichungen von der Resonanz-Frequenz.

Von den zahlreichen prinzipiell ähnlichen Methoden seien hier besonders die Untersuchungen von Ide[1] genannt, der ebenfalls mit erzwungenen longitudinalen Schwingungen gearbeitet hat und eine Genauigkeit von einigen Zehntel Prozent angibt. Die gut reproduzierbaren Laufzeitmessungen von Thyssen sind bei Gesteinen auf 1%, bei Metallen auf 0,1% genau angegeben.

Tabelle 4. *Messung der Geschwindigkeit $V_d$ der Stabdehnwelle.*

Die Werte sind nach Rayleigh korrigiert. $\varnothing$ = Durchmesser, $l$ = Länge, $f_0$ = Eigenfrequenz der Probe. Nach von Thyssen.

| Gesteinsbezeichnung und Tiefenlage | $\varnothing$ (mm) | $l$ (mm) | $f_0$ (Hz) | $V_d$ (m/sec) |
|---|---|---|---|---|
| 1. Salz (434 m) . . . . . . . . . . . | 50 | 153 | 15000 | 4695 |
| 2. Anhydrit, Lenne (1800 m) . . . . . | 110 | 195 | 13130 | 5415 |
| 3. Wealden-Sandstein (830 m) . . . . | 78 | 150 | 10170 | 3200 |
| 4. Unter-Eozän, ausgetrocknet (239 m) . | 50 | 108 | 6060 | 1385 |
| 5. Plattendolomit Westfalen (1270 m) . | 95 | 153 | 11900 | 3895 |
| 6. Gigaskonglomerat . . . . . . . . . | 95 | 136 | 14580 | 4315 |
| 7. Tonstein, unterer Buntsandstein Westfalen (1400 m) . . . . . . . . | 95 | 113 | 11500 | 2930 |
| 8. Sandiger Tonstein Indien (1700 m) . | 82 | 178 | 9050 | 3345 |

Neuerdings berichten Gassmann, Weber und Vögtli[2] über Messungen von longitudinalen Schwingungen verschiedener Gesteine. Die stabförmigen Proben werden an zwei dünnen Fäden aufgehängt, um Energieverluste zu vermeiden. Dann wird das eine Stabende mit Hilfe eines Tauchspulensystems mit konstanter Kraft sinusförmig angeregt, während die Geschwindigkeit des anderen Stabendes — ebenfalls über ein Tauchspulensystem — gemessen wird. Die Frequenz der eingeprägten Kraft kann von 30 bis 15000 Hz fast stetig verändert werden.

---

[1] John M. Ide: Some Dynamic Methods for Determination of Young's Modulus. Rev. Sci. Instrum. 6, No. 10 (1935).

[2] F. Gassmann, M. Weber u. K. Vögtli: Über Longitudinalschwingungen von Gesteinsstäben. Mitt. Inst. Geophys. E. T. H. Zürich 1952, Nr. 22.

Dabei beträgt die Amplitude der Schwingung höchstens 5 µ. Die Anordnung ist unempfindlich gegenüber unerwünschten Querschwingungen und kleinen Translationsbewegungen senkrecht zur Stabachse.

Außer der Grundschwingung werden auch Oberschwingungen gemessen, die oft an Stelle einer einzigen Resonanzüberhöhung zwei nahe beieinanderliegende Maxima aufweisen. Meßergebnisse werden in Tabelle 5 mitgeteilt, einen Ausschnitt aus einer Resonanzkurve zeigt Fig. 9.

FÖRTSCH[1] benutzte seine auf freistehende Brückenpfeiler angewandte Meßmethoden im Laboratorium bei der Untersuchung einseitig eingeklemmter Stäbe von 50 bis 100 cm Länge aus Holz, Glas und Metall. Der Elastizitätsmodul wurde erstens aus der Eigenfrequenz, die nach der Balkentheorie berechnet wurde, und zweitens aus der Dämpfung, dem Resonanzausschlag und der Exzenterkraft bestimmt. Der Einfluß der Klemmung wurde hierbei rechnerisch

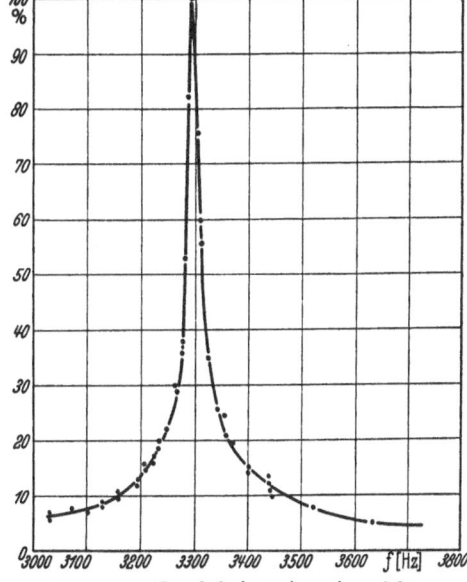

Fig. 9. Ausschnitt (Grundschwingung) aus einer mit konstanter Kraftamplitude (7,2 Pond) gemessenen Resonanzkurve (Gneiss, Tessin). (Nach GASSMANN, WEBER und VÖGTLI.)

Tabelle 5. *Elastizitätsmoduln aus den Eigenfrequenzen longitudinaler Grund- und Oberschwingungen für einige Gesteinsproben.* Nach GASSMANN, WEBER und VÖGTLI.

| Gestein | Schwingung | Frequenz (Hz) | $E$ $10^5 \cdot$ kp/cm² | |
|---|---|---|---|---|
| Schwedengranit (Soervik) | Grundschwingung | 3065 | 5,5 | |
| | 1. Oberschwingung | 6140 | 5,5 | |
| | 2. Oberschwingung aufgespalten | { 8830 / 9420 | 5,1 / 5,8 | 5,4 |
| Tessiner Gneis (Maggia) | Grundschwingung | 3290 | 3,1 | |
| | 1. Oberschwingung | 6850 | 3,4 | |
| Kalk (Roc-Argent, Lourdes) | Grundschwingung | 4680 | 5,9 | |
| | 1. Oberschwingung aufgespalten | { 8760 / 9800 | 5,2 / 6,5 | 5,8 |
| Kalk (Laufen) | Grundschwingung | 3290 | 5,5 | |
| | 1. Oberschwingung | 6670 | 5,6 | |
| | 2. Oberschwingung aufgespalten | { 9150 / 10320 | 4,7 / 6.0 | 5,3 |
| Antikorrodal | Grundschwingung | 4670 | 6,0 | |

berücksichtigt. Es ergab sich, daß die Größe des Resonanzausschlages unabhängig von Einklemmung und Kopflast ist.

FÖRTSCH bestimmte die Eigenfrequenz, indem er die ausklingende freie Schwingung über ein Spiegelsystem photographisch registrierte. Bei der zweiten Methode wurde am Stabende ein Lagerbock angebracht, in dem eine Scheibe mit exzentrischer Masse rotierte. Außerdem wurde die Probe auf zwei Schneiden gelegt und $E$ ermittelt aus den Querschwingungen in dieser Lage und aus der statischen Durchbiegung bei Belastung in der Mitte. Die

---

[1] O. FÖRTSCH: Das Verhalten noch freistehender Brückenpfeiler bei Schwingungen und deren Abhängigkeit vom Untergrund. Z. Geophys. **14**, 173 (1938).

kleinsten Werte für $E$ berechneten sich aus den Eigenfrequenzen, die größten aus Dämpfung und Resonanz. Korrektionen wurden angebracht.

Regula[1] setzte die Untersuchungen an einseitig einzementierten Stäben aus Messing, Marmor, Schiefer und Granit von 20 cm Länge und $1 \times 1$ cm² Querschnitt fort. In ähnlicher Weise wurden die Stäbe in freie Biegeschwingungen und freie und erzwungene Torsionsschwingungen versetzt. Die daraus gewonnenen Werte für $E$ und $G$ wurden ferner durch statische Biege- und Torsionsmessungen ergänzt und daraus auch $V_d$, $V_t$ und $\sigma$ berechnet. Einige Werte, die mit den verschiedenartigen Verfahren gewonnen wurden und untereinander erheblich streuen, sind in Tabelle 6 wiedergegeben.

Tabelle 6. *Meßwerte nach* Regula.

Es wurden bestimmt: Der Elastizitätsmodul $E_1$ und der Scherungsmodul $G_1$ aus statischer Biegung bzw. Torsion, $E_2$ und $G_2$ aus freien Biege- bzw. Torsionsschwingungen, $G_3$ aus Resonanzausschlag und Dämpfung bei erzwungenen Torsionsschwingungen. $E$ und $G$ sind korrigierte Werte aus $E_1$ und $E_2$ bzw. $G_1$ und $G_2$ unter Berücksichtigung der Einspannung. $V_d = \sqrt{E/\varrho}$, $V_t = \sqrt{G/\varrho}$, $\sigma = (E/2G) - 1$, $\beta^2 = $ Dämpfung ($\beta = $ Verhältnis zweier aufeinanderfolgender Ausschläge nach verschiedenen Seiten). Elastizitäts- und Scherungsmoduln in $10^{11}$ dyn/cm², $V_d$ und $V_t$ in km/sec.

| Probe | $E_1$ | $E_2$ | $E$ | $G_1$ | $G_2$ | $G$ | $G_3$ | $V_d$ | $V_t$ | $\sigma$ | $\beta^2$ |
|---|---|---|---|---|---|---|---|---|---|---|---|
| Messing . . . . | 7,51 | 6,71 | 8,09 | 4,45 | 3,45 | 2,82 | 3,03 | 3,09 | 1,82 | 0,43 | 1,005 |
| Grauer Marmor | 5,4 | 4,78 | 5,87 | 3,45 | 2,76 | 2,29 | 5,74 | 4,62 | 2,9 | 0,28 | 1,023 |
| Schiefer . . . | 5,24 | 5,19 | 5,27 | 3,3 | 2,91 | 2,6 | 4,74 | 4,32 | 3,04 | 0,015 | 1,017 |
| Granit . . . . | 7,59 | 6,79 | 8,17 | 3,99 | 3,62 | 3,31 | 3,44 | 5,22 | 3,32 | 0,24 | 1,03 |
| Weißer Marmor | 3,2 | 2,64 | 3,78 | 2,07 | 1,87 | 1,7 | 1,29 | 3,7 | 2,48 | 0.11 | 1,029 |

**10. Einiges über das elastische Verhalten von Gesteinen.** $\alpha$) *Über die Streuung des Elastizitätsmoduls innerhalb eines Gesteinstyps.* Es interessiert die Frage, wie

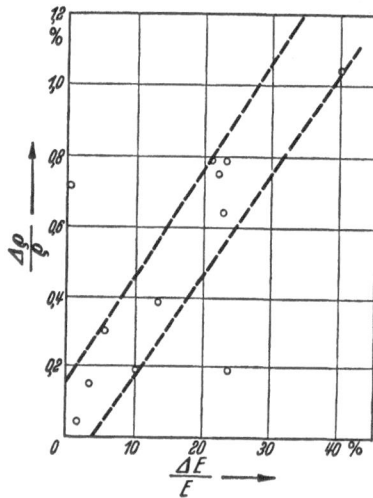

Fig. 10. Zusammenhang zwischen der Änderung der Gesteinsdichte und des $E$-Moduls für 12 Granitproben. Aufgetragen sind die relativen Abweichungen vom Mittelwert ohne Rücksicht auf das Vorzeichen. (Nach Ide.)

stark die elastischen Konstanten innerhalb eines einheitlichen Gesteinstyps streuen können. Ide[2] entnahm zu diesem Zweck dem Quincy- und dem Rockportgranit 12 große und 8 kleine Proben, ferner dem Sudbury-Norit 16 Proben und bestimmte den Elastizitätsmodul jeder Probe auf dynamische Art. Hierbei ergab sich, daß die großen Proben im Durchschnitt um 16%, die kleinen um 11% vom mittleren $E$-Modul abweichen. Dies sind erhebliche Beträge, wenn man an den weitgehend gleichmäßigen Aufbau und die praktisch unveränderte mineralogische Zusammensetzung des Granites denkt. Auch beim kompakten Norit streuen die $E$-Werte der Einzelproben, im Mittel jedoch nur um 3,3%.

Wenn man außer der Schwankung des $E$-Moduls auch die Änderung der Dichte $\varrho$ mißt, so ergibt sich im allgemeinen ein Zusammenhang zwischen beiden Größen in dem Sinne, daß eine Zunahme von $E$ mit einem Anwachsen von $\varrho$ verbunden ist. Fig. 10, in der $\Delta\varrho/\varrho$ als Funktion von $\Delta E/E$ dargestellt ist, veranschaulicht diese

[1] W. Regula: Untersuchungen elastischer Eigenschaften von Gesteinsstäben. Z. Geophys. **16**, 40 (1940).
[2] Ide: a. a. O.

Korrelation für einige Granitproben. Auch Norit verhält sich ähnlich: Bei den Proben mit der stärksten Abweichung vom Mittelwert gehören zu $\Delta E/E = 14{,}1$, 9,2 und 5,2% die Dichteänderungen $\Delta\varrho/\varrho = 3{,}2$, 2,4 und 2,2% [1].

*β) Abhängigkeit der Meßwerte von den Untersuchungsmethoden.* Die von verschiedenen Autoren veröffentlichten Meßergebnisse an gleichartigen Gesteinen weichen meist erheblich voneinander ab. Wenn man von Fehlern aller Art absieht, so beruht ein Teil dieser Unterschiede auf dem verschiedenartigen physikalischen Zustand des Gesteins, ein anderer Teil auf den experimentellen Bedingungen, denen die Proben unterworfen werden (Druck, Temperatur, Klemmung usw.). Schließlich hängen die Ergebnisse von der Wahl der Meßmethode ab, worüber hier berichtet werden soll.

Bei ideal elastischen Körpern liefern sämtliche Verfahren stets dieselben Elastizitätsgrößen. Gesteine sind nur unvollständig elastisch und außerdem porös. Beides hat zur Folge, daß die statisch bestimmten Konstanten von den dynamisch gemessenen abzuweichen pflegen. Beispiele dieser Art veranschaulicht Tabelle 6. Im Gegensatz zu den

Tabelle 7. *Elastizitätsmoduln in $10^{11}\,dyn/cm^2$, statisch, dynamisch und seismisch gemessen.* Nach ZISMAN, IDE und LEET.

| Gestein | $E_{stat}$ | $E_{dyn}$ | $E_{seism}$ |
|---|---|---|---|
| Sudbury Norit . . | 8,36 | 8,90 | 8,82 |
| Quincy Granit . . | 3,51 | 3,92 | 4,3 |
|  | 4,25 | 4,64 |  |
| Rockport Granit . | 3,89 | 4,51 | 5,0 |

dort angeführten Ergebnissen erhält man bei *dynamischen* Untersuchungen meist größere Maßzahlen für $E$ und $G$, als bei *statischen* Messungen. Dies zeigt auch Tabelle 7 für drei bereits mehrfach genannte Gesteine, die im Laboratorium statisch von ZISMAN[2], dynamisch von IDE[3] und in situ seismisch von LEET[4] untersucht worden sind. Die statischen Werte sind durchweg kleiner als die dynamischen, die mit den seismisch gemessenen verhältnismäßig gut übereinstimmen[5].

Die physikalisch zuverlässigsten Ergebnisse liefern die im Abschnitt C beschriebenen Methoden, die von manchen Nachteilen der bisher behandelten Verfahren frei sind, bei denen unmittelbar die Gruppengeschwindigkeit hochfrequenter Wellen gemessen wird.

*γ) Zur Frequenzabhängigkeit des Elastizitätsmoduls (Dispersion).* Die meisten dynamischen Elastizitätsmessungen werden bei hohen Frequenzen ausgeführt und es entsteht die Frage, ob diese Werte exakt auf den niederfrequenten Bereich der seismischen Praxis übertragen werden können, d.h., ob die elastischen Konstanten frequenzabhängig sind. Dazu beobachteten BIRCH und BANCROFT[6] an einer Granitprobe longitudinale, Torsions- und Biegeschwingungen im Frequenzbereich von 140 bis 4500 Hz. Innerhalb einer Meßgenauigkeit von 1% wurde bei den elastischen Konstanten *keine Frequenzabhängigkeit* gefunden. Es tritt hier keine Dispersion auf, und sie ist auch bei höheren Frequenzen nicht zu erwarten.

Nun sind die bei höheren Frequenzen dynamisch bestimmten $E$-Moduln von Gesteinen meist um etwa 20% größer als die statisch gemessenen. Daher

[1] Ähnliche Ergebnisse erhielt E. MÜLLER, vgl. Ziff. 15, Tabelle 11.

[2] ZISMAN: a. a. O.

[3] J. M. IDE: An Experimental Study of the Elastic Properties of Rocks. Geophysics 1, 347 (1936) u. a. a. O.

[4] L. D. LEET: Velocity of Elastic Waves in Granite and Norite. Physics 4, 375 (1933).

[5] Über einen Vergleich von seismischen Feld- und dynamischen Laboratoriumsmessungen an Sedimentgesteinen berichtet C. W. OLIPHANT in Geol. Soc. Amer. Bull. 61, No. 7, 759 (1950).

[6] F. BIRCH u. D. BANCROFT: Bull. Seism. Soc. Amer. 28, 243 (1938).

liegt die Vermutung nahe, daß sich beide Werte bei niedrigen Frequenzen annähern und somit ein Übergang vom dynamischen zum statischen Fall vorhanden ist. Die Frage, ob $E$ bei Frequenzverminderung abnimmt, wurde im Bereich von 120 bis 40 Hz von Bruckshaw und Mahanta[1] an einseitig eingeklemmten Proben von $6{,}5 \times 2{,}5 \times 1$ cm mit Hilfe induktiver Geber und Empfänger untersucht. Bei den gemessenen Proben aus Granit, Dolerit, Kalk- und Sandstein nimmt $E$ in der Tat ab, wie in Fig. 11 dargestellt ist. Jedoch beträgt diese Abnahme hier nur 2,1 bis 2,6% und wird um so kleiner, je höher die Frequenzen

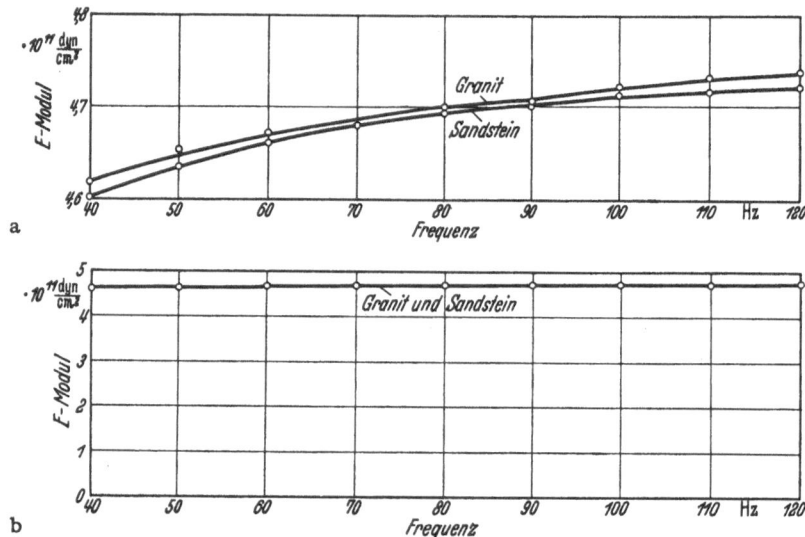

Fig. 11 a u. b. a Der $E$-Modul in Abhängigkeit von der Frequenz für zwei Proben aus Granit und Sandstein. b Ausschnitt aus a. $E$ ändert sich zwischen 40 und 120 Hz geringfügig um 2,1 bis 2,6%. (Nach Bruckshaw und Mahanta.)

sind. Die meist größere Differenz zwischen statischen und dynamischen Messungen wird dadurch nicht erklärt, es sei denn, man nimmt starke Änderungen unterhalb 40 Hz an, was jedoch nach allgemeiner Erfahrung nicht zu erwarten ist. Für die Prospektionsseismik, die üblicherweise mit Frequenzen zwischen 40 und 60 Hz arbeitet, spielt die geringe Abnahme von $E$ und die damit verbundene Verringerung der Wellengeschwindigkeiten um etwa 1% keine nennenswerte Rolle.

## C. Messung der Wellenausbreitung in Gesteinsproben.

**11. Allgemeines.** Die Erkenntnisse und Fortschritte auf den Gebieten des *Ultraschalls*[2] und der elektronischen Verfahren zur Erzeugung und Messung von kurzzeitigen Impulsen sind in den letzten Jahren in zunehmendem Maße für die Untersuchung der Wellenausbreitung in Gesteinsproben nutzbar gemacht worden. Hierbei werden die Ausbreitungsgeschwindigkeiten longitudinaler und transversaler Wellen gemessen und daraus die elastischen Konstanten berechnet. Ist die Wellenlänge $\lambda$ klein gegenüber den Abmessungen der Proben, so gelten

[1] J. M. Bruckshaw u. P. C. Mahanta: The Variation of the Elastic Constants of Rocks with Frequency. Vortragsmanuskript von der Tagung der Europ. Assoc. of Expl. Geophys., London 1952.

[2] L. Bergmann: Der Ultraschall und seine Anwendung in Wissenschaft und Technik. Stuttgart: S. Hirzel 1954.

für ein elastisches und isotropes Medium die Beziehungen[1]

$$V_l = \sqrt{\frac{E}{\varrho} \cdot \frac{1-\sigma}{(1+\sigma)(1-2\sigma)}}, \quad V_t = \sqrt{\frac{E}{\varrho} \cdot \frac{1}{2(1+\sigma)}}, \quad \sigma = \frac{V_l^2 - 2V_t^2}{2(V_l^2 - V_t^2)} = \frac{E}{2G} - 1,$$

$$G = V_t^2 \varrho, \quad E = 2G(1+\sigma) \quad \text{und} \quad \varkappa = \frac{3(1-2\sigma)}{E}.$$

Für $\lambda$ groß gegenüber dem Durchmesser einer stabförmigen Probe wird die Geschwindigkeit $V_d$ der *Dehnwelle* nach $V_d = \sqrt{E/\varrho}$ gemessen.

Die Ausbreitungsgeschwindigkeit der Wellen läßt sich auf verschiedene Weise bestimmen. Werden die Laufzeiten von Ultraschall-Impulsen nur für den Weg zwischen dem am Anfang einer Gesteinsprobe angebrachten Sender und dem am Ende der Probe befestigten Empfänger gemessen, so spricht man von *Durchstrahlungs-Verfahren*. HUGHES und JONES[2], HUGHES und CROSS[3], BUCHHEIM[4], RENTSCH[5] und RÖSLER[6] haben nach dieser Methode die Wellengeschwindigkeiten in verschiedenen Gesteinen bestimmt. Die Proben haben gewöhnlich einen Querschnitt von einigen Quadratzentimetern und Längen bis zu rund 10 cm, gegebenenfalls auch bis zu 20 cm. Sie lassen sich in Druckpressen und Druckgefäßen — KÁRMÁN-Gerät — bei verschiedenen Längs- und Manteldrucken und bei verschiedenen Temperaturen untersuchen.

Nach dem *Reflexionsverfahren*, bei dem die Laufzeit für Hin- und Rückweg in einer Probe gemessen wird, hat LUTSCH[7] mit einem für die Werkstoffprüfung gedachten Ultraschall-Echogerät sowohl $V_l$ als auch $V_t$ gemessen.

BAULE[8] ermittelt die Geschwindigkeit der Dehnwelle $V_d$ aus einer Laufzeitkurve für zahlreiche Meßpunkte entlang der stabförmigen Gesteinsprobe, die bis zu 2 m lang sein kann (Bohrkerne). BACHER[9] beobachtet die DEBYE-Beugungsbilder für $V_l$ und $V_t$, die beim schrägen Durchtritt von Ultraschall durch planparallele Gesteinsplättchen von wenigen Zentimeter Dicke auftreten.

Mit Hilfe der *Funkenphotographie* von Kopfwellen hat E. MÜLLER[10] nach der Schlierenmethode $V_l$ und $V_t$ aus den Winkeln ihrer Kopfwellenfronten zur ebenen Oberfläche der Gesteinsprobe bestimmt, die rund 3 bis 10 cm lang sein kann.

Alle genannten Verfahren sind vornehmlich für Messungen im Laboratorium gedacht. Am einfachsten in der Handhabung und mit dem geringsten Aufwand

---

[1] Die Beanspruchung des Festkörpers muß dabei so klein sein, daß Spannungen und Dehnungen linear zusammenhängen. Diese Voraussetzung ist bei Laboratoriumsuntersuchungen praktisch immer erfüllt. HÄNSEL und SCHARDIN berichteten auf der Tagung „Schall und Schwingungen in Festkörpern" (Göttingen, April 1955), daß selbst bei ungewöhnlich starken Explosionen an der Oberfläche von Metallproben keine Vergrößerung der Wellengeschwindigkeiten beobachtet werden konnte.

[2] HUGHES u. JONES: Variation of Elastic Moduli of Igneous Rocks with Pressure and Temperature. Bull. Geol. Soc. Amer. 61, 843 (1950). — Elastic Wave Velocities in Sedimentary Rocks. Trans. Amer. Geophys. Un. 32, Nr. 2 (1951).

[3] HUGHES u. CROSS: Elastic Wave Velocities in Rocks at High Pressures and Temperatures. Geophysics 16, No. 4 (1951).

[4] W. BUCHHEIM: Zum Problem der Drucksondierung in Gesteinen auf akustischer Basis. Bergakademie (Freiberg) 1, (1953).

[5] W. RENTSCH: Anwendung neuerer elektronischer Meßmethoden im Bergbau, I und II. Nachrichtentechnik 3, Nr. 5/7 (1953).

[6] R. RÖSLER: Experimentelle Untersuchungen zur Abhängigkeit der Schallgeschwindigkeit von der Druckbeanspruchung bei Gesteinen. Freiberger Forschungshefte C 12, 1954.

[7] A. LUTSCH: Eine einfache Methode zur Messung der elastischen Konstanten mit Hilfe von Ultraschallimpulsen. Z. angew. Phys. 4 (1952).

[8] H. BAULE: Laufzeitmessungen an Bohrkernen und Gesteinsproben mit elektronischen Mitteln. Geophysical Prospecting 1, No. 2 (1953).

[9] K. BACHER: Über die Bestimmung des elastischen Konstanten von Gesteinen mit Ultraschall. Erdöl u. Kohle 2, Nr. 4 (1949).

[10] E. MÜLLER: Experimente über Wellenausbreitung in Gesteinsproben. Geol. Jb. 70, 127 (1954). — Diss. Math.-Nat. Fak. Göttingen 1953.

an Auswertearbeit verbunden sind die direkten Methoden der Laufzeitmessung mit Hilfe des Braunschen Rohres. Die an kleinen Proben gewonnenen Ergebnisse können natürlich nur auf die Umgebung übertragen werden, wenn die Proben dafür repräsentativ sind. Besondere Beachtung ist in Zukunft Versuchen zu schenken, mit solchen Meßapparaturen die elastischen Konstanten des Gebirges *in situ* zu ermitteln, wie es von Buchheim[1] angedeutet worden ist.

**12. Methodik der Ultraschallmessungen.** In die Gesteinsproben werden mit piezoelektrischen oder magnetostriktiven Sendern kurze Ultraschall-Impulse hineingesendet und nach Durchlaufen eines abgemessenen Weges mit entsprechend hoch abgestimmten Empfängern — *Geophonen* — aufgenommen. Für kontinuierliche Laufzeitmessungen in der Gesteinswand von Bohrlöchern hat Vogel[2] einen Knallfunkensender beschrieben.

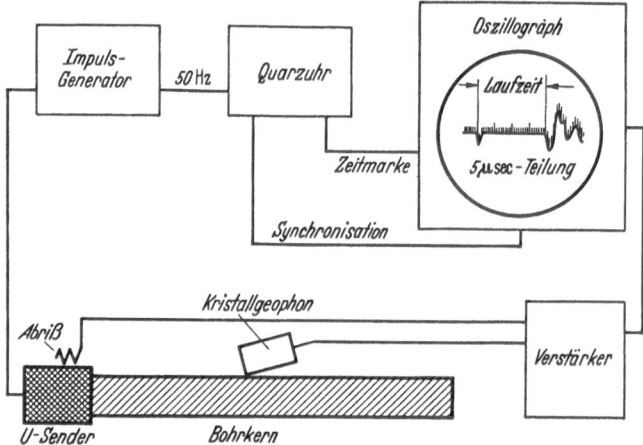

Fig. 12. Blockschaltbild zur elektronischen Laufzeitmessung. (Nach Baule.)

Die Frequenz der erzeugten Wellen beträgt bei $V_l$- und $V_t$-Messungen meist einige 100 kHz bis zu mehreren MHz, bei $V_d$-Messungen bis zu 25 kHz. Die Ankoppelung der Sender und Empfänger an die geglätteten Gesteinsproben geschieht mechanisch, oft unter Verwendung einer dünnen Zwischenschicht, z.B. Maschinenöl oder Fensterkitt. Der Sende-Impuls und das am Empfänger ankommende Signal werden, nach entsprechender Verstärkung und Synchronisierung, auf dem Schirm einer Braunschen Röhre zusammen mit einer meist quarzgesteuerten Zeitmarkierung sichtbar gemacht. Die hohe Zeitauflösung der Elektronenstrahloszillographen gestattet nach Hughes und Cross[3] eine Ablesung der Laufzeit bis auf 0,02 μsec für eine Probenlänge von rund 7 cm. Nach Bergmann[4] gibt Bradfield[5] einen guten Überblick über die bei solchen Verfahren allgemein zu beachtenden Punkte für die Art und Anbringung von Sender und Empfänger an die Probe und beschreibt eine Impulsanordnung, mit der sich noch Laufzeiten von 0,01 μsec messen lassen.

In Fig. 12 ist als Beispiel für eine Ultraschall-Laufzeitmessung das Prinzip der von Baule[6] beschriebenen Anordnung für $V_d$- und E-Modul-Bestimmungen

---

[1] Buchheim: a. a. O.
[2] C. B. Vogel: A Seismic Velocity Logging Method. Geophysics **17**, No. 3 (1952).
[3] Hughes und Cross: a. a. O.
[4] Bergmann: a. a. O., S. 643.
[5] G. Bradfield: Precise Measurement of Velocity and Attenuation using Ultrasonic Waves. (Kongreßbericht der Ultraschalltagung in Rom.) Nuovo Cim. **7**, Suppl. 2, 162—181 (1950).
[6] Baule: a. a. O.

wiedergegeben, die sich bei Serienmessungen an vielen hundert Bohrkernen bewährt hat. Mit dem magnetostriktiven Ultraschallsender wird in ein Ende des Bohrkerns 50mal pro sec ein Ultraschall-Impuls von 22 kHz hineingesendet

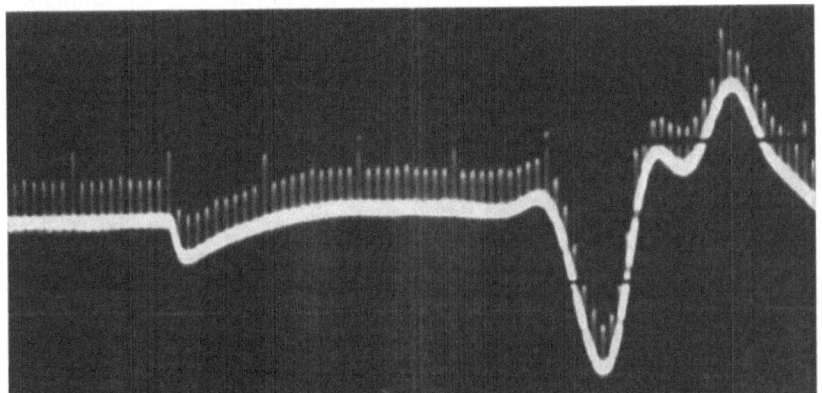

Fig. 13. Ultraschall-Oszillogramm eines Sandsteinkerns. Zeitmarken von 5 μsec Abstand. (Nach BAULE.)

und an vielen, verschieden weit vom Sender entfernten Meßpunkten am Bohrkern entlang mit dem Kristallgeophon empfangen. Die Entfernung der Meßpunkte vom Sender wird auf einem Metallmeterstab auf mindestens 1 mm genau abgelesen. Der Sende-Impuls — Abriß — und das am Geophon ankommende

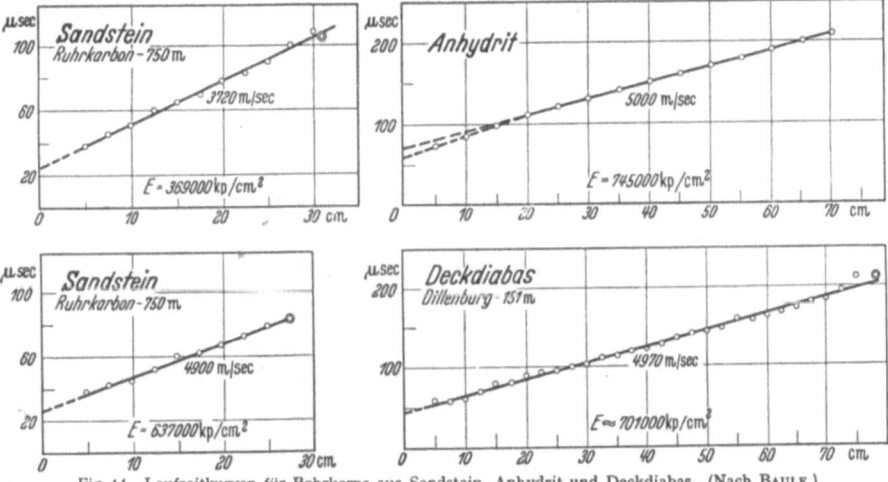

Fig. 14. Laufzeitkurven für Bohrkerne aus Sandstein, Anhydrit und Deckdiabas. (Nach BAULE.)

Signal werden zusammen mit den Zeitmarken einer Quarzuhr als stillstehendes Bild sichtbar gemacht und die Laufzeit abgemessen. Der Zeitmarkenabstand beträgt 5 μsec, jede 10. Marke ist durch Strichverlängerung herausgehoben. Ein Oszillogramm für einen Meßpunkt ist in Fig. 13 wiedergegeben.

Mit den Laufzeiten vieler Punkte wird die Laufzeitkurve gezeichnet, aus der sich die Geschwindigkeit ergibt. Einige Beispiele sind in Fig. 14 dargestellt.

In Fig. 15 sind die Dehnwellen-Geschwindigkeit $V_d$ und der $E$-Modul für 41 Kerne einer Gesteinsfolge in einem Bohrloch durch Sandstein und Schieferton im Ruhrkarbon wiedergegeben.

Der Vorteil dieser Methode gegenüber den Durchstrahlungsmessungen besteht darin, daß durch die punktweise Abtastung der Laufzeiten entlang der Meßstrecke plötzliche Geschwindigkeitsänderungen mit erfaßt werden und sich die Streuung bzw. Abweichung der Meßwerte von der mittleren durchschnittlichen Geschwindigkeitsgeraden gut kontrollieren läßt. Ein Nachteil dieses Verfahrens ist darin zu erblicken, daß die POISSONsche Konstante $\sigma$ sich nicht bestimmen läßt; dazu werden zusätzlich die Methoden von LUTSCH[1] und MÜLLER[2] herangezogen.

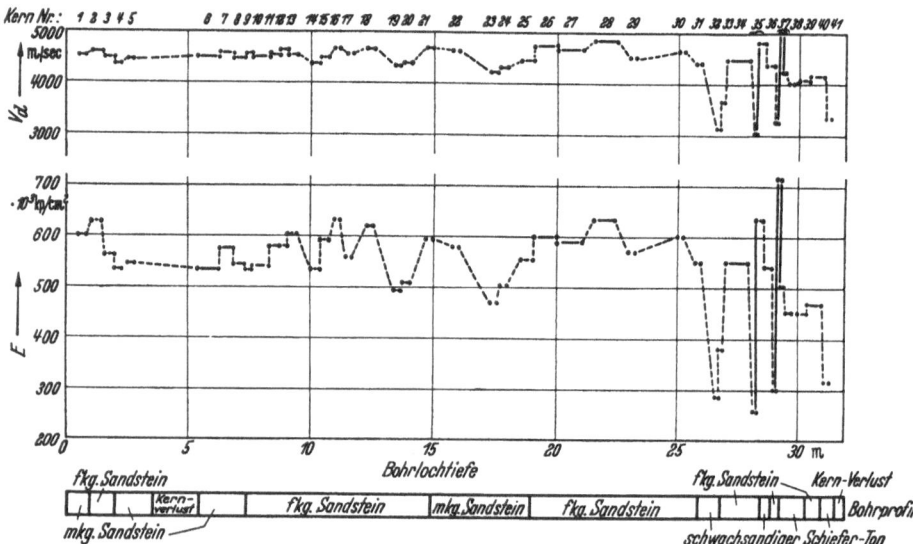

Fig. 15. Geschwindigkeit $V_d$ und E-Modul $E$ in Abhängigkeit von der Gesteinsart der Kerne aus einem Bohrloch des Ruhrkarbons. (Nach BAULE.)

## 13. Einfluß des äußeren Druckes und der Temperatur auf die Wellenausbreitung.

Allgemein wird angenommen, daß die Schallgeschwindigkeit in einem Gestein mit zunehmender Tiefe, d.h. mit zunehmendem Druck anwächst. Die wichtigsten Untersuchungen an Proben zu dieser Frage sind von HUGHES und JONES[3] und von HUGHES und CROSS[4] ausgeführt worden. Sie haben nach dem Durchstrahlungsverfahren mit Ultraschall-Impulsen die longitudinalen und transversalen Geschwindigkeiten in ummantelten[5] Gesteinsproben bei allseitigen Flüssigkeitsdrucken von 1 bis 5000 kp/cm² und bei Temperaturen von 25 bis 300° C in einem Druckzylinder untersucht. Die zylindrischen Gesteinsproben von etwa 2,5 cm Durchmesser und 4 bis 8 cm Länge waren dabei mit einem Mantel aus Kupferfolie oder Neoprene umgeben und die Enden, an denen der Kristallsender und Kristallempfänger angebracht war, mit dünnen Kupferkappen überzogen, um das Eindringen der Druckflüssigkeit in das Gestein zu verhindern. Die aus den Geschwindigkeiten berechneten POISSONschen Zahlen hatten für stark quarzitisches Gestein niedrige Werte von 0,13 bis 0,20, während sich für die meisten Gesteine Werte zwischen 0,26 bis 0,33 ergaben.

Tabelle 8 von HUGHES und CROSS zeigt für trockenen Sandstein die Zunahme von $V_l$ und $V_t$ mit wachsendem Druck und die geringe Abnahme mit zunehmender Temperatur.

---

[1] LUTSCH: a. a. O.
[2] MÜLLER: a. a. O.
[3] HUGHES u. JONES: a. a. O.
[4] HUGHES u. CROSS: a. a. O.
[5] „ummantelt" = bedeckt im Sinne von Ziff. 5.

Tabelle 8.

*$V_l$ und $V_t$ eines trockenen Sandsteins in m/sec bei verschiedenen Drucken und Temperaturen (nach* HUGHES *und* CROSS*).* Dichte 2,543 g/cm³, Porosität 5,1%. Aus etwa 2000 m Tiefe.

| T °C | 27 | | 100 | | 200 | |
|---|---|---|---|---|---|---|
| P(bar) | $V_l$ | $V_t$ | $V_l$ | $V_t$ | $V_l$ | $V_t$ |
| 0 | 3672 | | | | | |
| 50 | 3936 | 2483 | 3601 | | 3077 | |
| 100 | 4042 | 2514 | 3819 | | 3295 | |
| 250 | 4367 | 2594 | 4089 | 2472 | 3801 | |
| 500 | 4577 | 2700 | 4382 | | 4144 | 2600 |
| 750 | 4758 | 2807 | 4587 | 2688 | 4411 | 2637 |
| 1000 | 4870 | 2847 | 4737 | 2751 | 4575 | 2675 |
| 1500 | 5020 | 2884 | 4929 | 2833 | 4852 | 2797 |
| 2000 | 5110 | 2925 | 5061 | 2876 | 4992 | 2846 |
| 2500 | 5167 | 2944 | 5151 | 2922 | 5102 | 2879 |
| 3000 | 5127 | 2956 | 5208 | 2966 | 5157 | 2922 |
| 4000 | 5279 | 2975 | 5295 | 2969 | 5225 | 2944 |
| 5000 | 5363 | 2971 | 5345 | 2973 | 5312 | 2975 |

Die Tabelle 9 gilt für die mit Wasser gesättigte Probe, und Fig. 16 zeigt für beide Proben die Longitudinalgeschwindigkeit in Abhängigkeit vom Druck bei Zimmertemperatur. Bei niedrigen Drucken ist $V_l$ der trockenen Probe kleiner

Tabelle 9. *$V_l$ und $V_t$ eines mit Wasser gesättigten Sandsteins bei verschiedenen Drucken und Temperaturen (nach* HUGHES *und* CROSS*).* Dichte 2,606 g/cm³, Porosität 5,1%.

| T °C | 27 | | 100 | 200 |
|---|---|---|---|---|
| P (bar) | $V_l$ | $V_t$ | $V_l$ | $V_l$ |
| 0 | 4453 | 2319 | | |
| 50 | 4548 | 2432 | 4213 | |
| 100 | 4462 | 2687 | 4196 | 3789 |
| 250 | 4469 | 2761 | 4217 | 3802 |
| 500 | 4505 | 2896 | 4233 | 3873 |
| 750 | 4537 | 3008 | 4268 | 3932 |
| 1000 | 4553 | 3104 | 4307 | 3960 |
| 1500 | 4608 | 3263 | 4324 | 4074 |
| 2000 | 4651 | 3377 | 4373 | 4180 |
| 2500 | 4689 | 3468 | 4411 | 4295 |
| 3000 | 4706 | 3598 | 4442 | 4397 |
| 4000 | 4782 | 3651 | 4535 | 4600 |
| 5000 | 4887 | 3945 | 4660 | 4796 |

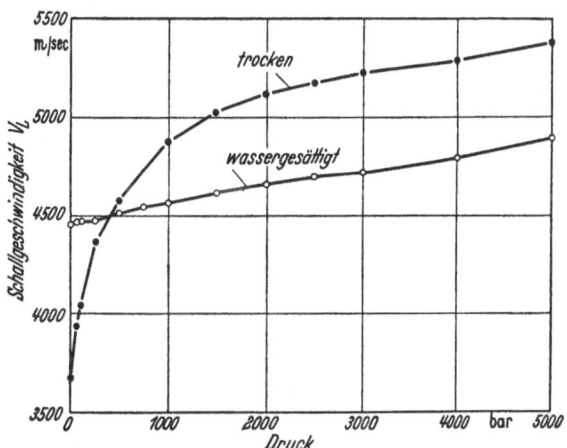

Fig. 16. Geschwindigkeit $V_l$ eines trockenen und eines wassergesättigten Sandsteins in Abhängigkeit vom Druck. (Nach HUGHES und JONES.)

als bei der feuchten; $V_l$ wächst bei der trockenen Probe mit zunehmendem Druck schneller und erreicht höhere Werte als bei der feuchten Probe.

Eine besonders starke Zunahme der Schallgeschwindigkeit mit steigendem Druck hat BUCHHEIM[1] bei Steinkohle festgestellt. Bereits bei 50 kp/cm² Längsdruck einer ummantelten Probe betrug die Geschwindigkeitszunahme 60%.

**14. Dynamische Messung der elastischen Hysteresis.** Mit dem elektronischen Durchstrahlungsverfahren haben BUCHHEIM[1] und RÖSLER[2] verschiedene Gesteinsarten sowohl in Druckrichtung als auch quer dazu bei steigenden und fallenden Drucken zwischen 0 und etwa 150 kp/cm² untersucht. Sie benutzten für ihre Messungen kleine magnetostriktive Sender aus Nickelblech und piezoelektrische

---

[1] BUCHHEIM: a. a. O.

[2] RÖSLER: a. a. O.

Empfänger, die in einem Abstand von rund 15 bis 25 cm in Vertiefungen der Gesteinsproben eingelassen waren. Die Proben wurden, mit und ohne Ummantelung aus Stahl, bei einseitiger, langsam wechselnder Druckbeanspruchung bis etwa 200 kp/cm² in einer Schopper-Presse untersucht[1].

Auf diese Weise ergab sich, daß die Werte der Ultraschallgeschwindigkeit bei steigendem Druck nicht mit denen bei fallendem Druck übereinstimmen, sondern eine Nachwirkung, die *elastische Hysteresis*, besteht. Wie aus Fig. 17 zu ersehen ist, nimmt bei einem nicht ummantelten Sandstein die Breite der

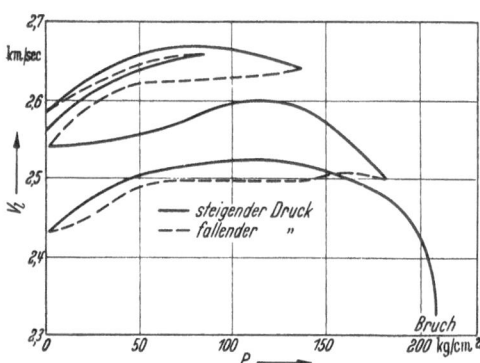
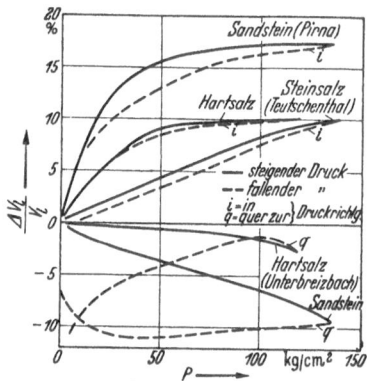

Fig. 17. Verlauf von $V_l$ in einer nassen Sandsteinprobe ohne Ummantelung bei wechselnder Be- und Entlastung bis zum Scherungsbruch. (Nach Buchheim.)

Fig. 18. Prozentuale Änderung $\Delta V_l/V_l$ in Druckrichtung und quer dazu bei nicht ummantelten Proben von Sandstein, Hartsalz und Carnallit. (Nach Buchheim.)

Hystereseschleife für die Geschwindigkeit *quer* zur Druckrichtung mit wachsendem Maximalwert des Druckes stark zu. Außerdem ist die äußerst starke Verminderung von $V_l$ bei Annäherung an den Bruch der Probe zu beachten. Schmale Hystereseschleifen ergeben sich für die in Richtung des *Längsdruckes* gemessenen Geschwindigkeiten, deren prozentuale Änderungen $\Delta V_l/V_l$ für einige nicht ummantelte Proben unter anderem in Fig. 18 wiedergegeben sind.

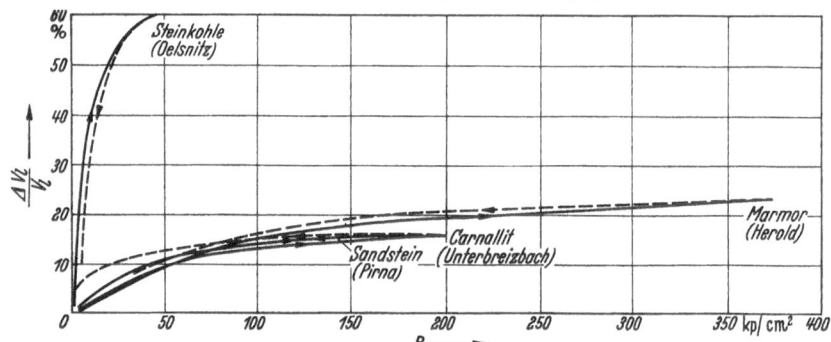

Fig. 19. Prozentuale Änderung $\Delta V_l/V_l$ in Druckrichtung bei ummantelten Proben von Sandstein, Carnallit, Marmor und Steinkohle. (Nach Buchheim.)

Eine weitere Verengung der Hystereseschleife wird nach Buchheim[2] dann erzielt, wenn die Druckänderungen nur sehr langsam, etwa im Verlaufe von Tagen oder Wochen erfolgen.

Die Hystereseschleifen für einige mit Stahlmantel versehene Gesteinsproben bei in Druckrichtung gemessener Geschwindigkeit werden in Fig. 19 gezeigt. Mit

[1] Einzelheiten bei Rösler: a. a. O.
[2] Buchheim: a. a. O.

steigendem Druck nähert sich die Schallgeschwindigkeit asymptotisch einem Grenzwert, der im allgemeinen 10 bis 20% höher als der Normalwert liegt. Dieser Grenzwert kann bei etwa 200 kp/cm² Längsdruck und rund 100 kp/cm² Manteldruck als praktisch erreicht gelten. Eine Ausnahme hiervon bildet Steinkohle, wie bereits oben erwähnt wurde.

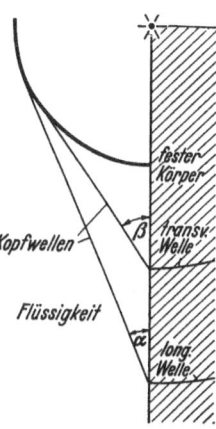

**15. Funkenphotographie von Kopfwellen.** Eine besonders anschauliche Meßmethode zur Bestimmung der Wellengeschwindigkeiten $V_l$ und $V_t$ in sehr kleinen Gesteinsproben bei Zimmertemperatur und atmosphärischem Druck ist von ERICH MÜLLER[1] beschrieben worden, der auf Arbeiten von MACH-CRANZ-SCHARDIN[2] und von SCHMIDT[3] zurückgreift. Man taucht die kleinen Gesteinsproben von 3 bis 10 cm Länge in eine mit Flüssigkeit gefüllte Küvette und photographiert die in der Flüssigkeit auftretenden Kopfwellen (MINTROP[4]-Wellen) mit Hilfe eines Funkenblitzes nach der TÖPLERschen Schlierenmethode. Die Winkel $\alpha$ und $\beta$ — vgl. Fig. 20 — zwischen den geradlinig verlaufenden Kopfwellen und der ebenen Gesteinsoberfläche werden gemessen und daraus die Wellengeschwindigkeiten nach den Beziehungen

Fig. 20. Kugel- und Kopfwellen in der Umgebung der Grenzfläche einer Gesteinsprobe-Flüssigkeit nach Zündung des Knallfunkens. (Nach E.MÜLLER.)

$$V_l = \frac{c}{\sin \alpha} \quad \text{und} \quad V_t = \frac{c}{\sin \beta}$$

berechnet, wobei $c$ die Schallgeschwindigkeit in der Flüssigkeit ist. Aus $V_l$, $V_t$ und der Gesteinsdichte $\varrho$ lassen sich dann alle anderen elastischen Konstanten bestimmen.

Auf den Knallfunken $F_1$ in der Küvette $K$ soll nach etwa $10^{-5}$ sec (durch Induktivität $L$ regelbar) der helle Photoblitz $F_2$ von $10^{-7}$ sec Dauer folgen (s. Fig. 21). Dazu werden die

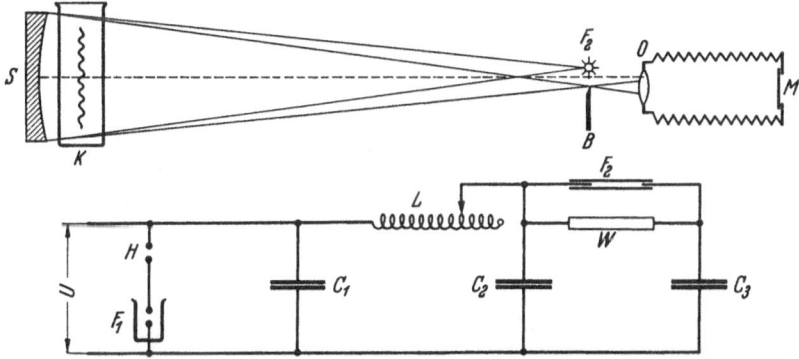

Fig. 21. Schema der optischen und elektrischen Versuchsanordnung (TÖPLERsche Schlierenmethode und MACHsche Schaltung). (Nach E. MÜLLER.)

Kondensatoren $C_1$ bis $C_3$ mit einer Spannung $U$ von 20 bis 30 kV aufgeladen, bis $F_1$ und die Hilfsfunkenstrecke $H$ zünden, wobei $C_3$ zunächst durch den hohen Widerstand $W$ isoliert wird, während sich $C_1$ und $C_2$ oszillierend entladen. Nach einer Halbschwingung des Kreises $L C_2$ tritt zwischen den Kondensatorplatten von $C_2$ und $C_3$, die sich vorher auf gleichem

[1] E. MÜLLER: a. a. O.

[2] H. SCHARDIN: Das TÖPLERsche Schlierenverfahren. VDI-Forsch.-Heft, Ausg. B **5**, 367 (1934).

[3] O. v. SCHMIDT: Über Knallwellenausbreitung in Flüssigkeiten und festen Körpern. Phys. Z. **39**, 868 (1938). — Über Kopfwellen in der Seismik. Z. Geophys. **15**, 141 (1939).

[4] L. MINTROP: 100 Jahre physikalische Erdbebenforschung und Sprengseismik. Naturwiss. **34**, 257, 289 (1947). — Über Anwendungen des seismischen Verfahrens im Erdölbergbau und ihre wirtschaftlichen und wissenschaftlichen Auswirkungen. Öl und Kohle **1943**, 269.

Potential befanden, eine Spannung von annähernd $2U$ auf und jetzt entlädt sich auch $C_3$ über die scharf berandete Funkenstrecke $F_2$, die von einem Hohlspiegel $S$ auf die Schneide der Schlierenblende $B$ abgebildet wird. Das Photoobjektiv $O$ entwirft von $K$ ein Bild auf der Mattscheibe $M$. Etwas Licht wird von den Kopfwellen nach der Seite gebrochen, fällt an $B$ vorbei und macht die Schlieren auf der Mattscheibe sichtbar[1].

Die Gesteinsproben können beliebige Gestalt besitzen, benötigt wird nur *eine* ebene Fläche. An dieser wird ein dünner Draht befestigt, auf den der Knallfunken überschlagen kann. Änderungen der elastischen Eigenschaften haben eine Deformation der geradlinigen Kopfwellen zur Folge und machen sich dadurch bemerkbar. Genauigkeitsangaben finden sich am Schluß von Ziff. 16.

In Fig. 22 sind die Schlierenaufnahmen für eine Tonschiefer-, Sandstein- und Lamprophyrprobe wiedergegeben. Man erkennt (vgl. Fig. 20) die unterschiedlichen Neigungen der longitudinalen und transversalen Kopfwellen in der Flüssigkeit.

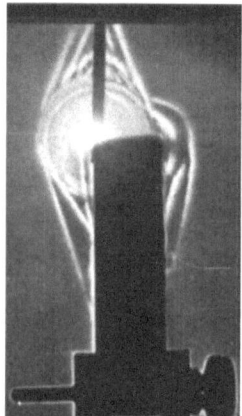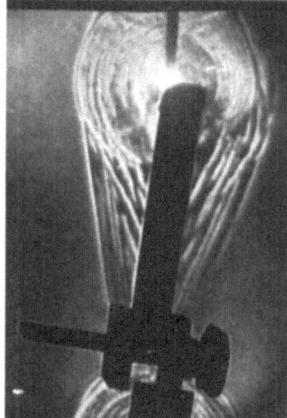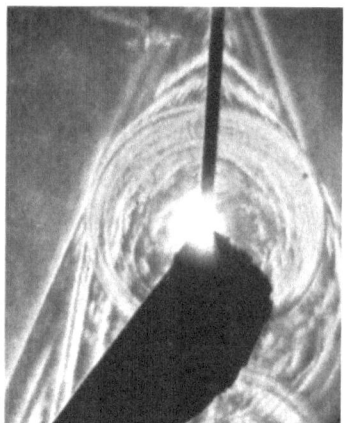

Fig. 22. Schlierenaufnahmen mit den Kopfwellenfronten von $V_l$ und $V_t$ (links im Bilde) bei Proben aus Tonschiefer, Sandstein und Lamprophyr. (Nach E. MÜLLER.)

Auf den Bildern sind die vertikal- oder schiefstehenden Proben zu sehen. Zwischen einem Metallstift und der linken ebenen Gesteinsoberfläche zündet der Knallfunken (heller Fleck). Die Momentaufnahmen zeigen die Kugelwelle im Wasser und — links auf den Bildern — die Kopfwellen. Auch der Metallstift ist von Kopfwellen umgeben.

In den Tabellen 10 und 11 sind einige Ergebnisse für verschiedene Gesteine nach Messungen von MÜLLER zusammengestellt. Die kleinsten Geschwindigkeitswerte, die noch gemessen werden können, werden bestimmt durch die Schallgeschwindigkeit im flüssigen Medium. Bei Verwendung von Tetrachlorkohlenstoff liegt diese untere Grenze bei 950 m/sec.

Mit einer prinzipiell gleichen Versuchsanordnung untersucht HELBIG[2] die Wellenausbreitung in *anisotropen Festkörpern*. Geschichtete Medien verhalten sich gegenüber elastischen Wellen anisotrop, sofern die Schichtdicken genügend klein zur Wellenlänge sind. Die Ausbreitungsverhältnisse in solchen Medien sind von RUDZKI[3], MATUZAWA[4], STONELEY[5] und RIZNICHENKO[6] untersucht worden.

---

[1] E. MÜLLER: Naturwiss. Rdsch. **1954**, H. 1, 29.

[2] K. HELBIG: Diss. Univ. Göttingen 1955.

[3] M. P. RUDZKI: Parametrische Darstellung der elastischen Welle in anisotropen Medien. Anz. Akad. Wiss. Krakau **1911**, 503.

[4] T. MATUZAWA: Elastische Wellen in einem anisotropen Medium. Tokio Imp. Univ. Earthqu. Res. Inst. Bull. **21**, 231 (1943).

[5] R. STONELEY: The Seismological Implications of Aeolotropy in Continental Structure. Mon. Not. Roy. Astronom. Soc., Geophys. Suppl. **5**, 343 (1949). — Polarisation of the S-Phase of Seismograms. Ann. Geofisica, Roma **4**, 3 (1951).

[6] J. V. RIZNICHENKO: Seismic Quasi-Anisotropy. Izv. Akad. Nauk SSSR., Ser. geogr. e geofiz. **13**, Nr. 6, 518 (1949).

Tabelle 10. *Einige Meßergebnisse nach* E. MÜLLER.

| Gestein, Herkunft und Tiefe | $V_l$ m/sec | $V_t$ m/sec | $\sigma$ | $E$ $10^{11}$ dyn/cm² | $G$ $10^{11}$ dyn/cm² | $\varkappa$ $10^{-11}$ cm²/dyn |
|---|---|---|---|---|---|---|
| Granit, Selb, Bayern . . . . . . . | 5200 | 3000 | 0,25 | 6,0 | 2,4 | 2,5 |
| Pechstein, Polenz, Meißen. . . . . | 5400 | 3100 | 0,25 | 5,5 | 2,2 | 2,7 |
| Basalt, Backenberg. . . . . . . . | 6400 | 3200 | 0,33 | 7,8 | 2,9 | 1,3 |
| Eklogit, Silberbach, Fichtelgebirge . | 8000 | 4300 | 0,30 | 16,3 | 6,3 | 0,74 |
| Anhydrit, Rehden | | | | | | |
| 2174 m, ∥ zur Schicht . . . . . | 5500 | 2400 | 0,38 | 4,7 | 1,7 | 1,5 |
| ⊥ zur Schicht . . . . . | 5150 | 2600 | 0,33 | 5,3 | 2,0 | 1,9 |
| 2180 m, ∥ zur Schicht . . . . . | 5700 | 2700 | 0,36 | 5,8 | 2,1 | 1,4 |
| ⊥ zur Schicht . . . . . | 5500 | 2600 | 0,36 | 5,4 | 2,0 | 1,6 |
| Anhydrit, Mittelwert von 12 Proben | | | | | | |
| ∥ zur Schicht . . . . . . . . | 5610 | 2690 | 0,352 | 5,7 | 2,11 | 1,56 |
| Plänerkalk, Sénon-Emscher, 1460 m | 2400 | 1350 | 0,27 | 1,1 | 0,45 | 12 |
| Mergelton, 1590 m . . . . . . . | 3100 | 1750 | 0,27 | 1,6 | 0,63 | 9 |
| Kalksandstein, 1721 m . . . . . | 3250 | 2000 | 0,20 | 2,3 | 0,97 | 8 |
| Feinkörniger Sandstein, 2170 m . . | 3500 | 2150 | 0,19 | 2,8 | 1,2 | 7 |

Tabelle 11. *Zusammenhang zwischen Gesteinsdichte* $\varrho$ *und den elastischen Konstanten für sechs Buntsandsteinproben*[1]. *Nach* E. MÜLLER.

| Buntsandstein Bohrung Rehden 5 Tiefe in m | $\varrho$ g/cm³ | $V_l$ m/sec | $V_t$ m/sec | $\sigma$ | $E$ $10^{11}$ dyn/cm² |
|---|---|---|---|---|---|
| 1676—1677 | 2,58 | 4400 | 2200 | 0,33 | 3,3 |
| 2028—2030 | 2,66 | 4700 | 2500 | 0,30 | 4,3 |
| 1794—1797 | 2,66 | 4700 | 2600 | 0,28 | 4,6 |
| 1925—1927 | 2,71 | 4900 | 2600 | 0,30 | 4,8 |
| 1532—1534 | 2,79 | 5200 | 2900 | 0,27 | 6,0 |
| 1532—1542 | 2,82 | 5600 | 2900 | 0,32 | 6,3 |

Danach sind in „transversal-isotropen" Medien drei verschiedene Wellentypen möglich: Der erste Typ besitzt vorwiegend longitudinale, der zweite und dritte transversale Bewegungsanteile. Die beiden letzten unterscheiden sich durch ihre Polarisation. Eine Trennung in divergenzfreie und rotationsfreie Wellen — wie im isotropen Medium — tritt nur in Sonderfällen auf. Die Wellennormale fällt im allgemeinen nicht mit der Ausbreitungsrichtung zusammen, daher ist zwischen der Geschwindigkeit in Richtung der Normalen (Phasengeschwindigkeit) und der „Strahlgeschwindigkeit" (Gruppen- oder Signalgeschwindigkeit) zu unterscheiden. Die beiden Geschwindigkeiten sind auch dann verschieden, wenn keine Dispersion vorliegt. Nur für den Fall, daß die Wellenfront senkrecht auf der Ausbreitungsrichtung steht, ist die „Normalengeschwindigkeit" gleich der Strahlgeschwindigkeit. In transversal-isotropen Medien ist dies der Fall bei Ausbreitung in Richtung der Symmetrieachse und parallel zur Schichtung. Ein typisches transversalisotropes Gestein ist z. B. Schiefer. Zum experimentellen Studium der Ausbreitungsverhältnisse im Labor eignet sich Schiefer nicht, da man wegen der relativ großen Plattendicke sehr große Wellenlängen benötigen würde.

HELBIG untersucht Hartgewebe und Hartpapiere, die sich durch geringe Schichtdicken auszeichnen. Ein Beispiel veranschaulicht Fig. 23. Gemessen werden die unterschiedlichen Strahlenwege entlang der Oberfläche und durch das Innere der Probe (Strahlenweg $OB$ kleiner als $OA$). Dargestellt ist nur die erste (longitudinale) Raumwelle. Im vorliegenden

---

[1] Vgl. Ziff. 10a.

Fall wurde eine 20 mm starke Hartgewebeplatte untersucht, die horizontal geschichtet ist. In c) sind die Strahlgeschwindigkeiten in Abhängigkeit von der Strahlrichtung für dieselbe Probe aus einer Reihe von Aufnahmen aufgetragen.

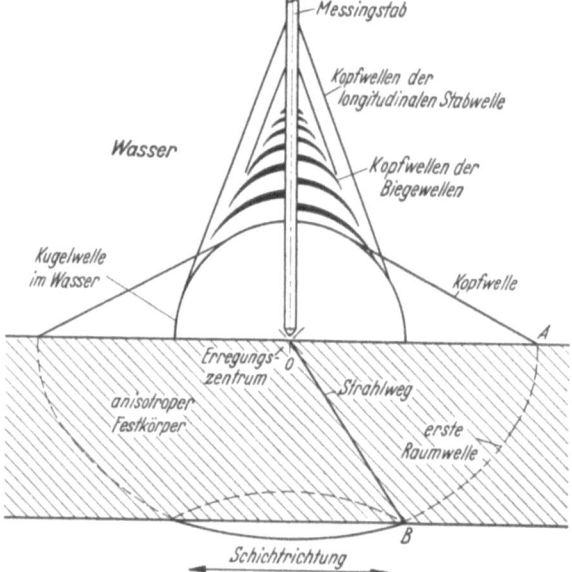

Auf den Fig. 22 und 23 sind auch Kopfwellen in der Umgebung des Messingstabes zu sehen, darunter solche, die zu den Biegewellen im Stab gehören. Biegewellen treten auf, da der Stabdurchmesser in der Größenordnung der Wellenlänge liegt. Die Krümmung und Vielzahl der Kopfwellen (schematische Darstellung s. Fig. 23a) rührt von der Dispersion der Biegewellen her; die Enveloppe dieser Biegewellen liefert die Richtung der „transversalen" Kopfwelle, die zu den transversalen Wellen im Stab gehört.

### 16. Ergebnisse von Vergleichsmessungen.

Bei einem Vergleich der Ergebnisse mehrere Autoren, die verschiedene Untersuchungsverfahren benutzt

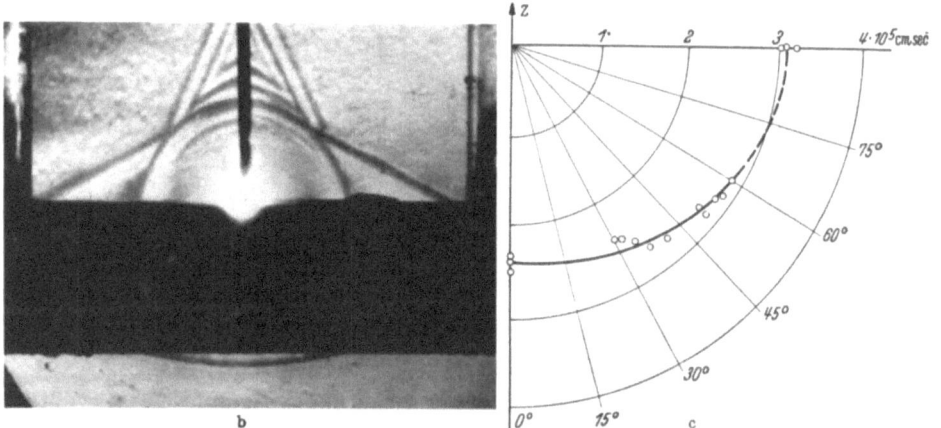

Fig. 23a—c. Elastische Anisotropie geschichteter Medien. a Schematische Darstellung. Wegen der Anisotropie sind die Strahlenwege $OA$ und $OB$ verschieden lang. b Eine Aufnahme des Wellenfeldes. c Strahlgeschwindigkeit in Abhängigkeit von der Strahlrichtung für Novotex-Hartgewebe. (Nach K. Helbig.)

haben, läßt sich, mit einigen Einschränkungen, feststellen, daß eine relativ gute Übereinstimmung der im Laboratorium an Proben ermittelten Geschwindigkeiten sowohl untereinander als auch mit denen der seismischen Praxis gefunden wurde.

Wie Tabelle 12 zeigt, haben Hughes und Cross[1] eine ganz besonders gute Übereinstimmung ihrer nach dem Durchstrahlungs-Verfahren ermittelten *Transversal*-Geschwindigkeiten mit denjenigen erzielt, die Birch[2] nach einer Resonanz-

---

[1] Hughes u. Cross: a. a. O.
[2] F. Birch: Handbook of Physical Constants, Geological Society of America Special Papers, No. 36, 1942.

methode ermittelte. Die im Laboratorium gemessenen Longitudinal-Geschwindigkeiten waren für einen Vergleich nicht geeignet.

Ein — sehr erwünschter — exakter Vergleich der nach verschiedenen Verfahren erhaltenen Ergebnisse ist natürlich nur möglich, wenn die verschiedenartigen Messungen an der *gleichen* Gesteinsprobe, soweit das überhaupt technisch möglich ist, ausgeführt werden. Aus diesem Grunde haben BAULE und MÜLLER mit Vergleichsmessungen begonnen, die folgendermaßen durchgeführt wurden: An zahlreichen Bohrkernen von mehreren dm Länge

Tabelle 12. *Transversalgeschwindigkeit $V_t$ in Abhängigkeit vom Druck.* Nach Messungen von BIRCH und HUGHES u. CROSS.

| Druck in bar | Solenhofener Schiefer $V_t$ in m/sec nach | |
|---|---|---|
|  | BIRCH | HUGHES u. CROSS |
| 1 | 2750 | 2854 |
| 500 | 2910 | 3008 |
| 4000 | 3080 | 3066 |

wurden von BAULE mit dem Laufzeitkurvenverfahren die Dehnwellengeschwindigkeit $V_d$ und der E-Modul bestimmt. Anschließend wurde von jedem so untersuchten Kern ein kurzes Probestück von 7 cm Länge abgesägt und dafür von MÜLLER nach der Schlierenmethode $V_l$ und $V_t$ ermittelt und unter anderem $\sigma$, $E$ und $V_d$ berechnet. Über ein Teilergebnis der bisherigen Vergleichsmessungen wird nachfolgend berichtet.

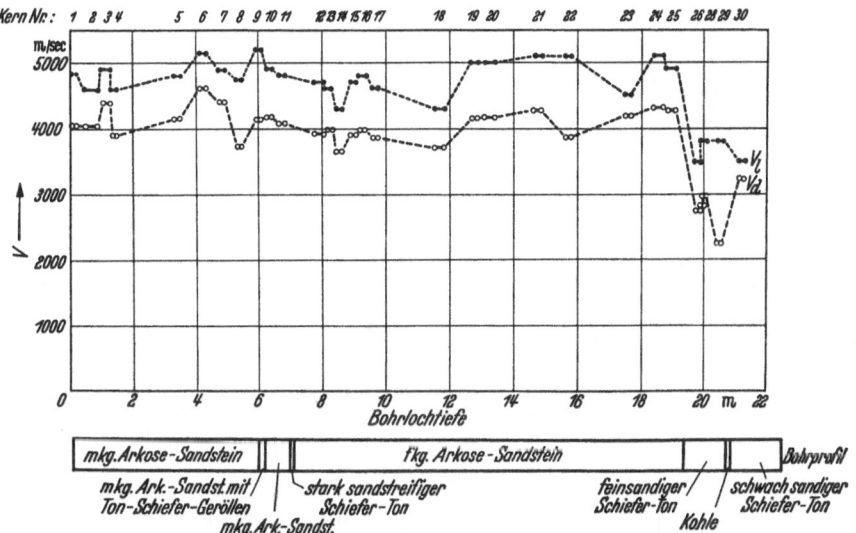

Fig. 24. Vergleich von $V_l$ (nach MÜLLER) und $V_d$ (nach BAULE) in Abhängigkeit von der Gesteinsart der Kerne eines Bohrloches aus dem Ruhrkarbon. ● Meßwerte nach MÜLLER. o Meßwerte nach BAULE.

In Fig. 24 sind über dem geologischen Bohrprofil[1] eines 21 m tiefen Bohrloches in Abhängigkeit von der Gesteinsart und der Bohrlochtiefe die für 30 Kerne von MÜLLER gemessenen Longitudinal-Geschwindigkeiten $V_l$ und die von BAULE gemessenen Dehnwellen-Geschwindigkeiten $V_d$ eingetragen. Die Kurven laufen fast parallel zueinander und zeigen eine gute Übereinstimmung insofern, als, der Theorie entsprechend, die $V_l$-Werte größer als die $V_d$-Werte sind. Durchschnittlich sind die $V_l$-Werte für die Kernstücke Nr. 10 bis 21 aus feinkörnigem Arkose-Sandstein um 19% höher als die $V_d$-Werte. Ein Vergleich der aus $V_l$

---

[1] Die petrographische Bestimmung der Gesteine wurde von Dr. F. W. HÜNERMANN, Bochum, vorgenommen.

und $V_l$ von Müller *berechneten* $V_d$-Werte nach der Beziehung

$$V_l = V_d \sqrt{\frac{1 - \sigma}{(1 + \sigma)(1 - 2\sigma)}}$$

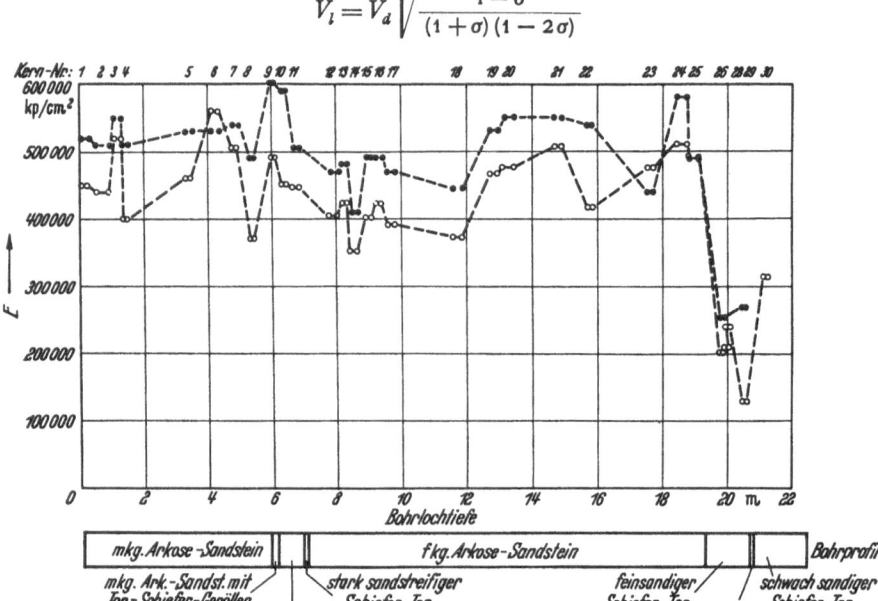

Fig. 25. Vergleich der $E$-Modulwerte an den Bohrkernen entsprechend Fig. 24. o Meßwerte nach Baule; ● Meßwerte nach Müller.

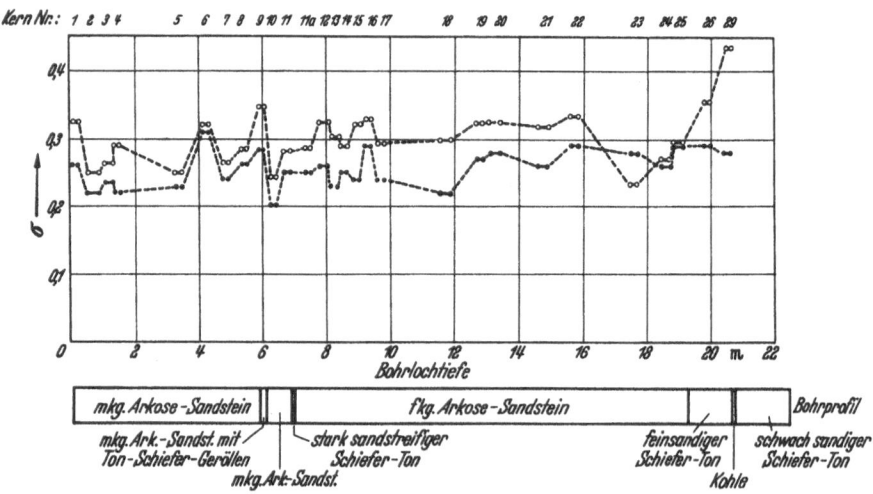

Fig. 26. Poisson-Zahl $\sigma$ der Bohrkerne entsprechend Fig. 24 und 25 aus $V_l$ und $V_t$ und aus $V_l$ und $V_d$ berechnet. (Nach Messungen von Müller und Baule.)

mit den von Baule *gemessenen* $V_d$-Werten zeigt allerdings keine gute Überein-stimmung. Ebenso ist bei einem Vergleich der für die gleichen Kerne nach bei-den Verfahren ermittelten $E$-Modulwerte — vgl. Fig. 25 — festzustellen, daß die von Müller gefundenen Werte systematisch um rund 17% für die Kerne Nr. 11 bis 18 höher als die von Baule ermittelten liegen.

In Fig. 26 ist die POISSONsche Konstante für die gleiche Bohrkernfolge in Abhängigkeit von der Gesteinsart und Bohrlochtiefe dargestellt. Die POISSON-Zahlen der unteren Kurve sind allein nach den Schlierenmessungen von MÜLLER aus $V_l$ und $V_t$ berechnet. Dagegen wurden die $\sigma$-Werte der oberen Kurve, die für die Kerne Nr. 11 bis 22 rund 21% höher liegen, aus den von BAULE gemessenen $V_d$- und den von MÜLLER gemessenen $V_t$-Werten berechnet. Die Kurven zeigen — abgesehen von den starken Abweichungen ab Kern Nr. 23 — fast parallelen Verlauf. Die Erwartung, daß sie übereinstimmen bzw. sich decken würden, ist nicht erfüllt worden.

Zur weiteren Klärung wurden sorgfältige Messungen an Metallen, Kunststoffen und Gesteinen durchgeführt. Bei mehreren Proben wurde der Elastizitätsmodul auch mit Hilfe eines sog. Hochfrequenzpulsators der Firma Amsler[1] gemessen. Die drei Verfahren arbeiteten im Frequenzbereich von etwa 200 Hz (AMSLER), 22 kHz (BAULE) und 2 MHz (MÜLLER).

Die Amsler Hochfrequenz-Pulsatoren für Materialprüfung arbeiten nach dem Resonanzprinzip, wobei die Betriebsfrequenz mit der Eigenfrequenz der Maschine zusammenfällt. Diese Frequenz wird durch die elastischen Eigenschaften der Maschine und des Probekörpers bestimmt. Der $E$-Modul eines Probestabes kann gemessen werden, wenn derjenige eines Vergleichsstabes bekannt ist. Gemessen werden hierbei die Frequenzen der Maschine mit Prüfling, mit Vergleichsstab und bei kurzgeschlossener Maschine. Die Proben werden dabei einer eindimensionalen Vorspannung von z. B. 300 kp (für die meisten Proben der Tabelle 13) und einer Wechsellast von 50 oder 100 kp ausgesetzt. Die Genauigkeit der Messung wird für die $E$-Moduln im Bereich von $(0,2 - 30) \cdot 10^5$ kp/cm² mit $\pm 3\%$ angegeben.

Tabelle 13. *Vergleichsmessungen.* $V_d$ und $E_B$ nach BAULE, $V_l$, $V_t$, $E_M$ und $\sigma_M$ nach MÜLLER, $\sigma_{BM}$ nach BAULE und MÜLLER, $E_A$ nach AMSLER (Vorspannung $P$ meist 300 kp). Genauigkeit von $V_l$ und $V_t$ bei den Metallproben 1 bis 2%, sonst 1,5 bis 2,5%. Weitere Genauigkeitsangaben im Text. $V_l$-Werte in m/sec, $E$-Werte in $10^5$ kp/cm².

| Probe | $V_l$ | $V_d$ | $V_t$ | $E_B$ | $E_M$ | $E_A$ | $\sigma_M$ | $\sigma_{BM}$ |
|---|---|---|---|---|---|---|---|---|
| Rundstahl, gezogen DIN 668 . . . . | 5850 | 5070 | 3000 | 20,5 | 20,6 | 22,0 | 0,290 | 0,295 |
| Silberstahl, legiert S 212 . . . . . . | 5850 | 5060 | 3070 | 20,3 | 19,6 | 22,7 | 0,310 | 0,296 |
| Stahldraht, federhart gezogen, DIN 671 . | 5920 | 5260 | 3000 | 22,4 | 21,4 | | 0,328 | 0,277 |
| Kupfer, elektrolyt hart . . . . . . | 4690 | 3820 | 2100 | 13,2 | 10,9 | 10,1 | 0,375 | 0,335 |
| Bondur (Duraluminium) . | 5900 | 5030 | 3000 | 7,4 | 7,0 | 6,7 | 0,326 | 0,306 |
| Plexiglas. . . . . . | 2770 | 2180 | 1350 | 0,57 | 0,59 | 0,6 | 0,344 | 0,350 |
| harte quarzitische Ruhrkarbon-Sandsteine Nr. 1 . . . . . . . | 5050 | 4350 | 2750 | 5,2 | 5,3 | 4,4 | 0,29 | 0,30 |
| 2 . . . . . . . | 4750 | 4250 | 2700 | 4,9 | 5,0 | 3,93 bei $P$ = 150 kp<br>4,28     $P$ = 300 kp<br>4,76     $P$ = 600 kp | 0,26 | 0,27 |
| 3 . . . . . . . | 4800 | 4540 | 2920 | 5,6 | 5,6 | 5,5 | 0,21 | 0,21 |
| 4 . . . . . . . | 4600 | 4300 | 2700 | 4,8 | 4,8 | 4,9 | 0,24 | 0,22 |
| 5 . . . . . . . | 4500 | 4070 | 2700 | 4,6 | 4,9 | 4,7 | 0,22 | 0,26 |
| 6 . . . . . . . | 4750 | 4160 | 2830 | 4,7 | 5,4 | 5,5 | 0,23 | 0,29 |
| 7 . . . . . . . | 4150 | 4000 | 2420 | 4,4 | 4,0 | | 0,24 | 0,17 |
| 8 . . . . . . . | 4500 | 4250 | 2550 | 4,9 | 4,5 | | 0,26 | 0,21 |
| 9 . . . . . . . | 4380 | 4270 | 2560 | 4,8 | 4,4 | | 0,24 | 0,14 |
| 10 . . . . . . . | 4360 | 3790 | 2400 | 3,8 | 3,9 | | 0,28 | 0,29 |
| 11 . . . . . . . | 3870 | 3380 | 2130 | 3,2 | 3,2 | | 0,28 | 0,29 |

[1] Alfred J. Amsler & Co., Schaffhausen.

Einige Ergebnisse, die mit Hilfe der drei verschiedenen Verfahren an den gleichen Probekörpern gewonnen worden sind, wurden in Tabelle 13 wiedergegeben. Die Daten von Amsler und auch die meisten Messungen von Baule beziehen sich hierbei auf das kurze Teilstück, welches von Müller untersucht worden ist. Außer $V_l$ und $V_t$ von Müller, $V_d$ von Baule und den drei verschiedenen Elastizitätsmoduln sind die Poisson-Zahlen $\sigma_M$ aus $V_l$ und $V_t$ und $\sigma_{BM}$ aus $V_l$ und $V_d$ angeführt. Die Elastizitätsmoduln stimmen untereinander relativ gut bis auf wenige Prozent überein. Irgendwelche systematischen Abweichungen der Meßergebnisse eines der drei Verfahren gegenüber den anderen sind nicht festzustellen. Stärker differieren auch hier die beiden $\sigma$-Werte, jedoch ist weder bei ihnen, noch bei den $E$-Moduln ein so systematisches Verhalten zu finden, wie in den Fig. 25 und 26.

Grundsätzlich kann angenommen werden, daß der von Baule bestimmte $E$-Modul genauer ist, als derjenige von Müller, da das $E$ von Müller mit Hilfe des weniger genauen $\sigma$ berechnet werden muß. Bei einem mittleren Fehler von 2% für $V_l$ und $V_t$ beträgt $\Delta\sigma/\sigma$ bei $\sigma = 0,20, 0,25, 0,30$ und $0,35$: $16, 8,5, 5,3$ und $3\%$. Der $E$-Modul läßt sich dann bei Metallen und einigen Gesteinen mit hohem $\sigma$ auf 3 bis 4%, bei normalen Sedimentgesteinen auf 4 bis 8% genau berechnen. Hingegen wird der Scherungsmodul $G$ bei Metallen auf 2 bis 3% und bei Sedimentgesteinen auf 3 bis 5% genau bestimmt. Von wesentlicher Bedeutung sind jedoch bei Müller und anderen Autoren die unmittelbaren Messungen der für die Praxis wichtigen longitudinalen und transversalen Geschwindigkeiten für den unendlich ausgedehnten Raum.

**17. Anwendungen für Seismik und Gebirgsdruckforschung.** Von praktischer Bedeutung sind die an Gesteinsproben ermittelten elastischen Konstanten für die Seismik und Gebirgsdruckforschung im Untertage-Bergbau. Insbesondere ergeben die an Proben gemessenen Geschwindigkeiten einen wichtigen Hinweis auf die bei reflexionsseismischen Untersuchungen in Bergwerken zu erwartenden Geschwindigkeitsverhältnisse. Inwieweit sich die an Bohrkernen aus großen Aufschlußbohrungen ermittelten Geschwindigkeiten in nützlicher Weise mit seismischen Bohrlochversenkmessungen in Beziehung bringen lassen, bedarf noch weiterer Klärung.

Auf dem Gebiet der Gebirgsdruck-Forschung wird seit langem versucht, $E$ und $\sigma$ *in situ* zu ermitteln und die Unterschiede der Wellengeschwindigkeiten in einem bestimmten Gestein als Maß für den unterschiedlichen Spannungszustand bzw. den Gebirgsdruck heranzuziehen. Heinrich[1] und Dixon[2] haben solche Messungen, die in der Durchführung oft sehr schwierig sind, im Karbon vorgenommen, bisher aber nur geringe Erfolge erzielt.

Seit 1936 hat sich besonders das US-Bureau of Mines und darin an hervorragender Stelle L. Obert[3] mit Druckuntersuchungen durch Geschwindigkeitsmessungen an großen Gesteinspfeilern in Erzminen befaßt. Obert hat sich — leider auch ohne Erfolg — bemüht, aus der unter Tage gemessenen Geschwindigkeit und der Geschwindigkeit, die er an Proben aus dem gleichen Pfeiler bei verschiedenen Drucken im Laboratorium ermittelte, auf den Druck im Pfeiler selbst zu schließen.

---

[1] Heinrich: Bestimmung von Fortpflanzungsgeschwindigkeiten elastischer Wellen im oberschlesischen Karbon. Z. Berg-, Hütt.- u. Salinenw. **1936**.

[2] J. K. Dixon: The Sonic Location of Pressure Zones around Mining Excavations. University School of Mines, King's College, Newcastle upon Tyne, 1953.

[3] L. Obert: Measurement of Rock Pressures in Underground Mines. Part I. U.S. Dep. of the Int. Bureau of Mines, Rep. of Invest. Nr. 3444, 1939. do. Part II. a. a. O. R. I. Nr. 3521, 1940. — L. Obert, S. L. Windes u. W. Duvall: Standardized Tests for Determining the Physical Properties of Mine Rock. a. a. O. R. I. Nr. 3891 (1946).

# Literatur.

BERGMANN, L.: Der Ultraschall und seine Anwendung in Wissenschaft und Technik. Stuttgart: S. Hirzel 1954.

Handbuch der Geophysik (Herausgeber B. GUTENBERG), Bd. VI, 1931 und Bd. II, 1933. Berlin: Gebrüder Borntraeger.

Internal Constitution of the Earth, 2nd edit, (Ed. by B. GUTENBERG), Dover Publications. New York 1951.

Handbuch der Physik, herausgeg. von GEIGER-SCHEEL, Bd. VIII (Akustik). Berlin: Springer 1927.

Handbuch der Physik, herausgeg. von S. FLÜGGE, Bde. VI und XI. Berlin: Springer 1956/57.

LORD RAYLEIGH: Theory of Sound, 2. Aufl., Bd. 1. London: MacMillan & Co. 1926; New York: Dover Publications 1945.

MOOS, A. v., u. F. DE QUERVAIN: Technische Gesteinskunde. Basel: Birkhäuser 1948.

REICH, H.: Über elastische Eigenschaften von Gesteinen usw., Gerlands Beitr. Geophys. **17**, 86, 432 (1927).

— Über Gesteinselastizität. Z. dtsch. geol. Ges. 79 (1927).

— u. R. v. ZWERGER: Taschenbuch der angewandten Geophysik. Leipzig: Akademische Verlagsgesellschaft 1943.

RINT, C.: Handbuch für Hochfrequenz- und Elektrotechniker, Bd. I—II Berlin: Verlag für Radio-, Foto-Kinotechnik 1954.

SOMMERFELD, A.: Vorlesungen über theoretische Physik, Bd. II (Mechanik deformierbarer Medien). Wiesbaden: Dieterichsche Verlagsbuchhandlung 1945.

VILBIG, F: Lehrbuch der Hochfrequenztechnik, Bd. I/II. Leipzig: Akademische Verlagsgesellschaft 1942.

WIEN-HARMS: Handbuch der Experimentalphysik, Bd. XXV, 3. Teil (Geophysik) 1931 und Bd. XVII, 2.—3. Teil (Technische Akustik). Leipzig: Akademische Verlagsgesellschaft 1934.

# Gravity and Isostasy.

By

## G. D. GARLAND.

With 20 Figures.

## I. Introduction.

**1. The Scope of the Chapter.** The relation between the variation of gravity over the earth and the earth's figure will be discussed by K. JUNG. In the present chapter, the problem of *measuring* the gravitational field, and the *analysis* of this field in terms of the mass distribution within the earth, will be treated.

A difficulty arises at the beginning, in that the absolute determination of gravity at any one point on the earth, to a useful accuracy of one part in one million, is a problem of great difficulty, and no one determination has yet provided a completely satisfactory result. Fortunately for the geophysicist, it is the *variation* in gravity over the earth which is important to him, and this variation may be established through *relative* measurements, or comparisons between points on the earth, which are less exacting experimentally. The *absolute* value is important in certain applications, namely *metrology*, so that the early sections of the chapter are devoted to an evaluation of recent determinations.

In discussing the *instruments* available for relative gravity measurements, emphasis has been placed on the physical principles involved rather than on details of construction. The torsion balance, for example, has been treated in some length because of the insight it provides on the nature of the gravitational field, notwithstanding the fact that the instrument itself is not in extensive use at the present time.

In discussing the *interpretation* of the measurements in terms of underground density variations, the aim has been to show the general nature of the approach, and limitations involved in reaching conclusions. The problems which have to be treated range from broad variations of the field from its normal value, indicative of mass anomalies of continental extent, to extremely local distortions of the field produced by concentrations of mass a few hundred feet in extent. The latter problem arises most frequently in the commercial application of gravity measurements to the location of economic minerals, but so much of the recent development, both of instruments and of techniques of analysis, is due to this incentive that one cannot in justice omit the applied phase of gravity measurements.

**2. Fundamental Concepts.** The acceleration of gravity varies over the surface of the earth between the limits of about 977 and 983 cm./sec.$^2$. At any particular point on the earth, the acceleration, or its numerical equivalent, the force acting on unit mass, is the resultant of the acceleration due to the attraction exerted by all particles of the earth, that due to the attraction of the sun, moon[1] and other planets, and that due to the earth's rotation. The broad-scale variation over the

---

[1] Tidal forces and their effects will be discussed in Vol. XLVIII of this Encyclopedia.

earth is intimately related to the earth's figure, but the more local departures from this smoothly varying field are related to the inhomogeneous density of the earth's crust. A concentration of material of higher than normal density near a particular station results in an excess of mass, proportional to the difference between the actual density and the normal density of the earth's crust, and this mass results in a local increase in the gravitational attraction.

As will be shown later, the range in density encountered is rather restricted, and the effects of structures in the earth's crust even of continental proportions amount to only a small proportion of the total gravitational field. To obtain useful information it is in fact necessary to measure *differences* in the acceleration of gravity of the order of one part in $10^6$ of the normal value. The unit of acceleration, that is, one centimetre per second, has been termed the *gal*, in honour of GALILEO. A convenient unit for geophysical purposes is 0.001 cm./sec.², or *milligal*, although in the case of local surveys of high relative precision, values may be expressed in units of 0.01 milligal.

### General References.

HEISKANEN, W.: Beobachtungen der Schwerkraft. Das Problem der Isostasie. In Handbuch der Geophysik, Bd. 1, S. 731—951. Berlin: Gebrüder Bornträger 1935.

BERROTH, A.: Schweremessungen. In GEIGER-SCHEELS, Handbuch der Physik, Bd. 2, S. 416—486. Berlin: Springer 1926.

SCHMEHL, H., u. K. JUNG: Schwere, Figur und Massenverteilung der Erde. In WIEN-HARMS' Handbuch der Experimentalphysik, Bd. 25, Teil 2, S. 141—357. Leipzig: Akademische Verlagsgesellschaft 1931.

LEJAY, P.: Développements modernes de la gravimétrie. Paris 1947.

See further discussion and references in the chapters on The Earth's Figure (by K. JUNG), Forces in the Earth's Crust (by A. E. SCHEIDEGGER) and Structure of the Earth's Crust (by M. EWING and F. PRESS) in this Volume.

# II. The Absolute Measurements of Gravity.

## a) Pendulum Methods.

**3. The Reversible Pendulum.** An account of a number of early absolute measurements of gravity has been given by SCHMEHL[1], but at present we need consider only the determinations made in this century at Potsdam, Washington, and Teddington. In each of these, some type of the *reversible pendulum* was employed. The principle of the method is that for a compound pendulum with knife edges on opposite sides of the mass centre, at distances $h_1$ and $h_2$ from the mass centre, the period $T$ of a simple pendulum of length $h_1 + h_2$ is given by

$$T^2 = \frac{T_1^2 h_1 - T_2^2 h_2}{h_1 - h_2}, \qquad (3.1)$$

where $T_1$ and $T_2$ are the periods of oscillation about the respective knife edges. In practice, $T_1$ and $T_2$ are adjusted to be very nearly equal, so that

$$T = \frac{T_1 + T_2}{2} + \frac{T_1 - T_2}{2} \cdot \frac{h_1 + h_2}{h_1 - h_2}. \qquad (3.2)$$

The second term becoming small[2] as the equality of periods is approached, the critical length measurement for the determination of $g$ is $l = h_1 + h_2$, the distance between knife edges. Then by the simple formula

$$g = \pi^2 l / T^2. \qquad (3.3)$$

---

[1] H. SCHMEHL: Handbuch der Experimental-Physik, Bd. XXV, Teil 2, S. 192. 1931.

[2] The pendulum is so constructed that its mass centre is asymmetrical between the knife edges, with $h_1$ about equal $2 h_2$.

The above principle was apparently discovered independently by three workers within about thirty years, but of these only Kater[1] actually constructed a reversible pendulum and attempted a measurement with it. His value, for a site in London, is about 35 parts in $10^6$ (i.e. 35 milligals[2]) higher than the mean of modern determinations. Considering the limitations of his apparatus, and the difficulties in the method found by later observers, it is remarkable that Kater's value is not more in error.

In gravity measurements, it is usual to call period $T$ the time between opposite extremes, which in other periodic phenomena would be called a half oscillation.

**4. Measurements at Potsdam, Washington and Teddington.** The absolute measurement at Potsdam was made in the Pendelsaal of the Geodetic Institute by Kühnen and Furtwängler[3], working under the direction of Helmert. They used five pendulums, varying in mass from 2.86 kg. to 6.23 kg., and a variety of methods of support. Great care was taken to eliminate all known sources of error, including the yielding of the knife edges under the weight of the pendulums, and the value finally adopted was

$$g\ Potsdam = 981.274\ \text{cm./sec.}^2.$$

This value remained unchallenged as the absolute basis for all gravity determinations until the completion of the series of measurements by Heyl and Cook[4] at the National Bureau of Standards in Washington. For the latter work, fused-silica pendulums in the form of tubes were employed. Each pendulum carried two planes, of silica or stellite, one near each end, while the knife edges were fixed to the support. The periods of a pendulum in the two positions were adjusted to equality by grinding one end of the pendulum.

After consideration of various sources of error, Heyl and Cook give the value
$$g\ Washington = 980.08 \pm 0.003\ \text{cm./sec.}^2\ \text{(average deviation)}.$$

The absolute determination at Teddington by Clark[5] was made with a single pendulum, in the form of an alloy stem of I section, to the ends of which were bolted massive bobs. Two plane surfaces, chromium plated, were provided on the pendulum, while the hard steel knife edges were fixed to the massive cast iron support. Clark's final values was

$$g\ Teddington = 981.1815 \pm 0.0016\ \text{cm./sec.}^2\ \text{(average deviation)}.$$

When differences in gravity between the sites of the above determinations were measured by the methods to be described later, it became apparent that there were discrepancies between the Potsdam and more recent measurements of 10 to 20 parts in $10^6$, and between the latter themselves of about 5 parts in $10^6$. A re-examination of the various corrections applied in each case was therefore desirable.

**5. Revisions of the Potsdam, Washington and Teddington values.** Dryden[6] and Jeffreys[7] have criticized the treatment adopted by the Potsdam observers for an apparent error systematic with the mass of the pendulum. The original argument was that yielding of the knife edge, and of the support, would tend to make the pendulum rotate about a point above the knife edge, and thereby

[1] H. Kater: Phil. Trans. Roy. Soc. Lond. **108**, 32 (1818).
[2] 1 milligal = 0.001 cm./sec².
[3] F. Kühnen and P. Furtwängler: Veröff. preuß. Geodät. Inst. **27** (1906).
[4] Paul R. Heyl and Guy S. Cook: J. Res. Nat. Bur. Stand. **17**, 805 (1936).
[5] J. S. Clark: Phil. Trans. Roy. Soc. Lond., Ser. A **238**, 65 (1940).
[6] H. L. Dryden: J. Res. Nat. Bur. Stand. **29**, 303 (1942).
[7] H. Jeffreys: Mon. Not. Roy. Astronom. Soc., Geophys. Suppl. **5**, 219 (1948/49).

increase the effective length of the pendulum. Kühnen and Furtwängler therefore extrapolated their observations to apply to a pendulum of zero mass. On the other hand, Heyl's observations indicated no such systematic variation with the mass of the pendulum, and Dryden and Jeffreys, through different but straightforward methods of weighting the many sets of observations, reduce the Potsdam value by some 12 parts in $10^6$.

In the case of the Teddington determination, the chief uncertainty has been in the correction applied by Clark for the bending of the pendulum as it swings. Jeffreys[1] has determined the correction through an application of Rayleigh's method for the frequency of a constrained vibrating system. His result is given in the form

$$\gamma^2 - \alpha^2 = \frac{1}{I\vartheta^2} \int \frac{M^2}{E\,S\,C} \, dx, \qquad (5.1)$$

where $\alpha$ and $\gamma$ are the frequencies of the actual pendulum and a rigid pendulum of the same dimension, $E$ is Young's modulus, $SC$ the quadratic moment of cross-section of the pendulum, $I$ the moment of inertia about the knife edge, $\vartheta$ the angular displacement, and $M$ is the bending moment required to prevent bending. The integration is to be along the length $x$ of the pendulum. When this formula is applied to Clark's case, the correction for bending is of opposite sign to that deduced by Clark, and the Teddington value is increased by 1.7 parts in $10^6$.

Jeffreys also recomputed the errors due to bending, stretching and other causes for the Washington determination of Heyl, and while no single correction was found to be greatly in error, consideration of a number of small effects originally neglected led to an increase in the final value of 1.6 parts in $10^6$.

The adjusted absolute values may be compared as in Table 1, which utilizes recent relative determinations for the reduction of each determination to Potsdam. It will be seen that even the recomputed values differ by up to 0.007 cm./sec.[2] of which not more than 0.001 cm./sec.[2] is likely to be due to the relative measurements used in the comparison. This fact has led to the search for other methods for the absolute determination of $g$.

Table 1. *Comparisons of Absolute Determinations of Gravity.*

| Observer | Original value cm./sec.[2] | Revised value cm./sec.[2] | Revised value Reduced to Potsdam cm./sec.[2] |
|---|---|---|---|
| Kühnen and Furtwängler[2] . . | 981.274 | 981.2633 | 981.2633 |
| Clark . . . . . . . . . . | 981.1815 | 981.1832 | 981.2612 |
| Heyl and Cook . . . . . . | 980.08 | 980.0816 | 981.2565 |

## b) Other Absolute Methods.

**6. Falling Body Experiments.** The direct measurement of gravity by means of observations of a freely falling body became possible with the perfection of methods for the precise measurement of short time intervals. Volet[3] first described an arrangement for such a determination, at the Bureau International des Poids et Mesures. A graduated rule was dropped in an evacuated chamber, and photographed at regular intervals by means of a spark discharge. A time scale, controlled by a quartz crystal oscillator, was placed on the same film. By examination of the photographs, the position of the rule relative to fixed

---

[1] H. Jeffreys: Mon. Not. Roy. Astronom. Soc., Geophys. Suppl. 5, 398 (1948/49).

[2] Values in the Table as given by Jeffreys; A. Berroth (Bull. Géodés., N. S. No. 12, p. 183, 1949) gives the revised value 981.2613.

[3] Ch. Volet: C. R. Acad. Sci. Paris 222, 373 (1946); 235, 442 (1952).

reference marks at successive times may be determined. An equation for the motion of the rule may then be determined by least squares.

The result of the first series of observations was given by VOLET at the Brussels meeting of the International Union of Geodesy and Geophysics, 1951. This indicated that the Potsdam value was 0.024 cm./sec.$^2$, high, and was therefore considerably lower than CLARK's or HEYL and COOK's value. The figure quoted,

$$g \ (Pavillon \ de \ Breteuil) = 980.916 \ \text{cm./sec.}^2$$

was based on 18 drops made with rules of invar and bronze, at pressures between 1 and 5 mm. of mercury.

Further work on this method of determining $g$ is in progress at the Bureau International des Poids et Mesures, the National Research Council of Canada, Ottawa, the National Physical Laboratory, Teddington, and the Physikalisch-Technische Bundesanstalt, Brunswick, Germany. Certain factors, such as the viscosity effect of the residual air in the chamber, the magnetic forces active on invar rules, and the elastic vibrations of rules as they are released, require more complete evaluation. When the results of three or four independent determinations can be shown by intercomparison to be consistent to within 0.001 cm./sec.$^2$ or so, a suitable absolute basis for all gravity measurements will be available. This problem is perhaps of more concern in the field of metrology (e.g. the determination of standards of pressure and of electric current) than in geodesy or geophysics, where relative values over the earth are the features of interest.

# III. Relative Gravity Measurements.

## a) Pendulum Measurement.

**7. Invariable Pendulums.** The measurement of *differences* in gravity by means of pendulum observations is much less exacting than the determination of the absolute value. Regardless of the actual form of the pendulum employed, it is convenient to speak of its *"length"*, referring to the length of the *equivalent* simple pendulum of the same period. If the properties of the pendulum are sufficiently stable, this length remains invariable, and the ratio of the values of gravity at two stations is inversely proportional to the squares of the observed periods.

The history of the pendulum dates from the experiments of HUYGENS, which led to its use for the regulation of clocks, and the subsequent discovery that the rates of pendulum clocks vary with location on the earth. Among the very early attempts to measure differences in gravity was the work of BOUGUER[1] in the mountains of Peru. The first observations approaching present standards of accuracy were those of KATER[2] whose work has also been mentioned in connection with absolute determinations. KATER observed a line of several stations from the Isle of Wight to the Orkneys, the relative values of which have been found to be correct within about 0.003 cm./sec.$^2$. BULLARD and JOLY[3] have remarked that no work of comparable accuracy was done in England during the remainder of the century.

The pendulums used for the very early determinations usually had a half-period $T$ of about one second, and therefore were about one metre long. VON STERNECK[4] developed a much more portable type by simply making the half-period close to one half second, and the length one quarter metre. In the

---

[1] P. BOUGUER: La Figure De La Terre. Paris 1749.

[2] H. KATER: Phil. Trans. Roy. Soc. Lond. **109**, 337 (1819).

[3] E. C. BULLARD and H. L. P. JOLY: Mon. Not. Roy. Astronom. Soc. Geophys. Suppl. **3**, 443 (1936).

[4] R. V. STERNECK: Z. Instrumentenkde. **8**, 157 (1888).

classical form, the VON STERNECK pendulum has a rather thin stem, with the knife edge fastened near the top and a massive bob cast on, or riveted to, the lower end. However, if the pendulum is made in the form of a uniform cylinder, it is possible for it to have the interesting *minimum period* characteristic.

To investigate this property, let $l$ be the equivalent simple pendulum length, so that the half period is given by

$$T = \pi \sqrt{\frac{l}{g}}. \qquad (7.1)$$

The quantity $l$ in this formula may be expressed in terms of two characteristic lengths $k$ and $h$ by the relation

$$l = \frac{k^2 + h^2}{h} \qquad (7.2)$$

where $k$ is the radius of gyration of the pendulum about an axis through the mass centre parallel to the plane of the knife edge, and $h$ is the distance from the knife edge to the mass centre.

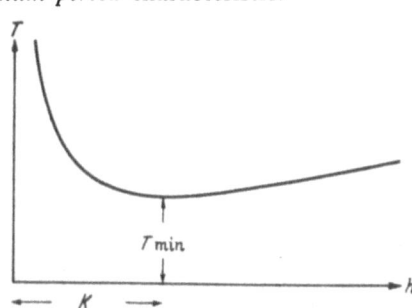

Fig. 1. Relation between the half-period $T$ of a pendulum and the distance $h$ from knife-edge to mass-centre.

The variation of $T$ as a function of $h$ is shown in Fig. 1. By differentiation of equation (7.1), with $l$ replaced by (7.2), it is readily seen that the minimum value of $T$ occurs when $h$ is made equal to $k$. The practical importance of this

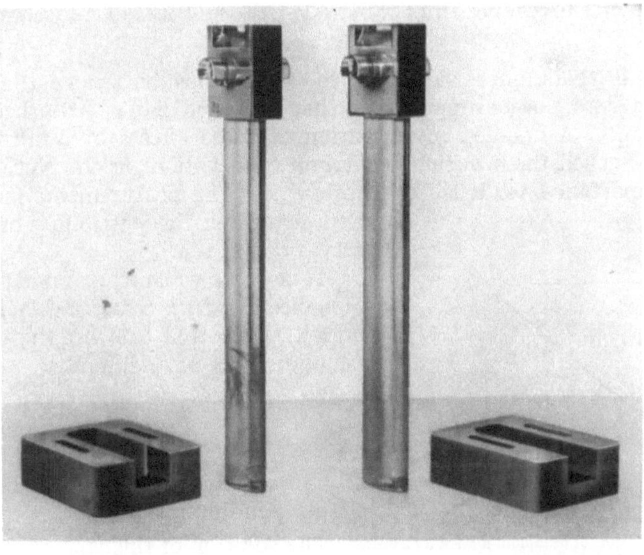

Fig. 2. Fused quartz minimum pendulums, with quartz knife edges and Pyrex bearing planes, as designed by the Gulf Research and Development Company.

result is that, near $h = k$, the half-period $T$ is independent of small changes in $h$, which could arise through wearing of the knife edge[1]. Pendulums of this type have been constructed in the form of circular or elliptical cylinders, by simply calculating the correct value $k$ (which is a simple function of length and radius) for the distance from the knife edge to the mass centre. Recent examples of this type are the beautifully constructed quartz pendulums of the GULF apparatus (Fig. 2).

[1] M. SCHULER: Z. techn. Phys. **10**, 392 (1929).

The choice of the *material* for the pendulum, and for the knife edge, has been the subject of much discussion. Brass, bronze, invar and quartz have probably been used the most for the pendulum itself. The obvious disadvantage of the first two is their large *temperature coefficient*, which is many times that of invar or quartz. Invar pendulums have been shown to have very stable characteristics, but because of their magnetic properties, they must be isolated from the earth's magnetic field during observations. Quartz might appear to be the ideal material, but it is capable of possessing an electrostatic charge, and also the pendulums are more subject to physical damage during transport. For knife edges, agate and also hard metallic alloys such as stellite have been used.

**8. Single Pendulum Apparatus.** For reasons to be discussed shortly, few pendulum instruments in use at the present time employ a single pendulum swinging at a time. An exception is the BROWN apparatus of the United States Coast and Geodetic Survey. In it, a pendulum of the VON STERNECK type is swung in a vacuum chamber, the motions of the pendulum being recorded by means of a light beam reflected from a mirror on the pendulum to a photo-electric cell. The output of the cell is amplified and recorded on a chronograph, together with marks controlled by radio time signals. A swing of eight hours duration is required if the period is to be determined to 1 part in $10^7$, but the chronograph record need be taken for only a few minutes at the beginning and end of the swing. The apparatus is essentially an evolution from an older arrangement designed by MENDENHALL, in which light flashes from the mirror on the pendulum were observed with a telescope, and compared with the beats of a mechanical chronometer.

The single pendulum is subject to the two serious limitations, that it produces a motion or flexure of its support, and that its period may be disturbed by motions of the ground. A detailed investigation of these effects by VENING MEINESZ[1] who had observed them during pendulum observations in the Netherlands, was of great importance, for it led to improvements in measurements made on land, and suggested the possibility of gravity observations at sea.

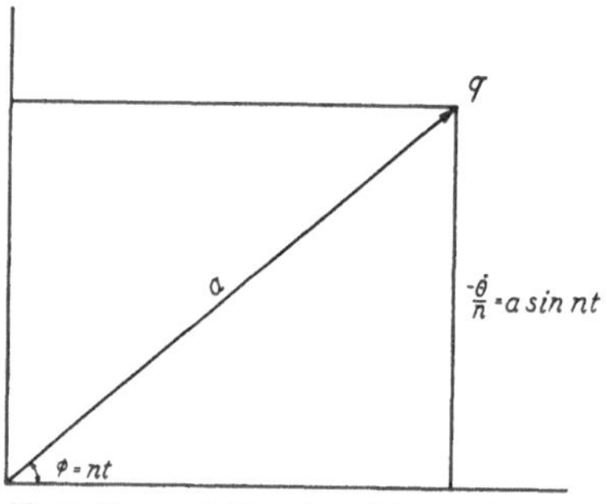

Fig. 3. Representation of pendulum motion by a rotating vector $q$.

It is worth while to examine the equations of *motion of a disturbed pendulum* in some little detail. As is well known, the equation for an undisturbed pendulum is

$$\ddot{\Theta} + n^2 \Theta = 0 \qquad (8.1)$$

where $\Theta$ is the angular displacement and $n^2 = g/l$, $l$ being the length of the equivalent simple pendulum.

The solution of this is

$$\Theta = a \cos n t \qquad (8.2)$$

where $a$ is the constant amplitude of the pendulum. The motion in this case may be represented by the rotation of a vector $q$ of constant length $a$ (Fig. 3), where $q$ is the complex quantity given by

$$q = \Theta - \frac{i}{n} \dot{\Theta}. \qquad (8.3)$$

---

[1] F. A. VENING MEINESZ: Observations de Pendule dans les Pavs-Bas. Delft 1923.

If the pendulum is under the influence of disturbing forces expressed by $S$, the equation of motion is of the form

$$\ddot{\Theta} + n^2 \Theta + S = 0. \tag{8.4}$$

In terms of the quantity $q$ the equation is

$$\dot{q} - i n q - i S/n = 0. \tag{8.5}$$

The solution is easily found to be

$$q = (q_0 + \delta q) e^{i n t} \tag{8.6}$$

where $q_0$ is the value at $t = 0$ and

$$\delta q = \frac{i}{n} \int_0^t S e^{-i n t} dt. \tag{8.7}$$

The vector $q$ may still be thought of as representing the motion, but this vector no longer rotates at constant velocity, nor is it of constant length. In fact, it is from the behaviour of $q$ that the effects of the disturbances on the amplitude and period of the pendulum are deduced. The velocity of the extremity of the vector $q$ is $\dot{q}$, and from equation (8.5) it is seen that the contribution to this velocity from the disturbance $S$ is $S/n$, parallel to the imaginary axis. The effect on the *angular* velocity of the vector at any time $t$ is

$$\frac{S \cos \Phi}{n a}$$

where $\Phi$ is the phase angle (Fig. 3), and therefore the total disturbance to the motion during time $T$ is

$$\Delta \Phi = \int_0^T \frac{S \cos \Phi \, dt}{n a}. \tag{8.8}$$

From this, the disturbance of the period, $\Delta T$, follows without difficulty.

In this equation $S$ may represent disturbances resulting from *ground movement*, or from movements of the *support* caused by the pendulum itself. For determinations with a single pendulum apparatus, it is usual to correct for the latter effect by measurements of the displacement of the case during a swing with an interferometer; but the former may be difficult to evaluate. However, if *two* pendulums of equal period are swung with equal amplitudes and phase difference $\pi$ on the same support, in the same vertical plane, both effects may be eliminated. For in this case, equation (8.8) shows that $\Delta T$ for each pendulum is equal but opposite in sign, and the mean period of the two pendulum is unaffected by the ground movements. Also, the forces exerted by the pendulums on the support are equal and opposite at all times, so that the tendency of the support to yield is eliminated. The recognition of these facts has led to the design of multiple pendulum apparatus.

**9. Multiple Pendulum Apparatus.** Instruments with provision for the swinging of two, three or four pendulums have been designed, but for most observations on land, apparatus employing two pendulums is in general sufficient, and somewhat more portable. One of the most precise instruments was that developed by the Gulf Oil Company[1], for use in the exploration for petroleum. In this apparatus two well-matched quartz minimum pendulums of half-period about 0.9 second are swung on Pyrex glass planes. Originally, when the apparatus was used to make a number of observations within a limited area, the periods of the field pendulums were compared with those of a similar set swung simultaneously at a base station. Photographic records were made of the pendulum motion at the two locations, and by registering any arbitrary radio signals on each record, the difference in periods between the base and field sets could be determined. The field pendulums were swung at the base station before and after each circuit of observations, so that finally the ratio of periods at the different field stations could be determined, without reference to absolute time. With the perfection of crystal chronometers, the periods of the pendulums may be directly determined with sufficient accuracy from a swing of about one hour, so that the base set of pendulums is not required. With this modification the apparatus is not restricted to a limited area.

As used by the University of Wisconsin for measurements in various parts of the world, a 100 Kc. quartz crystal oscillator supplies a 1000 cps. signal for the operation of a synchronous motor. The latter operates a slotted disc in such a way as to supply time marks on the photographic record of the pendulum motion. The crystal chronometer may be rated by beating suitable harmonics of the fundamental against standard radio frequencies, such as are broadcast by station WWV of the National Bureau of Standards, Washington.

The second pendulum apparatus which has been used in many parts of the world is that of Cambridge University, constructed originally by Sir GERALD LENNOX-CONYNGHAM[2], and modified by E. C. BULLARD and B. C. BROWNE. The pendulums, swung in matched pairs, are of invar, with knife edges of stellite. Timing of the pendulums is accomplished by a method essentially the same as that outlined above. Effects due to the magnetic properties of the pendulums may be virtually eliminated if the earth's field is annulled by HELMHOLTZ coils. The most important attribute of the apparatus is the stability of the pendulums themselves, which have changed in period by only minor amounts during twenty-five years, and will often reproduce their periods after journeys of several thousand miles to within 1 part in $10^7$.

**10. Gravity Measurements at Sea.** The theory developed by MEINESZ for the elimination of disturbances from the observations of pendulums swinging on an unstable support led him[3] to adapt the apparatus for measurements in a *submarine*, submerged below the region affected by surface water waves. However, even at sufficient depth, account must be taken of vertical and horizontal accelerations, and of rotations. For two perfectly isochronous pendulums of equivalent length $l$, the effect of a horizontal acceleration $\ddot{y}$ may be eliminated, to the first order, by simply recording the *difference* of the angular displacements. For if the equations of motion are

$$l\ddot{\vartheta}_1 + g\,\vartheta_1 + \ddot{y} = 0,\tag{10.1}$$

$$l\ddot{\vartheta}_2 + g\,\vartheta_2 + \ddot{y} = 0,\tag{10.2}$$

then

$$l\,(\ddot{\vartheta}_1 - \ddot{\vartheta}_2) + g\,(\vartheta_1 - \vartheta_2) = 0,\tag{10.3}$$

is the equation of motion of an undisturbed "*fictitious*" *pendulum*. The motion of the fictitious pendulum is recorded directly by reflecting a beam of light

---

[1] MALCOLM W. GAY: Geophysics 5, 176 (1940). See our Fig. 2, p. 207.
[2] Sir G. P. LENNOX-CONYNGHAM: Geogr. J. 73, 326 (1929).
[3] F. A. VENING MEINESZ: Theory and Practice of Pendulum Observations at Sea. Delft 1929.

from mirrors on the two pendulums in turn, so that the deflection of a trace on the photographic record is proportional to $(\vartheta_1 - \vartheta_2)$.

In the MEINESZ apparatus three pendulums are used, the central one being initially at rest and the other ones started in opposite phase. Records are made of the motion of two fictitious pendulums each formed by the central and one outer pendulum, and also of the motion of the central pendulum alone, which is required for a correction if not all pendulums are isochronous. A chronometer interrupts the light beam producing these traces at regular intervals, and from the pattern of time marks which results the period may be determined, by selection of corresponding points near the beginning and end of a swing.

The effect of *vertical* accelerations, $\ddot{x}$, cannot be separated from gravity, so that the quantity measured is $g + \bar{\bar{x}}$, where $\bar{\bar{x}}$ is the mean value over the period of observation. We have

$$\bar{\bar{x}} = \frac{1}{t} \int_0^t \ddot{x}\, dt = \frac{1}{t} [\dot{x}_t - \dot{x}_0]. \tag{10.4}$$

To the first order, the effect of vertical accelerations may thus be made negligibly small, if $t$ is sufficiently long. In practice, a swing of one half hour or more is used.

The effect of *rotations* is reduced by mounting the pendulum case in gimbals, placed at the same level as the knife edges. Rotation about a horizontal axis in the plane of oscillation is most serious, for besides imparting a centrifugal acceleration to the pendulums, it results in the measurement of a component of gravity. MEINESZ employed a damped pendulum, swinging perpendicular to the principal pendulums, to measure this rotation.

There has been considerable discussion of the perturbations introduced through neglect of second-order terms in the above considerations. BROWNE[1] pointed out that in the case where the period of the disturbance is much longer than that of the pendulums, it may be assumed that the pendulum measures the resultant $g\left(1 + \frac{(\ddot{y}^2 + \ddot{z}^2)}{2g^2}\right)$, where $\ddot{y}$ and $\ddot{z}$ are *horizontal* accelerations. This effect would always tend to make the observed value high. However, as the relation between period and deduced gravity is not linear, there is in addition a term $-\ddot{x}^2/4g$, due to *vertical* accelerations, which does not vanish even for an extended period of observations. The horizontal accelerations cannot be measured directly in a submarine, but variations in the direction of the resultant vector may be obtained by recording the angle between a short-period damped pendulum and a gyroscopically maintained vertical. Variations in the vertical acceleration may be estimated from a suitable accelerometer, or from irregularities in the actual pendulum records. The chief problem, however, is the limitation of BROWNE's analysis as the period of the disturbance approaches that of the pendulum. Recently, Miss WORSLEY[2] has solved the general equation by electronic integration for a number of cases, which indicate that the former correction for horizontal accelerations should be doubled even when the ratio of periods is fifteen to one. MEINESZ[3], has re-examined the problem with the conclusion that BROWNE's correction is adequate after all, and BROWNE and COOPER[4] found

---

[1] B. C. BROWNE: Mon. Not. Roy. Astronom. Soc., Geophys. Suppl. **4**, 271 (1937).
[2] Miss B. H. WORSLEY: Proc. Cambridge Phil. Soc. **48**, 718 (1952).
[3] F. A. VENING MEINESZ: Proc. Kon. Ned. Akad. Wetensch. B **56**, 218 (1953).
[4] B. C. BROWNE and R. I. B. COOPER: Phil. Trans. Roy. Soc. Lond., Ser. A **242**, 243 (1950).

by experiment that observations made at different depths at the same site would be more consistent with no correction for the second-order terms. It may be concluded that further investigation into the problem is desirable.

Most of the effects are fundamental to the situation, and would not be removed by substitution of a different method of measurement. In many respects, however, it may be more convenient in future to employ the string gravimeter of GILBERT[1], in which the resonant frequency of a stretched wire is measured and related to the tension in the wire. Such an instrument is easily adaptable to recording over an interval of time for the first-order removal of vertical accelerations.

Finally, it must be pointed out that one of the largest corrections involved in measurements made from any moving support is the EÖTVÖS *effect*. That is, any east-west component of motion of a body on the earth alters its centrifugal acceleration and therefore its effective weight. If $dg$ is the change in gravity due to this motion,

$$dg = da \cos \varphi \tag{10.5}$$

where $da$ is the change in centrifugal acceleration (normal to the earth's axis of rotation) and $\varphi$ is the geocentric latitude.

But

$$da = 2\omega r\, d\omega = 2\omega v \tag{10.6}$$

where $\omega$ is the angular velocity of the earth, $r$ the distance from the axis of rotation, and $v$ is the east-west component of velocity of the observer.

Finally

$$dg = 2\omega v \cos \varphi = 7.5 v \cos \varphi$$

where $dg$ is expressed in milligals and $v$ in miles per hour ($=$ knots). Clearly, therefore, the determination of gravity to an accuracy of 1 milligal in the middle latitudes requires the measurement of the submarine's velocity to 0.2 miles per hour or better.

**11. Correction of Pendulum Observations.** Regardless of the type of pendulum instrument employed, the prime necessity for the determinations of differences in gravity is a measure of the difference in period of the pendulums between two stations, under the same conditions of pressure, temperature and amplitude at the two locations. As the latter factors cannot in general be held constant at all stations, it is usual to measure them and apply corrections based on experimentally determined coefficients. The temperature coefficient, which reflects the change in length of the pendulum with changing temperature, is obviously a direct function of the coefficient of expansion of the pendulum material. A number of effects enter into the variation of period with pressure, the chief of these being the reduction in weight of the pendulum due to buoyancy, and the viscosity of the air.

The variation may usually be expressed as

$$\Delta S = k_1 \varepsilon + k_2 \sqrt{\varepsilon}, \quad \text{with} \quad \varepsilon = \frac{p}{760 (1 + t/273)}, \tag{11.1}$$

where $\Delta S$ is the change in period, $p$ and $t$ are pressure in mm. of Hg and temperature in °C. respectively, and $k_1$ and $k_2$ are determined by experiment. The observed period is also corrected for finite amplitude of swing, on the basis of the

[1] R. L. G. GILBERT: Proc. Phys. Soc. Lond. B **62**, 445 (1949).

series expansion

$$S_{\alpha=0} = S_\alpha \left( 1 - \frac{\alpha^2}{16} + \cdots \right) \tag{11.2}$$

where $\alpha$ represents a mean value of the amplitude (in arc) during the swing. With the introduction of crystal clocks for timing and the consequent reduction in the time necessary for a swing to about one hour, the diminution in $\alpha$ through the swing is usually sufficiently small for the simple arithmetic mean to be used in this expression.

When the observed periods have been corrected for pressure, temperature and amplitude, as well as for clock rate, they are analysed for changes in the length of the pendulums between repeat observations at some base station. If these are sufficiently small for a mean value for the base period to be adopted, differences in gravity from the base are computed by the inverse square relation.

## b) Static Methods of Measuring Gravity.

**12. Gravity Meters.** The difficulties encountered with pendulum measurements during the nineteenth century led to a demand for a more rapid, static method of determining differences in gravity. This is reflected in the development by Baron Eötvös of the *torsion balance* for the measurement of the spacial derivatives of the gravitational field, and in the request, circulated in 1886 by the British Association for the Advancement of Science, for designs of an instrument for rapid gravity determinations. However, the development of precise, portable gravity meters really began about 1930, and came as a direct result of the application of gravity observations to exploration for petroleum.

Fundamentally, gravity meters balance the force on a fixed mass against the elastic stresses in a system of springs or torsion fibres[1]. The chief problem in their design was the achievement of *sensitivity*, for to be useful in geophysical exploration, it was necessary that differences in gravity of 0.0001 cm./sec.[2] or less be detectable. Two general trends in design have been followed, depending upon the choice of a *stable* or unstable *(astatic)* mechanical system.

In the former case, the small differences in gravity to be measured produce only a minute displacement of the mass, but this is made detectable by extreme electrical or optical magnification. Examples of this type are the BOLIDEN instrument, in which the suspended mass forms one plate of a condenser in a sensitive tuned circuit, and the GULF gravimeter, in which the unwinding of the mainspring, which accompanies its extension, is detected by means of a long optical lever.

In the astatic instruments, an additional force is supplied to the system by a spring or tension fibre, in such a way as to increase any small displacement from the central position. Many instruments make use also of the principle of the LACOSTE[2] seismograph, by suspending a horizontal beam by means of an inclined spring whose tension is directly proportional to its physical length ("zero-length spring"). In such a system the forces opposing displacement of the beam may be made very small. It is usual to provide the astatic system with a restoring element, by means of which the operator may bring the suspended mass to its neutral position at each location. The instrument reading is thus a measure of the restoring force required for this operation. The materials for the mechanical system may be metallic, as in the case of most instruments which

---

[1] Text books on geophysical exploration give numerous examples. See, for example J. JAKOSKY, Exploration Geophysics, pp. 369—386. Los Angeles 1949, and the references given in the chapter on Seismic Prospecting by M. EWING and F. PRESS, p. 153.

[2] L. J. B. LACOSTE: Physics 5, 178 (1943). See also p. 49.

have been designed, or fused quartz, as in the Worden gravimeter (Fig. 4, in the foreground).

A prime consideration in any gravity meter is the elimination of the effect of temperature on the dimensions and elastic properties. In most instruments the measuring system is kept at a constant temperature, above that of the surroundings, by means of heating coils and sensitive thermostats. Otherwise, by a suitable combination of materials with different coefficients of expansion, an additional force must be applied to the suspension to compensate for the effects of temperature. The latter method has the advantage that the source of power

Fig. 4. Three types of gravity meters in current use. The two instruments in the background are provided with thermostats and heating circuits, while the small meter in the foreground is compensated against changes in temperature.

for heating coils may be eliminated, and the instrument made more portable, as in the case of the Worden instrument.

Apart from the effect of temperature, the elastic properties of the materials used in the measuring system of any gravimeter change with time, and this property, combined with the tidal variation of gravity itself, causes the instrument reading to vary or "*drift*" at any fixed location. For the difference between two fundamental stations to be determined with high precision, the instrument must be transported back and forth within a few hours, so that from the repeat readings time-variation or drift curves are obtained for each station, and the mean difference in reading may be computed with allowance for drift. The readings for subsidiary stations may then be determined from a single closed circuit based on a fundamental station.

All measurements of the differences in gravity obtained with gravity meters are expressed in the first instance in terms of instrumental divisions, whose value in absolute units of acceleration depends upon the elastic and geometrical properties of the system. In general it is not possible to compute the *scale constant* from these properties, and it must be obtained experimentally. This can sometimes be done in the laboratory by adding a small known mass to the suspension, or by tilting the instrument by known amounts so that the differences between successive components of gravity are measured. However, it is usually more satisfactory to obtain the *calibration factor* by a suitable combination of pendulum and gravity meter observations at a number of stations, as will be discussed in Sect. 14.

**13. The Torsion Balance.** A discussion of gravimetric instruments would not be complete without mention of the torsion balance, as developed by Baron ROLAND VON EÖTVÖS[1]. A study of this instrument provides a useful insight to the distortion of the gravitational field by local mass anomalies. In essence, it consists of a fine torsion wire (Fig. 5), carrying a mirror and a light horizontal beam. The beam carries a mass $m$ at one end, and a similar mass, suspended by a second light wire of length about 60 cms., at the other. It is this vertical displacement of the two masses which differentiates the EÖTVÖS balance from the otherwise similar apparatus (with the two masses at equal height), employed by CAVENDISH[2] for the determination of the constant of gravitation.

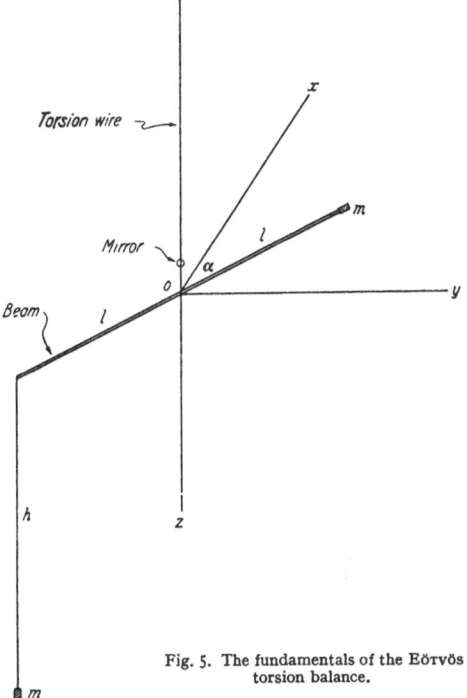

Fig. 5. The fundamentals of the Eötvös torsion balance.

To consider the forces acting on the suspension, we take axes as shown in Fig. 5, the origin being at the mid point of the rod, and the $z$-axis in the direction of the vertical at this point. In the presence of local gravity anomalies the direction of the vertical (i.e. the direction of the total gravity vector) will not necessarily be the same at other points of the suspension, because of the relatively rapid space-variation of the forces due to local effects, and it is precisely this fact which is utilized in the torsion balance.

In developing the equations, it is convenient to consider first the CAVENDISH balance, with both masses attached to the beam. If the components of gravity at $O$ are taken as $X_0 = 0$, $Y_0 = 0$, $Z_0 = g$, those at a point $(x, y)$ of the beam would be

$$
\left.
\begin{aligned}
X &= x\frac{\partial X}{\partial x} + y\frac{\partial X}{\partial y}, \\[6pt]
Y &= x\frac{\partial Y}{\partial x} + y\frac{\partial Y}{\partial y}, \\[6pt]
Z &= Z_0 + x\frac{\partial Z}{\partial x} + y\frac{\partial Z}{\partial y}.
\end{aligned}
\right\}
\tag{13.1}
$$

---

[1] R. v. EÖTVÖS: Verh. der XV. allgemeinen Konferenz der Internationalen Erdmessung in Budapest, 1906.

[2] H. CAVENDISH: Phil. Trans. Roy. Soc. Lond. **88** (1798).

The moment about the $Z$-axis of the forces acting on an element $dm$ of the beam is $dm(-yX + xY)$, or

$$dm\left(-xy\frac{\partial X}{\partial x} - y^2\frac{\partial X}{\partial y} + x^2\frac{\partial Y}{\partial x} + xy\frac{\partial Y}{\partial y}\right). \tag{13.2}$$

It will be recognized that the components $X, Y, Z$ are derivable from the gravitational potential $U$ through the equations

$$X = \frac{\partial U}{\partial x}, \qquad Y = \frac{\partial U}{\partial y}, \qquad Z = g = \frac{\partial U}{\partial z}, \tag{13.3}$$

and the potential function furnishes relations of the form

$$\frac{\partial X}{\partial y} = \frac{\partial^2 U}{\partial y\,\partial x} = \frac{\partial^2 U}{\partial x\,\partial y} = \frac{\partial Y}{\partial x}. \tag{13.4}$$

In terms of the potential, and in cylindrical polar coordinates $(r, \alpha, z)$, the couple acting on the element is

$$dT' = r^2\,dm\left[\frac{\partial^2 U}{\partial x\,\partial y}\cos 2\alpha + \left(\frac{\partial^2 U}{\partial y^2} - \frac{\partial^2 U}{\partial x^2}\right)\frac{\sin 2\alpha}{2}\right]. \tag{13.5}$$

For the entire beam,

$$T' = K\left[\frac{\partial^2 U}{\partial x\,\partial y}\cos 2\alpha + \left(\frac{\partial^2 U}{\partial y^2} - \frac{\partial^2 U}{\partial x^2}\right)\frac{\sin 2\alpha}{2}\right] \tag{13.6}$$

where $K$ is the moment of inertia of the suspension about the $z$-axis.

If one mass is now placed in the lower position, the only change will be to add forces on this mass of the form

$$\frac{\partial X}{\partial z}\cdot hm \quad \text{and} \quad \frac{\partial Y}{\partial z}\cdot hm$$

where $h$ is the vertical displacement of the mass. The equation of the couple acting on the system is then

$$\left.\begin{aligned} T = K\frac{\partial^2 U}{\partial x\,\partial y}\cos 2\alpha + K\left(\frac{\partial^2 U}{\partial y^2} - \frac{\partial^2 U}{\partial x^2}\right)\frac{\sin 2\alpha}{2} + \\ + mhl\frac{\partial^2 U}{\partial y\,\partial z}\cos\alpha - mhl\frac{\partial^2 U}{\partial x\,\partial z}\sin\alpha. \end{aligned}\right\} \tag{13.7}$$

If the torsion constant of the suspension is $\tau$, the effect of this couple will be to rotate the system through an angle $\vartheta$ from its equilibrium position, where

$$\tau\vartheta = T. \tag{13.8}$$

The recordings are in practice made by photographing the image of a spot of light reflected from the mirror carried on the torsion wire. If the photographic plate is at a distance $D$ from the mirror, and if $n_0$ and $n_a$ are the measurements on the plate, from some arbitrary origin, corresponding to the equilibrium position and the actual position of the suspension, then

$$\vartheta = \frac{n_\alpha - n_0}{2D}. \tag{13.9}$$

Finally, substituting in equation (13.7) and (13.8)

$$\left.\begin{aligned} n_\alpha - n_0 = C_1\left(\frac{1}{2}\left[\frac{\partial^2 U}{\partial y^2} - \frac{\partial^2 U}{\partial x^2}\right]\sin 2\alpha + \frac{\partial^2 U}{\partial x\,\partial y}\cos 2\alpha\right) - \\ - C_2\left(\frac{\partial^2 U}{\partial x\,\partial z}\sin\alpha - \frac{\partial^2 U}{\partial y\,\partial z}\cos\alpha\right) \end{aligned}\right\} \tag{13.10}$$

where

$$C_1 = \frac{2KD}{\tau} \quad \text{and} \quad C_2 = \frac{2Dmhl}{\tau} \qquad (13.11)$$

are instrumental constants to be determined by measurement or by timing vibrations of the system. Of the other quantities in equation (13.10), $n_\alpha$ is the actual measurement on the photographic plate, for an initial setting of the beam at an angle $\alpha$ from the $x$-axis, while the unknowns are $n_0$ (the reading which would have been obtained if no couple were acting on the system), and the second derivatives $\frac{\partial^2 U}{\partial x \partial z}$, $\frac{\partial^2 U}{\partial x \partial y}$, $\frac{\partial^2 U}{\partial y \partial z}$ and $\left(\frac{\partial^2 U}{\partial y^2} - \frac{\partial^2 U}{\partial x^2}\right)$. It is seen that if readings were taken for five values of $\alpha$, five equations could be set up which would be capable of solution. However, the time required for an observation is materially shortened by having two similar systems side by side, but with upper and lower masses transposed. If such an instrument is read in three equally-spaced azimuths, six equations result, which may be solved for the unknowns, which now number six (the additional unknown being $n_0$ for the second suspension). Very beautiful instruments were developed, which automatically set the beams in the three correct azimuths, released them, and made exposures on a photographic plate, all with a minimum of attention from the observer. Complete details of instrumental technique are included in some of the older references[1, 2, 3].

It is of more interest here to consider the application of the quantities determined. The two second derivatives $\frac{\partial^2 U}{\partial x \partial z}$ and $\frac{\partial^2 U}{\partial y \partial z}$ are of course *horizontal gradients of gravity* in the north and east directions. Their determination individually defines the value of the total horizontal gradient, and its direction. Vectors representing this gradient at each station may then be plotted on a map of the area surveyed (Fig. 7), the pattern of vectors serving to outline regions of positive or negative gravity anomaly. In addition, actual differences of gravity may be determined, by a graphical or mechanical integration of the gradient between stations. The anomaly contours in Fig. 7a were obtained in this fashion, and these may be compared with the contours in Fig. 7b which were obtained from gravimeter readings some years later.

The dimensions of the gradient of gravity are $LT^{-2}L^{-1}$ or $T^{-2}$, and since the values usually encountered range from 0 to $200 \times 10^{-9}$ sec.$^{-2}$, the unit $10^{-9}$ sec.$^{-2}$ has been adopted and named the Eötvös *unit* (abbreviation $E.$). The gradients which are actually plotted are strictly not those measured, but are the values after *correction* for the effects of local terrain (which even in regions of gentle topography may range up to tens of $E.$), and for the normal gradient due to the variation of gravity with latitude (7 $E.$ at latitude 30°).

The remaining second derivatives of the potential, namely $\frac{\partial^2 U}{\partial y^2}$, $\frac{\partial^2 U}{\partial x^2}$ and $\frac{\partial^2 U}{\partial x \partial y}$ are closely related to the *curvature* of the equi-potential surface which passes through the centre of the beam. In order to examine the relationship, let us take as the equation of the surface which passes through the origin

$$\tfrac{1}{2}A x^2 + \tfrac{1}{2}B y^2 + \tfrac{1}{2}C z^2 + H x y + G x z + F y z + g z = 0 \qquad (13.12)$$

in which terms of order higher than the second have been neglected [as was in fact assumed in the derivation of equation (13.1)], and the coefficients are

---

[1] D. C. BARTON: Amer. Inst. Min. a. Metallurg. Engr., Techn. Publ. **50** (1928).

[2] H. SHAW and E. LANCASTER-JONES: Proc. Phys. Soc. Lond. **35**, 151 (1923).

[3] K. JUNG: Gravimetrische Methoden der Angewandten Geophysik. In Handbuch der Experimentalphysik, Bd. 25, Teil 3, S. 49. Leipzig: Akademische Verlagsgesellschaft 1930.

given by

$$A = \frac{\partial^2 U}{\partial x^2}, \qquad B = \frac{\partial^2 U}{\partial y^2}, \qquad C = \frac{\partial^2 U}{\partial z^2},$$
$$F = \frac{\partial^2 U}{\partial y \partial z}, \qquad G = \frac{\partial^2 U}{\partial x \partial z}, \qquad H = \frac{\partial^2 U}{\partial x \partial y}. \qquad \right\} \qquad (13.13)$$

To examine the form of the surface near the origin, consider its intersection with the plane $z = h$ (Fig. 6). The curve of intersection (the *indicatrix*) is the ellipse[1]

$$A x^2 + B y^2 + 2H x y + 2G x h + 2F y h = -(2g h + C h^2) \qquad (13.14)$$

whose major axis makes an angle $\lambda$ with the $x$-axis, where

$$\tan 2\lambda = \frac{2H}{A - B} = \frac{2\dfrac{\partial^2 U}{\partial x \partial y}}{\dfrac{\partial^2 U}{\partial x^2} - \dfrac{\partial^2 U}{\partial y^2}}. \qquad (13.15)$$

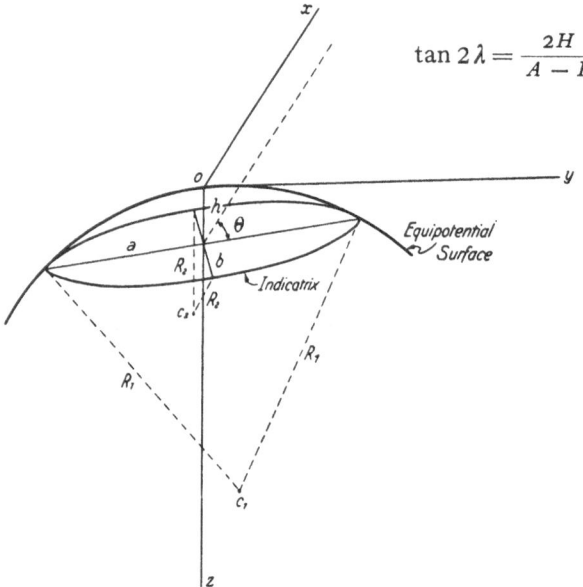

If $a$ and $b$ are the semi-axes of this ellipse, the greatest and least radii of curvature (which lie in perpendicular planes) of the equipotential surface at $O$ are given by

$$R_1 = \frac{a^2}{2h},$$
$$R_2 = \frac{b^2}{2h} \qquad \right\} \qquad (13.16)$$
$$\text{as} \quad h \to 0.$$

Substituting for the values of the semi-axes from equation (13.14) and (13.15) and allowing $h$ to approach zero, we find the convenient result

Fig. 6. Relationship between the principal curvatures of an equipotential surface and the indicatrix ellipse.

$$g\left(\frac{1}{R_2} - \frac{1}{R_1}\right) = \left(\frac{\partial^2 U}{\partial x^2} - \frac{\partial^2 U}{\partial y^2}\right) \sec 2\lambda. \qquad (13.17)$$

In other words, the quantities determined by the torsion balance define the difference between the principal curvatures of the surface, and the azimuths of the planes containing these radii. The difference $\left(\dfrac{1}{R_1} - \dfrac{1}{R_2}\right)$ really indicates the departure of the equipotential surface from a spherical form. For example, over a massive body of considerable extent in one direction, the equipotentials have minimum algebraic curvature (or greatest radius of curvature) in a plane parallel to this direction, and maximum curvature at right angles to it. In the case of an elongated structure which is deficient in density, the curvature of the surfaces is of opposite sign, and the plane of least algebraic curvature is at right angles to the structure. This is the case in Fig. 7, where the direction of minimum curvature, and the value of the curvature difference, are indicated at each torsion balance station, by a line without arrow.

---

[1] The indicatrix may also be a hyperbola, or degenerate into straight lines. See the detailed discussion in K. Jung's article cited above.

The Eötvös balance was undoubtedly one of the most elegant instruments ever designed for investigating gravitational fields. It suffered the practical limitations of low output of stations (1 or 2 per day, compared to 50 or more for a gravimeter), and of extreme sensitivity to nearby density inhomogeneities, so that today it is little used in routine surveys.

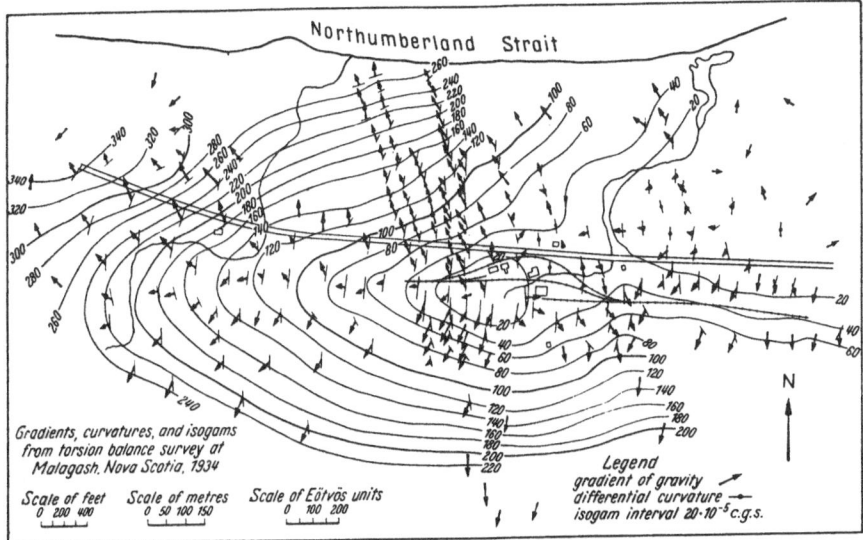

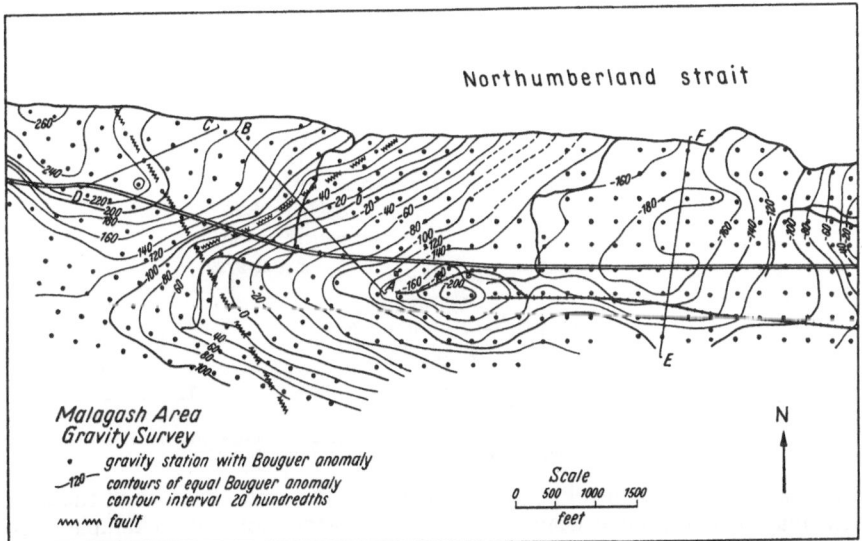

Fig. 7a and b. Gravity anomaly maps over a salt structure, as obtained from measurements with (a) torsion balance and (b) gravimeter. The torsion balance map shows vectors representing the gradient of gravity, and lines representing the direction of minimum curvature, in addition to the anomaly contours. (Torsion balance map after MILLER.)

## c) Gravity Networks.

**14. The Adjustment of Observations.** For studies of the gravitational field over the surface of the earth as a whole, or major portions of it, it is most important that the gravity observations form a self-consistent set, otherwise discontinuities in the data between different areas may be interpreted as having physical significance.

Such discontinuities may arise from errors in the connection of fundamental reference stations to the world base at Potsdam, or from inaccuracies in the calibration of gravity meters. If the maximum advantage is to be taken of the sensitivity of modern instruments, the observations of any survey should be made between stations of an interlocking network, which includes also accurate pendulum stations. In this way, the observed gravity differences around adjoining closed figures may be adjusted to give the minimum closing errors, as is done

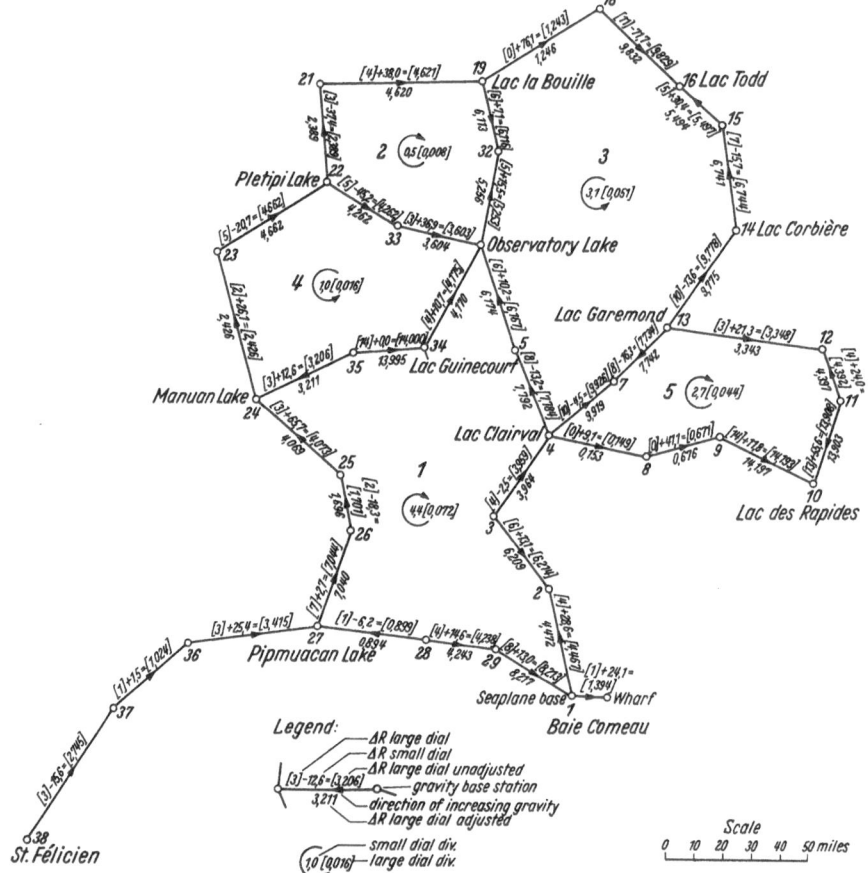

Fig. 8. Part of a network of stations observed with a gravimeter, indicating the closure error of various circuits in scale divisions, equivalent to about 0.1 milligal. (After INNES.)

in the case of geodetic levelling, and in addition the scale constant of the instrument may be determined from the pendulum observations. An example of such a network in Canada is shown in Fig. 8, which indicates the closure of circuits before and after the simultaneous adjustment. In this case, all gravity meter observations were made with one instrument, but Cook[1] has performed a most thorough adjustment of the principal gravity observations of Great Britain, in which the scale constants of four gravity meters were obtained by the inclusion of seven first-order pendulum stations in the network.

The response of gravity meters cannot be assumed to be linear over very large changes in gravity, and in a strong network there should be a distribution

[1] A. H. COOK: Mon. Not. Roy. Astronom.. Soc., Geophys. Suppl. **6**, 494 (1953).

of pendulum stations over the maximum range of gravity encountered. In other words, *the gravity meter should not be used to extrapolate values from a limited calibration range.*

An important limitation until recent years was the dearth of suitably located, dependable pendulum stations for calibration purposes. Efforts are now being made to establish such calibration lines in North America, Europe, and elsewhere.

**15. The World Gravity Network.** Even with the establishment of strong networks in various parts of the world, much remains to be done in the unification of the world gravity network as a whole, as was stressed by MORELLI[1]. Many of the connections of national reference stations to Potsdam (quoted to 0.001 cm./sec.[2]) depend upon uncertain pendulum observations, and in some cases where repeat observations have been made serious discordances have been found. The spread of values relative to Potsdam at Copenhagen (0.007cm./sec.[2]) and Dehra Dun (0.031 cm./sec.[2]) have often been pointed out. Recently investigators such as MARTIN[2], MORELLI[3], and WOOLLARD[4], have made numerous measurements with gravity meters between the important stations of the world, travelling by air. However, while these observations will undoubtedly improve the situation, the use of the gravity meter in this respect is not without difficulties. Apart from the problem of calibration, it is usually not possible to repeat observations for drift control. Instead, the observer must establish drift rates by observations made before and after a flight, then assume an average rate of drift for the time spent in transit. The final aim of a uniform distribution of stations over the world, the relative values of which are reliable to 0.001 cm./sec.[2], will only be obtained through a combination of more first-order pendulum stations and strong gravimeter connections, properly weighted into a single network. This question has been discussed in some detail by LEJAY[5].

# IV. Reduction of Gravity Observations.

**16. Variation with Latitude.** Before any information regarding the internal mass distribution of the earth may be deduced from gravity observations, the much larger variations due to *position* on the earth's surface must be removed. This is usually accomplished by comparing the measured value at a station with a *"normal"* value for the given position and elevation, the difference between the two quantities being known as the *gravity anomaly.*

In the chapter of this volume on "The Figure of the Earth", the derivation is discussed of a formula for the variation of gravity with latitude on the spheroid of reference. The International formula currently in use may be expressed as

$$\gamma_0 = 978.049 \left[1 + 0.0052884 \sin^2 \varphi - 0.0000059 \sin^2 2\varphi\right] \text{cm./sec.}^2 \qquad (16.1)$$

where $\gamma_0$ is *normal gravity* on the International Spheroid at geographic latitude $\varphi$. From this formula, $\gamma_0$ corresponding to the latitude of a gravity station may be computed[6]. The difficulty is that gravity is not measured on the spheroid, but usually on the earth's surface, at some measured height $h$ above the geoid.

---

[1] C. MORELLI: Geofisica Pura e Applicata **8**, 81 (1946).

[2] JEAN MARTIN: C. R. Acad. Sci. Paris **228**, 658; **229**, 18 (1949).

[3] C. MORELLI: Osserv. Geofis. Trieste n.s. **31** 1954.

[4] G. P. WOOLLARD: Geophysics **15**, 1 (1950) and Woods Hole Oceanog. Inst. Ref. 52—59, 1952.

[5] R. P. LEJAY: Bull. Gèod. **30**, 339 (1953).

[6] See tables given by K. JUNG in LANDOLT-BÖRNSTEIN, Zahlenwerte und Funktionen aus Physik usw., Bd. III, S. 265. Berlin: Springer 1952.

A *reduction* must therefore be applied to either the observed or theoretical value before these can be compared, and different types of anomaly are possible depending upon the manner in which this correction is computed.

**17. Elevation Reductions.** If we consider first the *reduction to the geoid*, through the height $h$, and neglect the effect of matter between the geoid and the earth's surface, the correction is readily obtainable from the inverse square principle. For a spherical earth of mass $M$, radius $a$, the pure attraction (neglecting the centrifugal acceleration) is

$$g = \frac{fM}{r^2} \tag{17.1}$$

where $f$ is the gravitational constant.

$$\frac{\partial g}{\partial r} = - \frac{2g}{r} = - \frac{2g}{a} = - 0.3086 \text{ mgal. per meter} \tag{17.2}$$

at the surface of the earth.

Thus, if the observed values for a height $h$ are multiplied by the factor $(1 + 2h/a)$, the corresponding quantities on the geoid are obtained. This is the *free air reduction*, and anomalies obtained through it are known as *free air anomalies*.

A more complete expression for the vertical gradient, taking account of the departure of the figure from a sphere, may be expressed as

$$\frac{\partial g}{\partial r} = - 0.30855 - 0.00022 \cos 2\varphi + 0.000144\, h \tag{17.3}$$

in mgals. per metre, where $\varphi$ is geographic latitude and $h$ is height in kilometres.

The variations with latitude are seen to be sufficiently small for the simple formula (17.2) to be useful.—Even in the more accurate formula (17.3), it is not necessary to take into account the change of the centrifugal acceleration[1].

The vertical gradient will be disturbed by any anomalous masses, above or below sea level, which distort the gravitational field. In view of this, together with the obvious disregard of mass above sea level, it might be questioned whether the free air reduction above has a practical value. Actually, it has considerable significance in studies of the external field from the point of view either of the earth's figure or of the corresponding large-scale mass variations. Its importance was stressed by Jeffreys[2] whose argument, valid to the first order in $h$, may be traced as follows.

The potential $U_p$ due to the earth's attraction at an external point $P$ may be expressed (by using Green's theorem) in terms of distributions over the earth's surface in the form

$$4\pi U_p = \iint \left[ \frac{\partial U}{\partial n} \cdot \frac{1}{R} - U \frac{\partial}{\partial n} \left( \frac{1}{R} \right) \right] dS \tag{17.4}$$

where $R$ is the distance from $P$ to $dS$, and $n$ is the normal to the surface, inward to the earth. The geopotential $\Psi$ is the sum of the potentials due to rotation and attraction, or

$$\Psi = U + \tfrac{1}{2}\,\omega^2\, r^2 \sin^2 \vartheta \tag{17.5}$$

in polar coordinates.

---

[1] E. A. Ansel cites Bruns in B. Gutenberg: Handbuch der Geophysik, Bd. 1, S. 687. Berlin: Gebrüder Bornträger 1936.

[2] H. Jeffreys: Gerlands Beitr. **31**, 378 (1931).

In terms of $g$, apart from a small, nearly constant, term in $\omega^2$,

$$4\pi U_p = \iint \left[ g\cos\varPhi \cdot \frac{1}{R} + gh\frac{\partial}{\partial n}\left(\frac{1}{R}\right) \right] dS \qquad (17.6)$$

where $\varPhi$ is the slope of the earth's surface at $dS$.

This may be shown to be equivalent to a distribution at a depth $h$ below the surface, and therefore to a distribution over the geoid, except that $dS$ is an element of the actual earth's surface. The corresponding element on the geoid is

$$dS_0 = \left(\frac{a}{a+h}\right)^2 \cos\varPhi\, dS \qquad (17.7)$$

where $a$ is the earth's radius, and therefore the surface density corresponding to the distribution is not $\left(\frac{g}{2\pi f}\right)\cos\varPhi$, where $f$ is the gravitational constant, but $\left(\frac{g}{4\pi f}\right)\left(\frac{a+h}{a}\right)^2$ or $\frac{g}{4\pi f}\left(1+\frac{2h}{a}\right)$. In other words $g\left(1+\frac{2h}{a}\right)$, which corresponds to *gravity computed on the geoid by the free air reduction, is simply related to an equivalent surface distribution on the geoid which would reproduce the external field at all points.*

However, for the more specific study of mass anomalies *below* sea level, particularly in a limited region of the earth's surface, it is convenient to correct for the attraction of material between sea level and the earth's surface. If this material is approximated by an infinite horizontal slab, its attraction is

$$\varDelta g = 2\pi f\sigma h \qquad (17.8)$$

where $\sigma$ is the density of the slab and $f$ the gravitational constant.

The effect of this attraction is to increase the value of gravity on the surface, so that the above quantity is subtracted from the observed value to reduce it to sea level. For a commonly employed value of $\sigma$, 2.67 grams per cubic centimetre, the correction is

$$\varDelta g = 0.1119 \text{ mgal. per metre}.$$

This method of reduction was first suggested by the French geodesist BOUGUER, and the anomalies which are obtained by the application of it in combination with the free air reduction are known as BOUGUER *anomalies*. It is evident that the BOUGUER *anomaly* depends upon the density assumed for the material above sea level, and the value adopted for this may involve a personal factor. The figure 2.67 is often quoted as a mean density for continental rocks, but in the case of an investigation of a limited area, it may be desirable to employ a value closer to the actual measured density of the underlying rocks.

For sea stations (observations in a submarine), the free air anomaly is obtained by application of the normal reduction to correct upward to the surface of the sea, together with a correction for the attraction of a layer of water above the point of observation. In the case of the BOUGUER anomaly, allowance is made for the deficiency in density of sea water as compared to crustal rock, the computation being made for a slab of thickness equal to the measured sea depth beneath the station.

If the topography surrounding a station departs appreciably from a horizontal plane, the observed value of gravity will be reduced, as both mass above the level of the station, and the deficiency of mass below, produce a negative effect. A *terrain reduction* must therefore be added to the observed value. This is often computed analytically, by dividing the region surrounding the station into standard zones, bounded by concentric circles and radii, estimating the mean difference in elevation of each zone from that of the station, and applying a correction for each zone, based on this difference and on the density of the surface material.

Two systems which have been widely used in the construction of templates are those of Hayford[1] and Hammer[2]. The outer radii of zones, and the number of compartments in each zone, are given in Table 2. Hayford's divisions are

Table 2. *Standard zones for terrain corrections.*

| Hayford Zones | | | Hammer Zones | | |
|---|---|---|---|---|---|
| Designation of Zone | Outer radius-metres | No. of Compts. | Designation of Zone | Outer radius-feet | No. of Compts. |
| A | 2 | 1 | A | 6.56 | 1 |
| B | 68 | 4 | B | 54.6 | 4 |
| C | 230 | 4 | C | 175 | 6 |
| D | 590 | 6 | D | 558 | 6 |
| E | 1280 | 8 | E | 1280 | 8 |
| F | 2290 | 10 | F | 2936 | 8 |
| G | 3520 | 12 | G | 5018 | 12 |
| H | 5240 | 16 | H | 8578 | 12 |
| I | 8440 | 20 | I | 14662 | 12 |
| J | 12400 | 16 | J | 21826 | 16 |
| K | 18800 | 20 | K | 32490 | 16 |
| L | 28800 | 24 | L | 48365 | 16 |
| M | 58800 | 14 | M | 71996 | 16 |
| N | 99000 | 16 | | | |
| O | 166700 | 28 | | | |

actually part of a world-wide system, which was adopted for broad-scale studies, and which will be referred to later (Sect. 21) in the discussion of isostatic anomalies. Factors for computing the effect of uneven topography out to the limits of Hayford's zone O have been listed by Bullard[3]. For certain applications, such as the detailed coverage of a restricted area, the effects of more distant features may be neglected, but a more detailed evaluation of local terrain is required. Hammer's system of zones was designed with this in view. In either case, transparent templates are constructed, according to the scales of the topographical maps available. The mean elevations of all compartments surrounding a gravity station are estimated from the contours of the map, multiplied by the appropriate factors, and summed to give the total correction. The inner zones of the systems are too small to be estimated from standard maps, so that either a sketch must be made of the topography surrounding each station, or, preferably, the stations must be judicously selected to minimize the effect of very local features.

In cases where the topography consists of elongated forms such as scarps or ridges, considerable time may be saved by the use of profiles calculated for simple shapes of infinite length[4]. For example, Fig. 9 illustrates the effect of a vertical scarp, and from this diagram the correction for a station at a given distance from the top or bottom of the scarp could be quickly estimated. It

[1] John F. Hayford and William Bowie: U.S. Coast and Geodetic Survey Spec. Publ. 10, 1912.
[2] Sigmund Hammer: Geophysics **4**, 184 (1939).
[3] E. C. Bullard: Phil. Trans. Roy. Soc. Lond., Ser. A **235**, 445 (1936).
[4] M. King Hubbert: Geophysics **13**, 226 (1948).

is of interest to note in this case how rapidly the correction falls to negligible values with increasing distance from the topographical feature.

In the case of either the free air or BOUGUER reductions, the observed values of gravity are reduced to the surface of the geoid, while the theoretical values $\gamma_0$ refer to the spheroid. The difference, known as the *indirect effect*, is a relatively small, slowly varying, quantity, which is often omitted in the study of local areas. It is important in large scale investigations, however, since the geoid is systematically elevated over continents and depressed over oceans.

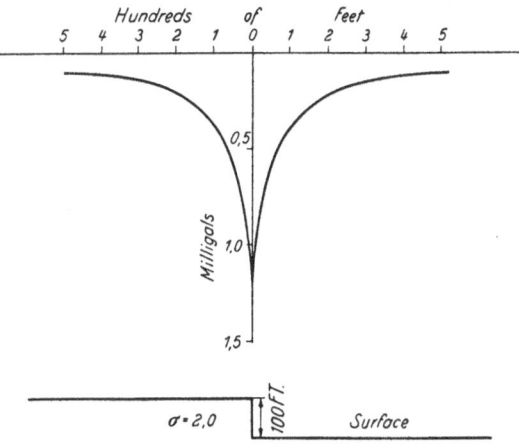

Fig. 9. The effect on gravity of a vertical scarp. (After HUBBERT.)

The free air and BOUGUER anomalies, which may be computed from the observed gravity and the latitude and elevation of the station in a relatively straightforward manner, form the starting point for investigations on the internal mass distribution. It is true that a number of other types of anomaly are possible, if allowance is made for supposed systematic mass distributions according to different hypotheses. These will be considered in a subsequent section, after the hypotheses themselves have been introduced.

## V. The Mathematical Interpretation of Gravity Anomalies.

**18. Limitations to the Interpretation of Potential Fields.** Before any specific results relating to the structure of the earth from gravity measurements are discussed, it would be well to outline the methods in use for the physical interpretation of the anomalies in terms of mass distributions. It is a well known result from the theory of potential that no interpretation can be *unique*, for an infinite number of mass distribution can always be found which will give the same gravitational field at all external points. This follows from GREEN'S theorem, or from solutions to the DIRICHLET problem.

Consider, for example, a mass anomaly of limited horizontal extent within the earth's crust, so that in the region of detectable gravitational effect the earth's surface may be taken as a horizontal plane. Then the observed potential field at all points above the region will be reproduced by a surface distribution $\Delta g/2\pi f$ over any horizontal plane between the anomalous mass and the earth's surface. It is thus never possible to place a minimum depth on the source of an anomaly, but it may be possible to determine a *maximum depth*, beyond which negative densities would be required to reproduce the observed effect.

Since a number of mass distributions will yield the same potential field at all external points, it is obvious that no advantage would be gained by *measuring* gravity at different heights above the earth, if such were possible, or more than one derivative of the potential. Earlier writers on the EÖTVÖS torsion balance, for example, were sometimes misled by the latter possibility.—But considerations of higher derivatives have proved useful (see Sect. 20), in other respects.

**19. Determination of Anomalous Masses.** In spite of these limitations, a number of very useful direct interpretation methods have been devised whereby the mass distribution corresponding to a given gravitational field may be computed, provided certain assumptions are made regarding the general form or position of this distribution. For example, if it is thought that the source of the anomaly may be approximated by a surface distribution at a depth $z$, the field itself $g_z$, may be computed from the surface field, and the corresponding surface density determined from

$$\sigma = \frac{g_z}{2\pi f}. \tag{19.1}$$

To determine $g_z$, the method of EVJEN[1] may be followed, in which use is made of the MACLAURIN expansion

$$g_z = g_0 + \left(\frac{\partial g}{\partial z}\right)_0 z + \left(\frac{\partial^2 g}{\partial z^2}\right)_0 \frac{z^2}{2!} + \cdots \tag{19.2}$$

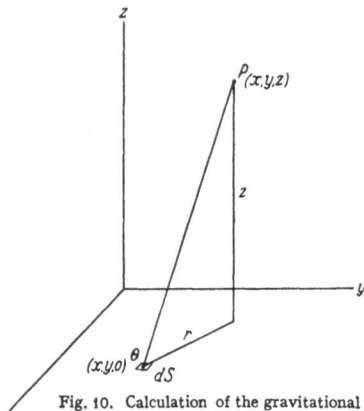

Fig. 10. Calculation of the gravitational field at points above the surface.

where $(\partial g/\partial z)_0$, $(\partial^2 g/\partial z^2)_0$, etc. are the successive vertical derivatives of gravity at the surface.

To compute the required derivatives, it is necessary to derive an expression for the field at points above the surface, in terms of the values measured on the surface. In other words, we are led to the NEUMANN and DIRICHLET problems of potential theory[2]. Briefly, the principle is that the given field is unchanged if the original masses are removed and replaced by a surface distribution of density $\Delta g/2\pi f$ over $S$ where $\Delta g$ is the measured anomaly on $S$. At a point $P$ above $S$ therefore (Fig. 10), the vertical component of the attraction of an element $dS$ about the point $Q$ is

$$\partial g_P = \frac{z \cdot \Delta g_Q \cdot dS}{2\pi (r^2 + z^2)^{\frac{3}{2}}} \tag{19.3}$$

and the total anomaly at $P$ due to the surface distribution is

$$\Delta g_P = \int_S \partial g_P \, dS. \tag{19.4}$$

In other words, the field in the region above $S$ is completely determined by the measured values on $S$. By means of differentiation under the integral sign, similar expressions may be set up for the vertical derivatives of $g$, and these may be evaluated numerically (with due regard for the singularity at the origin in the case of points on the surface) for use in the MACLAURIN expansion.

In a method due to TSUBOI[3], the surface field is analysed to give a double FOURIER series

$$g = \sum \sum c_{mn} \frac{\cos}{\sin} m x \frac{\cos}{\sin} n y. \tag{19.5}$$

It follows that the surface distribution at depth $h$ has the series form

$$\sigma = \sum \sum \frac{c_{mn}}{2\pi f} e^{\sqrt{m^2+n^2}\, h} \frac{\cos}{\sin} m x \frac{\cos}{\sin} n y. \tag{19.6}$$

Thus, when $h$ is given, the coefficients in the series are immediately determined.

[1] H. M. EVJEN: Geophysics **1**, 127 (1936).
[2] B. KOSBAHN: Geofisica Pura e Applicata **11**, 27 (1949).
[3] C. TSUBOI: Proc. Imp. Acad. Jap. **14**, 170 (1938).

Bullard and Cooper[1] point out that the arithmetic involved in analyzing the surface field may be considerable, and prefer the Bessel-Fourier integral expression

$$g_h(r, \varphi) = \frac{1}{2\pi} \sum_{n=-\infty}^{+\infty} \int_0^\infty \int_0^{2\pi} \int_0^\infty g_0(r_1, \varphi_1) e^{ph} J_n(p\, r_1) J_n(p\, r) \times \\ \times \cos n(\varphi_1 - \varphi)\, p\, r_1\, dr_1\, d\varphi\, dp \qquad (19.7)$$

where $(r, \varphi, z)$ are cylindrical coordinates. Providing $g_0$ (the measured field on the surface) is a sufficiently smooth function, the integral may be solved for $g_h$ (the field at a depth $h$), and the equivalent mass distribution found. Bullard and Cooper tabulate functions which may be used for the smoothing and the integration.

In many cases, however, it may not be convenient to deal with a surface distribution of mass, for example, in the case of a relatively shallow structure of significant vertical relief. Jung[2] showed that for simple geometrical forms, (spheres, elliptical cylinders, etc.) parameters such as depth could be determined from characteristics of the anomaly profiles. In practice, such an approach may be very dangerous, for the result is entirely dependent on the form assumed. However, Jung did emphasize a most useful theorem of Gauss which relates the integrated anomaly over a horizontal plane to the total anomalous mass $M$

$$M = \frac{1}{2\pi f} \int_{-\infty}^{\infty} \int_{-\infty}^{\infty} g(x, y)\, dx\, dy. \qquad (19.8)$$

The integration is carried to the limits of the detectable gravity anomaly $g(x, y)$. In other words, the total disturbing mass is uniquely determined by the surface gravity field; only the distribution of the mass is indeterminate.

The proof of (19.8) is easy: It is readily proved for a point mass $dm$ and will therefore hold for any mass $M = \int dm$.

One of the most elegant methods developed for direct interpretation is that of Grant[3]. It is well known that the expression for the potential of an attracting mass may be expressed in a series of inverse powers of the distance from the point of observation to a fixed point in the mass. Provided no dimension of the body is large compared to this distance, the series converges fairly rapidly. The coefficients in the series involve succeeding multipole moments of the body, and are therefore closely related to the symmetry. Grant divides structures into the usual crystal classes (monoclinic, orthorhombic, tetragonal and cubic), each class being characterized by certain non-vanishing moments. Equations may be set up relating these moments to various derivatives of the surface field, so that if these derivatives are calculated, and a general form is assumed for the unknown structure, its most probable symmetry classification, and the orientation of its axes, may be determined.

Apart from these direct methods of approach, resort may always be had to an indirect procedure, whereby the effects of simple hypothetic bodies are calculated, and the bodies are adjusted until a fit with the observed gravitational field is obtained. In many cases this is sufficient to indicate the order of magnitude of the attracting body, and indeed it may be all that is justified if the observations are sparse or of uncertain quality. As an example, Fig. 11 shows an observed anomaly profile and the calculated profiles due to three quite different structures, in the relative positions indicated. The size, density and position

---

[1] E. C. Bullard and R. I. B. Cooper: Proc. Roy. Soc. Lond. A **194**, 332 (1948).
[2] K. Jung: Z. Geophysik **13**, 45 (1937).
[3] Fraser Grant: Geophysics **51**, 344, 756 (1952).

of the assumed structures has been adjusted to give a satisfactory agreement with the observed anomaly, but the decision as to which, if any, approximates the actual geological conditions must be based on other lines of evidence.

Useful formulae and curves for estimating the effects of hypothetical structures have been given for simple cases by NETTLETON[1], for a wider variety of forms by ANSEL[2], and for quite general structures, given by depth contours, by BARANOV[3]. For the interpretation of torsion balance surveys, the extensive series of calculated profiles of the gradient and curvature by SHAW[4] is available.

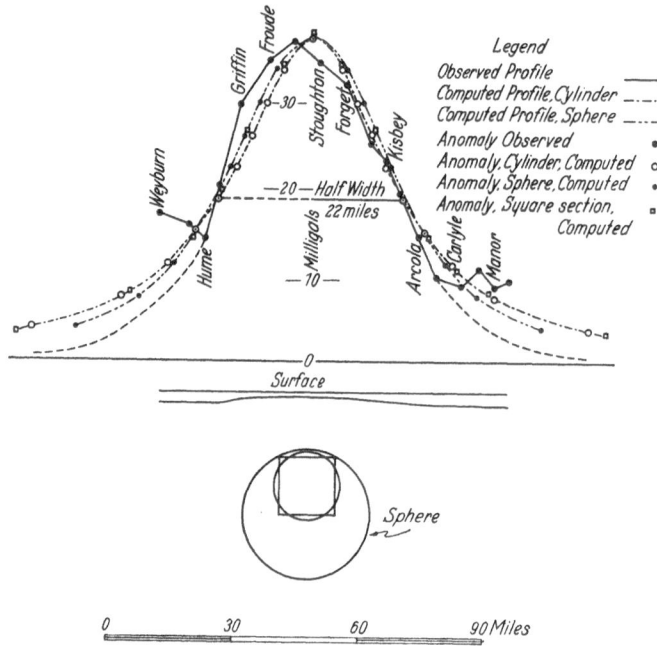

Fig. 11. Analysis of an observed anomaly profile by comparison with the calculated effects of simple forms. (After MILLER.)

In general, the question of calculating the attraction of type bodies may be divided into two cases, depending upon whether the body involved is finite in all dimensions, or is elongated in one dimension, with uniform cross-section. Structures of the latter type are spoken of as *"two-dimensional"*. In the first case, the fundamental formula is the vertical component of the attraction of a point mass, which is to be integrated over the limits of the body. With the notation of Fig. 12a, the effect at $O$ of a point mass $m$ at $M$ is:

$$\Delta g = f\, m\, z/r^3 .$$ (19.9)

As a specific case, the attraction of the sphere (which equals the attraction of its total mass concentrated at its center) shown in Fig. 12a, may be written as

$$\Delta g = \tfrac{4}{3}\,(\sigma_2 - \sigma_1)\,\pi f z\, a^3/r^3$$ (19.10)

where $(\sigma_2 - \sigma_1)$ is the density contrast between the sphere and the surroundings.

[1] L. L. NETTLETON: Geophysical Prospecting for Oil, pp. 102—115. New York 1940.
[2] ANSEL, E. A.: Beitr. angew. Geophys. 5, 763 (1936); 6, 141 (1937).
[3] V. BARANOV: Geophysical Prospecting 1, 36 (1953).
[4] H. SHAW: Trans. Amer. Inst. Min. a. Metallurg. Engr. 97, 271 (1932).

In practice, many geological bodies are elongated in one direction (the "strike" of the structure), and profiles calculated for the two-dimensional case may be used to approximate their attraction. In this case, the fundamental formula is the attraction of a line element parallel to the $y$-axis in Fig. 12b:

$$\Delta g = 2f\,\lambda z/r^2 \qquad (19.11)$$

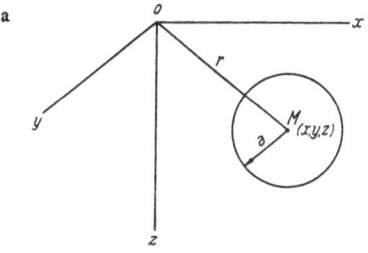

where $\lambda$ is the mass per unit length of the element. The attraction in other cases is obtained from an integration over the cross-section of the body in the $x$-$z$ plane. For the horizontal cylinder Fig. 12c we have

$$\Delta g = 2\pi f(\sigma_2 - \sigma_1)\,z\,\frac{a^2}{r^2}. \qquad (19.12)$$

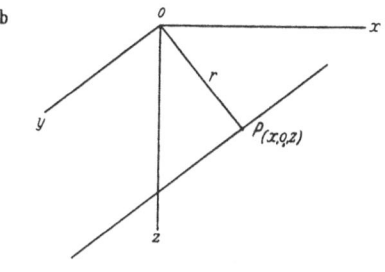

For two-dimensional structures of irregular cross-section, a variety of charts is available for the estimation of the gravitational effect[1].

It is customary for the geophysical interpreter to have at hand a collection of these calculated anomaly profiles, showing the effects of idealized structures of various sizes in different positions, to facilitate the comparison with the observed effect. The chief danger in this procedure is that the interpreter may be led into the belief that satisfactory agreement between a certain theoretical calculation and the observed field is sufficient to explain the anomalous conditions, whereas in fact it merely indicates the most likely dimensions of the particular type body he has selected for the process. It is best to qualify the conclusions relating to the interpretation by stating the assumptions made, or to restrict them to general limits, such as the maximum possible depth, or order of magnitude, of the unknown structure.

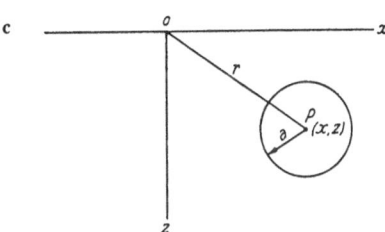

Fig. 12a—c. Notation employed in calculating the attractions of (a) point-mass or sphere (b) horizontal line and (c) horizontal cylinder.

**20. Isolation of Anomalies.** In all of the methods which have been discussed it is implied that the field under consideration is that due to a single, isolated body. Actually, the gravity anomalies at each station reflect the integrated attractions of *all* mass irregularities within the range of detection, and the separation of these effects may present a major problem. In the study of broad-scale effects, continental or sub-continental in scale, the question is one of smoothing out very local anomalies. Conversely, in the detailed study of a local area, a regional trend must often be removed before the smaller anomalies can be studied. Methods are available for the mathematical smoothing of the anomaly field, if the data are available in the form of a rectangular grid. A number of matrix equations may then be set up, which when solved, give the equation of the "anomaly surface" to the desired order.

---

[1] F. Breyer: Beitr. angew. Geophys. **7**, 317 (1938).

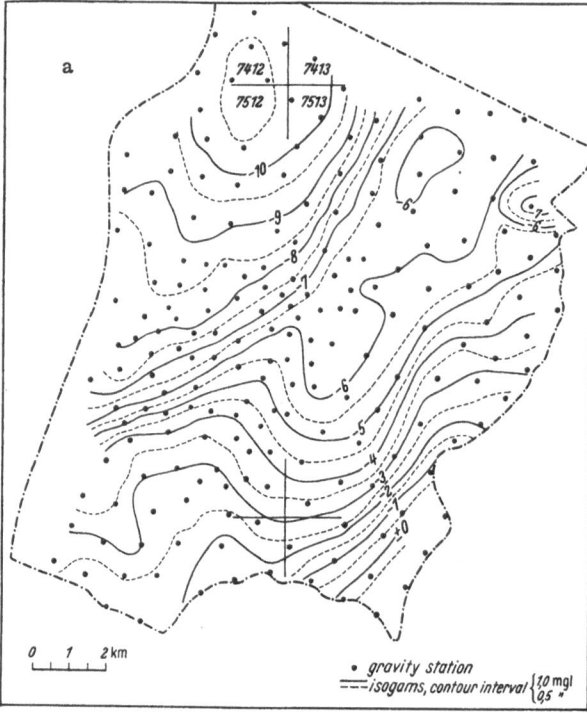

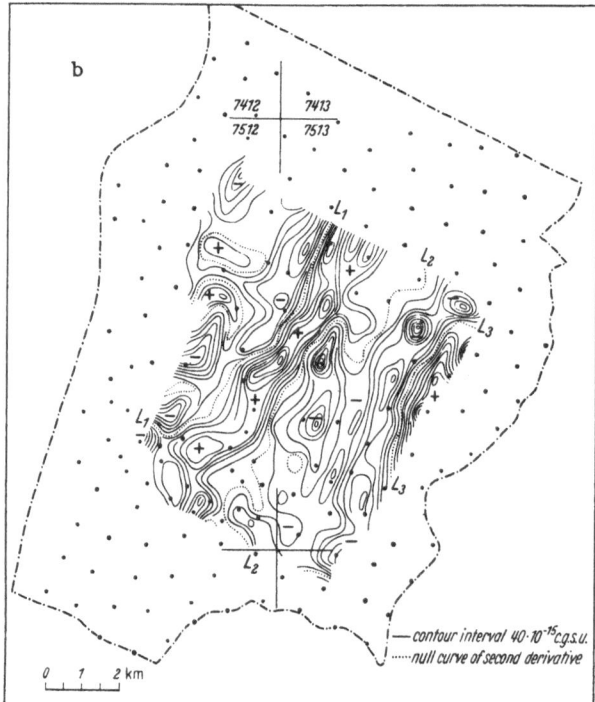

Fig. 13a u. b. BOUGUER anomaly and second vertical derivative maps of the same area. (After ROSENBACH.)

In geophysical prospecting, considerable use has recently been made of the first few *vertical derivatives* of the gravitational field[1, 2, 3] for the purpose of accentuating very local effects. The second vertical derivative in particular may be thought of as representing the *curvature* of the original" anomaly surface". Thus, local flexures in the BOUGUER anomaly field, which may be almost undetectable to the eye, may give very large values of the second derivative, while broad regional trends of considerable absolute magnitude on the BOUGUER map contribute practically nothing to the derivative. Some beautiful applications have been published, of which Figs. 13 and 14 are examples. It will be noted that the smooth regional increase of the BOUGUER anomaly map has been replaced by a succession of intense positive and negative centres, upon which further investigation may be concentrated. The interpretation of the individual features on the second-derivative map can be assisted by computing second-derivative profiles for hypothetical bodies and comparing these with the anomalies in hand.

[1] H. M. EVJEN: Geophysics **1**, 127 (1936).

[2] T. A. ELKINS: Geophysics **16**, 29 (1951).

[3] OTTO ROSENBACH: Geophysical Prospecting, II, 1 and 128, 1954.

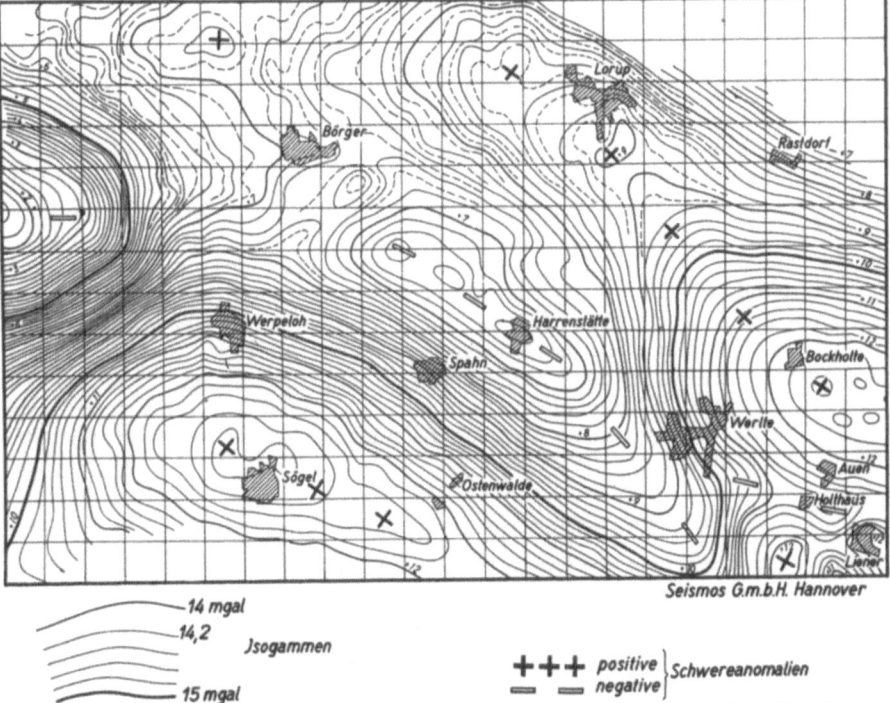

—————— 14 mgal
—————— 14,2  Jsogammen
—————— 15 mgal

**+ + +** positive } Schwereanomalien
**— — —** negative

Fig. 14a. Bouguer anomaly $\Delta g''$, expressed by isogams in steps of 0.2 milligal, for a region in North-West Germany.

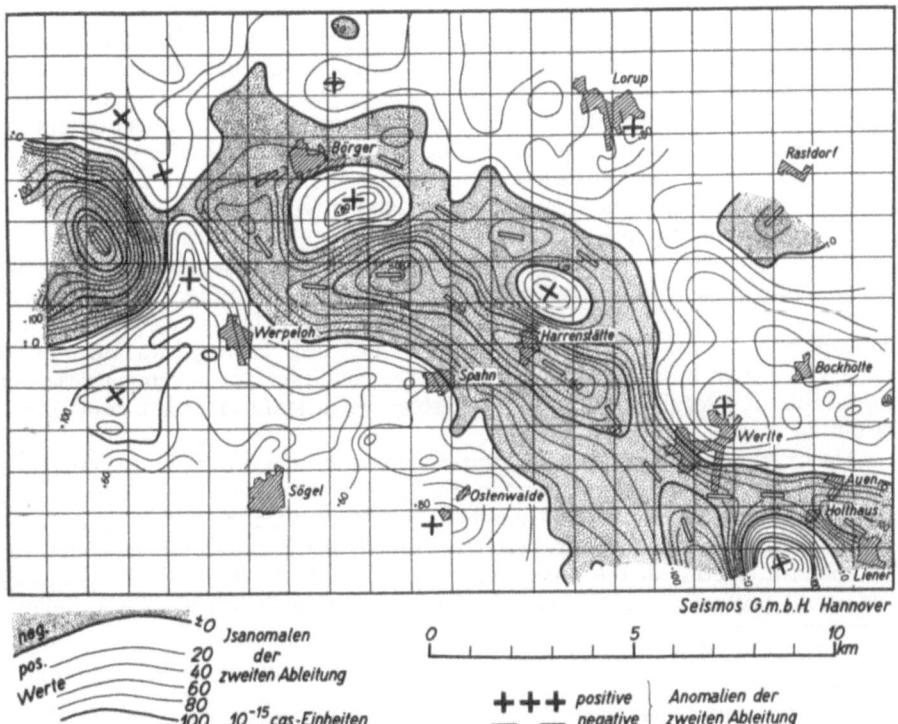

neg. ———— ±0  Jsanomalen
pos. ———— 20  der
———— 40  zweiten Ableitung
Werte ———— 60
———— 80
———— 100  $10^{-15}$ cgs-Einheiten

0          5          10
|_|_|_|_|_|_|_|_|_|_|  km

**+ + +** positive } Anomalien der
**— — —** negative  zweiten Ableitung

Seismos G.m.b.H. Hannover

Fig. 14b. Second derivative $\partial^2 g/\partial z^2$, expressed by isanomals in steps of $20 \cdot 10^{-15}$ cgs-units, for the same region as in Fig. 14a, computed according to H. Linsser's method, described at the Munich Meeting, Deutsche Geophysikalische Gesellschaft, 1955 (Courtesy Seismos-Hannover).

However, it must not be thought that the method is without limitations. The most serious of these is the extreme *magnification of errors* in the original gravity data, by the very nature of the differentiation process. The fundamental formula used in the calculation of any derivatives is equation (19.3), which may be differentiated with respect to $z$, but the best means of evaluating the surface integral has been a question of some discussion[1]. In particular, difficulties arise when the gravity stations of the original survey are not located on a regular grid, and interpolation must be carried out before graphical integration[2]. An additional disadvantage is the loss of one of the most desirable features of the BOUGUER gravity map, that is, the close relationship between anomaly magnitude and total mass of the attracting body, for the actual peak values of the second derivative map have relatively minor significance.

With this brief outline of the procedures available for the interpretation of force fields, we may turn to the evaluation of gravitational methods for determining the structure of the earth.

# VI. The Meaning of Gravity Anomalies.

In the previous section, the formal relationships between anomalies of gravity and the responsible mass distributions were discussed. For the mass distributions to have geophysical significance, it is necessary to relate them to structural conditions within the earth. It is the intention in this chapter to outline some of the more important contributions to our knowledge of the earth's interior from studies of the gravitational field. The applications to the determination of the earth's *figure* are treated separately[3].

**21. The Notion of Isostasy.** The suggestion that topographical irregularities in the earth's surface are somehow compensated by a deficiency in mass beneath elevated regions was made in the eighteenth century or earlier. The development of idealized mechanisms to explain in quantitative terms the deficiency beneath such mountains as the Himalayas came about the middle of the nineteenth century, as a result of the study of deflections of the plumb line measured during the triangulation of India, and before the era of extensive gravity surveys. At that time, theories were put forward by AIRY[4] and PRATT[5] which, with modifications, have formed the basis of many later studies.

AIRY suggested the presence of a light, but strong, crust floating on a denser substratum, in such a way that the topography is compensated by *roots* of the lighter material at the base of this crust. PRATT on the other hand proposed a crust extending to a *uniform depth* below sea level, but of *varying density*, depending on the height of the topography. The word *"isostasy"* was introduced some years later by DUTTON[6] to describe the condition of the compensation of topographical irregularities, and the existence of a state of hydrostatic equilibrium beneath a certain depth.

The existence of some form of isostatic compensation is proven by the universal tendency of BOUGUER anomalies toward large negative values with increasing elevation. In other words, the observed and theoretical values of gravity at mountain stations would be in better agreement if the BOUGUER correction for the direct attraction of the topography were *omitted*. As JEFFREYS[7] has noted, one would have to choose between the conclusions that the topography is hollow,

---

[1] See last two references.
[2] H. LINSSER's method, of which Fig. 14b gives an example, does not need a regular grid.
[3] K. JUNG: Last Chapter in this volume.
[4] Sir G. B. AIRY: Phil Trans. Roy. Soc. Lond. **145**, 101 (1855).
[5] J. H. PRATT: Phil. Trans. Roy. Soc. Lond. **149**, 745 (1859).
[6] CLARENCE E. DUTTON: Bull. Phil. Soc. Washington **11**, 51 (1892).
[7] H. JEFFREYS: Earthquakes and Mountains, 2nd. edit, p. 75. London 1950.

or that it is compensated by a mass deficiency at depth. For the testing of the completeness of compensation according to the different theories, it has been the practice to compute *isostatic anomalies*, in which allowance is made for the compensating masses. If isostasy is complete, these anomalies should be nearly zero for stations at all elevations, in contrast to the BOUGUER anomalies.

For computing the effect of the compensation, the surface of the earth is divided into the HAYFORD zones and compartments, which have been mentioned above (Sect. 17), in connection with terrain corrections. In table 2 the HAYFORD zones were listed to a distance of 166.7 kilometres from the station; eighteen additional zones (Numbered 18 to 1) extend to the antipodes[1]. The problem is then to calculate the effect, at the central station, of the mass defficiency

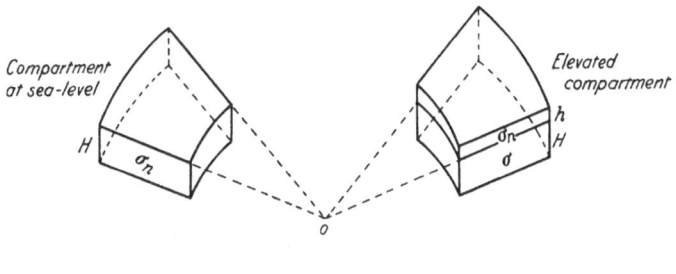

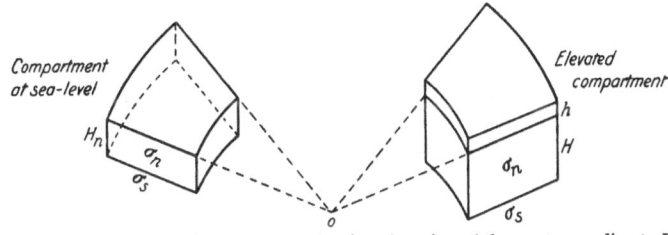

Fig. 15. Factors involved in the calculation of the compensation of a prism of the crust, according to PRATT and AIRY.

necessary to compensate the average surface elevation of each compartment.

On the PRATT theory, this mass deficiency consists of an anomaly in density extending to a *constant* depth $H$, (Fig. 15), the actual density in the column beneath any compartment being

$$\sigma = \frac{H}{H+h} \cdot \sigma_n$$

where $h$ is the mean surface elevation of the zone and $\sigma_n$ is the normal crustal density (usually assumed to be 2.67 grams per cubic centimetre). Beneath the ocean the density $\sigma$ of the column is greater than normal, beneath elevated land areas it is smaller. Now the attraction of such a prism of material at the central station is easily calculated, and tables have been prepared by HAYFORD and BOWIE giving the effect of the mass deficiency for various surface elevations of each of the standard zones.

For the AIRY theory, the density is assumed to be constant $= \sigma_n$, but the depth extent $H$ of a compartment varies with the surface elevation, as indicated in Fig. 15. In this case, the attraction of columns of varying length must be

---

[1] JOHN HAYFORD and W. BOWIE: U. S. Coast and Geodetic Survey Spec. Publ. 10, 1912.

calculated, after the adoption of the normal thickness of the crust for segments at sea level, $H_n$, and of the constant density difference between the crust and substratum $(\sigma_s - \sigma_n)$. Tables giving this effect for various parameters on the AIRY hypothesis have been published by HEISKANEN[1].

In computing isostatic anomalies, therefore, one must determine the elevations of each of the HAYFORD zones surrounding each station, and extending, theoretically at least, to the antipodes. The corrections for the compensation of each zone are then found from tables, so that the total correction to be applied to the observed gravity to obtain the *isostatic anomaly* may be summarized as follows:

(i) *Free Air Correction* as discussed in Sect. 17.

(ii) BOUGUER *Correction,* assuming the topography to be an infinite slab.

(iii) *Terrain Correction* for the departure of the topographic surface surrounding the station from a plane (usually computed to the limit of zone *O*).

(iv) A small *curvature correction*[2] to correct (ii) to the attraction of a cap of material extending to the limit of zone *O*.

(v) The *correction for compensation,* according to the assumed theory, for all zones from the station to zone *O*.

(vi) The *combined correction* for topography and compensation for the more distant zones. The combined effect for the most distant zones is rather slowly varying from place to place, and in many parts of the world the correction may be interpolated from previous calculations, or read from special charts[3].

In recent years, HEISKANEN and his students[4] have made investigations of the AIRY theory in many parts of the world. VENING MEINESZ[5] has suggested a modification of the AIRY hypothesis, in which the compensation of a elemental prism of the crust is not concentrated beneath that prism, but is spread out radially, which would be the case if the crust yielded elastically, in the manner of a beam, to the applied load.

The usual method of procedure has been to compute isostatic anomalies for a number of depths of compensation, to determine which hypothesis gives the smallest mean anomaly. Alternatively, the correlation factor between the anomalies and height may be found, and the depth adopted which gives the least correlation. HEISKANEN[6], in summarizing a number of studies has shown that a depth of the AIRY "crust" of between 30 and 40 kilometres beneath region at sea level (which would correspond approximately with the crust as defined seismologically) generally gives anomalies most independent of height. In practically all regions of the world the BOUGUER anomaly goes to extremely large negative values with increasing height, so that some form of compensation must be present. However, since the interpretation of the gravitational field is not capable of a unique solution, it is apparent that the criterion of determining smallest residuals for some particular idealization in a particular area is not in itself evidence that the hypothesis is close to the truth. The question of isostatic compensation must be examined from a more fundamental standpoint, especially from the point of view of broad scale variations in gravity which would demand strength at depths within the earth greater than the usual depth of compensation. The question of imperfect compensation of more local features, say less

---

[1] W. HEISKANEN: Publ. Isostatic Inst., **2** (1938).

[2] E. C. BULLARD: Phil. Trans. Roy. Soc. Lond., Ser. A **235**, 445 (1936).

[3] W. HEISKANEN and U. NUOTIO: Publ. Isostatic Inst. **3** (1938).

[4] Largely in the important series Publications of the Isostatic Institute of the International Association of Geodesy, Helsinki.

[5] F. A. VENING MEINESZ: Bull. Géod. **29**, 1 (1931); **63**, 1 (1939).

[6] W. HEISKANEN: Trans. Amer. Geophys. Un. **34**, 11 (1953).

than 50 kilometres in extent, is less vital. Such loads can be supported by a system of allowable stresses within the crust[1], without making demands upon the sub-crustal material. (See also pp. 253 ff.)

**22. The Analysis of the External Field.** The observations of gravity indicate that, to a first approximation, there is an *equality of mass per unit area, above a certain level within the earth.* A state of perfect isostasy would therefore be equivalent to purely hydrostatic stress beneath this level. However, broad scale departures from isostasy would imply shearing stresses within this region, and it is therefore of considerable interest to investigate these effects.

JEFFREYS[2] has taken the earth's external field, as represented by free air anomalies, and analysed it for harmonics[3] up to degree 4 which may be present in addition to the main ellipticity term. That is, a solution for the variation of gravity over the earth's surface as a function of both latitude $\varphi$ and longitude $\lambda$ is sought in the form

$$g = \sum_n \sum_m a_{nm} P_n^m (\varphi) \cos m\lambda + \sum \sum b_{nm} P_n^m (\varphi) \sin m\lambda. \qquad (22.1)$$

The International Formula (Eq. 16.1) expressed in this form would be (in milligals)

$$\gamma_0 = 979770 + 3446 \cdot 0\, P_2 (\varphi) + 5.3\, P_4 (\varphi).$$

JEFFREYS' method was to choose one station in each 1° "square" of the earth's surface, and use these to obtain a mean value for each 10° square, corresponding to the mean height of the latter. These in turn were combined into 30°-squares, the means of which were subjected to harmonic analysis by a least squares procedure. The final result is an equation of the form

$$\left.\begin{aligned}
\gamma_0 = (979772.5 \pm 1.9) + (3439.9 \pm 5.0)\, P_2(\varphi) + 5.3\, P_4(\varphi) +\\
+ (4.0 \pm 1.4)\, P_2^2(\varphi) \cos 2\lambda + (1.30 \pm 0.68)\, P_3^2(\varphi) \cos 2\lambda +\\
+ (4.2 \pm 2.4)\, P_3^1(\varphi) \cos \lambda + (0.46 \pm 0.26)\, P_3^3(\varphi) \sin 3\lambda.
\end{aligned}\right\} \qquad (22.2)$$

The uncertainties in the coefficients are standard errors. The fourth harmonic was not determined by the analysis, but is included because it is required by the second order theory of the earth's figure[4]. Due to the scarcity of data in some regions the uncertainties of the longitude terms are rather high, but the presence of the terms seems fairly well established. In addition, harmonics of degrees 20 and 6 with the considerable amplitudes of 20 and 12 milligals were determined. Since even a completely compensated topographical feature produces an effect on free-air gravity, the above variations could be construed as resulting from harmonics in the surface elevation. But it is found that the actual topography bears no relation to what would be required in this case, and in fact many of the harmonics in the surface elevation have the opposite sign. In other words, the variations established by JEFFREYS' analysis indicate *uncompensated loads within the earth,* extending over very considerable regions.

The strength required to support these loads depends upon the depth to which the rocks are assumed to have such strength. If some strength is assumed to the core of the earth, a stress difference of $1.5 \times 10^8$ dynes/cm.$^2$ is indicated

---

[1] H. JEFFREYS: The Earth. Third edit. Chapter VI. Cambridge: University Press 1952.
[2] H. JEFFREYS: Mon. Not. Roy. Astronom. Soc., Geophys. Suppl. 5, 1 (1941); 5, 55 (1943).
[3] As to the definition of the spherical harmonics $P_n^m$, see, e.g., JEFFREYS, The Earth (l. c.), p. 128; or the last chapter in Vol. XLIX of this Encyclopedia.
[4] K. JUNG: Last Chapter in this volume.

for the upper 600 kilometres. If it is assumed that there are no stress differences below 600 kilometres, which is about the limit of deep focus earthquakes, then a stress difference of $3.3 \times 10^8$ dynes/cm.$^2$ is required above this depth. The strength of the material in this region could then be about one third that of crustal rocks. The presence of such considerable strength beneath the usual depth of compensation has often been overlooked; and it has been implied that uncompensated features of the earth are tending toward more complete compensation. In fact, such features are being supported by the strong crust or by the weaker, but still appreciably strong, material beneath it.

We have so far discussed in a very broad way the theory of isostasy in the light of the probable conditions at depth within the earth. In order to come to conclusions regarding the actual density relationships closer to the earth's surface, it will be necessary to examine specific examples of the gravity anomalies observed over various features such as island arcs, mountain ranges, shields, etc.

**23. Island Arcs.** The origin of the forces which produce the deformation of the earth's crust remains an undecided question[1], but there is general agreement that the early stages in a cycle of mountain building are seen today in seismically-active island arcs. These features consist of arcuate patterns of volcanic and sedimentary islands, such as are found in the East and West Indies and around the Pacific, in which the tectonic activity is evidenced by active volcanoes and earthquakes. The latter are found to occur at increasing depths as one proceeds toward the continents, away from the island arcs. Very deep ocean trenches are known to be associated with these features, as are also *intense, but narrow, negative gravity anomalies* (Fig. 16[2]). An explanation for these anomalies, as put forth by MEINESZ and extended by HESS[3], is that they result from a down-buckling of the lighter earth's crust in the substratum, in the form of an immense root. In this connection, it should be noted that the method of reducing the gravity observations does not appreciably alter the form of the curves; the effect is so intense that free-air or isostatic anomalies yield similar pictures. Under the hypothesis of downbuckling, the root is assumed eventually to rise, so that the folded rocks above become a new mountain range, compensated after the theory of AIRY. The amount of rise for complete compensation may be easily estimated. For example, if buckling of a crust 30 kilometres thick of density 2.7 grams per cubic centimetre produces a root structure of an additional 30 kilometres protruding into the subcrustal material of density 3.3, the surface would rise 5.5 kilometres, leaving a compensating root of 24.5 kilometres. Such a thickening of the crust would imply a crustal shortening in the ratio of 2:1, and it is from a similar argument that JEFFREYS[4] estimates a shortening of 1.66:1 in the case of the Alps or Rockies. This estimate is important in the theory of the contraction of the earth[5].

In recent years, it has become apparent that an alternative hypothesis to that of a simple downbuckle of light crustal material may be required to explain the anomalies associated with island arcs. Strips of negative anomaly are found in areas, such as the West Indies, where the lower density crust is known from seismic investigations[6] to be thin or lacking. Also, the investigations of SCHEIDEGGER

[1] A. SCHEIDEGGER: Forces in the Earth's Crust, Chapter in this volume.
[2] F. A. VENING MEINESZ: Gravity Expeditions at Sea. Publ. Ned. Geod. Comm., Vol. 111. 1941.
[3] H. H. HESS: Proc. Amer. Phil. Soc. **79** (1938).
[4] H. JEFFREYS: The Earth, p. 307.
[5] A. SCHEIDEGGER: Chapter in this volume.
[6] M. EWING and F. PRESS: Structure of the Earth's Crust, Chapter in this volume.

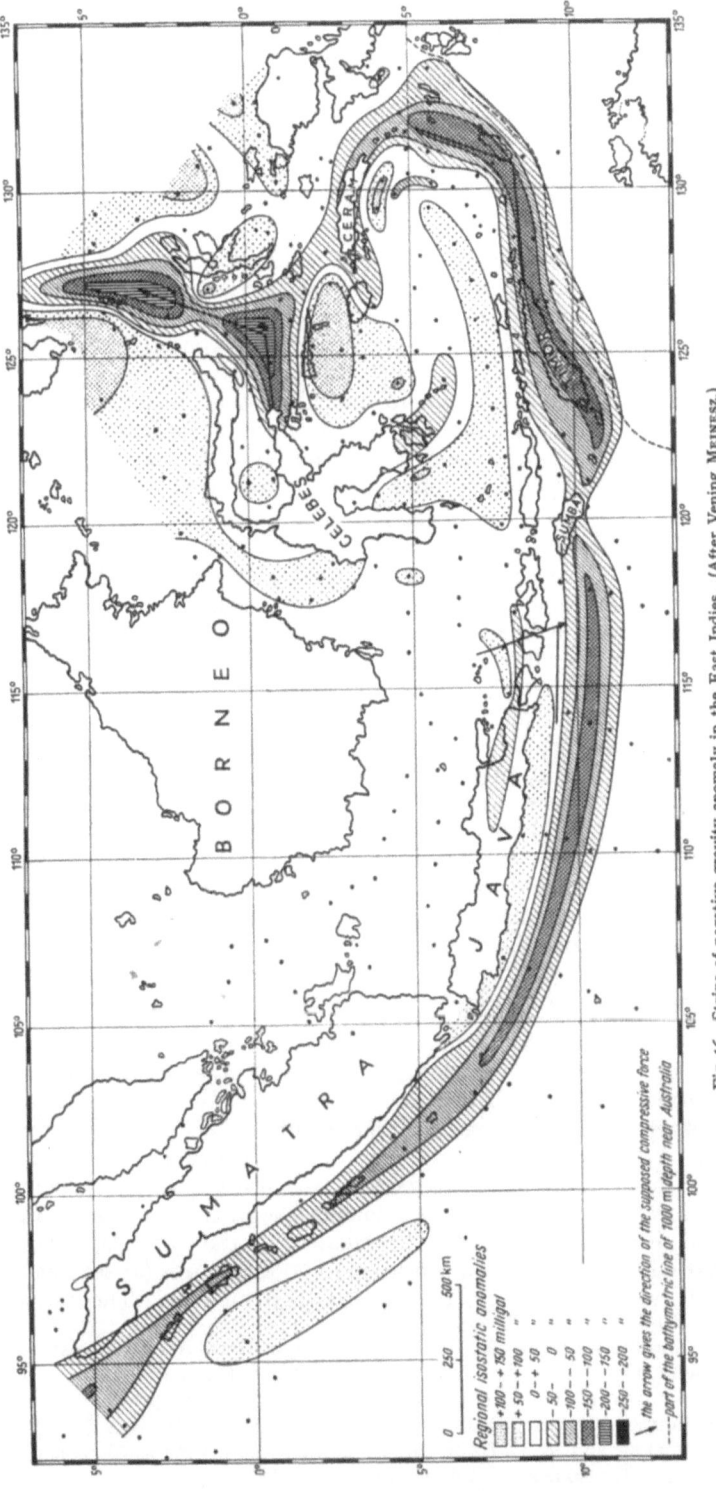

Fig. 16. Strips of negative gravity anomaly in the East Indies. (After Vening Meinesz.)

and WILSON[1] on the failure of the crust suggest that fracture along conical surfaces dipping toward the continents may play a more important part than is suggested by the buckling theory. The findings of EWING and WORZEL[2] in the vicinity of the trench north of Puerto Rico in the Caribbean Sea suggest that a large portion of the observed anomaly of 300 milligals is due to an accumulation of low density sediments in a crustal depression. At least, a thickness of low-velocity material of the correct order of magnitude was indicated by the seismic refraction profiles.

In the East Indies, the strips of negative anomaly often do not coincide exactly with the ocean deeps, but are displaced toward the continental margin. On the above explanation of the anomalies, this would indicate that the deepest part of the original trench has been filled with material transported from the continental side. From the point of view of isostasy, the importance of this explanation for the strips of negative anomaly is that, if the trenches do become island arcs and eventually mountain ranges, the explanation of the compensation must be somewhat changed, since under the buckling hypothesis, the root was assumed to form first, from a pre-existing lighter crust.

It should be pointed out in passing that there is a fundamental difference between the island arc type of structure and the mid-oceanic groups such as the *Hawaiian* chain. The latter, which apparently represent mountains formed by the outpouring of lava on top of the ocean floor without major crustal dislocations, are *not* accompanied by intense negative anomalies. In fact, the Hawaiian islands, which have been the subject of numerous investigations[3] show *positive* isostatic anomalies. The interpretation of these is complicated by the variations in density of the lavas. VENING MEINESZ[4] showed that the islands could be approximately compensated according to his regional hypothesis and still show positive anomalies if a mean density of somewhat over 2.9 grams per cubic centimetre was assumed for the volcanic material. WOOLLARD reports that the density of the lavas, if vesicular, could be as low as 2.3 grams per cubic centimetre, and if this is the case, incomplete compensation is implied. There are, however, volcanic throats containing much denser rocks, and the contribution of these to the overall effect is difficult to evaluate. There is in any case general agreement among the investigators that the island masses are not compensated *locally* according to the classic theories, but whatever compensation is present is provided by a *regional* crustal flexure. — (See also Fig. 11, p. 255.)

**24. Mountain Ranges.** In contrast to island arcs, present day mountains appear to be quite well compensated. At least, the residuals on an isostatic theory in such areas as the Alps or the Cordillera of North America are no larger than in other continental regions. For example, the mean without regard to sign, of the AIRY isostatic anomaly, for a depth of compensation of 40 kilometres, at stations in the mountainous regions of western Canada is 13 milligals, while for all of Canada, including many stations in the interior and on the Precambrian shield, the mean is 14 milligals[5]. Similarly, in the East Alps, HOLOPAINEN[6] found that with a depth of compensation of 20 kilometres (at sea level) the mean residual AIRY anomaly could be reduced to 6 milligals. The actual range in

[1] A. E. SCHEIDEGGER and J. TUZO WILSON: Proc. Geol. Assoc. Canada **3**, 167 (1950).
[2] M. EWING and J. WORZEL: Bull. Geol. Soc. Amer. **65**, 165 (1954).
[3] G. P. WOOLLARD: Trans. Amer. Geophys. Un. **32**, 358 (1951).
[4] F. A. VENING MEINESZ: Proc. Kon. Ned. Akad. Wetensch. **44**, 1 (1941).
[5] A. H. MILLER and W. G. HUGHSON: Publ. Dom. Obs. Ottawa **11**, No. 3 (1936).
[6] PAAVO E. HOLOPAINEN: Publ. Isostatic Inst. **16** (1947).

anomaly at individual stations was of course much larger, regardless of the hypothesis upon which the computation is based.

Accepting the implication that present day mountains are, for a good first approximation, compensated, the problem which remains is to determine the form and location of the compensating masses, or roots, and to relate them to the general process of mountain-building. This problem is difficult because the gravity field can really only indicate the maximum possible depth, while other lines of evidence, such as detailed seismological studies, are complicated by the terrain. BULLARD and COOPER[1] have investigated the gravity field of the Alps (Fig. 17) by their method of projecting the field to depth, and conclude that there must be major horizontal changes in density within 28 kilometres of the surface. If a greater depth is assumed for the compensating masses, improbable density contrasts are required to reproduce the form of observed field. The form of the BOUGUER profile across the Alps is asymmetrical, rising steeply on the south but continuing negative for some distance to the north of the mountains. BULLARD and COOPER point out that this could be the result of failure of the crust by overthrusting from the south, bringing denser material near the surface on the south side of the range and depressing the crust beneath, and to the north of, the mountains.

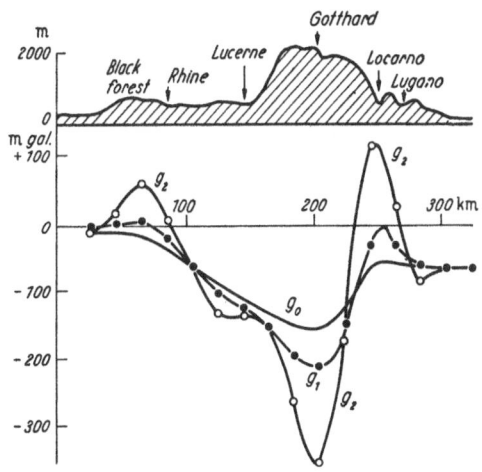

Fig. 17. Profile of BOUGUER gravity anomaly across Alps. (After BULLARD and COOPER.) $g_0$ is the observed profile, while $g_1$ and $g_2$ are the calculated fields at depths of 20 and 40 kilometres respectively.

In the Canadian section of the North American Cordillera, there would attempt to be a fundamental difference between the compensation of the folded sediment type of structure which forms the Rocky Mountains, and the Coast Range, with its core of granitic type rock. The results of a detailed gravimeter traverse by OLDHAM[2] across the northern Cordillera suggest that the mass deficiency in the latter case is directly connected with this intrusive core, which appears to be consistently less dense then the average rocks of the area. On the other hand, the axis of negative BOUGUER anomaly associated with the sedimentary Rocky Mountains does not always follow the line of maximum elevation, but in places lies to the west of it. The mass deficiency in this case may be related to crustal warping along the site of the original geosyncline, from which the mountains were thrust many miles to the east.

Seismological observations on the conditions in the crust beneath mountain ranges have been discussed by BYERLY[3] and GUTENBERG[4]. The most direct evidence for root structure is a delay of some seconds in the arrival times of waves traversing the Alps and the Sierra Nevada in California. This delay is observed for the wave travelling along the base of the layer in which the thickening occurs. Thus, under the Sierra Nevada, $P_n$, or the compressional wave

[1] E. C. BULLARD and R. I. B. COOPER: Proc. Roy. Soc. Lond. A **194**, 332 (1948).
[2] C. H. G. OLDHAM: Trans. Amer. Geophys. Un. **35**, 364 (1954) abstr. only.
[3] P. BYERLY: Bull. Geol. Soc. Amer. **48**, 2025 (1938).
[4] BENO GUTENBERG: Bull. Geol. Soc. Amer. **54**, 473 (1943).

travelling beneath the base of the crust, was found to be delayed, while the wave through the "intermediate layer", or lower portion of the crust, was not. In the case of the Alps, a delay was suggested in the case of the wave through the intermediate layer itself, which would imply a thickening of the "granitic layer", or upper portion of the crust. However, since the general question of layering within the crust remains somewhat a matter of interpretation, it would appear safer to say only that the seismological observations indicate a thickening of low velocity material beneath the mountains.

**25. Exposed Shields and Other Areas of Granitic Rocks.** The exposed Precambrian shields of the earth display the upper portion of the continental crust which is elsewhere concealed by later formations. As these areas also show evidence of former mountains now eroded to a fairly uniform elevation, they offer interesting possibilities for gravitational studies.

Recent investigations over the Canadian Precambrian shield[1, 2, 3] have shown rather surprising variations in anomaly for an old, stable area of uniform elevation. Changes in Bouguer anomaly from near zero to negative values of 90 milligals within a horizontal distance of a few miles are found.

Fig. 18 indicates the pattern of Bouguer anomalies over a portion of the shield, after allowance has been made for effects which are obviously related to known density variations, in the manner of the "*geological correction*" of Evans and Compton[4]. These writers suggest that after the usual factors have been applied in the calculation of the Bouguer anomalies, an additional correction be applied to remove the effects of all known geological structures. That is, if a certain formation of anomalous density is *known* to exist in the area under study, and if its thickness is *known* with some accuracy, the estimated effect of it is removed from the Bouguer anomaly. The gravity map, after application of the "geological correction", therefore is free from the influence of the known lithology, and the effect of previously unknown variations in density is more clearly shown. In the present case the attractions of known denser formations such as basic volcanic rocks on the surface of the shield have been eliminated by smoothing the anomalies over these features, so that the contours are believed to be those which would be obtained if only granite and gneiss were present. The isostatic correction, on any of the theories of compensation, is nearly constant over broad areas, so that the *range* of isostatic anomalies is nearly as great as that of the Bouguer anomalies.

As the number of observations has increased, it has become apparent that the most negative effects are distributed along structural trends within the Shield[5] and probably do not represent the aftermath of continental depression during the era of Pleistocene glaciation, as had been intimated on the basis of much less data[6]. As far as can be told, the negative trends follow the lines of Precambrian mountains, and may therefore represent a lack of adjustment since their disappearance, but it is not yet clear whether the anomalies are entirely due to density variations within the crust, or whether there is also the effect of varying crustal thickness. The results of a large number of density determi-

[1] G. D. Garland: Publ. Dom. Obs. Ottawa **16**, No. 1 (1950).

[2] M. J. S. Innes: Gravity in the Canadian Precambrian Shield. Paper presented at the ninth general assembly, Int. Union Geod. Geophys., Brussels 1951.

[3] C. H. G. Oldham: Geophysics **19**, 76 (1954).

[4] Percy Evans and Wilfred Compton: Quart. J. Geol. Soc. Lond. **102**, 211 (1946).

[5] J. Tuzo Wilson: Trans. Roy. Soc. Canada **43**, 157 (1949).

[6] See, for example R. A. Daly, Strength and Structure of the Earth, New York 1940 or Beno Gutenberg: Bull. Geol. Soc. Amer. **52**, 721 (1941).

nations suggest that the mean density of the gneissic rock, which is believed to be typical of the continental crust is close to 2.8 grams per cubic centimetre, rather than 2.67, as usually adopted. Many of the larger negative anomalies appear to result from bodies, which must extend to depths of 10 kilometres or more, of massive, lighter granite with a density of about 2.65. In other words, it would appear that "granite" is not the typical rock of the upper portion of the crust, as might be suggested by the term "granitic layer" as used in seismology, but is a lighter material of more restricted occurrence.

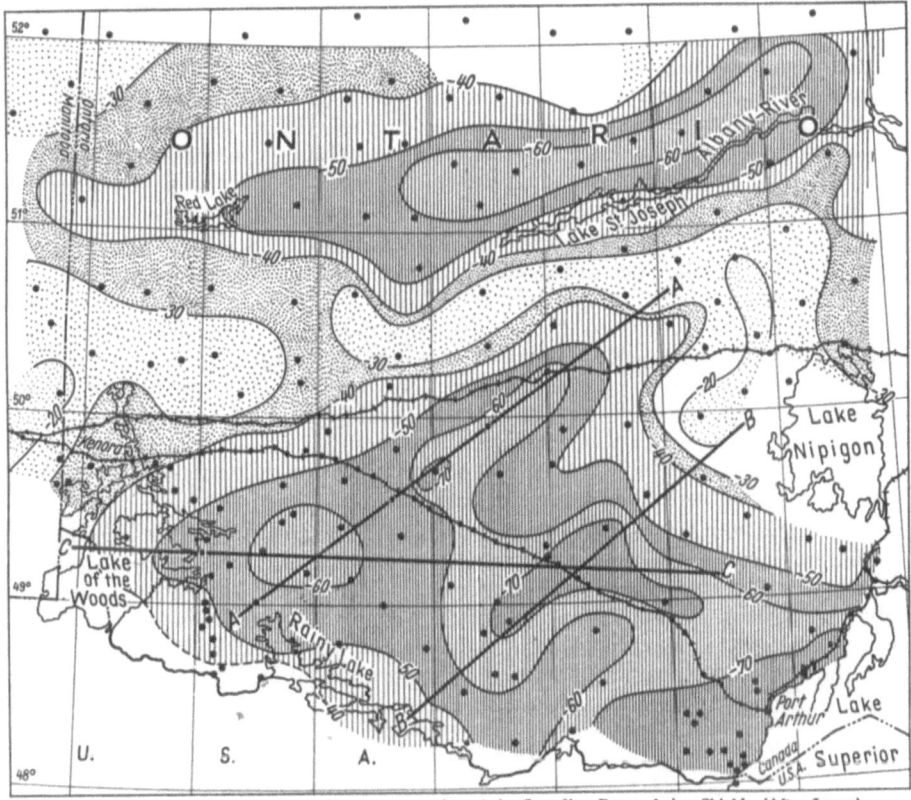

Fig. 18. Pattern of BOUGUER anomalies over a portion of the Canadian Precambrian Shield. (After INNES).

Similar large negative anomalies have been observed over the Rapakivi granites in the Fennoscandian Shield[1] and over many smaller bodies of granite, located outside shield areas in New England[2] and Eastern Canada[3], Ireland[4] and England[5]. However, in the Canadian Precambrian Shield at least, there remain other prominent variations in the gravity field which do not appear to be directly related to the measured densities, and these may indicate actual variations in crustal thickness, possibly dating from the time of the last orogeny.

It is outside the scope of the present chapter to draw conclusions on such questions as the origin of granite and the forces responsible for mountain building,

[1] H. REICH: Gerlands Beitr., Geophys. **2**, 1 (1932).
[2] G. P. WOOLLARD: Trans. Amer. Geophys. Un. **32**, 634 (1951).
[3] G. D. GARLAND: Publ. Dom. Obs. Ottawa **16**, No. 7 (1953).
[4] A. H. COOK and T. MURPHY: Geophys. Mem. Dub. Inst. Adv. Stud. **2**, Part 4 (1952).
[5] M. H. P. BOTT: Geol. Mag. **90**, 257 (1953).

but the purpose of outlining these investigations of island arcs, mountain ranges and shields has been to suggest the general processes involved. It is clearly not sufficient to limit isostatic studies to computations based on various idealizations. Apart from the major departures from isostasy analysed by JEFFREYS, some of the problems involved are the relative importance of crustal buckling and fracture in the root formation, the effect of low density oceanic sediments in trough areas, the possible role of a low density intrusive core in the case of certain ranges, and the preservation of the effects of old ranges in shield areas.

**26. Local Gravity Surveys.** The gravity effects which have been discussed have been those connected with the general problem of the compensation of the earth's surface features. However, gravity measurements are also used to investigate much more local structures in greater detail, and it would be worthwhile to investigate the density relationships which make such applications feasible. In particular, the application to commercial exploration for petroleum or minerals should not be overlooked, for it was as a result of this that portable, sensitive gravity meters were developed and certain techniques of interpretation devised.

The densities of the rocks and minerals comprising the earth's crust range from about 1.7 grams per cubic centimetre for unconsolidated soils to about 4.5 in the case of metallic ore minerals (Table 3). Among igneous rocks, the density is largely controlled by chemical composition, increasing toward the basic end of the scale. The density of sedimentary rocks is a function of both chemical composition and porosity, where the latter in turn is controlled by the original nature of the rock and the degree of compaction. Juxtaposition of any of these materials of varying density will in theory give rise to an anomaly in gravity, but the possibility of detecting and utilizing this anomaly is subject to the limitations mentioned above in Part C.

In exploration for petroleum the problem may range from the very broad one of proving the existence of a sedimentary basin suitable for more detailed investigation, to the more particular search for structural traps associated with petroleum accumulation. As a general rule unmetamorphosed sedimentary rocks are less dense than the crystalline basement rocks which underly the basins, and from the observed negative anomaly $\Delta g$, the approximate thickness $h$ of the basin may be estimated from the BOUGUER relation

$$\Delta g = 2\pi f \sigma h$$

where $\sigma$ is the density contrast between the basin rocks and the basement. The localization of structural conditions within the sedimentary column is a much more complex problem, for the desired effect is often only a flexure on the anomaly field due to some broader structure. For this reason, gravity surveys are now used chiefly in conjunction with detailed seismic reflection methods.

Early efforts in gravitational exploration were restricted to structures producing prominent effects, such as *salt domes*. These large cylindrical masses of salt, with a density less than that of most sedimentary rocks, are sometimes marked by circular negative anomalies of several milligals (Fig. 7). Petroleum accumulation may occur in strata around the dome or in the porous "cap rock" of the dome itself.

The next type of structure to be investigated was the basement uplift. An elevation of the surface of the igneous rocks beneath the sedimentary layers may be expected to produce a positive anomaly by bringing denser material closer to the surface and such a structure can be commercially important if the overlying

Table 3. *Densities of Common Rocks and Minerals.*

| Rock or Mineral | Density (gm./cc.) |
|---|---|
| **Sedimentary Rocks:** | |
| Unconsolidated Surface Material (drift, clay, etc.) | 1.1—2.0 |
| Shales, Gulf Coast, U.S.A., near surface | 1.9 |
| Shales, Gulf Coast, U.S.A., at depths over 8000 ft | 2.30 |
| Shales, Carboniferous, Eastern Canada, calcareous | 2.46 |
| Shales, Carboniferous, Eastern Canada, bituminous | 1.89 |
| Sandstone, Triassic, Europe | 2.35 |
| Sandstone, Carboniferous, Eastern Canada | 2.32 |
| Sandstone, Carboniferous, Scotland | 2.38 |
| Sandstone, Devonian, Eastern Canada | 2.62 |
| Limestone, Carboniferous, Eastern Canada | 2.66 |
| **Metamorphic Rocks:** | |
| Gneiss, Precambrian, Canadian Shield | 2.65—3.0 |
| Quartzite, Rocky Mountains | 2.6—2.7 |
| Schist, Rocky Mountains | 2.7—3.0 |
| Crystalline Limestone, Precambrian | 2.7 |
| **Igneous Rocks:** | |
| Granite, Devonian, Eastern Canada | 2.65 |
| Granite, Precambrian, Canadian Shield | 2.60 |
| Basalt, Hawaii | 2.95 |
| Peridotite | 3.0—3.4 |
| Greenstone, Precambrian | 3.0 |
| **Economic Minerals:** | |
| Salt | 2.2 |
| Gypsum | 2.4 |
| Potassium Salt (Germany) | 1.6 |
| Magnetite | 5.0 |
| Hematite, massive (density may be considerably less if porous) | 4.9—5.3 |
| Pyrite | 5.0 |
| Chalcopyrite | 4.1—4.3 |
| Chromite | 4.5 |
| Galena | 7.5 |

sedimentary beds conform to the basement surface. Even this problem is complicated by the difficulty in distinguishing between the gravitational effects of basement topography and of density variations within the igneous rocks themselves.

Finally, the smallest and most elusive anomalies are those caused by minor density variations within the sedimentary column, which, if associated with stratigraphic traps, may still be of great importance in the localization of petroleum.

As an example, Fig. 19 illustrates schematically some of the complicating factors which were involved in an actual survey to locate porous reefs. These structures, which are the prototype of present-day coral reefs, can form important reservoirs of petroleum or natural gas. Because of their high porosity their density is usually lower than that of the surrounding limestone and dolomite, a fact that would suggest the presence of a local gravity minimum over them. The diagram illustrates extraneous effects which would have to be recognized and separated from the effect of the reef itself, namely the variation in lithology within the basement rocks (density contrast 2.69 to 2.67); the uneven interface between the basement and the lowest sedimentary layer (density contrast 2.67 or 2.69 to 2.65); and the uneven surface of the rock beneath the unconsolidated surface material (density contrast 2.4 to 2.0). Even after regard for these factors, it was found that reefs were characterized by small local maxima, not minima, because of the thinning (presumably by compaction) of a layer of low-density salt immediately above them. In other words, the excess of mass produced in this way offset the deficiency in the reef itself, causing the observed anomaly to be quite different than might be predicted. The example emphasizes the need for careful geological consideration to accompany the physical interpretation of gravity anomalies.

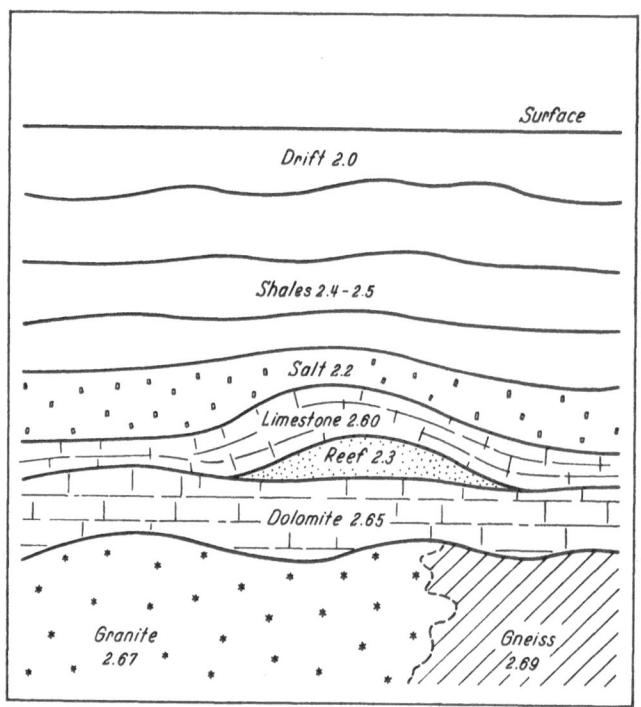

Fig. 19. Schematic diagram of densities relationships encountered in a gravity survey for limestone reefs.

In the case of exploration for *ore minerals* the problem is usually the direct location of the heavy ore itself. The important point usually is the smallness of the anomaly produced even by shallow, commercially important bodies of this type. For example, Fig. 20 shows the local effects related to two relatively large bodies of sulphide ore in the Canadian Precambrian Shield. The ore is about 0.7 grams per cubic centimetre denser than the surroundings. It will be noted that the anomaly over the westerly body, which is situated immediately beneath the surface, is only 0.6 milligals, while that related to the easterly body is considerably smaller. An anomaly of this magnitude is quite detectable instrumentally, but could easily be obscured by the effects of other local structures, or by incompletely corrected terrain effects. It will be observed that the anomaly over the easterly body is displaced from the surface exposure, apparently because of the dip of the body. The map also indicates a pronounced regional gradient toward the northwest, resulting from the contact of two rock types which crosses the area. However, in spite of its inherent limitations, one advantage of the gravitational method of exploration for ore is that the anomaly produced is directly related to the total mass (cf. Eq. 19.8), and this is the property of commercial interest also.

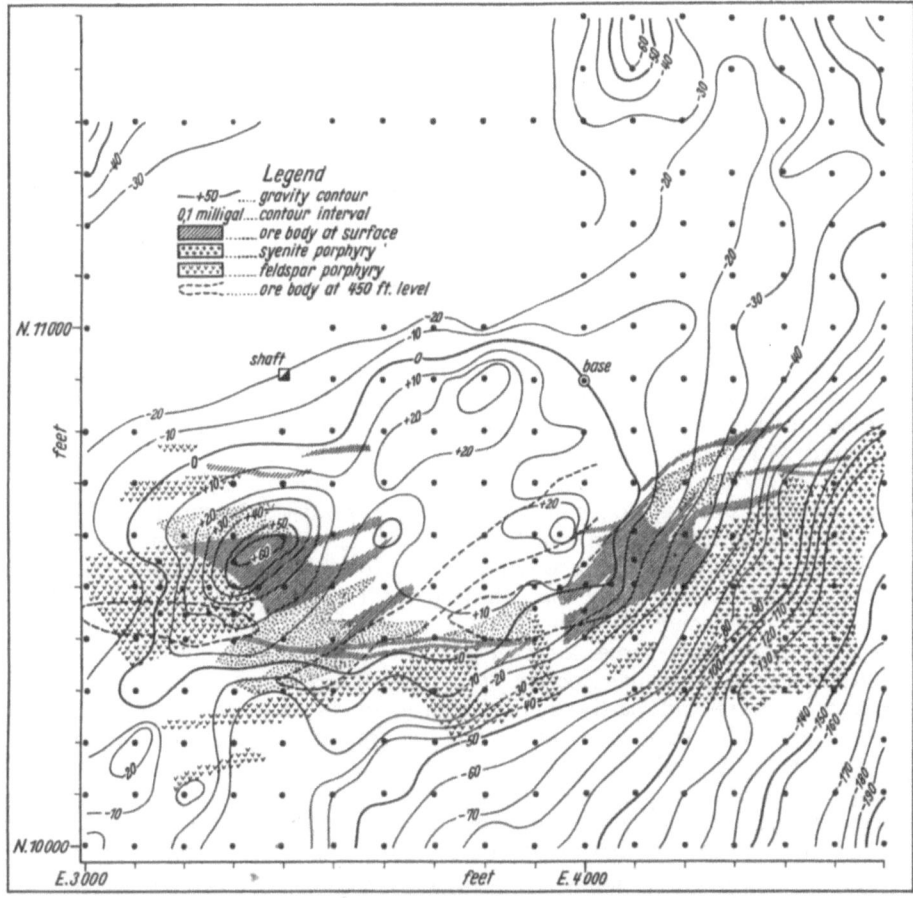

Fig. 20. BOUGUER gravity anomalies (in the unit 0.01 milligal) observed over two near-surface bodies of metallic sulphides. (After INNES.)

The table of densities (Table 3) will suggest other ore minerals most likely to exhibit a density contrast with normal crustal rocks. Among recent surveys for which published accounts are available are those for *chromite*[1], and *iron ore*[2], but in common with geophysical exploration as a whole, the activity in the mining field has been more restricted than in the search for petroleum.

[1] L. L. NETTLETON, W. K. HASTINGS and SIGMUND HAMMER: Geophysics **10**, 34 (1945).
[2] G. D. GARLAND: Trans. Canad. Inst. Min. a. Metallurgy **54**, 340 (1951).

# Structure of the earth's crust.

By

MAURICE EWING and FRANK PRESS.

With 13 Figures.

**1. Introduction.** The crust of the earth has often been defined in two different ways. The *silicic crust* is the part of the solid earth above the MOHOROVIČIĆ *discontinuity*, which separates it from the *mantle*. The existence of this discontinuity has been established by seismic refractions and reflections, which show an abrupt increase in seismic wave speeds to a value which appears to be the same all over the world. The silicic crust is about 5 km. thick under oceans and about 35 km thick under continents.

The *lithosphere* is that outer shell of the solid earth which has the shear strength without which topographic irregularities could not exist. Measurements of the acceleration of gravity, from which conclusions about density distribution in the lithosphere may be drawn, prove that the rocks beneath elevated areas of the globe must be less dense than those beneath depressed areas such as the ocean basins, as though the lithosphere were underlain by an *asthenosphere* in which only hydrostatic forces are exerted. The elevation of a given segment of the lithosphere is determined partly by vertical forces exerted on it by the adjacent lithosphere and partly by hydrostatic forces from the asthenosphere. For any area with dimensions of several hundred kilometers, the hydrostatic forces alone are adequate to determine the elevation, and the strength of the lithosphere becomes a negligible factor. This division of the earth into a lithosphere and an asthenosphere is the basis of the second definition of the crust. The evidence for determining the thickness of the lithosphere is extremely indefinite, but the demands of gravity are satisfied if its boundary is at the MOHOROVIČIĆ discontinuity or deeper[1].

We will be primarily concerned with the *silicic crust* as defined above. Recent geophysical investigations establish two basic crustal types which between them cover the greater part of the earth — continental and oceanic — and show that the constants of each type of crust may be defined within very narrow limits. Deviations from these simple crustal types occur in anomalous areas such as continental or oceanic mountain chains, island arcs and deep sea trenches, or as the zones of transition from continent to ocean. These all are long narrow belts which occupy only a small fraction of the earth's surface.

**2. Continents.** Geology describes the near-surface features of the continental crust. It tells us that beneath a thin veneer of sediments lies a thick section of crystalline silicic rock with average composition intermediate between that of granite and basalt. Studies on tectonic processes enable geologists to set broad limits for the thickness of the continental crust, but it remains for geophysicists to give quantitative information about it. The most effective methods of geophysics for this investigation are seismic refraction studies, earthquake surface wave dispersion, and gravitational surveys.

---

[1] A third definition of the crust has been given by BENIOFF, see end of Sect. 4.

α) *Seismic Refraction and Reflection Methods.* Seismic exploration gives the most detailed information about the crust and the underlying mantle. The sources of seismic waves may be either earthquakes, rock bursts or explosions. The basic measurement is the travel time of an elastic wave, and it is necessary to determine travel time as a function of distance, for several wave types, to a distance of about 500 km. For *explosions*, the field procedure consists of establishing a line of seismographs[1], preferably three-component instruments, from which wave types may be identified and travel times measured with accuracy of at least

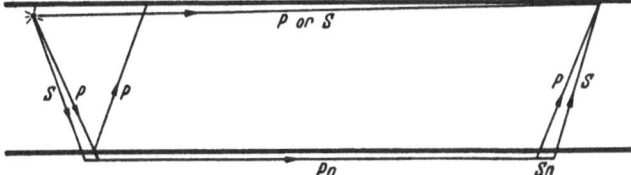

Fig. 1. Compressional (P) and shear (S) wave paths in crust.

0.1 sec. *Earthquakes* are usually studied by the use of networks of fixed stations, and can be used for only a few parts of the earth (usually anomalous parts) due to lack of adequate networks for determining epicenter location, focal depth and origin time upon which the travel time curve depends.

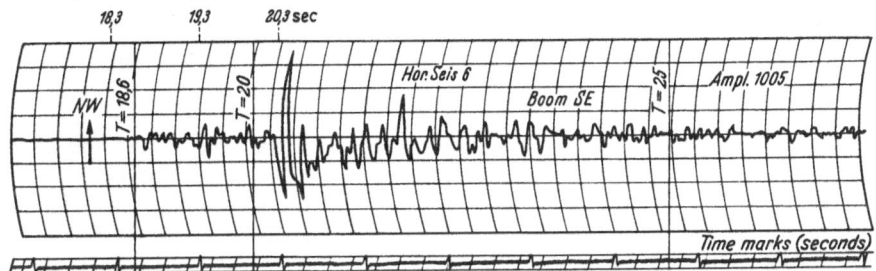

Fig. 2. Record of a horizontal one-second period velocity seismometer at a distance of 109.7 km. from an explosion of 2400 lbs. of *TNT*. Onset of ground motion at about 18.6 sec. is P. Large arrival at 20.4 sec. is critical reflection *PP*
(TATEL, ADAMS and TUVE).

Excellent results have been achieved with rock bursts in Canada[2] and South Africa[3]. In the latter case, an ingenious radio system was designed for triggering the field seismographs and telerecording at a central station.

An explosion or earthquake generates both compressional and distortional elastic waves which are radiated in all directions. At distant points the principal waves received are the direct P or S, the refracted $PPnP$ and $SSnS$ (usually called $Pn$ and $Sn$), and the reflected phase $PP$, the latter being important only for distances corresponding to reflection near the critical angle. In Fig. 1, the ray paths are depicted schematically. A seismogram obtained by TATEL, ADAMS and TUVE[4] at a distance of 109.7 km. from an explosion is shown in Fig. 2.

[1] For the typical Heligoland explosion of 18 April 1947, when 4000 tons of high explosive were fired, see P. L. WILLMORE [Phil. Trans. Roy. Soc. Lond., Ser. A **242**, 123 (1949)]. G. A. SCHULZE and O. FOERTSCH [Geolog. Jb. **64**, 204 (1950)]. H. REICH, O. FOERTSCH and G. A. SCHULZE [J. Geophys. Res. **56**, 147 (1951)]. The exact depth of the MOHOROVIČIĆ discontinuity was disclosed with unprecedented clarity.

[2] J. H. HODGSON: A seismic survey in the Canadian shield. Publ. Dominion Obs. Ottawa **16**, 113—163 (1953).

[3] P. L. WILLMORE, A. L. HALES and P. G. GANE: A seismic investigation of crustal structure in the Western Transvaal. Bull. Seism. Soc. Amer. **42**, 53—80 (1952).

[4] H. E. TATEL, L. H. ADAMS and M. A. TUVE: Studies of the Earth's Crust Using Waves from Explosions. Proc. Amer. Phil. Soc. **97**, 659—699 (1953).

A typical experimental travel-time curve for compressional waves from explosions observed in Pennsylvania[1] is presented in Fig. 3. For distances less than 180 km. the only waves generally identifiable are $P$ and $S$. From approximately 180 km. to 310 km. $Pn$ arrives before $P$ but both are identifiable. Corre-

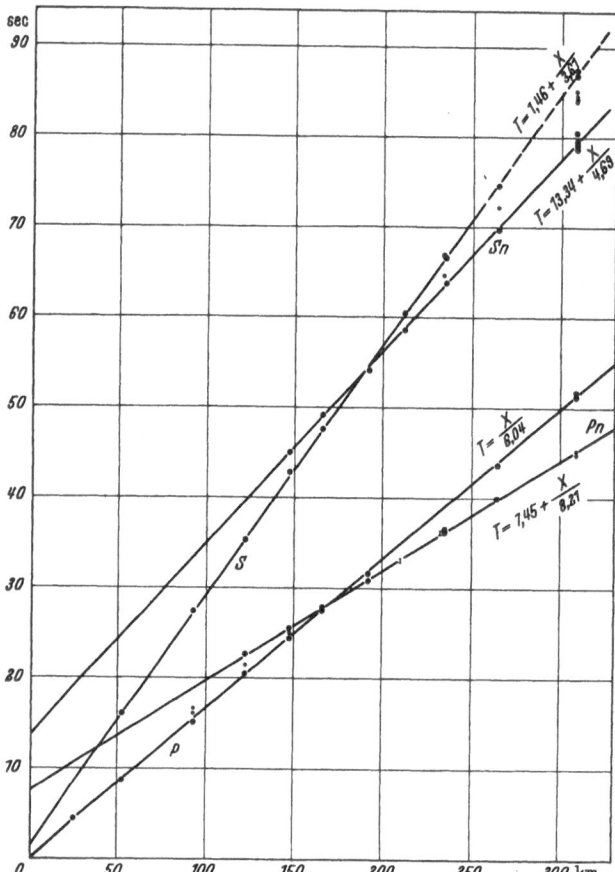

sponding relations in the shear waves may be found when near earthquakes, rockbursts, or certain quarry blasts are used as sources. The reflection $PP$ is readily identified by its characteristic appearance at about 100 km. with large amplitude over the narrow range of distances for which reflection occurs near the critical angle. Using the principles of interpreting seismic refraction and reflection travel time curves in which the crust is represented by one or more homogeneous layers[2], the elastic wave velocities in each layer and the depth to the interfaces may be calculated. A few investigators consider that a two-layered crust is required, but as data of high quality become available for various parts of the earth it becomes apparent that no choice of a double layer is applicable to *all* continents. A single layered crust, in which the velocity increases slightly with depth, accounts for almost all details of observed travel time curves.

Fig. 3. Time-distance curve showing $P$, $Pn$, $S$, $Sn$ observations from quarri blasts in Pennsylvania (Katz, unpublished).

Tuve, Tatel and Hart[3] have selected several velocity-depth functions which fit their travel-time curves equally well to illustrate the fact that data now available are inadequate for a unique solution. Their velocity functions are shown in Fig. 4. The possibility of a low velocity layer[4] cannot be ruled out on the basis of refraction travel-time curves alone. Selecting recent measure-

[1] S. Katz: Seismic Study of Crustal Structure in Pennsylvania and New York. Bull. Seism. Soc. Amer. In press.

[2] See the article Seismic Prospecting in this Volume, pp. 163 ff.

[3] M. A. Tuve, H. E. Tatel and P. J. Hart: Crustal Structure from Seismic Explosions. J. Geophys. Res. **59**, 415—422 (1954).

[4] B. Gutenberg: Crustal Layers of the Continents and Oceans. Bull. Geol. Soc. Amer. **62**, 427—440 (1951).

ments[1] made in areas free from obvious crustal anomalies, we find values of 6.0—6.2 and 3.5—3.7 km./sec. for the velocities of compressional and shear waves

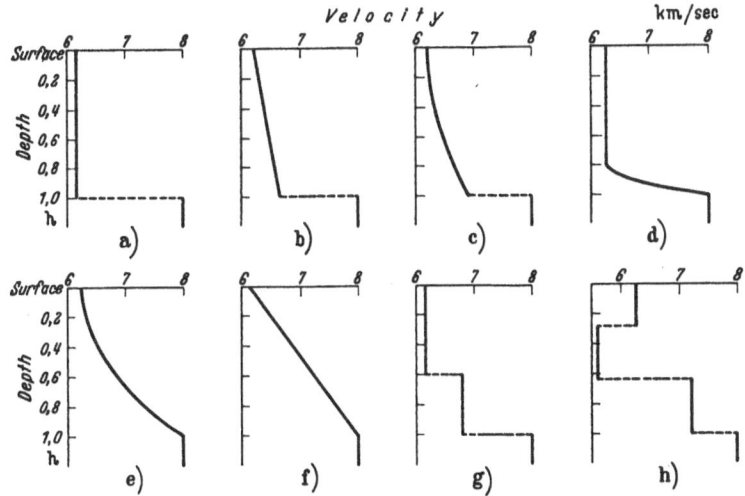

Fig. 4 a—h. Velocity-depth functions which fit travel time data. Only (b), (c), (d) fit all data equally well (Tuve, Tatel and Hart). h is depth to 8.0 km./sec. layer.

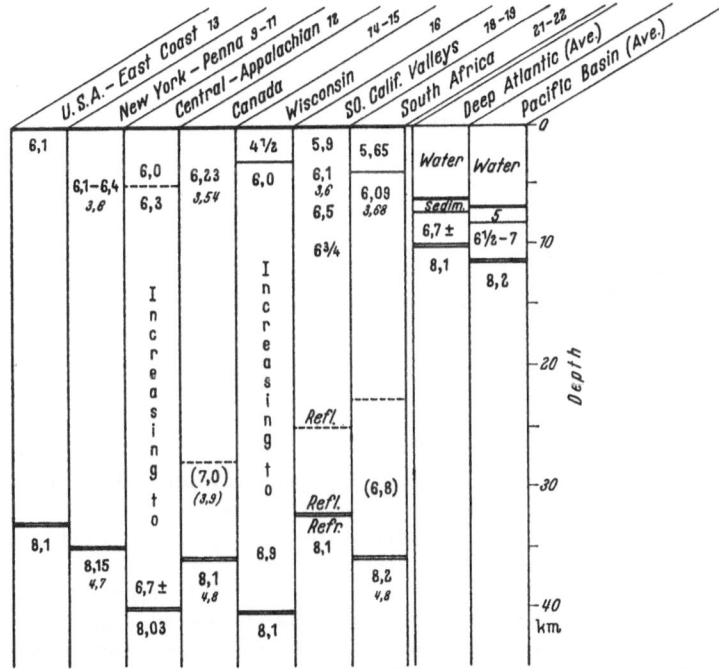

Fig. 5. Continental and oceanic crustal columns determined from seismic refraction measurements. $V_p$ and $V_s$ (written in italics below the value for $V_p$) are compressional and shear velocities in km./sec. respectively. (After Gutenberg.)

in the upper part of the silicic crust and 8.0—8.2 and 4.7—4.8 km./sec. for the upper part of the mantle. The crustal thickness is about 35 km. Fig. 5 summarizes

<hr>

[1] For references see B. Gutenberg: Wave Velocities in the Earth's Crust, The Crust of the Earth Symposium. Geol. Soc. Amer. 1955. In press.

the results of selected recent measurements. A comprehensive summary of world-wide results has been given by Reinhardt[1].

β) *Surface Wave Data.* Seismic measurements can give the average crustal structure along a line about 300 km. in length, but surface wave dispersion studies give the average properties over distances of continental dimensions. Thus the two methods are supplementary. The theories for Love wave and Rayleigh wave dispersion in layered media have been given on pp. 120ff.[2]. Theoretical dispersion curves may be calculated for various models of the crust, and compared with the dispersion read directly from seismograms. Such comparisons have been made in the preceding Chapter, where it was shown that the observed surface wave dispersion was compatible with the crustal structure deduced from seismic refraction measurements.

As a result of the wide distribution of earthquakes and of seismograph stations, the study of surface waves is the most effective method available for establishing the fact of the identical nature of the continental crust the world over. In particular, Fig. 6, p. 128, compares the dispersion of Rayleigh waves across North America with that across Africa. It is striking that for most of the range of periods for which the data overlap the difference in group velocity is never greater than 0.1 km./sec.

Fig. 1, p. 122, shows that Love wave data for North America yield an average crustal thickness of 35 km., providing the effect of the mantle velocity gradient be included in the Love wave theory.

The $Lg$ phase is a short period surface wave which is readily recognized on seismograms (see p. 133). Many suitable earthquakes are available for its study because of the strength of the phase. On short period seismograms its arrival time may be read to within a few seconds, hence its velocity may be measured with high accuracy. The velocity of $Lg$ is 3.51 km./sec. for North America, and velocities for Eurasia and Africa vary by less than 1 % from this value. As discussed on p. 133, this velocity is the same as that for shear waves in the upper part of the silicic crust.

The fact that $Lg$ is propagated only along continental paths provides an easy and precise method for world-wide determination of the extent of continents. For example, it has been found that the Gulf of Mexico, the Mediterranean Sea and the Norwegian Sea are underlain by oceanic crust, whereas the Canadian Archipelago, Baffin Bay, the Barents Sea, and the Bering Strait transmit $Lg$ as continents do. It has been found without exception that continents extend oceanward to approximately the 1000 fm. depth contour[3].

γ) *Gravity.* The role of gravity in geophysics is discussed on pp. 225. Here we are interested only in those gravity anomalies which distinguish continents from oceans or determine major variations from the typical continental structure described above. The international gravity formula was established primarily from gravity surveys on continents and is intended to represent the gravity field which would exist if the land surface everywhere were at sea level. The formula expresses the value of gravity at the surface of the oceans almost equally well, establishing the fact that the obvious deficit in mass in the water layer is compensated by an excess of mass in the underlying rocks. In Fig. 8 typical

---

[1] H. G. Reinhardt: Steinbruchsprengungen zur Erforschung des tieferen Untergrundes. Berlin: Akademie-Verlag 1954. 91 S.
[2] In the article Surface Waves and Guided Waves in this Volume.
[3] J. Oliver, M. Ewing and F. Press: Crustal structure of the Arctic regions from the $Lg$ phase. Bull. Geol. Soc. Amer. 1955.

continental and oceanic columns are compared. Gravity results and seismic results are compatible in their indications of crustal structure.

**3. Oceanic Crust.** Information about the oceanic crust is dependent upon seismic, gravitational, and similar measurements to an even greater extent than for continental crust. As a consequence of the difficulties of making such measurements at sea, our knowledge about the oceanic crust lagged behind. The advance began about 1920 with the development of the echo sounder and the VENING MEINESZ marine pendulum apparatus. With the aid of earthquake surface wave observations, it was concluded soon thereafter that the oceanic crust was thinner than the continental crust and that differences between the oceans existed. But it was not until about 1950 that seismic refraction measurements and improved surface wave analysis firmly established the thickness at about 5 km. for all oceanic areas investigated. The only exceptional areas are obviously anomalous mountains and trenches.

*α) Seismic Refraction Measurements.* The lack of a suitable network of oceanic seismograph stations restricts the refraction method to the use of explosive sources and shipborne seismographs. As with refraction measurements on land, the basic data consist of the travel times of elastic waves as a function of distance, and the methods of interpretation

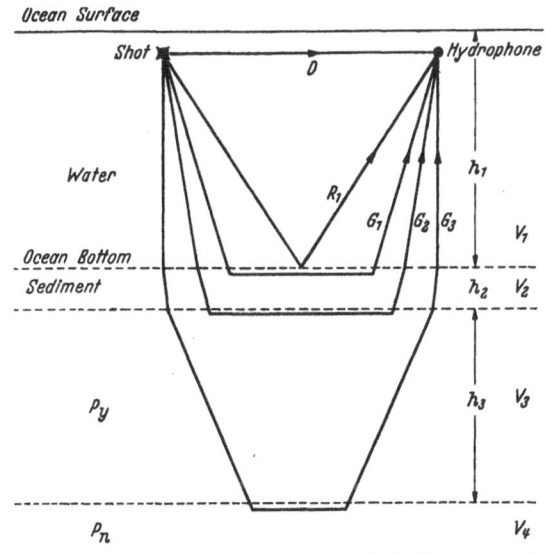

Fig. 6. Refraction and reflection paths for oceanic seismic measurements.

are the same. The elastic waves are detected with pressure-sensitive devices located at depths of the order of 50 ft. The explosive is detonated at a similar depth, and at distances ranging up to 60 miles, as measured from the travel time of the direct water wave. The paths of typical elastic waves observed are shown in Fig. 6. By the use of suitable filters in the amplifiers and by isolating the detectors from waves and ship disturbances, a favorable signal-to-noise ratio may de obtained.

A travel time curve typical of oceanic stations[1] is shown in Fig. 7. Three segments are usually found, corresponding to propagation of compressional waves in water (1.5 km./sec.), the silicic crust (6.3 to 7.0 km./sec.), and the mantle (7.9 to 8.3 km./sec.). The sedimentary layer is difficult to detect, but with detailed data it is found to have a velocity about 20% greater than that of water.

Again using the prospecting methods of interpretation (pp. 163 ff.), a typical oceanic crust is found to be that shown in Fig. 5. The depth of the MOHOROVIČIĆ discontinuity is only 10—12 km. beneath the sea surface in contrast to the 35 km. depth under the continents. The crystalline oceanic crust is only about 5 km. thick and is similar to rocks found near the bottom of the continental crust. With

[1] M. EWING, G. H. SUTTON and C. B. OFFICER, jr.: Seismic Refraction Measurements in the Atlantic Ocean, Part VI: Typical Deep Stations, North America Basin. Bull. Seism. Soc. Amer. **44**, 21—38 (1954).

the precision of measurement now available the mantle beneath continent and ocean is elastically indistinguishable. Variations from the typical oceanic section of Fig. 5 are minor, the most frequent ones being the addition of a layer with thickness of 1 to 2 km. and with velocity suggestive of consolidated sediments or volcanic rocks.

β) *Earthquake Surface Wave Dispersion.* Surface wave dispersion for oceanic paths is radically different from that for continental paths. For Rayleigh waves the water layer increases the dispersion at least fourfold, while for Love waves the very thin oceanic crust confines the strong dispersion to waves of

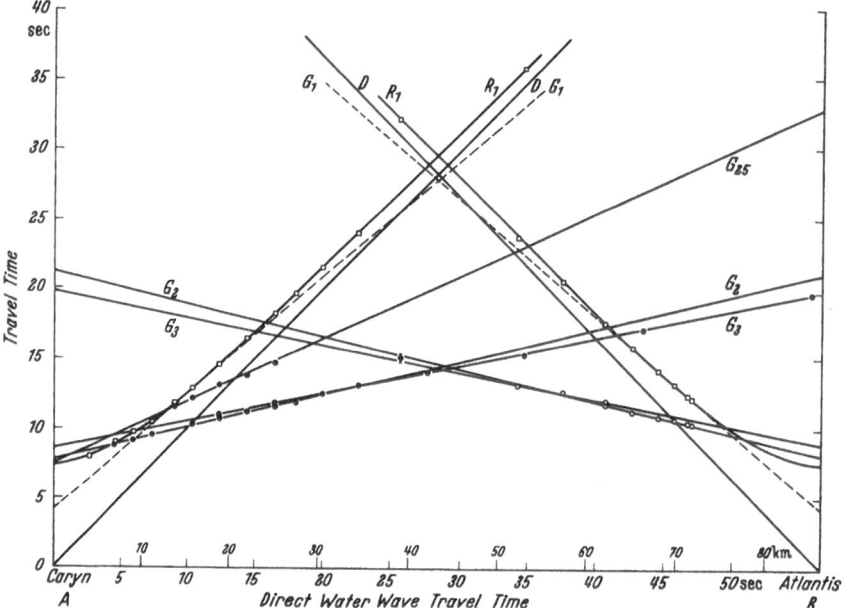

Fig. 7. Typical deep ocean refraction travel time curves. Atlantic Ocean, 26° N 64° W. The symbols used are explained by Fig. 6. $G_{18}$ (here erroneously marked $G_{38}$) is an $SV$ phase which is due to transformation of part of the compressional waves into vertically polarized shear waves at the contact between the sediment and the underlying rocks, with transformation back to compressional waves when the energy reenters the sediment.

period less than 15 sec. For longer Love waves (see Fig. 3, p. 124), which are usually the only ones received, the velocity change is less than 0.1 km./sec. over the entire range of periods, hence the seismograms show but little evidence of dispersion.

When the observed dispersion of Love and Rayleigh waves for oceanic paths is compared with the dispersion computed from crustal models chosen to fit seismic refraction results as in Fig. 3 (p. 124) and 8 (p. 130) excellent agreement, is obtained. Comparing surface wave dispersion for different oceans, and finding no difference[1], one must conclude that, except for obviously anomalous narrow belts, oceanic crust is everywhere the same.

γ) *Gravity Measurements at Sea.* Over 2000 measurements of gravity at sea are available and it is increasingly clear that the average deviation from isostatic equilibrium with the continents is only a few milligals. This fact being established, it is possible to use seismically determined layer thicknesses and

[1] For references see F. Press and M. Ewing: Earthquake Surface Waves and Crustal Structure, in: The Crust of the Earth. Geol. Soc. Amer. **1955**. In press.

demonstrate that the densities required for balance are those of the rock types expected from geological studies. Fig. 8, from the calculations of Worzel and Shurbet[1], show the densities adopted for silicic crust and the upper part of the mantle.

### 4. Continental Margins and Anomalous Areas.
These regions are long narrow belts, which include continental and submarine mountain chains and deep sea trenches. They are the most crucial regions of the earth's crust, but the least known. Investigations of them are usually very difficult because of thick sediments, rough topography and rapid lateral variations of crustal structure. The principal tools available for exploration of them are seismic refractions and gravity observations. So few data are available that we can only give brief descriptions of a few typical examples.

α) *East Coast of the United States of America.* This coast is a typical stable continental margin with a broad continental shelf. Fig. 9 is a structural section across the margin at Cape May, New Jersey, prepared by Worzel and Shurbet[2] from seismic refraction and gravity data. A sharp flexure in the Mohorovičić discontinuity is found at the margin of the continent. A rapid variation in the thickness of the crustal layer and great thickening of the sediments occur as the

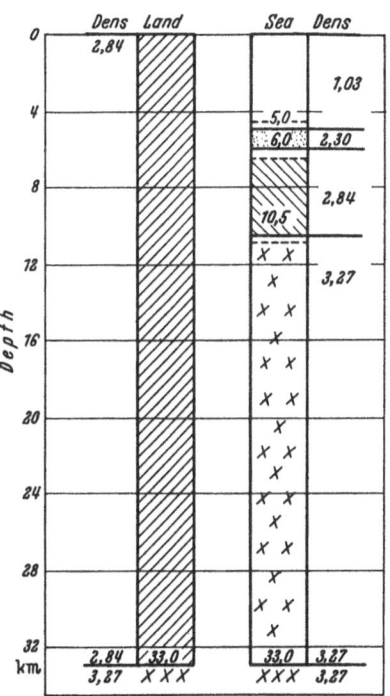

Fig. 8. Isostatically balanced continental and oceanic crustal columns consistent with seismic refractions.

margin is approached. In this manner the continental crust merges with the oceanic crust across a stable margin. In seismic velocity the oceanic crust resembles the lower part of the continental crust, but the details of the transition between them at the margin are not known.

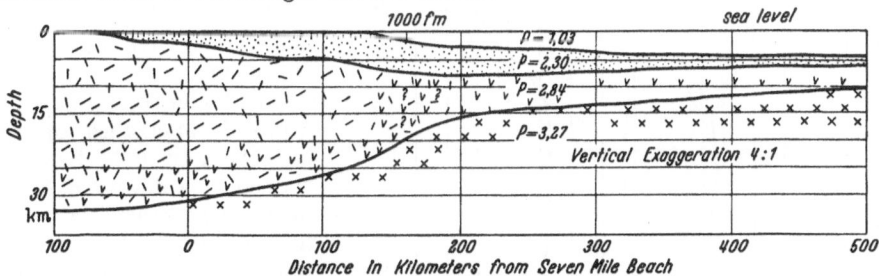

Fig. 9. Crustal section of continental margin off Cape May, New Jersey. In order of increasing density ϱ, layers are water, sediment, crust, mantle (Worzel and Shurbet).

β) *West Coast of South America.* This margin is characterized by seismic activity associated with a great mountain chain and an adjacent deep sea trench.

[1] J. L. Worzel and G. L. Shurbet: Gravity interpretation from Standard Oceanic and Continental Crustal Sections, in: The Crust of the Earth, loc. cit.

[2] J. L. Worzel and G. L. Shurbet, Gravity Anomalies at Continental Margins. Proc. Nat. Acad. Sci. USA. 1955. In press.

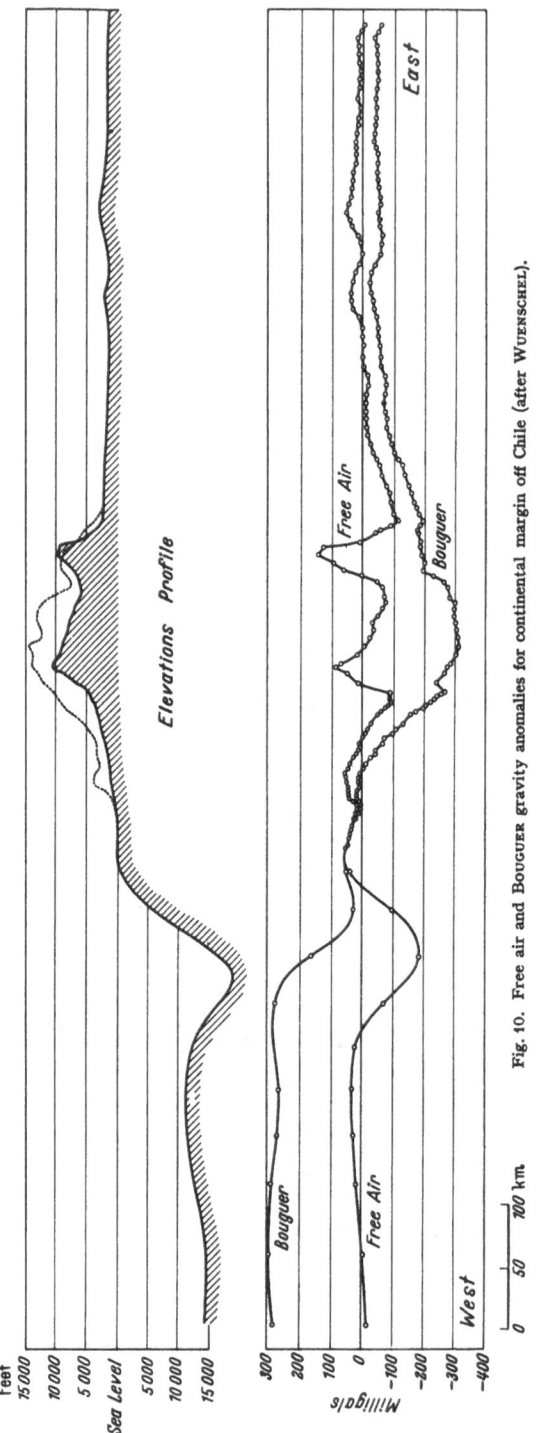

Fig. 10. Free air and Bouguer gravity anomalies for continental margin off Chile (after Wuenschel).

Geophysical information is available from studies of topography, gravity, seismicity, and strain release by earthquakes. A section[1] showing the topography and the Bouguer and free air gravity anomalies is shown in Fig. 10. Two striking features are evident. The first is a broad negative Bouguer anomaly of 300 to 400 milligals under the Andes, as broad as the mountain range, and explainable by the usual interpretation in which a root under the mountains provides the isostatic compensation. Although no direct seismic measurement giving the dimensions of the root of a mountain chain exists, Byerly[2] and Gutenberg believe delays in the seismic phase $Pn$ under the Sierra Nevada demonstrate a considerable downward displacement of the Mohorovičić discontinuity. If the upper half of the Andes root is less dense than the substratum by 0.39 gm./cc. and the lower half by 0.32 gm./cc. the root must extend to a depth of about 65 km. below sea level.

The second feature is the free air negative anomaly of 175 milligals in a zone only slightly wider than the trench. Although anomalies of this type have usually been explained as a downbuckle in a thick silicic crust, we prefer the view that this trench results from a depression of the crust,

[1] P. Wuenschel: Gravity Measurements and Their Interpretation in South America Between Latitudes 15° and 33° South. Geol. Soc. Amer. In press.

[2] P. Byerly: Comment on "The Sierra Nevada in the Light of Isostasy" by A. C. Lawson, Bull. Geol. Soc. Amer. **48**, 2025—2031 (1938).

in the marginal zone where the thickness is either oceanic or intermediate, the thickness remaining unchanged or diminishing[1].

*γ) Detached Island Arc.* Perhaps the most fruitful regions for study of crustal mechanics are the island arcs, of which the West Indies, the Aleutians, and the Indonesian Archipelago are examples. These arcs are long, narrow chains of islands, typically convex toward the ocean, in which the sedimentary rocks show late Cretaceous or Tertiary folding and faulting. They are characterized by great seismicity, association with deep sea trenches, volcanism, and very large gravity

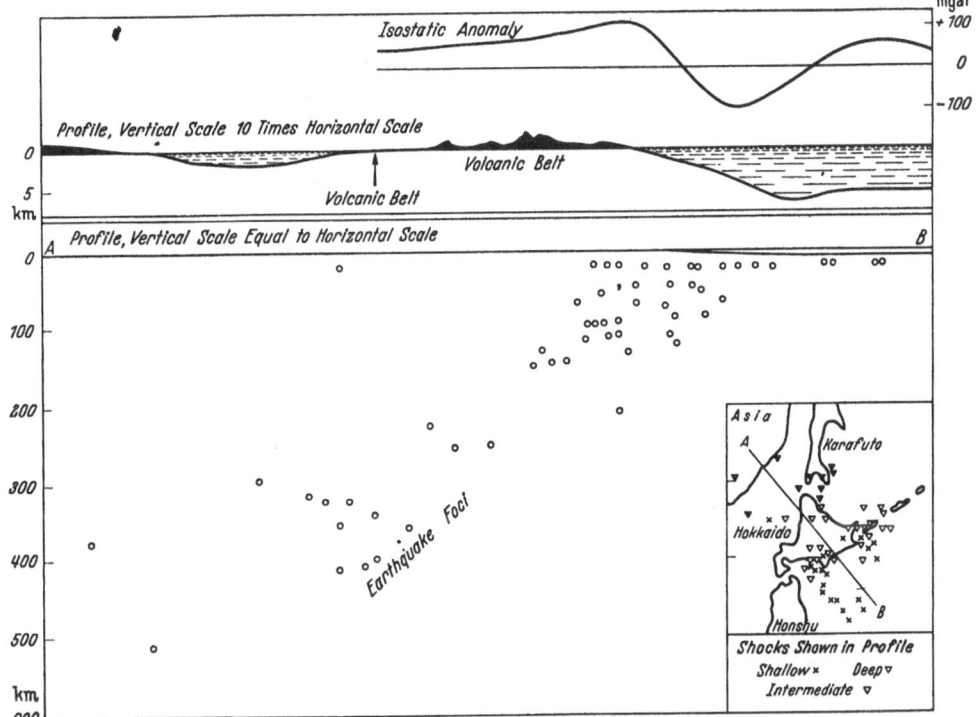

Fig. 11. Crustal section for typical island arc (GUTENBERG and RICHTER). In the map (right), deep foci are marked by black triangles.

anomalies. A typical cross section showing these features is presented in Fig. 11 from GUTENBERG and RICHTER[2]. The earthquake foci occur along a surface dipping away from the ocean. The surface of foci outcrops at the trench and extends downward to depths of 500 km. Volcanism occurs along a belt near the epicenters of the intermediate depth earthquakes. The greatest negative gravity anomaly (isostatic or free air) is either coincident with the axis of the trench or displaced slightly landward.

Seismic refraction and gravity profiles are available for a section crossing the West Indian Arc at Puerto Rico[1]. These are presented in Fig. 12 where Sect. a is based on topographic and seismic measurements, Sect. b gives the assumed structure and densities required to fit Sect. a and the gravity anomaly

---

[1] J. L. WORZEL and G. L. SHURBET: The Crust of the Earth, loc. cit.

[2] B. GUTENBERG and C. F. RICHTER: Seismicity of the Earth, p. 1—273. Princeton: University Press 1949.

in Sect. c. It is seen that the crust is about 25 km. thick under the island and that
the trench represents a down warp without thickening in the crust.

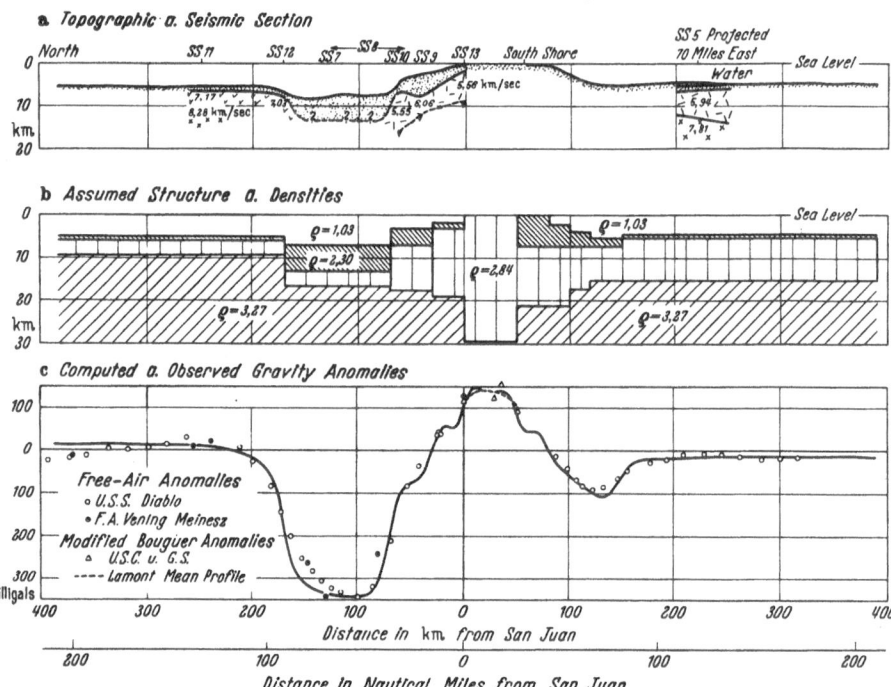

Fig. 12a—c. Crustal section of West Indian arc at Puerto Rico based on seismic and gravity data (Worzel and Shurbet).

Benioff[1] has studied the elastic strain- rebound characteristics of earthquakes together with the depth and geographic distribution of their foci. Fig. 13,

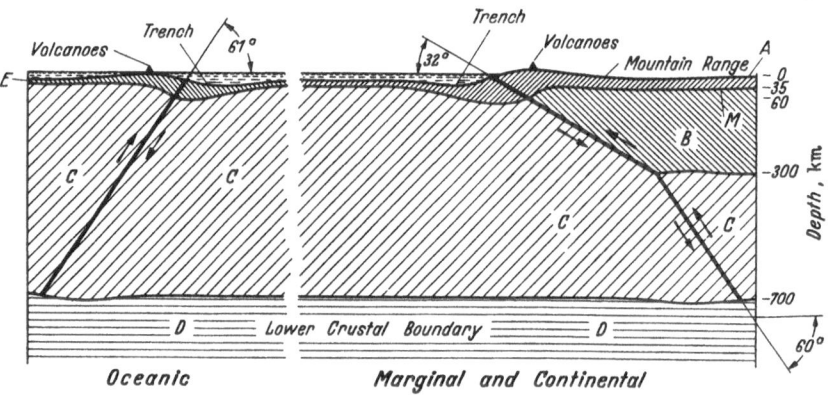

Fig. 13. Generalized crustal section for detached arc and continental margin based on earthquake foci (Benioff). — Note
exaggeration of surface relief.

taken from his paper, shows a continental margin structure (e.g. west coast of
South America) on the right and a detached island arc structure on the left. The

[1] H. Benioff: Orogenesis and Deep Crustal Structure — Additional Evidence from Seismology. Bull. Geol. Soc. Amer. 65, 385—400 (1954).

silicic crust is represented by $A$ and the Mohorovičić discontinuity by $M$. Benioff suggests that the principal orogenic feature of the margin or island arc is a complex reverse fault. The fault planes may be determined by the positions of earthquake foci and the indicated direction of slip is dictated by the geometry of the trenches and adjacent uplifts despite teleseismic studies which usually indicate strike-slip faulting. At the continental margin the discontinuity in angle of fault dip at 100 km. $\pm$ and the marked dissimilarity of strain-rebound characteristics of the shocks occuring above and below this level are indicative of a *major tectonic discontinuity between B and C* as shown in the figure. Only a single fault component (with the same dip as the deeper fault under the continents) occurs under the detached arcs, and it is assumed that both faults occur in a single continuous medium ($C$). Benioff defines the crust as the layer in which rigidity is sufficient to maintain elastic shearing strain for sufficient time necessary to accumulate the energy to generate earthquakes. Accordingly the layer extends down to 700 $\pm$ km., the level of the deepest earthquakes.

**5. Composition of the Crust.** The average composition of the continental crust determined by sampling of basement rock, is essentially that of an igneous rock intermediate between granite and basalt. The gradual increase of velocity with depth may be due either to the effect of pressure or to a variation toward basaltic composition or both. On the other hand, the $SiO_2$ percentage of igneous rocks from oceanic islands is much lower than for continents, and the higher velocity of the oceanic crust indicates basaltic composition.

Laboratory experiments on the velocity of rocks have shown that only three types have the values observed in the upper mantle—dunite (olivine), peridotite (olivine and pyroxene), and eclogite (garnet and pyroxene).

The mean thermal flux for continents has been given by Bullard[1] as $1.3 \times 10^{-6}$ cal./cm.$^2$ sec. and the average radiogenic heat production of granitic rocks as $53 \times 10^{-14}$ cal./cm.$^3$ sec. Using a conservative estimate of $0.3 \times 10^{-6}$ cal./cm.$^2$ sec. as the contribution to total flux from original heat, Bullard concludes that the mean radiogenic heat production for a 35 km. crust cannot exceed half the value estimated from surface rocks.

Recent observations of heat flow in the oceanic crust show values approximately equal with those for continents. Since granitic and other acidic rocks are far more radioactive than basaltic and ultrabasic rocks, it is difficult to account for the oceanic heat flow. One possibility is to assume that the upper mantle under the oceans differs from that under continents in having a higher radioactivity, but observed concentrations in oceanic crustal rocks are so low that it seems necessary to allow a layer at least 250 km. thick to produce the necessary heat. Although some theories of continental growth support this distribution of radioactivity in the upper mantle, it must be rejected as it implies temperatures reaching the melting point within a few hundred kilometers of the ocean floor. Another possibility is that convection in the mantle brings the necessary excess of original heat to the sea floor. This requires rising currents under the oceans, contrary to the direction usually assumed for mantle convection, but is open to all of the objections which have previously been raised against this process.

---

[1] E. Bullard: The Interior of the Earth, in: The Earth as a Planet. Ed. by G. P. Kuiper. University of Chicago Press 1954, 751 p.

# Forces in the Earth's Crust.

By

## ADRIAN E. SCHEIDEGGER.

With 12 Figures.

**1. The Earth's pertinent surface features.** The Earth is a rotating celestial body. Its matter is subject to gravitation. If stresses other than hydrostatic pressure were altogether absent, its surface would be perfectly smooth and exhibit a simple geometrical form. However, everyone knows that the surface of the Earth is not perfectly smooth but exhibits a great number of disturbances. This fact must be the result of forces other than hydrostatic compression in the Earth's crust and therefore requires a physical explanation.

The salient surface features of the Earth can be grouped into four categories: viz. faults, folds, mountain ranges, and continents. In such a grouping we shall begin with small-scale disturbances and proceed to larger-scale ones. From an aetiological standpoint it would be preferable to start with the large-scale disturbances; unfortunately, however, the more global the scale of a phenomenon becomes, the scantier the information about it gets. This is especially true if it comes to physical explanations, and therefore, for the sake of expanding upon the better known phenomena first, the order starting from small-scale phenomena and proceeding to large-scale ones will be preserved throughout this Chapter.

The first remarkable features upon the Earth's crust are faults. Such faults are dislocations that may be observed in many places on the Earth's surface. They occur in parallel or subparallel systems which have usually a wide lateral distribution. Geologists distinguish between three types of faults which they call *normal, transcurrent* and *reversed faults.*

*Normal faults,* which are the commonest ones, are characterized by an irregular strike, a steep dip of about 65—70°, and by the fact that the rocks appear to have slipped down on one side over the other, the inclination of the fault plane being such that the horizontal extent is *increased.* The relative vertical displacement may be anything from a few millimeters to one kilometer.

*Transcurrent faults* are always very straight and often stretch for several hundred kilometers. Their dip is almost vertical. The rocks appear to have moved in the direction of the strike of the fault, the motion on one side being opposite to that on the other.

Finally, *reversed faults* are similar to normal faults, except that the dip is much less and the movement is such that the slip *shortens* the horizontal extent.

Figs. 1—3 will serve to illustrate the geologists' conception of faulting.

The most conspicuous irregularities of the Earth's surface are undoubtedly *mountain ranges.* It appears that the strata in those mountain ranges have been contorted to a fabulous extent. It almost looks as if some supernatural giant took originally level strata, pulled them up and folded them over and over at his will. It also appears that a *shortening* of the Earth's crust has taken place where mountains have been folded up. Naturally, the action of water and wind will erode some of the folded materials and the physiographic appearance of a mountain range is therefore one of high peaks and deep valleys. Nevertheless, the continuity of the originally folded strata can be traced from peak to peak and the original (undisturbed by erosion) position of the layers can be reconstructed. The horizontal distances over which the strata are folded over may be up to several score kilometers. The Western Alps, in a profile from Lausanne over the Dent Blanche to the plain of Piemont, give a classical example.

If one examines the mountains somewhat more closely, a few more remarkable facts become apparent. It will of course be necessary to consider as *"mountains"* also such occurrences as *islands* in the sea; the latter are nothing but submerged mountains. Thus, we observe that mountains are not scattered at random over the Earth's surface.

*Firstly*, they occur in *ranges*. Each individual *mountain range* (or *island chain*) has an arcuate strike, close to being part of a circle, sometimes with a "tail"

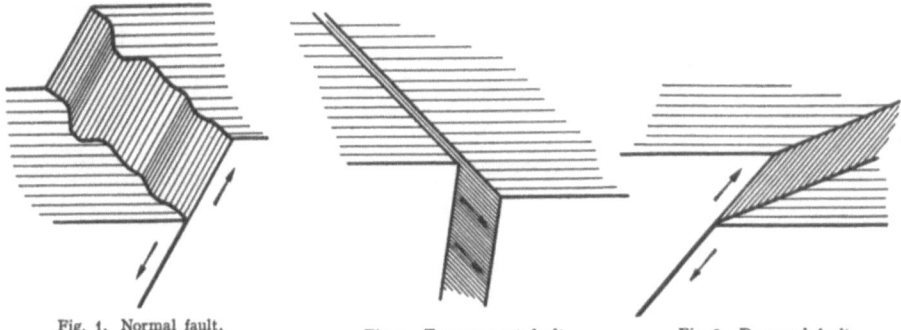

Fig. 1. Normal fault.          Fig. 2. Transcurrent fault.          Fig. 3. Reversed fault.

attached to it. This fact is most conspicuous on the island arcs in the North Pacific Ocean. Some people prefer to describe an "arc with tail" in terms of the top-part of an X-shaped feature rather than in the terms stated here. The cross-section of an island or mountain arc is quite specific. The main range is usually full of volcanic activity, followed on the convex side by a shallow foredeep, a second, purely sedimentary mountain range and a very deep ocean trench. Variations of this physiographic pattern are possible as some of the elements may be missing. If there is a continental block anywhere near the arcs, then the latter always point their concave side toward the continent. Furthermore, seismological investigations have shown that frequent earthquakes occur with hypocenters on a zone dipping down conically beneath the arcs toward their concave side.

*Secondly*, the island- and mountain-arcs themselves form chains, the centers of each arc of a chain lying almost on a great circle (cf. Fig. 4). One of the most systematic analyses of mountain ranges has been made by WILSON[1] who defines only four types of *junctions* between arcs of one chain, namely *linkages, fractured deflections, reversed arcs* and *capped deflections* (cf. Fig. 5). An example of a linkage may be found in the islands of Japan, of a fractured deflection in the Rocky Mountains of western North America, of a reversed arc in the Scotia arc (Antarctica), and of a capped deflection in the Appeninnian and Dalmatian Mountains capped by the Alps.

It must be noted that in the present systematization the Alps do not appear as a "fundamental" part of a chain of mountain arcs, but only as a *"cap range"*.

[1] J. T. WILSON: Proc. Geol. Assoc. Canada **3**, 141 (1950).

Such cap ranges are quite frequent and are always high. They do not exhibit any of the specific features discussed above (such as volcanism) of those mountain arcs that, physiographically, seem to form the "elements" of a chain of mountain arcs; therefore they have to be assigned the separate place as "cap range" in the system of chains of arcs.

*Thirdly,* the great-circle chains of mountain arcs themselves seem to form a system, in which two such great circles appear to be crossing each other at right angles. Geological age determinations show that mountain building occurs in singular *diastrophisms* of which there were about ten since the beginning of the geological (Precambrian) age. The diastrophisms were separated by long periods of acquiescence. Each diastrophism is believed to express itself in the formation of two great-circle systems of chains of

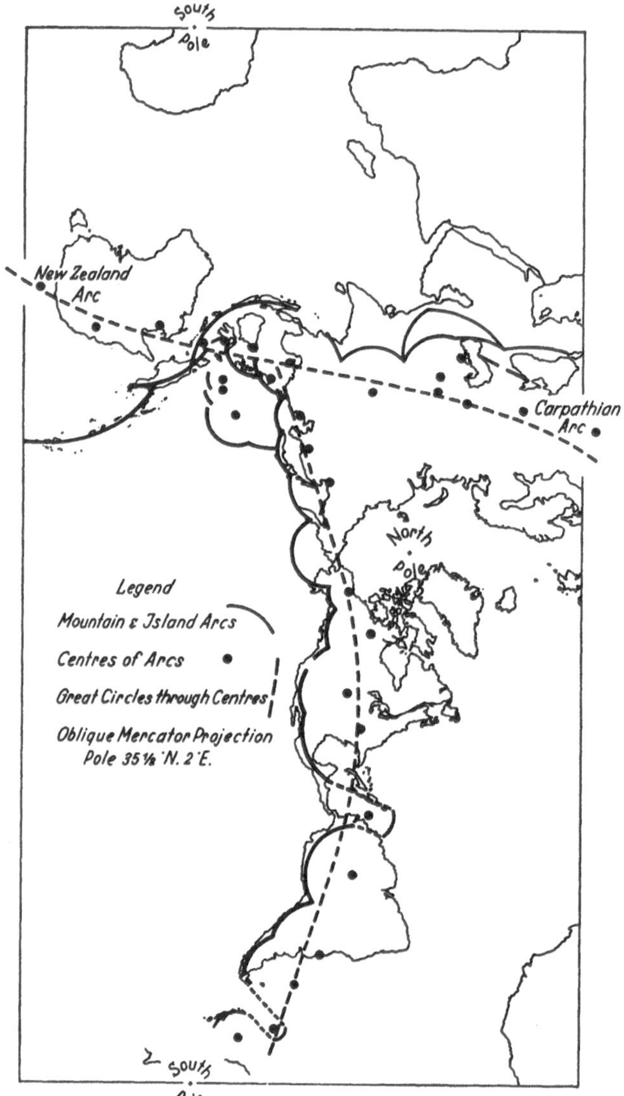

*New Zealand Arc*

*Carpathian Arc*

*Legend*

*Mountain & Island Arcs*

*Centres of Arcs* •

*Great Circles through Centres*

*Oblique Mercator Projection*
*Pole 35 ½ °N. 2 °E.*

Fig. 4. World map on an oblique mercator projection showing post-Triassic-mountain and island arcs, their poles or centers, and two great circles through their poles. After J. T. WILSON.

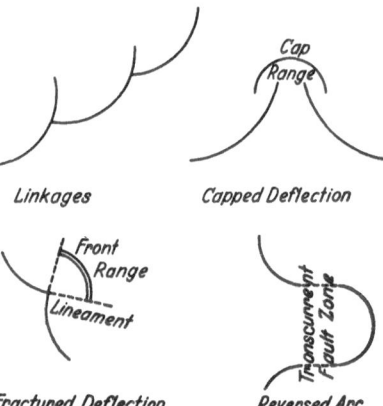

Fig. 5. The four types of junctions of arcs. After J. T. WILSON.

mountain arcs. Unfortunately, it becomes increasingly difficult to trace such mountain systems the older they are. It is quite easy to see the two great circles formed by the centers of the arcs of the youngest mountains (see Fig. 4), but for the older ones this is not so easy. Therefore, an alternative opinion exists assuming that the centers of the mountain arcs (including the newest) do not significantly lie on two great circles, but rather on lines following the margin of continents.

Indeed, concerning the newest mountains, the two postulated great circles are also lines following roughly the margin of continents. Obviously, more geological data about the older mountain ranges will be necessary to clear this pertinent question of physiography.

*Finally*, the last remarkable fact about the physiographic disturbances of the Earth's crust consists in the distribution of continents and oceans. Though quite irregular, it shows a few systematic features. The continents are nearly everywhere antipodic to oceans, and they are all roughly triangular, touching each other in the north and pointing southwards. Four old shields have their position, roughly speaking, at the corners of a tetrahedron. The level of the continents is about 5000 meters above that of the ocean floors, and the ratio of the area covered by continents to that occupied by oceans is roughly 1:2.

Summarizing, we may state that there are four categories of peculiarities upon the Earth's surface, as was outlined at the beginning of this Chapter. These categories may be termed

(i) faults,

(ii) folds,

(iii) the global distribution of mountains, and

(iv) the distribution of continents and oceans.

Owing to the physiographic regularities of these phenomena, it is hardly believable that they would be due to "chance". One is therefore prompted to look for a physical explanation.

**2. Principles of a physical explanation.** The explanation of the peculiarities of the Earth's surface must be based upon our knowledge of the laws of physics and upon such information as we may have of the Earth's constitution.

According to our present knowledge, the Earth consists of three principal layers, termed as *crust, mantle* and *core*[1].

The *crust*, as revealed by seismology, consists of at least two comparatively thin layers composed of material of low density. JEFFREYS[2] assumes in the crust an upper layer composed of granite, of mean thickness 11 km., and below this a layer, termed intermediate, of thickness 24 km., density 2.87. Its chemical composition is not definitely established. The lower limit of this layer is known in seismology as the MOHOROVIČIC *discontinuity* and is usually taken as the boundary of the *mantle*. The exact definition of all these layers, however, is still largely hypothetical.

The *mantle* reaches to a depth of 2800 km. where there is the boundary with the core. It is characterized by extremely high seismic velocities, for transverse and longitudinal waves alike. The pressures near its top are of the order of some 10,000 atmospheres which is almost within the experimentally accessible range. It can be shown that amongst the known minerals only a few silicates yield the observed high velocities. BRIDGMAN[3] has shown that these minerals have a lower compressibility than any other of the many substances investigated, elements or compounds.

---

[1] See also the chapters on the Earth's crust (EWING and PRESS) and the Earth's interior (JACOBS).

[2] H. JEFFREYS: Mon. Not. Roy. Astronom. Soc., Geophys. Suppl. **4**, 498, 537, 548, 594 (1939).

[3] P. W. BRIDGMAN: Phys. Rev. **17**, 3 (1945). — Rev. Mod. Phys. **18**, 1 (1946). — Proc. Amer. Acad. Sci. **76**, No. 1 (1945); No. 3 (1948).

Within the mantle, at a depth of about 900 km., there is probably a further discontinuity, involving a change of phase, or a chemical change, or both[1]. This has also been deduced from seismic data, but in connection with some investigations on the solid state. Thus, it has been observed that the variation of observed compressibility with pressure cannot be accounted for by the compression of a homogeneous substance between 200 and 800 km depth; compression alone accounts very well between 900 and 2800 km. depth. Therefore, there must be a discontinuity at 900 km. depth.

Finally, the *core* presumably consists of iron with some small admixture of nickel. Its density is around 11—12. It is quite possible that iron has such a high density under the pressures prevalent in the core. However, all that is really known is that the material has a great density, does not transmit transverse waves, and probably has a high electric conductivity in order to account for the magnetic field of the Earth. It is conceivable that the material would be chemically the same as in the mantle of the Earth, if a certain phase change would take place at the boundary.

The physical causes of the surface disturbances of the Earth will have to be sought in the upper layers of the Earth, i.e. in the mantle and crust. The core will, presumably, play only an indirect rôle as a heat supplier or the like. The distinction between "mantle" and "crust" as outlined above, is based essentially on *chemical* grounds. At one time it was thought that the crust only would be able to support (shear-) stresses to any appreciable extent, and therefore that the forces effecting the surface features of the "Earth" would be entirely located in the "crust" as defined above. However, since the discovery of deep-focus earthquakes with focal depths to about 700 km. below the surface, this hypothesis became untenable. From a mechanical standpoint, it seems that "crust" and "mantle" cannot be treated separately; both seem to behave mechanically in a similar way, at least to a depth of about 700 km. Both have enough rigidity to transmit transverse seismic waves, and both seem to be strong enough to support large (shear) stresses which, eventually, may become released in earthquakes. The latter fact, of course, does not preclude that plastic flow and creep may also take place in those layers.

Thus, the Earth's surface features must have their primary origin in some layer of the Earth which has its lower boundary at least as deep as the foci of the deepest earthquakes. This layer will be termed the "*orogenetic shell*" of the Earth. By saying "primary" origin it is understood that the orogenetic shell is that hypothetical layer whose mechanical behavior determines the occurrences on the Earth's surface. It is quite clear that there must be other physical causes (termed "fundamental") which make the orogenetic layer behave in the way it does, but it must be assumed that those causes are thermal or cosmic or such like. At any rate, they cannot be stresses themselves, as the assumption of "primeval" stresses would only be an *ad hoc* hypothesis and thus not furnish an explanation at all.

Several fundamental causes have been suggested that might make the orogenetic layer behave in the way it does, amongst them cosmic tidal forces, forces due to rotation of the Earth, and thermal effects. Most geophysicists agree nowadays that the *thermal effects* are probably the most important ones. Temperature gradients measured in mines etc. suggest strongly that the Earth is a heat engine, the heat supply being derived either from *primeval heat* (a remnant of a hypothetical gaseous, hot state of the Earth) by cooling, or else from the

---

[1] F. BIRCH: Trans. Amer. Geophys. Un. **32**, 533 (1951).

continued *radioactive* decay of fissionable material dispersed somewhere in the Earth. In either case there must be a mechanism by which the heat energy available is transformed into mechanical work evident in the upthrust of mountains, etc., and it is only natural to assume that this transformation occurs in a localizable zone of the Earth;—the latter is the hypothetical orogenetic shell. A similar argument can be put forth if one prefers to believe that the fundamental cause of mountain building etc. is other than thermal in origin. Whatever the fundamental effects, it must be assumed that there is a localizable zone in which these effects are transformed into mechanical work; and that zone can then be defined as the orogenetic shell.

As stated above, the actual surface features of the Earth may reflect the behavior of the orogenetic shell only in a secondary manner. Thus, the search for an explanation of the Earth's surface features will logically have to start with an analysis of the smallscale surface features in order to determine what sort of stresses and strains were necessary to produce the latter. Then one will proceed to infer what must have happened in the orogenetic shell to produce those stresses. Again, it would be much more satisfactory if one could start with the theory of the orogenetic shell and its behavior, then all further effects which were termed above as "secondary" would follow from the primary ones and could be explained by the deductive method. Unfortunately, the behavior of the orogenetic shell is largely hypothetical and can be riddled out only by induction. Thus it is, unfortunately, unavoidable to start with the small disturbances of the Earth's crust, of which an explanation is fairly easily constructed, and then to proceed to the more obscure, larger ones, later.

**3. Dynamics of faulting.** The faults which dissect the Earth's crust are relatively easily amenable to an investigation of their physical causes. This is due to the comparative smallness of the scale of the phenomenon. The principal features and classes of faults have been described in Sect. 1 of this Chapter.

From the physiographic appearance of the faults it seems likely that they are simply the expression of localized mechanical failure of the material of the Earth's crust. Using one of the theories of such mechanical failure, it should therefore be possible to reconstruct the field of stresses which must have been in existence when the fault was caused. If the prevalence of such a field of stresses can be rendered plausible from other considerations about the Earth's crust, the phenomenon of faulting will have been "explained".

The subject of *mechanical failure* has been extensively investigated in engineering literature. Although the theory is by no means complete, a substantial amount of information has been accumulated and several conditions of failure have been advanced. For any details of these theories the reader is referred to vol. VII, part 2 of this encyclopedia.

Although a great number of conditions of mechanical failure have been advanced, it seems that the classical condition of MOHR[1] is still adequate for most purposes. In brief terms it states that failure ocurs if the largest shearing stress reaches a limiting value determined empirically and represented graphically as envelopes in "MOHR's diagram". It further states that at each point in the material under stress there are two surfaces of probable rupture which contain the direction of the intermediate principal stress and which are inclined at an angle $\varphi \leq 45°$ toward the smallest principal stress (i.e. the greatest pressure;

---

[1] O. MOHR: Abhandlungen aus dem Gebiete der technischen Mechanik, 3. Aufl. Berlin: Wilh. Ernst & Sohn 1928.

*tension* is given the *positive* sign). This implies that the *magnitude* of the intermediate principal stress has no influence on the fracture of materials.—In case
of a pure tensile test the hypothesis of MOHR does not determine the surface of
rupture as the stress state is then degenerate. Thus, one has to add that in this
case the surface of probable rupture is perpendicular to the tensile stress.

Actually, MOHR's hypothesis is but a more scientific restatement and generalization of earlier principles advanced by NAVIER and COULOMB. The latter
postulated the theory of "shear fracture" which states that fracture occurs if
the greatest shearing stress reaches a definite value and that the fracture will
be in a plane bisecting the angle between greatest and smallest principal stress;
the former accounted for the fact that the actual fracture does not exactly occur in
those planes by introducing a "coëfficient of internal friction". The coëfficient
of internal friction has the effect to make the angle $\varphi$ (as defined above) smaller
than 45°. MOHR simply restated this fact by making the coëfficient variable.

The idea of applying mechanical failure theory to an explanation of faulting
is due to ANDERSON[1, 2]. ANDERSON observed that an undisturbed (or "standard")
state of stress in the Earth's crust cannot be entirely arbitrary. First, it must
be near the breaking point as is evidenced by the frequency of earthquakes.
Second, there can be no pressure or tension perpendicular to the surface and no
shearing force parallel to it. The latter condition implies that the normal to
the surface is one of the principal directions of stress at or near the surface. Thus,
except in strongly folded areas, one principal direction of stress is nearly vertical,
the other two are horizontal.

Now, assuming comparatively small disturbances of this standard stress
state in the Earth's crust, mechanical failure may occur which could appear as
a fault. ANDERSON[1, 2] arrived, based on NAVIER's (or MOHR's) theory of mechanical failure, at the following explanation of the three types of faults observed by geologists:

(i) *Normal faults.* Assuming that there is relief of pressure in all horizontal
directions, the greatest pressure is the vertical pressure which is due to gravity.
In general, the horizontal stresses will not be equal such that the greatest tension
will prevail in a certain direction. Thus, the intermediate stress will also be
horizontal, but at right angles to the direction of greatest tension (whichever
this be) and, if failure occurs, this will happen along a plane containing the
intermediate stress and inclined at an angle $\varphi \leq 45°$ toward the vertical, which
is the direction of smallest principal stress. One thus obtains the characteristics
of a normal fault. From the geometrical pattern of the stresses it is obvious
that the motion of the two parts must be such that the horizontal extent is
increased.

(ii) *Transcurrent faults.* Assume that there is an increase of pressure in one
horizontal direction and a relief of pressure in a horizontal direction at right
angles to it. The smallest principal stress is then horizontal and the intermediate
one is vertical. Now, if fracture occurs, this must happen according to MOHR's
theory in a vertical plane inclined at an angle $\varphi \leq 45°$ toward the greatest pressure.
One obtains thus a fault with a vertical dip, the motion of the two parts being
essentially horizontal. This is the characteristic pattern of a transcurrent fault.

(iii) *Reversed faults.* Assume that there is an increase of pressure in all horizontal directions. In general, one horizontal direction will be characterized by

[1] E. M. ANDERSON: Trans. Edinburgh Geol. Soc. **8**, part 3, 387 (1905).
[2] E. M. ANDERSON: The Dynamics of Faulting and Dyke Formation with Applications to
Britain. Edinburgh: Oliver & Boyd 1942.

the fact that along it the pressure will be greatest. Thus the minimum pressure will be vertical and the intermediate principal stress will be in a horizontal direction, at right angles to the greatest pressure. If conditions are such that failure occurs, this will happen according to Mohr's theory along a plane inclined at an angle $\varphi \leq 45°$ toward one horizontal direction, the motion of the two parts being toward each other. One thus obtains the characteristics of a reversed fault: the dip is shallow, and the motion is such that the horizontal extent is shortened.

Apart from the faulting phenomena discussed in Sect. 1 of this Chapter, there are other phenomena which appear related to faults: The occurence of *dykes* and *rift valleys*.

*Dykes* are, in the main, nearly vertical fissures between 3 and 30 meters wide that have been infilled with some intrusive material. The two sides of a dyke appear to have moved apart in a direction normal to the fissure such that there is neither a lateral nor a vertical dislocation. Dykes are much less common than faults and they are also somewhat restricted in distribution.

Rift valleys are giant trenches that appear as wide, deep valleys which may stretch for thousands of kilometers. The best known example is the system of rift valleys in East Africa.

The idea of Anderson to consider mechanical failure as a major geological agent, also provides for an easy explanation of dykes. Mohr's theory of fault planes does not apply in the case of a degenerate stress state. In the latter case, fracture is normal to the tensile stress. This is often referred to as "*tension fracture*" as opposed to "*shear fracture*" which was used to explain faults; but, in fact, "tension fracture" and "shear fracture" must be only different cases of any *consistent* mechanical fracture theory.

The occurrence of dykes is thus explained by the remark that they may be considered as the evidence of tension fracture. This remark is also due to Anderson[1] who elaborated somewhat more on the phenomenon, showing that also the intrusion of magmatic material into the dykes can be rendered plausible if tension fracture is taken as cause of dyke formation,—at least from a geological standpoint.

The cause of rift valleys is somewhat less certain,—mainly because not too much is known about them geologically. Thus, evidence of possible faulting is obscured as the sides may have caved in, their bottom is possibly filled with material from their sides etc.

In view of these uncertainties it seems quite reasonable to fit the rift valleys too, into the systems of Anderson's mechanical failure theory; either by assuming that they are double reversed faults, the center part being depressed by pressure from the sides[2] or else the expression of transcurrent faulting. The straightness of the rift valleys seems to speak in favor of transcurrent faulting, but it is difficult to see why they should be in the form of such remarkable depressions. Assumption of a double normal fault as causing a rift valley is out of the question because it has been found[3] that there are negative gravity anomalies across same; the bottom of the latter should therefore *rise* according to the theory of isostasy[4]. Double reversed faulting, however, as assumed above, would be able to keep the bottom of a rift valley "down", as is required.

---

[1] In Chapter 3 of his book cited above.
[2] Cf. H. Jeffreys: The Earth, 3d. edit., p. 355. Cambridge 1952.
[3] E. C. Bullard: Phil. Trans. Roy. Soc., Ser. A **237**, 237 (1938).
[4] Cf. Garland's Chapter in this volume, p. 232.

ANDERSON's theory, as outlined in the previous paragraphs, seems to account quite satisfactorily for the types of faults and related phenomena that have been observed by geologists. Although it is based on a very elementary theory of mechanical failure, its success seems to indicate that at least the fundamental ideas are correct which it employs. Naturally, it does not make any statements as to where the deviations from the standard stress state should have come from,—that is left to the speculations about orogenesis to be treated later in this Chapter.

The theory of ANDERSON, thus, leaves very little to be desired, and later modifications are more in the line of refinements. Thus, SEIGEL[1] devised a theory which encompasses shear and tension fracture,—as were treated as separate cases by ANDERSON, and shows that the several types of faults are the consequence of it. Actually, SEIGEL's theory of fracture is equivalent to MOHR's, except that the statements are clad into a mathematical scheme making use of probability concepts, instead of into a graphical scheme using "stress circles". In fact, both theories are equally heuristic (in that the angle $\varphi$ is determined heuristically and not theoretically) and the application of SEIGEL's theory to the general principles of faulting leads therefore duly to ANDERSON's results.

WALLACE[2] calculated and plotted stress patterns that seemed to be reasonable to occur in the Earth's crust. He then showed that the actually observed faults are the outcome of reasonable assumptions as to the "preferred directions of failure" in the stress states. The assumptions are equivalent to MOHR's except that allowance is made for a greater latitude of $\varphi$, due to the possibility of incidental effects such as anisotropy of the material. The resulting patterns are analyzed in relation to actually observed faults.

Finally, HAFNER[3] reviewed the subject of stress theory as related to faulting. He gave an analytical presentation of ANDERSON's standard state. He then proceeded to calculate analytically such deviations from this standard state as would seem reasonable and which could produce faulting. The faulting patterns to be expected were then also calculated upon the assumption of MOHR's criterion of failure. The results were compared with geologically observed facts. The improvement of HAFNER's theory over the theory of ANDERSON is that ANDERSON actually discussed only the elements of a fault, whereas HAFNER showed that whole fault systems and changes of characteristics (such as dip) can well be explained by the assumption of reasonable stress states.

For instance, ANDERSON's standard state can be expressed as a *two-dimensional stress state*. Thus, let $x$ and $y$ be two CARTESIAN co-ordinates, $x$ horizontal, $y$ downward, in the direction of gravity ($y = 0$ surface), and let the components of the stress tensor be $\sigma_x$, $\sigma_y$, $\tau_{xy}$. It is then convenient to express the stresses by means of AIRY's stress function $\Phi$ such that

$$\sigma_x = \partial^2 \Phi/\partial y^2, \quad \sigma_y = \partial^2 \Phi/\partial x^2 - \varrho g y, \quad \tau_{xy} = - \partial^2 \Phi/\partial x \partial y \qquad (3.1)$$

if gravity is the only body-force and $\varrho$ = density. The stress function, furthermore, has to satisfy the following equation

$$\frac{\partial^4 \Phi}{\partial x^4} + 2 \frac{\partial^4 \Phi}{\partial x^2 \partial y^2} + \frac{\partial^4 \Phi}{\partial y^4} = 0. \qquad (3.2)$$

Then, ANDERSON's standard state can be expressed by the following choice of the stress function:

$$\Phi = - \tfrac{1}{6} \varrho g y^3 \qquad (3.3)$$

which yields

$$\left. \begin{array}{l} \sigma_x = \sigma_y = - \varrho g y, \\ \tau_{xy} = 0. \end{array} \right\} \qquad (3.4)$$

[1] H. O. SEIGEL: Trans. Amer. Geophys. Un. **31**, 611 (1950).
[2] R. E. WALLACE: J. Geology **59**, 118 (1951).
[3] W. HAFNER: Bull. Geol. Soc. Amer. **62**, 373 (1951).

As an example of a practical stress state, one can assume the presence of a supplementary horizontal component in addition to the standard stress state, but the absence of an associated supplementary vertical component. This is expressed by: (no body force for supplementary stress)

$$\sigma_y = \frac{\partial^2 \Phi}{\partial x^2} = 0 \quad \text{for all values of } y. \tag{3.5}$$

Integrating, one obtains the stress function

$$\Phi = c f_1(y)\, x + a x + b f_2(y) + d. \tag{3.6}$$

To satisfy Eq. (3.2), the fourth order derivatives of $f_1$ and $f_2$ must vanish. Hence the second order derivatives may be either linear functions of $y$, constants, or zero. The stress components then are

$$\sigma_x = c f_1''(y)\, x + b f_2''(y); \quad \sigma_y = 0; \quad \tau_{xy} = -c f_1'(y). \tag{3.7}$$

The boundary conditions at the surface require that $f_1'(y) = 0$ for $y = 0$. Keeping within the limits of the above restrictions, one can set up the following subgroups:

(a) $\qquad f_1'(y) = 0; \quad f_2''(y) = y + d,$ $\qquad\qquad$ (3.8a)

$$\sigma_x = b(y + d); \quad \sigma_y = 0; \quad \tau_{xy} = 0. \tag{3.8b}$$

(b) $\qquad\qquad f_1' = y; \quad f_2''(y) = 0,$ $\qquad\qquad$ (3.9a)

$$\sigma_x = cx; \quad \sigma_y = 0; \quad \tau_{xy} = -cy. \tag{3.9b}$$

(c) $\qquad\qquad f_1'(y) = \tfrac{1}{2} y^2; \quad f_2''(y) = 0,$ $\qquad\qquad$ (3.10a)

$$\sigma_x = cxy; \quad \sigma_y = 0; \quad \tau_{xy} = -\frac{c}{2} y^2. \tag{3.10b}$$

The most general expression for the stress systems satisfying the assumption of absence of a vertical stress component is given by the superposition of Eqq. (3.8) to (3.10). It is seen that the stipulation $\sigma_y = 0$ is associated with two additional general properties of the internal stress system: (i) that the shearing stress is a function of $y$ only, i.e. constant in all horizontal planes, and (ii) that $\sigma_x$ has linear gradients in both the horizontal and vertical directions.

Of practical importance are the stress systems of the first two subgroups. The combination of (3.8), (3.9) with the standard stress state yields (with $d = 0$)

$$\sigma_x = cx + by - ay; \quad \sigma_y = -ay; \quad \tau_{xy} = -cy; \tag{3.11}$$

where $a = \varrho g$. An analysis of these expressions yields that the trajectories of maximum principal pressure are curved lines, dipping downward away from the area of maximum compression. The curvature is stronger if the vertical gradient of $\sigma_x$ is small. From these trajectories, the potential fault surfaces can be calculated according to MOHR's criterion. The potential faults obviously belong into the class of *thrust faults*. The set dipping towards the area of maximum pressure is slightly concave upwards, the complementary set concave downwards. Thrust faults of the former type are very common in nature and the theoretically deduced curvature is frequently observed. The latter type appears to occur only very rarely.

In view of the above results it appears that a physically reasonable theory of the phenomenon of faulting is in existence which seems to be satisfactory even concerning minor geological details. In fact, faulting has been reduced to the assumption of reasonable stress systems. The primary cause of those stress systems will be discussed in Sect. 5 on orogenesis.

**4. Dynamics of folding.** *a) Facts.* The most conspicuous irregularities of the Earth's surface, namely mountains, are the product of a process called "folding", as described in Sect. 1. It seems that mountains appear concurrently with a "crustal shortening" taking place in their neighborhood, due to a hypothetical lateral compression.

If the crustal shortening is assumed to be the cause of the mountains, then one can estimate the amount of such shortening necessary to produce the latter. This can be done simply by a direct measurement of the extent of the strata in the great mountain systems. If one assumes that in a normal cross section of a mountain range, the length of the section of a stratum (which is a curved line) is equal to the length of that section before it was folded, i.e. when it was flat on the surface, one can determine how much shortening must have taken place. This, of course, assumes that the strata have undergone no deformation of area but only one of shape. Estimates of shortening obtained in this manner are quoted to be 50 to 80 km. in the Appalachians, 40 km. in the Rocky Mountains of Canada and 17 km. in the Coast Range of California[1]. For the Alps, HEIM[2] quotes 240—320 km. Compared with the assumed original (unfolded) cross section of a mountain range, these values represent a shortening of up to 4 to 1.

However, it is by no means certain that the strata have not undergone other deformations than only bending. It may be observed that the present shape of the great mountain ranges is certainly not due to lateral compression alone such that much of the observed folding may be due to other causes than crustal shortening. This argument can be substantiated by observing that in most great mountain systems, the theory of isostasy gives a fair approximation to the facts[3]. If one supposes that a region of average structure is halved in area and therefore doubled in thickness (i.e. a crustal shortening of 1:2), the theory of isostasy[3] yields that equilibrium is restored if the outer layers have a residual height of 5,6 km.[4]. This is much more than the average height of most mountain systems. Yet, HEIM assumes crustal shortening for the Alps of 4:1 which makes matters even worse.

The theory of isostasy gives also a means to calculate what the crustal shortening must have been across a mountain range in order to produce the average height that is actually observed. JEFFREYS[5] find that it will be about 1.66:1. One is therefore faced with the difficulty of reconciling the standpoint of the geologists who find that the extent of the strata requires crustal shortening of up to 4:1, with that of the isostasists who do not allow for more than 1.66:1. The only possible answer to this problem is that the extent of the nappes as seen by the geologists is not entirely due to crustal shortening. In order to justify this answer, the mechanics of folding has to be studied.

The problem of folding has been treated by several approaches, some of them consisting of analytical methods, the others of experimental methods.

*b) Analytical methods.* The analytical methods face the difficulty of having to cope with a geometrically very complex phenomenon which is difficult to encompass by mathematical equations. There is also the problem of identifying the rheological behavior of the real material of which the mountains are composed, with that of some idealized body which would be amenable to analytical description.

Therefore, JARDETZKY[6] proposed to make use of the equations of elasticity and, after the deformation of the body were computed, to replace the displacement by the velocities of the particles. He states that this is naturally only an approximation to the actual facts, but that it is very probable that the qualitative results thus obtained will not differ much from the true picture. According to JARDETZKY, the folding mechanism is thus the same as the buckling of thin elastic plates. On this basis, JARDETZKY[6] computed four types of folds which

---

[1] L. V. PIRSON and C. SCHUCHERT: Textbook of Geology. New York, N. Y.: Wiley 1920.
[2] A. HEIM: Geologie der Schweiz. Leipzig 1921.
[3] Cf. Chapter by GARLAND in this volume, p. 238.
[4] See JEFFREYS: loc. cit. (p. 265[2]), p. 307.
[5] loc. cit. (p. 265[2]), p. 308; see also G. D. GARLAND's Chapter in this volume.
[6] W. S. JARDETZKY: Trans. Amer. Geophys. Un. **31**, 901 (1950).

he classified as (i) Precambrian, (ii) coastal mountain ranges, (iii) intercontinental ranges, and (iv) Himalaya type. The different types of folds are obtained by assuming various thicknesses of the buckling layer, and various modes of application of the compressive force. The mathematics of the theory is quite lengthy and will therefore be described here only very qualitatively. As outlined above, it is in every case an application of the standard theory of elasticity.

Thus, the "*Precambrian type*" of folding is obtained by assuming that a thin elastic layer is underlain by a plastic one which subjects the upper layer to tangential forces. This yields a multitude of small undulations which, allegedly, represent Precambrian mountain ranges. The "*coastal mountains*" are obtained by considering a strip which is dragged against another body. To simplify the problem, it is assumed that the deformation of the strip corresponds to plane strain. The "*intercontinental ranges*" are obtained by assuming that a strip of matter (representing a geosyncline) is compressed by two shields or continents. The "*Himalaya type*" of folding, finally, is obtained by assuming that an elastic rectangular plate is fixed on two adjacent sides and that a force is applied diagonally to the free corner. This, according to JARDETZKY, produces a curved bulging in the plate.

The calculations of JARDETZKY, valuable as they are, seem to be open to the general objection that it is by no means certain that the assumption of an elastic deformation leads to results which "will not differ much from the true picture". There is certainly no indication that elasticity is a predominant feature as far as the structure of the deformations of the Earth's crust is concerned. The theory of buckling may correctly give the location of the occurrence of folding, but as to the internal structure of the folded ranges, it must be noted that from the general looks of the strata it is rather to be expected that some such phenomenon as plastic flow is the basic agent.

In this instance, DE SITTER[1] has tried to give a semi-analytical solution without specifying the mechanism of deformation as being elastic. He advanced the "principle of concentric folding" which is expressed by the fact that the surface of a folded layer is formed by three circles (see Fig. 6). According to DE SITTER this is the result of two fundamental laws: (i) the law of conservation of volume and (ii) the law that each infinitesimal layer undergoes only bending during the folding. As is seen, these laws do not allow for the possible change of area of a stratum.

DE SITTER, thus, arrived at the picture of folding illustrated by Fig. 6; in this Figure, unprimed letters $A, B, \ldots$ refer to the situation before folding, whereas primed ones $A', B', \ldots$ refer to the situation after folding. The folding has been caused by the compression of the amounts $2s$ of the original strata. It is easy to see that above the line through $B$, the two laws stipulated by DE SITTER are indeed satisfied, below that line this is, however, not the case. Thus, the material from the first shaded area must have been transferred into that of the second one by plastic flow or some such phenomenon. Moreover, even above the line through $B$, DE SITTER is certainly satisfying his two assumptions, but it is quite obvious that this is not the *only* solution satisfying those principles. It may be observed that the solution of DE SITTER is not a real "explanation" of folds as the cause of the latter is not reduced to a field of forces, nor is any attempt made at a rationalization what forces could produce the particular type of bending assumed by DE SITTER. That the two assumed laws are satisfied, is not sufficient to account for this, as they are only an expression of the conservation of area and matter.

---

[1] L. U. DE SITTER: Proc. Kon. Ned. Akad. Wetensch. **52**, No. 5 (1939).

The theory of DE SITTER has been modified by TIEDEMANN[1]. The latter author replaced the circles which make up the form of a concentric fold by sine-curves. As in DE SITTER's scheme, the two fundamental assumptions are adhered to. It is fairly easy to calculate the shapes of a series of sine-curves that make up the strata in a layer of the Earth, and one thus obtains a picture as shown in a layer of the Earth, and one thus obtains a picture as shown in Fig. 7. It will be observed that the lower boundary of possible folding is now not a surface below which one has to assume plastic deformation or such like, but rather a

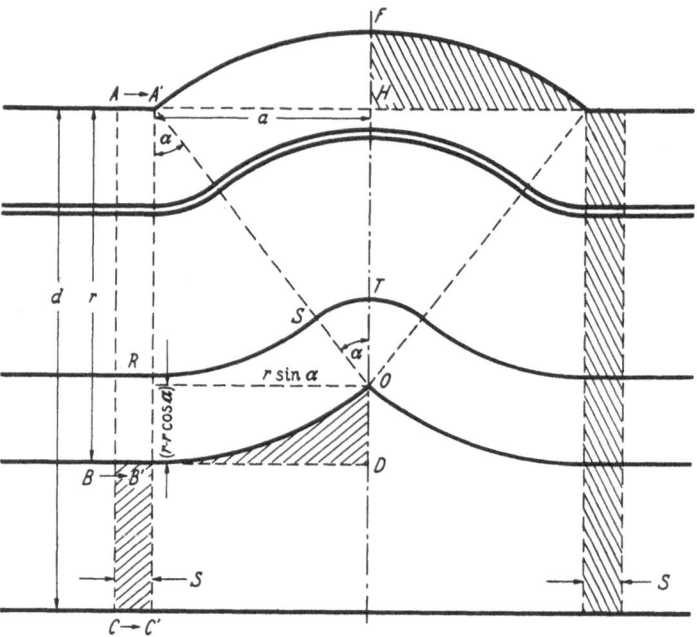

Fig. 6. DE SITTER's model of folding.

"shearing plane" above which a displacement takes place, and below which everything remains fixed.

Referring to Fig. 7, one has for a point on the curve:

$$y = n \sin x. \tag{4.1}$$

From the geometry apparent in Fig. 7, one can form the following equations which are based upon the fundamental assumptions:

$$\text{area } A'FH = \tfrac{1}{3} a h = s d; \tag{4.2}$$

$$\text{length } A' \text{ to } F = L = a + s. \tag{4.3}$$

However, the length $L$ of the sine curve can be calculated; one has:

$$d L = (1 + y'^2)^{\frac{1}{2}} d x = (1 + n^2)^{\frac{1}{2}} (1 - k^2 \sin^2 x)^{\frac{1}{2}} d x, \tag{4.4}$$

with

$$k^2 = n^2/(1 + n^2) \tag{4.5}$$

and hence:

$$L = 2 (1 + n^2)^{\frac{1}{2}} E(k), \tag{4.6}$$

where $E(k)$ is a standard elliptic integral:

$$E(k) = \int\limits_0^{\pi/2} (1 - k^2 \sin^2 x)^{\frac{1}{2}} d x. \tag{4.7}$$

---

[1] A. W. TIEDEMANN: Geologie en Mijnbouw 3, 199 (1941).

The integral $E(k)$ has been tabulated; using its values, one can calculate the values for $h$ under various assumptions for $s$ and $n$. It can be readily seen that the process of folding can be explained by a continuous movement, simply by adjusting the parameter $n$ for neighboring strata accordingly. One thus arrives at a *series* of folds as depicted in Fig. 7.

The theory of TIEDEMANN is open to the same criticisms as that of DE SITTER: The mechanism of folding is not uncovered as the sine-curve type of bending is nothing but an arbitrary shape satisfying two laws of continuity. Moreover, it is quite certain that the second law (which prohibits the areal extent of a stratum from being altered) is not fulfilled in nature due to the preponderance of plastic deformation, as has been mentioned in connection with the discussion of crustal shortening.

The analytical and semi-analytical attempts at solving the problem of folding seem thus to be open to the criticism of being based upon too restrictive assumptions. The problem is mainly one of practicability, as the calculations become unwieldy if such assumptions were made as would be more representative of the things in nature. However, let it be pointed out once more that the approach of JARDETZKY seems to give the correct location of the beginnings of the folds; the structure of such folds derived by him and by DE SITTER-TIEDEMANN, however, seems hardly creditable.

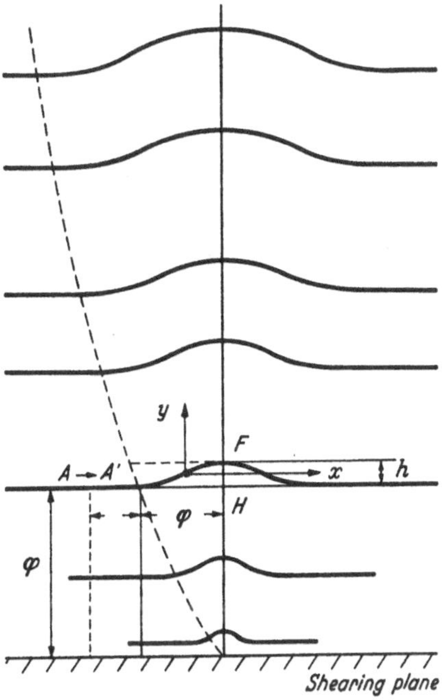

*c) Scale model experiments.* Therefore, the emphasis has been on *experimental* investigations, i.e. on investigations using *scale models*[1] to represent the actual phenomena.

Fig. 7. TIEDEMANN's model of folding.

The *theory* of scaling of phenomena concerned with the dynamics of the Earth's crust has been developed by HUBBERT[2]. It is based on the observation that the standard laws of mechanics must remain valid equally in the model experiment and in the process which the latter is supposed to represent.

One cannot give more than general principles by which the models have to be correctly scaled, the particular choice of scaling factors will be different from case to case. In general, one can choose 3 "fundamental" ratios independently, corresponding to the fact that there are 3 fundamental units in mechanics. If as *fundamental ratios* those of length $(L)$, mass $(M)$ and time $(T)$ are chosen, then all others are given by exactly such a combination of powers of $L, M, T$ as is indicated by their dimensional form in terms of the fundamental units "length", "mass" and "time".

Let us illustrate this on a specific example: Of particular importance for the understanding of the mechanics of folding is to know what the strength of the material of a small model of a mountain range would have to be. If this strength is known, one can proceed to investigate experimentally how a mountain range

---

[1] See e.g. H. CLOOS, Einführung in die Geologie, Gebr. Bornträger, Berlin 1936, and the papers by HUBBERT cited below.
[2] M. K. HUBBERT: Bull. Geol. Soc. Amer. **48**, 1459 (1937).

came into being. It is, then, also much easier to visualize what happened in the large scale phenomenon of which we do not have a good concept due to the unfamiliarity of the spans of time and space involved.

Thus, for illustrative purposes, let us imagine a large mass of granite. We wish to determine the properties of a dynamically similar model with length reduction $L = 5 \times 10^{-6}$ (i.e., 1 km. reduced to 5 mm.). The ratio of the gravitational force is given by reasons of practicality as equal to 1, as gravity is the same in the laboratory as in nature. Gravity being an acceleration, we have $LT^{-2} = 1$. Furthermore, any convenient model materials have a density which is not much less than that of granite, such that we may assume that the density ratio is $ML^{-3} = 0.5$.

By the assumption of these three ratios, all others are fixed. One immediately obtains: $M = 6.25 \times 10^{-17}$, $T = 2.24 \times 10^{-3}$. Most instructive is to find the *strength*

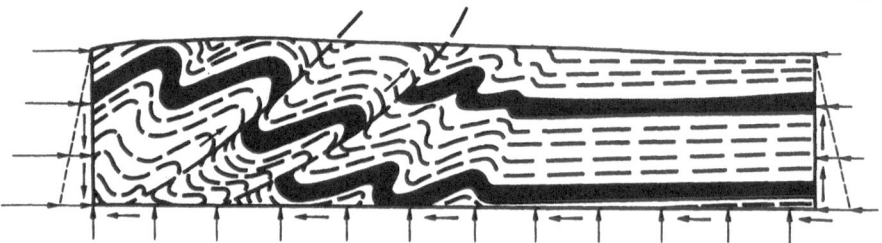

Fig. 8. Drawing of model experiment showing faulting and folding, after HUBBERT.

of the model-material. The strength is expressed in terms of stress (in the simplest theory as the limiting shearing stress) such that the strength ratio is equal to $ML^{-1}T^{-2}$. With the above values for $M$ etc. one finds a strength ratio of $2.5 \times 10^{-6}$. If the strength of granite is assumed as equal to $2 \times 10^{9}$ dynes/cm.², one finds that the model material must be a substance that starts being deformed under a stress of $5 \times 10^{3}$ dynes/cm.². This corresponds to a material so weak that a cube of it (density 1.5) larger than 3.3 cm. to the side could not support its own weight. Butter would be much too strong.

The foregoing remarks make it possible to visualize the process of folding with some more ease. At least, it is now no longer strange that violent distortion of the areal extent of the strata should have occurred, as plasticity and creep must be a major agent in so soft a material. Thus the discrepancy in crustal shortening as calculated from isostatic and geological reasoning can be resolved. Also, severe doubts are cast upon attempts of treating the structure of a fold by assuming that the fundamental mechanism be that of the theory of elasticity. The analogy with buckling plates, at best, can serve only to determine the location of the very first slight deformation.

The theory of scaling has not only been used for purposes of intuitive visualization of the processes involved, but also to perform actual experiments. HUBBERT[1] has given a review of such experiments where mainly soft clay is employed in the model. Later on, NETTLETON and ELKINS[2] and again HUBBERT[3] used granular materials such as sand. A drawing of the result of such an experiment, showing faulting as well as folding, is reproduced in Fig. 8 (after HUBBERT). In all these experiments it seems that structures much akin to folded strata were obtained.

[1] M. K. HUBBERT: Bull. Amer. Assoc. Petrol. Geol. 29, 1630 (1945). See also H. CLOOS l. c.
[2] L. L. NETTLETON and T. A. ELKINS: Trans. Amer. Geophys. Un. 28, 451 (1947).
[3] M. K. HUBBERT: Bull. Geol. Soc. Amer. 62, 355 (1951).

The employment of models to "explain" folding, of course, does not yield an actual "explanation", inasmuch as the mechanism of producing the folds is as little understood in the model as it is in the Earth. Nevertheless, it gives a means for duplication of large-scale phenomena on a small scale and experimentation in the laboratory which enable one, at least, to verify that the structures on the Earth are the very plausible outcome of very plausible processes, and not the result of some fantastic "catastrophies". In view of the difficulties of the mathematical theory of plasticity, experimental investigations of the problem of folding are probably the only means of a sensible attack for a long time to come yet.

**5. Physics of orogenesis.** *a) Introduction.* The analysis of the forces within the Earth's crust has, so far, been concerned with small-scale phenomena. Phenomena such as "faults" and "folds" have been "explained". When talking about such "explanations", it should be kept in mind that actually the mentioned small-scale phenomena have only been shown to be the plausible outcome of the action of hypothetical large-scale forces in the Earth's crust. The fundamental problem will therefore be to identify the process producing those large-scale forces. At the same time, this large-scale process will be expected to account for large-scale features on the Earth's surface such as were discussed in Sect. 1 of this Chapter. As outlined earlier, this hypothetical process is called "*orogenesis*". It has been shown that it is reasonable to assume that the process of orogenesis takes place in a rather thin layer which may be termed the "*orogenetic shell*" of the Earth.

Any theory of orogenesis, to be physically valid, must be based upon the mechanics of continuous matter of which the orogenetic shell is supposedly composed. Unfortunately, only very little is known about the mechanical behavior of those layers of the Earth's crust which might play the rôle of the orogenetic shell. Only for the outermost few kilometers do we have more or less definite data, and even there we do not know how the material behaves during prolonged time intervals. The theory of mechanical scaling discussed above, showed that for visualizing the behavior of a material even as strong as granite in large enough masses, one should properly think in terms of the mechanics of clay or butter of more customary dimensions. Thus, the orogenetic shell might have any mechanical properties ranging from those of a brittle shell to those of a layer of an ideal liquid.

In this instance, we shall recall briefly the principal ideal rheological bodies which an actual material such as that of which the orogenetic shell might be composed, may resemble. They are:
(i) the HOOKE solid or the ideally elastic body;
(ii) the NEWTONian or ideally viscous liquid;
(iii) the ST. VENANT or ideally plastic body;
(iv) composite bodies: the three fundamental bodies may be combined into a number of composite idealized bodies such as the KELVIN body, BINGHAM solid, MAXWELL liquid, etc. The particular characteristics of each of those bodies may be found in the chapter on rheology in Vol. 10.

The two fundamental ways in which an internal dislocation of the abovementioned rheological bodies (and therefore of the orogenetic shell) can take place, may be termed as "*failure*" and "*flow*". "*Failure*" is a localized, sudden internal movement releasing a built-up stress whereby neighboring mass elements may (but need not) be separated by a finite distance. "Flow" is a relatively slow, continuous movement. Accordingly, one will have to distinguish between two large classes of theories of orogenesis, according to whether they assume a "failure" or a "flow" phenomenon as basic occurrence in the orogenetic shell.

*b) Theories based on failure.* (i) *Physical background.* Turning first to the theories of orogenesis based upon failure phenomena, we may note that physically, they must be based upon the possible failure patterns of spherical shells. Such failure may occur by either buckling, or yielding, or fracture.

*Buckling* is specific to a purely elastic body. It has been shown[1,2] that the deformation of a buckling sphere is symmetrical about a diameter and that the deviations of the shape are given by a series of spherical harmonics along parallels of latitude associated with the diameter of symmetry.

*Yielding* is the type of failure which is specific to plastic bodies and pseudo-viscous (MAXWELLian) liquids. One of its characteristics is the occurrence of two families of *slip lines*[3]. Considering first bodies of the ST. VENANT type, SCHEIDEGGER and WILSON[4] have indicated how to obtain a qualitative idea of the plastic flow of a spherical shell. The general case is very difficult to treat, but the combination of results of NADAI[5] and HENCKY[6] suggests[4] that in the case of a spherical shell under tension (or compression) with a "weak region", the slip of the material occurs along surfaces whose traces on the shell are two families of logarithmic spirals with dip of about 45°. — Yielding may also occur in MAXWELL liquids. The loading conditions to produce yielding are somewhat different in this case, but the general characteristics are the same as in the ST. VENANT body.

*Fracture*, finally, is that type of failure commonly associated with the notion of a "solid" body. It is governed by the already mentioned theory of MOHR. According to that theory, the fractures expectable in a symmetrical stress state of a spherical shell are described in Table 1. A symmetrical stress state (with respect to one point) is the result of a "weak point" in an otherwise homogeneously stressed shell, — it is thus a stress state which could reasonably be expected to occur in the orogenetic shell.

Table 1.
*Fractures expectable in a symmetric stress state of a spherical shell.*

| Greatest Pressure | Intermediate Stress | | |
|---|---|---|---|
| | Vertical | Radial | Tangential |
| Vertical | | straight dip < 45° | circular dip < 45° |
| Radial | spiral dip vertical | | circular dip > 45° |
| Tangential | spiral dip vertical | straight dip > 45° | |

(ii) *Geological implications.* The geological implications of the various modes of failure lead to the so-called CONTRACTION HYPOTHESIS[7]. This hypothesis assumes that the Earth has been a hot, liquid celestial body at one stage of its life. It solidified quickly and continued to cool by radiation. JEFFREYS' theory[7] is that the silicate mantle solidified from its base at the top of the liquid iron core

[1] R. ZOELLY: Über ein Knickungsproblem an der Kugelschale, E. T. H., Zürich, Dissertation, 1915.

[2] W. LEUTERT: Die erste und zweite Randwertaufgabe der linearen Elastizitätstheorie für die Kugelschale, E. T. H. Zürich, Dissertation, 1948.

[3] See Chapter by the author in Vol. IX of this Encyclopedia.

[4] A. E. SCHEIDEGGER and J. T. WILSON: Proc. Geol. Assoc. Canada 3, 167 (1950).

[5] A. NADAI: Z. Physik 30, 106 (1924).

[6] H. HENCKY: Z. angew. Math. Mechan. 3, 241 (1923).

[7] Cf. e. g. JEFFREYS: loc. cit. p. 265[2], p. 303.

outwards and has since been cooling by conduction without convection currents. From the center of the Earth to within about 700 km. of the surface, there has not been time since the Earth solidified for any appreciable cooling nor for any significant change in volume to have taken place. Within the region from about 700 to 100 km., cooling by conduction is taking place and hence this shell is contracting and being stretched about the unchanging interior. Hence it is in a state of internal tension. This layer is the orogenetic shell (see Fig. 9). Near the surface the rocks have already largely cooled so that at the surface they are in thermal equilibrium with the heat of solar radiation. They are therefore not changing in volume and the cooling and contraction of the layer below puts the outermost shell into a state of internal compression. In between the shell under tension and that under compression there is a level of no strain at about 100 km. depth. The outermost layer is thus being folded up upon the shrinking interior.

According to the thus outlined contraction hypothesis, the primary orogenetic process is taking place in the cooling shell which is under tension. The actual folding of mountains, then, is only a secondary consequence of this orogenetic process.

Owing to the different types of failure phenomena that are possible in continuous matter, one has to distinguish between theories that assume either buckling, or yielding, or else fracture as basic orogenetic phenomenon.

*Buckling* might occur in the outermost shell which is collapsing

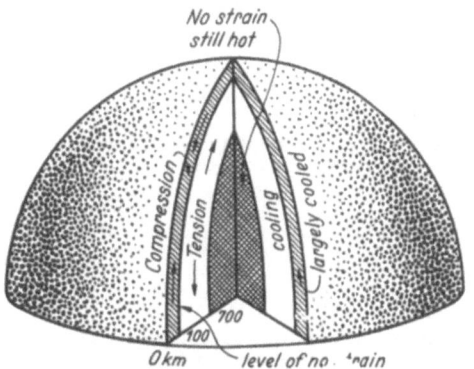

Fig. 9. Stresses in the Earth according to the contraction hypothesis.

upon the shrinking interior. However, the considerations of dyamical similarity outlined above make it extremely doubtful that buckling can be a major agent in determining the structure of the Earth's crust. It is most questionable whether there are any materials so strong as not to yield before any appreciable buckling deformation could take place. Nevertheless, the location (but not the structure) of any yielding may be determined by an almost infinitesimal buckling. However, no pattern resembling even only remotedly the buckling pattern of a sphere seems to be realized upon the Earth as the mountains are anywhere but on parallels of latitude associated with a single diameter.

Turning now to theories assuming *yielding*, one may note that the arcuate surface structure of the orogenetic belts can indeed be the result of plastic yielding in a simple stress state[1]. According to earlier remarks, the slip lines are logarithmic spirals in a symmetric stress state around a point of weakness. Underneath each of those lines would be a surface dipping down into the Earth at an angle. If the margins of continents are assumed to be weak zones, then island arcs might correspond to slip lines. They might be expected to be folded up because the "skin" of the Earth is becoming too large for the collapsing interior[2].

From a theoretical standpoint, there are objections to this theory. Slip lines form a double family of curves crossing each other, and there is no evidence that island arcs are crossing each other. Thus, theories postulating *fracture* might be more helpful. MOHR's theory gives the result that in a symmetrical

---

[1] Cf. SCHEIDEGGER and WILSON: loc. cit. p. 274[4].

[2] Cf. also P. P. BIJLAARD: Trans. Amer. Geophys. Un. **32**, 518 (1951).

stress state the fracture pattern may be as shown in Table 1. The case of the greatest pressure being vertical and the intermediate stress tangential might explain circular island arcs and mountain ranges, for it leads to conical fracture with a dip of about 45°.

This theory leads to a reasonable explanation of the principal features of the global distribution of mountain systems, at least if newer evidence from seismology (see below) is neglected. It also seems to explain many details of the orogenetic features quite satisfactorily. Thus, COULOMB[1] has shown that the occurrence of a constriction furrow in the orogenetic layer might explain the existence of a geosyncline below a mountain system and thus yield a simultaneous explanation of arcs and gravity anomalies. SCHEIDEGGER[2] has shown that a thin surface layer of a sphere will fold in a smaller pattern than a thick one—which fits the geological observations in connection with early (Precambrian) orogenesis.

In detail, the calculation to prove this, runs as follows: Let the volume of the surface layer be $V$ and the inner radius of the layer be $a$ and its thickness $b$. If the surface layer retains its volume during any short period of contraction brought about by cooling in the larger and hotter interior, one has:

$$V = 4\pi a^2 b. \tag{5.1}$$

If the radius of the interior sphere is changed by the amount $da$ and the thickness of the surface layer by $db$, then one obtains:

$$dV = 4\pi a^2\, db + 8\pi a b\, da. \tag{5.2}$$

However, $dV$ must vanish, as stated above; hence:

$$2\, da\, b + a\, db = 0, \tag{5.3}$$

or

$$db = -2b\, da/a. \tag{5.4}$$

As mentioned above, this equation indeed indicates that, for a shrinking of the interior by the amount $da$, the amount of material of the surface layer that has to be moved in order to make it continue to fit the interior, is proportional to its thickness. If one assumes that the shrinking of the interior occurs at a constant rate $\dot{a}$ in time, then the last equation becomes:

$$db/dt = -(2b/a)\,\dot{a}, \tag{5.5}$$

which indicates that the material in the surface layer that has to be moved around per unit time is just proportional to the thickness of that layer.

Furthermore, SCHEIDEGGER[3] has also shown that the four types of junctions of arcs (see Fig. 5) observed by WILSON[4] are the direct outcome of conditions of geometrical continuity that must be maintained during the development of a failure pattern. WILSON[5] has extended the failure theory to obtain a coherent theory of growth of terrestrial features. He shows that the geology, the volcanism, and the whole pattern of existing active mountain ranges can be explained in terms of this theory.

The criticisms against the failure theory come from two sides. Firstly, many fault plane determinations of earthquakes near island arcs seem to show that the relative motion therein is essentially transcurrent[6]. This does not agree with the concept of the contraction theory which requires that such faulting be essentially normal (below the level of no strain) or reversed (above the level of no strain). Secondly, the assumption of the Earth being in a state of cooling, is not definitely established. SLICHTER[7] investigated this question quite thoroughly

[1] J. COULOMB: Ann. Géophys. 1, 244 (1945).
[2] A. E. SCHEIDEGGER: Canad. J. Phys. 30, 14 (1952).
[3] A. E. SCHEIDEGGER: Canad. J. Phys. 31, 1148 (1953).
[4] J. T. WILSON: Proc. Geol. Assoc. Canada 3, 141 (1950).
[5] J. T. WILSON: Papers a. Proc. Roy. Soc. Tasmania 1950, 85.
[6] J. H. HODGSON, R. S. STOREY and P. C. BREMNER: Bull. Geol. Soc. Amer. 63, 1354 (1952).
[7] L. B. SLICHTER: Bull. Geol. Soc. Amer. 52, 561 (1941).

(except for the time variation of radioactivity) with very uncertain results. The heat flow through the accessible surface of the continents appears as higher than could be maintained by thermal conduction, but that could be due to an accumulation of radioactive materials near the surface of continents. No reliable data about heat flow through the ocean bottoms are available such that actually nothing certain is known about the true heat flow from the interior of the Earth to the surface.

Nevertheless, it is possible that the orogenetic shell is not at all in a state of cooling, but maybe in a state of heating up. In that case, the contraction theory would fall with the possibility of cooling. In order for an orogenetic phenomenon to take place, some hydrodynamic phenomenon due to an expansion must then be postulated as a fracture phenomenon due to an expansion could never produce any such crustal shortening as seems to be required for folding.

c) *Theories based on flow.*
(i) *Physical background.* Thus, turning now to the theories based upon flow phenomena, we may note that physically, they must be based upon the theory of flow of matter and heat. We separate the types of flow that might be of importance with respect to the Earth, according to the rheological behavior of certain idealized materials. One might start with the consideration of flow in PERFECT, INVISCOUS LIQUIDS.

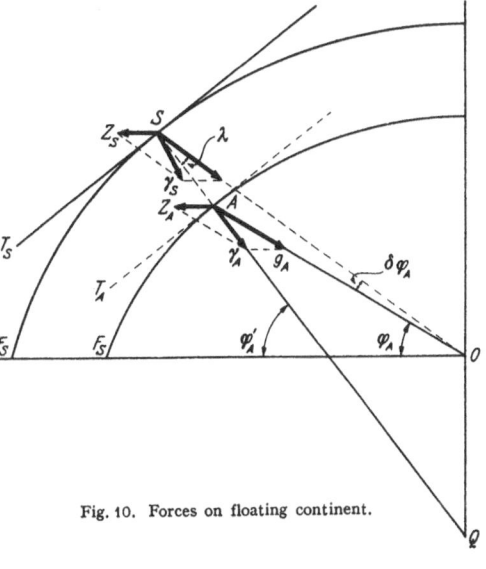

Fig. 10. Forces on floating continent.

Of special interest is the case where a liquid layer upon a sphere, in the gravitational field of the latter, is *oscillating*. VAN BEMMELEN and BERLAGE[1] have investigated this problem mathematically and found that such oscillations are indeed possible for certain frequencies.

Of similar interest is the case of a rotating *liquid sphere with musses floating upon it.* The liquid sphere is held together by its own gravitational forces. EÖTVÖS[2] and ERTEL[3] calculated that such floating masses will tend to drift towards the equator of gyration, which could produce a mechanism for orogenesis.

Thus, consider a "continent" floating upon a liquid sphere. Denote the center of the buoyant force by $A$, the center of gravity by $S$ (see Fig. 10). Furthermore, let $A$ and $g$ denote the vectors of the buoyant force and of the gravitational force, respectively, and $F_A$ and $F_S$ the equipotential surfaces passing through $A$ and $S$, respectively. The latter will diverge toward the equator. The line through $A$ and $S$ is assumed to be normal to $F_A$. Hence, the center of the buoyant force may be shifted into $S$ according to a well-known theorem in mechanics. In $S$, the buoyant force $A$ may be resolved into its components normal $(A')$ and tangential $(P_1)$ to $F_S$. The condition of floating for the continent then yields:

$$A' + g = 0 \tag{5.6}$$

whence it follows that $P_1$ is the force effecting the drift of the continent towards the equator. This, however, presumes that $S$ is moving in parallel to $F_S$. If the latter is not the case and

[1] R. W. VAN BEMMELEN and H. P. BERLAGE: Gerlands Beitr. **43**, 19 (1935).

[2] R. v. EÖTVÖS: C. R. 16me Conf. Assoc. Géod. Int. **1**, 38 (1910).

[3] H. ERTEL: Gerlands Beitr. **32**, 38 (1931); **43**, 327 (1935).

$S$ is assumed to move in parallel to $F_A$, then it is possible to resolve $\mathbf{g}$ into components normal ($g'$) and tangential ($P_2$) to $F_A$. The condition of floating yields in this case:

$$A + g' = 0, \tag{5.7}$$

and thus $P_2$ appears as the force effecting the drift towards the equator. In reality, $S$ will move neither in parallel to $F_A$ nor in parallel to $F_S$ but somehow in between. However, the resulting tangential forces in the two limiting cases will differ only in higher order terms than will be considered here; up to the significant order, the true drifting force may be identified with either $P_1$ or $P_2$. It is easier to calculate $P_2$ which may be done as follows:

Referring to Fig. 10, denote the *geographical* latitude of $A$ by $\varphi'_A$, the *geocentric* latitude by $\varphi_A$. Furthermore, $\mathbf{g}_A$ and $\mathbf{Z}_A$ are the accelerations due to gravity and the centrifugal force in $A$ which, combined, yield the total acceleration $\gamma_A$. If one sets $QA = a$, one has:

$$Z_A = \omega^2 a \cos \varphi'_A, \tag{5.8}$$

where $\omega$ is the angular velocity of the Earth.

The total acceleration $\gamma_A$ in $A$ has no component parallel to the tangential plane $T_A$, since $\gamma_A$ is normal to the equipotential surface $F_A$. This is expressed by the equation:

$$Z_A \sin \varphi'_A - g_A (\varphi'_A - \varphi_A) = 0. \tag{5.9}$$

In $S$, one has the following value for the tangential (to $T_A$) component $b_\varphi$ of $\gamma_S$:

$$b_\varphi = Z_S \sin \varphi'_A - g_S (\varphi'_A - (\varphi_A + \delta \varphi_A)); \tag{5.10}$$

for, when passing from $A$ into $S$ along the normal $n$ of $F_A$, $Z_A$ changes into $Z_S = Z_A + \delta Z_A$, $g_A$ into $g_S = g_A + \delta g_A$, $\varphi_A$ into $\varphi_S = \varphi_A + \delta \varphi_A$, whereas $\varphi'_A$ remains unchanged. Thus, one can rewrite equation (5.10) as follows:

$$b_\varphi = \delta Z_A \sin \varphi'_A + g_A \delta \varphi_A - \delta g_A (\varphi'_A - \varphi_A). \tag{5.11}$$

If one sets

$$\text{arc } QSO = \lambda, \qquad OA = r, \qquad AS = \delta n, \tag{5.12}$$

one has in view of $\lambda = \varphi'_A - (\varphi_A + \delta \varphi_A)$:

$$\delta n (\varphi'_A - \varphi_A - \delta \varphi_A) = r \delta \varphi_A; \tag{5.13}$$

or

$$(r + \delta n) \delta \varphi_A = \delta n (\varphi'_A - \varphi_A). \tag{5.14}$$

As $\delta n$ is much smaller than $r$, this is sufficiently approximated by:

$$\delta \varphi_A = \delta n (\varphi'_A - \varphi_A)/r. \tag{5.15}$$

Substituting (5.15) into (5.11), one has:

$$b_\varphi = \delta Z_A \sin \varphi'_A + (g_A/r - \partial g_A/\partial n) \delta n (Z_A/g_A) \sin \varphi'_A. \tag{5.16}$$

The expression $(g_A/r - \partial g_A/\partial n)$ can be rewritten by means of the gravitational potential $\Phi$. Firstly, one can set:

$$g_A = - \partial \Phi/\partial n, \tag{5.17}$$

and secondly use the fact that

$$\Delta \Phi = - 4 \pi \varkappa \varrho, \tag{5.18}$$

where $\varkappa$ is the gravitational constant and $\varrho$ the density. In polar coordinates, and with sufficient accuracy for the present calculation, one can rewrite this as follows:

$$\frac{\partial^2 \Phi}{\partial n^2} + \frac{\partial \Phi}{\partial n} \frac{2}{r} = - 4 \pi \varkappa \varrho, \tag{5.19}$$

which yields for $g_A$:

$$- \partial g_A/\partial n = 2 g_A/r - 4 \pi \varkappa \varrho. \tag{5.20}$$

Therefore:

$$g_A/r - \partial g_A/\partial n = 3 g_A/r - 4 \pi \varkappa \varrho; \tag{5.21}$$

and hence

$$b_\varphi = \delta Z_A \sin \varphi'_A + \left( \frac{3 g_A}{r} - 4 \pi \varkappa \varrho \right) \delta n \frac{Z_A}{g_A} \sin \varphi'_A. \tag{5.22}$$

From (5.8) one has

$$\delta Z_A = \omega^2 \cos \varphi_A' \, \delta n ,  \qquad (5.23)$$

and also:

$$Z_A = \omega^2 r \cos \varphi_A .  \qquad (5.24)$$

If all this be substituted into (5.22), one obtains:

$$b_\varphi = (\omega^2/2) \, \delta n \sin (2\varphi_A') + (3/2) \, (1 - 4\pi\varkappa\varrho \, r/(3 g_A)) \, \omega^2 \, \delta n \, 2 \cos \varphi_A \sin \varphi_A' .  \qquad (5.25)$$

Here, the small difference between $\varphi_A$ and $\varphi_A'$ may be neglected. Thus:

$$2 \cos \varphi_A \sin \varphi_A' = \sin (2\varphi_A') ,  \qquad (5.26)$$

whence:

$$b_\varphi = 2 \omega^2 \sin (2\varphi_A') \, \delta n \, (1 - \pi\varkappa\varrho \, r/g_A) .  \qquad (5.27)$$

If a mean density $\bar{\varrho}_m$ is defined by:

$$g_A = \tfrac{4}{3} \pi\varkappa\bar{\varrho}_m r ,  \qquad (5.28)$$

then it is seen that this cannot be very much different from the mean density $\varrho_m$ of the Earth which, therefore, may be used en lieu of $\bar{\varrho}_m$. Hence:

$$b_\varphi = 2 \omega^2 \, \delta n \sin (2\varphi_A') \left( 1 - \frac{3}{4} \frac{\varrho}{\varrho_m} \right).  \qquad (5.29)$$

Finally, if $M$ is the mass of the floating continent, one has for the *drifting force:*

$$K_\varphi = 2 M \omega^2 \, \delta n \sin (2\varphi_A') \left( 1 - \frac{3}{4} \frac{\varrho}{\varrho_m} \right).  \qquad (5.30)$$

The above deduction of the drifting force,—which is that proposed by ERTEL[1]— has been questioned by PREY[2], principally because the physical model under- lying ERTEL's calculations is somewhat of an oversimplification. It is not certain that the pressure of the substratum on the continent has its center on $QS$, shifted by the (unknown!) quantity $\delta n$. In particular, if certain types of increases of density with depth in the substratum are assumed, one does not obtain a drifting force toward the equator at all. The actual existence of the drifting force, there- fore, depends on how accurate the model underlying ERTEL's calculations ac- tually is[3].

Turning now to the (thermo-) hydrodynamics of VISCOUS FLUIDS, the problem of free thermal *convection currents* in such fluids is of importance. In this instance, JEFFREYS[4] has deduced that no stable thermal convection currents can occur unless a certain dimensionless parameter (the product of GRASHOFF and PRANDTL numbers) has a certain minimum value, correspond-- ing to a minimum temperature gradient. The convective motion between two shells (one supplying heat and the other abducting it) occurs then for a range of the temperature gradient, in a cellular pattern, each cell containing a vortex tube which is closed within the cell. This is in agreement with certain kinematical properties of vorticity according to which a vortex tube has to be closed. JEFFREYS' results have been extended and tested experimentally by others and appear to be very well established. For very high temperature gra- dients, the cellular patterns disappear and give way to irregular eddies.

The thermohydrodynamics of PLASTIC SUBSTANCES is also of importance in connection with orogenesis. Unfortunately, not much is known about heat

---

[1] J. T. WILSON: Proc. Geol. Assoc. Canada 3, 141 (1951).

[2] A. PREY: Gerlands Beitr. 48, 349 (1936).

[3] See also the discussion on *polar wandering* at the end of RUNCORN's chapter on "Magne- tism of the Rocks" in this Volume.

[4] H. JEFFREYS: Phil. Mag. [7] 2, 833 (1926). — Proc. Roy. Soc. Lond., Ser. A 118, 195 (1928).

transfer and the possibility of *convection currents* in such materials. In particular, no analytical solutions seem to be available. Thus, in order to obtain some idea about the thermomechanics of plastic flow, more qualitative arguments have to be used. From analogy with viscous flow[1] and from everybody's experience with the breakfast-porridge pot, it may be inferred that a high temperature gradient must be present for convection to set in. Under ordinary circumstances, the porridge simply chars at the bottom without getting into convective motion. This predicament seems to be the more pronounced, the "stiffer" the porridge is, i.e. the higher its yield strength. It is even doubtful whether thermal convection is at all possible in a medium where most of the apparent "viscosity" is due to its yield strength, such as toothpaste or an assembly of sand grains. It will be remembered that the latter has been successfully used to represent folding in small scale experiments. If convection currents are at all possible in such media, the temperature gradient must certainly be very much larger than in a corresponding viscous one.

(ii) *Geological applications.* We turn now to the geological applications of the mechanics of fluids.

Starting with the *oscillatory behavior* of a perfect liquid in its own gravitational field, we may note that such a phenomenon has been made responsible by HAARMANN[2] for orogenesis. In the rhythmic occurrence of cosmic influences, HAARMANN's "UNDATION THEORY" assumes that the figure of a geoid is disturbed and has to be restored to equilibrium by the displacement of large masses, especially of subcrustal magma. The displacement of these masses effects a vertical movement of the crust which shows up as the formation of mountains and geosynclines. Sliding of material from the momentary peaks of the oscillations provides for the contortion evident in the nappes.

The undation theory has generally not been taken very seriously. It is very questionable whether the obviously small cosmic perturbations could suffice to initiate large-scale mass displacements unless some consideration about resonance frequencies were made. Such investigations, however, do not seem to have been undertaken.

Proceeding with the next possibility, viz. that of *floating continents* drifting towards the equator under the EÖTVÖS force as especially advanced by WEGENER[3] we may note that it, too, is now almost abandoned. A most serious difficulty is that all DRIFT THEORIES can explain only *one* orogenetic diastrophism. If the continents were drifting towards the equator, they would proceed thither until they would collide, and that would be the end of it. The repeatedness of orogenetic diastrophisms remains thus totally unaccounted for.

If one wants to maintain that large *continental shifts* are taking place, with the understanding that "collisions" give rise to mountains, one will have to look for another formalism than that provided by the EÖTVÖS force. The search for such effects led HOLMES[4] to the postulate of subcrustal convection currents. However, the old objection that only single orogenetic diastrophisms could be explained in this manner, still holds. The idea of continental collisions was therefore abandoned and the claim made that CONVECTION CURRENTS

---

[1] Cf. J. G. OLDROYD: Proc. Cambridge Phil. Soc. **43**, 396, 521 (1947).

[2] E. HAARMANN: Die Oszillationstheorie, Stuttgart: Ferdinand Enke 1930.

[3] A. WEGENER: The origin of continents and oceans, transl. from 3d German ed. by J. Skerl. London: Methuen 1924.

[4] A. HOLMES: Mining Mag. **40**, 205, 286, 340 (1929).

alone would suffice to explain the orogenetic cycles,—as especially advanced by GRIGGS[1] and VENING MEINESZ[2].

GRIGGS assumed that convection takes place in what was formerly termed the mantle of the Earth due to a temperature gradient in the latter. To demonstrate the effect of these postulated currents in relation to orogenesis, a model experiment was performed. The vortex tubes of convection cells were represented by rotating drums. GRIGGS in his experiment took account of HUBBERT's theory of dynamical scaling, but only as far as mechanical properties are concerned. No attempt at scaling the thermal effects has been made. Furthermore, the plan and the three dimensional arrangements of convection currents in existing island and mountain arcs have been completely neglected. Also, the convection currents of GRIGGS are laminar and steady (as are the theoretical ones of JEFFREYS) and thus do not reflect the fact that orogenesis occurs in single diastrophisms and not steadily.

GRIGGS was himself aware of this last difficulty and therefore assumed a rather vague mechanism by postulating rather inadequately defined properties as to the rheology of the orogenetic shell, which are supposed to make the convection currents intermittent. It is to the credit of VENING MEINESZ to have stated in plain terms that he expects a *plastic substance* to give such INTERMITTENT CYCLES. His arguments, however, are largely phenomenological and not based upon sound rheological reasoning[3]. We have seen above that rocks are represented in small-scale models by plastic substances of very low strength: Sand has been successfully used to reproduce folding. Flour would be another possibility. It seems to the writer quite inconceivable that convection currents would start in a batch of flour when heated from below, and much less that they would be intermittent. However, the argument is somewhat incomplete as no proper attempt at scaling the thermal parameters appears ever to have been made.

The chief criticism of convection currents, however, comes from the recent observation by BIRCH[4] which was already mentioned in Sect. 2: Namely that it is very probable that either a chemical or at least a phase-discontinuity exists at about 900 km. depth. Such a discontinuity, if chemical, would make convection currents outright impossible, and if it were only a phase transition, would make such currents at least utterly doubtful. It does not seem possible that convection can occur in a medium which is other than homogeneous. Furthermore, convection currents are assumed to rotate in vertical planes and it is therefore difficult to see how they could give rise to horizontal differential displacements as encountered in fault plane studies of earthquakes.

Apart from those criticisms, convection currents would be able to provide for a mechanism producing such phenomena as geosynclines, mountains, etc. GRIGGS and VENING MEINESZ followed the consequences of such convection currents into great detail in the papers cited above. The question whether convection currents can provide a satisfactory mechanism of orogenesis is thus still a largely open one. More fundamental studies on the very subject of convection currents, especially in plastic bodies etc., will have to be undertaken before a final decision can be made.

[1] D. GRIGGS: Amer. J. Sci. **237**, 611 (1939). — Magmatic currents have also been discussed by G. KIRSCH, Geomechanik. Leipzig: Johann Ambrosius Barth 1938.

[2] F. A. VENING MEINESZ: Proc. Kon. Ned. Akad. Wetensch. **37**, 37 (1934); **50**, 237 (1947). Versl. Kon. Ned. Akad. Wetensch. **53**, No. 4 (1944). — Quart. J. Geol. Soc. Lond. **103**, 191 (1948).

[3] Cf. the discussion by A. E. SCHEIDEGGER: Bull. Geol. Soc. Amer. **64**, 127 (1953).

[4] F. BIRCH: Trans. Amer. Geophys. Un. **32**, 533 (1951).

*d) Conclusion.* The chief competitors as the true theory of orogenesis are thus the *contraction hypothesis* (with sliding fracture along arcs) and the hypothesis of *thermal convection currents*; — as all the other efforts mentioned above are pretty well ruled out for stated reasons.

Both the two principal theories seem to be able to account reasonably well for some of the observed phenomena and less adequately for others. Both theories assume essentially displacements which occur in a vertical direction. However, during the past 20 years, many studies of the displacements which are taking place during the occurrence of earthquakes in the foci have been published by various authors[1]. If the results of these studies are collected and their bearing upon the physics of orogenesis is analysed, it is seen that, in spite of the observed crustal shortening, most of the adjustments in earthquake-foci seem to be of a transcurrent nature. This is quite in contrast with what had been expected as it had always been thought that normal or reversed faulting would be the common case.

The occurrence of so much transcurrent faulting within the mantle and crust of the Earth presents a severe obstacle to accepting as correct either of the two theories of orogenesis advanced above as "likely". Both these theories are based upon the sole occurrence of normal or reversed faulting.

In a renewed inspection of the possible mechanics of failure in the Earth's mantle and crust, it appears that there is seemingly no previously considered material which could entertain displacements as they seem to be indicated from earthquake fault-plane studies. It might, therefore, be necessary to postulate a new type of model-material not heretofore considered, unless it either could be shown that a misinterpretation has occurred with fault plane studies of earthquakes, or else all connection between orogenesis and earthquakes were abandoned.

In any case, the ultimate understanding of orogenesis depends upon the proper understanding of the mechanics within the Earth's mantle and crust which obviously has not yet been achieved.

**6. The distribution of continents and oceans.** *a) The problem.* The final task is to give an explanation of the distribution of continents and oceans over the surface of the Earth consistent with the previously outlined theories of orogenesis. As pointed out in Sect. 1 of this Chapter, there are several morphological facts which have to be accounted for, namely (i) the appearance of two levels on the Earth, the continental 900 m. above and the oceanic 4000 m. below sea level; (ii) the ratio of the areas of "continental" and "oceanic" plateaus which is $1:2$; and (iii) the peculiar tetrahedral distribution of continents and oceans.

The theories of the origin of the continents reach all the way from those that assume a "catastrophism" to those that assume some hydrodynamic principle.

*b) Birth of the moon.* Of the theories assuming a catastrophy, those that ascribe the distribution of continents to the *birth of the moon* from the Earth merit being mentioned. The case for this theory has been stated recently by ESCHER[2]. This author supposes that the moon was separated from the Earth soon after solidification of its "top" layer. The position of the moon on the Earth would have been where there is now the Pacific Ocean. At the moment of the removal of the moon, there originated thus a wound in the Earth which

---

[1] Cf. e.g. H. HONDA, A. MASATUKA: Sci. Rep. Tôhôku Univ. (5), Geophys. 4, No. 1 (1952); J. H. HODGSON, R. S. STOREY: Bull. Seismol. Soc. Amer. 44, 57—83 (1954), where other references may be found. See also A. E. SCHEIDEGGER, Trans. Roy. Soc. Canad. 1955. Sect. 4 for a summary of published fault plane determinations of earthquakes.

[2] B. G. ESCHER: Bull. Geol. Soc. Amer. 60, 352 (1949).

caused a suction upon the remaining crust. The suction tore the crust to pieces which moved, in form of floes, towards the wound. This would explain eg. the fact that South America fits with its eastern shore into the western shore of Africa.

The physical aspects of such a theory should follow from a consideration of the mechanics of the supposed separation of the moon from the Earth. Such a mechanism could be provided only by a resonance of the oscillations of the Earth with the tidal forces exerted by the sun. JEFFREYS[1] discusses this possibility and shows that the argument although it might appear reasonable prima facie, is actually quite untenable for three reasons. Firstly, there would be a considerable discontinuity of velocity at the core boundary during the oscillations. This would lead to considerable loss of energy by friction which would restrict the amplitude of the oscillations to a maximum which is much less than the critical value for disruption. Secondly, nonlinear terms should not be neglected in the mathematical analysis of large oscillations of the Earth which would again restrict the maximally possible amplitude to a value much too small to cause the danger of disruption. This is much in contrast with what would happen if the oscillations were linear (harmonic). Thirdly, a celestial body ejected from another must of necessity return to the first one. Thus, any theory claiming that the Pacific is the remnant of a wound sustained by the Earth on the occasion of the separation of the moon, must fall with the possibility of such a separation.

A similar effect as by the theories postulating the birth of the moon from the Earth, is produced by a speculation of MATSCHINSKI's[2]. According to MATSCHINSKI, immediately after solidification of the top layer on an originally hot Earth, the interior must heat up disrupting the solid crust into a series of continents. This would again explain why some of the coasts opposite each other on an ocean seem to fit into each other. However, it is not clear why, then, the continents should be so few and far between.

*c) Tetrahedral shrinkage.* Another theory to account for the morphological facts about continents and oceans which, at the same time, aims at an explanation of the tetrahedral form of this distribution, is a theory assuming a particular type of *shrinkage*. It is based upon the assumption that the tendency exists for a contracting sphere to shrink tetrahedrally, simply because the tetrahedron has minimum volume for a given surface of all regular bodies. The case for this theory has recently been re-stated by WOOLNOUGH[3].

Thus, if it be assumed that the Earth had at one time cooled enough such that the outermost layer had become a solid skin incapable of changing its area, and the interior would have proceeded to cool and thereby to shrink, then a tetrahedral shape might be considered as the logical outcome of such a process. The corners of the tetrahedron would correspond to the continents, the faces to the ocean basins. A proper arrangement concerning the size of the tetrahedron would also explain the ratio 1:2 occupied by continents and by oceans.

The chief criticism of such a theory is that the "shell" of the Earth seems much too "soft" to retain its size upon a shrinking interior. The evidence of folding seems to show that adjustment of an outer shell to a collapsing interior takes place continually or in a rapid sequence of diastrophisms rather than in a slow settling to the form of a tetrahedron. Furthermore, the theories of deformation of such an outer shell outlined earlier seem to indicate that buckling

---

[1] loc. cit. p. 234.
[2] M. MATSCHINSKI: Cahiers Géol. de Thoiry **1951**, Nr. 7, 59. — C. R. Acad. Sci. Paris **230**, 1882 (1950).
[3] W. G. WOOLNOUGH: Bull. Amer. Assoc. Petrol. Geol. **30**, 1981 (1946).

would be the mechanism determining the adjustment of a *rigid* shell to a collapsing interior which would not be a tetrahedral adjustment as postulated above. At any rate, that an actual tetrahedron could form, seems out of the question as failure of the outer shell would take place long before such a shape could develop,— for the same reasons as buckling cannot determine the *shape* of folds.

*d) Octahedral convection.* To avoid such difficulties, VENING MEINESZ[1] proposed a hydrodynamic speculation, based upon the presence of convection

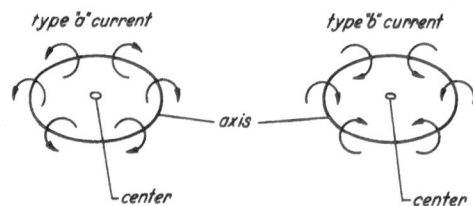

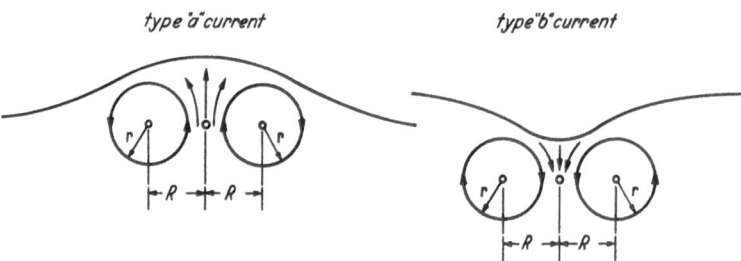

Fig. 11. Types of convection currents, after SCHEIDEGGER.

currents as also assumed in his mountain building theory. Thus, the tetrahedral distribution of the continents is postulated to be the outcome of an octahedral arrangement of convection currents at one time of the Earth's life. In a single convection cell, the motion of the fluid would be toroidal either outward or inwards ("type *a*" and "type *b*" currents, see Fig. 11). Currents in adjacent cells are of course of opposite type. It is now well possible that "type *a*" currents would give rise to continents, "type *b*" currents to oceans (Fig. 11). Thus, an octahedral arrangement of convection currents would give rise to a tetrahedral distribution of continents. Fig. 12 shows the system of currents as postulated by VENING MEINESZ: the axes of "type *a*" currents are represented by solid lines, of "type *b*" currents by broken lines. The creation of this system could be made plausible by the remark that a regular pattern is most likely to occur. The octahedron is the only regular surface in which an even number of sides touches in one corner, and this is a necessary condition for a convection current distribution.

It is rather difficult to rationalize the above intuitive arguments. PEKERIS[2] tried to calculate the system of convection currents to be expected in a cooling Earth. The result was that two continents should form at the poles with a ring-shaped ocean around the equator. Unfortunately, this bears not the least resem-

---

[1] F. A. VENING MEINESZ: Versl. Kon. Ned. Akad. Wetensch. **53**, No. 4 (1944).
[2] C. L. PEKERIS: Mon. Not. Roy. Astronom. Soc., Geophys. Suppl. **3**, 343 (1935).

blance with what is actually observed on the Earth. HILLS[1] tried to save the situation by assuming that the two polar continents formed at an early stage of the Earth's life were broken up and wandered towards the equator due to the Eötvös force. The complete asymmetry between the northern and the southern hemispheres, to say least, seems hardly understandable upon this basis.

Nevertheless, the octahedral convection current theory of VENING MEINESZ seems still attractive. A pertinent mathematical solution of sufficient complexity to bear any remote resemblance with anything observed on the Earth has obviously not yet been attained. Until proven otherwise, an octahedral distribution

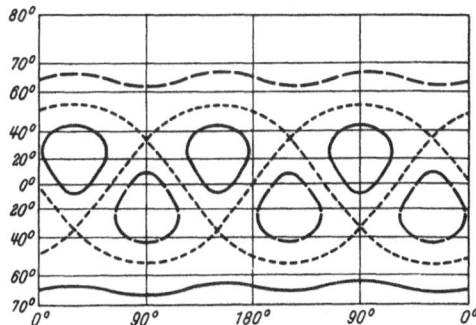

Fig. 12. Octahedral arrangement of convection currents after VENING MEINESZ.

of convection currents must therefore be regarded as a possibility in a cooling Earth. Furthermore, SCHEIDEGGER[2] has shown that there is no difficulty in assuming *slow* convection currents on a continental scale,—even in a medium of breaking strength and "viscosity" of limestone. It should be observed, however, that such convection currents would be very different from those postulated by GRIGGS and followers in their theory of *orogenesis*. Firstly, on a continental scale, there is no problem of intermittency,—the uplift of the continents may well have been due to steady flow of long duration. Secondly, the convection currents of such a continental scale may have been very much slower than those supposed causing orogenetic movements. With the heat available, it was calculated that 2 mm. per year would be quite sufficient which is about 1/10 of what is commonly thought necessary for orogenesis.

In detail, this can be shown as follows. One can calculate the rotating speed of a convection current as depicted in Fig. 11, which would be necessary to give the observed heat flow of about $6 \times 10^{-7}$ cal/cm.$^2$ sec.[3]. For, one may note that the heat $H$ transported by a revolving torus, corresponding to a convection current, is

$$H = 2\pi^2 r^2 R n \varDelta T/t.\tag{6.1}$$

The volume of the torus is $2\pi^2 r^2 R$, the temperature difference between the bottom and top level of the torus is $\varDelta T$, the specific heat per unit volume is $n$, and $t$ is the time of revolution. Thus, the heat flow $h$ per unit area is

$$h = H/\pi (R+r)^2 = 2\pi [r^2 R/(R+r)^2] n \varDelta T/t\tag{6.2}$$

and hence:

$$t = [2\pi r^2 R/(R+r)^2] n \varDelta T/h.\tag{6.3}$$

[1] G. F. S. HILLS: The Formation of the Continents by Convection. London: Arnold 1947.
[2] A. E. SCHEIDEGGER: Trans. Amer. Geophys. Un. **33**, 585 (1952).
[3] This value has been postulated by E. C. BULLARD [Mon. Not. Roy. Astronom. Soc., Geophys. Suppl. **6**, 36 (1950)] in order to maintain the Earth's magnetic field.

One must now make reasonable assumptions regarding the quantities occurring in the last equation. It will be noted that for an approximation only powers of ten will matter. Thus, one may assume that $r$ is of the order of 100 km., $R$ of 1000 km, $n$ of the order of unity and $\Delta T$ of the order of 1000° C. Thus one obtains that $t$ is of the order of $10^8$ years. The speed of the material where it rotates fastest is $2\pi r/t$ which yields approximately 2 mm. per year as stated above.

In view of the above results, Scheidegger[1] has therefore suggested that such continental currents might even subsist to the present day and be responsible for slow phenomena on a continental scale; cooling and fracture being responsible for the "rapid" phenomena of mountain building. To this, it should be remarked, however, that all the old shields show the structure of eroded mountains rather than of uplifted blocks of magma. It seems therefore not likely that the mechanism causing the original nucleation of continents would be persisting to-day.

*e) Growth of continents.* If it is accepted that the nucleation of continents took place by some mechanism which is now dead, then Wilson[2] provided an explanation for the growth of such continents by postulating the process of continuous orogenesis on the margin of such continents, the latter being a "weak" zone around which orogenesis by sliding fracture would be centered, the latter providing channels for additional magma to come forth and be added to the continental masses. The inner part of such continents would meanwhile be progressively eroded and thus present the picture of old shields consisting of mountain roots. The thus outlined theory has been followed up by Wilson into great detail and seems to provide for a reasonable explanation not only of continental growth, but also of many features about the global distribution of mountains.

*f) Polar wandering.* A further possible mechanism that has been considered responsible for the existence of continents is the effect of polar wandering. The case for polar wandering has been recently restated by Gold[3]. Accordingly, a movement of the entire Earth relative to the poles of rotation can be achieved by internal effects. If one has a *stationary* sphere, with exactly spherical symmetry, resembling the Earth in size and moments of inertia, a beetle weighing one gram could turn it over by walking $10^{27}$ times around it. On a *rotating* sphere, the beetle could cause any desired angle of movement of the poles relative to the sphere by merely sitting in the right place. It is thus seen that rather minor causes could be held responsible for a shift of the poles with respect to a perfectly spherical Earth.

In the case of the actual Earth, on a spheroid spinning around a principal axis of inertia of greater moment than the other two, no significant amount of polar wander could be expected if the Earth were perfectly *rigid*. However, since the Earth has a certain *plasticity*, polar wander may occur. Gold has given a detailed account of the various possibilities.

A shift of the poles, in itself, does not yet cause continents. However, Vening Meinesz[4] made calculations of shear patterns that would be induced by a suitable polar shift. He claimed that principal shear zones would coincide with actual margins of continents. One can therefore speculate that the shape of the continents was caused by a suitable polar wandering.

One thus has again, as in the last section, a series of alternative opinions as to the origin of the peculiar distribution of the continents. In contrast to the

[1] A. E. Scheidegger: Trans. Amer. Geophys. Un. **35**, 585 (1952).
[2] J. T. Wilson: Papers a. Proc. Roy. Soc. Tasmania **1950**, 85.
[3] T. Gold: Nature, Lond. **175**, 526 (1955); also **176**, 422 (1955).
[4] F. A. Vening Meinesz: Trans. Amer. Geophys. Un. **28**, 1 (1947).

theories of orogenesis, there is, however, only one theory that is at all widely accepted: that of the *formation of continents by convection*. It is the later development of those continents about which the opinions differ: Did they grow in the manner postulated by WILSON? Did they get broken up and drifted over the Earth's surface? The ultimate answer to none of these questions is known, and only a certain degree of probability can be assigned to each hypothesis;—which the writer tried to do without adding too much of his own bias.

**7. Conclusion.** In reviewing the physics of the forces in the Earth's crust one is faced with the peculiar condition that no coherent theory is known from which all the details would follow. In contrast to most physical theories where a fundamental equation is postulated whose consequences more or less agree with (and thereby "explain") the facts of nature which they concern, the subject of orogenesis is still much in the dark. The only attack to the problem is therefore by the inductive method. This procedure is made very difficult by the fact that it is near to impossible to express the results of geological exploration by a series of numbers which would be accessible to mathematical methods. Such an accessability to mathematical methods would considerably lighten the task of logical reasoning and making sensible deductions from the observed facts. The consequence is that the research on the physics of geological facts is literally haunted by wild conjectures falling little short of actual postulation of the supernatural. Hypotheses are lightly postulated without any regard to physical possibilities. Most of these hypotheses have been ignored in the present review as they would only add to the bulkiness of this volume. However, it is in most cases not easy to convince the public that such hypotheses are actually too far fetched to be regarded seriously.

Therefore, little has been said in this review of such attempts at "explaining" the forces in the Earth's crust by more or less obviously fantastic postulates. Nevertheless, it is hoped that the present compilation will serve as a suitable basis for further research into the dark subject with which it is concerned.

# Radioactivity and Age of Minerals.

By

J. T. WILSON, R. D. RUSSELL and R. M. FARQUHAR.

With 9 Figures.

## A. Introduction.

1. This chapter is chiefly devoted to methods of dating rocks and minerals. Although geologists have long known much about the *relative* order in which rocks were formed, *absolute* methods of geochronology have only been developed this century. These are all based upon the decay of naturally occuring radioactive isotopes, of which at least small quantities occur in practically all rocks and minerals. This has prompted an interest in measurements of the amounts of radioactivity present in rocks and minerals, to which subject part D of this chapter is devoted. This is also important in other branches of physics of the earth. For example, much of the earth's heat is generated by its radioactive elements.

The principles of radioactive decay are well known and they can be applied to many problems in geophysics by straight-forward experiments. Most radioactive age determinations depend, for example, upon the production at a known rate of a daughter isotope from its radioactive parent since the mineral was deposited in pure form. The application of this principle would seem to be a simple matter. It is true that some of the experimental techniques are delicate and require precise measurements of small quantities, but this is not uncommon in physics and few of the methods employed are novel. Some of the physical constants are not yet well established, but this merely requires more careful measurements.

Thus the study of the radioactivity of rocks and minerals appears in some respects to be simpler than many other branches of physics, but it is important that physicists be not misled by this apparent ease. There are less obvious features which are exceedingly difficult and which have caused progress to be disappointingly slow and uncertain. Although the first radioactive age determinations were made nearly fifty years ago, there is still no complete, absolute time scale for the history of the earth.

Before such a time scale can be achieved there will need to be a much closer union of geological and physical work than has yet been obtained. The contributions of these two branches of science and the bases for their different points of view can best be indicated in the historical review (part B). In essence the argument is this: As recently as the eighteenth century most men believed that the earth had been formed in a cataclysm only a few thousand years previously, although some had already begun to realise that the geological evidence pointed to a *slow* evolution of the earth's surface features. The approximate methods which were then devised for estimating the time required for such an evolution showed it to be at least hundreds of millions of years. This view was strongly contested by physicists who, until the discovery of radioactivity, could see no source of energy to keep the earth and sun hot for more than a few tens of millions of years. At the beginning of this century the geologists adopted a geological time scale which was a compromise between the two different views. For the younger and fossiliferous part of the geological record, upon which most of their work had been done, they retained a complex classification, while they fitted the older rocks, about which their knowledge was uncertain, into an abbreviated and simple scale to suit the physicists' view of a rapidly cooling earth.

Within the last fifty years a succession of radioactive methods of dating rocks have been developed and estimates of the age of the earth and its oldest rocks have been pushed progressively farther back. It is now apparent that the unfossiliferous or Precambrian rocks do not represent just a brief and simple introduction to the geological record, but rather they represent not less than five-sixths of the history of the earth. The time scale generally used for the Precambrian can now be seen to be telescoped and much in error, but there are difficult problems to be solved before a more correct time scale can be established.

These problems are now being faced but there are as yet too few good age determinations. Many of the earlier determinations were based on faulty assumptions or inadequate knowledge and have been shown to be wrong. It is often difficult to relate age determinations made on a single crystal to the age of a region. The old classification of the Precambrian, though invalid for the world as a whole, was based upon a sensible, if erroneous, interpretation and for many limited areas serves well so that there is a natural reluctance to abandon it, at least until a replacement has been agreed upon.

Thus it should be borne in mind while reading this chapter that the problem of providing the proper scale for geological time has only begun to be tackled. This subject, so deceptively simple in appearance, is in reality elusive and full of challenge for future work. The endeavour has been made to indicate in what direction that work should lie.

# B. Historical review.

**2.** Men have always wished to know the nature and history of the universe and of the earth on which they live. Information on these subjects has been sought by astronomers, physicists and geologists alike, but until the discovery of radioactivity by BECQUEREL in 1896, no proper understanding of the physical aspects of these matters was possible. This did not handicap geologists who by looking at the rocks during the eighteenth and nineteenth centuries compiled an account of the later part of the history of the face of the earth which is still largely valid today. F. D. ADAMS [1] has given a good account of discoveries by which they gradually established the relative time scale of successive periods represented by sedimentary rocks. Only the main trends will be mentioned here.

**3.** A century ago most people literally believed the biblical accounts of the creation and of the flood which carried the implication that the universe and all things in it had been formed in awful cataclysms only a few thousand years ago. From a study of the Bible, Archbishop UssHER fixed 4004 B.C. as the year of creation. This date was long defended with religious zeal.

As HOLMES ([2a], pp. 6—9) has said:

"From the days of earliest tradition down to the close of the eighteenth century, it was commonly believed that the spectacular features of landscapes, such as great mountains and deep gorges, were the result of sudden and violent catastrophes. The terrifying and destructive phenomena of earthquakes, volcanoes and floods suggested exaggerated ideas of still more terrible convulsions of Nature ... In Europe, where theological influence was dominant, the Catastrophists, with those conceptions of great world convulsions, had the apparent support of the Scriptures."

In the eighteenth century JEAN GUETTARD noticed the *gradual* changes in the landscape wrought by rain and by wind, by rivers and by frost. He and his supporters, who were called *Uniformitarians*, held that the processes of erosion and change which can be seen in action today had operated in the past and that it was these processes and not catastrophic events which had shaped the earth's surface features. To such ideas about geomorphological change were later added the biological concept of the evolution of living plants and creatures from primitive forms. The Uniformitarians could not admit that the whole of geological history could be compressed into any period as brief as six thousand years and, consequently, they were bitterly opposed by the *Catastrophists* as heretics and enemies of religion. Nevertheless, during the nineteenth century their view gradually prevailed and man first began to comprehend the true vastness of geological time and to fashion methods of measuring it.

**4.** The early geologists worked in Europe and the eastern United States in areas underlain for the most part by fossiliferous rocks. As they examined these they gradually discovered and applied three principles upon which the geological classification of time depends. First, HUTTON distinguished the igneous rocks which had been emplaced in a hot and molten state from the sedimentary rocks which had been deposited in water. WERNER and SMITH both independently realized that the sedimentary rocks had been laid down in succession so that in any undisturbed sequence the oldest were at the bottom. They also found that particular *fossils* were indicative of particular periods in the earth's history. They showed that the same living forms had not been repeated and that index fossils could thus be used to correlate rocks of the same age occuring in widely separated areas. During the nineteenth century work based on these principles was carried on all over the world and it was gradually established that all the fossiliferous sedimentary rocks everywhere could be placed in a uniform succession which was for convenience divided into three eras, 11 periods and about 55 stages all marked by characteristic fossils. These divisions are still used and the principal ones are shown in Table 1. The accuracy of this relative method of geochronology is surprisingly good. The average length of the 55 stages is less than ten million years each. In many of the stages much smaller subdivisions can be identified over wide areas of the earth so that relative correlation by some

Table 1. *Principle subdivisions of geological time as deduced from paleontology.*

| Era | | Period | Fossils |
|---|---|---|---|
| Cenozoic or Cainozoic | | Quarternary<br>Tertiary | Abundant |
| Mesozoic | | Cretaceous<br>Jurassic<br>Triassic | Abundant |
| Paleozoic or Palaeozoic | | Permian<br>Carboniferous<br>Devonian<br>Silurian<br>Ordovician<br>Cambrian | Abundant |
| Precambrian | Proterozoic | | Very few |
| | Archean or Archaeozoic | | None |

marine fossils, for example graptolites, is possible with an error of not more than a million years.

In some places at the base of this fossiliferous succession the early geologists found sedimentary rocks with few and uncertain fossils which they assigned to a *Proterozoic era*. In all places where exposure was deep the oldest rocks were found to be profoundly altered and to contain no fossils. These rocks were called the *fundamental complex* or *basement rocks* and they were classed together in the *Archean era*. The absence of any fossils by which these early rocks could be correlated prevented the establishment of any world wide classification other than this division into altered and unaltered types. Together they were called the pre-Cambrian or Precambrian because they lay below the earliest fossiliferous period.

At first men did not know what period of time the geological succession represented, but they realized that it must be vastly long and soon develped methods for estimating its length. As early as the beginning of the eighteenth century, HALLEY had suggested that the salt in the oceans had been derived from the land and that the rate at which the rivers carried more salt to the oceans might provide a method of dating them. Study of this suggested that the oceans were very old, but it was found that this method and the analogous one of calculating the time taken to accumulate all of the sedimentary rocks were only approximate. The rates of contributions of salts and of sedimentary materials have fluctuated, material has been recycled, and at best these methods only give *minimum ages* of a few hundreds of millions of years for the age of oceans and sedimentary rocks.

Estimates based upon the rate of evolution were later suggested and these estimates, which might appear to have been little more than guesses, oddly enough proved to be more accurate than the earlier approaches. In 1867 CHARLES LYELL[1] established a chronology based upon estimated *"cycles of evolution"*. For the oldest Cambrian fossiliferous rocks he found an age of 240 million years which is just half that of the latest measurements. For the younger rocks his estimates were even better. He also allowed time for that part of the geological record represented by the older and unfossiliferous Precambrian rocks. The great majority of geologists firmly believed that they had established the great age of the earth and that they had shown that for some hundreds of millions of years past its surface conditions were much like those of today.

**5.** No sooner had the geologists established the Uniformitarian principle of the earth's history, than they were attacked on the grounds that the time they required was impossibly long. In 1883 W. THOMSON[2] (later Lord KELVIN) pointed out that no source of the earth's heat was known except original heat and that the earth must be cooling from an earlier hot and molten state. He calculated that this could not have taken more than 400 million years nor less than 20 million, depending upon the assumptions used, but his final estimate[3] was 20 to 40 million years for the most probable age of the earth. The age of the oceans and of sedimentary rocks would be even less and the oldest rocks should show evidence of their hot and molten beginning.

Lord KELVIN's estimate received support from HELMHOLTZ who investigated the sources of energy available to the sun and found that all sources then known could not have maintained its heat for more than a few millions of years ([3],

---

[1] C. LYELL: Principles of Geology, 10th edit., Vol. 1, p. 300. London 1867.

[2] W. THOMSON and P. G. TAIT: Treatise on Natural Philosophy, 2nd edit., Vol. 1/2, p. 468. Cambridge 1883.

[3] Lord KELVIN: Phil. Mag. **47**, 66 (1899).

page 252). These views had the unanimous support of physicists, but the majority of geologists continued to hold out doggedly for a long and uniform history of the earth or at least for that part represented by the fossiliferous succession.

**6.** Precambrian rocks are not well represented in Western Europe and Eastern United States where geological work began, but there are large areas of exposed basement rocks in Canada which were first studied by WILLIAM LOGAN. Between 1842 and 1863 he found that it was possible to make a local classification of them. Similar work elsewhere enabled other local classifications to be established, but without fossils or age determinations, it was impossible to correlate Precambrian rocks satisfactorily and endless arguments and confusion resulted.

In 1902 to help to clarify the nomenclature and correlation of the Precambrian rocks in the Lake Superior region, a joint committee of the United States and Canadian Geological Surveys was appointed which, after examining some key areas, made a report in 1905[1]. The rocks were grouped into four principal divisions. Those in the two older divisions (*Laurentian* and *Keewatin*) were found to be principally volcanic, igneous and metamorphic rocks with smaller amounts of highly altered and deformed sedimentary rocks. Those in the younger division (*Huronian* and *Keweenawan*) were predominantly undisturbed rocks.

This report was only concerned with one small area, but it was taken by others and applied until it became the basis of a world wide chronology. Thus CHAMBERLIN and SALISBURY[2] referred the Laurentian and Keewatin divisions to the Archeozoic or Archean era lying in age between the original formation of the earth and the oldest predominantly sedimentary rocks. They said of them that "the rocks of no later era are so largely igneous, so notably deformed or so highly metamorphic". They placed the Huronian and Keweenawan divisions into the Proterozoic or Algonkian era described as "a time when igneous activity was still rather pronounced, though by no means so overwhelmingly as in Archean times. Sedimentation had become, for the first time, the leading process in the formation of the geological record". This classification of the Precambrian rocks into an older altered group and in a younger unaltered group was applied all over the world.

Here is a strong echo of KELVIN's view that the earth's history had been short and that it had been cooling rapidly. There is no doubt that a greatly foreshortened view was taken and that the Precambrian with its turbulent Archean beginning was regarded as an introduction to the important part of geological time which was represented by the fossiliferous rocks. We can now see how unfortunate it was that this basis of geochronology was established before the first rocks were dated in an absolute way by radioactive methods. The discovery of radioactivity with the realization that radioactive elements occur in all rocks and that they generate heat, destroyed the basis of Lord KELVIN's arguments. The earth's history is now known to be even longer than the geologists had thought and a hundred times as old as KELVIN had maintained. The question as to whether the earth is now cooling or whether it is warming up slowly has yet to be settled.

**7.** In 1907 BOLTWOOD published the first estimates of the age of radioactive minerals calculated from chemical analyses for uranium, thorium and lead [4]. He found the oldest fossiliferous rocks to be about 500 million years old, while some Precambrain rocks gave ages of over 2000 million years. While BOLTWOOD's ages may have been inaccurate in detail, they were sufficiently good to show

[1] C. R. VAN HISE: J. Geol. **13**, 89 (1905).
[2] T. C. CHAMBERLIN and R. D. SALISBURY: Geology, 2nd edit., Vol. 2, p. 140 and 163. New York 1905.

that Precambrian time, so far from being a brief and turbulent beginning, did in fact constitute the greater part of geological time. All subsequent work on radioactive age determinations has supported this view and the basis for the old and simple subdivision of the Precambrian has been destroyed. It might be supposed that this discovery would have led to its immediate revision, but this was not the case. For several reasons, radioactive age determinations did not at first prove very useful to geologists.

In the first place, although correct in order of magnitude for absolute time, BOLTWOOD's methods were not precise. For this reason, when the early age determinations were compared with the sedimentary record, it was found that the relative age could locally be determined much more accurately by the use of fossils and the order of succession than by radioactive methods. This is still true. Wherever index fossils occur they provide a much closer correlation to the accepted relative time scale than do radioactive age determinations. The latter have only served to place approximate absolute values upon the scale. Age determinations have proved most useful for dating unfossiliferous rocks, especially those of Precambrian age.

A second difficulty proved to be that some of the methods which were at first greeted with enthusiasm, particularly the helium method, were later found to be entirely unreliable[1]. Thirdly, accurate determinations are only made with difficulty so that few of them are available even now, and there are not yet nearly enough to provide an adequate geochronology for the Precambrian in all parts of the world. Fourthly, it is true that in many single localities the Precambrian does seem to have a dual division into altered and unaltered rocks and the idea that the two divisions are not everywhere of the same age has proved difficult to accept. For these and other reasons, age determinations have not to date been much relied on by geologists and they have only been used as adjuncts when no other methods were available.

Nevertheless methods of age determination have been steadily improved. BOLTWOOD's analyses and those made for thirty years thereafter depended upon *chemical* and *gas* analyses of the uranium, thorium, lead and helium content of both radioactive ores and rocks. Many of the early ages were incorrect because of inaccurate analyses, the use of wrong decay constants and the failure in the early years to understand that there were two uranium decay series. None of the results could be entirely relied upon because there was no method of estimating the extent of possible leaching or the addition of extraneous matter—particularly common or ore lead. The use of atomic weight determinations to correct for the presence of common lead met with indifferent success. After a great deal of effort, the measurement of helium was largely abandoned because helium was not retained sufficiently well by the rocks. The early work has been well reviewed by HOLMES [2a], HAHN [2b] and KNOPF [5].

8. A great advance was made when ASTON[2] started to make *isotopic* analyses to supplement the chemical ones. Between 1938 and 1941, NIER [6] to [8] published a series of 50 isotopic determinations of lead ores and radiogenic leads which for ten years remained the chief group of accurate isotopic lead abundances. In the past three or four years, several other laboratories have started to make many more isotopic analyses of radiogenic and ore leads [9] to [13]. A start has also been made to date ordinary rocks from the microscopic quantities of uranium and thorium occuring in them[3]. New methods also applicable to more common minerals are being developed using the decay of radioactive isotopes of potassium [14] to [16] and rubidium [17], [18], while organic material formed within the last 100,000 years is being dated by carbon-14 methods [19]. Other new methods propose the use of lead ores [20] and the study of lead-210 [21], [22] in radioactive ores.

---

[1] N. B. KEEVIL: Univ. Toronto Stud., Geol. Ser. (46) 39 (1941).

[2] F. W. ASTON: Nature, Lond. 137, 613 (1936).

[3] G. R. TILTON et al.: Bull. Geol. Soc. Amer. 66, 1131 (1955). — G. TILTON, L. ALDRICH and M. INGHRAM: Anal. Chem. 26, 894 (1954). — G. W. MOREY: Carnegie Inst. Wash., Year Book 52, 78 (1953).

Reviews of the more modern work on age determinations have been published by EVANS, GOODMAN and KEEVIL [23], BULLARD [24], ZEUNER [25], KNOPF [26], BURLING [27], KULP [13], RANKAMA [28], KOHMAN [29] and FAUL [30]. The whole literature is thoroughly chronicled in the *Reports of the Committee on the Measurement of Geologic Time* which have been written annually by A. C. LANE, from 1924 to 1946 and by J. P. MARBLE from 1947 to 1954 [31]. These reports are invaluable to anyone studying this subject.

**9.** In 1937 HUBBLE[1] placed what was in effect an upper limit upon age determinations. He estimated that the rate of expansion of the observable galaxies was such that if the rate had always remained constant the galaxies would have been together $1.86 \times 10^9$ years ago. As the rate appeared to be slowing down he suggested that the maximum probable *"age of the universe"* was only $1.5 \times 10^9$ years. This estimate was in conflict with the great age already found for some Precambrian rocks.

In 1946 HOLMES suggested that NIER's analyses of 25 common leads could be used to estimate the age of the earth [32]. In 1947 HOUTERMANS [33] independently made another similar suggestion. These methods have been modified and used with additional analyses to obtain the latest value of $(4.3 \pm 0.4) \times 10^9$ years [34], [35]. This made the contradiction between the apparent ages of the earth and the universe more marked and the problem remained unsolved until 1952 when BAADE reported that the Cepheid variables upon whose period-luminosity relations the apparent *"age of the universe"* depends, were divisable into two classes with different properties[2]. As a consequence the *"age of the universe"* is being revised and is now considered by astronomers to be $(5 \pm 1) \times 10^9$ years, which is in accord with the estimated age of the earth and the age of the oldest dated rocks.

**10.** We can now see that as a result of fifty years of work the problem of obtaining accurate ages of minerals and an absolute time scale for the earth are only really beginning to be understood. Only recently have the major contradictions that formerly existed between the findings of astronomers, geologists and physicists been removed. Accurate age determinations have been few and they have been published along with many that have later been shown to be unreliable. Geologists who were at first elated by the promise of age determinations have become cautious about accepting them and they have not altered the geochronology adopted fifty or more years ago.

At this time an account of the age of rocks and minerals should not therefore be a summary of past achievements so much as a statement of opportunities offered and a description of the methods now available for solving the problems which so clearly await settlement. The situation is typical of a borderline branch of science which has been neglected by physicists and geologists alike, although both may have much to gain by its solution.

Two great problems awaiting solution are the establishment of an accurate, absolute time scale in years for the younger and fossiliferous part of the succession which has at present only an accurate relative geochronology, and the devising of an adequate and proper time scale for the earlier or Precambrian part of geological time which is now known to be five-sixths of the whole. It is unfortunate that the system of subdivision now generally employed for the Precambrian can only be described as being confused, telescoped and misleading.

---

[1] E. HUBBLE: The observational approach to cosmology. Oxford 1937.
[2] O. STRUVE: Sky and Telescope, Vol. 12, p. 203 and 238. 1953.

It does not do justice to the excellent geological field work on which it was based and it disguises those problems of Precambrian time which it should illuminate. Fortunately the means now appear to be available to provide solutions for both these problems within the next few years.

## C. Physical considerations.

### I. Equations of radioactive decay.

**11.** RUTHERFORD and SODDY[1] were the first to put forward a general theory of radioactive decay. They assumed that from time to time a radioactive isotope spontaneously emits a beta particle or alpha particle with the release of a considerable amount of energy, and with a change in atomic mass and atomic number so that mass and electric charge are conserved. It is now known that other types of decay exist and that there is a small net change in mass corresponding to the energy released during the radioactive transformation. RUTHERFORD and SODDY postulated that the disintegration of the parent atom is a purely random event, and that the rate of decay of radioactive atoms should be proportional only to the number of atoms present. The constant of proportionality is called the *decay constant*, and is characteristic of the particular radioactive isotope. The various equations used for determining mineral ages and the production of radioactive heat are based on these assumptions.

**12.** The *simplest form of radioactive decay* is exemplified by the decay of rubidium-87 to form strontium-87. According to the assumptions of RUTHERFORD and SODDY:

$$d/dt \, (\mathrm{Rb}^{87}) = \lambda \, \mathrm{Rb}^{87}. \tag{12.1}$$

Here the chemical symbol, $\mathrm{Rb}^{87}$, is used to denote the *number of atoms* of that isotope. *This convention is used throughout this chapter.* Where it is convenient to use a symbol for *mass* rather than number of atoms, this is done by enclosing the chemical symbol in square brackets: $[\mathrm{Rb}^{87}]$. *Decay constants* are always represented by the symbol $\lambda$. Where it is necessary to distinguish between different decay constants this is done by using either superscripts or subscripts.

By integrating (12.1), with the boundary condition that the number of atoms of rubidium-87 present initially was $\mathrm{Rb}_0^{87}$,

$$\mathrm{Rb}^{87} = \mathrm{Rb}_0^{87} \, e^{-\lambda t}. \tag{12.2}$$

The number of radiogenic strontium-87 atoms must equal the number of rubidium atoms that have decayed. Therefore:

$$\mathrm{Sr}^{87} = \mathrm{Rb}_0^{87} - \mathrm{Rb}^{87} = \mathrm{Rb}_0^{87} (1 - e^{-\lambda t}). \tag{12.3}$$

Geophysicists usually do not adopt equation (12.3) *directly* for age determination studies because laboratory measurements give $\mathrm{Rb}^{87}$ = the number of atoms of rubidium-87 atoms *now* present, rather than $\mathrm{Rb}_0^{87}$, the number of atoms *initially* present.

The equation usually used is obtained by combining equation (12.2) and (12.3) to give

$$\mathrm{Sr}^{87} = \mathrm{Rb}^{87} (e^{\lambda t} - 1). \tag{12.4}$$

---

[1] E. RUTHERFORD and F. SODDY: Trans. Chem. Soc. **81**, 321, 837 (1902). — Phil. Mag. **4** 370, 569 (1902); **5**, 441, 446, 576 (1903).

$Sr^{87}$ is the number of atoms of radiogenic strontium that are found in a mineral of age $t$ having *at present* $Rb^{87}$ atoms of rubidium-87. Unless this convention is clearly understood, confusion can result in the application of standard physical formulae to geophysical problems.

Equations analogous to equation (12.4) apply to all cases where the decay scheme consists of a radioactive parent decaying directly into a stable daughter. It applies, for example, to carbon-14 and tritium. It also applies to decay schemes where a long-lived parent is separated from a stable end product by a series of intermediate radioactive isotopes which are short-lived in comparison with the age of the mineral. In particular, it applies to the accumulation of lead isotopes in geologically old uranium and thorium minerals provided that radioactive equilibrium has been maintained.

**13.** For age determinations based on the uranium and thorium series it is possible to use the *number of atoms of an intermediate, radioactive member of the decay series* rather than the number of atoms of the *parent*. If equilibrium can be assumed, the measurement of the intermediate member gives an immediate measure of atoms of the parent present. Suppose that a parent, $P$, decays through a series of *radioactive* intermediate members, $I_1$, $I_2$, etc., to form a stable daughter $D$; viz;

$$P \xrightarrow{\lambda_1} I_1 \xrightarrow{\lambda_2} I_2 \xrightarrow{\lambda_3} \cdots \xrightarrow{\lambda_n} I_n \xrightarrow{\lambda_{n+1}} \cdots \xrightarrow{\lambda_p} D.$$

A simple extension of the principles outlined above enables us to relate any of the intermediate products to the parent.

$$\frac{dP}{dt} = -\lambda_1 P.$$

Therefore

$$P = P_0 e^{-\lambda_1 t}, \tag{13.1}$$

$$\left. \begin{aligned} \frac{dI_1}{dt} &= \lambda_1 P - \lambda_2 I_1 \\ &= \lambda_1 P_0 e^{-\lambda_1 t} - \lambda_2 I_1. \end{aligned} \right\} \tag{13.2}$$

Put $I_1 = a e^{-\lambda_1 t} + b e^{-\lambda_2 t}$, where $a$ and $b$ are to be determined, then

$$\frac{dI_1}{dt} = -a \lambda_1 e^{-\lambda_1 t} - b \lambda_2 e^{-\lambda_2 t}.$$

When these expressions for $I_1$ and $dI_1/dt$ are substituted in (13.2), the following equation results:

$$a (\lambda_2 - \lambda_1) = \lambda_1 P_0.$$

Therefore

$$I_1 = \frac{\lambda_1 P_0}{(\lambda_2 - \lambda_1)} e^{-\lambda_1 t} + b e^{-\lambda_2 t}.$$

If $\lambda_2 > \lambda_1$, then for $\lambda_2 t \gg 1$

$$I_1 = \frac{\lambda_1 P_0}{(\lambda_2 - \lambda_1)} e^{-\lambda_1 t}. \tag{13.3}$$

For the second intermediate member

$$\left. \begin{aligned} \frac{dI_2}{dt} &= \lambda_2 I_1 - \lambda_3 I_2 \\ &= \frac{\lambda_2 \lambda_1 P_0 e^{-\lambda_1 t}}{(\lambda_2 - \lambda_1)} - \lambda_3 I_2. \end{aligned} \right\} \tag{13.4}$$

Since (13.4) is identical with (13.2) except for the constant coefficients, its solution can be obtained by changing the same constants in (13.3), i.e.

$$I_2 = \frac{\lambda_2 \lambda_1 P_0 e^{-\lambda_1 t}}{(\lambda_3 - \lambda_1)(\lambda_2 - \lambda_1)} \quad \text{where} \quad \lambda_3 t \gg 1, \quad \lambda_2 > \lambda_1 \quad \text{and} \quad \lambda_3 > \lambda_1.$$

Continuing in this way, we find that

$$I_n = \frac{\lambda_n \lambda_{n-1} \cdots \lambda_1}{(\lambda_{n+1} - \lambda_1)(\lambda_n - \lambda_1) \cdots (\lambda_2 - \lambda_1)} \cdot P_0 e^{-\lambda_1 t},$$

and therefore

$$\frac{I_n}{P} = \frac{\lambda_n \lambda_{n-1} \cdots \lambda_3 \lambda_2 \lambda_1}{(\lambda_{n+1} - \lambda_1)(\lambda_n - 1) \cdots (\lambda_2 - \lambda_1)}. \tag{13.5}$$

Equation (13.5) is only true where $\lambda_2, \lambda_3, \ldots \lambda_{n+1}$, are all greater than $\lambda_1$, and where all of $\lambda_2 t, \lambda_3 t, \ldots \lambda_{n+1} t$ greatly exceed unity. The abundances of both parent and intermediate member must be expressed in *numbers of atoms* or *moles*.

The presence of radium in a uranium mineral provides a good example of the application of equation (13.1). In this case the decay constant of uranium-238, $\lambda_1$, is negligibly small in comparison with the other decay constants in the series. Then, for equilibrium, equation (13.5) simplifies to

$$\begin{aligned}
\frac{Ra^{226}}{U^{238}} &= \frac{\lambda_1 \lambda_2 \lambda_3 \lambda_4 \lambda_5}{\lambda_2 \lambda_3 \lambda_4 \lambda_5 \lambda_6} \\
&= \frac{\lambda_1}{\lambda_6} \\
&= \frac{(\lambda_{U^{238}})}{(\lambda_{Ra^{226}})} \\
&= \frac{(1.54 \times 10^{-10})}{(4.28 \times 10^{-4})} \\
&= 3.60 \times 10^{-7}.
\end{aligned}$$

Thus measuring the amount of radium in a uranium mineral may be equivalent to measuring the amount of uranium, if radioactive equilibrium exists. Such a measurement, when combined with appropriate lead analyses can be used for determining mineral ages.

**14.** A *third type of radioactive decay* which is important for the determination of geological ages is the decay of a *single parent* into *two alternate stable decay products*. The decay of potassium-40 to both calcium-40 and argon-40 is such a case. Equations for this type of decay follow directly from the assumption that the decay of the parent to either end product is a purely random process. The rate of decay to each decay product is then proportional to the number of parent atoms and is independent of any other physical or chemical condition. There will be two constants of proportionality, one for each mode of decay. For potassium-40:

for decay to calcium by $\beta$-ray emission

$$\frac{d}{dt}(K^{40}) = -\lambda_\beta K^{40}, \tag{14.1}$$

for decay to argon by orbital electron capture,

$$\frac{d}{dt}(K^{40}) = -\lambda_e K^{40}, \tag{14.2}$$

for the total rate of decay,

$$\frac{d}{dt}(K^{40}) = -(\lambda_\beta + \lambda_e)\,K^{40} \atop = -\lambda\,K^{40}.$$ (14.3)

By integrating (14.3):

$$K^{40} = K_0^{40}\,e^{-\lambda t}.$$ (14.4)

Substituting (14.4) into (14.1) and (14.2) and integrating, we obtain in the notation commonly used:

$$\frac{Ca^{40}}{K^{40}} = \left(\frac{\lambda_\beta}{\lambda}\right)(e^{\lambda t} - 1),$$ (14.5)

$$\frac{A^{40}}{K^{40}} = \left(\frac{\lambda_e}{\lambda}\right)(e^{\lambda t} - 1).$$ (14.6)

The *branching ratio* of potassium-40 is defined as the ratio $\lambda_e/\lambda_\beta$. Owing to the dual decay, two decay constants are necessary to completely describe the radioactivity of potassium-40. The branching ratio and either $\lambda_\beta$ or $\lambda$ are commonly used.

**15.** The *half-life*, $T^{\frac{1}{2}}$, of a radioactive species is defined as the time in which half of the amount initially present will decay. From equation (12.2) for the case of rubidium-87

$$\frac{Rb^{87}}{Rb_0^{87}} = e^{-\lambda T^{\frac{1}{2}}} = \frac{1}{2}.$$

Therefore

$$T^{\frac{1}{2}} = \frac{\log_e 2}{\lambda} = \frac{0,6931}{\lambda}.$$ (15.1)

# II. Natural radioactive isotopes.

**16.** The most important class of radioactive isotopes contains those which have always existed and whose half-lives are sufficiently long so that they have not entirely vanished by decay during geological time. The isotopes in this class which have been used for making geological age determinations are listed with some of their properties in Table 2. Those which have not yet been so used (although some of them may be in the future) are given in Table 3.

Uranium-238, uranium-235 and thorium-232 are the best known long-lived, radioactive isotopes. The decay scheme of each of these is complex, their members forming a series of radioactive isotopes which have half-lives short compared with the age of the earth. Tables 4, 5 and 6 (taken from KULP, BATE and BROECKER [13]) list these isotopes and give their chief properties.

Table 2. *Naturally occurring radioactive isotopes which have been used for geological age determinations.*

| Isotope | Type of decay | Particle energy | End product | Half-life[1] |
|---------|---------------|-----------------|-------------|--------------|
| $_{19}K^{40}$ | Beta | 1.36 Mev | $_{20}Ca^{40}$ | $1.3 \times 10^9$ years |
| | Orbital electron capture | — | $_{18}A^{40}$ | |
| $_{37}Rb^{87}$ | Beta | 0.275 Mev | $_{38}Sr^{87}$ | $6.1 \times 10^{10}$ years |
| $_{90}Th^{232}$ | (complex cf. Table 4) | — | $_{82}Pb^{208}$ | $1.4 \times 10^{10}$ years |
| $_{92}U^{235}$ | (complex cf. Table 5) | — | $_{82}Pb^{207}$ | $7.1 \times 10^8$ years |
| $_{92}U^{238}$ | (complex cf. Table 6) | — | $_{82}Pb^{206}$ | $4.5 \times 10^9$ years |

[1] For references see the following pages.

Table 3. *Naturally occuring radioactive elements not yet used for age determinations* [1].

| Element | Isotope · | Type of decay | Particle energy | Probable half-life (years) | Probable end product |
|---------|-----------|---------------|-----------------|----------------------------|----------------------|
| Sm | 147 ( ?) | $\alpha$ | 2.0   Mev. | $1.4 \times 10^{11}$ | Nd |
| Lu | 176 | $\beta$ | 0.215 Mev. | $7.3 \times 10^{10}$ | $Hf^{176}$ |
| Re | 187 | $\beta$ | $\sim$8 kev. | $\sim 10^{11}$ | $Os^{187}$ |
| Bi | 209 | $\alpha$ | ? | very large | $Tl^{205}$ |
| W | 178 or 180 | $\alpha$ | ? | very large | Hf |
| Nd | 144 | $\alpha$ | ? | very large | $Ce^{140}$ |

[1] With the exception of samarium and lutecium the radioactivities of these elements are not indisputably established, nor are details of the decay known. KOHMAN and SAITO [29] give references to unsuccesful searches for activities in $Ca^{48}$, $V^{50}$, $Zr^{96}$, $Cd^{113}$, $In^{113}$, $Sb^{123}$, $Te^{123}$, $Nd^{150}$, and Mo. They also give references for the activities of the elements listed in this table.

Table 4. *The uranium* $(U^{238})$ *series.*

| Isotope | | Particle emitted | Particle energy (Mev) | Half-life | Reference for half-life |
|---------|--|------------------|-----------------------|-----------|-------------------------|
| $_{92}U^{238}$ | (UI) | $\alpha$ | 4.18 | $4.49 \pm 0.01 \times 10^9$ years | 1 |
| $_{90}Th^{234}$ | $(UX_1)$ | $\beta$ $\Big\{$ | 0.205 / 0.111 | $\Big\}$ $24.101 \pm 0.025$ days | 2 |
| $_{91}Pa^{234}$ | $(UX_2)$ | $\beta$ $\Big\{$ | 2.32, 1.50 / 0.60 | $\Big\}$ $1.175 \pm 0.003$ min. | 3 |
| $_{92}U^{234}$ | (UII) | $\alpha$ | 4.763 | $2.475 \pm 0.016 \times 10^5$ years | 4 |
| $_{90}Th^{230}$ | (Io) | $\alpha$ | 4.68, 4.61 | $8.0 \ \pm 0.3 \ \times 10^4$ years | 5 |
| $_{88}Ra^{226}$ | (Ra) | $\alpha$ | 4.77 | $1622 \ \pm 1$ years | 6 |
| $_{86}Rn^{222}$ | (Rn) | $\alpha$ | 5.486 | $3.825 \pm 0.005$ days | 7 |
| $_{84}Po^{218}$ | (RaA) | $\alpha$ | 5.998 | $3.050 \pm 0.009$ min. | 8 |
| $_{82}Pb^{214}$ | (RaB) | $\beta$ | 0.65 | $26.8 \ \pm 0.1$ min. | 9 |
| $_{83}Bi^{214}$ | (RaC) | $\alpha$ 0.04% / $\beta$ 99.96% | 5.46 / 1.65, 3.17 | $19.72 \pm 0.04$ min. | 10 |
| $_{84}Po^{214}$ | (RaC') | $\alpha$ | 7.680 | $163.7 \ \pm 0.2$ microsec | 11 |
| $_{81}Tl^{210}$ | (RaC'') | $\beta$ | 1.8 | $1.32 \ \pm 0.01$ min. | 12 |
| $_{82}Pb^{210}$ | (RaD) | $\beta$ | 0.018 | $22.5 \ \pm 0.4$ years | 13 |
| $_{83}Bi^{210}$ | (RaE) | $\beta$ | 1.17 | $4.989 \pm 0.013$ days | 14 |
| $_{84}Po^{210}$ | (RaF) | $\alpha$ | 5.298 | $138.374 \pm 0.032$ days | 15 |
| $_{82}Pb^{206}$ | (RaG) | | | Stable | |

[1] E. II. FLEMING jr., A. GHIORSO and B. B. CUNNINGHAM: Phys. Rev. **82**, 967 (1951). — C. A. KIENBERGER: Phys. Rev. **76**, 1561 (1949) — $4.498 \times 10^9$ Yr.

[2] G. B. KNIGHT and R. L. MACKLIN: Phys. Rev. **74**, 1540 (1948).

[3] F. BARENDREGT and SJ. TOM: Physica, Haag **17**, 817 (1951).

[4] E. H. FLEMING jr., A. GHIORSO and B. B. CUNNINGHAM: Phys. Rev. **88**, 642 (1952). — C. A. KIENBERGER: Phys. Rev. **87**, 520 (1952) — $2.522 \pm 0.008 \times 10^5$ Yr.

[5] SEABORG, KATZ and MANNING (edit.): The Transuranium Elements, National Nuclear Energy Series, IV–14B, 1435, 1949.

[6] B. H. KETTELLE: Oak Ridge National Laboratory, Oak Ridge, Tenn., ORNL–286, 1949.

[7] J. TOBAILEM: C. R. Acad. Sci Paris **233**, 1360 (1951).

[8] M. BLAU: Mitt. Ra. Inst. 161, Wien. Ber. IIa **133**, 17 (1924).

[9] E. RUTHERFORD: Radioactive Substances and Their Radiations, Chap. XIV, p. 489. 1913.

[10] P. BRACELIN: Proc. Cambridge Phil. Soc. **23**, 150 (1926).

[11] G. v. DARDEL: Phys. Rev. **79**, 734 (1950).

[12] E. ALBRECHT: Mitt. Ra. Inst. 123, Wien. Ber. IIa **128**, 925 (1919).

[13] E. SCHWEIDLER: Mitt. Ra. Inst. Wien. Ber. IIa **138**, 743 (1929). — I. JOLIOT-CURIE and M. CURIE: J. de Phys. **10**, 385, 388 (1929) — 19.5 to 23 Yr. — F. WAGNER, jr.: Argonne National Laboratory 4490, 5 (1950) — 25 Yr.

[14] E. E. LOCKETT and R. H. THOMAS: Nucleonics **11**, 14 (1953).

[15] M. L. CURTIS: Mound Lab. Atomic Energy Commission Declassified 3536, MLM–575, 8 pages, 1953.

Table 5. *The actinium* ($U^{235}$) *series.*

| Isotope | Particle emitted | Particle energy (Mev) | Half-life | Reference for half-life |
|---|---|---|---|---|
| $_{92}U^{235}$ (AcU) | $\alpha$ | 4.40, 4.58 4.20, 4.47 | 7.13 $\pm$ 0.16 × 10⁸ years | 1 |
| $_{90}Th^{231}$ (Uy) | $\beta$ | 0.094, 0.302 0.216 | 25.6 $\pm$ 0.1 hours | 2 |
| $_{91}Pa^{231}$ (Pa) | $\alpha$ | 5.00 | 3.43 $\pm$ 0.03 × 10⁴ years | 3 |
| $_{89}Ac^{227}$ (Ac) | $\beta$ | 0.04 | 22.0 $\pm$ 0.3 years | 4 |
| $_{90}Th^{227}$ (RdAc) | $\alpha$ | 6.00 | 18.6 $\pm$ 0.1 hours | 5 |
| $_{88}Ra^{223}$ (AcX) | $\alpha$ | 5.70 | 11.2 $\pm$ 0.2 days | 6 |
| $_{86}Rn^{219}$ (An) | $\alpha$ | 6.82 | 3.917 $\pm$ 0.015 sec. | 7 |
| $_{84}Po^{215}$ (AcA) | $\alpha$ | 7.37 | 1.83 $\pm$ 0.04 × 10⁻³ sec. | 8 |
| $_{82}Pb^{211}$ (AcB) | $\beta$ | 1.39, 0.50 | 36.1 $\pm$ 0.2 min. | 9 |
| $_{83}Bi^{211}$ (AcC) | $\alpha$ $\beta$ | 6.62 — | 2.16 $\pm$ 0.03 min. | 10 |
| $_{84}Po^{211}$ (AcC') | $\alpha$ | 7.43 | 0.52 $\pm$ 0.02 sec. | 11 |
| $_{81}Tl^{207}$ (AcC'') | $\beta$ | 1.44 | 4.79 $\pm$ 0.02 min. | 12 |
| $_{82}Pb^{207}$ (AcD) | | | Stable | |

[1] E. H. FLEMING jr., A. GHIORSO and B. B. CUNNINGHAM: Phys. Rev. **88**, 642 (1952).
[2] A. H. JAFFREY, J. LERNER and S. WARSHAW: Phys. Rev. **82**, 498 (1951). — G. B. KNIGHT and R. L. MACKLIN: Phys. Rev. **75**, 34 (1948) give 25.51 ± 0.23 Hr.
[3] Q. VAN WINKLE, R. G. LARSON and L. I. KATZIN: J. Amer. Chem. Soc. **71**, 2585 (1949).
[4] J. M. HOLLANDER and R. F. LEININGER: Phys. Rev. **80**, 915 (1950). F. WAGNER jr.: Argonne National Laboratory–4490, 5 (1950) — 27.7 Yr.
[5] SEABORG, KATZ and MANNING (edit.): The Transuranium Elements, National Nuclear Energy Serics, IV–14B, 1424, 1949. (S. PETERSON and A. GHIORSO Paper 19.12.)
[6] S. MEYER and F. PANETH: Mitt. Ra. Inst. 104, Wien. Ber. IIa **127**, 147 (1918).
[7] R. SCHMID: Mitt. Ra. Inst. 103, Wien. Ber. IIa **126**, 1065 (1917). — M. LESLIE: Phil. Mag. (6) **24** (1921) and P. B. PERKINS: Phil. Mag. (6) **27**, 720 (1914) — 3.92 ± 0.004.
[8] A. G. WARD: Proc. Roy. Soc. Lond., Ser. A **181**, 183 (1942).
[9] B. W. SARGENT: Canad. J. Res. A **17**, 103 (1939).
[10] S. MEYER and F. PANETH: Mitt. Ra. Inst. 104, Wien. Ber. IIa **127**, 147 (1918).
[11] R. L. LEININGER, E. SEGRÉ and E. N. SPIESS: Phys. Rev. **82**, 334 (1951).
[12] B. W. SARGENT, L. YAFFE and A. P. GRAY: Canad. J. Phys. **31**, 235 (1953).

Table 6. *The thorium series.*

| Isotope | Particle emitted | Particle energy (Mev) | Half-life | Reference for half-life |
|---|---|---|---|---|
| $_{90}Th^{232}$ (Th) | $\alpha$ | 3.98 | 1.39 $\pm$ 0.02 × 10¹⁰ years | 1 |
| $_{88}Ra^{228}$ (MsTh₁) | $\beta$ | 0.012 | 6.7 $\pm$ 0.1 years | 2 |
| $_{89}Ac^{228}$ (MsTh₂) | $\beta$ | 1.15 | 6.13 $\pm$ 0.03 hours | 3 |
| $_{90}Th^{228}$ (RdTh) | $\alpha$ | 5.42 | 1.90 $\pm$ 0.01 years | 4 |
| $_{88}Ra^{224}$ (ThX) | $\alpha$ | 5.68 | 3.64 $\pm$ 0.01 days | 5 |
| $_{86}Rn^{220}$ (Tn) | $\alpha$ | 6.28 | 54.53 $\pm$ 0.04 sec. | 6 |
| $_{84}Po^{216}$ (ThA) | $\alpha$ | 6.77 | 0.158 $\pm$ 0.008 sec. | 7 |
| $_{82}Pb^{212}$ (ThB) | $\beta$ | 0.355 | 10.67 $\pm$ 0.05 hours | 8 |
| $_{83}Bi^{212}$ (ThC) | $\alpha$ $\beta$ | 6.05 2.25 | 60.48 $\pm$ 0.04 min. | 9 |
| $_{84}Po^{212}$ (ThC') | $\alpha$ | 8.78 | 0.29 $\pm$ 0.01 microsec. | 10 |
| $_{81}Tl^{208}$ (ThC'') | $\beta$ | 1.79 | 3.1 $\pm$ 0.1 min. | 11 |
| $_{82}Pb^{208}$ (ThD) | | | Stable | |

[1] ALOIS F. KOVARIK and N. I. ADAMS: J. Phys. Rev. **54**, 413 (1938).
[2] M. CURIE, A. DEBIERNE, A. S. EVE, H. GEIGER, O. HAHN, S. C. LIND, S. MEYER, E. RUTHERFORD and E. SCHWEIDLER: Rev. Mod. Phys. **3**, 427 (1931).
[3] O. HAHN and O. ERBACHER: Phys. Z. **27**, 531 (1926).
[4] S. MEYER and F. PANETH: Mitt. Ra. Inst. 96, Wien. Ber. IIa **125**, 1253 (1916). — B. WALTER: Phys. Z. **18**, 584 (1917). — L. MEITNER: Phys. Z. **19**, 257 (1918). — In close agreement.

[5] M. Curie, A. Debierne, A. S. Eve, H. Geiger, O. Hahn, S. C. Lind, S. Meyer, E. Rutherford and E. Schweidler: Rev. Mod. Phys. **3**, 427 (1931).

[6] P. B. Perkins: Phil. Mag. (6) **27**, 720 (1914). — R. Schmid: Mitt. Ra. Inst. **103**, Wien. Ber. IIa **126**, 1065 (1917) — 54.50 ± 0.03 Sec.

[7] A. G. Ward: Proc. Roy. Soc. Lond., Ser. A **181**, 183 (1942).

[8] H. v. Buttlar: Naturwiss. **39**, 574 (1952).

[9] F. v. Lerch: Mitt. Ra. Inst. Wien. Ber. IIa **123**, 699 (1914).

[10] T. Hajashi, Y. Ishizaki and L. Kumabe: J. Phys. Soc. Japan **8**, 110 (1953). — D. E. Bunyan, A. Lundby and D. Walker: Proc. Phys. Soc. Lond., Ser. A **62**, 253 (1949) — 0.304 ± 0.004 Microsec.

[11] Same as [5]).

A third class consists of those naturally occurring, radioactive isotopes with short lives which are continually being formed in the atmosphere by cosmic ray bombardment. There are only two of these known[1], tritium and carbon-14, and only the latter has a half life long enough to make it useful for measuring even the most recent part of geological time. Some properties of these isotopes are given in Table 7.

Table 7. *Short-lived radioactive isotopes produced by cosmic rays.*

| | Isotope | Decay scheme | Particle energy | Half-life (years) | End product |
|---|---|---|---|---|---|
| Tritium | $_1H^3$ | Beta emission | 19 kev. | 12.5 | $_2He^3$ |
| Carbon-14 | $_6C^{14}$ | Beta emission | 155 kev. | 5568 ± 30 | $_7N^{14}$ |

It is true that other radioactive isotopes do occur naturally due to spontaneous fission of some of the heaviest elements or due to neutron capture, but the quantities of these isotopes are very small and they have not been used for determining the ages of any minerals [*29*].

# III. Decay schemes and decay constants.

## a) Uranium and thorium.

**17.** The discovery of radioactivity by Becquerel in 1896 followed closely after the discovery of x-rays by Röntgen and the observation that the glass walls of the x-ray tube fluoresced. Only a few months later Becquerel examined a variety of fluorescent materials for emission of x-rays by placing the materials in close contact with photographic plates and searching for any darkening that could be attributed to the new rays. One of the materials tested was a double sulphate of uranium and potassium which fluoresced brilliantly under ultra-violet light, and which Becquerel found to blacken photographic film even through black paper. Further experiments showed that the phenomenon had no relation to the fluorescence, but was characteristic of all uranium salts and uranium metal.

Rutherford analysed the rays emitted by uranium with a magnetic field and showed that they consisted of two types, one having short ranges of penetration which he called "alpha rays" and another having much greater powers pentration which he called "beta rays". Later even more penetrating rays were discovered by Villard who named them "gamma rays".

It was soon found that the radiation from uranium (as well as x-rays) increased the electrical conductivity of gases and hence could be detected with an electroscope. Making use of this fact Mme Curie tested many elements for radioactivity.

[1] *Note added in proof:* Be[7] is also produced by cosmic rays, cf. J. R. Arnold and H. Ali Al-Salih: Science, Lancaster, Pa. **121**, 451 (1955).

SCHMIDT and Mme CURIE soon discovered that thorium exhibited the same radioactive properties as uranium.

In 1898 Mme CURIE discovered that several uranium minerals, in particular pitchblende from Bohemia, were many times as active as pure uranium metal. This prompted her to carry out careful chemical separations of the elements in this pitchblende and to test them individually for radioactivity, which in turn led to the discovery of the radioactive elements polonium and radium. In a similar manner actinium was discovered by DEBIERNE and GIESEL, radiolead (uranium $D$ or lead-210) by HOFFMAN and STRAUSS, and ionium by BOLTWOOD. In 1900 the radioactive emanations radon, actinon and thoron were discovered by RUTHERFORD.

The explanation of the presence of these and other radioactive substances in uranium minerals was difficult to find, particularly in view of the fact that most of them had half-lives so short that they should have largely decayed in geological time. RUTHERFORD and SODDY (cf. Sect. 11, footnote 1) in 1903 proposed a theory of radioactive decay that accounted for the presence of the short-lived elements in uranium and thorium minerals by their production by uranium and thorium. The details of the decay schemes of uranium-238, uranium-235 and thorium-232, are now well understood. They are shown in Tables 4, 5 and 6 which are taken from KULP et al [13]. Uranium-235, the parent of the actinium series, was discovered as recently as 1935 by A. J. DEMPSTER.

The decay constants of the parent isotopes of the uranium, actinium and thorium series are of interest in age determination studies because they govern the rate at which the stable lead end-products are produced. The various determinations of the half-lives of these elements are given in Table 8. The half-life of uranium-238 seems to be most accurately known, the uncertainty being only a small fraction of one percent. The values of the half life of uranium-235 are much more widely scattered. The most recent value of $7.13 \times 10^8$ years is the one most usually accepted.

The half-life of thorium-232 has been determined only once, with the result shown in the table.

Table 8. *Half-lives of uranium-238, uranium-235 and thorium-232.*

| Isotope | Reference | Half-life (years) |
|---|---|---|
| $_{92}U^{238}$ | H. M. NEWMAN and I. PERLMAN: Unpublished Data (see Circular No. 499 of the National Bureau of Standards, USA). | $4.51 \times 10^9$ |
| | A. F. KOVARIK and N. E. ADAMS jr.: J. Appl. Phys. **12**, 296 (1941). | $4.51 \times 10^9$ |
| | F. L. CLARK, H. J. SPENCER-PALMER and R. N. WOODWARD: British Atomic Energy Projects Reports BR–521 and 522 (1944). | $4.498 \times 10^9$ |
| | C. A. KEINBERGER: Phys. Rev. **76**, 1561 (1949). | $4.498 \times 10^9$ |
| | E. H. FLEMING jr., A. GHIORSO and B. B. CUNNINGHAM: Phys. Rev. **82**, 967 (1951). | $4.49 \pm 0.01 \times 10^9$ |
| $U_{92}^{235}$ | A. O. NIER: Phys. Rev. **55**, 150 (1939). | $7.13 \times 10^8$ |
| | A. F. KOVARIK and N. I. ADAMS jr.: J. Appl. Phys. **12**, 296 (1941). | $7.07 \times 10^8$ |
| | F. L. CLARK et al.: See above. | $8.91 \times 10^8$ |
| | C. A. KEINBERGER: Phys. Rev. **76**, 1561 (1949). | $8.8 \times 10^8$ |
| | E. H. FLEMING jr., A. GHIORSO and B. B. CUNNINGHAM: Phys. Rev. **82**, 967 (1951). | $7.07 \times 10^8$ |
| | E. H. FLEMING jr., A. GHIORSO and B. B. CUNNINGHAM: Phys. Rev. **88**, 642 (1952). | $7.13 \pm 0.16 \times 10^8$ |
| $_{90}Th^{232}$ | A. F. KOVARIK and N. E. ADAMS. jr: Phys. Rev. **54**, 413 (1938). | $1.39 \times 10^{10}$ |

## b) The alkali metals.

**18.** Following the discovery of the radioactivity of uranium and thorium and their decay products at the end of the nineteenth century, there was a strong suspicion that radioactivity might be an atomic property characteristic of *all* the elements. This suspicion was strengthened by the observation that all material appeared to give off some quantity of ionizing radiation, although it was very difficult to discover whether the activity was due to contamination by decay products from the uranium and thorium series. Many experiments were soon reported which attempted to prove or disprove suspected radioactivities of the lighter elements.

In 1905 THOMSON showed that both rubidium and sodium-potassium alloys differed from many other light metals in emitting charged particles in the dark as well as when illuminated. This work was soon followed by the researches of CAMPBELL[1] which showed that all potassium salts emitted an ionizing radiation in proportion to their potassium content, and thus proved the radioactivity of potassium. He further demonstrated that the particles emitted by potassium had the same electric charge as the beta particles from uranium, and also showed that rubidium salts were radioactive. Other early experiments verifying the radioactivity of these elements are reviewed by RUTHERFORD, CHADWICK and ELLIS [36]. KOLHÖRSTER[2] first observed the gamma radiations that are emitted from potassium in addition to the beta particles.

Although the activities of both potassium and rubidium are weak, their beta activities were soon measured to obtain estimates of the decay constants of these elements. HOLMES and LAWSON[3] reviewed the early measurements, and found half lives of $10^{11}$ and $1.5 \times 10^{12}$ years for rubidium and potassium respectively, assuming that all atoms of these elements were radioactive. Making use of more recent discoveries that the radioactive isotopes are $Rb^{87}$ forming 27.2% of common rubidium and $K^{40}$ forming 0.0119% of common potassium the half-lives given by HOLMES and LAWSON become $3 \times 10^{10}$ years and $1.6 \times 10^9$ years, respectively.

**19.** The decay rate of rubidium has been of primary interest in the determination of geological ages by the *rubidium-strontium method*. The first of such age determinations was carried out chemically in 1942 by STEVENS and SCHALLER[4] who also determined a value for the half-life of $Rb^{87}$ from the rubidium-strontium ratio of a lepidolite from a pegmatite in southeastern Manitoba, Canada. An age of 2200 million years was assigned to the sample on the basis of lead-uranium ages determined for another nearby pegmatite, and the half-life value obtained was $6.1 \times 10^{10}$ years.

The decay constant has now been determined several times by beta-particle counting experiments, with the results listed in Table 9. Despite the fact that these values are reasonably consistent it has recently been suggested that the values may be too high, possibly due to the existence of a bound beta decay or a long-lived nuclear isomer. These suggestions were prompted by the observation that rubidium-strontium ages are often much larger than the corresponding uranium-lead values. This *discrepancy*, which is discussed more fully in section 58, is the only basis for arguing an incorrect value for the half-life, and may actually have another explanation.

[1] N. R. CAMPBELL: Proc. Cambridge Phil. Soc. **14**, 211 (1907). — N. R. CAMPBELL and A. WOOD: Proc. Cambridge Phil. Soc. **14**, 15 (1907).

[2] W. KOLHÖRSTER: Z. Geophys. **6**, 341 (1930).

[3] A. HOLMES and R. W. LAWSON: Phil. Mag. **2**, 1218 (1926).

[4] R. E. STEVENS and W. T. SCHALLER: Amer. Mineral. **27**, 525 (1942).

Table 9. *Determinations of the half-life of rubidium-87.*

| Reference | Half-life × 10⁻¹⁰ years |
|---|---|
| O. HAHN and M. ROTHENBACH: Phys. Z. **20**, 194 (1919). | 7.5 |
| A. HOLMES and R. W. LAWSON: Phil. Mag. **2**, 1218 (1926). | 2.8 |
| W. MÜHLHOFF: Ann. Phys. **7**, 205 (1930). | 12 |
| G. ORBAN: Akad. Wiss. Wien. Ber. **140**, 121 (1931). | 6.3 |
| F. STRASSMAN and E. WALLING: Ber. dtsch. chem. Ges. B **71**, 1 (1938). | 6.3 |
| S. EKLUND: Ark. Mat., Astronom. Fys., Ser. A **33**, No 14 (1946). | 5.8 |
| O. HAXEL, F. G. HOUTERMANS and M. KEMMERICK: Phys. Rev. **74**, 1886 (1948). — M. KEMMERICK: Z. Physik **126**, 399 (1949). | 6.0 |
| S. C. CURRAN, D. DIXON and H. W. WILSON: Phys. Rev. **84**, 151 (1951); recalculated by G. CHARPAK and F. SUZOR: C. R. Acad. Sci. Paris **233**, 1356 (1951). | 7.6 ±0.4 |
| S. C. CURRAN, D. DIXON and H. W. WILSON: Phil. Mag. **43**, 82 (1952). | 6.15 |
| M. H. MACGREGOR and M. L. WIEDENBECK: Phys. Rev. **86**, 420 (1952). | 6.23±0.3 |
| G. M. LEWIS: Phil. Mag. **43**, 1070 (1952). | 5.90±0.3 |
| J. FLINTA and S. EKLUND: Ark. Fysik **7**, 401 (1954). | 6.1 ±0.2 |

**20.** The decay of potassium is accompanied by the emission of beta particles, gamma rays, x-rays and AUGER electrons. Counting experiments have failed to give very conclusive results except for the number of beta particles emitted (Table 10). MÜHLHOFF[1] reported a figure of 23 beta particles per gram of potassium per second. ORBAN[2] reported 75 for the same constant. BRAMLEY[3] compared the terrestrial abundances of calcium and argon and concluded that the ratio of orbital electron capture to beta decays should lie in the range 1/100 to 1/700. He compared this rough estimate with measurements of KOLHÖRSTER and MÜHLHOFF (cf. Table 11) who had observed 6 and 4 gamma rays per 100 beta particles respectively, and concluded that the gamma radiation was associated with the orbital electron capture. This conclusion, which was disputed for more than fifteen years, now appears to be correct.

Table 10. *Rate of beta emission by potassium.*

| Authority | β's/sec. gm. K |
|---|---|
| L. B. BORST and J. J. FLOYD: Phys. Rev. **74**, 989 (1948). | 23 ±2[4] |
| T. GRÅF: Phys. Rev. **74**, 831 (1948). | 26.8±1.2 |
| O. HIRZEL and H. WÄFFLER: Phys. Rev. **74**, 1553 (1948). | 34 ±4 |
| J. J. FLOYD and L. B. BORST: Phys. Rev. **75**, 1106 (1949). | 25 ±2 |
| R. W. STOUT: Phys. Rev. **75**, 1107 (1949). | 30.6±2.0 |
| W. R. FAUST: Phys. Rev. **78**, 624 (1950). | 31.2±3.0 |
| F. G. HOUTERMANS, O. HAXEL and J. HEINTZ: Z. Physik **128**, 657 (1950). | 27.1 |
| G. A. SAWYER and M. L. WIEDENBECK: Phys. Rev. **79**, 490 (1950) | 28.3±1 Ordinary potassium |
| G. A. SAWYER and M. L. WIEDENBECK: Phys. Rev. **79**, 490 (1950). | 30.9±1.7 Enriched to 0.4% K⁴⁰ |
| B. SMALLER, J. MAY and M. FREEDMAN: Phys. Rev. **79**, 940 (1950). | 23[4] |
| F. W. SPIERS: Nature, Lond. **165**, 356 (1950). | 30.5 |
| C. F. G. DELANEY: Phys. Rev. **81**, 158 (1951). | 32.0±3.0 |
| M. L. GOOD: Phys. Rev. **83**, 1054 (1951). | 27.4 |

Mean ± standard deviation = 29.4±2.7

[1] W. MÜHLHOFF: Ann. Physik **7**, 205 (1930).
[2] G. ORBAN: Sitzgsber. Akad. Wiss. Wien **140**, 121 (1931).
[3] A. BRAMLEY: Science, Lancaster, Pa. **86**, 424 (1937).
[4] These results were omitted in calculating the mean.

Table 11. *Rate of gamma ray emission by potassium.*

| Authority | Quantity measured | | | Branching ratio |
|---|---|---|---|---|
| | Radium-C equivalent curies/gm. K | $\gamma$'s/sec. gm. K | $\lambda_\gamma/\lambda_\beta$ | |
| KOLHÖRSTER[1] . . . . . . . . . . | $5 \times 10^{-11}$ | | | 0.061 (0.049) [†] |
| MUHLHOFF[2] . . . . . . . . . | $3.3 \times 10^{-11}$ | | | 0.040 (0.032) [†] |
| BĔHOUNEK[3] . . . . . . . . . | $1.3 \times 10^{-10}$ | | | 0.16   (0.13) [†] |
| GRAY and TARRANT[4] . . . . . . | $1.6 \times 10^{-11}$ | | | 0.02   (0.02) [†] |
| AHRENS and EVANS[5] . . . . . . | | 3.3 | | 0.112 |
| GRÀF[6] . . . . . . . . . . . | | $3.4 \pm 0.08$ | | $0.115 \pm 0.025$ |
| HESS and ROLL[7] . . . . . . . . | $8.9 \times 10^{-11}$ | | | 0.109 (0.087) [†] |
| HIRZEL and WÄFFLER[8] . . . . . . | | | $0.087 \pm 0.012$ | $0.087 \pm 0.012$ |
| FLOYD and BORST[9] . . . . . . . | | | $0.05 \pm 0.01$ | $0.05 \pm 0.01$ |
| SAWYER and WIEDENBECK[10] . . . . | | $2.9 \pm 0.3$ [††] | | $0.098 \pm 0.010$ |
| FAUST[11] . . . . . . . . . . . | | $3.6 \pm 0.4$ | | $0.122 \pm 0.014$ |
| HOUTERMANS, HAXEL, and HEINTZ[12] | | $3.1 \pm 0.3$ | | $0.105 \pm 0.010$ |
| SMALLER, MAY and FREEDMAN[13] . . | | | $0.05 \pm 0.01$ | $0.05 \pm 0.01$ |
| SPIERS[14] . . . . . . . . . . . | | 3.0 | | 0.102 |
| BURCH[15] . . . . . . . . . . . | | $3.5 \pm 0.1$ | | $0.119 \pm 0.006$ |

Mean branching ratio $\pm$ standard deviation $= 0.090 \pm 0.038$

[†] These values are calculated assuming one gamma ray per disintegration of radium-C. The bracketed values are calculated assuming gamma rays in only 80% of the disintegrations. The mean was determined using the higher values.

[††] This value has been recalculated from the original data using a more recent value for the gamma activity of potassium-42 [B. KAHN and W. S. LYON: Phys. Rev. **91**, 1212 (1953)].

[1] W. KOLHÖRSTER: Z. Geophys. **6**, 341 (1930).
[2] A. MÜHLHOFF: Ann. Physik. **7**, 205 (1930).
[3] F. BĔHOUNEK: Physik. **69**, 654 (1931).
[4] L. H. GRAY and G. T. P. TARRANT: Proc. Roy. Soc. Lond., Ser. A **143**, 681 (1934).
[5] L. H. AHRENS and R. D. EVANS: Phys. Rev. **74**, 279 (1948).
[6] T. GRÀF: Phys. Rev. **74**, 1199 (1948).
[7] V. F. HESS and J. D. ROLL: Phys. Rev. **73**, 916 (1948).
[8] O. HIRZEL and H. WÄFFLER: Phys. Rev. **74**, 1553 (1948).
[9] J. J. FLOYD and L. B. BORST: Phys. Rev. **75**, 1106 (1949).
[10] G. A. SAWYER and M. L. WIEDENBECK: Phys. Rev. **76**, 1535 (1949).
[11] W. R. FAUST: Phys. Rev. **78**, 624 (1950).
[12] F. G. HOUTERMANS, O. HAXEL and J. HEINTZ: Z. Physik **128**, 657 (1950).
[13] B. SMALLER, J. MAY and M. FREEDMAN: Phys. Rev. **79**, 940 (1950).
[14] F. W. SPIERS: Nature, Lond. **165**, 356 (1950).
[15] P. R. J. BURCH: Nature, Lond. **172**, 361 (1953).

However, BRAMLEY's estimate seemed to be disproved by the counting experiments of THOMPSON and ROWLANDS[1] which indicated that the ratio of x-rays to beta particles was about 3 or 4. BLEULER and GABRIEL[2] repeated the measurements and found $1.9 \pm 0.4$ for the same ratio. The latter authors also assigned a new and high value to the rate of beta decay with the net result that the half-life of potassium-40 was put at 240 million years. This low half-life value was immediately attacked on geophysical grounds. GLEDITSCH and GRÀF[3] pointed out that such a low half-life would mean that the heat generated by potassium-40 would have been intolerably large in the early part of the earth's

[1] F. C. THOMPSON and S. ROWLANDS: Nature, Lond. **152**, 103 (1943).
[2] E. BLEULER and M. GABRIEL: Helv. phys. Acta **20**, 67 (1947).
[3] E. GLEDITSCH and T. GRÀF: Phys. Rev. **72**, 640 (1947).

history. Their arguments were expanded by BIRCH[1], POOLE[2], VERHOOGEN[3] and JEFFREYS[4] who also concluded that such a short half-life was not possible.

Meanwhile, several workers had concentrated on both beta and gamma measurements. The beta measurements, which were relatively straightforward, are summarized in Table 10. The mean of these results is 29.4 betas/gm. K. sec., with a standard deviation of about 9%.

The gamma measurements proved to be much more difficult (cf. Table 11). KOLHÖRSTER had obtained 1.5 gammas/gm. K. sec., and MÜHLHOFF had obtained 1 gamma/gm. K. sec., both by comparison of the ionization from potassium with that from a radium standard. BĚHOUNEK obtained 4.8 for the same constant from a similar experiment, while GRAY and TARRANT obtained 1.2. The latter measurement appeared to settle the matter and no further measurements were reported for thirteen years.

Since the war, the problem has been taken up with renewed interest. GLEDITSCH and GRÀF reported $3.4\pm0.8$ for the number of gammas/gm. K. sec., and HESS and ROLL reported 2.6, both by comparing the gamma radiation from potassium with that from a radium standard. AHRENS and EVANS compared the rate of emission of gamma ray energy from potassium with that from cobalt-60, and obtained 3.3 gammas/gm. K. sec.

All the above values resulted from attempts to determine the absolute number of gamma rays emitted. The first direct measurement of the ratio of gamma rays to beta particles was made in 1948 by HIRZEL and WÄFFLER who gave a result of $0.087\pm0.012$. Multiplying this figure by the average rate of beta decay from Table 10 gives 2.6 gammas/gm. K. sec. FLOYD and BORST carried out a similar experiment, but used potassium enriched 100-fold in potassium-40 and obtained a gamma-ray to beta-particle ratio of $0.05\pm0.01$ (1.5 gammas/gm. K. sec.). The latter value was supported by SMALLER, MAY and FREEDMAN who investigated the fluorescent properties of potassium iodide which are directly dependent on the gamma to beta ratio of potassium-40. In addition to these values, absolute counting experiments of SAWYER and WIEDENBECK, FAUST, and SPIERS gave values of $2.9\pm0.3$, $3.6\pm0.4$ and 3.0 gammas/gm. K. sec. respectively.

The branching ratio of potassium is defined as the ratio of orbital electron captures to beta-particle emissions. The ratio of gamma rays to beta particles cannot be considered to be equal to the branching ratio unless it is known that there is exactly one gamma ray per electron capture. In 1950 this was far from certain, for the only value for the x-ray to beta-particle ratio was $1.9\pm0.4$—over ten times larger than any gamma-ray to beta-particle ratio that had been measured. Then CECCARELLI, QUARCINI and ROSTANGI reported a ratio of x-rays to beta particles of less than 0.07, which cast great doubt on the earlier x-ray measurements[5]. SAWYER and WIEDENBECK[6] used the fact that AUGER electrons are produced by the internal conversion of $88\%$ of the x-rays and they measured the ratio of AUGER electrons to beta particles in order to obtain a value for the branching ratio of potassium. The value obtained was $0.135\pm0.04$. The most recent x-ray measurement was that of GRÀF[7] who found the ratio of electron captures to beta-particle emissions to be less than 0.2.

[1] F. BIRCH: Phys. Rev. 72, 1128 (1947).
[2] J. H. J. POOLE: Nature, Lond. 162, 775 (1948).
[3] J. VERHOOGEN: Nature, Lond. 164, 72 (1949).
[4] H. JEFFREYS: Ann. Géophys. 6, 10 (1950).
[5] M. CECCARELLI, G. QUARCINI and A. ROSTANGI: Phys. Rev. 80, 909 (1950).
[6] G. A. SAWYER and M. L. WIEDENBECK: Phys. Rev. 79, 490 (1950).
[7] T. GRÀF: Ark. Fysik 3, 171 (1951).

The "accepted" branching ratio, as indicated from x-ray and AUGER electron measurements, has been decreased by about a factor of 30 in the past ten years, from 3.5 to 0.1. Gamma-ray measurements have generally been much more consistent. The ratio of gamma rays to beta particles certainly lies between 0.03 and 0.14 and is probably not too far from the centre of the range. The general convergence of the gamma-ray and x-ray values leads one to suspect that there is one gamma ray emitted per electron capture. This was predicted from theory by MORRISON[1] and is strengthened by precision mass measurements of JOHNSON[2] which show that only one gamma can be associated with each electron capture, and that none can be associated with a beta emission. The decay scheme shown in Fig. 1 which was first proposed by SUESS[3], seems to be on a firm basis.

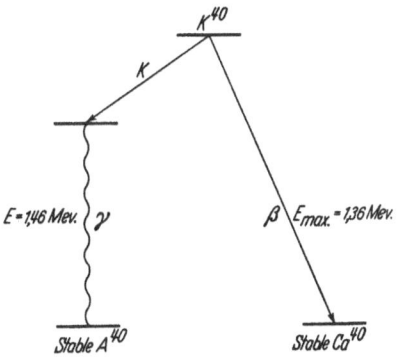
Fig. 1. Decay scheme of potassium-40.

The branching ratio has also been determined by INGHRAM, BROWN, PATTERSON and HESS[4] who found the ratio of argon to calcium in an ancient sylvite to be $0.126 \pm 0.003$. More recently SHILLIBEER, RUSSELL, FARQUHAR and JONES[5] have used potassium argon ratios obtained by themselves and by WASSERBURG and HAYDEN[6] for dated minerals to obtain a branching ratio of 0.089 while WASSERBURG and HAYDEN have used 0.085. All these values are well within the range of the counting experiment results.

**21.** The energy of the beta particles emitted by rubidium-87 has little direct importance in geophysics, for rubidium contributes a negligible amount to the heat generated by the radioactive materials within the earth. Early values for the end-point of the beta spectrum were widely scattered, but more recent results have been in better agreement, indicating a value of between 250 and 275 kev. The spectrum has a forbidden shape, tending to increase the proportion of beta particles of low energy. The average energy is approximately 45 kev. References for the various energy determinations are in general the same as those given in Table 9.

BIRCH [37] has reviewed the various determinations of the gamma and beta energies for potassium-40. These values are of greater interest because of the importance of the radioactive heat produced by potassium in the earth. The energy of the gamma rays is 1.46 Mev. and the beta particles have a maximum energy of 1.35 Mev. The mean energy from ordinary potassium is $26 \pm 2$ cals./ gm. K. year.

## c) Carbon and hydrogen.

**22.** Wherever a neutron flux is present there is a possibility that it will produce transmutations of the atoms under bombardment. In the upper atmosphere there is a continual flux of neutrons produced by cosmic rays. LIBBY[7] suggested

[1] P. MORRISON: Phys. Rev. 82, 209 (1951).
[2] W. H. JOHNSON jr.: Phys. Rev. 88, 1213 (1952).
[3] H. E. SUESS: Phys. Rev. 73, 1209 (1948).
[4] M. G. INGHRAM, H. BROWN, C. PATTERSON and D. C. HESS: Phys. Rev. 80, 916 (1950).
[5] H. A. SHILLIBEER, R. D. RUSSELL, R. M. FARQUHAR and E. A. W. JONES: Phys. Rev. 94, 1793 (1954).
[6] G. J. WASSERBURG and R. J. HAYDEN: Phys. Rev. 93, 645,(1954).
[7] W. F. LIBBY: Phys. Rev. 69, 671 (1946).

that these neutrons might react with the nitrogen-14 in the air to produce radioactive carbon-14. This prompted a search that resulted in the discovery of
radiocarbon in living matter[1].

This discovery and the subsequent development of the carbon-14 method
of age determination led to several accurate measurements of the carbon-14
half-life. A review of some of the recent values are shown in Table 12. The
half-life, approximately 5800 years, is so short that this isotope is of limited use
for dating geological formations. The abundance of carbon-14 is so low that the
amount of heat generated by it in the earth is insignificant.

Table 12. *Half-lives of tritium and carbon-14[2].*

| Isotope | Reference | Half-life in years |
|---------|-----------|--------------------|
| $_1H^3$ | M. GOLDBLATT, E. S. ROBINSON and R. W. SPENCE: Phys. Rev. **72**, 973 (1947). | 10.7 |
| | A. NOVICH: Phys. Rev. **72**, 972 (1947). | 12.1 |
| | G. H. JENKS, J. A. GHORMLEY and F. H. SWEETON: Phys. Rev. **75**, 701 (1949). | 12.5 |
| $_6C^{14}$ | R. C. HAWKINGS, R. F. HUNTER, W. B. MANN and W. H. STEVENS: Canad. J. Res. B **27**, 555 (1949). | 6100±200 or 6400±200 |
| | A. G. ENGELKEMEIR and W. F. LIBBY: Rev. Sci. Instrum. **21**, 550 (1950). | 5580±45 |
| | W. M. JONES: Phys. Rev. **76**, 885 (1949). | 5589±75 |
| | W. W. MILLER, R. BALLENTINE, W. BERNSTEIN, L. FRIEDMAN, A. O. NIER and R. D. EVANS: Phys. Rev. **77**, 714 (1950) | 5513 |
| | G. G. MANOV and L. F. CURTIS: Abstract, 118th Meeting, Amer. Chem Soc., Chicago Sept. 1950. | 5360 |
| | R. S. COSWELL, J. M. BRABANT and A. SCHWEBEL: J. Res. Nat. Bur. Stand. **53**, 27 (1954). | 5900±250 |

Helium-3, which is the decay product of tritium, is observed in the helium in
the atmosphere and in helium from well gases. KAUFMAN and LIBBY[3] show
that it is reasonable that in both cases the helium-3 has been produced from
tritium, which in turn is produced by cosmic ray neutrons or neutrons from
spontaneous fission of uranium.

Table 12 lists half-life determinations for tritium.

# D. Radioactivity of rocks.

## I. Introduction.

**23.** The distribution and intensity of the radioactivity of rocks are subjects
of great importance to geophysicists, in particular for the determination of
absolute geological ages (Part E) and for calculation of the production of
heat within the earth[4]. These subjects are also the concern of the applied
geophysicist, both in searching for economically important deposits of radioactive
minerals, and in utilizing induced and natural radioactive properties of rocks
to prospect for oil and oil bearing strata. While the stimulus of both academic
and economic interests has produced important contributions to our knowledge

---

[1] E. C. ANDERSON, W. F. LIBBY, S. WEINHOUSE, A. F. REID, A. D. KIRSCHENBAUM and
A. V. GROSSE: Science, Lancaster, Pa. **105**, 576 (1947).

[2] For additional references see LIBBY, "Radiocarbon Dating", University of Chicago
Press 1952, p. 34, from which most of these have been taken.

[3] S. KAUFMAN and W. F. LIBBY: Phys. Rev. **93**, 1337 (1954).

[4] See the chapter by J. A. JACOBS in this volume, pp. 389ff.

of rock radioactivity, finer aspects of the subject have not yet been thoroughly investigated. From the available data, is has been possible to formulate only rather general principles concerning the distribution and concentration of those radioactive isotopes which occur in nature.

Measurements of the radioactive properties of all natural materials were commenced shortly after the discovery of radioactivity itself. The extremely sensitive methods which were developed by the early workers in this field for detecting radiation indicated that a low level of activity was present in most rocks and minerals. It was soon shown that this activity was due almost entirely to traces of uranium and thorium and their radioactive decay products. Further investigations led to the discovery of the radioactivity of the alkali metals potassium and rubidium (Part C). As techniques for detecting radioactivity and for isolating chemical elements improved, other naturally occuring radioactive isotopes were discovered. These isotopes, which are listed in Tables 3 and 7, and also rubidium-87, are of such rare occurrence or are so long-lived that their contribution to the overall activity of rocks may be justifiably neglected.

**24.** Since the uranium and thorium decay series cause the major part of the observed activities of crustal rocks, an extensive literature regarding the distribution of these elements has accumulated over the past fifty years [38]. Many of the recent measurements suggest that the results of early determinations were too high, probably because incorrect radioactive standards were used, or because insufficient care was taken to avoid contamination. It is usual therefore to place more stress on those determinations which have been made since 1936 [39]. Nevertheless, the recent results have emphasized the general findings of earlier workers that the trace quantities of uranium and thorium in rocks are not distributed uniformly among the various mineral phases but tend to be concentrated in a few accessory minerals[1]. It has also been found that considerable quantities of these elements may be held interstitially among the different crystalline phases and along grain boundaries[2]. Because of this heterogeneous distribution, considerable variations in the uranium and thorium contents of rocks are to be expected, and any conclusions regarding the occurrence of these elements are necessarily restricted to a few cautious generalizations.

In brief, data available for igneous rocks indicate that the content of uranium and thorium increases with the silica content of the rocks. Thus granites are generally more radioactive than intermediate rocks, which in turn are more active than basic and ultrabasic rocks. Regional surveys of uranium and thorium concentrations in igneous rocks suggest that there are certain areas which are more radioactive than others[3]. These regional measurements also suggest that the average ratio of thorium to uranium in rocks is approximately constant although large variations are apparent in individual samples.

**25.** In any discussion of the radioactivity of rocks, it is important to include some mention of the abundance and distribution of potassium. The decay of potassium-40 contributes significantly to the radioactivity of the earth as a whole, and therefore to the generation of heat in the earth's interior. By

---

[1] C. S. Piggot: J. Amer. Chem. Soc. **50**, 2910 (1928). — N. B. Keevil: Amer. J. Sci. **36** 406 (1938). — E. Picciotto: Bull. Soc. belge Geol., Paleontol. et Hydrol. **59**, 170 (1950). — G. R. Tilton, C. C. Patterson, H. Brown, M. Inghram, R. Hayden, D. Hess, and E. S. Larsen jr.: Bull. Geol. Soc. Amer. **66**, 1131 (1955).

[2] P. M. Hurley: Bull. Geol. Soc. Amer. **61**, 1 (1950). — H. Brown, W. J. Blake, A. Chodos, R. Kowalkowski, C. R. McKinney, G. J. Nuerburg, L. T. Silver and A. Uchiyama: Geol. Soc. Amer., Program of Annual Meeting, 32, 1953. (Abstract.)

[3] F. E. Senftle and N. B. Keevil: Trans. Amer. Geophys. Un. **28**, 732 (1947).

comparison with uranium and thorium, potassium is relatively abundant in both granitic and basic rocks, and the limits of its distribution among the different rock types has been fairly well determined. Section 33 therefore includes some specific analytical data on the distribution and occurrence of this element in rocks together with those given for uranium and thorium.

## II. Methods and measurement.

### a) Uranium and thorium.

**26.** The concentrations of the small quantities of uranium and thorium in rocks have been determined by both direct and indirect methods. By far the greatest proportion of the published data has been obtained by physical methods, in which the radioactivity of the samples has been measured with an ionization chamber, Geiger-Mueller counter, scintillation counter or photographic emulsion.

The most direct method consists of determining the total alpha activity of the rock specimen by placing the crushed sample in an ionization chamber and counting the ionization pulses produced by alpha particles ejected from the uranium, thorium and their daughter products. The true alpha activity of the sample may then be estimated from alpha counting theory, thereby avoiding any possible error due to incorrect radioactive standards[1]. The results of a single measurement are only good to about 20% but this uncertainty may be reduced by making several additional determinations.

Since the method does not provide separate estimates of the uranium and thorium content of the sample, it is usual to combine the result with a separate measurement of either uranium or thorium. Uranium is best estimated by measuring the amount of radon which is in equilibrium with the radium in the sample. Assuming that the radium is in radioactive equilibrium with the parent uranium, the uranium content can be calculated as shown in Sect. 13. For fresh igneous rocks the assumption of radium equilibrium appears to be valid[2]. Radon-222 is a gaseous member of the uranium-238 decay chain (Table 4) having a half-life of 3.8 days. Its chemical properties are similar to those of the inert gases helium, argon, krypton and xenon, and it is therefore unreactive and easy to purify. Radon may be removed from a sample of rock by direct fusion, and its alpha activity then determined in an ionization chamber. By comparing the results with a standard sample of radon, the amount of radon and hence the radium in the rock sample can be estimated.

A similar procedure can be used to determine the thorium in rocks by measuring their thoron (radon-220) content. The thoron is liberated by the same fusion procedure used in releasing radon-222. However, the short half-life of thoron (54 seconds) makes it necessary to use a flow method for determining the activity of the sample. The gas from the fusion is passed through the ionization chamber at an optimum rate determined by the volume of the flow system and the half-life of the thoron. From the level of the observed alpha activity, the thoron, and hence the thorium can be estimated. The half-lives of thoron and radon are sufficiently different so that neither isotope interferes in the determination of the other. Some difficulty in the thoron flow method has been encountered due to distillation of the thoron parent, radium-224, out of the fused sample onto the cooler parts of the apparatus. Many of the thorium results in the literature have therefore been calculated indirectly by subtracting the uranium

---

[1] C. Goodman and R. D. Evans: Bull. Geol. Soc. Amer. **52**, 491 (1941).
[2] R. D. Evans and C. Goodman: Bull Geol. Soc. Amer. **52**, 459 (1941).

content determined by the radon-222 method from the total uranium plus thorium as determined from the alpha activity of the solid sample.

**27.** Recent improvements in the design and analysis techniques of solid source mass spectrometry have made possible the application of an *isotope dilution method* to the direct determination of both uranium and thorium in rocks. This method consists of equilibrating a measured quantity of the element to be determined with that in the sample to be analysed. A portion of the mixture is then extracted and isotopically analysed with a mass spectrometer. The concentration of the element in the sample can be determined provided that the added material has a known and different isotopic composition. In making uranium determinations on rock and mineral samples the usual procedure is to use a known quantity of the separated isotope uranium-235 as the diluant. The percentage concentration of uranium in the sample can then be calculated from the uranium-238/uranium-235 ratio in the mixture.

Thorium has been determined in a similar manner[1]. Since thorium isotopes other than the common thorium-232 are not as yet readily available, the diluant used has been prepared from natural sources, by extracting traces of thorium from large samples of the highly uraniferous ores of the Colorado Plateau area. Because of the high ratio of uranium-238/thorium-232 in the ores, the mole ratio of thorium-230/thorium-232 in the resultant extract is 3.9, nearly $8 \times 10^7$ times the value found in a rock having an average thorium-uranium ratio of 3.5. Such material therefore conveniently serves as an isotopic diluant in making thorium analyses of ordinary rocks.

These methods have not as yet been widely applied, but some uranium and thorium analyses have been made on rocks and mineral separates from rocks[2]. The details of the techniques and the necessary chemical processing of the samples preliminary to isotopic analysis have been fully described in these papers. A portion of the results obtained are included in the following section, since they indicate the manner in which uranium and thorium are distributed among the mineral phases of one particular granite.

**28.** Extremely small quantities of uranium may be determined directly by means of a fluorimetric method which has been developed for routine analyses of rocks by GRIMALDI[3,4]. Uranium salts when fused with sodium fluoride fluoresce an intense yellow green under ultraviolet light, the intensity of the fluorescence being proportional to the uranium concentration over fairly wide limits. The method is specific for uranium if the wavelength of the exciting radiation is about 3650 Å, but cobalt, chromium, nickel, manganese, silver, platinum, gold, lead and the rare earths must either be removed from the sample or their concentrations must be diluted since they tend to quench the fluorescence due to the uranium.

GRIMALDI et al[4] state that the limit of detection of this procedure is about $10^{-10}$ grams of uranium compared to limits of $10^{-5}$ grams for the polarographic and colorimetric methods.

**29.** Thorium has proved a difficult element to determine chemically at low concentrations. Attempts to apply methods for determining relatively large

---

[1] G. R. TILTON, L. T. ALDRICH and M. G. INGHRAM: Anal. Chem. **26**, 894 (1954).

[2] G. R. TILTON, C. C. PATTERSON, H. BROWN, M. INGHRAM, R. HAYDEN, D. HESS, and E. S. LARSEN jr.: Bull. Geol. Soc. Amer. **66**, 1131 (1955).

[3] F. S. GRIMALDI, I. MAY and M. S. FLETCHER: U.S. Geol. Surv. Circ. 199, 1952.

[4] F. S. GRIMALDI, I. MAY, M. S. FLETCHER and J. TITCOMB: U.S. Geol. Surv. Bull. 1006, 1954.

quantities of this element to the amounts of thorium in rocks have not been entirely successful. Difficult analytical problems arise in treating the elements with which thorium is often associated, and in selectively separating the thorium from such mixtures. A colorimetric method has recently been developed and at present gives an accuracy of $\pm 2$ to $3\,\%$[1]. The results compare favourably with those obtained by the isotope dilution method.

Measurements of natural total radioactivities have been made on rocks in situ, by adaptation of the well-known bore hole logging methods utilized in geophysical prospecting. The relative gamma-ray activities of rock formations are often measured by lowering geiger tubes or ionization chambers down cased wells. It has been found that many shales are much more radioactive than other sedimentary formations, and such beds of shale may often be identified and traced by means of their comparatively high radioactivities.

### b) Potassium and rubidium.

**30.** Determinations of the distribution of potassium in rocks have in the past been made almost entirely by means of standard gravimetric procedures based on the precipitation of potassium chloroplatinate from ethanol solution. The classic method due to J. Laurence Smith[2] depends on the separation of the alkali metals by a carbonate fusion followed by a selective water leaching of the readily soluble alkali metal salts. Alternatively the mineral specimens may be decomposed by the repeated action of hydrofluoric acid in a sulphuric acid medium. Both methods give trustworthy results for potassium concentrations of a few percent and greater.

For quantities of potassium less than one percent emission spectroscopic methods have been used with success[3] although the accuracy obtained was $+15\%$ to $10\%$. These investigators found that the small amounts of potassium in ultramafic rocks and meteorites had been apparently overestimated in many previous chemical analyses on these materials. The concentration of potassium in meteorites has also been determined by the isotopic dilution method, using potassium-41 as the diluant[4]. The results obtained by this method are in good agreement with those determinations made using a new extraction procedure coupled with a flame photometer analysis[5]. Flame photometer analyses have in general proved reliable throughout the entire range of potassium concentrations found in terrestrial and meteoritic material.

A study of the rubidium contents of rocks was undertaken by Ahrens[3] using the same samples on which potassium determinations were made. Previous rubidium analyses reported in the literature gave comparable results to those obtained by these workers.

## III. Results.

### a) Uranium and thorium in igneous rocks.

**31.** The results of measurements of uranium and thorium concentrations in igneous rocks show the statistical scatter inherent in trace element distributions in natural materials. This scatter is apparent in the measurements and estimates of various authors as given in Table 13a. The values shown in the table have

---

[1] G. R. Tilton, L. T. Aldrich and M. G. Inghram: Anal. Chem. **26**, 894 (1954).
[2] J. Laurence Smith: Amer. J. Sci. **1**, 269 (1871).
[3] L. H. Ahrens, W. H. Pinson and M. M. Kearns: Geochim. et Cosmochim. Acta **2**, 229 (1952).
[4] G. J. Wasserburg and R. J. Hayden: Phys. Rev. **97**, 86 (1955).
[5] G. Edwards and H. C. Urey: Geochim. et Cosmochim. Acta **7**, 154 (1955).

Table 13 a. *Average uranium and thorium contents of igneous rocks.*

| Rock type | No. of samples | Uranium (p.p.m.) [*] | Thorium (p.p.m.) | Reference |
|---|---|---|---|---|
| Granites (North America) . . . . . | 9 | 3.8 ±0.4 | 10.3 ± 1.7 | [1] |
| Acidic (North America) . . . . . . | 12 | 3.9 ±0.3 | 13.4 ± 1.6 | [1] |
| Acidic (weighted average) . . . . . | 26 | 3.0 ±0.3 | 13 ± 2 | [1] |
| Acidic (range of averages for Canadian and United States regions) . | 1257 | 3.84 to 4.02 | 13.1 to 13.5 | [2] |
| Acidic (North America, Greenland, Iceland, Ireland and Japan) . . | | 4.65±0.35[**] | 8.1 ± 08 | [3] |
| Acidic (Finland) . . . . . . . . | | 13.6 ±1.17[**] | 28.0 ± 2.4 | [3] |
| Acidic (Alps) . . . . . . . . . | | 12.9 ±1.98[**] | 33.0 ± 5.0 | [3] |
| Acidic (South Africa) . . . . . . | | 6.89±0.47[**] | — | [3] |
| Intermediate (North America) . . . | 6 | 1.4 ±0.2 | 4.4 ± 1.2 | [1] |
| Intermediate (range of averages for Canadian and United States regions) . . . . . . . . . . . . | 297 | 3.03 to 2.27 | 9.28 to 10.5 | [2] |
| Basalts (North America) . . . . . | 8 | 0.83±0.15 | 5.0 ± 0.3 | [1] |
| Diabase . . . . . . . . . . . . | 4 | 0.83±0.15 | 2.0 ± 0.2 | [1] |
| Trap . . . . . . . . . . . . . | 5 | 1.3 ±0.2 | 4.2 ± 0.2 | [1] |
| Basic rocks . . . . . . . . . . | 27 | 0.95±0.09 | 3.8 ± 0.5 | [1] |
| Basic rocks (weighted average) . . . | 34 | 0.96±0.11 | 3.9 ± 0.6 | [1] |
| Basalts (North America, Greenland, Iceland, Scotland, Ireland and Japan) . . . . . . . . . . . . | | 2.8 ±0.2[**] | 9.8 ± 0.8 | [3] |
| Basalts (England, Germany, France and Hungary) . . . . . . . . | | 3.80±0.8[**] | 8.8 ± 1.0 | [3] |
| Plateau basalts . . . . . . . . | | 2.1 ±0.1[**] | 5.2 ± 0.2 | [3] |
| Oceanic Island basalts . . . . . . | | 2.6 ±0.1[**] | 4.6 ± 0.3 | [3] |
| Dunites (world) . . . . . . . . . | | 1.2 ±0.2[**] | 3.3 ± 0.3 | [3] |
| Basic lavas . . . . . . . . . . | | 0.6 to 1.1 | — | [4] |
| Ultrabasic rocks . . . . . . . . | | 0.03 | — | [5] |

[*] p. p. m. is parts per million by weight.
[**] Calculated assuming a U/Ra ratio of $2.92 \times 10^6$ gm./gm.

[1] R. D. Evans and C. Goodman: Bull. Geol. Soc. Amer. **52**, 459 (1941).
[2] F. E. Senftle and N. B. Keevil: Trans. Amer. Geophys. Un. **28**, 732 (1947).
[3] H. Jeffreys: The Earth, 3rd edit. Cambridge University Press 1952.
[4] J. A. S. Adams: Nuclear Geology, edit. H. Faul, pp. 89—98. New York: John Wiley & Sons 1954.
[5] G. L. Davis: Amer. J. Sci. **245**, 677 (1947).

been arrived at by averaging determinations made on various classes of rocks from many different geological and geographical localities.

It is evident that the general terms granitic, intermediate and basic are too broad for the purpose of comparison of uranium and thorium contents. Furthermore regional surveys of radioactivity have shown areas of igneous rocks of higher than average radioactivity in North America and Africa[1], and on a smaller scale it has been shown that the radioactivity associated with single batholiths tends to be concentrated toward the periphery of the intrusive[2].

Recent work has been directed towards studies of the radioactivity in rocks from a single magmatic series in relation to the overall composition or major constituents[3].

[1] F. E. Senftle and N. B. Keevil: Trans. Amer. Geophys. Un. **28**, 732 (1947). — C. F. Davidson: Min. Mag., Lond. **85**, 329 (1951).
[2] H. A. Slack and K. Whitham: Trans. Amer. Geophys. Un. **32**, 44 (1951).
[3] E. S. Larsen jr. and G. Phair: Nuclear Geology, edit. H. Faul, pp. 75—89. New York: John Wiley & Sons 1954. — J. A. S. Adams: Nuclear Geology, edit. H. Faul, pp. 89—98. New York: John Wiley & Sons 1954.

Table 13b. *Distribution of uranium and thorium in igneous rocks.*

| Mineral constituent | Woodson granodiorite California[1] Uranium (p.p.m.) | Rattlesnake calc-alkaline granite California[1] Uranium (p.p.m.) | Tory Hill granite Ontario[2] | |
|---|---|---|---|---|
| | | | Uranium (p.p.m.) | Thorium (p.p.m.) |
| Quartz . . . . . . . . . . . | 2.2 | 2.4 | 0.13 | |
| Orthoclase . . . . . . . . . | | 1.2 | | |
| Perthite . . . . . . . . . . | 8.2 | | 0.22 | 0.41 |
| Plagioclase . . . . . . . . | 6.0 | 1.9 | 0.20 | |
| Biotite. . . . . . . . . . . | 2.6 | 5.2 | | |
| Hornblende . . . . . . . . | 2.8 | | | |
| Magnetite . . . . . . . . . | 3.5 | | 2.57 | |
| Sphene . . . . . . . . . . | | | 303 | 2205 |
| Garnet. . . . . . . . . . . | 7.5 | 5 | | |
| Zircon . . . . . . . . . . . | 1750 | 4600 | 2650 | 2180 |
| Monazite . . . . . . . . . . | 820 | | | |
| Apatite . . . . . . . . . . | 62 | 47 | 90.5 | |
| Xenotime . . . . . . . . . | 12700 | 360 | | |
| Muscovite . . . . . . . . . | | 8.1 | | |

[1] E. S. LARSEN jr. and G. PHAIR: In Nuclear Geology, edit. H. FAUL, p. 85. New York: John Wiley & Sons 1954.

[2] G. R. TILTON, C. C. PATTERSON, H. BROWN, M. INGHRAM, R. HAYDEN, D. HESS and E. S. LARSEN jr.: Bull. Geol. Soc. Amer. **66**, 1131 (1955).

The results of such investigations support the general conclusions which may be drawn from the data given in Table 13a. The radioactivity of rocks is highest in granitic rocks and decreases as the silica content decreases, to reach very low values in the basic and ultrabasic types. Other specialized studies have revealed a number of interesting facts about the concentration of uranium and thorium with respect to their distribution among the various phases and mineral species of igneous rock specimens. As the results in Table 13b indicate, these elements tend to be concentrated largely in the accessory minerals such as zircon, sphene and apatite, and between the grains and mineral boundaries of the rock components. This particular distribution can be understood when the general *geochemical factors* controlling the concentrations of uranium and thorium are considered:

The ionic radius and charges of uranium and thorium ions in a silicate melt are such as to prevent their inclusion during crystallization in the lattice structures of the common rock forming minerals such as quartz, plagioclase and perthite. They may however replace zirconium ions in zircon, or $Ca^{++}$-ions in apatite and sphene. In the process of crystallization of a rock, uranium and thorium will tend to be concentrated in the residual liquid, eventually being incorporated in minerals like the complex rare earth silicates, uraninite, monazite and uranothorite, or perhaps remaining as films among the previously solidified crystals. Uranium and thorium are treated similarly throughout these chemical processes, and on the average the ratio of thorium to uranium would be expected to remain the same in all types of igneous rocks. This description is oversimplified, but explains quantitatively the observed distribution of these elements in igneous rocks.

### b) Sedimentary rocks.

**32.** The uranium and thorium contents of sediments and sedimentary rocks are subject to even wider variations than their concentration in igneous rocks. However investigations have shown that broad concentration limits may be defined for different types of sedimentary rocks. A large percentage of the uranium and thorium which enter sediments are derived from the erosion of pre-existing igneous, metamorphic and sedimentary material. Under the physical and chemical processes which act during the weathering of rocks, these elements may follow several different courses. The uranium and thorium within the lattices of resistant minerals will be deposited with these minerals in placer deposits and sands and form the major contribution to the radioactivity of such clastic sediments. Accessory minerals such as apatite, monazite and xenotime which may contain considerable percentages of the uranium and thorium in igneous rocks are relatively impervious to chemical weathering but can be broken up by attrition. In placers which have been deposited close to the point of origin of the source material, these minerals will predominate. Zircon is resistant to both chemical and mechanical weathering and occurs over much wider areas. Because thorium-rich minerals are on the whole more resistant that uranium minerals, the average ratio of thorium to uranium in clastic sediments is considerably higher than the ratio in igneous rocks.

The remainder of these elements which occur in the more easily weathered minerals and between the mineral boundaries are dissolved by erosional processes and either remain in solution or are precipitated with marine sediments. Under proper chemical conditions uranium is adsorbed in considerable concentrations by certain forms of carbonaceous sediments, some types of clays, phosphatic sediments of marine origin, and to a smaller extent by gelatinous compounds of elements such as iron, manganese, or silicon[1]. In particular, many black, bituminous marine shales contain relatively large amount of uranium sometimes as much as 0.01 or 0.02 percent, the amount of uranium in general increasing with the carbon content. On the other hand, sediments consisting mainly of marine carbonates are extremely poor in uranium, since carbonate ion blocks both the precipitation and adsorption of uranium from solution[2].

The distribution of uranium in the oceans and deep sea sediments has also been determined. The concentration of dissolved uranium is a function of depth, but averages about $1.3 \times 10^{-9}$ gms. per ml. of sea water while deep sea sediments such as red clay contain 1 or 2 parts per million[3].

Uranium may often be introduced into sediments after they have been consolidated, being carried by hydrothermal solutions and deposited in cracks and porous zones under favourable chemical conditions. These conditions and the solution constituents determine the types of uranium mineral that is formed. Some of the world's largest uranium deposits are sedimentary beds carrying 0.1 or more percent of uranium and there has been much debate as to whether the uranium was deposited during sedimentation and recrystallized later or whether the uranium was introduced into the sedimentary beds by the subsequent passage of hydrothermal solutions[4].

Relatively little is as yet known about the distribution of thorium in sediments, mainly because of the difficulty in making rapid, accurate analyses for small quantities of this element. The portion of the thorium which is dissolved during rock erosion can be easily hydrolysed and subsequently precipitated with

[1] Yu. M. Tolmachev: Bull. Acad. Sci. USSR., Cl. Sci. Chim. no. 1, 28 (1943). — R. F. Beers and C. Goodman: Bull. Geol. Soc. Amer. 55, 1229 (1944). — V. E. McKelvey and J. M. Nelson: Econ. Geol. 45, 35 (1950).

[2] C. S. Piggot and W. D. Urry: Amer. J. Sci. 239, 81 (1941).

[3] G. Koczy: Öst. Akad. Wiss., Math.-naturwiss. Kl., Sitzgsber., Abt. IIa 158, 113 (1950). W. D. Urry and C. S. Piggot: Amer. J. Sci. 240, 93 (1942).

[4] C. F. Davidson: Trans. Instn. Min. Metal. 63, 244 (1954). — J. D. Louw: Trans. Geol. Soc. S. Africa (in press).

hydrolyzate sediments[1]. Hydrothermal solutions apparently carry very little thorium since the ratio of thorium to uranium is extremely low in deposits formed from such solutions.

Representative results of the uranium and thorium contents of various sedimentary rocks are presented in Table 14. Only wide limits of concentration can be assigned to any particular type of sedimentary formation and it is apparent

Table 14. *Average uranium and thorium contents of sedimentary rocks.*

| Rock type | Average Uranium p.p.m. * | Average Thorium p.p.m. * | Reference |
|---|---|---|---|
| Average in placers . . . . . . | 2 | 60 | 1 |
| Average of all sediments, exclusive of limestones . . . . | 1.2 | 3.3 | 2 |
| | — | 5 to 12 | 3 |
| Limestones . . . . . . . . . . | 1.3 | 1.1 | 2 |
| Shales . . . . . . . . . . . . | — | 10 | 4 |
| Carbonaceous sediments (shales). | up to 100 | — | 1 |

* p. p. m. is parts per million by weight.
[1] K. G. BELL: In Nuclear Geology, edit. H. FAUL, pp. 98–114. New York: John Wiley & Sons 1954. — K. G. BELL, C. GOODMAN and W. L. WHITEHEAD: Bull. Amer. Assoc. Pet. Geol. 24, 1529 (1940).
[2] R. D. EVANS and C. GOODMAN: Bull. Geol. Soc. Amer. 52, 459 (1941).
[3] J. J. JOLY: Phil. Mag. 20, 353 (1910).
[4] E. MINAMI: Nachr. Ges. Wiss. Göttingen, Math.-phys. Kl., Fachgruppe IV 1, 155 (1935).

that the different chemical properties of uranium and thorium and the different physical properties of the minerals containing them may result in considerable differentiation of these elements in sedimentary rocks.

The radioactivity of *recent marine sediments* is due in part to the presence of ionium and radium and their subsequent decay products, which have been adsorbed or precipitated through chemical action with the sediments[2]. In the deep oceans, the activity of the surface layers down to a depth of 40 to 80 cm. is due to such elements. Below this depth, the excess ionium and radium have entirely decayed and the remainder present is in radioactive equilibrium with the small quantities of uranium and thorium parents in the sediments.

### c) Potassium and rubidium.

33. Potassium is a relatively abundant element in most granitic rocks. Consequently potassium contents have been determined in the courses of most rock analyses, and the limits of distribution of this element in granitic type of surface rocks are well known. As presented in Table 15, DALY (1933)[3] has given an average of 2.9% for the potassium content of granites, based on a large number of analyses. Rocks of basaltic composition are subject to wider variations than granitic types, because of the marked effects of geochemical fractionation of potassium. AHRENS, PINSON and KEARNS[4] have found that the potassium content of a number of successive flows in the Columbia River basalts vary from

[1] KALERVO RANKAMA and T. G. SAHAMA: Geochemistry. University of Chicago Press 1950.
[2] H. PETTERSSON: Nuclear Geology, edit. H. FAUL, pp. 115–120. New York: John Wiley & Sons 1954.
[3] R. A. DALY: Igneous Rocks and the Depths of the Earth. New York: McGraw-Hill Book Co. 1933.
[4] L. H. AHRENS, W. H. PINSON and M. M. KEARNS: Geochim. et Cosmochim. Acta 2, 229 (1952).

0.65 to 1.4%. AHRENS[1] tabulation of average abundances of basaltic and similar material are given in Table 15.

It has recently been shown that the potassium contents of ultramafic rocks such as dunite, serpentine and perioditite are much lower than was previously believed[2]. The average value obtained by these authors is included in Table 15.

Table 15. *Average potassium contents of igneous and sedimentary rocks.*

| Rock type | % Potassium | Reference |
|---|---|---|
| Granite . . . . . . . . . . . . . . . . . | 2.9 | 1 |
| Plateau basalt . . . . . . . . . . . . . . | 0.65 | 1 |
| Oceanite . . . . . . . . . . . . . . . | 0.37 | 1 |
| Basalt . . . . . . . . . . . . . . . . . | 0.15 to 2.0 | 2 |
| Eclogite . . . . . . . . . . . . . . . . | 0.05 to 0.75 av. 0.24 | 2 |
| Dunite, serpentine and peridotite (type commonly associated with orogenic belts) . . | 0.001 | 3 |
| Shales . . . . . . . . . . . . . . . . . | 3.0 | 2 |
| Sandstones . . . . . . . . . . . . . . | 1.0 | 2 |
| Limestones . . . . . . . . . . . . . . | 0.1 to 0.3 | 2 |

[1] R. A. DALY: Igneous Rocks and the Depths of the Earth. New York: McGraw-Hill Book Co. 1933.

[2] L. H. AHRENS: Nuclear Geology, edit. H. FAUL pp. 128—132. New York: John Wiley & Sons 1954.

[3] L. H. AHRENS, W. H. PINSON and M. M. KEARNS: Geochim. et Cosmochim. Acta 2, 229 (1952).

Chemical fractionation determines to a great extent the abundance of potassium in sedimentary rocks. The general ranges of magnitude of the potassium contents of argillaceous sediments, sandstones and limestones are also included in Table 15, together with the results for igneous rocks. The amounts of potassium in these classes are extremely variable even for samples from single sedimentary formations.

A careful study of the abundance of *rubidium* in all types of igneous rocks and meteorites, and the relation of this element to potassium in these same rocks has been made by AHRENS, PINSON and KEARNS[3]. The similar chemical properties and ionic radii of the potassium and rubidium ions result in an extremely close relationship between these two elements in natural chemical processes, even throughout those which result in such large variations in the absolute abundances in basaltic rocks. The average value for the potassium/rubidium determined for all rock types is 90 gms/gm. EDWARDS and UREY[4] have recently found a K/Rb ratio of 147 gms/gm.

## d) Meteoritic abundances of uranium, thorium, potassium and rubidium.

**34.** The concentration of radioactive elements in meteorites has been closely linked in the past with the problem of the heat generated within the earth[5]. The levels of concentration of uranium and thorium in meteorites are extremely low,

[1] L. H. AHRENS: In Nuclear Geology, edit, H. FAUL, pp. 128—132. New York: John Wiley & Sons 1954.

[2] W. K. HOLYK and L. H. AHRENS: Geochim. et Cosmochim. Acta 4, 241 (1953).

[3] L. H. AHRENS, W. H. PINSON and M. M. KEARNS: Geochim. et Cosmochim. Acta 2, 229 (1952).

[4] G. EDWARDS and H. C. UREY: Geochim. et Cosmochim. Acta 7, 154 (1955).

[5] J. A. JACOBS: The Earth's Interior, Chapter in this volume, pp. 389 ff.

and have required most specialized techniques for their determination. For this reason the distribution of uranium and thorium among the different phases of meteoritic material is known only approximately. Some of the most recent estimates are given in Table 16, together with the concentrations of potassium and rubidium. It is evident that if meteoritic material is assumed typical of that of the interior of the earth, surface rocks are on the average considerably enriched in the radioactive elements[1].

Table 16. *Uranium, thorium, potassium and rubidium contents of meteorites.*

| Type of meteorite | No. of samples | U×10⁻⁸ gm./gm. | Th×10⁻⁸ gm./gm. | K p.p.m. | Rb p.p.m. | Reference |
|---|---|---|---|---|---|---|
| Iron (octahedrites and hexahedrites) . . . . . . . | 6 | 0.7 | 4.0 | | | 1 |
| | 7 | 0.33 | | | | 2 |
| Iron (based on combined U—Th method of determination . | 15 | 0.05 to 0.75 | ≤0.6 to 10.5 | | | 3 |
| Stone . . . . . . . . . . . | 1 | 1.0 | | | | 2 |
| | 2 | 1.1 | | | | 4 |
| Stone (chondrite) . . . . . | 1 | 10.6 | 33.5 | | | 5 |
| Stone (achondrite) . . . . . | 1 | 4.8 | | | | 2 |
| Pallasite (metal phase) . . . | 3 | 1.23 | | | | 2 |
| Pallasite (olivine phase) . . . | 2 | <0.11 | | | | 2 |
| Stone (chondrite) . . . . . | 21 | | | 900 | 9 | 6 |
| | 21 | | | 850 | | 7 |
| | 7 | | | | 5 | 7 |
| Stone (carbonaceous chondrite) | 8 | | | 300 to 600 | | 7 |
| Stone (achondrite) . . . . . | 4 | | | 8 to 830 | | 7 |
| Pallasite and chladnite . . . | 2 | | | <10 | <2 | 6 |

[1] W. J. ARROL, R. B. JACOBI and F. A. PANETH: Nature, Lond. 149, 235 (1942).
[2] G. L. DAVIS: Amer. J. Sci. 248, 107 (1950).
[3] J. C. DALTON, J. GOLDEN, G. R. MARTIN, E. R. MERCER and S. J. THOMSON: Geochim. et Cosmochim. Acta 3, 272 (1953).
[4] C. PATTERSON, H. BROWN, G. TILTON and M. G. INGHRAM: Phys. Rev. 92, 1234 (1953).
[5] K. F. CHACKETT, J. GOLDEN, E. R. MERCER, F. A. PANETH and P. REASBECK: Geochim. et Cosmochim. Acta 1, 3 (1950).
[6] L. H. AHRENS, W. PINSON and M. KEARNS: Geochim. et Cosmochim. Acta 2, 229 (1952).
[7] G. EDWARDS and H. C. UREY: Geochim. et Cosmochim. Acta 7, 154 (1955).

# IV. Conclusion.

**35.** Part D has outlined in a general manner the amounts of radioactivity which occur in rocks and meteorites, and some of the methods which have been applied to the determination of the low concentrations of the radioactive elements in natural material. From the variability of the results presented in the previous sections it is apparent that data can only be applied on a regional basis. Fuller accounts of the distribution of radioactivity in rocks are given in a recent book "Nuclear Geology", edited by H. FAUL [30] and a review "Radioactivity in Geology and Cosmology", by KOHMAN and SAITO [29]. Present interest in this subject has accelerated work on the techniques of determining small quantities of uranium and thorium with improved accuracy. There should shortly be available more precise methods with a considerably greater fund of data on the radioactive properties of natural materials in general. No attempt has been made here to include data on major deposits of uranium and thorium. These deposits are relatively rare occurences in nature, and are unimportant in considerations of the earth's internal sources of heat.

[1] A. HOLMES: Geol. Mag. 2, 64 (1915).

# E. Methods of age determination.

## I. Age determinations based on the decay of uranium and thorium to helium and lead.

### a) Methods for uranium minerals.

**36.** Among the naturally occurring radioactive isotopes, the two isotopes of uranium (uranium-238 and uranium-235) and the only abundant isotope of thorium (thorium-232) form a group because they occur together and all the end products of their decay are isotopes of helium and lead. They have been used as the standards in age determination work for several excellent reasons. There are no uranium and thorium isotopes which are not radioactive so that large amounts of decay products are produced. Both elements form major constituents of certain minerals, e.g. uraninite. All their isotopes disintegrate at rates which have been fairly accurately measured and the details of the decay systems have been well studied (cf. Part C III and Tables 4, 5 and 6).

Determinations of the quantity of *helium* formed by the decay of these elements was at one time thought to give reliable ages. Subsequent work showed that in many cases much of the helium generated in a rock or mineral escapes, so that now usually only *lead* is used for age determinations. In some special cases where the mineral is very dense, and the concentration of uranium and thorium low, reliable ages may perhaps be obtained by the helium method, but it is hard to be sure which of these ages are reliable.

The study of the lead end-product has therefore become the standard method. Fortunately, each of the three parent uranium and thorium isotopes produces a different lead isotope and these can be identified and the relative amounts of each measured with a mass spectrometer. Common lead contains the same three isotopes but it also contains lead-204 which serves as a measure of contamination of *radiogenic lead* by *common lead*. The amounts of uranium and thorium isotopes can also be determined and thus all three systems can be completely distinguished. It is possible to study the results of decay in these three systems in different ways which give rise to several separate methods of age determination.

**37.** Originally age determinations were based only on *chemical analyses* for uranium, thorium and lead. The time of deposition of the mineral may be calculated approximately from the formula:

$$t = \frac{7.37 \times 10^9 \, [\mathrm{Pb}]}{[\mathrm{U}] + 0.35 \, [\mathrm{Th}]} \text{ years} \tag{37.1}$$

where $t$ is the age of the mineral in years and [Pb], [U], and [Th] are the proportions by weight of the corresponding elements as determined by chemical analysis.

This formula was modified from time to time as the decay constants became more accurately known. While it is a good approximation for comparatively young ages a small correction is needed for very old minerals[1]. This method gives only one age determination for each specimen and thus provides no checks. At the time of much of the earlier work uranium-235 had not been recognized and incorrect decay constants were used. In addition, contamination by common lead cannot be detected and corrections for it cannot be made. Fortunately, some of the early determinations including many of those by ELLSWORTH [40], HILLEBRAND [4], and HOLMES [2a] were made on uraninite crystals which were

---

[1] A. HOLMES and R. W. LAWSON: Amer. J. Sci. (5) **13**, 327 (1927).

pure and unaltered. Some of these analyses gave fairly good results but others were found to be in error when checked by other methods. WICKMAN [41] has published curves which can give crude ages more accurately than the above simple formula.

Due to the uncertainty in the extent of common lead contamination, chemical analyses alone can no longer be recommended.

**38.** When, in addition to chemical analyses for lead, uranium and thorium, an *isotopic analysis* of the lead has been made, all three decay systems can be

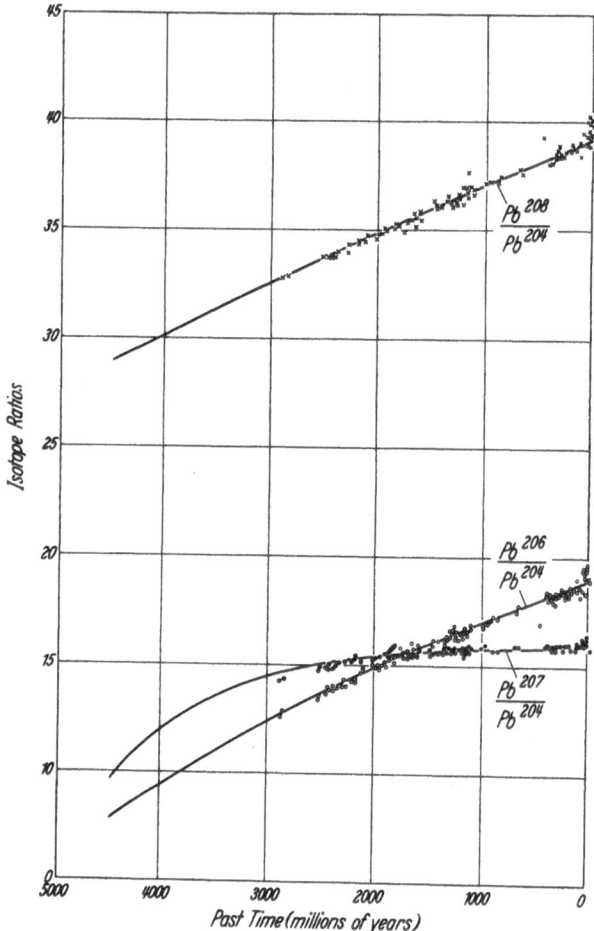

Fig. 2. Common lead abundance-time curves. (After G. L. CUMMING: Ph. D. Thesis, University of Toronto, 1955.)

treated separately using the knowledge that the ratio of uranium-235 to uranium-238 is always found to be $1:137.8$[1]. Common lead can be detected by the presence of lead-204, a correction made for it, and three independent ages determined, according to formulae of the type

$$Pb^{206} = U^{238} (e^{\lambda t} - 1) \tag{38.1}$$

---

[1] M. G. INGHRAM: Manhattan Project Technical Series (Div. 2, Gaseous Diffusion Project), Vol. 14, Chap. V, p. 35. 1947.

where $Pb^{206}$, etc., are numbers of atoms of these isotopes and $\lambda$ is the decay constant. Introducing the atomic weights and decay constants these formulae give

$$t_{206} = 6.50 \times 10^9 \log_e\left(1 + 1.163 \frac{[Pb^{206}]}{[U]}\right), \tag{38.2}$$

$$t_{207} = 1.03 \times 10^9 \log_e\left(1 + 159.2 \frac{[Pb^{207}]}{[U]}\right), \tag{38.3}$$

$$t_{208} = 20.0 \times 10^9 \log_e\left(1 + 1.126 \frac{[Pb^{208}]}{[Th]}\right), \tag{38.4}$$

where $t$ is the age in years and $[Pb^{206}]$ etc. are weights in grams. As some thorium is usually present in uranium minerals and vice versa, three independent ages can usually be obtained in this way. If these all agree it is most unlikely that any losses of any isotopes could have occurred and then this method provides the surest possible value for an age determination. An excellent example of such agreement has just been reported in considerable detail by HOLMES for a sample from Bikita, Southern Rhodesia (Table 17), while others have also been published [6], [7], [8], [12], [13].

If the isotopic analysis reveals that lead-204 is absent, or almost absent and if the ages agree, then this method presents no problems. Unfortunately, this is not usually the case. If lead-204 is present the question arises as to what isotopic ratios are to be used for the *common lead* correction, because common or ore leads vary widely in isotopic composition. Most common leads are what are called *ordinary* in which case the variation in composition is a function of age[1]. Variation in composition with age for these leads are illustrated by the curves drawn in Fig. 2. Unfortunately, some leads are *anomalous* with compositions quite different from ordinary leads (Tables 19 and 20) and these are much more variable. In making a correction for common lead it is usual to use the composition of a normal lead of the same age as the radiogenic lead but this clearly involves an assumption which could at times lead to large errors. BATE and ECKELMANN[2] hold that such large errors are in fact unlikely, but it is best to avoid contaminated specimens.

Table 17. *Typical age determinations of uranium minerals[3].*

| Location | Mineral | U | Th | Pb | 204 | 206 | 207 | 208 | Age (millions of years) |
|---|---|---|---|---|---|---|---|---|---|
| | | % by weight of mineral | | | atom percent of total lead | | | | |
| Huron Claim, Manitoba | Uraninite | 53.50 | 12.46 | 15.44 | 0.018 | 81.50 | 13.18 | 5.31 | 2490 (a) |
| | | (ELLSWORTH and DeLURY) (best analysis) | | | (NIER No 14) | | | | 1370 (b) 1980 (c) 1270 (d) |
| Huron Claim, Manitoba | Uraninite | — | — | — | 0.045 | 79.83 | 14.02 | 6.11 | 2565 (a) |
| | | | | | (University of Toronto No 504) | | | | |
| Huron Claim, Manitoba | Uraninite | — | — | — | 0.054 | 79.23 | 13.65 | 7.10 | 2550 (a) |
| | | | | | (University of Toronto No 845) | | | | |

[1] D. W. ALLAN, R. M. FARQUHAR and R. D. RUSSELL: Science, Lancaster, Pa. 118, 486 (1953).
[2] G. L. BATE and W. R. ECKELMANN: Bull. Geol. Soc. Amer. 64 (1953), abstract only.
[3] For NIER's samples see references [6] and [8]. For HOLMES' samples see A. HOLMES, Nature, Lond. 173, 612 (1954). For Toronto samples see reference [46]. For KULP et al. samples see J. L. KULP, W. R. ECKELMANN, H. R. OWEN and G. L. BATE: United States Atomic Energy Commission, NYO—6199, 1953. For WASSERBURG's samples see Geochim. et Cosmochim. Acta 7, 51 (1955).

Table 17. (Continued.)

| Location | Mineral | U | Th | Pb | 204 | 206 | 207 | 208 | Age (millions of years) |
|---|---|---|---|---|---|---|---|---|---|
| | | % by weight of mineral | | | atom percent of total lead | | | | |
| Huron Claim, Manitoba | Monazite | 0.281 | 15.63 (MUENCH) | 1.524 | 0.0097 | 10.2 (NIER No 28) | 1.86 | 87.93 | 2600 (a) 3200 (b) 2850 (c) 1860 (d) |
| Bikita, Southern Rhodesia | Monazite | 0.074 | 2.39 | 0.34 | 0.0066 | 9.49 (HOLMES) | 1.71 | 88.80 | 2680 (a) 2675 (b) 2680 (c) 2645 (d) |
| Irumi Hills, Broken Hill, Northern Rhodesia | Monazite | 0.09 | 8.15 | 0.58 | 0.071 | 5.76 (HOLMES) | 1.85 | 92.32 | 2620 (a) 2040 (b) 2330 (c) 1390 (d) |
| Salisbury, Southern Rhodesia | Monazite | 0.19 | 9.36 | 0.95 | 0.035 | 7.53 (HOLMES) | 1.75 | 90.68 | 2650 (a) 2260 (b) 2470 (c) 1940 (d) |
| Beaverlodge Lake, Saskatchewan YY Concession | Pitchblende | 0.916 | 0.0018 (National Chemical Laboratory, Teddington, England) | 0.25 | -- | 93.14 | 6.55 (University of Toronto No 79) | 0.33 | 940 (a) 190 (b) 260 (c) |
| Beaverlodge Lake, Saskatchewan Gil Group | Pitchblende | 8.21 | 0.0097 (National Chemical Laboratory, Teddington, England) | 1.02 | 0.01 | 93.29 | 6.52 (University of Toronto No 117) | 0.19 | 930 (a) 822 (b) 850 (c) |
| Beaverlodge Lake, Saskatchewan 50—AA—14 | Pitchblende | 32.02 | 0.0009 (National Chemical Laboratory, Teddington, England) | 2.83 | — | 92.97 | 6.53 (University of Toronte No 40) | 0.53 | 920 (a) 592 (b) 664 (c) |
| Eldorado Mine, Great Bear Lake, N.W.T. | Pitchblende | 52.32 | 0.004 (MARBLE) | 10.51 | 0.055 | 89.02 | 8.68 (NIER No 10) | 2.25 | 1425 (a) 1218 (b) 1290 (c) |
| Eldorado Mine, Great Bear Lake, N.W.T. | Pitchblende | — | — | — | 0.024 | 91.37 | 8.21 (University of Toronto No 173) | 0.43 | 1410 (a) |
| Wilberforce, Ontario | Uraninite | 53.52 | 10.37 (WELLS) | 9.26 | 0.010 | 87.98 | 6.59 (NIER No 15) | 5.42 | 1035 (a) 1060 (b) 1040 (c) 990 (d) |
| Wilberforce, Ontario | Uraninite | 56.57 | 10.96 (HECHT) | 9.90 | — | — | — | — | crude age = 1120 |
| Wilberforce, Ontario | Uraninite | — | — | — | — | 89.64 | 6.63 (University of Toronto No 122) | 3.74 | 1030 (a) |
| Conger Twp., Ontario | Uraninite | 69.78 | 2.83 (ELLSWORTH) | 10.83 | 0.0057 | 91.81 | 6.79 (NIER No 22) | 1.40 | 1030 (a) 1010 (b) 1015 (c) 1010 (d) |
| Blackstone Lake, Pit, Conger Twp., Ontario | Uraninite | 69.3 | 2.99 | 10.72 | 0.0069 | 91.96 | 6.658 (WASSERBURG and HAYDEN) | 1.38 | 990 (a) 994 (b) 993 (c) 897 (d) |
| Spruce Pine, North Carolina | Samarskite | 13.06 | 1.11 | 0.60 | <0.02 | 91.2 | 5.60 (KULP, ECKELMANN, OWEN and BATE) | 3.18 | 699 (a) ± 150 312 (b) 362 (c) 354 (d) 325 (e) |

(a) Indicates lead ratio age ($Pb^{207}/Pb^{206}$). (b) Indicates lead age ($Pb^{206}$). (c) Indicates lead age ($Pb^{207}$). (d) Indicates lead age ($Pb^{208}$). (e) Indicates lead ratio age ($Pb^{206}/Pb^{210}$).

WICKMAN has pointed out that all three decay systems involve transmutation through some gaseous isotope which may be lost by diffusion and thereby cause errors in the calculated age values. For instance, loss of radon-222 which is a member of the $U^{238} - Pb^{206}$ chain would lower the apparent lead-206 content of the mineral and affect the $Pb^{206}/U^{238}$ ratio. Since the rate of radon leakage is temperature dependent, compensation for such loss cannot be made without knowing the thermal history of the mineral. However, work at Columbia University[1] has shown that radon leakage is generally small for samarskites and fresh primary pitchblendes and uraninites. WICKMAN [41], BULLARD [24] and HOLMES [42] have discussed the effects of losses of uranium and lead by leaching.

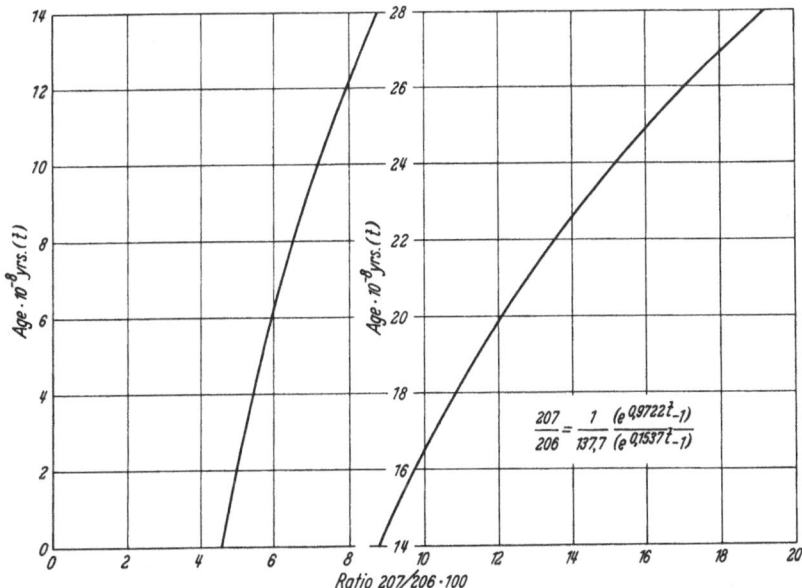

Fig. 3. Curve for dating uranium minerals by the lead-ratio method.

**39.** NIER utilized the fact that the two equations describing the decay of uranium-235 and uranium-238 could be combined to give the formula

$$\frac{Pb^{207}}{Pb^{206}} = \frac{U^{235}}{U^{238}} \cdot \left( \frac{e^{\lambda' t} - 1}{e^{\lambda t} - 1} \right). \tag{39.1}$$

It is evident that through the use of this formula the age of a uranium ore can be determined by isotopic analysis of the extracted lead without any chemical analyses. Although only one determination of age for each mineral can be made in this manner, this method has two advantages over the crude chemical method. Common lead can be detected and corrected for, and, if recent leaching has occurred, the effect upon the ratio of the two lead isotopes will be small compared with the differential effect of leaching upon two different elements.

Equation (39.1) cannot be solved directly for $t$, so that in practice the ages are taken from curves such as those shown in Fig. 3 or from nomographs published by KULP [13]. The small dependence of the $Pb^{207}/Pb^{206}$ ratio on time for ages less

[1] G. L. BATE, B. J. GILETTI and J. L. KULP: Bull. Geol. Soc. Amer. 63, 1233 (1952). — J. L. KULP, G. L. BATE and W. S. BROECKER: Amer. J. Sci. 252, 345 (1954).

than 500 m.y. suggests that this method cannot be expected to give reliable ages for minerals much younger than Precambrian. STIEFF[1] has quite rightly condemned its use for Mesozoic ores.

There has been some difference of opinion as to the relative merits of the *lead-ratio* and *uranium-lead* methods. It has been contended by some [13] that radon loss is the most serious cause of error and that consequently the age based on the $Pb^{207}/U^{235}$ ratio should be most reliable. Others [12] contend that the effects of radon loss are generally small in comparison with the effects of leaching, and that the most reliable age for all but the youngest minerals is that based on the ratio $Pb^{207}/Pb^{206}$ which is not seriously altered by loss of lead or uranium. Some minerals, when measured in the laboratory, are found to lose appreciable amounts of radon. However, it is doubtful whether this loss is representative of that which occurred before the minerals were exposed to recent surface weathering.

It is the writers' belief that the age determinations themselves indicate best which are the most reliable ages. Some typical results are shown in Table 17. Samples from Southern Rhodesia, Huron Claim and Beaverlodge Lake particularly show that the $Pb^{207}/Pb^{206}$ ages are by far the most consistent. For the Rhodesian samples the scatter of the $Pb^{207}/Pb^{206}$ ages is one-sixth of that for the $Pb^{207}/U^{235}$ ages, one-tenth that for the $Pb^{206}/U^{238}$ ages and one-twentieth that for the $Pb^{208}/Th^{232}$ ages. The majority of workers now favour the lead-ratio method for older specimens where it can be applied accurately.

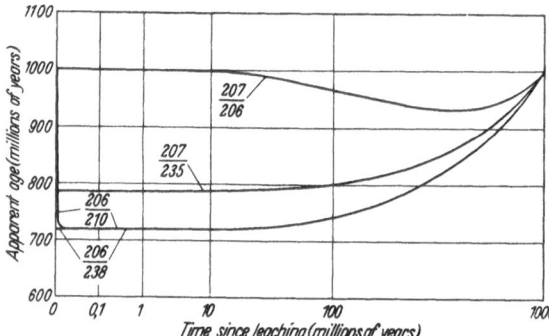

Fig. 4. The effect on apparent ages of leaching thirty percent of lead at various times from a uranium mineral 1000 million years old.

**40.** In theory the age of a uranium or thorium mineral can be estimated by measuring the ratio of a lead isotope daughter either to the uranium or thorium parent or to any of the intermediate members of the decay family assuming the latter are in radioactive equilibrium with the parent. In practice it would be difficult to measure most of the intermediate members, since their half-lives are small and hence their quantitative separation difficult. However, at least one useful method has been proposed. The determination of the ratio $Pb^{206}/Pb^{210}$ in a uranium-bearing mineral as a means of determining ages was first suggested by HOUTERMANS. Measurements by BEGEMAN et al [21] on pitchblende from the Belgian Congo indicated that the ratio gave results in good agreement with the chemical values. Recently, KULP et al [22] have shown that $Pb^{206}/Pb^{210}$ ages are consistent with the best determinations by other methods over a range of ages from 60 to 1400 million years.

Measurement of the ratio $Pb^{206}/Pb^{210}$ has several advantages.

(1) The ratio may be determined on any fraction of the total lead in the mineral, and hence, it is unnecessary to carry out a quantitative chemical analysis. The proportion of lead-206 present in a specimen of radiogenic lead is determined by means of a mass spectrometer.

---

[1] L. R. STIEFF, T. W. STERN and R. G. MILKEY: U.S. Geol. Surv. Cir. 271, 1953.

(2) Lead-210 has a half-life of about 22 years and is easily measured by radio-active counting methods, although the amount would always be too small to measure mass-spectrometrically.

(3) The danger of contamination is slight since lead can be readily separated by chemical means from other members of the radioactive family.

Although it has been claimed that $Pb^{206}/Pb^{210}$ ages are independent of recent leaching of uranium or lead [21], [22], loss at most times during the life of a uranium minerals would affect the $Pb^{206}/Pb^{210}$ age just as seriously as it would affect the $Pb^{206}/U^{238}$ age. This is shown in Fig. 4 and 5 which show the relative effects of leaching on the various methods. The $Pb^{206}/Pb^{210}$ method is independent of radon loss if such loss has been constant during the entire lifetime of the mi-neral, a circumstance rarely to be expected. This method, although a recent develop-ment, shows promise of being a valuable aid to the other established methods. Few de-terminations have yet been made by it.

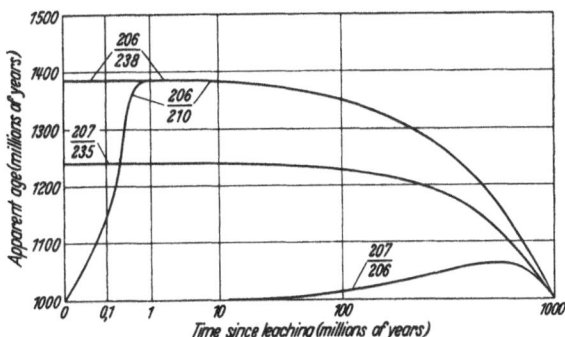

Fig. 5. The effect on apparent ages of leaching thirty percent of uranium at various times from a uranium mineral 1000 million years old. This has been calculated assuming no thorium leaching. If thorium is lost the $Pb^{206}/Pb^{210}$ curve lies closer to the $Pb^{206}/U^{238}$ curve.

**41.** It can be seen that there is no lack of methods for determining the age of uranium and thorium ores. Beyond the necessity of mak-ing an isotopic analysis of the extracted lead in all cases and chemical analyses or a deter-mination of lead-210 for young ores, there is no agreement upon the best method to use. In all cases it is desirable to have as much information as possible, but for old Precambrian ores the $Pb^{207}/Pb^{206}$ ratio seems to give the most satis-factory ages. More comparisons and measurements on the same material by different laboratories are needed and are being undertaken.

Everyone would probably agree that the best age determination possible in the light of our present knowledge would be one made on a fresh crystal of urani-nite, by chemical and isotopic analysis and by counting lead-210 activity. If radon loss were measured and found to be small, if little or no lead-204 were pre-sent and if ages by all the methods agreed, then the age would probably be as correct as possible. It is doubtful whether these ideal conditions will be found for many minerals.

### b) Methods for dating common leads.

**42.** When SODDY and HYMAN in 1914 showed that lead from a Ceylon thorite had an atomic weight approximately one unit greater than that for common lead, they prompted a series of investigations of the atomic weight of this element. Additional measurements of atomic weights of lead from uranium and thorium minerals also supported SODDY's prediction that lead produced by the decay of thorium should have atomic weight 208 while uranium-lead should have ato-mic weight 206. Later ROSE and STRANATHAN[1], by studying the hyperfine structure of the spectra of lead compounds, and ASTON[2], with a mass spectro-graph, also showed the great differences between common and radiogenic leads and

[1] J. L. ROSE and R. K. STRANATHAN: Phys. Rev. **49**, 916 (1936).
[2] Cf. first footnote in Sect. 8.

between different radiogenic leads. However, its was not until 1938 that NIER reported the first systematic investigation of common lead abundances.

In a paper published that year [6], and in a subsequent paper published in 1941 [8], he listed abundances for 25 common lead samples which varied greatly in geological age and geographical origin. In addition, many of the samples had been used previously for chemical atomic weight determinations. These measurements showed for the first time the enormous variations in the isotopic constitution of common leads. These variations had been previously hidden by the coincidence that they are of such a nature as to cause no appreciable changes in atomic weight.

It was immediately presumed that the variations resulted from different amounts of thorium and uranium leads in the common lead samples. Since common lead minerals (such as galenas) contain negligible uranium and thorium, the addition of radiogenic lead must have taken place before the lead was concentrated from the material of the outer part of the earth to form the mineral. Despite the general agreement on the cause of common lead abundance variations, a mathematical representation for the process is not easily decided upon. Since the growth of the continental masses and the formation of mineral

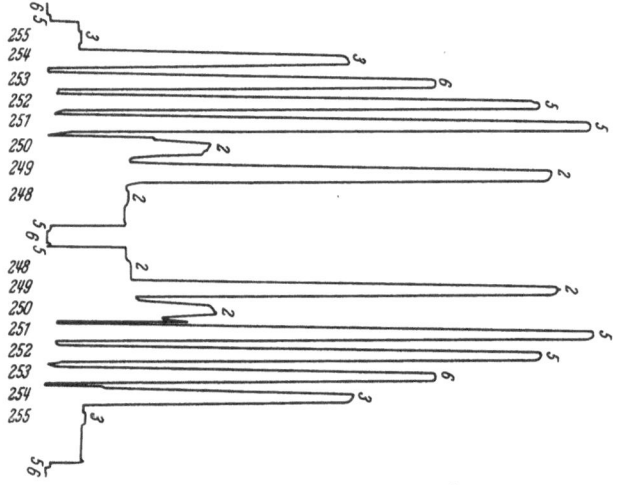

Fig. 6. Mass spectrogram showing the Pb(CH$_3$)$_3^+$ group for a typical common lead.

deposits are unquestionably very complex, any useable mathematical formula defining a relationship between the isotopic abundances of lead samples and their geological age must be approximate. It then follows that no common lead age determination method will be exact. However, in many cases the ages obtained are in striking agreement with other methods of determination and with known geological relationships. This fact lends support to their use in areas which cannot readily be dated by other methods. Moreover, "galena ages" can often give valuable information about the ages of periods of sulphide mineralization. This information cannot at present be obtained in any other way.

Two principal methods have been suggested for dating common lead minerals. These methods will be described separately below.

**43.** HOLMES, in 1946 [32], and HOUTERMANS, in 1947 [33], independently suggested a simple theory to explain the abundances of common leads. They suggested that at an early time in its history the earth was fluid and well mixed by rapid convection currents. At that time the uranium, thorium and lead were uniformly distributed throughout the earth's mantle, and the isotopic constitution of the lead was everywhere the same. This lead is usually referred to as *primeval lead*. We will call its abundances $a_0$ for Pb$^{206}$/Pb$^{204}$, and $b_0$ for Pb$^{207}$/Pb$^{204}$. Then at some time, $t_0$, the earth became sufficiently rigid to preserve inhomogeneities in its uranium and lead distribution. In each locality the ratio of lead to uranium differed slightly, but in any particular locality it did not change except for the steady decay of uranium to lead. At a later time $t$ a portion of lead was concentrated to form a lead mineral which is essentially free from uranium. The abundances of the lead isotopes in such a mineral will be given by the equations:

$$\frac{Pb^{206}}{Pb^{204}} = x = a_0 + \alpha V_m (e^{\lambda t_0} - e^{\lambda t})$$

and

$$\frac{Pb^{207}}{Pb^{204}} = y = b_0 + V_m (e^{\lambda' t_0} - e^{\lambda' t}). \tag{43.1}$$

Here $V_m$ is used to represent the ratio $U^{235}/Pb^{204}$ referred to the present time, and $\alpha$ is used to represent $U^{238}/U^{235}$, also referred to the present. $\lambda$ and $\lambda'$ are the decay constants of $U^{238}$ and $U^{235}$ respectively. A similar equation can be derived for $Pb^{208}/Pb^{204}$ but it is not necessary for this method of age determination.

By combining the equations 43.1, it is found that

$$\frac{y - b_0}{x - a_0} = \frac{1}{\alpha} \frac{(e^{\lambda' t_0} - e^{\lambda' t})}{(e^{\lambda t_0} - e^{\lambda t})}. \tag{43.2}$$

To determine the age of a lead ore, values for the lead isotope ratios $x$ and $y$ are substituted in this equation which is then solved for $t$. Values for $a_0$, $b_0$ and $t_0$ must be assumed, and the solution obtained graphically.

In their early discussions of lead ore abundances, both HOLMES and HOUTER-MANS used equation (43.2) with NIERS' 25 lead isotope analyses to obtain values for the parameters $a_0$, $b_0$ and $t_0$, which would give ages in best accord with esti-mated geological ages of the samples. At that time there were too few common lead isotope analyses to test the physical assumptions involved or to set up a general procedure for common lead dating.

In 1953, DAMON[1] carried out a more extensive study of the possibilities of this method of geological dating. In addition to NIER'S analyses, he was able to use analyses reported by COLLINS, FARQUHAR and RUSSELL[2], and by VINO-GRADOV, ZADOROZHNYI and ZYKOV [9], which more than doubled the available experimental data. He used values for the age of the earth and for the primeval abundances which had been determined by HOLMES [32]. Many of the samples gave ages in reasonable accord with the supposed ages of the deposits, where such ages were known. However, this check is not entirely valid in view of the fact many of the samples had been assigned ages by HOLMES in order to obtain the parameters in the equation.

More recently HOUTERMANS[3] has used the isotopic analyses of meteoric lead by PATTERSON, BROWN, TILTON and INGHRAM[4] for the primeval lead abun-dances. This assumption may be justified in view of the very minute amounts of uranium and thorium in the meteorites analysed, and the fact that the equality of meteoritic and terrestial isotopic abundances has been established for many other elements. By using these primeval abundances and abundances of young lead samples he calculated the age of the earth by means of equation (43.2). The set of $a_0$, $b_0$ and $t_0$ obtained in this way, which differ radically from that used by DAMON, were used by HOUTERMANS with equation (43.2) to give "model"

[1] P. E. DAMON: Trans. Amer. Geophys. Un. **34**, 906 (1953).
[2] C. B. COLLINS, R. M. FARQUHAR and R. D. RUSSELL: Phys. Rev. **88**, 1275 (1952).
[3] F. G. HOUTERMANS: Nuovo Cim. **10**, 1623 (1953).
[4] C. PATTERSON, H. BROWN, G. TILTON and M. INGHRAM: Phys. Rev. **92**, 1234 (1953).

ages for common lead minerals. Typical ages obtained with the parameters used by DAMON and HOUTERMANS are shown in Table 18.

There are numerous sources of possible error in this method. In the first place, the model assumed for the history of the lead ores is probably greatly oversimplified. If the lead in some part of the earth experiences geological disturbances of a nature which would change the relative proportions of uranium and lead present, equation (43.1) would be in error. This would in turn cause an error in equation (43.2) and an error in the ages obtained for lead ores in that locality. Such failures in the basic assumptions do occur, as is shown by the anomalous abundances of minerals from Sudbury, Ontario (Table 19), and

Table 18. *Comparison of common lead dating methods.*

(a) Greenstone areas which may be continental nuclei. Uranium-lead ages, where such exist, indicate that these areas are between 2000 and 3000 million years old.

| Sample and isotopic abundances | No.[1] | HOLMES-HOUTERMAN's Method | | RUSSELL's method [2] |
|---|---|---|---|---|
| | | DAMON's curve | HOUTERMAN's curve | |
| *Yellowknife geological province, Canadian Shield* | | | | |
| Vein north of Ptarmigan Mine (13.93, 15.05, 33.99) | T-205 | 2420 | 2900 | 2360 ± 140 |
| Vein north of Ptarmigan Mine (13.95, 15.04, 34.10) | T-450 | 2380 | 2940 | 2340 ± 140 |
| Con Mine 800' level (14.21, 15.25, 34.47) | T-444 | 2510 | 2890 | 2200 ± 150 |
| Con Mine, 1250' level (14.25, 15.26, 34.42) | T-446 | 2480 | 2850 | 2200 ± 150 |
| Negus Mine, 1900' level (14.27, 15.29, 34.50) | T-443 | 2530 | 2870 | 2190 ± 150 |
| Negus Mine, 2000' level (14.28, 15.31, 34.42) | T-445 | 2540 | 2870 | 2200 ± 150 |
| *Keewatin geological province, Canadian Shield* | | | | |
| Timmins, Ontario (13.75, 14.87, 33.72) | T-472 | 2230 | 2980 | 2450 ± 130 |
| Steeprock Lake, Ontario (13.94, 14.83, 33.75) | T-404 | 1950 | 2800 | 2410 ± 130 |
| New Norzone Property, Arntfield, Quebec (13.96, 14.96, 33.95) | T-582 | 2250 | 2860 | 2360 ± 140 |
| Alcona Property, Sioux Lookout, Ontario (14.03, 14.89, 33.78) | T-149 | 2020 | 2760 | 2400 ± 130 |
| Lake Shore Mine, Kirkland Lake, Ontario (14.37, 15.02, 33.99) | T-519 | 1970 | 2610 | 2280 ± 140 |
| *Other greenstone areas* | | | | |
| Rosetta Mine, Barberton District, South Africa (12.65, 14.27, 32.78) | T-193 | 1950 | 3370 | 2890 ± 100 |
| Kokosho Querry, near Kulu Matundu, Uele, N. Congo (12.81, 14.39, 32.84) | T-616 | 2160 | 3300 | 2840 ± 110 |
| Kalgoorlie, West. Astralia, 1700' level, west stope (13.89, 15.05, 33.88) | T-503 | 2450 | 2940 | 2400 ± 130 |

[1] All results but one are Toronto analyses. In this way we hope to minimize the effect of small consistent mass spectrometer errors which might make a comparison of methods more difficult. One of NIER's results is included to show the effect of such mass spectrometer differences.

[2] For method see reference [20]. The constants used are those found empirically by G. L. CUMMING. They are $a = 18.80$, $b = 15.85$, $c = 39.10$, $V = 0.080$, and $W = 40.8$.

*Table 18.* (Continued.)

(b) Samples from the Grenville geological province of the Canadian Shield, the accepted age of which is 800 to 1100 million years. Best dated localities are Conger Township and Fission Mines near Wilberforce, Ontario. Both are between 1000 and 1050 million years old.

| Sample and isotopic abundances | No.[1] | HOLMES-HOUTERMAN's Method | | RUSSELL's method[2] |
| --- | --- | --- | --- | --- |
| | | DAMON's curve | HOUTER-MAN's curve | |
| Anacon Mine, Quebec 423 stope (16.55, 15.64, 36.46) | T-559 | 1420 | 1570 | 1240 ± 190 |
| Anacon Mine, Quebec 617 to 660 stope (16.53, 15.63, 36.42) | T-560 | 1420 | 1570 | 1240 ± 190 |
| Anacon Mine, Quebec 826 to 859 stope (16.49, 15.58, 36.33) | T-561 | 1400 | 1580 | 1280 ± 190 |
| Ciglen Claims, Lake Baskatong, Quebec (16.59, 15.64, 36.71) | T-520 | 1430 | 1570 | 1160 ± 190 |

(c) Effect of mass spectrometer errors. The following two samples came from the same deposit but were analysed at different laboratories.

| Sample and isotopic abundances | No.[1] | HOLMES-HOUTERMAN'sMethod | | RUSSELL's method[2] |
| --- | --- | --- | --- | --- |
| | | DAMON's curve | HOUTER-MAN's curve | |
| Ivigtut, Greenland (14.89, 14.98, 35.17) | T-202 | 1330 | 2180 | 1925 ± 150 |
| Ivigtut, Greenland . (14.65, 14.65, 34.47) | N-13 | 620 | 2070 | 2110 ± 150 |

Table 19. *Lead isotope abundances of Sudbury, Ontario galenas.*

| Location | Toronto number | $Pb^{206}/Pb^{204}$ | $Pb^{207}/Pb^{204}$ | $Pb^{208}/Pb^{204}$ |
| --- | --- | --- | --- | --- |
| McKim Mine (1000' level) | 311 | 23.03 | 16.69 | 45.19 |
| McKim Mine (960' level) | 312 | 22.89 | 16.72 | 44.92 |
| McKim Mine[3] (800' level) | 310 | 16.30 | 15.85 | 36.97 |
| McKim Mine[3] (600' level) | 309 | 16.43 | 15.96 | 36.93 |
| Falconbridge Mine (3300' level) | 308 | 24.29 | 17.04 | 45.70 |
| Falconbridge Mine (2400' level) | 305 | 23.70 | 16.92 | 45.35 |
| Falconbridge Mine (1700' level) | 306 | 24.20 | 16.95 | 45.58 |
| Frood Mine[3] (disseminated in wall) | 232 | 15.99 | 15.84 | 36.56 |
| Frood Mine (post-ore slip) | 233 | 23.01 | 16.90 | 45.07 |
| Garson Mine | 217 | 22.95 | 16.69 | 44.79 |
| Garson Mine (post-ore slip) | 234 | 23.00 | 16.61 | 44.78 |
| Worthington Mine | 211 | 26.00 | 16.94 | 52.21 |
| Hardy Mine (750' level) | 307 | 23.23 | 16.77 | 52.64 |
| Fairbank Township[3] (in deformed micropegmatite) | 235 | 16.20 | 15.76 | 36.99 |
| Treadwell Yukon Mine[3] (centre of basin) | 518 | 16.15 | 15.60 | 35.94 |

the Tri-State mines near Joplin, Missouri (Table 20). The true significance of these anomalous leads has yet to be established, although it now appears that such leads have been recently exposed to an environment much richer in uranium and thorium than is the case for most lead ores.

A second source of error is the uncertainty in the value of the decay constant of uranium-235, which is still uncertain by at least one percent. An error of this

[1] See footnote 1, p. 328.

[2] See footnote 2, p. 328.

[3] All samples but these are anomalous.

Table 20. *Isotopic abundances of lead from Tri-State Mines, USA.*

| No.[1] | Mine | Bed | Lead isotope ratios | | |
|---|---|---|---|---|---|
| | | | $Pb^{206}/Pb^{204}$ | $Pb^{207}/Pb^{204}$ | $Pb^{208}/Pb^{204}$ |
| 316 | Diamond Joe | Chester Limestone | 21.38 | 16.16 | 41.03 |
| 318 | Webber-Westside | D—E | 22.15 | 16.21 | 41.63 |
| 320 | Howe (av. of 2 analyses) | E | 22.29 | 16.27 | 41.78 |
| 315 | Federal-Jarrett | G—H | 22.70 | 16.25 | 42.16 |
| 319 | Kitty | G—H | 22.73 | 16.31 | 42.07 |
| 314 | Webber-Westside | G—H | 22.12 | 16.22 | 41.63 |
| 317 | Grace B | J | 22.77 | 16.29 | 42.25 |
| 321 | Otis White | K | 22.77 | 16.28 | 42.13 |
| 323 | Blue Goose No 1 | M | 22.07 | 16.24 | 41.87 |
| 231 | Westside | M | 22.16 | 16.31 | 41.98 |
| 322 | Blue Goose No 2 (av. of 2 analyses) | O | 21.83 | 16.14 | 41.35 |

amount would cause an error of about 300 million years in the age of a young ore calculated by HOUTERMANS' method.

More serious is the uncertainty in the value for the age of the earth. The uncertainty of the value used by DAMON is at least 300 million years. HOUTERMANS states an uncertainty of 300 million years for his value. The fact that these two values differ by 1200 million years suggests that neither age may be as precise as these errors indicate. An uncertainty in $t_0$ of 300 million years results in uncertainties of 1400 million years and 1100 million years with the constants used by HOUTERMANS and DAMON respectively. This is a very serious objection when it is considered that the "age of the earth" used in these calculations refers to the time when the present uranium-lead distribution in the earth was established, for this time might differ at various places in the earth.

A final source of error which could change the values calculated by these methods by an enormous amount, is the assignment of the primeval abundances. In HOUTERMANS' calculations this will remain a serious cause for concern until it has been better established that all meteorites have the same isotopic abundances and that these abundances can be identified with the primeval lead abundances involved in these calculations.

While these various sources of uncertainty could each individually cause enormous errors in common lead ages calculated in this way, in practice the errors largely cancel. This arises from the fact that the method for assigning a value to $t_0$ forces the abundances of young leads to fit approximately to equation (43.2). Thus, the methods of DAMON and HOUTERMANS are basically contrived to give ages of the correct order of magnitude for young samples. The method has the advantage that it is independent of local differences in uranium and lead abundances, provided that the lead has not been subjected to multiple orogenies which change the uranium-lead ratios. There appears to be strong indication that this may be a powerful method, with the correct choice of parameters, but before it can be considered established much work will have to be done to investigate the possible effects of failures in the basic assumptions.

**44.** In 1951, ALPHER and HERMAN [43] proposed that the distribution of uranium, thorium and lead in the outer part of the earth from which the lead ores were derived is approximately homogeneous. If this assumption is valid, the isotopic abundances of common lead samples should satisfy approximately

---

[1] This number refers to the University of Toronto sample number.

equations of the type:

$$x = \frac{Pb^{206}}{Pb^{204}} = a - \alpha V(e^{\lambda t} - 1),$$

$$y = \frac{Pb^{207}}{Pb^{204}} = b - V(e^{\lambda' t} - 1),$$

$$z = \frac{Pb^{208}}{Pb^{204}} = c - W(e^{\lambda'' t} - 1).$$

$$(44.1)$$

Here $a$, $b$ and $c$ are the $Pb^{206}/Pb^{204}$, $Pb^{207}/Pb^{204}$ and $Pb^{208}/Pb^{204}$ ratios for very young lead samples, $W$ is the present average value of the ratio $Th^{232}/Pb^{204}$, $V$ is the present average value of the ratio $U^{235}/Pb^{204}$, and $\lambda''$ is the lead decay constant of $Th^{232}$. $z$ is the ratio $Pb^{208}/Pb^{204}$ for lead in a lead mineral of age $t$, and all other symbols are the same as those previously defined in Sect. 43.

ALPHER and HERMAN used isotopic analyses of 25 common lead samples reported by NIER, with ages assigned by NIER and by HOLMES, and calculated the parameters in equations (44.1). McCRADY[1], using somewhat different assumptions, obtained different values for the parameters from NIER's abundances. COLLINS, RUSSELL and FARQUHAR[2] used additional isotopic analyses and revised ages for the samples and calculated still another set of values for the parameters. ALLAN, FARQUHAR and RUSSELL[3] showed that the parameters could be obtained without assuming ages for the isotopically analysed common lead samples, and obtained values for the parameters which were essentially the same as those found by COLLINS, RUSSELL and FARQUHAR. The overall results of these various investigations was a surprising support for the basic assumptions of a homogeneous distribution of uranium, thorium and lead in the material from which the ores have been derived.

Since the continental masses are composed of rocks of a very heterogeneous nature, this apparent homogeneity is hard to understand if the lead has been derived from crustal rocks. However it has now been well established that the continents are not primary subdivisions of the original earth, but have grown during geological time. It has been suggested that material forming the earth's crust has been selectively derived from the material in the top 700 kilometers of the mantle. If this is true, and if periods of mineralization are closely related to periods of orogeny, it is possible that the parent material for most lead ores is the upper part of the mantle which may be far more homogeneous than the continents. Anomalous leads, which will be discussed in detail in later paragraphs, may be derived in part from crustal rocks.

In view of the apparent general validity of equations (44.1), it was a natural step to propose the use of these equations as a dating method. The first step in this direction was taken by McCRADY[1] in 1952 who proposed that all common lead abundances should obey equations (44.1) even if the ages obtained should differ from the ages of the geological regions in which the leads had occurred. He obtained the constants in the equations by assuming ages of 2780 million years and 100 million years for the lead ores from Ivigtut, Greenland and Joplin, Missouri. Both of these were unfortunate choices, for the age of Ivigtut had never been established and may be quite different from the value assumed, and leads from Joplin are known to be anomalous leads for which equations (44.1)

[1] E. McCRADY: Trans. Amer. Geophys. Un. 33, 156 (1952).

[2] C. B. COLLINS, R. D. RUSSELL and R. M. FARQUHAR: Canad. J. Phys. 31, 402 (1953).

[3] D. W. ALLAN, R. M. FARQUHAR and R. D. RUSSELL: Science, Lancaster, Pa. 118, 486 (1953).

do not apply. As a result the ages obtained by McCRADY bore no relationship to known geological ages.

The next step was taken by RUSSELL, FARQUHAR, CUMMING and WILSON [20] who based a dating method on common lead abundance-time curves calculated by COLLINS, RUSSELL and FARQUHAR[1] following the method of ALPHER and HERMAN [43]. In contrast to McCRADY's assumption of an exact fit between measured abundances and equations (44.1) they pointed out that these expressions could only be used for geological dating if due allowances were made for failures of the basic assumption of a homogeneous earth. They calculated possible errors from this cause and showed that for ages older than about 800 million years the errors were small enough so that useful ages could be obtained.

Further ages calculated by this method have been based on the constants obtained by G. L. CUMMING[2] which appear to be more satisfactory. Ages calculated in this way are shown in Table 18 where they are compared with the values calculated by the HOLMES-HOUTERMANS method.

The main objections to this method are two. First, the values for the parameters in the equations are not easily established. Their accurate determination involves the use of dated lead samples, of which there are very few. Secondly, heterogeneities in the distribution of uranium, thorium and lead in the source material for the ore leads do exist and do cause uncertainties in the ages obtained. If the uncertainties estimated by RUSSELL are correct, the second objection may not be very serious when old leads are to be dated. Ages younger than about 800 million years by this method have relatively enormous possible errors.

The advantages of this method are chiefly that it provides two ages, from the $Pb^{208}/Pb^{204}$ ratios and the $Pb^{206}/Pb^{204}$ ratios, and that it takes some account of the possible errors which may arise through failure of the basic assumptions.

Mass spectrometric errors are less important in the ALPHER-RUSSELL method than in the HOLMES-HOUTERMANS method, although they would cause errors in either case. The results of mass spectrometer intercalibrations show that the only serious mass differences occur in the smaller $Pb^{204}$ abundance [34].

**45.** Early in the study of common lead isotope abundance variations it became apparent that certain leads could not be fitted to any of the above theories. The Joplin, Missouri ores, which are of this type, were finally rejected by HOLMES for his determination of the earth's age. They had to be rejected by BULLARD and STANLEY [44] before their age of the earth calculations would yield any sensible result, and they could not be used by ALPHER and HERMAN in the development of their common lead abundance-time curves.

A second suite of anomalous leads was discovered in the region of Sudbury, Ontario. The Sudbury leads have $Pb^{206}/Pb^{204}$ and $Pb^{208}/Pb^{204}$ ratios that appear even more anomalous than those for Joplin leads. The fact that the $Pb^{207}/Pb^{204}$ ratios are not so anomalously large indicates that the excess radiogenic lead must have been generated in a relatively late stage of the earth's history when uranium-235 was relatively rare. Suites of samples from the Sudbury area and the Joplin-Tri-State area have been analysed at Toronto with the results shown in Tables 19 and 20.

The Sudbury ores show a remarkable variation of isotopic constitution even for samples from the same mine. Such a variations is not found in the case of deposits which do not contain anomalous leads. This is borne out by the analyses of leads from other mines, (e.g. the Anacon samples in Table 18), which are appa-

---

[1] See p. 331, footnote 2.
[2] G. L. CUMMING: Ph. D. Thesis, University of Toronto 1955.

rently not anomalous, which come from a greater range of depths than the Sudbury leads, and which show no measurable variations in isotopic constitution.

The large abundance variations of the Sudbury galenas have been used by RUSSELL, FARQUHAR, CUMMING and WILSON [20] to obtain information about their age. They postulated that these leads were mixtures in varying proportions of a single radiogenic lead and a single common lead (of the non-anomalous or "ordinary" type). By applying a simple least squares calculation they determined that the ratio $Pb^{207}/Pb^{206}$ in the radiogenic lead was $0.134 \pm 0.007$. It can be calculated that uranium was generating lead isotopes in these proportions 1300 million years ago. It follows that the lead in the Sudbury galenas must have been in contact with uranium at least as recently as 1300 million years ago. This age is then an upper limit to the age of these minerals.

Four of the samples do not appear to be anomalous. These can be dated from the curves of G. L. CUMMING[1]. The ages obtained are in excellent agreement, and the mean age of 1260 million years agrees within the limits of error with the age of 1040 million years found for the older dated pegmatites of the Grenville orogeny. The Sudbury ores appear to be associated with a series of faults which start at the northern border of the Grenville geological province and extend into the older Keewatin province to the north.

It is also possible to calculate the ratio of $Pb^{208}$ to $Pb^{206}$ in the radiogenic component of the Sudbury galenas. This was done and found to be 1.54. Such lead would be produced 1200 million years ago by a mixture of 5.2 parts thorium to one part of uranium. This figure is appreciably higher than the value 3.5 usually quoted for the average ratio of these elements in crustal rocks. However,

Table 21. *Thorium and uranium contents of rocks from Sudbury, Ontario[2].*

| Rock | Location | $Th \times 10^6$ (gms./gm.) | $U \times 10^6$ (gms./gm.) | Th/U |
|---|---|---|---|---|
| Granite | Creighton Mine, Ontario | $17 \pm 4$ | $3.5 \pm 0.7$ | 4.8 |
| Olivine Diabase | Worthington, Ontario | $15 \pm 3$ | $2.2 \pm 0.3$ | 6.8 |
| Norite | Outcrop, Creighton, Ontario | $4.3 \pm 0.8$ | $1.5 \pm 0.4$ | 2.9 |
| Olivine Diabase | Creighton Mine, Ontario 20' level | $2.4 \pm 0.3$ | $0.97 \pm 0.2$ | 2.4 |
| Norite | Creighton Mine, Ontario 40' level | $2.1 \pm 0.4$ | $0.71 \pm 0.14$ | 3.0 |

measurements by SENFTLE and KEEVIL, reproduced in Table 21, show that those rocks in the Sudbury area which are richer in uranium and thorium have larger thorium to uranium ratios.

The variation of the Tri-State abundances is not sufficient to apply successfully least squares calculations of the above type. However, variations do exist and do seem to be related to the position of the ore in the stratigraphic column. No theory has yet been put forward to explain the variation of the Tri-State lead isotope abundances.

### c) The age of the earth.

**46.** Reference has already been made to the use of lead isotope abundances to obtain information about the age of the earth (sects. 43—45). The mathematical equations used for this purpose are identical with those previosly given [equa-

---

[1] Cf. p. 332, footnote 2.
[2] N. B. KEEVIL: Amer. J. Sci. **242**, 309 (1944).

tions (43.1), (43.2), (44.1)] and their application to this problem has already been described, so that little further description need be given here. However the basic ideas behind these calculations are not simple and care is required in interpreting the results.

Age of the earth calculations of this type are made possible by the fact that the outer part of the earth is sufficiently rich in uranium and thorium and sufficiently poor in lead for the addition of radiogenic lead isotopes to play a large part in determining the isotopic abundances of terrestrial lead samples. The methods used depend also on the existence of inhomogeneities in the lead-uranium-thorium distribution in the outer part of the earth, which permanently affect the isotopic constitution of the lead present in each locality.

Leads in lead minerals of a similar geological age are found to differ in their proportions of lead-206, lead-207, and lead-208. These variations are thought to be the result of association of the lead with uranium and thorium when the lead was disseminated throughout a portion of the outer part of the earth, and before the mineral was formed. Association of leads with different amounts of uranium and thorium within the past billion years would cause important variations in the abundances of lead-206 and lead-208, but much smaller variations in lead-207, for in such recent times radiogenic contributions from the relatively rare uranium-235 have been much smaller than the corresponding contributions from the more abundant uranium-238 isotope. Lead ores from the Sudbury basin and from the Joplin Tri-State mines appear to be of this type. However, the vast majority of common leads in a single age group show variations in the lead-207 abundance quite comparable with the lead-206 variations. This is conclusive evidence that a large part of these variations were produced at a time when uranium-235 was much more abundant than at present, and hence shows that the lead found in lead ores must have existed at least a few billion years ago in some body relatively rich in uranium and capable of preserving heterogeneities in its composition.

Such calculations as those of HOLMES [32], HOUTERMANS [33] and BULLARD and STANLEY [44] etc., attempt to put these qualitative considerations into a quantitative mathematical form by assuming some simplified history for the ore leads. If this history is a reasonable approximation to the actual history of these ores, the age obtained will be accurate. Otherwise, it may be in error. However, the qualitative considerations outlined above are not dependent on any particular earth model, and the general magnitude of the result obtained is fortunately not greatly affected by the particular assumption used. This is seen from the values found in Table 22 which is taken from the paper of RUSSELL and ALLAN [34]. These were also calculated for other earth models as well as by two methods of calculation. The reader is referred to the original paper for detailed descriptions of the models and methods and for additional ages.

The physical event to which the *"age of the earth* $(t_0)$*"* refers deserves serious consideration. All the calculations assume that there is a time $t_0$ prior to which the lead, uranium and thorium isotopes were distributed uniformily (with respect to lead-204) throughout the material of the earth. At time $t_0$ some process of differentiation produced variations in the concentrations of the uranium and thorium isotopes with respect to lead-204, but none in the lead isotopes. This is a reasonable assumption, since the ratios of the lead isotopes to one another would not be affected by natural chemical processes, whereas uranium and thorium have different properties from lead and therefore variations from place to place may well have been produced in their concentrations with respect to lead-204. If the earth had a molten origin, $t_0$ would be practically the same as the time of

Table 22. *Age of the earth.* Unit $10^9$ *years.*

| Group of samples | No. of samples | Age from slope of straight-line graphs | | Age from standard deviations | |
|---|---|---|---|---|---|
| | | from $Pb^{207}$-$Pb^{206}$ graph | from $Pb^{207}$-$Pb^{208}$ graph | Uncorrected age | Age corrected for variations in time of deposition |
| Vinogradov . . . . . . . . . | 28 | $4.4 \pm 0.2^1$ | 4.5 | 4.6 | 4.7 |
| Nier . . . . . . . . . . . . | 13 | $3.1 \pm 0.3$ | 3.3 | 3.3 | 3.4 |
| Geiss . . . . . . . . . . . | 16 | $3.5 \pm 0.4$ | 3.1 | 4.4 | 5.0 |
| Farquhar . . . . . . . . . | 23 | $2.8 \pm 0.5$ | 3.4 | 3.8 | 3.9 |
| Nier, Geiss, and Farquhar . . | 52 | $3.6 \pm 0.2$ | 3.9 | 4.0 | 4.1 |
| All . . . . . . . . . . . . . | 80 | $4.1 \pm 0.1$ | 4.2 | 4.4 | 4.4 |
| Asian samples . . . . . . . | 24 | $4.5 \pm 0.2$ | 4.6 | 4.6 | 4.7 |
| European samples . . . . . . | 35 | $4.2 \pm 0.2$ | 4.5 | 4.6 | 4.7 |
| North American samples . . . . | 15 | $3.2 \pm 0.3$ | 3.5 | 3.4 | 3.5 |

solidification of the entire earth, since the latter process would be quite rapid. If the earth was formed by accretion of solid particles, it may have become molten and then solidified again, or it may have had a still more complicated history, but $t_0$ would still represent the time when the upper part of the earth's mantle become stable. At the time Holmes first wrote, it was generally believed that the continents had formed at the same time as the earth solidified, and he therefore called $t_0$ *"the age of the crust"*. In view of recent theories that the continents probably have grown during geologic time, it seems wiser to call $t_0$ *"the age of the upper mantle"*. If ores have, in general, a deep-seated origin, this is certainly what $t_0$ represents.

Finally, it should be remarked that, although these calculations depend on the presence of heterogeneities in the earth, these variations are sufficiently small to allow other calculations [43] which assume approximately homogeneous distributions of lead, thorium, and uranium to provide useful methods of dating lead ores, especially older ones. That the mantle should at once be sufficiently heterogeneous to enable one to date it and sufficiently homogeneous to enable one to date lead ores is almost suspiciously provident, but there seems at present no reason to doubt this particular piece of scientific good fortune.

Another age value obtained by Houtermans[2] does not depend on heterogeneities in the earth's lead-uranium ratios, but rather depends on the assumption that certain meteoritic lead can be identified with the earth's primeval lead. If this assumption is true, Houtermans' determination should be the more accurate. Fortunately the two methods give essentially the same results. In all cases care should be taken when comparing results obtained by different methods to make sure that the methods date the same event.

## II. The potassium methods.

### a) Introduction.

**47.** While the above methods have been enormously useful for dating periods of mineralization and orogenesis, they suffer from the comparative rarity of uranium and thorium. Some areas which are of considerable geological interest, for example the Lewisian province in northwestern Scotland, have not been dated because suitable radioactive material has never been found there. There has been much work carried out in an effort to establish dating methods which

---

[1] These uncertainties are standard deviations.
[2] F. G. Houtermans: Nuovo Cim. **10**, 1623 (1953).

depend on radioactive elements more common than uranium and thorium. Of all the elements occurring in nature which have long-lived radioactive isotopes, potassium is by far the most abundant. It is found in feldspars and micas which are among the commonest minerals of igneous rocks and of pegmatites, in sylvite, carnallite and other minerals of salt deposits, and in slates and shales among the sedimentary rocks. The widespread occurrence of potassium minerals and the fact that the radioactive isotope has a half-life comparable with the age of Precambrian minerals, make it a logical choice for age determination purposes.

The radioactive isotope of natural potassium is potassium-40 which forms 0.0119 atom percent of the whole. Its radioactivity was discussed in Sect. 30. It has two stable daughter isotopes: argon-40 produced by orbital electron capture and calcium-40 produced by beta-particle emission. Age determinations can, in principle, be based on the accumulation of either of these isotopes in a potassium-bearing mineral. In either case the usefulness of the determination depends on the sensitivity and accuracy with which the daughter element can be measured, and the ease with which it can be distinguished from non-radiogenic contamination. The occurrence of common calcium in minerals is so widespread, and calcium-40 forms such a large proportion of common calcium (97%) that the detection of radiogenic calcium is usually very difficult. On the other hand, only very small amounts of common argon are found in minerals and rocks so that despite the fact that 99.6% of common argon is argon-40, the correction for non-radiogenic argon presents no serious problems. For this reason all *potassium ages* that have yet been determined have been based on the *decay to argon*, although the possibility of *calcium ages* in the future cannot be ruled out. Adequate techniques now exist for the measurement of potassium, argon and calcium in minerals and rocks in the amounts encountered in age determination studies.

Unlike the uranium-thorium-lead methods which have been the subject of intensive study for approximately fifty years, the potassium-argon method has been under development for only eight years and has given positive signs of success within only the past two or three years. The writers are acutely aware of the fact that no final appraisal can be made of a method that is still so new. It is almost certain that many new experimental results will be reported before this account finally appears and these results may alter the aspect of the potassium-argon method. In the meantime the present summary will describe the great advances already made and give some of the recent results of potassium-argon dating which presage a bright future for the method.

## b) History and status.

**48.** Although the radioactivity of potassium was detected as early as 1905, it was not until 1937 that VON WEIZSÄCKER suggested the possibility of a dual decay forming argon as well as calcium as end products. This suggestion was very attractive, for it explained the phenomenally large abundance of argon-40 in the earth's atmosphere which contains only small amounts of argon-36 and -38 and of the other inert gases. These are believed to be far more abundant than argon-40 in the atmospheres of the sun and stars. The production of argon-40 by the great amount of potassium known to exist on the surface of the earth provided a simple explanation for the large abundance of argon-40 in its atmosphere.

The physical methods applied to prove the dual activity of potassium and to determine its two decay constants have already been described in section 20. Among the first to take a geophysical approach to the problem were HARTECK and SUESS[1] who attempted to detect argon in potassium minerals of known age. Although they failed to detect argon they were able to show that it was

---

[1] P. HARTECK and H. E. SUESS: Naturwiss. **34**, 214 (1947).

present in much smaller quantities than predicted by the decay constants then accepted. HARTECK and SUESS correctly interpreted their results as indicating that the values for the decay constants were greatly in error, although the possibility of gross loss of argon from the minerals could not be disproved.

A few years later FARRAR and CADY[1] reported results of argon analyses carried out on potassium minerals. They found no relationship between the argon content, the potassium content and the supposed age of the samples. No mass spectrometer analyses were carried out so that it was impossible to distinguish between non-radiogenic contamination and radiogenic argon produced by the potassium decay. At about the same time ALDRICH and NIER[2] reported argon measurements, including isotopic analyses, for five potassium minerals of different ages. These measurements showed for the first time a distinct correlation between argon content and the age of the samples. It was not possible to determine whether the relationship was exact, for the ages of the samples were never shown to be precise and the potassium contents were not measured, but were calculated from the accepted chemical formulae of the minerals.

The argon samples extracted by ALDRICH and NIER varied in radiogenic purity from 14 to 81%. A few years later MOUSUF[3] in an attempt to establish the rate of production of argon by potassium, extracted argon from minerals from the Grenville geological province of the Canadian shield and found it to be as high as 99.9% radiogenic. The possibility of extracting argon of such high radiogenic purity provided additional incentive to those engaged in developing the method.

**49.** Starting in 1950 GENTNER and his co-workers published a series of papers in which they reported upon the potassium-argon ratios from the Buggingen deposits in Germany. The results show a marked decrease in relative argon content for the more finely grained specimens. The observed effect, which is apparently quite regular, has been explained by the authors as caused by the diffusion of argon from the crystals. In the most recent paper [45] they report that the fit between the theoretical diffusion curves and the observed argon-potassium ratios is improved by allowing for a continual cooling of the deposits since their time of formation.

If this effect is real and if the assumed explanation is correct, this study paves the way for the development of a powerful tool to investigate the thermal history of these and other salt deposits. Unfortunately sylvites are very soluble in water and it remains possible that the phenomena observed by GENTNER is in some way related to solution and redeposition of the salts, and not due to diffusion. That this might be the case is suggested by the argon measurements which have in the meantime been carried out by several laboratories on insoluble feldspars and micas. Sylvite is a close-packed cubic structure through which diffusion might be expected to be slow in comparison with more open structures like feldspars and micas. However if the diffusion constant of $10^{-19}$ cm.$^2$ sec.$^{-1}$ found for the sylvites is assumed to be similar to that for the other less compact mineral structures, a loss of most of the argon from feldspars and micas 2500 million years old would be expected. Potassium-argon ratios for minerals in this age range have been determined and such loss seems impossible, even allowing for possible errors in decay constants and estimated geological ages. Admittedly the arguments are somewhat nebulous and no clear-cut conclusions can be reached

[1] R. L. FARRAR jr. and G. H. CADY: J. Amer. Chem. Soc. **71**, 472 (1949).
[2] L. T. ALDRICH and A. O. NIER: Phys. Rev. **74**, 876 (1948).
[3] A. K. MOUSUF: Phys. Rev. **88**, 150 (1952).

at the present time. Whatever the outcome, the variations observed by GENTNER are of fundamental importance and their study cannot help but add much to our understanding of the potassium-argon method.

50. The rate of argon production by potassium which MOUSUF had determined seemed small in comparison with that predicted from decay constants found by many counting experiments. In an effort to determine whether argon loss was the cause of this, RUSSELL, SHILLIBEER, FARQUHAR and MOUSUF[1] extracted argon from five potassium feldspar samples differing in age by as much as a factor of five. These specimens were all dated precisely by uranium minerals associated with the feldspars. No indications that argon had diffused from any of the samples were found. The result obtained for the decay constant of potassium-40, like MOUSUF's, was invalidated when WASSERBURG and HAYDEN[2] found an $A^{40}/K^{40}$ ratio that was approximately 35% higher for feldspar from one of the same deposits. Further investigation, reported by SHILLIBEER,

Table 23. *Potassium-argon ages.*

(a) Ages taken from "$A^{40}$—$K^{40}$ dating" by G. J. WASSERBURG and R. J. HAYDEN, Geochim. et Cosmochim. Acta 7, 51 (1955). The comparison ages are in each case $Pb^{206}/U^{238}$, $Pb^{207}/U^{235}$ and $Pb^{208}/Th^{232}$ ages for uraninite from the same pegmatite. The ages are reproduced with the permission of the authors. Decay constants used are $\lambda = 0.55 \times 10^{-9}$ years$^{-1}$ and $\lambda_e/\lambda_\beta = 0.085$.

| Mineral and location (Feldspar unless noted) | % K | $A^{40}/K^{40}$ | Age (m. y.) | Comparison age (m. y.) |
|---|---|---|---|---|
| Besner Mine, Lot 5, Con B, Henvey Twp., Ontario (40 to 80 mesh) | 11.00 | $0.0565 \pm 0.0014$ | 990 | (Uraninites give ages from 750 to 950) |
| Besner Mine, (18 to 40 mesh) { | 10.92 | $0.0573 \pm 0.0014$ | 1000 | |
| | 10.92 | $0.0567 \pm 0.0014$ | 990 | |
| Besner Mine (80 to 118 mesh) | 10.43 | $0.0542 \pm 0.0014$ | 960 | |
| Strickland Quarry, (Collins Hill) Portland, Conn | $10.51 \pm 0.02$ | 0.01282 | 275 | { 268 266 239 |
| Blackstone Lake Pit, Conger Twp., Parry Sound, Ontario | $11.98 \pm 0.02$ | 0.0556 | 975 | { 994 993 897 |
| | $11.98 \pm 0.02$ | 0.0543 | 957 | |
| Blackstone Lake Pit | $1.32 \pm 0.03$ | 0.0456 | 834 | |
| Blackstone Lake Pit | $10.62 \pm 0.02$ | 0.0592 | 1020 | |
| Concession 16, Cardiff Twp., Ontario | $3.02 \pm 0.06$ | 0.0629 | 1070 | { 1000 1020 870 |
| Concession 16, Cardiff Twp., Ontario | $4.79 \pm 0.05$ | 0.0587 | 1020 | |
| Tory Hill, Ontario | $6.15 \pm 0.02$ | 0.0540 | 953 | |
| Tory Hill, Ontario | $6.86 \pm 0.04$ | 0.0542 | 955 | |
| Viking Lake Pegmatite Beaverlodge Lake, Saskatchewan | $11.68 \pm 0.02$ | 0.1502 | 1950 | { 1850 1880 1670 |
| Beardsley I (meteorite) | 0.101 | $7.13 \times 10^{-5}$ cc. $A^{40}$/gm. | $4820 \pm 200$ | |
| Beardsley II (meteorite) | 0.101 | $5.87 \times 10^{-5}$ cc. $A^{40}$/gm. | $4500 \pm 200$ | |
| Forest City (meteorite) | 0.0831 | $5.15 \times 10^{-5}$ cc. $A^{40}$/gm. | $4570 \pm 200$ | |

---

[1] R. D. RUSSELL, H. A. SHILLIBEER, R. M. FARQUHAR and A. K. MOUSUF: Phys. Rev. 91, 1223 (1953).

[2] G. J. WASSERBURG and R. J. HAYDEN: Phys. Rev. 93, 645 (1954).

*Table 23.* (Continued.)

(b) Taken from SHILLIBEER and RUSSELL, Canad. J. Phys. **32**, 681 (1954), and SHILLIBEER and WATSON, Science, Lancaster, Pa. **121**, 33 (1955) and other unpublished University of Toronto measurements. Comparison ages assigned by original authors. Decay constants used are $\lambda = 0.548 \times 10^{-9}$ years$^{-1}$ and $\lambda_e/\lambda_\beta = 0.089$.

| Mineral and location (Feldspar unless noted) | % K$_2$O | A$^{40}$/K$^{40}$ | Age (m. y.) | Comparison age (m. y.) |
|---|---|---|---|---|
| Besner Quarry, Lot 5, Con B, Henvey Twp., Ontario. No 1178-A | 13.2 | 0.0514 | 890 ± 70 | 940 |
| Dill Twp., Ontario. No 1107-A | 12.6 | 0.052 | 900 ± 70 | Grenville (900 to 1200) |
| Dill Twp., Ontario. (mica) No 1109-A | 9.6(6) | 0.049 | 860 ± 60 | |
| Conger Twp., Lot 7, Con X, Ontario. No 1223-A | 13.1 | 0.055 | 940 ± 70 | cf. Blackstone Lake Pit, (a) part of this table |
| Conger Twp., Lot 7, Con X, Ontario. (mica) No 1124-A | 8.7(6) | 0.089 | 1040 ± 80 | cf. Blackstone Lake pit, (a) part of this table |
| Lee Lake, Lac la Ronge, Saskatchewan. 1061-A | 12.0 | 0.130 | 1730 ± 120 | 1740 (207/206 age of uraninite from same pegmatite) |
| Charlebois Lake, Saskatchewan. No 1062-A | 10.2 | 0.138 | 1800 ± 130 | 1790 (207/206 age of uraninite from same pegmatite) |
| Middle Foster Lake, Saskatchewan. No 1140-A | 11.6 | 0.135 | 1730 ± 120 | |
| Foster Lake, Saskatchewan. No 1145-A | 9.0(9) | 0.147 | 1880 ± 130 | |
| Viking Lake Pegmatite, Beaverlodge Lake, Saskatchewan. No 1086-A | 13.7 | 0.138 | 1800 ± 130 | cf. (a) part of this table. |
| Moose Dyke, Great Slave Lake, Northwest Territories. No 1069-A | 13.1 | 0.183 | 2150 ± 140 | 2230 (mean of eight galena ages) |
| Silver Leaf Mine, Winnipeg River, Manitoba. (mica) No 1080-A | 10.1 | 0.260 | 2600 ± 150 | 2550 (mean of 4 207/206 ages for Huron Claim pegmatite) |
| Ussusa, Sultan Hamud, Kenya, South Africa. No 1169-A | 12.9 | 0.0291 | 550 ± 50 | Mozambique belt |
| Kinyikio, Kenya, South Africa. No 1170-A | 12.7(5) | 0.0251 | 485 ± 50 | Mozambique belt |
| Tsaro, Nairobi, Kenya, South Africa. No 1171-A | 11.7 | 0.0289 | 560 ± 50 | Mozambique belt |
| Popes Claim, Southern Rhodesia, South Africa. No 1084-A | 8.6 | 0.268 | 2660 ± 150 | 2610 (mean of 4 207/206 ages from same region) |

(c) Ages taken from "Potassium-argon ages of lepidolites", G. W. WETHERILL, L. T. ALDRICH and G. L. DAVIS, Programme of the 1954 Thanksgiving meeting of the American Physical Society. Comparison ages are rubidium-strontium ages. Decay constants used are $\lambda = 0.552 \times 10^{-9}$ years$^{-1}$ and $\lambda_e/\lambda_\beta = 0.090$.

| Mineral and location (Lepidolite) | | K-A age (m. y.) | | Rb-Sr ages (m. y.) |
|---|---|---|---|---|
| Bikita, Southern Rhodesia | | 2750 | | 3400 |
| Bonneville, Wyoming, USA | | 2570 | | 3020 |
| Bagdad, Arizona, USA | | 1610 | | 1900 |
| Ohio City, Colorado, USA | | | | |
| White lepidolite | | 1500 | | 1740 |
| Coarse lepidolite | | 1540 | | 1750 |

Table 23. (Continued.)

(d) Ages taken from GENTNER, GOEBEL and PRÄG, Geochim. et Cosmochim. Acta 5, 124 (1954) and GENTNER, JENSEN and MEHNERT, Note in Z. Naturforsch., 9a, 176 (1954). Ages have been recalculated for a total decay constant of $0.548 \times 10^{-9}$ years$^{-1}$ and a branching ratio of 0.090. Ages obtained by the original authors are shown in brackets. This recalculation may be slightly in error for the Buggingen samples, for the diffusion correction applied will depend in part on the decay constants chosen. Comparison ages were assigned by the original authors.

| Mineral and location | % K$_2$O | A$^{40}$/K$^{40}$ | Age (m.y.) | Comparison age (m. y.) |
|---|---|---|---|---|
| Buggingen, Germany (sylvite) | — | — | 24 (20) | Oligocene |
| Waldkirch, Elztal, Black Forest, Germany (feldspar) | 13.02 | 0.0137 | 281 (223) | 200—250 |

(e) Age taken from FRITZE and STRASSMANN, Naturwiss. 39, 522 (1952). Age has been recalculated using a total decay constant of $0.548 \times 10^{-9}$ years$^{-1}$ and a branching ratio of 0.090. There was no mass spectrometer analyses of the argon extracted. There appears to be an error in arithmetic in the age calculated by the original authors. This age is given in brackets. Comparison age was assigned by the present authors.

| Mineral and location | % K$_2$O | A$^{40}$/K$^{40}$ | Age (m.y.) | Comparison age (m. y.) |
|---|---|---|---|---|
| Varutrask, Finland (feldspar) | 12.8 | 0.130 | 1730 (1880) | 1700 (uraninite) 1740 (lepidolite) |

RUSSELL, FARQUHAR and JONES[1] showed that the gas extractions made at Toronto by heating with metallic sodium were inadequate for releasing all of the argon from the minerals and has led to general agreement between University of Chicago and University of Toronto measurements. Thus the latest results, using improved techniques, have given values for the decay constants in much better agreement with those measured by counting experiments.

Besides these preliminary potassium-argon ages which have already been published, further studies of the method are known to be in progress at Massachusetts Institute of Technology, the Carnegie Institute of Washington, the University of California and at universities in Germany (FRITZE and STRASSMANN) and in Russia (GERLING). Some of the available potassium-argon ages have been summarized in Table 23.

The age determinations themselves, which are at present our only guide, give no indication of serious errors which could be the result of argon loss[2]. It seems likely that either such loss is entirely negligible, or that it is at least small enough to cause no serious relative errors in potassium-argon dating. Gross loss by diffusion has been strongly suspected by many because the loss of helium from uranium and thorium minerals has been shown to be so severe as to invalidate the helium method which also depends on the collection of a gas in a mineral. Helium ages, which were once widely accepted, have been proved to be almost always low because of the escape of large fractions of the helium. This comparison with helium is not valid because six to eight helium atoms and one lead atom are produced by the decay of each atom of uranium or thorium, and centres of great pressure within the minerals result. The helium nuclei are released as alpha particles with energies in the order of 5 Mev. capable of seriously damaging

---

[1] H. A. SHILLIBEER, R. D. RUSSELL, R. M. FARQUHAR and E. A. W. JONES: Phys. Rev. 94, 1793 (1954).

[2] Note added in proof: It now appears that some feldspars do lose argon, but that all micas retain it. This phenomenon is still being studied.

the crystal structure. All of these considerations favour the loss of helium. On the other hand, only one atom is produced by a decaying potassium-40 atom, and each argon atom is produced with a recoil energy of only about 30 ev. No pressure is built up, no damage to the crystal structure occurs, and the bulkier argon atom is less likely to diffuse. Moreover, the rarity of the radioactive isotope, even in minerals of high potassium content, minimizes the chances of accumulation and escape of the argon.

Thus it seems likely that the argon may be almost entirely retained. On this point the whole future of the method depends and there is a strong tendency to be optimistic. The special problems and special techniques necessary for the satisfactory application of the helium, common lead, radiogenic lead and rubidium-strontium methods are only now beginning to be realized after the completion of hundreds of age determinations. It is too much to hope that the potassium-argon method will not also have its pitfalls, or that the best procedures for its use will be clear after the analysis of only a handful of samples.

### c) Experimental procedures.

**51.** Potassium-argon ages are calculated by the use of the formula (14.6)

$$t = \left(\frac{1}{\lambda}\right) \log_e \left(1 + \frac{\lambda \, A^{40}}{\lambda_e \, K^{40}}\right) \tag{51.1}$$

where $t$ is the age of the potassium mineral, $\lambda$ is the total decay constant and $\lambda_e$ the decay constant for orbital electron capture. Argon-40 and potassium-40 represent the masses of these isotopes found in the mineral specimen. The mass of argon-40 must first be corrected for the presence of non-radiogenic argon.

It follows from equation (51.1) that the determination of the age of a potassium mineral requires the accurate measurement of its potassium and argon contents. Neither of these measurements are experimentally simple, and several methods exist for carrying them out. These methods will be described briefly in the paragraphs that follow to illustrate in a general way the problems involved. More detailed description of them can be found in many papers in the literature.

**52.** Chemical analysis for potassium has always been very difficult, due to the very great similarity in chemical behaviour of the alkali metals. Classically the potassium content of minerals is determined by the J. LAWRENCE-SMITH method or one of its modifications. This involves breaking down the mineral structure with a sodium carbonate fusion, leaching the alkalis from the resulting solid mass, and selectively precipitating the potassium as the chloroplatinate. This method has been widely used and yields very good results in the hands of a careful and experienced chemist. It is less satisfactory for minerals containing small amounts of potassium.

The detection and measurements of potassium by a flame photometer is also a recognized method of determining this element. This method has been widely used in biological studies, but is somewhat less well known than the LAWRENCE-SMITH method for mineral and rock analyses. It has the advantage of great simplicity when used on a routine basis. The writers have heard widely varying reports as to the relative merits of the flame photometer and wet chemical methods for potassium analyses, but if well done, they are probably about equally good. We have found the flame photometer best for our work, particularly for minerals containing less than 5% potassium.

The determination of potassium by isotopic dilution with a mass spectrometer is possible and should be very accurate. Since potassium forms no stable volatile compounds it must be used in a solid-source mass spectrometer, and hence this method requires specialized mass spectrometer techniques and costly equipment. The isotopic dilution method may shortly become a standard against which the other methods will be compared. It has recently been applied by WASSERBURG and HAYDEN[1] to determine the potassium content of meteorites.

A fourth method for potassium analyses makes use of its natural radioactivity. Determining potassium by counting techniques has been suggested and various experimental arrangements have been described. It has the advantage that it would determine the particular isotope responsible for the decay products, and the disadvantage that radium in very small amounts could interfere badly. The method has not yet been used for age work and probably would be less accurate than the others.

UREY and EDWARDS[2] have recently developed a method particularly suited to meteorites and rocks having very small potassium content. The potassium is distilled from the mineral at 1300° C after the addition of pure calcium chloride. The method has been checked against wet chemical methods and isotopic dilution measurements and appears to yield very good results. The method has the advantage that the equipment required is much less expensive than that for the isotopic dilution method.

**53.** Methods for measuring *argon* are divided into two basic types depending on whether the argon is extracted, and purified quantitatively and measured with a McLEOD gauge, or alternatively mixed with an isotopic tracer and measured mass spectrometrically. To release the argon the mineral must be melted, with or without a flux. If the volumetric measurement is intended, the whole of the argon must be released from the rock into the high vacuum system; if the tracer technique is to be used the tracer must be completely mixed with the argon from the melted rock. Neither of these conditions are easily tested in practice, for no standard minerals containing specific amounts of argon exist, and their preparation would be extremely difficult, if not impossible. The only present indication of the efficiency of the measurements is by comparing results obtained on identical specimens at different laboratories and with different methods of measurement. Too few such checks have been made, but indications are that present techniques by either method are adequate for age determination purposes.

In all cases potassium-argon ages depend on an accurate correction for the presence of non-radiogenic argon. This is accomplished by determining the ratio of argon-40 to argon-36 in the extracted sample by means of a mass spectrometer analysis, and comparing it with the corresponding ratio for atmospheric argon. It can easily be shown that the proportion of radiogenic argon is given by the relationship

$$\% \text{ Radiogenic Argon} = \frac{(A^{40}/A^{36}) \text{ sample} - (A^{40}/A^{36}) \text{ air}}{(A^{40}/A^{36}) \text{ sample}} \times 100\%. \qquad (53.1)$$

NIER[3] has found the ratio $A^{40}/A^{36}$ in air to be 296 atoms/atom.

#### d) Choice of constants.

**54.** In Sect. 20, it was shown that the decay constants of potassium are not accurately known. For the purpose of compiling a table of ages it was necessary to make a consistent choice of decay constants. We have used the mean values of the decay constants determined physically by gamma ray and beta particle

[1] G. J. WASSERBURG and R. J. HAYDEN: Phys. Rev. **97**, 86 (1955).
[2] G. EDWARDS and H. C. UREY: Geochim. et Cosmochim. Acta **7**, 154 (1955).
[3] A. O. NIER: Phys. Rev. **77**, 789 (1950).

counting experiments. The corresponding value of the branching ratio agrees to within 1% with the most recent geophysical value[1]. These means were as follows:

$$\lambda_\beta = 0.503 \times 10^{-9}\,\text{yrs}^{-1}, \quad \text{Std. dev.} = \pm\, 0.045 \times 10^{-9}\,\text{yrs}^{-1},$$

$$\lambda_e = \text{Branching Ratio} \times \lambda_\beta$$

$$= 0.090 \times 0.503 \times 10^{-9}$$

$$= 0.045 \times 10^{-9}\,\text{yrs}^{-1}, \quad \text{Std. dev.} = \pm\, 0.015 \times 10^{-9}\,\text{yrs}^{-1}.$$

It is also necessary in calculating ages from formula (53.1) to know the abundance of the isotope of potassium of mass 40. NIER[2] found this abundance to be 0.0119 atom percent. A second measurement by REUTERSWARD[3] gave 0.0118 atom percent, in excellent agreement with NIER's value. The values for the decay constants of potassium were calculated with NIER's value for the isotopic abundance, and the same value has been used in the age calculations. Any small error in the abundance of potassium-40 would largely cancel in the two calculations and produce negligible error in the final ages.

### e) Experimental results.

**55.** In order to give an indication of the work that has been done determining potassium-argon ages and to enable the results to be compared as far as possible with the other age determinations, some of these ages have been listed in Table 23. All calculations have been based on the decay constants just discussed, unless otherwise indicated, but $A^{40}/K^{40}$ ratios are given in all cases so that other decay constants can be used to recalculate the ages if the reader prefers. Comparison ages are included wherever such ages existed. In many cases the correlation between the dated mineral and the comparison age is not very close, but the agreement in ages is remarkable and augers well for the potassium-argon method.

## III. The rubidium methods.

**56.** The radioactivity of rubidium was discovered in 1905 by THOMSON and verified in 1907 by CAMPBELL and WOOD, as already outlined in Sect. 19. HEMMENDINGER and SMYTHE[4], in 1937 proved that the radioactivity was associated with the isotope of mass 87, which forms 27.2 % of ordinary rubidium[5]. The only other naturally occurring rubidium isotope is rubidium-85, which does not appear to be radioactive.

GOLDSCHMIDT[6] and HAHN and WALLING[7] suggested that the natural beta decay of rubidium to strontium could be used to determine the age of minerals. Although there are no true rubidium minerals, this element occurs in lithium micas, the most common of which often contains one or two percent rubidium. It is also found in lesser amounts in other common micas, such as biotite. Chemically it is very similar to potassium, and minerals rich in potassium are often also rich in rubidium.

The decay of rubidium-87 is given by equation (12.4).

[1] H. A. SHILLIBEER et al.: 1954, see footnote in Sect. 50.

[2] Footnote in Sect. 53.

[3] C. REUTERSWARD: Ark. Fysik **4**, 203 (1951).

[4] A. HEMMENDINGER and W. R. SMYTHE: Phys. Rev. **51**, 1052 (1937).

[5] A. O. NIER: Phys. Rev. **50**, 1041 (1936).

[6] V. M. GOLDSCHMIDT: Skrift. norske Vid.-Akad., Math.-naturwiss. Kl. **1937**, No. 4.

[7] O. HAHN and E. WALLING: Z. anorg. allg. Chem. **236**, 78 (1938).

The half-life of rubidium-87 is long with respect to the age of most minerals and only a small error is introduced if the exponential is approximated by the first two terms of its Taylor's expansion. Inserting the numerical value of $\lambda$, (12.4) becomes

$$t = 8.8 \times 10^{10} \times \frac{Sr^{87}}{Rb^{87}}.$$    (56.1)

For accurate age determinations in addition to precise chemical analyses a mass spectrometer analysis is necessary to correct for the presence of non-radiogenic strontium. This correction, which is small for lepidolites but may be very large for other minerals, is possible because common strontium has, in addition to strontium-87, isotopes of masses 86 and 88. Variations in the abundances of the isotopes of common strontium, now denied, have been reported recently[1] and suggest difficulties in applying accurately a correction for the presence of large amounts of non-radiogenic strontium.

**57.** Rubidium and strontium can both be determined chemically, and this method was the first employed[2]. The techniques are difficult because the amounts of radiogenic strontium are so small and because analysis for rubidium is difficult.

Ahrens[3] has described a method by which the ratio of strontium to rubidium can be determined spectroscopically in a single operation. In any spectrochemical measurement the intensity of the spectral line is given by the formula

$$I = K \cdot C^n$$    (57.1)

where $I$ is the intensity of the line, $C$ is the concentration of the element within the source of excitation, $K$ is a constant, and $n$ is the emission factor which is dependent on several energy conditions, but which is frequently close to unity. Ahrens, after investigating the subject carefully, used rubidium line 4202 A and strontium line 4078 A. If the emission factor is close to unity for these lines, the ratio of the concentrations of the elements is proportional to the ratio of the intensities of the two lines. A synthetic standard was used to determine the constant of proportionality.

Using this technique Ahrens and his co-workers determined a number of rubidium-strontium ages, some of which are reproduced in Table 24. The reproducibility of the intensity ratio on a series of arcings of the same lepidolite showed an approximate standard deviation from the mean of about 30 to 40 %. Since each analysis was carried out in quadruplicate, the error was reduced to 10 to 15 %, which is the uncertainty of the calculated age. In addition to the question of the reproducibility of analyses, the difficulty in preparing suitable synthetic standards is an additional source of possible error in any spectrochemical analysis. Such an error is difficult to assess.

An entirely new approach to the problem was taken by Aldrich and his co-workers[4] and by Tomlinson[5]. They developed methods which avoid completely spectroscopic determinations of rubidium and strontium, but make use of

---

[1] L. T. Aldrich, L. F. Herzog, J. B. Doak and G. L. Davies: Trans. Amer. Geophys. Un. **34**, 457 (1953). — L. T. Aldrich, L. F. Herzog, W. H. Pinsen jr., and G. L. Davies, ibid. **36**, 875 (1955).

[2] R. E. Stevens and W. T. Schaller: Am. Mineral. **27**, 525 (1942). — F. Strassman and E. Walling: Ber. dtsch. chem. Ges. **71**, 1 (1938).

[3] L. H. Ahrens: Bull. Geol. Soc. Amer. **60**, 217 (1949).

[4] L. T. Aldrich, C. R. Tilton, G. L. Davies and L. O. Nicolaysen: Carnegie Inst. Wash. Year Book **1952**, No. 51, 70; see also footnote[1].

[5] R. H. Tomlinson and A. K. Das Gupta: Canad. J. Chem. **31**, 909 (1953).

mass spectrometric analyses and the isotopic dilution method. In this method small known amounts of separated isotopes are added to the mineral samples as tracers, and are carried through all the chemical manipulations. The mass spectrometer then determines the abundances of the other isotopes in terms of the tracer. Since the important isotopes of both strontium and rubidium have the same mass, a complete separation of the rubidium from the smaller amount of radiogenic strontium is very important. This separation is accomplished by the use of ion exchange columns. The measurements carried out in this way are accurate to about 3 %, which is appreciable better than the accuracy usually obtained with optical spectroscopy. Ages calculated by the isotopic dilution technique are also shown in Table 24.

Table 24. *Rubidium-strontium ages of lepidolites* [1].

| Sample | Age by spectro-chemical analysis | Age by isotopic dilution | Approximate lead age [2] |
|---|---|---|---|
| Silver Leaf Mine, Winnipeg River, Manitoba | 2400 m.y. | | 2500 m.y. |
| Winnipeg River, Manitoba . . . . . . . | | 3500 m.y. | 2500 m.y. |
| Dixon, New Mexico . . . . . . . . . . | 800 m.y. | | |
| Dixon, New Mexico . . . . . . . . . . | | 3450 ± 400 m.y. | |
| Black Hills, South Dakota . . . . . . . | 900 m.y. | | |
| Black Hills, South Dakota . . . . . . . | 1500 m.y. | | |
| South Dakota . . . . . . . . . . . . . | | 2100 m.y. | |
| Popes Claim, Southern Rhodesia . . . . . | | 3470 m.y. | 2700 m.y. |
| Madagascar . . . . . . . . . . . . . . | | 690 m.y. | 650 m.y. |

**58.** A comparison of rubidium-strontium ages with uranium-lead ages revealed the disturbing fact that ages calculated by the "refined" isotopic dilution technique seem to be very high, while those calculated from the less accurate spectrochemical analyses seemed to be more reasonable. This anomaly has partly been explained by analytical errors in the early isotopic dilution results, but the latest rubidium-strontium ages are still approximately 20% higher than uranium-lead and potassium-argon ages. On one occasion ALDRICH has said [3] "All the ages given . . . appear high, but they are independent measurements in which systematic errors of some previous determinations have been eliminated, and at least they make it a matter of interest to redetermine the decay constants of $Rb^{87}$".

The oldest uranium minerals yet dated (which come from South Africa; see Table 17) are approximately 2700 million years old. The oldest common lead minerals appear to be about 2900 million years old, and the oldest potassium minerals have been dated at about 2600 million years. Of these, the age determined for the uranium minerals are the most reliable. The occurrence of many rubidium-strontium ages older than 3000 million years and some as old as 3800 million years is surprising. If they are in error, the fault may lie with the decay constant of rubidium-87, loss of rubidium from the minerals, the presence of a nuclear isomer of rubidium-87, or errors in the analytical methods. If the rubidium ages are correct, they necessitate a major revision of our ideas of geochronology. Too few rubidium-strontium ages are yet available to enable the difficulty to be settled.

---

[1] For spectroscopic values see reference [17]. For isotopic dilution values see the Annual Reports of the Director of the Department of Terrestrial Magnetism, Carnegie Institute of Washington. See also Table 23c for other rubidium-strontium ages.

[2] Assigned by the writers of this chapter.

[3] L. T. ALDRICH: Annual Report of the Carnegie Inst. Wash. **1951/52**.

## IV. The radiocarbon and tritium methods.

**59.** The remaining methods to be considered in detail here are those based on the natural radioactivities of tritium and carbon-14, isotopes which are now known to be present in all specimens of present day hydrogen and carbon. Unlike other radioactive isotopes which have been used for age determination purposes, both tritium and carbon-14 have half-lives extremely short compared with geological time, and consequently it is certain that these isotopes are being continually renewed by some natural processes. Their presence is explained by the neutron flux maintained in the upper atmosphere of the earth by cosmic ray bombardment. This flux continually creates, by nuclear transmutation, atoms of these two short-lived isotopes which are then rapidly mixed with the other carbon and hydrogen of the earth. So far as it is possible to tell, the processes of decay and production are in equilibrium so that the proportion of the radioactive isotopes in these elements are not changing with time.

The other age determination methods were based on the accumulation of daughter products in a mineral containing a radioactive parent. The radiocarbon and tritium methods use a different approach. If the assumptions of a steady state with thorough mixing are valid, it follows that all living matter will contain the same proportion of radioactive carbon-14. After death the radioactive isotope will cease to be replenished and will decay. Then the time interval since death can be determined from the difference between the radiocarbon content and that measured for identical living matter. Similarly, circulating waters maintain their tritium activity by the continual addition and mixing of rain. In isolated water, the tritium content decays exponentially.

The half-life of tritium is 12.5 years; the decay product is helium-3. With such a short half-life this isotope can be expected to be useful for dating only events of the last half-century. It seems likely that the method will be valuable for studying the mechanisms by which rain water is transported, and for determining the rate of circulation of underground waters. So far such work is only in its preliminary stages.

The half-life of carbon-14 is $5568 \pm 30$ years; its daughter isotope is nitrogen-14. This half-life, much longer than that for tritium, has permitted the determination of ages as old as 30,000 years by the radiocarbon method. It seems likely that this limit can be extended to 50,000 years, while maximum limits as great as 100,000 years have been envisioned by some workers. The best present estimate of the age of the earth is over 4300 million years. That is, present radiocarbon ages give information about only one hundred thousandth of the history of the earth, and consequently it is of no use for dating events throughout the greater part of geological time. On the other hand they have proved their usefulness for studying certain recent geological processes and they have been valuable to archeologists for dating events during the history of mankind. It is in the latter field, rather than in geophysical applications, that radiocarbon dating has earned a reputation as a major discovery of modern science.

Both the tritium and radiocarbon methods were pioneered by LIBBY and his co-workers, who have played a large part in establishing the validity of the carbon-14 method and who have determined a large number of valuable carbon-14 ages. A recent book by W. F. LIBBY [19] describes the principles of the method and gives full instructions for determining radiocarbon dates with technical information for the construction of an adequate counter. It is from this book that much of the background material presented here has been taken[1].

Radiocarbon ages are calculated from the ratio of the carbon-14 abundance of the carbon in the specimen to the carbon-14 abundance of the carbon in similar

---

[1] It has recently been agreed to publish summaries of $C^{14}$ ages in Nature, London and Science, Lancaster, Pa.

living material. These are assumed to be related by the formula

$$\frac{C^{14}_{\text{sample}}}{C^{14}_{\text{living carbon}}} = e^{-\lambda t}. \tag{59.1}$$

From this equation the time $t$ since the "death" of the specimen can be calculated, since the decay constant $\lambda$ is known. In all carbon-14 dating the time of death is assumed to be equivalent to the time at which the living matter ceased to exchange its carbon with the air.

The use of equation (59.1) involves two important physical assumptions. The first is that the radiocarbon content of carbon in modern living material is the same as the radiocarbon content of carbon in living material during the past few tens of thousands of years. This will only be true if the cosmic ray flux, which produces carbon-14 has been constant during that time. This assumption appears to be well justified by the excellent agreement between radiocarbon ages and known ages of materials which can be dated historically and of tree rings. Another factor which may influence the radiocarbon content of natural carbon is the production of enormous quantities of carbon dioxide in the air by modern industry and the resultant dilution of atmospheric radiocarbon. Experiments to detect this effect have not yet yielded conclusive results.

The second assumption is that the carbon found now in the specimen is the same carbon that was present when it was living. This is not always the case, for some samples can exchange carbon readily with their surroundings. Inorganic carbon compounds, which are of a chalky nature, are particularly suspect. This source of error can be avoided by choosing material which is likely to have preserved its original carbon.

Most radiocarbon ages have been determined by counting with a screen-wall counter, carbon carefully prepared from a specimen. Other counting techniques are being developed which may increase the precision of the results and which may allow older objects to be dated. One of the methods now being tested is the use of acetylene formed of the sample carbon, dissolved under pressure in an organic solvent which contains a fluorescing substance. Another makes use of benzene prepared from the carbon in the specimen. A fluorescing material is added to the benzene and the beta particles are counted by the scintillations. Whatever type of arrangement is used, elaborated precautions must be taken to minimize the cosmic ray background and the contamination of apparatus and materials by other radioactive substances.

# V. Accessory mineral methods and rock analyses.

**60.** The problem of *dating rocks* is closely related to that of dating specific minerals. While the same methods may in principle be applied in both cases, there are particular difficulties which complicate the former problem. In particular the smaller concentrations of radioactive parents necessitate special measuring techniques, and the small grain size of the mineral particles cause concern over possible loss of parent and daugther isotopes by diffusion or leaching.

While the difficulties encountered in dating rocks are greater than those in dating minerals, the importance of this work justifies the expenditure of a great deal of effort. The most reliable age determinations have been made on pegmatitic uraninites, and there are now several pegmatites to which very reliable ages can be assigned. If it can be assumed that pegmatites represent final stages of crystallization of rock magma, then these pegmatite ages can safely be equated to the time of formation of the associated igneous country rock. Often this assumption seems reasonable, but it could clearly lead to serious errors in some cases. The correlation of pegmatite ages with the fossiliferous time scale of sedimentary rocks is still more difficult. The interpretation of absolute age determinations of metamorphic rocks is very hazardous until the precise effect of the metamorphic processes is known. A proper understanding of many of the problems of geochronology awaits the establishment of methods for dating the various rock types.

**61.** HOLMES[1] was among the first to realize that the helium method, largely abandoned because of the proven loss of helium from uranium minerals, might be applied with better success to dating rocks. Helium, since it is a gas, can be separated and accurately measured in far smaller concentrations than is possible for lead by any of the chemical methods. Uranium and thorium can be detected by the radioactivity of the parent and intermediate members of the series. The very small concentrations of uranium and thorium in most rock minerals lessens the likelihood of loss of helium following radiation damage to the crystal structures and the formation of high pressures of helium within the lattices.

Subsequent work, including researches of PANETH, URRY, EVANS, KEEVIL, GOODMAN, LARSEN, GERLING and others, showed that helium loss from rock minerals was in general much less than from uranium minerals, but that such loss was sufficient to result in erratic helium ages. Close packed structures, such as corundum, rutile and many sulphide minerals, lost less helium than most rock-forming silicates. Recent work in this field has been done at the Massachusetts Institute of Technology, USA, by P. M. HURLEY and his associates who have contributed much to our understanding of the mechanisms for the loss of helium. They have found it possible to relate the loss of helium to the age and alpha activity of the sample[2]. Moreover it appears that all the observed discrepancies between helium ages and lead-uranium ages cannot be ascribed to helium diffusion alone, but that metamorphism, the solubility of the atmospheric helium in rocks and the existance of uranium interstitially between the rock grains may play important roles. It still remains possible that certain rock minerals do not lose more than negligible helium by diffusion.

**62.** While all minerals contain small concentrations of uranium and thorium, the radioactivity in common igneous rocks is largely concentrated in the common *accessory minerals*, zircon, sphene and apatite which may contain one-half of the total uranium and thorium. It is now known that alpha particles passing through these minerals use a small fraction of their energy in the displacement of atoms from their proper lattice sites. Such dislocations can produce permanent changes in such physical properties as coloration, thermoluminescence, lattice dimensions, and density. KERR and HOLLAND[3] and KULP et al[4] have suggested that the heat of recrystallization of rare earth oxides can be used to estimate the total lattice disarrangement, and hence the age of a specimen of known uranium and thorium contents. If accurate, LARSEN's suggestion for applying the chemical uranium-lead method to zircons will provide an easy method [51].

**63.** The first precise rock ages were determined by TILTON and co-workers[5] who used sensitive mass spectrometer techniques and micro-chemical separations to determine the amounts and isotopic constitutions of uranium, thorium and lead in minerals in a granite from Essonville, Ontario. Several independent ages were obtained, all very close to 1000 million years. This age is in excellent agreement with the well established age of 1030 million years for the nearby

[1] A. HOLMES and H. F. HARWOOD: Miner. Mag. **21**, 493 (1928). — V. S. DUBEY and A. HOLMES: Nature, Lond. **123**, 794 (1929). — V. S. DUBEY: Nature, Lond. **126**, 807 (1930).

[2] P. M. HURLEY: Trans. Amer. Geophys. Un. **33**, 174 (1952). — P. M. HURLEY: In Nuclear Geology, H. FAUL edit., p. 321 et seq. New York: John Wiley & Sons 1954.

[3] P. F. KERR and H. D. HOLLAND: Amer. Mineral. **36**, 563 (1951).

[4] J. L. KULP, H. L. VOLCHOK and H. D. HOLLAND: Amer. Mineral. **37**, 709 (1952).

[5] G. R. TILTON, C. C. PATTERSON, H. BROWN, M. INGHRAM, R. HAYDEN, D. HESS and E. S. LARSEN jr.: Bull. Geol. Soc. Amer. **66**, 1131 (1955).

Fission Mines at Wilberforce, Ontario, and consequently supports the supposition that ages found for pegmatites may also date associated country rock. TILTON also analysed isotopically lead extracted from pegmatitic feldspar from Tory Hill, Ontario and found it to be very similar in isotopic constitution to Grenville lead ores.

**64.** Because of its ubiquitous nature potassium should be an ideal element for the determination of ages of rocks. Little has yet been attempted in this regard, but some preliminary experiments have yielded promising results. A potassium-argon age was determined by TILTON et al[1] for the Essonville, Ontario granite. Although the age obtained was some thirty percent lower than those obtained by the uranium-lead method for the same rock, the difference may be explained by uncertainty in the decay constants (cf. part C), or by incomplete argon extraction.

A suite of aplites, syenites and granites from Wisconsin was dated by SHILLIBEER with the results given in Table 25. The agreement obtained for the various rocks is excellent. Some preliminary experiments have been carried out at Toronto on dating shales by the potassium-argon method. These shales contained index fossils that accurately located them in the sedimentary succession. Some ages obtained were apparently correct, while others were too large. More work is being done on this problem.

Table 25[2]. *Potassium-argon ages for ten rocks from the Wassau area of Wisconsin.*

| Toronto number | Rock | $\% \times 10^8$ by Wt. Rad. $A^{40}$ | $\%$ by Wt. $K_2O$ | $A^{40}/K^{40}$ | K-A age m. y. |
|---|---|---|---|---|---|
| 1207 | Aplite | $4.64 \pm 0.08$ | $5.21^3$ | 0.088 | $1330 \pm 90$ |
| 1208 | Syenite | $4.80 \pm 0.20$ | 5.30 | 0.090 | $1360 \pm 90$ |
| 1209 | Aplite | $4.35 \pm 0.07$ | 4.76 | 0.091 | $1360 \pm 90$ |
| 1210 | Syenite | $4.63 \pm 0.10$ | 5.96 | 0.077 | $1210 \pm 80$ |
| 1211 | Aplite | $2.37 \pm 0.08$ | 3.29 | 0.071 | $1140 \pm 80$ |
| 1212 | Syenite | $5.68 \pm 0.35$ | 5.62 | 0.100 | $1460 \pm 100$ |
| 1213 | Aplite | $5.15 \pm 0.15$ | 5.98 | 0.085 | $1300 \pm 90$ |
| 1214 | Syenite | $4.80 \pm 0.11$ | 5.35 | 0.089 | $1340 \pm 90$ |
| 1215 | Nepheline Syenite | $2.42 \pm 0.10$ | 2.84 | 0.086 | $1310 \pm 90$ |
| 1216 | Granite | $3.95 \pm 0.22$ | 4.28 | 0.089 | $1340 \pm 90$ |

Omitting No 1211, Mean age ± standard deviation = 1330 ± 70.
Including No 1211, Mean age ± standard deviation = 1310 ± 90.

# VI. Choice of methods.

**65.** An important practical consideration is the choice of a method for determining ages. Few laboratories will wish to try them all. Factors governing the choice are the availability of suitable minerals or rocks, the relative accuracy of the methods including the possibility of errors arising from leaching or alteration, and the technical difficulties involved in making the physical measurements.

Materials for the *potassium* method are by far the most abundant. Potassium forms at least several percent of most feldspars and micas which are common minerals in most igneous rocks and pegmatites. Among sedimentary rocks it is present in shales and also in some salt deposits as sylvite or carnallite.

The most satisfactory *uranium* mineral for age determinations is uraninite which occurs in pegmatites. So do samarskite and ellesworthite and also mona-

---

[1] See footnote 5, p. 348.
[2] This table has been taken from H. A. SHILLIBEER, Ph. D. Thesis, Department of Physics, University of Toronto 1955.
[3] All potassium analyses made by K. WATSON.

zite, the chief thorium mineral, but these do not generally give such satisfactory ages. Pitchblende, which is the form of uraninite occurring in veins, has in many cases been dissolved and redeposited, and consequently gives younger ages than would be found for the material from which it was derived. Other secondary and vein minerals are also undesirable if accurate dates of pegmatites or country rocks are to be obtained.

For the common lead method, *galena* is the commonest mineral, although others may do. This method has the advantages of making it possible to date many sulphide ore deposits. Few other minerals which can be dated occur in these deposits. As mentioned before, it is doubtful whether the galena method will ever be as precise as some other methods.

The mineral usually dated by the *rubidium* method is the mica lepidolite which occurs in some pegmatites. It is chosen because of its relatively large rubidium content. Other micas, such as biotite, contain much less rubidium, but can be used with special techniques. The interpretation of rubidium ages is still somewhat obscure (cf. Sect. 58), but this should be clarified by future work.

Concerning errors which may occur due to alteration and losses of isotopes, some general principles can be stated.. It has usually been supposed that gases would be lost. It seems that this is true for helium, and that radon may be lost from badly altered minerals, but not from fresh ones. The evidence so far suggests that argon is well retained by many potassium minerals. For solids, methods which only involve two or more isotopes of the same element are less likely to be affected by leaching and analytical errors than those involving different elements. Examples of such methods are the lead-ratio method (Sect. 39) and the lead-210 method (Sect. 40) for uranium minerals, and the common lead method by which old lead ores can be dated (Sect. 43).

Pegmatites and the minerals in them usually seem to be better crystallized and less susceptible to alteration and redeposition than minerals in veins. Most rubidium compounds are so soluble that they must come under some suspicion until it can be shown that rubidium ages are reliable.

The choice of methods will be determined in many laboratories by the type of *mass spectrometer* available. No very complete age work can be done without one, except by the carbon-14 method, although most of the work so far published by the rubidium method was carried out by AHRENS using a spectrograph, and excellent results were obtained. More recent work has been done with a mass spectrometer.

The crude chemical method for uranium ores is no longer regarded as accurate without isotopic analyses. All other uranium-thorium-lead methods require a mass spectrometer with a resolution of one mass unit in 250 or greater. Such instruments are commercially available, but they are expensive. They are not easy to build, and advice from someone experienced in their construction will help to avoid many pitfalls. With such an instrument it is possible to use several lead methods with only simple auxillary equipment.

For the potassium and rubidium methods a mass spectrometer with lower resolution is sufficient, in which case the less expensive radio frequency mass spectrometers offer a possible alternative to the conventional types. To analyse for argon, skill is required in high vacuum techniques and it is extremely easy to make errors in potassium analyses. It seems that the isotopic dilution method with a mass spectrometer, can be used for most of the analyses required for age studies.

It will be of help to laboratories if they can determine ages by more than one method, and arrange for intercalibrations with other groups.

# F. Geological considerations.

## I. The application of physical measurements to geological studies.

**66.** In making determinations of the age of minerals and rocks the very different natures of geological and physical studies must be taken into account. Physicists are accustomed to dealing with relatively simple systems which may be isolated, subjected to controlled experiment and measured precisely. In age determinations such techniques can be successfully used to determine the decay constants of radioactive isotopes. When these results are applied to rocks, physicists must realize that they are dealing with more complex systems whose history has been vastly long. These systems have not been under control nor observation except during the last brief period of their history. They may have been subjected to outside influences which are no longer easily detected. These are the usual handicaps to geological studies and account for the lack of precision in them and for the reliance placed by geologists upon judgement and intuition. On this account, physicists have in some cases underestimated the work of geologists, but it should be remembered that during the nineteenth century geologists discovered (without aid from physical methods) most of the facts known today about the earth's history. Their long-held views upon the great length of the earth's history have now been firmly established although they were strongly attacked by KELVIN before the discovery of radioactivity, and although they were to a less extent in conflict with early views about the age of an expanding universe.

These examples show the great errors which can be introduced into age determinations by dependence upon faulty assumptions. Since we cannot reproduce the great spans of geological time we can only avoid error by diligent search for checks and by comparisons of different methods. On the other hand geologists have not kept pace with the advances in physics. Their early hopes for radioactive methods of age determination have been dampened by the collapse of some promising methods and by the delay of fifty years required to develop others. With a few notable exceptions like HOLMES, LANE, MARBLE and WICKMAN the development of methods of age determinations has been made by physicists. The time is ripe to bring these branches of science together. Now physical methods are available to provide an absolute geochronology for the whole history of the rocks of the earth. This would span most of the history of the universe and therefore be its most complete record. The elucidation of this problem would therefore be useful for all science.

## II. The ages of minerals, rocks and formations and their use in establishing a time scale.

**67.** It is convenient to divide geological time into three or four parts as shown in Fig. 7. The most recent of these parts is best preserved and has been most studied. For these reasons it has always been given a prominence out of proportion to its length. It is the part since the beginning of CAMBRIAN time represented by sedimentary rocks containing an abundance of fossils. The details of its chronology are given in Tables 1 and 26.

The next oldest part and one which is about five times as long is represented by rocks called *pre-Cambrian* (or Precambrian), in which only occasional poor fossils are found. It is clear that life existed throughout most of this time, but that the creatures had not developed hard parts which could be readily preserved

as fossils. Traditionally the Precambrian has been divided into a younger Proterozoic part of less altered rocks and an older, more altered part, but it is now
known that this method of division is not a chronological one. It is beginning
to be replaced by a subdivision into a younger part in which the rocks, regardless
of their degree of alteration, were originally much like those of today, and an
older part which had a much higher
proportion of volcanic rocks. In most
parts of the world the Precambrian
rocks are covered by a veneer of younger
deposits, but they crop out in the central parts of continents in areas known
as *shields*.

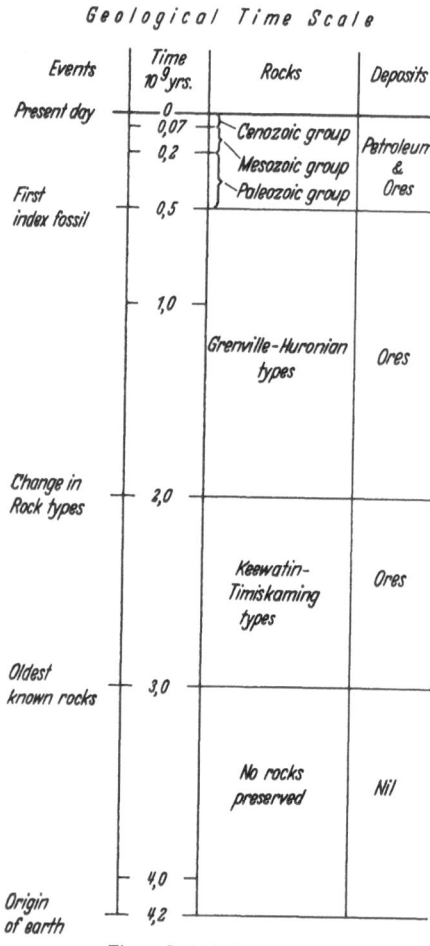

Fig. 7. Geological time scale.

The oldest part of the earth's history
is not represented by any known rocks
but is the period between the oldest rocks
and the time of origin calculated for the
earth. At that time the earth was in
an unstable and perhaps partly molten
condition.

It was shown in part B that for the
youngest part of geological time there is
a *paleontological time scale* based upon cor-
relating sedimentary strata in different
places into a complete succession in which
the evolution of fossil life can be traced.
The fossils which are preserved in these
younger sedimentary rocks represent animals which evolved and spread almost
simultaneously to all parts of the world.
Index fossils are those used for correlation because they changed rapidly from
one identifiable form to another.

In attempting to fix *absolute ages* to
this relative paleontological scale the
problem is encountered that age determinations, which are necessarily made
upon small samples of mineral or rock,
are of little use unless they can be related
to the sedimentary succession or tied to
the geology of the area in some other
manner. Unfortunately few sedimentary
rocks and few minerals occurring in them can be directly dated. Exceptions
are the uraniferous Kolm formation of Sweden and a few shales, evaporites
and other special sedimentary formations but the results to date have been
somewhat inconclusive.

Most of the best age determinations have been made upon crystals obtained
from pegmatites. Pegmatites are tabular bodies of coarsely crystalline minerals
which are considered by geologists to have been formed at high temperatures
within pre-existing rocks around the margins of larger, intrusive bodies of igneous
rocks. Thus dating a mineral in a pegmatite is considered to date the pegmatite
and the associated igneous rock. This rock may underlie a considerable area and
so assist in correlation.

Table 26. *Absolute dates suggested for periods since the Precambrian.*

| Era | Period | Date of start of period (millions of years) | | | | |
|---|---|---|---|---|---|---|
| | | BULLARD | HOLMES A | HOLMES B | KNOPF | MARBLE |
| Cenozoic | Quaternary | (5) | 1 | 1 | 1 | 1 |
| | Tertiary | 79 | 68 | 58 | 70 | 60 |
| Mesozoic | Cretaceous | 155 | 140 | 127 | | 130 |
| | Jurassic | 178 | 167 | 152 | | 155 |
| | Triassic | 208 | 196 | 182 | 200 | 185 |
| Paleozoic | Permian | 223 | 220 | 203 | | 210 |
| | Carboniferous | 270 | 275 | 255 | | 265 |
| | Devonian | 314 | 318 | 313 | | 320 |
| | Silurian | 332 | 350 | 350 | | 360 |
| | Ordovician | 379 | 430 | 430 | | 440 |
| | Cambrian | 426 | 510 | 510 | 500 | 520 |

*References*

E. C. BULLARD: Mem. and Proc. Manchester Lit. and Phil. Soc. 86 (1944).

A. HOLMES: Trans. Geol. Soc. Glasgow 32, 1, 117 (1947).

A. KNOPF: In Genetics, Paleontology and Evolution, edit. E. L. JEPSEN et al., Chap. 1, Princeton University Press 1949.

J. P. MARBLE: Rept. Ctee. Measur. Geol. Time 1949/50, 18 (1950).

In many cases a pegmatite or igneous rock will cut some older sedimentary formations and will be overlain by younger sedimentary formations, and thus an age is obtained which lies somewhere in the gap between two formations of the sedimentary succession. Unfortunately igneous activity usually accompanies mountain building and this disturbs sedimentation so that the gaps in which igneous rocks occur are usually long ones. Furthermore the activity often causes metamorphism of the older rocks and destroys the fossils.

For these reasons it is rarely possible to fix an absolute age in the sedimentary column precisely, even with a mineral which has been accurately dated. As an example, RODGERS[1] has pointed out that several dozen age determinations in the whole Appalachian region have so far failed to fix a single date in the sedimentary column. When enough determinations have been made this difficulty may be overcome provided the gaps have occurred at different times in different places.

In the meantime the progress of giving absolute dates to the fossiliferous succession is neither satisfactory nor precise. Few age determinations are accurate enough to discriminate between formations as well as can be done by fossils. Of the well-measured dates, few can be placed precisely in the sedimentary column. Such results as have been obtained for an absolute time scale of the fossiliferous part of the geological succession are summarized in part B and Table 26.

# III. Special problems of Precambrian geochronology.

## a) Length of Precambrian time.

**68.** Most attention has been given by geologists to the fossiliferous succession because it has been the most easily studied and because these rocks underlie the most densely populated parts of the world. Now that it is known that Precambrian time represents five-sixths of the whole, more attention is being given to it and it has been found to present perplexing problems which are by no means completely solved.

---

[1] J. RODGERS: Amer. J. Sci. 250, 411 (1952).

**69.** The vast length of geological time is hard to visualize or understand. The ages of the earth and of the oldest minerals are measured in thousands of millions of years and constitute a large fraction of the whole age of the universe (see Fig. 7). This was not realized until recently. Sixty years ago in the debate on the age of the earth which immediately preceded and overlapped the discovery of radioactivity, KELVIN and most physicists held that the earth was only a few tens of millions of years old, while the geologists believed its age was a few hundreds of millions of years. One group underestimated the age by a factor of a hundred, the other by a factor of ten.

The geologists were nearly correct in their estimate of the age of the Cambrian and other fossiliferous rocks, but they greatly underestimated the length of the Precambrian and telescoped together many rocks which appeared similar, but which really differed greatly in age. Unfortunately they firmly adopted this telescoped and short chronology and it is still widely followed today. Much confusion will continue until sufficient age determinations have been made to form an entirely new classification of these older rocks and until it is realized by geologists and physicists alike that they represent a history much longer and more complex than most existing accounts suggest.

### b) Proportions of metamorphic rocks and of unaltered rocks.

**70.** Another serious impediment to the study of Precambrian rocks is the small proportion of them which are unmetamorphosed. HIGGS[1] has pointed out that in the United States the proportion of sedimentary rocks preserved is less in each preceding period back to the Cambrian, and the same is probably also true for successively older sedimentary rocks in the Precambrian. Metamorphism and erosion are constantly reducing the volume of old unaltered sedimentary rocks.

The converse, that the older the period the larger the proportion of metamorphic and igneous rocks, is also true. This is particularly evident in North America where ancient metamorphic rocks form half the surface of Canada whereas younger metamorphic rocks are only exposed in much smaller areas in the Appalachian and Cordilleran regions.

Because old unaltered rocks are now found only in isolated basins which are no longer interconnected and because they have no fossils, the precise and convenient methods for establising the succession in the sedimentary column which formed the basis of the time scale used for the younger rocks, cannot be applied to the Precambrian. In the Precambrian, since the metamorphic and igneous rocks are much the most abundant and contain most of the minerals which can be dated, the time scale must be based on them.

Of course, within individual basins the succession of Precambrian sedimentary rocks provides a satisfactory local chronology but no means has yet been found for uniting these into a universal time scale. There is much evidence that the sediments in different basins may be of different ages. The idea that all unaltered Precambrian sedimentary rocks belong to just one world-wide Proterozoic era and that all metamorphic Precambrian rocks belong to another world-wide Archean area is now disproved [46], [47], [48].

### c) The ages of regions or geological provinces.

**71.** The dating of pegmatites and their use in establishing a time scale have been discussed. It has often been supposed that in old regions pegmatites of many ages would be found representing repeated periods of mountain building,

---

[1] D. V. HIGGS: Amer. J. Sci. **247**, 575 (1949).

but this has not been found to be the case either in Africa or in North America which are the continents most studied (Fig. 8). In all regions so far investigated it has been found that there was only one range of time, a few hundreds of millions of years long, during which all the pegmatites in that region were formed [12], [46], [47]. In different areas different spans of time were involved. Thus in the Appalachian region all the accurately dated pegmatites were formed during the period from $7 \times 10^8$ to $2 \times 10^8$ years ago and a few ages of doubtful validity would only extend the period back to $8 \times 10^8$ years. On the other hand in the Grenville and Adirondack region north of the St. Lawrence River all the

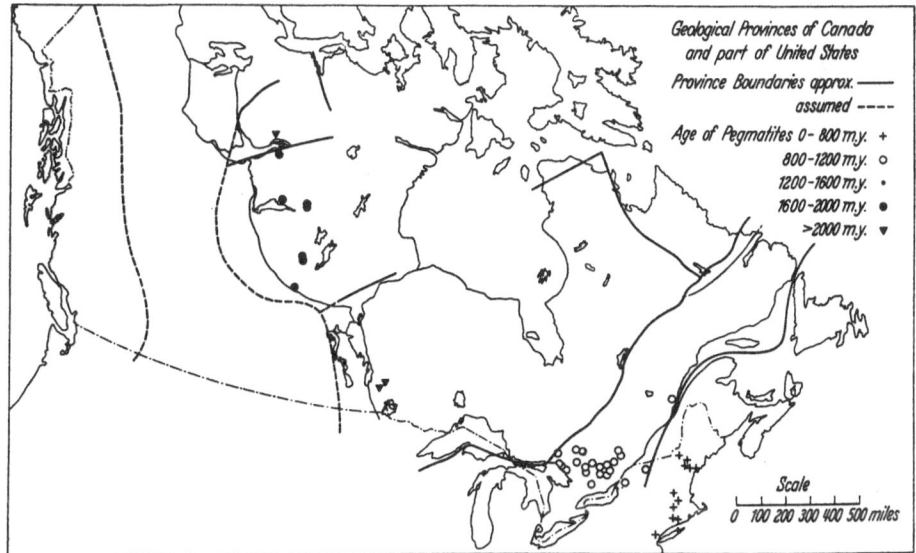

Fig. 8. Ages of North American pegmatites.

pegmatites were formed in a period from $13 \times 10^8$ to $8 \times 10^8$ years ago, while in the Keewatin province which lies further north across Ontario and Quebec all the pegmatites were formed in a period from 26 to $20 \times 10^8$ years ago. Thus each region has had one period during which all its pegmatites were formed and these periods were different for different regions.

It is sometimes stated that an unfossiliferous sedimentary formation cut by a dated pegmatite may be of any older age, for example that some formations in the Appalachian region may be as old as those in the Keewatin region, but there is no evidence to support this view and the trend of evidence is against it[1]. On the contrary, reasons have now been suggested for believing that every region has undergone only one period of activity during which sedimentation, igneous intrusion, formation of pegmatites and mountain building all occurred. If this is true one may speak of the *age of a region*, but the age would be a broad one covering a period of a few hundreds of millions of years during which that region was active. This is a very different concept from the age of a formation or of an individual pegmatite which would have been formed in one million years or less. Thus within the larger period there may have been several shorter times of igneous activity  For example the fact that both Ordovician and Devonian

---

[1] J. T. WILSON: In: The Earth as a Planet, edit. G. P. KUIPER, Chap. 4. University of Chicago Press 1954.

granites are known in the Appalachian region is no bar to speaking of Appalachian activity as a single longer period embracing both these ages.

This concept enables us to introduce a new and useful idea for Precambrian geology. In the fossiliferous sedimentary succession time is already divided into stages averaging perhaps ten million years apiece, into periods, like the Cambrian or Ordovician, each consisting of two or three stages and into three eras, each consisting of several periods (see Table 26). If we consider Precambrian time, which spans several thousands of millions of years, we have as yet no means of dividing it so well but we can distinguish regions like the Grenville and the Keewatin, each of which seems to have been formed during one interval several hundred million years long. Thus the best that can yet be done for most of the Precambrian is to subdivide and date shield areas in the manner shown for North America in the first part of Table 27 or as done by Holmes and Cahen [47] for Africa.

Table 27. *Time scale for growth of North America.*

| Primary province | Limits of known orogenetic activity Unit $10^8$ years | Secondary province |
|---|---|---|
| Keewatin (or Superior nucleus) . . | 26—22 | Timiskaming basins |
| Yellowknife nucleus . . . . . . . | 24—21 | Yellowknife sedimentary basins |
| Athabaska (or Churchill) . . . . . | 19—16 | East arm, Great Slave Lake |
| Labrador Coast . . . . . . . . . | c. 15 | Labrador Trough |
| Great Bear Lake (or Snare) . . . . | c. 14 | Unaltered part of Snare group |
| Grenville . . . . . . . . . . . | 12—9 | Mistassini, Huronian and Keweenawan basins |
| Crystalline part of Appalachians (Piedmont, New England coast) . | 7—2 | Folded Appalachians (Valley and Ridge Province) |
| Innuitian (or North Ellesmere Island) | c. 3 | Arctic Islands basin |
| Coast Ranges part of Cordillera . . | 3—0 | Rocky Mountain part of Cordillera |

### d) The ages of veins.

**72.** In order to avoid confusion it is well to point out that there is an important difference between dating veins and pegmatites (see Figs. 8 and 9).

Pegmatites seem to have been formed only in conjunction with igneous activity and mountain building, but veins, which are also tabular bodies of a different geological appearance, seem to have been formed both with mountain building and also at later dates after igneous activity had ceased in a region. Thus dating a vein gives only an upper limit to the age of a region. For example in the Appalachian region only veins of $7 \times 10^8$ years and less are found, but in the Keewatin region veins were found at many different times between $25 \times 10^8$ years ago and recent time. This interpretation is in conflict with the view which has been widely held that veins only accompany episodes of major igneous activity, but the conflict becomes apparent only in old regions.

For these reasons, and also because veins are very subject to alteration, their dates are not as useful in establishing a time scale as are those of pegmatites.

### e) Continental growth.

**73.** When the time scales formerly used for Precambrian rocks were being drawn up it was believed that the continental blocks had been formed complete early in the earth's history. Such a view was natural before the full length of the Precambrian time was recognized and before it was realized that the basement rocks cover a great range in time and that different regions or provinces are of different ages. Now it has been discovered that the provinces are arranged in zones that get progressively younger outwards from continental nuclei (see Figs. 8 and 9).

Many geologists believe that this is due to continental growth and that each province represents another addition and another stage in the growth. The earliest rocks to form were the continental nuclei. They have certain characteristics which will now be discussed.

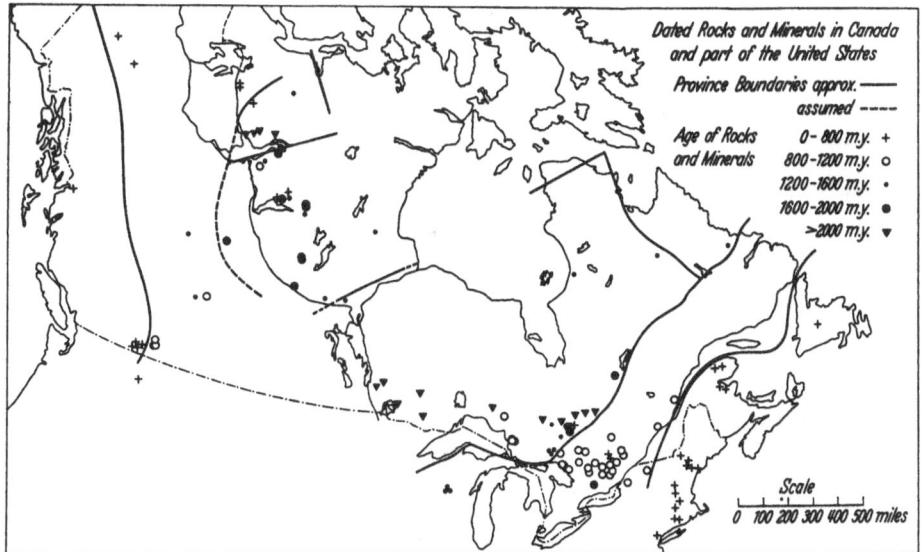

Fig. 9. Ages of North American pegmatites and veins.

### f) Continental nuclei and Keewatin-Timiskaming rock types.

**74.** Enough geological descriptions and age determinations are now available to identify at least eleven *continental nuclei* which are all older than two billion years (see Table 28). There must be others, for some continents are multiple and have more than one nucleus. In every case these nuclei have several characteristics that separate them clearly from later rocks (see Table 29). They are those which in North America are associated with Keewatin and Timiskaming rocks.

Table 28. *Continental nuclei.* (Others undoubtedly exist but have not yet been identified.)

| Continent | Name and location | Approximate age (m. y.) |
|---|---|---|
| North America | Keewatin (South and East of Hudson Bay) | 2600—2200 |
| | Yellowknife | 2400—2100 |
| South America | Guiana Highlands | ? |
| Europe | Sveco—Fennian ? | 1800+ |
| Africa | Swaziland—Transvaal | 3000—2100 |
| | Southern Rhodesia | 2700—2100 |
| | Kenya—Uganda | c. 2200 |
| | North Belgian Congo | c. 2700 |
| | Sierra Leone—Gold Coast | 2500—2100 |
| Asia | Mysore, India | c. 2300 |
| Australia | Kalgoorlie | c. 2300 |

Table 29. *Characteristics of continental nuclei.*

| | |
|---|---|
| *Age:* | Pegmatites and gold-quartz veins older than 2000 million years. |
| *Structure:* | Narrow sinuous synclines of Keewatin type volcanics overlain by Timiskaming type sediments and cut by ovoid granitic batholiths. |
| *Nature of volcanics:* | Chiefly andesite and basalt flows with much pillow lava. |
| *Nature of sediments:* | Poorly sorted greywackes, conglomerates and tuffs. Sandstones, limestones and arkoses lacking. |
| *Mineralization:* | Gold-quartz veins relatively common. Some old base metal deposits and iron formations. |

We do not yet have enough accurate information to subdivide and date the parts of the continental nuclei at all accurately. Pettijohn[1] has suggested that in northwestern Ontario and in the adjacent part of the United States the Keewatin province consists of multiple belts each of which represents a separate orogeny. This suggests that each nucleus was formed by a number of separate orogenies due to a mountain building process during the early history of the earth quite different from any which has operated since.

It is therefore believed that about 2000 million years ago a major change occurred in the method of mountain formation. Before that time mountain building processes produced many small sinuous belts characterized by a high proportion of basic volcanics and by poorly differentiated sediments. Since then the belts of mountains have been very much larger, mountain building has been less frequent and it has produced abundant gneisses from well assorted sediments. This change at the close of Keewatin time appears to be the most important tectonic event in geological history. The only comparable marker, the beginning of Cambrian time, is of great biological significance but does not represent an important tectonic event.

### g) Periodicity of mountain building.

**75.** Although the paelontological time scale was primarily based upon biological evolution the attempt was made to relate its major divisions, or eras, to times of uplift of mountain belts. This has some justification in Europe and eastern North America where the scale was developed but it does not apply at all well to other places like Japan or Africa. Generally speaking, it is now becoming realized that mountain building does not consist of single, brief episodes separated by long periods of calm but rather that it has been a continuous process sometimes operating in one part of the world and sometimes in another. It is very much open to question whether major orogenetic events in different parts of the world should be held to have been contemporaneous[2]. At least it is unwise to make this assumption except in cases where it can be proved. It will also be well to avoid making assumptions which may turn out to be too simple, such as the suggestion which was made without proof and once widely held that there were only two ages of granitic intrusion in the Precambrian. Rather for each part of the world it seems wise to try to establish the sequence of events and relate them to absolute dates with little reference to events elsewhere.

### h) Archean and Proterozoic types of rocks.

**76.** It is certainly true that in Precambrian areas there are two quite easily distinguished types of rocks; little-deformed rocks of Proterozoic type in basins

---

[1] F. J. Pettijohn: Bull. Geol. Soc. Amer. **54**, 925 (1943).
[2] J. Gilluly: Bull. Geol. Soc. Amer. **60**, 561 (1949). — L. M. R. Rutten: Bull. Geol. Soc. Amer. **60**, 1755 (1949).

resting upon highly altered rocks of Archean type. When it was believed that the continents had been formed as complete blocks early in the history of a rapidly cooling earth, it was natural to suppose that the Archean rocks everywhere represented rocks formed simultaneously in that early turbulent period and that the Proterozoic rocks above them belonged to another quieter era. Such a simple interpretation of history was not like that for the younger rocks whose history is known to have been far more complex, but it was perhaps a justifiable view as long as the Precambrian was considered to have been a short period with rapid cooling. Before the development of methods of age determinations there was in any case no way of making any other subdivision.

We have seen that the discovery of radioactivity and the first age determinations destroyed the justification for this short time scale and simple subdivision, but for some time there were too few age determinations to enable any other to be suggested. As early as 1935 LEITH[1] pointed out that although in any one region Proterozoic type of rocks always overlie and are younger than Archean type of rocks (and indeed the reverse would be impossible), nevertheless Proterozoic type of rocks are probably not all of the same age and some of them in one place may be older than some Archean type of rocks elsewhere. He suggested that these names should be confined to types of rocks and should not be used for periods of time.

Many age determinations have now shown that view to be correct. Thus 14 age determinations for pitchblende cutting unaltered Athabasca sandstones in the Martin Lake Mine, Saskatchewan, show that these rocks of Proterozoic type are at least $16 \times 10^8$ years old[2]. The underlying gneisses of Archean type contain pegmatites about $18 \times 10^8$ years old. On the other hand the Grenville rocks of Ontario which are also of Archean type have been shown by scores of age determinations to be much younger, for none of them are more than $12 \times 10^8$ years old and the majority are about $10 \times 10^8$ years [46].

Other cases could be cited in which Proterozoic type of rocks in one area are older than Archean type of rocks elsewhere. As has been said "the trend is clearly towards the recognition of more than two eras ... In the absence of fossil evidence of age, radioactive methods are the only real methods of estimating time"[3].

**77.** It is suggested that the explanation of the difference between these two types is this. Once continental nuclei had been formed successive periods of mountain building each produced a pair of belts of different rock types but of the same age. Within the primary part of the belt, which was being most actively mountain built and added to the continent, the rocks were altered and of Archean type, but on adjacent parts of the pre-existing continents some less disturbed beds were usually laid down at the same time and these were of Proterozoic type. These were often folded into a secondary belt of mountains[4].

Thus it is believed that after the Keewatin continental nucleus had been formed in central Canada between 25 and $20 \times 10^8$ years ago, the Grenville belt of Archean type rocks was added further southeast about 12 to $8 \times 10^8$ years ago, while at about the same time the Huronian and Keweenawan rocks of Proterozoic type were laid down resting on the old Keewatin on the inner or north-

[1] C. K. LEITH, R. J. LUND and A. LEITH: U.S. Geol. Surv. Prof. Paper **184**, 10 (1935).

[2] A. M. CHRISTIE: Geol. Surv. Canada Mem. **269**, 105 (1953).

[3] F. F. GROUT, J. W. GRUNER, G. M. SCHWARTZ and G. A. THIEL: Bull. Geol. Soc. Amer. **62**, 1017 (1951).

[4] J. T. WILSON: In: The earth as a planet, edit. G. P. KUIPER, Chap. 4. University of Chicago Press 1954.

western side of the Grenville. Later, while the crystalline Piedmont part of the Appalachian mountains was being added 7 to $2 \times 10^8$ years ago, gently folded and undisturbed Appalachian rocks of the same age were being laid down in the Valley and Ridge province on the western or inside to rest on the older Grenville. Although these rocks were mostly Paleozoic in age many of them are similar in appearance to Proterozoic types of rocks.

Thus, as geologists had correctly noted, Proterozoic types of rocks were laid down on older regions of Archean types. Confusion and error only arose when several regions were considered together and different pairs of Archean and Proterozoic rocks of diverse ages were equated together. To avoid further confusion it would seem best to abandon these terms.

## IV. Principles for establishing a Precambrian geochronology.

**78.** The ideas discussed above can be summarized in these principles:

(i) Rocks of Proterozoic type and rocks of Archean type can be readily recognized and distinguished, but both types were formed at many overlapping times in the past. The subdivision of Precambrian time into an older Archean era and a younger Proterozoic era is misleading and should be abandoned. As HOLMES has said "Obviously the time has come to liberate Precambrian geology from the tyranny of a telescoped classification in which the Archean ... is regarded as a single era of worldwide distribution. Five "Archean" cycles are already known and there has probably been time enough in the geological past for double that number" ([48], page 254).

(ii) Metamorphic rocks form most of the crust; they are the most widespread of Precambrian rocks and they contain most of the radioactive minerals. They form the natural means for subdividing Precambrian time.

(iii) In each continent the metamorphic rocks can be divided into provinces representing old mountain ranges formed at various times.

(iv) Those provinces with pegmatites more than $20 \times 10^8$ years old tend to be centrally situated in continents. They contain a high proportion of Keewatin-type volcanic rocks overlain by Timiskaming type sedimentary rocks. These are the continental nuclei. Their complex structures have nowhere been thoroughly disentangled.

(v) The younger provinces composed largely of gneisses and plutonic igneous rocks form belts about these nuclei. Each contains pegmatites formed only during one period of a few hundreds of millions of years. The youngest such provinces are considered to form the present primary mountain belts like the Appalachian Piedmont and the western ranges of the Cordillera. It has generally proved difficult to subdivide these primary belts, even where they are known to be contemporaneous with fossiliferous rocks.

(vi) On the inner or continental side of these primary provinces there are secondary basins of folded sedimentary rocks which have been little altered and which are of approximately the same age as the primary provinces. Thus the secondary basins of the Valley and Ridge province lie inside the associated primary crystalline Appalachians and the secondary basins of Huronian and Keweenawan sediments lie inside the related primary Grenville province. Each of these groups of sediments of Proterozoic type rest on metamorphic rocks of an earlier cycle. In any one basin the sedimentary succession can be worked out precisely, but there is no reason to believe that all the unfossiliferous Proterozoic type rocks in different basins are all of the same age.

(vii) The change in type of mountain building which occurred about $20 \times 10^8$ years ago appears to have been the most important physical event in all geological time. The only event of comparable importance is biological and occurred when animals began to form hard parts and to be preserved as fossils $5 \times 10^8$ years ago at the beginning of the Cambrian period and of the Paleozoic era.

(viii) Pegmatites were only formed during the period of active formation of primary provinces and serve to date them. Veins on the other hand may be many hundreds of millions of years younger than the country rock in which they occur.

(ix) There are at present no ways in which to apply age determinations to Precambrian rocks to give a chronology as detailed as that provided by fossils in the younger rocks. The first objective must be to date provinces in the basement and to establish the broad features of the time scale correctly for each continent. It will be impossible to do this until the idea is abandoned that all Proterozoic type rocks are younger than all Archean type rocks.

# V. Geological time scales.

## a) For Cambrian to recent time.

**79.** The most recent papers which have discussed the problem of giving absolute dates to the fossiliferous succession and which made recommendations for a time scale include those by BULLARD [24], HOLMES[1], KNOPF [26] and MARBLE[2] (see Table 26).

In constructing such a scale the best data is provided by the relative order of the fossiliferous rocks in the sedimentary column. Only a very few radioactive age determinations are available which are believed to be accurate, and unfortunately most of these can only be placed inexactly in the sedimentary column. A larger number of doubtful determinations have been made. It is therefore a matter of opinion which determinations should be used. Thus KNOPF used only three different dates, those for the Cambrian Swedish Kolm, for two Permian minerals from Norway and from Czechoslovakia and for Tertiary minerals from Colorado, whereas HOLMES selected only two of those dates and added three more to give fives dates altogether, while BULLARD used a method for averaging the results from a larger number of admittedly doubtful determinations.

To fill in the ages for the parts of the column not directly dated, HOLMES compiled the maximum thicknesses of sediments for each period known anywhere in the world and assumed that the length of each period was proportionate to this maximum thickness. KNOPF, on the other hand, claimed that the data was of little value, but made no alternative suggestion except for the youngest rocks for which he used estimates of the rate of evolution.

Clearly a great deal more needs to be known before the stratigraphic column can be considered to be accurately dated. On the other hand, there is no reason to doubt that the scales are correct in a broad way.

MARBLE's time scale is the most recent and is a compilation. He was in a position to have the latest data and his classification follows HOLMES "B" scale which HOLMES preferred and which ZEUNER has also adopted. There is really no means of choosing which is the most accurate scale until more dates are available.

---

[1] A. HOLMES: Trans. Geol. Soc. Glasgow **32**, 1, 117 (1947).
[2] J. P. MARBLE: Rept. Ctee. Measure. Geol. Time **1949/50**, 18 (1950).

## b) For Precambrian time.

**80.** The fact that no general stratigraphic column can yet be devised for Precambrian time has been pointed out. It has also been stated that there is no physical proof nor necessity for periods of activity to be contemporaneous in different parts of the world. All that can be done as yet is to define provinces of different ages for each continent. For no old province can any date of initiation of activity be given precisely as the evidence would have been destroyed. For some provinces like the Appalachian, a definite date of termination of activity can be given.

The secondary basins of folded rocks which were contemporaneous with and related to particular primary provinces can be named. What we know for North America is given in Table 27. The basis of this subdivision and the list of age determinations on which it is based are given elsewhere [46]. HOLMES and CAHEN [47] have just reviewed all the data for Africa and they have revised the earlier work of HOLMES [48], CAHEN [49] and GRETENER et al [50].

For other continents the data is still fragmentary. HOLMES[1] has suggested a similar subdivision for India while HILLS[2] has discussed the geology of the Australia Precambrian.

## Bibliography.

[1] ADAMS, F. D.: The birth and development of the geological sciences. Baltimore: Williams & Wilkins Co. 1938.
[2a] HOLMES, A.: The age of the earth. New York: Harper and Brothers 1927.
[2b] HAHN, O.: Was lehrt uns die Radioaktivität über die Geschichte der Erde. Berlin: Springer 1926.
[3] JEFFREYS, H.: The Earth. Cambridge University Press 1952.
[4] BOLTWOOD, B. B.: Amer. J. Sci. 23, 77 (1907).
[5] KNOPF, A. (editor): The age of the earth, Nat. Res. Council, Washington 1931.
[6] NIER, A. O.: J. Amer. Chem. Soc. 60, 1571 (1938).
[7] NIER, A. O.: Phys. Rev. 55, 150, 153 (1939).
[8] NIER, A. O., R. W. THOMPSON and B. F. MURPHEY: Phys. Rev. 60, 112 (1941).
[9] VINOGRADOV, A. P., I. K. ZADOROZHNYI and S. I. ZYKOV: Dokl. Akad. Nauk. SSSR. 87, 1107 (1952).
[10] HOUTERMANS, F. G.: Nouvo Cim. 10, 1623 (1953).
[11] GEISS, J.: Z. Naturforsch. 9a, 218 (1954).
[12] COLLINS, C. B., R. M. FARQUHAR and R. D. RUSSELL: Bull. Geol. Soc. Amer. 65, 1 (1954).
[13] KULP, J. L., G. L. BATE and W. S. BROECKER: Amer. J. Sci. 252, 345 (1954).
[14] GENTER, W., F. JENSEN and K. R. MEHNERT: Z. Naturforsch. 9a, 176 (1954).
[15] WASSERBURG, G. J., and R. J. HAYDEN: Geochim. et Cosmochim. Acta 7, 51 (1955).
[16] SHILLIBEER, H. A., and R. D. RUSSELL: Canad. J. Phys. 32, 681 (1954).
[17] AHRENS, L. H.: Bull. Geol. Soc. Amer. 60, 217 (1949).
[18] ALDRICH, L. T., G. R. TILTON, G. L. DAVIS, L. O. NICOLAYSEN and C. C. PATTERSON: Proc. Geol. Assoc. Canad. 7, pt. 2, 7 (1955).
[19] LIBBY, W. F.: Radiocarbon dating. University of Chicago Press 1952.
[20] RUSSELL, R. D., R. M. FARQUHAR, G. L. CUMMING and J. T. WILSON: Trans. Amer. Geophys. Un. 35, 301 (1954).
[21] BEGEMANN, F., H. V. BUTTLAR, F. G. HOUTERMANS, N. ISAAC and E. PICCIOTTO: Geochim. et Cosmochim. Acta 4, 21 (1953).
[22] KULP, J. L., W. S. BROECKER and W. L. ECKELMANN: Nucleonics 11, 19 (1953).
[23] EVANS, R. D., C. GOODMAN and N. B. KEEVIL: In Handbook of Physical Constants, Geol. Soc. Amer. Spec. Paper 36, sect. 18, p. 267. 1942.
[24] BULLARD, E. C.: Mem. and. Proc. Manchester Lit. Phil. Soc. 86, 55 (1944).
[25] ZEUNER, F. E.: Dating the past, an introduction to geochronology, 3rd ed. London: Methuen & Co. 1952.

---

[1] A. HOLMES, W. T. LELAND and A. O. NIER: Amer. Mineral. 35, 19 (1950).
[2] E. S. HILLS: J. Proc. Roy. Soc. N. S. Wales 79, 67 (1946).

[26] KNOPF, A.: In Genetics, paleontology and evolution, edit. G. L. JEPSEN, Chap. 1. Princeton University Press 1949.
[27] BURLING, R. L.: Determination of geologic time. Nucleonics 10, 30 (1952).
[28] RANKAMA, K.: Isotope geology. New York: McGraw-Hill Book Co. 1954.
[29] KOHMAN, T. P., and N. SAITO: Ann. Rev. Nuclear Sci. 4, 401 (1954).
[30] FAUL, H. (editor): Nuclear geology. New York: John Wiley & Sons 1954.
[31] Annual Report of the Committee on the Measurement of Geologic Time, Nat. Res. Council, Washington 1924—1954.
[32] HOLMES, A.: Nature, Lond. 157, 680 (1946).
[33] HOUTERMANS, F. G.: Z. Naturforsch. 2a, 322 (1947).
[34] RUSSELL, R. D., and D. W. ALLAN: Mon. Not. Roy. Astronom Soc., Geophys. Suppl. (in press).
[35] PATTERSON, C., G. TILTON and M. INGHRAM: Science, (Lancaster, Pa.) 121, 69 (1955).
[36] RUTHERFORD, E., J. CHADWICK and C. D. ELLIS: Radiations from radioactive substances. Cambridge University Press 1930.
[37] BIRCH, F.: J. Geophys. Res. 56, 107 (1951).
[38] BIRCH, F.: In Nuclear geology, edit. H. FAUL, Chap. 5. New York: John Wiley & Sons 1954.
[39] EVANS, R. D., and C. GOODMAN: Bull. Geol. Soc. Amer. 52, 459 (1941).
[40] ELLSWORTH, H. V.: Rare-element minerals of Canada. Geol. Surv. Canada, Econ. Geol. Ser. No. 11 1932.
[41] WICKMAN, F. E.: Sveriges geol. undersökn, Arsbok 33 (1939).
[42] HOLMES, A.: Trans. Edinburgh Geol. Soc. 14, Pt. 11, 176 (1948).
[43] ALPHER, R. A., and R. C. HERMAN: Phys. Rev. 84, 1111 (1951).
[44] BULLARD, E. C., and J. P. STANLEY: Suom. Geod. Lait. Julk. Ver. des Finn. Geod. Inst. 36, 33 (1949).
[45] GENTNER, W., K. GOEBEL and R. PRÄG: Geochim. et Cosmochim. Acta 5, 124 (1954).
[46] CUMMING, G. L., J. T. WILSON, R. M. FARQUHAR and R. D. RUSSELL: Proc. Geol. Soc. Canada 7, pt. 2. 27 (1955).
[47] HOLMES, A., and L. CAHEN: Col. Geol. and Min. Resources 5 (1), 3 (1955).
[48] HOLMES, A.: Internat. Geol. Cong., Rep. 18th Sess., Pt. 14, p. 254, 1948.
[49] CAHEN, L.: Ann. Soc. Geol. Belgique, Liege 1954.
[50] GRETENER, P. E. F., R. M. FARQUHAR and J. T. WILSON: Trans. Roy. Soc. Canada Sect. IV, Ser. III 48, 17 (1954). — WILSON, J. T., R. M. FARQUHAR, P. E. F. GRETENER, R. D. RUSSELL and H. A. SHILLIBEER: Nature, Lond. 174, 1006 (1954).
[51] LARSEN, jr., E. S., N. B. KEEVILL and H. C. HARRISON: Bull. Geol. Soc. Amer. 63, 1045 (1952).

*Acknowledgements.* The authors are indebted to Mrs. R. S. WADDINGTON for helping prepare the manuscript of this chapter and to Miss M. CLARK for typing it.

Many helpful suggestions were given by Professor W. H. WATSON, Dr. D. A. KEYES, Dr. G. L. CUMMING, Dr. H. A. SHILLIBEER, Dr. J. P. MARBLE and Mr. J. E. HOGG.

# The Earth's Interior.

By

## J. A. JACOBS.

With 10 figures.

**1. Introduction.** A major obstacle encountered in a study of the constitution of the Earth's interior is the difficulty of obtaining *experimental data*. It is impossible in the majority of cases to obtain direct measurements of physical quantities at depth. The only alternative is to simulate in the laboratory the conditions that prevail in the Earth's interior. In this connection important pioneering work has been carried out by BRIDGMAN, whose work will be discussed in more detail later. Even so, the greatest pressure obtained by him viz. 100,000 bars, corresponds to a depth of only 300 km. in the Earth. Thus extreme extrapolation is needed to reproduce conditions deep within the Earth—extrapolations which in many cases may not be justifiable. There are, however, two means of obtaining information about the Earth's interior. These are the study of *seismic waves* and the analysis of the Earth's *magnetic field* and its secular variation. Both these fundamental subjects will be considered in detail later.

A difficulty of a different nature arises over the question of *terminology*. The top-most layers of the Earth are called the *crust* and are not considered in this chapter. The thickness of the crust is not constant but varies between about 30 to 40 km. under the continents, whilst it is probably not much more than 5 or 6 km. thick under the oceans. That part of the Earth between the crust and a depth of approximately 2900 km. is called the *mantle*, and for the remaining part inside the mantle, the word *core* is used. These three main divisions of the Earth into the crust, the mantle and the core are based mainly on seismic evidence. Some authors use different terms to denote various subdivisions within the Earth. Such terms as *lithosphere* and *asthenosphere* (indicating shells of large and relatively small strength respectively) and *sial* and *sima* (which refer to the composition) will be avoided.

A more subtle difficulty in terminology arises from a certain looseness in the use of physical terms by some authors. For example, at the surface of the Earth if a material melts, it becomes *fluid* and the word molten implies fluid, and it is often assumed that under all circumstances a material is fluid above its melting point. The *melting point* defines the transition from a crystalline to a non-crystalline state. The transition from solid to liquid is a transition from relatively large rigidity and viscosity to much smaller values. These are two fundamentally different processes, although the one is usually connected with the other.

Finally any examination of the physical properties existing within the Earth must, if it is to be of any value, be correlated with the surface features of the Earth and the chemical constitution of the Earth. Thus a study of the *mechanical*, *thermal*, and *electrical* properties within the Earth must be integrated with the allied subjects of *geology* and *geochemistry*. The geological and chemical aspects will not be discussed in detail in this chapter, although they will be considered when evaluating the merit of any physical theory.

# A. Mechanical Properties.

## a) Analysis of Seismic Data.

The Earth is continually undergoing deformation due to the stresses that are set up within it. If the stresses are not too great, elastic or plastic deformation will occur. However, if the stresses continue to build up over a long time so that the principal stress difference exceeds the strength of the material, fracture will take place. This involves a sudden release of stress and the disturbance will give rise to elastic waves which will travel through the Earth. If the disturbance is large enough, it may be picked up by any of the seismological observatories that lie scattered across the world.

**2. Theory of Elastic Waves.** Consider a homogeneous, isotropic, perfectly elastic medium. Using the infinitesimal strain theory of elasticity, the vector equation of motion satisfied by the displacement vector $D$ is

$$(\lambda + \mu) \operatorname{grad} \operatorname{div} D + \mu \nabla^2 D + \varrho (F - f) = 0 \tag{2.1}$$

where $\lambda$, $\mu$ are the LAMÉ stress constants, $\varrho$ the density, $F$ the external force per unit mass, and $f$ the acceleration. Since the displacement and velocity are small, $f = d^2 D/dt^2$ may be replaced by $\partial^2 D/\partial t^2$. Then with $F = 0$, equation (2.1) becomes

$$(\lambda + \mu) \operatorname{grad} \operatorname{div} D + \mu \nabla^2 D - \varrho \frac{\partial^2 D}{\partial t^2} = 0. \tag{2.2}$$

Taking the divergence of equation (2.2),

$$(\lambda + 2\mu) \nabla^2 \operatorname{div} D - \varrho \frac{\partial^2}{\partial t^2} \operatorname{div} D = 0. \tag{2.3}$$

Thus the cubical dilatation $\delta = \operatorname{div} D$ satisfies the wave equation

$$\left\{ \nabla^2 - \frac{\varrho}{\lambda + 2\mu} \cdot \frac{\partial^2}{\partial t^2} \right\} \delta = 0 \tag{2.4}$$

and the disturbance is transmitted with speed

$$\alpha = \sqrt{\frac{\lambda + 2\mu}{\varrho}} \,. \tag{2.5}$$

Again taking the curl of equation (2.2),

$$\mu \nabla^2 \operatorname{curl} D - \varrho \frac{\partial^2}{\partial t^2} \operatorname{curl} D = 0 \tag{2.6}$$

which shows that the rotation $\omega = \frac{1}{2} \operatorname{curl} D$ satisfies the wave equation

$$\left\{ \nabla^2 - \frac{\varrho}{\mu} \cdot \frac{\partial^2}{\partial t^2} \right\} \omega = 0 \tag{2.7}$$

the disturbance being transmitted with speed

$$\beta = \sqrt{\frac{\mu}{\varrho}} \,. \tag{2.8}$$

In seismology these two waves are called $P$ and $S$ waves respectively. (For further details, see K. E. BULLEN's Chapter.) The first is a condensation-rarefaction wave involving change of volume. The second is a shear wave in which there is distortion without change of volume. Both wave velocities depend only on the elastic parameters and the density of the medium. In particular if the rigidity $\mu$ is zero, then $\beta$ is zero i.e. shear waves cannot be transmitted through a material of zero rigidity. Since the *incompressibility* or *bulk modulus* $k$ is given by

$$k = \lambda + \tfrac{2}{3}\mu \tag{2.9}$$

it follows from (2.5) and (2.8) that

$$\alpha^2 - \frac{4}{3}\beta^2 = \frac{k}{\varrho}. \tag{2.10}$$

This quantity will be denoted by $\varphi$. It must be pointed out that the basic equation (2.2) has been considerably simplified by the assumptions of uniformity and of absence of initial stress. How serious these assumptions are, will be discussed later.

**3. Travel-Time tables and Velocity-Depth curves.** The times at which signals from the same seismic shock arrive at different stations can be recorded so that

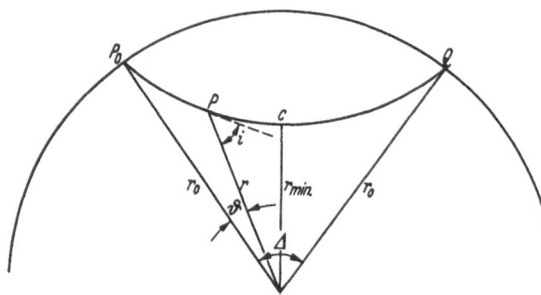

Fig. 1. Seismic ray $P_0 PCQ$.

it is possible to determine the travel time of the disturbance as a function of the distance. Tables have been constructed since the beginning of the century giving the travel-time $T$ as a function of the distance $\varDelta$ in angular measure along a great circle arc connecting the focus (assumed at the surface) with the point of observation. Tables constructed in more recent years by different investigators are in good agreement with one another. The most comprehensive data have been compiled by GUTENBERG and RICHTER and by JEFFREYS and BULLEN[1]. From the travel-time tables it is possible to obtain velocity-depth curves. JEFFREYS has applied a thorough statistical analysis to the data and claims that his results have an intrinsic accuracy of one half percent. With this degree of accuracy it is necessary to take into account the actual depth of the focus of the earthquake and the ellipticity of the Earth. Many calculations of the relation between velocity and depth have been published from 1910 onwards, the early pioneers being WIECHERT, GEIGER, ZOEPPRITZ, and GUTENBERG. An excellent account of this early work with a full bibliography has been given by MACELWANE[2]. An account of the method of obtaining the velocity-depth curves from the travel time tables will now be given. The theory is due in the first place to HERGLOTZ[3] and BATEMAN[4].

Let $(r, \vartheta)$ be the polar coordinates of any point $P$ of a seismic ray which originated at the surface at a point $P_0$ as indicated in Fig. 1. Let $v$ be the velocity of the ray at $P$, $v$ being a function of the distance $r$ only. The time taken to reach $P$ is

$$T = \int \frac{ds}{v} = \int \frac{1}{v} \left\{ \left(\frac{dr}{d\vartheta}\right)^2 + r^2 \right\}^{\frac{1}{2}} d\vartheta \tag{3.1}$$

[1] For a complete bibliography, see H. JEFFREYS, Reports on Progress in Physics **10**, 52 (1945).

[2] J. B. MACELWANE: Internal constitution of the Earth, Chapter X. Dover, New York: 1951.

[3] G. HERGLOTZ: Phys. Z. **8**, 145 (1907).

[4] H. BATEMAN: Phys. Z. **11**, 96 (1910).

where $ds$ is an element of length along the path. The actual path is such that the integral is stationary for small variations of the path. Writing $V$ for the integrand in (3.1) and $r'$ for $dr/d\vartheta$, it follows from the calculus of variations that $r$ must satisfy the differential equation

$$\frac{d}{d\vartheta}\left(\frac{\partial V}{\partial r'}\right) - \frac{\partial V}{\partial r} = 0 \tag{3.2}$$

a first integral of which is

$$V - r'\frac{\partial V}{\partial r'} = p \tag{3.3}$$

where $p$ is a constant for a given ray.

Substituting for $V$, we have

$$\frac{r^2}{v} = p\,(r^2 + r'^2)^{\frac{1}{2}}. \tag{3.4}$$

Hence, writing $\eta = r/v$,     $d\vartheta = \pm\,p\,r^{-1}(\eta^2 - p^2)^{-\frac{1}{2}}\,dr.$     (3.5)

If $\vartheta$ begins by increasing, then since $r$ will begin by decreasing, the negative sign must be taken.

The parameter $p$ has a simple physical interpretation. If $i$ is the angle that the ray makes with the radius $OP$, then for small displacements along the ray,

$$r\,d\vartheta = \sin i\,ds = \sin i\,\{r^2 + r'^2\}^{\frac{1}{2}}\,d\vartheta \tag{3.6}$$

so that from (3.4),

$$p = \frac{r\sin i}{v} \tag{3.7}$$

which is a generalized form of SNELL's law. When the ray reaches its deepest point $C$ where $r = r_{\min}$, $r'$ will vanish. The ray then bends upwards, remaining symmetrical about $OC$ and reaching the surface again at the point $Q(r_0, \varDelta)$. In particular, it follows from (3.7) that

$$p = \frac{r_{\min}}{v_{r_{\min}}} = \eta_{r_{\min}}. \tag{3.8}$$

There is another interpretation of the parameter $p$. Let the ray emerge at $Q$ at time $T$, and let $Q'$ be a neighbouring point $(r_0, \varDelta + d\varDelta)$ in the same plane and let $T + dT$

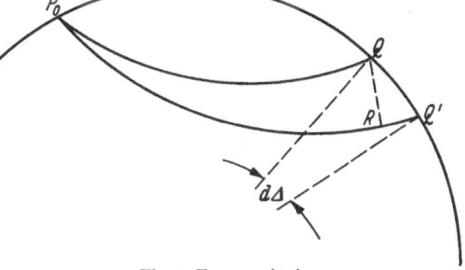

Fig. 2. To prove (3.9).

be the time required to reach $Q'$. Draw $QR$ perpendicular to the ray that reaches $Q'$ (see Fig. 2). Then to the first order of small quantities

$$RQ' = v\,dT \quad\text{and}\quad QQ' = r_0\,d\varDelta$$

and hence, with the values of $i$ and $v$ at the surface in $Q$,

$$\sin i = \frac{v\,dT}{r_0\,d\varDelta}. \tag{3.9}$$

Thus from (3.7), the ray parameter $p$ may be identified as $dT/d\varDelta$. Integrating equation (3.5) between $P_0$ and $C$,

$$\tfrac{1}{2}\varDelta = \int\limits_{r_{\min}}^{r_0} p\,r^{-1}(\eta^2 - p^2)^{-\frac{1}{2}}\,dr \tag{3.10}$$

i.e. using (3.8),

$$\Delta = \int_{p}^{\eta_0} 2p\,r^{-1}(\eta^2 - p^2)^{-\frac{1}{2}} \frac{dr}{d\eta}\,d\eta \tag{3.11}$$

which may be regarded as an integral equation whose solution determines $\eta$, and hence $v$, as a function of $r$.

The following solution due to G. Rasch, was given to Jeffreys in a private communication. Let the subscript 1 denote values of the variables at the level $r_1$, and let $\Delta_1$ be the value of $\Delta$ for the ray whose deepest point is at the level $r_1$. Multiply both sides of equation (3.11) by $(p^2 - \eta_1^2)^{-\frac{1}{2}}$ and integrate with respect to $p$ over the range $\eta_1$ to $\eta_0$. Then

$$\int_{\eta_1}^{\eta_0} \Delta\,(p^2 - \eta_1^2)^{-\frac{1}{2}}\,dp = \int_{\eta_1}^{\eta_0} dp \int_{p}^{\eta_0} 2p\,r^{-1}\{(\eta^2 - p^2)\,(p^2 - \eta_1^2)\}^{-\frac{1}{2}} \frac{dr}{d\eta}\,d\eta. \tag{3.12}$$

Change the order in the double integration. The limits for $p$ become $\eta_1$ to $\eta$ and those for $\eta$ run from $\eta_1$ to $\eta_0$. But if $\eta > \eta_1$,

$$\int_{\eta_1}^{\eta} \frac{p\,dp}{\sqrt{(\eta^2 - p^2)\,(p^2 - \eta_1^2)}} = \frac{\pi}{2} \tag{3.13}$$

so that the right hand side of equation (3.12) becomes

$$\int_{\eta_1}^{\eta_0} \pi\,r^{-1} \frac{dr}{d\eta}\,d\eta = \pi \log\left(\frac{r_0}{r_1}\right). \tag{3.14}$$

Integrating the left hand side of (3.12) by parts, leads to

$$\left[\Delta\,\mathrm{Ar\,Cos}\left(\frac{p}{\eta_1}\right)\right]_{\eta_1}^{\eta_0} - \int_{\eta_1}^{\eta_0} \frac{d\Delta}{dp} \cdot \mathrm{Ar\,Cos}\left(\frac{p}{\eta_1}\right)dp = \int_{0}^{\Delta_1} \mathrm{Ar\,Cos}\left(\frac{p}{\eta_1}\right)d\Delta. \tag{3.15}$$

Thus equation (3.12) finally reduces to

$$\int_{0}^{\Delta_1} \mathrm{Ar\,Cos}\left(\frac{p}{\eta_1}\right)d\Delta = \pi \log\left(\frac{r_0}{r_1}\right). \tag{3.16}$$

$p$ is a known function of $\Delta$, whilst $\eta_1$ is the known value of $dT/d\Delta$ at $\Delta_1$. Hence $r_1$ is determined by equation (3.16) in terms of $\Delta_1$, and thus in terms of $\eta_1$ or $r_1/v_1$. It is thus possible to determine numerically the velocity as a function of the depth below the surface.

The details of exceptional cases such as when there is a true surface of discontinuity will be omitted here. They may be found in works devoted to the subject[1].

**4. Velocity-depth curves and the Structure of the Earth.** Table 1 gives the velocities of the longitudinal and transverse waves at different depths in the Earth. These are the values given by Jeffreys[2] (1939) who takes the boundary of the crust as 33 km. More recently, Gutenberg[3] (1948) has given an alternative estimate of the velocities in the upper part of the mantle; they do not

---

[1] See, for example K. E. Bullen: Introduction to the Theory of Seismology. Cambridge 1953.

[2] H. Jeffreys: Mon. Not. Roy. Astronom. Soc., Geophys. Suppl. 4, 498, 594 (1939).

[3] B. Gutenberg: Bull. Seism. Soc. Amer. 38, 121 (1948).

Table 1. *Velocities of P and S waves at different depths in the Earth.*

| Mantle | | | Core | |
|---|---|---|---|---|
| Depth km. | Velocity of longitudinal (P) waves km./sec. | Velocity of transverse (S) waves km./sec. | Depth km. | Velocity of longitudinal (P) waves km./sec. |
| 33 | 7.75 | 4.35 | 2898 | 8.10 |
| 100 | 7.95 | 4.45 | 3000 | 8.22 |
| 200 | 8.26 | 4.60 | 3200 | 8.47 |
| 300 | 8.58 | 4.76 | 3400 | 8.76 |
| 413 | 8.97 | 4.96 | 3600 | 9.04 |
| 600 | 10.25 | 5.66 | 3800 | 9.28 |
| 800 | 11.00 | 6.13 | 4000 | 9.51 |
| 1000 | 11.42 | 6.36 | 4200 | 9.70 |
| 1200 | 11.71 | 6.50 | 4400 | 9.88 |
| 1400 | 11.99 | 6.62 | 4600 | 10.06 |
| 1600 | 12.26 | 6.73 | 4800 | 10.25 |
| 1800 | 12.53 | 6.83 | 4892 | 10.44 |
| 2000 | 12.79 | 6.93 | 5121 | (9.7) |
| 2200 | 13.03 | 7.02 | 5121 | 11.16 |
| 2400 | 13.27 | 7.12 | 5700 | 11.26 |
| 2600 | 13.50 | 7.21 | 6371 | 11.31 |
| 2800 | 13.64 | 7.30 | | |
| 2898 | 13.64 | 7.30 | | |

differ appreciably from those given by JEFFREYS. Fig. 3 shows the distribution of seismic velocities in both the mantle and core according to JEFFREYS and GUTENBERG. Apart from the major discontinuity at 2898 km. which marks the boundary between the core and mantle, there is another uncertainty at a depth of about 5000 km., JEFFREYS postulating another strong discontinuity. GUTENBERG[1] (1951), on the other hand, concludes that there is no decrease of velocity nor a strictly vertical segment, although there is a sharp change of slope in the velocity-depth curve. That part of the Earth below this discontinuity will be referred to as the *inner core*.

The extent to which individual judgment and choice of material have affected the construction of travel-time tables and hence velocity depth curves is well illustrated in a figure given by MACELWANE[2] in which half a dozen different solutions are plotted on the same graph. There is general agreement upon the major features of the curves, the differences for the most part being concerned

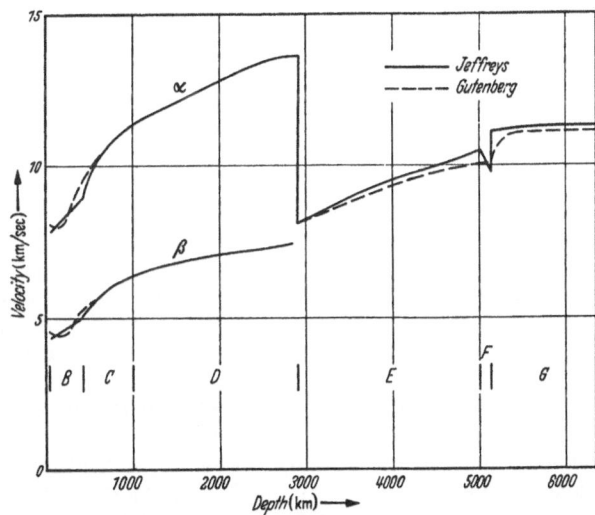

Fig. 3. Seismic velocities α (longitudinal) and β (transverse) as function of depth. [After F. BIRCH: J. Geophys. Res. 57, 231 (1952).]

[1] B. GUTENBERG: Trans. Amer. Geophys. Un. **32**, 373 (1951).
[2] J. B. MACELWANE: Internal Constitution of the Earth, p. 276. Dover, New York: 1951.

with minor inflections. In the latest work of JEFFREYS and GUTENBERG discussed above, most of these inflections have disappeared and it is unlikely that the velocities will be further improved.

The uppermost few tens of kilometers of the Earth, below which high seismic velocities are first encountered has been called the *crust*. The crust consists of at least two thin layers but comprises an insignificant fraction of the Earth's total mass, and for a discussion of the properties of the interior of the Earth, it may be represented conventionally by a thin layer of uniform thickness (33 km.) and density. Seismic data pertaining to the deeper parts of the Earth are insensitive to changes in the characteristics of the crustal layers.

An analysis of the velocity-depth curves indicates a number of sub-divisions of the interior of the Earth below the base of the crust which is known as the MOHOROVIČIĆ discontinuity. The subdivisions of Table 2 are taken from BULLEN[1]

Table 2. *Dimensions and descriptions of the internal layers.*

| Layer | Depth to boundaries km. | Fraction of volume | Features of region |
|---|---|---|---|
| Crust    A | 0 | 0.0155 | Conditions fairly heterogeneous |
|  | — 33 — | | |
| B | | 0.1667 | Probably homogeneous |
|  | 413 | | |
| Mantle  C | | 0.2131 | Transition region |
|  | 984 | | |
| D | | 0.4428 | Probably homogeneous |
|  | — 2898 — | | |
| E | | 0.1516 | Homogeneous fluid |
|  | 4982 | | |
| Core    F | | 0.0028 | Transition layer |
|  | 5121 | | |
| G | | 0.0076 | Inner Core (solid ?) |
|  | 6371 | | |

whose analysis is based on the velocity-depth curves of JEFFREYS. It is clear that the principal layers are $B, C, D$ and $E$ which together make up over 97% of the whole Earth. The existence of layer $F$ has not yet definitely been settled, whilst the nature and exact position of the division between $B$ and $C$ is also in doubt. In fact a more complete knowledge of conditions in the transition region $C$ would go a long way towards solving many of the geophysical problems of today.

The passage of transverse waves through the core has never been observed, from which it was deduced early on that the core is *liquid*. There is additional evidence for the conclusion, however. In a study of the refraction of elastic waves at the interface of two solids, JEFFREYS[2] pointed out that if the velocities of the longitudinal and transverse waves are in about the same ratio, then in all cases that were worked out an incident transverse wave gave a very small transmitted longitudinal wave and conversely. Now an *SKS* wave has undergone two such transformations, and yet observations show that its intensity is comparable to that of a transverse wave that has travelled through the mantle only. The observed intensity can readily be explained if the matter below the mantle is a liquid.

[1] See footnote 1, p. 368.
[2] See footnote 1, p. 366.

It has been suggested that the core might consist of a solid viscous medium of considerably lower rigidity than the mantle, so that transverse waves could be propagated, but would be very rapidly damped out. JEFFREYS (and EUCKEN[1]) have shown on the other hand that the existence of strongly damped transverse waves at the same time with virtually undamped longitudinal waves leads to impossible values of the elastic constants. Finally there is independent evidence obtained from an analysis of the bodily tide of the Earth (JEFFREYS[2], TAKEUCHI[3]), and from the nature of the secular variation of the Earth's magnetic field (BULLARD[4], ELSASSER[5]).

Although all this evidence is overwhelmingly in favour of a *liquid* core (region $E$), the nature of the *inner* core (region $G$) is still in doubt. There are strong reasons for believing that it may be *solid:* this question will be taken up again in section eleven.

**5. Variation of Density within the Earth.** Since the mean density of the Earth is about 5.5 gm./cm.$^3$, the material deep within the Earth must have a density considerably greater than that of typical surface rocks. Let $\varrho$ be the density, $p$ the pressure, $g$ the gravitational attraction, and $k$ the adiabatic incompressibility, defined by the equation

$$\frac{1}{k} = \frac{1}{\varrho}\frac{\partial \varrho}{\partial p}. \tag{5.1}$$

If $m$ is the mass of the material within a sphere of radius $r$, then, since the stress in the Earth's interior is essentially equivalent to a hydrostatic pressure,

$$\frac{dp}{dr} = -g\varrho \tag{5.2}$$

where

$$g = \frac{\gamma m}{r^2}. \tag{5.3}$$

$\gamma$ being the gravitational constant. (The assumption of hydrostatic stress would be a poor approximation for the crust where the strength of the rocks is of the same order as the mean pressure. But the mean pressure steadily increases with depth, whilst the strength, or maximum stress difference decreases, so that a depth is soon reached where the approximation is satisfactory.)

Consider a *homogeneous layer* in which the temperature varies along the adiabat as the pressure increases with depth. Then from (5.1) and (5.2),

$$\frac{d\varrho}{dr} = \frac{d\varrho}{dp} \cdot \frac{dp}{dr} = \frac{-g\varrho^2}{k}. \tag{5.4}$$

But from equation (2.10),

$$\varphi = \frac{k}{\varrho} = \alpha^2 - \frac{4\beta^2}{3}$$

where $\alpha$ and $\beta$ are the longitudinal and transverse wave velocities at the level $r$. Hence

$$\frac{d\varrho}{dr} = \frac{-g\varrho}{\varphi} = -\frac{\gamma m \varrho}{r^2 \varphi}. \tag{5.5}$$

[1] A. EUCKEN: Naturwiss. **32**, 112 (1944).
[2] H. JEFFREYS: The Earth. Cambridge 1952.
[3] H. TAKEUCHI: Trans. Amer. Geophys. Un. **31**, 651 (1950).
[4] E. C. BULLARD: Observatory **70**, 139 (1950).
[5] W. M. ELSASSER: Rev. Mod. Phys. **22**, 1 (1949).

This equation, first obtained by ADAMS and WILLIAMSON[1], has been extensively used by BULLEN to obtain by numerical integration the density variation within the Earth.

Any determination of the density distribution must satisfy two conditions. The integrated density must give the correct *total mass* of the Earth viz $597.7 \times 10^{25}$ gm. (OLCZAK[2]) and also the correct *moment of inertia* of the Earth about its axis viz. $81.04 \times 10^{43}$ gm. cm.[2] as determined from the precession of the equinoxes. Expressing the moment of inertia of a sphere of mass $m$ and radius $r$ as $z m r^2$, then for a homogeneous sphere $z = 0.4$, whilst for the Earth $z = 0.3341$. This indicates a large increase of density towards the centre. If the density distribution is determined by integrating from the outside inwards, then at any radius $r$, the residual mass $m$ can be calculated, and a reasonable value (slightly less than 0.4) must be obtained for the parameter $z$ at that radius. BULLEN[3] found that these restrictions imposed severe limits on admissible density distributions. A brief account of his work is now given.

The mass of the crustal layers is first substracted from the mass of the Earth. Equation (5.5) is then applied to the region $B$ (see Table 2), where the smoothness of the $P$ and $S$ velocity distributions indicate that the material is fairly homogeneous. The density at the top of layer $B$ which enters as a constant of integration is taken as 3.32 gm./cm.[3], this being the density of olivine at that pressure. BULLEN points out that an error of 0.1 gm./cm.[3] in this figure affects the estimation of the densities below the layer $B$ by less than $\frac{1}{4}$%. If the density distribution in the entire mantle is obtained by integration of equation (5.5), the impossibly high value of 0.57 for $z$ is obtained for the core. This gives further (independent) proof that the mantle is not homogeneous. In order to obtain the density distribution in the layers $C$ and $D$, BULLEN uses simple polynomials with undetermined coefficients. As a result of various trial distributions of density within the core, BULLEN found that $z$ for the core must be between 0.375 and 0.395. Assuming the value 0.38 and making the assumptions that there is no discontinuity of $\varrho$ in the mantle and that the upper part of layer $D$ is sufficiently homogeneous so that equation (5.5) can be applied to it, he obtains the density distribution in layers $C$ and $D$. The density distribution for the entire mantle is given in Table 3.

Table 3. *Density, Gravity and Pressure distribution in the mantle.*

| Depth km. | Density gm./cm.[3] | Gravity cm./sec.[2] | Pressure $\times 10^{12}$ dynes/cm.[2] | Depth km. | Density gm./cm.[3] | Gravity cm./sec.[2] | Pressure $\times 10^{12}$ dynes/cm.[2] |
|---|---|---|---|---|---|---|---|
| 33 | 3.32 | 985 | 0.009 | 1000 | 4.68 | 995 | 0.392 |
| 100 | 3.38 | 989 | 0.031 | 1200 | 4.80 | 991 | 0.49 |
| 200 | 3.47 | 992 | 0.065 | 1400 | 4.91 | 988 | 0.58 |
| 300 | 3.55 | 995 | 0.100 | 1600 | 5.03 | 986 | 0.68 |
| 400 | 3.63 | 997 | 0.136 | 1800 | 5.13 | 985 | 0.78 |
| 413 | 3.64 | 998 | 0.141 | 2000 | 5.24 | 986 | 0.88 |
| 500 | 3.89 | 1000 | 0.173 | 2200 | 5.34 | 990 | 0.99 |
| 600 | 4.13 | 1001 | 0.213 | 2400 | 5.44 | 998 | 1.09 |
| 700 | 4.33 | 1000 | 0.256 | 2600 | 5.54 | 1009 | 1.20 |
| 800 | 4.49 | 999 | 0.300 | 2800 | 5.63 | 1026 | 1.32 |
| 900 | 4.60 | 997 | 0.346 | 2898 | 5.68 | 1037 | 1.37 |

[1] E. D. WILLIAMSON and L. H. ADAMS: J. Wash. Acad. Sci. **13**, 413 (1923).
[2] T. OLCZAK: Acta astronomica **3**, 81 (1938).
[3] K. E. BULLEN: Bull. Seism. Soc. Amer. **30**, 235 (1940); **32**, 19 (1942).

In the core, layer $E$ is almost certainly a homogeneous fluid to which equation (5.5) is applicable. For the layers $F$ and $G$, BULLEN[1] gives two density distributions, based on extreme hypotheses. In the first case he assumes that the density varies smoothly throughout the entire core i.e. throughout the whole of regions $E$, $F$ and $G$. This hypothesis gives an approximate lower limit to the density at the centre of the Earth of 12.3 gm./cm.[3]. In the second case he assumes that the density at the centre of the Earth is greater than this figure by 10 gm./cm.[3].

Table 4. *Limiting densities.*

| Region | Depth km. | Density gm./cm.[3] | |
|---|---|---|---|
| | | Hypothesis (1) | Hypothesis (2) |
| B | | | |
| | 413 | 3.64 | |
| | 500 | 3.88 | 3.90 |
| C | 600 | 4.11 | 4.14 |
| | 800 | 4.46 | 4.52 |
| | 1000 | 4.65 | 4.71 |
| | 1400 | 4.88 | 4.95 |
| | 1800 | 5.10 | 5.17 |
| D | 2200 | 5.31 | 5.37 |
| | 2600 | 5.51 | 5.57 |
| | 2898 | 5.66 | 5.72 |
| | 2898 | 9.7 | 9.1 |
| | 3000 | 9.9 | 9.2 |
| | 3500 | 10.5 | 9.8 |
| E | 4000 | 11.1 | 10.3 |
| | 4500 | 11.6 | 10.8 |
| | 4982 | 11.9 | 11.1 |
| F | | | |
| | 5121 | 12.0 | |
| G | | | |
| | 6371 | 12.3 | 22.3 |

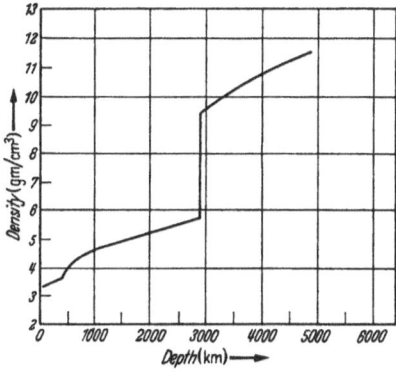

Fig. 4. Variation of density in the Earth's interior. (After K. E. BULLEN.)

On this hypothesis there would be a density jump of approximately 10 gm./cm.[3] between the bottom of layer $E$ and the top of layer $G$. These two extreme density distributions are given in Table 4.

The distribution given in Table 5 is obtained by an arbitrary assumption regarding the variation of density between the bottom of layer $E$ and the top of layer $G$ giving values about half way between the two extreme cases. For the sake of consistency these values will be used in all future calculations. Since the regions $F$ and $G$ constitute only about 1% of the Earth's total volume it

Table 5. *Density, Gravity and Pressure distribution in the core.*

| Depth km. | Density gm./cm.[3] | Gravity cm./sec.[2] | Pressure × 10[12] dynes/cm.[2] | Depth km. | Density gm./cm.[3] | Gravity cm./sec.[2] | Pressure × 10[12] dynes/cm.[2] |
|---|---|---|---|---|---|---|---|
| 2898 | 9.43 | 1037 | 1.37 | 4600 | 11.27 | 677 | 2.88 |
| 3000 | 9.57 | 1019 | 1.47 | 4800 | 11.41 | 646 | 3.03 |
| 3200 | 9.85 | 979 | 1.67 | 4982 | 11.54 | 626 | 3.17 |
| 3400 | 10.11 | 936 | 1.85 | 5121 | (14.2) | 585 | 3.27 |
| 3600 | 10.35 | 892 | 2.04 | 5121 | (16.8) | | |
| 3800 | 10.56 | 848 | 2.22 | 5400 | | 460 | 3.41 |
| 4000 | 10.76 | 803 | 2.40 | 5700 | | 320 | 3.53 |
| 4200 | 10.94 | 758 | 2.57 | 6000 | | 177 | 3.60 |
| 4400 | 11.11 | 716 | 2.73 | 6371 | (17.2) | 0 | 3.64 |

[1] See footnote 1, p. 368.

follows that in spite of uncertainties of the density variations in these regions, the density distribution within $E$ can be estimated fairly precisely. BULLEN shows that an uncertainty of 1 gm./cm.³ in the mean density of the innermost region $G$ affects the density at the top of layer $E$ by only 0.07 gm./cm.³. The density distribution in Tables 3 and 5 is shown graphically in Fig. 4. BULLEN believes that these density values are accurate to within 1% down to a depth of 2700 km. and to within about 3% below that depth.

**6. Pressure distribution and variation of the acceleration due to gravity within the Earth.** From equations (5.2) and (5.3), it follows that

$$\frac{dp}{dr} = -\frac{\gamma m \varrho}{r^2}. \tag{6.1}$$

Hence by numerical integration, the pressure distribution may be obtained once the density distribution has been determined. Since the density is only used to determine the pressure gradient, the pressure distribution is insensitive to small changes in the density distribution and may be determined quite accurately. The results are given in Tables 3 and 5 and graphically in Fig. 5. The accuracy in the values for the regions $F$ and $G$ is of the order of 3% and for all other regions probably less than 1%.

The variation of $g$ can be calculated from equation (5.3). Its values are also given in Tables 3 and 5. The value of $g$ does not differ by more than one percent from the value 990 cm./sec.² until a depth of over 2400 km. is reached. On the other hand the values of $g$ deep within the Earth are sensitive to changes in density and values below 4000 km. cannot be given with an accuracy greater than 5%.

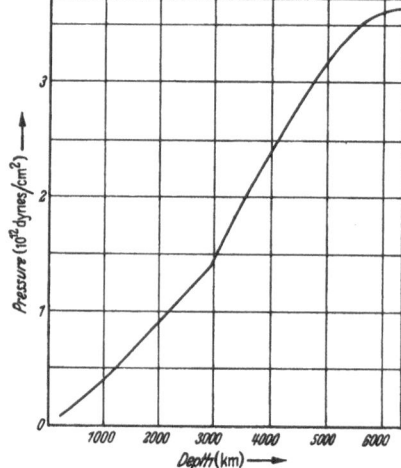

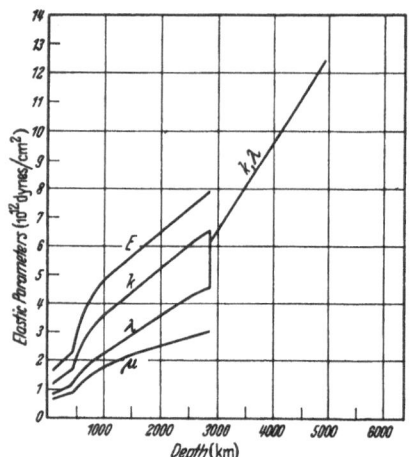

Fig. 5. Pressure distribution in the Earth's interior. (After K. E. BULLEN.)          Fig. 6. Variation of elastic parameters in the Earth's interior. (After K. E. BULLEN.)

**7. Values of the elastic constants within the Earth.** From a knowledge of the density distribution, it is easy to compute values of the elastic constants $\lambda, \mu$ and $k$ using equations (2.5), (2.8) and (2.10). Thus

$$\lambda = \varrho\,(\alpha^2 - 2\beta^2), \tag{7.1}$$

$$\mu = \varrho\beta^2 \tag{7.2}$$

and

$$k = \varrho\,(\alpha^2 - \tfrac{4}{3}\beta^2).$$ (7.3)

From the known relations between the elastic constants it is then possible to compute values of YOUNG's Modulus $E$ and POISSON's ratio $\sigma$. Thus

$$E = \frac{\mu\,(3\lambda + 2\mu)}{(\lambda + \mu)} = \frac{\varrho\,\beta^2\,(3\alpha^2 - 4\beta^2)}{(\alpha^2 - \beta^2)}$$ (7.4)

and

$$\sigma = \frac{\lambda}{2(\lambda + \mu)} = \frac{\alpha^2 - 2\beta^2}{2(\alpha^2 - \beta^2)}.$$ (7.5)

$\sigma$ is thus independent of the density $\varrho$. Values of these elastic constants in the regions $B, C, D$ and $E$ are given in Table 6 and shown graphically in Fig. 6. There is little change in the incompressibility $k$ across the boundary between the mantle and the core. The values of $\lambda$ and $k$ in the inner core are not included since they are subject to much uncertainty.

Table 6. *Values of the elastic constants.*

| Region | Depth km. | $\lambda$ | $\mu$ | $k$ | $E$ | $\sigma$ |
|---|---|---|---|---|---|---|
| | | | | (unit $10^{12}$ dynes/cm.$^2$) | | |
| | 33 | 0.74 | 0.63 | 1.16 | 1.60 | 0.269 |
| | 100 | 0.80 | 0.67 | 1.24 | 1.70 | 0.272 |
| B | 200 | 0.90 | 0.74 | 1.38 | 1.89 | 0.275 |
| | 300 | 1.01 | 0.81 | 1.54 | 2.07 | 0.277 |
| | 413 | 1.14 | 0.90 | 1.73 | 2.30 | 0.280 |
| | 500 | 1.42 | 1.10 | 2.15 | 2.82 | 0.283 |
| C | 600 | 1.69 | 1.32 | 2.57 | 3.38 | 0.282 |
| | 800 | 2.06 | 1.69 | 3.19 | 4.31 | 0.275 |
| | 1000 | 2.33 | 1.89 | 3.59 | 4.82 | 0.276 |
| | 1400 | 2.76 | 2.15 | 4.20 | 5.51 | 0.281 |
| | 1800 | 3.27 | 2.39 | 4.87 | 6.16 | 0.288 |
| D | 2200 | 3.81 | 2.63 | 5.57 | 6.81 | 0.295 |
| | 2600 | 4.32 | 2.88 | 6.23 | 7.49 | 0.300 |
| | 2898 | 4.49 | 3.03 | 6.51 | 7.87 | 0.300 |
| | 2898 | 6.2 | | 6.2 | | 0.5 |
| | 3000 | 6.5 | | 6.5 | | 0.5 |
| | 3500 | 8.1 | | 8.1 | | 0.5 |
| E | 4000 | 9.7 | | 9.7 | | 0.5 |
| | 4500 | 11.1 | | 11.1 | | 0.5 |
| | 4982 | 12.6 | | 12.6 | | 0.5 |

## b) Composition of the mantle and core.

8. **The work of BULLEN, BIRCH and VERHOOGEN.** BULLEN[1] has pointed out that the smallness of the change in the incompressibility $k$ across the boundary between the core and the mantle is less than the uncertainty of the determination. The difference is in fact les than 5% (6.5 and 6.2$\times 10^{12}$ dynes/cm.$^2$—see Table 6). He also found that there is no significant difference in the value of $dk/dp$ above and below the core boundary. On the other hand the core boundary represents a very sharp division between materials of very different rigidity and density. These observations led him to formulate what he has called the *compressibility pressure hypothesis* viz that at the high pressures that prevail deep within the Earth (below about 1000 km.) the compressibility of a substance is virtually independent of its chemical composition. On this assumption both $k$ and $dk/dp$

---

[1] K. E. BULLEN: Mon. Not. Roy. Astronom. Soc., Geophys. Suppl. 5, 355 (1949).

would be fairly smooth functions of $p$ throughout the Earth below a depth of about 1000 km. If the hypothesis is correct, it would lead to an increased precision in the values obtained for the density distribution, since an extra condition would be provided. In particular it would follow that the mean density of the region $G$ must be greater than 14 gm./cm.[3]. An implication[1] of the theory is the rejection of equation (5.5) to calculate the density distribution in the region $2700 \leq d \leq 2900$, where, however, it can be estimated independently. BULLEN finds that the density increase in this region is just over three times that indicated by the formal application of equation (5.5). The compressibility-pressure hypothesis thus entails a modification of his previous Earth model, indicating an accumulation of denser materials near the base of the Earth's mantle.

The hypothesis also throws some light on the constitution of the inner core. The velocity-depth curves of JEFFREYS show a diminution in the longitudinal velocity $\alpha$ from 10.44 km./sec. at the top of layer $F$ to 9.7 km./sec. at the base, followed by a discontinuous increase from 9.7 to 11.16 km./sec. across the boundary between $F$ and $G$. Assuming that the regions $E$ and $F$ are both liquid, $\alpha^2 = k/\varrho$ and it follows that either $dk/dp$ is negative in $F$ or else $\varrho$ increases by at least 16% through $F$. If the formula $\alpha^2 = k/\varrho$ holds approximately in the inner core $G$, $k$ would have to increase across the boundary between $F$ and $G$ by at least 32%, excluding the highly improbable case that the density decreases with depth. On the other hand if the inner core $G$ is solid and hence capable of transmitting transverse waves, the relation $\alpha^2 = k/\varrho$ is replaced by $\alpha^2 - \tfrac{4}{3}\beta^2 = k/\varrho$ in $G$. The discontinuity in the value of $k$ across the boundary between $F$ and $G$ can then be avoided if at the top of $G$, $\tfrac{4}{3}\beta^2$ is at least equal to $(11.16^2 - 9.7^2)$ km.$^2$/sec.$^2$ i.e. $\beta$ is about 4.8 km./sec. or greater inside $G$. Thus the compressibility pressure hypothesis ($dk/dp > 0$ and $k$ continuous throughout the entire Earth below about 1000 km.) indicates an increase in density in $F$ and a solid inner core capable of transmitting shear waves.

BULLEN[2] further shows that the inner core $G$ cannot be chemically homogeneous, and that his hypothesis indicates an accumulation of denser materials in the inner core in the same way that it indicated an accumulation of denser rocks near the base of the mantle. The density at the Earth's centre is found to lie between 16 and 20 gm./cm.[3].

The ADAMS-WILLIAMSON method of obtaining the density distribution described in Sect. 5 and leading to equation (5.5) is based on two assumptions

(1) the density changes only as a result of compression with no change of intrinsic density, as by introduction of a new material,

(2) the compression is adiabatic i.e. the rate of change of temperature with depth is taken to be just the change corresponding to adiabatic compression.

When BULLEN first made an estimate of the density variation within the Earth, he found that the mantle could not be uniform in composition with its density determined by compression alone as given by the ADAMS-WILLIAMSON method and the velocity depth curves now accepted. BIRCH[3] has more recently carried out a long and exhaustive study on the constitution of the Earth's interior and has examined in particular this question of the uniformity of the mantle.

---

[1] See footnote 1, p. 375.
[2] K. E. BULLEN: Mon. Not. Roy. Astronom. Soc., Geophys. Suppl. **6**, 50 (1950). (Tables of the distribution of $\varrho$, $g$ and $p$ for an Earth model based on the compressibility-pressure hypothesis are given on pages 56—57.)
[3] F. BIRCH: J. Geophys. Res. **57**, 227 (1952).

Here uniformity or homogeneity is taken to mean that the variations of quantities such as $\varrho$, $k$ etc. are caused by compression alone, as contrasted to variations resulting from chemical or phase change. He considers variations of the quantity $\varphi = k/\varrho$ which was defined in equation (2.10).

If the pressure distribution is approximately hydrostatic, (5.2) gives $dp = -g\varrho\,dr$, whilst for a material of uniform composition, (5.1) gives $dp = \dfrac{k\,d\varrho}{\varrho} = \varphi\,d\varrho$. Hence it follows from $k = \varphi\varrho$ that

$$\frac{dk}{dp} = \frac{(\varphi\,d\varrho + \varrho\,d\varphi)}{\varphi\,d\varrho} = 1 + \varrho\,\frac{d\varphi}{dp} = 1 - g^{-1}\frac{d\varphi}{dr}. \tag{8.1}$$

Since $g$ is very nearly constant throughout the mantle, the quantity on the right hand side of equation (8.1) can be calculated directly from a knowledge of the behaviour of $\varphi$, which, as equation (2.10) shows, may be obtained from seismic data ($\alpha$ and $\beta$) alone. On the other hand, Birch has used an equation of state which is in excellent agreement with Bridgman's experimental work up to pressures of $10^5$ kgm./cm.$^2$, to obtain a theoretical curve of the expression $1 - g^{-1}\dfrac{d\varphi}{dr}$ as a function of depth. This theoretical curve is in good agreement with the values calculated from seismic evidence between about 900 km. and the base of the mantle, but there are wide discrepancies between depths of about 200 to 900 km. Birch thus deduces that the lower part of the mantle is substantially uniform whilst the upper part is non-uniform i.e. compression of a homogeneous material cannot by itself account for the variation of $dk/dp$ between about 300 and 800 km. In addition Birch has calculated $\varphi_0$, the value of $\varphi$ at zero pressure. He finds that the mean value of $\varphi_0$ is 51 km.$^2$/sec.$^2$ and that all points between 700 and 2800 km. lie within about 1% of this value, again demonstrating the uniformity of the deeper layers of the mantle. This figure for $\varphi_0$ is about 50% higher than the corresponding figure for forsterite—in fact the values of $\varphi_0$ for all the silicates are too low. No known material with a plausible composition has also the required elasticity. Thus if the lower part of the mantle is a ferromagnesian silicate, it must exist as a denser, more tightly bound structure than olivine and may represent a high pressure form of an olivine pyroxene composition. Birch thus suggests that the region between about 300 and 800 km. is a *region of transition to the high pressure form*, probably with a change of chemical composition as well.

The composition of the mantle is thus envisaged as consisting of three layers. The uppermost layer, approximately 300 to 400 km. thick consisting perhaps of eclogitic material, followed by a transitional layer down to a depth of about 900 km. In this layer the elastic properties vary rapidly with depth — due to high-pressure phase changes and/or a change in chemical composition. Below 900 km. the mantle is homogeneous in composition with elastic properties that are found only in certain oxides such as corundum and rutile.

Verhoogen[1] has raised a number of objections to this interpretation of the structure of the mantle. He shows that the equation of state used by Birch, although apparently adequate to represent experimental data for the alkali metals, is not likely to be applicable to silicate minerals and is inconsistent with the assumption that the velocity-depth curve is continuous. He shows further that it is extremely unlikely that silicates would occur in a corundum or rutile type of structure. There is also the objection that notable phase changes would produce sharp and not gradual changes in elastic properties, although small

---

[1] J. Verhoogen: J. Geophys. Res. **58**, 337 (1953).

phase transitions that do not correspond with any major reorganization of the crystal lattices involved are quite likely. It is not impossible that the composition of the mantle might *gradually* change with depth—with, for example, a gradual increase in the amount of metallic iron.

Although Verhoogen's criticisms are in the main justifiable, it is well to remember that the conclusion that the mantle is not homogeneous in both phase and composition does not depend upon a particular equation of state. Birch did not set out to prove that the upper part of the mantle is not uniform—this fact had been amply demonstrated by the earlier work of Bullen—rather he attempted to analyse and explain the complex structure of the Earth's interior.

It must be confessed that there is still no clear picture of the physical constitution of the mantle, and it is not surprising that there should be conflicting opinions on this subject in view of the very meagre amount of data on which any theory can be based. The existence, depth and suddenness of the Mohorovičić discontinuity which marks the boundary between the crust and mantle, are not in doubt, but opinions on the nature of the material in the mantle are necessarily speculative. The range of choice is severely limited since any material must

(1) be petrologically likely i.e. it must occur as an intrusive rock and not be excessively rare

(2) be composed of reasonably common elements and

(3) give the right seismic velocity at the appropriate temperature and pressure.

These three requirements seem best met by peridotite. Peridotite also satisfies less certain geochemical criteria such as a low concentration of alkali metals and uranium and has a density substantially greater than most of the rocks of the granitic layer.

**9. Ramsey's Hypothesis.** It is generally assumed that the core of the Earth consists of iron or an iron nickel alloy. The main reasons in support of this viewpoint are

(1) the mean density of the Earth is considerably larger than would be expected from the materials found on or near the surface. Thus the core must be composed of a heavy material and iron is the only abundant heavy element;

(2) a ferromagnetic core may account for terrestrial magnetism, and

(3) the existence of iron meteorites.

Ramsey[1,2], however, has put forward the suggestion that the large increase of density at the boundary of the Earth's core is due to a *pressure transition from the molecular to a metallic phase*, rather than to the appearance of a new material such as an alloy of iron and nickel. He thus assumes that the Earth is of uniform chemical composition (below the crustal layers) which he identifies as *olivine*— a mixture of 90% magnesium orthosilicate and 10% iron orthosilicate.

Originally Ramsey put forward his hypothesis to account for the densities of the terrestrial planets. On his hypothesis the pressure at the boundary of the core will be characteristic of the chemical composition of the material which he assumes is the same for all the terrestrial planets. However Kuiper[3] has pointed out that the dimensions and masses of the terrestrial planets as now known

---

[1] W. H. Ramsey: Mon. Not. Roy. Astronom. Soc. **108**, 406 (1948).
[2] W. H. Ramsey: Mon. Not. Roy. Astronom. Soc., Geophys. Suppl. **5**, 409 (1949).
[3] G. P. Kuiper: The Atmospheres of the Earth and Planets, 2nd Edit., Chicago 1952 (see, in particular, pages 308 and 339).

show that this hypothesis is untenable. RABE's[1] work on the orbit of Eros and the more accurate values of the masses of Mercury and Venus rule it out on astronomical grounds. RAMSEY was undoubtedly influenced when he first put forward his hypothesis by the explanation given by BERNAL[2] for the 20° discontinuity at a depth of about 413 km. in the Earth's mantle. BERNAL accounts for this as a transition to a crystal structure which is more stable at high pressures i.e. it occurs at a pressure which is characteristic of the substance concerned.

One of the arguments put forward in favour of an iron core theory is the immiscibility of liquid iron and the silicates. RAMSEY attacks this on the grounds of the work of KUHN and RITTMANN[3], who have shown that even under the most favourable conditions, the time required for such a chemical separation is much larger than the age of the Earth. However, as KUIPER[4] has pointed out, the KUHN and RITTMAN objections to the separation of the iron and silicate phases cannot, on empirical grounds, apply to planets. These phases are often separated in large meteorites which almost certainly derived from asteroidal bodies of roughly 500 km. in diameter.

ELSASSER[5] has attempted to show that RAMSEY's hypothesis conflicts with the data on the Earth interpreted in the light of theoretical physics. He compares the estimates of the densities and compressibilities of a large number of elements and compounds found experimentally by BRIDGMAN[6], for pressures up to $10^5$ atmospheres, with the limiting computed values at pressures of the order of $10^7$ atmospheres and above obtained from quantum mechanical theory. These theoretical values are based on a THOMAS-FERMI-DIRAC model of the electronic density in a closest packed, cubic, monatomic lattice and are obtained chiefly from the results of the computations of FEYNMAN, METROPOLIS and TELLER[7]. ELSASSER has interpolated in the gap between $10^5$ and $10^7$ atmospheres and states that the densities of all elements can be determined as functions of the pressure in this range, with a maximum error of at most 15 to 20%. In comparing the density variation within the Earth with his interpolated curves, ELSASSER finds strong support to the theory that the mantle consists mainly of silicates and the core of iron, thus excluding the possibility of RAMSEY's hypothesis. ELSASSER also finds some discrepancies between his extrapolated curves and geophysical data.

As is the case with the mantle, however, complete agreement has not been reached on the question of the constitution of the core. BULLEN[8] has analysed the above data of BRIDGMAN, FEYNMAN, METROPOLIS and TELLER and disagrees with some of ELSASSER's findings. BULLEN's work not only fits the data used by ELSASSER but is also consistent with geophysical data. His calculations imply that the atomic number to be associated with the material of the outer core i.e. the region $E$ (2900 to 4980 km.) should be at least 6 units less than the value derived using simple extrapolations from quantum mechanical calculations for high pressures. If the reduction in the atomic number is no more than six units, the most probable composition of the region $E$ would still be nickel-iron, and to this extent ELSASSER's main conclusion is supported. But other aspects of

[1] E. RABE: Astrophys. J. **55** ,112 (1950).

[2] J. D. BERNAL: Observatory **59**, 268 (1936).

[3] W. KUHN and A. RITTMANN: Geol. Rundschau **32**, 215 (1941).

[4] G. P. KUIPER: Astrophys. J. **55**, 164 (1950). See also A. EUCKEN: Naturwiss. **32**, 112 (1944); **33**, 311 (1946). — Nachr. Akad. Göttingen, Math.-Phys. Kl. 1944.

[5] W. M. ELSASSER: Science, Lancaster, Pa. **113**, 105 (1951).

[6] P. W. BRIDGMAN: Proc. Amer. Acad. Arts Sci. **76**, No. 1 (1945); No. 3, **1948**.

[7] R. P. FEYNMAN, N. METROPOLIS and E. TELLER: Phys. Rev. **75**, 1561 (1949).

[8] K. E. BULLEN: Mon. Not. Roy. Astronom. Soc., Geophys. Suppl. **6**, 383 (1952).

BULLEN's new calculations, which are accepted by ELSASSER, suggest that the needed reduction may be greater than six units indicating that the region $E$ may consist of a modification of ultra-basic rock[1].

UREY[2] has put forward a new theory for the development of the planets which rests in large part on arguments from physical chemistry. He concludes on several grounds that the Earth's core has an iron composition. BULLEN[3] questions a number of his arguments, although agreeing that the inner core is chemically distinct from the rest of the Earth and of nickel-iron composition. It is extremely difficult to be definite about the composition of the outer part of the core. The two extreme viewpoints may not be mutually exclusive, however, and a composition of about 80% iron and 20% modified ultrabasic rock may be the answer.

## B. Thermal Properties.

**10. Introduction.** Of all the problems connected with the constitution of the Earth's interior, its thermal properties are the least well understood, and a more complete understanding of them would go a long way towards solving many of the outstanding problems in geophysics. In any investigation of the thermal properties of the Earth, sooner or later the question of the initial temperature of the Earth and the mode in which it has cooled (or heated) since then, is bound to arise. This in its turn leads to the question of the origin of the Earth and to other deeper and more far reaching astrophysical problems. Since such questions are in their very nature bound to be extremely controversial, they will be avoided as far as possible, although some discussion of them is inevitable.

Most theories of the origin of the Earth agree that it passed through a *molten stage* before finally separating into different layers. URRY[4] has shown that even an Earth formed from the cold accretion of dust particles would melt by the accumulation of heat from radioactive elements and afterwards solidify as the amount of radioactive material decreased with increasing age of the Earth. BIRCH[5] has repeated URRY's calculations and come to the same conclusion. He estimates that the time required for melting would probably be of the same order as the rather indefinite time of accumulation[6]. HOYLE[7] has also shown that accretion should lead to a liquid planet. Thus the cooling of the Earth from a molten state will be considered.

So long as the Earth remained fluid, the process was comparatively straight-forward. The heavy material of the core settled to the centre and stayed there. In general liquids, as they cool, contract and become denser. Thus the matter at the surface of the Earth cooled by radiation into space and then sank through the hotter liquid below. This would set up irregular convection currents and ensure a continued supply of heat to the surface. The increase in temperature $dT$ for a reversible adiabatic increase of pressure $dp$ is given by

$$dT = \frac{T\alpha'}{\varrho c_p} \cdot dp \tag{10.1}$$

---

[1] A quantum statistical method for estimating the densities of compounds at high pressures, with applications to the Earth, has been given by L. KNOPOFF and R. J. UFFEN: J. Geophys. Res. **59**, 471 (1954).

[2] H. C. UREY: The Planets, their origin and development. Yale University Press 1952.

[3] K. E. BULLEN: Nature, Lond. **170**, 363 (1952).

[4] W. D. URRY: Trans. Amer. Geophys. Un. **30**, 171 (1949).

[5] F. BIRCH: J. Geophys. Res. **56**, 107 (1951).

[6] It must be remarked, however, that the concentration of radioactive elements is very uncertain and the values used by URRY and BIRCH may be too high. If a cold origin for the Earth is postulated, some doubt must remain as to whether it could have melted in any reasonable time.

[7] F. HOYLE: Mon. Not. Roy. Astronom. Soc. **106**, 406 (1946).

where $\alpha'$ is the volume coefficient of thermal expansion, $\varrho$ the density, and $c_p$ the specific heat at constant pressure. This gives the rate of increase of temperature on adiabatic compression, which condition is satisfied with a high degree of accuracy in a fluid cooling by convection. Assuming hydrostatic equilibrium, the variation of pressure with depth $z$ is given by

$$dp = g \varrho \, dz \tag{10.2}$$

Hence from (10.1) and (10.2), the *adiabatic temperature gradient* is given by

$$\frac{dT}{dz} = \frac{g \alpha' T}{c_p} . \tag{10.3}$$

Taking $g = 981$ cm./sec.$^2$, $\alpha' = 2 \times 10^{-5}$ per degree, $T = 1400°$, $c_p = 0.2$ cal./gm. degree, JEFFREYS[1] obtained an approximate estimate of the gradient of 0.3°/km. This temperature gradient would be maintained as long as the Earth was fluid. If the gradient were exceeded, convection currents would increase in strength and redistribute the temperature adiabatically. On the other hand if the gradient became less, convection currents would be damped down by viscosity, cooling at the top would become more rapid and the gradient would increase again.

The *melting point* of the material depends on the pressure and hence on depth. The effect of pressure on the melting point is given by the CLAUSIUS-CLAPEYRON equation

$$\frac{dT_m}{dp} = \frac{T_m}{L} \left( \frac{1}{\varrho_1} - \frac{1}{\varrho_2} \right) \tag{10.4}$$

where $T_m$ is the melting temperature, $L$ the latent heat of fusion, and $\varrho_1, \varrho_2$ the densities in the liquid and solid states. Thus, using equation (10.2), it follows that in a liquid layer,

$$\frac{dT_m}{dz} = \frac{g T_m}{L} \left( 1 - \frac{\varrho_1}{\varrho_2} \right) . \tag{10.5}$$

Taking fairly typical values for silicate rocks viz $T_m = 1300°$, $L = 100$ cal./gm. $\varrho_1/\varrho_2 = 0.9$, JEFFREYS obtained a gradient of approximately 3°/km. Although the numerical estimates obtained for equations (10.3) and (10.5) can only be considered approximate, they do show that, whilst the adiabatic temperature and the melting point both increase with depth, the melting point increases the faster. Refinements in the evaluation of these gradients is unlikely to alter this result. Thus as the liquid cooled, the *melting point is first reached at the bottom*, solidification proceeding from the bottom upwards. This result was first pointed out by ADAMS[2].

The above arguments have supposed that the material of the mantle was originally thoroughly mixed and remained so, that there was no internal generation of heat due to radioactivity and that the gradients given by equations (10.3) and (10.5) were constant. This last assumption is almost certainly not realized. Thus if the numerical value of the melting point gradient is extrapolated to 2900 km., it leads to a value of about 10,000° as the melting point of the mantle at the core boundary. The validity of such extrapolations is extremely doubtful. Most of the data on the Earth's interior discussed in the preceding sections has been obtained from the interpretation of seismic data by physical laws the truth of which has been well established. It is natural to ask whether seismic evidence can yield any information on the thermal properties of the Earth. Considerable headway has been made in this direction by

---

[1] See footnote 2, p. 371.
[2] L. H. ADAMS: J. Wash. Acad. Sci. **14**, 459 (1924).

combining the seismic data with the theory of the solid state. This will be discussed in detail in later sections where more accurate estimates are obtained of the melting point and adiabatic gradients.

**11. The Earth's inner core.** All evidence indicates that at least part of the Earth's core is liquid, whilst the mantle is solid (see Sect. 4). Any theory of the thermal history of the Earth must at least satisfy these two conditions. It has been suggested, however, that the core contains an inner core beginning at a depth of approximately 5000 km. which is solid, although no definite proof of this has ever been given.

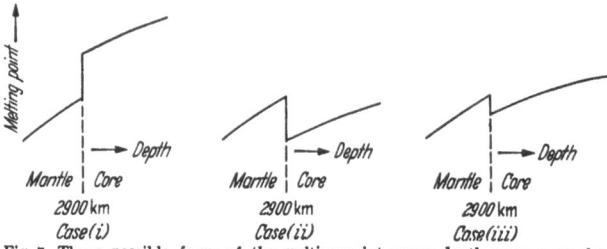

Fig. 7. Three possible forms of the melting point *versus* depth curve near the boundary between mantle and core.

BULLEN[1] has shown, however, that the rise in the velocity of longitudinal waves at this depth can be explained by assuming it to be solid and of the same composition as the rest of the core. BIRCH[2] has also come to the conclusion that the inner core is most probably crystalline iron and the outer part liquid iron, perhaps alloyed with a small fraction of lighter elements. Assuming the core to consist mainly of iron and the mantle of silicates, JACOBS[3, 4] has put forward the following physical explanation for the existence of a solid inner core.

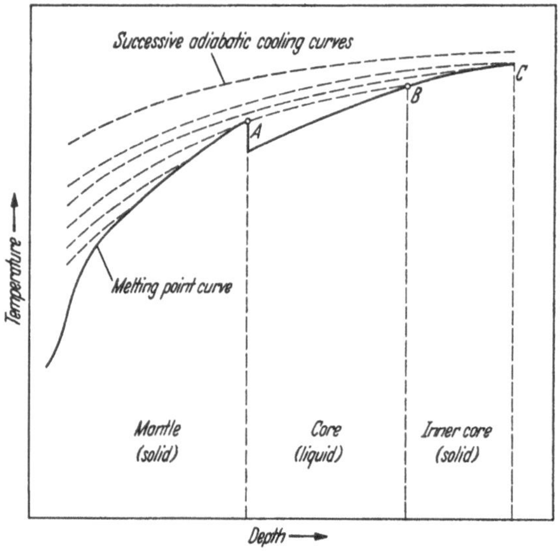

At the boundary between the silicate mantle and the iron core there must be a discontinuity in the melting point-depth curve, although the actual temperature must be continuous across the boundary. The form of this discontinuity could, mathematically, take any of the three cases shown in Fig. 7. Case (i) in which the melting point curve in the core is always above that in the mantle is impossible; for the actual temperature curve must lie below the melting point curve in the mantle, above the melting point curve in the core, and yet be continuous across the boundary. Cases (ii) and (iii) are both possible. In case (ii) the melting point curve in the core never rises above the value of the melting point in the mantle at the core boundary, whilst in case (iii) it exceeds this value for part of the core. Considering case (iii), the melting point curve will be of the general shape shown in Fig. 8.

Fig. 8. Cooling of the Earth from the molten state.

[1] See footnote 1, p. 368.
[2] See footnote 3, p. 376.
[3] J. A. JACOBS: Nature, Lond. **172**, 297 (1953).
[4] J. A. JACOBS: Nature, Lond. **173**, 258 (1954).

As the earth cooled from its molten state, the temperature gradient would be essentially adiabatic, there being strong convection currents and rapid cooling at the surface. Solidification would commence at that depth at which the curve representing the adiabatic temperature first intersected the curve representing the melting point temperature. It is suggested, therefore, that *solidification began at the centre of the Earth*, and not at the boundary of the core and mantle as has been supposed. A solid inner core would continue to grow until a curve representing the adiabatic temperature intersected the melting-point curve twice, once at $A$, the boundary of the core and mantle, and again at $B$, as shown in the figure. As the Earth cooled still further, the mantle would begin to solidify from the bottom upwards. The liquid layer between $A$ and $B$ would thus be trapped. The mantle would cool at a relatively rapid rate, leaving this liquid layer essentially at its original temperature, insulated above by a rapidly thickening shell of silicates and below by the already solid (iron) inner core.

In the above argument no specific values of the temperatures are postulated, and the behaviour of both the adiabatic and melting-point curves need not be known exactly. If they vary qualitatively as shown however, then the above argument does give a physical reason for the existence of a solid inner core.

It remains to consider the possibility envisaged in case (ii) of Fig. 7. It follows by similar reasoning to that given above that in this case as the Earth cooled from a molten state, the entire core would be left liquid. Reasons will be given in a later section, however, for believing that case (iii) is far more likely than case (ii). Finally if the Earth is still heating up, then either case (ii) or case (iii) could lead to a liquid core with a solid inner core.

**12. The melting point gradient.** Using the theory of the solid state, Uffen[1] has computed from seismic data the ratio of the melting point at various depths in the mantle to that at a depth of 100 km. The basis of his method is the computation of the characteristic frequency of vibration of the crystal lattice as given in the works of Einstein and Debye using the measured values of the velocities $\alpha$, $\beta$ of the longitudinal and transverse waves. The melting point is then computed using Lindemann's theory of fusion which relates the characteristic frequency to the melting point. Debye[2] treated a solid as a continuous elastic medium and took into account the discontinuous (molecular) structure of the solid by assuming that the allowable frequencies of vibration stop abruptly at an upper limit $\nu_{max}$ such that the total number of independent vibrations equals the $3N$ vibrational degrees of freedom of the $N$ molecules. He obtained the result[3]

$$\nu_{max} = \left(\frac{9\varrho N_0 n}{4\pi M}\right)^{\frac{1}{3}} \left(\frac{1}{\alpha^3} + \frac{2}{\beta^3}\right)^{-\frac{1}{3}} \tag{12.1}$$

where $\varrho$ is the density, $N_0$ is Avogadro's number, $M$ the molecular weight and $n$ the number of atoms per molecule. It has been assumed that the material of the mantle may be treated as a monatomic solid with a "mean atomic weight" equal to $M/n$. For olivine the mean atomic weight is 20 and Birch[4] has shown that most of the rock forming silicates do not deviate much from this value. Einstein[5] derived an expression for a characteristic frequency $\nu$ on the assumption that the compressibility $\chi$ is directly proportional to the mean distance

[1] R. J. Uffen: Trans. Amer. Geophys. Un. **33**, 893 (1952).
[2] P. Debye: Ann. Phys., (4) **39**, 784 (1912).
[3] See this Encyclopedia, Vol. VII/1, p. 327.
[4] See footnote 3, p. 376.
[5] A. Einstein: Ann. Phys., (4) **34**, 170 (1911).

between molecules and inversely proportional to the force per unit displacement acting on the molecules. For molecular crystals he obtained the result

$$\nu = C_1 V_M^{\frac{1}{3}} (\chi M)^{-\frac{1}{2}} \tag{12.2}$$

where $V_M$ is the volume per mole and $C_1$ a constant.

When a crystal is heated to a sufficiently high temperature, the lattice breaks down and the crystal melts. LINDEMANN[1] assumed that this occurs when the amplitude of oscillation becomes so great that "direct collisions" take place between neighbouring molecules. He also assumed that the molecules of the solid are arranged in a simple cubic lattice and that the melting point $\tau$ is large compared with the DEBYE characteristic temperature (defined as $h\nu_{max}/k$, where $h$ is PLANCK's constant and $k$ is BOLTZMAN's constant), so that the energy of an oscillator is given by its classical value $3k\tau$.

UFFEN[2] has verified that this is the case for the mantle. LINDEMANN obtained the result

$$\nu = C_2 M^{-\frac{1}{2}} V_M^{-\frac{1}{3}} \tau^{\frac{1}{2}} \tag{12.3}$$

where $C_2$ is a function of the lattice spacing and the diameter of the "effective sphere of action" of a molecule. LINDEMANN assumed that $C_2$ was constant. Using the known value of $\nu$ for platinum, obtained from specific heat measurements, he found good agreement for a number of substances between the frequencies determined from specific heat measurements and those obtained from melting point data. Similar agreement was found for the DEBYE characteristic frequency which is the justification for identifying the DEBYE $\nu_{max}$ with the

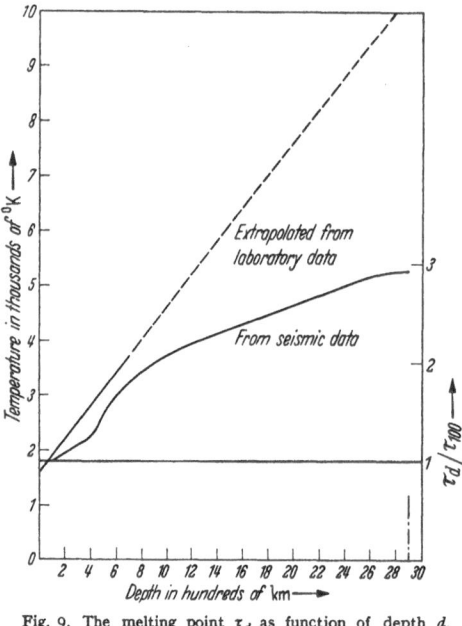

Fig. 9. The melting point $\tau_d$ as function of depth $d$. (After R. J. UFFEN[2].)

Table 7. Values of $\tau_d/\tau_{100}$ for various depths.

| Depth km. | From Eq. (12.4) | From Eq. (12.5) | Mean |
|---|---|---|---|
| 100 | 1.00 | 1.00 | 1.00 |
| 200 | 1.08 | 1.07 | 1.08 |
| 300 | 1.18 | 1.14 | 1.16 |
| 413 | 1.30 | 1.24 | 1.27 |
| 500 | 1.51 | 1.42 | 1.47 |
| 600 | 1.70 | 1.62 | 1.66 |
| 800 | 1.95 | 1.90 | 1.92 |
| 1000 | 2.10 | 2.04 | 2.07 |
| 1400 | 2.34 | 2.22 | 2.28 |
| 1800 | 2.60 | 2.36 | 2.48 |
| 2200 | 2.86 | 2.50 | 2.68 |
| 2600 | 3.08 | 2.65 | 2.86 |
| 2898 | 3.13 | 2.71 | 2.92 |

LINDEMANN $\nu$. Eliminating $\nu$ from (12.2) and (12.3), UFFEN[2] obtained the ratio of the melting point at depth $d$ in the mantle to the melting point at depth 100 km. Thus

$$\frac{\tau_d}{\tau_{100}} = \frac{(\chi \varrho)_{100}}{(\chi \varrho)_d} . \tag{12.4}$$

[1] F. A. LINDEMANN: Phys. Z. 11, 609 (1910).
[2] See footnote 1, p. 383.

Alternatively from (12.1) and (12.3),

$$\frac{\tau_d}{\tau_{100}} = \left[\frac{(\beta^3 + 2\alpha^3)_{100}}{(\beta^3 + 2\alpha^3)_d}\right]^{\frac{1}{3}} \left[\frac{(\alpha\beta)_d}{(\alpha\beta)_{100}}\right]^2. \tag{12.5}$$

Values of $\tau_d/\tau_{100}$ as given by equations (12.4) and (12.5) are given in Table 7, and represented graphically in Fig. 9 which also includes JEFFREY's temperature curve, extrapolated from laboratory data. It is seen that the melting point gradient decreases with depth. If the mantle consists of olivine, taking $\tau_{100} = 1800°$ K (GUTENBERG[1]) the value of $\tau_{2900}$, the melting point of the mantle at the core boundary is $5300°$ K. The most serious assumptions in UFFEN's work are that the molecular weight remains constant throughout the mantle and that $C_2$ in equation (12.3) is the same at all pressures.

**13. The adiabatic gradient.** Several writers [VERHOOGEN[2], UFFEN[3], VALLE[4], JACOBS[5]], have sought to obtain a better estimate of the adiabatic temperature gradient in the mantle which is given [see equation (10.3)] by

$$\frac{d}{dz} \log T = \frac{g\alpha'}{c_p}. \tag{13.1}$$

It follows that the temperature at any depth can be computed if the temperature at any assigned depth is known and the value of the ratio $\alpha'/c_p$ is known throughout the mantle. VERHOOGEN[2] obtained values of the ratio $\alpha'/c_p$ at any depth using seismic data and BRILLOUIN's[6] expression for the internal pressure corresponding to the radiation pressure of the elastic waves. UFFEN[3] has obtained an alternative expression based on the GRÜNEISEN-DEBYE equation of state for solids. His equation for the adiabatic temperature gradient is

$$\frac{dT}{dz} = g\varrho\gamma T\chi_s \tag{13.2}$$

where $\chi_s$ is the *adiabatic* compressibility and $\gamma$ is GRÜNEISEN's constant defined by the equation

$$\gamma = \frac{\varrho}{v} \cdot \frac{dv}{d\varrho} \qquad \text{(GRÜNEISEN equation of state)}[7]$$

or

$$\gamma = \frac{\varrho}{v_{max}} \cdot \frac{dv_{max}}{d\varrho} \qquad \text{(DEBYE equation of state)}.$$

From (13.1) and (13.2) it follows that

$$\frac{\alpha'}{c_p} = \varrho\gamma\chi_s. \tag{13.3}$$

Assuming that $\gamma$ is independent of the temperature and differentiating equation (12.1) with respect to $\varrho$, UFFEN obtained the variation of $\gamma$ with depth. No assumptions about the chemical composition of the Earth's interior are necessary for the evaluation of $\gamma$. Since $\chi_s$ can be obtained direct from the measurement of the velocity of seismic waves, equation (13.3) then gives the variation of $\alpha'/c_p$ with depth. A comparison of the results of VERHOOGEN and UFFEN is given in Table 8.

---

[1] B. GUTENBERG: Internal Constitution of the Earth, p. 162. Dover, New York: 1951.
[2] J. VERHOOGEN: Trans. Amer. Geophys. Un. **32**, 41 (1951).
[3] R. J. UFFEN: Ph. D. Thesis, Univ. of Western Ontario, Canada 1952.
[4] P. E. VALLE: Ann. Geofisica, Roma **5**, 41 (1952).
[5] J. A. JACOBS: Canad. J. Phys. **31**, 370 (1953).
[6] L. BRILLOUIN: Les Tenseurs en Mécanique et en Élasticité. Paris: Masson & Cie. 1938.
[7] See this Encyclopedia, Vol. VII/1, p. 276.

Equation (13.2) can be integrated to give the ratio of the temperature (adiabatic) at any depth to the temperature at any assigned depth. Thus

$$\log\left(\frac{T_d}{T_{200}}\right) = \int_{200}^{d} g\varrho\gamma\chi_s\,dz = \int_{200}^{d} \frac{g\alpha'}{c_p}\,dz$$

which can be evaluated graphically. Both Verhoogen's and Uffen's values of $T_d/T_{200}$ are included in Table 8.

Uffen has extended his work to obtain estimates of both the coefficient of thermal expansion $\alpha'$ and the thermal conductivity $K$ with depth. He first showed that the value of $c_v$, the specific heat at constant volume, is not likely to vary much throughout the mantle, and he obtained a value of $c_v$ of $1.2 \times 10^7$ ergs/gm. deg. The specific heat at constant pressure $c_p$ is related to $c_v$ by the equation,

$$c_p - c_v = \frac{T\alpha'^2}{\varrho\chi_T} \qquad (13.4)$$

where $\chi_T$ is the *isothermal compressibility*.

Since

$$\gamma = \frac{\alpha'}{\varrho c_p \chi_s} = \frac{\alpha'}{\varrho c_v \chi_T} \qquad (13.5)$$

it follows from equation (13.4) that

$$c_p = \frac{c_v}{1 - \left(\frac{\alpha'}{c_p}\right)\gamma T c_v}. \qquad (13.6)$$

Table 8. *Values of $\alpha'/c_p$ and $T_d/T_{200}$ for various depths in the mantle. ($\alpha' =$ volume coefficient of thermal expansion, $c_p =$ specific heat, $T_d$ and $T_{200} =$ temperatures at depths $d$ and 200 km.)*

| Depth | $\alpha'/c_p \times 10^{11}$ gm./erg. | | $T_d/T_{200}$ | |
|---|---|---|---|---|
| km. | Verhoogen | Uffen | Verhoogen | Uffen |
| 200 | 3.61 | 4.47 | 1.00 | 1.00 |
| 600 | 2.05 | 2.12 | 1.11 | 1.11 |
| 1000 | 1.54 | 1.54 | 1.19 | 1.20 |
| 1400 | 1.27 | 1.29 | 1.25 | 1.26 |
| 1800 | 1.08 | 1.12 | 1.31 | 1.32 |
| 2200 | 0.92 | 0.99 | 1.37 | 1.38 |
| 2600 | 0.81 | 0.82 | 1.41 | 1.43 |

Using his values of $\alpha'/c_p$ and $\gamma$, Uffen obtained values of $c_p$. For $T$ he used his melting point temperature, which will be too large above approximately 600 km. due to cooling. He found that $c_p$ was approximately constant throughout the mantle, and thus the decrease with depth in the value of $\alpha'/c_p$ is essentially due to the change in $\alpha'$. Finally using Debye's theory of lattice conductivity and seismic data, Uffen has obtained the ratio of the thermal conductivity at depth to that at 100 km. Taking $K_{100} = 0.007$ cal/cm. sec. deg., values of $K_d$ can be obtained. Values of $c_p$, $\alpha'$ and $K_d/K_{100}$ are given in Table 9. In view of the limitations of the theory and the uncertainty of many of the quantities involved, the numerical values (especially of $K_d/K_{100}$) can only be regarded as approximate. However, in the absence of any other data, the results are of value in indicating the trend of the variation of the quantities concerned.

Table 9. *Values of $c_p$, $\alpha'$ and $K_d/K_{100}$ at various depths in the mantle ($K_d$ and $K_{100} =$ thermal conductivity).*

| Depth km. | $c_p$ erg./gm. degree | $\alpha'$ (degree)$^{-1}$ | $K_d/K_{100}$ |
|---|---|---|---|
| 100 | $1.40 \times 10^7$ | $65 \times 10^{-6}$(?) | 1.0 |
| 600 | 1.33 | 28 | 1.9 |
| 1000 | 1.31 | 20 | 2.8 |
| 1400 | 1.29 | 17 | 3.4 |
| 1800 | 1.28 | 14 | 4.0 |
| 2200 | 1.26 | 12 | 4.6 |
| 2600 | 1.25 | 10 | 5.9 |
| 2900 | 1.24 | 9 | 7.7 |

**14. Temperature-pressure hypothesis.** In view of Bullen's compressibility-pressure hypothesis (Sect. 8) that the reciprocal of the compressibility i.e. the in compressibility $k$ is a linear function of the pressure $p$ both in the core and the mantle below a depth of about 1000 km., JACOBS[1, 2, 3] has proposed that a similar relation may exist between the volume coefficient of thermal expansion $\alpha'$ and the pressure $p$. Thus he assumed that

$$\frac{1}{\alpha'} = C + b\,p \tag{14.1}$$

where $C$ and $b$ are constants. Support for this hypothesis was obtained by two independent methods. Firstly it was found that UFFEN's results for $\alpha'$ (given in Table 9) were in excellent agreement with a relationship of the form given by equation (14.1). The values of the constants $C$ and $b$ were found to be

$$C = 2.4 \times 10^4 \text{ }^\circ\text{K}, \qquad b = 6.2 \times 10^{-8} \text{ }^\circ\text{K cm.}^2/\text{dyne}. \tag{14.2}$$

Corroboration of this result was obtained from an investigation based on MUR-NAGHAN's[4] theory of finite strain. The theory gives the changes in $\alpha'$ due to compression alone and ignores the effect of temperature changes—these will however be small in comparison with those due to pressure. Again excellent agreement was found with the linear hypothesis (14.1), the values of the constants $C$ and $b$ this time being

$$C = 2.78 \times 10^4 \text{ }^\circ\text{K}, \qquad b = 5.45 \times 10^{-8} \text{ }^\circ\text{K cm.}^2/\text{dyne}. \tag{14.3}$$

More important is the fact that the values of the coefficients $C$ and $b$ in equation (14.1) as obtained by these two methods, agree to within 10%. Thus two independent lines of attack not only support the general hypothesis, but also agree amongst themselves as to the numerical value of the coefficients.

The adiabatic temperature gradient $dT/dp$, given by equation (10.1), may now be evaluated, using equation (14.1). Thus

$$\frac{dT}{dp} = \frac{T}{\varrho\,c_p\,(C + b\,p)} \tag{14.4}$$

which integrates to give the (adiabatic) temperature

$$\log T = \int \frac{dp}{\varrho\,c_p(C + b\,p)}. \tag{14.5}$$

Both VERHOOGEN[5] and UFFEN[6] found that the values of $c_p$ were approximately constant (in agreement with BRIDGMAN's observations at relatively low pressures). Taking a constant value for $c_p$, the integral in equation (14.5) can be evaluated numerically, once the value of $T$ is known at any assigned depth. Assuming the relationship (14.1) to be valid in the core also, as BULLEN postulated his compressibility-pressure hypothesis to be, JACOBS evaluated the integral in equation (14.5) throughout the core as well. The increase in temperature throughout the core was found to be only 500°. Values of the temperature gradient are given in Table 10 for three different assumed values of $T_{1000}$. UFFEN

[1] See footnote 5, p. 385.
[2] J. A. JACOBS: Nature, Lond. 170, 838 (1952).
[3] J. A. JACOBS: Nature, Lond. 171, 835 (1953).
[4] F. D. MURNAGHAN: Amer. J. Math. 59, 235 (1937). See also footnote 3 on page 376.
[5] See footnote 2, p. 385.
[6] See footnote 3, p. 385.

gives the melting point at 1000 km. as 3600° K, and this is taken as the upper limit for $T_{1000}$[1].

VALLE[2] has also obtained values of the ratio $\alpha'/c_p$ in the mantle using seismic data and has obtained the ratio of the adiabatic temperature at any depth in the mantle to the temperature at 33 km. He has extended his theory to the core to obtain values of the ratio of the temperature at any depth in the core to its value at the core-mantle boundary. His adiabatic temperature gradient is steeper than that obtained by VERHOOGEN, UFFEN and JACOBS. In particular he obtains over 30% increase at a point halfway across the core—such a rise, however, seems very excessive.

Table 10. *Adiabatic temperature gradient at various depths in the Earth.*

| Depth km. | Adiabatic temperature gradient | | |
|---|---|---|---|
| | $T_{1000} = 2400° K$ | $T_{1000} = 3000° K$ | $T_{1000} = 3600° K$ |
| 1000 | 0.38 | 0.47 (5) | 0.57 |
| 1400 | 0.32 | 0.40 (5) | 0.48 |
| 1800 | 0.27 (5) | 0.35 | 0.41 (5) |
| 2200 | 0.24 (5) | 0.31 | 0.37 |
| 2600 | 0.22 (5) | 0.28 | 0.33 (5) |
| 2900 | 0.21 (5) | 0.26 (5) | 0.31 (5) |
| 3000 | 0.20 | 0.25 | 0.29 (5) |
| 3400 | 0.15 (5) | 0.19 (5) | 0.23 (5) |
| 3800 | 0.12 (5) | 0.15 (5) | 0.18 (5) |
| 4200 | 0.10 | 0.12 (5) | 0.15 |
| 4600 | 0.08 | 0.10 | 0.12 |
| 4982 | 0.07 | 0.08 (5) | 0.10 (5) |
| 5121 | 0.06 (5) | 0.08 | 0.09 (5) |

**15. Temperature distribution in the core.** In a study of the effect of pressure on the melting point, SIMON[3] has put forward the formula

$$\frac{p}{\alpha} = \left(\frac{T}{T_0}\right)^c - 1 \qquad (15.1)$$

where $\alpha$ is a constant related to the internal pressure, $T_0$ is the normal melting temperature and $c$ a numerical constant. This formula was found to be in good agreement with experimental results on low melting point substances which have been raised to very high relative temperatures—helium, for example, was solidified at temperatures more than ten times the critical temperature. The formula has also been given a theoretical foundation which, although not accounting for all details, confirms the general form. Applying the equation to iron, SIMON arrived at the result

$$p_{\text{atm.}} = 150,000 \left[\left(\frac{T}{1805}\right)^4 - 1\right]. \qquad (15.2)$$

BULLARD[4] has also applied equation (15.1) to iron and obtained similar results. Equation (15.2) rests on the assumption that the body-centred cubic $\delta$-modification of iron which is stable in the region of the normal melting point continues to be so at higher temperatures and that no critical point between the solid and fluid phase intervenes. The equation would also cease to be valid if a change in the configuration of the electrons in the outer shell of the iron atom should occur at very high pressures. Table 11 gives the melting temperatures calculated from equation (15.2). The table also gives values calculated with different (extreme) values of $c$ and $\alpha$ in order to indicate possible limits of error.

Assuming the existence of a solid inner core (see Sect. 11) it is now possible to calculate the actual temperature at the boundary of the core and mantle

---

[1] J. VERHOOGEN [Trans. Amer. Geophys. Un. **36**, 866 (1955)] found that the empirical relation $\alpha' T = b (T/\Theta)^{1.5}$ where $b$ is a constant and $\Theta$ the DEBYE temperature (see this Encyclopedia, Vol. VII/1, p. 248 and 328) represents satisfactorily the variation of $\alpha'$ with temperature $T$. He concludes that the temperature at the core boundary is not likely to exceed 2700° K, confirming a result he had obtained earlier [Trans. Amer. Geophys. Un. **35**, 85 (1954)] on petrological grounds on the assumption of convection in the mantle.

[2] See footnote 4, p. 385.

[3] F. E. SIMON: Nature, Lond. **172**, 746 (1953).

[4] E. C. BULLARD: The Solar System. Vol. 2, Chapter 3. Chicago: University Press 1954.

using the above melting point curve of SIMON and the adiabatic temperature curve for the core as obtained by JACOBS[1]. Since the boundary $B$ (see Fig. 8) of the inner core is the point of transition between the liquid and solid state in the core, the melting point at $B$ must be the actual temperature at that level. Hence, by extending the adiabatic curve back from this point to $A$, the boundary between the core and mantle, we can obtain the actual temperature at $A$. The actual temperature at $A$ must be less than the melting point of the silicates that compose the mantle and above the melting point of iron. Taking SIMON's value of the melting temperature at $B$ as 3900° K., the actual temperature at $A$ is 3600° K.

Finally, it is possible to obtain an upper limit for the melting point of the mantle at the core boundary by calculating the adiabatic curve that passes through $C$, the melting point at the centre of the Earth. The melting-point curve of the mantle must be below this curve, or a solid inner core could not have formed. SIMON's value for the melting point at $C$ is 4040° K., and this leads to the upper limit to the melting point of the mantle at 2900 km. as 3680° K. SIMON places an upper limit of 4380° K. for his temperature at $C$, and this in turn would raise the upper limit to the melting point of the mantle at the core boundary to 4000° K.

Table 11. *Variation of the melting point of iron with pressure.*

| Pressure <br> × (10⁶ atm.) | $T°$ K <br> $c = 3.5$ <br> $\alpha = 170,000$ atm. | $T°$ K <br> $c = 4$ <br> $\alpha = 150,000$ atm. | $T°$ K <br> $c = 4.5$ <br> $\alpha = 133,000$ atm. |
|---|---|---|---|
| 0.1 | 2060 | 2050 | 2040 |
| 0.2 | 2250 | 2230 | 2210 |
| 0.5 | 2670 | 2600 | 2550 |
| 1.0 | 3130 | 3000 | 2900 |
| 1.5 | 3460 | 3290 | 3150 |
| 2.0 | 3730 | 3510 | 3340 |
| 2.5 | 3970 | 3700 | 3500 |
| 3.0 | 4170 | 3860 | 3640 |
| 3.5 | 4340 | 4010 | 3760 |
| 4.0 | 4500 | 4140 | 3870 |

The above arguments of JACOBS[1] are based on the assumption of a solid inner core. If UFFEN's[2] results are used for the melting point curve for the mantle, the melting point at the bottom of the mantle is considerably greater than that given by SIMON at the centre of the Earth. Unless there has been considerable cooling of the mantle this would favour case (ii) of Fig. 7 (see Sect. 11) i.e. an all liquid core, solidification beginning at the core boundary. However, such a large difference between the two melting points (about 2000°) seems unlikely.

**16. Radioactivity and the thermal history of the Earth.** In a study of the thermal problems of the Earth most investigators have considered the initial, the present, and, in some cases, the final equilibrium thermal state of the Earth. Few writers have investigated the thermal *history* of the Earth, although URRY[3] has stressed this very point. In a discussion of the thermal properties within the Earth, it has been the custom to assume that a steady equilibrium state has been reached. In practice this is likely to be far from the truth. Temperatures at the core boundary, for example, probably fluctuate, not only in space, but also in time— especially if convection currents are present. The discovery of natural radioactivity at the turn of the century completely upset the cosmological consequences that KELVIN and others had drawn from the computed rate of cooling of the Earth based on the assumption of thermal conduction without internal sources of heat. In examining the significance of radioactivity in geophysical problems, most investigators have assumed (for simplicity) a constant rate of

---

[1] See footnote 5, p. 385.
[2] R. J. UFFEN: Nature, Lond. **173**, 259 (1954).
[3] See footnote 3, p. 380.

heat generation, yet the decrease in heat generation during the Earth's lifetime is approximately 50%. Thus in an exhaustive study of the cooling of the Earth, SLICHTER[1] considered the rate of generation of heat $H$ from radioactive sources as a function only of $r$ the distance from the Earth's centre. In all his Earth models, he has neglected any change in $H$ with time. SLICHTER's paper, however, remains a valuable contribution to the subject. One of the difficulties of the problem is the unknown distribution and concentration of radioactive substances in rocks. Present knowledge is expanding rapidly but even by 1940 most results were unreliable and have been shown to be too high.

The equation of heat conduction for a radioactive Earth is

$$\varrho c \frac{\partial T}{\partial t} = \frac{1}{r^2} \frac{\partial}{\partial r} \left( K r^2 \frac{\partial T}{\partial r} \right) + H(r, t) \tag{16.1}$$

where $H(r, t)$ is the rate of production of heat by radioactivity per unit time and volume, $c$ is the specific heat, and $K$ the thermal conductivity. The solution of (16.1) presents considerable mathematical difficulties and the assumptions made in obtaining a solution will now be discussed. It will be shown later that the practice of using the solution for a semi-infinite plane as an approximation can lead to considerable errors. Only two types of radioactive distribution are of practical importance. The first is one in which the concentration of radioactivity falls off exponentially with depth, and the other is one in which the radioactive substances are distributed in spherical shells in each of which their concentrations are constant. The second type of distribution is preferred since it can be made to fit any given distribution fairly well by taking a sufficient number of shells. It also gives rise to a solution well suited to calculation with different sets of values of the radioactive concentrations in the various shells. Moreover the first type of distribution leads to an extremely slowly convergent series—VAN OSTRAND[2] using LOWAN's[3] solution had to sum over 6000 terms for just one depth.

The Earth is considered radially symmetric as regards all physical properties and taken as solid throughout. The effect of the liquid part of the core will be to equalize the temperature in the liquid, causing the temperature gradient to become approximately the adiabatic and the error produced by treating the entire Earth as solid will be negligible in the greater part of the mantle. The temperature at the surface of the Earth is determined, not by the amount of heat arriving from the interior, but by equilibrium between heat received from the sun and that radiated back into space. The effect of diurnal and annual variations in surface temperature dies out in a few tens of feet, and the effect of climatic changes such as the ice ages is also negligible when considered over time intervals of billions of years. Provided that there has been no great change in the amount of heat radiated by the sun—a fact which is substantiated on astronomical grounds—it is safe to assume that the surface temperature of the Earth has remained practically constant throughout the whole of geologic time and an average value of 0° C has been chosen. It should be noted that even under the most extreme hypotheses as to the extent of vulcanism in the past, the transport of heat to the surface by volcanic processes is an order of magnitude less than the loss by conduction for the whole Earth and will be neglected (see LOTZE[4] and VERHOOGEN[5]).

[1] L. B. SLICHTER: Geol. Soc. Amer. Bull. **52**, 561 (1941).
[2] C. E. VAN OSTRAND: Geophysics **5**, 57 (1940).
[3] A. N. LOWAN: Phys. Rev. **44**, 769 (1933).
[4] F. LOTZE: Nachr. Ges. Wiss. Göttingen, Math.-Phys. Kl. **75** (1927).
[5] J. VERHOOGEN: Amer. J. Sci. **244**, 744 (1946).

Before a solution of (16.1) is given, there remains a discussion of the values of the thermal conductivity $K$ and the thermal diffusivity $k$ defined by the equation

$$k = \frac{K}{\varrho c}. \qquad (16.2)$$

The heat capacity $\varrho c$ with which heat conduction is concerned is essentially $\varrho c_v$. For reasons given in the next section, both $K$ and $k$ are treated as constant. $K$ varies fairly widely for common rocks, and UFFEN [1] using seismic data and the theory of solids finds a considerable increase in $K$ with depth. However URRY [2] using the experimental results of BIRCH and CLARK [3], together with thermo-dynamical theory, has constructed a semi-empirical theory of conductivity of rocks. His results show that the diffusivity is remarkably constant down to a depth of 600 km. at least and that a value of $k = 0.007$ cm.$^2$/sec. should be reasonably accurate for the mantle. The effects of these assumptions will be discussed further in the next section.

**17. The solution of the equation of heat conduction for a radioactive Earth.** The rate of production of heat due to radioactivity $H(r, t)$ may be written

$$H(r, t) = \sum_i h_i(r) e^{-\lambda_i t} \qquad (17.1)$$

where $h_i(r)$ is the initial heat production per unit volume and $\lambda_i$ the decay constant of the $i^{\text{th}}$ radioactive substance. The time $t$ is measured from the time of solidification $t_0$ years ago.  Equation (16.1) may then be written

$$\varrho c \frac{\partial T}{\partial t} = \frac{\partial K}{\partial r} \cdot \frac{\partial T}{\partial r} + K\left(\frac{2}{r} \cdot \frac{\partial T}{\partial r} + \frac{\partial^2 T}{\partial r^2}\right) + \sum_i h_i(r) e^{-\lambda_i t} \qquad (17.2)$$

The boundary conditions for the temperature $T = T(r, t)$ are

$$T(a, t) = 0 \qquad (17.3)$$

where $a$ is the radius of the Earth, and

$$T(r, 0) = f(r) \qquad (17.4)$$

where $f(r)$ is the initial temperature distribution. If $T = T' + T''$, where $T'$ satisfies the equation

$$\varrho c \frac{\partial T}{\partial t} = \frac{\partial K}{\partial r} \cdot \frac{\partial T}{\partial r} + K\left(\frac{2}{r} \frac{\partial T}{\partial r} + \frac{\partial^2 T}{\partial r^2}\right) \qquad (17.5)$$

with boundary condition (17.3) and

$$T'(r, 0) = f(r), \qquad (17.6)$$

then $T''$ must satisfy (17.2) with the boundary condition (17.3) and

$$T''(r, 0) = 0. \qquad (17.7)$$

Thus the solution for a radioactive Earth with a given initial temperature distribution is the sum of the solutions for a non-radioactive Earth cooling from the given initial temperature distribution and for a radioactive Earth heating up from zero temperature. The solution to the first problem is well known and will be given in Sect. 18. Consider then the latter problem and drop the

---

[1] See footnote 3, p. 385.
[2] W. D. URRY and G. COMENETZ: Thermal History of the Earth (unpublished) 1947.
[3] BIRCH and CLARK: Amer. J. Sci. **238**, 529, 613 (1940).

superscript " from $T$. It is easy to shaw that the effect of each radioactive substance may be evaluated separately and that their sum will give the total effect. Thus in discussing the solution, the last sum in (17.2) will be replaced by $h(r)\, e^{-\lambda t}$.

Assuming that the radioactive substances are distributed in shells in each of which their concentrations are constant (see Sect. 16), write

$$h(r) = \begin{cases} H_0, & (0 \leq r \leq r_1), \\ H_1, & (r_1 \leq r \leq r_2), \\ \dots\dots\dots\dots\dots \\ H_n, & (r_n \leq r \leq a) \end{cases} \tag{17.8}$$

and

$$\begin{aligned} Q_0 &= H_0, \\ Q_1 &= H_1 - H_0, \\ &\dots\dots\dots\dots \\ Q_n &= H_n - H_{n-1}. \end{aligned} \tag{17.9}$$

If $T_m(r, t)$ is the solution of

$$\varrho\, c\, \frac{\partial T_m}{\partial t} = \frac{\partial K}{\partial r} \cdot \frac{\partial T_m}{\partial r} + K\left(\frac{2}{r}\,\frac{\partial T_m}{\partial r} + \frac{\partial^2 T_m}{\partial r^2}\right) + \begin{cases} Q_m\, e^{-\lambda t}, & (r_m \leq r \leq a) \\ 0, & (0 \leq r \leq r_m) \end{cases} \tag{17.10}$$

then it is easy to show that the temperature $T_j$ in the $j^{\text{th}}$ shell is given by

$$T_j = \sum_{m=0}^{j} T_m \tag{17.11}$$

i.e. the general solution can be built up from a number of solutions of one type. This cannot be done if $K$ and $\varrho\, c$ are not continuous across the shell boundaries i.e. the general solution for the temperature cannot be found by the superposition of a number of simple solutions unless $K$ is treated as a *constant*. The actual value of $K$ however is not needed, only the value of the diffusivity $k$ which has been shown to be remarkably constant throughout the upper part of the mantle. With this simplification, the general solution can be built up from the solution of the equation

$$\varrho\, c\, \frac{\partial T}{\partial t} = K\left(\frac{\partial^2 T}{\partial r^2} + \frac{2}{r}\,\frac{\partial T}{\partial r}\right) + \begin{cases} Q\, e^{-\lambda t}, & (r_1 \leq r \leq a) \\ 0, & (0 \leq r \leq r_1) \end{cases} \tag{17.12}$$

with boundary conditions

$$T(a, t) = 0, \quad T(r, 0) = 0, \quad T(0, t) < \infty. \tag{17.13}$$

Writing $u = r\,T$, equation (17.12) reduces to

$$\frac{\partial u}{\partial t} = k\, \frac{\partial^2 u}{\partial r^2} + \begin{cases} \dfrac{Q}{\varrho\, c}\, r\, e^{-\lambda t} & (r_1 \leq r \leq a) \\ 0, & (0 \leq r \leq r_1) \end{cases} \tag{17.14}$$

where

$$u(a, t) = 0, \quad u(0, t) = 0, \quad u(r, 0) = 0. \tag{17.15}$$

Consider a solution of the form

$$\bar{u} = R(r)\, e^{-\lambda t}. \tag{17.16}$$

Then $R(r)$ must satisfy

$$\frac{d^2 R}{dr^2} + \eta^2 R + \begin{cases} \dfrac{Q\, r}{K} \\ 0 \end{cases} = 0 \tag{17.17}$$

where $\eta^2 = \lambda/k$ and

$$\left.\begin{array}{c} R_1(0) = 0 = R_2(a), \\ R_1(r_1) = R_2(r_1), \\ \frac{d}{dr} R_1(r_1) = \frac{d}{dr} R_2(r_1). \end{array}\right\} \tag{17.18}$$

The subscripts 1 and 2 refer to the regions $0 \leq r \leq r_1$ and $r_1 \leq r \leq a$ respectively. The last two conditions ensure that the temperature and heat flow will be continuous across the shell boundary. The solutions of equations (17.17) with the boundary conditions (17.18) are

$$\left.\begin{array}{l} R_1 = c \sin \eta r, \\ R_2 = c_1 \sin \eta r + c_2 \cos \eta r - \dfrac{Q r}{\eta^2 K} \end{array}\right\} \tag{17.19}$$

where the constants $c$, $c_1$, and $c_2$ are determined from the equations

$$\left.\begin{array}{l} c_1 \sin \eta a + c_2 \cos \eta a - \dfrac{Q a}{\eta^2 K} = 0, \\[2mm] c_1 \sin \eta r_1 + c_2 \cos \eta r_1 - \dfrac{Q r_1}{\eta^2 K} = c \sin \eta r_1, \\[2mm] \eta \left\{ c_1 \cos \eta r_1 - c_2 \sin \eta r_1 \right\} - \dfrac{Q}{\eta^2 K} = c \eta \cos \eta r_1. \end{array}\right\} \tag{17.20}$$

Solving for the constants $c$, $c_1$ and $c_2$, the corresponding solutions for $\bar{u}$ are found to be

$$\left.\begin{array}{l} \bar{u}_1 = \dfrac{Q \sin \eta r \, e^{-\lambda t}}{\eta^2 K \sin \eta a} \left\{ a - r_1 \cos \eta \, (a - r_1) - \dfrac{1}{\eta} \cdot \sin \eta \, (a - r_1) \right\} \\[3mm] \bar{u}_2 = \dfrac{Q e^{-\lambda t}}{\eta^2 K \sin \eta a} \left\{ a \sin \eta r - r \sin \eta a + \left( r_1 \cos \eta r_1 - \dfrac{1}{\eta} \sin \eta r_1 \right) \sin \eta \, (a - r) \right\}. \end{array}\right\} \tag{17.21}$$

Now write $u = \bar{u} + U$. Then in order to obtain the complete solution of equations (17.14) and (17.15), $U$ must satisfy

$$\frac{\partial U}{\partial t} = k \frac{\partial^2 U}{\partial r^2} \tag{17.22}$$

where $U(a, t) = 0$, $U(0, t) = 0$ and

$$\left.\begin{array}{l} U(r, 0) = \dfrac{-a Q}{K \eta^2 \sin \eta a} \times \\[3mm] \times \left[ \begin{array}{ll} \sin \eta r \left\{ 1 - \dfrac{1}{\eta a} \sin \eta \, (a - r_1) - \dfrac{r_1}{a} \cos \eta \, (a - r_1) \right\}, & 0 \leq r \leq r_1 \\[3mm] \sin \eta r - \dfrac{r}{a} \sin \eta a + \left( \dfrac{r_1}{a} \cos \eta r_1 - \dfrac{1}{\eta a} \sin \eta r_1 \right) \sin \eta \, (a - r), & r_1 \leq r \leq a \end{array} \right]. \end{array}\right\} \tag{17.23}$$

The solution is

$$U = \sum_{m=1}^{\infty} A_m e^{-\frac{\pi^2 m^2 k t}{a^2}} \sin \frac{m \pi r}{a}, \tag{17.24}$$

where

$$\left.\begin{array}{l} A_m = \dfrac{2}{a} \displaystyle\int_0^a U(r, 0) \sin \dfrac{m \pi r}{a} \, dr \\[4mm] \quad = \dfrac{-2 a^3 Q}{K \pi} \cdot \dfrac{(-1)^m + \dfrac{1}{\pi m} \sin \dfrac{m \pi r_1}{a} - \dfrac{r_1}{a} \cos \dfrac{m \pi r_1}{a}}{m \, (a^2 \eta^2 - m^2 \pi^2)}. \end{array}\right\} \tag{17.25}$$

Since $T = \dfrac{u}{r} = \dfrac{\bar{u}}{r} + \dfrac{U}{r}$, the final solution is given by

$$
\frac{T}{Q} = \frac{a e^{-\lambda t}}{K r \eta^2 \sin \eta a} \times
$$

$$
\times \left\{
\begin{array}{ll}
\sin \eta r \left[ 1 - \dfrac{1}{\eta a} \sin \eta (a - r_1) - \dfrac{r_1}{a} \cos \eta (a - r_1) \right], & 0 \leq r \leq r_1 \\[2ex]
\sin \eta r - \dfrac{r}{a} \sin \eta a + \left( \dfrac{r_1}{a} \cos \eta r_1 - \dfrac{1}{\eta a} \sin \eta r_1 \right) \sin \eta (a - r), & r_1 \leq r \leq a
\end{array}
\right\} -
$$

$$
- \frac{2 a^3}{\pi K r} \sum_{m=1}^{\infty} e^{-\frac{\pi^2 m^2 k t}{a^2}} \sin \frac{m \pi r}{a} \left[ \frac{(-1)^m + \dfrac{1}{\pi m} \sin \dfrac{m \pi r_1}{a} - \dfrac{r_1}{a} \cos \dfrac{m \pi r_1}{a}}{m (a^2 \eta^2 - m^2 \pi^2)} \right]. \tag{17.2}
$$

Equation (17.26) may be further simplified and written in the alternative form

$$
\frac{T}{Q} = \frac{2 a^3}{\pi K r} \sum_{m=1}^{\infty} \left( e^{-\lambda t} - e^{-\frac{\pi^2 m^2 k t}{a^2}} \right) \sin \frac{m \pi r}{a} \left[ \frac{(-1)^m + \dfrac{1}{\pi m} \sin \dfrac{m \pi r_1}{a} - \dfrac{r_1}{a} \cos \dfrac{m \pi r_1}{a}}{m (a^2 \eta^2 - m^2 \pi^2)} \right]. \tag{17.2}
$$

The heat flow is given by

$$
F(r, t) = - K \frac{\partial T}{\partial r} \tag{17.27}
$$

and may be obtained direct from (17.26) by differentiation.

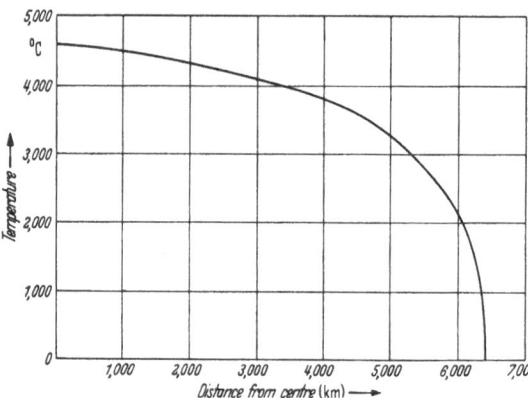

Fig. 10. Initial temperature distribution in the Earth.

**18. The cooling of a non-radioactive Earth.** To complete the thermal picture of the Earth, the solution for the cooling of a non-radioactive Earth is required. The numerical results will depend of course upon the values adopted for the initial temperature distribution, which has been assumed to be everywhere the melting point of the substance concerned — with a slight correction near the surface where the temperature is assumed to be zero. A smooth curve (see Fig. 10) was drawn using the melting point values of DALY[1] for the mantle and SIMON'S[2] values of the melting point of iron at high pressures in the core. The equation of heat conduction (16.1) with $H(r, t)$ zero and the conductivity $K$ constant reduces to

$$
\frac{\partial T}{\partial t} = k \left( \frac{\partial^2 T}{\partial r^2} + \frac{2}{r} \frac{\partial T}{\partial r} \right) \tag{18.1}
$$

with boundary conditions

$$
T(a, t) = 0, \quad T(r, 0) = f(r), \quad T(0, t) < \infty. \tag{18.2}
$$

[1] R. A. DALY: Geol. Soc. Amer. Bull. **54**, 401 (1943).
[2] See footnote 2, p. 388.

The solution, first given by FOURIER, is

$$T(r, t) = \frac{1}{r} \sum_{m=1}^{\infty} A_m e^{-\frac{m^2 k^2 \pi^2 t}{a^2}} \sin \frac{m \pi r}{a} \tag{18.3}$$

where

$$A_m = \frac{2}{a} \int_0^a r f(r) \sin \frac{m \pi r}{a} dr. \tag{18.4}$$

The heat flow, defined by equation (17.27) follows at once from (18.3). The $A_m$ can be regarded as the coefficients in the expansion of the function $r f(r)$ in a FOURIER sine series. Hence these coefficients can be found to any desired degree of accuracy by a harmonic analysis of the function $r f(r)$. For the present calculation a sine series for $r f(r)$ was evaluated to 44 terms.

**19. Numerical computation of the temperature and heat flow for a Number of Earth Models.** The naturally radioactive substances of long period which occur in sufficiently high concentrations in rocks to have a decided thermal effect are $U^{238}$, $U^{235}$, $Th^{232}$ and $K^{40}$. The numerical evaluation of the solution given in the last section requires a knowledge of the decay constant $\lambda$ of each of these four radioactive substances, together with their rates of heat production $H$ and their concentrations in the shells of whatever Earth model is selected. The values used for $\lambda$ and $H$ are given in Table 12. The source of information

Table 12. *Radioactive constants and heat production.*

| Substance | Decay constant $10^{-9}$/year | | | Heat Production cal./gm. yr. | |
|---|---|---|---|---|---|
| | Value and possible error | Value used in calculation | Reference | BIRCH's[1] (1954) value | Value used in calculation |
| $U^{238}$ ..... | $0.154 \pm 0.0003$ | 0.154 | 2, 3 | 0.71 | 0.72 |
| $U^{235}$ ..... | $0.972 \pm 0.021$ | 0.980 | 3, 4, 5 | 4.3 | 4.7 |
| $Th^{232}$ .... | 0.0499 | 0.0499 | 6 | 0.20 | 0.21 |
| $K^{40}$...... | $0.55 \pm 0.03$ | 0.536 | 7 | 0.22 | 0.21 |
| K........ | . . . | . . . | . . . | $27 \times 10^{-6}$ | $26 \times 10^{-6}$ |

for the values of $\lambda$ is indicated. The values of $H$ for each of the four radioactive substances were found by adding up the energies of the individual radiations of their decay systems, allowing for the energy of the neutrinos. Values were taken from the Handbook of Physical Constants (1953 edition) for the uranium and thorium series and agree well with those calculated by BIRCH (1954). The information on $K^{40}$ is the least certain—until quite recently even the decay scheme was in doubt. It has now been established with near certainty that $K^{40}$ decays to $Ca^{40}$ by a beta-decay, and to $A^{40}$ by a K capture followed by a gamma ray emission. A discussion of the values used for $\lambda$ and $H$ for $K^{40}$ has been given by ALLAN[8] who also estimated the effects of uncertainties in the values of $\lambda$.

---

[1] F. BIRCH: Nuclear Geology. Chapter 5. New York, Wiley, 1954.

[2] C. A. KIENBERGER: Phys. Rev. **76**, 1561 (1949).

[3] E. H. FLEMING Jr., A. GHIORSO and B. B. CUNNINGHAM: Phys. Rev. **88**, 642 (1952).

[4] A. O. NIER: Phys. Rev. **55**, 153 (1939).

[5] A. F. KOVARIK and N. I. ADAMS Jr.: J. Appl. Phys. **12**, 296 (1941).

[6] A. F. KOVARIK and N. I. ADAMS Jr.: Phys. Rev. **54**, 413 (1938).

[7] R. D. RUSSELL, H. A. SHILLIBEER, R. M. FARQUHAR and A. K. MOUSUF: Phys. Rev. **91**, 1223 (1953).

[8] D. W. ALLAN: M. A. Thesis, University of Toronto, Canada 1954. See also the chapter by J. T. WILSON et al. in this volume.

Table 13. *Radioactive contents of different rock types.*

| Rocktype | Radioactive concentration in gm./gm. | | | | |
|---|---|---|---|---|---|
| | K | $K^{40}$ | $U^{238}$ | $U^{235}$ | $Th^{232}$ |
| Granite-granodiorite . | 0.026 | $3.1 \times 10^{-6}$ | $2.90 \times 10^{-6}$ | $2.10 \times 10^{-8}$ | $10.0 \times 10^{-6}$ |
| Intermediate ........ | 0.011 | $1.3 \times 10^{-6}$ | $1.40 \times 10^{-6}$ | $1.00 \times 10^{-8}$ | $4.3 \times 10^{-6}$ |
| Basaltic ........... | 0.009 | $1.1 \times 10^{-6}$ | $0.90 \times 10^{-6}$ | $0.65 \times 10^{-8}$ | $3.2 \times 10^{-6}$ |
| Dunite ............ | $1.0 \times 10^{-5}$ | $1.2 \times 10^{-9}$ | $1.30 \times 10^{-8}$ | $0.96 \times 10^{-10}$ | $4.6 \times 10^{-8}$ |
| Pallasite ........... | $0.8 \times 10^{-5}$ | $1.0 \times 10^{-9}$ | $0.64 \times 10^{-8}$ | $0.47 \times 10^{-10}$ | $2.2 \times 10^{-8}$ |
| Iron meteorite ...... | $0.3 \times 10^{-5}$ | $0.4 \times 10^{-9}$ | $0.32 \times 10^{-8}$ | $0.23 \times 10^{-10}$ | $1.1 \times 10^{-8}$ |

As has been pointed out in Sect. 16, knowledge of the radioactive content of the different rocks and the distribution of radioactive substances throughout the Earth is very limited, although the situation has improved considerably during the last few years. Allan[1] has given a critical review of existing knowledge, and the values of the radioactive content of the different rock types used in the numerical computations are given in Table 13. The most striking feature of the distribution of radioactivity within the Earth is the extreme rapidity with which it must decrease with depth. If the amount of radioactivity found in surface rocks was at all representative of the mantle as a whole, the heating that would result would cause the entire Earth to become molten. Since seismic evidence indicates that the Earth is solid to at least a depth of 2900 km., the radioactive content of the rocks that constitute the mantle must be extremely small. This coupled with the evidence obtained from seismic data imposes severe restrictions on the possible composition of the mantle. It most probably consists of dunite or peridotite.

Because of the absence of granitic rocks, with their relatively high radioactive content, in the crust under the oceans, it has been suggested that the heat flow through the deep sea floor should be considerably smaller than that observed through the continents. The few measurements of surface heat flow that have been made at sea do not support this theory, however. The heat flow as observed through the floor of the North Pacific (Revelle and Maxwell[2]) and through the Atlantic (Bullard[3]) indicates that the average heat flow from unit areas of the Earth is the same on the continents and in the oceans. There are two possible explanations for this. Either the rocks in the first 100 km. or so beneath the oceans are much more radioactive than has been assumed or heat is transported from deep in the mantle by convection. If the radioactivity is distributed through a greater range of depth under the oceans than under the continents, the temperatures under the oceans will be higher and thus there will be a tendency for convection currents in the mantle (if they exist) to rise under the oceans and sink under the continents. This is the reverse of what has usually been assumed[4]. The first explanation viz that the rocks beneath the oceans have the same total amount of radioactivity as those beneath the continents but spread through a greater range of depth is adopted as the basis for one of the Earth models. It is a natural explanation if the continents are continuously expanding by a process of differentiation in which radioactive material is concentrated vertically[5].

---

[1] See footnote 8, p. 395.
[2] R. Revelle and A. E. Maxwell: Nature, Lond. **170**, 199 (1952).
[3] E. C. Bullard: Proc. Roy. Soc. Lond., Ser. A **222**, 408 (1954).
[4] For further information on the question of convection in the mantle, see Chapter "Forces in the Earth's Crust".
[5] J. T. Wilson: Trans. Roy. Soc. Canada **43**, 157 (1949).

The most important feature of any Earth model is that it should be capable of portraying as faithfully as possible the physical properties it is desired to investigate. Thus in a discussion of the properties of the mantle it is legitimate to neglect the detailed complex structure of the crust and to represent it by a thin shell of uniform thickness and composition. This does not imply that such a crude representation of the crust is considered actually to exist, but such a representation is perfectly adequate for the investigation in hand. If a model can satisfy the geological and chemical facts as well, it is of course to be preferred, but there is no harm in constructing an abstract model to investigate specific physical phenomena as long as it is realized that there will be no attempt to treat the structure in detail and identify each shell with a specific rock type.

Four models will be considered here. The first is a simplified one which does not attempt to fit the details of crustal structure, but which should nevertheless yield useful information on the thermal properties of the interior. This simplified model consists of three shells—a thin crustal layer 20 km. thick of granite-granodiorite composition, a uniform mantle of dunite, and a dense core of the composition of iron meteorites. Apart from the simplified crustal layer, the model is the same as that proposed by BULLEN[1]. The second model, which was proposed by ADAMS and WILLIAMSON[2] and later slightly revised by ADAMS[3] is considerably more detailed, and was the one URRY[4] considered best in his 1947 study. This model will show whether changes in the crustal layers have much thermal effect at depth. The properties of these first two models which are required in the computations are summarized in Table 14.

Table 14. *Summary of data for two Earth models.*

| Depth (km.) | Shell | Density | Initial heat production $H_0$ (10⁻⁶ cals/cm.³ yr.) | | | | $H_0/\varrho c$ (10⁻⁶ degrees/yr.) | | | |
|---|---|---|---|---|---|---|---|---|---|---|
| | | | K | $U^{238}$ | $U^{235}$ | Th | K | $U^{238}$ | $U^{235}$ | Th |
| Model I: | | | | | | | | | | |
| 0–20 | granite-granodiorite | 2.76 | 15 | 11 | 14 | 7.1 | 24 | 18 | 23 | 12 |
| 20–2900 | dunite | 4.77 | 0.0099 | 0.083 | 0.11 | 0.056 | 0.0071 | 0.059 | 0.079 | 0.040 |
| 2900–6371 | iron (meteorite) | 10.70 | 0.0081 | 0.046 | 0.060 | 0.031 | 0.0074 | 0.042 | 0.055 | 0.028 |
| Model II: | | | | | | | | | | |
| 0–15 | granite-granodiorite | 2.8 | 15 | 11 | 14 | 7.1 | 24 | 18 | 23 | 11 |
| 15–30 | changing from | 2.9 | 6.9 | 5.4 | 7.1 | 3.1 | 8.8 | 6.9 | 9.1 | 4.0 |
| 30–45 | acidic to | 3.1 | 6.1 | 3.7 | 4.8 | 2.3 | 6.1 | 3.7 | 4.8 | 2.3 |
| 45–60 | basic | 3.3 | 3.0 | 2.6 | 3.2 | 1.6 | 3.1 | 2.7 | 3.3 | 1.7 |
| 60–1600 | dunite | 3.9 | 0.0085 | 0.069 | 0.091 | 0.046 | 0.0057 | 0.046 | 0.061 | 0.031 |
| 1600–3000 | pallasite | 7 | 0.013 | 0.059 | 0.076 | 0.039 | 0.0087 | 0.039 | 0.051 | 0.026 |
| 3000–6371 | iron (meteorite) | 10 | 0.0072 | 0.043 | 0.055 | 0.028 | 0.0072 | 0.043 | 0.055 | 0.028 |

Computation of the temperature and heat flows using the solutions given in equations (17.26), (17.27) and (18.3) was carried out on *Ferut*, the electronic computer at the University of Toronto. The actual concentration and distribution of radioactivity does not appear in the series solution (although the decay constant $\lambda$ does), so that once these series have been summed and tabulated, the results may be used for any desired distribution of radioactivity. Tables have thus been drawn up allowing for any desired concentration of the four radioactive substances in nine different shells. These tables enable the temperature and heat flow

---

[1] See footnote 1, p. 368.
[2] L. H. ADAMS and E. D. WILLIAMSON: Smithsonian Report for 1923, 241 (1925).
[3] L. H. ADAMS: Smithsonian Report for 1937, 255 (1938).
[4] See footnote 2, p. 391.

at six depths within the Earth and at six times in the Earth's history to be readily evaluated for any Earth model. The construction of these tables involved summing $2 \times 4 \times 9 \times 6 \times 6 = 2592$ series, many of them to more than 100 terms. Such an undertaking would have been impossible by hand computation. It is not possible to publish the complete set of tables in this chapter, but the results for the first two Earth models are given in Tables 15 and 16. It is not implied that the temperatures and heat flows are known to the accuracy given in these tables. The extra figures have been retained in the computation in order to facilitate comparison of the results at different depths and times in the Earth's history.

Table 15. *Temperatures in the two Earth models.*

| Time (10⁹ yr.) | Depth (km.) | | | | | |
|---|---|---|---|---|---|---|
| | 50 | 100 | 500 | 1000 | 2000 | 2900 |
| Temperatures for Model I (°C): | | | | | | |
| 0 | 665 | 1220 | 2310 | 2856 | 3615 | 3958 |
| 0.25 | 905 | 1216 | 2339 | 2894 | 3656 | 3992 |
| 0.5 | 861 | 1166 | 2361 | 2930 | 3693 | 4024 |
| 1 | 763 | 1056 | 2382 | 2988 | 3753 | 4077 |
| 2 | 606 | 892 | 2358 | 3072 | 3844 | 4162 |
| 3 | 521 | 785 | 2308 | 3124 | 3913 | 4225 |
| 4 | 461 | 714 | 2248 | 3163 | 3970 | 4282 |
| Temperatures for Model II (°C): | | | | | | |
| 0 | 665 | 1220 | 2310 | 2856 | 3615 | 3958 |
| 0.25 | 1447 | 1520 | 2330 | 2885 | 3643 | 3985 |
| 0.5 | 1424 | 1576 | 2346 | 2910 | 3667 | 4010 |
| 1 | 1267 | 1481 | 2369 | 2960 | 3705 | 4052 |
| 2 | 976 | 1226 | 2379 | 3017 | 3763 | 4117 |
| 3 | 825 | 1073 | 2347 | 3057 | 3808 | 4168 |
| 4 | 712 | 934 | 2298 | 3086 | 3841 | 4209 |

Table 16. *Heat-flows in the two Earth models.*

| Time (10⁹ yr.) | Depth (km.) | | | | | |
|---|---|---|---|---|---|---|
| | 50 | 100 | 500 | 1000 | 2000 | 2900 |
| Heat-flows for Model I (cals/cm.² yr.): | | | | | | |
| 0.25 | 15.1 | 14.5 | 5.3 | 5.8 | 4.8 | 3.3 |
| 0.5 | 15.2 | 14.5 | 5.7 | 5.8 | 4.9 | 3.2 |
| 1 | 14.2 | 13.8 | 7.0 | 5.9 | 5.0 | 3.2 |
| 2 | 13.1 | 13.0 | 9.1 | 6.1 | 5.0 | 3.3 |
| 3 | 12.3 | 12.5 | 10.4 | 6.7 | 5.1 | 3.5 |
| 4 | 11.6 | 11.8 | 11.1 | 7.4 | 5.1 | 3.6 |
| Heat-flows for Model II (cals/cm.² yr.): | | | | | | |
| 0.25 | 12.0 | 5.4 | 5.3 | 5.8 | 4.8 | 3.7 |
| 0.5 | 14.8 | 7.1 | 5.6 | 5.8 | 4.9 | 3.7 |
| 1 | 16.3 | 9.5 | 6.6 | 5.9 | 4.9 | 3.8 |
| 2 | 15.8 | 11.1 | 8.2 | 6.1 | 5.0 | 3.9 |
| 3 | 14.7 | 11.2 | 9.3 | 6.6 | 5.0 | 4.0 |
| 4 | 13.7 | 11.0 | 10.9 | 7.2 | 5.1 | 4.2 |

A number of interesting conclusions can be drawn from Tables 15 and 16. Thus, in spite of the extremely low concentrations of radioactivity in the deep mantle and core, there is still a heating up there of two or three hundred degrees. Near-surface conditions appear to have been greatly different in the far past from those existing at present, and, indeed, for the second model, in which the region of high radioactivity penetrates to 60 km. depth, the results indicate that there may even have been remelting of the material at depths of from

50 to 100 km. during the first billion or so years. If so, this could explain the now fairly well established fact that the Earth as a whole is about $4\frac{1}{2}$ billion years old, whereas no rocks have been found much older than about 3 billion years. However, this brief temperature rise at the very beginning soon ceases and cooling commences. The rate of cooling, for both models, was greater in the past than it is now, and this suggests that orogenetic activity may have decreased with time and perhaps have been caused by different processes in the far past. The present temperatures at depths of 50 to 100 km. differ by as much as 200° for the two models, and this suggests that similar differences may hold beneath the oceans and continents. Interpretation of the heat flow results is not so obvious, but the present values near the core boundary are of the order of magnitude required by BULLARD's dynamo theory of the origin of the Earth's magnetic field, although a little smaller in actual numerical value (see Sect. 21). The present surface heat flow for the two models can readily be calculated. The value obtained for Model I agrees well with observed values, while that for Model II is somewhat higher, which suggests that too much radioactivity has been included in the surface layers of this model.

This work has now been greatly extended and a large number of different Earth models have been investigated covering a wide range of possible Earth structures. It has also been shown how different initial conditions, and hence different theories of the origin of the Earth, affect its future thermal history.

Thus for the third model the amount of radioactivity in the surface layers is reduced, a structure of two fifteen km. layers being taken for the crust, instead of the four of Model II. Thus the layer of dunite in Model III extends from 30 to 1600 km. instead of from 60 to 1600 km. Finally, for the sake of comparison, Model IV was constructed which allows for a uniform distribution of radioactivity. Although showing the same continual increase in temperature with time at depth, the rapid rise followed by a decrease near the surface is absent in this model. This reflects the removal of the high concentration of radioactivity in the crustal layers which is a feature of the other models[1].

All these calculations have been repeated with different initial temperature distributions. The initial temperature depends on the origin of the Earth. If the Earth had a hot origin, its thermal history is traced from the time of solidification, the initial temperature being the melting point temperature. In addition to the initial temperature curve already considered (Fig. 10), the melting point curves of UFFEN[2] and BULLARD[3], which give higher values, have been used. The temperature at the centre of the Earth is about 5500° C in UFFEN's curve, and about 8000° C in BULLARD's as compared to about 4500° C in the distribution of Fig. 10. If the Earth was formed from the cold accretion of dust particles, then, quite apart from radioactivity, it would heat up due to compression, although it may not become completely molten. There would not be much variation of temperature with depth, and a constant value of 2000° C was chosen as the initial temperature distribution for an Earth with a cold origin. The cooling of an Earth from any one of these initial temperature distributions shows very similar features and demonstrates how slowly temperatures change at great depths. At these depths thermal conditions (appart from the effects of radioactivity) are not greatly different from their initial values. Near the surface, on the other hand, cooling is considerable, and the surface heat flow for

---

[1] For further details see J. A. JACOBS, Proc. Assoc. Seism., I.U.G.G. Xth Assembly 1954.
[2] Reference UFFEN, page 383.
[3] Reference BULLARD, page 388.

$4 \times 10^9$ years is but little affected by the values taken for the initial temperature distribution in the first few tens of km.

To obtain the actual temperature in any Earth model, the heating up due to radioactivity must be added to the cooling from the initial temperature i.e. the cooling from any one of the initial states considered must be combined with any one of Models I, II, III or IV. It is found that the thermal history for an Earth with a cold origin is very different. Although the temperature increases with time at depth (as was the case for an Earth with a hot origin), there is no initial rise near the surface. Two points stand out from this investigation: firstly all models of the Earth indicate that the temperature at depth has continued to increase throughout geologic time, the temperature at the core-mantle boundary having increased most probably by about 300° C. Secondly, although radio-activity plays a major role in the thermal history of the Earth, near-surface conditions during the first billion years or so are dominated by the initial temperature distribution i.e. by the conditions under which the Earth came into being. The cosmological issues which this involves cannot unfortunately be avoided.

# C. Geomagnetism and the Earth's Interior.

**20. Theories of terrestrial magnetism.** A magnetic field may be produced by *permanent magnetization*, by the *motion of charges*, by *rotation*, or by *electric currents*. The Earth's magnetic field will be considered in the light of these four possible causes[1].

A temperature of the order of the Curie point for iron, 750° C is reached at a depth of about 25 km., so that at greater depths all ferromagnetic substances will have lost their magnetic properties unless the Curie point is higher under the higher pressures. Experiments (at relatively low pressures) do not indicate any significant change of the Curie point with pressure. Thus in order to account for the Earth's magnetic moment, a degree of magnetization in the Earth's crust of from 2 to 8 gauss would be required—which is impossible from experimental evidence. Finally an explanation of the Earth's magnetic field based on permanent magnetization fails to account for the magnetic poles being near the geographical poles and for the secular variation.

The second possible cause is easily seen to be inadequate. If the Earth contained a distribution of negative charge in its outer parts and an equal positive charge deep in its interior, a magnetic field would be produced by the motion of the charges as the Earth rotates about its axis. But to explain the observed magnetic field, the charges must be so great that they would set up a vertical electrical potential gradient $10^8$ times as large as that found by atmospheric observations which is of the order of 1 volt/cm. Such potential gradients are far beyond what can be sustained by any known substance.

If all the magnetic elements in a body are alike, rotating it at an angular velocity of $\omega$ r.p.s. will produce the same intensity of magnetization as placing it in a field of strength $2\pi\varrho\omega$ gauss, where $\varrho$ is the gyromagnetic ratio of a magnetic element i.e. the ratio of the angular momentum of the element to its magnetic moment. For the Earth this gives a field in the right direction but too small by an order of $10^{10}$. As a result of Babcock's[2] discovery of large magnetic fields in rotating stars, Blackett[3] proposed a general law for the magnetic fields of massive rotating bodies. Blackett's theory suggested that the cause

---

[1] See also the chapter "Magnetism of the Earth's body" by Runcorn in this Volume, and D. R. Inglis, Rev. Mod. Phys. **27**, 212 (1955).

[2] H. W. Babcock: Astrophys. J. **105**, 105 (1947); **108**, 191 (1948).

[3] P. M. S. Blackett: Nature, Lond. **159**, 658 (1947). — Phil. Mag. **40**, 125 (1949).

of the field is distributed through the whole bulk of the Earth so that on descending a mine, part of the cause will be above. This can be shown to lead to a decrease of the horizontal field with depth. RUNCORN[1] has carried out measurements in coal mines to determine the radial variation of the geomagnetic field— the experimental values obtained were significantly different from those predicted by such a theory. BLACKETT[2], in a final extensive memoir, described this experiment relating to magnetism and the Earth's rotation as "negative".

There remains *electric currents* as a possible cause for the Earth's magnetic field. These could arise in a number of ways. The Earth might act as an electromagnet, the field being due to internal currents which must have a magnitude of 1000 million amperes. Since the currents must decay owing to electrical resistance, the original electrical system must have been more intense. Calculations show that prohibitively large currents would have been needed in the past[3]. A more promising suggestion, due to ELSASSER[4], is that the Earth's magnetic field is due to *thermoelectric currents* produced by the migration of electrons in the metallic interior of the Earth. These currents owe their existence to inhomogeneities continually created by turbulent convective motions. In order to obtain a non-vanishing resultant angular momentum of the currents around the Earth's axis, the current system must exhibit a particular asymmetry, which is produced by the Coriolis force upon the convective motions. RUNCORN[5] has also suggested that thermo-electric currents may account for the Earth's main field and its secular variation.

The most promising explanation of the Earth's magnetic field is due originally to a suggestion by LARMOR[6] that the magnetic field of the sun might be maintained in a way analogous to that of a self-exciting dynamo. ELSASSER[7,8,9] and BULLARD[10–15] have followed up this suggestion and an account of their work is given in the next section. It is because the causes of the Earth's magnetic field are to be found in the interior of the Earth that geomagnetism is discussed in this Chapter.

**21. The dynamo theory of the Earth's magnetic field.** The major features of the Earth's magnetic field can be approximately represented by a dipole located at the centre of the Earth (see RUNCORN's chapter, this Volume, pp. 498). Sufficient data is also available to show that the secular variation, like the main field has its origin within the Earth. BULLARD[10] has shown that the secular variation can be explained by assuming that conducting material in the core moves across the Earth's magnetic field and thus produces electric currents. These electric

[1] S. K. RUNCORN, A. C. BENSON, A. F. MOORE and D. H. GRIFFITHS: Phil. Mag. **41**, 783 (1950).
[2] P. M. S. BLACKETT: Phil. Trans. Roy. Soc. Lond., Ser. A **245**, 309—370 (1952).
[3] H. LAMB: Phil. Trans. Roy. Soc. Lond., Ser. A **174**, 519 (1883); **180**, 513 (1889).
[4] W. M. ELSASSER: Phys. Rev. **55**, 489 (1939).
[5] S. K. RUNCORN: Trans. Amer. Geophys. Un. **35**, 49 (1954).
[6] Sir J. LARMOR: Brit. Assoc. Report, Bournemouth, 159, 1919.
[7] W. ELSASSER: Phys. Rev. **69**, 106 (1946).
[8] W. ELSASSER: Phys. Rev. **70**, 202 (1946).
[9] W. ELSASSER: Phys. Rev. **72**, 821 (1947).
[10] E. C. BULLARD: Mon. Not. Roy. Astronom. Soc., Geophys. Suppl. **5**, 248 (1948).
[11] E. C. BULLARD: Proc. Roy. Soc. Lond., Ser. A **197**, 433 (1949).
[12] E. C. BULLARD: Proc. Roy. Soc. Lond., Ser. A **199**, 413 (1949).
[13] E. C. BULLARD: Mon. Not. Roy. Astronom. Soc., Geophys. Suppl. **6**, 36 (1950).
[14] E. C. BULLARD, C. FREEDMAN, H. GELLMAN and J. NIXON: Phil. Trans. Roy. Soc. Lond. (A) **243**, 67 (1950).
[15] E. C. BULLARD and H. GELLMAN: Phil. Trans. Roy. Soc. Lond., Ser. A **247**, 213—278 (1954).

currents in turn produce a further magnetic field and the changes in this field constitute the secular change.

Two possible causes of motion in the Earth's liquid core have been put forward. The first is ascribed to the variation of the Earth's angular velocity of rotation as observed astronomically and attributed to tidal friction. The second is based on a sufficient temperature gradient in the core to produce thermal convection. These two mechanisms have been studied in detail by BULLARD[1,2] who found that only thermal convection can produce the required motion, the first effect being extremely small. The conditions necessary for radioactive heating in the core to produce convection have been considered by BULLARD[2] and lead to certain difficulties. These difficulties have been reduced by a more careful consideration of the quantities involved by JACOBS[3,4].

The dynamo theory of the Earth's magnetic field suggests that the field is ultimately produced and maintained by a mechanism of induction, the magnetic energy being drawn from the kinetic energy of the fluid motions in the core. This could be achieved if a "feed-back" mechanism could be devised whereby a part of the magnetic energy of the toroidal field is returned to the original dipo e field in such a way that it reinforces the latter. A solution of the complete problem would be extremely difficult, if not impossible. ELSASSER and BULLARD separlate the elctromagnetic and hydrodynamic problems and attempt to solve the former one alone. They start with an assumed system of motions in the Earth's core together with an assumed magnetic field and calculate the electromagnetic interaction occuring within this system of motions and fields. The question that has to be answered is "do there exist motions of a simply connected, symmetrical fluid body which is homogeneous and isotropic, which will cause it to act as a self-exciting dynamo and to produce a magnetic field in the absence of any sustaining field from an external source?" BULLARD calls such dynamos *homogeneous* to distinguish them from the dynamos of the electrical engineer. In the engineering dynamos, the coil has the symmetry of a clock face in which the two directions of rotation are not equivalent. It is this feature that causes the current to traverse the coil in such a direction that it produces a field which reinforces the initial field. A simple body such as a sphere does not have this property—asymmetry can only be in the motion. This fundamental difference is the crux of the problem of whether asymmetry of motion can suffice for a dynamo, or whether asymmetry of structure is necessary as well. COWLING[5] has shown that a motion symmetrical about an axis cannot maintain a steady field in a body which is also symmetrical about the axis.

In looking for solutions of the required type, the equations to be satisfied are, in the first place, MAXWELL's equations viz

$$\operatorname{curl} \boldsymbol{H} = 4\pi \boldsymbol{I} = 4\pi \varkappa (\boldsymbol{E} + \boldsymbol{v} \times \boldsymbol{H}) , \tag{21.1}$$

$$\operatorname{curl} \boldsymbol{E} = - \frac{\partial \boldsymbol{H}}{\partial t} , \tag{21.2}$$

$$\operatorname{div} \boldsymbol{H} = 0 \tag{21.3}$$

where $\boldsymbol{H}$, $\boldsymbol{E}$ are the magnetic and electric fields, $\boldsymbol{I}$ the current, and $\varkappa$ the electrical conductivity, all measured in electromagnetic units. $\boldsymbol{v}$ is the velocity of the fluid in cm./sec. The displacement current, the Hall current, and the magnetic field due to the motion of charge have been omitted. $\varkappa$ is assumed constant through-

[1] See footnote 11, p. 401.
[2] See footnote 13, p. 401.
[3] See footnote 5, p. 385.
[4] J. A. JACOBS: Trans. Amer. Geophys. Un. **35**, 161 (1954).
[5] T. G. COWLING: Mon. Not. Roy. Astronom. Soc. **94**, 39 (1934).

out the core and the permeability taken as unity. The fields $E$ and $H$ must have no singularities either inside or outside the body. At the boundary of the core, all components of $H$ and the tangential components of $E$ must be continuous, and the normal component of $I$ must vanish. The normal component of $E$ need not be continuous. Outside the core, $H$ must be derivable from a potential and $E$ contain a part derivable from a potential and a part connected with $\partial H/\partial t$ by equation (21.2). In a dynamo problem all fields must vanish at infinity at least like $1/r^3$.

Equations (21.1), (21.2), (21.3) with the above boundary conditions determine the fields once $v$ is specified. Since they are linear and homogeneous, any solution may be multiplied by a constant factor. This indeterminateness may be removed by including the *hydrodynamic equations* with specified forces, and using them to determine the velocities. The inclusion of the hydrodynamic equations increases the difficulty of the problem enormously, since they contain a force $I \times H$ which is quadratic in the fields. BULLARD[1] therefore leaves $v$ as arbitrary, subject only to the equation of continuity

$$\operatorname{div} \boldsymbol{v} = 0, \tag{21.4}$$

i.e. the medium is treated as incompressible.

Eliminating $E$ from equations (21.1) and (21.2) gives

$$\nabla^2 \boldsymbol{H} = 4\pi\varkappa \left[ \frac{\partial \boldsymbol{H}}{\partial t} - \operatorname{curl}(\boldsymbol{v} \times \boldsymbol{H}) \right]. \tag{21.5}$$

Thus analytically the problem of the existence of homogeneous dynamos is to determine whether there are any vector fields $v$ satisfying (21.4), for which (21.3) and (21.5) have solutions without singularities which satisfy the boundary conditions and which do not decrease to zero with the time. BATCHELOR[2] has pointed out that equation (21.5) which connects the velocity and the field is identical with the relation connecting velocity and vorticity, except that the latter has the kinematic viscosity $\nu$ in place of $\dfrac{1}{4\pi\varkappa}$. It is known both experimentally and in some simple cases theoretically that turbulent motions are possible in which the vorticity does not decrease indefinitely, so that it seems highly probable that solutions of the required type do exist of equations (21.3) and (21.5) for some types of motion.

It is not practicable to try and solve equations (21.3) and (21.5) by a direct arithmetical method. This is partly due to the complexity of (21.5) which is a vector equation being equivalent to three simultaneous scalar partial differential equations in four independent variables, and partly to the difficulty of satisfying the boundary conditions. The only practicable method is to expand the solutions in a series of spherical harmonics. An electric or magnetic field or a current system satisfying MAXWELL's equations can be derived from the scalar functions.

$$\left.\begin{aligned} \psi_n^{m\,c} &= R(r)\, P_n^m(\cos\vartheta)\cos m\,\varphi\,, \\ \psi_n^{m\,s} &= R(r)\, P_n^m(\cos\vartheta)\sin m\,\varphi \end{aligned}\right\} \tag{21.6}$$

where $\vartheta$ is the co-latitude, $\varphi$ the longitude, $R(r)$ a function of $r$, and $P_n^m(\cos\vartheta)$ the semi-normalized associated LEGENDRE functions. The fields derived from these functions are of two kinds called *toroidal* and *poloidal*. The toroidal field $T = \operatorname{curl} \boldsymbol{r}\,\psi = \operatorname{grad} \psi \times \boldsymbol{r}$, and the poloidal field $S = \operatorname{curl} T$. This symbolism was used by ELSASSER and retained by BULLARD. His paper[3] contains diagrams

---

[1] See footnote 15, p. 401.

[2] G. K. BATCHELOR: Proc. Roy. Soc. Lond., Ser. A **201**, 405 (1950).

[3] Footnote 2, p. 401.

of the general form of the fields for $n = 1,2$. The $S$ fields are more difficult to visualize than the $T$'s since they have a radial component. The same scheme can be used to classify motions within a sphere. A system of convection currents in a stationary sphere might involve motions of any of the $S$ types—in a rotating sphere these would be accompanied by motion of $T$ type. BULLARD and ELSASSER have drawn up tables showing what interactions are possible between these motions and a given field. The field outside the Earth being predominantly $S_1$ it is natural to start with it and consider the network of possible interactions that stems from it. It is found that the combinations of motions $T_1 S_1$ and $T_1 S_2$ do not produce a closed chain returning to $S_1$ and therefore cannot maintain a field. The combinations $T_1 S_2^{2c}$, $T_1 S_2^c$ and $T_1 S_1^c$ do produce closed chains—as do the corresponding motions with $s$ in place of $c$ in the upper index. The combination $T_1 S_2^{2c}$ was the one selected for detailed investigation. ELSASSER used a series of spherical BESSEL functions to represent the radial variation of the velocities and fields. His analysis was extremely complicated and his results not very successful. BULLARD and GELLMAN[1] met with more success by assuming a perfectly general form for the radial functions. The differential equations and their boundary conditions were replaced by the finite difference approximation to them, resulting in a set of linear homogeneous algebraic equations. These were solved by evaluating the relevant matrices using an electronic computer (*Ferut* at the University of Toronto, and *ACE* at the National Physical Laboratory, England). The amount of work involved was however still considerable, but they did succeed in showing that the $T_1 S_2^{2c}$ system could act as a dynamo and hence that there was no general theorem prohibiting them. It is natural to enquire what other systems can act as dynamos, and whether it is possible to decide if a given system is a dynamo without so much labour. TAKEUCHI and SHIMAZU[2] also conceived the idea of leaving the form of the radial velocity arbitrary, but their analysis was not so complete and did not yield any conclusive results. However in a later paper[3], they conclude that the self-exciting dynamo is possible by a variety of fluid motions.

**22. Further considerations of the Earth's Magnetic Field.** The only tenable explanation for the secular variation lies in the field produced by electromagnetic induction in material moving near the surface of the core (ELSASSER[4], BULLARD[5]). The rapidity of the change in field and the restricted size of the centres of rapid change require that the motions should not be more than a few hundred kilometres below the surface of the core—if they were deeper, their effects would be much reduced by screening of overlying conducting material and would be more widespread. It is natural to suppose that the non-dipole field is largely the integrated result of the secular variation and that its cause lies in the shallow depths in the core. BULLARD gives an explanation of the westward drift of the Earth's magnetic field based on the differential rotation of the core. If there is convection in the core it will be radial, the motion being outwards in some areas and inwards in others. In a rotating sphere such a motion contradicts the conservation of angular momentum and must be combined with a radial variation of angular velocity—the material near the outside of the core rotating with a smaller angular velocity than that inside. Thus the minor features of

---

[1] See footnote 15, p. 401.
[2] H. TAKEUCHI and Y, SHIMAZU: J. Phys. Earth **1**, 1, 57 (1952).
[3] H. TAKEUCHI and Y. SHIMAZU: J. Geophys. Res. **58**, 497 (1953).
[4] See footnote 8, p. 401.
[5] See footnote 10, p. 401.

the field and the secular variation may be expected to drift westward relative to the inner part of the core. BULLARD shows that owing to the relatively low conductivity of the silicates of the mantle, the electromagnetic forces between the core and mantle will be much smaller than those between different parts of the core, but far greater than the viscous forces. The electromagnetic forces thus provide a coupling causing the mantle to follow not the outer part of the core, but some weighted average of the whole core. Such a coupling allows the outer part of the core (and with it the nondipole field and the secular variation) to drift westward relative to the mantle.

The interactions between fluid motion and magnetic fields can be formulated in terms of the differential equations of classical physics (see for example LUND-QUIST[1]). The complete solution of these magneto-hydrodynamic equations in the core is extremely difficult and has not so far been attempted, although CHANDRASEKHAR has made valuable contributions. BULLARD separated the electromagnetic and hydrodynamic problems and attempted to solve the former assuming an arbitrary velocity distribution. CHANDRASEKHAR, on the other hand, has approached the whole question by considering the hydrodynamical problems first. In 1952[2] he discussed the magnetic inhibition of convection in a horizontal layer of fluid heated below and subject to rotation and hoped to extend his work to the case of a sphere with a view to applying his results to the Earth's core. As a preliminary investigation he considered the thermal instability of a non-rotating fluid sphere heated within in the absence of a magnetic field[3] and the onset of convection by thermal instability in spherical shells[4]. He has not yet published, however, any results for the broader and far more complex problem of convection in a fluid sphere in the presence of a magnetic field.

ELSASSER[5] has also considered a dynamo model of the Earth's magnetic field based on the statistical theory of turbulence. In a turbulent fluid the energy will gradually be distributed into smaller and smaller eddies before it is finally dissipated by viscosity. As a result of this process, stray magnetic fields in the fluid may be amplified and as a consequence of this amplificatory and de-amplificatory process, an average statistical distribution of the energy results in which a certain fraction of the total energy has become transformed into electromagnetic field energy. Thus the presence of a magnetic field is attributed to the natural statistical tendency of the energy to spread as widely and as evenly as possible. A dynamical theory of these phenomena has not yet been established, although CHANDRASEKHAR[6] has carried out a detailed kinematical analysis which is essentially an extension to the magneto-hydrodynamical case of the well-known VON KARMAN-HOWARTH theory of turbulent correlations. In this turbulent generator model, the magnetic fields in the core are closely correlated to the individual eddies, and, in some form or other, follow their patterns. This would account for the irregular magnetic field observed, but does nothing to explain the origin of the dipole field.

**23. The electrical conductivity.** The only evidence as to the variation of the electrical conductivity within the Earth is that provided by the short period fluctuations of the geomagnetic field. In this connection most work[7,8] has been

[1] S. LUNDQUIST: Ark. Fysik **5**, No. 15, 297 (1952).
[2] S. CHANDRASEKHAR: Phil. Mag. **7**, 43, 501 (1952), Proc. Roy. Soc. Lond. (A) **225**, 173 (1954).
[3] S. CHANDRASEKHER: Phil. Mag. **7**, 43, 1317 (1952).
[4] S. CHANDRASEKHAR: Phil. Mag. **7**, 44, 233 (1953); 44, 1129 (1953).
[5] W. M. ELSASSER: Trans. Amer. Geophys. Un. **35**, 73 (1954).
[6] S. CHANDRASEKHAR: Proc. Roy. Soc. Lond., Ser. A **204**, 435 (1951).
[7] S. CHAPMAN and A. T. PRICE: Phil. Trans. Roy. Soc. Lond., Ser. A **229**, 427 (1930).
[8] B. N. LAHIRI and A. T. PRICE: Phil. Trans. Roy. Soc. Lond., Ser. A **237**, 509 (1938).

done on the daily variations and the storm-time variations. Both these are primarily of external origin, being due to varying electric currents in the ionosphere. However, these ionospheric currents induce other currents within the Earth, and a comparison of the magnetic fields of the inducing and induced currents can yield information about the electrical conductivity of the region in which the induced currents flow. To account for the observations it is necessary to assume that the conductivity increases rapidly with depth, from about $10^{-7}$ ohm$^{-1}$ cm$^{-1}$ or less for dry surface rocks to a value of at least $10^{-2}$ ohm$^{-1}$ cm$^{-1}$ at a depth of 900 km. No direct information on the conductivity at greater depths can be obtained from variations of the magnetic field, since the corresponding induced currents do not penetrate appreciably farther than this. This increase in conductivity could be due to a change in composition or to the increase of temperature with depth. From the earlier discussion of the constitution of the mantle, the second cause is far more likely. The material of the mantle is probably some form of olivine and almost certainly a semi-conductor, in which the conducticity $\sigma$ and temperature $T$ satisfy a relationship of the form

$$\sigma = \sigma_0 \exp\left(\frac{-A}{k\,T}\right) \tag{23.1}$$

where $k$ is BOLTZMANN'S constant. The coefficients $\sigma_0$ and $A$ may depend on pressure and temperature, but the important changes in $\sigma$ will be due to the exponential factor.

HUGHES[1] has made laboratory measurements of the conductivity of various olivines at temperatures up to 1500° C. These experiments extend the earlier work of COSTER[2] and indicate a very large increase of $\sigma$ with temperature. It also appears that most of the materials examined behave like ionic semi-conductors at sufficiently high temperatures. HUGHES has also obtained values of the coefficients $\sigma_0$ and $A$ for the Earth's mantle by comparing the distribution of conductivity, as derived from geomagnetic data, with recent estimates of the distribution of temperature within the Earth. His values, $\sigma_0 = 2 \times 10^6$ ohm$^{-1}$ cm$^{-1}$, $A = 3.5$ eV, are of the same order of magnitude as those derived from his laboratory studies, and he thus uses equation (23.1) to extrapolate for $\sigma$ throughout the mantle. He obtains a value of 30 ohm$^{-1}$ cm$^{-1}$ for $\sigma$ at the boundary of the core and mantle, which is somewhat higher than earlier estimates. This would imply a tighter electromagnetic coupling between the mantle and the core.

Little is known with certainty about the electrical conductivity of the core. Assuming the core to have a composition similar to that of iron meteorites, BULLARD[3] obtained a value of 1000 ohm$^{-1}$ cm$^{-1}$ for $\sigma$. He obtained this value from empirically determined pressure and temperature laws. He increased $\sigma$ by a factor of two to allow for the effect of pressure, although the latest work of BRIDGMAN indicates that the conductivity is more likely to decrease. ELSASSER[4], by a different method, obtained a value of 5000 ohm$^{-1}$ cm$^{-1}$ for $\sigma$, and BULLARD[5] in a later paper suggested that a mean of his previous estimate and ELSASSER'S viz 3000 ohm$^{-1}$ cm$^{-1}$ might be nearer the truth. In the light of the most recent work, BULLARD'S earlier estimate of 1000 ohm$^{-1}$ cm.$^{-1}$ is to be preferred, although as in so many branches of geophysics, it is difficult to be more definite about a subject on which there is so little direct evidence.

---

[1] H. HUGHES: Ph. D. thesis, Univ. Cambridge 1954.
[2] H. P. COSTER: Mon. Not. Roy. Astronom. Soc., Geophys. Suppl. **5**, 193 (1948).
[3] See footnote 10, p. 401.
[4] See footnote 8, p. 401.
[5] See footnote 13, p. 401.

# Électricité tellurique.

Par

L. CAGNIARD.

Avec 27 figures.

## A. Introduction.

**1. Les précurseurs.** AMPERE, DAVY et beaucoup de physiciens de la première moitié du dix-neuvième siècle pensaient généralement que le champ magnétique terrestre était dû à des courants, des courants continus, à peu près constants, circulant d'Est en Ouest dans l'intérieur du Globe. Cette vieille hypothèse d'AMPERE n'a été jusqu'à présent ni confirmée ni démentie par l'expérience. Il est fort possible que, dans les régions très profondes de la Terre, hors d'atteinte d'une expérimentation directe, il circule des courants de ce genre. Mais *en surface*, ceux qui se manifestent et qu'on appelle *courants telluriques* n'ont pas du tout le caractère de courants continus, du moins si l'on fait abstraction de certains courants d'origine très locale qu'on se refuse à considérer comme étant des courants telluriques à proprement parler. Le courant tellurique est essentiellement un courant variable dont l'intensité, la direction, le sens même varient d'une façon très capricieuse en fonction du temps. Il est possible qu'il présente une composante continue, mais en tout cas très petite, dont la réalité même est considérée aujourd'hui comme douteuse.

FARADAY, après qu'il eût découvert l'induction électromagnétique, ne pouvait douter qu'il existât des courants dans le sol conducteur, courants qui devaient être induits par les variations du champ magnétique terrestre. Cependant les galvanomètres encore bien rudimentaires de l'époque (1831) ne permirent pas à l'illustre physicien de démontrer expérimentalement l'existence de ces courants. Dès l'année précédente pourtant, l'Anglais R. W. Fox[1] avait pu faire dévier un galvanomètre en le reliant simplement à deux prises de terre, mais il s'agissait d'un phénomène électrochimique sans aucun rapport avec les courants telluriques véritables.

Plus tard, dès que fut inventée la télégraphie électrique, les ingénieurs s'aperçurent de la possibilité de substituer, à l'emploi du fil de retour, une mise à la terre des extrémités de la ligne. Ils constatèrent que la ligne était alors parcourue par des courants très irréguliers. Ces courants se manifestèrent avec une intensité exceptionnelle lors de la grande aurore boréale observée le 17 novembre 1847 en Europe occidentale. A l'ingénieur anglais BARLOW revient d'avoir fait, en 1847/48, la toute première étude systématique de ces courants telluriques[2]. Mais ce n'est guère qu'une dizaine d'années plus tard que les milieux scientifiques commencèrent à leur porter un réel intérêt. La raison en fut que d'énormes perturbations, provoquant une interruption presque complète du service, s'étaient produites sur toutes les lignes d'Europe et d'Amérique lors d'une autre grande

---

[1] R. W. Fox: On the electromagnetic properties of metalliferous veins in the mines of Cornwall. Phil. Trans. Roy. Soc. Lond. **120**, 399 (1830).
[2] W. H. BARLOW: Phil. Trans. Roy. Soc. Lond. **139**, 61 (1849).

aurore boréale, qui eut lieu du 29 août au 3 septembre 1859 et qui put être observée jusqu'à la Martinique, à la très basse latitude de 14° N. En France par exemple, des forces électromotrices allant jusqu'à 900 Volts se manifestèrent sur les lignes N—S, produisant des étincelles dans les postes récepteurs [1].

On a vraiment quelque peine aujourd'hui à se figurer les théories bizarres que proposaient les savants de cette époque. De la Rive [2] par exemple, qui faisait autorité, pensait que la vapeur d'eau, formée en abondance dans les régions équatoriales, se chargeait d'électricité positive lors de l'évaporation, abandonnant des charges négatives sur la Terre. Les vents de la haute atmosphère entraînaient cette vapeur dans les régions polaires où elle se condensait. L'aurore polaire n'était pas autre chose que l'espèce de décharge provenant de la recombinaison des électricités au moment de la condensation. Quant aux courants telluriques, ils figuraient la contrepartie, dans le sol, du transport d'électricité ayant lieu dans la haute atmosphère, etc. ... Il ne faut pourtant pas sous-estimer l'œuvre scientifique accomplie par les Ingénieurs des Télégraphes, de toutes nationalités, pendant cette seconde moitié de siècle. Dès avant 1900, si l'origine des courants telluriques demeurait encore bien mystérieuse, les circonstances même du phénomène, dans ce qu'elles ont d'essentiel, étaient connues grâce à ces ingénieurs.

**2. Les premiers observatoires telluriques.** Malheureusement, l'exploitation des lignes télégraphiques ne permet que des enregistrements sporadiques, de sorte que la création d'observatoires, équipés de lignes autonomes, ne tarda pas à s'imposer. Le premier fut créé dès 1859, à Monaco, par Lamont [3]. Les lignes étaient bien trop courtes (200 m.) et les électrodes, fortement polarisables, étaient médiocres. Matteucci [4] perfectionna la technique de Lamont, adoptant des lignes de 6 km. et des électrodes impolarisables. Puis ce fut Airy qui, en 1862, avec des lignes de 13 et 16 km., équipa à Greenwich un Observatoire qui put fonctionner de façon ininterrompue à partir de 1865 [5]. Sur le modèle de l'observatoire anglais ne tardèrent pas à s'en installer plusieurs autres, les uns temporaires, les autres permanents. Parmi ces derniers, on citera ceux de Berlin, de Paris-Parc St Maur et de Pawlowsk.

Ces observatoires ne connurent, les uns comme les autres, qu'une vie très éphémère. Car la construction de lignes de tramways et de chemin de fer électriques, engendrant des courants vagabonds, occasionna des perturbations telles, que tous ces observatoires furent contraints d'interrompre définitivement leurs travaux les uns après les autres. Cela coïncidait d'ailleurs — il est juste de le reconnaître — avec une certaine lassitude et un certain découragement des géophysiciens pour ce qui concernait les études telluriques. Car, en regard de la masse des résultats de haute valeur obtenus par le moyen des lignes télégraphiques, la contribution des observatoires n'avait pas été en proportion des efforts déployés. Il faut signaler cependant la reprise des observations telluriques, en 1910, à Tortosa (Espagne) par les Pères Jésuites de l'Observatoire de l'Èbre. Les observations de Tortosa, ininterrompues de 1910 à 1938, ont constitué jusqu'à ces derniers temps la plus longue série continue de mesures telluriques.

**3. La prospection électrique et le renouveau des études telluriques.** C'est par le biais de la prospection qu'allait se manifester un regain des études telluriques. Le premier brevet de prospection électrique est de 1900 (Brown et Mc Clatchey). La première prospection est réussie en 1906 par Petersson. Mais c'est incontestablement le Français Conrad Schlumberger qui, à partir de 1912, est le véritable créateur de toutes les méthodes de prospection pratiques et correctes

---

[1] K. T. Clement: Das große Nordlicht in der Nacht zum 29. August 1859 und die Telegraphenverwirrung in Nord-Amerika und Europa. Hamburg 1860.
[2] De la Rive: C. R. Acad. Sci. Paris 49, 424 (1859).
[3] J. Lamont: Der Erdstrom und der Zusammenhang desselben mit dem Magnetismus der Erde. Leipzig 1862.
[4] Matteucci: C. R. Acad. Sci. Paris 53, 942 (1864); 54, 511 (1864).
[5] G. B. Airy: Phil. Trans. Roy. Soc. Lond. 158, 465 (1868); 160, 215 (1870).

utilisant un courant continu qu'on injecte dans le sol. C'est également à lui qu'on doit l'explication des phénomènes découverts comme il fut dit, par Fox, ainsi que l'utilisation de ces phénomènes pour une nouvelle et amusante technique de prospection (Polarisation spontanée). Et durant que les Français se cantonnent prudemment dans les investigations par courants continus, toutes sortes de techniques, fondées sur l'emploi du courant alternatif, et plus ou moins heureuses d'ailleurs, sont proposées ou expérimentées dans le Monde.

Quand on lit les Mémoires telluriques des Géophysiciens du siècle écoulé, on n'a guère l'impression que ces savants se souciaient sérieusement de la structure, de la nature, de la résistivité du sous-sol sur lequel était bâti leur observatoire et dans lequel circulaient les courants étudiés par eux. Pour les prospecteurs au contraire, la Géologie est le but, la Physique n'est que le moyen d'atteindre ce but. La prospection électrique en particulier, attira donc fort opportunément l'attention des Géophysiciens d'Observatoires sur le rôle certainement fondamental que joue le sous-sol de leurs établissements dans les phénomènes qu'on y étudie. À l'encontre des Observateurs, les Prospecteurs ne demeurent pas non plus éternellement rivés dans le même lieu. Par la force des choses, ils se trouvent amenés à découvrir, donc à interpréter, toutes sortes de phénomènes variés qu'ils rencontrent sur leur route et qu'ils ne se font pas faute de considérer d'abord comme «perturbateurs» tant qu'ils n'ont pas réussi à les domestiquer pour s'en servir.

Parmi les perturbations en question, vont figurer en bonne place, précisément, les *variations telluriques*. Et dès 1921 les prospecteurs s'avisent — en l'espèce, c'est toujours de Conrad Schlumberger qu'il s'agit — que la comparaison des variations telluriques simultanées, en divers points d'une région, ne peut manquer d'apporter des renseignements très précieux sur la structure du sous-sol de ladite région[1]. Tels furent donc les débuts de la prospection tellurique, qui connaît actuellement un plein essor. Les phénomènes telluriques ne sont pas locaux. Ils sont à l'échelle du Globe. Et cela, les Ingénieurs des Télégraphes, dont les réseaux de lignes couvraient parfois toute l'étendue d'une grande nation, l'avaient compris depuis toujours. La création des vieux observatoires telluriques avait eu un inconvénient, celui de restreindre l'horizon des Observateurs, de les inciter à ne considérer que l'aspect strictement local du phénomène. Les Prospecteurs, amenés à dresser des «cartes telluriques» régionales, devaient faire craquer ces frontières étriquées de l'antique Géophysique, replacer les études telluriques dans leur cadre véritable.

À la condition de ne pas perdre de vue ces aspects modernes et vivants des études telluriques, il n'en demeure pas moins que beaucoup d'études de détail ne peuvent être menées à bien que dans des Observatoires permanents. Encore faut-il que lesdits observatoires soient parfaitement équipés, capables de faire des mesures vraiment correctes auxquelles on puisse accorder une confiance sans réserve. C'est à cette nécessité qu'a répondu la création d'observatoires modernes et modèles comme ceux de Watheroo (Australie) en 1923, de Huancayo (Pérou) en 1926, ou de Tucson (Arizona) en 1931, sous les auspices de la «Carnegie Institution of Washington».

## B. Bases théoriques de la distribution des courants électriques dans le sol.

**4. Conductibilité électrique du sol.** La circulation des courants électriques dans le sol n'est pas régie par d'autres lois que les lois classiques de l'Electromagnétisme, que la loi d'Ohm en particulier quand il s'agit des *courants continus*

---

[1] E. G. Leonardon: Terr. Magn. **33**, 91 (1928).

qui vont être envisagés en premier lieu. En vertu de la loi d'Ohm, le vecteur «densité de courant» $i$, en un point, est proportionnel au champ électrique $e$, lequel champ dérive d'un potentiel $V$:

$$i = \sigma\, e = \frac{1}{\varrho}\, e = -\,\sigma\, \mathrm{grad}\, V = -\,\frac{1}{\varrho}\, \mathrm{grad}\, V\,. \tag{4.1}$$

$\sigma$ désigne la *conductibilité électrique* du sol au point considéré. Son inverse $\varrho$ s'appelle la *résistivité*.

En Géophysique on utilise les unités pratiques électromagnétiques C. G. S. pour évaluer l'intensité des courants (Ampère), les différences de potentiel (Volt) et les résistances (Ohm), mais on préfère mesurer les longueurs en mètres plutôt qu'en centimètres. Dans ce système électromagnétique à l'usage des Géophysiciens, l'unité de résistivité n'est donc plus l'Ohm. cm., mais une unité 100 fois plus grande. l'Ohm.mètre ($\Omega$.m.), qu'on peut se représenter comme étant la résistivité d'un cube ayant 1 m. de côté, qui serait traversé par un courant de 1 A lorsqu'on établit une différence de potentiel de 1 V entre deux faces opposées.

Quelques minerais peu nombreux, des oxydes comme la magnétite, des sulfures tels que les pyrites de fer et de cuivre, la galène, également les anthracites et le graphite, présentent une *conductibilité de type électronique* (ou *métallique*). La majorité des minerais ainsi que la totalité des roches ont au contraire une *conductibilité ionique* (ou *électrolytique*) et deviennent en conséquence extrêmement résistantes dès qu'on les prive par dessication de l'eau qui les imprègne. De là résulte que la résistivité des roches, surtout des roches poreuses, dépend très étroitement de la nature du liquide d'imbibition: la même roche réservoir d'un gîte pétrolifère, extrêmement résistante quand elle renferme l'huile, devient très conductrice quand elle contient de l'eau salée.

Pour fixer cependant les idées par quelques ordres de grandeur, on peut dire que les roches les plus conductrices sont en principe les marnes, dont la résistivité est habituellement inférieure à 20 $\Omega$.m. Cette résistivité peut même s'abaisser jusqu'à 0,5 dans le cas des marnes salées des terrains pétrolifères. Les argiles et les schistes (50 à 500) sont normalement plus conducteurs que les calcaires (100 à 5000). Les roches éruptives, intrusives ou d'épanchement, les roches métamorphiques, les roches quartzeuses ou fortement silicifiées sont extrêmement résistantes (1000 à 10000 ou davantage encore) à condition qu'elles ne soient ni broyées, ni altérées. La résistivité des eaux de mer, très variable suivant la température et la salinité, est de l'ordre moyen de 0,3 tandis que celle des eaux normales de rivière, non saumâtres, est de l'ordre de 50.

La résistivité des minerais à conductibilité électronique est beaucoup plus petite que celle des roches. Celle de la pyrite n'est par exemple que de l'ordre de $10^{-3}$, celle du graphite de $3 \cdot 10^{-4}$. De telles résistivités, considérées par le Géophysicien comme extrêmement petites, sont pourtant énormes si on les compare à celle du cuivre ($1,6 \cdot 10^{-8}$). A propos des minerais, on remarquera qu'ils peuvent comporter une minéralisation conductrice, enrobée dans une gangue isolante, comme du quartz ou de la fluorine. S'il n'y a pas de continuité entre les imprégnations minéralisées, le minerai pris en bloc se comporte comme un isolant et non comme un conducteur.

De ce qui précède résulte aussi qu'il est impossible de mesurer correctement la résistivité d'une roche en opérant au laboratoire sur des échantillons ou des carottes de sondage ramenés du terrain. Les mesures de résistivités doivent être faites *«in situ»* par les procédés même que les prospecteurs emploient pour déterminer les «résistivités apparentes», procédés dont le principe sera indiqué un peu plus loin. On trouve en outre dans cette façon d'opérer l'avantage de faire intervenir un grand volume de terrain, d'obtenir par conséquent une *résistivité moyenne* dépouillée de fluctuations, d'irrégularités locales qui n'ont pas plus d'intérêt pour le géophysicien que pour le géologue.

Les roches ne sont pas seulement hétérogènes, mais encore *anisotropes*. On remarquera cependant que le Géophysicien, au contraire du Minéralogiste, ne s'occupe pas d'échantillons minuscules. A l'échelle où il doit se placer, les phénomènes ne se présentent pas sous le même aspect: un grain de sable est peut-être anisotrope, mais le tas de sable est isotrope. La juxtaposition désordonnée d'éléments anisotropes permet donc assez souvent de considérer la roche, en bloc, comme isotrope.

Tout de même, il serait imprudent de s'exagérer le caractère désordonné de l'assemblage des éléments qui constituent un terrain: la schistosité et la stratification par sédimentation sont des facteurs qui imposent une régularité relative dans la structure. Qu'on imagine par exemple un flysh qui serait constitué par une alternance, indéfiniment et périodiquement répétée sur des centaines de mètres d'épaisseur, de couches minces calcaires et d'autres plus marneuses. Soit 10 cm. l'épaisseur des premières, 2 cm. celle des secondes, 500 la résistivité des premières, 20 celle des secondes. Lorsqu'un courant s'écoule à travers cet empilement de flysh, les lignes de courant subissent d'innombrables réfractions sur chacune des surfaces de discontinuité. Pour le Géophysicien cependant, qui doit faire abstraction de ces infimes détails, tout se passe comme si le flysh était un terrain unique, homogène mais anisotrope. La résistivité est plus grande ($\varrho = 420$) si le courant s'écoule normalement aux strates que s'il leur est parallèle ($\varrho = 100$). Il s'agit là d'une fausse anisotropie, d'une *macro-anisotropie*, très différente de l'anisotropie vraie, laquelle est une micro-anisotropie[1, 2]. Cependant, bien que le phénomène en cause soit très loin d'être dénué d'importance pratique, on supposera toujours désormais, dans la suite de cet exposé, que les terrains sont *isotropes*.

**5. Courant continu.** On fera abstraction pour le moment des forces électromotrices internes de polarisation, d'électrofiltration, thermiques, etc. La distribution du courant continu est alors régie par la relation (4.1). Le courant continu étant conservatif par définition, le flux de $i$ à travers une surface fermée quelconque est nul, de sorte que:

$$\operatorname{div} \boldsymbol{i} = 0 \qquad\qquad (5.1)$$

ou:

$$\Delta V = \frac{\partial^2 V}{\partial x^2} + \frac{\partial^2 V}{\partial y^2} + \frac{\partial^2 V}{\partial z^2} = 0. \qquad\qquad (5.2)$$

$V$ varie de façon continue quand on franchit une surface qui sépare deux milieux différents. A cause de la conservativité du courant, il en est de même pour $\frac{1}{\varrho}\frac{\partial V}{\partial \nu}$, quand on désigne par $\nu$ la normale à ladite surface; et, en particulier, les surfaces équipotentielles sont normales à la surface du sol. Ces conditions régissent comme on sait la réfraction des lignes de courant et des surfaces équipotentielles à la traversée des surfaces de discontinuité de la résistivité. Comme le rapport des résistivités des milieux au contact est souvent énorme, la réfraction des lignes de courant apparaît comme étant un très gros phénomène. Dès que le rapport est très grand ou très petit, tout se passe comme si l'un des deux milieux au contact était infiniment conducteur par rapport à l'autre.

[1] R. Maillet et H. G. Doll: Erg.-Hefte angew. Geophys. **3**, 109 (1932).

[2] L. Cagniard: Ann. Inst. Phys. Globe Strasbourg, 3ième Partie, **4**, 3 (1948).

Pour injecter le courant dans le sol, les prospecteurs se servent de *prises de terre* ou *«électrodes»*. Une prise est constituée soit par un unique pieu d'acier fiché en terre, soit par plusieurs pieux de ce genre, reliés électriquement et groupés dans un espace qui occupe par exemple quelques dizaines de mètres carrés. Les dimensions d'une telle prise sont malgré tout habituellement petites en regard de sa distance aux autres électrodes, dont les unes concourrent à l'injection du courant et dont les autres servent aux mesures. De sorte que, sous cette réserve, il est légitime d'assimiler la prise en question à l'électrode «ponctuelle» et superficielle chère au Mathématicien. Si $I$ est l'intensité du courant qui y pénètre dans le sol, le potentiel $V$ d'un point $M$ devient infini comme $\frac{\varrho I}{2\pi} \cdot \frac{1}{r}$ quand la distance $r$ de $M$ à la source tend vers zéro. $\varrho$ désigne ici la résistivité du sol, à la source. Avec l'équation indéfinie (5.2), avec les conditions aux limites sur les surfaces de discontinuité, enfin avec les conditions initiales traduisant l'existence des sources, le problème analytique de la distribution des potentiels (et des courants) se trouve complètement et correctement posé. La solution est unique, abstraction faite d'une constante additive arbitraire, qui disparaît si l'on adopte pour zéro des potentiels le potentiel à l'infini.

De là résulte que si le sol est homogène, si sa surface est plane, si $A$ et $B$ sont les deux électrodes, reliées aux pôles d'un générateur de telle sorte que le courant $I$ pénètre en $A$ et sorte en $B$, le potentiel de tout point $M$ s'exprime par:

$$V = \frac{\varrho I}{2\pi} \left( \frac{1}{\overline{MA}} - \frac{1}{\overline{MB}} \right). \tag{5.3}$$

On trouve dans tous les traités de prospection géophysique [1] le dessin des surfaces équipotentielles et des lignes de courant qui correspondent à cette distribution, qu'on qualifie de normale. En vertu de (5.3), la différence de potentiel $\delta V$ entre deux points $M$ et $N$ de la surface du sol est alors:

$$\delta V = V_M - V_N = \frac{\varrho I}{2\pi} \left( \frac{1}{\overline{MA}} - \frac{1}{\overline{MB}} - \frac{1}{\overline{NA}} + \frac{1}{\overline{NB}} \right). \tag{5.4}$$

Pour mesurer «in situ» la résistivité d'un terrain, on placera donc un quadripôle $ABMN$ sur un affleurement de ce terrain, assez étendu en tous sens et suffisamment enraciné pour que les lignes de courant, s'écoulant de $A$ en $B$, ne sortent pratiquement pas du terrain en question. On mesurera $I$ et $\delta V$, et l'on résoudra l'équation (5.4) par rapport à $\varrho$.

**6. Equations de Maxwell.** Le comportement d'un courant variable est régi par les équations de Maxwell:

$$\left. \begin{aligned} -\mu \dot{\boldsymbol{h}} &= \operatorname{rot} \boldsymbol{e}, \\ 4\pi\sigma \boldsymbol{e} + \frac{\dot{\boldsymbol{e}}}{\mu \Omega^2} &= \operatorname{rot} \boldsymbol{h}, \\ \operatorname{div} \boldsymbol{e} = \operatorname{div} \boldsymbol{h} &= 0. \end{aligned} \right\} \tag{6.1}$$

Le point sur $\dot{\boldsymbol{e}}$ ou $\dot{\boldsymbol{h}}$ désigne, suivant l'usage, la dérivée par rapport au temps $t$ du champ électrique $\boldsymbol{e}$ ou du champ magnétique $\boldsymbol{h}$. On pose ici que le milieu est homogène et isotrope. Sa conductibilité est $\sigma$. On désigne par $c$ la vitesse de la lumière dans le vide et par

$$\Omega = \frac{c}{\sqrt{\varepsilon\mu}}$$

la vitesse de phase des ondes électromagnétiques planes qui, comme le montre précisément l'intégration de (6.1), peuvent se propager dans le milieu considéré. $\varepsilon$ et $\mu$ sont respectivement la constante diélectrique et la perméabilité du milieu relativement au vide. A l'exception de $\varepsilon$ et de $\mu$, toutes les grandeurs qui figurent dans (6.1) sont exprimées en unités électromagnétiques. Enfin les axes de coordonnées sont des axes « à droite ».

Les constantes diélectriques $\varepsilon$ des roches atteignent souvent plusieurs unités, avec un maximum de 80 dans le cas de l'eau. Par contre, le cas de la magnétite étant bien entendu exclu, $\mu$ est toujours extrêmement voisin de l'unité. Dorénavant on posera donc $\mu = 1$.

Pour être en mesure de déterminer complètement la distribution dans le sol d'un courant variable, il faut adjoindre aux équations indéfinies (6.1) les conditions aux limites qui traduisent la présence des surfaces de discontinuité: les composantes tangentielles de $e$ et de $h$ doivent être continues. Enfin les conditions initiales traduiront l'existence des sources, électrodes, câbles inducteurs, etc. en exprimant que les champs présentent des singularités convenables dans le voisinage de ces sources. On reviendra plus loin sur les singularités en question.

**7. Vecteur de HERTZ.** Plutôt que d'intégrer directement les équations de MAXWELL, il est préférable d'opérer un changement de fonctions préalable. Celui dont il va être fait usage au cours du présent exposé est certainement le plus commode à manier quand on porte plus spécialement son attention sur l'aspect *électrique* du phénomène électromagnétique. C'est un changement de fonctions différent qu'utilisent avec raison CHAPMAN et BARTELS [2] au chapître XXII de leur très classique ouvrage « Geomagnetism », du fait que ces auteurs s'intéressent surtout au phénomène *magnétique*. Pour les questions de théorie soulevées au cours du présent chapître, il n'y a d'ailleurs qu'intérêt à se reporter à n'importe quel traité d'électromagnétisme. Rien ne surpassera jamais en clarté le magistral exposé rédigé par SOMMERFELD dans le traité de FRANK et VON MISES [3].

On posera donc ici:

$$h = \mathrm{rot}\left(4\pi\sigma p + \frac{\dot{p}}{\Omega^2}\right), \\ e = \mathrm{grad\ div}\, p - \Delta p, \quad \right\} \tag{7.1}$$

$p$ est le vecteur de HERTZ. Il joue, dans la théorie des courants variables, un rôle similaire à celui du potentiel scalaire $V$ dans la théorie du courant continu. A partir du vecteur de HERTZ on calcule à la fois $e$ et $h$ par de simples dérivations, de même que, connaissant $V$, on calculait par dérivation les composantes de $e$. Le vecteur de HERTZ satisfait à l'équation indéfinie:

$$\Delta p - 4\pi\sigma\dot{p} - \frac{\ddot{p}}{\Omega^2} = 0 \tag{7.2}$$

car on vérifie sans peine que cette équation équivaut à (6.1).

Grâce à l'introduction du vecteur de HERTZ, non seulement réalise-t-on une très substantielle simplification des calculs, mais encore fait-on beaucoup mieux ressortir l'unité du phénomène électromagnétique, au sein duquel champ électrique et champ magnétique sont indissolublement unis.

À l'égard des Géophysiciens qui, dans un passé récent, l'ont parfois oublié, on ne soulignera jamais assez que le champ électrique ne dérive pas d'un potentiel. Car il ne peut pas s'exprimer à l'aide d'un seul gradient du fait que son rotationnel n'est pas nul, d'après la première des relations (6.1). *Parler de potentiel électrique dans le cas d'un régime variable n'a fondamentalement pas de sens.* Certes tout champ électrique, qu'il soit d'induction, coulombien ou autre, possède les mêmes «dimensions» physiques qu'un gradient de potentiel électrique, en raison de sa définition même. Il peut donc toujours être évalué en millivolts par km. si on le désire. Mais, s'il advient qu'à l'aide d'une ligne tellurique d'un kilomètre de long on ait mesuré un champ électrique de $n$ mV/km., il ne s'en suit pas pour autant qu'il existe entre les deux électrodes terminales une différence de potentiel de $n$ Millivolts.

**8. Courants alternatifs.** Pour étudier le comportement d'un courant alternatif de période $T$, de fréquence $n$, de pulsation $\omega = 2\pi/T$, autrement dit pour faire la théorie particulière d'un état de régime harmonique, il est commode d'utiliser la notation imaginaire $(i = \sqrt{-1})$ et de poser que toutes les grandeurs électromagnétiques dépendent du temps $t$ par l'intermédiaire du facteur $e^{-i\omega t}$. En fin de calcul on retrouve la siginification physique des grandeurs en question lorsqu'on ne conserve, par exemple, que la partie réelle des expressions imaginaires obtenues. Soit donc:

$$h = H e^{-i\omega t}; \quad e = E e^{-i\omega t}; \quad p = P e^{-i\omega t} \tag{8.1}$$

où les vecteurs $H$, $E$, $P$ ne dépendent plus que des variables d'espace. Les équations (7.1) deviennent:

$$\left. \begin{aligned} H &= \left(4\pi\sigma - \frac{i\omega}{\Omega^2}\right) \operatorname{rot} P, \\ E &= \operatorname{grad} \operatorname{div} P - \varDelta P \end{aligned} \right\} \tag{8.2}$$

et la relation (7.2) se transforme en:

$$\varDelta P + i\omega \left(4\pi\sigma - \frac{i\omega}{\Omega^2}\right) P = 0. \tag{8.3}$$

**9. Simplifications. Conséquences.** Pour ce qui touche plus spécialement à la Géophysique, $4\pi\sigma$ est toujours très grand par rapport à $\omega/\Omega^2$. Afin de le montrer on se placera dans des conditions particulièrement défavorables; on supposera une fréquence très élevée, de 1000 par exemple. Car, même en Géophysique Appliquée, de telles fréquences sont à l'extrême limite de celles qui sont pratiquement utilisables, du fait que la pénétration dans le sous-sol de courants de cette fréquence est très réduite. Pour la démonstration, on doit choisir aussi un terrain particulièrement résistant, de 5000 $\Omega$.m. par exemple, ce qui correspond à $\sigma = 2 \cdot 10^{-15}$ *em* C.G.S. $\Omega$ ne peut guère non plus devenir plus petit que $10^{10}$ C.G.S. Or, dans cette hypothèse, le terme $4\pi\sigma$ est quand même 400 fois plus grand que $\omega/\Omega^2$. Il est habituellement beaucoup plus grand encore. Puisqu'on est donc autorisé à négliger $\omega/\Omega^2$ devant $4\pi\sigma$, les équations de la Sect. 8 se simplifient et deviennent:

$$\left. \begin{aligned} H &= 4\pi\sigma \operatorname{rot} P, \\ E &= \operatorname{grad} \operatorname{div} P - \varDelta P, \end{aligned} \right\} \tag{9.1}$$

$$\varDelta P + 4\pi\sigma\omega i P = 0. \tag{9.2}$$

Il résulte d'abord de là que $\varepsilon$ ne joue pratiquement pas un rôle plus grand que $\mu$ dans le comportement des courants alternatifs au sein du sol, puisque ce paramètre n'intervient que par l'intermédiaire de $\Omega$, qui a disparu désormais. On remarquera en outre qu'un des coefficients de l'équation (9.2) est imaginaire. D'où résulte qu'en général on ne peut pas trouver de solutions réelles de cette équation. Les composantes du vecteur de Hertz, des champs électrique et magnétique, de la densité de courant, ont des expressions imaginaires. En langage de physicien, cela veut dire que toutes ces grandeurs sont d'habitude décalées en phase les unes par rapport aux autres. On a par exemple:

$$\left.\begin{aligned}
h_x &= X \cos(\omega t - \varphi_x), \\
h_y &= Y \cos(\omega t - \varphi_y), \\
h_z &= Z \cos(\omega t - \varphi_z)
\end{aligned}\right\} \tag{9.3}$$

où $\varphi_x$, $\varphi_y$, $\varphi_z$ sont normalement différents. Donc l'extrémité du vecteur $h$, comme de n'importe quel autre vecteur électromagnétique, au point $M$, décrit périodiquement une ellipse de centre $M$, et non un segment de droite.

**10. Cas limite des très basses fréquences.** Quand $\omega$ tend vers zéro, le second terme de (9.2) devient lui aussi négligeable, et les équations se simplifient encore en:

$$\left.\begin{aligned}
H &= 4\pi\,\sigma\,\mathrm{rot}\,P, \\
E &= \mathrm{grad}\,\mathrm{div}\,P,
\end{aligned}\right\} \tag{10.1}$$

$$\Delta P = 0. \tag{10.2}$$

Il advient donc, *dans ce cas limite*, que le champ électrique dérive du potentiel $V$ défini par:

$$V = -\,\mathrm{div}\,P \tag{10.3}$$

et satisfaisant, en conséquence, à l'équation de LAPLACE.

Ainsi, comme on devait le prévoir, un courant alternatif *de fréquence suffisamment basse* se comporte dans le sol exactement comme ferait un courant continu. Un exemple illustrera ce qu'il faut entendre par là: on supposera que, dans une première expérience, un courant continu d'intensité $I$ soit injecté dans le sol. En un point $M$ du sol, le potentiel est $V$, le champ électrique $e$, etc. À la pile située dans le circuit d'injection, on substitue désormais un alternateur de fréquence très basse, débitant un courant d'intensité $I \cos \omega t$. Au même point $M$ que tout à l'heure, le potentiel est maintenant $V \cos \omega t$, le champ électrique $e \cos \omega t$, etc. Dans ce cas limite, toute différence de phase a disparu entre n'importe lesquelles des grandeurs qu'on considère.

La conductibilité des roches est incomparablement plus petite que celle du cuivre qu'utilise l'ingénieur électricien. De plus, en géophysique, la fréquence des variations électromagnétiques est toujours très petite. Elle devient parfois extraordinairement petite, quand on se trouve par exemple amené à étudier la variation diurne des courants telluriques. Ces deux circonstances contribuent l'une et l'autre à diminuer le coefficient $4\pi\,\sigma\,\omega$. Et pourtant, qu'il s'agisse de prospection ou de Physique du Globe, il n'est jamais permis de consentir aux simplifications envisagées dans le présent paragraphe, d'assimiler la distribution des courants telluriques à celle de courants continus. Une affirmation aussi tranchée, et qui surprend de prime abord, ne se trouvera justifiée que lorsqu'on pourra, dans la suite de l'exposé, apprécier à sa véritable importance le rôle du *skin-effect*. Elle paraîtra pourtant moins paradoxale si l'on veut tenir compte de la disparité d'échelle des phénomènes en Électrotechnique d'une part, en Géophysique d'autre part.

Soient en effet deux milieux conducteurs, électriquement et géométriquement semblables. Les deux milieux se correspondent point par point. $M'$ est le point homologue de $M$. Les conductibilités en $M$ et $M'$ sont respectivement $\sigma$ et $\sigma'$; les vecteurs de HERTZ sont $P$ et $P'$. Deux longueurs homologues sont $l$ et $l'$.

Enfin $T$ et $T'$ sont les périodes des courants alternatifs utilisés dans l'un et l'autre cas. Soit:

$$l' = K_l\, l; \quad \sigma' = K_\sigma\, \sigma; \quad T' = K_T\, T; \quad K_l, K_\sigma, K_T: \text{Constantes.} \tag{10.4}$$

Si l'on veut que la similitude s'étende à l'ensemble des deux phénomènes électro-magnétiques, il faut, comme on le montre aisément:

$$K_\sigma\, K_l^2 = K_T. \tag{10.5}$$

Or les dimensions géométriques des conducteurs sont d'un tout autre ordre de grandeur en Géophysique qu'en Électrotechnique. Il est rare, dans ce dernier cas, d'utiliser des conducteurs dont le diamètre excède le centimètre, mais la Géophysique la plus modeste s'inquiète du comportement du courant jusqu'à 100 m. de profondeur au minimum. Le rapport $K_l$ de similitude géométrique dépasse $10^4$ et contribue donc, en ce qui le concerne, à accroître, par un facteur de plus de $10^8$, l'importance du second terme de (9.2) relativement au premier.

**11. Skin-effect dans une nappe tellurique harmonique uniforme. Profondeur de pénétration.** A grande distance des sources, le champ électromagnétique tend à devenir uniforme, du moins si la structure du sol permet qu'il en soit ainsi. Pour fixer les idées, on considérera seulement ici un sol homogène. L'uniformité qu'on suppose ici au champ électromagnétique ne peut avoir lieu que par rapport aux coordonnées horizontales $x$ et $y$, car s'il s'agissait d'une uniformité complète, $\Delta P$ serait nul, et $P$ le serait également en vertu de (9.2) puisque ni $\sigma$, ni $\omega$ ne le sont. $P$ est donc une fonction de $z$ qui satisfait à:

$$\frac{d^2 P}{d z^2} + 4\pi\, \sigma\, \omega\, i\, P = 0 \tag{11.1}$$

et qui, en conséquence, est de la forme:

$$P = A\, e^{z\sqrt{-4\pi\sigma\omega i}} + B\, e^{-z\sqrt{-4\pi\sigma\omega i}}; \quad A, B: \text{Vecteurs constants.} \tag{11.2}$$

On adoptera d'autre part pour axe $Oz$ la verticale descendante et l'on désignera par

$$\sqrt{-4\pi\sigma\omega i} = (1 - i)\sqrt{2\pi\sigma\omega}$$

celle des deux déterminations du radical dont la partie réelle est positive. Dans (11.2), il faut donc poser $A = 0$ car le premier terme devient infini aux grandes profondeurs. En définitive, la solution s'écrit:

$$P = P_0\, e^{(i-1)z\sqrt{2\pi\sigma\omega}}. \tag{11.3}$$

D'où:

$$p = P\, e^{-i\omega t} = P_0\, e^{-i(\omega t - z\sqrt{2\pi\sigma\omega}) - z\sqrt{2\pi\sigma\omega}}. \tag{11.4}$$

Ce qui signifie qu'une grandeur électromagnétique quelconque $N$, exprimée cette fois en termes réels, est de la forme:

$$N = N_0\, e^{-z\sqrt{2\pi\sigma\omega}} \cos\left(\omega t - z\sqrt{2\pi\sigma\omega} - \varphi\right); \quad N_0, \varphi: \text{Constantes.} \tag{11.5}$$

Les formules équivalentes (11.3), (11.4), (11.5) traduisent ce qu'on appelle le skin-effect. Quand $z$ augmente, on constate un amortissement exponentiel par rapport à $z$ en même temps qu'un retard de phase progressivement croissant. Sous le nom conventionnel de «profondeur de pénétration» (en sous-entendant qu'il s'agit d'un terrain de conductibilité $\sigma$ et d'une nappe de période $T$), on désignera la profondeur $\zeta$ où l'amplitude est réduite à la fraction $1/e$ de ce qu'elle est en surface:

$$\zeta = \frac{1}{\sqrt{2\pi\,\sigma\,\omega}} = \frac{1}{2\pi}\sqrt{\frac{T}{\sigma}} = \frac{1}{2\pi}\sqrt{\varrho\,T}\,. \qquad (11.6)$$

Quant à la phase, elle retarde d'un radian supplémentaire chaque fois que $z$ augmente de $\zeta$.

**12. Conditions initiales.** Pour achever de mettre en équations la distribution d'un courant variable, il reste à exprimer les conditions initiales devant être satisfaites au voisinage des sources. Or, au voisinage d'une source *idéalisée*, c'est à dire d'une électrode qu'on considère comme un point, d'un câble qu'on assimile à une ligne géométrique, etc., les champs, le vecteur $p$ de HERTZ ainsi que $\Delta p$ deviennent infinis, et il est clair que le rapport de $p$ à $\Delta p$ tend vers zéro. Dans le voisinage infinitésimal des sources, les conditions de la Sect. 10 sont rigoureusement satisfaites, quel que soit l'ordre de grandeur de $\sigma$ et de $T$, et le courant variable s'y comporte exactement comme un courant continu, au sens de la Sect. 10. Il n'y a donc jamais de difficulté, dans chaque cas particulier, à formuler les conditions initiales convenables.

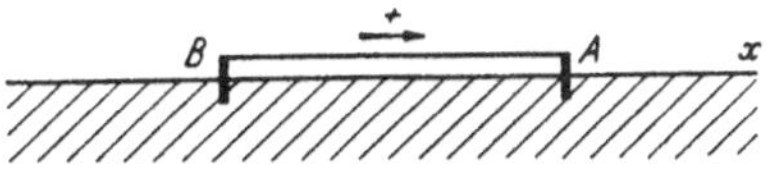

Au long de deux exposés qui comportent malheureusement quelques incorrections mathématiques, STEFANESCU[1] a excellemment souligné l'importance théorique,

Fig. 1. Dipôle T.P.S.

en Géophysique, de ce qu'il a appelé le dipôle T.P.S., ces trois lettres étant une abréviation du terme de «*Télégraphie par le sol*» sous lequel on désignait en 1914—1918 un système de transmissions utilisé dans la guerre de tranchées. Le dipôle T.P.S., tel que le Mathématicien l'idéalise, n'est autre chose (Fig. 1) qu'une ligne infiniment courte, posée sur le sol, isolée, de longueur $dx$, qui est traversée par un courant variable, harmonique ou non, dont l'intensité $I(t)$ est rapportée au sens positif $Ox$. La ligne se termine par deux électrodes $A$, $B$ qui sont supposées ponctuelles. En vertu de (5.3), la composante $p_x$ de $p$ devient infinie comme

$$\frac{\varrho\,(I\,dx)}{2\pi}\cdot\frac{1}{r} \qquad (12.1)$$

lorsque tend vers zéro la distance $r$ du point considéré au dipôle.

Si le dipôle est harmonique et le sol homogène, il est facile d'intégrer le problème jusqu'au bout. Cependant, malgré la grande simplicité de ce cas, la solution ne peut s'exprimer que sous forme d'intégrales définies, d'un maniement bien moins commode que la relation (5.3) qui lui équivaut quand il s'agit de courant continu.

À partir du dipôle T.P.S., STEFANESCU construit la solution générale pour une ligne d'excitation $AB$, de longueur et de forme quelconques, posée à terre et terminée par les deux électrodes $A$, $B$. On suppose qu'un alternateur, par exemple, est inséré sur la ligne. C'est là un dispositif d'excitation couramment utilisé par les prospecteurs. On supposera en outre, pour la rigueur du raisonnement qui suit, que la «durée d'établissement»[2] du courant dans la ligne est petite par

---

[1] S. STEFANESCU: Etudes sur la prospection électrique du sous-sol. Bucarest. 1. Série (1929), 2. Série (1932).

[2] Lorsqu'on insère instantanément une force électromotrice dans le câble, et qu'on maintient constante ladite force électromotrice à partir de cet instant, le courant de régime exige une durée illimitée pour s'établir asymptotiquement. Par «durée d'établissement», on désigne le temps nécessaire pour que le courant de régime soit «pratiquement» atteint, avec une approximation fixée de façon plus ou moins arbitraire. La durée d'établissement dépend d'ailleurs de l'emplacement occupé par la source sur la ligne. Peu importe quand il ne s'agit, comme ici, que d'ordres de grandeur.

rapport à la période du courant alternatif utilisé. Un tel câble équivaut à une suite de dipôles T.P.S. en nombre infini, placés bout à bout le long de $BA$ depuis $B$ jusqu'en $A$ et qui débiteraient tous le même courant $I \cos \omega t$. L'électrode terminale d'un de ces dipôles coïncide en effet avec l'électrode initiale du dipôle qui lui fait suite, de sorte que l'électrode dont il s'agit peut être supprimée sans que le phénomène électromagnétique s'en trouve altéré. En d'autres termes, lorsqu'on sait intégrer le problème pour un dipôle excitateur, il suffit d'une intégration curviligne supplémentaire effectuée le long du câble depuis $B$ jusqu'en $A$ pour exprimer la solution dans le cas d'un émetteur de longueur et de forme quelconques. S'il s'agit de courant continu, la solution ne dépend que des électrodes terminales. S'il s'agit de courant alternatif, faut-il faire remarquer que la forme du câble joue un rôle essentiel? Il doit en effet exister en ce cas une singularité du champ électrique tout le long du câble. Enfin rien n'interdit de faire coïncider l'électrode finale $A$ avec l'électrode initiale $B$. On développe ainsi la théorie générale d'un émetteur purement inductif, formé d'un câble isolé dont les deux bouts sont reliés aux bornes de l'alternateur.

**13. Phénomènes transitoires. Régimes variables.** Les prospecteurs qui se servent du courant continu doivent s'intéresser au mode d'établissement du phénomène électromagnétique dans le sol, quand on ferme brusquement l'interrupteur de la ligne $AB$. Il a même été proposé naguère de fonder une méthode régulière de prospection sur l'étude expérimentale de ces phénomènes transitoires très embrouillés. Des injections brusques de courant se produisent également sur les lignes de traction électrique, engendrant des courants telluriques artificiels, parasites, dénommés «vagabonds». Les problèmes théoriques soulevés par l'étude de ces phénomènes, et qui sortiraient du cadre de cet exposé, gagnent à être traités par les élégantes méthodes du calcul opérationnel.

D'autre part, les courants telluriques naturels manifestant un comportement d'une complexité extrême, il sera le plus souvent commode de *se figurer* qu'on a exprimé sous forme d'intégrales de Fourier l'évolution en fonction du temps, évolution dont l'allure est si tourmentée, des composantes telluriques. Autrement dit on supposera ces composantes telluriques décomposées en spectres plus ou moins continus de constituants harmoniques de toutes fréquences. Par cet artifice on se trouve donc ramené aux régimes harmoniques.

**14. Théorie de la mesure du champ tellurique.** D'après une opinion très répandue, mais qui conduit presque fatalement à des conclusions et formules erronées, il semble admis que la mesure du champ électrique est une opération qui «va de soi». D'où le paragraphe actuel, qui s'efforce d'introduire la rigueur indispensable à qui veut aborder correctement l'étude des variations rapides.

Ce qu'on appelle, d'un nom commode, champ tellurique, c'est plus précisément la composante tangentielle, à la surface du sol, du champ électrique. D'après les lois fondamentales de l'électromagnétisme, cette composante tangentielle n'éprouve aucune discontinuité lorsqu'on franchit la surface du sol. Peu importe donc, pour la théorie, que la ligne de mesure, la ligne tellurique, soit posée sur le sol, qu'elle soit souterraine, enterrée en tranchée ou qu'elle soit aérienne, accrochée à une file de poteaux. Ce que l'expérience, bien entendu, n'a pas manqué de confirmer.

Comme il a été dit au paragraphe précédent, le comportement de toute nappe tellurique est conditionné en dernière analyse par le comportement individuel de chacun de ses constituants harmoniques. On va donc supposer ici qu'on a affaire à un état de *régime harmonique*.

Soit une ligne tellurique constituée par un fil isolé $BA$ reliant deux électrodes $A$ et $B$. L'appareil de mesure est inséré sur la ligne en $G$ (Fig. 2). C'est habituellement un galvanomètre. Cela peut être n'importe quoi. On supposera en tous cas, expressément, que les caractéristiques de cet appareil de mesure sont linéaires. Plus généralement, tout ce qu'il peut convenir au physicien d'insérer aussi sur la ligne, des bobines de self-induction, des transformateurs, des appareils électroniques … doit posséder des *caractéristiques linéaires*. On exclut par exemple, ou du moins l'on n'admet qu'avec la plus grande méfiance, des selfs ou des transformateurs dont le noyau serait en fer ou en alliages à haute perméabilité. Ces réserves faites le phénomène tellurique considéré, qui est sinusoïdal et de période $T$ avant la mise en place de la ligne tellurique, demeure sinusoïdal; l'appareillage utilisé perturbe assurément le phénomène tellurique primitif, mais il n'engendre pas d'harmoniques de fréquences double, triple, etc.

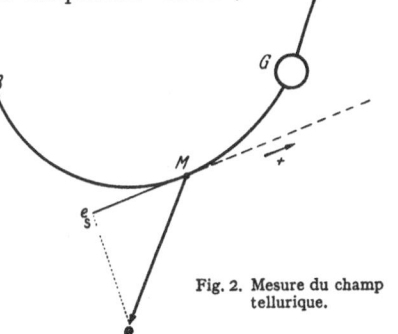

Fig. 2. Mesure du champ tellurique.

Le but que poursuit l'expérimentateur est bien entendu *de mesurer le champ tellurique primitif, tel qu'il serait si la ligne n'existait pas*. Quand elle existe, elle est parcourue par un courant alternatif si, comme on doit le supposer pour la généralité nécessaire des raisonnements, son impédance n'est pas infinie. C'est ce courant qui perturbe le phénomène, et la perturbation dont il s'agit est fréquemment *énorme*. Cela n'exclut d'ailleurs pas, comme on va le montrer, la possibilité de faire quand même une mesure correcte.

On comptera positivement de $B$ vers $A$ l'abscisse curviligne $s$ sur la ligne. Ce même sens positif servira à évaluer algébriquement l'intensité du courant et les forces électromotrices qu'on va supposer insérées sur la ligne, pour le besoin de la démonstration. En un point $M$ de la ligne, on figure par $e$ le champ tellurique instantané *primitif*, non perturbé, dont on rappellera que l'extrémité décrit périodiquement une ellipse. La composante $e_s$ de $e$ est sinusoïdale, ainsi que l'intensité du courant en un point de la ligne et que la déviation de l'appareil de mesure. Dans la notation imaginaire, le facteur $e^{-i\omega t}$ étant sous-entendu, ces grandeurs sont désignées par $E_s, I, D$.

Pour que la présence de la ligne tellurique n'apporte aucune perturbation électromagnétique, pour que cette ligne ne soit parcourue par aucun courant bien qu'elle se trouve soumise à l'action d'un champ électrique, le mathématicien imaginera que des alternateurs élémentaires de force électromotrice $-E_s\,ds$ et d'impédance nulle sont intercalés sur chaque tronçon de longueur $ds$. À vrai dire on supposera toujours que la durée d'établissement dans la ligne est très courte par rapport à la période, de sorte que les choses se simplifient notablement: $I$ est le même partout, $D$ ne dépend pas de l'emplacement de $G$ sur la ligne, enfin l'on peut remplacer tous ces alternateurs élémentaires par un alternateur unique dont la force électromotrice est:

$$\Phi = - \int_B^A E_s\,ds. \qquad (14.1)$$

Du fait que les équations de l'électromagnétisme sont linéaires, que la somme de deux solutions est aussi une solution, que l'insertion de $\Phi$ a pour effet d'annuler le courant de ligne, on peut conclure que l'insertion de $\Phi$ engendrerait un courant

$-I$ dans la ligne et une déviation $-D$ dans l'appareil de mesure si le phénomène tellurique qu'on étudie présentement n'existait pas. La méthode correcte pour étalonner la ligne tellurique résulte aussitôt de cette remarque. Le phénomène tellurique étant supposé ne pas exister quand on procède à cet étalonnage, champs et courants étant nuls initialement lors de l'opération, on insère une force électromotrice $\Phi_0$ de même période que celle supposée plus haut[1]. Elle engendre un courant $I_0$, une déviation $D_0$. Or l'on peut écrire:

$$\frac{\Phi}{\Phi_0} = \frac{(-I)}{I_0} = \frac{(-D)}{D_0} \tag{14.2}$$

ou:

$$\int_B^A E_s \, ds = D \cdot \left(\frac{\Phi_0}{D_0}\right) = I \cdot \left(\frac{\Phi_0}{I_0}\right). \tag{14.3}$$

On en conclut que la mesure de la déviation $D$ fait connaître l'intégrale du premier membre en grandeur et en phase.

La valeur de cette intégrale dépend non seulement de la position des électrodes $A$ et $B$, mais aussi de la forme du circuit. Le raisonnement ci-dessus s'applique sans modification au cas d'un circuit fermé: l'intégrale n'est pas nulle, elle mesure la dérivée du flux d'induction dans la boucle. Mais quand cette boucle devient très petite, quand sa longueur est assimilée à un infiniment petit du premier ordre, l'intégrale est infiniment petite du second ordre, en même temps que l'aire de la boucle. Ce résultat peut s'interpréter en d'autres termes: dans un domaine géométrique suffisamment petit, le champ électrique, au même titre d'ailleurs que n'importe quel vecteur à distribution continue, peut être considéré à la limite comme uniforme. Le Géophysicien, tant qu'il ne s'intéresse pas à ce qui se passe en dehors de ce petit domaine, conserve le droit de dire que *le champ tellurique*, comme n'importe quel vecteur uniforme, *dérive d'un potentiel*.

Quand la période $T$ devient suffisamment grande, les rapports $\Phi_0/D_0$ ou $\Phi_0/I_0$ deviennent, à la limite, indépendants de $T$. On peut donc dans ce cas, comme il est de règle dans l'étude expérimentale des variations telluriques très lentes, remplacer la force électromotrice alternative d'étalonnage $\Phi_0$ par une force électromotrice ou une chute de potentiel continues $\Phi_c$, lesquelles engendreront un courant continu $I_c$. D'après la loi d'OHM, on a d'ailleurs

$$\Phi_c = (R + R_A + R_B) \cdot I_c \tag{14.4}$$

où $R$ est la résistance totale de la ligne entre $A$ et $B$, tandis que $R_A$ et $R_B$ sont ce qu'on appelle les résistances des terres $A$ et $B$:

$$\int_B^A E_s \, ds = \frac{I}{R + R_A + R_B}. \tag{14.5}$$

Dans ce cas limite, intensité du courant et déviation sont en concordance de phase avec l'intégrale du premier membre.

Il faut beaucoup de prudence pour juger de la validité de (14.5) car il ne suffit pas que, dans le domaine de fréquences étudié, la self-induction des appareils en ligne soit négligeable, ni que les caractéristiques de ces appareils soient pratiquement des caractéristiques statiques. En effet, en régime alternatif, les phénomènes d'induction interviennent encore, et cela de deux manières différentes, pour fausser

---

[1] Il est clair qu'on ne peut pas supprimer le phénomène tellurique. Mais on peut toujours, par exemple, utiliser pour l'étalonnage une force électromotrice $\Phi_0$ assez grande pour qu'on puisse alors négliger le phénomène naturel superposé.

les raisonnements qui ont conduit à (14.5). D'une part le courant de retour par le sol, au lieu de se disperser rapidement, à la manière du courant continu, dans tout le volume offert, tend à se localiser dans les parties du sol situées au voisinage immédiat du câble. D'autre part, comme la densité de courant dans le sol est relativement élevée au voisinage de la ligne tellurique, il en résulte que le champ électrique y est aussi particulièrement intense. Tout se passe comme s'il apparaissait dans la ligne une force contre-électromotrice supplémentaire et comme si l'impédance de la ligne (résistance et self-induction) s'en trouvait modifiée.

# C. La prospection électrique.

**15. Remarque liminaire.** Les théories et les phénomènes dont, chacun de leur côté, s'occupent le Prospecteur sur le terrain et le Physicien du Globe dans son Observatoire, s'interpénètrent à un point tel qu'un exposé consacré à l'Electricité tellurique serait bancal s'il prétendait ignorer la prospection électrique. Mais on ne retiendra ici de ce qui touche à la prospection, en se limitant strictement à l'essentiel, que les principes théoriques, l'esprit des méthodes, les résultats d'ordre très général, en bref l'aspect proprement scientifique de la question. Pour les aspects plus particulièrement techniques, géologiques ou économiques qui concernent l'application des divers procédés de prospection, on consultera les traités spécialisés [1] ou les Mémoires des Ingénieurs.

**16. Prospection par courant continu. Technique des mesures.** La distribution du courant qu'on injecte dans le sol à l'aide de deux électrodes dépend des inégalités de conductibilité des roches et minerais du sous-sol. La mesure des différences de potentiel entre divers points de la surface fournit donc des indications sur la constitution du sol sous-jacent. Tel est le principe général de la prospection par courant continu. Il existe de nombreuses variantes d'application, entre lesquelles il faut faire un choix suivant la nature du problème géologique posé. Mais, dans les tous cas, une mesure individuelle se fait à l'aide d'un *quadripôle ABMN*. Comme il fut dit à la Sect. 5, les électrodes $A$ et $B$ servent à injecter le courant d'intensité $I$. Les électrodes $M$ et $N$, habituellement constituées par de «petits» piquets métalliques, servent à mesurer la différence de potentiel $\delta V$ entre ces deux points.

Si simple que paraisse la mesure d'une différence de potentiel, on y rencontre pourtant certaines difficultés qui doivent être signalées. Et d'abord, avant même qu'on ait injecté de courant, une différence de potentiel préexiste entre $M$ et $N$. Elle est due pour une part aux forces électromotrices telluriques les plus diverses dont le sol est le siège et pour une autre part, de beaucoup la plus importante en général, aux différences de potentiel existant au contact des piquets $M$ et $N$ avec le sol. On sait que les différences de potentiel au contact d'un métal et d'un électrolyte peuvent atteindre ou dépasser le volt: qu'elles ne sont guère supérieures à quelques millivolts au contact de deux électrolytes ou de deux métaux. Dans le cas actuel, bien que $M$ et $N$ soient de même métal, les électrolytes du sol dans le voisinage de ces deux électrodes ne sont jamais absolument identiques et il résulte de là que les piquets $M$ et $N$, avec la ligne de mesure qui les connecte, sont l'équivalent d'une pile dont la force électromotrice est souvent de l'ordre du volt. Or, à moins de lignes $AB$ très courtes, il est difficile en pratique d'engendrer des $\delta V$, dus au courant injecté, supérieurs à quelques Millivolts.

D'après une opinion courante, mais erronée, le remède consiste à utiliser en $M$ et $N$ des «*électrodes impolarisables*» au lieu de piquets métalliques. On fabrique facilement une électrode impolarisable avec un vase poreux, empli d'une solution saturée de sulfate de cuivre et de cristaux en excès, au sein duquel plonge une tige de cuivre. Quand on emploie de telles électrodes, la dissymétrie des connexions sol-électrode ne dépend plus de celle de contacts métal-électrolyte mais de celle, beaucoup plus faible, de contacts électrolyte-électrolyte qui se font au travers de la paroi poreuse. Car, de cette manière, les tiges de cuivre de $M$ et $N$ sont en contact avec des solutions à peu près identiques de sulfate de cuivre. Si les électrodes ne sont pas à la même température, la dissymétrie qui en résulte est de l'ordre du millivolt par degré d'écart. On constate effectivement, avec de telles électrodes bien construites, que la force électromotrice de la «pile» $MN$ ne dépasse guère quelques millivolts. La dissymétrie se trouve donc très fortement réduite, mais pas assez toutefois, tant s'en faut, pour qu'il soit permis de la négliger quand on mesure $\delta V$.

En fait, si l'on a recours à une technique appropriée, il importe peu que cette dissymétrie soit petite ou soit grande. Ce qui compte est qu'elle soit stable; et l'expérience montre qu'en prospection courante des piquets de métaux bien choisis sont à peu près aussi «stables» que des vases poreux. Ils sont aussi d'un emploi plus agréable. Qu'il s'agisse ou non d'électrodes impolarisables, on peut remarquer de reste que les «quantités d'électricité» mises en jeu lors d'une mesure sont tellement infimes

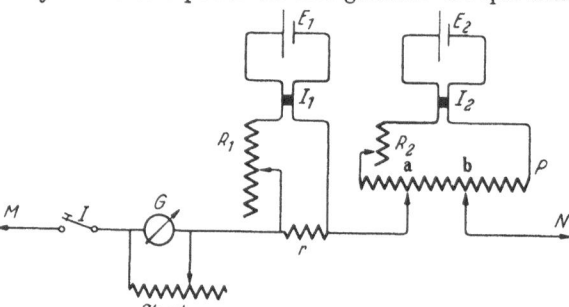

Fig. 3. Schéma d'un potentiomètre de prospection.

qu'elles ne risquent pas de faire varier la polarisation des électrodes. La technique va consister à insérer dans la ligne $MN$, par le jeu d'un potentiomètre continu à réglage progressif, une force électromotrice de «compensation» égale et opposée à la différence de potentiel préexistante, qu'il est d'ailleurs sans intérêt de chercher à déterminer. Pour les mesures courantes, avec de courtes lignes $AB$, mesures dont la précision n'a jamais besoin d'atteindre 1%, il est commode d'utiliser un robuste et léger «potentiomètre de prospection» plus ou moins analogue à celui dont la Fig. 3 indique le schéma et dont la Fig. 4 montre l'aspect extérieur. Les éléments de piles sèches de 1,5 V, $E_1$ et $E_2$, sont insérés le premier dans le dispositif potentiométrique de compensation, le second dans celui de mesure. Les deux dispositifs en question sont d'ailleurs identiques, si ce n'est que le premier comporte des rhéostats à curseur, le second des rhéostats à plots. Au début de la mesure, les inverseurs $I_1$ et $I_2$ sont ouverts. Avant d'injecter un courant dans $AB$, on commence par régler la compensation: le galvanoscope $G$ dévie sous l'effet de la polarisation de $M$ et $N$ quand on ferme l'interrupteur de ligne $I$, et le réglage initial consiste à fermer l'inverseur $I_1$ dans le sens convenable puis à manœuvrer le rhéostat $R_1$ jusqu'à annuler la déviation de $G$. Désormais l'on ne touche plus au circuit de compensation ainsi équilibré et l'on injecte le courant dans $AB$. A nouveau $G$ dévie, et l'on compense encore cette déviation en agissant sur le circuit potentiométrique de mesure de la même façon qu'on a fait précédemment sur le circuit de compensation. Ce réglage obtenu, il suffit de lire directement sur les rhéostats à décades du circuit potentiométrique de mesure la valeur de $\delta V$ en millivolts.

Pas davantage que par la polarisation des électrodes, la précision ni la sensibilité de la mesure ne sont limitées par l'existence de forces électromotrices telluriques naturelles, qui sont constantes dans le temps ou ne varient que lentement.

Il en va tout autrement quand se manifestent les fluctuations incessantes et capricieuses des courants telluriques proprement dits, et c'est pourquoi il ne servirait à rien de vouloir équiper les potentiomètres de terrain avec des galvanomètres vraiment sensibles.   Quand les fluctuations telluriques commencent à agiter l'aiguille du galvanomètre, la précision de la mesure ne peut être accrue que si l'on parvient à augmenter $\delta V$ sans augmenter pour autant les perturbations telluriques dans le même rapport.   De sorte que le seul remède est d'accroître, *si on peut*, l'intensité du courant injecté dans $AB$.

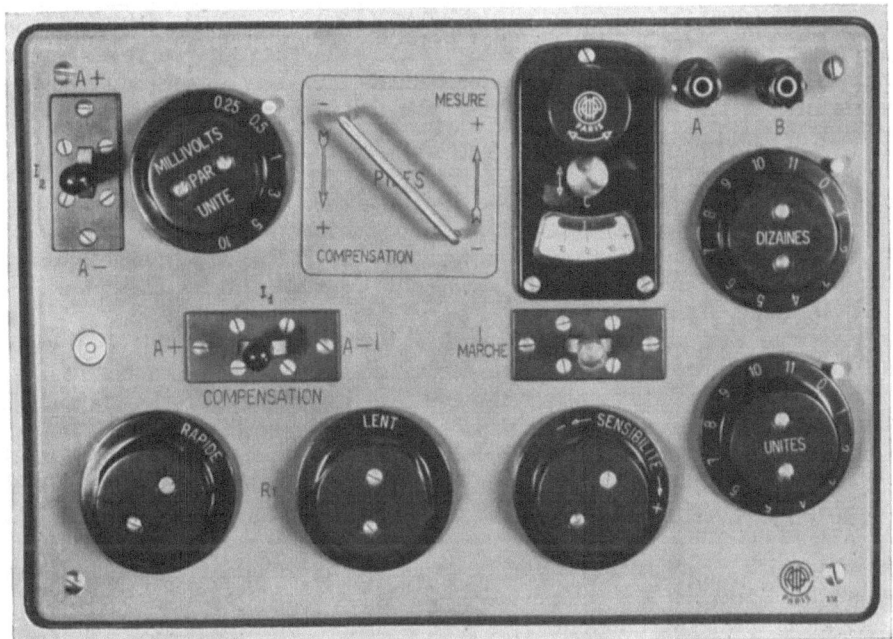

Fig. 4. Aspect extérieur d'un potentiomètre de prospection.

Lorsque les lignes $AB$ sont longues de plusieurs km. et que, pour des raisons pratiques, il ne faut plus songer à accroître l'intensité du courant injecté, on doit renoncer au potentiomètre et le remplacer par un galvanomètre *enregistreur* suffisamment sensible.   On emploie usuellement les mêmes enregistreurs que ceux qui servent à la prospection tellurique.   A proprement parler, le dispositif de mesure n'est alors pas autre chose qu'une «ligne tellurique» classique.   Les fluctuations telluriques rapides s'inscrivent photographiquement sur une bande de papier qu'on déroule à une vitesse de l'ordre du cm par minute.   On fait un étalonnage «statique» suivant la méthode exposée avec plus de généralité à la Sect. 14. Pour préciser, on insère dans $MN$ une force électromotrice continue donnée $\Phi_c$ de 1 mV. par exemple, ce qui entraine un déplacement $d$ du spot, ayant le caractère d'une discontinuité.   Plus tard, on injecte brusquement le courant dans la ligne $AB$; le spot éprouve alors une discontinuité $D$ et l'on a, rigoureusement:

$$\delta V = \Phi_c \cdot \frac{D}{d} . \tag{16.1}$$

Lorsqu'on emploie cette méthode de déviation galvanométrique au lieu de la méthode d'opposition et de zéro utilisée avec le potentiomètre, la différence de potentiel existant entre $M$ et $N$ n'est pas égale, bien entendu, à $\delta V$ et lui est même très inférieure si la résistance du galvanomètre est petite en comparaison de celle des terres $M$ et $N$. Le régime des courants dans le sol se trouve alors profondément perturbé par l'existence même du dispositif de mesure, ce qui n'empêche pourtant pas la mesure d'être correcte.

Cette méthode de déviation ne permet quand même pas d'augmenter indéfiniment la sensibilité des mesures, car l'établissement d'un courant dans une ligne n'est pas instantané, ainsi qu'on l'a remarqué à la Sect. 12. La durée d'établissement est d'autant plus grande que la ligne est plus longue et le sol plus conducteur. Si $AB$ est long d'une dizaine de km. et si le sol est marneux, l'état de régime électromagnétique n'est pas encore atteint au bout d'une minute. Le spot n'éprouve donc plus de discontinuité, mais un déplacement progressif de plus en plus difficile à évaluer ou même à reconnaître au milieu de l'agitation tellurique incessante. Il faut alors envisager des procédés de mesure plus complexes dont il ne sera pas question ici.

**17. Carte des potentiels.** Les premiers prospecteurs s'en tenaient à la « carte des potentiels ». Dans cette méthode, les électrodes $A$ et $B$ du quadripôle sont fixes, et l'on maintient constante l'intensité du courant injecté. Ce sont les électrodes $M$ et $N$ qu'on déplace systématiquement afin d'aboutir à dresser une carte des potentiels de la surface, sous forme d'un réseau de lignes équipotentielles graduées dont le zéro est à l'infini.

Les potentiels ainsi mesurés ne sont pas les potentiels normaux (5.3) qu'on aurait mesurés sur un sol homogène. On constate l'existence d'«anomalies» qui attirent l'attention du prospecteur et doivent lui fournir des indications précieuses. Par exemple, la frontière entre deux affleurements inégalement résistants se traduit sur la carte des potentiels par une ligne tout le long de laquelle les équipotentielles sont réfractées et présentent un point anguleux. Rien n'est pratiquement changé, sinon que le point anguleux fait place à un coude plus ou moins brusque, lorsque les affleurements sont masqués par un recouvrement qui n'est pas trop épais. De même une zone de formations conductrices se signale par un espacement anormal des équipotentielles.

L'idéal serait pourtant d'obtenir des informations plus précises, qui seraient numériques, quantitatives. On se rend compte d'ailleurs que le processus d'une telle interprétation quantitative ne pourrait pas être autre chose que le suivant : imaginer une structure vraisemblable du sous-sol en tenant compte de ce que la géologie permet de présumer et de ce que la carte géophysique autorise à soupçonner ; représenter cette structure hypothétique sous forme d'une épure des diverses formations, de telle ou telle résistivité connue ou supposée ; calculer les anomalies théoriques correspondant à ladite structure hypothétique ; comparer anomalies calculées et observées ; retoucher l'épure primitive en vue d'atténuer les divergences constatées ; continuer ainsi par approximations successives jusqu'à accord satisfaisant entre les anomalies théoriques et expérimentales.

Ce pourrait être dangereux de s'illusionner trop sur la portée véritable d'interprétations quantitatives qu'on parviendrait à mener de la sorte. Car, indépendamment d'une certaine marge inévitable d'imprécision, il est clair que le problème analytique qui consiste à déterminer la structure du sol en se donnant la répartition des potentiels superficiels comporte de très larges indéterminations. Malgré tout, une interprétation pouvant se justifier par des nombres sera toujours beaucoup plus sûre qu'une interprétation exclusivement qualitative. Et par surcroît, il ne faut pas oublier que la géologie impose des limitations très strictes à l'imagination mathématicienne, de sorte que les indéterminations signalées ne sont quand même pas si redoutables qu'il peut paraître au premier abord. Sans le moindre doute possible, si la géophysique des pionniers fut exclusivement une géophysique qualitative, la géophysique de demain sera de plus en plus une géophysique quantitative.

C'est précisément là qu'apparaît le véritable «drame» de la prospection électrique. Car à moins d'envisager des structures tellement simples et schématiques qu'elles n'ont pas d'intérêt pratique, on ne peut que s'avouer incapable d'intégrer les équations de la Sect. 5. On ne sait pas faire d'interprétation quantitative, à moins d'expérimenter sur modèles réduits. Ce qui n'est pas du tout facile en pratique et n'est certes pas un moyen qui fut beaucoup utilisé jusqu'ici en prospection.

**18. Carte des résistivités apparentes. Profondeur d'investigation.** On trouve souvent avantage à changer de tactique et à dresser une «carte des résistivités apparentes». Il faut d'abord définir ce qu'on entend par résistivité apparente. On a observé déjà qu'en résolvant (5.4) on peut déterminer la résistivité $\varrho$ du sol supposé homogène. Quand le sol n'est pas homogène, quand sa structure est absolument quelconque et même quand sa surface n'est pas plane, il demeure possible d'implanter un quadripôle, de mesurer $I$ et $\delta V$, puis de résoudre (5.4) par rapport à $\varrho$ exactement comme si le sol était homogène. Le nombre $\varrho_a$ qu'on calcule ainsi est ce qu'on appelle «*résistivité apparente*». La résistivité·apparente dépend non seulement de l'emplacement du quadripôle, mais aussi de sa configuration, de ses dimensions, de son orientation. On réduit d'ordinaire le nombre de paramètres arbitraires en n'utilisant qu'un quadripôle de configuration et de dimensions fixes qu'on transporte tout d'une pièce d'une station à l'autre en lui conservant une orientation invariable, laquelle est habituellement celle de la direction tectonique principale.

Les quadripôles employés en pratique sont toujours symétriques: leurs quatre électrodes sont en ligne droite, le milieu $O$ de $AB$ est le même que celui de $MN$, et c'est au point $O$ qu'on attribue, quand on dresse la carte de résistivités, la résistivité apparente qu'on a mesurée dans ces conditions. On utilise surtout le quadripôle de WENNER, défini par $AM = MN = NB$, et le quadripôle de SCHLUMBERGER où $MN$ est théoriquement un infiniment petit. Dans le quadripôle de SCHLUMBERGER, $MN$ doit être assez petit en effet par rapport à $AB$, le dixième au plus de $AB$, pour que la mesure de $\delta V$ soit équivalente à celle du champ en $O$.

En vertu de ce qu'on vient de dire et de (5.4), la résistivité apparente $\varrho_a$ qui, d'une manière tout à fait générale, a pour expression

$$\varrho_a = 2\pi \cdot \frac{\delta V}{I} \cdot \frac{1}{\dfrac{1}{\overline{MA}} - \dfrac{1}{\overline{MB}} - \dfrac{1}{\overline{NA}} + \dfrac{1}{\overline{NB}}} \tag{18.1}$$

devient, quand le quadripôle est symétrique

$$\varrho_a = \frac{\pi}{4} \cdot \frac{\delta V}{I} \cdot \frac{\overline{AB^2} - \overline{MN^2}}{\overline{MN}}. \tag{18.2}$$

S'il s'agit du quadripôle WENNER, on a donc

$$\varrho_a = \frac{2\pi}{3} \cdot \frac{\delta V}{I} \cdot \overline{AB} \tag{18.3}$$

tandis que la formule qui s'applique au quadripôle SCHLUMBERGER s'écrit

$$\varrho_a = \frac{\pi}{4} \cdot \frac{\delta V}{I} \cdot \frac{\overline{AB^2}}{\overline{MN}}. \tag{18.4}$$

Ce n'est pas ici le lieu de discuter les mérites respectifs des deux types de quadripôles, mais il est bon de remarquer quand même:

1. Que les «résistivités SCHLUMBERGER» ne sont pas égales aux «résistivités WENNER». Il faut se garder d'interpréter les premières avec des abaques WENNER, et vice-versa.

2. Que $\delta V$ est beaucoup plus grand dans un quadripôle Wenner que dans un quadripôle Schlumberger, mais qu'il serait faux d'en inférer que la mesure est plus facile et plus précise dans le premier cas que dans le second. Car il ne faut pas oublier que les perturbations telluriques sont proportionnelles à la longueur de $MN$.

Les Ingénieurs affectionnent les formules lapidaires, qui font image mais trahissent parfois la réalité. C'est ainsi qu'ils déclarent que la «profondeur d'investigation» est égale au quart de $\overline{AB}$. D'autre part la résistivité apparente serait une moyenne des résistivités (vraies) des divers terrains contenus dans une tranche d'épaisseur $\overline{AB}/4$. Il n'y a dans ces affirmations tranchantes que ce qu'il faut de vrai pour justifier dans une certaine mesure l'usage qu'on veut faire des cartes de résistivités.

La notion de *profondeur d'investigation* répond à la préoccupation dominante du prospecteur: quelle longueur de quadripôle doit-il adopter quand il présume que la masse perturbatrice à déceler se trouve à une certaine profondeur? La

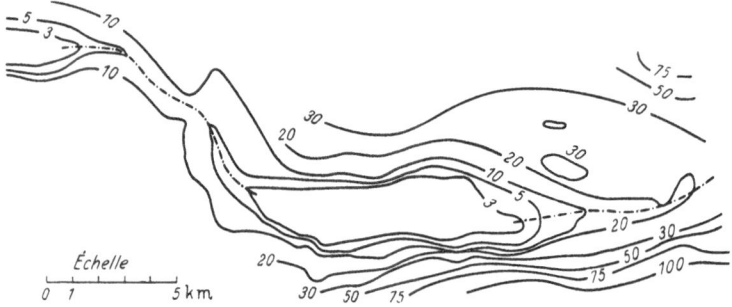

Fig. 5. Carte de résistivités apparentes (anticlinal de Grosny, U.R.S.S.). D'apres C. Schlumberger.

profondeur d'investigation est donc par définition la profondeur maxima au-dessous de laquelle la masse perturbatrice ne doit pas se trouver pour qu'on ait chance de la découvrir. Que la définition soit forcément arbitraire, cela résulte de ce que la présence d'une masse perturbatrice se manifeste «en théorie» à n'importe quelle profondeur où elle puisse se situer. Mais ce n'est pas une raison valable pour dénier au prospecteur le droit de définir très vaguement un para-mètre qui a tant d'importance et de signification pratiques pour lui.

De quelque manière qu'on veuille la définir, la profondeur d'investigation est proportionnelle à $\overline{AB}$, dans la mesure où l'on traduit par là une simple loi de similitude. Mais quand il s'agit de préciser davantage, il faudrait que la réponse fût beaucoup plus nuancée. Car la profondeur d'investigation ne peut pas dépendre uniquement de $\overline{AB}$. Elle dépend aussi, énormément, de la configuration de la masse perturbatrice, de ses dimensions et de la valeur du contraste de résistivités. On se doute qu'à profondeur égale il soit plus facile de découvrir une masse volumineuse qu'une masse minuscule. Il est aussi plus facile de rechercher une masse conductrice sous un recouvrement résistant que de faire l'inverse, etc. Quant à la règle des prospecteurs, règle de $\overline{AB}/4$, elle péche toujours, comme il se doit, par un très large excès d'optimisme. Dans les circonstances les plus favorables, une longueur $\overline{AB}$ de $4h$ permet tout juste de commencer à déceler la présence de quelque chose à la profondeur $h$, mais si l'on désire un minimum de précision et de sécurité dans l'interprétation, il est beaucoup plus prudent d'adopter une longueur $\overline{AB}$ de 20 à 100 fois $h$. Il apparaît ainsi un handicap nouveau et très regrettable des méthodes électriques, savoir leur profondeur d'investigation très limitée. Ces méthodes font souvent merveille en prospection minière, dans les études de Génie Civil comme dans les recherches hydrologiques. Il ne faut guère compter sur elles pour l'étude des grandes structures, du genre des structures pétrolifères, en dehors de quelques exceptions rarissimes. La Fig. 5 est précisément une de ces exceptions. C'est, d'après un travail «quasi-historique» de SCHLUMBERGER, un fragment de la carte des résistivités de l'anti-clinal pétrolifère de Grosny (Russie). Les lignes $AB$ n'avaient pourtant qu'un km. de long. Les résistivités apparentes sont inscrites en $\Omega$.m. sur les lignes d'équirésistivité. La résistivité $3\,\Omega$.m. est celle du sarmatien qui affleure en deux endroits et qui s'enfonce ailleurs sous des formations plus récentes et plus résistantes. Il n'est pas besoin de plus amples commentaires pour montrer combien une carte de résistivités est plus suggestive pour un Géologue qu'une carte de potentiels, pourtant équivalente, mais où le Physicien lui-même ne se reconnaît pas toujours.

**19. Sondages électriques.** L'opérateur peut encore stationner au même endroit, au centre fixe $O$ de son dispositif dont il accroît progressivement la longueur en passant d'une mesure à la suivante. Quand $\overline{AB}$ augmente, des terrains de plus en plus profonds interviennent dans la définition de la résistivité apparente et si l'on construit le graphique qui représente les variations de la résistivité apparente en fonction de la longueur $\overline{AB}$, on se trouve par là-même informé, dans une certaine mesure, de la variation de résistivité du sol en fonction de la pro-fondeur. Il y a, dans cette opération, quelque chose qui rappelle le sondage mécanique. C'est un «sondage électrique».

Le domaine idéal pour appliquer les sondages électriques est celui des *struc-tures tabulaires* (ou approximativement telles), où le sol est formé de couches horizontales superposées, en nombre quelconque, d'épaisseurs et de résistivités quelconques. Au contraire des méthodes électriques, les méthodes gravimétri-ques et magnétiques ne sont pas bien adaptées à ce genre d'études, car les ano-malies gravimétriques ou magnétiques sont nulles quand les structures sont rigoureusement tabulaires. Le cas de telles structures est d'ailleurs extrêmement important en prospection pratique puisque les grands bassins sédimentaires peu plissés le réalisent en première approximation. Et c'est aussi à peu près le seul cas véritablement intéressant qu'il soit donné au théoricien d'aborder par le calcul avec succès.

Le premier calcul fut fait par HUMMEL[1], par la méthode des images électri-ques. Plus tard, STEFANESCU et SCHLUMBERGER[2] appliquèrent à ce problème la méthode générale d'intégration de RIEMANN. HUMMEL trouve le résultat sous forme de séries. La méthode de RIEMANN, bien plus élégante, l'exprime sous forme d'intégrales définies. Ce sont d'ailleurs les séries de HUMMEL qui se prê-tent le mieux aux calculs numériques mais, pour aboutir en fin de compte à ces séries, il vaut tout de même mieux passer par l'intermédiaire des intégrales.

Pour l'Ingénieur, l'important n'est pas de calculer un sondage électrique quand il connaît le sous-sol. C'est le problème inverse: déterminer les épais-seurs et résistivités des couches, à partir du sondage électrique. A la condition expresse de supposer a priori que le sol est isotrope et, bien entendu, tabulaire, il ne semble pas exclu que la solution du problème inverse, si elle existe, soit unique. Mais les prétendues démonstrations d'unicité apportées jusqu'ici ne paraissent guère convaincantes; et il est certain d'autre part que deux struc-tures tabulaires très différentes peuvent avoir des sondages électriques, sinon

[1] J. N. HUMMEL: Z. Geophys. **5**, 89, 228 (1929).
[2] S. STEFANESCU et C. et M. SCHLUMBERGER: J. de Phys. **1**, 132 (1930).

absolument identiques, du moins indiscernables en pratique[1]. C'est pourquoi, dans la méthode des sondages comme dans toutes les autres méthodes électriques, c'est toujours aux approximations successives qu'on doit en revenir en dernière analyse.

La méthode des approximations successives n'est pourtant pratiquement applicable que si le calcul a priori d'une structure hypothétique donnée est

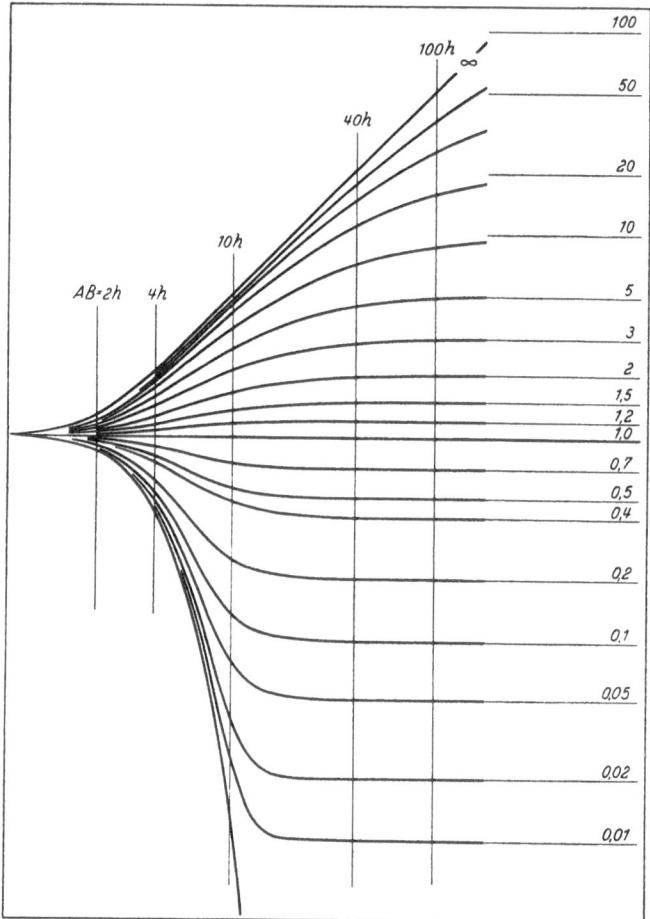

Fig. 6. Abaque bi-logarithmique de sondages électriques «2 terrains».

relativement rapide. C'est loin d'être le cas pour un sondage électrique un peu complexe, qui peut exiger facilement une bonne semaine de laborieux calculs. De sorte qu'on se borne en fait à dresser des catalogues de sondages théoriques dans l'espoir, hélas toujours déçu, d'y trouver précalculée la courbe théorique permettant d'interpréter toute courbe expérimentale. Et ces déceptions font abandonner également de plus en plus ces inutiles et encombrants catalogues en faveur de procédés semi-empiriques d'interprétation qui reposent, tout compte fait, sur l'utilisation des abaques «2 terrains»[2].

[1] L. CAGNIARD: Ann. Inst. Phys. Globe Strasbourg, 3. Partie **4**, 3 (1948).
[2] L. CAGNIARD: Congrès Ankara de 1952, Hydrologie zone aride, U.N.E.S.C.O. Paris, (1953), p. 184.

Les sondages «*2 terrains*» constituent une famille de courbes à 3 paramètres, savoir les deux résistivités en présence et l'épaisseur du premier terrain. Des considérations élémentaires de similitude électrique ou géométrique permettent de ramener le cas général au cas particulier d'un premier terrain de résistivité unité et d'épaisseur unité, de telle sorte que la totalité des sondages «2 terrains» ne représente plus qu'un réseau de courbes à un seul paramètre. Cela revient à utiliser des coordonnées bi-logarithmiques où les abscisses sont les logarithmes du rapport de $\overline{AB}/2$ à l'épaisseur du premier terrain, où les ordonnées sont les logarithmes du rapport de la résistivité apparente à la résistivité du premier terrain. C'est ainsi qu'est construit l'abaque de la Fig. 6. Les nombres inscrits sur les asymptotes sont les rapports de la résistivité du second terrain à celle du premier. On constate donc que la résistivité apparente (ici c'est une résistivité SCHLUMBERGER), égale à la résistivité du premier terrain quand $\overline{AB}$ est très court, tend vers la résistivité du second quand $\overline{AB}$ croît indéfiniment.

Les abaques d'interprétation ne comportent aucune graduation d'abscisses ou d'ordonnées car ce sont les sondages expérimentaux qu'on dessine sur un calque bi-logarithmique du commerce, de mêmes modules évidemment que sur les abaques. On regarde l'abaque théorique par transparence au travers du calque expérimental en cherchant à faire coïncider la courbe expérimentale qu'on présume du type «2 terrains», avec l'une des courbes théoriques. Si la coïncidence peut être réalisée, au prix d'une translation convenable par rapport à l'axe des abscisses et d'une autre translation par rapport à l'axe des ordonnées, on connaît du même coup l'épaisseur du premier terrain et les deux résistivités. Pour rendre la Fig. 6 plus clairement intelligible, on a quand même jugé bon d'y surimprimer une graduation en abscisses $\overline{AB} = 2h$, $4h$, $10h$, etc., où $h$ représente l'épaisseur du premier terrain. On méditera sur les énormes longueurs de lignes qui sont nécessaires, surtout quand le substratum est résistant, pour ne réaliser en fin de compte que des investigations de profondeur relativement médiocre.

**20. Courants alternatifs ou variables.** Dans les débuts de la prospection électrique, quand on préconisait l'emploi des courants alternatifs au lieu et place du courant continu, c'était manifestement surtout pour supprimer les difficultés provenant de la polarisation des électrodes. Il est pourtant bien facile, comme on a vu, de s'affranchir de ces difficultés sans renoncer pour autant à l'emploi si pratique du courant continu.

Par ailleurs, à l'époque héroïque des pionniers, on avait grand tort de s'imaginer que les courants alternatifs utilisés, dont la fréquence s'élevait parfois jusqu'à 1000, suivaient encore la loi d'OHM. On était souvent loin de compte. La plupart de ces mesures anciennes effectuées en courant alternatif ne sont donc pas correctes. Et il va de soi qu'à ce point de vue l'on ne saurait établir de distinction entre les procédés qui utilisent un véritable courant alternatif et ceux dans lesquels on commute un courant continu, à une fréquence plus ou moins rapide, à l'aide d'un commutateur rotatif.

On a montré au chapître **B** comment la distribution d'un courant alternatif peut ou pourrait être traitée correctement. Cela entraînerait bien entendu à des calculs plus laborieux qu'en continu, mais ces calculs demeurent praticables quand il s'agit de structures tabulaires, c'est à dire dans le seul cas vraiment important où l'on en vient à bout quand le courant est continu.

Mais le skin-effect ne se borne malheureusement pas à rendre les phénomènes plus complexes et les calculs plus laborieux. Sa conséquence de beaucoup la plus fâcheuse, et souvent irrémédiable, est d'imposer une limite infranchissable à la profondeur utile des investigations. Avec le courant continu, la profondeur d'investigation peut toujours théoriquement être accrue à volonté quand on augmente $AB$. Avec l'alternatif, quoi qu'on fasse, on ne dépassera jamais la profondeur d'investigation permise par le skin-effect.

C'est donc une aberration scientifique que de songer à employer l'alternatif pour les recherches de pétrole. Quand il s'agit de détecter des mines explosives, il est possible, rationnel et pratique de s'adresser aux fréquences radioélectriques, mais l'emploi de telles fréquences en prospection minière serait une autre aberration. Si l'utilisation de l'alternatif présente occasionnellement quelque avantage, c'est en définitive parce qu'il autorise l'emploi de dispositifs purement inductifs, ne comportant aucune *prise de terre*. Cela peut constituer un avantage décisif dans des terrains arides ou des rocs desséchés, quand des fréquences acoustiques de 100 à 500 permettent une profondeur d'investigation qui suffit à l'objet des recherches. Le courant alternatif circule alors dans un câble isolé posé à terre et relié aux bornes de l'alternateur. Au lieu de mesurer le champ électrique, on détermine le champ magnétique en se servant d'un cadre de petites dimensions, orientable, qui comporte un grand nombre de spires de fil fin, dans lequel les variations de flux d'induction engendrent une force électromotrice.

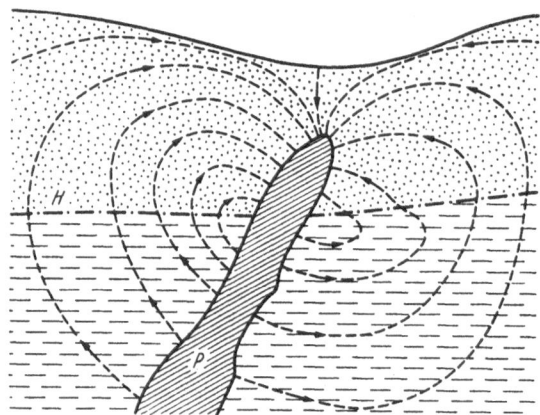

Fig. 7. Phénomènes de polarisation spontanée provoqués par un filon.

**21. Polarisation spontanée.** Dans certaines conditions, tout à fait exceptionnelles, peuvent prendre naissance des courants telluriques naturels locaux, dont la direction et l'intensité sont invariables dans le temps. Ce sont les courants découverts par Fox, ainsi qu'il fut dit à la Sect. 1. Schlumberger expliqua, avec de jolies expériences de laboratoire à l'appui, qu'il s'agissait d'un phénomène de polarisation spontanée, en abrégé P. S., de certains gîtes miniers soumis à oxydation. Le phénomène se produit quand il existe en profondeur une masse oxydable de substances qui présentent le type métallique de conductibilité. Il s'agit surtout des pyrites et chalcopyrites, des associations B. P. G. de blende, pyrite et galène, des anthracites, du graphite, mais aussi ... des ferrailles et canalisations de fonte ou d'acier enfouies. La Fig. 7 représente une masse pyriteuse $P$ allongée dans le sens vertical, enracinée au-dessous du niveau hydrostatique $H$. Les parties superficielles, soumises à l'action de l'oxygène, atmosphérique ou dissous dans les eaux d'infiltration, sont dans un milieu beaucoup plus oxydant que les parties profondes. La différence de potentiel au contact du minerai et de l'électrolyte du sol atteint l'ordre du volt, et elle n'est pas la même suivant qu'il s'agit des parties superficielles ou profondes. C'est grâce à la dissymétrie chimique due à l'oxydation que le gisement se comporte comme une immense pile dont la force électromotrice atteint couramment plusieurs centaines de millivolts, et dont le circuit se trouve fermé par le sol environnant. Le courant suit le sens des flèches de la figure, apportant par conséquent des ions hydrogène réducteurs dans les parties superficielles du gîte et tendant à y atténuer les phénomènes d'oxydation. En raison du sens du courant, le potentiel à la surface du sol est minimum à l'aplomb du gisement.

Pour mesurer les différences de potentiel de P. S., on se sert du potentiomètre de prospection décrit plus haut, sans que le circuit de compensation soit alors utile car on utilise, bien entendu, des électrodes impolarisables. La mesure ne soulève aucune difficulté car les champs telluriques normaux, variables en fonction du temps, sont d'un ordre de grandeur très inférieur aux champs telluriques de P. S. Il convient cependant d'être en garde envers certaines causes d'erreur provenant de champs telluriques d'électrofiltration engendrés par le ruissellement, l'infiltration ou l'évaporation des eaux. Ce sont d'ailleurs là des phénomènes sur lesquels il faudra revenir plus loin car ils peuvent perturber toute mesure tellurique.

La Fig. 8 est un exemple de prospection par P. S. sur un gîte de sulfures métalliques. Les équipotentielles sont graduées en millivolts. Les parties superficielles du gîte correspondent à la zone d'anomalies négatives, avec trois apophyses qui donnent lieu à des minima voisins de —150 millivolts. Les anomalies positives de la partie gauche reflètent l'influence des parties profondes du gisement, dont le pendage général est orienté vers la gauche de la figure.

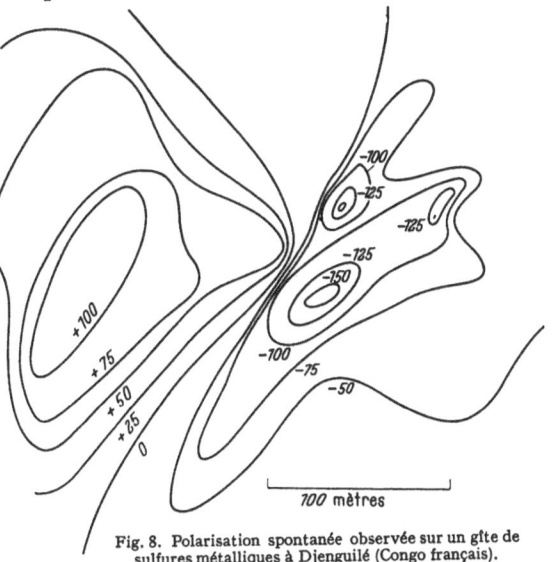

Fig. 8. Polarisation spontanée observée sur un gîte de sulfures métalliques à Djenguilé (Congo français).

# D. Dispositifs expérimentaux pour l'étude des courants telluriques. Perturbations diverses.

**22. Champs telluriques proprement dits et champs perturbateurs.** Les champs électriques en général, les champs telluriques en particulier, relèvent des causes les plus diverses. Certaines sont manifestement locales, et ne sont même parfois qu'instrumentales, quand il s'agit par exemple de la polarisation des électrodes de mesure. Le Géophysicien se refuse à bon droit à considérer tous ces champs d'origine locale comme de vrais champs telluriques, et il les traite comme des perturbations qu'il faut dépister soigneusement pour faire en sorte que les mesures n'en soient affectées qu'au minimum. Alors, et alors seulement, on peut dire qu'on enregistre le champ tellurique véritable dans toute sa pureté, et l'on se trouve en présence d'un phénomène géophysique à l'échelle de la planète, directement apparenté à d'autres grands phénomènes géophysiques ou cosmiques, aux variations du champ magnétique terrestre, à l'activité solaire, à l'activité aurorale.

Parmi les phénomènes perturbateurs figurent les courants vagabonds provoqués par les lignes télégraphiques, par les mises à la terre occasionnelles ou systématiques des installations électriques, par les effets d'induction des lignes électriques ou par leurs effets de capacité, mais surtout par les défauts d'isolement des rails conducteurs sur les lignes de chemin de fer électrifiées en continu. Ce sont ces courants vagabonds qui, depuis longtemps déjà, interdisent tout enregistrement tellurique valable dans les pays de civilisation industrielle. En raison

de leurs méfaits variés, qui ne sont pas seulement d'ordre géophysique, ces vagabonds ont souvent donné lieu, de la part d'Ingénieurs, à des études dont certaines présentent un réel intérêt géophysique[1, 2].

La *foudre*, en tombant à proximité de l'Observatoire, provoque des perturbations telluriques aussi bien que magnétiques, mais qui ne sont guère gênantes en raison de leur caractère sporadique et de leur brièveté.

Personne ne songerait à installer une station tellurique à proximité de gîtes
à P. S., mais il faut se méfier de la P. S. à laquelle donnent lieu les pylônes,
canalisations, ferrailles diverses qui s'oxydent dans le sol. D'autre part toute
canalisation métallique, pour amener l'eau par exemple, à supposer qu'elle
n'engendre pas de P. S., peut quand même perturber, par sa seule présence en
terre, le régime de circulation des courants telluriques. Les canalisations modernes
en matériaux plastiques sont certainement plus recommandables.

Un observatoire tellurique doit être situé loin de la mer, car on verra que
les enregistrements telluriques effectués au voisinage des côtes sont très fortement perturbés par des phénomènes variés en rapport avec la marée.

Une ligne tellurique aérienne que le vent fait osciller dans le champ magnétique terrestre est le siège de forces électromotrices d'induction indésirables.

Au surplus, chaque fois que le sol présente un défaut d'uniformité de quelque
nature que ce soit, chaque fois qu'il y a un gradient thermique, ou un gradient
de concentration ionique, ou un gradient de pression, etc., il faut s'attendre
à l'existence de champs perturbateurs. Exception faite de la polarisation et
de la P. S., il semble heureusement que tous ces champs électriques de non-
homogénéité demeurent très faibles. Comme on peut espérer d'autre part qu'ils
ne varient guère en fonction du temps, ils sont peu gênants lorsqu'on porte attention principalement aux *variations* telluriques pas trop lentes.

Enfin la circulation souterraine des eaux, l'infiltration, l'évaporation peuvent
être aussi l'origine d'importantes perturbations.

Comme l'essentiel de ce qui concerne la P. S. a été signalé à la Sect. 21, on
va se borner ci-après à évoquer les phénomènes d'électrofiltration et l'influence
des marées océaniques.

**23. Phénomènes d'électrofiltration.** Quand on crée un gradient de pression
pour forcer un électrolyte à filtrer au travers d'un diaphragme ou d'un tube
capillaire, il naît une différence de potentiel entre deux électrodes immergées
dans les portions de liquide que sépare la cloison poreuse ou le tube. Si le circuit
électrique est fermé, il naît un courant électrique. Autrement dit, du fait même
de la circulation du liquide, une force électromotrice, la *«force électromotrice d'électrofiltration»*, se manifeste dans la traversée du diaphragme. Tout se passe comme
s'il y avait un champ électrique, le «champ d'électrofiltration» en chaque point
situé dans l'épaisseur de la cloison.

Réciproquement, si l'on crée une différence de potentiel entre les deux com-
partiments du liquide, on force l'électrolyte à filtrer au travers du diaphragme
ou à s'écouler dans le tube. En cela consiste *l'osmose électrique*. Les deux phénomènes, électrofiltration et osmose électrique, sont d'autant plus intenses que
le liquide est *moins* conducteur.

Ces phénomènes sont connus depuis un siècle. Ils ont été étudiés remar-
quablement dans le passé, par QUINCKE en particulier, dont les expériences
sur le sujet sont classiques. HELMHOLTZ les avait déjà expliqués, pour ce qui

---

[1] G. GIROUSSE: Courants telluriques d'origine artificielle, p. 499. Dans E. MATHIAS, Traité
d'électricité atmosph. et tell. Paris: Presses Univ. 1924.
[2] G. DUPOUY: Ann. Géophys. **6**, 18 (1950).

est d'essentiel. L'illustre savant donc, admet qu'il se forme une «double couche» d'électricité au contact de la paroi des canalicules capillaires traversant la cloison. Dans un langage plus moderne, on préfère dire aujourd'hui que les ions du liquide sont plus ou moins adsorbés par la paroi, et que l'adsorption est plus prononcée pour les ions d'un signe que pour les autres. La constitution de la double couche dépend à la fois de la nature du liquide et de celle de la paroi. Le plus souvent d'ailleurs, cependant pas toujours, ce sont les ions négatifs qu'on trouve contre la paroi. Les ions positifs, avec leur cortège de molécules neutres qui leur sont plus ou moins liées, conservent une certaine mobilité et sont susceptibles d'être entraînés, soit par l'effet d'un gradient de pression (électrofiltration), soit par l'effet d'un champ électrique (osmose électrique).

C'est encore aux prospecteurs qu'on doit d'avoir reconnu que les phénomènes d'électrofiltration sont très fréquents dans la Nature et qu'ils y présentent une réelle importance géophysique. Bien que les gradients de pression soient en général petits, les forces électromotrices peuvent atteindre plusieurs volts parce que la conductibilité des eaux filtrantes est faible et parce que la filtration peut s'opérer sur de grandes hauteurs. On va donner des exemples variés de ces phénomènes.

Soit une couche sableuse, relativement perméable, dont le pendage est prononcé et qui s'intercale entre deux formations argileuses. La couche sableuse peut, si les conditions topographiques et météorologiques le permettent, constituer une zone privilégiée d'infiltration. Elle se signale alors par un minimum du potentiel mesuré en surface suivant la technique habituelle des mesures de P. S.

On rencontre fréquemment des cuvettes topographiques où se rassemblent les eaux de pluie. Le potentiel y est plus petit que dans les environs. Mais s'il advient qu'à la période sèche la cuvette se transforme en une zone privilégiée d'évaporation, le signe de l'anomalie de potentiel s'inverse.

De même les parties fraîchement labourées d'une prairie sont positives, de plusieurs millivolts, par rapport au reste du terrain.

De ces phénomènes, SCHLUMBERGER fit aussi une application industrielle de toute première importance. Afin d'identifier, sans prélever mécaniquement de «carotte», les diverses couches géologiques recoupées par un sondage, il avait inventé en premier lieu la technique du «carottage électrique», laquelle connut un succès extraordinaire puisqu'à l'heure actuelle on n'entreprend nulle part au Monde un sondage au pétrole sans le carotter électriquement. Avant de tuber le trou, et pendant qu'il est encore plein de boue, on y descend un «quadripôle» de minimes dimensions, analogue aux quadripôles de la prospection de surface, afin de pouvoir enregistrer sous forme d'un diagramme la variation de résistivité apparente en fonction de la profondeur. Les opérations de carottage furent évidemment d'abord effectuées en continu, car on ne s'attendait certes pas à être particulièrement gêné par l'électrofiltration, réputée pour n'être qu'une petite curiosité de laboratoire de l'antique Physique.

Or la pression exercée par la boue dans un sondage est énorme, de sorte que toute variation brusque de pression au cours des opérations de forage effectuées dans le voisinage se traduit par des variations brutales du régime de filtration, donc aussi de l'intensité des courants «telluriques» d'électrofiltration. Dans le sondage lui-même, où l'opérateur est en train de carotter électriquement, il ne règne pas non plus un équilibre hydrostatique. En principe, la pression de la boue est supérieure à la pression naturelle qui existe dans les formations perméables recoupées par la sonde. D'où un minimum du potentiel à la traversée des couches poreuses. Pour n'être pas gêné par l'électrofiltration quand on mesure les résistivités, il n'est que d'alimenter le quadripôle en alternatif ou, ce qui revient au même, en continu périodiquement inversé par un commutateur rotatif; on peut noter à ce propos qu'il y a cette fois équivalence complète entre continu et alternatif, au sens de la Sect. 10, en raison des très petites dimensions du quadripôle.

Du même coup, l'emploi d'alternatif autorise à enregistrer *simultanément* la différence moyenne de potentiel entre une quelconque des électrodes mobiles du quadripôle de résistivités et une électrode fixe plantée en surface. La différence de potentiel en question présente des minima bien localisés quand l'électrode mobile franchit des zones perméables. En réalité le phénomène est complexe: l'électrofiltration y joue le rôle le plus important, mais les forts gradients locaux de pression, de température, de concentration et de composition interviennent certainement aussi pour une part non négligeable.

Les données du carottage d'électrofiltration, que les techniciens appellent aussi carottage de porosité ou même, improprement, carottage P. S., complètent très opportunément celles du carottage de résistivités. Une résistivité très élevée peut être le fait d'un calcaire imperméable comme d'une couche réservoir imprégnée de pétrole. Une faible résistivité caractérise une marne imperméable comme une formation aquifère. Dans l'un et l'autre cas, le diagramme d'électrofiltration permet de lever le doute.

L'électrofiltration semble suffire à expliquer pourquoi les sommets des collines sont habituellement à un potentiel moindre que les régions de plus basse altitude. POLDINI [4] cite un exemple typique qu'il a observé en Serbie sur une colline d'un quartz anormalement spongieux. Une galerie percée vers le bas de la colline a permis de faire aussi quelques mesures sous terre. Comme on le voit sur la Fig. 9, le potentiel du sommet est de l'ordre de −300 millivolts quand on

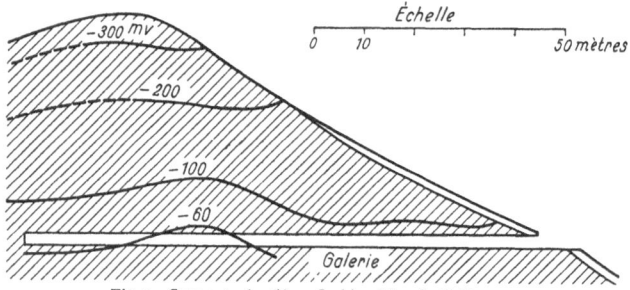

Fig. 9. Sommet négatif en Serbie. D'après E. POLDINI.

prend pour zéro celui de la vallée voisine. Les équipotentielles épousent approximativement les lignes de niveau topographiques. Les prospecteurs sont assez fréquemment trompés par des phénomènes de ce genre, s'imaginant à tort qu'ils ont découvert quelque beau gîte pyriteux.

Bien avant que les prospecteurs ne s'en occupassent, ces phénomènes étaient connus des Géophysiciens qui aimaient y voir la preuve de l'existence de «courants telluriques verticaux», dont la circulation s'effectuerait de bas en haut. On cite par exemple des observations très nettes faites naguère en deux observatoires français de montagne, par MARCHAND au Pic du Midi et par BRUNHES au Puy de Dôme. Cependant, au Ben Nevis, montagne d'Ecosse, DICKSON avait signalé que le sens de la différence de potentiel dépend essentiellement des conditions météorologiques (pluie, nuages, neige, ciel clair) régnant au sommet. Dans le même ordre d'idées, au Vésuve, PALMIERI montrait l'influence de l'activité fumerollienne.

On citera particulièrement un Mémoire plus récent de KOENIGSBERGER et HECKER[1]. Les auteurs se sont livré à des observations méthodiques en divers points d'une montagne de Suisse, non seulement en surface mais aussi dans les galeries d'une mine creusée en cette montagne. Ils ont fait un travail similaire dans les galeries d'une autre mine située, cette fois, en plaine. Ils constatent, eux aussi, que sauf en de rares points exceptionnels le potentiel décroît quand l'altitude augmente, à raison de 70 mV. en moyenne pour 100 m. de dénivellation. Connaissant la résistivité moyenne des roches de la montagne et admettant que les différences de potentiel observées correspondent à l'écoulement d'un courant vertical, les auteurs calculent la densité de ce courant et la trouvent du même ordre de grandeur que celle prévue par BAUER[2] pour ces hypothétiques et mystérieux «courants électriques ascendants». Il est utile de rappeler très sommairement ici que le «travail», ou circulation, du champ magnétique terrestre le long d'un contour fermé tracé à la surface du Globe devrait être nul si aucun courant ne traversait l'aire limitée par le contour en question. Lorsqu'on tente le calcul, on ne trouve pas que le résultat est nul. De là BAUER croyait pouvoir conclure à la réalité des courants verticaux, dirigés vers le haut ou vers le bas suivant qu'il s'agit de telle ou telle région du Globe. Quant à la densité de ces courants, elle apparaissait comme d'un ordre de grandeur très supérieur à celle des courants transportés par les ions de l'Atmosphère. La majorité des Géophysiciens d'aujourd'hui se montre à bon droit très sceptique, car la précision des levés magnétiques n'est pas telle, qu'on puisse avoir une confiance totale dans le résultat de ces calculs d'intégration[3].

Pour interpréter les observations faites en montagne ou dans les mines, il n'est certes pas besoin d'invoquer ces hypothétiques courants verticaux. On doit d'ailleurs remarquer que l'existence d'une composante verticale du *champ* tellurique n'implique pas obligatoirement un écoulement corrélatif d'électricité, c'est à dire un *courant*. Dans la majorité des cas, il semble qu'il n'y ait pas lieu d'expliquer les phénomènes observés autrement que par l'électrofiltration due à la rupture de l'équilibre hydrostatique, par l'existence même de la montagne ou des puits et galeries de la mine. Accessoirement d'autres phénomènes qui n'ont rien de mystérieux non plus peuvent aussi jouer leur rôle, notamment la différence de température entre les deux électrodes terminales de la ligne tellurique, ou encore certaines manifestations de P.S. dues à la proximité de minerai, à d'infimes intercalations graphiteuses, quand ce n'est pas à quelque ferraille enfouie dans la mine.

**24. Perturbations dues aux marées océaniques.** Les Ingénieurs chargés d'exploiter les câbles sous-marins ont eu parfois la possibilité d'imiter leurs collègues des lignes terrestres, et de procéder à des enregistrements «telluriques» de quelque durée. Ils constatèrent généralement que les telluriques ainsi enregistrés éprouvaient une variation diurne *lunaire*, autrement dit qu'ils suivaient le rythme de la marée océanique, alors que les telluriques enregistrés sur lignes terrestres suivent le rythme *solaire*. On peut citer les observations faites par SAUNDERS en 1880 sur le trajet Suez-Aden, ainsi que celles de DRESING, à la même date, sur le câble Ecosse-Norvège dans lequel le courant s'inverse toutes les 6 heures environ, selon que la mer monte ou descend. Cela ne faisait d'ailleurs que confirmer bien d'autres observations antérieures, effectuées sur des tronçons de câble après une rupture survenue au large. Les Ingénieurs ne manquent jamais, du reste, de mettre à profit ces accidents d'exploitation pour se

[1] J. KOENIGSBERGER et O. HECKER: Z. Geophys. 1, 152 (1924/25).
[2] L. A. BAUER: Terr. Magn. 28, 1 (1923).
[3] AD. SCHMIDT et J. BARTELS: Gerlands Beitr. Geophys. 55, 292 (1939).

livrer à des observations instructives. C'est ainsi par exemple qu'en 1938, le câble transatlantique français se rompit à 500 km. au large de Brest par un fond de 4400 m.[1]. La ligne «tellurique» fut constituée avec ce tronçon associé à une «prise de terre», située en fait au fond de l'eau à plusieurs km. de la côte. Les observations furent poursuivies durant plus de trois mois et montrèrent que le courant «tellurique» s'inverse à peu près en même temps que le flot, qu'il demeure pratiquement nul pendant les 15 à 20 minutes que dure l'étale de basse ou de pleine mer à Brest et que son amplitude, pouvant atteindre 1500 mV, est à peu près proportionnelle à l'amplitude de la marée locale.

Ces faits sont à rapprocher de ceux constatés dans l'île de Jersey par le R. P. Dechevrens[2], qui se contentait d'abord de relier à travers un galvanomètre les canalisations d'eau et de gaz de la petite ville de St Hélier, puis préféra se servir d'une très courte ligne tellurique, n'ayant que 13 m. de long, terminée par deux petites plaques de cuivre enfouies à 80 cm. de profondeur. Le courant tellurique ainsi observé suit aussi, fidèlement, le rythme de la marée, avec un retard de l'ordre de 2 heures. Mais déjà dès 1879 les Ingénieurs des Télégraphes s'étaient trouvé à même de constater des phénomènes semblables sur la ligne Londres-Cardiff, qui est voisine de la côte, et eux aussi avaient noté ce retard de 2 heures. Wollaston également, sur une ligne qui longeait la Severn, avait fait des observations similaires.

Tous les phénomènes qui viennent d'être évoqués semblent pouvoir être attribués sans erreur à deux causes principales, auxquelles s'adjoignent accessoirement d'autres causes secondaires. L'une de ces causes est encore l'électrofiltration. En particulier, les masses d'eau douce contenues dans les nappes souterraines et aboutissant à la mer se trouvent alternativement refoulées vers les terres ou aspirées vers la mer, et cela jusqu'à de grandes distances du rivage, au gré des variations de pression occasionnées par les fluctuations du niveau marin.

Le second de ces phénomènes n'est autre que l'induction électromagnétique. Les courants de marée, que ce soient des courants marins ou des courants d'estuaire, consistent en déplacements de masses d'eau conductrice dans le champ magnétique terrestre, de sorte que ces masses en mouvement sont le siège de champs électriques induits. Faraday avait prévu le phénomène dès 1832 et tenté, sans succès d'ailleurs, de le mettre en évidence à l'aide d'électrodes immergées dans la Tamise à Londres. L'étude expérimentale et théorique de ces courants induits fut effectuée pour la première fois en 1918 sous les auspices de l'Amirauté Britannique[3]. Elle fut d'ailleurs reprise un certain nombre de fois par d'autres. Chapman et Bartels s'étendent avec quelque détail sur cette question au cours du chapître XIII de «Geomagnetism» [2] qui est consacré aux courants telluriques.

Les diverses observations ou expériences relatées ci-dessus, évidemment très intéressantes en soi, ne sont quand même pas bien significatives au point de vue de la Géophysique. Leurs circonstances sont toujours plus ou moins complexes, de sorte qu'il est malaisé, d'un cas particulier à l'autre, de faire une discrimination exacte entre les causes diverses qui peuvent les expliquer. On connaît toujours mal en effet, et il arrive souvent qu'on ne connaît pas du tout, la constitution locale du sous-sol. On ne sait pas très bien non plus dans quelle mesure les courants peuvent se fermer, et comment ils le font. Ce sont là, comme il advient le plus souvent quand la Géologie est en cause, de bien médiocres circonstances expérimentales pour espérer parvenir à des conclusions très nettes.

**25. Dispositifs d'observation.** Ingénieurs des Télégraphes comme Prospecteurs électriciens possèdent une grande expérience des prises de terre. Les Télégra-

---

[1] M. Bernard: Onde électr. **17**, 465 (1938).

[2] M. Dechevrens: Terr. Magn. **23**, 37, 145 (1918); **24**, 33, 175 (1919).

[3] F. B. Young, H. Gerrard et W. Jevons: Phil. Mag. **40**, 149 (1920).

phistes français utilisent habituellement des plaques de fer galvanisé, planes, dont la surface est de l'ordre du m², et qu'ils enterrent verticalement. Ils estiment que la «terre» est bonne quand sa résistance est inférieure à 10 Ω, c'est à dire lorsqu'elle est du même ordre de grandeur que celle d'un Kilomètre de ligne. Les lignes aériennes le plus communément utilisées sont en fer galvanisé. Les deux prises terminales d'une ligne télégraphique ne sont jamais symétriques et ne présentent jamais, quoi qu'on fasse, des différences de potentiel au contact qui soient parfaitement équilibrées. La ligne est donc le siège d'une force électromotrice parasite atteignant habituellement quelques centaines de mV. Quand la ligne est en service depuis longtemps, cette force électromotrice paraît se stabiliser, sans qu'on ait pour autant l'impression qu'elle tende vers zéro.

Les prospecteurs n'utilisent les peu pratiques électrodes impolarisables que lorsqu'ils ne peuvent faire autrement. Deux piquets métalliques de même métal, enfoncés dans ce qu'on pourrait croire être exactement la même terre, mis à côté l'un de l'autre si l'on veut, n'en présentent pas moins presque toujours une forte dissymétrie. SCHLUMBERGER avait adopté le cuivre, en dépit de fort médiocres qualités mécaniques qui ne recommandent guère ce métal très mou pour en façonner des piquets. Les différences de potentiel au contact ne sont ni plus petites ni plus symétriques avec le cuivre qu'avec d'autres métaux, mais sont «peut-être» plus stables. Sur le choix optimum du métal, chacun prône son opinion. Tel recommande le fer, tel autre le cadmium, un troisième l'acier inoxydable, un quatrième le plomb, pendant qu'un cinquième assure qu'il faut des électrodes de plomb électrolysées au préalable durant de longues heures, en courant alternatif, dans une solution de chlorure de sodium. En fait la stabilité d'une électrode métallique dépend au premier chef de la nature du sol et de la configuration du terrain. Une terre argileuse est favorable, ainsi qu'un sol plat. Le vent, la pluie, le gel, l'insolation produisent, cela va sans dire, des effets désastreux. En prospection tellurique, où l'on a besoin d'une stabilité «prolongée», on préfère habituellement en revenir aux électrodes impolarisables.

Dans les Observatoires, les prises·de terre·ont toujours été établies avec le plus grand soin. Celles du Parc St Maur étaient faites de quatre spirales de fil de fer galvanisé, de 30 m. chacune, enfouies les unes au-dessus des autres tous les 50 cm., échelonnées entre 1 m. et 2,5 m. de profondeur. L'intervalle entre les spirales était comblé par des couches alternées de terre et de charbon de bois[1]. A l'Observatoire de l'Èbre, on avait d'abord adopté le système du Parc St Maur, puis on préféra des plaques horizontales de plomb, ayant 1 m² de surface, entourées de coke et enfouies à 2 m. de profondeur[2]. Dans les stations plus modernes de Huancayo et de Watheroo, chaque électrode comporte quatre boucles circulaires concentriques, placées dans un même plan horizontal, faites de lames de plomb connectées entre elles et enfouies à 1 ou 2 m. de profondeur, l'ensemble ayant 6 m. de diamètre extérieur. Quel que soit le système adopté, ce type de prises de terre présente généralement une résistance inférieure à 100 Ω, laquelle ne demeure d'ailleurs pas absolument constante [5], [6].

Les variations éventuelles de la résistance des prises de terre sont beaucoup moins graves que les fluctuations de la différence de potentiel au contact. Elles n'interviennent d'ailleurs absolument pas quand la mesure est faite par la méthode d'opposition. Ainsi opère-t-on à Watheroo [5] et à Huancayo, où l'on utilise un potentiomètre enregistreur automatique qui comporte un commutateur rotatif entraîné à vitesse uniforme par un moteur, de telle façon que, toutes les

---

[1] T. MOUREAUX: Ann. Bur. Cent. Météor. France 1, 25 (1893).
[2] S. J. GARCIA MOLLA: La section tellurique. Mém. Observ. de l'Ebre 4, 95 (1910).

50 sec. une mesure puisse être faite successivement sur les multiples lignes de l'Observatoire. Lorsque par exemple 8 lignes sont en service, un point s'inscrit toutes les 6 min. 40 sec.

Les potentiomètres enregistreurs ont certes de grands avantages mais ils ne permettent que des tracés discontinus et n'autorisent donc pas l'étude des fluctuations telluriques relativement rapides. Par contre, qu'il s'agisse de variations rapides ou de variations très lentes, les galvanomètres d'Arsonval à cadre mobile fournissent toujours d'excellents résultats. Par le moyen de résistances en série ou en dérivation, on peut régler à volonté sensibilité comme amortissement. Par ailleurs, il est toujours avantageux d'utiliser un galvanomètre suffisamment sensible pour qu'on puisse insérer sur la ligne une résistance supplémentaire de quelques milliers d'ohms. En la choisissant grande, on rend négligeable l'effet des minimes fluctuations éventuelles de la résistance des lignes et des prises. Elle doit cependant demeurer petite en regard de la résistance d'isolement de la ligne.

A une époque qui n'est pas encore bien lointaine, les Observateurs, en raison précisément des soins méticuleux apportés à la construction des prises de terre, se leurraient volontiers quant au rôle des effets d'électrode. Non seulement la stabilité leur semblait suffisante pour ne pas introduire d'erreur appréciable dans l'étude des variations diurnes, mais encore ils n'hésitaient guère à admettre que les courants telluriques présentent une composante *continue*, c'est à dire constante ou très lentement variable, qu'il est légitime d'évaluer en se fiant à la valeur moyenne du courant dans le galvanomètre. Très rares étaient ceux qui, comme Weinstein [7] se déclaraient convaincus que la soi-disant partie constante du courant tellurique n'est due qu'aux effets d'électrode et ne présente aucune espèce d'intérêt.

Beaucoup de faits, pourtant, qui semblent évidents aujourd'hui, auraient pu mettre en garde. Puisqu'on constatait une évolution dans le temps de la résistance des terres, il était assez invraisemblable que la différence de potentiel au contact dût faire exception. S'il advenait que, dans un Observatoire, une électrode dût être réparée, modifiée ou déplacée, il en résultait toujours une modification radicale du champ tellurique moyen. Quand on comparait un observatoire à l'autre, Berlin à Paris ou Paris à l'Èbre, la soi-disant composante continue se modifiait curieusement en grandeur et en direction. Avec des observations comme celles dont on dispose aujourd'hui à Watheroo, où de nombreuses électrodes alignées dans la même direction permettent toutes sortes de combinaisons et de comparaisons, il n'est plus possible de conserver le moindre doute à ce sujet. Car le champ moyen différerait au même lieu et à la même époque, en grandeur comme en direction et même en signe, suivant qu'on se référerait à telle combinaison d'électrodes plutôt qu'à telle autre [5]. D'où résulte en résumé que, si l'on ne saurait assurer aujourd'hui encore qu'il y a ou qu'il n'y a pas de composante continue, il est absolument certain en tous cas qu'elle ne peut être que bien inférieure à tout ce qu'on a cru avoir observé naguère. Il n'est pas inconcevable, loin de là, que l'existence d'une composante continue puisse être démontrée un jour. Mais il faudra sans doute, après avoir choisi l'emplacement avec grand soin, se résoudre à installer quelques lignes très longues, équipées avec des électrodes impolarisables, lesquelles seront minutieusement contrôlées et très profondément enfouies.

Les inconvénients d'une réduction de la longueur des lignes sont d'autant moins à craindre qu'on se préoccupe davantage de l'étude des variations rapides que de celle des variations lentes, et qu'on dispose d'électrodes plus stables. En prospection, où l'on ne s'est guère intéressé jusqu'ici qu'aux périodes échelonnées depuis quelques secondes jusqu'à quelques minutes, on s'accommode fort bien de lignes n'ayant que quelques centaines de mètres, même si on ne les équipe pas avec des électrodes impolarisables. Les électrodes métalliques soignées des observatoires suffisent assurément pour étudier la variation diurne de façon correcte. Il n'est pas nécessaire pour cela d'utiliser des lignes de 120 et 262 km. comme faisait Weinstein [7] à Berlin (Berlin-Dresden et Berlin-Thorn), ni des lignes de 56,8 et 93,9 km. comme à Tucson. Les 15,6 et 25 km. des lignes de Greenwich, les 14,8 et 14,8 km. des lignes du Parc St Maur sont peut-être un peu exagérés. Mais les lignes de l'Èbre (1,3 et 1,4 km.), celles de College Fairbanks, de Chesterfield, même de Huancayo, dont la longueur n'est guère plus grande que celle des lignes de l'Èbre, sont tout de même bien courtes, et les lignes doublées de Watheroo (3,4 et 9,9 — 2,0 et 5,6 km.) sont sans doute préférables.

Les lignes aériennes présentent divers inconvénients, qu'il ne faut pourtant pas exagérer: elles risquent d'être rompues de temps en temps; il faut les surveiller, et contrôler périodiquement leur bon isolement; elles imposent des parafoudres pour la protection des appareils enregistreurs. Mais leur principal inconvénient est d'osciller dans le vent, ce qui engendre des forces électromotrices induites parasites et interdit leur emploi pour les variations rapides. On peut pourtant s'en contenter pour les lignes de quelques km., recommandables pour l'étude des variations lentes. L'action du vent ne gêne pas le prospecteur quand il peut poser ses lignes directement sur le sol, sans avoir à les suspendre aux branches des arbustes ni à les poser sur de hautes herbes. Dans les observatoires au contraire, il est très désirable que les lignes pour variations rapides, lignes qui sans inconvénient peuvent être courtes, soient souterraines comme à Watheroo. Cependant leur isolement doit être impeccable, sans quoi le remède serait pire que le mal. De plus, si les câbles possèdent une armature extérieure métallique de protection, il faut que cette armature soit elle-même bien isolée.

Le champ tellurique est défini par ses composantes suivant deux axes horizontaux arbitraires et l'on peut donc orienter les deux lignes telluriques d'une station comme on veut. Pour faciliter les comparaisons entre observatoires différents, l'usage est d'adopter des lignes orthogonales suivant les *axes géographiques*, quoique l'Observatoire de l'Èbre fasse exception avec ses lignes orientées suivant les axes *magnétiques*. Le Congrès géophysique d'Edinburgh de 1936 a décidé en outre que les composantes telluriques seraient réputées positives vers le Nord ou vers l'Est. Or on montrera par la suite que les composantes magnétiques et telluriques sont étroitement apparentées deux à deux. S'il s'agit de la composante tellurique suivant une direction donnée qu'on prend pour axe $Ox$ (Fig. 10), c'est $Oy$, normal à $Ox$, qui porte la composante magnétique lui faisant «pendant», et l'on verra de plus que le sens positif logique à adopter sur $Oy$ est celui de la Fig. 10, c'est à dire la droite d'un observateur $O$ ayant les pieds à terre et regardant dans la direction $Ox$.

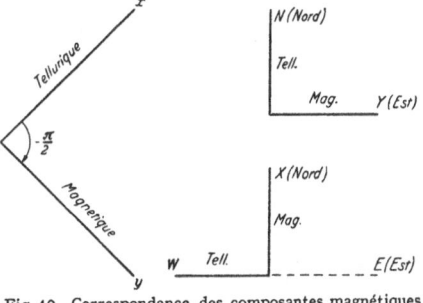

Fig. 10. Correspondance des composantes magnétiques et telluriques.

La convention d'Edinburgh est donc vicieuse, en ce sens qu'elle ne correspond pas aux conventions antérieures des spécialistes du Magnétisme. Comme on le voit sur la Fig. 10 à droite, la magnétique $Y$ (vers l'Est) fait bien pendant à la tellurique $N$, mais à la magnétique $X$ (vers le Nord) fait pendant la tellurique $W$, et non la tellurique $E$ du Congrès d'Edinburgh.

Malgré l'exemple de l'Èbre, il vaut sans doute mieux conserver l'habitude d'orienter les lignes telluriques suivant les *axes géographiques* car, autrement, l'on serait forcé de temps en temps, à cause des variations séculaires magnétiques, de modifier l'orientation des lignes. Pour les variations lentes, l'orientation géographique des lignes ne présente pas beaucoup d'inconvénients pratiques.

Par contre, la comparaison des variations magnétiques et telluriques rapides serait certainement facilitée si les variomètres magnétiques «rapides» enregistraient directement les composantes $X$ et $Y$ plutôt que la déclinaison et la composante horizontale.

Pour clore le présent chapître, on dira qu'aujourd'hui les Géophysiciens rompent avec les errements du passé, et trouveraient peu raisonnable d'implanter un nouvel observatoire sans une sérieuse étude préalable du sous-sol. Ils ne se font donc pas faute — et il faut les en louer — de faire une prospection électrique. Par malheur les nappes de courants telluriques pénètrent jusqu'à des profondeurs tellement hors de proportion avec la portée utile de ces petites investigations électriques courantes qu'au point de vue tellurique, autant vaudrait ne rien faire du tout. Outre les sondages électriques ultra-profonds, il faudrait employer précisément les méthodes de prospection qui vont être décrites plus loin sous le nom de méthodes telluriques.

# E. Résultats généraux d'expérience.

**26. Évidence d'une corrélation magnéto-tellurique.** Depuis un siècle, depuis qu'on s'intéresse aux courants telluriques, on sait qu'il y a une ressemblance frappante entre les enregistrements telluriques et les enregistrements magnétiques obtenus en même temps. Par situation magnétique calme, les telluriques sont calmes, et inversement. S'il y a agitation magnétique, il y a aussi agitation tellurique, et réciproquement. Une tempête tellurique correspond à une tempête magnétique, et les termes de cette proposition peuvent être retournés. Perturbations magnétiques et telluriques commencent et finissent en même temps. Si les fluctuations magnétiques sont rapides, il en est de même des fluctuations magnétiques, et quand les unes sont lentes, les autres le sont aussi. À condition que les perturbations signalées au chapitre précédent n'aient pas affecté les mesures, il n'est pas un phénomène, pas un accident constaté sur les enregistrements telluriques qui n'ait son «pendant» sur les enregistrements magnétiques. Il n'est pas non plus d'accident constaté sur l'enregistrement magnétique qui ne trouve son «pendant» sur le tellurique.

Les ressemblances et similitudes dont il s'agit sont déjà manifestes quand on compare les variations de n'importe quelle composante tellurique à celles de n'importe quelle autre composante magnétique. Mais elles deviennent plus frappantes

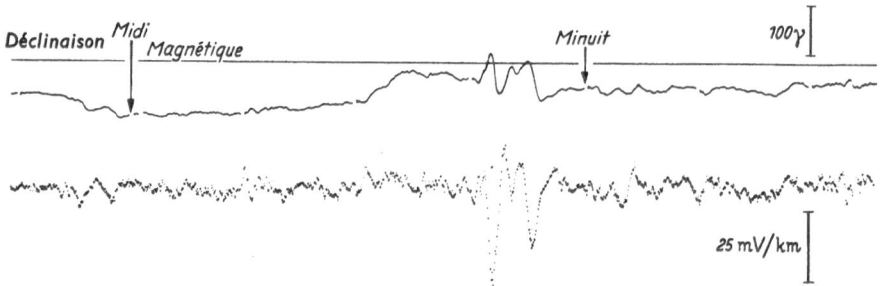

Fig. 11. Enregistrements tellurique et magnétique lents.

et plus évidentes encore, elles prennent l'allure de ces «airs de famille» étonnants qu'on ne reconnaît jamais qu'aux personnes de même sang, lorsqu'on compare une composante tellurique à la composante magnétique horizontale qui lui est perpendiculaire. Il y a aussi un siècle qu'on l'avait remarqué, et qu'on l'avait remarqué sans surprise en pensant qu'il devait bien y avoir là quelque aspect particulier de la classique règle d'électromagnétisme où l'on invoque le fameux Bonhomme d'AMPÈRE.

Qu'il existe une connexion, une corrélation entre les variations magnétiques et telluriques, c'est dont personne n'a jamais douté. Qu'il y ait entre elles une véritable relation, susceptible d'être explicitée, c'est beaucoup moins évident. On arriverait même parfois à n'en plus être bien convaincu quand les phénomènes sont complexes, quand il existe un mélange de périodes très diverses, comme c'est un peu le cas des Fig. 11 et 12. Il est alors besoin du «Fil d'Ariane» de la théorie pour retrouver malgré tout, dans ces apparences embrouillées, cet air de famille dont on pourrait parfois se croire autorisé à douter. Les deux enregistrements de la Fig. 11 sont des enregistrements «lents» (1 cm. par heure sur les originaux) effectués durant un intervalle de 24 heures à l'Observatoire de Chambon-la-Forêt. La courbe du haut est celle de la déclinaison magnétique. En bas se trouve celle de la composante tellurique suivant le méridien magnétique. Les enregistrements de la Fig. 12 sont «rapides» (1 cm. par minute sur la bande originale). Ils furent

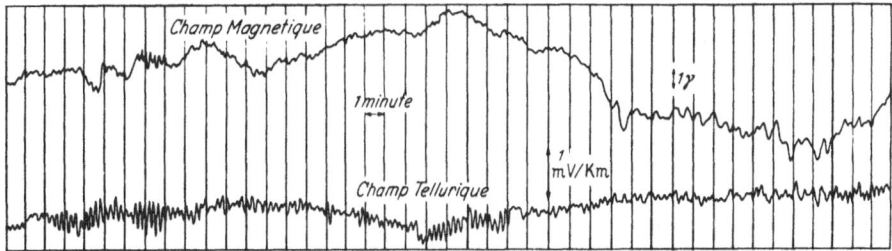

Fig. 12. Enregistrements tellurique et magnétique rapides.

obtenus avec un magnétomètre et des galvanomètres apériodiques de courte période. Les deux composantes qui y sont enregistrées sont encore perpendiculaires l'une à l'autre, mais leur orientation n'offre rien de particulier. Il s'agit en ce cas de mesures faites en rase campagne, pour lesquelles il fut commode de poser la ligne tellurique à terre, le long d'un chemin. Il convient de signaler que, pour être reproduit, le document a dû non seulement être réduit en dimensions, mais encore redessiné avec plus ou moins de bonheur, de sorte qu'il est loin de conserver les finesses de l'original.

On a vu naguère les Géophysiciens se partager en deux clans. Pour les uns, les variations magnétiques n'étaient pas autre chose que celles du champ de BIOT et SAVART des courants telluriques. En vertu d'un raisonnement beaucoup trop sommaire pour être bien exact, ils pensaient qu'il existe une proportionnalité pure et simple entre les deux composantes respectives. Les autres voulaient voir dans les courants telluriques des courants qui seraient «induits» par les variations magnétiques et, se fondant sur des raisonnements non moins sommaires que ceux des premiers nommés, en concluaient que la composante tellurique devait être proportionnelle à la dérivée par rapport au temps de la composante magnétique.

A condition de n'y pas regarder de trop près, certains enregistrements ont quelque apparence en effet de donner raison à l'une des deux thèses en présence. Mais d'autres enregistrements peuvent également être invoqués en faveur de l'opinion adverse, avec un pouvoir de persuasion qui n'est ni pire, ni meilleur. En fait, quand on examine sans parti pris n'importe quel enregistrement, force est de constater que le champ tellurique n'est proportionnel, ni au champ magnétique, ni à la dérivée du champ magnétique. Et la conclusion est encore plus évidente si l'on accepte d'examiner des enregistrements d'allure un peu complexe comme ceux des Fig. 11 et 12, plutôt que des fragments triés en raison de la simplicité occasionnelle de leur allure.

Ce qui apparaît généralement quand on y regarde de façon attentive, ce qui apparaît beaucoup mieux encore, bien entendu, quand on peut faire une véritable analyse harmonique, ce sont les correspondances approximatives que voici:

1. En un lieu donné, à une époque donnée, les variations telluriques et magnétiques ont même période quand elles ont toutes deux une allure quasi-harmonique. Elles comportent les mêmes superpositions de périodes dans les cas plus complexes.

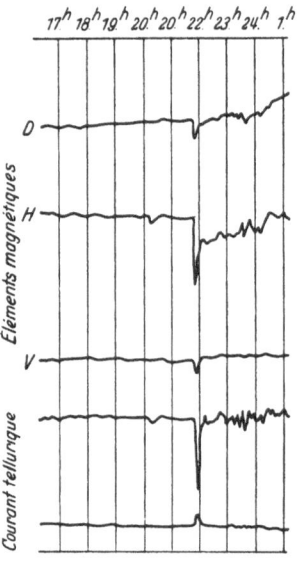

Fig. 13. Simultanéité des variations brusques sur les enregistrements magnétiques et telluriques. D'après J. Bosler.

2. En un point donné, le rapport des amplitudes telluriques et magnétiques au même instant ne dépend que de la période. Il ne change pas d'un instant à l'autre, d'un jour à l'autre. Il ne se modifie guère non plus, en général, quand on oriente différemment la ligne tellurique. Le rapport de l'amplitude tellurique à l'amplitude magnétique croît d'habitude quand la période devient plus petite. Il croît cependant généralement moins vite que s'il était en raison inverse de la période, comme ce serait le cas si la composante tellurique était proportionnelle à la dérivée de la composante magnétique.

3. Des variations très rapides, des débuts ou des changements brusques, des discontinuités apparaissent au même instant sur les deux types d'enregistrements. La Fig. 13, empruntée à Bosler [8], illustre cette proposition.

4. En un point donné, la différence de phase des composantes tellurique et magnétique à la même époque ne dépend que de la période. Elle ne change pas d'un moment à l'autre. Elle ne se modifie guère non plus, d'habitude, quand on change l'orientation de la ligne tellurique. Lorsqu'on adopte la disposition d'axes de la Fig. 10, la composante magnétique est normalement en retard de phase par rapport à la composante tellurique, ce retard étant généralement compris entre 0 et $\pi/2$, quelconque entre lesdites limites. Le retard serait nul si les deux composantes étaient proportionnelles l'une à l'autre. Il serait $\pi/2$ si la composante tellurique était proportionnelle à la dérivée de la composante magnétique.

Les règles qui précèdent reviennent à dire que les enregistrements telluriques et magnétiques expriment à peu près la même chose, bien que ce soit en des langages différents. Un enregistrement tellurique équivaudrait en gros à un enregistrement magnétique qu'on ferait avec un magnétomètre de caractéristique spéciale ayant, comme disent les radioélectriciens, une «courbe de réponse» particulière. Et cette courbe de réponse dépendrait de la nature et de la structure du sous-sol local.

Après avoir fait ces constatations, il n'apparaît plus très utile d'insister, autant que font la plupart des exposés d'électricité tellurique, sur divers autres aspects de cette corrélation magnéto-tellurique. Car une corrélation qui se révèle ainsi jusque dans les détails ne peut pas manquer non plus de se manifester dans l'allure générale. On se bornera donc à rappeler brièvement les autres correspondances que voici:

1. Les fluctuations de l'activité tellurique épousent très fidèlement celles de l'activité magnétique, comme en fait foi la Fig. 14, empruntée à Rooney [6].

Il est bien connu par ailleurs, comme le rappelle la Fig. 14, que l'activité magnétique est en relation très étroite avec l'activité solaire, le nombre de taches solaires. La même proposition vaut donc pour l'activité tellurique. Il existe aussi les mêmes relations entre l'activité tellurique et l'activité aurorale qu'entre l'activité magnétique et cette dernière.

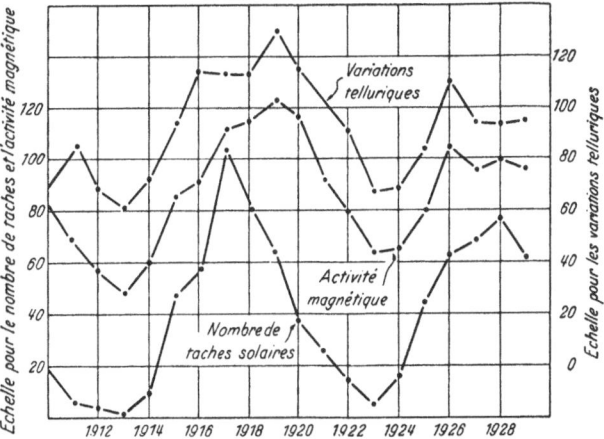

Fig. 14. Parallélisme de la marche des activités tellurique, magnétique et solaire. D'après Rooney.

2. La période de récurrence de 27 jours, caractéristique du magnétisme, se retrouve aussi dans l'activité tellurique. Les prospecteurs qui font de la prospection tellurique utilisent même couramment cette loi de récurrence pour choisir d'avance les journées le plus propices à une expérimentation efficace[1].

**27. Variation diurne du champ tellurique.** Comme la plupart des phénomènes géophysiques, le champ tellurique présente une variation diurne extrêmement nette, très facile à mettre en évidence dans les moyennes horaires. Si l'on en croit même Weinstein [7], «il n'est guère sous nos latitudes d'autre phénomène naturel qui présente une oscillation diurne aussi régulière. Ce qu'Al. von Humboldt a dit des oscillations de l'aiguille aimantée sous les tropiques s'applique totalement aux phénomènes telluriques diurnes et à leur

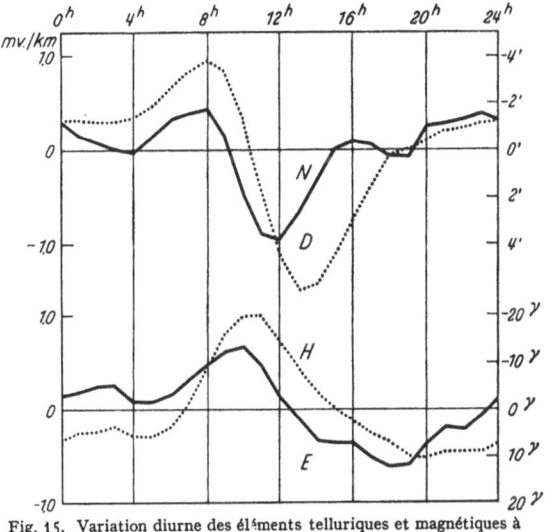

Fig. 15. Variation diurne des éléments telluriques et magnétiques à Paris (Parc Saint Maur). D'après E. Rougerie.

évolution au cours de l'année». C'est d'ailleurs Weinstein qui a publié les plus remarquables courbes de variation diurne tellurique, dont les ordonnées ne sont malheureusement graduées qu'en «unités arbitraires». Soit dit en passant, Weinstein n'a pu émettre une opinion aussi tranchée qu'en raison de la longueur démesurée de ses lignes.

La Fig. 15 représente, d'après Rougerie [9], la variation diurne des composantes telluriques $N$ et $E$ au Parc St Maur (Paris), pour un intervalle de 27 mois.

[1] G. Kunetz: Congrès de l'Assoc. Franç. Avanc. des Sci. Luxembourg. 1953. — Actes du Congrès, Luxembourg 1955.

Sur la figure sont représentées aussi, pour le même laps de temps, les variations diurnes de la déclinaison $D$ et de la composante horizontale magnétique $H$. Le parallélisme, qui est de règle dans l'allure des variations magnétiques et telluriques, apparaît là comme tout aussi frappant que dans le cas des variations plus rapides. Suivant la règle, la variation tellurique précède la variation magnétique, plus exactement elle est en avance de phase par rapport à elle.

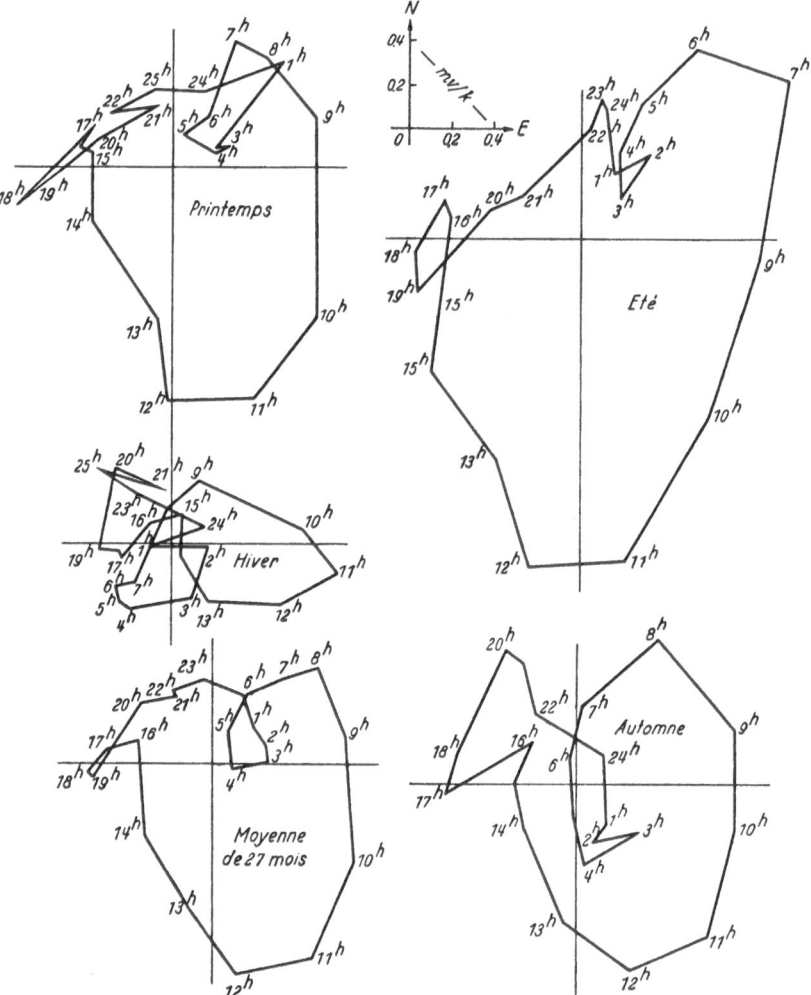

Fig. 16. Diagrammes polaires de la variation diurne à Paris (Parc Saint Maur). D'après P. Rougerie.

Il n'est pas sans intérêt non plus de représenter la variation diurne par un diagramme polaire en forme de boucle fermée, ladite boucle étant le lieu géométrique de l'extrémité du vecteur qui représente le champ tellurique aux diverses heures de la journée. La Fig. 16, empruntée à Rougerie, représente les boucles de Paris. Comme on voit, la boucle se déforme progressivement au cours de l'année. Indépendamment des lentes déformations séculaires, on constate une oscillation de période annuelle, fort nette. Tout cela rappelle encore étroitement la manière dont se comporte le Magnétisme. Les boucles de Paris sont «ventrues». En d'autres stations, à Stockholm par exemple, elles sont beaucoup plus étroites, bien plus allongées suivant une direction privilégiée. On le voit sur la Fig. 17, d'après STENQUIST[1]. A l'Observatoire de l'Ebre, le champ tellurique conserve même une direction presque invariable au cours de la journée. Les variations diurnes y sont presque exclusivement des variations d'intensité et de sens.

On sait que les variations diurnes du magnétisme sont approximativement les mêmes en toutes les stations de même latitude géographique, à condition de

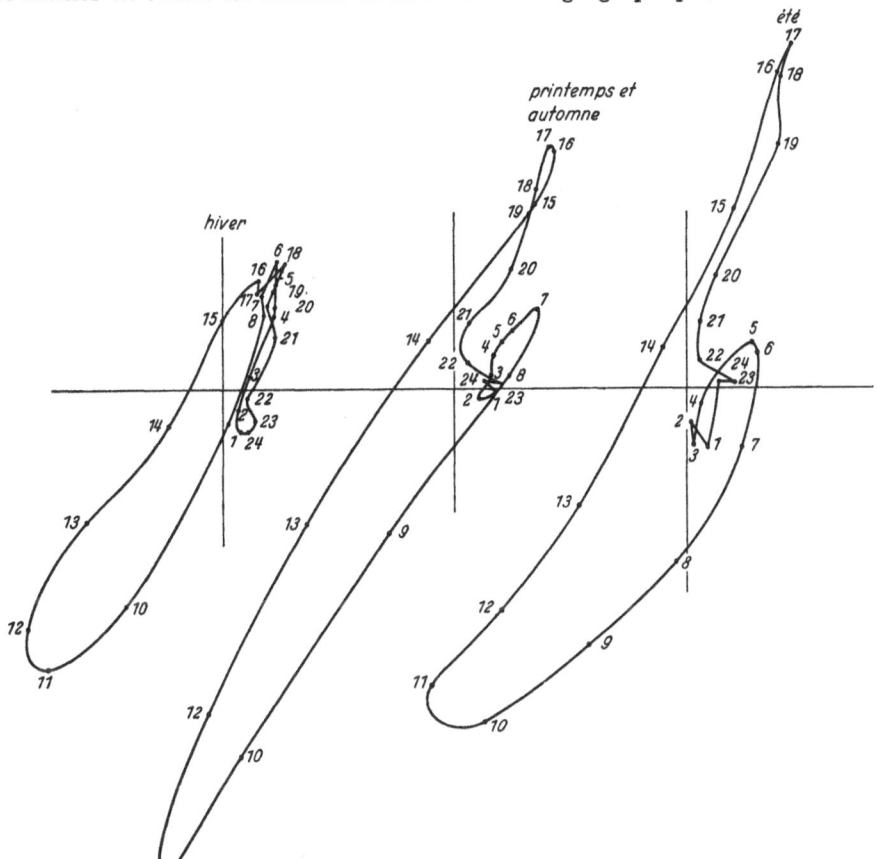

Fig. 17. Diagrammes polaires de la variation diurne à Stockholm. D'après D. STENQUIST.

les exprimer en fonction de l'heure șolaire locale. S'il s'agit des variations telluriques, la même proposition appelle de sérieuses réserves. Elle reste encore à peu près valable, en première et grossière approximation, tant qu'on se borne à envisager les variations de direction et de sens du champ, mais elle devient grossièrement erronée si l'on veut prendre aussi en considération la grandeur de ce champ. Un exemple typique est celui de l'Ebre (41° N.), ou l'amplitude du tellurique est environ 18 fois celle observée pas tellement loin de là, à Paris (49° N.).

En tenant pour vraie la première partie de la proposition, celle qui concerne seulement l'orientation du champ, il devient alors possible de représenter, sous une forme claire et suggestive, le système mondial des lignes de courant correspondant à la variation tellurique diurne. Car il suffit de figurer ce système tel qu'il se présente à une heure déterminée. Pour imaginer la configuration du système à une heure autre que celle-là, il suffit de translater en bloc sur la planisphère tout le système de courants de l'Est à l'Ouest, à raison de 15° de longitude par heure.

C'est ce que Gish et Rooney ont fait dans un travail justement classique [10] La Fig. 18 reproduit le système des *tubes de courants* telluriques mondiaux, tel que ces auteurs l'ont publié. On a pensé toutefois qu'il serait plus intéressant d'imiter Rougerie [9], qui surimprima sur cette carte les vecteurs telluriques

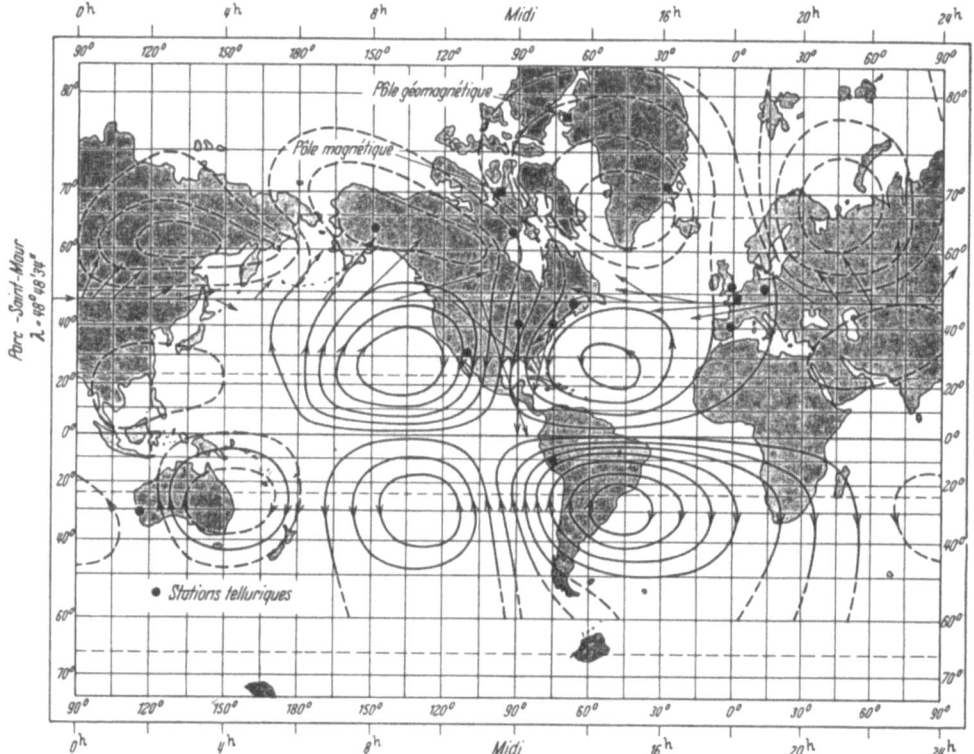

Fig. 18. Variation diurne des courants telluriques: système mondial des lignes de courant à 18 h T. M. G. D'après Gish, Rooney et Rougerie.

du Parc St. Maur, vecteurs qui ne furent calculés, à partir des observations de Mou-reaux, que postérieurement au travail de Gish et de Rooney. De prime abord, on est quelque peu surpris de la beauté du résultat, obtenu en partant d'un nombre telle-ment infime de stations telluriques réparties sur le Globe. Mais il ne faut pas oublier qu'en raison de l'hypothèse faite, la connaissance de la variation diurne en une seule station fait connaître, à l'heure déterminée choisie (18 h. T.M.G.), la direction du champ en tous les points du parallèle de cette station. On peut même assurer, non sans quelque malice, que pour dessiner des courbes régulières et cohérentes, il vaut mieux avoir moins de points expérimentaux. Gish et Rooney ont admis par surcroît que les lignes de courant doivent se fermer sur elles-mêmes en forme de «tubes de courants». Qu'il en soit approximativement ainsi, que les «spirales de courants» se ferment à peu près exactement, c'est assez vraisemblable, mais n'a guère de chance, semble-t-il, d'être rigoureusement exact. Quoi qu'il en soit au juste, les auteurs ont désiré que chacun de leurs «tubes de courants» correspondît à une même «intensité» du courant tellurique. Ils avouent d'ailleurs qu'en ce qui concerne plus spécialement cet aspect quantitatif de leur schéma, ils ont rencontré des difficultés telles, qu'ils ont dû se résoudre à ne tenir compte, pour l'ensemble du Globe, que de trois stations. On les croira volontiers; pour évaluer l'«intensité», en effet, la seule connaissance du champ tellurique, en surface, ne saurait suffire, quand même on connaîtrait aussi la résistivité superficielle. Il faudrait savoir la distribution du champ électrique et de la résistivité sur une épaisseur d'au moins 100 km. La résistivité des premiers kilomètres de l'Ecorce terrestre compte assurément fort peu en la circonstance. La preuve en est, si l'on en pouvait douter, que la configuration des grands océans, des vastes bassins sédimentaires ou des énormes boucliers de roches archéennes n'influe que très peu sur la direction des lignes de courant qui correspondent aux variations telluriques diurnes.

C'est précisément pour servir d'illustration aux commentaires qu'on vient de lire qu'on a tenu à représenter les vecteurs de ROUGERIE. Lesdits vecteurs s'accordent passablement avec les lignes de courant de GISH et ROONEY tant qu'il ne s'agit que de l'Atlantique, de l'Amérique du Nord et du Pacifique. Mais ils deviennent franchement aberrants sur l'Europe et l'Asie, où ils présentent parfois un sens *opposé* à celui qu'ils devraient avoir. A tel point qu'il est bien évident qu'un système de courants qui serait compatible avec les vecteurs de ROUGERIE n'aurait guère de ressemblance avec celui qu'ont figuré les auteurs.

**28. Caractère régional de la circulation des courants telluriques.** Si le travail de GISH et ROONEY met en lumière le caractère «planétaire» de la variation diurne, les variations irrégulières, elles aussi, présentent un caractère de très large régionalité.

La preuve en fut donnée depuis longtemps en ce qui concerne les variations «lentes», c'est à dire celles qu'on étudie commodément avec des vitesses d'inscription de l'ordre du centimètre à l'heure. Les résultats les plus significatifs à ce point de vue semblent avoir été ceux de BLAVIER [*11*]. Il s'est trouvé que, vers les années 1880, la construction de lignes télégraphiques souterraines en France avait libéré plusieurs lignes aériennes qui purent être judicieusement utilisées à des études telluriques. Semblable occasion ne se représentera plus jamais, maintenant que le sous-sol des pays industrialisés a été «empoisonné» de courants vagabonds. Les expériences de BLAVIER furent aussi favorisées admirablement par la constitution particulière du réseau français, où Paris se trouve inséré comme au centre d'une vaste toile d'araignée. C'est ce qui fait que le très ancien Mémoire de BLAVIER a beaucoup moins vieilli que bien d'autres plus récents, et pourquoi il mérite encore aujourd'hui d'être lu et relu avec le plus grand intérêt.

Parmi beaucoup d'autres résultats qui seraient dignes d'être relevés, BLAVIER publie des enregistrements simultanés effectués entre Paris et Nancy (280 km. en ligne droite) sur deux lignes différentes reliant les mêmes électrodes terminales. L'une des lignes est souterraine et passe par Reims, tandis que l'autre, aérienne, traverse Châlons sur Marne à 40 km. au sud de la première. On constate que les deux enregistrements obtenus sont «pratiquement» superposables.

BLAVIER utilise maintenant des lignes de longueurs très différentes, ayant à peu près la même direction, telles que Paris-Nancy, Paris-Bar le Duc, Paris-Châlons ou Paris-Reims. Cette fois encore, les enregistrements demeurent à peu près identiques, à condition de réduire les ordonnées de chaque courbe au prorata de la longueur de ligne. Même résultat encore quand on compare trois autres lignes de même direction: Paris-Châlons (150 km.), Paris-Vincennes (7 km. seulement) et enfin un simple fil de 1800 m. tendu en plein coeur de la capitale! Quelle époque bénie pour les études telluriques!

Mais un résultat beaucoup plus significatif encore est le suivant : de son bureau de Paris, Blavier enregistre sur la même bande Paris-Lille (212 km.), Paris-Juvisy (20 km.) et Nancy-Dijon (172 km.). Il s'agit encore de trois directions parallèles, mais le fait nouveau est qu'elles ne sont plus alignées. La parallèle Nancy-Dijon est à 270 km. de distance de la parallèle Lille, Paris, Juvisy. Et néanmoins les trois enregistrements sont encore à peu près les mêmes, à la réduction près des ordonnées au prorata des longueurs de ligne.

Il apparaîtra pourtant dans la suite de cet exposé que ce serait assurément quelque peu exagéré, de conclure de là à une quasi-uniformité des champs telluriques sur des distances de plusieurs centaines de kilomètres en tous sens. Les enregistrements telluriques «rapides», que les prospecteurs pratiquent depuis quelques années, et dont il va être question au paragraphe suivant, permettent de serrer le détail des phénomènes beaucoup mieux que les expériences de Blavier. En fait l'uniformité constatée par ce dernier est plus apparente que réelle, et résulte surtout de deux circonstances particulières. D'abord, il ne s'agit que de variations lentes, beaucoup moins sensibles que les variations rapides à l'influence des structures géologiques de surface. De plus le réseau des lignes télégraphiques de Blavier ne débordait guère des limites du grand bassin sédimentaire parisien dont la tectonique est relativement régulière. En tout cas, quoi qu'il en soit au juste de cette *uniformité*, les expériences de Blavier n'en suffisent pas moins pour faire éclater l'évidence, qu'à l'échelle des variations «lentes» les nappes telluriques conservent une *large régularité d'ensemble* sur des territoires au moins aussi vastes que la France entière.

**29. Variations rapides étudiées par les prospecteurs.** Il est fatal que la distribution des courants telluriques dépende de la résistivité des diverses formations géologiques, de sorte que l'étude de cette distribution, en son détail, doit apporter des renseignements utiles sur la structure du sol. De cette remarque devait naître la méthode tellurique de prospection, dont C. Schlumberger fut l'initiateur dès 1921[1], encore que la mise au point et l'application sur une grande échelle ne datent que de la fin de la dernière guerre mondiale. La prospection tellurique a été jusqu'ici une spécialité de la Compagnie générale de Géophysique, dont les principales publications sur le sujet, du moins celles qui présentent un intérêt d'ordre scientifique, sont citées dans la bibliographie [*12*], sans compter celles signalées en bas de pages.

Au point de vue technique, la prospection tellurique présente l'avantage de dispenser à la fois des générateurs comme des câbles de grande longueur, nécessaires en prospection classique. Mais l'intérêt essentiel de la nouvelle méthode est de permettre l'investigation de structures bien plus profondes que celles accessibles aux procédés courants, en raison du fait qu'il est a priori certain que les courants telluriques pénètrent dans le sous-sol à de très grandes profondeurs.

Étant donné que le champ tellurique, en un point donné, varie sans cesse en fonction du temps d'une manière compliquée et désordonnée, la prospection tellurique exige que ce champ soit enregistré simultanément en deux points. La méthode en effet ne peut être raisonnablement fondée que sur une comparaison, la comparaison des deux enregistrements en question. En pratique les prospecteurs commencent par installer un enregistreur permanent dans une station fixe qu'ils appellent «Base». Pendant que l'enregistrement est en cours à la Base, des opérateurs installent des stations temporaires d'enregistrement, lesquelles vont fonctionner chacune pendant une demi-heure par exemple et qui vont être successivement réparties, en nombre suffisant, dans un rayon de quelques dizaines

---

[1] E. G. Leonardon: Terr. Magn. **33**, 91 (1928).

de kilomètres autour de la Base. Les enregistrements photographiques de courte durée obtenus par les opérateurs de ces stations volantes sont destinés à être comparés par la suite avec l'enregistrement qui se trouve effectué à la Base dans le même temps.

A la Base comme à la station volante, on installe deux lignes telluriques courtes, qui sont simplement posées à terre, et dont la longueur n'a généralement pas besoin d'excéder 1 km. Ce n'est pas absolument nécessaire, mais c'est en tout cas bien commode, pour des raisons qu'on verra, que les lignes de chaque station soient orthogonales, et qu'elles soient deux à deux parallèles. Comme on a dit plus haut, on préfère les électrodes impolarisables, pour éviter des dérives intempestives, telles qu'il s'en produit avec des piquets de métal lorsque, par exemple, un rayon de soleil vient tomber sur l'un d'eux. Les enregistreurs utilisés à la Base et à la station volante sont identiques. La vitesse de déroulement du papier est de l'ordre du cm. ou de quelques cm. à la minute. Les galvanomètres sont du type Schlumberger-Picard: on sait qu'en dépit d'une robustesse à toute épreuve ces galvanomètres peuvent être très sensibles ($10^{-9}$ A. par mm. de déviation à 1 m.). Ils sont suramortis et leur période propre dépasse de peu la seconde. La technique opératoire adoptée se prête donc tout particulièrement à l'étude des variations dites «rapides» dont les périodes s'échelonnent depuis quelques secondes jusqu'à quelques minutes.

**30. Caractères particuliers des variations rapides.** Divers types caractéristiques d'enregistrements, obtenus par la Compagnie générale de Géophysique, sont reproduits sur la Fig. 19. Quelques commentaires préalables ne sont pas inutiles pour sa bonne intelligence:

Sur chaque enregistrement on distingue les deux tracés qui correspondent à chacune des deux lignes orthogonales de la station. Pour éviter les confusions, l'un des spots est simple, l'autre est dédoublé: ce résultat est obtenu en formant l'image d'un diaphragme éclairé, percé d'un seul trou ou de deux trous voisins, suivant le cas. En outre, on rencontre dans ce genre d'enregistrements la difficulté technique que voici: aux variations rapides qu'on désire enregistrer s'en superposent d'autres, plus lentes, moins intéressantes, et dont les amplitudes sont beaucoup plus grandes. De sorte que si l'on adopte la sensibilité qui convient bien à l'étude des variations rapides auxquelles on s'intéresse, en fait, exclusivement, il devient impossible de garder bien longtemps le spot à l'intérieur des 20 cm. de large de la bande photographique. C'est pourquoi les enregistreurs de la Compagnie générale de Géophysique sont pourvus de plusieurs miroirs latéraux, pour former des spots multiples chargés de se relayer les uns les autres. Quand, sous l'effet des dérives ou des variations lentes de grande amplitude, le spot qu'on est en train d'enregistrer vient à approcher dangereusement du bord droit du papier, l'un des miroirs latéraux de droite commence à intercepter une fraction du faisceau lumineux en la réfléchissant sur le bord gauche du papier. C'est ce qui explique pourquoi l'on voit parfois sur la Fig. 19 deux ou même trois tracés parallèles équivalents, qui sont fournis par les spots démultipliés d'un même galvanomètre.

G. Kunetz et M. Schlumberger[1] ont eu la possibilité de classer et de comparer la masse énorme de documents accumulés en sept années de prospection tellurique. Les enregistrements étudiés ont été obtenus surtout en France, mais aussi en Amérique du Nord, en Afrique du Nord et Afrique équatoriale, ainsi qu'à Madagascar. Ils représentent au total 50 000 heures, soit plus de 100 km. de bande photographique. Cette étude statistique d'une réelle ampleur a permis aux auteurs de distinguer les principaux types que voici:

1. Des *oscillations* qu'ils qualifient de «normales» (Fig. 19b). Elles sont très régulières, quasi-sinusoïdales. Leur période peut varier entre 15 et 30 sec. mais, au cours d'un même train d'ondes, d'une centaine d'oscillations par exemple, la période reste stable à quelques secondes près.

---

[1] M. Schlumberger et G. Kunetz: Observations sur les variations rapides des courants telluriques. Paris: Dorel 1948.

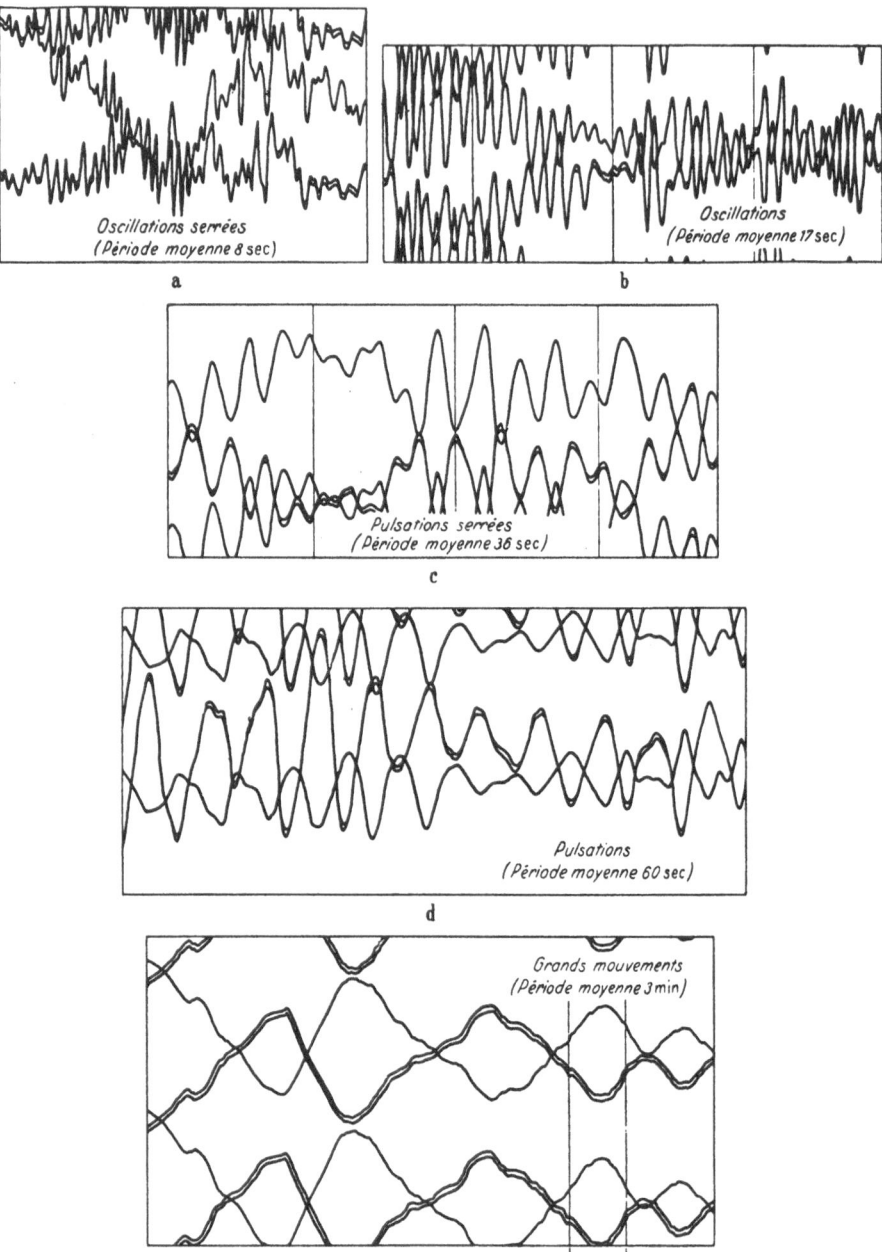

Fig. 19. Différents types de variations telluriques rapides. D'après M. Schlumberger et G. Kunetz.

2. Des *oscillations serrées* (Fig. 19a). Leur période peut descendre jusqu'à 6 sec. Ce régime est beaucoup moins fréquent.

3. Des *pulsations* (Fig. 19d). La période est plus longue, voisine de la minute. Quant à la régularité des mouvements du spot, elle est bien moindre. D'une

oscillation à l'autre, on constate de fortes variations dans les amplitudes comme dans les pseudo-périodes.

4. Des *pulsations serrées* (Fig. 19c). Les périodes y vont de 30 sec. à 1 min., formant transition entre les oscillations et les pulsations.

5. Des *grands mouvements* (Fig. 19e). Les périodes varient de quelques minutes à quelques dizaines de minutes, restant entendu que le mode d'enregistrement utilisé se prête fort mal à mettre en évidence des phénomènes de périodes encore plus longues.

Les phénomènes relativement purs sont d'ailleurs rares. On se trouve habituellement en présence de régimes complexes, avec superposition de phénomènes divers, d'amplitudes et de périodes disparates. C'est ce que les auteurs appellent *«régime moyen»*.

Il est extrêmement difficile de donner une idée précise de l'amplitude des variations telluriques rapides car cette amplitude dépend, non seulement de l'époque d'observation, comme c'est le cas en Magnétisme, mais aussi, et dans une très large mesure, du lieu où l'on opère. Les indications données quand même ci-après concernent d'ailleurs exclusivement les variations rapides dont la période ne dépasse pas quelques minutes et elles s'entendent des amplitudes moyennes, envisagées durant des intervalles de quelques heures. Cela dit, c'est en Afrique du Nord que les prospecteurs ont rencontré les valeurs les plus faibles: elles ne dépassaient pas, certains jours et en certains lieux, une vingtaine de microvolts par km. En France, les amplitudes sont rarement descendues au-dessous de 1000 microvolts par km. Quant aux valeurs habituelles, elles sont, en gros, cinq fois plus fortes. Par temps d'orage magnétique, il s'agit d'amplitudes qui sont dix fois, cent fois supérieures aux valeurs habituelles, parfois même bien davantage encore.

A un autre point de vue, il n'est pas rare de constater qu'en deux stations dont la distance n'excède guère le kilomètre, le rapport des amplitudes moyennes aux mêmes époques peut atteindre 1000.

Il semble que l'activité tellurique (pour les variations rapides s'entend) soit plus grande le jour que la nuit, et qu'elle soit maxima aux premières heures de la matinée, après le lever du soleil. Mais on observera que les enregistrements des prospecteurs ont lieu normalement de jour, entre 8 h. et 18 h. C'est seulement d'après un trop petit nombre d'enregistrements exceptionnels, qui n'ont pas été interrompus de nuit, que les auteurs ont pu exprimer cette opinion.

Il est amusant de chercher à construire des diagrammes polaires, analogues à ceux qu'on a considérés (Fig. 16 et 17) à propos des variations diurnes. Les courbes obtenues sont d'ordinaire plus irrégulières, plus tourmentées, plus capricieuses que n'en pourrait jamais dessiner sur un tableau noir, avec un morceau de craie, le plus facétieux des singes.

Dans ce domaine des variations rapides qui est essentiellement le leur, les Prospecteurs démontrent aussi, chaque jour comme partout, que les telluriques offrent ce même caractère de large régionalité qu'un BLAVIER avait pu prouver pour les variations lentes. C'est précisément sur ce postulat de régionalité qu'est fondée la méthode tellurique, de sorte que ce postulat trouve sa justification a posteriori dans les succès indiscutables de la méthode tellurique. On verra que la vérification va même présenter un certain caractère quantitatif, puisqu'on déduira du postulat que les composantes telluriques, enregistrées en deux stations pas trop éloignées l'une de l'autre, doivent être liées par des relations linéaires dont l'existence sera effectivement vérifiée par l'expérience. Ce postulat de régionalité ou d'uniformité, la Compagnie générale de Géophysique l'énonce ainsi: à condition de ne considérer qu'un territoire de dimensions assez restreintes, dont la superficie ne dépasse guère quelques milliers de km², le champ tellurique instantané présenterait une distribution pratiquement uniforme si la structure géologique était elle-même uniforme, plus précisément si elle était tabulaire. C'est seulement quand il circule des courants vagabonds provenant de sources industrielles trop proches que la superficie de la zone d'uniformité risque de se trouver restreinte.

Quand la distance entre stations devient exagérée, lorsqu'elle dépasse par exemple 50 à 100 km., le postulat d'uniformité se vérifie plus mal, et les relations linéaires dont il a été question deviennent moins bonnes. Cependant les corrélations

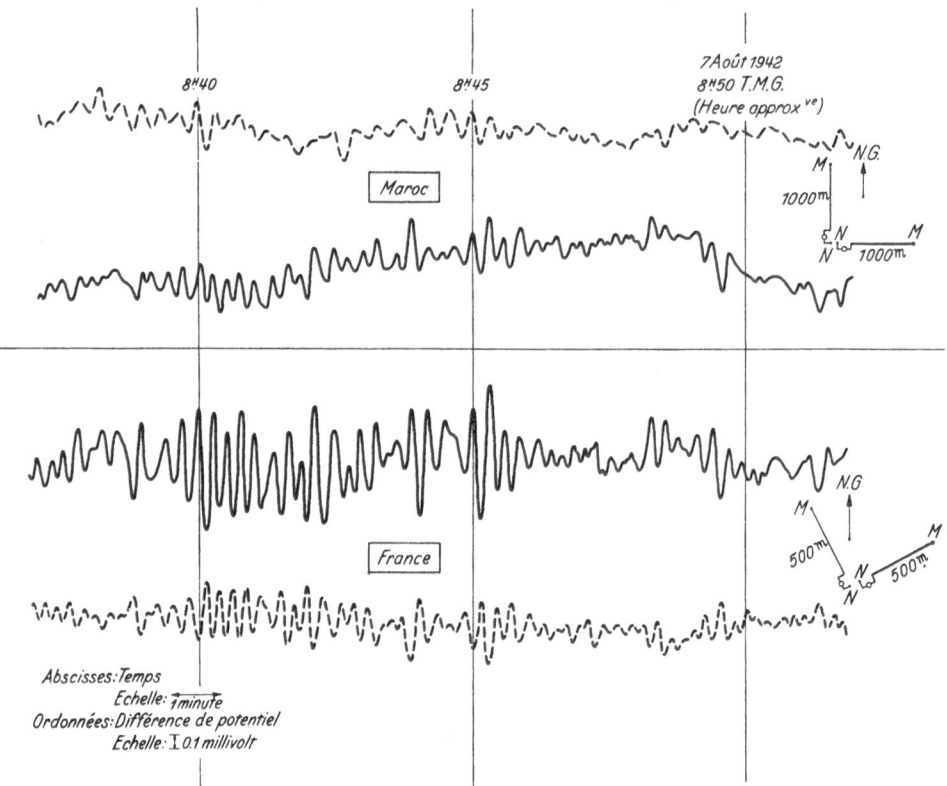

Fig. 20. Variations telluriques simultanément observées au Maroc et en France.
D'après M. SCHLUMBERGER et G. KUNETZ.

qualitatives des composantes telluriques s'étendent à des distances beaucoup plus grandes. Ainsi des phénomènes oscillatoires concordent encore, souvent jusque dans le moindre détail, en des stations aussi éloignées que France et Maroc (2000 km.), ainsi qu'on peut voir sur la Fig. 20. Des corrélations de ce genre sont même parfois encore assez nettes lorsqu'on passe de France au Gabon (6000 km.) ou à Madagascar (9000 km.)[1].

Comme on verra plus loin, les relations linéaires auxquelles il n'a encore été fait qu'allusion, permettent de tracer les lignes de force instantanées du champ tellurique, autrement dit les lignes de courant superficielles, à un instant arbitraire donné. Le tracé en question n'est d'abord obtenu que dans les limites du domaine d'uniformité mais, en déplaçant la Base quand il le faut, on peut prolonger ce tracé, de proche en proche, sur des territoires très vastes. C'est ainsi que la Fig. 21,

---

[1] M. SCHLUMBERGER et G. KUNETZ: C. R. Acad. Sci. Paris **223**, 551 (1946).

d'après un document de la Compagnie générale de Géophysique, concerne une grande partie du Dauphiné français. Rien ne peut démontrer de façon plus lumineuse la régionalité du phénomène. On remarquera comme la distorsion des lignes de courant est vraiment peu marquée. Encore convient-il d'observer qu'il s'agit en l'occurence d'une région géologiquement fort complexe, et que les dis-

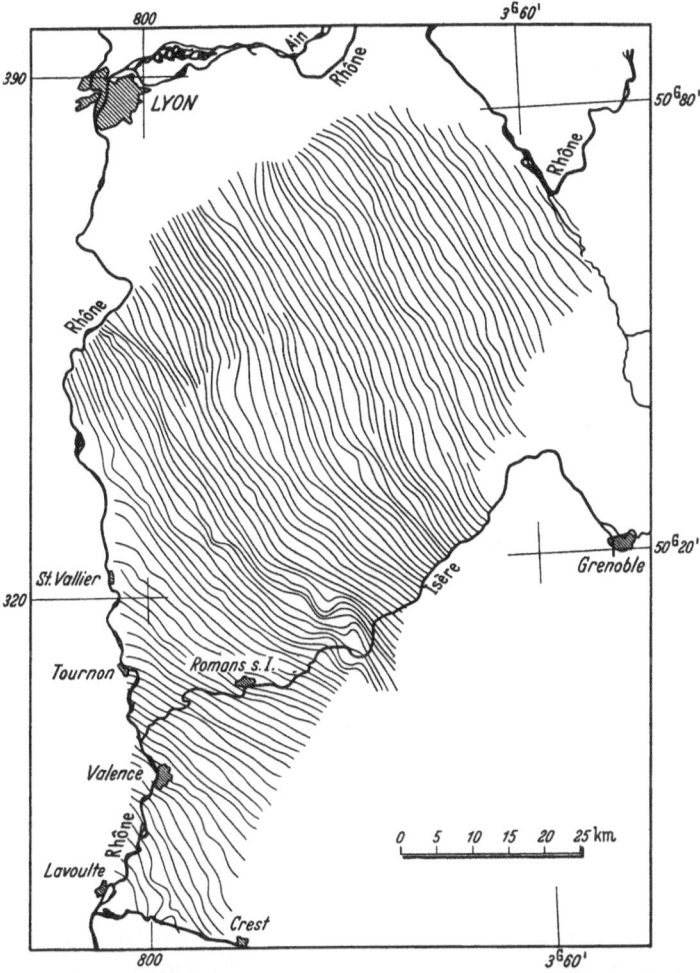

Fig. 21. Système de lignes de courant instantanées (Dauphiné, Bassin du Rhône, France). Une graduation de ces «tubes de courant» n'aurait pas de sens. D'après la Compagnie Générale de Géophysique.

torsions y sont dues sûrement bien davantage à l'hétérogénéité du sous-sol qu'à un manque intrinsèque d'uniformité du phénomène tellurique.

**31. Les relations linéaires de la prospection tellurique.** Au postulat d'uniformité, les Prospecteurs en adjoignent implicitement un autre, beaucoup plus discutable, savoir que les variations telluriques ont des périodes si grandes qu'on peut légitimement *négliger le skin-effect*. La nappe tellurique se comporterait donc à la manière d'une nappe artificielle de courant continu. A supposer du moins qu'on puisse en engendrer de telles, à l'aide d'électrodes d'injection placées très loin en dehors des frontières de la zone dite d'uniformité. A supposer en outre qu'on puisse, de minute en minute, changer non seulement l'emplacement des électrodes d'injection, mais aussi l'intensité et la polarité du courant injecté. Aussi admet-on que le champ tellurique instantané est complètement déterminé partout dans la zone d'uniformité dès qu'on se donne le champ en un seul point quelconque de ladite zone[1].

A la Base $B$ et à la station $S$ (Fig. 22), on considérera des axes de coordonnées $Bx$, $By$ et $SX$, $SY$ coïncidant avec les directions correspondantes des lignes telluriques. Ces directions sont en principe complètement arbitraires, bien qu'en pratique elles soient rectangulaires et deux à deux parallèles. On désignera par $x$ et $y$ les composantes du champ instantané $e$ à un instant donné, à la Base $B$. De même $X$ et $Y$ seront celles du champ $E$ au même instant en $S$. On appellera $e_1$

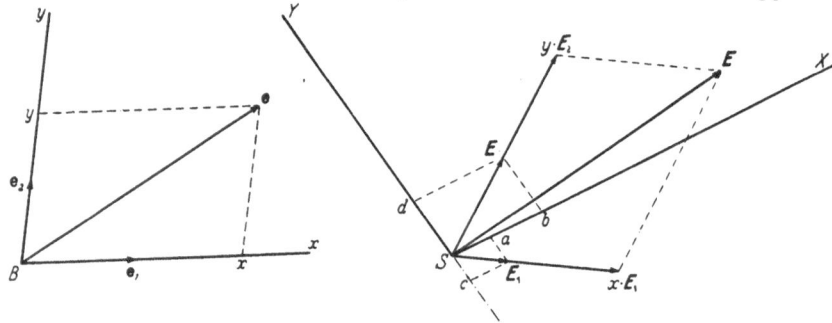

Fig. 22. Corrélation tellurique entre la station $S$ et la Base $B$.

et $e_2$ les champs unitaires en $B$ qui sont portés par les axes. À $e_1$ et $e_2$ correspondent respectivement, en $S$, les champs $E_1$ et $E_2$, qui ne sont pas unitaires et ne sont pas portés par les axes en $S$, du moins en général. Or on a par définition:

$$e = x\,e_1 + y\,e_2 \qquad (31.1)$$

et comme les équations régissant la circulation de courants continus sont linéaires, il s'en suit:

$$E = x\,E_1 + y\,E_2 \qquad (31.2)$$

ce qui, en projetant $E$ sur $SX$ et $SY$, peut encore s'écrire:

$$\left.\begin{array}{l} X = a\,x + b\,y, \\ Y = c\,x + d\,y \end{array}\right\} \qquad (31.3)$$

quand on désigne respectivement par $a$ et $b$ les projections de $E_1$ et $E_2$ sur $SX$, par $c$ et $d$ leurs projections sur $SY$.

La comparaison des enregistrements effectués en $S$ et en $B$, faite en vue de déterminer expérimentalement les valeurs des coefficients $a$, $b$, $c$, $d$ des relations (31.3), se trouve facilitée quand la vitesse de déroulement du papier est la même en $S$ qu'en $B$. Un marquage des temps sur les deux bandes est utile, mais n'exige pourtant pas un synchronisme précis. En effet, la parenté d'allure[2] des deux courbes est toujours telle, que c'est elle qui assure ipso facto le meilleur repérage

---

[1] L'énoncé de cette proposition devrait en réalité s'accompagner de quelques réserves, car il est facile d'imaginer des cas particuliers où elle tombe en défaut. Par exemple, en tout point du contour d'un affleurement infiniment résistant, le champ est tangent à ce contour. On ne peut donc pas se donner arbitrairement le champ en un tel point et, une fois qu'on a pu se le donner valablement, on n'a pas déterminé pour autant le champ tellurique ailleurs qu'en ce point.

[2] Dans la prospection pétrolière des bassins sédimentaires, dont les couches ne présentent souvent que de faibles pendages, la distorsion des lignes de courant est très peu marquée. C'est pour cette raison qu'on adopte des directions fixes pour les lignes telluriques. Comme on le voit sur la Fig. 23, $X$ et $x$, $Y$ et $y$ sont alors presque proportionnels deux à deux. Les nombres $b$ et $c$ sont petits en comparaison de $a$ et $d$ respectivement.

de synchronisme qui soit: deux accidents homologues peuvent être considérés comme synchrones avec une rigueur d'autant plus parfaite qu'ils se rapportent à des variations d'un caractère plus brusque, qu'ils ressemblent davantage à des discontinuités cinématiques affectant le mouvement des spots.

On ne connaît naturellement pas $x$, $y$, $X$, $Y$, autrement dit les grandeurs même des quatre composantes telluriques, mais seulement les variations $\delta x$, $\delta y$, $\delta X$, $\delta Y$ qu'elles éprouvent dans un intervalle de temps donné. Cela n'a pas d'inconvénient car les variations satisfont aux mêmes relations linéaires que les composantes elles-mêmes. En pratique, on choisit l'intervalle de temps qui correspond à une oscillation simple de grande amplitude, entre un maximum et un minimum d'élongation consécutifs de l'une des quatre composantes (Fig. 23). Il est absolument nécessaire de procéder ainsi, non seulement pour obtenir une bonne précision numérique, mais aussi pour réduire au minimum l'intervention des variations telluriques de grande période ou encore celle des effets de dérive. Il est clair en effet que les relations linéaires qu'on cherche à mettre en évidence n'ont rien à voir avec les caprices de la polarisation des électrodes. Quant aux variations lentes, il est peu probable, comme on verra plus loin, qu'elles soient régies par des relations ayant les mêmes coefficients numériques que celles qui valent pour les variations rapides.

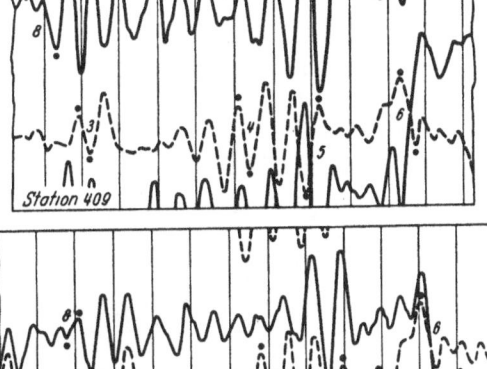

Fig. 23. Comparaison des enregistrements telluriques effectués à la station 509 et à la Base de Roybon (Noter que les tracés en trait continu, à la Base et à la station, sont inversés par suite d'une connexion erronée de l'un des galvanomètres). D'après L. MIGAUX.

En fait, par la façon même dont il conduit le dépouillement de ses enregistrements, le prospecteur opère une véritable sélection. De l'ensemble complexe des variations telluriques, il ne retient pour son calcul que certaines d'entre elles, dont les pseudo-périodes sont en somme très voisines les unes des autres. On verra qu'en fin de compte c'est ce qui assure la réussite de son calcul, que c'est ce qui permet une compatibilité suffisante des équations linéaires surabondantes dont il dispose pour faire ce calcul.

La connaissance des coefficients $a$, $b$, $c$, $d$, équivaut donc complètement à celle des champs $E_1$ et $E_2$ qui correspondent aux vecteurs $e_1$ et $e_2$ de la Base. Le prospecteur parvient ainsi en définitive à connaître le comportement de deux nappes telluriques distinctes et indépendantes. La première est définie par $e_1$ à la Base, la seconde par $e_2$. A partir de ces deux nappes, on peut prévoir le comportement de n'importe quelle autre. Ce sont précisément les lignes de courant d'une nappe tellurique de ce genre qui furent représentées sur la Fig. 21.

Les difficultés d'interprétation de semblables cartes de distribution du champ tellurique sont le plus souvent inextricables. Elles apparaissent comme du même ordre que celles soulevées par l'interprétation des cartes de potentiel dans la méthode de prospection par courant continu. C'est pourquoi l'on préfère présenter les résultats d'une prospection tellurique sous une forme différente, en fait nécessairement équivalente, mais qui a l'avantage d'être beaucoup plus suggestive.

L'artifice employé est du même ordre que celui auquel on a recours dans la prospection par courant continu lorsqu'on substitue aux cartes de potentiels celles de résistivités apparentes.

Les relations (31.3) établissent une correspondance réciproque bi-univoque entre tout point $m$ du plan des $x\,y$ et tout point $M$ du plan des $XY$. Quand $m$ décrit une courbe fermée $c$, qui délimite une aire $s$, le point $M$ décrit une autre courbe fermée $C$, laquelle circonscrit une surface $S$. Le rapport $S/s$, qui est indépendant de la forme des courbes $c$ et $C$, est égal à $a\,d - b\,c$[1]. On convient alors de poser $s = 1$ en un point arbitraire de la zone prospectée, de sorte que chaque

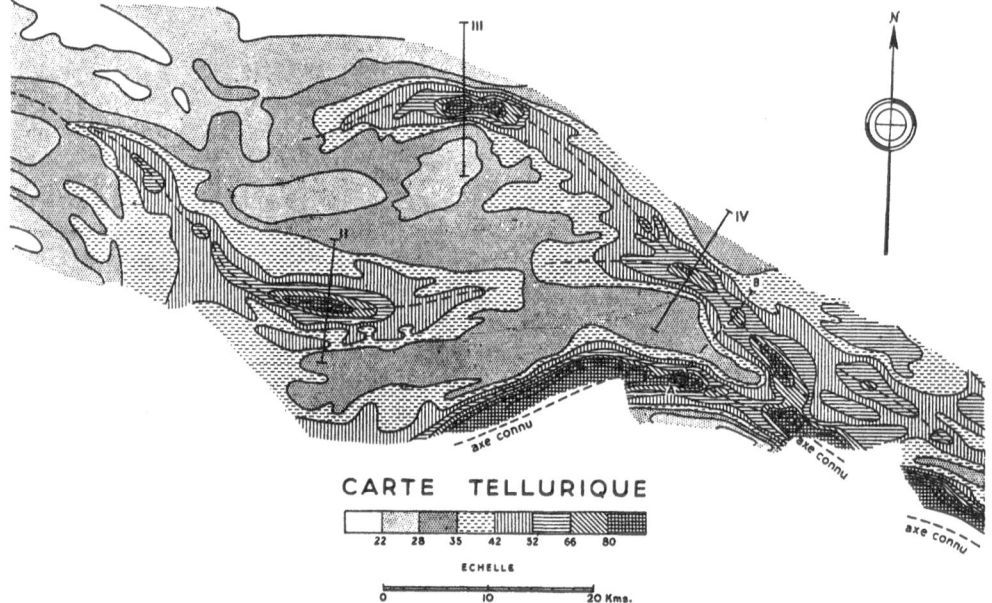

Fig. 24. Prospection pétrolière d'Aquitaine centrale (France S-W): carte d'iso-aires telluriques. D'après L. MIGAUX.

station se trouve affectée du nombre $a\,d - b\,c$, qu'on appelle «*Aire*» et qui n'est en fait qu'un rapport de deux aires. On réunit par une courbe toutes les stations caractérisées par une même valeur de l'«Aire», ce qui conduit à tracer une carte d'«iso-aires», comme celle que représente la Fig. 24.

Sur de telles cartes d'iso-aires, les axes anticlinaux se signalent d'habitude par un maximum de l'Aire. Un exemple extrêmement schématique va faire comprendre pourquoi il en est ainsi. On va supposer que le sous-sol ne comporte que deux terrains; que la résistivité du terrain superficiel est uniforme, celle du substratum infinie; que la surface du sol est plane et que le toit du substratum a forme de cylindre. On va adopter des axes $B\,x\,y$ et $S\,XY$ rectangulaires, avec $B\,x$ et $S\,X$ normaux aux génératrices du cylindre, autrement dit situés dans le plan de la «coupe géologique». Les coefficients $b$ et $c$ sont alors nuls par raison de symétrie. Quant à $d$, il est égal à l'unité car une nappe de courant *continu* qui s'écoule pa-

---

[1] Quand on tente d'établir des corrélations, lesquelles ne peuvent être au demeurant que fort médiocres, entre des stations très éloignées, entre France et Madagascar ou Gabon par exemple, il arrive que les sens de parcours sur $c$ et $C$ soient inverses. Cela n'arrive jamais en prospection, où le déterminant fonctionnel $\dfrac{D(X,Y)}{D(x,y)} = a\,d - b\,c$ est toujours positif, quand on adopte les dispositions d'axes qui sont usuelles et qu'on a indiquées plus haut.

rallèlement aux génératrices d'un cylindre est caractérisée par une densité uniforme. Par contre, pour ce qui est des lignes de courant normales aux génératrices, la plus superficielle suit la surface du sol, tandis que la plus profonde suit le toit du substratum. La densité de courant est donc maxima dans les régions de resserrement des lignes de courant, là où la profondeur $h$ du substratum est minima. Si par surcroît les rayons de courbure du cylindre sont grands en comparaison des profondeurs $h$ du substratum, il est clair que la densité de courant est en raison inverse de $h$. On a donc alors:

$$\frac{S}{s} = a\,d = d = \frac{K}{h}; \quad K: \text{Constante.} \tag{31.4}$$

Dans l'interprétation des résultats de prospections telluriques, on ne va malheureusement guère plus loin que ces considérations qualitatives, lesquelles apparaissent tout de même comme bien sommaires en regard de réalités géologiques qui sont tellement plus complexes. La méthode tellurique ne peut donc avoir, et d'ailleurs n'a pas, l'ambition d'apporter des renseignements d'ordre quantitatif. C'est essentiellement une excellente méthode de première reconnaissance, au même titre que la Gravimétrie. Elle sert pour obtenir à relativement peu de frais une première et très vague esquisse du sous-sol, avant de faire appel au procédé de séismique-réflexion, beaucoup plus précis, mais aussi beaucoup plus coûteux.

Malgré les réserves qui viennent d'être faites, le succès industriel de la méthode tellurique n'est pas niable. Cette constatation n'entraîne pourtant pas, sur le plan proprement scientifique, que tous les postulats sur lesquels elle repose, celui d'uniformité mis à part, s'en trouvent automatiquement validés. Les courants telluriques, puisque ce sont des courants variables, ne peuvent pas être des courants continus. Ils sont régis par les équations de Maxwell; ils ne le sont pas par la loi d'Ohm. Le champ tellurique $E$, instantané, à la station $S$, n'est certainement pas déterminé d'une façon complète dès qu'on se donne $e$ à la Base $B$. Car les équations de Maxwell, s'il est bien vrai qu'elles soient linéaires, ne font pas intervenir uniquement les grandeurs des champs: elles dépendent aussi des dérivées de ces champs par rapport au temps. Lorsqu'on accepte de ne considérer que des régimes harmoniques, la variable temps disparaît des équations, comme on a vu au chapître B, mais les coefficients de ces équations cessent en même temps d'être réels, et de plus ils dépendent désormais de la période. De sorte que, même quand on se limite aux régimes harmoniques, les choses ne sont pas si simples qu'on a dit: les coefficients $a, b, c, d$, ainsi que l'«Aire» $a\,d - b\,c$, sont en réalité des fonctions imaginaires de la période et non pas des constantes réelles.

Il y a donc intérêt à discuter le comportement de nappes telluriques d'une façon moins sommaire et plus correcte qu'il n'a été fait au cours du présent chapître. Cela sera examiné au chapître suivant.

## F. Relations entre variations magnétiques et telluriques.

**32. Origine des courants telluriques.** D'un point de vue exclusivement expérimental, et plus qualitatif à vrai dire que quantitatif, la Sect. 26 a déjà souligné l'évidence de la corrélation magnéto-tellurique. Au contraire le présent chapître reprend-il la question d'un point de vue surtout théorique et quantitatif.

Si les lois de l'électromagnétisme laissaient assurément prévoir qu'un enregistrement tellurique et un enregistrement magnétique obtenus en un même lieu ne pouvaient pas être complètement indépendants l'un de l'autre, on ne pouvait pourtant pas considérer comme évidente a priori la si parfaite réciprocité constatée expérimentalement. Chaque détail, a-t-on dit, de l'enregistrement tellurique trouve sa contrepartie dans un détail correspondant de l'enregistrement magnétique, et *réciproquement*. Constater cela, c'est prouver que l'enregistrement tellurique réalisé est celui d'un phénomène *pur*, et que ce phénomène pur est exclusivement de nature *électromagnétique*.

On a déjà fait allusion à cette pauvre controverse qui, dans un passé assez récent, partageait encore les Géophysiciens en deux clans. La grande question, croyait-on, était de savoir si les variations magnétiques étaient la *cause* des variations telluriques ou si c'était l'inverse. La question ainsi posée était pourtant le type parfait des questions fausses, des questions dépourvues de toute espèce de signification.

Une métaphysique sommaire enseigne effectivement que lorsque deux phénomènes sont liés, l'un d'eux est forcément une cause pendant que l'autre est un effet. Mais, d'un point de vue scientifique strict, force est de reconnaître que la classification des phénomènes en causes et en effets n'est, le plus souvent, pour ne pas dire toujours, qu'affaire de convention ou de définition. Quand par exemple un courant continu circule dans un conducteur, n'est-il pas vain de se demander si le courant existe parce que le potentiel n'est pas uniforme, ou s'il existe des différences de potentiel parce qu'il y a courant? Dans le cas actuel, c'est assurément la même chose: le phénomène électromagnétique est «un», quoiqu'il revête deux aspects, à la manière du dieu Janus qui avait, paraît-il, deux visages. Dans certaines conditions expérimentales, c'est exclusivement l'aspect magnétique qui intervient. Dans d'autres conditions d'expérience, c'est l'aspect électrique seul qui se révèle. Dans d'autres conditions encore, phénomènes électriques et magnétiques sont imbriqués simultanément sans qu'il soit possible de les dissocier expérimentalement.

C'était toujours la même grande question qui, sous une forme apparemment moins naïve, mais identique au fond et exactement aussi vaine, se trouvait parfois posée dans les termes suivants: les courants telluriques sont-ils des courants induits? Ou le champ magnétique, au contraire, n'est-il pas le *champ de* BIOT *et* SAVART associé aux courants telluriques? Assurément, dans une pédagogie élémentaire, il est indispensable d'établir des classifications. OERSTED ferme l'interrupteur d'un circuit, ce qui *cause* la déviation de l'aiguille, tandis que FARADAY déplace un aimant, ce qui *cause* le courant induit. Mais si tout le monde tombe «à peu près» d'accord quand il s'agit de qualifier de cause la volonté d'OERSTED qui anima sa main ou celle de FARADAY qui anima son bras, absolument rien n'autorise à considérer dans le premier cas que le champ magnétique est l'effet du courant, ni à admettre dans le second que le courant est l'effet d'une variation de flux. Lorsqu'on avance dans l'étude de la science, et qu'on a affaire à des phénomènes moins schématiques que ceux des expériences de cours, il n'y a généralement qu'avantage à bannir du langage des termes abusivement évocateurs tels que ceux de courants *induits*, de *cause*, d'*effet*, qui souvent ne veulent plus rien dire du tout.

Quant aux arguments invoqués en faveur d'une thèse ou de la thèse contraire, ils étaient aussi vains que la question elle-même. On constatait par exemple que le terme magnétique diurne présentait un *retard* de phase par rapport au terme tellurique correspondant. Cela prouvait, disait-on que le courant tellurique n'était pas un courant induit, c'est à dire un effet de la variation magnétique, puisqu'un effet ne saurait précéder sa cause. Mais on oubliait qu'il n'y a de différences de phase que dans les phénomènes périodiques, c'est à dire éternels, sans commencement ni fin, et qu'il est illusoire de décider, entre deux phénomènes éternels, lequel a débuté. On oubliait aussi qu'il suffit au mathématicien d'un changement de signe, d'une différenciation, etc., pour convertir un retard de phase en une avance, ou pour faire l'opération inverse, sans avoir altéré pour autant la réalité physique que traduit toujours arbitrairement tel ou tel symbolisme conventionnel.

Dès qu'il ne s'agit plus de phénomènes éternels, mais de perturbations ayant un début, l'expérience a montré (Fig. 13) que les débuts sont synchrones sur les enregistrements magnétiques et les enregistrements telluriques. Car il n'est, en fait, qu'un seul et unique début, celui de la perturbation électromagnétique. Et, pour la même raison, tout ce qu'on sait concernant l'*origine* des variations et des perturbations magnétiques vaut, identiquement, quand il s'agit de préciser l'origine des variations et des perturbations telluriques.

C'est un argument différent, méritant qu'on s'y arrête, qu'invoquait récemment BONDARENKO[1] pour prouver que les courants telluriques sont bien «des courants induits». L'auteur fait la remarque judicieuse que l'une des équations de MAXWELL:

$$-\frac{\partial H_z}{\partial t} = \frac{\partial E_y}{\partial x} - \frac{\partial E_x}{\partial y} \tag{32.1}$$

peut être facilement soumise au contrôle de l'expérience. Il suffit en effet de deux lignes telluriques voisines, parallèles à $Oy$, pour mesurer $\partial E_y/\partial x$, de deux autres lignes pour mesurer $\partial E_x/\partial y$, enfin d'une boucle horizontale pour mesurer $\partial H_z/\partial t$. L'expérience, sur laquelle on aimerait avoir plus de détails que n'en donne l'article, a été faite, et elle est concluante. Mais la conclusion à en tirer n'est assurément pas celle de BONDARENKO. Cette expérience n'est, ni plus ni moins, qu'une vérification des équations de MAXWELL. Il était assez improbable de les prendre en défaut au cours d'une expérience somme toute assez grossière. Rien n'a autorisé jusqu'à présent à les mettre en doute, pas plus lorsqu'il s'agit de courants telluriques naturels, que de courants telluriques vagabonds ou de tous autres courants électriques.

### 33. Théories modernes.

Les équations de MAXWELL ne cessant vraisemblablement pas de garder leur validité dès qu'il s'agit de Géophysique, il n'est donc que de les intégrer si on peut, afin de comparer les résultats expérimentaux avec les solutions obtenues. C'est uniquement dans cet accord de l'expérience avec la théorie qu'il faut chercher, comme toujours, le seul test valable des hypothèses émises.

C'est LAMB[2] qui eut pour la première fois la réelle audace de traiter un problème d'électromagnétisme à l'échelle du Globe, problème d'ailleurs en dehors du sujet du présent exposé. Pour ce qui est de la variation diurne, c'est à CHAPMAN que revient le mérite de l'avoir abordée par la théorie. Il fut l'initiateur de toute une série de beaux problèmes de physique mathématique, évoqués avec assez de détails et avec références bibliographiques dans le chapître XXII de «Geomagnetism», par CHAPMAN et BARTELS [2]. Ces problèmes, situés aux confins de l'Electricité tellurique et du Magnétisme, ressortissent davantage au Magnétisme qu'à l'Electricité et ne seront donc envisagés ici que très superficiellement.

Le calcul suppose qu'on a d'abord fait une analyse harmonique sphérique, sous forme d'un petit nombre de termes, de la partie périodique du potentiel magnétique, telle qu'elle est connue par les observations de la variation magnétique diurne à la surface de la Terre. Cette analyse, proposée par C. F. GAUSS et exécutée la première fois par SCHUSTER[3] pour ce qui est de la variation diurne solaire, fut reprise et améliorée depuis par d'autres Géophysiciens, notamment

[1] A. P. BONDARENKO: Dokl. Akad. Nauk. SSSR. **89**, 443 (1953).
[2] H. LAMB: Phil. Trans. Roy. Soc. Lond. **174**, 519 (1883); **180**, 513 (1889).
[3] A. SCHUSTER: Phil. Trans. Roy. Soc. Lond. **180**, 467 (1889).

par Chapman[1]. On admet toujours que la variation diurne, rapportée à l'heure locale, est la même tout le long d'un parallèle géographique. Chapman, qui fit l'analyse la plus précise, ne disposait quand même que des données de 21 Observatoires. L'estimation des coefficients du développement comporte donc forcément une certaine marge d'incertitude et l'accord des divers calculateurs, sur ces données expérimentales de base, n'est pas parfait.

Pour incertaines qu'elles soient, ces analyses gaussiennes présentent dès l'abord un intérêt intrinsèque certain. Elles permettent en effet de distinguer deux parties dans le potentiel magnétique évalué sur la surface de la Terre. La première partie se rapporte à des sources, telles que les courants électriques de l'ionosphère, situées à l'extérieur de la surface terrestre. La seconde partie se rapporte aux sources internes qui, en la circonstance, ne peuvent guère être autre chose que les courants telluriques. Le résultat est que le quotient $r$ d'un terme du premier groupe par le terme de même période du second groupe, qu'il s'agisse du terme diurne proprement dit ou d'un des harmoniques, qu'il s'agisse de variation diurne solaire ou lunaire, est de l'ordre de 2 à 3. La différence de phase $f$ entre les deux termes de même période est petite, et excède rarement 20°.

Quand il s'agit, pour comparer, d'intégrer les équations de Maxwell, on se trouve malheureusement dans l'obligation d'adopter des hypothèses très simples sur la constitution du sol, beaucoup trop simples pour qu'elles aient quelque chance de ressembler à la réalité. On suppose d'abord, bien entendu, que la structure physique du Globe présente la symétrie de la sphère, ce qui est certainement très inexact quand il s'agit de l'Écorce, de l'Écorce où circulent précisément les courants telluriques. En fait, on utilise un «modèle» de Globe à deux paramètres seulement: l'Écorce, d'épaisseur constante $l$ et de résistivité infinie, recouvre un substratum de résistivité $\varrho_s$ uniforme.

Le principe du calcul réside en ce que n'importe quel modèle de Globe ne peut être compatible avec les valeurs numériques expérimentales des rapports $r$ et des différences de phase $f$. Si par exemple le Globe était infiniment résistant dans son ensemble, il n'y circulerait pas de courants, de sorte que $r$ serait infini. On cherche donc en définitive une épaisseur d'Ecorce $l$ et une résistivité $\varrho_s$ de substratum approximativement compatibles avec les valeurs admises pour les $r$ et les $f$. Un modèle de Globe qui donne un résultat satisfaisant correspond par exemple à $l = 250$ km. et à $\varrho_s = 28\ \Omega.$ m. On remarquera que cette résistivité de 28 $\Omega$. m. trouvée pour le *substratum* n'est que du même ordre de grandeur que celle des roches sédimentaires, et qu'elle est bien supérieure à celle des eaux océaniques. Ces résultats sont assez déroutants, mais il n'est pas exclu, comme le dit Chapman, que des modèles de Globe bien différents de celui-là puissent se révéler tout aussi satisfaisants. On évoquera par exemple, au cours du prochain paragraphe, des calculs de Tikhonov et de Lipskaiia[2] qui sont d'ailleurs d'un type très différent de ceux de Chapman. Ces auteurs parviennent eux aussi à mettre en accord la théorie et l'expérience en partant d'un modèle qui diffère radicalement de celui de Chapman: une Écorce, d'épaisseur variable $l$ et de résistivité variable $\varrho_0$, reposant sur un substratum infiniment conducteur. A Tucson (Arizona), $l$ serait de l'ordre du millier de kilomètres et $\varrho_0$ de l'ordre de 100 $\Omega$. m. A Toyohara (Sud de Sakhaline), $l$ serait un peu plus grand (1200 à 1400 km.) tandis que $\varrho_0$ serait environ moitié moindre. Enfin à Zouy (Province d'Irkoutsk), $l$ tomberait à une centaine de km. et $\varrho_0$ à environ 1 $\Omega$. m.

[1] S. Chapman: Phil. Trans. Roy. Soc. Lond. **218**, 1 (1919).

[2] A. N. Tikhonov et N. V. Lipskaiia: Dokl. Akad. Nauk. SSSR. **87**, 547 (1952) — N. V. Lipskaiia: Izv. Akad. Nauk SSSR. Sér. Géophys. **1953**, No. I, 41.

Les formules de CHAPMAN donnent bien entendu l'expression des champs électrique et magnétique partout dans le Globe et, en particulier, celle du champ tellurique. Sur ce point, la comparaison entre théorie et expérience s'avère vraiment médiocre. C'est ainsi qu'à Tortosa les amplitudes théoriques sont environ cinq fois plus petites que les amplitudes observées. A Watheroo, elles sont par contre six fois plus grandes. Tout cela n'est pas très surprenant: la structure locale de l'Écorce joue manifestement un rôle déterminant dans le comportement des courants telluriques, et les modèles de Globe à symétrie sphérique ne peuvent satisfaire vraiment ni les Géologues, ni les Géophysiciens.

**34. Comparaison des composantes horizontales magnétiques et telluriques dans le cas des variations rapides.** Ce sujet a fait récemment l'objet de divers déve-

loppements théoriques d'inspiration voisine. On citera entre autres KATO et KIKUCHI[1], RIKITAKE[2], TIKHONOV[3], CAGNIARD [13]. On prend ici pour point de départ le postulat d'uniformité de la prospection tellurique. Il ne s'agit donc que de théories approchées, qu'il n'est d'ailleurs pas interdit de pouvoir perfectionner. On admet que si le sol est homogène ou, plus généralement, s'il est tabulaire, la nappe tellurique est uniforme.

Le cas du sol homogène est particulièrement simple et mérite d'être discuté en premier lieu. A la Sect. 11, dont on reprend ici les notations, le vecteur de HERTZ correspondant

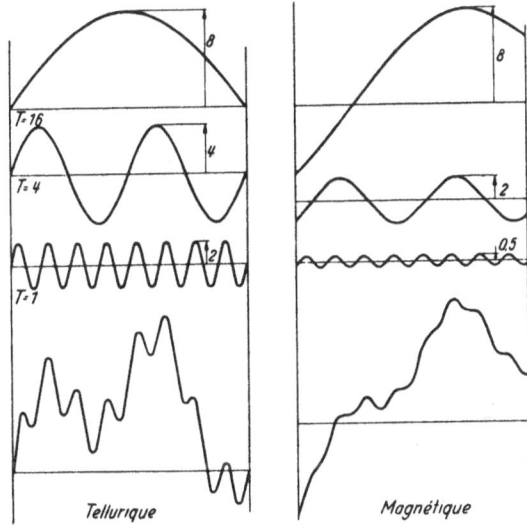

Fig. 25. Champs tellurique et magnétique dans une nappe tellurique imaginée comportant trois constituants harmoniques superposés.

à ce cas a déjà été exprimé de façon explicite (11.4). Conformément à (9.1), de simples dérivations permettent de calculer les champs. Quand on suppose que la nappe s'écoule suivant $Ox$, seules les les composantes $E_x$ et $H_y$ ne sont pas nulles, et quand on exprime le champ électrique à la surface du sol, autrement dit le champ tellurique, sous la forme

$$E_x = \cos \omega t \qquad (34.1)$$

le champ magnétique, à la surface du sol également, s'écrit:

$$H_y = 2 \sqrt{\frac{\pi \sigma}{\omega}} \cos\left(\omega t - \frac{\pi}{4}\right). \qquad (34.2)$$

*En conclusion, le quotient des amplitudes des champs tellurique et magnétique est égal à:*

$$\frac{E_x}{H_y} = \frac{1}{2} \sqrt{\frac{\omega}{\pi \sigma}} = \sqrt{\frac{\varrho}{2T}} \qquad (34.3)$$

*et, de plus, le champ magnétique retarde de $\pi/4$ par rapport au champ tellurique.*

[1] Y. KATO et KIKUCHI: Sci. Rep. Tôhoku Univ., Ser. V Geophys. **2**, 139 (1950).
[2] T. RIKITAKE: Bull. Earthq. Res. Inst. **28**, 45, 219 (1950); **29**, 61, 271 (1951).
[3] A. N. TIKHONOV: Dokl. Akad. Nauk SSSR. **73**, 295 (1950).

Pour illustrer les remarques qui précèdent, on a imaginé une nappe tellurique en milieu homogène qui résulterait de la superposition de trois constituants harmoniques ayant pour périodes 16; 4; 1 et auxquelles correspondraient des amplitudes telluriques respectives 8; 4; 2 (Fig. 25). Sur la droite de la figure sont représentées les variations des champs magnétiques associés, auxquelles correspondent donc des amplitudes qui sont entre elles comme les nombres 8; 2; 0,5. Les retards de phase sont tous $\pi/4$ et se traduisent par des décalages de temps proportionnels aux périodes, qui sont donc entre eux comme 16; 4; 1. Cet exemple schématique permet de mieux comprendre le mécanisme de la corrélation magnéto-tellurique tel qu'il se présente sur les courbes expérimentales des Fig. 11 ou 12.

Quand il s'agit d'une nappe tellurique naturelle, qui évolue de façon quelconque par rapport au temps, les résultats ci-dessus se généralisent facilement par l'emploi du calcul opérationnel:

$$E_x(t) = \frac{1}{2\pi\sqrt{\sigma}} \int_{-\infty}^{t} \frac{d[H_y(u)]}{\sqrt{t-u}}. \tag{34.4}$$

On obtient ainsi une formule élégante, moins suggestive cependant que la décomposition en constituants harmoniques.

**35. Retour sur le postulat d'uniformité.** A supposer l'uniformité du sous-sol, l'uniformité dans l'écoulement de la nappe tellurique ne peut provenir que d'une uniformité dans le mécanisme d'excitation. Cette uniformité de l'excitation est due en premier lieu aux grandes distances auxquelles se situent les sources des perturbations électromagnétiques. A ce propos, il n'est pas inutile de souligner combien sont énormes les longueurs d'onde de ces perturbations. Pour $T = 30$ sec., la longueur d'onde atteint par exemple 9 millions de km., soit 1400 fois le rayon terrestre, ce qui fait jouer un rôle primordial à la diffraction dans le comportement du phénomène électromagnétique. Lorsqu'une perturbation est enregistrée quelque part, elle n'est donc pas due uniquement à des sources situées au-dessus de l'horizon du lieu, mais à la totalité des courants ionosphériques, auxquels il faut ajouter la totalité des courants telluriques du Globe. Par surcroît, et au fond pour la même raison, la source des perturbations ne peut pas résider dans une oscillation, au caractère très local, de quelques particules ionosphériques seulement, en d'autres termes dans ce que le Mathématicien appelle volontiers une excitation par dipôle. Les courants électriques de l'ionosphère constituent, eux également, de véritables nappes, et cela contribue aussi à assurer l'uniformité de l'excitation.

D'autre part, dans un sous-sol homogène ou tabulaire, la circulation des courants telluriques ne peut se faire qu'approximativement suivant l'horizontale. Du fait que les courants de déplacement sont négligeables. l'horizontalité en question est rigoureuse à la surface même du sol, et elle reste approximative en profondeur parce que la nappe tellurique, en raison du skin effect, est relativement mince. L'uniformité et l'horizontalité sont donc a priori d'autant mieux réalisées que la profondeur $\zeta$ de pénétration est plus faible, c'est à dire que $T$ et $\varrho$ sont plus petits. Les ordres de grandeur sont utiles à considérer ici. Ils sont donnés par le tableau 1. Lorsqu'on exprime $\zeta$ en km., $\varrho$ en $\Omega$. m. et $T$ en secondes, on a

$$\zeta = \frac{1}{2\pi}\sqrt{10\,\varrho\,T}. \tag{35.1}$$

La relation (34.3) cesse d'être exacte quand l'excitation manque d'uniformité. A ce propos Wait[1] se livre à un calcul intéressant, toujours dans l'hypothèse

---

[1] J. R. Wait: Geophysics **19**, 281 (1954).

Tableau 1. *Profondeurs de pénétration exprimées en kilomètres.*

| $\varrho$ \ $T$ | 1 sec. | 3 sec. | 10 sec. | 30 sec. | 1 min. | 2 min. | 5 min. | 10 min. | 30 min. |
|---|---|---|---|---|---|---|---|---|---|
| 0,2 | 0,225 | 0,390 | 0,712 | 1,23 | 1,74 | 2,47 | 3,90 | 5,51 | 9,54 |
| 1 | 0,503 | 0,872 | 1,59 | 2,76 | 3,90 | 5,51 | 8,72 | 12,3 | 21,4 |
| 5 | 1,13 | 1,95 | 3,56 | 6,16 | 8,72 | 12,3 | 19,5 | 27,6 | 47,7 |
| 10 | 1,59 | 2,76 | 5,03 | 8,72 | 12,3 | 17,4 | 27,6 | 39,0 | 67,5 |
| 50 | 3,56 | 6,16 | 11,3 | 19,5 | 27,6 | 39,0 | 61,6 | 87,2 | 151 |
| 250 | 7,95 | 13,8 | 25,2 | 43,6 | 61,6 | 87,2 | 138 | 195 | 338 |
| 1000 | 15,9 | 27,6 | 50,3 | 87,2 | 123 | 174 | 276 | 390 | 675 |
| 5000 | 35,6 | 61,6 | 113 | 195 | 276 | 390 | 616 | 872 | 1510 |

d'un sol homogène. A cette relation, il substitue une formule rigoureuse qui se présente sous forme d'un développement en série, à convergence généralement rapide, dont le premier terme correspond évidemment à (34.3). L'auteur fait l'application numérique suivante: en supposant $\varrho = 10$ et admettant que la source de la perturbation est un «dipôle» situé au zénith, à 100 km. d'altitude, le second terme du développement cesse d'être négligeable quand $T$ dépasse une dizaine de secondes. En fait, le domaine de validité est considérablement plus étendu, car les sources ne sont pas des dipôles qui rayonnent au-dessus de nos têtes.

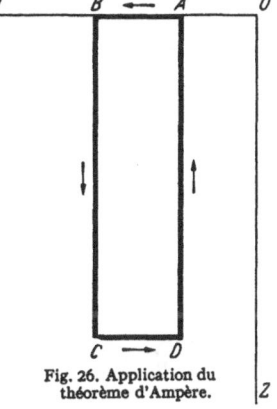

Fig. 26. Application du théorème d'Ampère.

L'application du théorème d'AMPERE donne lieu aussi à quelques remarques intéressantes. Le raisonnement qui suit suppose seulement l'uniformité de la nappe et pas nécessairement l'homogénéité du sol. Sur la Fig. 26 est représenté un rectangle $ABCD$, dans lequel $AB$ est à la surface tandis qu'on imaginera que $CD$ est rejeté à une profondeur infinie. Comme on suppose expressément que la conductibilité du sol ne s'annule pas aux grandes profondeurs, l'existence du skin-effect entraine que le champ magnétique est nul pour $z$ infini. De sorte que le calcul du travail du champ magnétique le long de $ABCD$ conduit à l'expression:

$$H_y = 4\pi \int_0^\infty i_x(z) \cdot dz \qquad (35.2)$$

où $i_x(z)$ désigne la densité de courant à la profondeur $z$. On voit donc que le champ magnétique $H_y$ mesure, au facteur $4\pi$ près, l'intensité du courant tellurique qui traverse une bande rectangulaire telle que $ABCD$, et de largeur unité. Et c'est parce que le courant est conservatif, et parce qu'il faut des accidents considérables dans la tectonique du sous-sol pour créer une distorsion des lignes de courant, que les variations magnétiques, contrairement à ce qui se passe pour les variations telluriques, demeurent toujours à peu de chose près les mêmes en deux stations pas trop éloignées l'une de l'autre.

Au demeurant, il est bien connu que le champ magnétique de BIOT et SAVART accompagnant une nappe tellurique, parallèle à $Ox$ et définie par la densité $i_x(z)$, s'exprime sous la forme:

$$h_y = 2\pi \int_0^\infty i_x(z) \cdot dz = \frac{H_y}{2}. \qquad (35.3)$$

La relation (35.3) diffère donc de (35.2) par la substitution du facteur $2\pi$ au facteur $4\pi$, de sorte que $h_y$ n'est que la moitié de $H_y$. Cela peut sembler paradoxal, mais s'explique pourtant aisément. Suivant un type de raisonnement qu'on fait assez souvent en Physique, soit en effet $h_1$, vecteur qu'on n'a pas jugé utile de représenter sur la Fig. 26, le champ de Biot et Savart associé à la totalité des courants lointains, aussi bien aux courants telluriques lointains qu'à l'ensemble des courants ionosphériques. Ce champ $h_1$ est pratiquement le même en $A$, *au-dessus* de la nappe tellurique locale, qu'en $C$, à une profondeur suffisante pour qu'on puisse considérer que $C$ est «pratiquement» *au-dessous* de la nappe. Le champ magnétique total est la résultante de $h_1$ et du champ $h_2$ associé aux portions de nappe tellurique voisines du point d'observation. C'est d'ailleurs ce champ $h_2$, au point $A$, qu'exprime (35.3). Or en $C$, le champ associé aux portions locales de nappe est $- h_2$, le champ total en $C$ est $h_1 - h_2$, et l'on sait de reste que ce champ total doit être nul.

$h_1$ est donc égal à $h_2$, et le paradoxe est éclairci. A ce point du raisonnement, on se trouve tenté d'établir un rapprochement avec les résultats de l'analyse gaussienne évoquée plus haut, qui permettait de dissocier le champ en deux portions, l'une associée à des sources externes, l'autre à des sources internes. Mais le rapprochement ne va pas plus loin, car si $h_2$ est indiscutablement associé à des sources internes, $h_1$ se trouve associé à la fois à des sources internes et à des sources externes.

Le raisonnement qui vient d'être fait contribue lui aussi à montrer pourquoi l'approximation (34.3) devient plus médiocre quand la profondeur de pénétration devient plus grande, car lorsque $C$ est très profond, on n'a plus le droit de poser que $h_1$ est pratiquement le même en $A$ et en $C$.

Ces nappes telluriques qu'on a envisagées ici ne comportent que des champs horizontaux, magnétiques comme telluriques. Les hypothèses qui furent faites excluent l'existence de variations affectant la composante verticale du champ magnétique. C'est d'ailleurs ce que montre très clairement l'équation (32.1) du système des équations de Maxwell, laquelle fait dépendre l'existence de $H_z$ de celle de variations, dans le plan horizontal, des composantes telluriques, en d'autres termes du manque d'uniformité du champ tellurique. Ce manque d'uniformité peut tenir au sous-sol, ou à l'excitation, ou aux deux à la fois. Dans tous les cas cependant, c'est la petitesse relative des variations de la composante verticale magnétique qui constitue le meilleur critère pour juger du degré de validité des approximations qui sont faites en ce moment.

**36. La prospection magnéto-tellurique.** La relation (34.3) fait apparaître le rapport $E_x/H_y$ comme une fonction de $\varrho$ et de $T$. Il est utile d'acquérir une notion des ordres de grandeur. C'est pourquoi le tableau 2 donne les amplitudes

Tableau 2. *Amplitudes du champ magnétique exprimées en $\gamma$.*

| $T$ \ $\varrho$ | 1 sec. | 3 sec. | 10 sec. | 30 sec. | 1 min. | 2 min. | 5 min. | 10 min. | 30 min. |
|---|---|---|---|---|---|---|---|---|---|
| 0,2 | 1 | 1,73 | 3,16 | 5,48 | 7,75 | 11,0 | 17,3 | 24,5 | 42,4 |
| 1 | 0,447 | 0,775 | 1,41 | 2,45 | 3,46 | 4,90 | 7,75 | 11 | 19,0 |
| 5 | 0,2 | 0,346 | 0,632 | 1,10 | 1,55 | 2,19 | 3,46 | 4,90 | 8,49 |
| 10 | 0,141 | 0,245 | 0,447 | 0,775 | 1,10 | 1,55 | 2,45 | 3,46 | 6 |
| 50 | 0,0632 | 0,110 | 0,2 | 0,346 | 0,490 | 0,693 | 1,10 | 1,55 | 2,68 |
| 250 | 0,0283 | 0,049 | 0,0894 | 0,155 | 0,219 | 0,310 | 0,490 | 0,693 | 1,2 |
| 1000 | 0,0141 | 0,0245 | 0,0447 | 0,0775 | 0,110 | 0,155 | 0,245 | 0,346 | 0,6 |
| 5000 | 0,00632 | 0,0110 | 0,0200 | 0,0346 | 0,0490 | 0,0693 | 0,110 | 0,155 | 0,268 |

en $\gamma$ du champ magnétique associé à un champ tellurique de 1 mV/km. suivant diverses valeurs de $\varrho$ et de $T$, les résistivités étant exprimées en $\Omega$. m. On remarquera combien le rapport des amplitudes des champs varie dans des proportions très importantes, de 1 à 700 dans les limites du tableau 2.

Inversement, si l'on peut savoir d'avance que le sol est homogène, il suffira de mesurer le rapport des amplitudes des champs pour une période $T$ déterminée, et de résoudre (34.3) par rapport à $\varrho$ pour connaître la résistivité du sol. Dans le système d'unités qui vient d'être adopté, $E_x$ en mV/km., $H_y$ en $\gamma$, $\varrho$ en $\Omega$. m. et $T$ en sec., on a:

$$\varrho = 0.2\, T \left(\frac{E_x}{H_y}\right)^2. \tag{36.1}$$

On peut d'ailleurs s'assurer que le sol est effectivement homogène, car la valeur ainsi calculée ne doit dépendre ni de $T$ ni de l'orientation de la ligne tellurique. De plus le décalage de phase des champs doit toujours être $\pi/4$.

Mais le sol n'est pas toujours homogène, ni même tabulaire. On s'en aperçoit d'abord à ce que le décalage de phase devient alors une fonction de $T$ ne présentant qu'exceptionnellement la valeur $\pi/4$. Mais il reste toujours possible néanmoins d'effectuer le calcul exprimé par (36.1). Le nombre qu'on obtient n'est pas la résistivité du sol puisque ce sol n'est pas homogène et n'est pas défini par une résistivité unique. Ce nombre a quand même les dimensions physiques d'une résistivité et peut être appelé *«résistivité apparente»*. Ladite résistivité apparente $\varrho_a$ dépend de la période $T$, et même aussi de l'orientation de la ligne quand le sol n'est pas tabulaire. Elle représente une sorte de moyenne des résistivités (vraies) des diverses formations géologiques présentes dans le volume de sol pratiquement intéressé par la circulation de la nappe tellurique de période $T$. En raison du skin effect, plus $T$ augmente, et plus devient grande l'épaisseur de la tranche de sol à laquelle s'applique la moyenne en question.

La comparaison magnéto-tellurique permet donc d'envisager la prospection du sous-sol, soit par une méthode de résistivités apparentes où, d'une station à l'autre, on n'envisagera jamais que la même période $T$, soit par une méthode de sondages où, dans la même station, on évaluera $\varrho_a$ en fonction de $T$.

On a remarqué sans doute que le raisonnement et le mode d'exposition adoptés au cours du présent paragraphe font exactement le «pendant» de ceux des Sect. 18 et 19 où furent décrites les méthodes classiques de prospection électrique. L'utilisation de la comparaison magnéto-tellurique en vue de la prospection constitue essentiellement le principe de la méthode magnéto-tellurique récemment proposée par CAGNIARD [13]. Cette méthode magnéto-tellurique, dont les applications industrielles n'ont pas encore débuté, bénéficie a priori de tous les avantages inhérents aux méthodes exclusivement telluriques, et elle permet par surcroît d'envisager, comme on va voir, des investigations «quantitatives».

Le parallélisme se poursuit en effet entre méthodes classiques et méthode magnéto-tellurique quand on envisage le cas d'un sous-sol tabulaire. Car on montre que le calcul théorique d'un sondage magnéto-tellurique, exactement comme celui d'un sondage électrique, est possible quels que soient le nombre, l'épaisseur et la résistivité des couches[1]. Malheureusement, quand il s'agit de sondages électriques, on ne pouvait guère profiter de cette possibilité, du fait qu'elle demeurait plus théorique que pratique. Dans un sondage magnéto-tellurique au contraire, il n'intervient que des calculs fort élémentaires, sans

---

[1] En supposant expressément qu'il s'agit d'un sous-sol tabulaire, l'un des résultats de calcul qu'on peut signaler est le suivant: le retard de phase de $H_y$ par rapport à $E_x$ est toujours compris entre 0 et $\pi/2$.

intégrales ni séries, qui peuvent même être encore facilités par l'adoption de procédés graphiques. Cela permet assurément d'augurer que, du point de vue quantitatif, les sondages magnéto-telluriques pourront tenir les promesses auxquelles ont failli les sondages électriques. Et ils garderont aussi sur eux l'avantage d'une profondeur d'investigation incomparable, bien faible sans doute en regard du rayon terrestre pour répondre totalement aux aspirations des Physiciens du Globe, suffisante quand même pour combler les voeux des Géologues les plus exigeants.

La Fig. 27 reproduit l'un des deux abaques, celui des résistivités apparentes, qui sert à l'interprétation des sondages magnéto-telluriques «2 terrains». Elle fait un exact «pendant» à la Fig. 6 et est construite sur les mêmes principes: coordonnées bi-logarithmiques, résistivités apparentes en ordonnées, périodes en abscisses, rapport des résistivités des deux terrains inscrit sur les diverses courbes du réseau. On utilise ces abaques de la même manière que ceux de la prospection électrique. Destinés à être regardés par transparence au travers d'un calque qui, lui, est gradué, les abaques d'interprétation originaux ne comportent pas la moindre graduation d'abscisses ou d'ordonnées. Mais pour rendre la Fig. 27 plus intelligible et plus instructive, on a surimprimé une graduation des axes, valable pour le cas où le terrain superficiel est épais de 5 km. et présente une résistivité de 10 $\Omega$. m.

En matière de sondage électrique, la résistivité apparente, égale à la résistivité du premier terrain quand la ligne $AB$ est courte, tend vers la résistivité du second quand $\overline{AB}$ croît indéfiniment. Ici les mêmes phénomènes se produisent quand la période est respectivement très petite ou très grande: c'est la période, qui fait «pendant» à la longueur de ligne. L'analogie gagne même à être précisée d'une manière quasi-numérique. Car on peut constater que la profondeur d' «investigation» est tout à fait du même ordre de grandeur, soit quand la longueur de ligne est $\overline{AB}$ dans un sondage électrique, soit quand la profondeur de «pénétration» $\zeta$ (35.1) est moitié moindre, dans un sondage magnéto-tellurique. Plus concisément, $\zeta$ est, en gros, comparable à $\overline{AB}/2$.

Dans le cas particulier qu'on vient d'imaginer, celui pour lequel sont valables les graduations de la Fig. 27, on constate qu'avec des périodes allant jusqu'à quelques minutes on pourra toujours déterminer correctement la profondeur du substratum, mais qu'il faudrait presque atteindre des périodes de quelques heures pour pouvoir apprécier valablement la résistivité du second terrain. On constate donc là à quel point les profondeurs utiles d'*investigation* sont beaucoup plus petites que les profondeurs de *pénétration*. De même avait-on constaté plus haut combien les profondeurs d'investigation des sondages électriques sont hors de proportion avec les longueurs de lignes utilisées. Et c'est toujours pour la même raison que, dans la prospection tellurique proprement dite, on est déçu de constater à quel point les indications obtenues sont, jusqu'à un certain point, superficielles, bien plus superficielles en tout cas qu'on serait tenté de le croire pour des raisons a priori.

Quand il s'agit de si grandes périodes, il faut quand même se montrer circonspect dans l'extrapolation de raisonnements qui ne valent que pour les variations suffisamment rapides. Les variations lentes de la composante verticale, en matière de variation diurne en particulier, atteignent des amplitudes qu'on ne saurait juger négligeables en comparaison de celles des composantes horizontales, et sont là par conséquent pour inciter à la prudence. Quelle confiance faut-il donc accorder à des résistivités apparentes qu'on continuerait alors à calculer d'après la définition (36.1)? En attendant que l'avenir ne réponde, il demeure que ces résistivités, à supposer qu'elles soient erronées de 50% ou davantage, possèdent néanmoins une valeur indicative précieuse. Au Parc St Maur, où ROUGERIE [9] a dépouillé les vieux enregistrements de MOUREAUX, on trouve ainsi une résis-

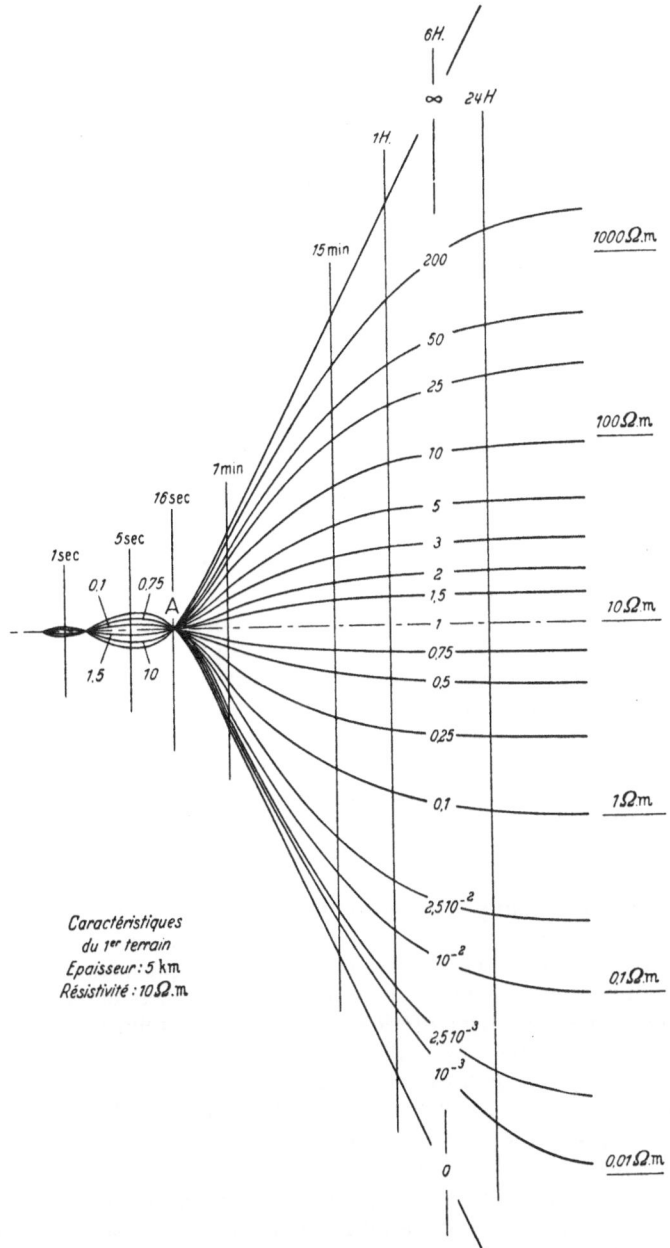

Fig. 27. Abaque bi-logarithmique de sondages magnéto-telluriques «2 terrains».

tivité apparente de 12,5 Ω. m. quand il s'agit de l'onde diurne, de 10,2 Ω. m. pour la semi-diurne, et ces résultats sont à peu de chose près les mêmes quand on considère la ligne $N - S$ ou la ligne $E - W$, quand on utilise dans le calcul l'une ou l'autre des moyennes saisonnières ou la moyenne générale. Les diffé-rences de phase sont, comme le veut la règle, des retards de phase du champ magnétique relativement au champ tellurique, et sont, elles aussi, très cohérentes, environ 69° pour l'onde diurne, 37° pour la semi-diurne.

Les résistivités apparentes ainsi calculées sont, de toute manière, bien dif-férentes de celles d'un granite, du moins d'un granite d'affleurement. Elles se laissent d'ailleurs comparer, pour l'ordre de grandeur, aux résistivités envisagées soit par Chapman, soit par Tikhonov-Lipskaia, et dont il fut fait état à la Sect. 33. Il y a là, sans nul doute, les tout premiers débuts, encore hésitants, d'un mode nouveau d'investigation ultra-profonde de l'Écorce terrestre, d'une Électroprospection profonde, comme dit Tikhonov, laquelle semble tout parti-culièrement prometteuse. Il n'est d'ailleurs pas exclu qu'on ne puisse améliorer la théorie, en lui apportant les retouches nécessaires qui permettraient de tenir compte de la non-uniformité. Wait a déjà montré la voie pour les variations rapides, pendant que Tikhonov et Lipskaia l'ont fait pour la variation diurne.

# G. Conclusion.

**37.** Le voile d'étrangeté et de mystère dont se paraient hier encore les courants telluriques est aujourd'hui tombé. Le champ tellurique n'est qu'un aspect par-ticulier du champ électromagnétique, et les enregistrements telluriques sont l'équivalent des enregistrements magnétométriques.

A l'heure présente, dans l'étude des variations et perturbations électroma-gnétiques rapides, il est plus facile de procéder à un enregistrement tellurique qu'à un enregistrement magnétique. Il ne s'agit pourtant, sans doute, que d'un état de choses provisoire, qui tient à ce que les Géophysiciens n'ont porté jusqu'ici bien sérieuse attention, ni aux variations magnétiques rapides, ni aux moyens expérimentaux propres à les enregistrer.

Moyennant quelques précautions expérimentales, une ligne tellurique et son galvanomètre sont l'équivalent d'un magnétomètre, mais d'un magnétomètre dont on ignore a priori les caractéristiques. Car ces caractéristiques sont déter-minées par la structure du sous-sol, structure qui est fort variable d'une station à l'autre, qui est toujours plus ou moins complexe, qui est toujours plus ou moins mal connue.

De nombreux aspects du phénomène tellurique restent à étudier, à découvrir, à préciser. On ne sait à peu près rien sur les variations plus lentes que la variation diurne, ni sur celles dont les périodes sont inférieures à quelques secondes. Bien des travaux qu'implique n'importe quel programme de recherches en ces domaines ne pourront être poursuivis qu'à l'intérieur d'Observatoires telluriques permanents, où l'observation magnétique devra aller de pair avec l'observation tellurique. Il ne semble pas d'ailleurs que ces Observatoires magnéto-telluriques aient besoin d'être bien nombreux s'ils sont convenablement répartis sur le Globe, s'ils sont judicieusement implantés, s'ils sont parfaitement outillés, et s'il règne entre eux une étroite et confiante collaboration permanente.

Selon toute vraisemblance pourtant, ce ne sont pas les Observatoires qui récolteront la plus belle moisson de résultats. Du point de vue scientifique comme du point de vue pratique, l'importance, l'importance exceptionnelle des études telluriques, réside en cet espoir magnifique qu'elles apportent de réaliser quelque jour la prospection ultra-profonde, et malgré tout relativement précise, de l'Écorce terrestre [14]. Cette réconciliation de la Géologie et de la Physique du Globe, de ces deux sœurs qui s'ignorent encore trop fréquemment, ne se fera pas dans des Observatoires permanents, mais dans des stations temporaires et mobiles. Elle impliquera l'usage de techniques relevant davantage de la Géophysique Appliquée que de la Géophysique Pure. Et il n'est pas interdit d'espérer qu'au cours des prochaines années, quelques-uns des Mystères de la Géologie profonde auront pu se trouver élucidés, tout au moins en partie.

## Bibliographie.

[1] Traités de prospection géophysique, faisant mention des procédés électriques: SCHLUMBERGER, C.: Etude sur la prospection électrique du sous-sol. Paris: Gauthier-Villars 1ère édit. 1920; 2ème édit. 1930. — NETTLETON, L. L.: Geophysical prospecting for oil. New York: McGraw-Hill 1940. — JAKOSKY, J. J.: Exploration Geophysics, 2ème édit. Los Angeles: Times Mirror 1950. — CAGNIARD, L.: La Prospection géophysique. Paris: Presses Universitaires 1950. — DOBRIN, M. B.: Introduction to Geophysical Prospecting. New York: McGraw-Hill 1952.

[2] CHAPMAN, S., and J. BARTELS: Geomagnetism. 2. Vol. Oxford: Clarendon·Press 1940.

[3] FRANK, P., u. R. VON MISES: Die Differential- und Integralgleichungen der Mechanik und Physik. Braunschweig: Vieweg & Sohn. 2. Aufl., Bd. 1 1930; Bd. 2 1935.

[4] POLDINI, E.: Les phénomènes de polarisation spontanée et leur application à la recherche des gîtes métallifères. Mém. Soc. Vaudoise Sci. Natur. 6, 1 (1938).

[5] GISH, O. H.: General description of the Earth-current measuring system at the Watheroo magnetic Observatory. Terr. Magn. 28, 89 (1923).

[6] ROONEY, W. J.: Earth-currents, Chapter VI of «Terrestrial Magnetism and Electricity», Physics of the Earth, VIII. New York: Dover Publications. 1ère édit. 1939; 2ème édit. 1949.

[7] WEINSTEIN, B.: Die Erdströme im deutschen Reichstelegraphengebiet und ihr Zusammenhang mit den erdmagnetischen Erscheinungen. Braunschweig: Vieweg & Sohn 1900.

[8] BOSLER, J.: Les courants telluriques. 3ème partie, chap. 1 du «Traité d'Electricité atmosphérique et tellurique», par E. MATHIAS. Paris: Presses Universitaires 1924.

[9] ROUGERIE, P.: Contribution à l'Etude des courants telluriques. Ann. Inst. Phys. Globe Paris 20, 60 (1942).

[10] GISH, O. H., and W. J. ROONEY: New aspects of Earth-current circulations revealed by polar-year data. Trans. Edinburgh Meeting of sept. 1936. Int. Union Geod. a. Geop. Copenhagen 1937.

[11] BLAVIER, E. E.: Etude des courants telluriques. Paris: Gauthier-Villars 1884.

[12] Publications de la Compagnie Générale de Géophysique sur la prospection tellurique: FRIEDEL, E., et J. GOGUEL: La prospection géophysique du Bas Dauphiné. Ann. Mines et Carb. 10, 417 (1944). — MIGAUX, L.: Une méthode nouvelle de Géophysique appliquée, la prospection par courants telluriques. Ann. Géophys. 2, 131 (1946). — BOISSONNAS, E., and E. LEONARDON: Geoph. Expl. by telluric currents, with special reference to a survey of the Haynesville salt-dome, Wood county, Texas. Geophysics 13,. 387 (1948). — MIGAUX, L.: Une méthode nouvelle de Géophysique appliquée, la prospection par courants telluriques. Evreux: Hérissey 1948.

[13] CAGNIARD, L.: Principe de la méthode magnéto-tellurique, nouvelle méthode de prospection géophysique. Ann. Géophys. 9, 95 (1953). — Basic theory of the magnetotelluric method of geophysical prospecting. Geophysics 18, 605 (1953).

[14] Note de l'éditeur: Les géomagnéticiens ont commencé d'utiliser les baies — variations avec des périodes une heure à-peu-près — pour cette prospection ultra-profonde. Signalons quelques travaux: T. RIKITAKE, I. YOKOYAMA and W. HISHIJAMA: The anomalous behaviour of geomagnetic variations of short period in Japan and its relations to the subterranean structure. Bull. Earthquake Res. Inst. Tokyo 30, 207 (1952); 31, 19, 89, 101, 119 (1953); 33, 297 (1955); summary in J. Geomag. a. Geoelectr., Kyoto 5, 59—65 (1953). — U. FLEISCHER: Ein Erdstrom im tieferen Untergrund Norddeutschlands während erdmagnetischer Baistörungen. Naturwiss. 41, 114 (1954). — Charakteristische erdmagnetische Baystörungen in Mitteleuropa und ihr innerer Anteil. Z. Geophys. 20, 120 (1954). — J. BARTELS: Erdmagnetisch erschließbare lokale Inhomogenitäten der elektrischen Leitfähigkeit im Untergrund. Nachr. Akad. Wiss. Göttingen, Phys.-math. Kl., Abt. II a 1954, 95. (Dort weitere Lit.) — W. KERTZ: Modelle für erdmagnetisch induzierte Ströme im Untergrund. Nachr. Akad. Wiss. Göttingen, Phys.-math. Kl., Abt. II a 1954, 101.

# Magnetization of Rocks.

By

## S. K. RUNCORN.

With 14 Figures.

The main topic of this chapter is the discussion of the natural remanent magnetization of rocks with the object of inferring the direction of the geomagnetic field in the geological past. The reader who is interested in other aspects of rock magnetism is referred to the additional references at the end.

## A. Mineralogical aspects.

The chief minerals occurring in rocks which account for their magnetic properties are those oxides of the system $FeO - Fe_2O_3 - TiO_2$ shown in Fig. 1. However, the iron sulphide of range of composition $Fe_{11}S_{12}$ to $Fe_6S_7$, called pyrrhotite,

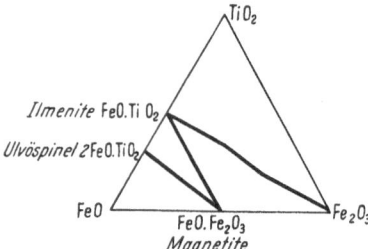

Fig. 1. Ternary diagram for the system $FeO-Fe_2O_3-TiO_2$.

is ferrimagnetic. Its Curie point is between 300 and 325° C, its saturation magnetization is of the order of 62 emu/cm.[3] and its coercive force is 15 to 20 gauss[1]. Its contribution to the magnetization of rocks is less well known than that of oxides and will not be discussed further here.

**1. Properties of end members of oxide system.** The properties of the more important minerals of this system will now be summarized: further numerical data are given by NAGATA [1]. Magnetite ($Fe_3O_4$ or $Fe^{++}Fe_2^{+++}O_4$) is a member of the spinel group of minerals, with cubic structure[2], the unit cell $a = 8.396$ Å, containing 32 oxygen ions and 24 cations, the latter distributed among 8$A$ sites or tetrahedral positions in 4 fold coordination with oxygen atoms and 16$B$ sites or octahedral positions in 6 fold coordination. Magnetite has a Curie point of 578° C, a saturation magnetization at room temperature of 92 to 93 emu/gram and a coercive force at room temperature of 50 gauss.

*Magnetite* is an example of a *ferrite*, the general formula of which is $Fe_2O_3MO$ when $M$ is a divalent metal. Its *ferromagnetism*, or *ferrimagnetism* as NÉEL has termed it, arises largely through the interaction between ions on $A$ sites and those on $B$ sites [2]. The exchange integral is such that at low temperatures the magnetic moments of ions on $A$ sites are parallel to one another and antiparallel to those on $B$ sites within any one domain; a fact recently verified by neutron diffraction studies[3]. Magnetite is an example of an inverse spinel in which in a unit cell the ferric ions are distributed equally between the $A$ and $B$ sites, the ferrous ions occupying the remaining $B$ sites. The spontaneous saturation magnetization per molecule is therefore equal to 4 Bohr magnetons, the magnetic moment of the ferrous ion. It is strong evidence for this scheme that the spontaneous magnetiza-

[1] T. NIRONE and N. TSUYA: Phys. Rev. 83, 1063 (1951).
[2] W. H. BRAGG: Phil. Mag. 30, 305 (1915).
[3] C. A. SHULL, E. O. WOLLAN and W. C. KOEHLER: Phys. Rev. 84, 912 (1951).

tion is equal to the magnetic moment of the divalent ions when the ferrous ion is replaced by other ions ($Mn^{++}$, $Co^{++}$, $Ni^{++}$, $Cu^{++}$ and $Zn^{++}$) which also form inverse spinels.

*Maghemite* ($\gamma Fe_2O_3$) is cubic (cell dimensions $a = 8.322$ Å [1]) and is an inverse spinel in which 1 in 9 of the iron ions are absent, probably from the $B$ positions. Its Curie point is estimated as $675°$ C and has a saturation magnetization at room temperature of 83.5 emu/gram.

It is ferrimagnetic but the theory of this has not been studied in detail, although it is clear that a net magnetic moment of the lattice is to be expected.

It may be produced by the low temperature oxidation of magnetite or the dehydration of lepidocrocite FeO(OH) but its occurrence has been rarely recognised in those lavas and sediments studied up to the present.

*Haematite* ($\alpha Fe_2O_3$) is rhombohedral with cell dimensions $a_R = 5.4271$ Å $\alpha 55°$ 15.8' for a rhombohedral unit cell [2]. Haematite appears to be ferrimagnetic [3] with a Curie point of $675°$ C and saturation magnetization equal to 0.5 emu/gram and coercive force 7600 gauss. NÉEL [2] attributes this ferromagnetism to the presence of small ferromagnetic domains in a paramagnetic lattice; the latter arises because the ferric ions are distributed on two equal sublattices of opposite magnetization. The ferromagnetic domains are thought to result from slight unbalances in this system of equal sub-lattices, probably arising from dislocations in the lattice. NÉEL [3] had previously suggested that the ferromagnetism was the result of a magnetite impurity of about 1%, interleaved into the haematite structure.

*Ilmenite* (FeTiO$_3$) is rhombohedral with cell dimensions $a_R = 5.523$ Å $\alpha 54°$ 51' for a rhombohedral unit cell. Ilmenite appears to be antiferromagnetic but some evidence has been produced that an ilmenite from Japan has a saturation magnetization of 0.2 emu/gram and a Curie point of 100 to $150°$ C.

*Ulvöspinel* (Fe$_2$TiO$_4$) is cubic of inverse spinel structure in which 2Fe$^{+++}$ is replaced by Fe$^{++}$Ti$^{++++}$. It is paramagnetic and occurs naturally as intergrowths with magnetite, as a result of exsolution from solid solution. It is not found naturally except in a multicomponent-system but it is possible to synthesize it.

**2. Properties of solid solutions of the iron oxides.** Solid solution between Fe$_3$O$_4$ and Fe$_2$TiO$_4$ is complete at high temperatures. POUILLARD [4] has shown

Table 1. *Variation of cell divisions and Curie point in series FeTiO—FeO.*

| Cell dimension $a$ in Å | 8.413 | 8.420 | 8.428 | 8.43 | 8.44 | 8.46 | 8.46 | 8.534 |
|---|---|---|---|---|---|---|---|---|
| Curie point in °C | 575 | 524 | 485 | 475 | 468 | 230 | 215 | — |
| Mol-% TiFeO | 0 | 8 | 10 | 18 | 25 | 42 | 42 | 100 |

that the Curie point varies with composition as shown in Table 1, the Curie point depending on composition by the following

$$T_C = 1245s \quad \text{where} \quad s = \frac{\text{Wt \% FeO}}{\text{Wt \% Fe}_2\text{O}_3 + \text{FeO}}. \tag{2.1}$$

If the solid solution involves a replacement of Fe$^{+++}$ in the $B$ sites by Ti$^{++++}$ and in the $A$ sites by Fe$^{++}$, the saturation magnetization should decrease with increasing content of Fe$_2$TiO$_4$ as is the case [4].

[1] G. HAGG: Z. phys. Chem., Abt. B **29**, 95 (1935).
[2] B. T. M. WILLIS and H. P. ROOKSBY: Proc. Phys. Soc. Lond., Ser. B, **65**, 950 (1952).
[3] J. ROQUET: C. R. Acad. Sci. Paris **224**, 1418 (1947).
[4] S. AKIMOTO: J. Geomag. a. Geoelect. **6**, 1 (1954).

At lower temperatures intermediate compositions exsolve on cooling into two separate phases; the exsolved component, usually the more titaniferous one, forms lamellae parallel to the cube faces (100) of the host crystal.

Solid solution is complete between haematite ($\alpha$ $Fe_2O_3$) and ilmenite ($FeTiO_3$) at temperatures above 1050° C. At normal temperatures haematite contains only up to about 20% ilmenite (in molecular proportions) and ilmenites only up to 6% $Fe_2O_3$. A small proportion of $TiO_2$ may be present in solid solution in any of this series at about 1000° C but it appears to exsolve as *rutile* at low temperature. Some cases are known[1] in which the intensity of saturation magnetization of these solid solutions is much greater than that of the end members. For example a solid solution having the molecule composition 26.9% $Fe_2O_3$, 66.4% $FeTiO_3$ and 6.7% $TiO_2$ gave a value of 20 emu/gram (with a Curie point of 250°C)[1]. No data are available on the variation of Curie point with composition in this series.

Néel [2] explains the high magnetization of the solid solutions by the following analysis. $\alpha$ $Fe_2O_3$ is antiferromagnetic because its structure is such that alternate layers of the ferric ions have opposite magnetizations. Ilmenite is also antiferromagnetic because every other layer of iron atoms is replaced by a layer of $Ti^{++++}$ ions and these are not magnetic. Thus the layers can be indicated as $(+Fe)$, $(-Fe)$, $(+Fe)$ for $\alpha$ $Fe_2O_3$ and $(+Fe)$, $(Ti)$, $(-Fe)$, $(Ti)$, $(+Fe)$ for $FeTiO_3$. Nagata suggests that the solid solution $Fe_4\,Ti_3\,O_3$ consists of two oppositely magnetized layers $A$ and $B$ of slightly different moments consisting of $+Fe_4^{++}Fe_3^{+++}Ti_4^{++++}$ and $-Fe_4^{++}Ti_3^{++++}$ respectively.

Very restricted solid solution is possible between magnetite ($Fe_3O_4$) and haematite ($\alpha Fe_2O_3$) at temperatures of igneous rock formation, but intergrowths of haematite and magnetite from natural sources are known.

Maghemites sometimes contain titanium. Such titanomaghemites, Nicholls[5] suggests, are likely to be solid solutions between $\gamma$ $Fe_2O_3$, $TiO_2$ and a cubic $Fe^{++}Ti^{++++}O_3$ or $\gamma$ $FeTiO_3$. Magnetic data on the titanomaghemites are not available.

Intergrowths of solid solutions of the $Fe_3O_4-Fe_2TiO_4$ series with ilmenite are commonly found in igneous rocks. Chevallier and Girard [6] have by synthesis prepared solid solutions of $FeTiO_3$ and $Fe_3O_4$. The variation of Curie point and intensity of magnetization is given in Table 2.

Table 2. *Mganetic properties of the* $Fe_3O_4$. $FeTiO_3$ *series of solid solutions.*

| % $Fe_3O_4$ | 100 | 90 | 85 | 78 | 71 | 63 |
|---|---|---|---|---|---|---|
| Curie point in °C | 580 | — | 540 | 500 | 450 | 405 |
| Saturation magnetization (rough measure of) gauss | 90 | 82 | 71 | 61.5 | 52.5 | 43.5 |

In the titaniferrous magnetites Chevallier, Mathieu and Vincent[2] show that the Curie point and saturation magnetization depend on the concentration of $Fe_3O_4$ in the material[3].

# B. Physical processes in the magnetization of rocks.

**3. Variation of saturation magnetization with temperature.** The variation of saturation magnetization of ferrimagnetics with temperature up to the Curie

---

[1] T. Nagata, S. Akimoto and S. Uyeda: J. Geomag. a. Geoelect. 5, 168 (1953).

[2] R. Chevallier, S. Mathier and E. A. Vincent: Cosmochim. Acta 6, 27 (1954).

[3] Many further details and further information will be found in a paper recently published (too late for inclusion here) by R. Chevallier, J. Bolfa, S. Mathieu: Bull. Soc. franç. Miner.-Crist. **78**, 307, 365 (1955).

point may or may not be similar to that of ferromagnetics in which the variation is rapid only just below the Curie point. E. W. GORTER [7] found in ferrites of general formula $Ni_{1.5-a}Mn_aFeTi_{0.5}O_4$ $(0.675 > a > 0.4)$ a maximum of spontaneous magnetization at a temperature between room temperature and the Curie point ($P$ type). In Fig. 2 NAGATA shows the contrasting variations of saturation magnetization of a magnetite (mineral $A$ of the Haruna dacite) and a highly titaniferous haematite (mineral $B$ of the same rock). The former has a similar variation to a ferromagnetic, the latter an almost linear variation, apparently rather typical of haematites. NÉEL [8] predicted that it was possible for the spontaneous magnetization of certain ferrites ($N$ type) to decrease with temperature, eventually reversing in sign, and, having reached a maximum, to decrease to zero at the Curie point. The reversal is caused by spontaneous magnetization of two sub-lattices possessing different temperature coefficients. E. W. GORTER [7] discovered this phenomenon in the mixed ferrites $Li_{0.5}Fe_{2.5-a}Cr_aO_4$ $(1.7 > a > 1)$ and GUIOT-GUILLAIN, PAUTHENET and FORESTIER[1] in the rare earth ferrites of general formula $Fe_2M_2O_6$ when $M$ is a trivalent rare earth element.

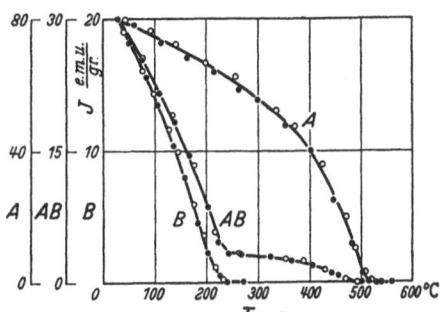

Fig. 2. Change in saturation magnetization $J$ with temperature $T$ of grains of the types $A$, $B$ and $AB$ from Haruna rock (after T. NAGATA).

**4. Mechanism of magnetization of igneous rocks.** The *natural remanent magnetization* of a rock specimen, that is, the magnetization remaining after the geomagnetic field has been nulled, may have been acquired in various ways, which it is desired to infer as far as possible with the aid of field evidence and by physical theory. There seem to be three ways in which the iron oxides can be magnetized without undergoing any chemical change. This group will be termed *physical magnetization*. The magnetizations are *thermo-remanent magnetization (T.R.M.)*, *isothermal remanent magnetization (I.R.M.)* and *anhysteretic magnetization*. The former is obtained by cooling the material in a weak field from above its Curie point, the two latter by applying a field at constant temperature for a certain time and removing it. *T.R.M.* is much greater and more stable than *I.R.M.*[2].

It was early recognized that igneous rocks such as dykes and lava flows, which had cooled in their present position, had become strongly magnetized and that this process of magnetization in a weak field could be reproduced in the laboratory. NÉEL [2] has explained *T.R.M.* and *I.R.M.* in terms of his theory of the magnetization of small particles. *I.R.M.* arises because a small proportion of the grains have a coercive force less than the applied field $H$ and are thus magnetized irreversibly upon its application. This proportion is of the order of $H/H_c$ when $H_c$ is the coercive force of the bulk material. In *T.R.M.* on the other hand the coercive force (being roughly proportional to the saturation magnetization) is very small just below the Curie point. Therefore the bulk material becomes magnetized irreversibly at this temperature and the magnetization rises with fall in temperature (except for those cases considered by NÉEL above as $P$ and $N$ types).

Of particular interest is the phenomenon of *partial thermoremanent magnetization (P.T.R.M.)* discovered by THELLIER [32] for bricks and baked clays and

---

[1] GUIOT-GUILLAIN, PAUTHENET and FORESTIER: C. R. Acad. Sci. Paris **239**, 155 (1954).
[2] J. G. KOENIGSBERGER was the first to demonstrate the importance of thermo-remanence in geomagnetism (1930). See his own report, with numerous references [22].

Nagata [1] for lavas. If in cooling through any temperature range (below the Curie point and above ordinary temperatures) the specimen acquires a moment from a certain field to which it is exposed over that temperature range and no other, the moment acquired has, as it were, a memory of the circumstances of its creation and will lose that memory only if heated up in zero field to the same temperature range. In order to explain *P.T.R.M.* it is necessary to go further into Néel's theory:

In a single domain grain of volume $v$ the height of the potential barrier between two orientations of minimum energy is proportional to $v$ and thus thermal fluctuations may cause the moment to change spontaneously from one position to another. Néel showed that if a remanent moment $M_0$ is given to each grain of an assemblage of grains of equal size, the moment $M_r$ after time $t$ in zero field will be given by

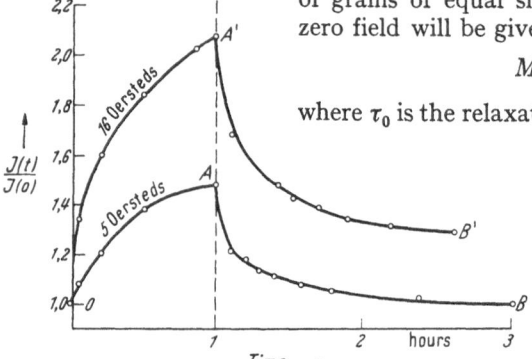

$$M_r = M_0 \exp\left(- t/\tau_0\right)$$

where $\tau_0$ is the relaxation time of the grain and is given by

$$1/\tau_0 = C \exp\left\{- v\, H_c\, J_s/2kT\right\}$$

and

$$C = (e\, H/2m)\, |3\, G\, \lambda + D\, J_s^2|\, \times \\ \times [2v/\pi\, G\, kT]^{\frac{1}{2}}$$

where $e$ and $m$ are the charge and mass of the electron $\lambda$ the magnetostriction constant, $G$ the shear modulus and $D$ a numerical constant depending on the shape of the grain equal to about 3.

Fig. 3. Isothermal remanent magnetization acquisition and decay in the Triassic red sandstones of Gt. Britain.

It can be shown that at a given temperature $T$ there is a well defined critical diameter $d_T$ above which the magnetic moment is stable in fields small compared to the coercive force $H_c$ and below which the thermodynamic equilibrium is reached. The blocking temperature $T_B$ is the temperature at which the diameter of the grain is critical.

In cooling in a weak magnetic field, grains of diameter $d_T$ will acquire their moment effectively at temperature $T$ and this will be stable below that temperature and unstable above it. It can therefore be effectively removed only by heating above that temperature. Thus most of the *T.R.M.* is very stable requiring a field of the order of $H_c$ to disturb it and it has extremely long relaxation times.

*I.R.M.* on the other hand can be removed by a field of the same order as that which produced it since only grains of critical diameter at ordinary temperatures have become magnetized and these have comparatively short relaxation times. Néel [2] shows that the *I.R.M.* $\sigma_r$ after time $t$ is related to its value $\sigma_{r_0}$ at time $t_0$, where time is measured from the instant just after the magnetizing field has been removed, by the expression

$$\sigma_r = \sigma_{r_0} - A\, T_0\, (\log t - \log t_0).$$

Fig. 3 shows results obtained by Clegg, Almond and Stubbs on some Triassic sandstones of England. These figures show the acquiring and decay of isothermal remanent magnetization, probably by the fine grained haematite which forms a coating of the quartz grains in such red sandstones.

It has been shown[1] that *A.C.* demagnetization removes *I.R.M.* more easily than it removes *T.R.M.*

In practice however igneous rocks, though still fine grained, have iron oxide particles which are larger grains and not single domains. Because any changes in magnetization require relatively little energy even in a stressed material with lattice defects such larger grains have a coercive force smaller than the fine grained haematite found in the red sandstones referred to above. The coercive force has been found to be proportional to the diameter of the grains, see for example NAGATA [1]. NÉEL [2] gives a theory applicable to larger grains.

**5. Laboratory experiments on thermoremanent magnetization.** Many experiments have been done to investigate the way in which thermoremanent magnetization is acquired when samples of igneous rock are cooled in the laboratory, often in a neutral atmosphere to minimize the effects of oxidation. Results are shown in Fig. 4 of some experiments

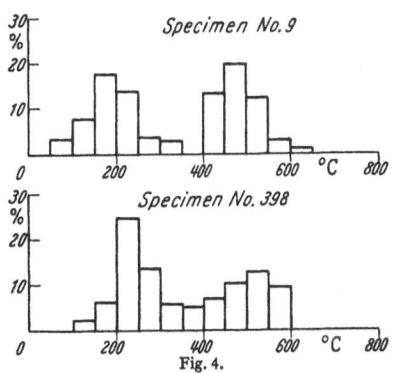

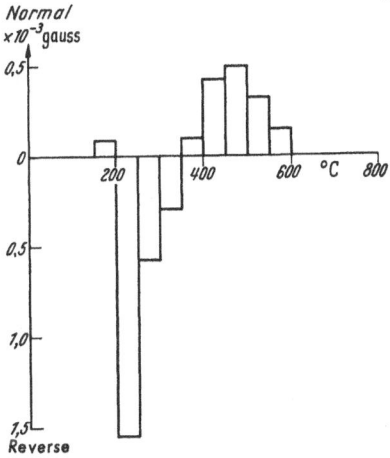

Fig. 4.

Fig. 4. Diagram of the partial thermo-remanent magnetization acquired by Icelandic specimens of Tertiary plateau basalt with reverse natural permanent magnetization. For easier comparison the intensities are expressed in percents of the sum. This sum is for specimen 9: $13.2 \times 10^{-3}$ gauss, and for specimen 398: $5.6 \times 10^{-3}$ gauss. The partial thermo-remanent magnetization is in the direction of the field for all temperature intervals. After HOSPERS [9].

Fig. 5. Diagram of the partial thermo-remanent magnetization acquired by a specimen of dacite pumice from Mt. Haruna, Japan. "Normal" denotes that the magnetization is in the direction of the external field, "Reverse" that it is in the opposite direction. After J. HOSPERS [9].

made on the Icelandic basalts by HOSPERS [9]. Magnetization is picked up mainly around 500 and 200° C. This indicates the probable presence in the rocks of two ferromagnetic constituents with different Curie points.

AKIMOTO has made an important distinction with reference to such experiments: in the process of heating, irreversible changes in the iron oxides may occur, such as exsolution, which considerably affect their magnetic properties. For example, CHEVALLIER and PIERRE[2] have found lava samples in which the variation of magnetization with temperature was the same in both heating and cooling as well as samples in which irreversible changes take place. In the latter case there are samples in which the magnetization disappears on heating for the first time around 300° C and on cooling again an appreciable proportion of the magnetization reappears just under 600° C. In these basalts it is reasonable to suppose that the ferromagnetic material is a solid solution of $Fe_3O_4$ and other constituents, which on heating in the laboratory liberates $Fe_3O_4$ as a separate phase.

---

[1] E. THELLIER and F. RIMBERT: C. R. Acad. Sci. Paris **239**, 1399 (1954).

[2] R. CHEVALLIER and J. PIERRE: Ann. Phys. **18**, 383 (1932).

An interesting and apparently anomalous type of thermoremanent magnetization has been discovered by Nagata[1], termed by him *reverse thermoremanent magnetization*. Lava from two volcanoes has been found which on cooling in the geomagnetic field acquired a remanent magnetization opposite to the field. The hypersthene hornblende dacite of Mt. Haruna, Japan, has a natural magnetization opposite to the present earth's field and of intensity $4.3 \times 10^{-3}$ emu. Fig. 5 shows the *P.T.R.M.* obtained by cooling a specimen of the dacite, by applying a field of 0.5 gauss over a range of 50° C during cooling. The abscissa give the temperatures at which the field was applied and the ordinates the intensity remaining after cooling to ordinary temperatures.

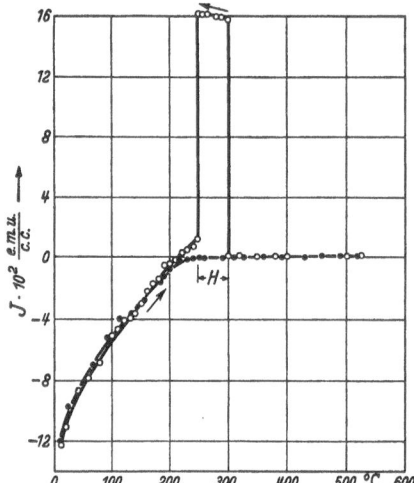

Fig. 6. Development of reverse magnetization with decrease in temperature when a magnetic field is applied only during cooling from 300 to 250° C. After Nagata[3].

The open circles of Fig. 6 show measurements of the moment of the rock obtained while the specimen was in the process of cooling for the case in which the field was applied only over the range 300 to 250° C. The black dots show the decay of this acquired moment when the material is heated in zero field. The strong effect of induction when the field is present will be noticed in the cooling curve. When the field is removed the remanent moment at 250° C is in the same direction as the field but it rapidly decreases with temperatures on cooling, reversing at 220° C and finally a considerable reversed moment is obtained. The ferromagnetic mineral grains in the Haruna rocks can be separated into three groups, $A$ having a Curie point between 500 and 530° C, $B$ having a Curie point below 300° C and the $AB$ grains composed of both types. The separation is effected by passing the grains through a tube in which the temperature can be maintained at any value up to 700°C. The tube is put into a strong non-uniform magnetic field which at any temperature can draw out those grains which have Curie points above that temperature. The saturation magnetization as a function of temperature has been given in Fig. 2.

The $AB$ grains, after being ground to particles between 3 to 5 $\mu$, could be separated magnetically into $A$ and $B$ material.

Nagata and his collaborators[2] found that both $A$ and $B$ grains showed normal thermoremanent magnetization and only the $AB$ type showed reverse thermoremanent magnetization. As a mixture of $A$ and $B$ did not show this anomalous effect, it was inferred that $AB$ represented an intimate association of $A$ and $B$ in definite geometrical arrangement, i.e. an intergrowth. Fig. 7 shows Nagata and Uyeda's[3] electron-microscopic photograph of surfaces of the minerals $AB$ in the Haruna rock.

Néel [8] had earlier suggested, among other possible ways in which a rock could receive reversed magnetization from a normal field, a *two constituent theory*.

[1] T. Nagata: Nature, Lond. **169**, 704 (1951).
[2] T. Nagata et al.: Nature, Lond. **172**, 630 (1953). — Naturwiss. **42**, 62 (1955).
[3] T. Nagata and S. Uyeda: Nature, Lond. **175**, 35 (1955). Later [Nature, Lond. **177**, 179 (1956)], the authors found that the reverse thermo-remanent magnetism of Haruna rock becomes more pronounced after a heat treatment of the ferromagnetic minerals (heated to 680° C, quenched to 20° C).

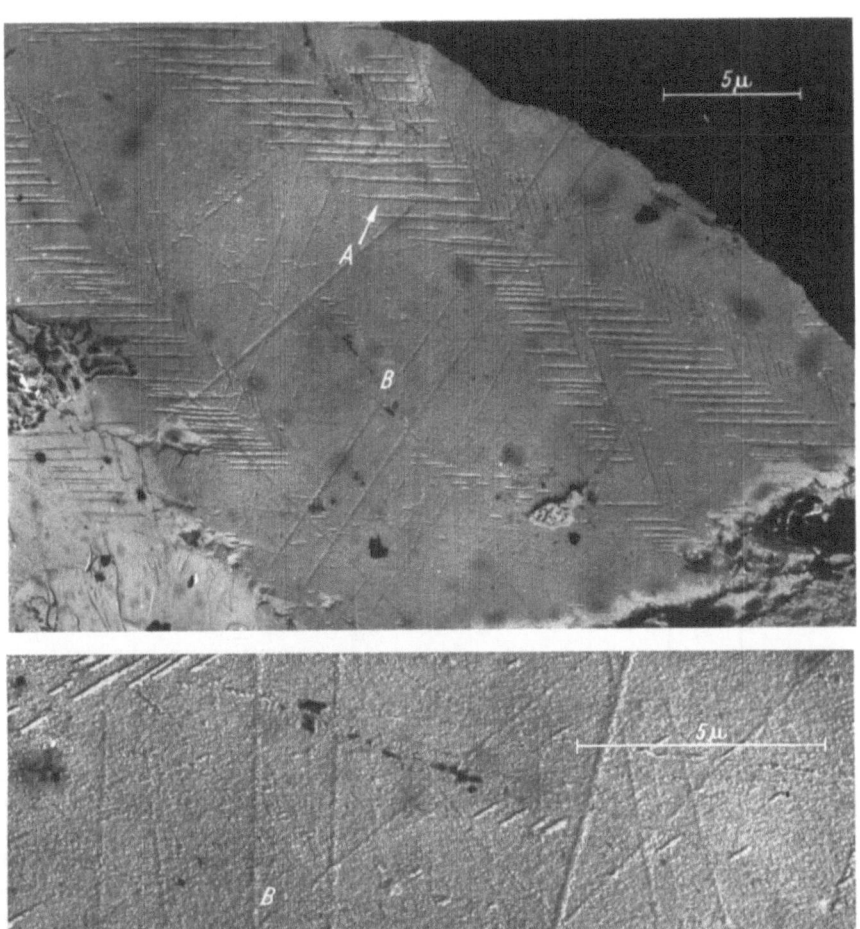

Fig. 7. Intergrowth of constituent $A$ in echelon with constituent $B$ in a Haruna ferromagnetic mineral (after Nagata and Uyeda).

This assumed the presence in the rock of two ferromagnetic minerals $A$ and $B$, the latter having the lower Curie point. On cooling from a temperature above both Curie points to that just above the lower Curie point, the mineral $A$ would be magnetized in the direction of the ambient magnetic field. The back field due to the magnetization of these grains would be opposed to the ambient field and

could in certain circumstances be greater than it. The constituent $B$ might therefore cool in a reverse field and become reversely magnetized. The final direction
of magnetization would depend on the temperature coefficients of the magnetization of materials $A$ and $B$. NAGATA and UYEDA therefore interpret the Haruna
rock as resulting from such a process. As can be seen by Fig. 7 constituent $A$
is in the form of strips of width about $0.2\,\mu$ and length 1 to $5\,\mu$. Where the
constituent $A$ has been magnetized in the cooling process the material $B$ will be
exposed to the back field of $A$.

$A$ is found to be a titanium poor titanomagnetite having a crystal structure
of an inverse spinel and a Curie point of $500°$ C. $B$ is a solid solution of ilmenite
and haematite (haemo-ilmenite) having rhombohedral crystal structure and a
Curie point of about $230°$ C.

Table 3. *Physical properties of minerals containing two iron oxide constituents.*

| Locality | Grain | Crystal Structure | Lattice Parameters | Curie Point |
|---|---|---|---|---|
| Haruna | $A$ | Cubic | $a_c = 8.403 \pm 0.002$ Å | $460°$ C |
| | $B$ | Rhombohedral | $A_{rh} = 5.480 \pm 0.002$ Å $\alpha_{rh} = 55°08' \pm 01'$ | $250°$ C |
| Asio | $A$ | Cubic | $a_c = 8.397 \pm 0.004$ Å | $460°$ C |
| | $B$ | Rhombohedral | $A_{rh} = 5.483 \pm 0.004$ Å $\alpha_{rh} = 55°02' \pm 03'$ | $230°$ C |
| Towada | $A$ | Cubic | $a_c = 8.412 \pm 0.003$ Å | $390°$ C |
| | $B$ | Rhombohedral | $A_{rh} = 5.491 \pm 0.001$ Å $\alpha_{rh} = 54°59' \pm 01'$ | $100°$ C |

While it is clear that a qualitative explanation has been obtained to explain
the anomalous behaviour of the Haruna rock, the quantitative treatment which
would enable an estimate to be made of the frequency of occurrence in natural
rocks of the circumstances necessary for this phenomenon is lacking. NAGATA,
AKIMOTO and UYEDA[1] show that two other igneous rocks from Asio and Towada
possess minerals $A$ and $B$, rather similar in properties to those of the Haruna
rock, see Table 3. However, most of the Asio rocks possess normal thermoremanent
magnetization when cooled to ordinary temperatures, although it was found that
mineral $B$ had become reversely magnetized in the process. In the Towada rocks,
on the other hand, the $T.R.M.$ is entirely normal, mineral $B$ being magnetized
in the same direction as mineral $A$. The presence of two distinct ferromagnetic
constituents therefore does not by any means necessarily imply spontaneous
reversals of magnetization.

**6. Magnetization of sedimentary rocks.** Sediments are derived from the
chemical or mechanical breakdown of other rocks and have been deposited after
transport by wind or water. Sediments which have been little altered since
deposition will tend to retain the original magnetization of the iron oxide minerals
if this was acquired as $T.R.M.$ in cooling during the formation of the parent
igneous or metamorphic rock. It is possible in some cases that the iron oxides
such as haematite have been formed in the sediment after deposition from iron
hydroxide which will fill the interstices of the quartz grains. It is possible that in
such circumstances the haematite would receive a stable magnetization in the
direction of the ambient magnetic field but such "chemical magnetization"
has not been reported. Thus if the sediment possesses a natural magnetization in

---

[1] T. NAGATA, S. AKIMOTO and S. UYEDA: J. Geomag. a. Geoelect. **1955.**

bulk, some method of orienting the magnetized iron oxide grains during the deposition of the sediment must be postulated. There seems to be plenty of evidence that this can result from the action of the earth's magnetic field while deposition is proceeding, provided the conditions are sufficiently quiet. Thus cases have been found where coarse grained sandstones, laid down in turbulent water have highly scattered directions of magnetization, whereas fine grained sandstones of similar age have consistent directions.

Varved clays are among the few sediments which may easily be dispersed and the redeposition of such material under various controlled conditions in the laboratory can therefore throw light on the origin of the magnetization of sediments.

JOHNSON, MURPHY and TORRESON [10] showed that the magnetization, acquired on redeposition of dispersed varved clay from New England in low magnetic fields, was accurate in declination but the inclination was appreciably less than that of the field. Their interpretation of these results postulated that flat particles would tend to settle horizontally whatever the field direction was.

Table 4. KING's *data on the bedding errors and inclination errors of the Swedish varved clays.*

| $I_F$ | δ | β |
|---|---|---|
| 0 | 2° | 8° |
| 10° N. | 3° | 10° |
| 25° N. | 9° | 8° |
| 45° N. | 21° | 15° |
| 65° N. | 27° | 17° |
| 90° | 10° | − 25° |
| 80° S. | 33° | − 17° |

KING[1] has carried out further experiments of this kind on the Swedish varves, in which deposition takes places not only on to a horizontal surface but on one inclined at 10° to the horizontal towards the field. He concludes that the ferromagnetic grains are either flattened discs or spheres, of which the former will tend to settle horizontally onto the bedding plane.

KING determines for various inclinations $(I_F)$ of applied field values of the bedding error β and the inclination error δ. If $I_0$ is the inclination of the clay magnetization on the horizontal bed and $I_{10}$ is that when deposition takes place on the 10° slope then $β = I_{10} - I_0$ and $δ = I_F - I_0$. Table 4 shows KING's data on these two errors.

GRIFFITHS[2] had found that measurements on two sets of contemporaneous varved clays from Sweden could be reconciled only if the 14° tilt of one series was assumed to be post depositional, whereas the geological evidence indicated that the deposition took place on an inclined surface. Table 4 shows how this difficulty can be resolved: At a value of $I_F$ of about 65° N. deposition on an inclined slope of 10° will cause the inclination of the magnetization acquired to be about 10° less than that of the field. Thus a correction for the "geological dip" of the beds will approximately give the correct angle of inclination.

In the absence of appreciable movement and shear in the fluid, the magnetic moments of the iron oxide grains will lie along the field during deposition. As the size of particle decreases, the orienting effect of the magnetic moment will become large compared to that of the moment of inertia. The orienting effect of the field will thus only become unimportant for large particles. Those particles not elongated or disc-shaped will consequently settle with their moments on the average along the field—some scatter will result from the process of settling in between the grains. Particles which are elongated or disc-shaped will settle predominantly with their long axes closer to the horizontal than the field, an effect which may become more marked during the first stages of compaction. The experimental fact that the inclination of the polarization is generally less than that of the field shows that the elongated particles are magnetized along

---

[1] R. F. KING: Mon. Not. Roy. Astronom. Soc., Geophys. Suppl. 7 (1955).
[2] D. H. GRIFFITHS: Nature, Lond. **172**, 539 (1953).

their length, as this component is more resistant to demagnetization. We may also deduce that if the field is horizontal or vertical no error in inclination is to

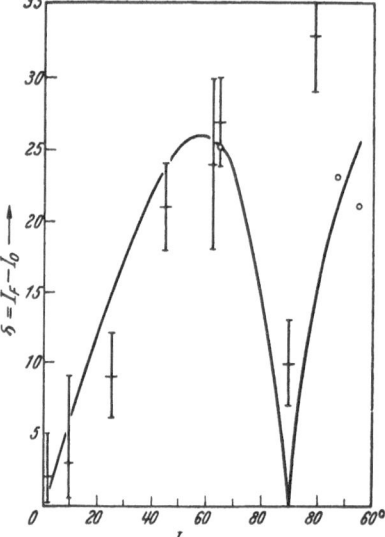

Fig. 8. Plot of the inclination error $\delta$ against the inclination $I_F$ for redisposed varved clays from Sweden (after R. F. KING).

be expected, in the latter case because there is an equal probability that the particles will rotate in either direction on settling.

These expectations are borne out by KING's results in Fig. 8. KING's simple considerations lead to a connection between the angle of inclination $(I_F)$ of the field and the resulting inclination $(I_0)$ of the polarization of the sediment as follows:

$$\tan I_0 = f \tan I_F \qquad (6.1)$$

where $f$ is a factor less than 1 which is proportional to the relative number of *round* grains and the probability with which *flat* grains take up a horizontal position on deposition. Fig. 8 for $\delta = I_F - I_0$ as a function of $I_F$ shows that this equation (represented by the curve) fits the data fairly well.

It is not known how frequently such inclination errors will occur in nature. A comparison of the difference between the magnetic inclination in various lithologies will throw light on this.

# C. Determination of the original magnetization of rocks.

**7. The measurement of magnetic polarization.** The measurement of a remnant magnetization of a rock formation is best accomplished by measuring the magnetization of a number of hand specimens, the orientation of which has been determined. It is convenient to cut the specimen into cubes or disks: their magnetization is measured either by rotating the specimen between pick-up coils or by using an astatic magnenometer.

The former method has been developed in the USA[1] the latter in England[2] and Japan[3]. The high sensitivity in the case of the astatic magnetometer is obtained either by very careful annulling of the earth's field and a high astaticism[2] or by obtaining resonance by rotating the sample at the natural period of oscillation of the magnetometer[3]. All these methods enable specimens with an intensity of polarization of $10^{-8}$ gauss to be measured accurately.

When directions of magnetization of a number of specimens from one formation have been determined, it is often necessary to assess the reliability of the mean direction. FISHER[4] has shown how this may be done. Around the mean direction may be described a "cone of confidence" of semi-angle $\alpha$ which to a probability of 95% will contain the true direction. $\varkappa$, a measure of the scatter of the direction, may also be calculated.

**8. Elimination of recently acquired magnetizations.** Before a measured polarization can be assumed to be the original magnetization of the rock, the

[1] E. A. JOHNSON and A. C. McNISH: Terr. Magn. **43**, 401 (1938).
[2] P. M. S. BLACKETT: Phil. Trans. Roy. Soc. Lond. **245**, 309 (1952).
[3] N. KUMAGAI and N. KAWAI: Mem. Coll. Sci. Univ. Kyoto B **20**, 307 (1953).
[4] R. A. FISHER: Proc. Roy. Soc. Lond., Ser. A **217**, 295 (1953).

effect of *I.R.M.* must be determined. THELLIER has developed laboratory methods of eliminating this unwanted magnetization—"magnetic cleaning". However, the determination of the extent to which a given rock strata is affected by *I.R.M.* or other magnetization acquired late in the history of the rock, is probably best studied by field tests. These were originated by GRAHAM [18].

If the magnetizations of rock strata which have been folded or tilted by different amounts, become parallel when corrected for geological tilt, it is evident that the strata have not been magnetized since folding. Similarly the magnetization of pebbles in a conglomerate bed will remain randomly oriented unless the material is unstable.

Where such tests can be carried out, the likelihood of the magnetism of undisturbed rocks of the same type being stable may be estimated. Where the same strata are tilted to different extents, the *I.R.M.* may be eliminated if the specimens have varying degrees of instability, for the magnetizations lie in a plane containing the present geomagnetic field and the stable magnetization of the rock and these planes are different for different geological dips. So far, only limited use has been made of this principle.

# D. Survey of observational data of remanent magnetization.

**9. Palaeomagnetism of recent varved clays.** JOHNSON, MURPHY and TORRESON [10] measured the magnetization of the varved clays of New England. Their results are shown in Fig. 9.

Fig. 10 shows GRIFFITHS'[1] results on the Swedish varved clays at Prästmon and Undrom. GRIFFITHS finds the scatter in the recent delta deposits to be greater than those of the Prästmon and Undrom polarizations and postulates that the effect of river currents must be important in the process of magnetization. GRIFFITHS has also found strong evidence for this in recent Icelandic varves, in which discordant magnetizations having no relation to the known field changes are found from localities 100 yards apart.

Tests of stability of the material are not easy because the original directions of magnetization are bound to be nearly coincident with the present field and the geological situation is such that GRAHAM's methods cannot be used. Consequently GRIFFITHS tested the varves for the presence of an appreciable component of isothermal remanent magnetization by remeasuring some of Prästmon samples after leaving them in the geomagnetic field in random orientations for over a year. No appreciable change in the direction of magnetization was noted. The natural remanent magnetization was not destroyed by 1000 gauss a.c. and although the varved clays could be magnetized in a field of 100 gauss this isothermal remanent magnetization could be removed by 100 gauss a.c.

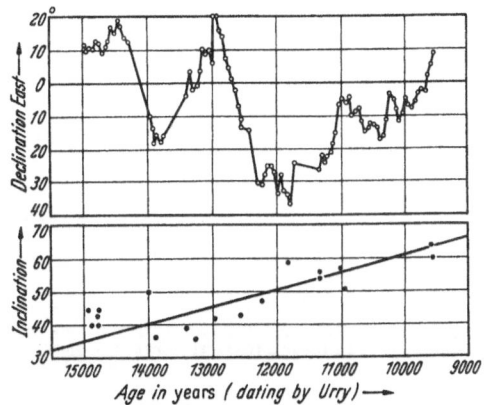

Fig. 9. Variation of declination and inclination of New England varved clays, 15000 to 9500 B.C. (after JOHNSON, MURPHY and TORRESON).

---

[1] D. H. GRIFFITHS: Mon. Not. Roy. Astronom. Soc. Geophys. Suppl. 7 (1955).

Thus it seems reasonable that the magnetizations of the iron oxide grains were thermoremanent in origin, the grains being orientated during deposition.

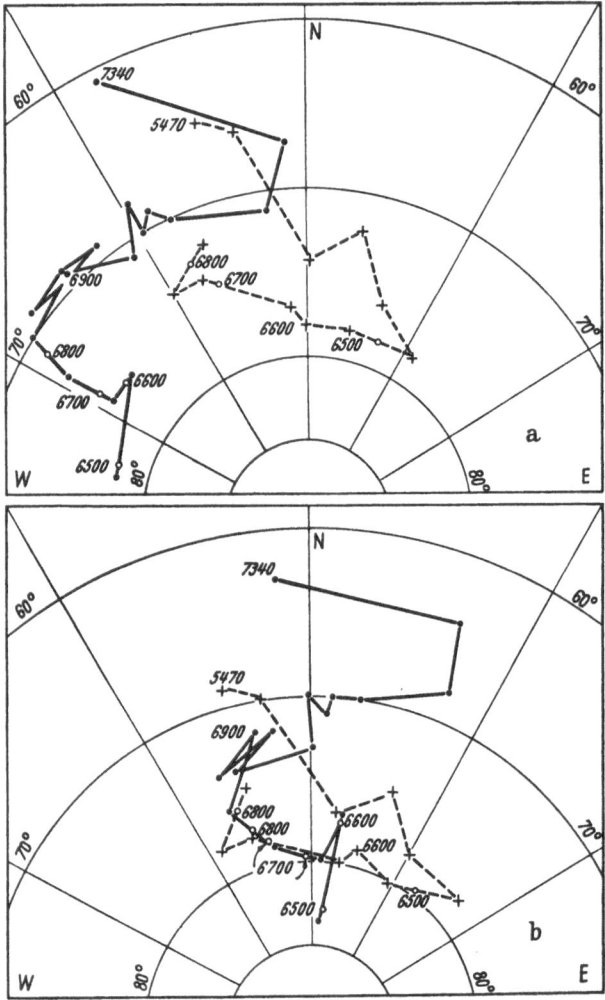

Fig. 10a and b. Directions of remanent magnetism of Swedish varved clays (a) Uncorrected and (b) Corrected for dip. Dates in years on Liden's time scale where 7602 equals 1000 A.D. Prästmom ●, Undrom +, Interpolated points ○ (after D. H. Griffiths).

**10. Palaeomagnetism of quaternary lava flows** (last million years). α) *Historic flows.* Chevallier [11] carried out a thorough study of the permanent magnetization of the dated lava flows of Mt. Etna. Chevallier was able to show that the known secular variation of the geomagnetic field agrees with these results, obtained on large specimens, within about 2° from 1600 onwards.

Hospers [9] examined flows near Mt. Hekla in Iceland of A.D. 1766, 1845, 1878, 1913 and 1947/48 and a flow at Eldhraun N.E. of Lake Myvatn of date A.D. 1729. The mean direction of magnetization, calculated from measurements on 5, 4, 5, 3, 9 and 6 samples respectively from each flow, is given in the first line of Table 5.

Table 5. *Directions of magnetization of quaternary lava flows in Iceland.*

| Flows | Mean declination | Mean inclination | Angle with present field | Angle with dipole field | Radius of [1] circle of confidence |
|---|---|---|---|---|---|
| Historic. . . . . . . | N.    4.3° E. | + 76.2° | 7° | 1° | 8.2° |
| Postglacial  . . . . . | N.    1.1° E. | + 73.8° | 7.5° | 2.5° | 8.2° |
| Postglacial, N. Iceland | N.   67.0° E. | + 86.4° | 13.5° | 11.5° | 15.2° |
| Palagonite formation . | N.    7.1° E. | + 75.5° | 6.1° | 2.0° | 8.2° |
| Early  Quaternary . . | N. 181.0° E. | − 75.2° | 180° − 6.3° | 180° − 1.5° | 7° |

MINAKAMI [13] showed that the Yoridai-Sawa lava flow, which erupted on July 12, 1940 in Japan, became magnetized in the direction of the local geomagnetic field.

β) *Prehistoric flows.* In S.W. Iceland an average of four samples was measured by HOSPERS [9] from each of eight flows younger than the end of the last ice age. The Palagonite formation (mid Quaternary in age) was also sampled. Table 5, lines 2 to 4, shows the mean directions for each series of flows. There is no recorded example of a flow younger than mid Quaternary times which is not magnetized in approximately the present direction of the field.

In W. and S.W. Iceland, there is an upper series of plateau basalts which are slightly different in appearance from the Tertiary flows below them and contain intercalated glacial beds which rest (at Snaefellsnes) on an early Quaternary sedimentary bed. HOSPERS[2] therefore argues that these flows are early Quaternary in age but some geologists consider them to be Tertiary. HOSPERS [9] demonstrates that they are all reversely magnetized (last line in Table 5) not only there but also in Hvalfjordhur and in Mt. Esja, 120 and 125 km. away from Snaefellsnes.

**11. Palaeomagnetism of the Tertiary (1 to 70 million years).** At Snaefellsnes and in other sections in Iceland the underlying Tertiary basalts were found by HOSPERS [9] to be magnetized in alternating zones of reversed and normal magnetization, each zone of roughly 10 to 40 flows. Mean directions of these different groups are given in Table 6, and are seen to agree closely with those directions to be expected if the field is on the whole an axial dipole one but with either polarity.

Table 6. *Mean angle of magnetization of Tertiary lava flows of Iceland.*

| | Angle of mean direction of magnetization with | |
|---|---|---|
| | present field | dipole field (axial) |
| Normal Tertiary, W. Iceland . . . | 11.0° | 12.0° |
| Reversed Tertiary, W. Iceland  . . | 180° − 16.0° | 180° − 13.6° |
| Normal Tertiary, N. Iceland . . . | 6.2° | 1.5° |
| Reversed Tertiary, N. Iceland  . . | 180° − 16.0° | 180° − 11.3° |

Mean of Tertiary: declination N. 1.5° E., inclination 77.8°.

EINARSSON and SIGURGEIRSSON[3] have surveyed, using an ordinary compass, the lava flows of Iceland. They examined a total thickness of sections of 21 000 metres of which they found 10 000 metres normally magnetized and 9000 metres reversely magnetized, the rest being irregular. They find at least three periods in which the field was reversed, the last one covering the Pliocene-Pleistocene

---

[1] See R. A. FISHER: Proc. Roy. Soc. Lond., Ser. A **217**, 295 (1953).
[2] J. HOSPERS: Geologie en Mijnbouw **16**, 491 (1954).
[3] T. EINARSSON and T. SIGURGEIRSSON: Nature, Lond. **175**, 892 (1955).

boundary as confirmed by ROCHE[1] in France. They invariably find the same magnetization in the country rock close to a dyke whether this is clay or basalt and also concordant magnetizations in the baked clays beneath basalt flows.

The Tertiary dykes in Great Britain are an interesting example of reversed magnetism. POOLE, WHETTON and TAYLOR[2] studied these dykes in the Northumberland coalfield with a view to the applicability of magnetic methods of tracing dykes, concealed from surface examination over long distances by glacial drift. They made traverses across the dykes with a Watts vertical force variometer. The Acklington dyke was examined in three localities at Broomhill, Swarland and Rothbury. Twenty-one, eighteen and eight traverses 200 to 400 ft. long were made at these three localities respectively at intervals along the dyke between 200 and 2000 ft. Seven traverses of 1500 ft. of the dyke at Broomhill gave pronounced negative anomalies up to 400 gammas in intensity, indicating that the magnetization was equivalent to a strong concentration of N. poles on the upper surface. About half the traverses of the Rothbury section and a third of the Swarland section also give rise to negative anomalies. The authors also found the Coley Hill dyke to give rise to negative anomalies. They clearly saw that a normally magnetized dyke, even of complicated geometry, could not give rise to the inverted anomalies, but the authors considered that in some way the weathering of the surface of the dyke was the cause.

BRUCKSHAW and ROBERTSON [14] later examined in detail three of the Tholeiite dykes of North England and showed them to be in general reversely magnetized at a number of places along their length in nearly the opposite direction to the field of the present day. They were among the first to point out that such magnetization might have been caused by the presence of an inverted magnetic field but left open the possibility that it arose quite locally. They found some sections of the dykes to be normally magnetized but noticed that these specimens when left in the laboratory for some months changed their directions of magnetization considerably. It seems likely that these specimens possessed isothermal remanent magnetization. More difficult to understand is that BRUCKSHAW and ROBERTSON found that the magnetization of the dykes was not uniform across the dyke but considerably scattered in direction. They suggested that a certain amount of flow had taken place within the dykes after magnetization had occurred, but it is unlikely that such basic material can be still fluid at temperatures as low as the Curie point of magnetite.

BRUCKSHAW and VINCENZ [21] have investigated the magnetization of the Tertiary lavas of the Isle of Mull, from which centre of igneous activity the Tholeiite dykes referred to above radiate. They found 16 reversely magnetized lavas, the mean declination being 173° and the mean inclination being — 79°.

VINCENZ [12] extended the work to the intrusives of the Isle of Mull and finds in these examples of both reversely and normally magnetized bodies.

The Lower Basalts of Northern Ireland are separated from the Upper Basalts by the so-called interbasaltic horizon and were considered to be Eocene but recently SIMPSON[3] has presented arguments in favour of a Miocene age. HOSPERS and CHARLESWORTH[4] sampled between 5 and 7 flows at four different localities. All the flows were found to be reversely magnetized. The mean direction for all

[1] A. ROCHE: C. R. Acad. Sci. Paris 236, 107 (1953). Similar results for the Miocene basalts of the Vogelsberg (Germany) will be reported by G. ANGENHEISTER, Nachr. Akad. Wiss. Göttingen, Math.-phys. Kl. 1956; G. ANGENHEISTER u. A. HAHN: Geol. Jb. 1956.
[2] G. POOLE, J. T. WHETTON and A. TAYLOR: Trans. Min. Met. Eng. 89, 34 (1935).
[3] F. B. SIMPSON: Adv. of Sci. (Rep. Brit. Assoc. Belfast Meeting) 9, 331 (1953).
[4] J. HOSPERS and H. A. K. CHARLESWORTH: M. N. R. A. S., Geophys. Suppl. 7, 32 (1954).

the 24 flows is shown below. A positive inclination means one in which the north seeking pole is below the horizontal.

BULLERWELL[1] has also shown by magnetic survey across the edge of lava flows in Northern Ireland that some flows are reversely magnetized. He also showed the presence of normally magnetized flows but it is not known whether their magnetization is T.R.M. or I.R.M. so that the evidence for normally magnetized flows in N. Ireland is lacking. Although reversed flows predominate, the number is so small that it is probable that the Northern Ireland results are no different from the Icelandic results except in one particular. The mean direction of magnetization found by HOSPERS and CHARLESWORTH implies a pole at 133° E. 76° N. and they interpret this as a possible movement of the pole of rotation (assuming the N. Ireland lava flows to be older than the Icelandic flows). This interpretation is doubtful as the 24 flows sampled may have erupted within a short space of time; thus the magnetizations could represent the deviation of the magnetic field due to the secular variation rather than any movement of the mean pole.

CAMPBELL and RUNCORN[2] have shown that the Columbia River basalts of Miocene age of Oregon and Washington, USA, show similar alternating zones of normal and reversed magnetization to those in the other regions of Tertiary basalts. The mean number of flows between successive reversals is 7 compared with 20 in Iceland. It is therefore possible that the average rate of eruption of lava flows was less than half as quick on the Columbia plateau as in Iceland. CAMPBELL and RUNCORN find 4 lava flows lying between a normally and reversely magnetized series which are magnetized in directions widely different from these predominant directions. They interpret this as evidence of a reversal of the field, for reversal due to the anomalous properties of the iron oxide minerals would not be expected to result in *oblique* directions of this kind.

ROCHE [15] reports work on the Tertiary lava flows of the Central Massif of France. He shows the presence of many reversely magnetized flows. ROCHE[15] and HOSPERS [9] examine the process by which lavas become reversely magnetized by studying contact zones. ROCHE finds that a number of clays and a marly limestone, baked by overlying reversely magnetized lava flows, were invariably reversely magnetized in the same direction, even when the clays were not derived from rocks of volcanic origin. HOSPERS found a reversely magnetized dyke, probably of Quaternary age, cutting a series of normally magnetized Tertiary lava flows. That thickness of the lavas, which had at the time of intrusion of the dyke been heated above the Curie point, was magnetized in the same direction as the dyke.

It is important to note that the *intensity of magnetization* ($I_r$) and the magnetic *susceptibilities* ($\varkappa$) are nearly the same for reversely ($R$) and normally ($N$) magnetized lavas as shown in Table 8.

HOSPERS estimates the length of time elapsing between geomagnetic reversals as between a few hundred thousand and a million years on the average. There is no reason to suppose that reversals take place at regular intervals.

Table 7. *Magnetization of the Tertiary lavas of N. Ireland.*

|  | Declination | Inclination | Angle, with mean direction |
|---|---|---|---|
| Mean direction. | N. 194° E. | − 60.2° | 0° |
| Present field . | N. 13° W. | + 70° | 180° − 15° |
| Dipole field . . | N. 0° | + 70.8° | 180° − 11.5° |

Radius of circle of confidence = 5.4°.

[1] W. BULLERWELL: Mem. Geol. Surv., Gt. Britain 6, 21 (1954).
[2] C. D. CAMPBELL and S. K. RUNCORN: To be published 1955.

Hospers found examples of reversely magnetized *sediments* in Iceland (other than those baked by volcanic action). A thick series of Pliocene sediments at Tjornes were found, normally magnetized at the top and reversely magnetized at the bottom.

On the other hand Torreson, Murphy and Graham [16], who have examined the magnetization of 99 samples of flat-lying, undisturbed sediments, mainly Tertiary but including some Jurassic, conclude that the direction of the geomagnetic field has not changed. They show that the directions of magnetization are grouped along the direction of the geocentric axial dipole field, in declination markedly and in inclination moderately different from the present direc-

Table 8. Roche *and* Hosper's *comparison of normally and reversely Magnetized Lavas.*

|  |  | Mean $I_r$ gauss | Mean $\varkappa$ |
|---|---|---|---|
| Miocene . . . . . | N | $1.20 \times 10^{-3}$ | $0.32 \times 10^{-3}$ |
| France . . . . . | R | $1.14 \times 10^{-3}$ | $0.31 \times 10^{-3}$ |
| Early Quaternary . | N | $4.99 \times 10^{-3}$ | $1.05 \times 10^{-3}$ |
| Iceland . . . . . | R | $4.93 \times 10^{-3}$ | $1.11 \times 10^{-3}$ |
| Tertiary . . . . . | N | $3.02 \times 10^{-3}$ | $1.18 \times 10^{-3}$ |
| Iceland . . . . . | R | $4.27 \times 10^{-3}$ | $1.14 \times 10^{-3}$ |

tion of the geomagnetic field. Eight specimens are reversed but there is no evidence of even a small peak opposite to the main one. The authors concluded that this evidence was consistent with the idea that for the past 50 million years or so the polarity of the earth's field has not changed but that its mean direction has been close to that of the field of an axial dipole. However, they were not able to provide evidence, either of a field or laboratory type, to demonstrate the stability of the sediments which they had studied.

Kawai[1] shows that, when allowance is made for $I.R.M.$, some Tertiary sediments of Japan are reversely magnetized.

## 12. Palaeomagnetism of the Mesozoic (70 to 190 million years). Clegg, Almond

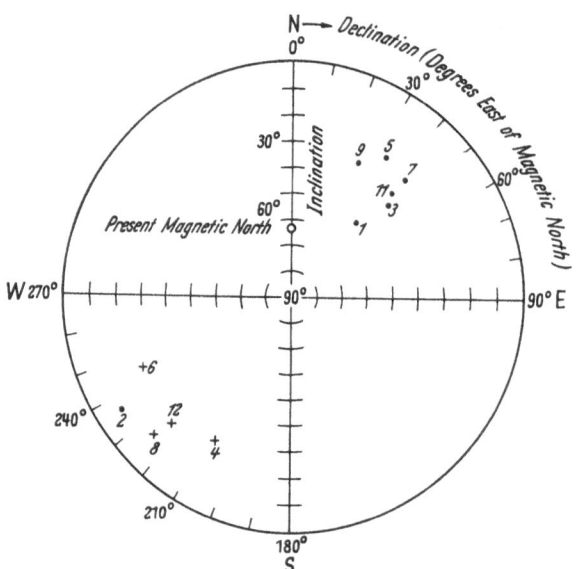

Fig. 11. Polar projection showing mean directions of magnetic polarization of specimens from various sites. Triassic sediments in England. After Clegg, Almond and Stubbs[2] · Downward dip, + Upward dip.

Table 9. *Direction of Magnetization in the Triassic Sandstones of England.* $\alpha$ = radius of circle of confidence, $\varkappa$ = dispersion.

| Declination | Inclination | $\alpha$ | $\varkappa$ |
|---|---|---|---|
| N 29° E | + 34° | 10° | 58 |
| N 39° E | − 16° | 27° | 13 |

and Stubbs[2] have measured the natural remanent magnetization of Triassic sediments in England, in nine sites over a wide area. They find the Keuper

[1] N. Kawai: J. Geomag. a. Geoelect. **6**, 208 (1954).
[2] Clegg, Almond and Stubbs: Phil. Mag. **45**, 583 (1954).

Marls to be magnetized in a North East—South West direction. The mean values are shown in table 9; the mean values at different sites are given in Fig. 11.

RUNCORN[1], from a consideration of the magnetization of the Maenkapi and Springdale sandstones of the Colorado plateau, USA, making certain allowances for *I.R.M.*, finds a mean declination of N 22° W and a mean inclination of +16° for Triassic times in the south western USA.

**13. Palaeomagnetism of the Palaeozoic (190 to 500 million years).** Important surveys of the magnetizations of Palaeozoic rocks have been made by GRAHAM[2], by CREER[3] and by GOUGH[4].

GRAHAM[2] reports measurements on 182 samples from the eastern United States of Ordovician to Permian age. The polarizations are somewhat scattered but are grouped about the present direction of the geomagnetic field. GRAHAM concluded tentatively that because the beds sampled are flat lying and undisturbed, the results could be taken as implying that for most of Palaeozoic time the direction and sense of the geomagnetic field were very much as they are today. No tests for stability were made.

RUNCORN[1] finds the Naco sandstones of Carboniferous (Pennsylvanian) age of Arizona, USA, to be magnetized, with a mean declination S 30.3° E and a mean inclination —3.4°.

He also finds the Supai formation of Arizona (Permian) to have a mean declination of S 47.3° E and a mean inclination of +22.9°. DOELL[5] finds the mean for the same formation to be S 34° E and +8°, respectively.

Table 10. *Mean Directions from 15 Sites in the Old Red Sandstone (Creer's data).*

| D | I | α | ϰ | |
|---|---|---|---|---|
| S. 20° W. | + 6 | 12° | 36 | flat lying |
| S. 14° W. | − 12 | 5° | 115 | |
| S. 14° W. | − 8 | 9° | 60 | |
| S. 15° W. | + 2 | 10° | 37 | |
| S. 16° W. | − 10 | 15° | 36 | |
| S. 28° W. | + 10 | 8° | 27 | |
| S. 22° W. | + 25 | 11° | 7 | |
| S. 19° W. | + 21 | 6° | 57 E. | younging to S. |
| S. 16° W. | − 16 | 7° | 61 SE. | |
| S. 19° W. | + 21 | 6° | 57 N. | |
| S. 32° W. | + 31 | 17° | 28 N. | |
| S. 55° W. | + 36 | 23° | 17 N. | |
| S. 29° W. | − 33 | 14° | 23 S. | |
| S. 25° W. | − 36 | 14° | 33 S. | |
| S. 42° W. | − 25 | 29° | 19 S. | |

CREER selected from Palaeozoic rocks in Britain those fine-grained red sandstones and siltstones which seemed likely to be magnetically stable. CREER sampled a stratigraphical thickness of 350 ft. of purple Caerbwdy sandstones at Caerbwdy Bay, S. Wales. He showed that the correction for the dip of the beds, about 50° towards the south, decreased the scatter of the magnetizations, thereby suggesting that they were stable. This he also confirmed by re-measuring the specimens after some months in the laboratory. CREER obtained a mean declination of S. 7° W. and a mean inclination of +39°, with a value of α of 8° and a value of ϰ of 32.4.

CREER also sampled extensively the Old Red Sandstone rocks of the Anglo-Welsh Cuvette. His results are summarized in Table 10.

Recently D. I. GOUGH[6] has made a thorough study of the directions of magnetization of the *Pilansberg dykes*, near Johannesburg, South Africa, the reversed magnetization of which has long been known from the magnetic surveys of

[1] S. K. RUNCORN: Bull. Geol. Soc. Amer. **1956** (in the press).
[2] J. W. GRAHAM: J. Geophys. Res. **59**, 215 (1954).
[3] K. M. CREER: To be published.
[4] D. I. GOUGH: (in the press).
[5] R. R. DOELL: Nature, Lond. **176**, 1167 (1955).
[6] D. I. GOUGH: Mon. Not. Roy. Astronom. Soc., Geophys. Suppl. (in the press).

Gelletich[1]. Two hundred specimens from outcrops of these dykes were measured and found to show random directions of magnetization and variations in intensity between $1 \times 10^{-3}$ and $570 \times 10^{-3}$ gauss. Magnetization by lightning or during weathering was suspected and so five of the dykes were sampled at depth in the gold mines of the Witwatersrand. The East Geduld dyke shows only 4 specimens magnetized in the direction of the present field, opposite to the other 58 specimens which were all reversely magnetized. Three of these four have a very low intensity of magnetization and therefore may have acquired sufficient I. R. M. in the present field to obscure their reversed magnetization.

The Goch dyke is composite and the inner part is much more weakly magnetized than the outer part and has a low inclination, which may possibly be due to isothermal remanent magnetization in the present geomagnetic field. The Pilansberg dykes are Palaeozoic or possibly Pre-Cambrian.

Table 11. *Magnetization of Pilansberg dykes* (D. I. Gough).

| Dyke | No. of samples | Declination | Inclination | Range of intensities (gauss) | Circle of confidence ($\alpha$) |
|---|---|---|---|---|---|
| Robinson . . . | 45 | 11.4° | +68.1° | $\begin{pmatrix} 0.86 \times 10^{-3} \\ 6.1 \times 10^{-3} \end{pmatrix}$ | 2.4° |
| E. Geduld. . . | 58 | 31.1° | +69.6° | $\begin{pmatrix} 0.039 \times 10^{-3} \\ 14.0 \times 10^{-3} \end{pmatrix}$ | |
| Goch . . . . . | | 9.4° | +58.8° | . . . | 2.8° 3.6° |

Belshé[2] also finds horizontal magnetizations in igneous rocks from Queen Maud's land in Antarctica. These have an uncertain Palaeozoic age.

Hallimond and Butler[3] concluded from magnetic survey that the Whin Sill of N. England, which was intruded in Carboniferous times, is horizontally magnetized and reversed with respect to the horizontal component of the present field.

J. C. Belshé[2] has measured samples of the Lower Carboniferous of Derbyshire and finds a mean declination of N. 26 E. and a mean positive inclination of 43° for the toadstones and 38° for the sandstones and siltstones.

Creer[4] also sampled five traps of the Permian lavas of Devon giving a mean declination of S. 9° W. and an inclination of −9°. The radius of confidence at the 5° level was 20° and the value of $\varkappa = 15$.

Early work by Mercanton[5] showed that a Permian lava between Keania and Wollongong in New South Wales was reversely magnetized with an angle of inclination of +87°.

**14. Palaeomagnetism of the Pre-Cambrian.** Irving[6] shows that the late Pre-Cambrian Torridonian Sandstones of Scotland are magnetized in significant directions. It appears that in the upper Applecross and Aultbea groups two predominant polarizations exist; a N. W. declination with a negative inclination and a S. E. declination with a positive inclination. Sandwiched between these were found, in some instances, oblique zones in which the directions were widely

[1] Gelletich, H.: Beitr. angew. Geophys. 6, 337 (1937).
[2] J. C. Belshé: Personal communication.
[3] A. F. Hallimond and A. J. Butler: Min. Mag., Lond. 80, 265 (1949).
[4] K. M. Creer: To be published.
[5] P. L. Mercanton: Terr. Magn. 31, 187 (1926).
[6] E. Irving: To be published 1955

scattered and different from the predominant directions. Statistical treatment shows that the N. W. and S. E. directions are nearly at 180°—the reversals being a similar phenomenon to that discovered in the Tertiary lava flows. CREER[1] shows the Longmyndian of Shropshire of late Pre-Cambrian age to contain red sandstones similarly magnetized to those of the Applecross and Aultbea groups of the Torridonia.

The lower group of the Torridonian, called the Diabaig, is likely to be much older than the upper two groups. This has been found by IRVING to be magnetised in a very different direction. Table 12 shows these results where $\alpha$ is the angle of the cone of confidence and $\varkappa$ the dispersion.

IRVING studied in detail the field evidence for the stability of these magnetizations, using the methods of GRAHAM [18]. He studied the dispersion of the present directions of polarization of pebbles of fine and very fine Torridonian

Table 12. *Comparison of directions of magnetization in Pre-Cambrian of Gt. Britain.*

|  | Declination | Inclination | $\alpha$ | $\varkappa$ |
|---|---|---|---|---|
| Applecross and Aultbea $\{$ | 127° | + 52° | 6° | 14.0 |
|  | 295° | − 34° | 9° | 11.8 |
| Longmyndian . . . . . | 294° | − 29° | 12° | 4.9 |
| Diabaig Group . . . . . | 307° | + 34° | 7° | 40.0 |

sandstones in conglomerate beds of New Red Sandstone Age. He finds $\varkappa = 1.1$ and 1.3 for beds of definite and presumed N. R. S. Age.

IRVING has examined a steep Caledonian fold in the fine and very fine Torridonian sandstones on the S. W. slopes of Ben Lioth Mhor. The $\varkappa$ for the mean directions at the eight different sites is 4.4 before correction and 25.5 after correction. This represents an extremely convincing example of the stability of sedimentary magnetization over a considerable proportion of geological time. IRVING has examined the magnetization of two zones of knotted folds with amplitudes between 0.1 and 10 cm, between which lie flat bedded layers, uniformly magnetized in a N. W. direction. One group has clearly been magnetized after slumping and one group before. Development of such studies will be of great importance in determining the exact time of origin of the magnetization of rocks, for it is possible, as has been noted by CLEGG, that if a sediment is very full of water, change of direction of the ambient field may succeed in reorienting the magnetic grains. Consequently in some sediments the magnetization may be finally determined some little time after deposition.

A particularly impressive proof that certain of the Torridonian Sandstones are highly stable has been found by IRVING, who has shown that the directions of magnetization of 10 blocks taken from an inter-formational conglomerate at Stoer in the Diabaig sequence are randomly orientated with $\varkappa = 1.6$.

RUNCORN[2] finds the Hakatai shales in the Grand Canyon, USA, to have a mean declination of S 87.7° W and a mean inclination of + 72.8°, while DOELL[3] finds the figures to be respectively S 66° W and + 72°.

DU BOIS[4] has determined the directions of magnetization for specimens of the Middle and Upper Keweenawan of N. Michigan. The Freda sandstone and Nonesuch Shales of the Upper Keweenawan have a mean declination of 75° west

[1] K. M. CREER: To be published 1955.
[2] S. K. RUNCORN: Bull. Geol. Soc. Amer. 1956 (in the press).
[3] R. R. DOELL: Nature, Lond. 176, 1167 (1955).
[4] P. M. DU BOIS: Nature, Lond. 176, 506 (1955).

of north and a mean inclination of 0°. From sandstones in the underlying Cooper Harbor Conglomerate of the Upper Keweenawan and the Portage Lake Lava series of the Middle Keweenawan a mean declination of 70° west of north and a mean positive inclination of 40° have been obtained.

**15. Palaeomagnetism of metamorphic rocks.** Field evidence which appears to associate the presence of reversed magnetization with the presence in the rocks of certain minerals has been given by Balsley and Buddington [19]. They are concerned with the explanation of large magnetic anomalies in the Adirondack

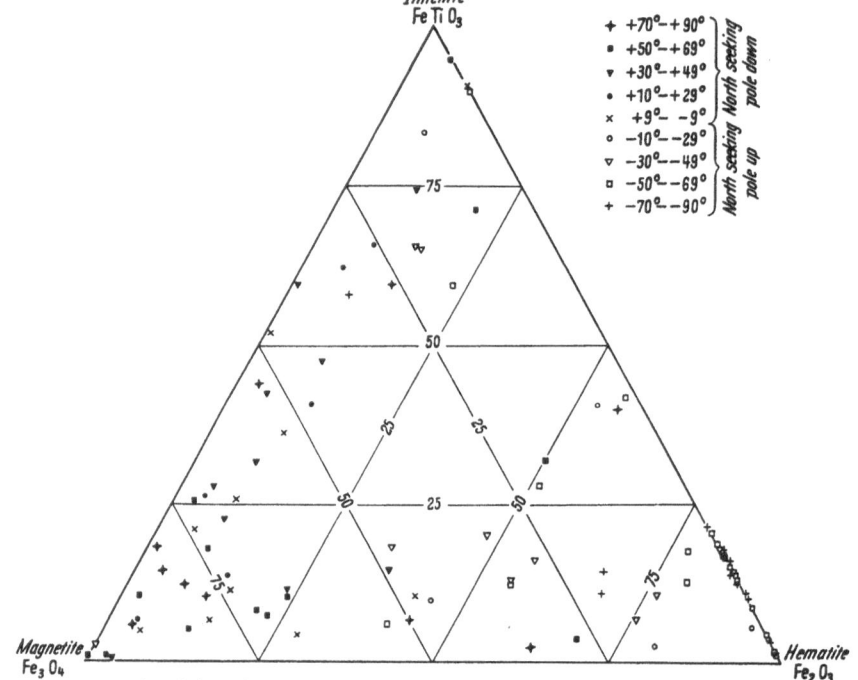

Fig. 12. Accessory minerals in rocks from the Adirondack mountains, with inclinations of the remanent magnetizations of the rocks shown in a Fe$_3$O$_4$—FeTiO$_3$—Fe$_2$O$_3$ ternary diagram (after Balsley and Buddington).

Mountains, New York, revealed by aeromagnetic survey. Two hundred specimens of varied igneous and metamorphic rocks were collected. Buddington and Balsley show that the rocks showing normal magnetization in the field contain as accessory minerals magnetites with exsolved intergrowths of ilmenite and ulvospinel—the Koenigsberger ratio[1] being not greater than 1. In rocks showing reversed magnetization they find the accessory minerals are exclusively members of the haematite—ilmenite—rutile series. Rocks in this group have a large Koenigsberger ratio and coercive force (the titanohaematites have a ratio 1700 and a coercive force of more than 2000 gauss). Buddington and Balsley find also a third group of rocks in which both the magnetite-ilmenite and the haematite-ilmenite-rutile series occur. This group has directions of magnetization transitional between those of the other two groups (reversed and normal). Only the values of magnetic inclination and total intensity are given by Buddington and Balsley. These results are shown in Figs. 12, 13, 14.

---

[1] The Koenigsberger ratio is that of the observed *remanent* magnetization of a rock to the *induced* magnetization $\varkappa F$ which the rock, due to its susceptibility $\varkappa$, would acquire if exposed, after preliminary demagnetization, to the present geomagnetic field $F$.

BUDDINGTON and BALSLEY conclude that the strong magnetization of the reversed group must represent the action of the haematite-ilmenite-rutile series

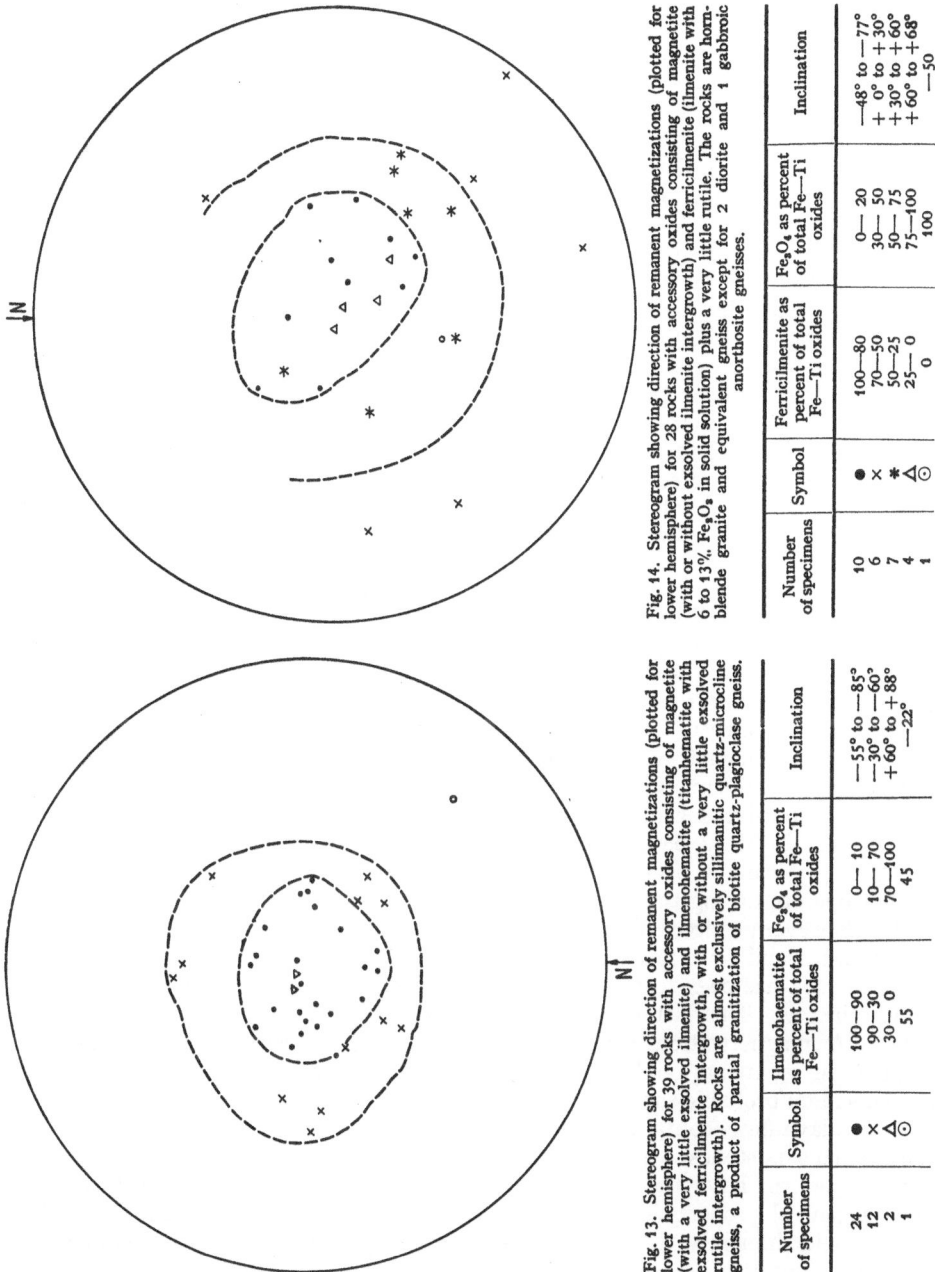

Fig. 14. Stereogram showing direction of remanent magnetizations (plotted for lower hemisphere) for 28 rocks with accessory oxides consisting of magnetite (with or without exsolved ilmenite intergrowth) and ferricilmenite (ilmenite with 6 to 13%, Fe$_3$O$_4$ in solid solution) plus a very little rutile. The rocks are hornblende granite and equivalent gneiss except for 2 diorite and 1 gabbroic anorthosite gneisses.

Fig. 13. Stereogram showing direction of remanent magnetizations (plotted for lower hemisphere) for 39 rocks with accessory oxides consisting of magnetite (with a very little exsolved ilmenite) and ilmenohaematite (titanhaematite with exsolved ferricilmenite intergrowth, with or without a very little exsolved rutile intergrowth). Rocks are almost exclusively sillimanitic quartz-microcline gneiss, a product of partial granitization of biotite quartz-plagioclase gneiss.

in the presence of a field not dissimilar to that existing at the present day. However, the magnetization found in the Adirondack rocks is reversed only in the sense that N. poles appear on the top surface of the rocks. In fact there seems some evidence that the directions of magnetization do not make an angle of 180° with the present geomagnetic field (see Figs. 13 and 14). Only if it is assumed that the magnetite-ilmenite group is magnetized along the direction of the present geomagnetic field, whether or not it is present in the rock alone or present with the haematite group in varying proportions, may the transitional magnetization be understood. In this case it may be inferred that the magnetite is always magnetized along the present direction of the geomagnetic field and the haematite along a direction more or less due W. and with a negative inclination of about 70°. Thus where both magnetite and haematite are present in one rock the direction of magnetization will be intermediate as BUDDINGTON and BALSLEY find.

RUNCORN[1] suggested that the dissimilar magnetization of the magnetite and haematite might be due either to the contrasting stabilities of these two magnetic oxides suggested by their different coercive forces or to the fact that the magnetization was acquired at totally different times due to the dissimilar Curie points. The former hypothesis would seem the more attractive, because the process of magnetization of the haematite would in view of the high KOENIGSBERGER ratio seem a vastly more efficient one. It is thus probably thermoremanent in origin. The low KOENIGSBERGER ratio of the magnetite would seem to be indicative of a less effective process of magnetization; even perhaps I. R. M. produced in the last few hundred thousand years. There are not sufficient results yet on this interesting problem to support this contention.

# E. Geophysical interpretation of results.

**16. Interpretation of palaeomagnetic results.** It remains to discuss whether NEEL's suggested processes of self reversal account for an appreciable fraction of the natural occurences of reversed remanence in the igneous and sedimentary rocks. The difficulty in deciding this lies in the fact that in the reproduction of the process of magnetization (reheating and cooling an igneous rock or even redispersing a sediment) the vital property of the oxide minerals may be destroyed. Thus the extensive reheating tests that have been carried out on the reversed and normal Tertiary basalts (BRUCKSHAW and VINCENZ [15], VINCENZ [16], HOSPERS [9]) cannot be regarded as decisive evidence that a NÉEL mechanism did not operate during the initial cooling of the rock.

It is however possible either that separate mechanisms account for the reversals in sediments and lava flows or that *reversals* mainly occur *spontaneously* long after the rock has been formed due to chemical or structural changes in the iron oxide minerals. This latter NÉEL [2] suggests and it is a suggestion not easy to test. As magnetizations of opposed sign seem to occur with roughly the same frequency in sediments and lavas, two separate mechanisms of reversal are unlikely. It seems more probable that reversals occur long after the formation of the rocks. However, lava flows contain the magnetite series as the ferromagnetic constituent and the red sandstones, which so far have provided most of the data on sedimentary magnetism have the haematite series as their ferromagnetic constituent. Thus a general property of *self-reversal* must be applicable to a wide range of the iron oxide minerals. It is also relevant[2] that no very recent lavas or sediments have been found to be reversely magnetized.

Quite apart from reversals of magnetization, which are frequent geologically speaking, there is also the slow wander in the mean direction of magnetization

---

[1] S. K. RUNCORN: Trans. Amer. Geophys. Un. **35**, 49 (1954).
[2] N. D. OPDYKE and S. K. RUNCORN: Sciene, Lancaster, Pa. Dec. **1955**.

through geological time, which has to be considered. If the coincidence of the magnetic axis with the axis of the earth's rotation is taken as a fundamental hypothesis, the poles of the earth cannot have been in the same relative position with respect to the land masses as at present. If there has been no relative movement between the land masses then there must have been movement of the whole crust at least, and more plausibly the whole earth, relative to the axis of rotation in space. This is known as the hypothesis of *polar wandering* (p. 286). Alternatively relative movement between the continents, with polar wandering as well, might be assumed. This is the theory of *continental drift*. Both are of interest in connection with the speculations of geologists concerning palaeoclimates and the world-wide distribution of related organisms.

CREER, IRVING and RUNCORN [20] display the palaeomagnetic results obtained in Great Britain in terms of the polar wandering hypothesis. Table 13 shows their suggested positions of the N. pole at various geological epochs, corresponding to the mean directions of magnetization of the strata which have been fully described above. Fields having the same sense with respect to the axis of rotation as the present one are termed negative.

Table 14 shows the corresponding results from the USA; the agreement in the pole positions for the same epochs is reasonably good.

It has been suggested that palaeomagnetism may throw light on the hypothesis of *continental drift*.

WEGENER[1] suggested that at the beginning of Carboniferous times N. America was joined to Eurasia while Africa, Australia, India and S. America formed a single continent, Gondwanaland. This hypothesis, with its modification

Table 13. *Inferred positions of the pole throughout geological time. British strata.*

| Geological Period | Averaged Directions | | Sign of Field | Position of Magn. Poles |
|---|---|---|---|---|
| | D | I | | |
| Pre-Cambrian . . | N. 56° W. | − 38° | − | N. Pole — Middle of Pacific 130° W. 0° |
| (Longmyndian) . | S. 69° E. | + 19° | + | S. Pole — Near Ethiopia 50° E. 0° |
| Cambrian . . . . | S. 7° W. | + 39° | + | N. Pole — Marshall Is. in Pacific 173° E. 15° N. |
| | | | | S. Pole — Middle of South Atlantic 7° W. 15° S. |
| Silurian . . . . | S. 25° W. | − 16° | + | N. Pole — 140° E. 40° N. Japan, N. Hondo |
| | | | | S. Pole — 40° W. 40° S. in S. Atlantic |
| Devonian . . . . | N. 34° E. | + 2° | − | N. Pole — In Pacific Nr. Kamchatka 156° E. 34° N. |
| | S. 19° W. | − 2° | + | S. Pole — S. Atlantic about 3000 miles S. E. of Buenos Aires 24° W. 34° S. |
| Permian . . . . | S. 9° W. | − 9° | + | N. Pole — In Pacific Nr. Kamchatka 168° E. 48° N. |
| | | | | S. Pole — 3000 miles S. E. of Buenos Aires 12° W. 48° S. |
| Triassic . . . . | N. 26° E. | + 26° | − | N. Pole — In Pacific Nr. Kamchatka 133° E. 47° N. |
| | S. 34° W. | − 26° | + | S. Pole — In S. Atlantic 47° W. 47° S. |
| Eocene . . . . . (Tertiary) | S. 14° W. | − 60° | + | N. Pole — In Arctic near New Siberian Is. 118° E. 75° N. |
| | | | | S. Pole — 62° W. 75° S. in Weddell Sea Antarctica |

[1] See chapter by A. E. SCHEIDEGGER, in this Volume, Sect. 5 and 6.

Table 14. *Inferred positions of the pole throughout geological time. American strata.*

| Geological Period | Averaged Directions | | Sign of Field | Position of Magn. Poles |
|---|---|---|---|---|
| | D | I | | |
| Pre-Cambrian . . . . . . . | S. 87.7° W. | + 72.8° | — | 31° N. 150° W. (RUNCORN) |
| (Hakatai shales) . . . . . | S. 66  ° W. | + 72.8° | — | 18° N. 144° W. (DOELL) |
| Pre-Cambrian or Cambrian  . | N. 75°    W. | 0° | — | 20° N. 165° E. |
| (Freda and Nonesuch shales) | | | | |
| Carboniferous . . . . . . . | S. 30.3° E. | − 3.4° | + | 49° N. 120° E. |
| Permian  . . . . . . . . . | S. 47.3° E. | + 22.9° | + | 26° N. 121° E. (RUNCORN) |
| | S. 34° E. | + 8° | + | 39° N. 115° E. (DOELL) |
| Triassic . . . . . . . . . | N. 22° W. | + 16° | + | 33° N. 107° E. |

by DU TOIT, purports to provide an explanation for the rapid migration of organisms throughout the world which is a feature of every stage of biological evolution. It is not clear how types with a continental habitat accomplish this and it was early postulated that extensive and transitory land bridges existed in every epoch. The theory of isostasy made it clear that the disappearance of land bridges into what is now deep ocean was unlikely. The forces which WEGENER postulated as a cause of continental drift are now known to have negligible effects, but the concept of the continents as a granitic scum (WEGENER's sial) floating in a denser basaltic medium which also forms the floor of the oceans suggests that such drift might be feasible. JEFFREYS[1] argues that however plastic the basaltic material below the continents may be because of the moderately higher temperature, the basaltic shell under the oceans must have greater strength than the continental blocks. Drifts of the continental masses through the floor of the oceans is therefore unimpressive as a geophysical hypothesis. JEFFREYS also points out that the similarity of shape of the west coast of Africa and the east coast of S. America, which figures prominently in popular expositions of continental drift, is a poor fit, even allowing that the continental shelf rather than the coast is the most permanent boundary of the continental masses.

Geological evidence of the distribution of former glaciations, and the palaeontological evidence of plants flourishing in regions of the world which are now climatically unsuitable for them, has been held to be strong evidence for continental drift[2]. The widespread glaciations of South Africa, India, Australia and South America in Permian and Carboniferous times when Laurasia enjoyed mild or tropical climates is not however explained by continental drift alone. The Dwyka tillite, or bounder clay, of South Africa can be traced from the Transvaal to the Cape and from South West Africa to Natal, in many places resting on a glaciated floor. In some cases successive glaciations, such as occurred in the Pleistocene ice age in Europe, may be recognized and it is possible to infer that the ice sheets moved towards the south, as no known mountain ranges existed to supply the glaciers. The glaciation is thought to be the result of great continental ice sheets. Thus the simplest hypothesis is that the pole must have wandered far from its present position. In India similar evidence of glaciation moving northwards is found in Orissa, the Central Provinces and the Punjab. Because this Permo-Carboniferous glaciation is contemporaneous in these four widely separated continents of the southern hemisphere, it might appear that polar wandering cannot alone be the explanation. However, geological correlation

---

[1] H. JEFFREYS: The Earth, 3rd edit. Cambridge: Univ. Press 1952.
[2] A. P. COLEMAN: Ice ages Recent and Ancient. New York: MacMillan 1926. — C. E. P. BROOKS: Climate through the ages. London: Ernest Benn 1926. — H. SHAPLEY: Climatic change. Harvard University Press 1953.

in the Palaeozoic may not be sufficiently accurate and the pole might have moved fast enough at this time to shift from one continent to another.

The problem of the *dynamics of polar wandering* was considered by Darwin[1], who concluded that surface movements had negligible effect on the stability of the axis of rotation. In any sphere, rigid except for a ring of material which is caused to move through an angle $\Theta_1$, a marked spot on the surface will move through an angle $\Theta_2$ relative to a fixed point in space. $\Theta$ is given by

$$\Theta_2 = \frac{I_1}{I_2} \Theta_1$$

by the conservation of angular momentum where $I_1$ is the moment of inertia of the ring about an axis perpendicular to its plane through its centre and $I_2$ the moment of inertia of the rest of the sphere about the same axis. The rapid rotation of the sphere about an axis perpendicular to this will not affect the situation but merely alter the reaction between the moving of matter and the sphere, so that the ring exerts on the sphere a torque about an axis perpendicular to the two mentioned above. This torque of course remains fixed relative to an axis in the body, and if the polar wandering is slow and the rotation is fast will hardly affect the direction of the latter in space. On the contrary the effect of an external torque is to cause precession of the axis of rotation of the body in space.

In the case of the earth it is necessary to consider the effect of the equatorial bulge. If the earth's crust were a continuous shell, a force to distort this to its new shape would be necessary, but a more realistic picture is that of a number of more or less loosely connected blocks floating on a plastic mantle. Thus the remodelling of the geoid consequent upon polar wandering must overcome only the friction between neighbouring blocks and the viscous resistance to flow in the mantle. This appears to be quite negligible.

Goguel[2] has considered the possibility that marine currents, such as the Gulf Stream, may produce considerable polar wandering over geological time. It is however likely that even if there is a net angular momentum in the upper parts of the ocean it is likely to be compensated by an opposite angular momentum in the deeper ocean which cannot be so easily detected.

Vening Meinesz[3] considers the possibility of polar wandering occurring as a result of widespread convection in the mantle. However, the latter is not definitely known to occur. Nevertheless it has been seen to be a very effective process. If the moment of inertia $(I_1)$ of the convecting system in the mantle were one-fiftieth of that of the earth, then convective overturn must occur about 14 times to cause 90° of polar wandering. In 500 million years this implies velocities in the mantle of a few tens of centimeters a year. Such convection on so long a time scale is a very attractive explanation of polar wandering.

Gold[4] has considered in some detail the instability of the earth's axis of rotation resulting from the torques produced by the redistribution of mass on the earth's surface associated with tectonics, in particular mountain building. The free or Eulerian nutation of the earth, determined by astronomical observations of the variations of latitude, is apparently excited by random impulses but damped by at least $1/e$ of its amplitude in 10 periods (of 420 days). Bondi and Gold show that, contrary to the view of Jeffreys (1952), the moment of inertia of the core is insufficient to cause this damping, from which it is therefore not possible to derive an upper limit for the viscosity of the core. They therefore

[1] G. H. Darwin: Phil. Trans. Roy. Soc. Lond. **167**, 1 (1877).
[2] J. Goguel: Ann. Géophys. **6**, 139 (1950).
[3] F. A. Vening, Meinesz: Gravity Expeditions at Sea, IV. Delft 1948.
[4] T. Gold: Nature, Lond. **175**, 526 (1955).

assume that the mantle is, for slowly varying strains, inelastic and the necessary dissipation of energy occurs within it.

Gold argues that the restricted areas on the earth's surface where isostatic adjustment is not complete and the very near equilibrium shape of the geoid in spite of changes in the rate of rotation of the earth due to tidal friction, point to an absence of permanent stiffness in the earth as a whole over geological time. If the earth were perfectly rigid the random impulses exciting the free nutation would build up its amplitude without limit.

Gold suggests that tectonic movements, representing the addition of an excess mass between the pole and equator, would be effective in producing polar wandering. Gold estimates that if a continent of the size of South America were suddenly raised by 30 metres, an angle of 0.01° would in consequence separate the axis of figure from the axis of rotation and due to plastic flow polar wandering at a rate of 0.001°/year would occur until the additional mass was situated on the equator. Gold estimates that large angles of polar wandering of the type envisaged above would occur in a million years.

## F. The magnetic susceptibility of rocks.

**17. The measurement of magnetic susceptibility[1].** The induced magnetization of rocks contributes in an important way to the magnetic anomalies observed over geological structures and is therefore of importance in applied geophysics. The susceptibility arises from the presence of the iron oxides and sulphides discussed in Part A and varies with the field strength. Consequently the standard laboratory method of the measurement of susceptibility in a strong magnetic field is of doubtful application in applied geophysics.

Measurement of the magnetic susceptibility of rock samples [1] may be made by a ballistic method or by the astatic magnetometer[2] or by an A.C. method[2, 3]. Mooney[4] has recently developed an A.C. method for use in measuring the susceptibility of rocks in situ.

**18. Interpretation of experimental values of susceptibilities of rocks.** A broad correlation between the value of susceptibility and the rock type is well attested. L. B. Slichter[5], for example, has shown that the range of susceptibilities of the basic effusives, basic plutonics, granites, gneisses, etc. and sedimentary rocks overlap but that the mean values decrease in this order, being lowest for the sediments. Mooney and Bleifuss[5] find, from examination of rocks from Minnesota, that the mean susceptibilities of acid intrusives, basic intrusives, and basic extrusives are in the ratio $1:2:7$.

In contrast with the remanent magnetization of rocks, only magnetite, of the oxides, appears to play an important role in determining the susceptibility. Slichter[6] quotes the susceptibility of magnetite as between 0.3 and 0.8, that of the other iron oxides between $1/10$ and $1/100$ of these. He states that the susceptibility of a rock is well estimated by multiplying a susceptibility value of 0.3 by the proportion of magnetite present in the rock. Mooney and Bleifuss[6] find a definite dependance of susceptibility on magnetite content, giving a value of the susceptibility of the magnetite near to that given by Slichter.

[1] See the more detailed discussion in Nagata's book [1], and typical hysteresis curves shown in Landolt-Börnstein [23].

[2] J. M. Bruckshaw and E. I. Robertson: J. Sci. Instrum. 25, 444 (1948).

[3] J. A. Sharpe: Proc. Geophys. Soc. Tulsa 1, 23 (1953).

[4] H. M. Mooney: Geophysics 17, 531 (1952).

[5] Handbook of Physical Constants; Geol. Soc. Amer. Special Papers No. 36, 1942.

[6] H. M. Mooney and R. Bleifuss: Geophysics 18, 383 (1953).

# References:

[1] NAGATA, T.: Rock Magnetism. Tokyo: Maruzen & Co. 1953. 230 pp.
[2] NÉEL, L.: Adv. Physics **4**, 191 (1955).
[3] NÉEL, L.: Ann. Phys. **4**, 249 (1949).
[4] POUILLARD, E.: Ann. Chimie **5**, 164 (1950).
[5] NICHOLLS, G. D.: Adv. Physics **4**, 113 (1955).
[6] CHEVALLIER, R., et J. GIRARD: Bull. Soc. chem. Fr. **17**, 576 (1950).
[7] GORTER, E. W.: Philips Res. Rep. **9**, 295. 321, 403 (1954).
[8] NÉEL, L.: Ann. Geophys. **7**, 90 (1951).
[9] HOSPERS, J.: Nature, Lond. **168**, IIII (1951). — Proc. Kon. Ned. Akad. Wetensch. **56**, 467, 477 (1954); **57**, 112 (1954).
[10] JOHNSON, E. A., T. MURPHY and O. W. TORRESON: Terr. Magn. **53**, 349 (1948).
[11] CHEVALLIER, R.: Ann. Phys. **4**, 5 (1925).
[12] VINCENZ, S. A.: M. N. R. A. S., Geophys. Suppl. **6**, 590 (1954).
[13] MINAKAMI, T.: Tokyo Imp. Univ. Earthquake. Res. Inst. Bull. **19**, 612 (1941).
[14] BRUCKSHAW, J. M., and E. I. ROBERTSON: Mon. Not. Roy. Astronom. Soc., Geophys. Suppl. **5**, 308 (1949).
[15] ROCHE, A.: C. R. Acad. Sci. Paris **230**, 113 (1950); **233**, 11'32 (1951).
[16] TORRESON, O. W., T. MURPHY and J. W. GRAHAM: J. Geophys. Res. **54**, 111 (1949).
[17] KAWAI, N.: J. Geophys. Res. **56**, 73 (1951).
[18] GRAHAM, J. W.: J. Geophys. Res. **54**, 131 (1949).
[19] BALSLEY, J. R., and A. F. BUDDINGTON: J. Geomag. a. Geoelect. **6**, 176 (1954).
[20] CREER, K. M., E. IRVING and S. K. RUNCORN: J. Geomag. a. Geoelect. **6**, 163 (1954).
[21] BRUCKSHAW, J. M., and S. A. VINCENZ: M. N. R. A. S., Geophys. Suppl. **6**, 579 (1954).

## Additional references.

[22] KOENIGSBERGER, J. G.: Terr. Magn. **43**, 119, 229 (1938). A voluminous report with numerous references to older literature.
[23] RÖSSIGER, M.: Sect. 32, 34 in LANDOLT-BÖRNSTEIN, Zahlenwerte aus Physik, 6. Aufl., Bd. 3, S. 331. Berlin: Springer 1952.
[24] THELLIER, E.: Ann. Inst. Phys. du Globe Paris **16**, 157 (1938). (A treatise on magnetization of bricks, with geophysical applications.)

# The Magnetism of the Earth's Body.

By

S. K. RUNCORN.

With 11 Figures.

## A. Main field and secular variation: Observations and analysis.

**1. Introduction.** The existence of a geomagnetic field was made use of in the Mariners' compass, long before its origin within the earth was recognized. ALEXANDER NECKAM[1] refers in a celebrated passage for the first time in European literature to the directional property of a magnetized needle and its use at sea. It was originally supposed that this attraction was due to the pole star and it was not until the work of GILBERT[2] that its true nature was understood. GILBERT was able to show that in the neighbourhood of a magnetized sphere of lodestone—a *terrella* as he termed it—the magnetic field was directed in a similar manner to the field observed at the surface of the earth. Thus by this simple model experiment GILBERT came to his well-known conclusion, "*Magnus magnes ipse est globus terrestris*".

The discovery of the declination, inclination and the secular variation and the construction of the early magnetic maps are described by CHAPMAN and BARTELS [1].

The study of geomagnetism has been the occasion of many notable advances in physics since GILBERT's time. For the absolute measurement of the intensity of the field, GAUSS developed the well-known method using the vibration and deflection magnetometers. His development of spherical harmonic theory of potential fields was carried out to analyse the magnetic surveys of the earth.

**2. The magnetic elements.** The geomagnetic field $F$ is a vector quantity, the variation of which in time and space has been studied, in gradually greater detail, through the last few centuries. At any point its component in a horizontal plane, $H$, has components in a north direction $X$ (positive northwards) and $Y$ in an east direction (positive eastwards). The component of $F$ in a vertical direction is known as $Z$ (measured positive downwards). The angle made by $H$ with the meridian is known as the *declination D* (or by mariners as the variation) and is reckoned positive if to the east. The angle of inclination (or dip) $I$ is the angle made by $F$ with the horizontal and is reckoned positive if the north seeking end of the vector $F$ points downward. $F$ is thus known if three suitably chosen elements are known. When *transient variations* of these elements with a time scale of a few days, associated with magnetic storms and with the daily variation, are averaged out, the *main geomagnetic field* is left. It is usual to take an annual mean based on quiet days as the value of the main field at any station but even this procedure does not eliminate entirely effects arising from flow of current in the ionosphere[3].

---

[1] A. NECKAM: De Naturis Rerum, St. Albans 1180.
[2] W. GILBERT: De Magnete, London 1600.
[3] See the "General survey of geomagnetic time-variations" in Vol. XLIX of this Encyclopedia.

**3. Measurement of the geomagnetic elements.** A distinction must be made between *absolute* and *relative* magnetic measurements and instruments[1]. An absolute measurement connects a magnetic field quantity with the basic standards of length, mass and time, while a relative measurement involves an accurate measurement of the difference between the unknown magnetic field and an arbitrary field or base line, which may be relied upon to remain constant over a sufficient time and which may from time to time be measured absolutely.

No difficulty is involved in measuring *declination* to a high accuracy by a freely suspended magnet and a meridian mark. Such an instrument is known as a declinometer. The angle of *inclination* was measured by a dip circle, but has largely been replaced by the dip inductor, in which a coil is rotated about an axis in its plane, and the inclination of the axis is changed until the induced electromotive force is reduced to zero. $H$ was the component first measured with high accuracy by the GAUSS magnetometer. A magnet is first suspended horizontally in the earth's field and its period of oscillation measured. It is then used with a deflection magnetometer, the field at a measured distance from the magnet being compared with that of the earth. The sine galvanometer and the SCHUSTER-SMITH coil magnetometer are now more generally used. In the SCHUSTER-SMITH coil magnetometer[2], $H$ is compared with the field at the centre of a HELMHOLTZ coil system of accurately known size carrying a current of known magnitude. The axis of the coil is rotated in a horizontal plane until the resultant of its field and that of the earth lies normal to the magnetic meridian. Similarly $Z$ may be determined[3] by nulling the component in the centre of a HELMHOLTZ coil system with a vertical axis, various magnetic detectors being used to register the null. The accuracy of these absolute instruments for $H$ and $Z$ depends on the accurate measurement of current and of their constants. The latter depend on knowledge of the dimensions of the windings of the HELMHOLTZ coil systems and of their changes with temperature.

In an observatory it is convenient and in field surveys essential to use the less cumbersome relative instruments, often self recording, which must be, however, standardized regularly against absolute instruments. Such variometers usually rely for their stability of base line on the constancy over some days of the moment of a carefully aged magnet.

Two variometers of exceptional stability were designed by the late Dr. LA COUR, the Quartz Horizontal Magnetometer[4] (Q.H.M.) for the measurement of $H$ and the $Z$ magnetometric balance[5] (B.M.Z.) for the measurement of $Z$. They are valuable for intercomparison of observatories as well as for regular observations in the observatory or field.

The Q.H.M. is an instrument in which a horizontal magnet is suspended by a fine quartz fibre. It is mounted on a divided circle which enables the instrument to be rotated about a vertical axis and its azimuth read. An autocollimating telescope with its axis horizontal is attached to the case and a mirror on the magnet system enables the horizontal circle reading of the magnetic meridian to be determined. The case is then rotated through an angle greater than one revolution until the magnet is again in the same position relative to the case. In this position there is exactly 360° of twist in the fibre, plus any initial twist in the fibre. The latter may be eliminated by rotation in the opposite direction. The stability

[1] See also V. LAURSEN and J. OLSEN in Vol. 49 of this Encyclopedia.
[2] F. E. SMITH: Phil. Trans. Roy. Soc. Lond., Ser. A **223**, 175 (1922).
[3] D. W. DYE: Proc. Roy. Soc. Lond., Ser. A **117**, 434 (1928).
[4] D. LA COUR: Comm. Magnétiques, Copenhagen, Dan. Met. Inst. No. 15. 1936.
[5] D. LA COUR,: Comm. Magnétiques, Copenhagen, Dan. Met. Inst. No. 19. 1942.

of the instrument depends on the constancy of the elastic properties of the quartz fibre and the moment of the magnet. The value of $H$ is calculated from a formula involving these, the readings of the divided circle and the temperature.

The B.M.Z. involves the cancelling of the vertical component by a fixed "field" magnet and a "turn" magnet, the contribution of the latter and weaker magnet being varied by rotation about a horizontal axis. The null detector is a horizontal "Monad" magnet, which is delicately balanced on knife-edges to be horizontal when the vertical component is nulled. The nulling is accomplished by observing the inclination of the magnet through an autocollimating telescope. Small changes in the moment of the Monad magnet over time do not affect the baseline of the instrument, and changes in the centre of gravity of the Monad magnet may be minimized by making the magnet from a single piece of steel with none of the variable parts necessary in ordinary variometers using the Lloyd balance principle.

*Surveys at sea* have played an important role in extending our knowledge of the magnetic field and early in this century the Carnegie Institution of Washington conducted extensive surveys of this kind. Unfortunately no non-magnetic vessel has been available for such surveys since the loss of the "Carnegie" in Apia harbour, Samoa, in 1929. The development of the *fluxgate magnetometer*[1] makes it likely that extensive surveys of the geomagnetic field may be done from the air. The fluxgate magnetometer cancels out the component of field in a certain direction with a solenoid, inside which is a null detector depending on the non-linear properties of a highly permeable material, such as mu metal. By mounting three such elements at right angles and rotating the system by servomechanisms it is possible to use two of the elements as null detectors maintaining the third parallel to the magnetic field. The response time of the instrument is such that an accuracy of 1 gamma in the measurement of total field is possible even in flight. The measurement of three components is more difficult and essentially less accurate, for the misalignment between an element and the component which it is desired to measure causes an error as large as the sine of the angle of misalignment multiplied by the field component perpendicular to the element, but where the element is aligned along the total field the error involved merely depends on the cosine of the error angle.

**4. Local anomalies in the geomagnetic field.** The main geomagnetic field when surveyed on a close net is seen to possess local anomalies, sometimes of the order of thousands of gammas or more. One of the greatest anomalies is that of Kursk, south of Moscow; along strips of 250 km. length, $Z$ is everywhere above normal, up to 1.9 gauss.

The origin of these anomalies resides in the induced and permanent magnetization of rocks of the earth's crust. The magnetization of rocks depends on the amount and nature of the iron oxide minerals contained in them and, if the original permanent magnetization has been retained, whether the latter is larger than the induced magnetization in the present earth's field (this is the case in all relevant anomalies). The intensity of magnetization of most rocks will not exceed $10^{-2}$ gauss and is likely to be larger than $10^{-6}$ gauss.

The magnetization of rocks (discussed in the following chapter of this volume) presumably decreases with depth and will vanish at a depth of around 25 km. where the temperature will exceed the Curie points of the iron oxide minerals. Thus the local anomalies of the main field are likely to have lateral extents of

---

[1] See J. R. BALSLEY in Vol. XLIX of this Encyclopedia.

some tens of kilometers. Because of their small scale the gradients arising from local anomalies are in places larger than that of the general dipole field.

DEEL[1] has commented on the absence of anomalies of a scale intermediate between these local ones and those of some thousands of kilometers observed

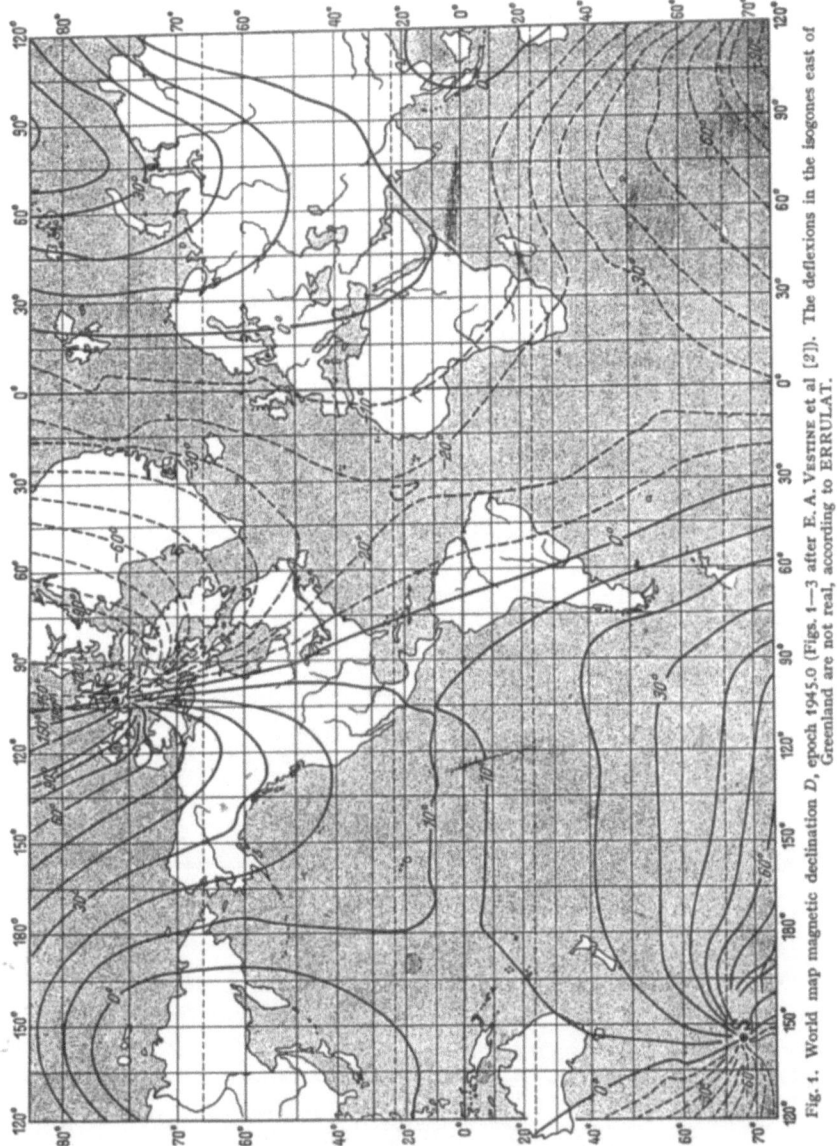

Fig. 1. World map magnetic declination $D$, epoch 1945.0 (Figs. 1—3 after E. A. VESTINE et al [2]). The deflexions in the isogones east of Greenland are not real, according to ERRULAT.

on world maps. It may be inferred from this that no appreciable magnetic sources lie between the Curie point isotherm and the depths of the mantle.

No comparison has yet been made of the scale of anomalies over the oceans with those over the continents. Systematic differences in the depth of the Curie

[1] S. A. DEEL: Pap. U. S. Coast Geod. Sur. No. 664, 1945.

point isotherm between the oceans and continents would be detectable in this way[1].

On a world wide scale such local anomalies contribute little or nothing to the main geomagnetic field and will be ignored in this account.

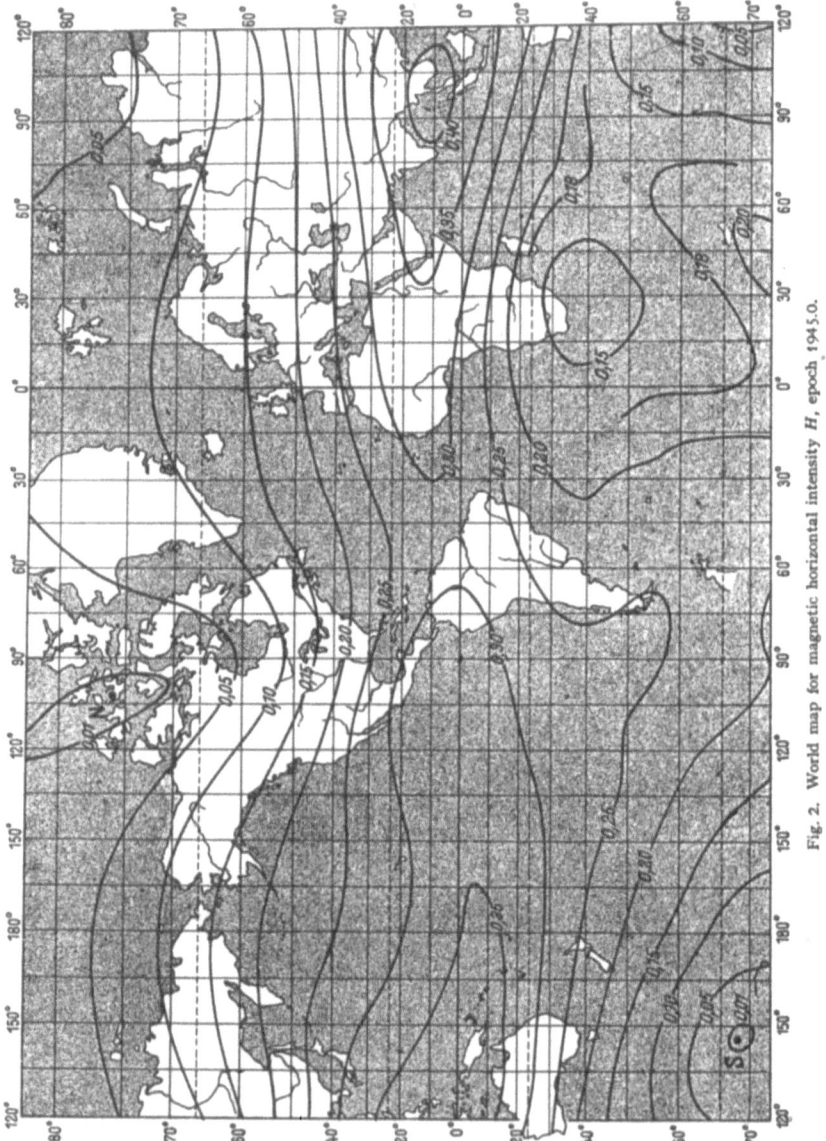

Fig. 2. World map for magnetic horizontal intensity $H$, epoch 1945.0.

**5. Maps of the geomagnetic field.** At various times the accumulation of magnetic observations has enabled world maps of the field to be compiled. Measurements at fixed stations or observatories or the reoccupation of repeat stations enable measurements during surveys to be corrected for the secular

[1] S. K. Runcorn: Nature, Lond. **173**, 281 (1954).

variation, which is assumed linear with time, varying smoothly over wide areas, and unaffected by local anomalies. Thus all measurements may be *reduced to a single epoch*. Maps of the main field, the *residual* or *non-dipole field*, i.e., the observed field minus that axial or inclined or eccentric dipole field which best

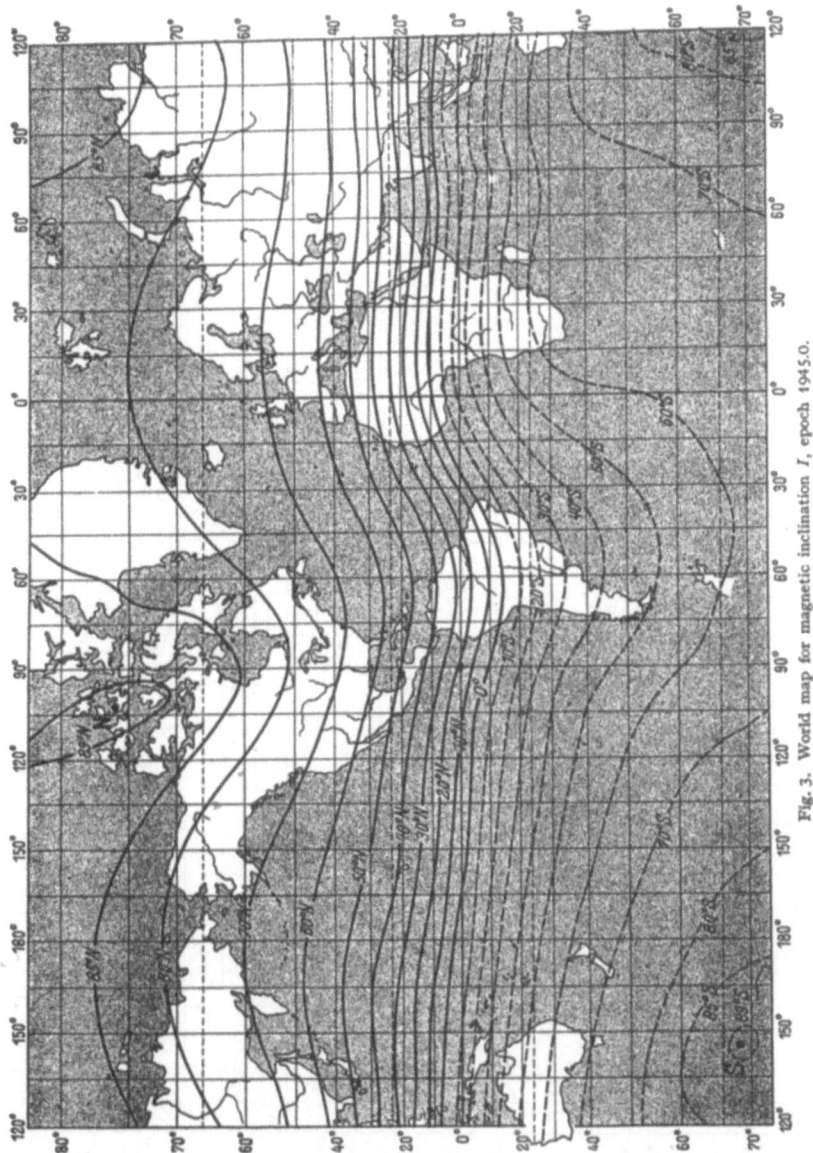

Fig. 3. World map for magnetic inclination $I$, epoch 1945.0.

describes it, and the secular variations have often been prepared. A compilation of early values of declination enabled VAN BEMMELEN [5] to draw isogonic charts for 1550 to 1700 at intervals of fifty years over much of Europe, Africa and the Atlantic. The British Admiralty and the US Hydrographic office have published at frequent intervals over a long period charts of equal declination, horizontal

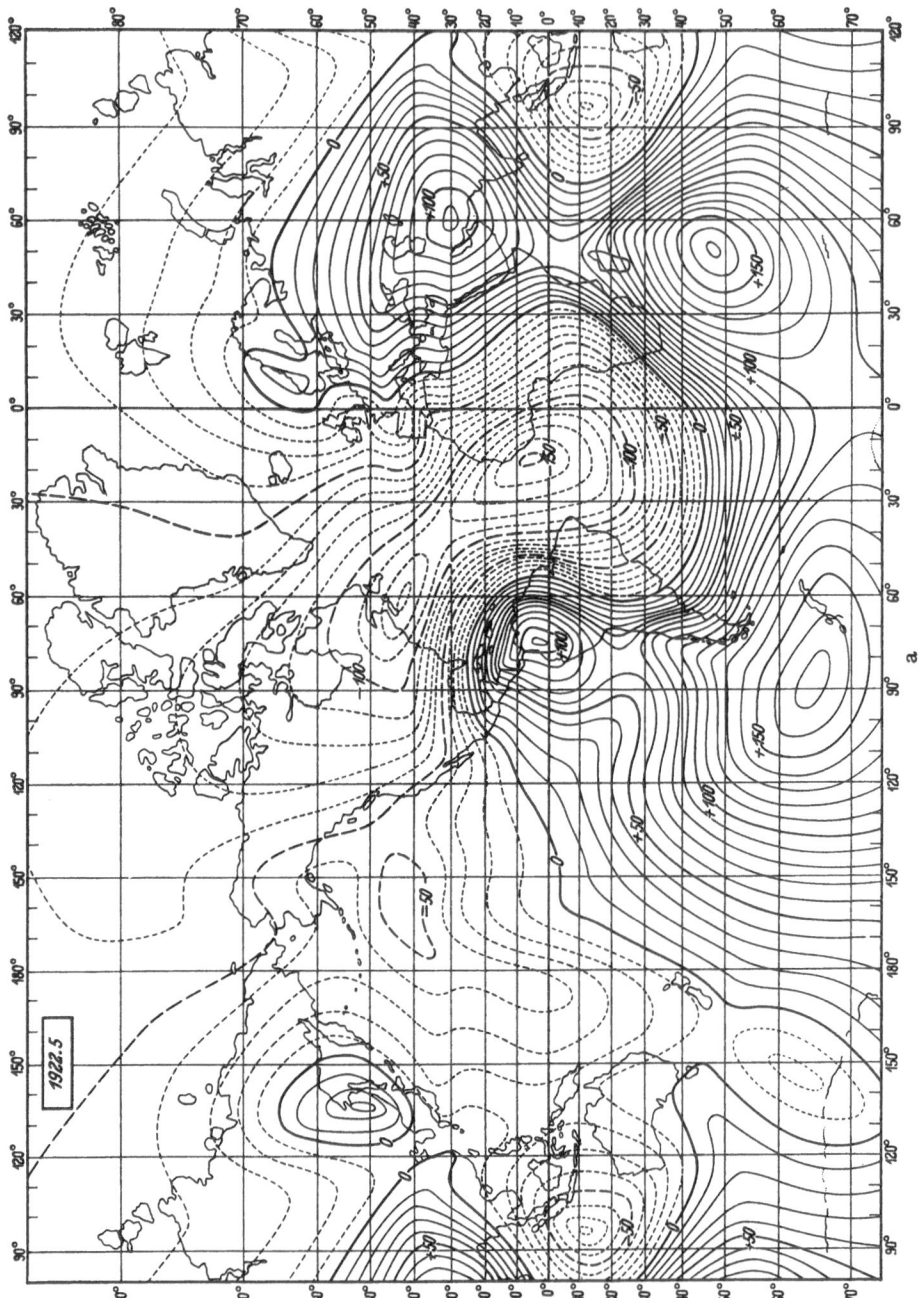

intensity and dip. The gap between VAN BEMMELEN's maps and the modern data is perhaps best filled by NEUMAYER's atlas [9], but see his preface.

The early charts all showed that, though the earth's field was approximately that of a dipole at the geocentre, the fit was better if the axis of this dipole was inclined to the geographical axis. Even so it was evident that there were regions

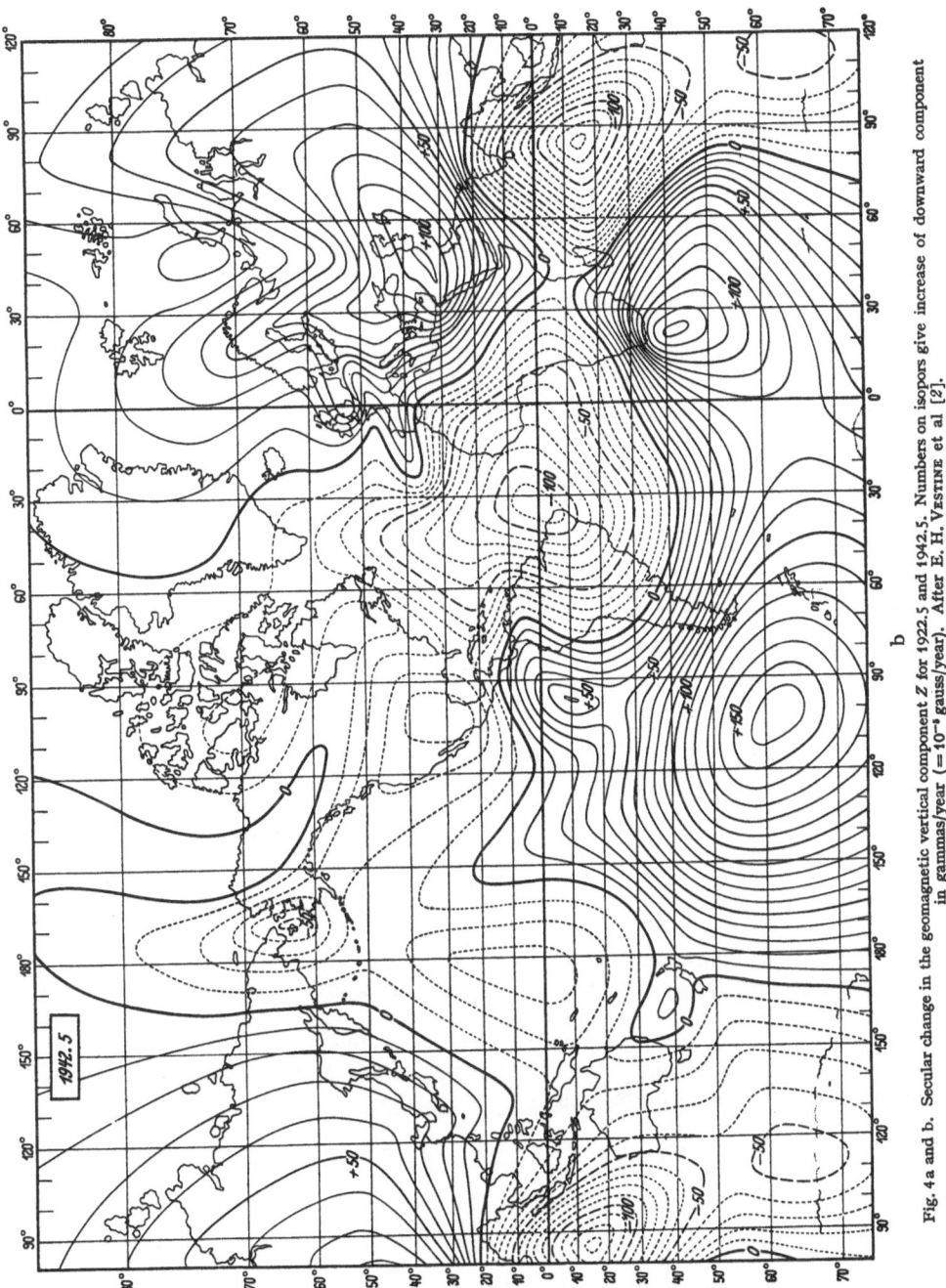

Fig. 4a and b. Secular change in the geomagnetic vertical component Z for 1922.5 and 1942.5. Numbers on isopors give increase of downward component in gammas/year (=10⁻⁵ gauss/year). After E. H. Vestine et al [2].

of continental size on the earth's surface over which the divergence from this simple model was considerable. The isomagnetic lines in such areas appear on these maps as a series of approximately concentric closed curves. The early maps of van Bemmelen show this quite clearly in Europe where large declinations were observed.

Much effort therefore was made early in this century, mainly on the initiative of the Carnegie Institution of Washington, to extend the magnetic measurements over the oceans and unexplored areas of the globe, to enable a proper description of the earth's field and and its secular change to be attempted.

The modern network of observatories and special surveys has enabled Vestine and his collaborators [2] to describe the main geomagnetic field and its secular variation thoroughly between the years 1912 and 1942. They give for each of the four epochs, 1912.5, 1922.5, 1932.5 and 1942.5 maps showing *isogonics* (lines of equal *D*), *isoclinics* (lines of equal *I*) and *isodynamics* (lines of equal components of force) for the main field and various *isopors* (lines of equal secular change in an element). Fig. 4 shows the *secular change* in *Z* for 1922.5 and 1942.5.

World maps for *Z, H, D, F* and *I* and the secular changes in these elements are to be published for the epoch 1955.0 and fixed intervals thereafter by the U.S. Navy Hydrographic Office, Washington, D.C.

The magnetic poles are the points on the earth's surface where the horizontal field vanishes; in general the compass needle does not point exactly to them. The positions of the poles have been found from time to time, and Table 1 gives these data.

Table 1a. *Coordinates of North Magnetic Pole.*

| Date or epoch | North latitude | West longitude | Observer |
|---|---|---|---|
| 1831.4 | 70° 15′ | 96° 45′ | J. C. Ross[1] |
| 1904.5 | 70° 30′ | 95° 30′ | R. Amundsen[1] |
| 1948.0 | 73° 00′ | 100° 00′ | P. H. Serson et al.[2] |

Table 1b. *Coordinates of South Magnetic Pole.*

| Date or epoch | South latitude | East longitude | Observer |
|---|---|---|---|
| 1841.1 | 75° 00′ | 153° 45′ | J. C. Ross |
| 1909.0 | 72° 25′ | 155° 16′ | D. Mawson |
| 1912.5 | 71° 12′ | 150° 42′ | E. N. Webb |
| 1952 | 68° 42′ | 143° 00′ | P. Mayaud[3] |

**6. Spherical harmonic analysis of the field.** A magnetic field known over a spherical surface may be analysed into parts of external and internal origin by a method first used by Gauss. Suppose in the region under study no electric currents flow; then the field $F$ may be written as the gradient of a scalar potential $V$. This potential obeys Laplace's equation and therefore may be expressed as a series of spherical harmonics

$$V = a \sum_{n=0}^{\infty} \sum_{m=0}^{n} P_n^m(\vartheta) \left[ \left\{ C_n^m \left(\frac{r}{a}\right)^n + (1 - C_n^m) \left(\frac{a}{r}\right)^{n+1} \right\} \underline{A}_n^m \cos m\varphi + \right. $$
$$\left. + \left\{ S_n^m \left(\frac{r}{a}\right)^n + (1 - S_n^m) \left(\frac{a}{r}\right)^{n+1} \right\} \underline{B}_n^m \sin m\varphi \right] \tag{6.1}$$

when $a$ is the radius of the earth and $(r, \vartheta, \varphi)$ spherical coordinates, $C_n^m, S_n^m$ are the portions of the parts of the harmonic terms of *external* origin. From this by

[1] A. S. Steen, N. Russeltvedt and K. F. Wasserfall: Geofysiske Publikasjoner, Oslo 7, part II (1933).
[2] R. G. Madill: Arctic 1, 8 (1948).
[3] P. Mayaud: Expéditions Polaires Françaises, Résultats Scientifiques No S IV 2, 1954.

differentiation the following expressions for $X$, $Y$ and $Z$ at the earth's surface are obtained, putting $n\underline{A}_n^m = A_n^m$ and $n\underline{B}_n^m = B_n^m$,

$$X = \sum_n \sum_m X_n^m \left(A_n^m \cos m\,\varphi + B_n^m \sin m\,\varphi\right), \tag{6.2}$$

$$Y = \sum_n \sum_m Y_n^m \left(A_n^m \sin m\,\varphi - B_n^m \cos m\,\varphi\right), \tag{6.3}$$

$$Z = \sum_n \sum_m P_n^m \left[\left\{n\,C_n^m - (n \dotplus 1)\,(1 - C_n^m)\right\}\left(\frac{A_n^m}{n}\right)\cos m\,\varphi \right.$$
$$\left. + \left\{n\,S_n^m - (n+1)\,(1 - S_n^m)\right\}\left(\frac{B_n^m}{n}\right)\sin m\,\varphi\right] \tag{6.4}$$

where

$$X_n^m = d\,P_n^m(\cos\vartheta)/n\,d\,\vartheta$$

and

$$Y_n^m = m\,P_n^m(\cos\vartheta)/n\sin\vartheta$$

which together with $P_n^m(\vartheta)$ have been extensively tabulated by SCHMIDT[1].

From the harmonic analysis of the observed values of $X$ and $Y$ around lines of latitude, the coefficients of $\cos m\,\varphi$ and $\sin m\,\varphi$ in (6.2) and (6.3) can be obtained for various values of longitude. Then for each value of $m$, using the tabulated values of $X_n^m$ and $Y_n^m$, the various coefficients $A_n^m$, $B_n^m$ can be calculated. Any difference between the values of the coefficients derived using $X$ or $Y$ is attributable either to inadequate data or to an earth-air current, the existence of which would make it impossible to express $V$ in terms of a scalar potential (see the discussion in Sect. 8). From the observed values of $Z$ and the previously determined values of $A_n^m$ and $B_n^m$ it is clear that a similar procedure will enable the quantities $C_n^m$ and $S_n^m$ to be calculated. These for the main field and the secular variation field are found to be approximately zero and thus the symbols $g_n^m$ and $h_n^m$ are commonly used in (6.1) for $\underline{A}_n^m$ and $\underline{B}_n^m$ in geomagnetic work. Corrections of the spherical harmonic analysis to take account of the ellipticity of the earth's surface have been discussed by AD. SCHMIDT[1].

**7. Analysis of the main field and value of the dipole moment.** The results of the principal analyses of the main field are given in Table 2a. The extent to which they may be compared has been discussed by MAUERSBERGER[2]. The axial dipole term $g_1^0$ predominates, the equatorial dipole being about 15% of the axial one. The diminution of the dipole moment over the last 100 years by about 7% is evident but recently [3,4] it has been suggested that evidence exists that the moment began to increase again about 1935. GAIBER-PUERTAS[3] compares the total dipole moment at 1922 derived from an analysis by BAUER with that at 1945 from VESTINE's analysis. He also compares maps of the surface fields in the decades 1933 to 1943 and 1913 to 1923. BULLARD[4] argues that the methods of extrapolating field values in the construction of magnetic maps may tend to prevent changes in the secular variation rates from becoming evident until some time after they have occurred. He shows from the records of 24 observatories during the periods 1928 to 1930, 1938 to 1940 and 1948 to 1950 that the "substitute" dipole moment, deduced from the values of $H$ and $Z$, shows a minimum in 1936.

[1] AD. SCHMIDT: Tafeln der normierten Kugelfunktionen. Gotha: Engelhard-Reyher 1935. — See also the chapter on "Analysis of geophysical functions" in Vol. XLIX of this Encyclopedia, and Chapters 17 und 18 in [I]. The definition of the "normalized" spherical harmonics is also given in Landolt-Börnstein [3], article 32 503.

[2] P. MAUERSBERGER: Geophys. Inst. Potsdam, Abh. No. 5, p. 5, 1952.

[3] C. GAIBER-PUERTAS: Geofisica Pura e Applicata 23, 6 (1952).

[4] E. C. BULLARD: J. Geophys. Res. 58, 277 (1953).

Table 2a. *The spherical harmonic analysis of the main field.* Units $10^{-4}$ gauss.

| Source | Epoch | $g_1^0$ | $g_1^1$ | $h_1^1$ | $g_2^0$ | $g_2^1$ | $h_2^1$ | $g_2^2$ | $h_2^2$ |
|---|---|---|---|---|---|---|---|---|---|
| Gauss . . . . . . | 1835 | − 3235 | − 311 | + 625 | + 51 | + 292 | + 12 | − 2 | + 157 |
| Erman-Petersen . | 1839 | − 3201 | − 284 | + 601 | − 8 | + 257 | − 4 | − 14 | + 146 |
| Adams . . . . . | 1845 | − 3219 | − 278 | + 578 | + 9 | + 284 | − 10 | + 4 | + 135 |
| Adams . . . . . | 1880 | − 3168 | − 243 | + 603 | − 49 | + 297 | − 75 | + 61 | + 149 |
| Fritsche . . . . | 1885 | − 3164 | − 241 | + 591 | − 35 | + 286 | − 75 | + 68 | + 142 |
| Schmidt . . . . . | 1885 | − 3168 | − 222 | + 595 | − 50 | + 278 | − 71 | + 65 | + 149 |
| Neumayer-Petersen | 1885 | − 3157 | − 248 | + 603 | − 53 | + 288 | − 75 | + 65 | + 146 |
| Dyson-Furner . . | 1922 | − 3095 | − 226 | + 592 | − 89 | + 299 | − 124 | + 144 | + 84 |
| Afanasieva . . . . | 1945 | − 3032 | − 229 | + 590 | − 125 | + 288 | − 146 | + 150 | + 48 |
| Vestine-Lange . . | 1945 | − 3057 | − 211 | + 581 | − 127 | + 296 | − 166 | + 164 | + 54 |

Third order coefficients.

| Source | Epoch | $g_3^0$ | $g_3^1$ | $h_3^1$ | $g_3^2$ | $h_3^2$ | $g_3^3$ | $h_3^3$ |
|---|---|---|---|---|---|---|---|---|
| Neumayer-Petersen | 1885 | + 98 | − 129 | − 24 | + 144 | + 2 | + 41 | + 70 |
| Vestine-Lange . . | 1945 | + 115 | − 173 | − 52 | + 121 | + 18 | + 88 | + 3 |

Macht[1] argues from a comparison of Vestine's analyses that such a minimum is not likely to have occurred before about 1940.

The higher harmonics of the main field are not large but if the field is extrapolated to the base of the mantle[2] the contribution of the higher harmonics becomes important and Elsasser [4] shows that because of the comparative uncertainty with which higher harmonics of the field are known the field cannot accurately be extrapolated to this depth, even if there are no electric currents flowing in the lower mantle.

For some purposes it is convenient to express positions on the earth's surface or above it in terms of spherical coordinates related to the dipole axis rather than to the axis of rotation. Convenient tables have been prepared by Vestine and Knapp[3] to enable this transformation to be made.

**8. Possible existence of external field.** Although the main geomagnetic field has been freed[4] as far as possible from the effects of the transient magnetic fields originating in the ionosphere or in ring currents encircling the earth, the existence of small components of *external* origin has sometimes been claimed. It is possible that small steady fields rotating with the earth could be detected by spherical harmonic analysis. The early analyses did in fact show such a field amounting to a few per cent of the main field but that this was due to defective data seems likely[5] and on the whole later analyses show smaller and smaller effects of this kind. Of more interest is the possible existence of fields which do *not* rotate with the earth. The surface field of the sun has never been claimed to be more than 50 gauss at the pole and assuming this decreases outside the sun according to an inverse cube law this field at the earth should be only $10^{-6}$ gauss. However,

---

[1] H. G. Macht: J. Geophys. Res. **59**, 369 (1954).

[2] See the interesting maps of current-systems at depths 0, 1000, 2000 and 3000 km. within the earth which reproduce the surface main field and its secular variation, given by Vestine et. al. [*14*], pp. 30ff., and 74ff.

[3] Table 511, p. 493, Smithsonian Physical Tables, Ninth Revised edition. Smithsonian Miscellaneous Collections, vol. 120, 1952.

[4] In the analyses of Sect. 7.

[5] The non-reliable "*exterior*" and "*non-potential*" parts found in the geomagnetic field by early analyses have been criticized by F. Dyson and H. Furner [Mon. Not. Roy. Astronom. Soc., Geophys. Suppl. **1**, 76 (1923)], and the fallacy of supposed air-earth currents invented to "explain" the non-potential part has been demonstrated by Ad. Schmidt and J. Bartels [Gerlands Beitr. Geophys. **55**, 292 (1939)].

various lines of astrophysical evidence[1,2,3] lead to the hypothesis that there is a general *galactic magnetic field* of the order of $10^{-5}$ gauss. Certain theories of the origin of cosmic rays and the necessity of finding an explanation of their nearly isotropic distribution as observed at the earth's surface involve general fields of this order. Such a galactic magnetic field if it existed might be larger or smaller by factors of 10 near the solar system, so that it is of interest to know whether there exists a uniform external field of approaching $10^{-4}$ gauss, or about 0.02% of the main field (to take a favourable case). To determine the component by this parallel to the axis of the earth by spherical harmonic analysis would require the geomagnetic field over the earth to be known at any one time in much greater detail than it is at present.

It would not be impossible to search for the component of such a supposed external field perpendicular to the axis of rotation of the earth in the data of the transient variations. Such a field would at any one station result in a discrete period of one sidereal day but would be masked even by the solar quiet day variation. The separation of such a period from the solar daily variation would be very difficult. The field during the night hours at a station is less affected by transient magnetic fields and CHAPMAN and BARTELS [1] suggest that the mean field shortly after midnight of a quiet day may be supposed appreciably free of transient field. Thus the study of the annual variation of this datum is probably most likely to reveal an external field if it exists. LEWIS, McINTOSH and R. A. WATSON[4] show that at Abinger this annual variation in $H$ is $5.5\,\gamma$. It is therefore certain that the external field is no larger than this though it is probable that seasonal effects in the ionosphere are the cause of this annual variation.

**9. The geomagnetic secular change: description.** The secular variation of the field was discovered in 1635 by GELLIBRAND who observed that the variation,

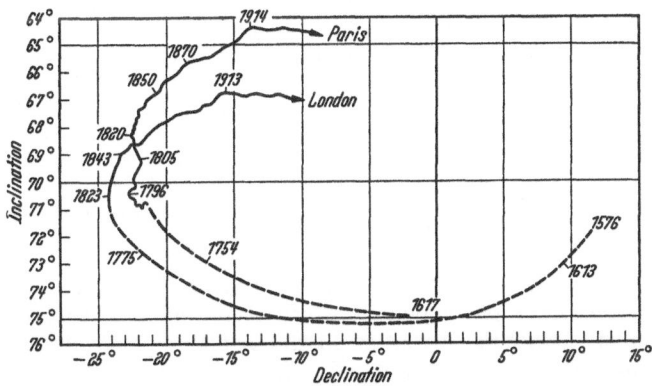

Fig. 5. Secular variation of the field direction in Paris and London. After GAIBAR-PUERTAS [10].

more usually called the declination, of the compass needle changed with time. Changes in the angle of dip and in the field intensity also occur. The data in declination and dip go back further than those of intensity at many places on the earth's surface. The change in the direction of the field vector in space at any place may be best represented by the motion of the point on a sphere cut by

[1] L. SPITZER and J. W. TUKEY: Astrophys. J. **114**, 187 (1951).
[2] E. FERMI: Phys. Rev. **75**, 1169 (1949).
[3] S. CHANDRASEKHAR and E. FERMI: Astrophys. J. **118**, 113 (1953).
[4] R. P. W. LEWIS, D. H. McINTOSH and R. A. WATSON: J. Geophys. Res., **60**, 71 (1955).

a unit vector drawn parallel to the direction of the field. This motion on a sphere may be transferred by the usual projections to a plane curve. Figs. 5 and 6 show examples of such curves over considerable periods.

From such results at any one station, it might be inferred that the magnetic poles precessed round the geographical poles. At London the period of such a precession would be about 480 years. However, the secular variation is not well represented by such a change in the direction of the dipole component, as is seen

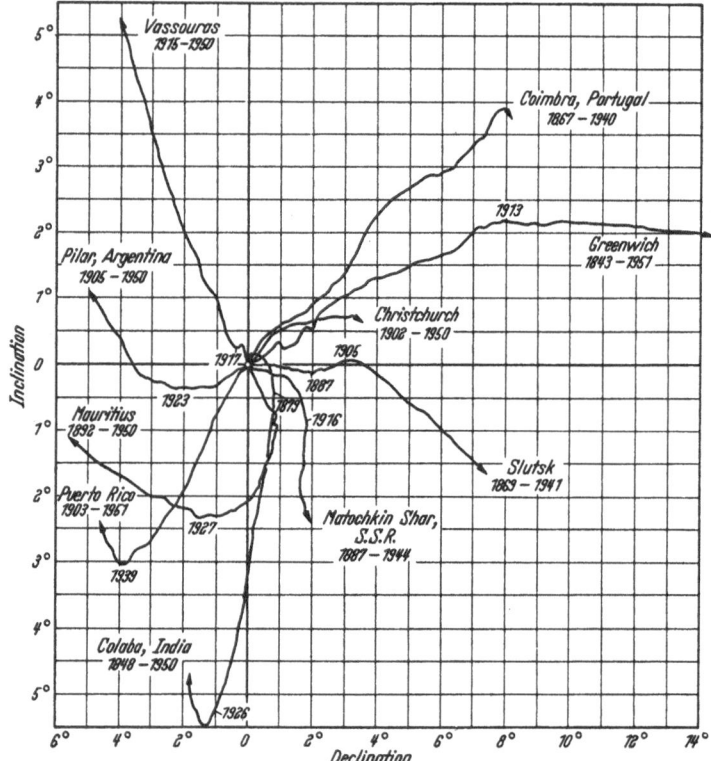

Fig. 6. Declination and inclination measured from arbitrary datum appropriate for each station.
After GAIBAR-PUERTAS [10].

by a comparison of the secular change at a number of stations. It is *regional* rather than a *planetary* phenomenon[1] which is particularly well illustrated in the discussion of the secular variation in the last fifty years by VESTINE, LA-PORTE, COOPER, LANGE and HENDRIX [2]. It is evident also from the spherical harmonic analyses of the secular variation field: no dominant first harmonic is present, see Table 2b.

It will be seen from Figs. 5 and 6 that there is a tendency for the north end of the magnetic vector to describe a *clockwise curve* in space (viewed along the vector). This important discovery was made by BAUER [6] and has again been emphasised in GAIBAR-PUERTAS's study of the geomagnetic secular variation [10], from which Figs. 5 and 6 have been compiled.

The isoporic lines appear to form sets of ovals centering on points of local maximum change, called *isoporic foci*. Considerable changes take place in the

---

[1] J. BARTELS: Preuss. Meteor. Inst. Berlin, 8, No. 2, p. 23, 1925.

Table 2b. *The spherical harmonic analysis of the secular change in $10^{-2}$ gauss/year.*

| Source | Epoch | $g_1^0$ | $g_1^1$ | $h_1^1$ | $g_2^0$ | $g_2^1$ | $h_2^1$ | $g_2^2$ | $h_2^2$ |
|---|---|---|---|---|---|---|---|---|---|
| DYSON-SCHMIDT. . | 1922—1885 | + 20 | − 1 | − 1 | − 10 | + 6 | − 14 | + 21 | − 18 |
| BARTELS . . . . . | 1920—1902 | + 42 | − 9 | + 12 | − 7 | + 8 | − 25 | + 13 | − 8 |
| CARLHEIM-GYLLENSKÖLD. . | 1920—1902 | 0 | + 13 | + 4 | 0 | − 4 | − 12 | + 13 | − 17 |
| VESTINE-LANGE . | 1912.5 | + 25 | + 1 | − 7 | − 7 | − 1 | − 9 | + 24 | − 17 |
| | 1922.5 | + 28 | + 4 | − 7 | − 10 | + 1 | − 14 | + 17 | − 17 |
| | 1932.5 | + 23 | + 1 | − 5 | − 14 | + 1 | − 18 | + 10 | − 14 |
| | 1942.5 | + 9 | + 2 | + 1 | − 18 | 0 | − 20 | + 2 | − 14 |

Third order terms.

| Source | Epoch | $g_3^0$ | $g_3^1$ | $h_3^1$ | $g_3^2$ | $h_3^2$ | $g_3^3$ | $h_3^3$ |
|---|---|---|---|---|---|---|---|---|
| VESTINE-LANGE | 1912.5 | + 6 | − 6 | − 6 | − 1 | + 3 | + 13 | − 11 |
| | 1922.5 | + 7 | − 7 | − 5 | − 2 | + 3 | + 7 | − 11 |
| | 1932.5 | + 4 | − 7 | − 5 | + 1 | + 4 | + 3 | − 13 |
| | 1942.5 | + 2 | − 7 | − 1 | + 3 | + 4 | 0 | − 10 |

general distribution of such lines within a few decades[1], the secular variation for epoch 1942.5 being markedly different from that of earlier epochs in this century (see Fig. 4). The centre of rapid secular change in South Africa seems to have arisen quite suddenly just before the beginning of the present century. It might be surmised that the lifetime of centres of secular changes may be only about a hundred years for in that time the field can change by the order of 10000 gammas which is the rough magnitude of the anomalies in the residual field.

An interesting relationship[2] exists at any point on the earth's surface between the gradient of $\dot{Z}$ (the annual rate of secular variation in $Z$) in any direction $l$ on the earth's surface and the secular change of the horizontal component in that direction $\dot{H}_l$. The relation is as follows

$$\frac{d\dot{Z}/dl}{\dot{H}_l} = \frac{1}{1400} \tag{9.1}$$

where $l$ is measured in kilometers. An explanation of this in terms of spherical harmonic analysis has been suggested[3]. Thus isoporic foci in $Z$ tend to occur near lines of zero change in $H$.

The isoporic foci in $Z$ tend to occur near the local maxima or minima of $H$ of the main field. This is made clear by comparing Fig. 4 and 2.

**10. Westward drift of the geomagnetic field.** Prominent among the characteristics of the secular change of the geomagnetic field is its *westward drift*. HALLEY[4] noted that certain non-axial features of the geomagnetic field drifted westward at a rate of 0.5° per year. VAN BEMMELEN'S [5] construction of isogonic charts from 1550 to 1750 at intervals of 50 years shows that a region of maximum declination existed, which moved from the Baltic to the east Atlantic or about 25° in 250 years. BAUER [6] constructed isapoclinic charts for 1780 and 1885 showing the differences between the observed values of inclination and those appropriate

---

[1] N. H. HECK gave the *D*-isopors for North America, which show a complete change of their pattern about 1910 (Reproduced in [*1*], pp. 124ff.).

[2] A. LUNDBAK: Unpublished dissertation, Copenhagen.

[3] S. K. RUNCORN: J. Geophys. Res. **60**, 231 (1955).

[4] E. HALLEY: Phil. Trans. Roy. Soc. Lond. **17**, 563 (1692).

to an axial dipole field. These show two "subsidiary dip poles", one of magnitude 23 to 24° at 8° N. 65° E. in 1780 and 4° N. 42° E. in 1885 and one of magnitude about 25° at 10° S. 30° W. in 1780 and 20° S. 42° W. in 1885. Carlheim-Gyllensköld[1] from analyses of the field suggested that the nondipole fields were rotating to the west at rates of 0.114°/year for the harmonic $n=1, m=1, 0.26°$/year for the harmonic $n=2$ $m=1$ and 0.795°/year for the harmonic $n=2$, $m=2$.

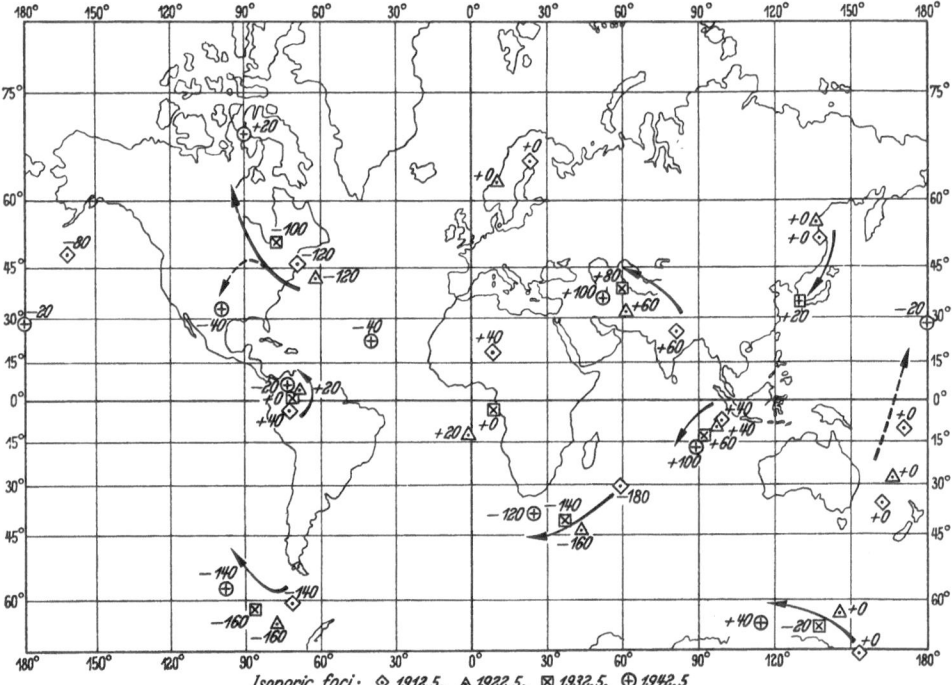

Isoporic foci: ◇ 1912.5, △ 1922.5, ⊠ 1932.5, ⊕ 1942.5

Fig. 7. Movement of isoporic foci in $Z$, 1912 to 1942. Approximate annual rates indicated in gammas. For 1922 and 1942, the foci may be inferred also from Figs. 4a and b. Arrows show general trends of respective foci, 1912 to 1942. After J. A. Fleming [7].

The secular variation charts for the epochs 1912.5, 1922.5, 1932.5 and 1942.5 of Vestine, Laporte, Lange, Cooper and Hendrix [2] show that the isoporic foci have an unmistakable tendency to drift west. This is shown clearly by the map of the isoporic foci in $Z$ due to Fleming [7] reproduced in Fig. 7. Elsasser [4] drew attention to the steady westward motion of the point of zero declination in the western hemisphere at the equator: in the last 400 years its rate has been 0.23°/year.

The secular variation and non-dipole fields for the epochs 1912.5 to 1942.5 have been analysed statistically by Bullard, Freedman, Gellman and Nixon [8] who have calculated, for each component of these fields, the displacement in longitude necessary to produce the best fit between fields at different epochs. They find the mean westward drift of the non-dipole field to be $0.18 \pm 0.015°$/year and of the secular variation field $0.32 \pm 0.067°$/year.

Vestine[2] [31] has shown, by calculating the position of the eccentric dipole between 1830 and 1950, that there is evidence that the velocity of the westward drift fluctuates by about 20% in 50 years (Fig. 8).

---

[1] V. Carlheim-Gyllensköld: Astron. Jakt. Stockholm 5, 1 (1896).
[2] E. H. Vestine: Proc. Nat. Acad. Sci. 38, 1030 (1952).

The secular change arises of course not only from this movement of the non-axial components of the main field with respect to the earth's surface but also from the field variation within the core: in fact, only about one third of the secular variation rates at any one time seem to be due to the former phenomenon. Nevertheless, the contribution of the westward drift to the secular variation causes the more persistent and steady change in the field observed at one place when results over several centuries are considered. Thus the clockwise rotation of the geomagnetic vector at any one station discussed above and shown in Fig. 5 and 6 can be explained as the long-term result of the westward drift as will be discussed in Sect. 14.

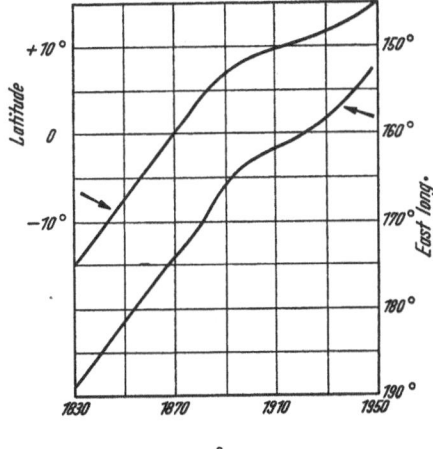

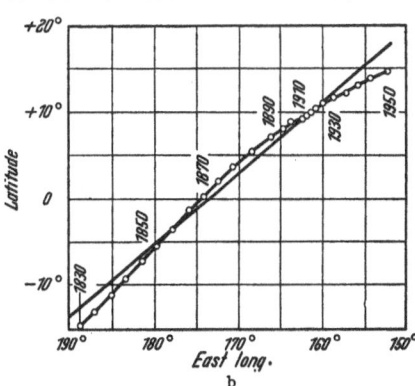

Fig. 8a and b. (a) Motion of eccentric dipole 1830—1950. (b) Shift of latitude and longitude of eccentric dipole, 1830 to 1950. After E. A. Vestine [31].

Vestine[1] [31] on discussing the motion of the eccentric dipole concluded that there is a steady northward drift since 1830 averaging about 0.25° per year (see Fig. 8) and interprets it as a rotation of the core relative to the mantle about an axis perpendicular to the axis of rotation. Such a drift does not, however, appear as a systematic effect in any of the isoporic or isodynamic maps, (see Fig. 4). It is thus somewhat doubtful that such an effect exists.

**11. The spectrum of the geomagnetic secular variation.** The estimation of the energy in different periodicities of the main geomagnetic field is hampered by lack of data. Nevertheless the energy spectrum is of great interest and an attempt has been made by Hughes and Moore[2] to determine it by a combination of data from various sources. Investigations of the modern records of observatories reveal fluctuations of the geomagnetic elements of time scale from a few years to a few decades. The longer records, mainly of declination available from old established observatories, provide periods up to a few hundred years. The presence of fluctuations of 30° in declination with periods of the order of thousands of years is revealed by the palaeomagnetic observations on recent varied clays[3]. From this Hughes and Moore have derived a spectrum and Runcorn [11] has shown that it may be approximated by

$$A = 40 \exp\left(- 13.2\,\omega^{\frac{1}{2}}\right) \tag{11.1}$$

where $A$ is the declination in degrees and $\omega$ is the angular frequency in (yrs)$^{-1}$.

[1] E. H. Vestine: Proc. Nat. Acad. Sci. **38**, 1030 (1952).

[2] H. Hughes and A. F. Moore: Mon. Not. Roy. Astronom. Soc., Geophys. Suppl. (in the press).

[3] See the preceding chapter in this Volume.

Of course each spherical harmonic coefficient of the geomagnetic field will possess its own spectrum and it must be assumed that equation (11.1) refers roughly to those harmonics $n = 3$, 4, which analyses of the secular variation and non-dipole fields show to be most prominent at the present time. Further, the observations of the palaeomagnetic deviations in series of Tertiary and Quaternary lava flows show that the mean over a number of consecutive flows approximates closely to that of an axial dipole field, and the scatter of deviations is not too great[1]. Consequently for long periods of the order of perhaps tens or hundreds of thousands of years, the spectrum must decay to a small value; thus equation (11.1) should be multiplied by a function which causes it to tail off at these periods.

The spectrum of the geomagnetic secular variation may eventually be of importance in the study of rock magnetism as well as in the discussion of the electrical conductivity of the mantle.

**12. The geomagnetic secular variation in the Pacific.** Defining the "Pacific hemisphere" as lying between 120° E. and 80° W., it is noticeable that the secular variation is markedly smaller in that hemisphere than over the rest of the earth's surface (see Fig. 4). Taking the mean without respect to sign of the secular variation rates we find a ratio of about 3:1 between the two hemispheres. Chapman and Bartels [1] carefully consider whether the low values of the secular change and the lack of well formed isoporic foci in the Pacific could be due to paucity of observations and they are inclined to regard the phenomenon as real. Unfortunately measurements do not go back sufficiently far to determine whether this was true in past centuries. Whether this is a permanent feature of the geomagnetic field or only a statistical fluctuation is difficult to establish, but the latest reoccupation of certain of the Carnegie Institution of Washington repeat stations in the South Pacific by New Zealand workers[2] shows the secular variation rates to be still only a few gammas a year. Palaeomagnetic measurements may eventually throw light on the question.

Assuming low values of a secular variation in the Pacific are a permanent feature of the geomagnetic field, it is out of the question that the field is affected by the broad differences in the structure of the crust which there must be between the essentially continental non-Pacific hemisphere and the oceanic Pacific hemisphere. Consequently these differences must be reflected at great depths within the mantle in some systematic difference of physical properties and the assumption, basic to much of the physics of the earth's interior, of an essentially spherically symmetrical earth, may require important modification.

It will be seen also from Bullard's Figure (reproduced in 32528 of Landolt-Börnstein [3]) that the residual field (deviations from the dipole field) is systematically less in the Pacific hemisphere; this was evident already in Bartels' charts (1, p. 660f.).

**13. Secular change impulses.** It is often assumed that the secular change rates remain essentially constant over many years and change gradually. Careful study, however, of observatory records shows that sudden changes in rate of the secular variation occur, which become apparent within a few years. Walker and O'Dea[3] term the phenomena "secular change impulses". Howe[4] noticed such a change in the records of the observatory at Cheltenham in about 1940. A. F. Moore[5] has shown the phenomenon to be frequent in the records of all

---

[1] See the following chapter in this Volume.
[2] A. L. Burrows: Unpublished.
[3] G. B. Walker and P. L. O'Dea: Trans. Amer. Geophys. Un. **33**, 797 (1952).
[4] H. H. Howe: J. Geophys. Res. **46**, 246 (1941).
[5] A. F. Moore: Diss. Cambridge 1955.

observatories. Such studies appear to indicate that there are important features of the secular change with a time scale of only a few years which correlate only over small distances of the earth's surface. It seems possible that the sources of such changes are higher in the mantle than was at one time thought possible.

The difficulty in studying this phenomenon is that studies of the secular variation must be based on annual means. However, there may be some residual effect of the transient variations appearing in the annual means and varying irregularly from year to year, depending on sun spot activity etc. Thus the finer detail of the secular variation field is not easily determined.

**14. Spatial variation of the geomagnetic field.** The determination of the variation of the main field with height above and depth below the earth's surface is of some importance. VESTINE, LAPORTE, LANGE, and SCOTT [14] have computed the main geomagnetic field and the secular change field at various heights in the atmosphere to 500 kms. for 1945. Apart from local anomalies the predominance of the dipole field is such that for practical purposes the elements should vary according to an inverse cube law from the geocentre and it is of theoretical importance to test whether this is so.

APPLETON[1] estimated from ionosphere measurements the magnitude of the magnetic field at a height of 200 kms. and showed that it had decreased by about 10%. More recently measurements of the total field by fluxgate magnetometers in rockets[2] and aircraft have given the expected variation.

The measurement of the *variation of the main field with depth* within the crust is more difficult because of the smaller differences which can be measured and the anomalous gradients arising from rocks. In five mines in Northern England it has been found possible to make measurements in passages sufficiently removed from man-made magnetic disturbances and to make accurate allowance for magnetic gradients arising from local anomalies [12]. Table 2c summarises the

Table 2c. *Results of determination of variation (excess of values at depth over surface values) of the geomagnetic horizontal and vertical force components, $\Delta H$ and $\Delta Z$, in five mines (collieries) in North England.*

| Colliery | depth (ft.) | theoretical | | | experimental | |
|---|---|---|---|---|---|---|
| | | $\Delta Z$ | $\Delta H$ (core) | $\Delta H$ (fundamental) | $\Delta Z$ | $\Delta H$ |
| Hickleton . . . . | 2500 | + 16 | + 7 | − 15 | + 12 ± 2.7 | + 17 ± 2.5 |
| Cadeby . . . . . | 2600 | + 16 | + 7 | − 15 | + 12 ± 2.3 | + 20 ± 2.2 |
| Brodsworth . . . | 2500 | + 16 | + 7 | − 15 | + 5 ± 4.6 | + 21 ± 2.9 |
| Astley Green . . . | 2900 | + 18 | + 8 | − 17 | + 17 ± 2.8 | + 9 ± 2.4 |
| Nook . . . . . . | 2900 | + 18 | + 8 | − 17 | + 12 ± 2.7 | + 8 ± 2.3 |
| Average (referred to 2900 ft.) | | + 18 | + 8 | − 17 | + 13 | + 16 |

results, $\Delta Z$ and $\Delta H$ being the values of the elements at depth $d$ minus the values at the surface. The elimination of the field gradient arising from local anomalies was accomplished by making careful surveys of the surface field in the region where the measurements at depth were made. It was found that the surface field gradients were not greatly different from those arising from the main dipole field. Consequently as the vertical gradients of the field components and the horizontal gradients are related by the fundamental laws of magnetostatics, an

---

[1] E. V. APPLETON: Nature, Lond. **133**, 793 (1934).
[2] E. MAPLE, W. A. BOWEN and S. F. SINGER: J. Geophys. Res. **55**, 115 (1950).

estimate can be made of the deviations produced in the depth measurements by the magnetic fields of local anomalies.

An attempt to determine the variation with depth of the main geomagnetic field by measurements in the ocean was made by Arley et al.[1] but the instrumental difficulties made the results difficult to interpret.

### 15. Physical representations of the residual and secular variation fields.

Spherical harmonic analysis expresses the field as arising from *multipoles* at the geocentre. The regional as opposed to planetary nature of the secular variation field has long been recognized and consequently some analysis which allows the field to be represented in terms of a number of sources remote from the geocentre is likely to be more valuable physically than spherical harmonic analysis.

McNish[2] saw that the non-dipole and the secular variation fields for 1920.5 could be approximated by 14 and 13 *dipoles* respectively directed *radially* on a surface of half the radius of the earth and with equal magnitudes of $10^{24}$ gauss cm.[3] and $1.37.10^{22}$ gauss cm.[3]/year respectively. Elsasser[3] showed that a random distribution of dipoles within the core would reproduce some of the characteristics of the secular variation.

Bullard[4] examined the region of rapid secular variation in South Africa using the data for epoch 1922.5 given by Vestine et al [14] and found that it could be well produced by a *horizontal* dipole 700 kms below the surface of the core. Lowes and Runcorn [13] examined the secular variation field for the same epoch on a world wide scale. The representation of the field in terms of a small number of dipoles depends on the fact that each region of the secular change is likely to be produced by a single current loop of diameter of the order of a thousand kilometers below the boundary of the core. Lowes and Runcorn show that the resulting magnetic field at the surface of the earth would be nearly the same as that produced by a dipole perpendicular to the plane of the loop and slightly displaced from it and that the fields of neighbouring current loops do not appreciably interfere at the surface. A simple graphical method of finding the positions of the dipoles is then possible. If the horizontal components of the secular change field are drawn on a gnomonic projection the lines along which the vectors lie parallel for a considerable distance represent diametral planes in which the sources lie. An examination of the distribution of the total secular variation along these so-called $S$ lines enables the source to be determined. Fig. 9 shows the sources found and the $S$ lines. Table 3 shows that all are *radial dipoles* at a depth of about 0.6 of the radius of the earth. Such dipoles are found to be equivalent to horizontal current loops at the surface of the core.

Lowes[5] has discussed briefly a similar analysis of the residual field. Macht[6] considers a model of the main field for 1945 consisting of two eccentric dipoles; a polar one $(M_p)$ parallel to the axis of rotation and displaced 410 km. from it in the meridian plane 151° E. and 230 km. north of the equatorial plane; and a transverse one $(M_g)$ in the meridian 268° E. and in the equatorial plane about 1030 km. from the geocentre. In the past 120 years $M_p$ has undergone small displacements but while the position of $M_g$ has migrated along nearly a quarter of an elliptical orbit the direction of $M_g$ remained fixed with respect to the earth's

[1] N. Arley, P. Andreasen, J. Espersen and J. Olsen: Nature, Lond. 171, 384 (1953).
[2] A. G. McNish: Trans. Amer. Geophys. Un. 21, 287 (1940).
[3] W. M. Elsasser: Phys. Rev. 60, 876 (1941).
[4] E. C. Bullard: Mon. Not. Roy. Astronom. Soc., Geophys. Suppl. 5, 248 (1948).
[5] F. J. Lowes: Ann. Géophys. 11, 91 (1955).
[6] H. G. Macht: Trans. Amer. Geophys. Un. 32, 555 (1951).

crust. This interesting representation of the field does not, however, suggest any clear physical interpretation.

That the westward drift of the *residual field* is only a third of the *secular variation pattern* at any one time is due to the presence of strong centres of change,

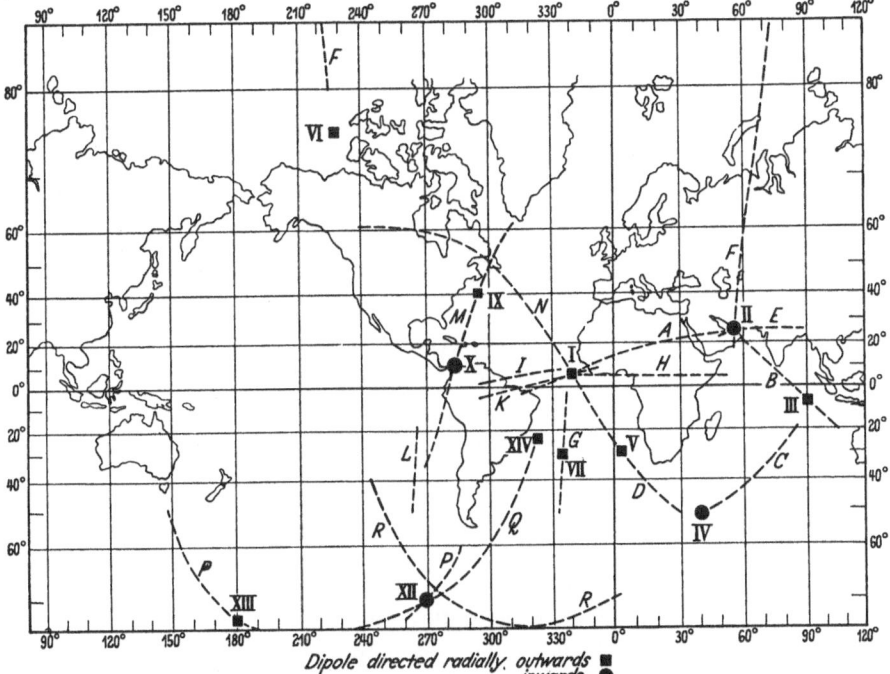

Dipole directed radially. outwards ■<br>inwards ●

Fig. 9. Dipole sources of secular change for epoch 1922.5 showing *S* lines.

such as are revealed in LOWES and RUNCORN's analysis. Many of these centres of change seem to have a comparatively short life of perhaps a hundred years. Consequently the *long* series of observations at certain observatories referred to above reflect more obviously the influence of the westward drift. It may

Table 3. *Sources of secular change for epoch 1922.5.*

| No. | Source | Depth as a fraction of $R$ | Position | Maximum surface field ($\gamma$/year) | |
|-----|--------|--------------------------|----------|------------------------------|---|
| I | vertical dipole | $0.6 \pm 0.05$ | $341°$ E., $5°$ N. $\pm 3°$ | $135 \pm 5$ | Definitely identified |
| II | | $0.6 \pm 0.05$ | $57°$ E., $23°$ N. $\pm 3°$ | $100 \pm 5$ | |
| III | | $0.6 \pm 0.05$ | $92°$ E., $7°$ S. $\pm 3°$ | $60 \pm 5$ | |
| IV | | $0.6 \pm 0.05$ | $46°$ E., $50°$ S. $\pm 3°$ | $160 \pm 10$ | |
| VI | vertical dipole | $0.6 \pm 0.1$ | $210°$ E., $75°$ N. $\pm 5°$ | $65 \pm 10$ | reasonable identification |
| IX | | $0.6 \pm 0.1$ | $295°$ E., $40°$ N. $\pm 5°$ | $125 \pm 10$ | |
| X | | $0.5 \pm 0.1$ | $284°$ E., $10°$ N. $\pm 5°$ | $120 \pm 10$ | |
| V | vertical dipole | $0.6 \pm 0.1$ | $4°$ E., $30°$ S. $\pm 3°$ | $60 \pm 5$ | existence as separate sources doubtful; |
| VII | | $0.6 \pm 0.1$ | $335°$ E., $30°$ S. $\pm 5°$ | $90 \pm 10$ | |
| VIII | | $0.6 \pm 0.1$ | $327°$ E., $75°$ S. $\pm 5°$ | $140 \pm 10$ | |
| XVI | | $0.55 \pm 0.1$ | $323°$ E., $25°$ S. $\pm 5°$ | $60 \pm 5$ | |
| XI | unidentified | — | — | — | |
| XII | unidentified | — | $270°$ E., $70°$ S. $\pm 3°$ | $170 \pm 10$ | in south polar region |
| XIII | vertical dipole | $0.55 \pm 0.1$ | $180°$ E., $73°$ S. $\pm 5°$ | $50 \pm 5$ | |

be simply shown[1] that the *westward* movement of a source, such as a dipole, below a fixed observatory results in a *clockwise* movement of the north pole of the magnetic field vector, and this is true whether the dipole is to the north or south of the station and is true whatever its polarity. BAUER's discovery [6] of this motion must therefore be regarded now as one of the most interesting manifestations of the westward drift.

**16. Energy of the geomagnetic field.** The magnetic energy of the geomagnetic field is small compared to those considered in other physical processes within the earth such as its thermal energy. The simplest case would be to consider the magnetic energy of a dipole field generated by a sinusoidal distribution of electric current over a spherical surface, such as the interface of the mantle and core. The uniform field $H_0$ inside the spherical surface is equal to the polar field. Thus the magnetic energy within the sphere is $V H_0^2/8\pi$, where $V$ is the volume enclosed by the spherical surface. The magnetostatic energy $E$ of a field $H$ derived from a scalar potential $W$, in a volume $\tau$ with a surface $\sigma$ is given by

$$ E = \int_\tau \frac{H \cdot H}{8\pi} \, d\tau = \int_\tau \frac{V(WH)}{8\pi} \, d\tau = \frac{1}{16\pi} \int_\sigma \frac{d}{dr} (W^2) \, d\sigma. $$

It is simple to calculate that the magnetic field energy outside the sphere is equal to twice this value; thus the total magnetic energy is $3 V H_0^2/8\pi$ ergs. Extrapolating the geomagnetic field downwards to the core boundary $H_0$ is 4 gauss; thus the total energy in the field would on this model be $3.4 \times 10^{26}$ ergs.

The Joule heat expended by the currents maintaining the geomagnetic fields is $J^2/\sigma$ ergs per c.c. which can be shown to involve a rise in temperature of the core of only $1°$ C in the whole of geological time.

As will be seen later, this is an underestimate, as we have failed to take into account the presence of larger fields inside the earth which do not emerge from the conducting core. However, the estimate above is simple in principle.

# B. The electrical conductivity within the earth.

**17. The electrical conductivity of the mantle determined from the transient variations.** By spherical harmonic analysis the quiet day daily variation of the geomagnetic field $S_q$ has been separated into an internal part $S_q^i$ and an external part $S_q^e$, the latter being three or four times as great as the former. It is natural to assume that the former is the result of the currents which flow in the earth's mantle due to the varying external field. Comparison of the amplitude and phase relationships of the separate periodicities, (24 hours, 12 hours, etc.) enables the conductivity of the earth to be inferred, assuming it to be zero down to a certain depth and uniform at greater depths. Using this model a value of $3.6 \times 10^{-4}$ ohm$^{-1}$ cm.$^{-1}$ below a depth of 160 miles fits the experimental results.

Taking into account the magnetic storms, each of which extends over a few days, it becomes possible to examine more carefully the variation of the conductivity with depth. LAHIRI and PRICE [15] have accomplished this and Fig. 10 shows their results. Curve a shows the uniform "core" distribution described above. Curves d and e are the only curves compatible with the storm time variations. In each there is a thin shell of relatively high conductivity near the surface of integrated conductivity $5.1 \times 10^3$ ohms$^{-1}$. The conductivity of this surface shell

---

[1] S. K. RUNCORN: Unpublished.

is not uniformly distributed over the earth's surface as it is assumed to arise from the oceans. A further development of the discussion to take this into account by considering induction in a non-uniformly conducting shell is needed. A. T. PRICE[1] (private communication) does not, however, think the general conclusions will be modified.

**18. Theory of the electrical conductivity of the mantle.** Insulators show three types of electrical conductivity[2] each of which is distinguished from metallic conductivity by increasing with rise of temperature.

*Ionic* or *electrolytic conduction* occurs by mobile ions moving through the crystal lattice as a result of defects in it. In the SCHOTTKY process two energies are of importance, half that required to produce a pair of oppositely charged lattice defects and the height of the potential barrier separating adjacent lattice sites occupied by the mobile ions. The excitation energy $E$ of the process is the sum of these two energies. The conductivity ($\sigma$) is related to the absolute temperature $T$ by the equation

$$\sigma = \sigma_0 \exp\left(-E/kT\right) \quad (18.1)$$

where $k$ is BOLTZMANN's constant. The constant $\sigma_0$ is proportional to the number of ion pairs per cm.[3]

*Impurity semi-conduction* arises by the thermal excitation of electrons occupying energy levels produced by impurities in solid solution into the unoccupied con-

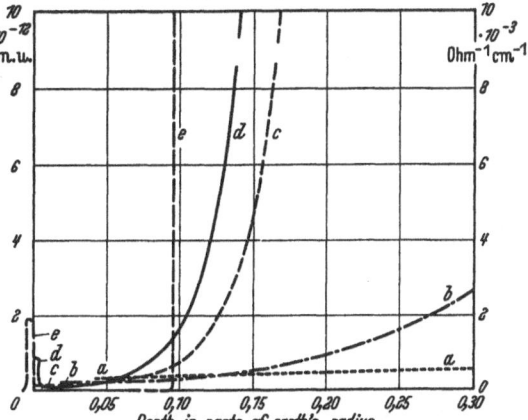

Fig. 10. The electric conductivity of the mantle. Electromagnetic units (1 e.m.u. = 10⁹ ohm⁻¹cm⁻¹). From CHAPMAN-BARTELS [1]. after LAHIRI and PRICE.

duction band of the crystal, or by the excitation of electrons from the occupied valence band of the crystal into unoccupied impurity levels. In the first case the current is carried by electrons; in the second case by positive holes. If the excitation energy between the levels between which the electrons are excited is $E$, the electrical conductivity is given by

$$\sigma = \sigma_0 \exp\left(-E/2kT\right) \quad (18.2)$$

where $\sigma_0$ is proportional to the number of electrons associated with the impurity levels per cm.[3]

*Intrinsic semi-conductors* are those in which the conductivity would be present in the pure state. The occupied valence band and the unoccupied conduction band are separated by a band of forbidden energy $E$. The thermal or optical excitation of electrons across this band renders the crystal conducting. The electrical conductivity is given by

$$\sigma = \sigma_0 \exp\left(-E/2kT\right) \quad (18.3)$$

where $\sigma_0$ is proportional to the mobility of the valence electrons and the square root of their number per cm.[3] Light of frequency $\nu$ greater than $E/h$, where $h$ is PLANCK's constant, is absorbed and should render the crystal photoconducting.

---

[1] See his chapter in Vol. XLIX of this Encyclopedia.
[2] N. F. MOTT and R. W. GURNEY: Electronic Processes in Ionic Crystals, 2nd edit Oxford 1948.

Coster[1] showed that rocks exhibit the typical dependance of conductivity on temperature to be expected from *insulators*. The mantle is thought to consist of a ferromagnesian silicate, the mineral *olivine*, and it is therefore of interest to investigate the variation of electrical conductivity of olivine with temperature and pressure. H. Hughes[2] shows that all three of the conduction mechanisms described above are present in olivine at ordinary pressure. The values of $\sigma_0$ and $E$ for each process are different as shown in Table 4 and consequently at

Table 4. Hughes's *data on the Conductivity of olivine.*

| | $\sigma_0$ | $E$ | Range of Importance |
|---|---|---|---|
| Impurity Semi-Conduction . . . . | $10^{-4}\,\Omega^{-1}\,cm^{-1}$ | 1  e.v. | 600° C |
| Intrinsic Semi-Conduction . . . . | $10^{-1}\,\Omega\,cm^{-1}$ | 3.3 e.v. | 600 to 1100° C |
| Ionic Conduction . . . . . . . . | $10^5\,\Omega^{-1}\,cm^{-1}$ | 3.0 e.v. | 1100° C |

low temperatures only impurity semi-conduction is important, while between 600 and 1100° C intrinsic semi-conduction becomes predominant and finally above 1100° C both these processes become negligible in comparison with ionic conduction with its high value of $\sigma_0$. The conduction mechanism above 1100° C is recognized to be ionic because if an iron electrode is used in contact with the magnesium orthosilicate, iron diffuses into the silicate replacing the magnesium, as can be shown chemically after the crystal has carried appreciable current. The proof that intrinsic semi-conduction is responsible for the conductivity between 600 and 1100° C is that the value of $E$ derived from the conductivity— temperature relation agrees with that derived from the value $\nu$ of the threshold of absorption in the ultraviolet (about 3100 Å).

The application of these ideas to the explanation of the distribution of electrical conductivity in the mantle are clear in a general way but difficulties appear when a quantitative account is attempted. Impurity semi-conduction can never be large enough, on account of the small value of $\sigma_0$, to be of importance except in the crust. Rikitake[3] assumes that ionic conduction accounts for the conductivity and calculates the temperatures throughout the mantle assuming values of $E$ and $\sigma_0$ derived from laboratory experiments. However, some estimate has to be made of the effect of pressure on the processes of conduction before it can be plausibly stated whether ionic conduction or intrinsic semi-conduction is the more important at any given temperature.

Hughes[4] has measured the change of conductivity of a specimen of olivine at temperatures within the ionic region up to pressures of 10000 atmospheres. Assuming the main effect of applying pressure is to alter $E$, he finds $E$ increases by 2% per 10000 atmospheres. Runcorn (in a paper to be published) has likewise measured the effect of pressure on the intrinsic semi-conduction of olivine by the indirect method of finding the effect of pressure on the threshold value $\nu$ of the absorption of ultraviolet light. He infers that $E$ decreases by 0.65% per 10000 atmospheres. These changes agree well with estimates based on simple theoretical ideas[5] relating the variation of $E$ and the lattice constant.

The temperature distribution in the mantle is estimated by assuming that the material of the mantle is everywhere at its melting point except in the outer

[1] H. P. Coster: Mon. Not. Roy. Astronom. Soc., Geophys. Suppl. 5, 193 (1948).
[2] H. Hughes: Thesis, Cambridge University 1953.
[3] T. Rikitake: Bull. Earth. Res. Inst. 29, 61 (1951).
[4] H. Hughes: J. Geophys. Res. 60, 187 (1955).
[5] S. K. Runcorn and D. C. Tozer: To be published.

700 kms (where cooling has been appreciable) or by assuming the temperature gradient is the adiabatic one. JEFFREYS and UFFEN have estimated the temperatures on the former assumption; JEFFREYS[1] by extropolating the labotatory value of the melting point gradient with pressure of olivine and UFFEN[2] by determining the melting point at any depth using the elastic constants of the material derived from the seismic velocities. VERHOOGEN[3] has estimated the temperatures on the assumption of an adiabatic gradient in the mantle. RUNCORN and TOZER[4] show that except for a region just below the crust the temperature is not likely to be sufficient for ionic conduction to play a role, when the effect of pressure is included. Assuming $\sigma_0 = 10\,\Omega^{-1}$ cm.$^{-1}$, RUNCORN [11] shows that the value of $E$ must be smaller than that of olivine under laboratory conditions, if the conductivity of the lower part of the mantle is to be about $1\,\Omega^{-1}$cm.$^{-1}$. The decrease of conductivity with increasing radius, as LAHIRI and PRICE's data show, can only be obtained (unless the temperature gradient is much greater than in any of the models discussed above), by assuming an increase of $E$ with increasing radius in that part of the mantle. Such an anomaly in this region is plausible in view of the rapid variation of seismic velocities between depths of 300 and 1000 km. The change in chemical composition, or phase change, to which seismologists attribute this "20° discontinuity" (p. 100), would also be expected to result in a change of excitation energy.

**19. The electrical conductivity of the core.** In spite of speculation to the contrary the balance of evidence[5] is that the core consists of iron or an ironnickel mixture. ELSASSER [4] attempts to predict its conductivity by writing its dependance on pressure and temperature in the form

$$\sigma = C\,\Theta^2/T \tag{19.1}$$

where $\Theta$ is the Debye temperature, $T$ is the absolute temperature and $C$ a factor depending on the pressure effect on the electronic wave functions. The Debye temperature is proportional to the velocity of sound which in the core is twice that in ordinary iron. Thus taking $T$ to be 9000° C, he finds $\sigma = 1.3 \times 10^4\,\Omega^{-1}$ cm.$^{-1}$ compared with $10^5\,\Omega^{-1}$ cm.$^{-1}$ at ordinary temperatures and pressures. The temperature of the core is more probably half the value ELSASSER took: if so, his estimate should be $3 \cdot 10^4\,\Omega^{-1}$ cm.$^{-1}$. BULLARD [16], allowing for a doubling of resistance on melting, arrives at a value of $10^3\,\Omega^{-1}$ cm.$^{-1}$ later revised to $3 \cdot 10^3\,\Omega$ cm.$^{-1}$.

It may be possible to infer the conductivity from the form of the secular variation spectrum (Sect. 11). Dimensional analysis shows that this spectrum depends on a function of a non-dimensional number, $x \cdot \omega^{\frac{1}{2}} \sigma^{\frac{1}{2}}$ when $x$ is the length scale of the eddies within the core associated with the secular variation. From equation (11.1), making a small correction for the screening of the mantle, we find $\sigma^{\frac{1}{2}} x = 6 \cdot 10^4$ c.g.s.e.m.u. LOWES and RUNCORN [13] have suggested that the current loops associated with secular variation are of the order of 3000 km. across. Putting $x$ equal to this, $\sigma$ is found to be $10\,\Omega^{-1}$ cm.$^{-1}$. In view of the general nature of the argument this value cannot be definitely said to conflict with the above estimates, though ELSASSER's value would seem to be too high.

---

[1] H. JEFFREYS: The earth, 3rd. edit. Cambridge 1952.

[2] R. J. UFFEN: Trans. Amer. Geophys. Un. **33**, 893 (1952).

[3] J. VERHOOGEN: Trans. Amer. Geophys. Un. **32**, 41 (1951).

[4] S. K. RUNCORN and D. C. TOZER: To be published.

[5] F. BIRCH: J. Geophys. Res. **57**, 227 (1952). — See also the chapter by J. A. JACOBS in this Volume, p. 378.

# C. Theories.

**20. Ferromagnetic theory of the main field.** The main dipole field of the earth would be explained if the earth were uniformly magnetized with an intensity of 0.075 gauss or if the crust were magnetized with an intensity of about 8 gauss. Even the former intensity is only rarely exceeded in surface rocks and there is sufficient knowledge of the rocks at the base of the crust to rule out the possibility that the content of iron oxides is likely to be high enough for the lower part of the crust to contribute appreciably to the field.

Although laboratory experiments point to a decrease in the Curie point with pressure, there is no knowledge of whether iron or iron oxides in the *inner* mantle could be ferromagnetic. However, if the dipole field were the result of a uniform magnetization of the inner mantle, the field within the core would be zero and the whole phenomenon of the secular variation would have no explanation.

It is unlikely that the *fluid core* could be ferromagnetic and even if it were the strong magnetic forces between different parts of the magnetized fluid would tend to cause the elementary domains to form closed loops. The inner body, however, is thought to be *solid* iron and under the extreme pressures might be ferromagnetic and it would be natural for its direction of magnetization to line itself up in time with the axis of rotation. Such a possibility cannot be entirely excluded as motions in the liquid core could be postulated to provide the electromagnetic induction, associated with the secular variation. The evidence that reversals in the geomagnetic field occur is, however, a greater stumbling block. The intensity of magnetization of the inner body would need to be about 9 gauss: the generation of electric modes in the fluid core would probably amplify the magnetic field resulting from this.

**21. The possibility that the geomagnetic field arises directly from rotation (Fundamental theories).** The most obvious cause of magnetization resulting from rotation occurs in a ferromagnetic body because the magnetic moments of the atoms are associated with their angular momenta. Thus *gyroscopic effects* cause a partial lining up of the elementary magnetic moments along the axis of rotation. Such a body is uniformly magnetized with an intensity proportional only to the angular velocity independent of the size of the body. Thus though such effects can just be demonstrated in the laboratory at high speeds of rotation, they are utterly *negligible* for the earth.

A possibility discussed by Sutherland[1] and Angenheister[2] is that the earth is negatively charged but Schuster [17] and Schlomka[3] showed that, while the rotation of this charge would result in a dipole field as observed by an observer *not* partaking in the earth's rotation, the field seen on the earth's surface would have an entirely different distribution.

Sutherland and Angenheister assumed that, though the earth was electrostatically neutral so that this difficulty would not appear, it possessed a surface negative charge and a positive volume charge, thus leading to a net dipole moment. This, however, is seen to involve electric fields of the order of $10^8$ volts per cm., considerably greater than the dielectric strength of most materials. Schlomka also examined the possibility of a slight difference between the electric forces between two protons, two electrons and an electron and proton, in the ratio of $(1+\alpha):(1+\beta):1$ respectively. Adjustment of the constants $\alpha$ and $\beta$ allowed the

---

[1] W. Sutherland: Terr. Magn. **5**, 73 (1900); **8**, 49 (1903); **9**, 167 (1904); **13**, 155 (1908)
[2] G. Angenheister: Phys. Z. **26**, 307 (1925).
[3] T. Schlomka: Gerlands Beitr. Geophys. **38**, 357 (1933).

forces of gravitation to be associated with electrostatic forces and also would cause slight electrical polarization which in a rotating body would lead to a magnetic field.

SWANN[1] modified the electromagnetic equations by the addition of small terms depending on the *acceleration* of electric charges as well as their velocity. The modification is such that new effects arise in rotation but not in translation —the latter being easily detectable in the laboratory. SWANN and LONGACRE[2] carried out an experiment based on this theory, in which the surface field varies as $\varrho\, \omega^4\, r^4$ when $\varrho$ is the density, $r$ the radius of the body and $\omega$ the angular velocity. They rotated a copper sphere of 10 cm. radius at 2000 r.p.s. and showed that the predicted field of $10^{-4}$ gauss near the body did *not* exist.

H. A. WILSON[3] suggested that a mass element of $m$ gram moving with velocity $v$ might produce a magnetic field given by

$$H = -\frac{G^{\frac{1}{2}} m}{c\, r^3}\, v \times r \tag{21.1}$$

($G$ = constant of gravitation) in analogy with the field due to a moving charge. On integrating this for a rotating sphere the magnetic moment is found to be given by

$$P = \left(\frac{G^{\frac{1}{2}}}{2\, c}\right) U \tag{21.2}$$

when $U$ is the angular momentum.

WILSON'S postulate was intended to apply to *translational* motion and this he disproved by swinging a heavy body near to a sensitive magnetic detector[4].

Thus the manifest inadequacy of suggested effects based on classical physics and the interest aroused by the success of the general theory of relativity and the attempts to provide a unified field theory embracing gravitational and electromagnetic phenomena, had led to an attempt to explain the geomagnetic field as a new fundamental property of rotating matter on the cosmic scale, an effect too small to have been noticed in the laboratory. SCHUSTER [17] himself summed up in favour of such a view after arguing from considerations of symmetry that suitable electromotive forces to maintain electric currents of the type required to produce an axial magnetic field were not likely to be found.

The discovery by BABCOCK [18] of the magnetic field of the A type star 78 Virginis led BLACKETT [19] and BABCOCK [20] to propose a proportionality between angular momentum and magnetic moment which was roughly obeyed in the case of 78 Virginis, the Sun and the Earth. BLACKETT further showed that the constant of proportionality was of such a magnitude that it might be equal to $\beta\, G^{\frac{1}{2}}/2c$, when $G$ is the constant of gravitation, $c$ the velocity of light and $\beta$ a constant of the order of 1. Doubt as to the value of the radius and angular velocity of 78 Virginis and the exact value of the surface field of the Sun made an exact test of the relationship BLACKETT proposed impossible. The discovery of the apparent reversals of stellar magnetic fields by BABCOCK [21] and the evidence of reversals of the geomagnetic field also threw further doubt on the fundamental character of these fields. It is possible that BABCOCK'S results on reversals of the field may be explained otherwise than by an actual large change of field with respect to the star, but then it is necessary to invoke an "oblique rotator theory" in which it is assumed that there is a considerable angle between

[1] W. F. G. SWANN: Phil. Mag. 3, 1088 (1927).
[2] W. F. G. SWANN and A. LONGACRE: J. Franklin Inst. 206, 421 (1928).
[3] H. A. WILSON: Proc. Roy. Soc. Lond., Ser. A 104, 451 (1923).
[4] H. A. WILSON: Proc. Roy. Soc. Lond. Ser. A 104, 451 (1923).

the axis of rotation and the dipole axis[1]. The evidence for fundamental theories is not therefore *prima facie* impressive, but such an effect if it existed is so important that further investigation is desirable.

Further difficulties arise when an attempt is made to construct an elementary relationship to describe how a particle of matter in a rotating body contributes to the field, supposing only that this relation involves the physical properties, such as the value of gravity, at that point[2].

Nevertheless a direct test of such a fundamental theory in the case of the earth is possible and has been carried out with negative results. The theory has been worked out by RUNCORN [22] and CHAPMAN [23], who compare the variation within the earth of the geomagnetic elements for the *fundamental theory* and for the *core theories*, which suppose no sources of field exist in the outer mantle of the earth. It is necessary to introduce a further assumption in addition to the proportionality between angular momentum and magnetic moment if the variation of the field with depth is to be calculated.

The only simple assumption, which does not evidently conflict with laboratory experience, is that a ring of material in the earth possesses the magnetic effects, though not the electrical effect, of a current of density

$$J_0 = \beta \, G^{\frac{1}{2}} \, \varrho \, a \, \omega \sin \vartheta / c \tag{21.3}$$

where $\varrho$ is the density at a point, $\omega$ the angular velocity of the earth and $a$ the radius.

At a small depth $d$, the values of the elements $Z_d$ and $H_d$ are related to those at the surface $Z_0$ and $H_0$ by the equation

$$Z_d = Z_0(1 + 3\,d/R), \tag{21.4}$$

$$\left. \begin{aligned} H_d &= H_0(1 + 3\,d/R) - 4\pi\,J_0\,d \\ &= H_0(1 - 6.5\,d/R) \end{aligned} \right\} \tag{21.5}$$

when $R$ is the radius of the earth assuming the density of the surface rocks to be 2.7.

For *core* theories on the other hand the relations are as follows

$$Z_d = Z_0(1 + 3\,d/R), \tag{21.6}$$

$$H_d = H_0(1 + 3\,d/R). \tag{21.7}$$

As was explained in Sect. 14, RUNCORN, BENSON, MOORE and GRIFFITHS [12] showed that there was strong evidence in favour of the variation expected by core theories.

A further consequence of the above interpretation of the fundamental core theory is that a small magnetic field should be produced by a dense body rotating with the earth, because of its equivalence to a virtual current element. BLACKETT [24] investigated this using a sensitive astatic magnetometer and proved that a field of the order of magnitude expected was not present near dense bodies at rest in the laboratory.

An inverse effect, which has not yet been tested, although this is experimentally possible, may be simply predicted from the above arguments. By equation (21.1) a ring of matter rotating about an axis perpendicular to a plane would be expected to produce a magnetic field similar to that of a ring of electric current.

---

[1] S. K. RUNCORN: Vistas in Astronomy. Pergamon Press 1955.
[2] T. GOLD: Nature, Lond. **163**, 513 (1949).

Thus if a unit north pole completes a circuit through the ring an amount of work equal to $4\pi J_0$ will have been done and this must appear as a change in the rotation of the ring. From this it can be deduced[1] that a ring of matter placed in a strong magnetic field perpendicular to its plane will experience a torque about an axis parallel to the magnetic field given by

$$T = \frac{G^{\frac{1}{2}}}{c} \varrho\, A\, r \frac{d\Phi}{dt}$$

where $A$ is the cross sectional area of the ring, $\Phi$ the magnetic flux linked and $r$ the radius.

## 22. Mathematical theory of magnetic fields in spheres or spherical shells.

A magnetic field $H$, being a solenoidal vector, may be expressed in terms of a vector potential $A$ by the equation

$$H = \operatorname{curl} A \tag{22.1}$$

$A$ may be expressed in terms of two scalar functions $\psi_E$ and $\psi_M$ known as generating functions.

$$A = -\,r\,\psi_E - r \times \operatorname{grad} \psi_M. \tag{22.2}$$

Thus there arise two different kinds of magnetic field $H_E$ and $H_M$ given by

$$H_E = r \times \operatorname{grad} \psi_E = -\operatorname{curl} r\, \psi_E \tag{22.3}$$

and

$$H_M = \operatorname{curl}^2 r\, \psi_M. \tag{22.4}$$

The electric currents associated with each of these fields $I_E$ and $I_M$ are given by

$$I_E = -\left(\frac{1}{4\pi}\right) \operatorname{curl}^2 r\, \psi_E \tag{22.5}$$

and

$$I_M = \left(\frac{1}{4\pi}\right) \operatorname{curl}^3 r\, \psi_M. \tag{22.6}$$

$I_E$ and $H_E$ are known as the electric mode and $I_M$ and $H_M$ as the magnetic mode. The vector field $H_E$ is known as a toroidal vector as it has no radial component and therefore each line of force lies entirely on a spherical surface. Such fields as $H_M$ are known as poloidal fields.

MAXWELL's equations for a medium of permeability 1 are

$$\operatorname{curl} H = 4\pi I = 4\pi\sigma\,(E + v \times H),$$

$$\operatorname{curl} E = -\frac{dH}{dt}$$

where $H$, $E$ and $I$ are measured in one frame of reference.

In the case where movements in a conducting fluid are being considered it is convenient to expand $\psi_E$ and $\psi_M$ in terms of solid harmonics.

$$T_{nm}(r)\, P_n^m(\vartheta) \frac{\sin m\varphi}{\cos m\varphi},$$

$$S_{nm}(r)\, P_n^m(\vartheta) \frac{\sin m\varphi}{\cos m\varphi},$$

---

[1] S. K. RUNCORN: Diss. Manchester 1949.

the radial functions being, in general, complicated ones. Another solenoidal vector field which can be treated in the same way is the velocity $v$ in an incompressible fluid. In the case of a solid conductor Maxwell's equations become

$$\nabla^2 H + 4\pi\sigma\frac{dH}{dt} = 0$$

and then the generating functions are solutions of the analogous scalar equation

$$\nabla^2\psi + 4\pi\sigma\frac{d\psi}{dt} = 0.$$

If $\psi$ is proportional to $\sin\omega t$ or $\cos\omega t$ then $\psi$ depends on the sums and differences of similar solutions of the equations

$$\nabla^2\psi + k^2\psi = 0$$

where

$$k^2 = \pm 4\pi\sigma i\omega.$$

Solutions of this equation are sums of the terms

$$\psi = j_n(kr)\, P_n^m(\vartheta)\frac{\sin m\varphi}{\cos m\varphi} \qquad \text{finite at origin},$$

$$\psi = n_n(kr)\, P_n^m(\vartheta)\frac{\sin m\varphi}{\cos m\varphi} \qquad \text{singularity at origin},$$

where $j_n$ and $n_n$ are spherical Bessel functions of the first and second order. In addition for a solid conductor equation (22.6) reduces to

$$I_M = \left(\frac{1}{4\pi}\right)k^2\operatorname{curl} r\,\psi_E.$$

Thus an interesting reciprocity exists between the electric and magnetic modes: $H_E$ and $I_M$ are toroidal vector fields and $I_E$ and $H_M$ are poloidal vector fields.

These mathematical methods are described in Stratton [25] and Chapman and Bartels [1] and were introduced into the study of the main geomagnetic field by Elsasser [26], [27]. [28]. Bullard [29] makes use of the method in discussion of induction problems in rotating spheres and spherical shells. Thus fields may exist in the interior of the earth which cannot be directly detected at the surface.

**23. Theory of the westward drift and the irregular fluctuations in the length of the day.** The westward drift of the geomagnetic field has generally been interpreted to imply that the outer core is moving more slowly than the mantle, on the ground that the time constant of free decay of currents in the core is very much longer than that in the mantle.

An explanation of the westward drift of the core has been proposed [8], relating it to the effect of the Coriolis forces on the motions within the core. It is suggested that the interchange of fluid between the outer and inner parts of the core through convection will cause material with smaller transverse velocity to rise to the surface of the core. Thus a differential angular velocity will be set up between the outer and inner parts of the core. Due to electromagnetic interaction between the core and mantle, it was shown that the mean angular velocity of the core will be about the same as that of the mantle and therefore the outer layers of the core will be rotating more slowly than that of the mantle.

Since this view was put forward MUNK and REVELLE[1], VESTINE [31] and RUNCORN [30] have suggested that the westward drift and the irregular fluctuations of the length of the day may be causally related. The latter phenomenon has been extensively studied, most recently by BROUWER[2]. Apparently every few decades the angular velocity of the earth changes by about 1 part in $10^8$, either increasing or decreasing. How sudden these changes are is a difficult matter to examine with the available data but that many of the changes are well established within three or four years seems certain. Doubtless smaller irregular fluctuations occur at much more frequent intervals. MUNK and REVELLE made a careful study of the possible changes at the surface of the earth which could explain these irregular fluctuations and came to the conclusion that none was sufficiently large. They therefore assumed that the cause resided in the turbulent core. VESTINE has found evidence, as was described above, that the westward drift is not constant with time but shows fluctuations of about 20% over the last 50 years. As shown in Fig. 11 the curve of change in the westward drift has a general similarity to that of the fluctuations in the earth's angular velocity.

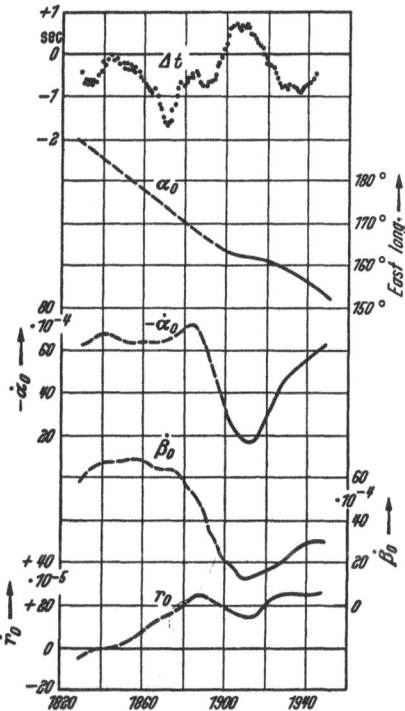

Fig. 11. Fluctuations in length of day $\Delta t$, compared with shift of the geomagnetic eccentric dipole: westward drift given by longitude $\alpha_0$ and its time derivative $\dot\alpha_0$, time derivatives of latitude $\dot\beta_0$ and radial distance $\dot r_0$.
After E. H. VESTINE [31].

Because of the short time in which the changes in the length of the day take place, RUNCORN [30] has suggested appreciable relative motion between parts of the core would not occur during these changes; the strong magnetic fields within the core would appear to prohibit sudden readjustments in its radial variation of angular velocity, such as are postulated by VESTINE [31]. Such a rigid core model leads, assuming conservation of angular momentum, to a relation between the change (increase) in the length of the day ($\delta t$ seconds) and the change (in crease) in the velocity of westward drift ($\delta\Omega$ degrees of longitude per year):

$$\delta T = -0.067\,\delta\Omega. \qquad (23.1)$$

The interpretation of the westward drift is too ambiguous for this relationship to be tested by the observational data, for different methods of computing the westward drift from the secular variation field and from the non-dipole field lead to rather different results. The increase in the length of the day of about 3 milliseconds, which occurred just before the beginning of this century and was thought by DE SITTER to have taken place in 1897, would correspond according to equation (23.1) to a decrease in the westward drift of about 20%. Thus if the fluctuations in the change of the length of the day have been random, as BROUWER urges, and assuming changes of this order of magnitude every 20 years, the westward drift might be expected to change by 100% in 500 years. Such

---

[1] W. MUNK and R. REVELLE: Mon. Not. Roy. Astronom. Soc., Geophys. Suppl. 6, 331 (1952).
[2] D. BROUWER: Astronom. J. 57, 125 (1952). — See the first chapter in this Volume.

considertions seem to imply that the westward drift is unlikely to have any fundamental significance and might very well change in sign in quite short times, merely reflecting the loose coupling of the core and mantle.

The coupling mechanism between the core and mantle has as yet been little investigated. Contrary to the view that electromagnetic coupling is more important, it has recently been suggested[1] that the fluctuating torques on the mantle arise from changes in a rapid radial variation of angular velocity in the outer part of the core, assuming a rather high viscosity of the fluid. However, it seems inevitable that some of the electric currents flow in the inner mantle; there will thus be varying electromagnetic torques exerted by the core on the mantle and vice-versa and these seem adequate to account for the phenomenon. For a uniformly conducting sphere or spherical shell, the electromagnetic torque arising from the mechanical forces exerted on material in which a current is flowing in the presence of a magnetic field can be calculated. Four interactions have to be considered: those between the current and magnetic field of the electric mode, the current and magnetic field of the magnetic mode, the current of the electric mode and the magnetic field of the magnetic mode and vice-versa. It may be proved that the first and fourth interactions lead to zero torque. The second interaction leads to zero torque if the magnetic field is entirely axial, or if it is steady where no magnetic mode currents can exist (in the absence within the conductor of electromotive forces). Further, such interaction exists only between currents and fields derived from the same harmonic of the generating function. It seems probable in the case of the inner mantle that the fluctuating non-axial components of the secular change are insufficient to cause appreciable torques. This argument consequently leads to the assumption of the presence within the inner mantle of considerable electric mode fields. This conclusion is tentative in view of the possibility that the phenomenon of the fluctuations of the length of the day may not be caused electromagnetically.

The existence of appreciable electric modes in the inner mantle implies a current system with radial components. The decrease of the electrical conductivity of the mantle with height causes the leakage current from this system near the surface of the earth to be small but not negligible. Potential differences between the equator and pole of the order of a volt at the bottom of the mantle involve *potential gradients* near the earth's surface of the order of a millivolt per kilometer. These are of the order of the fluctuating potentials due to the currents induced in the crust by the transient variations of the geomagnetic field. Such steady potential differences at the earth's surface, if present, would be of great interest in connection with the physical processes of the earth's interior. Whether they can be detected in the presence of extraneous effects is a matter for future experimental work.

**24. Hydrodynamics of the earth's core.** The recognition by ELSASSER [32] that the short time scale of the geomagnetic secular variation could be understood if the origin of the field was associated with motions in the earth's core was the starting point of the present theory. There seems no alternative but to assume that the motions result from thermal convection, the heat sources being a slight radioactivity in the core or inner body [16], [33]. For convection to occur the adiabatic gradient in the core must be exceeded: consequently the heat conducted out of the core under this gradient in addition to that convected must escape through the mantle. The lower thermal conductivity of the mantle creates some difficulty here and it may be that slow convection must be assumed to occur

---

[1] W. M. ELSASSER and H. TAKEUCHI: Trans. Amer. Geophys. Un. (in the press).

in the mantle in order that this should be so [16]. Jacobs[1] has, however, reviewed the problem and concluded that convection in the mantle is not necessary.

Frenkel[2] pointed out that for large scale motions in the core the viscous and inertial forces were of negligible importance compared with the Coriolis and the electromagnetic forces. The equation of motion of material in the core is therefore:

$$2\,\omega \times v + I \times H/\varrho = -\operatorname{grad} p/\varrho + g\,(1 + T\beta) \qquad (24.1)$$

where $v$ is the velocity vector, $\varrho$ the density, $p$ the pressure, $g$ gravity, $\beta$ the volume coefficient of expansion and $T$ the temperature. The Coriolis force does not absorb energy so that the energy released by the convection must be absorbed by the magnetic field.

Runcorn [30] put forward a physical argument that the effect of rotation will be to line up the axes of the fluid eddies parallel to the axis of the earth's rotation. This would supply a general explanation for the axial character of the geomagnetic field, a conclusion suggested by observations in historic times and confirmed by palaeomagnetic studies[3]. He also attempted to show that if the electromagnetic forces are left out of equation (24.1) a simple mathematical proof of this could be obtained by taking the curl of the equation. This proof would be true in the case of plane boundaries perpendicular to the axis of rotation. However, the presence of the spherical boundary of the core results in even more restricted motions than two dimensional ones[4] [34]. These motions are toroidal ones in which no radial motions occur. This is the natural result of leaving the electromagnetic forces out of account, for the Coriolis forces cannot do any work; thus the motions must be such that energy is not released through the density differences. Consequently there is still no mathematical treatment of the effects of the Coriolis force on radial motions within the core, yet this must be of cardinal importance.

Hide[5] has given a general account of the hydrodynamics of the core.

**25. Induction theories of the secular variation.** The short time scale of the secular variation suggests its explanation lies in the motions of the earth's core. That this field is due to a number of discrete sources within the outer parts of the core has been shown to be plausible. It is natural to suppose these dipole sources to be the result of electromagnetic induction caused by local eddies in the core in a primary field. Bullard[6] considered the centre of the secular variation in South Africa and showed that electromagnetic induction would not be a quantitative explanation of the secular variation unless fields about two orders of magnitude greater than the dipole field were present in the core. Lowes and Runcorn [13] in considering the pattern of the secular variation over the whole globe came to the same conclusion. Since that time the existence of strong electric mode fields in the core has become an essential feature in theories of geomagnetism. The simplest reasonable assumption about the nature of the electric mode magnetic field in the interior is that it vanishes at the equator [30], [34]. If it does, a serious difficulty for the induction theory of the secular variation emerges, as the secular variation foci would tend to avoid the equator. Lowes and Runcorn [13] show that in fact some of the strongest foci lie within

---

[1] J. A. Jacobs: Canad. J. Phys. **31**, 370 (1953). — See also his chapter in this Volume.
[2] J. Frenkel: R. Acad. Sci. USSR. **49**, 98 (1945).
[3] J. Hospers: J. Geology **63**, 59 (1955).
[4] E. C. Bullard, S. K. Runcorn: Trans. Amer. Geophys. Un. **36**, 491 1955.
[5] R. Hide: Physics and Chemistry of the earth. Pergamon Press 1955.
[6] E. C. Bullard: Mon. Not. Roy. Astronom. Soc., Geophys. Suppl. **5**, 248 (1948).

a few degrees of it. Thus physical models of the secular variation are still disappointingly vague.

Lowes and Runcorn [13] attempted the interpretation of another feature of the secular variation; the fact that the dipoles are all radial. They suggested that the electric currents associated with the smaller periods of the order of a hundred years would be confined by electromagnetic induction to a small depth near the surface of the core. This results from the well known skin effect, that the depth of penetration of a field varying with angular frequency $\omega$ in a medium of electrical conductivity $\sigma$ will be of the order of $1/\sqrt{\omega\sigma}$. It is thus inferred that the current loops associated with the secular variation are largely horizontal, and therefore their magnetic effects are nearly those of a radial dipole. It is not, however, clear that these considerations are appropriate in a fluid where motions may result in field disturbances being brought up to the surface of the core from depths considerably greater than the skin depth.

Lowes and Herzenberg[1] have considered the problems presented by induction in the core with great thoroughness. They have shown that the presence of the core boundary near an eddy makes the approximation used by Bullard of considering the induction of a rotating eddy surrounded by an infinite conducting fluid invalid. They have considered therefore the induced fields resulting from rotating a solid conducting cylinder in a conducting fluid asymmetrically distributed about it. They show that in general this leads to an augmented induced field. However, their work is mainly concerned with the steady state so that its application to the core is not without ambiguity.

**26. Dynamo theories.** Self excited dynamos were introduced into cosmical physics by Larmor[2] who suggested that circulatory motions in meridian planes within the Sun opposite in both hemispheres could, by interaction with the dipole magnetic field, generate currents circulating in the same direction in both hemispheres round the axis of rotation and in a sense tending to reinforce the dipole field. The analogy with the self excited dynamo of elementary text books is evident. Cowling[3] considered this suggestion in detail and showed that a contradiction appeared in the mathematical analysis. By symmetry the electric field is zero everywhere and thus Maxwell's equations reduce to

$$\operatorname{curl} \boldsymbol{H} = 4\pi\sigma\boldsymbol{v} \times \boldsymbol{H}. \tag{26.1}$$

There is a line of zero $H$ encircling the axis in the equatorial plane and if equation (26.1) is integrated over an area $\boldsymbol{da}$ (bordered by $\boldsymbol{dl}$) perpendicular to this line, Stokes's theorem given

$$\int_{dl} \boldsymbol{H}\cdot\boldsymbol{dl} = 4\pi\sigma\int_{da}\boldsymbol{v}\times\boldsymbol{H}\cdot\boldsymbol{da}. \tag{26.2}$$

As $da\to0$ it is clear that the R.H.S. of this equation becomes small more rapidly than the L.H.S. Thus unless $v$ is infinite the equation cannot be true along the line of zero $H$. Elsasser [26], [27], [28] and Bullard [16] and Bullard and Gellman [34] sought to circumvent this difficulty by considering motions and fields more complicated than those Larmor chose.

There are two possible approaches to a discussion of dynamo theories. One may seek to set up a model such as Larmor suggested (though a more complicated one) having a definite and suitably chosen set of motions. One then can examine

---

[1] F. J. Lowes and A. Herzenberg: Phil. Trans. Roy. Soc. Lond. (1956).
[2] J. Larmor: Rep. Brit. Assoc. 1919, 159.
[3] T. G. Cowling: Mon. Not. Roy. Astronom. Soc. 94, 39 (1934).

whether, with the appropriate boundary condition, the equation for a steady state dynamo

$$\operatorname{curl}^2 \boldsymbol{H} = 4\pi\sigma \operatorname{curl} \boldsymbol{v} \times \boldsymbol{H} \tag{26.3}$$

possesses solutions of $H$ with a field external to the sphere in which motions are occurring.

PARKER[1] has considered a plausible model of convection in the case in which he supposes the rising fluid motions to spiral about a radial direction under the influence of the Coriolis force and in opposite senses in the two hemispheres. These motions twist the lines of force of the toroidal field into loops in the meridian planes which, because the toroidal field is oppositely directed in the two hemispheres, eventually diffuse from the core as lines of force of the dipole magnetic field. PARKER gives mathematical arguments to support this intuitive approach.

It would be helpful if any general theorems could be found to throw light on the nature of the solutions $H$ given $v$ but none has been found. Consequently BULLARD and GELLMAN [34] have examined the problem by computation.

No simple theorems, such as COWLING'S, have yet been found of general application i.e. when there is no simple symmetry about the axis of rotation.

The other approach is to eschew any definite model considering the motions to be turbulent and to base the argument on the interchange of energy between a magnetic field and a turbulent conducting fluid [4].

ELSASSER [4] discusses in a general way the interchange of energy between mechanical motions of a conducting fluid and a magnetic field. ELSASSER and BATCHELOR[2] show that the fundamental equation to be considered is

$$\frac{d\boldsymbol{H}}{dt} = \frac{1}{4\pi\sigma} \nabla^2 \boldsymbol{H} - (\boldsymbol{v} \cdot \nabla) + (\boldsymbol{H} \cdot \nabla)\boldsymbol{v}.$$

The change in field at a point is the sum of those terms. The first term on the right hand side gives the effect of diffusion which is the smaller the higher the conductivity. The second term gives the change arising from the movement of the field with the fluid. The third term represents the rate of stretching of the lines of force which may be sufficient to overcome the dissipating effects of the first term.

BATCHELOR[2] argues that if the field vector $\boldsymbol{H}$ is replaced by the vorticity and $1/4\pi\sigma$ by the kinematic viscosity $\nu$ the hydrodynamic equation for an incompressible fluid is obtained. As it is known that turbulent motions are possible in which the vorticity is maintained, BATCHELOR argues that similar motions will maintain a magnetic field, for sufficiently high $\sigma$. BATCHELOR does not consider the effect of boundaries to the medium and it is not known how they would affect his argument in respect of fields outside the boundary. It is simple to see how if $\sigma$ is infinite the stretching of the lines of force in a moving fluid increases the magnetic field intensities within the fluid to any degree providing energy is available to maintain the motions. However, if the fluid is enclosed within a simple boundary such as a spherical one, the amplification of field occurring within the sphere does not create new lines of force outside it[3]. However, in the case of a finite conductivity some diffusion of lines of force generated by the motion of the fluid would probably occur.

---

[1] E. N. PARKER: Astrophys. J. **122**, 293 (1955).
[2] G. K. BATCHELOR: Proc. Roy. Soc. Lond., Ser. A **201**, 405 (1950)
[3] H. BONDI and T. GOLD: Mon. Not. Roy. Astronom. Soc. **110**, 607 (1951).

**27. Specific electromotive forces in the earth's interior.** Possible causes of electromotive forces within the deep interior of the earth are thermoelectric or chemical. A likely seat for either would be a major contact between dissimilar materials or between the same substances under markedly different physical conditions. Elsasser [32] tried to invoke the latter as a source of thermoelectric potential supposing that convection within the core would involve pressure differences within the liquid. The effects were not quantitatively important. The only surface discontinuities of the former kind known are the core-mantle boundary, the core inner body boundary and possibly that represented by the 20° discontinuity. Potential differences across such spherical surfaces with concentric shells of uniform conductivity on either side have been shown by Runcorn [30] to excite the electric mode but no magnetic mode. Consider the case of steady potential difference across the core-mantle boundary and assume the radius of the core is $a_1$ and the radius of the effectively conducting part of the inner mantle is $a_2$. The electric mode satisfies the boundary conditions. Let the generating function in the core be $\Psi_1$ and in the mantle $\Psi_2$. Then

$$\left[\frac{d}{dr}(r\Psi_1)\right]_{a_1} = \left[\frac{d}{dr}(r\Psi_2)\right]_{a_1} \quad \text{and} \quad \left[\frac{d}{dr}(r\Psi_2)\right]_{a_2} = 0.$$

The potential difference is given by $a(\Psi_1 - \Psi_2)$.

Runcorn shows that potential differences of the order of about 1 volt are sufficient to produce a large enough electric mode magnetic field. Any theory of this kind supposes that the generation of the magnetic mode occurs through the interaction of the convection motions in the earth's core. Quantitative consideration of these theories is difficult in the absence of satisfactory theories of the properties of matter at high pressure.

**28. The possibility of reversal of the dipole field of the earth.** No features of the thermoelectric and dynamo theories of the geomagnetic field suggest an obvious reason for the present polarity of the magnetic field

The study of palaeomagnetism (see the preceding chapter) has produced strong but not as yet decisive evidence in favour of frequent reversals in the polarity of the geomagnetic field through geological time and it is of interest to enquire to what extent such reversals might be expected on the theory of the field, as at present developed. It may be stated in general that there has appeared no fundamental reason why the field should have a definite polarity, though the important role played by the Coriolis force seems to indicate a sufficient reason for its axial character.

Runcorn[1] has considered the generation of a magnetic mode field from an electric mode field (supposed constant) by motion in the core. He shows that the radial variation of the motion plays a very large part in the direction of the dipole field and suggests that otherwise unimportant changes in the type of convection occurring in the core could reverse the polarity of the field. Bullard and Gellman [34] likewise show that the form of the radial functions is of importance in determining the sign of the magnetic field generated.

# References.

[1] Chapman, S., and J. Bartels: Geomagnetism. Oxford 1940.
[2] Vestine, E. H., L. Laporte, C. Cooper, I. Lange and W. C. Hendrix: Carnegie Instn. Publ. No. 578 **1947**.
[3] Landolt-Börnstein: Zahlenwerte, 6. Aufl., Bd. 3, Astronomie und Geophysik, pp. 396 to 425. Berlin; Springer 1952.

[1] S. K. Runcorn: Ann. Géophys. **11**, 73 (1955).

[4] ELSASSER, W. M.: Rev. Mod. Phys. **22**, 1 (1950).
[5] BEMMELEN, W. VAN: Batavia R. Magn. Met. Obs. Suppl. to vol. 21 of Observations, 1899.
[6] BAUER, L. A.: Amer. J. Sci. **50**, 109, 189, 314 (1895).
[7] FLEMING, J. A.: Trans Assoc. Terr. Mag. a. Electr. Inter. Un. Geod, Geophys., Oslo **1948**, 37.
[8] BULLARD, E. C., J. FREEDMAN, O. GELLMAN and M. NIXON: Phil. Trans. Roy. Soc. Lond., Ser. A **243**, 67 (1950).
[9] NEUMAYER, G.: Atlas des Erdmagnetismus. Gotha: Justus Perthes 1891.
[10] GAIBAR-PUERTAS, C.: Observ. del Ebro, Mem. No. 11 **1953**.
[11] RUNCORN, S. K.: Trans. Amer. Geophys. Un. **36**, 191 (1955).
[12] RUNCORN, S. K., A. C. BENSON, A. F. MOORE and D. H. GRIFFITHS: Phil. Trans. Roy. Soc. Lond., Ser. A **244**, 113 (1951).
[13] LOWES, F. J., and S. K. RUNCORN: Phil. Trans. Roy. Soc. Lond., Ser. A **243**, 525 (1951).
[14] VESTINE, E. H., L. LAPORTE, I. LANGE and W. E. SCOTT: Carnegie Instn. Publ. No. 580 **1947**.
[15] LAHIRI, B. N. and A. T. PRICE: Phil. Trans. Roy. Soc. Lond., Ser. A **237**, 509 (1939).
[16] BULLARD, E. C.: Proc. Roy. Soc. Lond., Ser. A **197**, 433 (1949).
[17] SCHUSTER, A.: Proc. Phys. Soc. Lond. **24**, 121 (1912).
[18] BABCOCK, H. W.: Astrophys. J. **105**, 105 (1947).
[19] BLACKETT, P. M. S.: Nature, Lond. **159**, 658 (1947).
[20] BABCOCK, H. W.: Publ. Astr. Soc. Pacif. **59**, 112 (1947).
[21] BABCOCK, H. W.: Nature, Lond. **166**, 249 (1950).
[22] RUNCORN, S. K.: Proc. Phys. Soc. Lond. **61**, 373 (1948).
[23] CHAPMAN, S:. Ann. Géophys. **4**, 109, (1948)
[24] BLACKETT, P. M. S.: Phil. Trans. Roy. Soc. Lond., Ser. A **245** 213, (1952).
[25] J. A. STRATTON: Electromagnetism. New York: McGraw Hill 1941.
[26] ELSASSER, W. M.: Phys. Rev. **69**, 106 (1946).
[27] ELSASSER, W. M.: Phys. Rev. **70**, 202 (1946).
[28] ELSASSER, W. M.: Phys. Rev. **72**, 821 (1947).
[29] BULLARD, E. C.: Proc. Roy. Soc. Lond., Ser. A **197**, 433 (1949).
[30] RUNCORN, S. K.: Trans. Amer. Geophys. Un. **35**, 49 (1954).
[31] VESTINE, E. H.: J. Geophys. Res. **58**, 127 (1953).
[32] ELSASSER, W. M.: Phys. Rev. **55**, 489 (1939).
[33] ELSASSER, W. M.: Trans. Amer. Geophys. Un. **31**, 454 (1950).
[34] BULLARD, E. C. and H. GELLMAN: Phil. Trans. Roy. Soc. Lond., Ser. **247**, 213 (1954).

# Figur der Erde.

Von

KARL JUNG.

Mit 57 Figuren.

## A. Grundbegriffe.

### a) Das Geoid als Figur der Erde.

**1. Erdkörper, Erdkruste, Erdoberfläche.** Mit *Erdkruste* bezeichnet man die Gesamtheit der sehr vielgestaltig aufgebauten, aus festem Material gebildeten äußeren Schichten des *Erdkörpers*. Abweichend vom üblichen Gebrauch sollen bei den folgenden Betrachtungen auch die Wassermassen der Ozeane der Erdkruste und somit dem Erdkörper zugerechnet werden. Dann ist unter *Erdoberfläche* die Grenzfläche zu verstehen, die den Erdkörper von der ihn umgebenden Atmosphäre trennt.

Es gibt verschiedene Definitionen der Erdkruste und verschiedene Angaben über die Tiefe, bis zu der sie herabreicht. Da das Schwerefeld der Erde, von dem viel die Rede sein wird, in enger Beziehung zur Massenverteilung steht, liegt es nahe, mit der Bezeichnung „Erdkruste" diejenigen Schichten zusammenzufassen, die in gleichem Niveau verschiedene Dichten haben. Es wird angenommen, daß man den tieferen Schichten einen ausgeglichenen, hydrostatischen Zustand zuschreiben kann, bei dem sich die Dichte nur in vertikaler Richtung ändert.

Der Erdkörper soll als starr angesehen werden. Periodische Massenverschiebungen, wie die der Gezeiten, werden gesondert betrachtet[1], ebenso die Rotationsgeschwindigkeit[2] und die Polbewegungen, die man bei astronomischen Ortsbestimmungen berücksichtigen muß[3] ([*22*], S. 257). Die Masse der Atmosphäre, etwa $10^{-6}$ der gesamten Erdmasse, ist so gering, daß sie vernachlässigt werden darf.

**2. Oberflächengestalt, Polyeder, Geoid.** Wie der Augenschein zeigt, ist die Oberfläche der Kontinente äußerst vielgestaltig. Sie folgt keinem einfachen mathematischen Gesetz. Man sucht sie zu erfassen, indem man, vom großen ins kleine arbeitend, geeignete Punkte markiert und ihre gegenseitige Lage mit den Methoden der Geodäsie bestimmt. Solche Punkte und ihre Verbindungsgeraden bilden ein *Polyeder*. Es gibt so viele Polyeder wie zusammenhängende geodätische Netze vorhanden sind. Erst wenn es gelingt, die Ozeane mit geodätischen Messungen zu überbrücken, kann ein einheitliches *Erdpolyeder* ausgemessen werden.

Bei der Darstellung von Vermessungsergebnissen pflegt man die Lotlinien der Polyederecken heranzuziehen. Die Richtungswinkel der Lotlinie, geo-

---

[1] J. BARTELS: Gezeitenkräfte. — R. TOMASCHEK: Tides of the solid earth. Dieses Handbuch, Bd. XLVIII.

[2] H. SPENCER-JONES: The rotation of the earth, S. 1 in diesem Band.

[3] Ziff. 39 bis 43.

graphische Breite und Länge, geben die Lage der Eckpunkte an, und der von einer Bezugsfläche und dem Eckpunkt begrenzte Lotlinienabschnitt gibt die Höhe. Letzten Endes bezieht man die Höhen auf das *Meeresniveau*, das als Niveaufläche des Schwerefeldes auf den Lotlinien senkrecht steht und ungefähr mit der wirklichen Meeresoberfläche zusammenfällt.

Auch wenn man von Meereswellen, Gezeiten und meteorologisch bedingten Schwankungen absieht, ist der wirkliche Meeresspiegel keine Niveaufläche. Die Coriolis-Beschleunigungen der Meeresströme und Dichteunterschiede im Meerwasser bewirken Abweichungen, die einige Dezimeter betragen und aus ozeanographischen Beobachtungen abgeschätzt werden können[1]. Jedoch genügen die vorhandenen Beobachtungen noch nicht, ein für die ganze Erde gültiges, einheitliches Meeresniveau festzulegen. Die praktische Geodäsie hilft sich damit aus, daß jedes Land durch Anschluß an gut vermarkte Punkte sein eigenes Bezugsniveau definiert und die Niveauunterschiede dieser Landeshorizonte nachträglich ermittelt werden, wenn aneinander grenzende Netze genügend gemeinsame Punkte haben, wobei die Feinnivellements von besonderer Bedeutung sind.

Dem üblichen Wortsinn entsprechend, müßte man das Erdpolyeder als *Figur der Erde* bezeichnen. Dem steht entgegen, daß man kein einheitliches Polyeder ausmessen kann, auch ist es für Betrachtungen, die sich nicht speziell auf seine Einzelheiten beziehen, viel zu unregelmäßig gestaltet. Es empfiehlt sich, eine einfachere Fläche, die der mathematischen Behandlung zugänglich ist und für die ganze Erde definiert werden kann, als Figur der Erde anzusehen. Es hat sich als zweckmäßig erwiesen, hierfür eine Niveaufläche des Schwerefeldes auszuwählen, die in der Nähe der Meeresoberfläche verläuft. Man nennt sie *Geoid*.

Das *Geoid* unterscheidet sich hinsichtlich seiner Eigenschaften in keiner Weise von den benachbarten Niveauflächen des Schwerefeldes. Es ist eine geschlossene, stetige Fläche ohne Ecken und Kanten mit stetiger Krümmung, wo sich die Dichte stetig ändert, und unstetiger Krümmung an Dichtesprüngen. Im allgemeinen ist es konvex gekrümmt, doch sind in kleinem Bereich konkave Krümmungen möglich, wenn Gesteine von verschiedener Dichte an stark gekrümmten Grenzflächen aneinanderstoßen.

Die unsichere Festlegung des Meeresniveaus und seiner Beziehungen zu den Landeshorizonten verhindert eine scharfe Festlegung des Geoids. Man muß sich damit abfinden, daß sein Abstand von den Landeshorizonten um etwa einen Meter unsicher ist und sein Abstand vom Zentrum der Erde dieselben Unsicherheiten aufweist wie der Erdradius[2].

Die Kräuselungen der Niveauflächen, die von der inhomogenen Verteilung der Erdkrustenmassen verursacht sind, nehmen ab, wenn man sich von der Erde entfernt. Nahe der Erdoberfläche ist diese Abnahme so gering, daß man das Geoid und die ihm benachbarten Niveauflächen als Parallelflächen ansehen kann[3]. Für die Bestimmung der Gestalt des Geoids sind die beschriebenen Unsicherheiten ohne Belang.

Daß der Erdkörper ungefähr die Gestalt einer Kugel hat, ist allgemein bekannt und bei jeder Mondfinsternis an dem stets kreisrunden Erdschatten anschaulich zu sehen. Man wird das Geoid in erster Annäherung als Kugel ansehen können. Eine bessere, sehr weitgehende Annäherung erhält man mit einem Rotationsellipsoid.

---

[1] Ozeanisches Nivellement. G. DIETRICH: Z. Geophys. **12**, 287 (1936).
[2] Ziff. 61.
[3] Man schätze $\partial \zeta / \partial l$ mit (33.5) und (33.7) ab.

## b) Bezugs- und Annäherungsflächen des Geoids.

**3. Schwerepotential, Normalschwere, Schwereanomalie.** Die Niveauflächen des Schwerefeldes, also auch das Geoid, sind Flächen gleichen *Schwerepotentials* $W$, und es kann das Geoid mit

$$W = W_0 = \text{const} \tag{3.1}$$

definiert werden. Hierbei ist $W$ das Potential einer auf die Masseneinheit bezogenen (,,spezifischen'') Kraft und hat im cgs-System die Einheit $\text{erg} \cdot \text{g}^{-1}$.

Die Schwere setzt sich aus der Gravitation der Erdmassen[1] und der Zentrifugalbeschleunigung der Erdrotation zusammen. Beide Felder haben Potentiale, und es gilt

$$W = V + Z \tag{3.2}$$

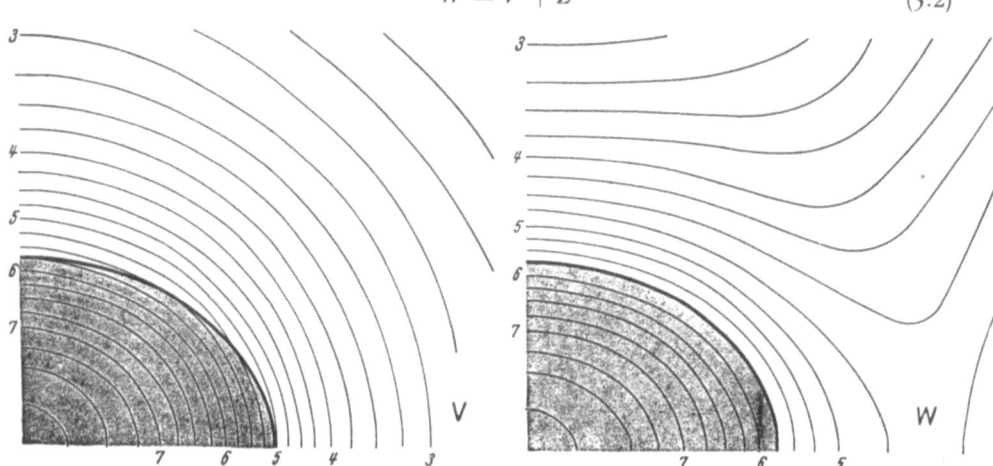

Fig. 1. Gravitationspotential $V$ eines homogenen Rotationsellipsoids in willkürlichen Einheiten.

Fig. 2. Schwerepotential $W$ des Rotationsellipsoids der Fig. 1 wenn es so schnell rotiert, daß seine Oberfläche eine Niveaufläche ist. Einheit willkürlich (wie in Fig. 1).

$V = $ Gravitationspotential, $Z = $ Potential der Zentrifugalbeschleunigung. Eine anschauliche Vorstellung dieser Potentiale vermitteln die Fig. 1 bis 3.

Es ist zweckmäßig, von $V$ das Potential der in der Erdkruste enthaltenen Inhomogenitäten abzusondern:

$$V = N + T \tag{3.3}$$

$N = $ Gravitationspotential der *normalen* Massenverteilung, $T = $ Gravitationspotential der *Störungsmassen*,

$$W = N + T + Z \tag{3.4}$$

und die Bestandteile $N$ und $Z$ zusammenzufassen:

$$N + Z = U \tag{3.5}$$

$$W = U + T \tag{3.6}$$

$U = $ Potential der *Normalschwere*.

Die Ausdrücke $W$ und $U$ beschreiben das Potential in Punkten, die mit der Erde rotieren. In Punkten, die nicht mitrotieren, gelten die Ausdrücke $V$ und $N$.

---

[1] Die übrigen Massen des Weltalls können hier außer Acht gelassen werden; der durchschnittliche Einfluß der Sonne und des Mondes auf die Erdfigur wird dem Gezeiten-Potential zugerechnet und durch die langperiodischen Terme ausgedrückt. Vgl. Bd. XLVIII dieses Handbuches.

Den Schwerevektor $g$ erhält man als Gradient des Schwerepotentials.

$$g = \operatorname{grad} W. \tag{3.7}$$

Der Gradient von $U$ wird *Normalschwere* genannt und oft mit $\gamma$ bezeichnet.

$$\gamma = \operatorname{grad} U. \tag{3.8}$$

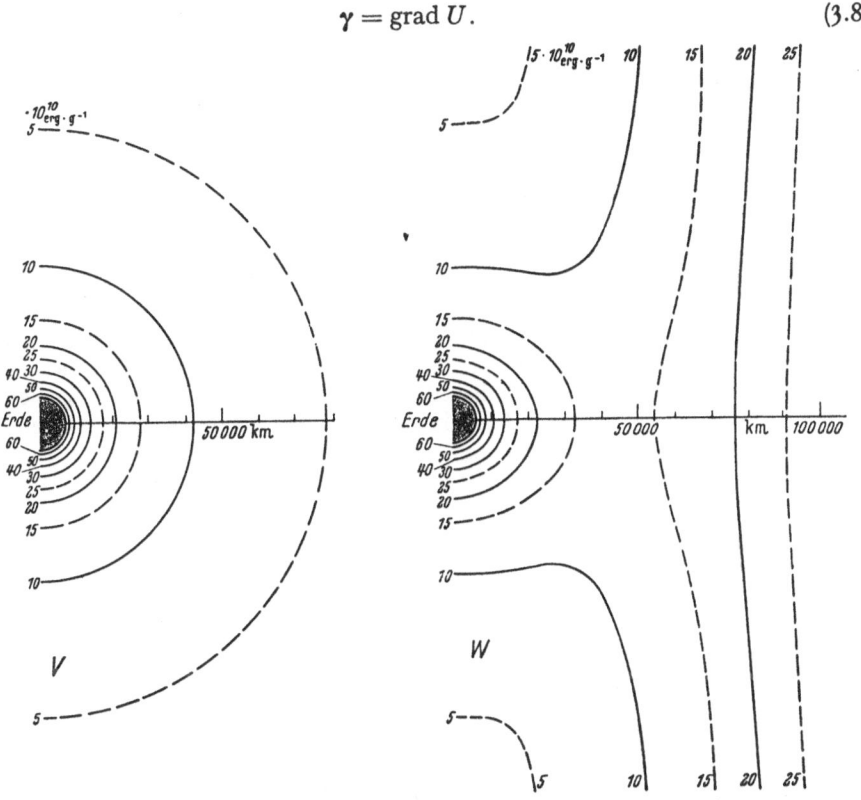

Fig. 3. Gravitationspotential $V$ und Schwerepotential $W$ der Erde in cgs-Einheiten.

Diese Vektoren stehen senkrecht auf den entsprechenden Äquipotentialflächen
— daher die Bezeichnung „Niveaufläche" —, und ihr Betrag ist dem Abstand
benachbarter Niveauflächen umgekehrt proportional.

Die Richtung von $g$ heißt *Lotrichtung*, der Betrag von $g$ wird *Schwereintensität*
oder kurz *Schwere* genannt. Entsprechend bezeichnet man Richtung und Betrag
von $\gamma$ als normale Lotrichtung und Normalschwere. Ausdrücke von der Form
$g - \gamma$ nennt man *Schwereanomalie*.

$g$ und $\gamma$ sind auf die Masseneinheit bezogene („spezifische") Kräfte und
haben die Dimension einer Beschleunigung. Ihre cgs-Einheit ist

$$1 \text{ cm} \cdot \text{sec}^{-2} = 1 \text{ Gal}$$

(Gal: Abkürzung von GALILEI). Betrachtet man Schwereunterschiede, z. B.
Schwereanomalien, so ist es zweckmäßig, mit der Einheit $10^{-3}$ Gal $= 1$ mgal
(Milligal) zu rechnen.

**4. Ellipsoid und Sphäroid, Erdkugel.** Wie man die Erdoberfläche und das
Polyeder auf das Geoid bezieht, kann man das Geoid auf eine Fläche beziehen,
die sich mit einem einfachen mathematischen Ausdruck definieren läßt. Eine

solche *Referenzfläche* kann nach rein mathematischen oder nach potential-theoretischen Gesichtspunkten ausgewählt werden.

Da homogene, langsam rotierende Himmelskörper die Gestalt eines ab-geplatteten Rotationsellipsoids annehmen, dessen kleine Achse mit der Rotations-achse zusammenfällt, liegt es nahe, *Referenzellipsoide* zu wählen, deren kleine Achsen dieselbe Richtung haben wie die Rotationsachse der Erde. Man wird ihre Abmessungen so bestimmen, daß sie das Geoid in dem betrachteten Gebiet mög-lichst gut annähern. Es gibt mindestens so viele Referenzellipsoide wie zusammen-hängende geodätische Netze. Ein einheitliches *Erdellipsoid* erhält man auf diese Weise noch nicht; jedoch dürfte man bei dem heutigen Stand der geodätischen Aufnahme den Abmessungen eines solchen Erdellipsoids recht nahe kommen, wenn man die Abmessungen der Referenzellipsoide in geeigneter Weise mittelt.

Handelt es sich nur darum, die geometrische Gestalt von Geoidstücken zu beschreiben, so kann man die Bedingung der Achsenrichtung fallen lassen und ganz allgemein danach fragen, welches Ellipsoid — das nun gekippt sein kann — sich dem Geoid am besten anpaßt[1].

Die rein geometrischen Verfahren geben keine Auskunft über die physi-kalischen Beziehungen, insbesondere ist die Lage des Schwerpunkts der Erde nicht erkennbar. Für geophysikalische Betrachtungen sind sie nicht geeignet, und es genüge daher, wegen der Einzelheiten auf die reichhaltige Literatur ([5 I],), [13], [15], [16] zu verweisen.

Bei der potentialtheoretischen Entwicklung bezieht man das Geoid $W = W_0$ auf die Niveaufläche $U = W_0$ von gleichem Potential. Wegen ihrer kugelähn-lichen Gestalt bezeichnet man eine solche Bezugsfläche als *Sphäroid*. Auch die Ausdrücke *Niveausphäroid*, *Referenzsphäroid* und *Normalsphäroid* sind gebräuch-lich. Je nach der mathematischen Form des Potentialausdrucks $U$ hat das Sphäroid eine mehr oder weniger einfache Gestalt. Immer ist es einfacher ge-staltet als das Geoid.

Die gebräuchlichen Sphäroide stimmen ganz oder bis auf wenige Meter mit den Rotationsellipsoiden von gleicher Achsenlänge überein. Meist ist man berechtigt, den Ausdruck „Ellipsoid" für „Sphäroid" zu setzen.

Das Sphäroid ist *potentialtheoretisch* und *nicht geologisch* definiert. Gleiches gilt von der Normalschwere. Bei der geologischen Auswertung von Schwere-messungen ist es üblich, die Normalschwere als diejenige Schwere anzusehen, die man auf einer eingeebneten, eine homogene Erdkruste begrenzenden Erd-oberfläche messen würde. Diese geologische Definition der Normalschwere ist nicht mit der potentialtheoretischen Definition identisch, und die Normalschwere, die der geologischen Definition entspricht, ist nicht genau gleich der potential-theoretischen Normalschwere. Der Betrag ihrer Abweichung läßt sich erst ab-schätzen, wenn man die Schwereverteilung auf der ganzen Erde hinreichend genau kennt. Bei dem heutigen Stand der Schweremessungen ist man noch nicht so weit.

In Ermangelung besserer Möglichkeiten bleibt bei erdumspannenden geo-logischen Betrachtungen vorerst nichts anderes übrig, als den Gradient von $U$ als Normalschwere anzunehmen[2]. Bei kleinräumigen Untersuchungen kann es zweckmäßig sein, das ungestörte Schwerefeld jeweils empirisch aus den Messungs-ergebnissen zu ermitteln[3].

---

[1] Jedoch ist jetzt nicht einzusehen, warum man sich nun noch auf Ellipsoide festlegt und nicht auch Stücke von Paraboloiden und zweischaligen Hyperboloiden zuläßt.

[2] Ein Versuch von H. HAALCK, eine der geologischen Definition näher kommende „nor-male Erdgestalt" anzugeben (Veröff. Geodät. Inst. Potsdam Nr. 4, 1950), befriedigt nach Ansicht des Referenten nicht: Ziff. 16.

[3] Diese Bestimmung des „regionalen Feldes" ist ein wichtiger Teil der Auswertung in der angewandten Gravimetrie.

Mit *Erdkugel* pflegt man diejenige Kugel zu bezeichnen, die den gleichen Rauminhalt hat wie das Sphäroid. Der Radius der Erdkugel beträgt ungefähr 6371,2 km. Das Erdellipsoid hat einen Äquatorradius $a$ von ungefähr 6378,4 km und einen Polradius $c$ von ungefähr 6356,9 km [1] (Fig. 4).

**5. Geoidundulation und Lotabweichung.** Das Lot in einem Geoidpunkt $P$ (Fig. 5) schneidet das Sphäroid im Punkt $Q$ [2]. Die Länge $\zeta$ dieses Lotes bezeichnet man als *Geoidundulation*. Sie ist positiv, wenn das Geoid über dem Sphäroid liegt [3]. Der Winkel $\varepsilon$, den das Lot mit der Normalschwere $\gamma$ bildet, wird *Lotabweichung* oder *Lotstörung* genannt.

Geoidundulationen und Lotabweichungen hängen von der Gestalt des Geoids und der gewählten

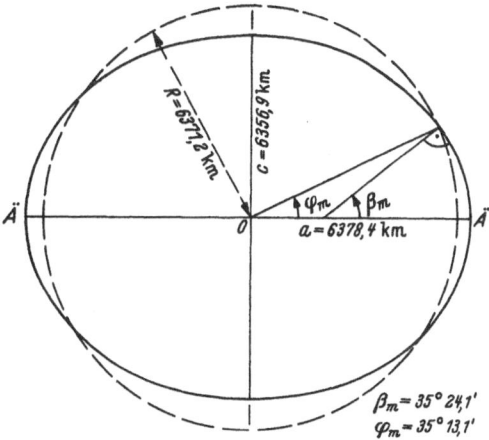

Fig. 4. Erdellipsoid und Erdkugel. $\beta_m$ mittlere geographische Breite, $\varphi_m$ mittlere geozentrische Breite.

Referenzfläche ab. Ist die Referenzfläche dem Geoid nur ungenügend angepaßt (Fig. 6), so haben Geoidundulationen und Lotabweichungen einen regionalen systematischen Gang, der sich nicht aus der Massenverteilung der Erdkruste erklären läßt. Mit Methoden der Ausgleichsrechnung [4] kann eine Verbesserung der Referenzfläche hergeleitet werden [5].

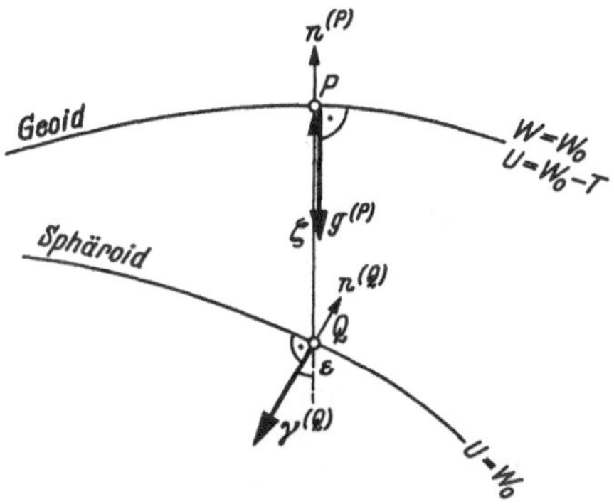

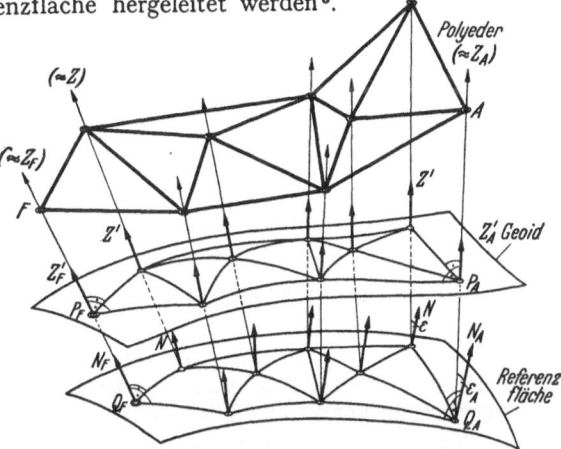

Fig. 5. Geoid und Sphäroid. $\zeta$ Geoidundulation, $g$ Schwere, $\gamma$ Normalschwere. $n^{(P)}$, $n^{(Q)}$ Einheitsvektoren senkrecht zu den Niveauflächen $W = W_0$ und $U = W_0$.

Fig. 6. $Z'$ Zenitrichtung am Geoid ($\approx$ Zenitrichtung am Polyeder), $N$ Ellipsoid-Normale, $\varepsilon$ Lotabweichung. Im Fundamentalpunkt $F$ fallen $Z$ und $N$ zusammen, $\varepsilon$ wächst mit der Entfernung von $F$.

---

[1] Genauere Zahlen in Ziff. 61, 62 und in [22].

[2] Diese Zuordnung findet sich auch in [16]. In der theoretischen Geodäsie wird oft die Zuordnung mit der Sphäroid-Normalen bevorzugt.

[3] In der Literatur kommen auch andere Bezeichnungen (z. B. $N$ und $u$) und das umgekehrte Vorzeichen vor.

[4] K. ARNOLD: Das Minimumprinzip für die Geoidundulationen bei der Bearbeitung astronomisch-geodätischer Netze. Veröff. Geodät. Inst. Potsdam **1955**, Nr. 8.

[5] Soweit es sich um rein geometrische Verfahren zur Bestimmung von Geoidstücken handelt, sei auf die geodätische Literatur verwiesen.

Grundlage der Berechnung der auf das potentialtheoretische Sphäroid bezogenen Geoidundulation ist die Zerlegung des Schwerepotentials

$$W^{(P)} = U^{(P)} + T^{(P)}.\tag{5.1}$$

Entwickelt man $U$ in eine TAYLORsche Reihe und bricht man nach dem linearen Glied ab, so gilt

$$W^{(P)} \approx U^{(Q)} + \mathrm{grad}^{(Q)}\, U \cdot \zeta + T^{(P)}$$

und wegen $U^{(Q)} = W^{(P)}$ ist

$$T^{(P)} \approx -\,\mathrm{grad}^{(Q)}\, U \cdot \zeta = \gamma^{(Q)}\, \zeta \cos \varepsilon,$$

$$\boxed{\zeta \approx \frac{T^{(P)}}{\gamma^{(Q)} \cdot \cos \varepsilon} \approx \frac{T^{(P)}}{\gamma^{(Q)}}.}\tag{5.2}$$

**6. Differentialgleichung der physikalischen Geodäsie, Term von BRUNS.** Um $T^{(P)}$ zu ermitteln, bildet man mit (5.1) den Gradient von $W$ in $P$ und multipliziert ihn skalar mit dem Gradient von $U$ in $Q$.

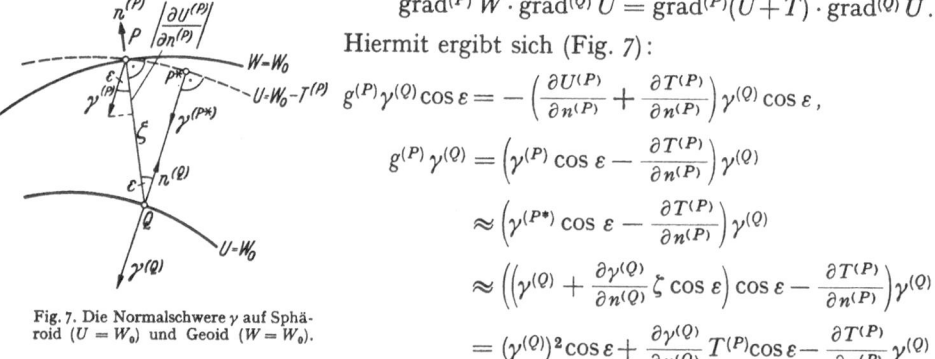

Fig. 7. Die Normalschwere $\gamma$ auf Sphäroid ($U = W_0$) und Geoid ($W = W_0$).

$$\mathrm{grad}^{(P)}\, W \cdot \mathrm{grad}^{(Q)}\, U = \mathrm{grad}^{(P)}(U + T) \cdot \mathrm{grad}^{(Q)}\, U.$$

Hiermit ergibt sich (Fig. 7):

$$g^{(P)} \gamma^{(Q)} \cos \varepsilon = -\left(\frac{\partial U^{(P)}}{\partial n^{(P)}} + \frac{\partial T^{(P)}}{\partial n^{(P)}}\right) \gamma^{(Q)} \cos \varepsilon,$$

$$g^{(P)}\, \gamma^{(Q)} = \left(\gamma^{(P)} \cos \varepsilon - \frac{\partial T^{(P)}}{\partial n^{(P)}}\right) \gamma^{(Q)}$$

$$\approx \left(\gamma^{(P*)} \cos \varepsilon - \frac{\partial T^{(P)}}{\partial n^{(P)}}\right) \gamma^{(Q)}$$

$$\approx \left(\left(\gamma^{(Q)} + \frac{\partial \gamma^{(Q)}}{\partial n^{(Q)}} \zeta \cos \varepsilon\right) \cos \varepsilon - \frac{\partial T^{(P)}}{\partial n^{(P)}}\right) \gamma^{(Q)}$$

$$= (\gamma^{(Q)})^2 \cos \varepsilon + \frac{\partial \gamma^{(Q)}}{\partial n^{(Q)}} T^{(P)} \cos \varepsilon - \frac{\partial T^{(P)}}{\partial n^{(P)}} \gamma^{(Q)}$$

und mit $\cos \varepsilon \approx 1$ erhält man die *Differentialgleichung der physikalischen Geodäsie*

$$\boxed{\frac{\partial T^{(P)}}{\partial n^{(P)}} - \frac{1}{\gamma^{(Q)}} \frac{\partial \gamma^{(Q)}}{\partial n^{(Q)}} T^{(P)} + (g^{(P)} - \gamma^{(Q)}) \approx 0.}\tag{6.1}$$

$g^{(P)} - \gamma^{(Q)}$ wird *scheinbare Schwereanomalie* genannt. Die für *geologische* Auswertungen maßgebliche *wahre Schwereanomalie* ist $g^{(P)} - \gamma^{(P)}$. Die *geodätische* Auswertung kann auf die scheinbare Schwereanomalie gegründet werden.

Die Reduktion der Messungsergebnisse[1] führt auf die scheinbare Anomalie. Die wahre Schwereanomalie muß aus der scheinbaren mit einer Zusatzreduktion berechnet werden. Hierzu hat man (Fig. 7)

$$\gamma^{(P)} \approx \gamma^{(P*)} \approx \gamma^{(Q)} + \frac{\partial \gamma^{(Q)}}{\partial n^{(Q)}} \zeta \cos \varepsilon$$

$$\approx \gamma^{(Q)} + \frac{\partial \gamma^{(Q)}}{\partial n^{(Q)}} \zeta,$$

$$\boxed{g^{(P)} - \gamma^{(P)} \approx (g^{(P)} - \gamma^{(Q)}) - \frac{\partial \gamma^{(Q)}}{\partial n^{(Q)}} \zeta.}\tag{6.2}$$

---

[1] Ziff. 31.

Das letzte Glied auf der rechten Seite wird *Term von* BRUNS genannt. Um ihn zu berechnen, muß die Geoidundulation $\zeta$ bekannt sein. Die Geoidundulation ist also nicht nur von geodätischer Bedeutung: auch für die geologische Auswertung der Schwereanomalien ist es wichtig, $\zeta$ zu kennen.

## B. Theorie der Geoidbestimmung.

### a) Die hypothesenfreie Bestimmung der Erdfigur nach BRUNS.

**7. Die Messungsverfahren und ihre Tragweite.** Wie H. BRUNS gezeigt hat [4], ist eine hypothesenfreie Bestimmung möglich, wenn folgende Messungen mit der erforderlichen Genauigkeit ausgeführt wurden:

1. Astronomische Ortsbestimmungen: geographische Breite (Polhöhe) und Länge einer Polyederecke, Azimut einer Polyederkante.

2. Triangulationen: Horizontalwinkel in den Polyederecken, eine Basislänge.

3. Trigonometrisches Nivellement[1]: Vertikalwinkel (Zenitdistanzen) in den Polyederecken.

4. Geometrisches Nivellement: Höhenunterschiede, ermittelt mit horizontalen Zielungen.

5. Bestimmungen der Schwereintensität.

Horizontalwinkel, Vertikalwinkel und die Basislänge legen Gestalt und Größe des Polyeders fest und geben zugleich die auf die Polyederkanten bezogenen Lotrichtungen in den Polyederecken (Fig. 8 oben). Mit der Längenmessung in einem Polyederpunkt wird dessen Meridianebene in Beziehung zum Nullmeridian gebracht, die Polhöhe und das Azimut vollenden die Orientierung des Polyeders, das nun die richtige Lage in bezug auf Erdachse und Nullmeridian erhält (Fig. 8 Mitte). Der Abstand von der Erdachse und die Entfernnng vom Erdschwerpunkt sind noch zu bestimmen. Hierzu muß man das Polyeder in das System der Niveauflächen einpassen (Fig. 8 unten) und braucht dazu

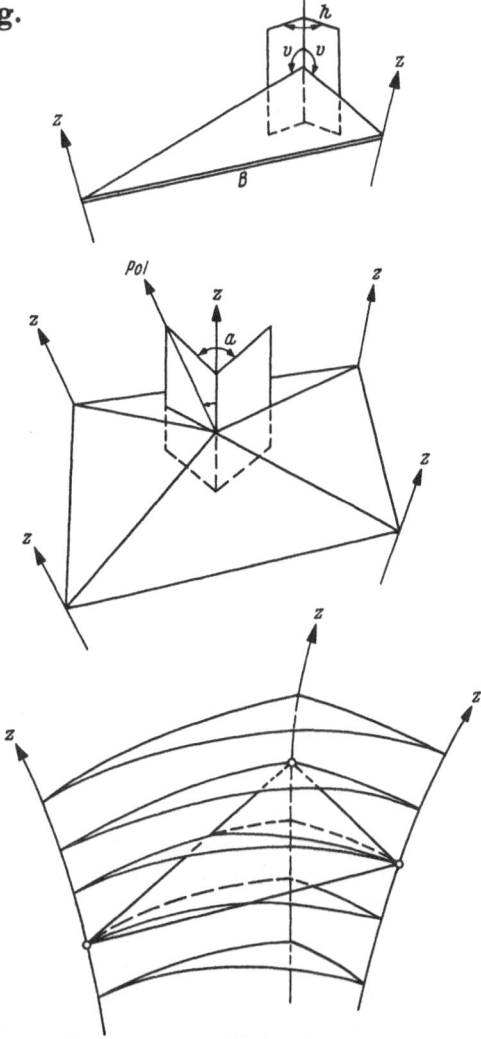

Fig. 8. Die hypothesenfreie Bestimmung der Erdfigur nach BRUNS. Oben: Mit einer Basismessung ($B$), Horizontal- und Vertikalwinkeln ($h$, $v$) sind Größe und Gestalt des Erdpolyeders gegeben. Mitte: Der Zenitabstand des Poles und ein Azimut ($a$) genügen zur Orientierung des Polyeders. Unten: Zur Einpassung in das System der Niveauflächen braucht man geometrische Nivellements und Schweremessungen. $Z$ Zenitrichtung.

die geometrischen Nivellements (Fig. 37, Ziff. 37). Schweremessungen ermöglichen die Berücksichtigung des veränderlichen Abstands der Niveauflächen.

---

[1] Die Bezeichnung „Nivellement" ist irreführend. Denn es handelt sich um rein geometrische Messungen ohne Beziehung auf die Niveauflächen.

Umfaßt das Polyeder die ganze Erde, so ist das Polyeder mit den Niveauflächen hypothesenfrei bestimmt. Vorläufig gibt es aber nur getrennte geodätische Netze, die jeweils einen größeren oder kleineren Teil der Kontinentoberflächen bedecken. Jedes dieser Netze ist in das jeweilige System von Niveauflächen eingepaßt; die Niveauflächensysteme sind aber noch nicht aufeinander bezogen, und auch die Beziehung zum Erdschwerpunkt ist noch nicht hergestellt.

**8. Praktische Grenzen, Notwendigkeit von Hypothesen und Näherungsverfahren.** Diese Unvollständigkeiten, die auf den Lücken des Beobachtungsmaterials beruhen, lassen sich vielleicht mit dem Fortschritt der geodätischen Meßtechnik[1] beseitigen. In der Praxis scheitert aber das BRUNSsche Verfahren an der atmosphärischen Strahlenbrechung, die die genaue Messung von Vertikalwinkeln verhindert. Zwar hat P. GAST gezeigt, daß man das trigonometrische Nivellement mit einer Kombination von Triangulationen, geometrischen Nivellements und astronomischen Ortsbestimmungen ersetzen kann[2] (Fig. 9) und im Prinzip wohl eine Möglichkeit zur Anwendung des BRUNSschen Verfahrens besteht. Bei der großen Zahl der astronomischen Messungen, die erforderlich sind, ist diese Methode schwerfällig und umständlich; sie hat auch keine praktische Bedeutung erlangt.

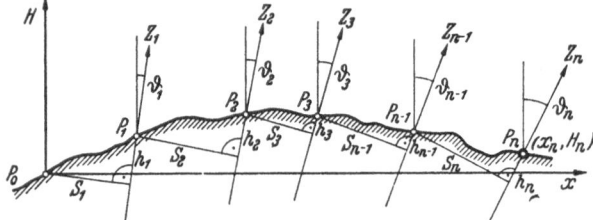

Fig. 9. Ersatz von trigonometrischen Höhenmessungen durch Triangulationen $(s_i)$, astronomische Ortsbestimmungen $(\vartheta_i)$ und geometrische Nivellements $(h_i)$. $Z_i$ Zenitrichtungen. $x_n = \sum_i (s_i \cos \vartheta_i + h_i \sin \vartheta_i)$. $H_n = \sum_i (-s_i \sin \vartheta_i + h_i \cos \vartheta_i)$.

So ist man gezwungen, andere Wege zu suchen, die nun nicht mehr ganz hypothesenfrei sind. Bei ihnen stehen die Schweremessungen und die Potentialtheorie des Schwerefeldes im Vordergrund, und die Beobachtung himmelsmechanischer Erscheinungen, die mit der Gestalt des Erdkörpers zusammenhängen, dürften mit wachsender Verfeinerung der Messungsverfahren noch aufschlußreicher werden. Die Aufstellung der mathematischen Beziehungen wird dadurch erleichtert, daß die Erde ungefähr die Gestalt einer Kugel hat. Einige Annahmen über die Dichteverteilung sind nötig, wenn man die Beziehungen zwischen den Trägheitsmomenten und der Abplattung der Erde entwickelt[3].

### b) Die Potentialtheorie des Schwerefeldes, Anwendung auf die Figur der Erde.

**9. Die Reihenentwicklung des Schwerepotentials.** Die bekannten Darstellungen des Schwerepotentials[4] [5], [8], [10], [13], [16], [18] beruhen auf einer Entwicklung des Gravitationspotentials nach Kugelfunktionen. Sie hat die Form

$$V = \sum_{n=0}^{\infty} \frac{l_0^{n+1}}{l^{n+1}} V_n(\vartheta, \lambda) \quad [l \geq l_0], \tag{9.1}$$

[1] Ziff. 50.
[2] P. GAST: Z. Geophys. **9**, 189 (1933).
[3] Ziff. 38.
[4] Ziff. 17, 33.

wobei

$$V_n(\vartheta, \lambda) = \sum_{\nu=0}^{n} V_{n\nu}(\vartheta, \lambda)$$

$$= \sum_{\nu=0}^{n} ({}_cV_{n\nu} \cos \nu \lambda + {}_sV_{n\nu} \sin \nu \lambda) P_{n\nu}(\cos \vartheta),$$

$l, \vartheta, \lambda$ Radiusvektor, Polabstand, Länge des Aufpunkts; $l_0$ Radius der um den Koordinatenanfang beschriebenen Kugel, die gerade alle Massen einschließt; $P_{n\nu}(\cos \vartheta)$ zugeordnete Kugelfunktionen erster Art, zweckmäßig normiert nach AD. SCHMIDT[1] [27]; ${}_cV_{n\nu}, {}_sV_{n\nu}$ Zahlen-Koeffizienten.

Die Reihe auf der rechten Seite von (9.1) konvergiert auf und außerhalb der Kugel $l = l_0$. In besonderen Fällen, z.B. beim homogenen, abgeplatteten

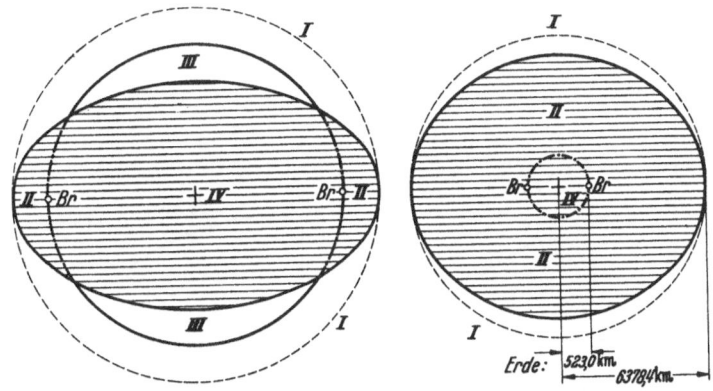

Fig. 10. Konvergenz und Gültigkeit der Entwicklung des Potentials nach Kugelfunktionen (Außenraumformel) beim homogenen, abgeplatteten Rotationsellipsoid. I: konvergent und gültig, II: konvergent und nicht gültig, III und IV: nicht gültig. Beim schwach abgeplatteten Rotationsellipsoid (rechts) fehlt das Gebiet III. *Br* Brennpunkte der Meridianellipsen.

Rotationsellipsoid[2] (Fig. 10), kann sich der Konvergenzbereich auch in das Innere dieser Kugel hinein erstrecken. (9.1) erfüllt die LAPLACEsche Gleichung[3] und stellt das Gravitationspotential im massenfreien Teil des Konvergenzbereichs dar. Im Innern der Massen, wo die POISSONsche Gleichung[4] gilt, ist (9.1) nicht anwendbar, auch dann nicht, wenn die Reihe konvergiert[5].

Denkt man sich die Erde *normalisiert*, die Erdoberfläche eingeebnet und die Unregelmäßigkeiten der Massenverteilung beseitigt, so ist sie nahezu von einem schwach abgeplatteten Rotationsellipsoid begrenzt, und die Flächen gleicher Dichte in ihrem Innern weichen nur wenig von Rotationsellipsoiden ab. Jedoch ist dieser Erdkörper nicht homogen: die Dichtezunahme mit der Tiefe ist beträchtlich (Fig. 39, Ziff. 38). Es ist nicht selbstverständlich, daß man die für $l \geq l_0$ gültigen

---

[1] $P_n(\xi) \equiv P_{n0}(\xi) = \dfrac{1}{2^n n!} \dfrac{d^n}{d\xi^n} (\xi^2 - 1)^n$;    $P_{n\nu}(\xi) = \sqrt{\dfrac{\varkappa_\nu (n-\nu)!}{(n+\nu)!}} \sqrt{1 - \xi^2}^\nu \dfrac{d^\nu}{d\xi^\nu} P_n(\xi).$

$\varkappa_0 = 1, \quad \varkappa_1 = \varkappa_2 = \cdots = \varkappa_n = 2.$    (Zahlenwerte in Anhang I.)

[2] K. JUNG: Gerlands Beitr. Geophys. **29**, 346 (1931).

[3] $\dfrac{\partial^2 V}{\partial x^2} + \dfrac{\partial^2 V}{\partial y^2} + \dfrac{\partial^2 V}{\partial z^2} = 0.$

[4] $\dfrac{\partial^2 V}{\partial x^2} + \dfrac{\partial^2 V}{\partial y^2} + \dfrac{\partial^2 V}{\partial z^2} = -4\pi f \sigma$ ($f$ = Gravitationskonstante, $\sigma$ = Gesteinsdichte).

[5] Vgl. Bemerkung zu Literaturangabe [14].

Entwicklungen bis zur Oberfläche anwenden kann; die Zulässigkeit einer solchen Vereinfachung muß nachgewiesen werden[1].

Hierzu ist es zweckmäßig, rotationssymmetrische Ellipsoid-Koordinaten $\varrho, \overline{\vartheta}, \lambda$ einzuführen[2], und zwar so, daß eine der ellipsoidischen Koordinatenflächen mit dem Erdellipsoid zusammenfällt[3]. Der Koordinatenanfang liegt, wie üblich, im Schwerpunkt der Erde. Fällt die $z$-Achse eines kartesischen Koordinatensystems mit der Rotationsachse zusammen, so gilt (Fig. 11)

$$\left.\begin{aligned} x &= \sqrt{\varrho^2 + e^2} \cdot \sin \overline{\vartheta} \cos \lambda, \\ y &= \sqrt{\varrho^2 + e^2} \cdot \sin \overline{\vartheta} \sin \lambda, \\ z &= \varrho \cdot \cos \overline{\vartheta}. \end{aligned}\right\} \tag{9.2}$$

Die Koordinatenflächen $\varrho = \text{const}$ sind konfokale, abgeplattete Rotationsellipsoide. $\varrho$ ist die kleine Achse des den Aufpunkt enthaltenden Koordinaten-

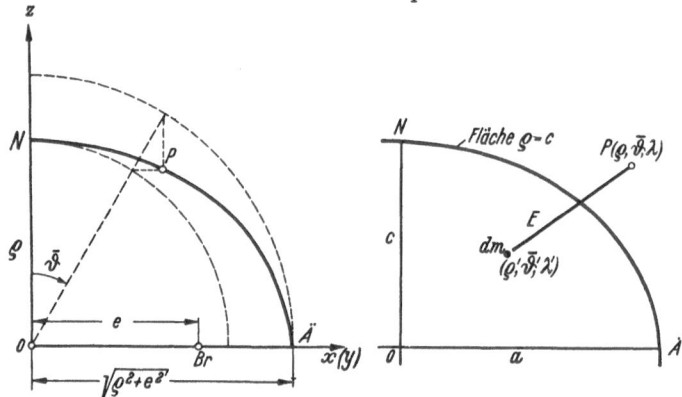

Fig. 11. Ellipsoidkoordinaten. $a$ große Achse der Meridianellipse, $\varrho$ kleine Achse, $e$ halber Brennpunktsabstand, $\overline{\vartheta}$ reduzierter Polabstand.

Ellipsoids, $\overline{\vartheta}$ der *reduzierte Polabstand* (Komplement der *reduzierten Breite*) und $\lambda$ die *Länge* des Aufpunkts. Das Volumenelement ist

$$dv = \frac{\partial(x, y, z)}{\partial(\varrho, \overline{\vartheta}, \lambda)} \, d\varrho \, d\overline{\vartheta} \, d\lambda = (\varrho^2 + e^2 \cos^2 \overline{\vartheta}) \sin \overline{\vartheta} \, d\varrho \, d\overline{\vartheta} \, d\lambda. \tag{9.3}$$

$\varrho = c$ ist das Erdellipsoid, $\varrho = c_0$ sei das Koordinaten-Ellipsoid, das gerade alle Massen der Erde einschließt[4]. Ist $E$ der Abstand des Massenelements $dm$ mit den Koordinaten $\varrho', \overline{\vartheta}', \lambda'$ vom Aufpunkt $(\varrho, \overline{\vartheta}, \lambda)$, so gilt für $\varrho \geq c_0$ [24], [26 II]

$$\frac{1}{E} = \frac{i}{e} \sum_{n=0}^{\infty} \sum_{\nu=0}^{n} (-1)^\nu C_{n\nu} Q_{n\nu}\left(\frac{i\varrho}{e}\right) \cdot P_{n\nu}\left(\frac{i\varrho'}{e}\right) \cdot P_{n\nu}(\cos \overline{\vartheta}) \cdot P_{n\nu}(\cos \overline{\vartheta}') \cdot \cos \nu (\lambda - \lambda') \tag{9.4}$$

---

[1] Die folgenden Ausführungen schließen sich in naheliegenden Verallgemeinerungen geläufigen Gedankengängen an. Sehr elegant sind diese Fragen in [19], Kapitel IV, behandelt.

[2] Anhang III, Ziff. 53.

[3] Nur diese Koordinatenfläche kann eine Niveaufläche der Normalschwere sein (Beispiel: internationales Ellipsoid und internationale Schwereformel).

[4] Bei der Erde ist $c = 6356,9$ km, $\sqrt{c^2 + e^2} = a = 6378,4$ km, $e = 523,0$ km, $e/c = 0,0823$, $(e/c)^2 = 0,00677 = 2/295$, $c_0 - c \approx 8,8$ km, $(c_0 - c)/c \approx 1,4 \cdot 10^{-3}$. Genauere Zahlen Ziff. 62.

$[Q_{n\nu} =$ zugeordnete Kugelfunktionen zweiter Art[1]; $i = \sqrt{-1}$; $C_{n0} = 2n + 1$,
$C_{n1} = C_{n2} = \cdots = C_{nn} = (2n + 1)/2]$ und es ist das Gravitationspotential:

$$V = f \cdot \int\limits_{\varrho'=0}^{c_0} \int\limits_{\bar\vartheta'=0}^{\pi} \int\limits_{\lambda'=0}^{2\pi} \sigma(\varrho', \bar\vartheta', \lambda') \cdot \frac{i}{e} \Bigl( \sum_{n=0}^{\infty} \sum_{\nu=0}^{n} (-1)^\nu C_{n\nu} \, Q_{n\nu}\Bigl(\frac{i\varrho}{e}\Bigr) \cdot P_{n\nu}\Bigl(\frac{i\varrho'}{e}\Bigr) \times$$
$$\times P_{n\nu}(\cos\bar\vartheta) \cdot P_{n\nu}(\cos\bar\vartheta') \cdot \cos\nu(\lambda - \lambda')\Bigr) \cdot (\varrho'^2 + e^2\cos^2\bar\vartheta') \sin\bar\vartheta' \, d\varrho' \, d\bar\vartheta' \, d\lambda' \quad\Biggr\} \quad (9.5)$$

$f =$ Gravitationskonstante ($6{,}67_0 \cdot 10^{-8}$ cgs-Einheiten); $\sigma =$ Gesteinsdichte; $\varrho \geq c_0$.
Mit

$$\cos\nu(\lambda - \lambda') = \cos\nu\lambda \cdot \cos\nu\lambda' + \sin\nu\lambda \cdot \sin\nu\lambda'$$

erhält man

$$V = \sum_{n=0}^{\infty} \sum_{\nu=0}^{n} Q_{n\nu}\Bigl(\frac{i\varrho}{e}\Bigr) \cdot V_{n\nu}(\bar\vartheta, \lambda),$$
$$V_{n\nu}(\bar\vartheta, \lambda) = ({}_cV_{n\nu} \cos\nu\lambda + {}_sV_{n\nu} \sin\nu\lambda) \cdot P_{n\nu}(\cos\bar\vartheta),$$
$$\left.\begin{matrix} {}_cV_{n\nu} \\ {}_sV_{n\nu} \end{matrix}\right\} = f \cdot \frac{i}{e}(-1)^\nu C_{n\nu} \int\limits_{\varrho'=0}^{c_0} \int\limits_{\bar\vartheta'=0}^{\pi} \int\limits_{\lambda'=0}^{2\pi} \sigma(\varrho', \bar\vartheta', \lambda') \begin{Bmatrix} \cos\nu\lambda' \\ \sin\nu\lambda' \end{Bmatrix} \cdot P_{n\nu}\Bigl(\frac{i\varrho'}{e}\Bigr) \times \qquad (9.6)$$
$$\times P_{n\nu}(\cos\bar\vartheta') \cdot (\varrho'^2 + e^2\cos^2\bar\vartheta') \sin\bar\vartheta' \, d\varrho' \, d\bar\vartheta' \, d\lambda'$$
$$C_{n0} = 2n + 1; \quad C_{n1} = C_{n2} = \cdots = C_{nn} = (2n + 1)/2; \quad [\varrho \geq c_0].$$

Führt man die Integrationen aus, so ergibt sich

$$Q_{00}V_{00} = fM \frac{1}{e} \arctan\frac{e}{\varrho},$$

$$M = \text{Gesamtmasse der Erde} = \int\limits_{\varrho'=0}^{c_0} \int\limits_{\bar\vartheta'=0}^{\pi} \int\limits_{\lambda'=0}^{2\pi} \sigma(\varrho', \bar\vartheta', \lambda') \cdot dv',$$

$$Q_{10}V_{10} = 0,$$
$$Q_{11}V_{11} = 0,$$
$$Q_{20}V_{20} = \frac{15}{2}f\frac{1}{e^3} Q_{20}\Bigl(\frac{i\varrho}{e}\Bigr) \cdot \Bigl(C - \frac{A+B}{2} - \frac{1}{3}e^2M\Bigr) \cdot P_{20}(\cos\bar\vartheta)$$

$$= -\frac{45}{4}\frac{f}{e^3}\Bigl(\frac{\varrho}{e}\Bigr)^2 \Bigl(\Bigl(1 + \frac{1}{3}\Bigl(\frac{e}{\varrho}\Bigr)^2\Bigr) \arctan\frac{e}{\varrho} - \frac{e}{\varrho}\Bigr)\Bigl(C - \frac{A+B}{2} - \frac{1}{3}e^2M\Bigr) \cdot P_{20}(\cos\bar\vartheta),$$

---

[1] $Q_n(\xi) \equiv Q_{n0}(\xi) = \frac{1}{2} P_n(\xi) \cdot \ln\frac{\xi+1}{\xi-1} - \frac{2n-1}{1 \cdot n} P_{n-1}(\xi) -$

$$- \frac{2n-5}{3(n-1)} P_{n-3}(\xi) - \frac{2n-9}{5(n-2)} P_{n-5}(\xi) - \cdots$$

[bricht nach dem Glied mit $P_1(\xi)$ oder $P_0$ ab].

$$Q_{n\nu}(\xi) = \sqrt{\frac{\varkappa_\nu(n-\nu)!}{(n+\nu)!}} \sqrt{1-\xi^2}^\nu \, \frac{d^\nu}{d\xi^\nu} Q_n(\xi).$$

Bei der Berechnung von $Q_n\Bigl(\frac{i\varrho}{e}\Bigr)$ beachte man, daß $\frac{1}{2}\ln\frac{i\alpha+1}{i\alpha-1} = -i\arctan\Bigl(\frac{1}{\alpha}\Bigr)$. Da $\frac{1}{\alpha}$
klein ist, kann man in TAYLORsche Reihen entwickeln. Siehe Anhang III, Ziff. 55 und 57.

wobei $A$, $B$, $C$ die Haupt-Trägheitsmomente der Erde bezeichnen ($C$ bezogen auf die *Figurenachse Z*). Hiermit wird

$$
V = f M \frac{1}{e} \arctan \frac{e}{\varrho} + \sum_{n=2}^{\infty} \sum_{\nu=0}^{n} Q_{n\nu}\left(\frac{i\varrho}{e}\right) \cdot \left({}_cV_{n\nu} \cos \nu \lambda + {}_sV_{n\nu} \sin \nu \lambda\right) \cdot P_{n\nu}(\cos \bar{\vartheta}),
$$

$$
{}_cV_{20} = \frac{15}{2} f \frac{i}{e^3}\left(C - \frac{A+B}{2} - \frac{1}{3} e^2 M\right).
$$

(9.7)

Mit dem Potential der Zentrifugalbeschleunigung,

$$
Z = \frac{\omega^2}{2}(x^2 + y^2) = \frac{\omega^2}{3}(\varrho^2 + e^2)\left(1 - P_{20}(\cos \bar{\vartheta})\right) \tag{9.8}
$$

[$\omega$ = Winkelbeschleunigung der Erdrotation ($2\pi/86164$ sec = $7{,}2921 \cdot 10^{-5}$ sec$^{-1}$)] erhält man das Schwerepotential für $\varrho \geq c_0$:

$$
\begin{aligned}
W &= V + Z \\
&= W_{00}(\varrho) + \sum_{n=2}^{\infty} \sum_{\nu=0}^{n} Q_{n\nu}\left(\frac{i\varrho}{e}\right) \cdot \left({}_cW_{n\nu} \cos \nu \lambda + {}_sW_{n\nu} \sin \nu \lambda\right) \cdot P_{n\nu}(\cos \bar{\vartheta}), \\
W_{00}(\varrho) &= V_{00}(\varrho) + \frac{\omega^2}{3}(\varrho^2 + e^2) \\
&= f M \frac{1}{e} \arctan \frac{e}{\varrho} + \frac{\omega^2}{3}(\varrho^2 + e^2), \\
Q_{20}\left(\frac{i\varrho}{e}\right) \cdot {}_cW_{20} &= Q_{20}\left(\frac{i\varrho}{e}\right) \cdot {}_cV_{20} - \frac{\omega^2}{3}(\varrho^2 + e^2) \\
&= \frac{15}{2} f \cdot \frac{i}{e^3} Q_{20}\left(\frac{i\varrho}{e}\right) \cdot \left(C - \frac{A+B}{2} - \frac{1}{3} e^2 M\right) - \frac{\omega^2}{3}(\varrho^2 + e^2), \\
\left.\begin{aligned} {}_cW_{n\nu} &= {}_cV_{n\nu} \\ {}_sW_{n\nu} &= {}_sV_{n\nu} \end{aligned}\right\} &\text{ für alle anderen } n \text{ und } \nu.
\end{aligned}
$$

(9.9)

**10. Das Potential der Normalschwere.** Zweckmäßig ist das Potential $U$ so zu definieren, daß die Niveaufläche $U = W_0$ einem abgeplatteten Rotationsellipsoid angepaßt werden kann. So wird man zwangsläufig auf die Form

$$
U = W_{00}(\varrho) + W_{20}(\varrho, \bar{\vartheta}) + W_{40}(\varrho, \bar{\vartheta}) \tag{10.1}
$$

geführt, und es ist nach (9.9):

$$
\begin{aligned}
U(\varrho, \bar{\vartheta}) &= f M \frac{1}{e} \arctan \frac{e}{\varrho} + \frac{\omega^2}{3}(\varrho^2 + e^2) + \\
&\quad + \left(Q_{20}\left(\frac{i\varrho}{e}\right) \cdot C_{20} - \frac{\omega^2}{3}(\varrho^2 + e^2)\right) \cdot P_{20}(\cos \bar{\vartheta}) + \\
&\quad + Q_{40}\left(\frac{i\varrho}{e}\right) \cdot C_{40} \cdot P_{40}(\cos \bar{\vartheta}); \\
C_{20} &= {}_cV_{20} = \frac{15}{2} f \frac{i}{e^3}\left(C - \frac{A+B}{2} - \frac{1}{3} e^2 M\right), \\
C_{40} &= {}_cW_{40} = \text{const.}
\end{aligned}
$$

(10.2)

Auf dem Niveausphäroid $U = W_0$ sei $\varrho = r(\overline{\vartheta})$. $e$ sei so gewählt, daß das Sphäroid dieselben Achsen hat wie das Ellipsoid $\varrho = c$ (Fig. 12). Dann gilt

$$r(0°) = c, \quad r(90°) = c,$$

$$U(c, 0°) - U(c, 90°) = 0,$$

und es ergibt sich[1]

$$C_{20} = \frac{1}{Q_{20}(ic/e)}\left(\frac{\omega^2}{3}(c^2 + e^2) - \frac{5}{12}Q_{40}(ic/e) \cdot C_{40}\right), \tag{10.3}$$

$$\boxed{\begin{aligned} U(\varrho, \overline{\vartheta}) = {}& fM\frac{1}{e}\arctan\frac{e}{\varrho} + \frac{\omega^2}{3}(\varrho^2 + e^2)\left(1 - P_{20}(\cos\overline{\vartheta})\right) + \\ & + \frac{Q_{20}(i\varrho/e)}{Q_{20}(ic/e)}\left(\frac{\omega^2}{3}(c^2 + e^2) - \frac{5}{12}Q_{40}(ic/e) \cdot C_{40}\right) \cdot P_{20}(\cos\overline{\vartheta}) + \\ & + Q_{40}(i\varrho/e) \cdot C_{40} \cdot P_{40}(\cos\overline{\vartheta}). \end{aligned}} \tag{10.4}$$

Ist $C_{40} = 0$, so ist $U(c, \overline{\vartheta})$ konstant und unabhängig von $\overline{\vartheta}$. In diesem Fall ist das Ellipsoid $\varrho = c$ eine Niveaufläche. $C_{40}$ steht in enger Beziehung zu der Abweichung des Niveausphäroids vom gleichachsigen Rotationsellipsoid[2].

Nach bekannten Sätzen ist[3]

$$\gamma = \mathrm{grad}\, U = \left\{\frac{1}{\sqrt{q_\varrho}} \cdot \frac{\partial U}{\partial \varrho}, \right.$$

$$\left. \frac{1}{\sqrt{q_{\overline{\vartheta}}}} \cdot \frac{\partial U}{\partial \overline{\vartheta}}, \quad 0\right\}$$

mit

$$q_\varrho = \left(\frac{\partial x}{\partial \varrho}\right)^2 + \left(\frac{\partial y}{\partial \varrho}\right)^2 + \left(\frac{\partial z}{\partial \varrho}\right)^2,$$

$$q_{\overline{\vartheta}} = \left(\frac{\partial x}{\partial \overline{\vartheta}}\right)^2 + \left(\frac{\partial y}{\partial \overline{\vartheta}}\right)^2 + \left(\frac{\partial z}{\partial \overline{\vartheta}}\right)^2.$$

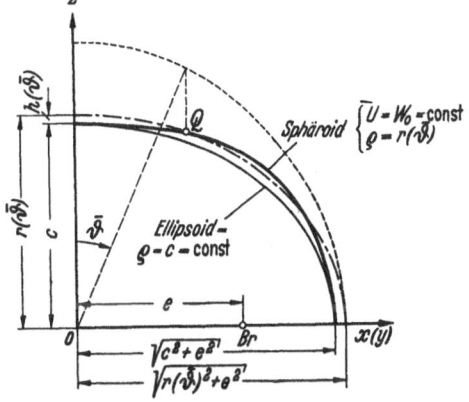

Fig. 12. Abweichung des Sphäroids vom achsengleichen Rotationsellipsoid in Ellipsoidkoordinaten. $Br$ Brennpunkt der Ellipsoid-Meridiane, $Q$ Sphäroidpunkt, $\overline{\vartheta}$ reduzierter Polabstand.

Hiermit erhält man

$$\gamma = \left\{\frac{\partial U}{\partial \varrho}\sqrt{\frac{\varrho^2 + e^2}{\varrho^2 + e^2\cos^2\overline{\vartheta}}}, \quad \frac{\partial U}{\partial \overline{\vartheta}}\frac{1}{\sqrt{\varrho^2 + e^2\cos^2\overline{\vartheta}}}, \quad 0\right\}$$

$$\gamma = \frac{1}{\sqrt{\varrho^2 + e^2\cos^2\overline{\vartheta}}}\sqrt{\left(\frac{\partial U}{\partial \varrho}\right)^2(\varrho^2 + e^2) + \left(\frac{\partial U}{\partial \overline{\vartheta}}\right)^2}.$$

Da sich das Sphäroid nur um wenige Meter vom gleichachsigen Ellipsoid unterscheidet[4], ist $\dfrac{\partial U}{\partial \overline{\vartheta}}$ im Verhältnis zu $\dfrac{\partial U}{\partial \varrho}\sqrt{\varrho^2 + e^2}$ sehr klein[5], und es gilt genau

---

[1] Es ist $P_{20}(\xi) = \frac{3}{2}\left(\xi^2 - \frac{1}{3}\right)$, $P_{40}(\xi) = \frac{35}{8}\left(\xi^4 - \frac{6}{7}\xi^2 + \frac{3}{35}\right)$; $P_{20}(0°) = 1$, $P_{20}(90°) = -1/2$, $P_{40}(0°) = 1$, $P_{40}(90°) = 3/8$.

[2] Gl. (12.10).

[3] [23], S. 181.

[4] Tabelle 11, Ziff. 63.

[5] Größenordnung: $10^{-6}$.

genug für den Absolutbetrag der Normalschwere

$$\gamma(\varrho,\bar{\vartheta}) = -\frac{\partial U}{\partial \varrho}\sqrt{\frac{\varrho^2 + e^2}{\varrho^2 + e^2\cos^2\bar{\vartheta}}},$$

mit

$$-\frac{\partial U}{\partial \varrho} = \frac{fM}{\varrho^2 + e^2} - \frac{2}{3}\omega^2\varrho\left(1 - P_{20}(\cos\bar{\vartheta})\right) -$$
$$-\frac{i}{e}\left\{\frac{Q'_{20}(i\varrho/e)}{Q_{20}(ic/e)}\left[\frac{\omega^2}{3}(c^2 + e^2) - \frac{5}{12}Q_{40}\left(\frac{ic}{e}\right)\cdot C_{40}\right]\right\}\cdot P_{20}(\cos\bar{\vartheta}) -$$
$$-\frac{i}{e}Q'_{40}(i\varrho/e)\cdot C_{40}\cdot P_{40}(\cos\bar{\vartheta}), \tag{10.5}$$

wobei $Q'_{n0}(\xi)$ für $\frac{d}{d\xi}Q_{n0}(\xi)$ gesetzt ist.

Auf dem Sphäroid $\varrho = r(\bar{\vartheta})$ sind einige Vereinfachungen gestattet. Vom zweiten Glied an kann auf den rechten Seiten $c$ für $r$ gesetzt werden, und in dem Glied mit $P_{40}(\cos\bar{\vartheta})$ setzt man die Wurzel genau genug gleich 1.

$$U(r,\bar{\vartheta}) = fM\frac{1}{e}\arctan\frac{e}{r} + \frac{\omega^2}{3}(c^2 + e^2) -$$
$$-\frac{5}{12}Q_{40}(ic/e)\,C_{40}\cdot P_{20}(\cos\bar{\vartheta}) +$$
$$+ Q_{40}(ic/e)\cdot C_{40}\cdot P_{40}(\cos\bar{\vartheta}), \tag{10.6}$$

$$\gamma(r,\bar{\vartheta}) = \frac{fM}{r^2 + e^2}\sqrt{\frac{r^2 + e^2}{r^2 + e^2\cos^2\bar{\vartheta}}} - \frac{2}{3}\omega^2 c\sqrt{\frac{c^2 + e^2}{c^2 + e^2\cos^2\bar{\vartheta}}}\left(1 - P_{20}(\cos\bar{\vartheta})\right) -$$
$$-\frac{i}{e}\frac{Q'_{20}(ic/e)}{Q_{20}(ic/e)}\left[\frac{\omega^2}{3}(c^2 + e^2) - \frac{5}{12}Q_{40}(ic/e)\,C_{40}\right]P_{20}(\cos\bar{\vartheta})\times$$
$$\times\sqrt{\frac{c^2 + e^2}{c^2 + e^2\cos^2\bar{\vartheta}}} - \frac{i}{e}Q'_{40}(ic/e)\cdot C_{40}\cdot P_{40}(\cos\bar{\vartheta}). \tag{10.7}$$

$C_{40}$ wird durch Vergleich mit dem $P_{40}$-Glied der Schwereformel bestimmt (12.10).

**11. Die Schwereformel** [1]. Nachdem die gemessenen Schwerewerte so reduziert sind, daß sie Randwerten auf dem Geoid entsprechen [2], werden die Koeffizienten einer *Schwereformel* berechnet, mit der man die Abhängigkeit der Sphäroidschwere von der Breite darstellt. Oft erscheint die Schwereformel in der Gestalt

$$\gamma = \gamma_{\ddot{a}}(1 + \mathfrak{b}\sin^2\beta - \mathfrak{b}'\sin^2 2\beta) \tag{11.1}$$

$\gamma =$ Schwere auf dem Niveausphäroid *(Normalschwere)*, $\gamma_{\ddot{a}} =$ Äquatorschwere, $\mathfrak{b}, \mathfrak{b}' =$ Zahlenkoeffizienten, $\beta =$ geographische Breite (Fig. 13). Häufig sind auch Formeln von der Form

$$\begin{rcases} \gamma = \gamma_{\ddot{a}}(1 + \mathfrak{b}_2\sin^2\beta + \mathfrak{b}_4\sin^4\beta), \\ \mathfrak{b}_2 = \mathfrak{b} - \mathfrak{b}_4, \\ \mathfrak{b}_4 = 4\mathfrak{b}' \end{rcases}, \tag{11.2}$$

---

[1] Zahlenwerte Anhang IV, Ziff. 64.
[2] Ziff. 31.

gelegentlich tritt die Schwereformel mit Kugelfunktionen auf[1]:

$$\left.\begin{aligned}
\gamma &= B_{00} + B_{20} \cdot P_{20}(\sin\beta) + B_{40} \cdot P_{40}(\sin\beta), \\
B_{00} &= \gamma_a \left(1 + \tfrac{1}{3}\mathfrak{b}_2 + \tfrac{1}{5}\mathfrak{b}_4\right), \\
B_{20} &= \gamma_a\left(\tfrac{2}{3}\mathfrak{b}_2 + \tfrac{4}{7}\mathfrak{b}_4\right), \\
B_{40} &= \gamma_a \tfrac{8}{35}\mathfrak{b}_4.
\end{aligned}\right\} \tag{11.3}$$

Mit

$$\cot\beta = \frac{\varrho}{\sqrt{\varrho^2 + e^2}}\tan\overline{\vartheta} \tag{11.4}$$

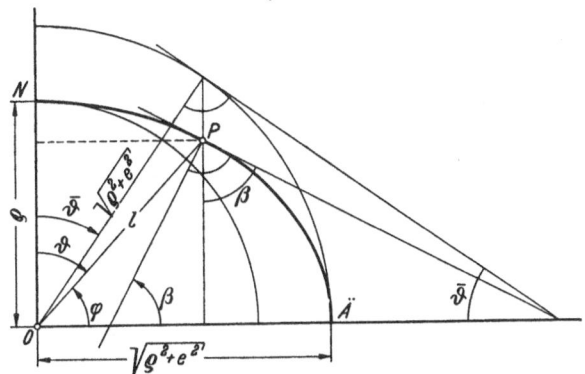

Fig. 13. Beziehungen zwischen Ellipsoidkoordinaten und Kugelkoordinaten.

kann der reduzierte Polabstand $\overline{\vartheta}$ und mit

$$\cot\beta = \frac{\varrho^2}{\varrho^2 + e^2}\tan\vartheta \tag{11.5}$$

der *geozentrische Polabstand* $\vartheta$ (Komplement der *geozentrischen Breite* $\varphi$) eingeführt werden.

Setzt man zur Abkürzung

$$\boxed{e/c = \varepsilon,} \tag{11.6}$$

und behält in Reihenentwicklungen die Glieder mit der zweiten Potenz von $\varepsilon$ bei, so findet man

$$\gamma = \gamma_a \left(\begin{aligned}\left(1 + \tfrac{1}{3}\mathfrak{b} - \tfrac{2}{15}(\mathfrak{b}_4 - \varepsilon^2\mathfrak{b})\right) + \left(\tfrac{2}{3}\mathfrak{b} - \tfrac{2}{21}(\mathfrak{b}_4 - \varepsilon^2\mathfrak{b})\right)P_{20}(\cos\overline{\vartheta}) + \\ + \tfrac{8}{35}(\mathfrak{b}_4 - \varepsilon^2\mathfrak{b})\,P_{40}(\cos\overline{\vartheta})\end{aligned}\right), \tag{11.7}$$

$$\gamma = \gamma_a \left(\begin{aligned}\left(1 + \tfrac{1}{3}\mathfrak{b} - \tfrac{2}{15}(\mathfrak{b}_4 - 2\varepsilon^2\mathfrak{b})\right) + \left(\tfrac{2}{3}\mathfrak{b} - \tfrac{2}{21}(\mathfrak{b}_4 - 2\varepsilon^2\mathfrak{b})\right)P_{20}(\cos\vartheta) + \\ + \tfrac{8}{35}(\mathfrak{b}_4 - 2\varepsilon^2\mathfrak{b})\,P_{40}(\cos\vartheta)\end{aligned}\right), \tag{11.8}$$

übersichtlich ist auch die Form[2]

$$\gamma = \gamma_a \left(\begin{aligned}1 + (\mathfrak{b} + 2\varepsilon^2\mathfrak{b} - \mathfrak{b}_4)\cos^2\vartheta - \\ - (2\varepsilon^2\mathfrak{b} - \mathfrak{b}_4)\cos^4\vartheta\end{aligned}\right). \tag{11.9}$$

---

[1] H. JEFFREYS in [21], S. 178, und Month. Not. Roy. Astronom. Soc., Geophys. Suppl. **5**, 65 (1949).

[2] Mit $\varepsilon^2 = 2\mathfrak{a}\left(\mathfrak{a} = \dfrac{a-c}{a} = \text{Abplattung des Sphäroids}\right)$ geht (11.9) in die von HELMERT entwickelte Form über ([5, II], S. 84).

**12. Reihenentwicklung des Schwerepotentials. Abstand des Sphäroids vom gleichachsigen Ellipsoid.** Setzt man (Fig. 12)

$$r(\bar{\vartheta}) = c + h(\bar{\vartheta}),\qquad(12.1)$$

so ist $h(\bar{\vartheta})$ sehr klein gegen $e$ und $c$. Man kann in Reihen entwickeln und bricht sie nach den Gliedern mit $\varepsilon^4$ und $(h/c)$ ab. Schreibt man noch

$$h(\bar{\vartheta}) = h_{00} + h_{20}P_{20}(\cos\bar{\vartheta}) + h_{40}P_{40}(\cos\bar{\vartheta}),\qquad(12.2)$$

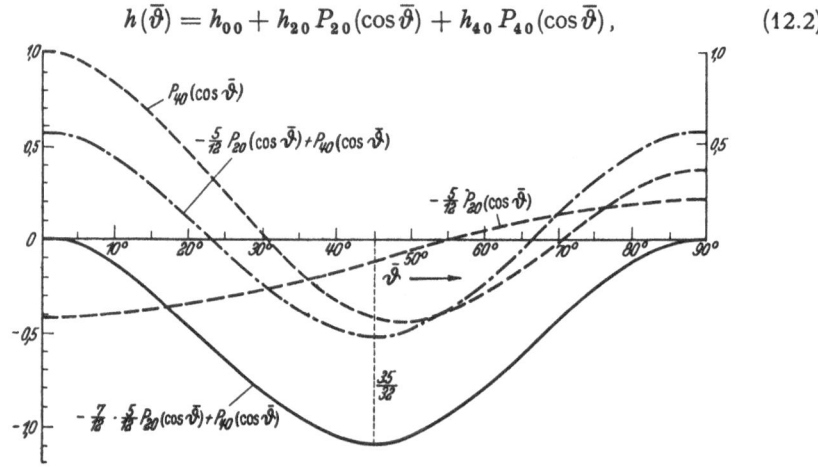

Fig. 14. Die Abweichung des Sphäroids vom achsengleichen Rotationsellipsoid ($h_{40} = 1$ gesetzt). $\bar{\vartheta}$ reduzierter Polabstand.

so erhält man aus (10.6)[1]

$$U(r,\bar{\vartheta}) = W_0 = \frac{fM}{c}\left(1 - \frac{1}{3}\varepsilon^2 + \frac{1}{5}\varepsilon^4 - \frac{h_{00}}{c}\right) + \frac{1}{3}\omega^2 c^2 (1 + \varepsilon^2) + \\ + \left(-\frac{fM}{c^2}h_{20} + i\frac{2}{189}\varepsilon^5 C_{40}\right)\cdot P_{20}(\cos\bar{\vartheta}) + \\ + \left(-\frac{fM}{c^2}h_{40} - i\frac{8}{315}\varepsilon^5 C_{40}\right)\cdot P_{40}(\cos\bar{\vartheta}).\qquad(12.3)$$

Für $\bar{\vartheta} = 0$ ist $h = 0$, $P_{20} = 1$, $P_{40} = 1$. Hiermit findet man

$$W_0 = \frac{fM}{c}\left(1 - \frac{1}{3}\varepsilon^2 + \frac{1}{5}\varepsilon^4\right) + \frac{1}{3}\omega^2 c^2 (1 + \varepsilon^2) + \\ + i\frac{2}{189}\varepsilon^5 C_{40} - i\frac{8}{315}\varepsilon^5 C_{40}.\qquad(12.4)$$

Man subtrahiert (12.4) von (12.3) und erhält (Fig. 14)

$$h_{00} = i\frac{c^2}{fM}\frac{2}{135}\varepsilon^5 C_{40},\\ h_{20} = i\frac{c^2}{fM}\frac{2}{189}\varepsilon^5 C_{40},\\ h_{40} = -i\cdot\frac{c^2}{fM}\frac{8}{215}\varepsilon^5 C_{40};\qquad(12.5)$$

$$h_{00} = -\tfrac{7}{12}h_{40},\qquad h_{20} = -\tfrac{5}{12}h_{40},\qquad(12.6)$$

$$h(\bar{\vartheta}) = \left(-\tfrac{7}{12} - \tfrac{5}{12}P_{20}(\cos\bar{\vartheta}) + P_{40}(\cos\bar{\vartheta})\right)h_{40} = -\tfrac{35}{32}h_{40}\sin^2 2\bar{\vartheta}.\qquad(12.7)$$

---

[1] Die wichtigsten der hier und bei ähnlichen Rechnungen vorkommenden Reihenentwicklungen sind ohne Ableitung im Anhang III angeführt (Ziff. 56).

Das Extremum von $h(\bar\vartheta)$, in der reduzierten Breite $\bar\vartheta = 45°$, beträgt

$$h_e = -\tfrac{35}{32}\,h_{40}, \tag{12.8}$$

und es ist

$$h_{00} = \tfrac{8}{15}\,h_e, \qquad h_{20} = \tfrac{8}{21}\,h_e, \qquad h_{40} = -\tfrac{32}{35}\,h_e, \tag{12.9}$$

und

$$\varepsilon^5 C_{40} = -i\,\frac{fM}{c^2}\cdot 36\,h_e. \tag{12.10}$$

Mit (10.3) ergibt sich

$$\varepsilon^3 C_{20} = -i\,\frac{5}{2}\left(\omega^2 c^2\left(1 + \frac{13}{7}\,\varepsilon^2\right) + \frac{8}{7}\,\frac{fM}{c^2}\,h_e\right) \tag{12.11}$$

und mit (10.2)

$$\boxed{\;\frac{f}{c^3}\left(C - \frac{A+B}{2} - \frac{1}{3}\,c^2\varepsilon^2 M\right) = -\frac{1}{3}\left(\omega^2 c^2\left(1 + \frac{13}{7}\,\varepsilon^2\right) + \frac{8}{7}\,\frac{fM}{c^2}\,h_e\right).\;} \tag{12.12}$$

Diese Beziehung wird bei der Bestimmung der Trägheitsmomente wichtig werden.

Der Abstand zweier übereinander liegender Punkte sei $H$. Dann ist $h$ der Achsenabschnitt zwischen den beiden diese Punkte enthaltenden Koordinaten-Ellipsoiden (Fig. 15). Zwischen $H$ und $h$ besteht die Beziehung[1]

$$\left.\begin{aligned} &H : h = dn : d\varrho, \\ &H = h\sqrt{\frac{\varrho^2 + e^2\cos^2\bar\vartheta}{\varrho^2 + e^2}}, \\ &H = h\left(1 - \frac{(\varepsilon)^2}{2}\sin^2\bar\vartheta\right), \end{aligned}\right\} \tag{12.13}$$

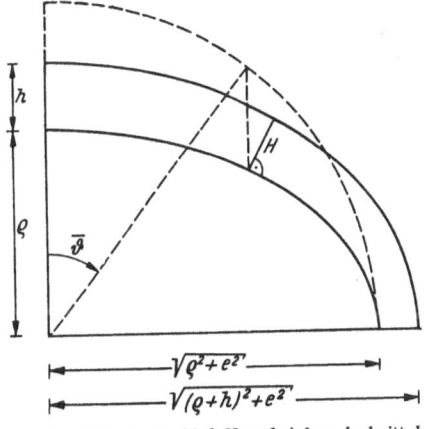

Fig. 15. Höhenunterschied $H$ und Achsenabschnitt $h$.

wobei $(\varepsilon) = e/\varrho$. Da $h/\varrho$ und $(\varepsilon)^2$ klein sind, genügt fast immer $H = h$.

### 13. Reihenentwicklung der Normalschwere. Vergleich mit der Schwereformel.
Führt man die Reihenentwicklung an (10.7) aus[2], so wird die Normalschwere bei Anwendung von (12.2), (12.5) und (12.10):

$$\left.\begin{aligned} \gamma(r,\bar\vartheta) = &\frac{fM}{c^2}\left(1 - \frac{2}{3}\,\varepsilon^2 + \frac{8}{15}\,\varepsilon^4\right) - \frac{2}{3}\,\omega^2 c\left(1 + \frac{1}{2}\,\varepsilon^2\right) - \frac{16}{15}\,\frac{fM}{c^3}\,h_e + \\ &+ \left(-\frac{1}{3}\,\frac{fM}{c^2}\,\varepsilon^2\left(1 - \frac{8}{7}\,\varepsilon^2\right) + \frac{5}{3}\,\omega^2 c\left(1 + \frac{32}{35}\,\varepsilon^2\right) + \frac{8}{21}\,\frac{fM}{c^3}\,h_e\right)\cdot P_{20}(\cos\bar\vartheta) + \\ &+ \left(\frac{3}{35}\,\frac{fM}{c^2}\,\varepsilon^4 - \frac{2}{7}\,\omega^2 c\,\varepsilon^2 - \frac{96}{35}\,\frac{fM}{c^3}\,h_e\right)\cdot P_{40}(\cos\bar\vartheta). \end{aligned}\right\} \tag{13.1}$$

Vergleicht man das $P_{40}$-Glied mit dem entsprechenden Glied der Schwereformel (11.7), so erhält man eine Beziehung zwischen $h_e$, $\varepsilon$ und den Beobachtungswerten:

$$\boxed{\;h_e = -\frac{1}{12}\,\frac{\gamma_ä c^3}{fM}\left(\mathfrak{b}_4 - \varepsilon^2\,\mathfrak{b}\right) + \frac{1}{32}\,c\,\varepsilon^4 - \frac{5}{48}\,\frac{\omega^2 c^4}{fM}\,\varepsilon^2.\;} \tag{13.2}$$

---

[1] Anhang III, Ziff. 53.
[2] Anhang III, Ziff. 56.

Vergleicht man das $P_{20}$-Glied mit dem entsprechenden Glied von (11.7), so ergibt sich eine Beziehung zwischen $\varepsilon$, $\gamma_{\ddot{a}}$ und $\mathfrak{b}$.

$$\gamma_{\ddot{a}}\,\mathfrak{b} = -\frac{1}{2}\frac{fM}{c^2}\,\varepsilon^2\left(1 - \frac{5}{4}\,\varepsilon^2\right) + \frac{5}{2}\,\omega^2 c\left(1 + \frac{39}{70}\,\varepsilon^2\right) - \frac{8}{7}\frac{fM}{c^3}\,h_e. \qquad (13.3)$$

Nach einfachen Umformungen[1] führt sie auf das CLAIRAUTsche Theorem. Der Vergleich des absoluten Glieds mit dem absoluten Glied von (11.7) ergibt Zuzusammenhänge, mit denen die Masse der Erde berechnet werden kann:

$$\begin{aligned}\frac{fM}{c^2} = \gamma_{\ddot{a}}&\left(1 + \frac{2}{3}\,\varepsilon^2 + \frac{1}{3}\,\mathfrak{b} - \frac{4}{15}\,\varepsilon^4 - \frac{2}{15}\,\mathfrak{b}_4 + \frac{16}{45}\,\varepsilon^2\mathfrak{b}\right) + \\ &+ \frac{2}{3}\,\omega^2 c\left(1 + \frac{7}{6}\,\varepsilon^2\right) + \frac{16}{15}\frac{fM}{c^3}\,h_e.\end{aligned} \qquad (13.4)$$

In den Formeln (13.2), (13.3), (13.4) und (12.12) ist die Theorie des Niveausphäroids im wesentlichen enthalten. Einfache Umformungen[2] führen auf verschiedene, für die Zahlenrechnung geeignete Formelsysteme, unter denen sich die bekannten Systeme von HELMERT [5 II] und JEFFREYS [21] befinden.

**14. Einführung des Äquatorradius a.** Meist bezieht man sich nicht auf den Polradius, sondern auf den Äquatorradius $a$ des Niveausphäroids.

$$a = \sqrt{c^2 + e^2} = c\sqrt{1 + \varepsilon^2}. \qquad (14.1)$$

Führt man außerdem das Verhältnis $\mathfrak{c}$ der Zentrifugalbeschleunigung am Äquator zur Äquatorschwere ein:

$$\mathfrak{c} = \frac{\omega^2 a}{\gamma_{\ddot{a}}}, \qquad (14.2)$$

so erhält man aus (13.2)

$$\frac{h_e}{a} = -\frac{1}{12}\left(\mathfrak{b}_4 - \varepsilon^2\mathfrak{b}\right) + \frac{1}{32}\,\varepsilon^4 - \frac{5}{48}\,\varepsilon^2\mathfrak{c} \qquad (14.3)$$

und aus (13.3)

$$\frac{1}{2}\,\varepsilon^2 + \mathfrak{b} = \frac{5}{2}\,\mathfrak{c} + \frac{7}{24}\,\varepsilon^4 - \frac{1}{6}\,\varepsilon^2\mathfrak{b} - \frac{4}{21}\,\varepsilon^2\mathfrak{c} - \frac{8}{7}\frac{h_e}{a}. \qquad (14.4)$$

Diese Beziehung ist die den rotationssymmetrischen Ellipsoid-Koordinaten entsprechende Formulierung des CLAIRAUTschen *Theorems*.

Die Glieder von der Größenordnung $\varepsilon^4$ sind in dieser Form für die Zahlenrechnung unbequem[3]. Mit dem CLAIRAUTschen Theorem, insbesondere seiner abgekürzten (ursprünglichen) Gestalt,

$$\tfrac{1}{2}\varepsilon^2 + \mathfrak{b} \approx \tfrac{5}{2}\mathfrak{c}, \qquad (14.5)$$

lassen sie sich erheblich vereinfachen. So erhält man aus (14.3) und (14.4) die übersichtlicheren Beziehungen

$$\frac{h_e}{a} = -\frac{1}{12}\,\mathfrak{b}_4 - \frac{1}{96}\,\varepsilon^4 + \frac{5}{48}\,\varepsilon^2\mathfrak{c} \qquad (14.6)$$

---

[1] Ziff. 14 und 15.
[2] Ziff. 15—23; 26.
[3] Bei den Zahlenrechnungen pflegt man mit (14.5) einen Näherungswert von $\varepsilon^2$ zu berechnen und ihn in die rechte Seite von (14.4) einzusetzen (Iterations-Verfahren). Es ist $\omega = 7,2921 \cdot 10^{-5}\ \mathrm{sec}^{-1}$, $\sqrt{c^2 + e^2} = 6378,4$ km, $c = 6356,9$ km, $\varepsilon^2 = 0,0067680$, $\mathfrak{c} = 0,0034678 = 1:288,37$, $\tfrac{1}{2}\mathfrak{c} = 0,0086695$. Ausführlichere Angaben im Anhang IV, Ziff. 62.

und

$$\frac{1}{2}\,\varepsilon^2 + \mathfrak{b} = \frac{5}{2}\,\mathfrak{c} + \frac{3}{8}\,\varepsilon^4 - \frac{17}{28}\,\varepsilon^2\mathfrak{c} - \frac{8}{7}\,\frac{h_e}{a}. \tag{14.7}$$

Aus (12.4) leitet man den Potentialwert $W_0$ des Niveausphäroides ab:

$$W_0 = \frac{fM}{a}\left(1 + \frac{1}{6}\,\varepsilon^2 + \frac{1}{3}\,\mathfrak{c} - \frac{11}{120}\,\varepsilon^4 + \frac{1}{9}\,\varepsilon^2\mathfrak{c} - \frac{1}{9}\,\mathfrak{b}\mathfrak{c} - \frac{2}{9}\,\mathfrak{c}^2 - \frac{8}{15}\,\frac{h_e}{a}\right), \tag{14.8}$$

während der Verlauf des Potentials auf dem gleichachsigen Ellipsoid mit

$$U(c,\overline{\vartheta}) = \frac{fM}{a}\left(\begin{array}{l}\left(1 + \frac{1}{6}\,\varepsilon^2 + \frac{1}{3}\,\mathfrak{c} - \frac{11}{120}\,\varepsilon^4 + \frac{1}{9}\,\varepsilon^2\mathfrak{c} - \frac{1}{9}\,\mathfrak{b}\mathfrak{c} - \frac{2}{9}\,\mathfrak{c}^2\right) + \\[2mm] + \frac{8}{21}\,\frac{h_e}{a}\,P_{20}(\cos\overline{\vartheta}) - \frac{32}{35}\,\frac{h_e}{a}\,P_{40}(\cos\overline{\vartheta})\end{array}\right) \tag{14.9}$$

gegeben ist, berechnet durch Reihenentwicklung aus (10.6) für $\varrho = c$. Die Sphäroidschwere findet man in ähnlicher Weise aus (10.7):

$$\gamma = \frac{fM}{a^2}\left(\begin{array}{l}\left(1 + \frac{1}{3}\,\varepsilon^2 - \frac{2}{3}\,\mathfrak{c} - \frac{2}{15}\,\varepsilon^4 - \frac{2}{9}\,\varepsilon^2\mathfrak{c} + \frac{2}{9}\,\mathfrak{b}\mathfrak{c} + \frac{4}{9}\,\mathfrak{c}^2 - \frac{16}{15}\,\frac{h_e}{a}\right) + \\[2mm] + \left(-\frac{1}{3}\,\varepsilon^2 + \frac{5}{3}\,\mathfrak{c} + \frac{1}{21}\,\varepsilon^4 + \frac{97}{126}\,\varepsilon^2\mathfrak{c} - \frac{5}{9}\,\mathfrak{b}\mathfrak{c} - \frac{10}{9}\,\mathfrak{c}^2 + \frac{8}{21}\,\frac{h_e}{a}\right)\cdot P_{20}(\cos\overline{\vartheta}) + \\[2mm] + \left(\frac{3}{35}\,\varepsilon^4 - \frac{2}{7}\,\varepsilon^2\mathfrak{c} - \frac{96}{35}\,\frac{h_e}{a}\right)\cdot P_{40}(\cos\overline{\vartheta})\end{array}\right). \tag{14.10}$$

Für Masse und Äquatorschwere ergibt sich

$$\frac{fM}{a^2} = \gamma_a\left(1 - \frac{1}{2}\,\varepsilon^2 + \frac{3}{2}\,\mathfrak{c} + \frac{3}{8}\,\varepsilon^4 - \frac{15}{28}\,\varepsilon^2\mathfrak{c} + \frac{16}{7}\,\frac{h_e}{a}\right), \tag{14.11}$$

$$\gamma_a = \frac{fM}{a^2}\left(1 + \frac{1}{2}\,\varepsilon^2 - \frac{3}{2}\,\mathfrak{c} - \frac{1}{8}\,\varepsilon^4 - \frac{5}{7}\,\varepsilon^2\mathfrak{c} + \frac{1}{2}\,\mathfrak{b}\mathfrak{c} + \mathfrak{c}^2 - \frac{16}{7}\,\frac{h_e}{a}\right). \tag{14.12}$$

Trägheitsmomente und Masse verbindet

$$C - \frac{A+B}{2} = \frac{1}{3}\,a^2 M\left(\begin{array}{l}\varepsilon^2 - \mathfrak{c} - \varepsilon^4 + \frac{1}{7}\,\varepsilon^2\mathfrak{c} \\[2mm] + \frac{3}{2}\,\mathfrak{c}^2 - \frac{8}{7}\,\frac{h_e}{a}\end{array}\right). \tag{14.13}$$

**15. Einführung der Abplattung α. Das Formelsystem von HELMERT.** Anschaulicher als $\varepsilon^2 = \dfrac{a^2 - c^2}{c^2}$ ist die *Abplattung* des Niveausphäroids[1]:

$$\alpha = \frac{a-c}{a}, \tag{15.1}$$

und es ist seit jeher[2] üblich, mit $\alpha$ statt mit $\varepsilon^2$ zu rechnen. Mit

$$\left.\begin{array}{l}\varepsilon^2 = 2\alpha + 3\alpha^2, \\[2mm] \varepsilon^4 = 4\alpha^2\end{array}\right\} \tag{15.2}$$

---

[1] Beim internationalen Ellipsoid ist $\alpha = 1:297 = 0{,}0033670$.
[2] P. CLAIRAUT 1743 [1].

und dem abgekürzten Clairautschen Theorem erhält man eine Fülle von Beziehungen, die für Zahlenrechnungen geeignet sind. Ein Teil von ihnen ist aus der klassischen Literatur bekannt, ein viel benutztes Formelsystem findet sich bei Helmert [5, II][1].

*Schwereformel:*

$$
\left.
\begin{aligned}
\gamma &= \gamma_{\ddot{a}}(1 + \mathfrak{b}\sin^2\beta - \mathfrak{b}'\sin^2 2\beta)\,, \\
\gamma &= \gamma_{\ddot{a}}(1 + \mathfrak{b}_2\sin^2\beta + \mathfrak{b}_4\sin^4\beta)\,, \quad \mathfrak{b}_4 = 4\,\mathfrak{b}' \quad [\text{H}]\,, \\
\gamma &= B_{00} + B_{20}P_{20}(\sin\beta) + B_{40}P_{40}(\sin\beta)\,.
\end{aligned}
\right\} \quad (15.3)
$$

*Clairautsches Theorem [2],[3]:*

$$
\left.
\begin{aligned}
\mathfrak{a} + \mathfrak{b} &= \frac{5}{2}\,\mathfrak{c} - \mathfrak{a}^2 - \frac{1}{2}\,\mathfrak{a}\mathfrak{c} + \frac{2}{7}\,\mathfrak{b} \quad [\text{H}]\,, \\
\mathfrak{a} + \mathfrak{b} &= \frac{5}{2}\,\mathfrak{c} - \frac{6}{5}\,\mathfrak{a}^2 - \frac{1}{5}\,\mathfrak{a}\mathfrak{b} + \frac{2}{7}\,\mathfrak{b}\,, \\
\mathfrak{a} + \frac{3}{2}\frac{B_{20}}{B_{00}} &= \frac{5}{2}\,\mathfrak{c} - \frac{5}{8}\frac{B_{40}}{B_{00}} - \frac{3}{4}\left(\frac{B_{20}}{B_{00}}\right)^2 - \mathfrak{a}^2 - \frac{1}{2}\,\mathfrak{a}\mathfrak{c} + \frac{2}{7}\,\mathfrak{b}\,;
\end{aligned}
\right\} \quad (15.4)
$$

$$
\left.
\begin{aligned}
\mathfrak{b} &= \frac{1}{3}\,(11\mathfrak{a}^2 - 10\mathfrak{a}\mathfrak{c} + \mathfrak{b}_4)\,, \\
\mathfrak{b} &= \frac{1}{3}\,(7\mathfrak{a}^2 - 4\mathfrak{a}\mathfrak{b} + \mathfrak{b}_4) \quad [\text{H}]\,, \\
\mathfrak{b} &= \frac{1}{3}\left(11\mathfrak{a}^2 - 10\mathfrak{a}\mathfrak{c} + \frac{35}{8}\frac{B_{40}}{B_{00}}\right). \\
\mathfrak{b} &= \frac{1}{3}\left(7\mathfrak{a}^2 - 6\mathfrak{a}\frac{B_{20}}{B_{00}} + \frac{35}{8}\frac{B_{40}}{B_{00}}\right), \\
\mathfrak{b} &= \frac{7}{2}\,\mathfrak{a}^2 - \frac{5}{2}\,\mathfrak{a}\mathfrak{c} - 4\,\frac{h_e}{a}\,, \\
\mathfrak{b} &= \frac{5}{2}\,\mathfrak{a}^2 - \mathfrak{a}\mathfrak{b} - 4\,\frac{h_e}{a}\,.
\end{aligned}
\right\} \quad (15.5)
$$

*Abstand des Niveau-Sphäroids vom achsengleichen Ellipsoid:*

$$
\left.
\begin{aligned}
\frac{h_e}{a} &= \frac{7}{8}\,\mathfrak{a}^2 - \frac{5}{8}\,\mathfrak{a}\mathfrak{c} - \frac{1}{4}\,\mathfrak{b}\,, \\
\frac{h_e}{a} &= \frac{5}{8}\,\mathfrak{a}^2 - \frac{1}{4}\,\mathfrak{a}\mathfrak{b} - \frac{1}{4}\,\mathfrak{b} \quad [\text{H}]\,, \\
\frac{h_e}{a} &= -\frac{1}{24}\,\mathfrak{a}^2 + \frac{5}{24}\,\mathfrak{a}\mathfrak{c} - \frac{1}{12}\,\mathfrak{b}_4\,, \\
\frac{h_e}{a} &= \frac{1}{24}\,\mathfrak{a}^2 + \frac{1}{12}\,\mathfrak{a}\mathfrak{b} - \frac{1}{12}\,\mathfrak{b}_4\,, \\
\frac{h_e}{a} &= -\frac{1}{24}\,\mathfrak{a}^2 + \frac{5}{24}\,\mathfrak{a}\mathfrak{c} - \frac{35}{96}\frac{B_{40}}{B_{00}}\,, \\
\frac{h_e}{a} &= \frac{1}{24}\,\mathfrak{a}^2 + \frac{1}{8}\,\mathfrak{a}\frac{B_{20}}{B_{00}} - \frac{35}{96}\frac{B_{40}}{B_{00}}\,.
\end{aligned}
\right\} \quad (15.6)
$$

---

[1] In den folgenden Aufstellungen werden die Helmertschen Formeln mit [H] kenntlich gemacht.

[2] Abgekürzte (ursprüngliche) Form: $\mathfrak{a} + \mathfrak{b} = \frac{5}{2}\mathfrak{c}$.

[3] Über die Bedeutung von $\mathfrak{b}$ siehe (17.2).

*Masse der Erde, Äquatorschwere:*

$$\frac{fM}{a^2} = \gamma_{\ddot{a}}\left(1 - \mathfrak{a} + \frac{3}{2}\mathfrak{c} - \frac{15}{14}\mathfrak{a}\mathfrak{c} + \frac{16}{7}\frac{h_e}{a}\right), \tag{15.7}$$

$$\left.\begin{aligned}\gamma_{\ddot{a}} &= \frac{fM}{a^2}\left(1 - \mathfrak{b} + \mathfrak{c} - \frac{8}{7}\mathfrak{a}\mathfrak{b} + \mathfrak{b}\mathfrak{c} - \frac{1}{4}\mathfrak{c}^2 + \frac{2}{7}\mathfrak{b}_4\right) \quad [H], \\ \gamma_{\ddot{a}} &= \frac{fM}{a^2}\left(1 - \mathfrak{b} + \mathfrak{c} - \mathfrak{a}\mathfrak{b} + \frac{9}{14}\mathfrak{b}\mathfrak{c} + \frac{9}{14}\mathfrak{c}^2 - \frac{24}{7}\frac{h_e}{a}\right).\end{aligned}\right\} \tag{15.8}$$

*Trägheitsmomente:*

$$\left.\begin{aligned}C - \frac{A+B}{2} &= \frac{1}{3}a^2 M\left(4\mathfrak{c} - 2\mathfrak{b} - 4\mathfrak{a}^2 + \frac{3}{2}\mathfrak{c}^2 + \frac{6}{7}\mathfrak{b}\right) \quad [H], \\ C - \frac{A+B}{2} &= \frac{1}{3}a^2 M\left(2\mathfrak{a} - \mathfrak{c} - \mathfrak{a}^2 + \frac{2}{7}\mathfrak{a}\mathfrak{c} + \frac{3}{2}\mathfrak{c}^2 - \frac{8}{7}\frac{h_e}{a}\right).\end{aligned}\right\} \tag{15.9}$$

*Potentialwert des Niveau-Sphäroids:*

$$W_0 = \frac{fM}{a}\left(1 + \frac{1}{3}\mathfrak{a} + \frac{1}{3}\mathfrak{c} + \frac{2}{15}\mathfrak{a}^2 + \frac{1}{3}\mathfrak{a}\mathfrak{c} - \frac{1}{2}\mathfrak{c}^2 - \frac{8}{15}\frac{h_e}{a}\right). \tag{15.10}$$

**16. Spezielle Sphäroide.** *Das Niveau-Ellipsoid* $(h_e = 0)$. Ist $h_e$ so klein, daß man es vernachlässigen kann, so fällt das Niveau-Sphäroid mit dem Ellipsoid $\varrho = c$ zusammen. Das Formelsystem vereinfacht sich nicht wesentlich.

$$\left.\begin{aligned}\mathfrak{a} + \mathfrak{b} &= \frac{5}{2}\mathfrak{c} - \frac{17}{14}\mathfrak{a}\mathfrak{c}, \\ \mathfrak{a} + \mathfrak{b} &= \frac{5}{2}\mathfrak{c} - \frac{17}{35}\mathfrak{a}^2 - \frac{17}{35}\mathfrak{a}\mathfrak{b},\end{aligned}\right\} \tag{16.1}$$

$$\left.\begin{aligned}\mathfrak{b} &= \frac{7}{2}\mathfrak{a}^2 - \frac{5}{2}\mathfrak{a}\mathfrak{c}, \\ \mathfrak{b} &= \frac{5}{2}\mathfrak{a}^2 - \mathfrak{a}\mathfrak{b},\end{aligned}\right\} \tag{16.2}$$

$$\left.\begin{aligned}\mathfrak{b}_4 &= -\frac{1}{2}\mathfrak{a}^2 + \frac{5}{2}\mathfrak{a}\mathfrak{c}, \\ \mathfrak{b}_4 &= \frac{1}{2}\mathfrak{a}^2 + \mathfrak{a}\mathfrak{b} \quad [H].\end{aligned}\right\} \tag{16.3}$$

*Das BRUNSsche Sphäroid.* Beim Sphäroid von BRUNS[1] [4], [9], [13] ist $\mathfrak{b} = 0$, und es treten entsprechende Vereinfachungen in den Formeln auf.

$$\left.\begin{aligned}\frac{1}{2}\varepsilon^2 + \mathfrak{b} &= \frac{5}{2}\mathfrak{c} + \frac{1}{8}\varepsilon^4 - \frac{1}{2}\varepsilon^2\mathfrak{c}, \\ \mathfrak{a} + \mathfrak{b} &= \frac{5}{2}\mathfrak{c} - \mathfrak{a}^2 - \mathfrak{a}\mathfrak{c},\end{aligned}\right\} \tag{16.4}$$

$$\left.\begin{aligned}\mathfrak{b}_4 &= -\frac{11}{4}\varepsilon^4 + 5\varepsilon^2\mathfrak{c}, \\ \mathfrak{b}_4 &= -11\mathfrak{a}^2 + 10\mathfrak{a}\mathfrak{c},\end{aligned}\right\} \tag{16.5}$$

$$\left.\begin{aligned}\frac{h_e}{a} &= \frac{7}{32}\varepsilon^4 - \frac{5}{16}\varepsilon^2\mathfrak{c}, \\ \frac{h_e}{a} &= \frac{7}{8}\mathfrak{a}^2 - \frac{5}{8}\mathfrak{a}\mathfrak{c}.\end{aligned}\right\} \tag{16.6}$$

---

[1] Ziff. 17.

*Die Normalform des Sphäroids nach* HELMERT [*5*, II].

$$\mathfrak{b}_4 = 0, \quad \mathfrak{b}_2 = \mathfrak{b}. \quad \gamma = \gamma_a (1 + \mathfrak{b} \sin^2 \beta), \tag{16.7}$$

$$\left.\begin{aligned} \mathfrak{b} &= \frac{1}{3}(11\,\mathfrak{a}^2 - 10\,\mathfrak{a}\mathfrak{c}), \\ \mathfrak{b} &= \frac{1}{3}(7\,\mathfrak{a}^2 - 4\,\mathfrak{a}\mathfrak{b}), \end{aligned}\right\} \tag{16.8}$$

$$\left.\begin{aligned} \frac{h_s}{a} &= -\frac{1}{24}\,\mathfrak{a}^2 + \frac{5}{24}\,\mathfrak{a}\mathfrak{c}, \\ \frac{h_e}{a} &= \frac{1}{24}\,\mathfrak{a}^2 + \frac{1}{12}\,\mathfrak{a}\mathfrak{b}. \end{aligned}\right\} \tag{16.9}$$

*Das* MCLAURIN*sche Ellipsoid.* Das MCLAURINsche Ellipsoid ist die rotationssymmetrische Gleichgewichtsfigur eines homogenen Körpers, der bei gleichem Äquatorradius und gleicher Äquatorschwere mit derselben Winkelgeschwindigkeit rotiert wie die Erde. Seine Oberfläche, die zugleich Niveaufläche ist, hat die Gestalt eines abgeplatteten Rotationsellipsoids [*14*], [*16*]. Man berechnet

$$C = \frac{2}{5}\,M\,a^2,$$

$$A + B + C = \frac{6}{5}\,M\,a^2\left(1 - \frac{1}{3}\,\varepsilon^2 + \frac{1}{3}\,\varepsilon^4\right)$$

und findet hieraus:

$$C - \frac{A+B}{2} = \frac{1}{5}\,M\,a^2\,\varepsilon^2(1 - \varepsilon^2) = \frac{2}{5}\,M\,a^2\,\mathfrak{a}\left(1 - \frac{1}{2}\,\mathfrak{a}\right). \tag{16.10}$$

Beim MCLAURINschen Ellipsoid ist, wie bei allen Niveau-Ellipsoiden, $h_s = 0$. Aus (15.9) erhält man dann

$$(\mathfrak{a})_{\mathrm{McL}} = \frac{5}{4}\,\mathfrak{c}\left(1 - \frac{69}{56}\,\mathfrak{c}\right) \tag{16.11}$$

und berechnet mit $\mathfrak{c} = 0{,}003468$:

$$(\mathfrak{a})_{\mathrm{McL}} = 0{,}004316 = 1/231{,}7.$$

*Das Sphäroid mit größter Massenkonzentration.* Das homogene MCLAURINsche Ellipsoid ist ein Sphäroid geringster Massenkonzentration[1]. Als entgegengesetzten Grenzfall erhält man das Sphäroid mit größter Massenkonzentration, wenn man eine Punktmasse $M$ im Zentrum anbringt und der übrigen Volumenausfüllung die Dichte Null zuschreibt. Für einen solchen Körper gilt

$$C - \frac{A+B}{2} = 0,$$

und man findet

$$(\mathfrak{a})_{\mathrm{konz}} = \frac{1}{2}\,\mathfrak{c}\left(1 - \frac{3}{2}\,\mathfrak{c}\right). \tag{16.12}$$

Mit dem $\mathfrak{c}$-Wert der Erde berechnet man

$$(\mathfrak{a})_{\mathrm{konz}} = 0{,}001725 = 1/579{,}7.$$

Der größte Abstand des Sphäroids vom gleichachsigen Ellipsoid beträgt

$$(h_e)_{\mathrm{konz}} = -\frac{3}{32}\,a\,\mathfrak{c}^2. \tag{16.13}$$

Die Abplattung der Erde liegt nahe bei 1/297. Die Erde ist einem MCLAURINschen Ellipsoid näher als einem Sphäroid größter Massenkonzentration.

*Das Normalsphäroid von* HAALCK[2, 3]. Um dem einer normalisierten Erdkruste entsprechenden Sphäroid möglichst nahe zu kommen, wird die Annahme gemacht, daß das Quadrat

---

[1] Wenn man voraussetzt, daß die Dichte bei zunehmender Tiefe nicht abnimmt.

[2] H. HAALCK: Veröff. Geodät. Inst. Potsdam Nr. 4, 1950.

[3] So begrüßenswert es ist, daß ein Versuch unternommen wird, das geologisch definierte Normalsphäroid der normalisierten Erde mit mathematischen Mitteln zu erfassen, glaubt der Referent, daß dieser Versuch noch nicht befriedigend gelungen ist. Der Betrag von $\mathfrak{a}$

der Abplattung dieses „*Normalsphäroids*" die Differenz $(\mathfrak{a})^2_{\text{McL}} - (\mathfrak{a})^2_{\text{konz}}$ in demselben Verhältnis teilt wie der Abstand $h_e$ die Differenz $(h_e)_{\text{McL}} - (h_e)_{\text{konz}}$. Der Ansatz lautet

$$\frac{(\mathfrak{a})^2_{\text{McL}} - \mathfrak{a}^2}{(\mathfrak{a})^2_{\text{McL}} - (\mathfrak{a})^2_{\text{konz}}} = \frac{(h_e)_{\text{McL}} - h_e}{(h_e)_{\text{McL}} - (h_e)_{\text{konz}}}. \tag{16.14}$$

$(h_e)_{\text{McL}}$ ist Null, ferner ist $\mathfrak{a}^2 = (\tfrac{5}{4}c - \mathfrak{b})^2$, und man findet mit (16.11), (16.13), (16.14):

$$h_e = \frac{1}{224}\,a\,(75\,c^2 - 80\,\mathfrak{b}c + 16\,\mathfrak{b}^2). \tag{16.15}$$

Für $c$ ist der oben gebrauchte Wert einzusetzen. $\mathfrak{b}$ kann noch frei gewählt werden; man wird es den Schwereformeln anpassen. Dann ist $h_e$ bestimmt, und die anderen Daten des Normalsphäroids sind festgelegt, auch der Koeffizient $\mathfrak{b}_4$ der Schwereformel. Mit $\mathfrak{b} = 0{,}005288$ wird $h_e = -3{,}3$ m.

**17. Transformation in Kugelkoordinaten.** Wie sich zeigen wird[1], beträgt $h_e$ nur einige Meter. Somit bestehen bei Ellipsoidkoordinaten keine Konvergenzbedenken, wenn man an Stelle des im Meeresniveau gelegenen Sphäroids eine Niveaufläche des Potentials $U$ untersucht, die einige Meter höher liegt und ganz im Außenraum derjenigen Koordinatenfläche verläuft, die das Niveausphäroid gerade einschließt. Es liegt auf der Hand, daß sich die Abplattung der höher gelegenen Niveaufläche von der Abplattung des Niveausphäroids so wenig unterscheidet, daß man die beiden Abplattungen einander gleichsetzen kann.

Drückt man, wie in (9.1), das Potential in Kugelkoordinaten aus, so ist nicht selbstverständlich, daß man die Reihe des Außenraumpotentials bis an die Oberfläche heran anwenden darf. Gelingt es jedoch, das Potential (10.2) und die Normalschwere (10.5) so in Kugelkoordinaten überzuführen, daß bei Beibehaltung kleiner Glieder bis zur Größenordnung $\varepsilon^4$ die bekannten Entwicklungen nach Kugelfunktionen entstehen, so folgt, daß man diese Ausdrücke zur Grundlage einer Theorie des Niveausphäroids machen darf.

Zu diesem Zweck werden der Radiusvektor $l$ und der geozentrische Polabstand $\vartheta$ eingeführt (Fig. 13).

$$l^2 = \varrho^2 + e^2\,(1 - \cos^2\overline{\vartheta}), \qquad \tan\vartheta = \frac{\sqrt{\varrho^2 + e^2}}{\varrho}\,\tan\overline{\vartheta}. \tag{17.1}$$

Man entwickelt in Reihen[2] und erhält als Ergebnis:

$$U = \frac{fM}{l}\left\{ 1 + \frac{1}{2l^2}\,\frac{C - \dfrac{A+B}{2}}{M}\,(1 - 3\cos^2\vartheta) + \right.$$
$$+ \frac{\omega^2 l^3}{2fM}\sin^2\vartheta +$$
$$\left. + \frac{D}{l^4}\left(\cos^4\vartheta - \frac{6}{7}\cos^2\vartheta + \frac{3}{35}\right)\right\} \tag{17.2}$$

mit
$$D = a^4\,\mathfrak{b}$$

beruht auf der Massenverteilung des tiefen Erdinnern, während $h_e$ von den Massen der Erdkruste maßgeblich mitbestimmt wird. (16.14) schafft eine enge Beziehung zwischen Größen, die nur in lockerem Zusammenhang stehen können, und muß daher als willkürlich angesehen werden. Zu den übrigen Ausführungen ist zu bemerken, daß das CLAIRAUTsche Theorem in der dort entwickelten Gestalt in die HELMERTsche Form übergeführt werden kann. Die Zahlenabschätzungen können sicherer gestaltet werden, wenn man von $c$ ausgeht und berücksichtigt, daß die Beziehungen, die $\mathfrak{b}$ mit $c$ verbinden, in beiden Grenzfällen verschieden sind.

[1] Tabelle 11, Ziff. 63.
[2] Anhang III, Ziff. 56.

und

$$
\begin{aligned}
\gamma = \frac{fM}{l^2}\Bigg\{ & 1 + \frac{3}{2}\frac{1}{l^2}\frac{C - \frac{A+B}{2}}{M}(1 - 3\cos^2\vartheta) - \frac{\omega^2 l^3}{fM}\sin^2\vartheta + \\
& + \frac{5D}{l^4}\left(\cos^4\vartheta - \frac{6}{7}\cos^2\vartheta + \frac{3}{35}\right) + \\
& + \frac{1}{2}\left(\frac{3}{l^2}\frac{C - \frac{A+B}{2}}{M} + \frac{\omega^2 l^3}{fM}\right)^2 (\cos^2\vartheta - \cos^4\vartheta)\Bigg\}.
\end{aligned}
\tag{17.3}
$$

(17.2) und (17.3) sind mit den HELMERTschen Formeln für $U$ und $\gamma$ identisch[1]. Die Anwendbarkeit der Kugelkoordinaten in üblicher Art ist, soweit sie das Niveausphäroid betrifft, erwiesen.

Setzt man in (17.2) $D = 0$, so erhält man den von BRUNS [4] ausführlich behandelten Potentialausdruck, auf den auch später mehrfach zurückgegriffen wurde [9] [13]. Mit $U = W_0$ erhält man die Gleichungen der Sphäroide:

$$D = 0 \quad \text{BRUNSsches Sphäroid,}$$
$$D \neq 0 \quad \text{HELMERTsches Sphäroid.}$$

In kartesischen Koordinaten $x$, $y$, $z$ ist das BRUNSsche Sphäroid eine algebraische Fläche 14. Ordnung, ein Umstand, der in keiner Beziehung zu den charakteristischen Eigenschaften der Sphäroide steht, aber seiner Einprägsamkeit wegen immer wieder erwähnt wird. Dabei pflegt man zu übersehen, daß das HELMERTsche Sphäroid sogar eine Fläche 22. Ordnung ist.

**18. Die Änderung der Normalschwere mit der Höhe im Außenraum.** Bei Betrachtung der Änderungen genügt es, in den Differentialquotienten bis zu den Gliedern der Größenordnung $\varepsilon^2$ und dem linearen Glied mit der Höhe $H$ zu gehen. Man braucht dann nicht mehr zwischen der reduzierten, der geozentrischen und der geographischen Breite zu unterscheiden.

Ist $n$ die äußere Normale des Niveausphäroids, so gilt[2]

$$
\frac{\partial\gamma}{\partial n} = \frac{\partial\gamma}{\partial\varrho}\sqrt{\frac{\varrho^2 + \varepsilon^2}{\varrho^2 + \varepsilon^2\cos^2\vartheta}}.
\tag{18.1}
$$

Setzt man

$$
\varrho = c + H(\vartheta, \lambda),
\tag{18.2}
$$

nimmt $\gamma$ aus (10.5), differenziert und entwickelt bis zu den Gliedern mit $\varepsilon^2$ und $H$, so erhält man

$$
\frac{\partial\gamma}{\partial n} = -\frac{2fM}{c^3}\left(1 - \frac{1}{2}\varepsilon^2 - \frac{3}{2}\varepsilon^2\cos^2\vartheta - \frac{1}{2}\mathfrak{c} + \frac{5}{2}\mathfrak{c}\cos^2\vartheta - 3\frac{H}{c}\right).
\tag{18.3}
$$

Aus (13.1) leitet man ab:

$$
\frac{2fM}{c^3} = \frac{2\gamma}{a}(1 - \varepsilon^2)^{-\frac{1}{2}}\left(1 - \frac{1}{2}\varepsilon^2 - \frac{1}{2}\varepsilon^2\cos^2\vartheta - \frac{3}{2}\mathfrak{c} + \frac{5}{2}\mathfrak{c}\cos^2\vartheta - 2\frac{H}{c}\right)^{-1},
$$

und eingesetzt in (18.3) ergibt sich

$$
\left.
\begin{aligned}
\frac{\partial\gamma}{\partial n} &= -\frac{2\gamma}{a}\left(1 + \frac{1}{2}\varepsilon^2 + \mathfrak{c} - \varepsilon^2\cos^2\vartheta - \frac{H}{a}\right), \\
\frac{\partial\gamma}{\partial n} &= -\frac{2\gamma}{a}\left(1 + \mathfrak{a} + \mathfrak{c} - 2\mathfrak{a}\cos^2\vartheta - \frac{H}{a}\right) \quad [\text{H}].
\end{aligned}
\right\}
\tag{18.4}
$$

---

[1] [5, II], S. 77 und S. 81.
[2] Anhang III, Ziff. 53.

Nun ist

$$
\left.\begin{aligned}
\gamma_H &= \gamma_0 + \left(\frac{\partial \gamma}{\partial n}\right)_0 H + \frac{1}{2}\left(\frac{\partial^2 \gamma}{\partial n^2}\right)_0 H^2, \\
\gamma_0 &= \gamma_H - \left(\frac{\partial \gamma}{\partial n}\right)_H H + \frac{1}{2}\left(\frac{\partial^2 \gamma}{\partial n^2}\right)_H H^2,
\end{aligned}\right\}
\tag{18.5}
$$

genau genug gilt

$$
\frac{\partial^2 \gamma}{\partial n^2} = \frac{\partial}{\partial n}\left(-\frac{2\gamma}{a}\right) = 6\frac{\gamma}{a^2} \quad [H] \;^1,
\tag{18.6}
$$

und es wird hiermit

$$
\left.\begin{aligned}
\gamma_H &= \gamma_0\left[1 - 2\frac{H}{a}\left(1 + \frac{1}{2}\varepsilon^2 + \mathfrak{c} - \varepsilon^2\cos^2\vartheta\right) + 3\frac{H^2}{a^2}\right], \\
\gamma_0 &= \gamma_H\left[1 + 2\frac{H}{a}\left(1 + \frac{1}{2}\varepsilon^2 + \mathfrak{c} - \varepsilon^2\cos^2\vartheta\right) + \frac{H^2}{a^2}\right],
\end{aligned}\right\}
\tag{18.7}
$$

$$
\boxed{\begin{aligned}
\gamma_H &= \gamma_0\left[1 - 2\frac{H}{a}\left(1 + \mathfrak{a} + \mathfrak{c} - 2\mathfrak{a}\cos^2\vartheta\right) + 3\frac{H^2}{a^2}\right] \quad [H], \\
\gamma_0 &= \gamma_H\left[1 + 2\frac{H}{a}\left(1 + \mathfrak{a} + \mathfrak{c} - 2\mathfrak{a}\cos^2\vartheta\right) + \frac{H^2}{a^2}\right].
\end{aligned}}
\tag{18.8}\;^2
$$

Das quadratische Glied wird selten berücksichtigt [3].

**19. Die Änderung der Abplattung mit der Höhe im Außenraum** [5, II]. Die Definition von $\varepsilon^2$

$$
\varepsilon^2 = \frac{a^2 - c^2}{c^2}
$$

führt auf

$$
\Delta(\varepsilon^2) = \frac{2a}{c^2}\Delta a - \frac{2a^2}{c^3}\Delta c.
$$

Mit

$$
\Delta a = \frac{\gamma_p}{\gamma_a}\Delta c = (1 + \mathfrak{b})\Delta c
$$

ergibt sich

$$
\left.\begin{aligned}
&\Delta(\varepsilon^2) = (2\mathfrak{b} - \varepsilon^2)\frac{\Delta a}{a}, \qquad \Delta(a - c) = \mathfrak{b}\,\Delta a, \\
&\Delta\mathfrak{a} = (\mathfrak{b} - \mathfrak{a})\frac{\Delta a}{a}, \qquad \frac{\Delta(\varepsilon^2)}{\varepsilon^2} = \left(\frac{2\mathfrak{b}}{\varepsilon^2} - 1\right)\frac{\Delta a}{a}, \\
&\qquad\qquad\qquad\qquad\;\; \frac{\Delta\mathfrak{a}}{\mathfrak{a}} = \left(\frac{\mathfrak{a}}{\mathfrak{b}} - 1\right)\frac{\Delta a}{a}.
\end{aligned}\right\}
\tag{19.1}
$$

Für $\Delta a/a$ kann $\Delta c/c$ gesetzt werden.

Nach den Schwereformeln ist $\mathfrak{b} \approx 0{,}00529$, $\mathfrak{a}$ ist ungefähr $0{,}00337$. Hiermit berechnet man

$$
\begin{aligned}
&\Delta(\varepsilon^2) = 0{,}00384\,\frac{\Delta a}{a}, \qquad \Delta(a - c) = 0{,}00529\,\Delta a, \\
&\Delta\mathfrak{a} = 0{,}00192\,\frac{\Delta a}{a}, \qquad\quad \frac{\Delta(\varepsilon^2)}{\varepsilon^2} = 0{,}571\,\frac{\Delta a}{a}, \\
&\qquad\qquad\qquad\qquad\qquad\;\; \frac{\Delta\mathfrak{a}}{\mathfrak{a}} = 0{,}571\,\frac{\Delta a}{a}.
\end{aligned}
$$

---

[1] Bei HELMERT ([5, II], S. 96) steht hier $2\gamma/a^2$. Man erhält diesen Ausdruck, wenn man die Abhängigkeit des Zählers von $a$ unberücksichtigt läßt. Der HELMERTsche Ausdruck findet sich auch in [16].

[2] Der Fußnote 1 entsprechend fehlt bei HELMERT und in [16] der Faktor 3 im quadratischen Glied. Die zweite Formel bringt auch JEFFREYS [Geophys. Suppl. (1954), S. 317/318 und [21], S. 375] und definiert die Höhe mit $(U_0 - U_H) : \frac{1}{2}(\gamma_0 + \gamma_H)$.

[3] Mit $\varphi = \begin{cases} 0° \\ 90° \end{cases}$, $\gamma_0 = \begin{cases} 978{,}0 \text{ Gal} \\ 983{,}2 \text{ Gal} \end{cases}$ wird $\dfrac{2\gamma_0}{a}(1 + \mathfrak{a} + \mathfrak{c} - 2\mathfrak{a}\cos^2\vartheta) = \begin{cases} 0{,}3088 \text{ mgal/m} \\ 0{,}3083 \text{ mgal/m} \end{cases}$ und $\dfrac{3\gamma_0}{a^2} = 0{,}072$ mgal/km².

**20. Die Änderung der Abweichung des Sphäroids vom Ellipsoid mit der Höhe im Außen-
raum [5, II].** Aus der letzten der Gln. (15.5) folgt

$$h_e = \frac{a}{4}\left(\frac{5}{2}\mathfrak{a}^2 - \mathfrak{a}\,\mathfrak{b} - \mathfrak{b}\right) \tag{20.1}$$

und

$$\frac{dh_e}{da} = \frac{1}{4}\left(\frac{5}{2}\mathfrak{a}^2 - \mathfrak{a}\,\mathfrak{b} - \mathfrak{b}\right) + \frac{a}{4}\left(5\mathfrak{a}\,\frac{d\mathfrak{a}}{da} - \mathfrak{b}\,\frac{d\mathfrak{a}}{da} - \mathfrak{a}\,\frac{d\mathfrak{b}}{da} - \frac{d\mathfrak{b}}{da}\right). \tag{20.2}$$

Nun ist

$$a\,\frac{d\mathfrak{a}}{da} = (\mathfrak{b} - \mathfrak{a}) \qquad \text{(Ziff. 19)},$$

$$a\,\frac{d\mathfrak{c}}{da} = 3\mathfrak{c}, \quad \text{da} \quad \mathfrak{c} = \frac{\omega^2 a^3}{fM};$$

$$a\,\frac{d\mathfrak{b}}{da} = \frac{5}{2}\,a\,\frac{d\mathfrak{c}}{da} - a\,\frac{d\mathfrak{a}}{da} = \frac{15}{2}\mathfrak{c} + \mathfrak{a} - \mathfrak{b} = 4\mathfrak{a} + 2\mathfrak{b},$$

$$a\,\frac{d\mathfrak{b}}{da} = -4\mathfrak{b}, \quad \text{da} \quad \mathfrak{b} = \frac{D}{a^4}.$$

Hiermit wird

$$\frac{dh_e}{da} = \frac{1}{4}\left(-\frac{13}{2}\mathfrak{a}^2 + 3\mathfrak{a}\,\mathfrak{b} - \mathfrak{b}^2 + 3\mathfrak{b}\right); \tag{20.3}$$

die Beobachtungen ergeben, daß

$$\mathfrak{b} \approx \frac{3}{2}\mathfrak{a} \tag{20.4}$$

und daher

$$\mathfrak{b} \approx \frac{1}{3}\left(7\mathfrak{a}^2 - 4\mathfrak{a}\,\mathfrak{b} + \mathfrak{b}_4\right) = \frac{1}{3}\mathfrak{a}^2\left(1 + \frac{\mathfrak{b}_4}{\mathfrak{a}^2}\right), \qquad \mathfrak{a}\,\mathfrak{b} \approx \frac{3}{2}\mathfrak{a}^2, \qquad \mathfrak{b}^2 \approx \frac{9}{4}\mathfrak{a}^2.$$

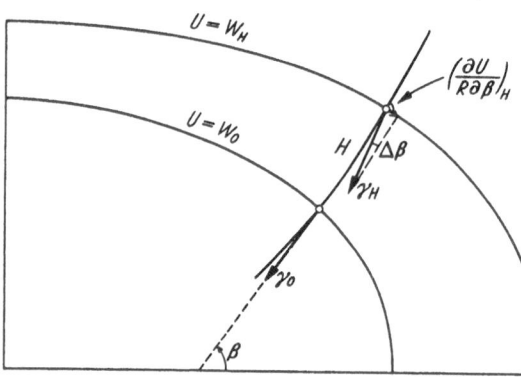

Fig. 16. Die Änderung der Breite mit der Höhe.

Eingesetzt in (20.3), erhält man

$$\frac{dh_e}{da} = -\left(\frac{13}{16} - \frac{\mathfrak{b}_4}{4\mathfrak{a}^2}\right)\mathfrak{a}^2. \tag{20.5}$$

In den gebräuchlichen Schwere-
formeln ist $\frac{1}{4}\mathfrak{b}_4 \approx 0{,}000006$. $\mathfrak{a}^2$ ist
ungefähr $0{,}000011$, also

$$\frac{\mathfrak{b}_4}{4\mathfrak{a}^2} \approx \frac{1}{2}, \qquad \frac{dh_e}{da} \approx -\frac{1}{2}\mathfrak{a}^2.$$

**21. Die Änderung der Breite mit
der Höhe im Außenraum.** Mit der
Krümmung der Lotlinien (Fig. 16)
ist eine Änderung der Breite verbunden. Aus der Figur liest man ab

$$\Delta\beta = \frac{1}{R}\left(\frac{\partial U}{\partial \beta}\right)_H \cdot \frac{1}{\gamma}. \tag{21.1}$$

Nun ist, da $(\partial U/\partial\beta)_0 = 0$,

$$\left(\frac{\partial U}{\partial \beta}\right)_H = \frac{\partial}{\partial n}\frac{\partial U}{\partial \beta} = -\frac{\partial \gamma}{\partial \beta} = -\gamma H 2\mathfrak{b}\sin\beta\cos\beta,$$

$$\Delta\beta = 2\frac{H}{R}\mathfrak{b}\sin\beta\cos\beta, \qquad \Delta\beta'' = \varrho''\frac{H}{R}\mathfrak{b}\sin 2\beta. \tag{21.2}$$

Mit $\mathfrak{b} \approx 0{,}00529$, $R = 6{,}371\cdot 10^3$ km, $\varrho'' = 206265''$/rad findet man

$$\Delta\beta'' = 0{,}171\sin 2\beta \cdot H_{km}'' \quad \text{[H]}.$$

**22. Oberfläche, Volumen, Radius der volumengleichen Kugel, mittlere Dichte.**
Die Oberfläche des Niveausphäroids unterscheidet sich nur sehr wenig von der

Oberfläche des gleichachsigen Ellipsoides; man kann den Unterschied vernachlässigen. Somit ist[1]

$$O = {}_E O = 2\pi \int_{\vartheta=0}^{\pi} \sqrt{c^2 + e^2 \cos^2 \overline{\vartheta}} \; \sqrt{c^2 + e^2} \; \sin \overline{\vartheta} \, d\overline{\vartheta} . \tag{22.1}$$

Die Integration ergibt

$$O = 2\pi c^2 \sqrt{1 + \varepsilon^2} \left( \frac{1}{\varepsilon} \operatorname{Ar\,Sin} \varepsilon + \sqrt{1 + \varepsilon^2} \right) . \tag{22.2}$$

Entwickelt man in eine Reihe und behält kleine Glieder bis zur Größenordnung $\varepsilon^4$ bei, so wird

$$\left. \begin{aligned} O &= 4\pi c^2 \left( 1 + \tfrac{2}{3} \varepsilon^2 - \tfrac{1}{15} \varepsilon^4 \right), \\ O &= 4\pi a^2 \left( 1 - \tfrac{2}{3} \mathfrak{a} + \tfrac{1}{15} \mathfrak{a}^2 \right). \end{aligned} \right\} \tag{22.3}$$

Um das Volumen zu bestimmen, denkt man sich die Erde aus einem vom Ellipsoid begrenzten Körper und der Restmasse zusammengesetzt.

$$\mathrm{Vol} = \tfrac{4}{3} \pi a^2 c + O \cdot h_{00} . \tag{22.4}$$

Mit (12.9) ergibt sich

$$\mathrm{Vol} = \frac{4}{3} \pi c^3 \left( 1 + \varepsilon^2 + \frac{8}{5} \frac{h_e}{c} \right), \qquad \mathrm{Vol} = \frac{4}{3} \pi a^3 \left( 1 - \mathfrak{a} + \frac{8}{5} \frac{h_e}{a} \right). \tag{22.5}$$

Hieraus folgt der Radius der volumengleichen Kugel *(Erdkugel)*:

$$\left. \begin{aligned} R &= c \left( 1 + \frac{1}{3} \varepsilon^2 - \frac{1}{9} \varepsilon^4 + \frac{8}{15} \frac{h_e}{c} \right), \\ R &= a \left( 1 - \frac{1}{3} \mathfrak{a} - \frac{1}{9} \mathfrak{a}^2 + \frac{8}{15} \frac{h_e}{a} \right). \end{aligned} \right\} \tag{22.6}$$

Dividiert man die Masse durch das Volumen, so erhält man die *mittlere Dichte*

$$\left. \begin{aligned} \sigma_m &= \frac{3\gamma_a}{4\pi f c} \left( 1 - \frac{1}{2} \varepsilon^2 + \frac{3}{2} \mathfrak{c} + \frac{3}{8} \varepsilon^4 - \frac{15}{28} \varepsilon^2 \mathfrak{c} + \frac{24}{35} \frac{h_e}{c} \right), \\ \sigma_m &= \frac{3\gamma_a}{4\pi f a} \left( 1 + \frac{3}{2} \mathfrak{c} + \frac{3}{7} \mathfrak{a} \mathfrak{c} + \frac{24}{35} \frac{h_e}{a} \right). \end{aligned} \right\} \tag{22.7}$$

**23. Mittlere Schwere, Gravitation auf der nicht rotierenden, volumengleichen Kugel; mittlere Breite.** Die mittlere Schwere auf dem Niveausphäroid erhält man am einfachsten, wenn man die mittlere Schwere für das gleichachsige Ellipsoid mit derselben Gesamtmasse berechnet und mit einer Höhenreduktion auf das Niveau $h_{00}$ übergeht.

$$\gamma_m = {}_E \gamma_m + \frac{\partial \gamma}{\partial n} \cdot h_{00}, \tag{23.1}$$

$$\gamma_m = {}_E \gamma_m - \frac{2\gamma}{R} h_{00}, \qquad \gamma_m = {}_E \gamma_m - \frac{16}{15} \frac{\gamma}{R} h_e, \tag{23.2}$$

wobei

$${}_E \gamma_m = \frac{1}{{}_E O} \int_{\overline{\vartheta}=0}^{\pi} \int_{\lambda=0}^{2\pi} {}_E \gamma \sqrt{c^2 + e^2 \cos^2 \overline{\vartheta}} \; \sqrt{c^2 + e^2} \; \sin \overline{\vartheta} \, d\overline{\vartheta} \, d\lambda .$$

${}_E \gamma$ entnimmt man (10.7) mit $C_{40} = 0$ und $r = c$. Das Glied mit $P_{20}(\cos \overline{\vartheta})$ verschwindet bei der Integration, der Rest wird in geschlossener Form integriert.

---

[1] Anhang III, Ziff. 53.

Man erhält

$$_E\gamma_m = \frac{4\pi c^2(1+\varepsilon^2)}{_EO}\left(\frac{fM}{c^2(1+\varepsilon^2)} - \frac{2}{3}\omega^2 c\right),\qquad(23.3)$$

$$\gamma_m = \frac{2\sqrt{1+\varepsilon^2}}{\frac{1}{\varepsilon}\text{Ar Sin }\varepsilon + \sqrt{1+\varepsilon^2}}\left(\frac{fM}{c^2(1+\varepsilon^2)} - \frac{2}{3}\omega^2 c\right) - \frac{16}{15}\frac{\gamma}{R}h_e.\qquad(23.4)$$

Schließlich kommt man mit Reihenentwicklungen auf die Form

$$\left.\begin{aligned}
\gamma_m &= \gamma_{\ddot a}\left(1 - \frac{1}{6}\varepsilon^2 + \frac{5}{6}\mathfrak{c} + \frac{19}{360}\varepsilon^4 + \frac{19}{252}\varepsilon^2\mathfrak{c} - \frac{16}{15}\frac{h_e}{c}\right),\\
\gamma_m &= \gamma_{\ddot a}\left(1 - \frac{1}{3}\mathfrak{a} + \frac{5}{6}\mathfrak{c} - \frac{13}{45}\mathfrak{a}^2 + \frac{19}{126}\mathfrak{a}\mathfrak{c} - \frac{16}{15}\frac{h_e}{a}\right),\\
\gamma_m &= \gamma_{\ddot a}\left(1 + \frac{1}{3}\mathfrak{b} - \frac{1}{60}\varepsilon^4 + \frac{1}{9}\varepsilon^2\mathfrak{b} - \frac{16}{15}\frac{h_e}{c}\right),\\
\gamma_m &= \gamma_{\ddot a}\left(1 + \frac{1}{3}\mathfrak{b} - \frac{1}{15}\mathfrak{a}^2 + \frac{2}{9}\mathfrak{a}\mathfrak{b} - \frac{16}{15}\frac{h_e}{a}\right).
\end{aligned}\right\}\qquad(23.5)$$

Die Gravitation auf der nicht rotierenden, volumengleichen Kugel ist

$$G = \frac{fM}{R^2} = \frac{fM}{a^2}\cdot\frac{a^2}{R^2}.\qquad(23.6)$$

Mit (14.11) und (22.6) findet man

$$\left.\begin{aligned}
G &= \gamma_{\ddot a}\left(1 - \frac{1}{6}\varepsilon^2 + \frac{3}{2}\mathfrak{c} + \frac{7}{72}\varepsilon^4 - \frac{1}{28}\varepsilon^2\mathfrak{c} + \frac{128}{105}\frac{h_e}{c}\right),\\
G &= \gamma_{\ddot a}\left(1 - \frac{1}{3}\mathfrak{a} + \frac{3}{2}\mathfrak{c} - \frac{1}{9}\mathfrak{a}^2 - \frac{1}{14}\mathfrak{a}\mathfrak{c} + \frac{128}{105}\frac{h_e}{a}\right).
\end{aligned}\right\}\qquad(23.7)$$

Diejenige Breite, in der die Normalschwere den Betrag $\gamma_m$ annimmt, kann man als *mittlere Breite* bezeichnen:

$$\gamma(\beta_m) = \gamma_m.\qquad(23.8)$$

Es sind aber auch andere Definitionen der mittleren Breite im Gebrauch: man setzt die Funktion $P_{20}$ der geozentrischen, reduzierten oder geographischen Breite gleich Null, oder den Radiusvektor gleich dem Radius der volumengleichen Kugel:

$$P_{20}(\cos\vartheta_1) = P_{20}(\sin\varphi_1) = 0,\qquad(23.9)$$

$$P_{20}(\cos\bar\vartheta_2) = P_{20}(\sin\psi_2) = 0,\qquad(23.10)$$

$$P_{20}(\sin\beta_3) = 0,\qquad(23.11)$$

$$l(\beta_m^{(R)}) = R.\qquad(23.12)$$

Diese verschieden definierten mittleren Breiten weichen etwas voneinander ab[1], bei genauen Untersuchungen müssen die Unterschiede berücksichtigt werden. Es gilt:

$$\varphi_1 = \psi_2 = \beta_3 = 35°\,15'\,52'',\qquad(23.13)$$

ferner ist

$$\cos^2{}_E\vartheta_m^{(R)} = \frac{1}{3} - \frac{1}{9}\varepsilon^2 + \frac{5}{81}\varepsilon^4,\qquad \cos^2{}_E\vartheta_m^{(R)} = \frac{1}{3} - \frac{2}{9}\mathfrak{a} - \frac{7}{81}\mathfrak{a}^2,\qquad(23.14)$$

$$\cos^2{}_E\bar\vartheta_m^{(R)} = \frac{1}{3} + \frac{1}{9}\varepsilon^2 - \frac{4}{81}\varepsilon^4,\qquad \cos^2{}_E\bar\vartheta_m^{(R)} = \frac{1}{3} + \frac{2}{9}\mathfrak{a} + \frac{11}{81}\mathfrak{a}^2,\qquad(23.15)$$

$$\sin^2{}_E\beta_m^{(R)} = \frac{1}{3} + \frac{1}{3}\varepsilon^2 - \frac{1}{81}\varepsilon^4,\qquad \sin^2{}_E\beta_m^{(R)} = \frac{1}{3} + \frac{2}{3}\mathfrak{a} + \frac{77}{81}\mathfrak{a}^2.\qquad(23.16)$$

Bei Einführung der mittleren Breiten werden einige Beziehungen erheblich einfacher.

---

[1] Eine Zusammenstellung, mit den zugehörigen Werten der Normalschwere, enthält Anhang IV in Ziff. 62.

**24. Größenordnungen.** Während sich einführende Darstellungen[1] oft damit begnügen, die Theorie des Niveausphäroids bis zur Größenordnung $\alpha$ zu entwickeln, ist seit langem entschieden, daß man die Größenordnung $\alpha^2$ berücksichtigen muß, wenn man der Beobachtungsgenauigkeit gerecht werden will[2] (Tabelle 1). Hierbei sollte man sich nicht, wie es gelegentlich geschehen ist[3], auf

Tabelle 1. *Größenordnungen.* $R = 6\,371\,221$ m; $G = 982\,037$ mgal; $\omega = 7{,}292\,115 \cdot 10^{-5}$ sec$^{-1}$

| | | | | | |
|---|---|---|---|---|---|
| $\varepsilon^2$ | 0,006 7682 | $R\varepsilon^2$ | 43 120 m | $G\varepsilon^2$ | 6646 mgal |
| $\alpha$ | 0,003 3670 | $R\alpha$ | 21 452 m | $G\alpha$ | 3 307 mgal |
| $c$ | 0,003 4678 | $R\varepsilon^4$ | 293 m | $Gc$ | 3406 mgal |
| $\varepsilon^4$ | 0,000 0458 | $R\alpha^2$ | 72 m | $R\omega^2$ | 3 388 mgal |
| $\alpha^2$ | 0,000 0113 | $R\alpha c$ | 74 m | $G\varepsilon^4$ | 45 mgal |
| $\alpha c$ | 0,000 0117 | $R\varepsilon^6$ | 2,0 m | $G\alpha^2$ | 11 mgal |
| | | $R\alpha^3$ | 0,2 m | $G\alpha c$ | 11 mgal |
| | | | | $G\varepsilon^6$ | 0,3 mgal |
| | | | | $G\alpha^3$ | 0,04 mgal |

das BRUNSsche Sphäroid[4] beziehen, in dessen Potential — wenn man es in Kugelkoordinaten ausdrückt — das $P_{40}$-Glied fehlt. Man soll dieses Glied beibehalten und zur Anpassung an das Beobachtungsmaterial oder an geeignete mathematische Festsetzungen[5] verwenden. Die für wissenschaftliche Untersuchungen brauchbaren Formelsysteme beruhen auf der HELMERTschen Darstellung (17.2) des Sphäroidpotentials oder einem geschlossenen Ausdruck für das Potential auf einem Niveauellipsoid[6].

Im allgemeinen nimmt man an, daß man den mittleren Erdradius mit einem mittleren Fehler von rund $\pm 50$ m kennt[7]. Dies entspricht der Größenordnung $R\alpha^2$. Schon hierdurch wird gefordert, solche Glieder in den Entwicklungen mitzunehmen.

Dies gilt besonders, wenn man die Abweichung des Geoids von der Normalfigur, die Geoidundulationen, betrachtet. Sie dürften wohl nirgends $\pm 100$ m überschreiten[8], und in kleinen Bereichen kann man Unterschiede der Geoidundulation mit einer Genauigkeit von etwa einem Meter ermitteln[9].

Von der Größenordnung einiger Meter ist auch der Abstand des Niveausphäroids vom gleichachsigen Ellipsoid. Sichere Aussagen über diesen Betrag werden erst möglich sein, wenn ein hinreichend dichtes Netz gleichmäßig verteilter Schwerestationen über die ganze Erde ausgebreitet sein wird. Vorläufig ist sogar das Vorzeichen noch unsicher[10].

**25. Schwereformeln. Abplattung des Niveausphäroids nach Schweremessungen.** Tabelle 11[11] gibt eine Zusammenstellung der wichtigsten, aus Schweremessungen abgeleiteten Schwereformeln. $\gamma_a$ und $\mathfrak{b}$ werden aus den auf das Meeresniveau

---

[1] Zum Beispiel [*18*] und B. GUTENBERG, Lehrbuch der Geophysik.

[2] Daß die ungleichmäßige Verteilung und die Lückenhaftigkeit der Schweremessungen es noch nicht gestatten, die Genauigkeit der Theorie auszuschöpfen, ist kein Grund, die Entwicklung der Auswertungsmethoden aufzuschieben. Es ist anzunehmen, daß die Lücken in nicht allzuferner Zeit im wesentlichen geschlossen werden.

[3] [*4*], [*9*], [*13*]. Ferner O. MEISSER: Praktische Geophysik. 1943.

[4] Ziff. 16 und 17.

[5] Zum Beispiel das internationale Ellipsoid. Anhang IV, Ziff. 62.

[6] Ziff. 27/28.

[7] Man beachte aber den abweichenden, aus umfassendem Beobachtungsmaterial abgeleiteten Wert von JEFFREYS. Anhang IV, Ziff. 61.

[8] Ziff. 34.

[9] Ziff. 29.

[10] Ziff. 25 und Anhang IV, Tabelle 11 (Ziff. 63).

[11] Anhang IV, Ziff. 63.

reduzierten Beobachtungswerten durch Ausgleichung ermittelt; der Berechnung von $\mathfrak{b}'$ pflegt man nicht die Beobachtungen, sondern naheliegende Annahmen über die Massenverteilung im Erdinneren oder die Gestalt des Niveausphäroids zugrunde zu legen. Man kann nicht anders, solange das notwendige Netz gleichmäßig verteilter Schweremessungen noch nicht vorliegt.

Der Gedanke HELMERTS, $\mathfrak{b}'$ gleich Null zu setzen[1], ein wenig fruchtbarer Verzicht, hat keinen Anklang gefunden. In den älteren Schwereformeln ist $\mathfrak{b}' = 0,000007$. Dieser Wert geht auf E. WIECHERT zurück[2] und beruht auf der Annahme einer zweischichtigen, nur aus Mantel und Kern bestehenden Erde in hydrostatischem Gleichgewicht. G. H. DARWIN[3], ausgehend von einer Dichteverteilung nach dem Dichtegesetz von ROCHE[4], kam auf $0,0000067$, also fast denselben Wert, und man pflegte aus dieser Übereinstimmung zu schließen, daß die Massenverteilung im tiefen Erdinnern nur eine geringe Wirkung auf die Gestalt des Niveausphäroids ausübt. Diese Ansicht läßt sich wohl nicht mehr halten, seit ANSEL feststellen mußte[5], daß der DARWINsche Wert durch einen kleinen Rechenfehler nicht unerheblich entstellt ist und in $-0,0000017$ abgeändert werden sollte. Der Einfluß dieser Abänderung ist nicht unbedeutend: wie Tabelle 11 zeigt, liegt das verbesserte Niveausphäroid außerhalb des gleichachsigen Ellipsoids, während der ursprüngliche Wert ein Niveausphäroid ergab, das vom gleichachsigen Ellipsoid eingeschlossen wird. Eine Neuberechnung von E. C. BULLARD[6], der die Dichteverteilung von BULLEN[7] zugrunde liegt, kommt auf $\mathfrak{b}' = 0,0000079$.

Nachdem sich die Internationale Union für Geodäsie und Geophysik (IUGG) entschieden hat, das HAYFORDsche Ellipsoid[8] mit abgerundeten Zahlenwerten als *Internationales Ellipsoid*[9] anzunehmen und zur allgemeinen Anwendung zu empfehlen[10], lag es nahe, eine Schwereformel zu berechnen, die dem internationalen Ellipsoid entspricht und die Schwere wiedergibt, die man auf ihm messen würde, wenn es eine Niveaufläche wäre und alle Massen einschlösse. Bei dieser *Internationalen Schwereformel*[11] ist nur $\gamma_a$ einer Ausgleichung entnommen (HEISKANEN 1928), $\mathfrak{b}$ und $\mathfrak{b}'$ wurden so berechnet[12], daß das Niveausphäroid $U = W_0$ mit dem internationalen Ellipsoid zusammenfällt. Die meist zitierte Form[13] ist

$$\boxed{\gamma = 978,0490 \cdot (1 + 0,0052884 \sin^2 \beta - 0,0000059 \sin^2 2\beta \text{ Gal.}}  \tag{25.1}$$

Als Vergleichs- und Ausgangswert hat sich die internationale Schwere gut bewährt. Bis jetzt liegt kein Anlaß vor, von der internationalen Schwereformel

---

[1] Ziff. 16.
[2] E. WIECHERT: Über die Massenverteilung im Innern der Erde. Göttinger Nachr. **1897**, 221. Dichte des Kerns 8,206, des Mantels 3,20 g/cm³, Radius des Kerns 0,7839 R.
[3] Month. Not. Roy. Astronom. Soc. **60**, 82 (1899).
[4] $\bar{\sigma} = \sigma_0 \left[ 1 - k \left( \dfrac{\varrho}{c} \right)^\lambda \right]^\mu$; $\sigma_0$ Mittelpunktsdichte; $\bar{\sigma}$ mittlere Dichte der von dem Ellipsoid mit der kleinen Achse $\varrho$ eingeschlossenen Masse; $k, \lambda, \mu$ Konstanten.
[5] [11], S. 668.
[6] Geophysics Suppl. **5**, 186 (1948).
[7] Bull. Seism. Soc. Amer. **30**, 235 (1940); **32**, 19 (1942).
[8] Anhang IV, Ziff. 61.
[9] Anhang IV, Ziff. 62. Definitionsgemäß hat das internationale Ellipsoid die große Achse $a = 6378,388$ km und die Abplattung $\mathfrak{a} = 1:297$ (genau). Tafeln und Literatur in [22].
[10] Madrid 1924. Bull. Géod. **1925**, Nr. 7, 552.
[11] Von der IUGG in Stockholm angenommen 1930. Bull. Géod. **1930**, Nr. 27, 239. Tafeln und Literatur in [22].
[12] Nach CASSINIS. Ziff. 28.
[13] Andere Darstellungen der internationalen Schwereformel im Anhang IV, Ziff. 64.

abzugehen, wenn sie auch nicht mehr die beste Annäherung an das gemessene Schwerefeld darstellt[1]. Fast alle neueren Schwereformeln haben den $b'$-Wert 0,000 005 9 von der internationalen Formel übernommen, oft übernimmt man auch den $b$-Wert 0,005 288 4.

Die Äquatorschwere 978,0490 bezieht sich auf die absolute Schwerebestimmung von KÜHNEN und FURTWÄNGLER, die dem *Potsdamer Schweresystem*[2] zugrunde liegt. Nach Vergleich mit neueren absoluten Schweremessungen ist sie wahrscheinlich etwa 12 bis 15 mgal zu hoch[3]. Eine entsprechende Abänderung der internationalen Schwereformel soll erst dann vorgenommen werden, wenn die begonnenen absoluten Schwerebestimmungen Klarheit erbracht haben. Es ist möglich, daß diese Verbesserung auch $b$ beeinflußt.

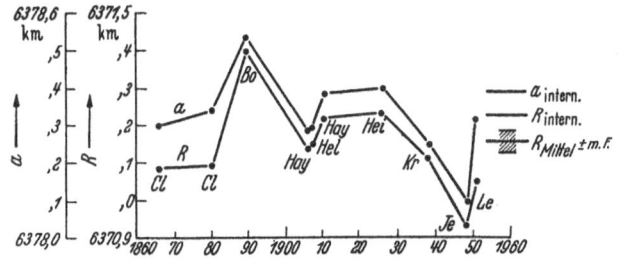

Vergleicht man die den neueren Ausgleichungen entnommenen Abplattungswerte[4] (Fig. 17), so sieht es aus, als sei die reziproke Abplattung mit einem mittleren Fehler von weniger als $\pm 1$ bekannt. Man wird in dieser Ansicht bestärkt, wenn man auch die aus Lotabweichungen ermittelten Abplattungen[6] heranzieht. Es ist jedoch möglich, daß die gute Übereinstimmung vorgetäuscht ist und auf einer gleichartigen Auswahl des Ausgangsmaterials beruht.

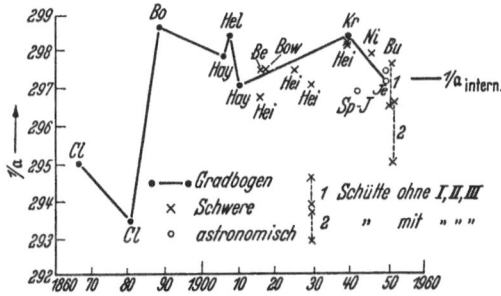

Fig. 17. Die wichtigsten Bestimmungen der Erdabmessungen, zeitlich geordnet. *Cl* CLARKE, *Bo* BONSDORFF, *Hay* HAYFORD, *Hel* HELMERT, *Hei* HEISKANEN, *Kr* KRASSOWSKI, *Je* JEFFREYS, *Le* LEDERSTEGER, *Bow* BOWIE, *Sp-J* SPENCER-JONES, *Ni* NISKANEN, *Bu* BULLARD[5].

Zu denken geben der Äquatorradius von JEFFREYS[7] und die Abplattungswerte, die H. SCHÜTTE[8] aus dem Schwereverzeichnis von N. F. ZHURAVLEV[9] berechnet hat.

ZHURAVLEV hat, in üblicher Weise, folgende Stationsgruppen von der Ausgleichung ausgeschlossen: 1. Höhenstationen mit Höhen von mehr als 2000 m,

---

[1] Einige neuere Ausgleichsergebnisse enthält die Zusammenstellung im Anhang IV, Ziff. 63.

[2] $g = 981,274$ Gal im Pendelsaal des Geodätischen Instituts, Potsdam: $\beta = 52° 22,86'$ N, $\lambda = 13° 04,06'$ E, $H = 87$ m.

[3] Siehe S. 204 f in diesem Bande.

[4] Anhang IV, Ziff. 63.

[5] Eine Änderung von $1/a$ um $+0,1$ bedeutet bei festgehaltenem $a$ eine Änderung von $c$ um $+7,2$ m und eine Änderung von $R$ um $+2,4$ m, bei festgehaltenem $R$ eine Änderung von $a$ um $-2,4$ m und eine Änderung von $c$ um $+4,8$ m.

[6] Anhang IV, Ziff. 61.

[7] Month. Not. Roy Astronom. Soc., Geophys. Suppl. **5**, 219 (1948). Ferner Anhang IV, Ziff. 61.

[8] Gerlands Beitr. Geophys. **62**, 9 (1950).

[9] N. F. ZHURAVLEV: Bestimmung der Abplattung des Erdsphäroids auf Grund gravimetrischer Messungen. [Russisch.] Moskau 1940. Enthält ein Verzeichnis von 10712 mit Pendeln gemessenen Schwerewerten.

2. Insel- und Meeresstationen mit Freiluftanomalien von mehr als 100 mgal,
3. Landstationen mit Freiluftanomalien von mehr als 100 mgal. SCHÜTTE hat
den Einfluß dieser Gruppen getrennt untersucht und fand: Die Hinzunahme
von Gruppe 1 erniedrigt 1/α um 0,3. Die Hinzunahme von Gruppe 2 erniedrigt
1/α um 1,6. Die Hinzunahme von Gruppe 3 erhöht 1/α um 0,3. Die Hinzunahme
aller Gruppen zugleich erniedrigt 1/α um höchstens 1,5. Besonders wirksam
erweist sich die Gruppe der Insel- und Meeresstationen. Das ist erklärlich. Denn
viele Inselstationen sind stark gestört (Fig. 18), und der weite Abstand der
Meßpunkte auf See bringt es
mit sich, daß der Einzelmessung ein größerer Einfluß
eingeräumt wird als in den
engmaschigen Netzen von
Landstationen. Eine engmaschige Vermessung der
Umgebung fraglicher Schwerestationen ist zu fordern;
anders wird die Bedeutung
herausfallender Einzelwerte
nicht richtig erkannt. Sie
einfach wegzulassen, ist ebenso bedenklich, wie sie kritiklos mitzunehmen.

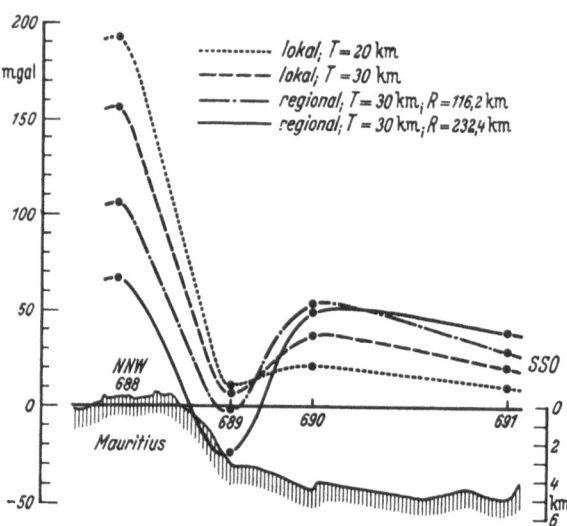

Fig. 18. Verlauf isostatischer Schwereanomalien (AIRY-HEISKANEN und
VENING-MEINESZ) in der Nähe einer Vulkaninsel.

**26. Das Niveauellipsoid.
Formeln von JEFFREYS.** In
allen Fällen ergab sich, daß
die Abweichung des Niveausphäroids vom gleichachsigen
Ellipsoid nur einige Meter
beträgt und oft so klein
herauskommt, daß sie mit den in den Reihenentwicklungen vernachlässigten
Größen vergleichbar ist. So liegt es nahe, mit einem *Niveauellipsoid* statt
dem allgemeineren Niveausphäroid zu rechnen und in den mitgeteilten Formelsystemen $h_e$ [definiert in (12.8)] gleich Null zu setzen. Man hat dann eine
Niveaufläche von geometrisch einfacher Gestalt, die man ohne weitere Vernachlässigungen den Referenzellipsoiden der geodätischen Netze oder dem internationalen Ellipsoid anpassen kann.

Setzt man nur $h_e$ gleich Null, so werden die mathematischen Beziehungen nicht wesentlich einfacher. Sie kürzen sich jedoch ab, wenn man statt c [definiert in (14.2)] die Größe

$$\bar{c} = \frac{\omega^2 a^2 c}{f M} = \frac{\omega^2 R^3}{f M} \qquad (26.1)$$

einführt. Ein solches Formelsystem hat H. JEFFREYS entwickelt und mitgeteilt[1].
Zur Umrechnung findet man

$$\bar{c} = c\left(1 - \frac{3}{2} c\right), \qquad c = \bar{c}\left(1 + \frac{3}{2} \bar{c}\right) \qquad (26.2)$$

---

[1] [21], S. 125ff. Formeln, die sich in gleicher oder sehr ähnlicher Form bei JEFFREYS
finden, werden mit [J] kenntlich gemacht. Die wichtigsten Bezeichnungen bei JEFFREYS:
$\Phi$ für $\beta$, $\Phi'$ für $\varphi$, $r$ für $l$, $e$ für $\alpha$, $m$ für $\bar{c}$, $U$ für $V$, $\Psi$ für $W$, $\Psi'$ für $W - f M/l$, $\Psi_1$ für $U - f M/l$,
$\Psi_2$ für $T$, $J$ für $\dfrac{2}{3}\dfrac{C - \dfrac{A+B}{2}}{M a^2}$, $C$ für $W_0$, $g$ für $\gamma$, $g_0$ für $\gamma_{\ddot{a}}$, $D$ für $\mathfrak{d}$, $h$ für $H$.

und erhält[1]:

$$\mathfrak{a} + \mathfrak{b} = \frac{5}{2}\,\bar{c} - \frac{17}{14}\,\mathfrak{a}\,\bar{c} + \frac{15}{14}\,\bar{c}^2 \qquad [J], \qquad (26.3)$$

$$\mathfrak{b} = \frac{7}{2}\,\mathfrak{a}^2 - \frac{5}{2}\,\mathfrak{a}\,\bar{c} \qquad [J], \qquad (26.4)$$

$$\mathfrak{b}' = -\frac{1}{8}\,\mathfrak{a}^2 + \frac{5}{8}\,\mathfrak{a}\,\bar{c} \qquad [J], \qquad (26.5)$$

$$\gamma_{\ddot{a}} = \frac{f\,M}{a^2}\left(1 + \mathfrak{a} - \frac{3}{2}\,\bar{c} + \mathfrak{a}^2 - \frac{27}{14}\,\mathfrak{a}\,\bar{c}\right) \qquad [J], \qquad (26.6)$$

$$\frac{3}{2}\,\frac{C - \dfrac{A+B}{2}}{M\,a^2} = \mathfrak{a} - \frac{1}{2}\,\bar{c} - \frac{1}{2}\,\mathfrak{a}^2 + \frac{1}{7}\,\mathfrak{a}\,\bar{c} \qquad [J], \qquad (26.7)$$

$$\frac{a}{l} = 1 + \mathfrak{a}\cos^2\vartheta + \frac{3}{2}\,\mathfrak{a}^2\cos^2\vartheta - \frac{1}{2}\,\mathfrak{a}^2\cos^4\vartheta \qquad [J], \qquad (26.8)$$

$$\gamma = \gamma_{\ddot{a}}\left\{1 + \left(\frac{5}{2}\,\bar{c} - \mathfrak{a} - \frac{17}{14}\,\mathfrak{a}\,\bar{c} + \frac{15}{4}\,\bar{c}^2\right)\cos^2\vartheta - \left(\frac{7}{8}\,\mathfrak{a}^2 - \frac{15}{8}\,\mathfrak{a}\,\bar{c}\right)\cos^2 2\vartheta\right\} \quad [J], \quad (26.9)$$

$$\gamma = \gamma_{\ddot{a}}\left\{1 + \left(\frac{5}{2}\,\bar{c} - \mathfrak{a} - \frac{17}{14}\,\mathfrak{a}\,\bar{c} + \frac{15}{4}\,\bar{c}^2\right)\cos^2\beta + \left(\frac{1}{8}\,\mathfrak{a}^2 - \frac{5}{8}\,\mathfrak{a}\,\bar{c}\right)\sin^2 2\beta\right\} \quad [J]. \quad (26.10)$$

Es sei $\varphi_1$, die „mittlere geozentrische Breite", so definiert, daß $P_{2\,0}(\sin\varphi_1) = 0$, und $l_1$ sei der Radiusvektor des Niveauellipsoids in der Breite $\varphi_1$. Man findet

$$\varphi_1 = 35°\,15'\,52'', \qquad (26.11)$$

$$l_1 = a\left(1 - \frac{1}{3}\,\mathfrak{a} - \frac{1}{3}\,\mathfrak{a}^2\right), \qquad l_1 = R\left(1 - \frac{2}{9}\,\mathfrak{a}^2\right), \qquad R = l_1\left(1 + \frac{2}{9}\,\mathfrak{a}^2\right). \qquad (26.12)$$

$\gamma_1$ sei die Normalschwere in der Breite $\varphi_1$. Mit ihr erhält man

$$\gamma_{\ddot{a}} = \gamma_1\left(1 + \frac{1}{3}\,\mathfrak{a} - \frac{5}{6}\,\bar{c} + \frac{8}{9}\,\mathfrak{a}^2 - \frac{229}{126}\,\mathfrak{a}\,\bar{c} - \frac{5}{9}\,\bar{c}^2\right), \qquad (26.13)$$

$$\gamma_1 = \frac{f\,M}{l_1^2}\left(1 - \frac{2}{3}\,\bar{c} - \frac{10}{9}\,\mathfrak{a}^2 + \frac{10}{9}\,\mathfrak{a}\,\bar{c}\right), \qquad (26.14)$$

$$\frac{\bar{c}}{1 - \frac{2}{3}\bar{c}} = \frac{\omega^2\,l_1}{\gamma_1}, \qquad [J] \qquad (26.15)$$

$$\bar{c} = \frac{\omega^2\,l_1}{\gamma_1}\left(1 + \frac{2}{3}\,\bar{c}\right), \qquad (26.16)$$

$$a = l_1\left(1 + \frac{1}{3}\,\mathfrak{a} + \frac{4}{9}\,\mathfrak{a}^2\right), \qquad (26.17)$$

$$\text{Oberfläche} = 4\pi\,l_1^2\left(1 + \frac{28}{45}\,\mathfrak{a}^2\right), \qquad (26.18)$$

$$\text{Volumen} = \frac{4}{3}\,\pi\,l_1^3\left(1 + \frac{2}{3}\,\mathfrak{a}^2\right), \qquad (26.19)$$

$$\sigma_m = \frac{3\gamma_{\ddot{a}}}{4\pi\,f\,l_1}\left(1 + \frac{2}{3}\,\bar{c} + \frac{4}{9}\,\mathfrak{a}^2 - \frac{10}{9}\,\mathfrak{a}\,\bar{c} + \frac{4}{9}\,\bar{c}^2\right), \qquad (26.20)$$

$$G = \gamma_1\left(1 + \frac{2}{3}\,\bar{c} + \frac{2}{3}\,\mathfrak{a}^2 - \frac{10}{9}\,\mathfrak{a}\,\bar{c} + \frac{4}{9}\,\bar{c}^2\right). \qquad (26.21)$$

[1] Da sich alle Größen auf das Niveauellipsoid beziehen, wird der Index $E$ weggelassen.

**27. Das Niveauellipsoid in geschlossener Form.** Während sich das allgemeine Niveausphäroid nur mit Reihenentwicklungen behandeln läßt, können die Beziehungen am Niveauellipsoid ohne Vernachlässigungen in geschlossener Form entwickelt werden. So erhält man die Schwere aus (10.5) mit $r = c$ und $C_{40} = 0$. Setzt man noch (11.6)

$$e/c = \varepsilon,$$

so gilt

$$\gamma(c, \bar{\vartheta}) = \left( \frac{fM}{c^2(1+\varepsilon^2)} - \frac{2}{3}\,\omega^2 c\,(1 - P_{20}(\cos \bar{\vartheta})) - \frac{i}{3}\frac{Q'_{20}(i/\varepsilon)}{Q_{20}(i/\varepsilon)}\,\omega^2 c\,\frac{1+\varepsilon^2}{\varepsilon}\,P_{20}(\cos \bar{\vartheta}) \right) \sqrt{\frac{1+\varepsilon^2}{1+\varepsilon^2\cos^2\bar{\vartheta}}}\,, \right\} \tag{27.1}$$

$$\gamma_{\ddot{a}} = \gamma(c, 90°) = \left( \frac{fM}{c^2(1+\varepsilon^2)} - \omega^2 c + \frac{i}{6}\frac{Q'_{20}(i/\varepsilon)}{Q_{20}(i/\varepsilon)}\,\omega^2 c\,\frac{1+\varepsilon^2}{\varepsilon} \right) \sqrt{1+\varepsilon^2}\,, \tag{27.2}$$

$$\gamma_p = \gamma(c, 0°) = \frac{fM}{c^2(1+\varepsilon^2)} - \frac{1}{3}\frac{Q'_{20}(i/\varepsilon)}{Q_{20}(i/\varepsilon)}\,\omega^2 c\,\frac{1+\varepsilon^2}{\varepsilon}\,, \tag{27.3}$$

$$\gamma_2 = \gamma(c, \bar{\vartheta}_2) = \left( \frac{fM}{c^2(1+\varepsilon^2)} - \frac{2}{3}\,\omega^2 c \right) \sqrt{\frac{1+\varepsilon^2}{1+\frac{1}{3}\varepsilon^2}}\,, \tag{27.4}$$

wobei $P_{20}(\cos \bar{\vartheta}_2) = 0$.

Führt man nach (14.2)

$$c = \frac{\omega_{\ddot{a}}^2\,a}{\gamma_{\ddot{a}}} = \frac{\omega^2 c\,\sqrt{1+\varepsilon^2}}{\gamma_{\ddot{a}}}$$

ein, so findet man

$$\mathfrak{b} = \frac{\gamma_p - \gamma_{\ddot{a}}}{\gamma_{\ddot{a}}} = -\frac{\gamma_2}{\gamma_{\ddot{a}}}\,\sqrt{\frac{1+\frac{1}{3}\varepsilon^2}{1+\varepsilon^2}}\,\left(\sqrt{1+\varepsilon^2} - 1\right) + \\ + \mathfrak{c}\left( \frac{1}{3\sqrt{1+\varepsilon^2}} - \frac{i}{6}\frac{Q'_{20}(i/\varepsilon)}{Q_{20}(i/\varepsilon)}\frac{\sqrt{1+\varepsilon^2}}{\varepsilon} \right)\left(2 + \sqrt{1+\varepsilon^2}\right), \right\} \tag{27.5}$$

$$fM = \gamma_2\,c^2\,\sqrt{1 + \frac{1}{3}\varepsilon^2}\,\sqrt{1+\varepsilon^2} + \frac{2}{3}\,\omega^2 c^3\,(1+\varepsilon^2), \tag{27.6}$$

und man erhält aus (10.2) und (10.3) mit $C_{40} \doteq 0$:

$$C - \frac{A+B}{2} = -i\frac{2}{45}\frac{\omega^2 c^5}{f}\,\frac{\varepsilon^3}{Q_{20}(i/\varepsilon)}\,(1+\varepsilon^2) + \frac{1}{3}\,c^2\,\varepsilon^2\,M. \tag{27.7}$$

(27.5), (27.6) und (27.7) enthalten die Theorie des Niveauellipsoids in geschlossener Form. (27.5) entspricht dem CLAIRAUTschen *Theorem* und bestimmt $\varepsilon$ aus $\mathfrak{b}$, $\mathfrak{c}$, $\gamma_{\ddot{a}}$ und $\gamma_2$, wenn die von $\varepsilon$ abhängigen Ausdrücke tabelliert sind, wozu — da der ungefähre Betrag von $\varepsilon$ bekannt ist — ein kleiner Bereich von $\varepsilon$ genügt. Alsdann führen (27.6) und (27.7) auf die Masse und den Unterschied der Hauptträgheitsmomente.

Für die Zahlenrechnung sind Reihenentwicklungen bequemer als (27.5).

**28. Die Normalschwere auf dem Niveauellipsoid. Formeln von SOMIGLIANA[1].** Setzt man

$$\frac{1}{\varepsilon^3}\,(\varepsilon - \arctan \varepsilon) = \Psi(\varepsilon), \tag{28.1}$$

---

[1] C. SOMIGLIANA: Teoria del campo gravitazionale dell'ellissoide di rotazione. Mem. Soc. astronom. ital. **4** (1929). C. SOMIGLIANA: Sul campo gravitazionale esterno del geoide ellissoidico. Rend. Accad. naz. Lincei. **1930**. Ausführlich wiedergegeben in [*16*].

so ergibt sich

$$-\frac{i}{3}\frac{Q'_{20}(i/\varepsilon)}{Q_{20}(i/\varepsilon)}\,\omega^2\,\frac{1+\varepsilon^2}{\varepsilon}=\frac{2}{3}\,\omega^2\,\frac{3(1+\varepsilon^2)\,\Psi(\varepsilon)-1}{(3+\varepsilon^2)\,\Psi(\varepsilon)-1}$$

und hiermit aus (27.1):

$$\left.\begin{aligned}\frac{\gamma}{c}=\left(\frac{4}{3}\,\pi f\,\sigma_m+\frac{2}{3}\,\frac{\omega^2}{(3+\varepsilon^2)\,\Psi(\varepsilon)-1}\right)\sqrt{\frac{1+\varepsilon^2}{1+\varepsilon^2\cos^2\bar\vartheta}}-\\-2\omega^2\,\frac{(1+\varepsilon^2)\,\Psi(\varepsilon)}{(3+\varepsilon^2)\,\Psi(\varepsilon)-1}\sqrt{\frac{1+\varepsilon^2\cos^2\bar\vartheta}{1+\varepsilon^2}}\,,\end{aligned}\right\}\tag{28.2}$$

wobei

$$\sigma_m=\frac{3}{4\pi}\,\frac{M}{c^3(1+\varepsilon^2)}$$

die mittlere Dichte bedeutet.

Man führt die geographische Breite $\beta$ ein:

$$1+\varepsilon^2\cos^2\bar\vartheta=\frac{1+\varepsilon^2}{1+\varepsilon^2\cos^2\beta}\,,$$

setzt

$$\begin{aligned}\frac{4}{3}\,\pi f\,\sigma_m+\frac{2}{3}\,\frac{\omega^2}{(3+\varepsilon^2)\,\Psi(\varepsilon)-1}&=A^*,\\-2\omega^2\,\frac{(1+\varepsilon^2)\,\Psi(\varepsilon)}{(3+\varepsilon^2)\,\Psi(\varepsilon)-1}&=B^*,\\\sqrt{1+\varepsilon^2\cos^2\beta}&=V\end{aligned}\tag{28.3}$$

und erhält die Schwere in der einfachen Form

$$\frac{\gamma}{c}=A^*\,V+\frac{B^*}{V}\,.\tag{28.4}$$

Zur Bestimmung von $\varepsilon$ genügt es, die Schwere in drei verschiedenen Breiten zu kennen. Dann hat man drei lineare Gleichungen, die in $(1/c)$, $A^*$ und $B^*$ homogen sind und nur dann eine nicht triviale Lösung haben, wenn

$$\begin{vmatrix}\gamma_{\mathrm{I}}&V_{\mathrm{I}}&\dfrac{1}{V_{\mathrm{I}}}\\[2mm]\gamma_{\mathrm{II}}&V_{\mathrm{II}}&\dfrac{1}{V_{\mathrm{II}}}\\[2mm]\gamma_{\mathrm{III}}&V_{\mathrm{III}}&\dfrac{1}{V_{\mathrm{III}}}\end{vmatrix}=0.\tag{28.5}$$

Durch Auflösen der Determinante erhält man

$$\begin{aligned}\gamma_{\mathrm{I}}V_{\mathrm{I}}(\cos^2\beta_{\mathrm{II}}-\cos^2\beta_{\mathrm{III}})+\gamma_{\mathrm{II}}V_{\mathrm{II}}(\cos^2\beta_{\mathrm{III}}-\cos^2\beta_{\mathrm{I}})+\\+\gamma_{\mathrm{III}}V_{\mathrm{III}}(\cos^2\beta_{\mathrm{I}}-\cos^2\beta_{\mathrm{II}})=0,\end{aligned}\tag{28.6}$$

eine Beziehung, die $\varepsilon$ als einzige Unbekannte enthält.

Im Spezialfall $\beta_{\mathrm{I}}=0^\circ$, $\beta_{\mathrm{II}}=90^\circ$, $\beta_{\mathrm{III}}=\beta$ (beliebig) erhält man

$$\gamma=\frac{(\gamma_\ddot{a}\sqrt{1+\varepsilon^2}-\gamma_p)\cos^2\beta+\gamma_p}{\sqrt{1+\varepsilon^2\cos^2\beta}}\,,\tag{28.7}$$

eine Schwereformel in strenger, geschlossener Form, die sich in

$$\gamma = \frac{\gamma_{\mathfrak{a}}\, a \cos^2\beta + \gamma_{\mathfrak{p}}\, c \sin^2\beta}{\sqrt{a^2 \cos^2\beta + c^2 \sin^2\beta}} \tag{28.8}$$

und

$$\gamma = \gamma_{\mathfrak{a}} \frac{1 + (\mathfrak{b} - \mathfrak{a} - \mathfrak{a}\,\mathfrak{b}) \sin^2\beta}{\sqrt{1 - (2\,\mathfrak{a} - \mathfrak{a}^2) \sin^2\beta}} \tag{28.9}$$

umformen läßt. Hat man $\varepsilon$ berechnet, so erhält man $B^*$ aus (28.1) und (28.3), findet $c$ aus

$$\frac{\gamma_i}{V_i} - \frac{\gamma_j}{V_j} = c\,B^*\left(\frac{1}{V_i^2} - \frac{1}{V_j^2}\right), \tag{28.10}$$

$A^*$ aus (28.4) und $\sigma_m$ aus (28.3).

Aus drei Werten der Normalschwere kann das Niveauellipsoid mit der von ihm eingeschlossenen Masse vollständig ermittelt werden. Die Bestimmung von $c$ ist jedoch nicht sehr genau, und es ist vorzuziehen, $c$ aus geodätischen Messungen abzuleiten, wie es auch·allgemein geschieht.

Für $\gamma_i = \gamma_{\mathfrak{p}}$, $\gamma_j = \gamma_{\mathfrak{a}}$ ergibt sich aus (28.10):

$$\frac{\gamma_{\mathfrak{p}}}{c} - \frac{\gamma_{\mathfrak{a}}}{c\,\sqrt{1 + \varepsilon^2}} = B^*\left(1 - \frac{1}{1 + \varepsilon^2}\right)$$

und hieraus nach einfachen Umformungen:

$$\mathfrak{a} + \mathfrak{b} = \frac{2\,c}{\sqrt{1 + \varepsilon^2}}\, \frac{\varepsilon^2\,\Psi(\varepsilon)}{1 - (3 + \varepsilon^2)\,\Psi(\varepsilon)}, \qquad \varepsilon^2 = \frac{1}{(1 - \mathfrak{a})^2} - 1 \tag{28.11}$$

eine strenge Form des CLAIRAUTschen Theorems.

Für die Zahlenrechnung eignen sich Reihenentwicklungen besser. Mit Entwicklung von $\Psi(\varepsilon)$ ergibt sich

$$\begin{aligned}
\mathfrak{a} + \mathfrak{b} &= \frac{5}{2}\,c\left(1 - \frac{17}{70}\,\varepsilon^2 + \frac{71}{392}\,\varepsilon^4 - \frac{45379}{301840}\,\varepsilon^6\right), \\
\mathfrak{a} + \mathfrak{b} &= \frac{5}{2}\,c\left(1 - \frac{17}{35}\,\mathfrak{a} - \frac{1}{245}\,\mathfrak{a}^2 - \frac{13}{18865}\,\mathfrak{a}^3\right),
\end{aligned} \tag{28.12}$$

und nach Entwicklung von $\dfrac{V}{\sqrt{1 + \varepsilon^2}}$ wird

$$\begin{aligned}
&\gamma = \mathfrak{b}_0 + \mathfrak{b}_2 \sin^2\beta + \mathfrak{b}_4 \sin^4\beta + \mathfrak{b}_6 \sin^6\beta + \cdots \\
&\text{mit} \\
&\mathfrak{b}_0 = c\left(A^* \sqrt{1 + \varepsilon^2} + \frac{B^*}{\sqrt{1 + \varepsilon^2}}\right), \qquad \mathfrak{b}_2 = -\frac{c}{2}\,\frac{\varepsilon^2}{1 + \varepsilon^2}\left(A^* \sqrt{1 + \varepsilon^2} - \frac{B^*}{\sqrt{1 + \varepsilon^2}}\right) \\
&\mathfrak{b}_{2n} = -\frac{1 \cdot 1 \cdot 3 \cdot 5 \ldots (2n - 3)}{2 \cdot 4 \cdot 6 \cdot 8 \ldots 2n}\,c\left(\frac{\varepsilon^2}{1 + \varepsilon^2}\right)^n\left(A^* \sqrt{1 + \varepsilon^2} - (2n - 1)\frac{B^*}{\sqrt{1 + \varepsilon^2}}\right).
\end{aligned} \tag{28.13}$$

Schließlich führt die Entwicklung von (28.9) nach längeren Umformungen auf

$$\gamma = \gamma_a \left(1 + \mathfrak{b} \sin^2 \beta - \mathfrak{b}' \sin^2 2\beta - \mathfrak{b}'' \sin^2 \beta \sin^2 2\beta - \mathfrak{b}''' \sin^4 \beta \sin^2 2\beta \ldots\right)$$

mit

$$\mathfrak{b}' = \frac{1}{8}\, \mathfrak{a}\,(\mathfrak{a} + 2\,\mathfrak{b}),$$

$$\mathfrak{b}'' = \frac{1}{8}\, \mathfrak{a}^2\,(2\,\mathfrak{a} + 3\,\mathfrak{b}) - \frac{1}{32}\, \mathfrak{a}^3\,(3\,\mathfrak{a} + 4\,\mathfrak{b}),$$

$$\mathfrak{b}''' = \frac{5}{32}\, \mathfrak{a}^3\,(3\,\mathfrak{a} + 4\,\mathfrak{b}) - \frac{3}{32}\, \mathfrak{a}^4\,(4\,\mathfrak{a} + 5\,\mathfrak{b}) + \frac{1}{64}\, \mathfrak{a}^5\,(5\,\mathfrak{a} + 6\,\mathfrak{b}),$$

$$\cdot \quad \cdot \quad \cdot \quad \cdot \quad \cdot \quad \cdot \quad \cdot \quad \cdot \quad \cdot \quad \cdot \quad \cdot \quad \cdot \quad \cdot \quad \cdot \quad \cdot \quad \cdot \quad \cdot \quad \cdot \quad \cdot$$

(28.14)

einen Ausdruck, der die Berechnung der Schwereformel gestattet, wenn $\gamma_a$ und $\mathfrak{a}$ gegeben sind. $\mathfrak{b}$ hängt über das CLAIRAUTsche Theorem mit $\gamma_a$ und $\mathfrak{a}$ zusammen.

**29. Die Beziehung zwischen der Geoidundulation und den Schwereanomalien.** *α) Entwicklung nach Kugelfunktionen.* Wie sich zeigen wird[1] bleiben die Geoidundulationen $\zeta$ im Bereich von $\pm 100$ m. Solange es nicht möglich ist, sie mit wesentlich größerer Genauigkeit als $\pm 1$ m zu bestimmen, kann schon die Größenordnung $\zeta \varepsilon^2$ vernachlässigt werden. Hiermit sind weitgehende Vereinfachungen möglich.

Grundlage ist die *Differentialgleichung der physikalischen Geodäsie* (6.1). Sie verbindet die Schwereanomalie

$$\Delta g = g^{(P)} - \gamma^{(Q)},$$ (29.1)

die man aus den auf das Meeresniveau reduzierten Schwerewerten unmittelbar erhält, mit der Normalschwere und dem Potential der Störungsmassen. Ist dieses Potential ermittelt, so sind mit (5.2) die Geoidundulationen gegeben.

$\gamma$ erhält man genau genug, wenn man in (10.5) nur das erste Glied der rechten Seite berücksichtigt und $e$ gegen $\varrho$ vernachlässigt. Dann ist

$$\gamma = \frac{f M}{\varrho^2},$$ (29.2)

$$\frac{\partial \gamma}{\partial n} = -\frac{2 f M}{\varrho^3},$$ (29.3)

$$\frac{1}{\gamma}\frac{\partial \gamma}{\partial n} = -\frac{2}{\varrho},$$ (29.4)

so daß (6.1) — mit Vernachlässigung des Unterschieds zwischen $\dfrac{\partial}{\partial n}$ und $\dfrac{\partial}{\partial \varrho}$ — in der Gestalt

$$\Delta g = -\frac{\partial T}{\partial \varrho} - \frac{2}{\varrho}\, T$$ (29.5)

erscheint. Hierbei kann auch vernachlässigt werden, daß sich $\dfrac{1}{\gamma}\dfrac{\partial \gamma}{\partial n}$ und $T$ eigentlich auf verschiedene Aufpunkte beziehen.

Vernachlässigt man $e$ gegen $\varrho$, so erhält man aus (9.9), wenn man auch die Unterschiede der verschiedenen Arten von Polabständen außer acht läßt,

$$T = \sum_{n=0}^{\infty}{}^{*} \sum_{\nu=0}^{n}{}^{*} Q_{n\nu}\left(\frac{i\varrho}{e}\right) W_{n\nu}(\vartheta, \lambda),$$ (29.6)

---

[1] Ziff. 34.

wobei

$$W_{n\nu}(\vartheta, \lambda) = ({}_cW_{n\nu}\cos\nu\lambda + {}_sW_{n\nu}\sin\nu\lambda)\, P_{n\nu}(\cos\vartheta) \tag{29.7}$$

eine LAPLACEsche Kugelfunktion bezeichnet. Die Sterne an den Summenzeichen deuten an, daß diejenigen Glieder weggelassen werden, die im Potential der Normalschwere enthalten sind.

Eingesetzt in (29.5) ergibt sich:

$$\Delta g = -\sum_{n=0}^{\infty}{}^{*} \sum_{\nu=0}^{n}{}^{*} \left(\frac{i}{e}\, Q'_{n\nu}\left(\frac{i\varrho}{e}\right) + \frac{2}{\varrho}\, Q_{n\nu}\left(\frac{i\varrho}{e}\right)\right) W_{n\nu}(\vartheta, \lambda). \tag{29.8}$$

$\Delta g$ ist eine bekannte Funktion von $\vartheta$ und $\lambda$ und wird nach Kugelfunktionen entwickelt[1]:

$$\boxed{\Delta g = \sum_{n=0}^{\infty}{}^{*} \sum_{\nu=0}^{n}{}^{*} G_{n\nu}(\vartheta, \lambda).} \tag{29.9}$$

Mit (29.8) ergibt gliedweiser Vergleich:

$$W_{n\nu}(\vartheta, \lambda) = -\frac{G_{n\nu}(\vartheta, \lambda)}{\dfrac{i}{e}\, Q'_{n\nu}\left(\dfrac{i\varrho}{e}\right) + \dfrac{2}{\varrho}\, Q_{n\nu}\left(\dfrac{i\varrho}{e}\right)}, \tag{29.10}$$

$$\boxed{T = -\sum_{n=0}^{\infty}{}^{*} \sum_{\nu=0}^{n}{}^{*} \frac{Q_{n\nu}\left(\dfrac{i\varrho}{e}\right)}{\dfrac{i}{e}\, Q'_{n\nu}\left(\dfrac{i\varrho}{e}\right) + \dfrac{2}{\varrho}\, Q_{n\nu}\left(\dfrac{i\varrho}{e}\right)}\, G_{n\nu}(\vartheta, \lambda).} \tag{29.11}$$

Nun gilt für $|x| \geq 1$

$$Q_{n\nu}(x) \sim \sqrt{1 - x^2}^{\nu}\, \frac{1}{x^{n+\nu+1}} + \cdots, \tag{29.12}$$ [2]

und bei hinreichend großem $x$ ist

$$Q_{n\nu}(x) \sim \frac{1}{x^{n+1}}, \qquad Q'_{n\nu}(x) \sim -\frac{n+1}{x^{n+2}} = -\frac{n+1}{x}\, Q_{n\nu}(x).$$

Hiernach gilt bei kleinem $e/\varrho$ [3]:

$$Q'_{n\nu}\left(\frac{i\varrho}{e}\right) \sim i\,\frac{e}{\varrho}\,(n+1)\, Q_{n\nu}\left(\frac{i\varrho}{e}\right). \tag{29.13}$$

Eingesetzt in (29.11), ergibt sich

$$T = \sum_{n=0}^{\infty}{}^{*} \sum_{\nu=0}^{n}{}^{*} \frac{\varrho}{n-1}\, G_{n\nu}(\vartheta, \lambda). \tag{29.14}$$

$\varrho$ ist variabel, weicht aber von den konstanten Größen $c$ und $R$ nur wenig ab. Somit gilt hinreichend genau:

$$\boxed{T = c \sum_{n=0}^{\infty}{}^{*} \sum_{\nu=0}^{n}{}^{*} \frac{G_{n\nu}(\vartheta, \lambda)}{n-1} \approx R \sum_{n=0}^{\infty}{}^{*} \sum_{\nu=0}^{n}{}^{*} \frac{G_{n\nu}(\vartheta, \lambda)}{n-1},} \tag{29.15}$$

---

[1] Die Konvergenz dieser Entwicklung und ihren Zusammenhang mit der Massenverteilung hat H. JEFFREYS untersucht [21], S. 171.
[2] Abgeleitet in den Lehr- und Handbüchern über Kugelfunktionen. Zum Beispiel [24], [26, II]. Anhang III, Ziff. 57.
[3] Im Anhang III, Ziff. 57, wird nachgewiesen, daß die relativen Vernachlässigungen die Größenordnung $\varepsilon^2 = (e/\varrho)^2$ haben.

und mit (5.2) wird[1]

$$\zeta = \frac{c}{\gamma} \sum_{n=0}^{\infty}{}^* \sum_{\nu=0}^{n}{}^* \frac{G_{n\nu}(\vartheta, \lambda)}{n-1} \approx \frac{R}{\gamma} \sum_{n=0}^{\infty}{}^* \sum_{\nu=0}^{n}{}^* \frac{G_{n\nu}(\vartheta, \lambda)}{n-1}.$$   (29.16)[2]

Man kann die Koeffizienten von $G_{n\nu}$ nur dann hinreichend genau bestimmen, wenn die Schwereanomalien in einem großen Bereich, am besten über die ganze Erde, bekannt sind. Es ist nicht möglich, die Geoidundulation $\zeta$ eines bestimmten Ortes zu berechnen, wenn man nur die Schwereanomalie in seiner nächsten Umgebung kennt.

β) Die STOKESsche Formel. Man kann die Reihe in (29.15) und (29.16) zu einem geschlossenen Ausdruck aufsummieren und erhält die als STOKESsche Formel bekannte Integraldarstellung der Geoidundulation [3], [5, II]. PIZETTI[3] hat gezeigt, wie man die STOKESsche Formel als Lösung einer Randwertaufgabe erhalten kann, ohne Reihenentwicklungen zu verwenden. Später haben IDELSON und MALKIN[4] eine ähnliche Ableitung gegeben.

Es werden Kugelkoordinaten eingeführt, und in (29.4) wird $\varrho$ durch $l$ ersetzt.

$$\frac{1}{\gamma} \frac{\partial \gamma}{\partial n} = -\frac{2}{l}.$$   (29.17)

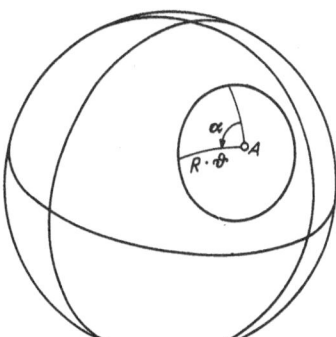

Hiermit wird aus (6.1):

$$\varDelta g = -\frac{\partial T}{\partial l} - \frac{2}{l} T,$$   (29.18)

$$l \varDelta g = -\frac{1}{l} \frac{\partial}{\partial l} (l^2 T).$$   (29.19)

Es läßt sich zeigen, daß die rechte Seite eine harmonische Funktion ist, d.h. die LAPLACEsche Gleichung[5] erfüllt. Ihre Randwerte auf der Kugel $l = R$ sind

Fig. 19. Zur Ableitung der STOKESschen Formel. $R$ Radius der Erdkugel, $\vartheta$ „Polabstand" bezogen auf den Aufpunkt $A$.

$$(l \varDelta g)_{l=R} = R \varDelta g (\vartheta, \alpha),$$   (29.20)

wobei $\vartheta$ den sphärischen Abstand vom Punkt $A$, dessen Geoidundulation bestimmt werden soll, bezeichnet (Fig. 19). $\alpha$ ist der Azimutwinkel im Koordinatensystem mit dem Zentrum $A$.

[1] Die Vereinfachungen sind beim Schluß von $\varDelta g$ auf $T$ und $\zeta$ solange unbedenklich, wie die Größenordnung $\zeta \varepsilon^2$ vernachlässigt werden darf. Beim umgekehrten Schluß von $\zeta$ oder $T$ auf $\varDelta g$ kann $\dfrac{n-1}{R}$ bei großem $n$ erheblich von $\dfrac{n-1}{\varrho}$ abweichen. Wesentlich kleiner sind die Fehler von $\dfrac{n-1}{c}$. Zur Klarstellung dieser Verhältnisse wurden die Ellipsoidkoordinaten bis zum Schluß beibehalten.

[2] Mit $R = 6370$ km, $\gamma = 981$ Gal wird $R/\gamma = 6{,}49$ m/mgal.

[3] P. PIZETTI: Sopra il calcolo teorico delle deviazioni del geoide dall'ellissoide. Atti di Torino **46** (1911). Die Bedeutung einer solchen Ableitung beruht darauf, daß alle mit Konvergenzfragen zusammenhängenden Schwierigkeiten vermieden werden. Von wesentlicher Bedeutung ist die erste GREENsche Funktion der Kugel. Da eine entsprechende Funktion des Rotationsellipsoids nicht angegeben werden kann, gibt es keine ähnliche Formel, die sich auf das Rotationsellipsoid bezieht. Die STOKESsche Formel gilt nur mit der Genauigkeit, mit der man die Erdkruste einer Erd*kugel* statt einem Erdellipsoid aufgesetzt denken kann. Nach VENING MEINESZ haben die Vernachlässigungen die Größenordnung $R \alpha^3 \approx \zeta \varepsilon^2$.

[4] Gerlands Beitr. Geophys. **29**, 156 (1931).

[5] Fußnote 3, S. 543.

Die Lösung von (29.19) mit den Randwerten (29.20) lautet [23], [24], [26, II]:

$$\frac{1}{l}\frac{\partial}{\partial l}(l^2 T) = \frac{R^2 - l^2}{4\pi R}\int\limits_{\vartheta=0}^{\pi}\int\limits_{\alpha=0}^{2\pi} R\,\Delta g\,(\vartheta, \alpha)\,\frac{df}{E^3}\,, \qquad (29.21)$$

wobei

$$E^2 = l^2 + R^2 - 2lR\cos\vartheta, \qquad df = R^2\sin\vartheta\,d\vartheta\,d\alpha.$$

Hiermit ergibt sich

$$\frac{\partial}{\partial l}(l^2 T) = \frac{R^2 l - l^3}{4\pi R}\int\limits_{\vartheta=0}^{\pi}\int\limits_{\alpha=0}^{2\pi} R^3\,\Delta g\,(\vartheta, \alpha)\,\frac{\sin\vartheta\,d\vartheta\,d\alpha}{E^3}\,, \qquad (29.22)$$

Integriert man beide Seiten über $l$ von $\infty$ bis $R$, so erhält man[1]

$$R^2 T = \frac{R^2}{4\pi}\int\limits_{\vartheta=0}^{\pi}\int\limits_{\alpha=0}^{2\pi} \Delta g\,(\vartheta, \alpha)\int\limits_{l=\infty}^{R}\frac{R^2 l - l^3}{E^3}\,dl\,\sin\vartheta\,d\vartheta\,d\alpha. \qquad (29.23)$$

Das Integral über $l$ ergibt

$$\int\limits_{l=\infty}^{R}\frac{R^2 l - l^3}{E^3}\,dl = \left(\frac{2l^2}{E} - 3E - 3R\cos\vartheta\ln\frac{l - R\cos\vartheta + E}{2R}\right)_{\infty}^{R}. \qquad (29.24)$$

Bei großem $l$ ist genau genug

$$\left.\begin{aligned}
\frac{1}{E} &= \frac{1}{l}\left(1 - \frac{2R}{l}\cos\vartheta\right)^{-\frac{1}{2}} = \frac{1}{l} + \frac{R\cos\vartheta}{l^2}\,, \\
\ln\frac{l - R\cos\vartheta + E}{2R} &= \ln\frac{l}{R}\,,
\end{aligned}\right\} \qquad (29.25)$$

und die eingeklammerte Funktion in (29.24) wird

$$-l + 5R\cos\vartheta - 3R\cos\vartheta\ln\frac{l}{R}\,. \qquad (29.26)$$

Auf der Kugel $l = R$ ist

$$E = 2R\sin\frac{\vartheta}{2}\,,$$

und die rechte Seite von (29.24) wird

$$\frac{R}{\sin\dfrac{\vartheta}{2}} - 6R\sin\frac{\vartheta}{2} - 3R\cos\vartheta\ln\left(\sin\frac{\vartheta}{2} + \sin^2\frac{\vartheta}{2}\right) \qquad (29.27)$$

Setzt man (29.24) mit (29.26) und (29.27) in (29.23) ein, so wird[1]

$$\left.\begin{aligned}
T &= \frac{R}{4\pi}\int\limits_{\vartheta=0}^{\pi}\int\limits_{\alpha=0}^{2\pi} \Delta g\,(\vartheta, \alpha)\,\overline{S}\,(\vartheta)\sin\vartheta\,d\vartheta\,d\alpha, \\
\text{wobei} \qquad & \\
\overline{S}\,(\vartheta) &= \frac{1}{\sin\dfrac{\vartheta}{2}} - 6\sin\frac{\vartheta}{2} - 3\cos\vartheta\ln\left(\sin\frac{\vartheta}{2} + \sin^2\frac{\vartheta}{2}\right) - 5\cos\vartheta.
\end{aligned}\right\} \qquad (29.28)$$

---

[1] In (29.23) ist vorausgesetzt, daß $\lim\limits_{l\to\infty} l^2 T = 0$, und (29.28) verlangt, daß die Integrale $\iint\Delta g\,df$ und $\iint\Delta g\cos\vartheta\,df$ verschwinden. Beides ist der Fall, wenn die Entwicklung von $\Delta g$ nach Kugelfunktionen keine Glieder der Ordnungen 0 und 1 enthält. Diese Glieder müssen vor Anwendung der STOKESschen Formel beseitigt werden. — R. A. HIRVONEN: The removal of spherical harmonics of first order from a field of observed gravity anomalies. Veröff. finn. Geodät. Inst. **1955**, Nr. 46, 59 (Publ. dedicated to W. A. HEISKANEN).

Meist schreibt man, mit (5.2)

$$\zeta = \frac{R}{4\pi\gamma} \int\limits_{\vartheta=0}^{\pi} \int\limits_{\alpha=0}^{2\pi} \Delta g(\vartheta, \alpha)\, S(\vartheta) \sin\vartheta\, d\vartheta\, d\alpha$$

mit

$$S(\vartheta) = \frac{1}{\sin\dfrac{\vartheta}{2}} - 6\sin\frac{\vartheta}{2} - 3\cos\vartheta \ln\left(\sin\frac{\vartheta}{2} + \sin^2\frac{\vartheta}{2}\right) + 1 - 5\cos\vartheta.$$

(29.29)

Man nennt $S(\vartheta)$ die STOKESsche *Funktion*[1] und (29.29) die STOKESsche *Formel*.

Da $S(\vartheta)$ bei kleinem $\vartheta$ sehr groß wird, empfiehlt es sich, die Schwereanomalie auf den Kreisen $\vartheta = \text{const}$ zu mitteln. Dann ist

$$\zeta = \frac{R}{\gamma} \int\limits_{\vartheta=0}^{\pi} \overline{\Delta g}(\vartheta)\, F(\vartheta)\, d\vartheta,$$

wobei

$$\overline{\Delta g}(\vartheta) = \frac{1}{2\pi} \int\limits_{\alpha=0}^{2\pi} \Delta g(\vartheta, \alpha)\, d\alpha \quad \text{und} \quad F(\vartheta) = \frac{1}{2}\sin\vartheta \cdot S(\vartheta).$$

(29.30)[2]

Der Verlauf von $S(\vartheta)$ und $F(\vartheta)$ ist in Fig. 20 dargestellt.

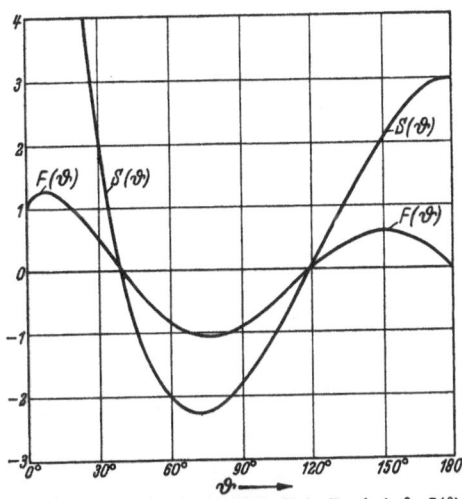

Fig. 20. Die STOKESsche Funktion $S(\vartheta)$. $F = \frac{1}{2}\sin\vartheta \cdot S(\vartheta)$.

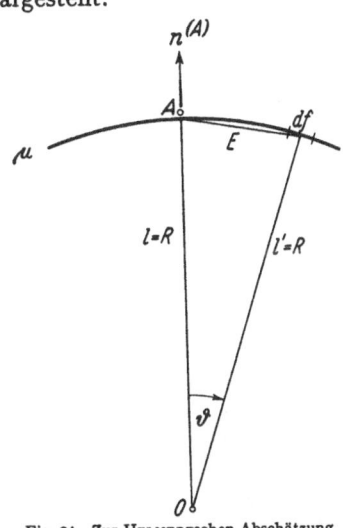

Fig. 21. Zur HELMERTschen Abschätzung.

Streng genommen, erfordert die STOKESsche Formel die Kenntnis der Schwereanomalien auf der ganzen Erde. In der Praxis kommt man aus, wenn die Schwereanomalie bis in Entfernungen von einigen hundert Kilometer gut bekannt ist und für die weiteren Gebiete abgeschätzt werden kann. Es gibt große Bereiche auf der Erde, in denen diese Bedingung nicht erfüllt ist[3].

---

[1] Das zugefügte Glied liefert keinen Beitrag zum Integral. Man findet auch $S(\vartheta)$ mit umgekehrtem Vorzeichen, z.B. in [16].

[2] LAMBERT, W. D., u. F. W. DARLING: Tables for determining the form of the geoid and its indirect effect on gravity. U.S. Coast and Geodetic Survey Spec. Publ. No. 199, 1936. Enthält $F$, $S/2$, $\int\limits_{0}^{\vartheta} F\, d\vartheta$.

[3] Ziff. 34. Schwierigkeiten, die mit der ungünstigen Verteilung der Schwerestationen zusammenhängen, untersucht H. JEFFREYS: The use of STOKES' formula in the adjustment of surveys. Bull. Géod. 1953, Nr. 30, 331.

*γ) Die* HELMERT*sche Abschätzung.* Eine Formel von BRUNS[1] scheint dagegen die Abschätzung der Geoidundulation zu gestatten, wenn die Schwereanomalie nur an dem betreffenden Ort gegeben ist. Schon HELMERT[2] hat gezeigt, daß die BRUNSsche Formel unvollständig ist[3].

Bei der HELMERTschen Abschätzung geht man von (29.18) aus und ersetzt die Störungsmassen durch eine Flächenbelegung von der Flächendichte $\mu$ auf der Erdkugel $l = R$ (Fig. 21). Nach bekannten Sätzen der Potentialtheorie[4] ist dann

$$T^{(A)} = f \iint \frac{\mu \, df}{E}, \qquad E = R \sqrt{2(1 - \cos\vartheta)},$$

$$\frac{\partial T^{(A)}}{\partial n^{(A)}} = -2 f \pi \mu^{(A)} + f \iint \mu \frac{\partial}{\partial n} \frac{1}{E} \, df.$$

Führt man die Integration aus, so ergibt sich

$$\frac{\partial T^{(A)}}{\partial n^{(A)}} = -2 f \pi \mu^{(A)} - \frac{T^{(A)}}{2R}, \qquad \Delta g^{(A)} = 2 f \pi \mu^{(A)} - \frac{3 T^{(A)}}{2R},$$

und mit (5.2) wird

$$\Delta g^{(A)} = 2 f \pi \mu^{(A)} - \frac{3\gamma}{2R} \zeta^{(A)},$$

$$\boxed{\zeta^{(A)} = \frac{2R}{3\gamma} \left(2\pi f \mu^{(A)} - \Delta g^{(A)}\right).} \qquad (29.31)$$

Setzt man[5]

$$\boxed{\mu^{(A)} = \sigma D^{(A)}}$$

und bedenkt[6], daß

$$\gamma = \frac{4}{3} f \pi \sigma_m R,$$

so erhält man mit $\sigma = \sigma_m/2$ die HELMERTsche Form von (29.31):

$$\boxed{D^{(A)} = 2 \left(\frac{2}{3} \frac{R}{\gamma} \Delta g^{(A)} + \zeta^{(A)}\right).} \qquad (29.32)$$

Bei der Abschätzung von Größenordnungen leisten (29.31) und (29.32) gute Dienste. Weitergehende Bedeutung haben sie nicht.

**30. Die Randwertaufgabe der physikalischen Geodäsie.** Formal wurde die Bestimmung der Geoidundulationen in (29.15) und (29.29) auf eine Art zweite Randwertaufgabe zurückgeführt, während (29.18) die Gestalt einer dritten Randwertaufgabe hat und die rechte Seite von (29.21) einen Ausdruck enthält, der einer ersten GREENschen Funktion nahe steht und sich somit auf eine erste Randwertaufgabe bezieht. Die Vielfältigkeit dieser Randwertaufgaben deutet darauf hin, daß die Randwertaufgabe der physikalischen Geodäsie enge Beziehungen zu denen der Potentialtheorie hat, aber nicht mit einer von ihnen identisch ist.

[1] [*4*], S. 27: $\dfrac{\Delta g}{\gamma} = \dfrac{3}{2} \dfrac{\zeta}{R}$; $\zeta = \dfrac{2}{3} R \dfrac{\Delta g}{\gamma}$.

[2] [*5*, II], S. 261.

[3] BRUNS legt den Aufpunkt $A$ (Fig. 21) von vornherein *in* die Flächenbelegung. Er hätte $A$ von außen an die Flächenbelegung heranführen müssen.

[4] Zum Beispiel [*23*], S. 164.

[5] $\sigma =$ Dichte der Störungsmasse, $D =$ Dicke der *„ideellen störenden Schicht"*.

[6] Die Substitution $2\pi f = \dfrac{3\gamma}{2\sigma_m R}$ ist auch bei Berechnung der Wirkung der BOUGUERschen Platte gebräuchlich. Sie bringt keine prinzipiellen Vorteile, höchstens einen geringen Nutzen bei zahlenmäßigen Abschätzungen, wenn $\sigma$ und $\sigma_m$ ein einfaches Verhältnis haben. Da sie die wahren Zusammenhänge verschleiert, sollte man auf sie verzichten. Es ist eine Täuschung, wenn man annimmt, man würde von der Gravitationskonstanten unabhängig. Nach einer Neubestimmung der Gravitationskonstanten muß $\sigma_m$ gleichfalls verbessert werden, während $\sigma$ seinen Wert beibehält.

In der Potentialtheorie (Fig. 22 unten) pflegt eine Fläche gegeben zu sein, auf der man das Potential, die zu der Fläche senkrechte Komponente des Feldvektors oder eine lineare Kombination beider Größen kennt. Die gegebene Fläche wird im allgemeinen keine Niveaufläche sein. In der Geodäsie dagegen ist die Fläche unbekannt, man weiß aber, daß sie eine Niveaufläche ist und die gemessenen Feldvektoren auf ihr senkrecht stehen. Man hat die Aufgabe, die gegebenen Flächenelemente zu einer einheitlichen, geschlossenen Fläche zusammenzusetzen (Fig. 22 oben).

Wie die ihr entsprechenden Randwertaufgaben der Potentialtheorie ist auch die Randwertaufgabe der physikalischen Geodäsie nur lösbar, wenn alle Massen im Innern der Fläche liegen. Für den bei der Erde vorliegenden Fall, daß die Fläche nur wenig von einer Kugel oder einem Ellipsoid abweicht, gibt es die beschriebenen Näherungsverfahren. Jedoch ist nicht bekannt, ob es nicht noch ganz anders geartete Flächen gibt, die gleichfalls als Lösungen auftreten könnten.

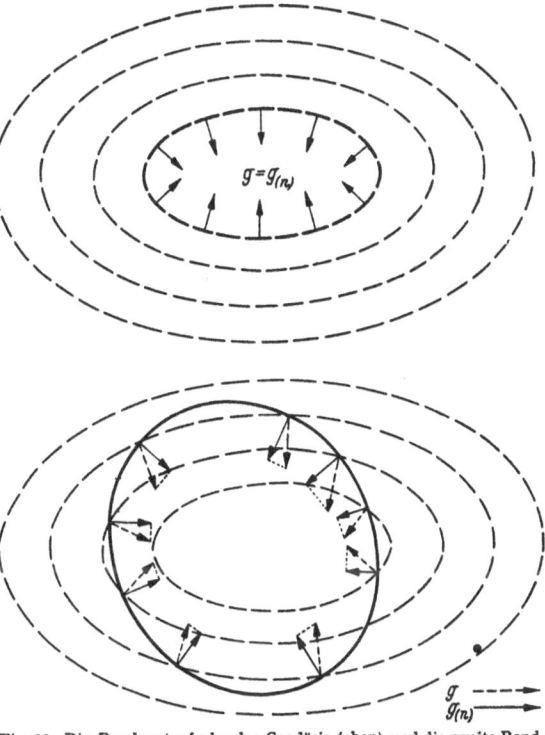

**31. Die Reduktion der Schweremessungen. Der indirekte Effekt.** Bei Anwendung der beschriebenen Verfahren zur Bestimmung des Niveausphäroids und der Geoidundulationen müssen die Schwereanomalien *Randwerte* sein, sie müssen sich auf eine Niveaufläche beziehen, die alle Massen einschließt.

Da sich die Kontinentmassen über das Geoid erheben und die Meßorte im allgemeinen nicht auf dem Geoid liegen, ist eine Umrechnung der gemessenen Schwerewerte in Randwerte nötig. Diese Umrechnung nennt man *Schwerereduktion*[1].

Bei der Schwerereduktion

Fig. 22. Die Randwertaufgabe der Geodäsie (oben) und die zweite Randwertaufgabe der Potentialtheorie (unten). $g_{(n)}$ die auf der Fläche senkrechte Schwerekomponente.

ist die außerhalb vom Geoid liegende Masse, die man auch *topografhische Masse* nennt, rechnerisch zu erfassen und zu beseitigen oder in das Innere des Geoids zu verschieben. Bei jeder Massenverschiebung ändert sich das Potentialfeld und verlagern sich die Niveauflächen. Nach einer solchen Reduktion ist das (ursprüngliche) Geoid[2] im allgemeinen keine Niveaufläche mehr, und die beschriebenen Methoden führen auf die Niveaufläche in ihrer verschobenen Lage.

Die bei massenverschiebenden Reduktionen hervorgerufenen Verlagerungen der Niveauflächen bezeichnet man als *indirekten Effekt*[3].

[1] Es wird vorausgesetzt, daß die wichtigsten Schwerereduktionen hinsichtlich ihrer Bedeutung und Ausführung bekannt sind. (Vgl. GARLAND, S. 222ff dieses Bandes.) Näheres hierüber enthalten alle geophysikalischen Lehr- und Handbücher, auch Tabellenwerke. Zum Beispiel [5, II], [7], [9], [11], [12], [16], [18], [21], [22].

[2] Auch "actual geoid" genannt.

[3] Auch als BOWIE-Effekt bezeichnet, da man sich auf Veranlassung von BOWIE mit ihm näher zu befassen begann. Ansätze, in etwas anderem Zusammenhang, finden sich schon bei BRUNS [4].

Grundsätzlich ist in folgender Weise vorzugehen:

1. Man beseitigt rechnerisch die außerhalb vom Geoid liegenden Massen, was auch durch Verschiebung ins Innere des Geoids geschehen kann.

2. Man verlegt die Schwerestation rechnerisch auf das Geoid.

3. Man berechnet den indirekten Effekt.

4. Man verlegt die Schwerestation auf die verlagerte Niveaufläche.

5. Man bestimmt die Gestalt der verlagerten Niveaufläche.

6. Man bestimmt die Gestalt des Geoids, indem man an der Gestalt der verlagerten Niveaufläche den indirekten Effekt mit umgekehrtem Vorzeichen anbringt.

Massen, die zwischen dem Geoid und der verlagerten Niveaufläche liegen, müssen ebenfalls beseitigt werden, wenn die verlagerte Niveaufläche unter dem Geoid verläuft. Hierbei tritt eine weitere Verlagerung der Niveaufläche auf, die mit einer nochmaligen Anwendung der Reduktionen (3) bis (6) berücksichtigt werden kann. In der Praxis zeigt sich, daß die einmalige Durchrechnung wohl in fast allen Fällen genügt.

Im Prinzip läßt sich bei der Geoidbestimmung jede Reduktionsmethode verwenden, mit der man Randwerte erhält und deren indirekter Effekt genau genug berechnet werden kann[1]. Aus praktischen Gründen wird man solche Methoden vorziehen,

bei denen der indirekte Effekt klein ist,

nach deren Anwendung zwischen den Schwerestationen gut interpoliert werden kann,

und die auf die Schwerewerte führen, die auch für andere Zwecke, insbesondere geologische Auswertungen, verwendet werden können.

Diese Forderungen schränken die praktisch brauchbaren Reduktionsverfahren erheblich ein[2].

Sehr gering ist der indirekte Effekt bei der Kondensation der topographischen Massen als Flächenbelegung in das Meeresniveau. Wenn es nicht auf besondere Feinheiten ankommt, können die Reduktionen 1, 3, 4 und 6 vernachlässigt werden, die Reduktion der Schwerewerte wird mit der „Freiluftformel" (18.8) ausgeführt. Man spricht daher von „Freiluftanomalien", obwohl die Bezeichnung „Kondensationsanomalien" zutreffender wäre. Aus den „Freiluftanomalien" erhält man unmittelbar die Gestalt des Geoids. In mehreren Veröffentlichungen hat JEFFREYS potentialtheoretisch nachgewiesen, daß die „Freiluftreduktion" diejenigen Schwerewerte liefert, die in die STOKESsche Formel (29.29; 29.30) einzusetzen sind[3].

Gegen die „Freiluftanomalien" wird angeführt, daß sie schwer interpolierbar sind, da sie oft recht sprunghaft auftreten. In dieser Beziehung sind die isostatischen Anomalien bequemer, und man nimmt bei ihnen in Kauf, daß der indirekte Effekt der isostatischen Reduktionen[4] von der Größenordnung der zu ermittelnden Geoidundulationen ist, also sorgfältig bestimmt und berücksichtigt werden muß. Das gilt auch von der Berechnung der Erdabplattung mit dem CLAIRAUTschen Theorem. Schwereformeln, die ohne Berücksichtigung des indirekten Effektes aus isostatisch reduzierten Schwerewerten abgeleitet sind, entsprechen einer regularisierten Erde[5] und mögen gute Vergleichswerte für die

---

[1] Angaben über Tafeln in [22].

[2] Ausführliche Untersuchungen über die verschiedenen Reduktionen und die Berücksichtigung des indirekten Effekts bringt L. TANNI (Ziff. 34).

[3] H. JEFFREYS: An Application of the Free-air Reduction of Gravity. Gerlands Beitr. Geophys. 31, 378 (1931). Ferner [21], S. 125ff.

[4] Sie regularisieren eine im Schwimmgleichgewicht befindliche Erdkruste.

[5] Die durch den indirekten Effekt isostatischer Reduktionen verlagerte Niveaufläche wird „regularisiertes Geoid" und „kompensiertes Geoid" — abgekürzt: „Co-Geoid" — genannt; LEJAY [20] nennt es „géoïde fictif".

geologische Auswertung darstellen. Sie entsprechen aber nicht dem Niveau-sphäroid, das sich dem (ursprünglichen) Geoid am besten anpaßt, und es ist eine Frage, wie weit die aus ihnen ermittelten Abplattungswerte als zuverlässig angesehen werden dürfen.

Kaum brauchbar ist die BOUGUER*sche Reduktion* [S. 223, Formel (17.8)], selbst wenn man sie dadurch verfeinert, daß man die wirkliche Geländegestalt und die Krümmung der Erdoberfläche berücksichtigt. Sie ist zu empfindlich gegen Fehler der Gesteinsdichte und verursacht, da es sich um eine Wegnahme der topographischen Massen handelt, eine große Verlagerung der Niveauflächen.

Die *Inversionsmethode von* RUDZKI scheint ideal zu sein, da sie die topographischen Massen in solcher Weise durch innerhalb des Geoids liegende Massen ersetzt, daß das Geoid nicht verändert wird. Sie führt ohne indirekten Effekt auf die Gestalt des Geoids, verändert aber alle anderen Niveauflächen so stark, daß man deren Gestalt nicht ohne weiteres richtig erhält. Gegen diese Reduktion spricht auch die verhältnismäßig mühsame Zahlenrechnung.

Die *Reduktion nach* PREY führt auf Schwerewerte, wie man sie im Innern der Kontinente auf dem Geoid messen würde, wenn man dorthin gelangen könnte. Sie gibt die wirkliche Geoidschwere, aber keine Randwerte. Ohne zusätzliche Berechnungen, mit denen man die unverlagerten topographischen Massen erfaßt, ist sie für die Geoidbestimmung nicht brauchbar. Sie ist wichtig bei der Reduktion von Nivellements[1].

### 32. Ermittlung der absoluten Lotabweichung aus den Schwereanomalien[2].

Die *absolute*, auf die Sphäroid-Normale bezogene *Lotabweichung* $\varepsilon$ ist so klein, daß man $\varepsilon = \mathrm{tg}\,\varepsilon$ und $\cos \varepsilon = 1$ setzen kann. Dann gilt (Fig. 23)

$$\varepsilon = - \operatorname{grad} \zeta. \qquad (32.1)$$

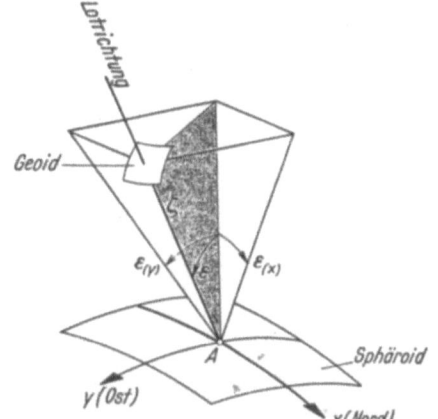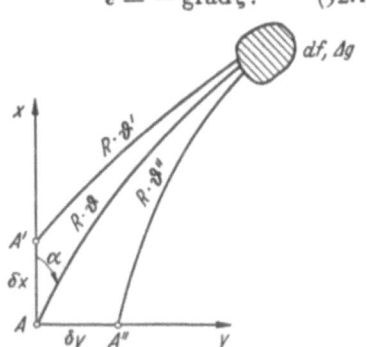

Fig. 23. Die absolute Lotabweichung $\varepsilon$ und ihre Komponenten $\varepsilon_{(x)}$, $\varepsilon_{(y)}$. $\zeta$ Geoidundulation.

Fig. 24. Zu den STOKESschen Formeln der absoluten Lotabweichung (Formeln von VENING MEINESZ).

Bezeichnen $x$ und $y$ die Nord- und Ostrichtung auf dem Sphäroid, so sind die entsprechenden Komponenten der absoluten Lotabweichung

$$\varepsilon_{(x)} = - \frac{\partial \zeta}{\partial x}, \qquad \varepsilon_{(y)} = - \frac{\partial \zeta}{\partial y}. \qquad (32.2)$$

Zur Berechnung von $\varepsilon_{(x)}$ und $\varepsilon_{(y)}$ schreibt man die STOKESsche Formel (29.29) in der Form (Fig. 24)

$$\zeta = \frac{1}{4 \pi \gamma R} \iint \Delta g\, S(\vartheta)\, df,$$

---

[1] Ziff. 37.

[2] F. A. VENING MEINESZ: A formula expressing the deflection of the plumb-line in the gravity anomalies and some formulae for the gravity-field and the gravity-potential outside the geoid. Proc. Kon. Akad. Wetensch., Amst. **31**, 315 (1928). — J. DE GRAAFF HUNTER: The figure of the earth from gravity observations and the precision obtainable. Phil. Trans. Roy. Soc. Lond., Ser. A **234**, 377 (1935).

wobei über die ganze Erde zu integrieren ist. Geht man auf die allgemeine Definition des Differentialquotienten zurück, so ist

$$\varepsilon_{,x} = -\lim_{\delta x \to 0} \frac{\zeta^{(A')} - \zeta^{(A)}}{\delta x} = -\lim_{\delta x \to 0} \frac{1}{4\pi\gamma R} \iint \Delta g \, \frac{S(\vartheta') - S(\vartheta)}{\delta x} \, df,$$

$$\varepsilon_{(y)} = -\lim_{\delta x \to 0} \frac{\zeta^{(A'')} - \zeta^{(A)}}{\delta y} = -\lim_{\delta y \to 0} \frac{1}{4\pi\gamma R} \iint \Delta g \, \frac{S(\vartheta'') - S(\vartheta)}{\delta y} \, df,$$

$$\left. \begin{aligned} \varepsilon_{(x)} &= -\frac{1}{4\pi\gamma R} \iint \Delta g \, \frac{\partial S(\vartheta)}{\partial x} \, df, \\ \varepsilon_{(y)} &= -\frac{1}{4\pi\gamma R} \iint \Delta g \, \frac{\partial S(\vartheta)}{\partial y} \, df. \end{aligned} \right\} \tag{32.3}$$

Nun ist

$$\frac{\partial S(\vartheta)}{\partial x} = \frac{dS(\vartheta)}{d\vartheta} \cdot \frac{\partial\vartheta}{\partial x} = -\frac{1}{R} \frac{dS(\vartheta)}{d\vartheta} \cdot \cos\alpha,$$

$$\frac{\partial S(\vartheta)}{\partial y} = \frac{dS(\vartheta)}{d\vartheta} \cdot \frac{\partial\vartheta}{\partial y} = -\frac{1}{R} \frac{dS(\vartheta)}{d\vartheta} \cdot \sin\alpha,$$

$$df = R^2 \sin\vartheta \, d\vartheta \, d\alpha,$$

und man erhält die Stokessche *Formeln für die absolute Lotabweichung (Formeln von* Vening Meinesz)[1]:

$$\varepsilon_{(x)} = \frac{1}{4\pi\gamma} \int_{\vartheta=0}^{\pi} \int_{\alpha=0}^{2\pi} \Delta g(\vartheta, \alpha) \, \frac{dS(\vartheta)}{d\vartheta} \sin\vartheta \cos\alpha \, d\vartheta \, d\alpha,$$

$$\varepsilon_{(y)} = \frac{1}{4\pi\gamma} \int_{\vartheta=0}^{\pi} \int_{\alpha=0}^{2\pi} \Delta g(\vartheta, \alpha) \, \frac{dS(\vartheta)}{d\vartheta} \sin\vartheta \sin\alpha \, d\vartheta \, d\alpha \tag{32.4}$$

mit

$$\frac{dS(\vartheta)}{d\vartheta} \sin\vartheta = -\frac{1}{\sin\frac{\vartheta}{2}} - 3 - 8\sin\frac{\vartheta}{2} + 32\sin^2\frac{\vartheta}{2} +$$

$$+ 12\sin^3\frac{\vartheta}{2} - 32\sin^4\frac{\vartheta}{2} + 3\sin^2\vartheta \ln\left(\sin\frac{\vartheta}{2} + \sin^2\frac{\vartheta}{2}\right).$$

Diese Formeln liefern $\varepsilon_{(x)}$ und $\varepsilon_{(y)}$ im Bogenmaß. Will man $\varepsilon_{(x)}$ und $\varepsilon_{(y)}$ in Winkelsekunden haben, so sind die rechten Seiten mit $\varrho'' = 206265''/\text{rad}$ zu multiplizieren.

Bei der Integration ist es zweckmäßig, die unmittelbare Umgebung des Aufpunkts $A$ für sich zu behandeln[2]. Mit $R \cdot \vartheta = s$ erhält man für kleines $\vartheta$:

$$\Delta g = \Delta g^{(A)} + \left(\frac{\partial \Delta g}{\partial x}\right)^{(A)} s \cos\alpha + \left(\frac{\partial \Delta g}{\partial y}\right)^{(A)} s \sin\alpha,$$

$$\frac{dS(\vartheta)}{d\vartheta} \sin\vartheta \, d\vartheta = -\left(\frac{2}{s} + \frac{3}{R}\right) ds,$$

---

[1] Oft wird der Winkel $\alpha$ von Süden aus im Uhrzeiger-Sinn gezählt; er ist dann um 180° größer als in (32.4) und (32.5), und die rechten Seiten haben das umgekehrte Vorzeichen. In der Literatur kommt auch das umgekehrte Vorzeichen bei $\varepsilon_{(x)}$ und $\varepsilon_{(y)}$ vor, und es wird $S$ für $_\tau S$ geschrieben.

[2] A. D. Sollins: Tables for the computation of the deflection of the vertical from gravity anomalies. Bull. Géod., N.S. 1947, Nr. 6. Die Tafeln enthalten

$$\frac{1}{2} \frac{dS(\vartheta)}{d\vartheta} \sin\vartheta \quad \text{und} \quad \frac{1}{2} \int\limits_{R \cdot \vartheta = 10\,\mathrm{m}}^{r} \frac{dS(\vartheta)}{d\vartheta} \sin\vartheta \, d\vartheta$$

und es wird für kleines $s_0/R$:

$$\int\limits_{\vartheta=0}^{s_0/R}\int\limits_{\alpha=0}^{2\pi}\Delta g\,\frac{dS(\vartheta)}{d\vartheta}\sin\vartheta\cos\alpha\,d\vartheta\,d\alpha=-2\pi\left(s_0+\frac{3}{4R}s_0^2\right)\left(\frac{\partial\Delta g}{\partial x}\right)^{(A)},$$

$$\int\limits_{\vartheta=0}^{s_0/R}\int\limits_{\alpha=0}^{2\pi}\Delta g\,\frac{dS(\vartheta)}{d\vartheta}\sin\vartheta\sin\alpha\,d\vartheta\,d\alpha=-2\pi\left(s_0+\frac{3}{4R}s_0^2\right)\left(\frac{\partial\Delta g}{\partial y}\right)^{(A)}$$

$$\boxed{\begin{aligned}\varepsilon_{(x)}&=-\frac{1}{2\gamma}\left(s_0+\frac{3}{4R}s_0^2\right)\left(\frac{\partial\Delta g}{\partial x}\right)^{(A)}+\frac{1}{4\pi\gamma}\int\limits_{\vartheta=s_0/R}^{\pi}\int\limits_{\alpha=0}^{2\pi}\Delta g\,(\vartheta,\alpha)\,\frac{dS(\vartheta)}{d\vartheta}\sin\vartheta\cos\alpha\,d\vartheta\,d\alpha,\\[2mm]\varepsilon_{(y)}&=-\frac{1}{2\gamma}\left(s_0+\frac{3}{4R}s_0^2\right)\left(\frac{\partial\Delta g}{\partial y}\right)^{(A)}+\frac{1}{4\pi\gamma}\int\limits_{\vartheta=s_0/R}^{\pi}\int\limits_{\alpha=0}^{2\pi}\Delta g\,(\vartheta,\alpha)\,\frac{dS(\vartheta)}{d\vartheta}\sin\vartheta\sin\alpha\,d\vartheta\,d\alpha.\end{aligned}}\qquad(32.5)$$

Die Zahlenrechnung wird mit ausführlichen Tabellen[1] erheblich erleichtert.

**33. Synthetische Untersuchungen über Schwere und Geoid.** *α) Bedeutung und allgemeine Ergebnisse.* Synthetische Berechnungen, bei denen man das Schwerefeld einer vorgegebenen Massenverteilung untersucht, dienen der Abschätzung von Größenordnungen und liefern Vergleichswerte für die geologische Auswertung der Schwereanomalien. Bei Fragen der Isostasie sind sie kaum zu entbehren.

Bereits Bruns[2] hat gezeigt, daß die Geoidundulationen einer nicht kompensierten Erdkruste Unterschiede von der Größenordnung eines Kilometers erreichen können. Genauere Berechnungen von Helmert[3] bestätigten dieses Ergebnis und zeigten zugleich, daß die Schwereunterschiede bei fehlender Kompensation in ausgedehnten Gebieten 150 mgal erreichen können, wobei die hohen Schwerewerte im wesentlichen auf den Kontinenten, die niedrigen Werte auf den Ozeanen zu finden sind. Eine sorgfältige Untersuchung von Prey[4] kommt auf Geoidhebungen von 1620 m in Zentralasien und Senken von 1260 m im Stillen Ozean mit Freiluftanomalien zwischen +38 mgal (Zentralasien) und −426 mgal (Nord-Polarmeer), und auf Freiluftanomalien, die in Europa ungefähr 200 mgal niedriger als im zentralen Teil von Nordamerika sind. Daß die Beobachtungen diesen Verhältnissen nicht entsprechen, ist bekannt. Der unkompensierte Zustand läßt sich mit den Beobachtungen nicht vereinbaren.

---

bis $r=5200$ m, sowie

$$\frac{1}{2}\,\frac{dS(\vartheta)}{d\vartheta}\qquad\text{und}\qquad\frac{1}{2}\int\limits_{\vartheta=0,01°}^{\vartheta}\frac{dS(\vartheta)}{d\vartheta}\sin\vartheta\,d\vartheta$$

bis $\vartheta=180°$.

[1] A. D. Sollins: Siehe die vorige Fußnote.
[2] [*4*], S. 22f.
[3] [*5*, II], S. 336ff.
[4] A. Prey: Zur Frage nach dem isostatischen Ausgleich der Erdrinde. Gerlands Beitr. Geophys. **29**, 201 (1931).

Abschätzungen für eine Erde mit kompensierter Kruste wurden mehrfach ausgeführt. PREY[1], LAMBERT[2], K. JUNG[3,4], u. a. fanden übereinstimmend, daß in diesem Fall die Geoidundulation um rund 70 m (Fig. 25) schwanken kann, wobei die hohen Werte über den großen Kontinenten, die niedrigen Werte über den weiten Ozeanen auftreten. Diese Größenordnung, aber nicht das Vorzeichen, wurde von den Beobachtungen bestätigt[5].

*β) Grundlage der Zahlenrechnung: Die Entwicklung von PREY.* Die neueren Untersuchungen verwenden meist eine Entwicklung nach Kugelfunktionen. Grundlage der Zahlenrechnung ist die Darstellung des Erdreliefs nach PREY[6], gelegentlich ergänzt mit einer Entwicklung des Quadrats seiner Höhen und Tiefen[4]. Bei rohen Abschätzungen genügt es, bis zur 7. Ordnung zu gehen; denn die allgemeinen Züge des Erdreliefs sind von der 7. Ordnung an bereits deutlich erkennbar (Fig. 26).

Bildet man aus den Koeffizienten $_cL_{n\nu}$ und $_sL_{n\nu}$ (Relief der Lithosphäre) die quadratischen Mittelwerte

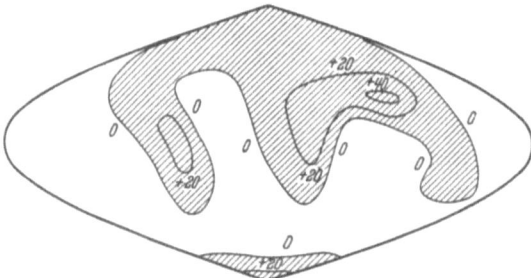

Fig. 25. Geoidundulation $\zeta$ bei Isostasie nach AIRY (in Metern). Entwicklung nach Kugelfunktionen bis zur 7. Ordnung.

$$\overline{L}_n = \sqrt{\sum_{\nu=0}^{n} \frac{_cL_{n\nu}^2 + {_s}L_{n\nu}^2}{2n+1}}, \quad (33.1)$$

so kann man beurteilen, wie weit die einzelnen Ordnungen $n$ an der Darstellung beteiligt sind (Fig. 27 und Tabelle 2). $\overline{L}_n$ nimmt mit wachsendem $n$ im allgemeinen ab. Jedoch heben sich die Ordnungen 1, 3, 4 und 5 recht deutlich hervor, ein Anlaß für interessante Theorien[7,8].

Bei genaueren Untersuchungen müssen Land und Meer mit ihren verschiedenen Dichten in getrennten Entwicklungen eingeführt werden. Bei Überschlagsrechnungen kann man die Wassermassen auf Kontinentdichte kondensiert denken und mit einer einheitlichen Erdkruste rechnen. Bezeichnet $L$ die Oberfläche der Lithosphäre und $M$ die Grundfläche der Hydrosphäre im Sinn der Entwicklungen von PREY, so ist ihre Oberfläche durch $L - \dfrac{\sigma_w}{\sigma_k} M$ gegeben, wenn $\sigma_w$ und $\sigma_k$ die Dichten des Meerwassers und der Kontinentmassen bezeichnen. Für $\sigma_w$ setzt man 1,03 g/cm³, für $\sigma_k$ wird oft 2,7 g/cm³ angesetzt.

*γ) Die mathematischen Grundlagen synthetischer Abschätzungen.* Mit Vernachlässigung von $e^2$ gegen $\varrho^2$ d. h. $\varepsilon^2$ gegen 1, gehen die Koordinaten-Ellipsoide in

[1] A. PREY: Neue Formeln zur Isostasie. Gerlands Beitr. Geophys. **18**, 185 (1927).

[2] W. D. LAMBERT: The form of the geoid on the hypothesis of complete isostatic compensation. Bull. Géod. **1930**, Nr. 26.

[3] K. JUNG: Schwere und Geoid bei Isostasie. Z. Geophys. **8**, 40 (1932).

[4] K. JUNG: Die rechnerische Behandlung der AIRYschen Isostasie mit einer Entwicklung des Quadrats der Meereshöhen nach Kugelfunktionen. Gerlands Beitr. Geophys. **62**, 39 (1950).

[5] Ziff. 34.

[6] A. PREY: Darstellung der Höhen- und Tiefenverhältnisse der Erde durch eine Entwicklung nach Kugelfunktionen bis zur 16. Ordnung. Abh. Ges. Wiss. Göttingen **11**, 1 (1922). Gibt eine Entwicklung des Reliefs der *Lithosphäre* (Kontinentoberfläche und Meeresboden) und der *Hydrosphäre* (auf den Kontinenten Null, auf den Ozeanen Meeresboden). Koeffizienten bis 8. Ordnung in Anhang II, Ziff. 52 (umgerechnet für nach AD. SCHMIDT normierte Kugelfunktionen).

[7] F. A. VENING MEINESZ: A remarkable feature of the earth's topography, origin of continents and oceans, I and II. Proc., Kon. nederl. Akad. te Wetensch., Ser. B **54**, 212 (1951).

[8] F. A. VENING MEINESZ: Convection-currents in the earth and the origin of the continents; I, II, III. Proc., Kon. nederl. Akad. Wetensch., Ser. B **55**, 527 (1952).

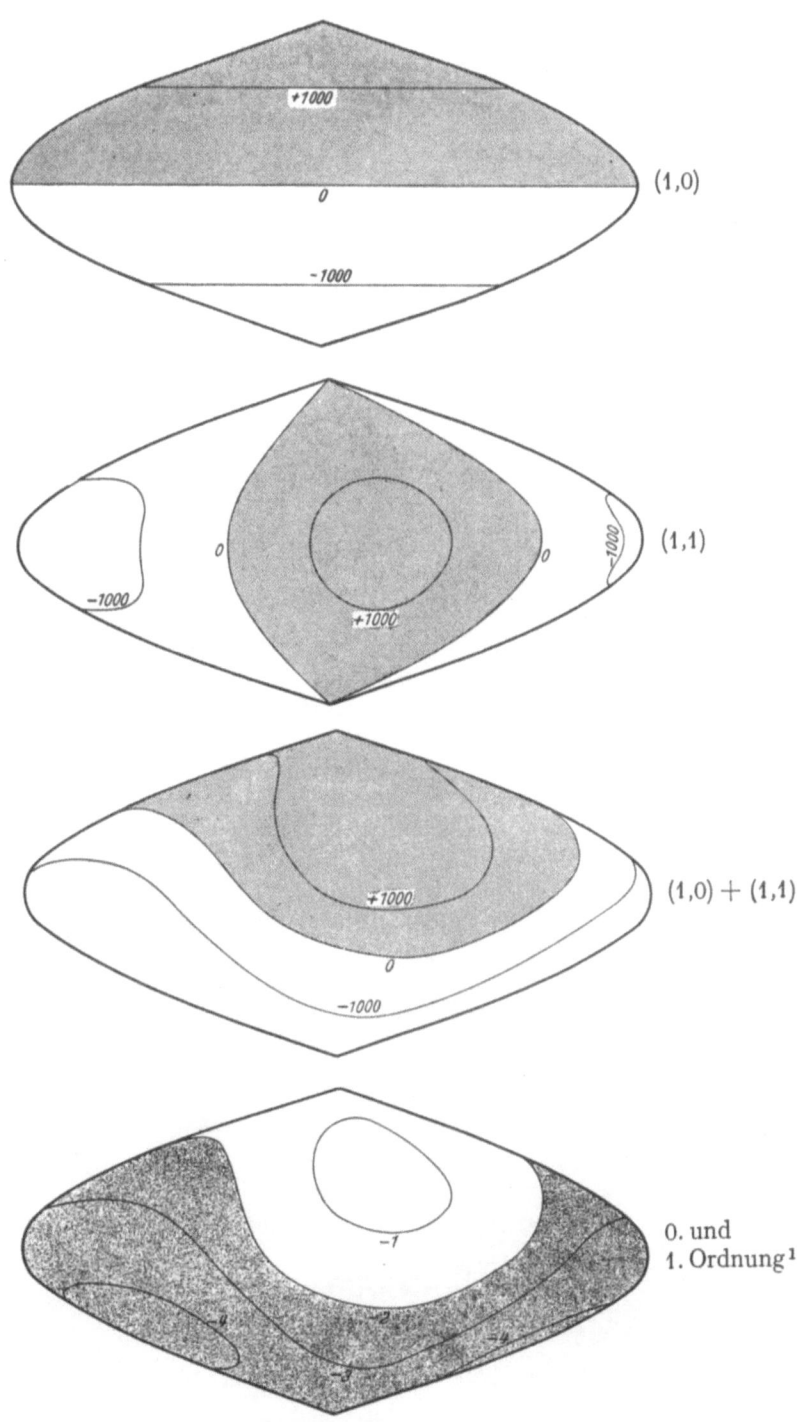

Fig. 26.

---

[1] 0. Ordnung: −2460 m.

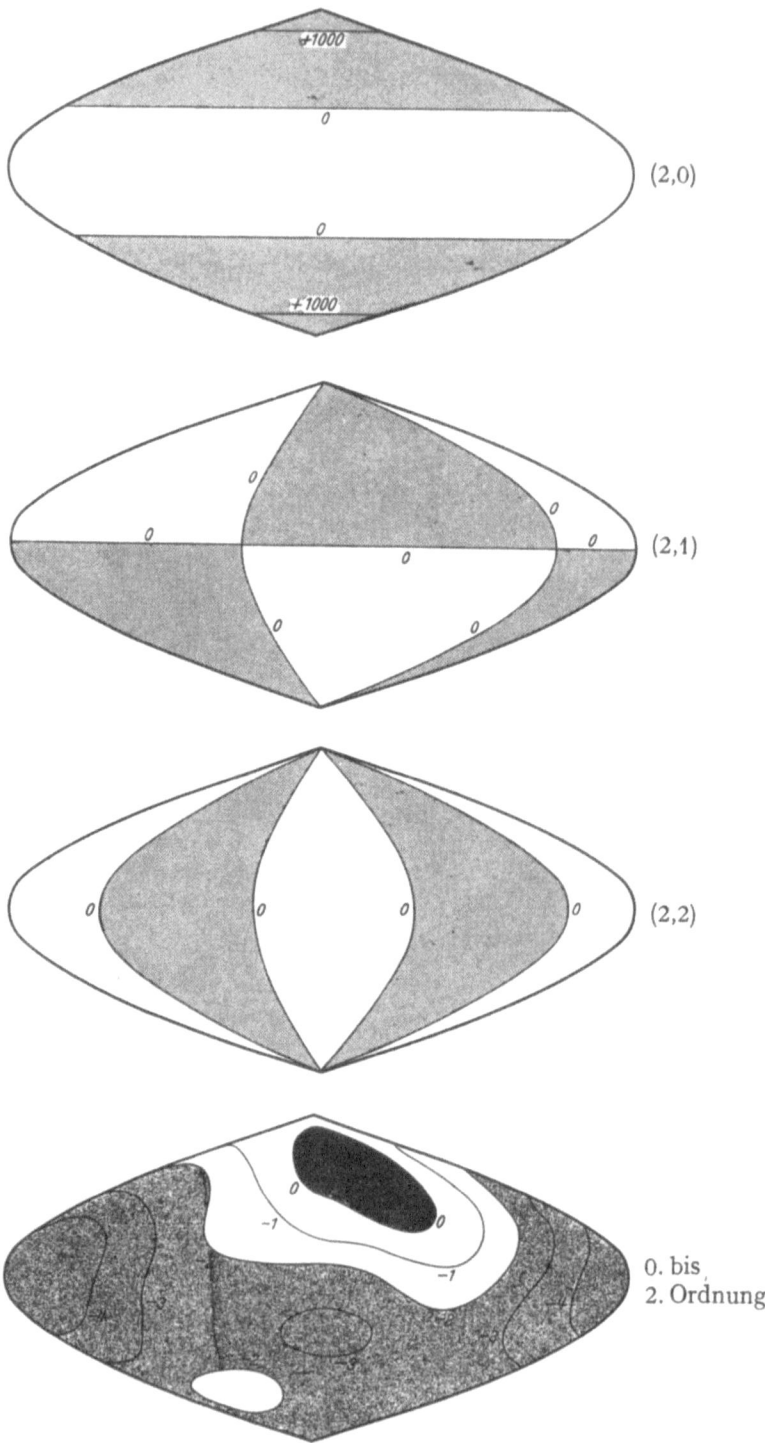

(2,0)

(2,1)

(2,2)

0. bis
2. Ordnung

Fig. 26.

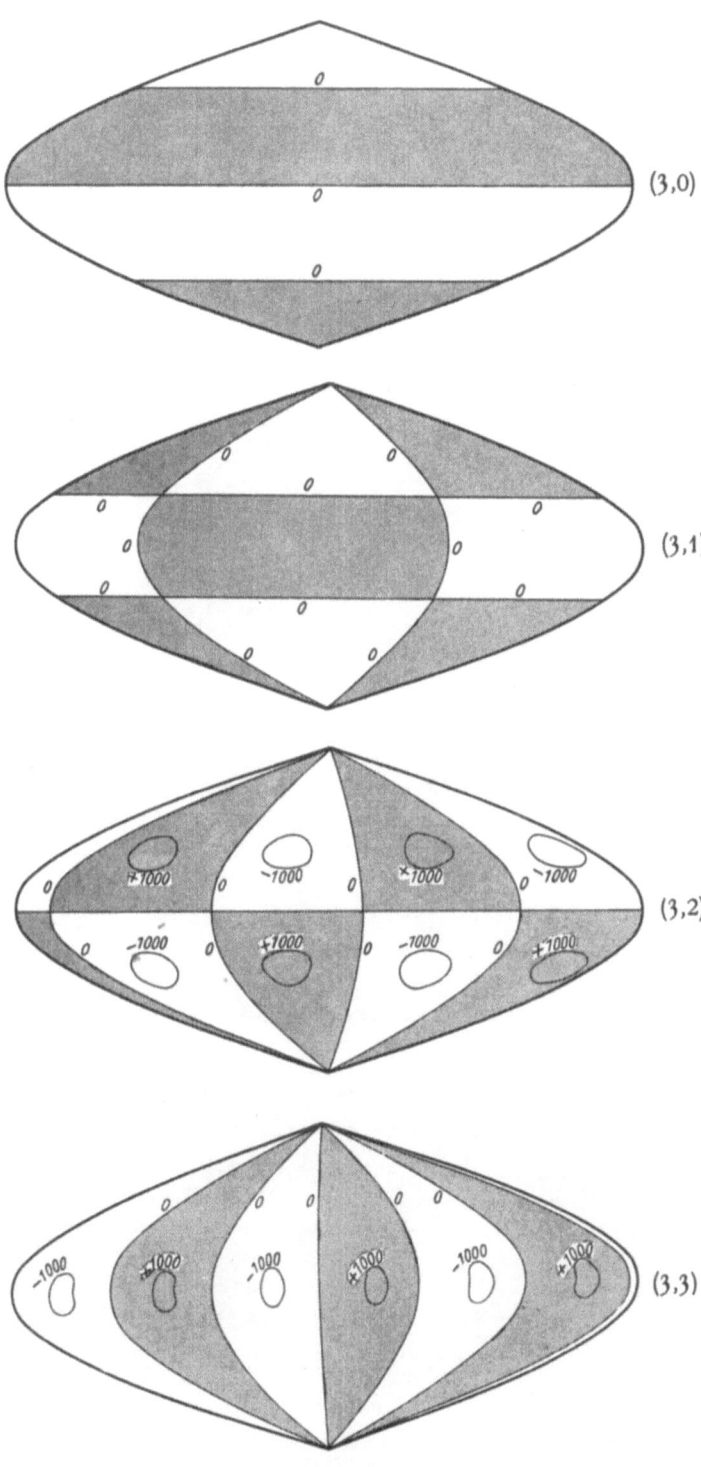

Fig. 26.

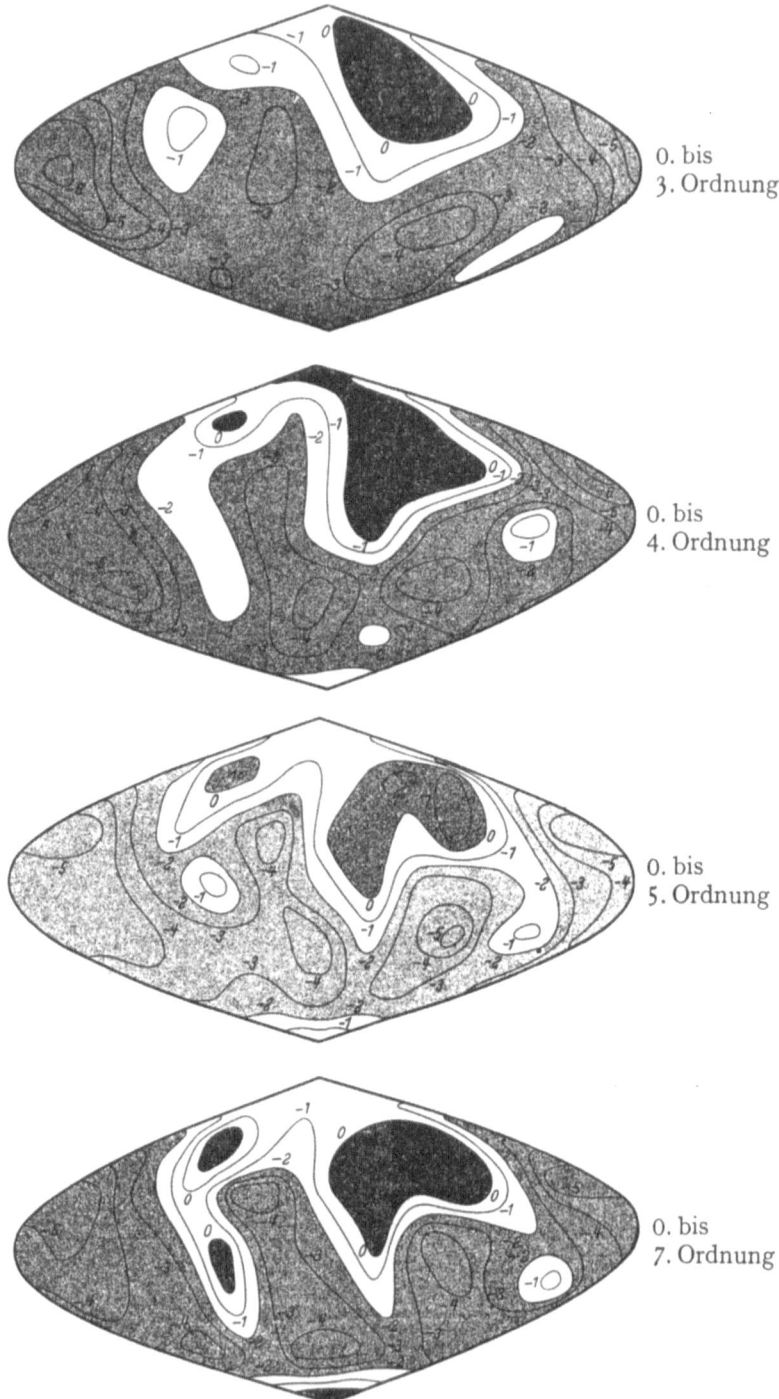

Fig. 26. Relief der Erde (Lithosphäre). Entwicklung nach Kugelfunktionen bis zur 7. Ordnung [28]. Flächentreuer Entwurf (Mercator-Sanson). Der Greenwich-Meridian liegt in der Mitte. Höhen- und Tiefenlinien im Abstand von 1000 m.

konzentrische Kugeln über. Die Funktionen $P_{n\nu}(i\varrho/e)$ und $Q_{n\nu}(i\varrho/e)$ werden proportional $l^n$ und $l^{-(n+1)}$.[1] $\varepsilon^2$ ist ungefähr $\frac{1}{150}$. Da die Geoidundulationen wohl kaum größere Beträge als 100 m erreichen und bis jetzt nicht mit größerer Genauigkeit als etwa $\pm 5$ m bestimmt werden können[2], sind Vernachlässigungen von dieser Größenordnung zulässig. Dann kann man auch die Rotation der Erde unbeachtet lassen und bei Berechnung von Geoidundulationen und Schwereanomalien so vorgehen, als seien die Unregelmäßigkeiten der Erdkruste einer nicht rotierenden Erdkruste aufgesetzt. Hiermit ergeben sich erhebliche Vereinfachungen.

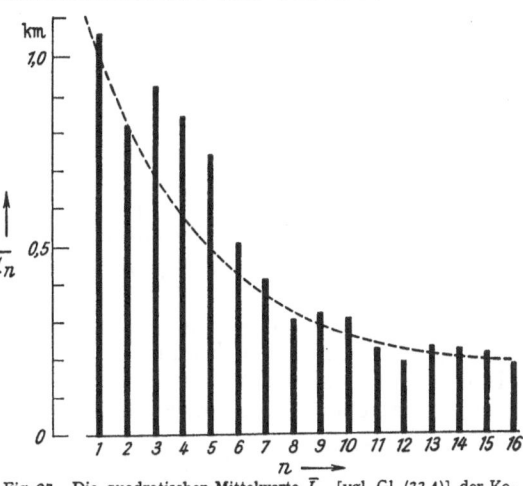

Fig. 27. Die quadratischen Mittelwerte $\bar{L}_n$ [vgl. Gl. (33.1)] der Koeffizienten der Entwicklung des Erdreliefs (Lithosphäre) nach Kugelfunktionen.

Tabelle 2. *Quadratische Mittelwerte der* PREY*schen Koeffizienten der Lithosphäre (nach* VENING MEINESZ)[3].

| $n$ | $\bar{L}_n$ km | $n$ | $\bar{L}_n$ km |
|---|---|---|---|
| 1 | 1,055 | 9 | 0,3287 |
| 2 | 0,822 | 10 | 0,3143 |
| 3 | 0,931 | 11 | 0,2359 |
| 4 | 0,850 | 12 | 0,2048 |
| 5 | 0,751 | 13 | 0,2350 |
| 6 | 0,510 | 14 | 0,2277 |
| 7 | 0,416 | 15 | 0,2179 |
| 8 | 0,309 | 16 | 0,1810 |

Es sei

$l$  der Abstand des Aufpunktes vom Erdmittelpunkt;
$l'$  der Abstand eines Massenpunktes vom Erdmittelpunkt;
$R$  der Radius einer Bezugskugel[4];
$R_a$  der Radius derjenigen Kugel, die gerade alle Massen der Kugel-Erde einschließt;
$\vartheta, \lambda$  Polabstand und Länge;
$f$  die Gravitationskonstante;
$\sigma$  die Gesteinsdichte;
$V$  das Schwerepotential (hier gleich dem Gravitationspotential);
$g$  die Schwere.

Dann gilt für $l \geqq R_a$:

$$V = V^{(a)} = \sum_{n=0}^{\infty} \sum_{\nu=0}^{n} \frac{R^{n+1}}{l^{n+1}} V_{n\nu}^{(a)}(\vartheta, \lambda),$$

$$g = g^{(a)} = \sum_{n=0}^{\infty} \sum_{\nu=0}^{n} (n+1) \frac{R^{n+1}}{l^{n+2}} V_{n\nu}^{(a)}(\vartheta, \lambda),$$

$$V_{n\nu}^{(a)} = (_cV_{n\nu}^{(a)} \cos \nu\lambda + _sV_{n\nu}^{(a)} \sin \nu\lambda) \cdot P_{n\nu}(\cos \vartheta),$$

$$\left.\begin{matrix} _cV_{n\nu}^{(a)} \\ _sV_{n\nu}^{(a)} \end{matrix}\right\} = \frac{f}{R^{n+1}} \int_{l'=0}^{l} \int_{\vartheta'=0}^{\pi} \int_{\lambda'=0}^{2\pi} \sigma(l', \vartheta', \lambda') \begin{Bmatrix} \cos \nu\lambda' \\ \sin \nu\lambda' \end{Bmatrix} \times$$

$$\times P_{n\nu}(\cos \vartheta') \cdot \sin \vartheta' \, l'^{n+2} \, dl' \, d\vartheta' \, d\lambda';$$

$$(33.2)[5]$$

---

[1] Anhang III, Ziff. 57.
[2] W. HEISKANEN: Science, Lancaster Pa. **121**, 50 (1955).
[3] F. A. VENING MEINESZ: A remarkable feature of the earth's topography, origin of continents and oceans, I and II. Proc., Kon. nederl. Akad. Wetensch., Ser. B **54**, 212 (1951).
[4] Meist setzt man $R = 6370$ km (Radius der dem Erdellipsoid volumengleichen Kugel).
[5] Für die obere Integrationsgrenze $l$ kann hier auch $R_a$ gesetzt werden, nicht aber in (33.3).

und für $l \leq R_a$ gilt

$$
\left.
\begin{aligned}
& V = V^{(a)} + V^{(i)}, \qquad g = g^{(a)} + g^{(i)}, \\[4pt]
& V^{(a)}, g^{(a)} \qquad \text{wie in (33.3)}, \\[4pt]
& V^{(i)} = \sum_{n=0}^{\infty} \sum_{\nu=0}^{n} \frac{l^n}{R^n} V_{n\nu}^{(i)}(\vartheta, \lambda), \\[4pt]
& g^{(i)} = -\sum_{n=0}^{\infty} \sum_{\nu=0}^{n} n \frac{l^{n-1}}{R^n} V_{n\nu}^{(i)}(\vartheta, \lambda), \\[4pt]
& V_{n\nu}^{(i)} = ({}_cV_{n\nu}^{(i)} \cos \nu\lambda + {}_sV_{n\nu}^{(i)} \sin \nu\lambda) \cdot P_{n\nu}(\cos\vartheta), \\[4pt]
& \left.{}_c V_{n\nu}^{(i)} \atop {}_s V_{n\nu}^{(i)}\right\} = f R^n \int_{l'=0}^{l} \int_{\vartheta'=0}^{\pi} \int_{\lambda'=0}^{2\pi} \sigma(l', \vartheta', \lambda') \begin{Bmatrix} \cos\nu\lambda' \\ \sin\nu\lambda' \end{Bmatrix} \times \\[4pt]
& \qquad\qquad \times P_{n\nu}(\cos\vartheta') \cdot \sin\vartheta' \, l'^{-(n-1)} \, dl' \, d\vartheta' \, d\lambda'.
\end{aligned}
\right\} \qquad (33.3)
$$

Im Innern der Erde und außerhalb des Erdkörpers zwischen seiner Oberfläche und der Kugel mit dem Radius $R_a$ ist das Formelsystem (33.3) zuständig. Im Innern ist es nicht zu umgehen. In dem massenfreien Raumteil ist es jedoch üblich, zur Vereinfachung das Formelsystem (33.2) anzuwenden.

Über die Zulässigkeit dieser Vereinfachung muß man sich Rechenschaft geben. Hierzu werde als Beispiel eine Erde mit homogener Kruste betrachtet, deren Oberfläche sich um $H(\vartheta, \lambda)$ über die Bezugskugel mit dem Radius $R$ erhebt. Man setzt

$$
H(\vartheta, \lambda) = -t(\vartheta, \lambda) + h(\vartheta, \lambda),
$$

wobei

$$
-t(\vartheta, \lambda) = \begin{cases} H(\vartheta, \lambda), & \text{wenn} \quad H(\vartheta, \lambda) \leq 0 \\ 0, & \text{wenn} \quad H(\vartheta, \lambda) \geq 0, \end{cases}
$$

$$
h(\vartheta, \lambda) = \begin{cases} 0, & \text{wenn} \quad H(\vartheta, \lambda) \leq 0 \\ H(\vartheta, \lambda), & \text{wenn} \quad H(\vartheta, \lambda) \geq 0. \end{cases}
$$

Der Aufpunkt liege auf der Bezugskugel, in ihm ist $h(\vartheta, \lambda) = 0$, $H(\vartheta, \lambda) \leq 0$.

Es ist zweckmäßig, $R = l$ zu setzen. $t$ und $h$ sind klein gegen $l$, man kann also $l - t(\vartheta, \lambda)$ und $l + h(\vartheta, \lambda)$ in TAYLOR-Reihen entwickeln. Bei den Integrationen macht man von dem Integralsatz der Kugelfunktionen[1] Gebrauch. So erhält man mit (33.3) den korrekten Potentialanteil $\delta V$ und den korrekten Schwereanteil $\delta g$ des Oberflächenreliefs:

$$
\begin{aligned}
\delta V = 4\pi f \sigma l^2 \sum_{n=0}^{\infty} \sum_{\nu=0}^{n} \frac{1}{2n+1} & \left\{ \left[ {}_c\!\left(\frac{H}{l}\right)_{n\nu} + \frac{n+2}{2}\, {}_c\!\left(\frac{t^2}{l^2}\right)_{n\nu} - \frac{n-1}{2}\, {}_c\!\left(\frac{h^2}{l^2}\right)_{n\nu} + \cdots \right] \cos\nu\lambda \; + \right. \\
& \left. + \left[ {}_s\!\left(\frac{H}{l}\right)_{n\nu} + \frac{n+2}{2}\, {}_s\!\left(\frac{t^2}{l^2}\right)_{n\nu} - \frac{n-1}{2}\, {}_s\!\left(\frac{h^2}{l^2}\right)_{n\nu} + \cdots \right] \sin\nu\lambda \right\} \cdot P_{n\nu}(\cos\vartheta),
\end{aligned}
$$

$$
\begin{aligned}
\delta g = 4\pi f \sigma l \sum_{n=0}^{\infty} \sum_{\nu=0}^{n} & \left\{ \left[ -(n+1)\, {}_c\!\left(\frac{t}{l}\right)_{n\nu} - n\, {}_c\!\left(\frac{h}{l}\right)_{n\nu} + \frac{(n+1)(n+2)}{2}\, {}_c\!\left(\frac{t^2}{l^2}\right)_{n\nu} - \right. \right. \\
& \left. - \frac{(n+1)(n-1)}{2}\, {}_c\!\left(\frac{h^2}{l^2}\right)_{n\nu} + \cdots \right] \cos\nu\lambda + \left[ -(n+1)\, {}_s\!\left(\frac{t}{l}\right)_{n\nu} - n\, {}_s\!\left(\frac{h}{l}\right)_{n\nu} + \right. \\
& \left. \left. + \frac{(n+1)(n+2)}{2}\, {}_s\!\left(\frac{t^2}{l^2}\right)_{n\nu} - \frac{(n+1)(n-1)}{2}\, {}_s\!\left(\frac{h^2}{l^2}\right)_{n\nu} + \cdots \right] \sin\nu\lambda \right\} \cdot P_{n\nu}(\cos\vartheta).
\end{aligned}
$$

---

[1] Anhang III, Ziff. 59.

In gleicher Weise erhält man mit (33.2) den vereinfachten Potentialanteil $(\delta V)$ und den vereinfachten Schwereanteil $(\delta g)$:

$$(\delta V) = 4\pi f \sigma l^2 \sum_{n=0}^{\infty} \sum_{\nu=0}^{n} \frac{1}{2n+1} \left\{ \left[ {}_c\!\left(\frac{H}{l}\right)_{n\nu} + \frac{n+2}{2}\, {}_c\!\left(\frac{H^2}{l^2}\right)_{n\nu} + \cdots \right] \cos\nu\lambda + \right.$$
$$\left. + \left[ {}_s\!\left(\frac{H}{l}\right)_{n\nu} + \frac{n+2}{2}\, {}_s\!\left(\frac{H^2}{l^2}\right)_{n\nu} + \cdots \right] \sin\nu\lambda \right\} \cdot P_{n\nu}(\cos\vartheta),$$

$$(\delta g) = 4\pi f \sigma l \sum_{n=0}^{\infty} \sum_{\nu=0}^{n} \left\{ \left[ (n+1)\, {}_c\!\left(\frac{H}{l}\right)_{n\nu} + \frac{(n+1)(n+2)}{2}\, {}_c\!\left(\frac{H^2}{l^2}\right)_{n\nu} + \cdots \right] \cos\nu\lambda + \right.$$
$$\left. + \left[ (n+1)\, {}_s\!\left(\frac{H}{l}\right)_{n\nu} + \frac{(n+1)(n+2)}{2}\, {}_s\!\left(\frac{H^2}{l^2}\right)_{n\nu} + \cdots \right] \sin\nu\lambda \right\} \cdot P_{n\nu}(\cos\vartheta)$$

und die relativen Vernachlässigungen [1]:

$$\frac{(\delta V) - \delta V}{\delta V} = \frac{\displaystyle\sum_{n=0}^{\infty} \sum_{\nu=0}^{n} \left\{ \frac{n(n+1)}{24}\, {}_c\!\left(\frac{h^4}{l^4}\right)_{n\nu} \cos\nu\lambda + \frac{n(n+1)}{24}\, {}_s\!\left(\frac{h^4}{l^4}\right)_{n\nu} \sin\nu\lambda + \cdots \right\} P_{n\nu}(\cos\vartheta)}{\displaystyle\sum_{n=0}^{\infty} \sum_{\nu=0}^{n} \frac{1}{2n+1} \left\{ \left[ {}_c\!\left(\frac{H}{l}\right)_{n\nu} + \cdots \right] \cos\nu\lambda + \left[ {}_s\!\left(\frac{H}{l}\right)_{n\nu} + \cdots \right] \sin\nu\lambda \right\} P_{n\nu}(\cos\vartheta)},$$

$$\frac{(\delta g) - \delta g}{\delta g} = \frac{\displaystyle\sum_{n=0}^{\infty} \sum_{\nu=0}^{n} \left\{ \left[ 2n\, {}_c\!\left(\frac{h}{l}\right)_{n\nu} + \cdots \right] \cos\nu\lambda + \left[ 2n\, {}_s\!\left(\frac{h}{l}\right)_{n\nu} + \cdots \right] \sin\nu\lambda \right\} P_{n\nu}(\cos\vartheta)}{\displaystyle\sum_{n=0}^{\infty} \sum_{\nu=0}^{n} \left\{ \left[ (n+1)\, {}_s\!\left(\frac{H}{l}\right)_{n\nu} + \cdots \right] \cos\nu\lambda + \left[ (n+1)\, {}_s\!\left(\frac{H}{l}\right)_{n\nu} + \cdots \right] \sin\nu\lambda \right\} P_{n\nu}(\cos\vartheta)}.$$

In den PREYSchen Tabellen sieht man, daß die Koeffizienten ${}_c(h/l)_{n\nu}$ und ${}_c(t/l)_{n\nu}$ nur selten Beträge von $\frac{1}{10000}$ erreichen. Daher sind die Vernachlässigungen, die bei der allgemeinen Verwendung von (33.2) im Außenraum entstehen, bis zur Ordnung $n = 35$ sicher unbedenklich. Bei Gliedern von wesentlich höherer Ordnung muß man vorsichtig sein und sich nötigenfalls von Glied zu Glied überzeugen, ob die Vernachlässigungen noch tragbar sind.

Weiterhin ist es üblich, räumlich verteilte Massen durch auf Kugelflächen ausgebreitete Flächenbelegungen zu ersetzen. Eine ähnliche Abschätzung führt zu dem Ergebnis, daß diese Vereinfachung bei den topographischen Massen, deren Reliefunterschiede in der Größenordnung von $\frac{1}{1000}$ des Erdradius liegen, bis zu Gliedern von hoher Ordnung unbedenklich sind. Bei den AIRYSchen Kompensationsmassen, deren Grundfläche Reliefunterschiede mit der Größenordnung von $\frac{1}{100}$ des Erdradius aufweist, können sie bedenklich sein [2].

*δ) Störungspotential, Undulationen der Niveaufläche, Niveauflächenschwere.* Das Störungspotential ist

$$T = \sum_{n=0}^{\infty}{}^* \sum_{\nu=0}^{n}{}^* \frac{R^{n+1}}{l^{n+1}} V_{n\nu}^{(a)}(\vartheta, \lambda), \tag{33.4}$$

wobei die Sterne andeuten, daß die der Normalschwere entsprechenden Glieder wegzulassen sind. Hieraus ergibt sich die Undulation der Niveaufläche [3]:

$$\zeta = \frac{1}{\gamma} \sum_{n=0}^{\infty}{}^* \sum_{\nu=0}^{n}{}^* \frac{R^{n+1}}{l^{n+1}} V_{n\nu}^{(a)}(\vartheta, \lambda). \tag{33.5}$$

---

[1] Für die erste Formel müssen die TAYLOR-Entwicklungen bis zur vierten Ordnung weitergeführt werden, da die Glieder bis zur dritten Ordnung einander aufheben. In der zweiten Formel erhält man rechts im Zähler zunächst $2n + 1$ statt $2n$; da aber im Aufpunkt $h = 0$ ist, heben sich die mit der 1 multiplizierten Glieder weg.

[2] In solchen Fällen verlangen auch räumlich verteilte Massen eine Berücksichtigung der quadratischen Glieder der TAYLOR-Entwicklungen. K. JUNG: Die rechnerische Behandlung der AIRYSchen Isostasie mit einer Entwicklung des Quadrats der Meereshöhen nach Kugelfunktionen. Gerlands Beitr. Geophys. **62**, 39 (1950). Siehe auch VENING MEINESZ: Bull. Géod. N.S. **1** (1946).

[3] (5.1) und (33.2). Dabei enthalten $T$ und $\zeta$ auch Glieder von den Ordnungen 0,0 und 2,0, wenn die Schwereformel nicht genau der angenommenen Massenverteilung entspricht.

Die Schwereänderung mit der Höhe ist genau genug[1]

$$\frac{\partial g}{\partial \zeta} = -\frac{2\gamma}{l},$$

so daß

$$\frac{\partial g}{\partial \zeta} \cdot \zeta = -2 \sum_{n=0}^{\infty}{}^* \sum_{\nu=0}^{n}{}^* \frac{R^{n+1}}{l^{n+2}} V_{n\nu}^{(a)}(\vartheta, \lambda) \tag{33.6}$$

den Schwereunterschied zwischen der Niveaufläche und dem die Niveaufläche annähernden Sphäroid darstellt. Addiert man (33.6) zu (33.2), so erhält man

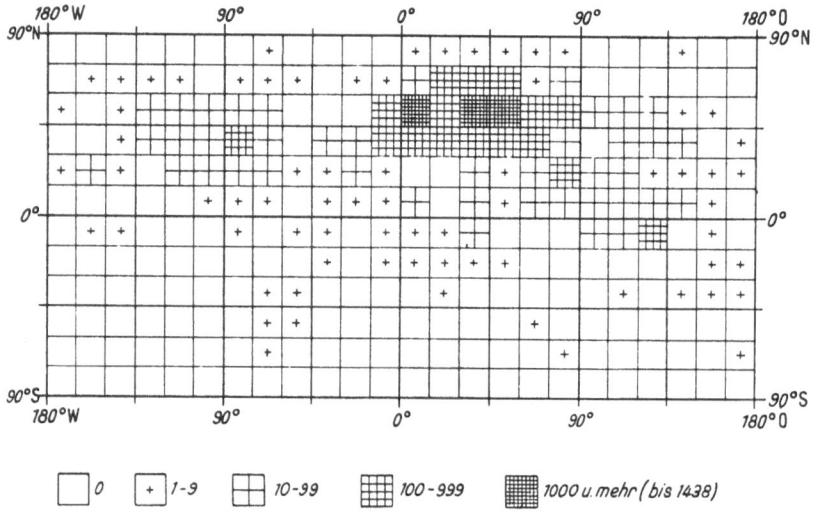

Fig. 28. Die Verteilung der Pendelstationen (bis 1936) nach dem Verzeichnis von ZHURAVLEV.

die Niveauflächenschwere $g_{\mathrm{Ni}}$. Die Schwereanomalie auf der Niveaufläche ist sodann

$$g_{\mathrm{Ni}} - \gamma = \sum_{n=0}^{\infty}{}^* \sum_{\nu=0}^{n}{}^* (n-1) \frac{R^{n+1}}{l^{n+2}} V_{n\nu}^{(a)}(\vartheta, \lambda). \tag{33.7}$$

Der Faktor $n-1$ zeigt an, daß diese Schwereanomalie kein Glied erster Ordnung enthalten kann[2]. Wenn die Analyse von Schwereanomalien auf Glieder erster Ordnung führt, so dürfte dieses Ergebnis von der Lückenhaftigkeit des Beobachtungsmaterials (Fig. 28) vorgetäuscht sein[3].

$\varepsilon$) *Isostatische Abschätzungen mit Flächenbelegungen.* Ist $\mu(\vartheta, \lambda)$ die Flächendichte einer auf der Kugel mit dem Radius $r$ ausgebreiteten Flächenbelegung, so gilt

für $l \geq r$

$$T^{(a)}(\vartheta, \lambda) = 4\pi f \sum_{n=0}^{\infty}{}^* \sum_{\nu=0}^{n}{}^* \frac{1}{2n+1} \frac{r^{n+2}}{l^{n+1}} \mu_{n\nu}(\vartheta, \lambda), \tag{33.8}$$

für $l \leq r$

$$T^{(i)}(\vartheta, \lambda) = 4\pi f \sum_{n=0}^{\infty}{}^* \sum_{\nu=0}^{n}{}^* \frac{1}{2n+1} \frac{l^{n}}{r^{n-1}} \mu_{n\nu}(\vartheta, \lambda). \tag{33.9}$$

---

[1] Ziff. 18.
[2] Dasselbe Ergebnis erhält man auch, wenn man (29.15) nach den $G_{n\nu}$ auflöst.
[3] Im Innern der Erde können Glieder erster Ordnung vorkommen. K. JUNG: Über das harmonische Glied erster Ordnung bei Isostasie. Publ. dedicated to W. A. HEISKANEN, Veröff. finn. Geodät. Inst. Helsinki **1955**, Nr. 46.

Als Anwendung sollen eine topographische Flächenbelegung $\mu(\vartheta, \lambda)$ auf der Kugel $r = R$ und ihre Kompensation $\bar{\mu}(\vartheta, \lambda)$ auf der Kugel $r = R - T$ mit drei verschiedenen Isostasiebeziehungen betrachtet werden:

a) Angenäherter Druckausgleich (übliche Isostasiebeziehung)

$$\bar{\mu}(\vartheta, \lambda) = -\mu(\vartheta, \lambda),$$

b) Massenausgleich

$$\bar{\mu}(\vartheta, \lambda) = -\frac{R^2}{(R - T)^2}\,\mu(\vartheta, \lambda),$$

c) Potentialausgleich unter der Ausgleichsfläche

$$\bar{\mu}_{n\,\nu}(\vartheta, \lambda) = -\frac{(R - T)^{n-1}}{R^{n-1}}\,\mu_{n\,\nu}(\vartheta, \lambda).$$

*Das Glied erster Ordnung und die Schwerpunktverlagerung bei der isostatischen Reduktion.*

In eine Kugel mit dem Radius $R$ sei eine Flächenbelegung

$$\mu_1(\vartheta, \lambda) = \mu_{10}\cos\vartheta + {}_c\mu_{11}\sin\vartheta\cos\lambda + {}_s\mu_{11}\sin\vartheta\sin\lambda$$

eingelagert. Die Kugel sei kugelsymmetrisch aufgebaut, ihre Gesamtmasse sei $m$. Der Radius der die Flächenbelegung tragenden, konzentrischen Kugel sei $r$. Dann hat das Maximum der Belegungsdichte die Koordinaten

$$\lambda_{\max} = \lambda_{11} = \arctan\frac{{}_s\mu_{11}}{{}_c\mu_{11}} \qquad (s\,g\cos\lambda_{11} = s\,g\,{}_c\mu_{11}, \qquad s\,g\sin\lambda_{11} = s\,g\,{}_s\mu_{11}),$$

$$\vartheta_{\max} = \arctan\frac{\mu_{11}}{\mu_{10}} = \frac{\sqrt{{}_c\mu_{11}^2 + {}_s\mu_{11}^2}}{\mu_{10}},$$

und es beträgt

$$(\mu_1)_{\max} = \sqrt{\mu_{10}^2 + {}_c\mu_{11}^2 + {}_s\mu_{11}^2}.$$

Der den Schwerpunkt der gesamten Massen enthaltende Radius ist zum Maximum gerichtet. Die Schwerpunktsentfernung vom Mittelpunkt sei $s$. Man berechnet:

$$s = \frac{4\pi}{3}r^3\frac{(\mu_1)_{\max}}{m}.$$

Wendet man diese Formel auf eine Kugel mit topographischer und kompensierender Flächenbelegung an, so erhält man die Schwerpunktlage

$$s = \frac{4\pi}{3}R^3\frac{(\mu_1)_{\max}}{m}\cdot S,$$

wobei

$$S = \begin{cases} 1 & \text{im unkompensierten Fall,} \\ +\dfrac{3\,T}{R} & \text{im isostatischen Fall (a),} \\ +\dfrac{T}{R} & \text{im Fall (b),} \\ +\dfrac{3\,T}{R} & \text{im Fall (c).} \end{cases}$$

Tabelle 3. $\sigma = 2{,}7$ g/cm³, $T/R = \frac{30}{6370}$, $m = 5{,}98 \cdot 10^{-27}$ g.

| | $L$ km | $M$ km | $L - \dfrac{1{,}03}{2{,}7}M$ km |
|---|---|---|---|
| $\dfrac{1}{\sigma}\mu_{10} =$ | $+1{,}264$ | $+1{,}112$ | $+0{,}840$ |
| $\dfrac{1}{\sigma}{}_c\mu_{11} =$ | $+1{,}130$ | $+1{,}006$ | $+0{,}746$ |
| $\dfrac{1}{\sigma}{}_s\mu_{11} =$ | $+0{,}667$ | $+0{,}564$ | $+0{,}449$ |
| $\dfrac{1}{\sigma}(\mu_1)_{\max} =$ | $1{,}821$ | $1{,}602$ | $1{,}210$ |
| $\lambda_{\max} =$ | | | $31{,}0°$ |
| $\vartheta_{\max} =$ | | | $46{,}0°$ |
| $s$ { unkomp. $=$ | | | $+592$ m |
| isost. (a) $=$ | | | $+\ 9{,}27$ m |
| isost. (b) $=$ | | | $+\ 2{,}79$ m |
| isost. (c) $=$ | | | $+\ 9{,}27$ m |

Tabelle 3 zeigt die aus der Entwicklung von PREY berechneten Zahlen. Um den gleichen Betrag, aber in umgekehrter Richtung, ist der Schwerpunkt des Co-Geoids gegen den Schwerpunkt des ursprünglichen Geoids verschoben.

*Das zonale Glied zweiter Ordnung und der Einfluß der isostatischen Reduktion auf die Bestimmung der Erdabplattung.*

Differenziert man (33.8), so erhält man den Schwereanteil $\delta g$ einer topographischen Belegung $\mu(\vartheta, \lambda)$. Ihr zonales Glied zweiter Ordnung lautet für $r = R$:

$$\delta g_{20}^{\text{top}} = \frac{12\pi}{5} f \frac{R^4}{l^4} \mu_{20} P_{20}(\cos\vartheta).$$

Für die Kompensationsbelegung auf der Kugel $r = R - T$ gilt

$$\delta g_{20}^{\text{komp}} = \frac{12\pi}{5} f \frac{(R-T)^4}{l^4} \bar{\mu}_{20} P_{20}(\cos\vartheta),$$

und der Einfluß der isostatischen Reduktion berechnet sich für $l = R$ zu

$$-\delta g_{20}^{\text{top}} - \delta g_{20}^{\text{komp}} = -\frac{12\pi}{5} f\left(1 + \frac{(R-T)^4}{R^4} \cdot F\right)\mu_{20} P_{20}(\cos\vartheta)$$

mit

$$F = \begin{cases} 0 & \text{im unkompensierten Fall,} \\[4pt] -1 & \text{im isostatischen Fall (a),} \\[4pt] -\dfrac{R^2}{(R-T)^2} & \text{im Fall (b),} \\[6pt] -\dfrac{R-T}{R}. & \text{im Fall (c).} \end{cases}$$

Dann beträgt der Einfluß auf den Koeffizient $\mathfrak{b}$ der Schwereformel

$$\delta\mathfrak{b} = -\frac{18\pi}{5} \frac{f}{\gamma} \mu_{20} B$$

mit

$$B = \begin{cases} 1 & \text{im unkompensierten Fall,} \\[4pt] \dfrac{4T}{R} & \text{im isostatischen Fall (a),} \\[6pt] \dfrac{2T}{R} & \text{im Fall (b),} \\[6pt] \dfrac{5T}{R} & \text{im Fall (c).} \end{cases}$$

Nach der einfachen Form des CLAIRAUTschen Theorems ist

$$\delta\mathfrak{a} = -\delta\mathfrak{b}, \qquad \delta\left(\frac{1}{\mathfrak{a}}\right) = -\frac{\delta\mathfrak{a}}{\mathfrak{a}^2}.$$

Nun ist $\mu_{20} = \sigma\left(L_{20} - \dfrac{1{,}03}{2{,}7} M_{20}\right)$, wobei $L_{20}$ und $M_{20}$ der PREYschen Entwicklung zu entnehmen sind[1]. Mit $\gamma = 981$ Gal, $\sigma = 2{,}7$ g/cm³, $T/R = \frac{30}{6370}$, $\mathfrak{a} = \frac{1}{297}$ erhält man

im unkompensierten Fall: $\quad \delta\mathfrak{a} = +1{,}53 \cdot 10^{-4}, \qquad \delta\left(\dfrac{1}{\mathfrak{a}}\right) = -13{,}53,$

im isostatischen Fall (a): $\quad \delta\mathfrak{a} = +2{,}89 \cdot 10^{-6}, \qquad \delta\left(\dfrac{1}{\mathfrak{a}}\right) = -0{,}255$

im Fall (b): $\qquad\qquad\quad \delta\mathfrak{a} = +1{,}44 \cdot 10^{-6}, \qquad \delta\left(\dfrac{1}{\mathfrak{a}}\right) = -0{,}127$

im Fall (c): $\qquad\qquad\quad \delta\mathfrak{a} = +3{,}61 \cdot 10^{-6}, \qquad \delta\left(\dfrac{1}{\mathfrak{a}}\right) = -0{,}318.$

---

[1] Anhang II, Ziff. 52: $L_{20} = +1{,}134$ km, $M_{20} = +1.037$ km.

Nach Schütte[1] ergab sich aus dem Schwereverzeichnis von Zhuravlev:

$$\delta\left(\frac{1}{a}\right) = +1 \text{ bis } +2,5.$$

Der Unterschied gegen die errechneten Werte dürfte von der ungleichmäßigen Verteilung der Schwerestationen verursacht sein.

**34. Einige Beispiele von Geoidbestimmungen mit Schwereanomalien.** Daß die Geoidundulationen verhältnismäßig klein sind und nicht entfernt die für eine unkompensierte Erdkruste errechneten Werte[2] erreichen, wurde früh erkannt. So findet Galle[3] in der Umgebung des Harzes einen von NNW nach SSO gerichteten Anstieg des Geoids von ungefähr 1 m auf 40 km, dem ein Buckel aufgesetzt ist, der in der Brockengegend den Betrag von 1 m erreicht. Der regionale Anstieg verschiebt das Maximum der Geoidhebung an den Südrand des Harzes in die Gegend von Bad Sachsa. Galle hat das Geoid aus Lotabweichungen bestimmt und die Schwereanomalien nur zur Berechnung einer die Krümmung der Lotlinien betreffenden Korrektur herangezogen.

Die erste verläßliche Geoidbestimmung mit der Stokesschen Formel hat Hirvonen ausgeführt[4]. Sie gründet sich auf Freiluftanomalien und läßt gut erkennen, daß die Geoidundulationen im allgemeinen nicht größer als ± 100 m

[1] Fußnote 8, S. 565.
[2] Ziff. 33α, S. 581.
[3] A. Galle: Das Geoid im Harz. Veröff. preuß. Geodät. Inst., N.F. **61** (1914). Wiedergegeben in Lehr- und Handbüchern.
[4] R. A. Hirvonen: Über die kontinentalen Undulationen des Geoids. Gerlands Beitr. Geophys. **40**, 18

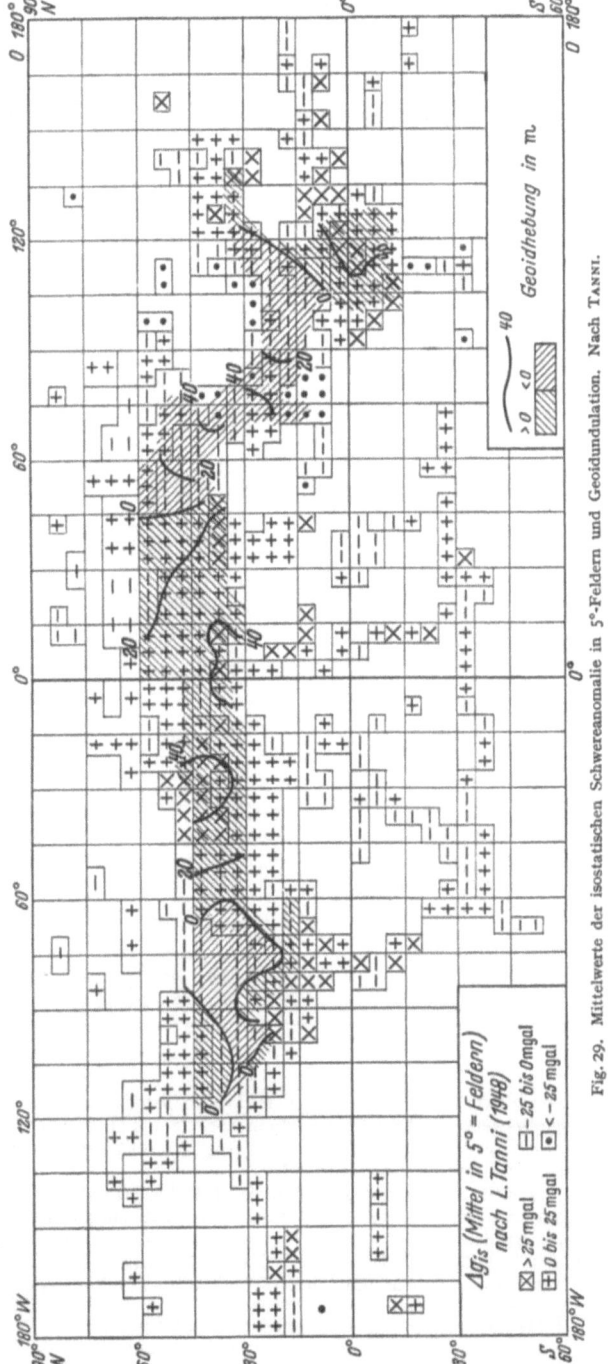

Fig. 29. Mittelwerte der isostatischen Schwereanomalie in 5°-Feldern und Geoidundulation. Nach Tanni.

(1933). — R. A. Hirvonen: The continental undulations of the geoid. Veröff. finn. Geodät. Inst., Helsinki **1934**, Nr. 19. — Die Ergebnisse sind auch wiedergegeben von K. Jung: Z. Vermessungswesen **44**, 558 (1935). — Vgl. auch R. A. Hirvonen: On the precision of the gravimetric determination of the geoid. Trans. Amer. Geophys. Un. **37**, 1 (1956).

werden. Sie brachte als überraschendes Ergebnis, daß die Geoidhebungen vorzugsweise auf den Ozeanen, die Geoidsenkungen in den Zentren der großen Kontinente liegen, gerade umgekehrt, als es für streng isostatischen Ausgleich zu erwarten war[1]. Als Erklärung wird man an eine geringe Überkompensation denken.

Wesentlich umfangreicheres Material stand TANNI zur Verfügung[2]. Nach ausführlichen Erörterungen über die zweckmäßigste Schwerereduktion verwendet er *isostatisch* reduzierte

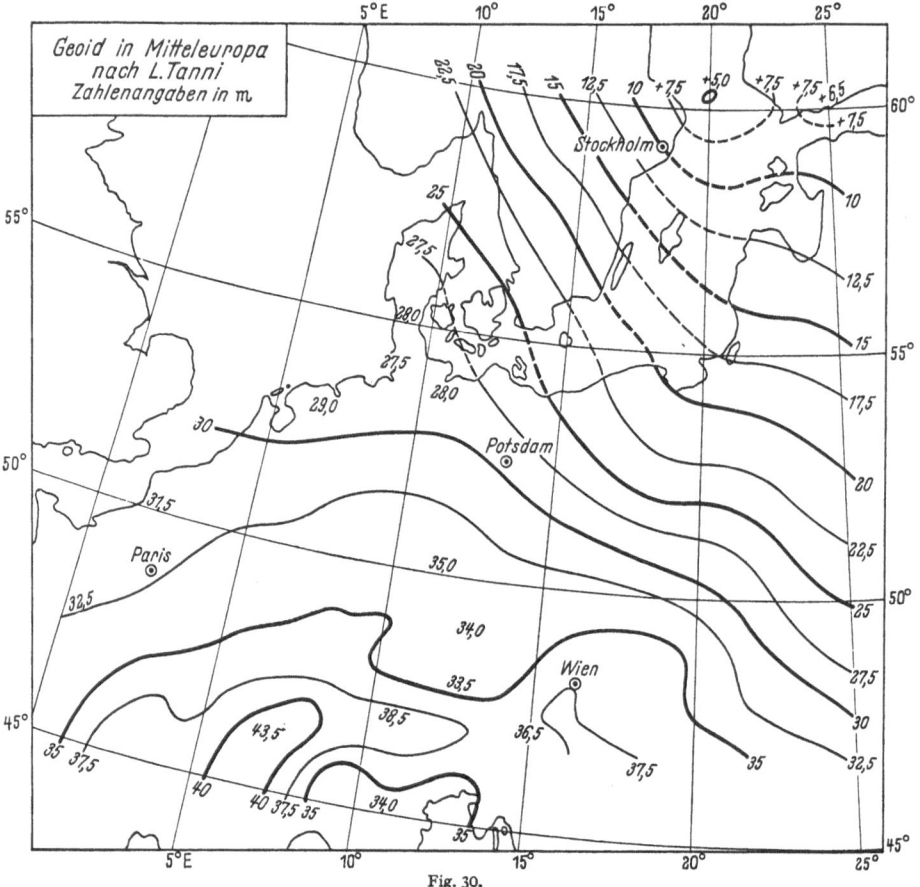

Fig. 30.

Schwerewerte. Der indirekte Effekt wird sorgfältig berücksichtigt. Die Verteilung seiner Ausgangswerte und die Höhenlinien des Geoids sind in Fig. 29 dargestellt. Sehr aufschluß-reich ist der Vergleich des Geoids (Fig. 30) mit dem Co-Geoid (Fig. 31) in Mitteleuropa, gleichfalls berechnet von TANNI[3].

Aus dem unerwartet geringen Einfluß des Himalaya-Gebirges auf die Lotabweichungen in Indien haben PRATT und AIRY ihre isostatischen Theorien entwickelt, und dort gehört Indien zu den Gebieten der Erde, die regional stark gestört sind (Fig. 32). Interessant ist der Einfluß, den die Wahl der Schwereformel auf das Aussehen der Anomalie ausübt, und die Glättung der Anomalie nach Rückberechnung aus der Geoidgestalt.

---

[1] Man vergleiche die Fig. 29 und 30 mit Fig. 25.
[2] L. TANNI: On the continental undulations of geoid as determined from the present gravity material. Publ. of the Isostatic Institute of the Intern. Ass. of Geodesy No. 18; Ann. Acad. Sci. fenn., Ser. A, III, Geologica-Geographica **16** (1948).
[3] L. TANNI: The regional rise of the geoid in Central Europe. Publ. of the Isostatic Institute of the Intern. Ass. of Geodesy No. 22. Ann. Acad. Sci. fenn., Ser. A, III, Geologica-Geographica **20** (1949).

Wie man mit der Stokesschen Formel und den Formeln von Vening Meinesz das Geoid vielfältig gestörter Gebiete bestimmen kann, zeigt de Vos van Steenwijk an dem Störungs-gebiet im Bereich der Sunda-Inseln[1]. Die von Vening Meinesz entdeckten Schwereanomalien sind in Fig. 33 dargestellt; Fig. 34 bringt eine vereinfachte Übersicht des Co-Geoids im ganzen Bereich, Fig. 35 einige Einzelheiten an einer besonders vielgestaltigen Stelle.

Eine ausführliche Bearbeitung amerikanischer Messungen[2] hat zwischen astronomisch-geodätisch und geophysikalisch ermittelten Lotabweichungen Übereinstimmungen von

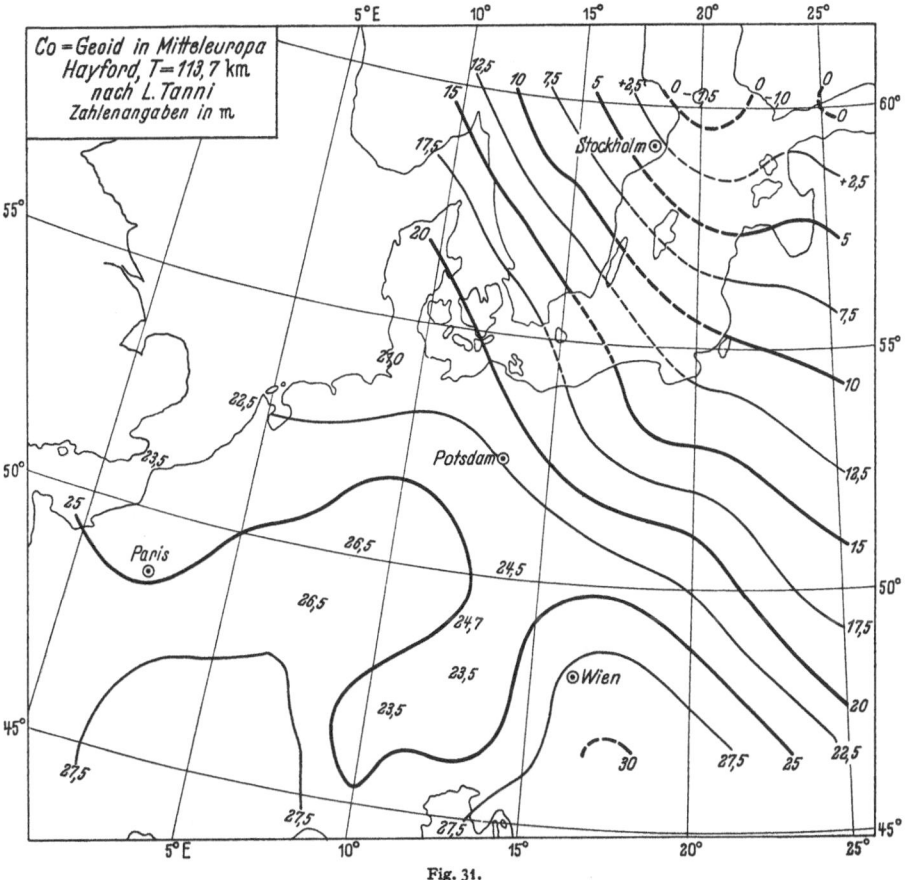

Fig. 31.

ungefähr 1″ erbracht. Sie ließen sich noch verbessern, wenn im Randgebiet noch mehr Schweremessungen vorhanden wären.

Bei Anwendung der Integralformeln[3] werden Geoidundulationen und Lotabweichungen punktweise ermittelt. Dagegen umfaßt die Koeffizientabelle der Kugelfunktionen die ganze Erde. Die Methode der Kugelfunktionen scheint für erdumfassende Betrachtungen im Vorteil zu sein. Dem steht aber als Nachteil gegenüber, daß man ohne weitgehende Fehlerbetrachtungen nicht in der Lage ist, die Verläßlichkeit der Ergebnisse abzuschätzen, während man bei der Ausrechnung der Integrale unmittelbar erkennt, wo die Beobachtungsdaten

[1] J. E. de Vos van Steenwijk: Plumb line deflectione and geoid in Eastern Indonesia. Publ. of the Netherlands Geodetic Comm. 1947.

[2] D. A. Rice: Deflections of the vertical from gravity anomalies. Bull. Géod. 1952, Nr. 25. 285.

[3] R. A. Hirvonen: On the precision obtainable for gravimetric determinations of the geoid. Techn. Paper of MCRL of Ohio State Univ. No. 171, 1952. — R. A. Hirvonen: Bull. Géod. 1955, Nr. 36, 62. — F. A. Vening Meinesz: Veröff. finn. Geodät. Inst. 1955, Nr. 46, 171 (Publ. dedicated to W. A. Heiskanen). — Vgl. auch das letzte Zitat auf S. 593.

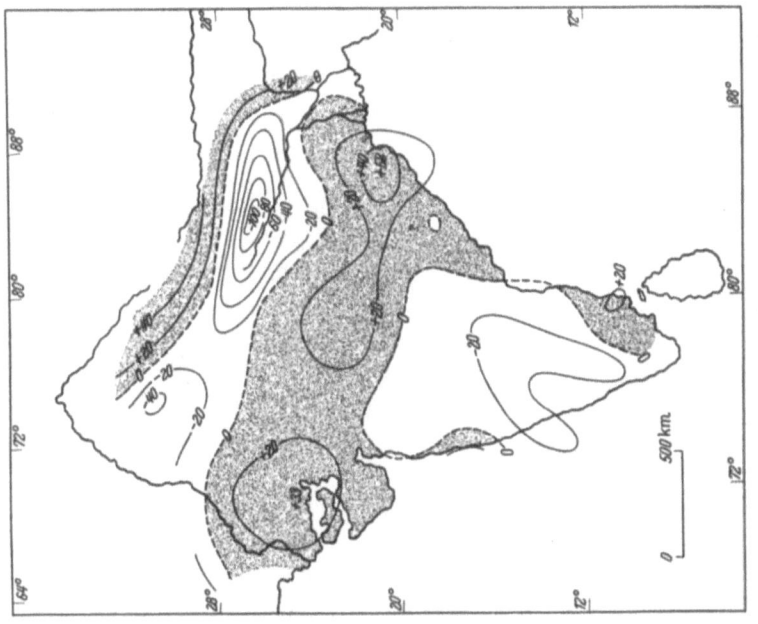

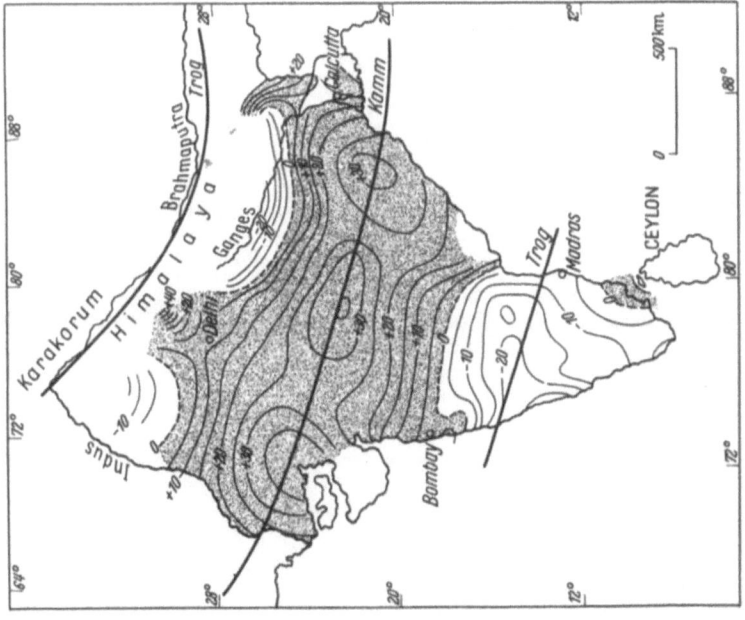

Fig. 32 a—d. Isostatische Schwereanomalien (nach Hayford) und Undulationen des Geoids in Indien (Survey of India). a Isostatische Anomalien, bezogen auf die Formel von Helmert 1901 (mgal). b Isostatische Anomalien, bezogen auf die internationale Schwereformel (mgal). c Geoidundulationen (in Fuß). d Aus den Geoidundulationen berechnete isostatische Anomalien (mgal).

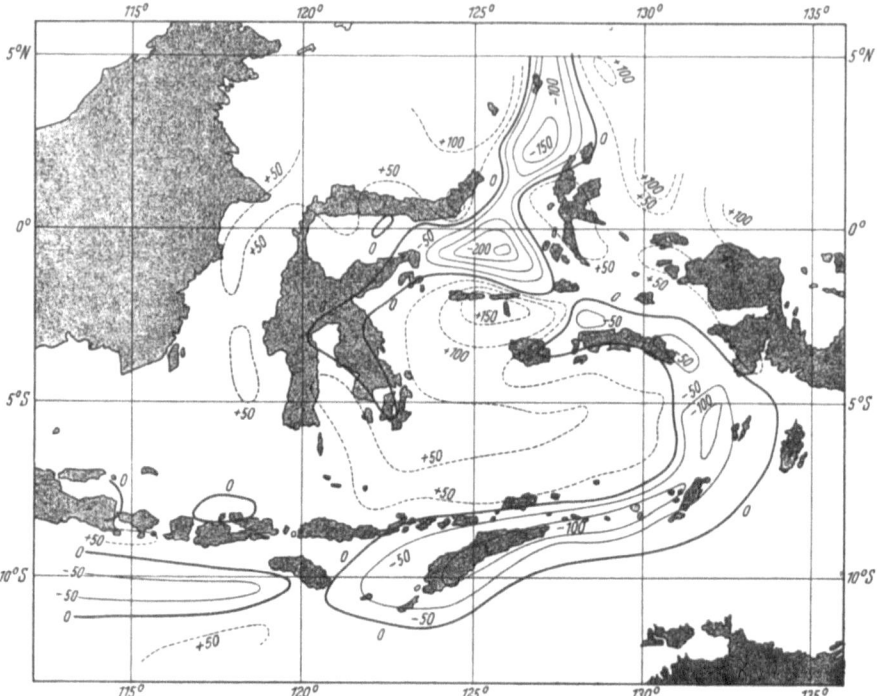

Fig. 33. Isostatische Schwereanomalien (HAYFORD) in mgal. Gezeichnet nach Tabellen von VENING-MEINESZ.

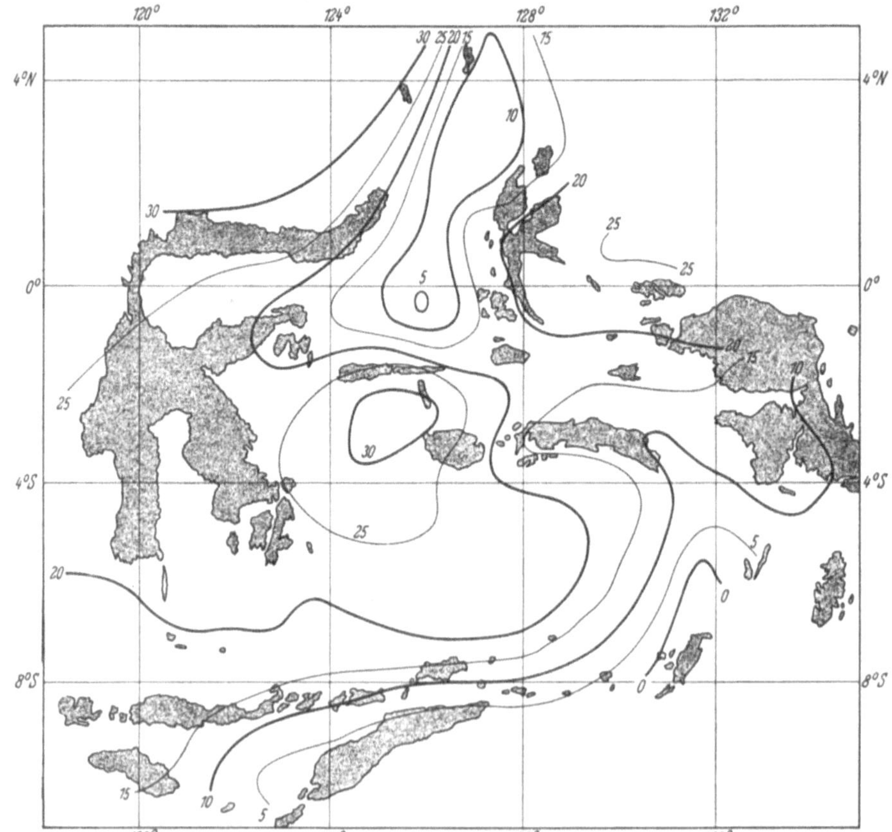

Fig. 34. Erhebung des Co-Geoids über dem internationalen Ellipsoid, abgeleitet aus den Schwereanomalien der Fig. 33. Nach DE VOS VAN STEENWIJK (vereinfacht). Zahlenangaben in Metern.

ausreichen oder zu lückenhaft sind. Die Entwicklung nach Kugelfunktionen kann erst dann mit Vorteil angewendet werden, wenn ein genügend dichtes, *gleichmäßiges* Netz repräsentativer Schwerestationen die ganze Erde überdeckt.

Nach JEFFREYS ist es möglich, einige Koeffizienten bis zur dritten Ordnung aus den Beobachtungsdaten zu errechnen. Seine Entwicklungen[1] beziehen sich auf Freiluftanomalien. Umgerechnet für normierte Kugelfunktionen[2] ergibt sich mit den Normalwerten der internationalen Schwereformel

$$g - \gamma = (2{,}5 \pm 1{,}9) -$$

$$- (6{,}1 \pm 5{,}0) \cdot P_{20}(\sin \beta) +$$

$$+ (13{,}8 \pm 4{,}8) \cos 2\lambda \cdot P_{22}(\sin \beta) +$$

$$+ (10{,}3 \pm 4{,}8) \cos \lambda \cdot P_{31}(\sin \beta) +$$

$$+ (10{,}1 \pm 5{,}3) \cos 2\lambda \cdot P_{32}(\sin \beta) +$$

$$+ (8{,}7 \pm 4{,}9) \sin 3\lambda \cdot P_{33}(\sin \beta) \text{ mgal.}$$

Hieraus berechnet man mit (29.16)[3]:

$$\zeta = 90 \cdot \cos 2\lambda \cdot P_{22}(\sin \beta) +$$

$$+ 34 \cdot \cos \lambda \cdot P_{31}(\sin \beta) +$$

$$+ 33 \cdot \cos 2\lambda \cdot P_{32}(\sin \beta) +$$

$$+ 28 \cdot \sin 3\lambda \cdot P_{33}(\sin \beta) \text{ m.}$$

**35. Das sektorielle Glied zweiter Ordnung und das dreiachsige Erdellipsoid.** Das sektorielle Glied zweiter Ordnung erhält man aus (9.6) mit $n = 2$, $\nu = 2$.

$$V_{22} = Q_{22}(i \varrho/e) ({}_c V_{22} \cos 2\lambda +$$

$$+ {}_s V_{22} \sin 2\lambda) \cdot P_{22}(\cos \bar{\vartheta}).$$

Man schreibt es auch in der Form

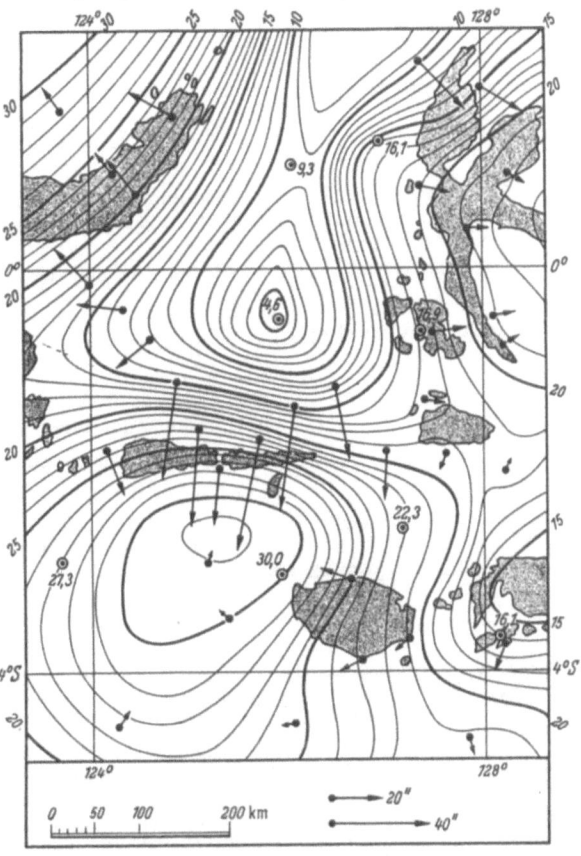

Fig. 35. Einzelheiten aus dem zentralen Gebiet der Fig. 34. Höhenlinien des Co-Geoids, bezogen auf das internationale Ellipsoid. Abstand der Höhenlinien 1 m. Aus den Schwereanomalien abgeleitete Geoidhebungen (Zahlenangaben in Metern) und Lotabweichungen (Pfeile). Die Lotabweichungspfeile sind die Gradienten der Höhen. Nach DE VOS VAN STEENWIJK.

$$V_{22} = Q_{22}(i \varrho/e) \cdot C_{22} \cos 2(\lambda - \lambda_{22}) \cdot P_{22}(\cos \bar{\vartheta}),$$

wobei

$$C_{22} = \sqrt{{}_c V_{22}^2 + {}_s V_{22}^2},$$

$$\tan 2\lambda_{22} = \frac{{}_s V_{22}}{{}_c V_{22}} \quad \begin{pmatrix} s\,g \cos 2\lambda_{22} = s\,g\,{}_c V_{22} \\ s\,g \sin 2\lambda_{22} = s\,g\,{}_s V_{22} \end{pmatrix},$$

$$\left.\begin{matrix} {}_c V_{22} \\ {}_s V_{22} \end{matrix}\right\} = f \cdot \frac{i}{e} \cdot \frac{5}{2} \int\limits_{\varrho'=0}^{c a} \int\limits_{\vartheta'=0}^{\pi} \int\limits_{\lambda'=0}^{2\pi} \sigma(\varrho', \vartheta', \lambda') \begin{Bmatrix} \cos 2\lambda' \\ \sin 2\lambda' \end{Bmatrix} P_{22}\left(\frac{i \varrho'}{e}\right) \times$$

$$\times P_{22}(\cos \bar{\vartheta}') \cdot (\varrho'^2 + e^2 \cos^2 \bar{\vartheta}') \sin \vartheta' \, d\varrho' \, d\bar{\vartheta}' \, d\lambda.$$

(35.1)

[1] H. JEFFREYS: The determination of the earth's gravitational field. Second Paper. MNRAS Geophys. Suppl. 5, 237 (1948) u. [21], S. 178.

[2] Die von JEFFREYS angegebenen Koeffizienten müssen mit $\sqrt{\dfrac{1}{\varkappa_\nu} \dfrac{(n+\nu)!}{(n-\nu)!}}$ multipliziert werden. Zahlenwerte Anhang I, Ziff. 51.

[3] Einen Vergleich dieser Werte mit den Geoidundulationen von TANNI (Fig. 29) bringt [22] auf S. 269.

Führt man die Integrationen aus, so ergibt sich, wenn $B$ und $A$ die Hauptträgheitsmomente im Äquator bezeichnen,

$$V_{22} = f \cdot \frac{18}{5} \cdot \frac{1}{e^3} (B - A) \cdot Q_{22}\left(\frac{i\varrho}{e}\right) \cdot \cos 2(\lambda - \lambda_{22}) \cdot P_{22}(\cos \bar{\vartheta}) \tag{35.2}$$

und mit Vernachlässigung von $(e/\varrho)^2$ gegen 1 erhält man den bekannten Ausdruck

$$V_{22} = \frac{3}{4} f \frac{B - A}{l^3} \sin \vartheta \cos 2(\lambda - \lambda_{22}). \tag{35.3}$$

Ein sektorielles Glied zweiter Ordnung zeigt also an, daß die Äquator-Trägheitsmomente verschieden sind. Hat es einen überragenden Betrag, so kann die Erde als dreiachsiges Ellipsoid angesehen werden.

Das entsprechende Glied der Schwereanomalie ist

$$G_{22} = c_{22} \sin^2 \vartheta \cdot \cos 2(\lambda - \lambda_{22}) \tag{35.4}$$

und man erhält aus (29.16)

$$\left. \begin{array}{rl} a - b = 2\dfrac{R}{\gamma} \cdot c_{22}, & \\[2mm] a = \text{größter Äquatorradius,} & \\[1mm] b = \text{kleinster Äquatorradius,} & \\[1mm] \lambda_{22} = \text{Länge der großen Äquatorachse.} & \end{array} \right\} \tag{35.5}$$

wobei

Ferner ist nach (29.15)

$$V_{22} = T_{22} = R \cdot G_{22},$$

und es folgt aus (35.3) für $l = R$:

$$B - A = \frac{4}{3} \frac{R^4}{f} \cdot c_{22} = \frac{2}{3} \frac{R^3 \gamma}{f} (a - b), \tag{35.6}$$

Mit $R = 6370$ km, $\gamma = 981$ Gal, $f = 6{,}67 \cdot 10^{-8}$ cgs-Einheiten erhält man $\dfrac{2R}{\gamma} = 1{,}274 \cdot 10^6 \mathrm{g}^{-1} \sec^2$, $\dfrac{4}{3} \dfrac{R^4}{f} = 3{,}297 \cdot 10^{42}$ cm sec$^2$, $\dfrac{2}{3} \dfrac{R^3 \gamma}{f} = 2{,}536 \cdot 10^{36}$ g cm.

Man hat mehrfach versucht, die Dreiachsigkeit der Erde nachzuweisen. Einige dieser Ergebnisse sind im Anhang III, Ziff. 62 angeführt. Sie überzeugen nicht. K. JUNG hat mit statistischen Methoden nachgewiesen[1], daß das sektorielle Glied zweiter Ordnung nicht gesichert ist. Seine Zufallswahrscheinlichkeit ergab sich zu rund 20%.

Es liegt nahe, die Abplattung einzelner Meridiane mit dem CLAIRAUTschen Theorem zu berechnen[2], ein Verfahren, das aber nur zulässig ist, wenn die Abweichungen des Geoids vom Rotationsellipsoid mit Gliedern zweiter Ordnung dargestellt werden können. In allen anderen Fällen wird die Schwereverteilung auf der *ganzen* Erde gebraucht, selbst wenn man den Geoidverlauf nur einen einzelnen Meridian untersuchen will.

Rotationsellipsoide und Rotationssphäroide von den Dimensionen des Erdkörpers können Gleichgewichtsfiguren rotierender Himmelskörper sein. Es gibt auch eine Reihe von dreiachsigen (JACOBIschen) Ellipsoiden, die als Gleichgewichtsfiguren rotierender homogener Himmelskörper auftreten können. Sie sind wesentlich anders geformt als die Erde. Das JACOBIsche Ellipsoid, das der Winkelgeschwindigkeit der Erdrotation entspricht, hat das Achsenverhältnis von $1:1{,}002:52{,}442$.

## 36. Zur Bestimmung der Gesteinsdichte für die topographische Reduktion.
Bei der Schwerereduktion kommt es darauf an, daß die außerhalb vom Geoid liegenden topographischen Massen vollkommen weggerechnet werden. Hierzu muß man die Gesteinsdichte gut kennen.

Meist werden die Dichtebestimmungen an Handstücken oder Bohrkernen im Laboratorium vorgenommen. Unter günstigen Verhältnissen kann man mit sog. NETTLETON-Profilen[3] die Dichte der für die Geländereduktion maßgeblichen Schichten bestimmen. Hierzu werden Schwereprofile über die Unebenheiten des Geländes gelegt und mit Annahme

---

[1] K. JUNG: Über das dreiachsige Erdellipsoid und seine Zufallswahrscheinlichkeit. Gerlands Beitr. Geophys. **59**, 331 (1943). — Betrachtungen zum dreiachsigen Erdellipsoid. Z. Geophys. **20**, 201 (1954).

[2] N. F. ZHURAVLEV: Fußnote 9, S. 565.

[3] L. L. NETTLETON: Determination of density for reductions of gravimeter observations. Geophysics **4**, 176 (1939).

verschiedener Dichte nach BOUGUER reduziert. Diejenige Dichte wird als die beste angesehen, deren Schwerekurve die geringsten Beziehungen zu dem Geländeprofil hat (Fig. 36). Das Verfahren läßt sich in mathematischer Form bringen, wenn man bedenkt, daß der Korrelationskoeffizient der BOUGUERschen Anomalie und der Beobachtungshöhe gleich Null sein

muß[1]. Man kann auch die Schwereanomalie als zufällige Fehler ansehen und mittels einer Ausgleichung die Beziehung zwischen der Schwere und der Höhe bestimmen[2]. Es ist notwendig, den regionalen Teil der Schwerestörung vorher von den Messungsergebnissen abzuziehen. Hierbei haben Formeln mit Gliedern erster und zweiter Ordnung gute Dienste getan[3]. Das Verfahren versagt, wenn eine Korrelation zwischen dem Geländeprofil und der Schwerestörung unbekannter Störungsmassen besteht.

Schweremessungen in verschiedenen Tiefen, nahe an Schächten oder in Bohrlöchern, haben sich gelegentlich bewährt[4]. Eine systematische Theorie hat die Möglichkeit noch günstigerer Anordnung der Meßpunkte erwiesen[5].

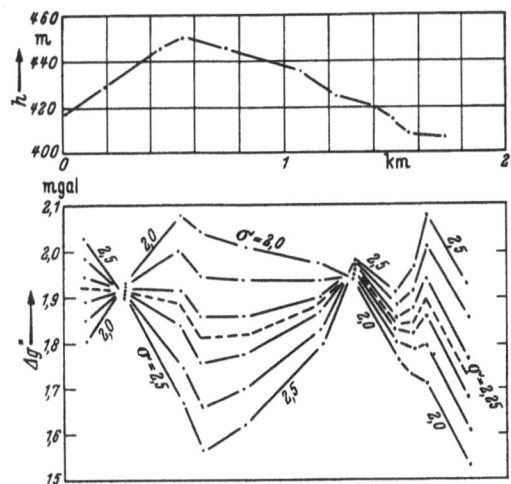

Fig. 36. Bestimmung der Gesteinsdichte σ nach NETTLETON. h Höhe, Δg″ BOUGUERsche Schwereanomalie. Beste Dichte: 2,25 g/cm³.

**37. Schwere und Nivellement.** Zwei Niveauflächen haben konstanten Potentialunterschied, aber nicht konstanten Abstand (Fig. 37). Ihr Abstand ist umgekehrt proportional der Schwere:

$$g_i h_i = g_i' h_i' = \text{const}.$$

Daher kommt es, daß die Schleifensummen der Nivellementshöhen im allgemeinen nicht gleich Null sind. Selbst bei fehlerfreiem Nivellement ergeben sich Schleifenschlußfehler, die vom Nivellementsweg abhängen. Es ist notwendig, die nivellierten Höhenunterschiede mit Hilfe der Schwerkraft zu reduzieren, d.h. in widerspruchsfreie Systeme umzuwandeln. Es sind verschieden definierte *reduzierte Höhen* im Gebrauch.

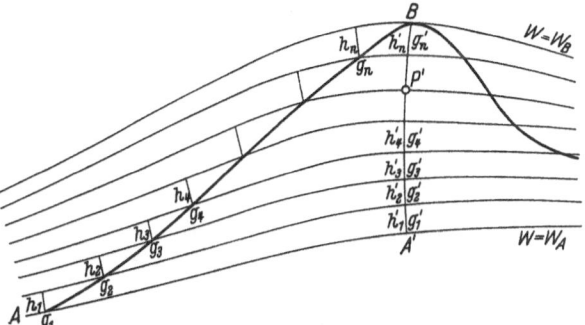

Fig. 37. Zur Reduktion des geometrischen Nivellements. Nivellementshöhe: $\sum_i h_i$, orthometrische Höhe: $\sum_i h_i'$, dynamische Höhe: $(W_B - W_A)/g_m$

($g_m$ ein Mittelwert der Schwere).

[1] K. JUNG: Über die Bestimmung der Bodendichte aus den Schweremessungen. Beitr. angew. Geophys. **10**, 154 (1943). — Zur Bestimmung der Bodendichte nach den NETTLETON-Verfahren. Z. Geophys. **19**, 54 (1953).

[2] D. S. PARASNIS: A study of the rock density in the English Midlands. Month. Nat. Roy. Astronom. Soc., Geophys. Suppl. **6**, 252 (1952).

[3] R. BORTFELD: Bemerkungen zur Dichtebestimmung nach dem NETTLETON-Verfahren. Erdöl und Kohle **7**, 353 (1954). — K. JUNG: Zur Dichtebestimmung nach dem NETTLETON-Verfahren. Erdöl u. Kohle **8**, 401 (1955).

[4] H. JUNG: Dichtebestimmungen im anstehenden Gestein durch Messung der Schwerebeschleunigung in verschiedenen Teufen unter Tage. Z. Geophys. **15**, 56 (1939). — S. HAMMER: Density determinations by underground gravity measurements. Geophysics **15**, 637 (1950).

[5] A. YARAMANCI: Eine allgemeine Methode zur gravimetrischen Gesteinsdichtebestimmung. Diss. Zürich Nr. 2122. Rev. Fak. Sci. Univ. Istanbul, Ser. A **18** (1952).

*Nivellementshöhen* (unreduziert).

$$\overset{B}{\underset{A}{H}} = \overset{B}{\underset{A}{\sum}} h_i.$$

Frei von Hypothesen über die Massenverteilung in der Erdkruste, aber nicht eindeutig, abhängig vom Weg. Punkte einer Niveaufläche erhalten verschiedene Höhen.

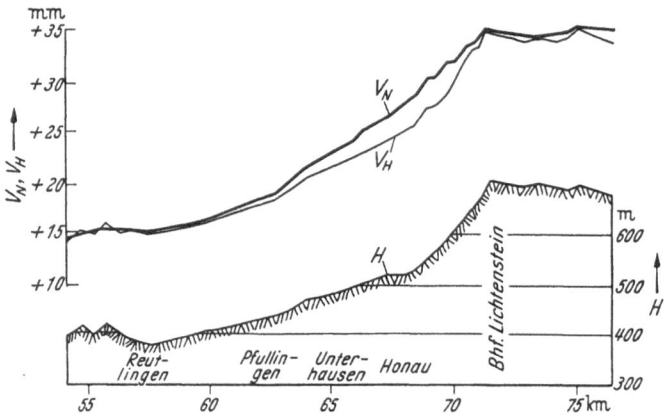

*Geodynamische Koten*[1].

$$\overset{B}{\underset{A}{K}} = \overset{B}{\underset{A}{\sum}} g_i h_i.$$

Potentialunterschiede. Einheit: meterkgal (1 meterkgal $\cong$ 1,02 m). Hypothesenfrei, widerspruchsfrei. Punkte einer Niveaufläche erhalten gleiche Koten.

*Dynamische Höhen.*

$$\overset{B}{\underset{A}{H_d}} = \overset{B}{\underset{A}{\sum}} \frac{g_i}{\gamma_m} h_i.$$

Fig. 38. Beispiel einer orthometrischen Reduktion eines Nivellements. $H$ Meereshöhe, $V_N$ orthometrische Reduktion nach Niethammer, $V_H$ Reduktion nach Helmert. Nach K. Ramsayer[3].

Hypothesenfrei, widerspruchsfrei, Punkte einer Niveaufläche erhalten gleiche Höhen. $\gamma_m$ ein Mittelwert der Schwere[2].

*Orthometrische Höhen (Seehöhen).*

$$\overset{B}{\underset{A}{H_0}} = \overset{B'}{\underset{A'}{\sum}} h_i' = \overset{A}{\underset{B}{\sum}} \frac{g_i}{g_i'} h_i.$$

Abhängig von Hypothesen über die Massenverteilung in der Erdkruste. Widerspruchsfrei. Punkte einer Niveaufläche erhalten verschiedene Höhen.

$$g_i'^{(P)} = g^{(B)} - \text{Top}^{(B)} - \text{E}^{(B)} - \text{Pl}^{(B)} - \overset{B}{\underset{P'}{\text{Ni}}} + \text{Pl}^{(P')} + \text{E}^{(P')} + \text{Top}^{(P')}.$$

$g^{(B)}$ = in $B$ gemessene Schwere,

$\text{Top}^{(B)}$ = Gelände*wirkung* in $B$ (die Gelände*reduktion* hat das umgekehrte Vorzeichen),

$\text{E}^{(B)}$ = Wirkung vorborgener Einbettungen in $B$,

$\text{Pl}^{(B)}$ = Wirkung der Bouguer-Platte $\left(\overset{B}{\underset{P'}{}}\right)$ in $B = 2\pi f \sigma \overset{B}{\underset{P'}{H_0}} \approx 2\pi f \sigma \overset{B}{\underset{P'}{H}}$

$\text{Pl}^{(P')}$ = Wirkung der Bouguer-Platte $\left(\overset{B}{\underset{P'}{}}\right)$ in $P' \approx -2\pi f \sigma \overset{B}{\underset{P'}{H}}$,

$\overset{B}{\underset{P'}{\text{Ni}}}$ = Schwereänderung mit der Höhe im Vakuum von $P'$ nach $B = -\frac{2g}{R} \overset{B}{\underset{P'}{H}}$.

Es handelt sich um eine Reduktion nach Prey.

Die Niethammer-Höhen vernachlässigen nur $\text{E}^{(B)}$ und $\text{E}^{(P')}$. Baeschlin gibt eine Methode an, mit der die isostatischen Kompensationsmassen berücksichtigt werden können[4].

Die Helmert-Höhen vernachlässigen auch $\text{Top}^{(B)}$ und $\text{Top}^{(P')}$. In sehr vielen Fällen reichen sie aus.

---

[1] War in der Meteorologie lange im Gebrauch. 1 „geodyn. Meter" = 1 mkgal. Heute wird gebraucht 1 geopotentielles Meter = 980 metergal.

[2] Oft $\gamma_{45°}$.

[3] Veröff. d. Deutschen Geodät. Komm. A. Nr. 6 (1953).

[4] C. F. Baeschlin: Ergänzung zur Berechnung der mittleren Schwere in einer Lotlinie nach Th. Niethammer, unter Berücksichtigung der Isostasie. Sitzgsber. bayr. Akad. Wiss., Math.-Nat. Kl. 1955, 109.

Die Größe der orthometrischen Reduktionen nach NIETHAMMER und HELMERT zeigt Fig. 38.

*Gebrauchshöhen.* Die beschriebenen Höhensysteme sind nicht voll befriedigend. Die Nivellementshöhen lassen sich wegen der vom Weg abhängigen Schleifenschlußfehler schwer kontrollieren und sind für Anschlüsse ungünstig. Die dynamischen und orthometrischen Höhen sind zwar widerspruchsfrei, weichen aber im Gebirge reichlich weit von den Nivellementshöhen ab.

Es gibt daher verschiedene Vorschläge zur Berechnung von Gebrauchshöhen, die ohne Widersprüche sind und sich nicht weit von den Nivellementshöhen entfernen. Über die Wahl der zweckmäßigsten Gebrauchshöhe ist man sich in Fachkreisen noch nicht einig[1].

**38. Theorie des Schwerefeldes im Erdinnern, Dichteverteilung, dynamische Abplattung**[2]. Die Flächen gleicher Dichte $\sigma(R)$ im Innern der Erde seien nahezu kugelförmig und mögen sich in der Form

$$\left.\begin{array}{l} l_{\sigma=\text{const}} = R\left(1 + \sum_{n=0}^{\infty} \sum_{\nu=0}^{n} \alpha_{n\nu}(R)\, Y_{n\nu}(\vartheta, \lambda)\right), \\[2mm] Y_{n\nu}(\vartheta, \lambda) \quad \text{eine Kugelflächenfunktion,} \quad \alpha_{n\nu}(R) \ll 1 \end{array}\right\} \tag{38.1}$$

darstellen lassen. Hier hat $\alpha_{n\nu}(R)$ die Größenordnung der Erdabplattung, Glieder von der Größenordnung $\mathfrak{a}^2 R$ sollen außer Acht bleiben[3]. An der Erdoberfläche sei $R = R_0$.

Maßgebend ist die CLAIRAUTsche Differentialgleichung

$$\sigma(R)\left(\frac{d^2\alpha_{n\nu}(R)}{dR^2} - \frac{n(n+1)\,\alpha_{n\nu}(R)}{R^2}\right) + 6\,\frac{\sigma(R)}{R}\left(\frac{d\alpha_{n\nu}(R)}{dR} + \frac{\alpha_{n\nu}(R)}{R}\right) = 0, \tag{38.2}$$

die man unter Annahme hydrostatischen Gleichgewichtes aus dem Graviationspotential ableiten kann. Sie gilt in dieser Form bei kontinuierlicher Dichtezunahme[4]. Im allgemeinen wird sie nur für $n=2$, $\nu=0$ angeführt.

Es ist nicht notwendig, von vornherein in dieser Weise zu spezialisieren. Setzt man voraus, daß die Dichte bei wachsender Tiefe nicht abnimmt und $\alpha_{n\nu}(R)$ sich im Erdmittelpunkt wie $R^p$ ($p > 0$) verhält, so kann man nachweisen, daß bei den vorliegenden Vernachlässigungen alle Glieder bis auf das Abplattungsglied $\binom{n=2}{\nu=0}$ verschwinden. Dann lautet die CLAIRAUTsche Differentialgleichung

$$\left.\begin{array}{l} 2\,\dfrac{d\sigma(R)}{dR}\cdot\alpha_{20}(R) + 6\,\sigma(R)\,\dfrac{d\alpha_{20}(R)}{dR} + R\,\sigma(R)\,\dfrac{d^2\alpha_{20}(R)}{dR^2} = 0 \\[3mm] \hspace{5cm} \alpha_{20}(R_0) = \mathfrak{a}. \end{array}\right\} \tag{38.3}$$

---

[1] Über Einzelheiten der Rechenverfahren, Fehlerabschätzungen, Vergleiche und Vorschläge muß auf die sehr reichhaltige geodätische Literatur verwiesen werden. Dort fast stets ausführliche Literaturangaben. Sogar die Wirkung zeitlicher Schwereänderungen (Gezeiten) wird erörtert [H. JENSEN: Bull. Géod. 1950, Nr. 17, 267].

[2] Unter sehr allgemeinen Voraussetzungen behandelt von H. JEFFREYS in [21].

[3] Wenn es sich, wie hier, nur um die Dichtezunahme mit der Tiefe handelt, genügt diese Annäherung. Sie genügt nicht, wenn der Koeffizient $\mathfrak{b}'$ der Schwereformel berechnet werden soll. Die Theorien von WIECHERT und DARWIN berücksichtigen noch die Größenordnung $R \cdot \mathfrak{a}^2$ und sind erheblich komplizierter [11]. Die einfachere Theorie von W. DE SITTER berücksichtigt ebenfalls die Größenordnung $R \cdot \mathfrak{a}^2$ [Bull. astron. Inst. Netherlds. **2**, 97 (1924)]. Anwendung E. C. BULLARD: Month. Not. Roy. Astrom. Soc. Geophys. Suppl. **5**, 186 (1948).

[4] Eine entsprechende Gleichung für Dichtesprünge gibt A. PREY an [9], [18].

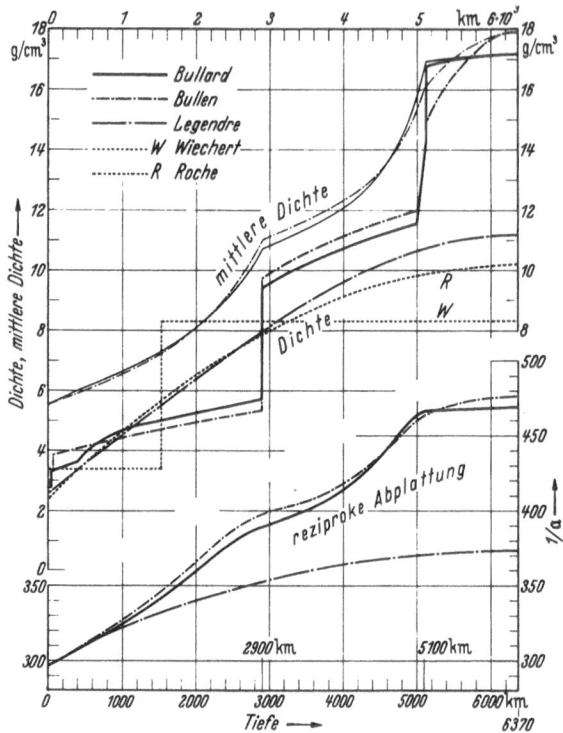

Fig. 39. Dichte im Innern der Erde, mittlere Dichte der unter den betreffenden Tiefen liegenden Massen, Abplattung der Niveauflächen in der Tiefe.

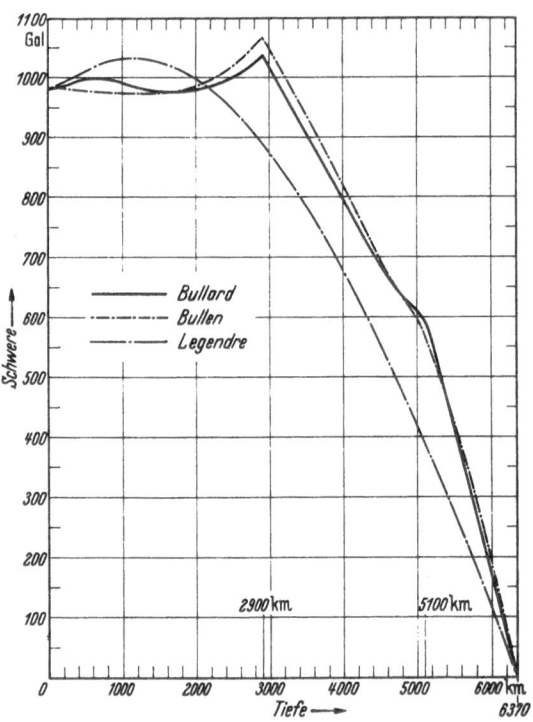

Fig. 40. Die Schwere im Innern der Erde.

Die Auflösung, die mit bekannten Kunstgriffen vorgenommen wird[1], führt auf

$$\frac{H}{c} = \frac{\dfrac{a}{c} - \dfrac{1}{2}}{1 - \dfrac{2}{5}\sqrt{\dfrac{5}{2}\dfrac{a}{c} - 1}},$$

wobei                                                                                                    (38.4)

$$c = \frac{\omega^2 R_0^3}{f M} = \frac{\omega^2}{\frac{4}{3}\pi f \sigma_m},$$

$$H = \frac{C - \frac{1}{2}(A + B)}{C} \quad \text{[dynamische Abplattung]}.$$

Während $\dfrac{C - \dfrac{A + B}{2}}{M}$ hypothesenfrei aus dem Schwerefeld ermittelt werden kann[2], enthält (38.4) die Bedingung des hydrostatischen Gleichgewichts und den Ansatz für das Verhalten der Dichte im Erdmittelpunkt, ist also nicht hypothesenfrei.

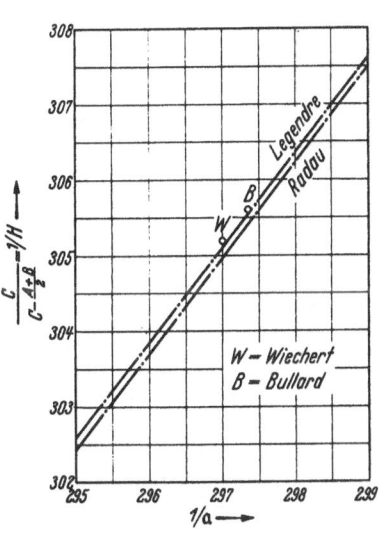

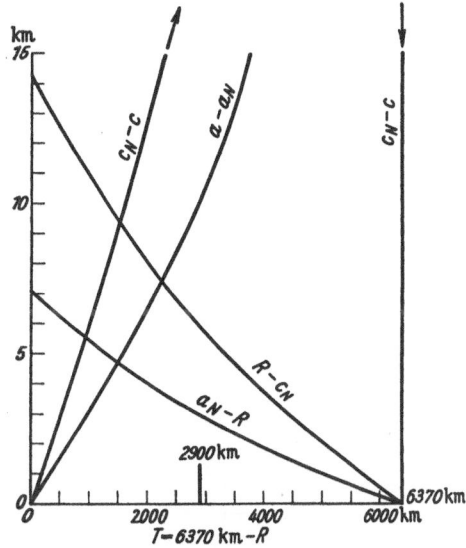

Fig. 41. Die Beziehung zwischen den Trägheitsmomenten und der Abplattung der Erde nach verschiedenen Theorien.

Fig. 42. Niveauflächen (Index: $N$) und volumengleiche Koordinatenellipsoide (ohne Index) im Erdinnern. $a$ Äquatorradien, $c$ Polradien, $R$ Radien der volumengleichen Kugeln.

Zur Auswertung von (38.4) berechnet man die rechte Seite für hinreichend viele Werte von $a/c$ und findet den einem gegebenen $H$-Wert entsprechenden Betrag von $a/c$ mit Interpolation. $H$ kann aus der Präzession ermittelt werden[3].

Fig. 39 zeigt einige Dichteverteilungen, die mit der CLAIRAUTschen Differentialgleichung verträglich sind, dazu die reziproke Abplattung der Niveauflächen in den verschiedenen Tiefen und die mittlere Dichte der jeweils von der Niveaufläche eingeschlossenen Masse. Aus der Dichte kann die Schwere berechnet werden (Fig. 40).

In Fig. 41 sind die Beziehungen zwischen $H$ und $a$ eingetragen, wie sie sich für verschiedene Dichteverteilungen ergeben. Es bestätigt sich, daß (38.4) trotz aller Vernachlässigungen mit beachtlicher Genauigkeit gilt, und man erkennt, daß es sehr präziser Untersuchungen bedarf, wenn man zwischen den verschiedenen Dichteverteilungen allein aus dem Schwerefeld

---

[1] Siehe z.B. [6], [10], [14], [16], [18], [21].
[2] Ziff. 15.
[3] Ziff. 46.

entscheiden will. Hier müssen andere Zweige der Geophysik, vor allem wohl die Seismik, klärend eingreifen.

Man kann nachweisen, daß bei Berücksichtigung der Größenordnung $R\,a^2$ die Niveauflächen im Innern der Erde keine genauen Ellipsoide sein können, wenn sie sich auch nur sehr wenig von Ellipsoiden entfernen. Die annähernden Ellipsoide sind anders geformt als die konfokalen Koordinatenellipsoide, was schon daraus hervorgeht, daß die Abplattung der Niveauflächen bei wachsender Tiefe abnimmt, die Abplattung der Koordinatenellipsoide aber größer wird. In Fig. 42 und Tabelle 10[1] sind die Abstände der Koordinatenellipsoide von den volumengleichen Niveauflächen und Kugeln zusammengestellt. In der Erdkruste können die Koordinatenellipsoide noch als gute Annäherungen an die Niveauflächen angesehen werden, und in diesen geringen Tiefen kann sich die Verwendung der Ellipsoidkoordinaten lohnen. Mit zunehmender Tiefe wird die Annäherung immer schlechter, und von 1500 km Tiefe an ist die Annäherung mit den Kugeln auf alle Fälle vorzuziehen.

# C. Bewegung der Erdachse.

## a) Polbewegung (Breitenschwankung).

**39. Die Bedeutung der Polbewegung für Geodäsie und Geophysik.** Das Koordinatensystem der Breiten und Längen hat seine Pole in den Punkten des Erdellipsoids, in denen es von der Rotationsachse durchstoßen wird. Verlagerungen der Pole machen sich bei Längen- und Breitenbestimmungen bemerkbar. In die Koordinatenverzeichnisse nimmt man die der *mittleren* Pollage entsprechenden Daten auf, die gemessenen Daten müssen dementsprechend reduziert werden. Diese Reduktion ist nur möglich, wenn man die augenblickliche Lage des Rotationspols kennt. Daher wird die Polbewegung fortlaufend von den Observatorien des *Internationalen Breitendienstes* verfolgt. Diese Observatorien liegen sämtlich in 39° 8' nördlicher Breite[2] und messen die Schwankungen ihrer Polhöhe.

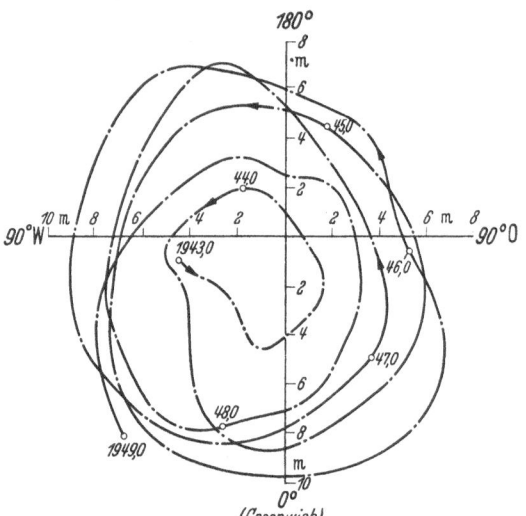

Fig. 43. Bahn des nördlichen Rotationspols $R$ auf der Erde um seine Mittellage. Nach Ergebnissen des internationalen Breitendienstes (Bull. géodésique 1948, Nr. 10, 1949, Nr. 13).

Man unterscheidet säkulare *Polwanderungen* und eine *periodische Polbewegung (Breitenschwankung)*. Während man bisher keine sicheren Schlüsse auf die Existenz, noch weniger auf Richtung und Betrag der Polwanderung (vgl. S. 286 und S. 493) schließen konnte, sind die Breitenschwankungen in ihren Einzelheiten recht gut bekannt (Fig. 43): Der Rotationspol beschreibt eine spiralige Bahn. Der Umlauf dauert etwa $1\frac{1}{5}$ Jahre, der Abstand des Momentanpols von der Mittellage überschreitet selten 10 m ($\frac{1}{3}''$). Man gewinnt den Eindruck, daß es sich wohl um eine im Lauf einiger Jahre abklingende Bewegung handelt, die in unregelmäßigen Zeitabständen neu erregt und von Störungen überlagert wird[3].

---

[1] Anhang IV, Ziff. 62, S. 634.
[2] Mizusawa, Japan (141° 8' E); Tschardjui, Turkestan (63° 29' E); Carloforte, Italien (kleine Insel bei Sardinien, 8° 19' E); Gaithersburg, Maryland USA (77° 12' W); Cincinati, Ohio, USA (84° 25' W); Ukiah, Californien, USA (123° 13' W).
[3] Vgl. den Artikel "The Rotation of the Earth" in diesem Band.

Die Periode der Polbewegung hängt von den Hauptträgheitsmomenten des Erdkörpers und von seiner Nachgiebigkeit ab. Man bestimmt die Trägheitsmomente aus der Schwereverteilung und der Präzession[1] und kann die Polbewegung zu Schlüssen über die Righeit des Erdkörpers heranziehen[2].

**40. Die EULERschen Gleichungen der deformierbaren Erde.** Die Massen der Erde sind gegeneinander verschiebbar. Ihre individuelle Bewegungsgeschwindigkeit sei $v$. Es wird eine *mittlere Drehung* so definiert, daß sie die Gesamtheit der auf den Schwerpunkt bezogenen Bewegungen am besten annähert[3]. Wie in der Fehlertheorie, möge auch hier das *Prinzip der kleinsten Quadratsumme* maßgebend sein.

$$\iiint_{\ddot{o}} (v - (\omega \times l))^2 \, dm = \text{min.}, \qquad (40.1)$$

$$\frac{d}{d\omega} \left\{ \iiint_{\ddot{o}} (v - (\omega \times l))^2 \, dm \right\} = 0. \qquad (40.2)$$

Nach kleinen Umformungen mit bekannten Vektor-Identitäten wird differenziert, und man erhält

$$- \iiint_{\ddot{o}} (l \times v) \, dm + \iiint_{\ddot{o}} \omega \, l^2 \, dm - \iiint_{\ddot{o}} (\omega l) \, l \, dm = 0.$$

Nun ist

$$\iiint_{\ddot{o}} (l \times v) \, dm = S \qquad (40.3)$$

der Vektor des *Drehimpulses* (Schwung), so daß

$$S = \iiint_{\ddot{o}} \omega \, l^2 \, dm - \iiint_{\ddot{o}} (\omega l) \, l \, dm. \qquad (40.4)$$

Man geht zu rechtwinkligen Koordinaten über und führt die auf die Koordinatenachsen bezogenen Trägheits- und Deviationsmomente ein, die nun veränderlich sind. Sie müssen also mitdifferenziert werden, wenn Differentialquotienten nach der Zeit zu bilden sind.

Zwischen dem Drehmoment $M$ der äußeren Kräfte und dem Drehimpuls besteht die Beziehung

$$M = \frac{dS}{dt} = \frac{d'S}{dt} + \omega \times S, \qquad (40.5)$$

wobei $d'S/dt$ die Relativgeschwindigkeit des Vektors $S$ in dem rechtwinkligen Koordinatensystem $x, y, z$ bezeichnet.

In Koordinaten erhält man

$$\left. \begin{aligned} A\dot{\omega}_x - F\dot{\omega}_y - E\dot{\omega}_z + \dot{A}\omega_x - \dot{F}\omega_y - \dot{E}\omega_z + (C - B)\,\omega_y\,\omega_z + \\ + F\omega_z\omega_x - E\omega_x\omega_y + D(\omega_z^2 - \omega_x^2) = M_x \end{aligned} \right\} \qquad (40.6)$$

mit

$$A = \iiint_{\ddot{o}} (y^2 + z^2) \, dm, \quad B = \iiint_{\ddot{o}} (z^2 + x^2) \, dm, \quad C = \iiint_{\ddot{o}} (x^2 + y^2) \, dm,$$

$$D = \iiint_{\ddot{o}} y z \, dm, \qquad E = \iiint_{\ddot{o}} z x \, dm, \qquad F = \iiint_{\ddot{o}} x y \, dm$$

und zwei weitere Beziehungen, die man mit cyclischer Vertauschung ableitet.

---

[1] Oder anderen astronomischen Bewegungen, die von der Erdabplattung beeinflußt sind.
[2] Zum Beispiel A. PREY: Über die Elastizitätskonstante der Erde. Gerlands Beitr. Geophys. **23**, 379 (1929). — H. JEFFREYS in [*21*].
[3] [*21*], S. 206ff.

Die Quadrate und Produkte der Winkelgeschwindigkeiten sind fast alle so klein, daß sie vernachlässigt werden dürfen; ebenso fast alle Produkte der Deviationsmomente und ihrer Ableitungen mit den Winkelgeschwindigkeiten und deren Ableitungen, wenn der Koordinatenursprung im Schwerpunkt liegt und der Winkel $(z\,\boldsymbol{\omega})$ klein ist. Übrig bleibt nur

$$\begin{aligned} A\dot{\omega}_x + (C-B)\,\omega_y\omega_z - \dot{E}\omega_z + D\omega_z^2 &= M_x,\\ B\dot{\omega}_y - (C-A)\,\omega_z\omega_x - \dot{D}\omega_z - E\omega_z^2 &= M_y,\\ C\dot{\omega}_z &= M_z \end{aligned} \qquad (40.7)$$

das System der EULERschen Gleichungen für eine deformierbare Erde.

Mond, Sonne und Planeten üben Drehmomente auf die abgeplattete Erde aus. Ihre Wirkung, die *Präzession*, wird später für sich betrachtet (Ziff. 44 bis 46). Der Präzession überlagern sich freie Achsenschwankungen, die als *Polbewegung* beobachtet werden. Ihre Amplitude ist im Vergleich zur fortschreitenden Präzession so gering, daß man sie als kleine Störungen betrachten und unabhängig von der Präzession behandeln kann, obwohl die EULERschen Gleichungen nicht linear sind. Bei der Polbewegung ist also

$$\boldsymbol{M} = 0, \quad \boldsymbol{S} = \mathrm{const}, \qquad (40.8)$$

der Drehimpulsvektor ist unveränderlich; man wird die Erdachsenschwankungen auf seine Richtung beziehen.

Mit (40.8) erhält man aus der dritten EULERschen Gleichung

$$\dot{\omega}_z = 0; \qquad \boxed{\omega_z = \mathrm{const} = \omega_0.} \qquad (40.9)$$

Der kleine, nicht einmal sicher nachgewiesene Unterschied zwischen $B$ und $A$ kann vernachlässigt werden, und es gilt genau genug:

$$\begin{aligned} A\dot{\omega}_x + (C-A)\,\omega_0\omega_y - \dot{E}\omega_0 + D\omega_0^2 &= 0,\\ A\dot{\omega}_y - (C-A)\,\omega_0\omega_x - \dot{D}\omega_0 + E\omega_0^2 &= 0. \end{aligned} \qquad (40.10)$$

**41. Der gleichmäßige Teil der Polbewegung.** $\alpha$) *Der Polhodiekegel.* Nimmt man an, daß sich der Erdkörper in gleichmäßiger Weise deformiert und keine Störungen eintreten, so läuft die Polbewegung gleichmäßig ab. Deformierend wirken die wechselnden Zentrifugalkräfte.

Das Potential der Zentrifugalbeschleunigung an der Spitze des Ortsvektors $\boldsymbol{l}$ ist

$$Z = \tfrac{1}{2}\,(\boldsymbol{\omega}\times\boldsymbol{l})^2 = \tfrac{1}{2}\,\big(\omega^2(x^2 + y^2 + z^2) - (\omega_x x + \omega_y y + \omega_z z)^2\big).$$

Vernachlässigt man Quadrate und Produkte von $\omega_x/\omega$ und $\omega_y/\omega$ so ist

$$Z = \tfrac{1}{2}\,\omega_0^2(x^2 + y^2) - \omega_0 z(\omega_x x + \omega_y y). \qquad (41.1)$$

Das erste Glied rechts ist konstant und liefert einen kleinen Beitrag zur Abplattung. Veränderlich ist nur das zweite Glied, das *deformierende Zentrifugalpotential*

$$Z_D = -\,\omega_0 z(\omega_x x + \omega_y y), \qquad (41.2)$$

Es wird angenommen, daß das *Gravitationspotential der Deformation* dem deformierenden Zentrifugalpotential proportional ist; die Proportionalitätskonstante sei mit $k$ bezeichnet[1].

$$V_D = - k \omega_0 z (\omega_x x + \omega_y y) \frac{R^5}{l^5}$$

$$R = \text{Erdradius}, \quad l = \text{Abstand vom Erdmittelpunkt}, \quad l \geq R. \qquad (41.3)$$

In der Entwicklung (33.2) des Gravitationspotentials lautet das entsprechende Glied, wenn man es in rechtwinkligen Koordinaten ausdrückt,

$$- f \cdot 3 \frac{E z x + D y z}{l^5} .$$

Durch Vergleich ergibt sich

$$E = \frac{k \omega_0 \omega_x R^5}{3 f}, \quad D = \frac{k \omega_0 \omega_y R^5}{3 f}, \qquad (41.4)$$

und man erhält mit (40.10):

$$\boxed{\begin{array}{l} \dot\omega_x + \Omega \omega_y = 0, \quad \dot\omega_y - \Omega_y = 0, \\[2mm] \text{mit} \quad \Omega = \omega_0 \dfrac{3 f (C - A) - k \omega_0^2 R^5}{3 f A + k \omega_0^2 R^5} . \end{array}} \qquad (41.5)$$

Die Lösung lautet

$$\boxed{\begin{array}{l} \omega_x = Q \cos \Omega (t - \tau), \\[1mm] \omega_y = Q \sin \Omega (t - \tau), \\[1mm] \text{mit} \quad Q^2 = \omega_x^2 + \omega_y^2 = \text{const} \end{array}} \qquad (41.6)$$

und es folgt unmittelbar

$$\boxed{\omega = \sqrt{\omega_x^2 + \omega_y^2 + \omega_z^2} = \text{const.}} \qquad (41.7)$$

Nach (41.6) beschreibt die Rotationsachse im System $x, y, z$ einen Kreiskegel mit dem halben Öffnungswinkel $\arctan (Q/\omega_0)$, den *Polhodiekegel*. Da $Q \ll \omega_0$ ist, fällt seine Achse fast mit der Achse des größten Hauptträgheitsmoments zusammen.

Wegen $E/D = \omega_x/\omega_y$ ist

$$\begin{vmatrix} \omega_x & \omega_y & \omega_0 \\ A \omega_x - E \omega_0 & A \omega_y - D \omega_0 & C \omega_0 \\ 0 & 0 & 1 \end{vmatrix} = 0. \qquad (41.8)$$

$\omega$, $S$ und die $Z$-Achse liegen in einer Ebene (Fig. 44).

*β) Die* CHANDLER*sche und die* EULER*sche Periode.* Die harmonische Analyse der Polbewegung ergab, daß vorzugsweise zwei Perioden in ihr auftreten: eine jährliche Periode, die wohl meteorologischen Massenverschiebungen zuzuschreiben

---

[1] In der deutschen Literatur wird diese Konstante auch mit $h$ bezeichnet und eine bei den Gezeiten auftretende Konstante $h$ mit $k$. Hier ist Vorsicht geboten.

ist, und die CHANDLERsche (oder NEWCOMBsche) *Periode* von 1,2 Jahren (Fig. 45)[1]. Nach JEFFREYS[2] beträgt sie

$$\frac{2\pi}{\Omega} = \begin{cases} 439{,}0 \pm 5{,}8 \text{ Sonnentage} \\ 440{,}2 \pm 5{,}9 \text{ Sterntage}. \end{cases}$$

Zusammen mit der Winkelgeschwindigkeit der Erdrotation ($\omega_0 = 7{,}2921 \times 10^{-5} \sec^{-1}$) und den (aus der Schwereverteilung und der Präzession errechneten) Trägheitsmomenten

$$A = 8{,}089 \cdot 10^{44} \text{ g cm}^2, \qquad \frac{C-A}{A} = 1/304{,}8$$

findet man aus (41.5)[3]

$$k = 0{,}290$$

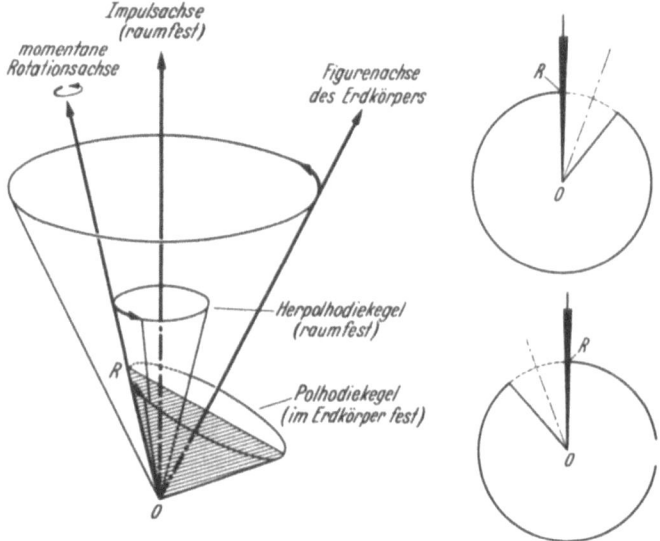

Fig. 44. Zur Kinematik der Polbewegung. Rechts die Lage von Polhodie- und Herpolhodiekegel im Abstand eines halben Tages.

Bei starrer Erde wäre $k = 0$, und die Periode der Polbewegung

$$\left(\frac{2\pi}{\Omega}\right)_{k=0} = \frac{2\pi}{\omega_0}\frac{A}{C-A} = 304{,}8 \text{ Sterntage (EULERsche Periode)}.$$

Die CHANDLERsche Periode kann als eine von der Nachgiebigkeit der Erde verlängerte EULERsche Periode angesehen werden.

So einfach die Polbewegung nach der geschilderten Theorie erscheint, so vielfältig erweist sie sich bei genauerer Betrachtung der Beobachtungsergebnisse[4]. Im wesentlichen dürfte es sich um eine gedämpfte[5], in unregelmäßigen Zeitabständen wieder auflebende Umlaufsbewegung handeln, bei der jede Massenverschiebung, die den Drehimpuls beeinflußt, einen störenden Einfluß ausübt. Es wird von Änderungen und Aufspaltungen der CHANDLER-

---

[1] Gerlands Beitr. Geophys. **16**, 108 (1927).
[2] [*21*], S. 210.
[3] JEFFREYS gibt an: 0,288 ± 0,008.
[4] Hier muß auf die — leider sehr zerstreute — Literatur hingewiesen werden. Zum Teil mitgeteilt in [*22*], S. 258.
[5] Relaxationszeit nach JEFFREYS ([*21*], S. 210): 15,1 ± 1,8 Jahre.

schen Periode berichtet[1], auch hat man Beziehungen zu anderen kosmischen Perioden gefunden[2]. Eine eigenartige Störung ist das sog. KIMURA-Glied. Es täuscht eine gleichzeitig jährliche Zu- und Abnahme der Breiten aller Stationen vor. Seine Ursachen sind in Refraktionsanomalien und der Parallaxe der beobachteten Sterne zu vermuten.

γ) *Der momentane Trägheitspol, Hauptträgheitsmomente.* Der momentane Trägheitspol fällt mit dem Maximum des Gravitationspotentials der abgeplatteten, von der Polbewegung deformierten Erde zusammen. Aus Symmetriegründen liegt er in der $Z\,S\,\omega$-Ebene. Für seinen Polabstand findet man

$$\vartheta_P = (\delta + \nu)\,\frac{A}{C - A}\cdot\frac{\dfrac{C - A}{A}\,\dfrac{\omega_0}{\Omega} - 1}{\dfrac{\omega_0}{\Omega} + 1}, \tag{41.9}$$

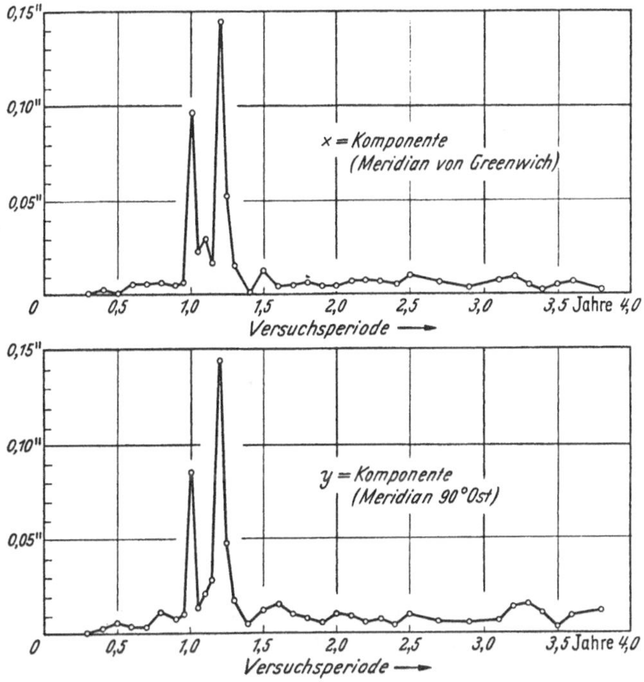

Fig. 45. Das Periodogramm der Polbewegung 1890,0 — 1924,1. Nach L. W. POLLAK[3]

wobei $\delta + \nu$ den halben Öffnungswinkel des Polhodiekegels bezeichnet (Fig. 46). In Zahlen erhält man:

$$\vartheta_P = 0{,}307 \cdot (\delta + \nu).$$

Denkt man sich das Gravitationspotential der Deformation von einer auf der Erde ausgebreiteten Flächenbelegung hervorgerufen, so kann man die Hauptträgheitsmomente der mit dieser Belegung behafteten Erde berechnen. Wird die Länge vorübergehend von der $Z\,S\,\omega$-Ebene aus gezählt, so ergibt sich

$$\left.\begin{aligned}
J(\vartheta_P,\,0°) &= C + \left(A - \frac{C}{2}\right)\vartheta_P^2 + 2A_\mu\vartheta_P,\\[4pt]
J(\vartheta_P + 90°,\,0°) &= A + \left(C - \frac{A}{2}\right)\vartheta_P^2 - 2A_\mu\vartheta_P,\\[4pt]
J(90°,\,90°) &= A,
\end{aligned}\right\} \tag{41.10}$$

[1] E. WAHL: Zusammenfassender Bericht. Zbl. Geophys. **4**, 1 (1939).
[2] K. LEDERSTEGER: Neue Analyse der CHANDLERschen Polbewegung. Bull. Géod. **1952**, Nr. 23, 67.
[3] L. W POLLAK: Gerlands Beitr. z. Geophysik **16**, 176 (1927).

und in Zahlen ist

$$\left.\begin{aligned}
J(\vartheta_P, 0°) &= C \cdot (1 + 0,047 \cdot (\delta + \nu)^2), \\
J(\vartheta_P + 90°, 0°) &= A \cdot (1 + 0,047 \cdot (\delta + \nu)^2), \\
J(\vartheta_P, 0°) - J(\vartheta_P + 90°, 0°) &= (C - A) \cdot (1 - 0,143 \cdot (\delta + \nu)^2)
\end{aligned}\right\} \quad (41.11)$$

$(\delta + \nu)^2$ hat die Größenordnung $10^{-12}$.

Schreibt man formal

$$\Omega = \omega_0 \frac{\overline{C} - \overline{A}}{\overline{A}} = \frac{\omega_0}{440,2},$$

$$\frac{\overline{C} - \overline{A}}{\overline{A}} = \frac{1}{440,2}, \qquad \frac{\overline{C}}{\overline{A}} = \frac{441,2}{440,2},$$

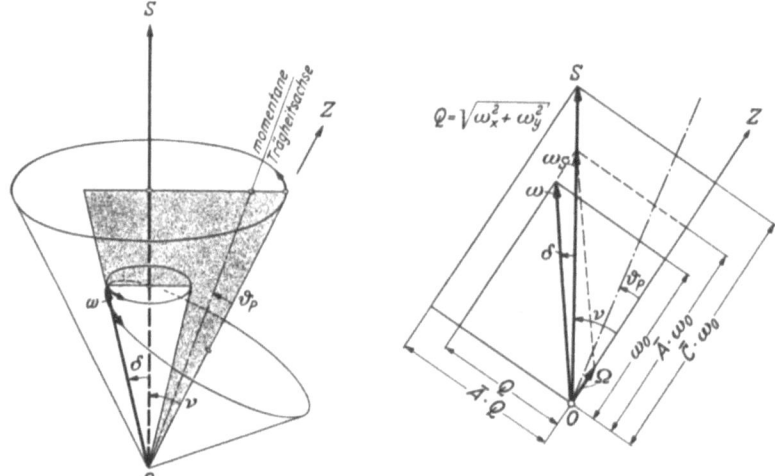

Fig. 46. Zur Theorie der Polbewegung: Winkel und Vektoren in der $Z\,S\,\omega$-Ebene.

so sind $\overline{C}$ und $\overline{A}$ nicht die Hauptträgheitsmomente der deformierten Erde, sondern diejenigen eines starren Körpers, der sich hinsichtlich Rotation und Polbewegung ebenso verhält wie die deformierte Erde.

$\delta$) *Einige Zahlen.* Rotationsperiode der Erde: $2\pi/\omega_0 \approx 1$ Sterntag [1].
Periode des Umlaufs auf der Polhodie: $2\pi/\Omega = 440,2$ Sterntage $= 439,0$ Sonnentage.
Periode des Umlaufs auf der Herpolhodie (Umlauf der $Z\,S\,\omega$-Ebene):

$$2\pi/\omega_S = \frac{440,2}{441,2} \text{ Sterntage} = \frac{439,0}{440,0} \text{ Sonnentage.}$$

Öffnungswinkel des Polhodiekegels: $\delta + \nu \lessgtr 0,3''$ ($\triangleq 10$ m).

Öffnungswinkel des Herpolhodiekegels:

$$\delta = \frac{1}{441,2} (\delta + \nu) \lessgtr \cdot 0,001'' \ (\triangleq 3 \text{ cm}).$$

Abstand des momentanen Trägheitspols von seiner Mittellage:

$$\vartheta_P = 0,307 \cdot (\delta + \nu) \lessgtr 0,1'' (\triangleq 3 \text{ m}).$$

---

[1] Genauer: 1 Sterntag $+ 0,009$ sec, da der Sterntag auf den rückläufig wandernden Frühlingspunkt bezogen ist.

*ε) Die Konstante k und die Abplattung der Erde.* Mit

$$c = \frac{\omega^2 R^3}{f M}, \qquad \frac{3}{2}\frac{C - A}{M R^2} = \alpha - \frac{1}{2}c, \qquad \frac{A}{C - A} = \frac{\omega_0}{\Omega_0} \left(\frac{2\pi}{\Omega_0} = \text{EULERsche Periode}\right)$$

kommt man auf

$$k = \left(2\frac{\alpha}{c} - 1\right)\frac{\Omega_0 - \Omega}{\omega_0 + \Omega} = \left(2\frac{\alpha}{c} - 1\right)\left(1 - \frac{\Omega}{\Omega_0}\right),$$

$$k = \left(\frac{2\alpha}{c} - 1\right)\left(1 - \frac{304,8}{440,2}\right).$$

*ζ) Die Hebung der Niveauflächen.* Dem Potential $Z_D + V_D$ entspricht eine Hebung der Niveauflächen vom Betrag $\dfrac{1}{\gamma} \cdot (Z_D + V_D)$. Für $(\delta + \nu) \lessgtr 0{,}3''$ ergibt sich

$$h_{\max} \lessgtr 4,1 \text{ cm}.$$

**42. Erzwungene Polbewegungen.** Bei Massenverschiebungen treten Polbewegungen auf. Man berechnet die Wirkung der Massenverlagerung auf $D$, $E$, $\dot D$, $\dot E$ und setzt sie in die Bewegungsgleichungen (40.10) ein.

Merklich sind Glieder von Jahresperiode. Kurzperiodische Bewegungen, wie die Gezeiten, machen wenig aus, konnten aber bei Bearbeitung einzelner Breitenstationen nachgewiesen werden[1].

**43. Säkulare Polwanderungen.** Säkulare Polwanderungen sind nicht mit Sicherheit nachgewiesen. Auf jeden Fall sind sie klein[2] und erreichen nicht die Beträge, die z.B. zur Erklärung von Eiszeiten gefordert werden. Es sind Anzeichen vorhanden, daß die Lage des mittleren Rotationspoles in enger Beziehung zum Mittelwasser der Ozeane steht und daß Beziehungen zwischen Amplitude und Periode der Polbewegung und der Amplitude des Gezeiteneinflusses bestehen[3].

## b) Präzession.

**44. Die Bedeutung der Präzession für Geodäsie und Geophysik.** Unter *Präzession* sei die Gesamtheit der Drehbewegungen verstanden, die der Erdachse

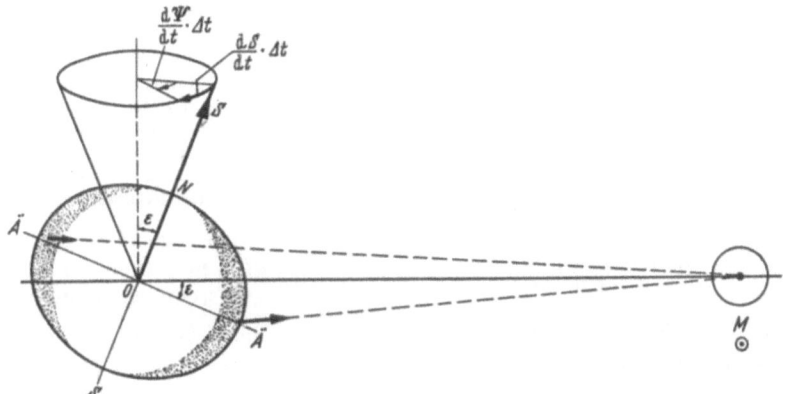

Fig. 47. Zur Theorie der solaren Präzession: Die Anziehungskräfte und der Drehimpulsvektor.

von äußeren Kräften aufgezwungen werden (Fig. 47). Wirksam sind die Gravitationskräfte der Sonne und des Mondes, wobei — wie bei den Gezeiten — der Mondeinfluß rund doppelt so groß ist wie der Einfluß der Sonne. Auch eine

---

[1] E. WAHL: Zusammenfassender Bericht. Zbl. Geophys. **4**, 4 (1939).

[2] B. WANACH: Eine fortschreitende Lagenänderung der Erdachse. Z. Geophys. **3**, 102 (1927). — Vgl. hierzu die Diskussionen über Polwanderungen: S. 286 u. S. 495 dieses Bandes.

[3] K. HOSOYAMA: On secular change of latitude. Trans. Amer. Geophys. Un. **33**, 345 (1952).

Wirkung der Planeten ist vorhanden, die ungefähr $\frac{1}{500}$ der luni-solaren Präzession beträgt[1].

Allgemeine Bedeutung hat die Präzession bei der Zeitbestimmung, da sich der Sterntag auf den wandernden Frühlingspunkt und nicht auf eine raumfeste Richtung bezieht. Die Geschwindigkeit der Präzessionsbewegung hängt von der Abplattung der Erde ab und kann zur Bestimmung der Abplattung herangezogen werden.

Über die Dichteverteilung im Innern der Erde sagt die Präzessionsgeschwindigkeit verhältnismäßig wenig aus; denn wie Fig. 41 erkennen läßt, gibt es sehr verschiedene Dichteverteilungen, die auf fast dieselbe Beziehung zwischen Präzessionsgeschwindigkeit und 'Erdabplattung führen.

In der beobachtenden Astronomie ist es seit jeher üblich, die Präzessionsbewegung formal in einen säkularen, regelmäßigen Anteil und den ihn überlagernden periodischen Glieder zu zerlegen, die man unter der Bezeichnung *Nutation* zusammenfaßt. So zweckmäßig diese Zerlegung für die Praxis der Zeitbestimmung ist, so wenig wird sie den physikalischen Zusammenhängen gerecht. Eine Erschwerung für das Verständnis bedeutet es auch, daß die Bezeichnungen „*Präzession*" und „*Nutation*" in anderem Sinn gebraucht werden als in der physikalischen Kreiseltheorie.

Die folgenden Betrachtungen sollen die physikalischen Beziehungen erläutern und ihre wesentlichen Züge wiedergeben. Da die Präzession von den Kräften hervorgerufen wird, die auch die Gezeiten erzeugen[2], sind die komplizierten Einzelheiten des Gezeitenablaufs auch im Ablauf der Präzession enthalten. Zum Teil wirken sie sich sogar stärker aus, weil ihre Wirkung sich im Lauf der Zeit aufsummieren und dann beobachtet werden können[3]. Eine Darstellung dieser Einzelheiten geht über den Rahmen eines physikalischen Handbuches hinaus. Weil sie am einfachsten darzustellen ist, soll zuerst nur die *solare* Präzession betrachtet werden; erst zum Schluß wird das Zusammenwirken der solaren und der lunaren Präzession beschrieben.

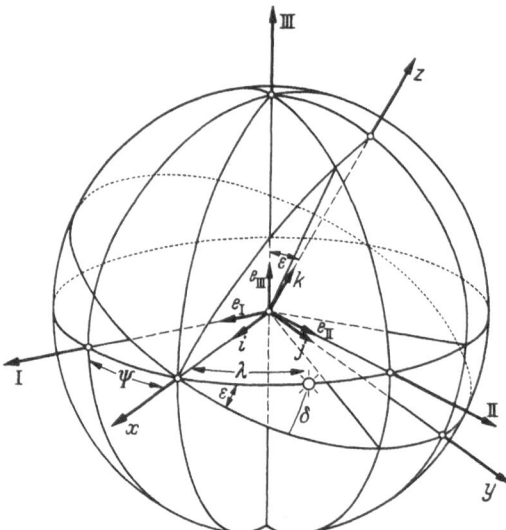

Fig. 48. Zur solaren Präzession: Koordinaten und Winkel. Das ekliptikale System (I, II, III) ist fest im Raum, die $X$-Achse des äquatorialen Systems ($x, y, z$) ist auf den Frühlingspunkt gerichtet. $\Psi$ Präzessionswinkel, $\varepsilon$ Schiefe der Ekliptik, $\lambda$ Länge der Sonne, $\delta$ Deklination der Sonne. Einheitsvektoren $e_I$, $e_{II}$, $e_{III}$; $i, j, k$.

**45. Die solare Präzession. α)** *Die Koordinatensysteme.* Es werden zwei Bezugssysteme eingeführt (Fig. 48):

ein ekliptikales System (I, II, III) mit den Einheitsvektoren $e_I$ $e_{II}$, $e_{III}$ und ein äquatoriales System ($x, y, z$) mit den Einheitsvektoren $i, j, k$.

Beide Systeme haben ihren Ursprung im Schwerpunkt der Erde. Das ekliptikale System hat raumfeste Richtungen. Das äquatoriale System dreht sich in der Erde und im Raum. Seine $Z$-Achse fällt mit der Figurenachse der Erde zusammen, seine $X$-Achse weist nach dem Frühlingspunkt[4].

---

[1] Genaue Zahlenangaben in der astronomischen Literatur, auch in [22].

[2] Vgl. den Beitrag „Gezeitenkräfte" in Bd. XLVIII dieses Handbuches.

[3] So kommt es, daß der tagesperiodische Gezeiteneinfluß der Planeten nicht beobachtet werden kann, während die planetarische Präzession berücksichtigt werden muß.

[4] In der Literatur kommen auch Systeme vor, deren $X$-Achse nach dem Herbstpunkt gerichtet ist.

Wichtig sind noch die EULERschen Winkel

$$\Psi = \sphericalangle (I, X),$$

$$\varepsilon = \text{Schiefe der Ekliptik} = \sphericalangle(III, z)$$

und

$\varphi$, ein Rotationswinkel um die $Z$-Achse, gezählt von der $X$-Achse aus.    Er beschreibt die Rotation der Erde.

$\dot{\Psi}$ ist die Präzessionsgeschwindigkeit.

*β) EULERsche Gleichungen, Drehmoment der äußeren Kräfte, gleichförmiger Teil der Präzessionsgeschwindigkeit.*  Der Erdkörper kann starr und als Rotationsellipsoid vorausgesetzt werden.  Die $Z$-Achse liegt in einer Hauptträgheitsrichtung, im Äquator herrscht Kreissymmetrie.  Dann verschwinden die Deviations-

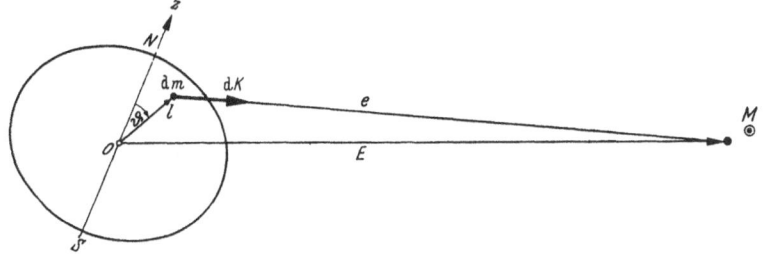

Fig. 49. Zur Theorie der solaren Präzession: Die Vektoren.

momente mit ihren Ableitungen; $A$, $B$ und $C$ werden Hauptträgheitsmomente, und $B$ wird gleich $A$. So vereinfacht sich (40.7) zu

$$\begin{aligned}
A\dot{\omega}_x + (C - A)\,\omega_y \omega_z &= M_x, \\
A\dot{\omega}_y - (C - A)\,\omega_x \omega_z &= M_y, \\
C\dot{\omega}_z &= M_z.
\end{aligned} \tag{45.1}$$

Man kann das Drehmoment $M$ aus dem Gezeitenpotential ableiten. Anschaulicher ist, es direkt zu bestimmen. Man denkt sich die Masse $M_\odot$ der Sonne, ihrer großen Entfernung wegen, im Sonnenschwerpunkt vereinigt (Fig. 49) und erhält

$$M = f M_\odot \iiint_{\ddot{O}} \left( l \times \frac{E - l}{e^3} \right) dm. \tag{45.2}$$

Mit bekannten Vektor-Identitäten und den Schwerpunktseigenschaften erhält man

$$M = \frac{3f\,M_\odot}{E^3} \iiint_{\ddot{O}} E\,l\,(l \times E)\,dm$$

und findet, wenn man zu Koordinaten übergeht,

$$M = \frac{3f\,M_\odot}{E^5} (C - A)\,(yZ\boldsymbol{i} - zX\boldsymbol{j}),$$

wobei

$$\boldsymbol{l} = x\boldsymbol{i} + y\boldsymbol{j} + z\boldsymbol{k},$$

$$\boldsymbol{E} = X\boldsymbol{i} + Y\boldsymbol{j} + Z\boldsymbol{k} \quad \text{und} \quad C > A.$$

Führt man die Länge $\lambda$ der Sonne und die Schiefe $\varepsilon$ der Ekliptik ein (Fig. 50), nimmt man zur Vereinfachung an, daß sich die Länge gleichförmig ändert

$$\lambda = \frac{2\pi}{T} t = \Omega t,$$

$T = 1$ tropisches Jahr,

$\Omega = $ (mittlere) Winkelgeschwindigkeit der Erde in ihrer Bahn,

und setzt man

$$\frac{3}{2} f \frac{M_\odot}{E^3} \frac{C-A}{A} = K, \tag{45.3}$$

so wird

$$
\begin{aligned}
& M = M_s + M_p, \\
& M_s = \text{säkularer Anteil von } M, \\
& M_p = \text{periodischer Anteil von } M, \\
& M_s = AK \sin \varepsilon \cdot \cos \varepsilon \cdot i, \\
& M_p = -AK \sin \varepsilon \left( \cos \varepsilon \cdot \cos 2\Omega t \cdot i + \sin 2\Omega t \cdot j \right).
\end{aligned}
\tag{45.4}
$$

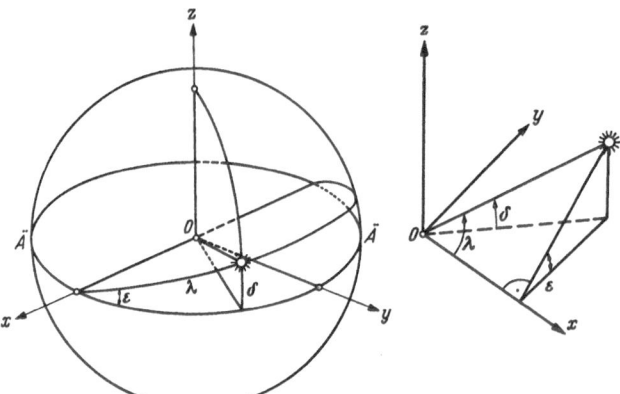

Fig. 50. Zur Theorie der solaren Präzession: Die Winkel. $\varepsilon$ Schiefe der Ekliptik, $\delta$ Deklination der Sonne, $\lambda$ Länge der Sonne.

Zur elementaren Abschätzung der Geschwindigkeit $\dot{\Psi}_s$, des gleichförmigen Teils der solaren Präzession, betrachtet man Fig. 47. Aus ihr folgt

$$\dot{\Psi}_s = \frac{\dot{S}_s}{S \cdot \sin \varepsilon} = \frac{M_s}{S \cdot \sin \varepsilon}. \tag{45.5}$$

Mit

$$\omega = 7,2921 \cdot 10^{-5} \, \text{sec}^{-1},$$

$$f = 6,670 \cdot 10^{-8} \, \text{cgs-Einheiten},$$

$$M_\odot = 1,19448 \cdot 10^{33} \, \text{g},$$

$$E = 1,485 \cdot 10^{13} \, \text{cm},$$

$$\varepsilon = 23° 26' 44,8'' \, (\text{für } 1950),$$

$$C - A = 1,327 \cdot 10^{42} \, \text{g cm}^2$$

erhält man

$$M_s = 5{,}758 \cdot 10^{28}\, i \quad \mathrm{g\,cm^2\,sec^{-2}},$$

$$M_p = (-\,5{,}758 \cdot \cos 2\Omega t \cdot i - 6{,}276 \cdot \sin 2\Omega t \cdot j) \cdot 10^{28} \quad \mathrm{g\,cm^2\,sec^{-2}}.$$

Fig. 51 zeigt den Verlauf von $M$ im Lauf eines Jahres und Fig. 52 die Bewegung der Spitze des Drehimpulsvektors unter dem Einfluß von $M$.

Die Länge des Drehimpulsvektors ist $S = C\omega = 5{,}918 \cdot 10^{40}\, \mathrm{g\,cm^2\,sec^{-1}}$.

Weiter berechnet man

$$\dot\Psi_s = \begin{cases} 2{,}445 \cdot 10^{-2}\,\mathrm{sec^{-1}}, \\ 5{,}043 \cdot 10^{-7}\,''/\mathrm{sec}, \\ 15{,}92\,''/\mathrm{Jahr}. \end{cases}$$

Diese Zahlen stimmen mit den Beobachtungen überein. Die wesentlichen Züge der Präzession werden von der elementaren Theorie erfaßt.

$\gamma)$ *Die Wirkung des säkularen und des periodischen Teils des Drehmoments.* Für

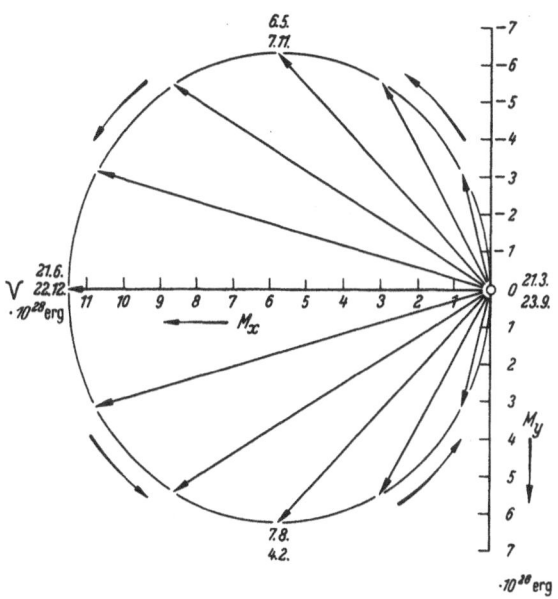

Fig. 51. Das Drehmoment der solaren Präzession im äquatorialen System Die $X$-Achse ist auf den Frühlingspunkt gerichtet.

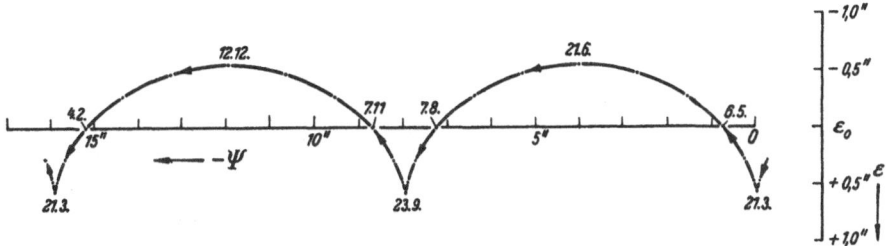

Fig. 52. Solare Präzession der Erde. Bahn der Spitze des Drehimpulsvektors im raumfesten System im Lauf eines Jahres unter Einfluß des Drehmoments der Fig. 51.

weitergehende Erkenntnisse reichen die elementaren Betrachtungen nicht aus. Einen Schritt vorwärts führen folgende Überlegungen[1].

---

[1] Man wird zuerst versuchen $M_s$ in die EULERschen Gleichungen einzusetzen. Macht man keine vereinfachenden Annahmen über $\varepsilon$, so kommt man bald auf unübersichtliche Beziehungen, die sich nur schwer auflösen lassen. Setzt man aber $\varepsilon$ konstant an, so zeigt sich, daß man nur einen Teil der zu bestimmenden Größen [nämlich $(S_x)_s = C\omega$, $\dot S_s$, $\dot\Psi$] richtig erhält, dagegen ergeben sich falsche Werte für $(S_y)_s$ und $(\omega_y)_s$. Der Grund liegt darin, daß sich der gleichförmigen Kegelbewegung mit ihrer Periode von 81400 Jahren kleine periodische Kräuselungen überlagern, deren Periode von der Länge eines Tages ist und bei denen sich $\varepsilon$ periodisch ändert. Obwohl die *Amplitude* dieser Kräuselungen unbeobachtbar klein ist, sind die *Geschwindigkeiten*, der kurzen Periode wegen, im Vergleich zur Geschwindigkeit des gleichförmigen Anteils der Präzession nicht klein.

Es ist $S_x \ll S$, $S_y \ll S$, also genau genug $S = C\,\omega_0$, und der Vektor $S$ bildet mit der $Z$-Achse einen sehr kleinen Winkel.

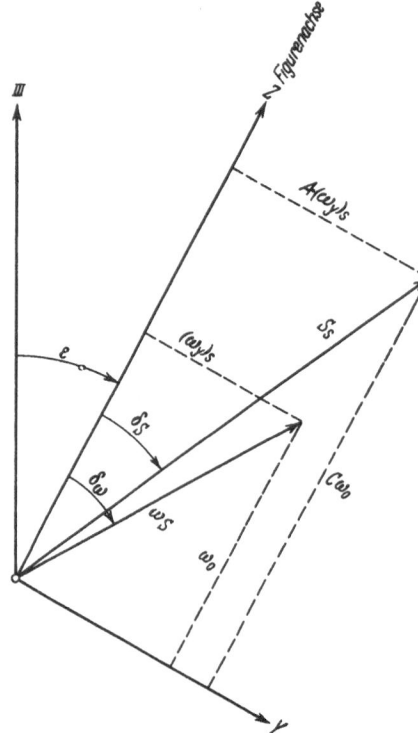

Fig. 53. Solare Präzession. Die Vektoren des Drehimpulses ($S$) und der Winkelgeschwindigkeit ($\omega$) mit ihren Komponenten, säkularer Anteil. $A$, $C$ Hauptträgheitsmomente der Erde.

Aus Fig. 48 entnimmt man

$$\dot{S} = |S| \cdot \dot{\varepsilon} \cdot j - |S| \sin \varepsilon \cdot \dot{\Psi} \cdot i,$$

$$M = C\,\omega_0 \cdot \dot{\varepsilon} \cdot j - C\,\omega_0 \sin \varepsilon \cdot \dot{\Psi} \cdot i,$$

und findet durch Vergleich mit (45.4):

$$\left.\begin{aligned}
\dot{\Psi}_s &= -\frac{A\,K}{C\,\omega_0}\cos \varepsilon, \\[4pt]
\dot{\Psi}_p &= +\frac{A\,K}{C\,\omega_0}\cos \varepsilon \cdot \cos 2\Omega t, \\[4pt]
\dot{\varepsilon}_s &= 0, \\[4pt]
\dot{\varepsilon}_p &= -\frac{A\,K}{C\,\omega_0}\sin \varepsilon \cdot \sin 2\Omega t.
\end{aligned}\right\} \quad (45.6)$$

$$\left.\begin{aligned}
(\omega_x)_s &= -\dot{\varepsilon}_s = 0, \\[4pt]
(\omega_x)_p &= -\dot{\varepsilon}_p \sin \varepsilon \\[4pt]
&= \frac{A\,K}{C\,\omega_0}\sin \varepsilon \cdot \sin 2\Omega t; \\[4pt]
(\omega_y)_s &= -\dot{\Psi}_s \sin \varepsilon \\[4pt]
&= \frac{A\,K}{C\,\omega_0}\sin \varepsilon \cdot \cos \varepsilon, \\[4pt]
(\omega_y)_p &= -\dot{\Psi}_p \sin \varepsilon \\[4pt]
&= -\frac{A\,K}{C\,\omega_0}\sin \varepsilon \cdot \cos \varepsilon \cdot \cos 2\Omega t,
\end{aligned}\right\} \quad (45.7)$$

$$\left.\begin{aligned}
(S_x)_s &= 0, \qquad (S_x)_p = \frac{A^2 K}{C\,\omega_0}\sin \varepsilon \cdot \sin 2\Omega t; \\[4pt]
(S_y)_s &= \frac{A^2 K}{C\,\omega_0}\sin \varepsilon \cdot \cos \varepsilon, \qquad (S_y)_p = -\frac{A^2 K}{C\,\omega_0}\sin \varepsilon \cdot \cos \varepsilon \cdot \cos 2\Omega t; \\[4pt]
\frac{(S_y)_s}{S_z} &= \frac{A^2 K}{C^2 \omega_0^2}\sin \varepsilon \cdot \cos \varepsilon
\end{aligned}\right\} \quad (45.8)$$

und in Zahlen:

$$\frac{A\,K}{C\,\omega_0} = 17{,}35\,''/\text{Jahr}; \qquad \frac{A^2 K}{C\,\omega_0} = 2{,}17 \cdot 10^{33}\,\text{g cm}^2\,\text{sec}^{-1}; \qquad \frac{A^2 K}{C^2 \omega_0^2} = 0{,}00750''.$$

Der Drehvektor $\omega$ und die Figurenachse bilden mit dem Impulsvektor $S$ unmeßbar kleine Winkel (Fig. 53).

$\delta$) *Die übliche Form der Präzessionsgeschwindigkeit.* Mit dem dritten Keplerschen Gesetz,

$$\frac{E^3}{T^2} = \frac{f}{4\,\pi^2}\,(M_\odot + M_\text{☾})$$

erhält man aus (45.6) und (45.3):

$$\dot{\Psi}_s = -\frac{3}{2}\frac{M_\odot}{M_\odot + M_\delta}\frac{\Omega^2}{\omega_0}\frac{C-A}{C}\cos\varepsilon.$$ (45.8)

Auf der rechten Seite sind alle Größen, bis auf $\dfrac{C-A}{C}$, gut bekannt. Aus $\dot{\Psi}_s$ kann dann $\dfrac{C-A}{C} = H$ bestimmt werden. Es ist Aufgabe der messenden Astronomie, $\dot{\Psi}_s$ durch Analyse der Beobachtungen zu ermitteln.

**46. Luni-solare Präzession.** $\alpha$) *Die geometrischen Beziehungen (Fig. 54).* Zur solaren Präzession, deren regelmäßiger Teil aus einem kegelförmigen Umlauf des Impulsvektors um eine zum Pol der Ekliptik gerichteten Achse besteht, kommt die vom Mond hervorgerufene, lunare Präzession mit einem ähnlichen Umlauf um eine senkrecht zur Mondbahn gerichtete

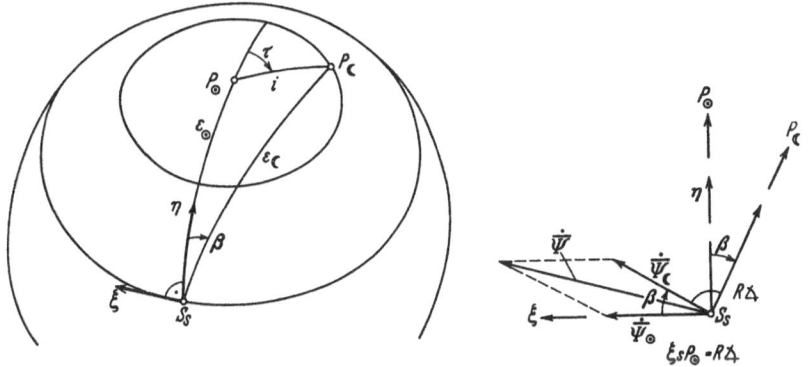

Fig. 54. Luni-solare Präzession. $P_\odot$ Pol der Ekliptik, $P_{\mathbb{C}}$ Pol der Mondbahn. $S_s$ Spitze des Drehimpulsvektors auf seiner
säkularen Bahn; $\dot{\overline{\Psi}}_{\odot,\mathbb{C}} = -\dot{\Psi}_{\odot,\mathbb{C}}\cdot\sin\varepsilon_{\odot,\mathbb{C}}.$  ($\dot{\Psi}_\odot$ solare Präzessionsgeschwindigkeit,
$\dot{\Psi}_{\mathbb{C}}$ lunare Präzessionsgeschwindigkeit.)

Achse. Anders ausgedrückt: der Himmelspol umkreist im sphärischen Abstand $\varepsilon_\odot$ den Pol der Ekliptik, zugleich aber auch im Abstand $\varepsilon_{\mathbb{C}}$ den Pol der Mondbahn, der seinerseits den Pol der Ekliptik im Abstand $i$ umläuft. Sämtliche Umläufe sind rückläufig. Die Umlaufsperiode des Mondes beträgt 18,6 Jahre; diese Bewegung kann auch als eine Art solarer Präzession aufgefaßt werden.

Bei der Zusammensetzung der Bewegungen muß man die Scheitel der Winkel $\Psi$ in den Mittelpunkt der Erde legen. Die Winkel seien dann mit $\overline{\Psi}$ bezeichnet.

$$\overline{\Psi} = -\Psi\cdot\sin\varepsilon.$$

Nun gilt (Fig. 54 rechts):

$$\dot{\xi} = \dot{\overline{\Psi}}_\odot + \dot{\overline{\Psi}}_{\mathbb{C}}\cos\beta,$$

$$\dot{\eta} = \dot{\overline{\Psi}}_{\mathbb{C}}\cdot\sin\beta$$

mit

$$\dot{\overline{\Psi}}_\odot = \frac{3}{2}\frac{M_\odot}{M_\odot + M_\delta}\frac{\Omega_\odot^2}{\omega}\frac{C-A}{C}\sin\varepsilon_\odot\cos\varepsilon_\odot,$$

$$\dot{\overline{\Psi}}_{\mathbb{C}} = \frac{3}{2}\frac{M_{\mathbb{C}}}{M_{\mathbb{C}} + M_\delta}\frac{\Omega_{\mathbb{C}}^2}{\omega}\frac{C-A}{C}\sin\varepsilon_{\mathbb{C}}\cos\varepsilon_{\mathbb{C}},$$ (46.1)

und nach den Regeln der sphärischen Trigonometrie erhält man

$$\dot{\xi} = \frac{3}{2} \frac{M_\odot}{M_\odot \ M_\oplus} \frac{\Omega_\odot^2}{\omega} \frac{C-A}{C} \times$$

$$\times \left\{ \sin \varepsilon_\odot \cos \varepsilon_\odot + \frac{M_\odot + M_\oplus}{M_\odot} \cdot \frac{M_\mathbb{C}}{M_\mathbb{C} + M_\oplus} \frac{\Omega_\mathbb{C}^2}{\Omega_\odot^2} \begin{pmatrix} \sin \varepsilon_\odot \cos \varepsilon_\odot \left( \frac{3}{2} \cos i - \frac{1}{2} \right) \\ + \frac{1}{4} \cos 2\varepsilon_\odot \sin 2i \cdot \cos \tau \\ - \frac{1}{4} \sin 2\varepsilon_\odot \cos^2 i \cdot \cos 2\tau \end{pmatrix} \right\}, \quad (46.2)$$

$$\dot{\eta} = \frac{3}{2} \frac{M_\mathbb{C}}{M_\mathbb{C} + M_\oplus} \frac{\Omega_\mathbb{C}^2}{\omega} \frac{C-A}{C} \left( \frac{1}{2} \cos \varepsilon_\odot \sin 2i \cdot \sin \tau - \frac{1}{2} \sin \varepsilon_\odot \sin^2 i \cdot \sin 2\tau \right)$$

mit

$$\tau = \frac{2\pi}{T_{\odot \mathbb{C}}}, \qquad T_{\odot \mathbb{C}} = 18,6 \text{ Jahre.}$$

$\beta$) *Die Bestimmung von* $\dfrac{C-A}{C} = H$. Der säkulare Anteil der luni-solaren Präzessions-geschwindigkeit beträgt

$$\dot{\Psi}_{s, \odot + \mathbb{C}} = \frac{3}{2} \frac{M_\odot}{M_\odot + M_\oplus} \frac{\Omega_\odot^2}{\omega} \frac{C-A}{C} \cos \varepsilon_\odot \times$$

$$\times \left( 1 + \frac{M_\odot + M_\oplus}{M_\odot} \frac{M_\mathbb{C}}{M_\mathbb{C} + M_\oplus} \frac{\Omega_\mathbb{C}^2}{\Omega_\odot^2} \left( \frac{3}{2} \cos^2 i - \frac{1}{2} \right) \right). \quad (46.3)$$

$\dot{\Psi}_{s, \odot + \mathbb{C}}$ ist aus Beobachtungen bekannt und beträgt 50,37''/Jahr[1]. Mit

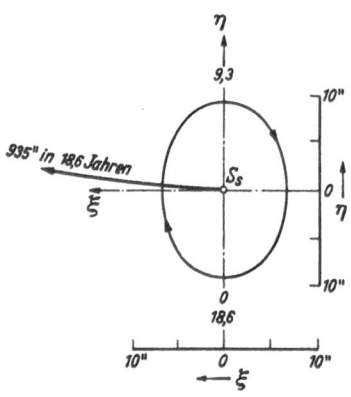

Fig. 55. Die Nutationsellipse.

$$\varepsilon_\odot = 23° 27'$$

$$i = 5° 9'$$

$$\frac{M_\odot}{M_\odot + M_\oplus} \approx 1,$$

$$\frac{2\pi}{\Omega_\odot} = 365,256 \text{ Tage (siderisches Jahr)},$$

$$\frac{M_\mathbb{C}}{M_\mathbb{C} + M_\odot} = \frac{1}{82,53},$$

$$\frac{2\pi}{\Omega_\mathbb{C}} = 27,322 \text{ Tage (siderischer Monat)}$$

erhält man $\dfrac{C-A}{C} = \dfrac{1}{304}$. Der genaue Wert ist $\dfrac{1}{305,8}$. Die hier entwickelte Theorie vermag die Beziehungen im wesentlichen zu beschreiben, bedarf aber der Ergänzung durch vernachlässigte Einzelheiten der Erd- und Mondbewegung, sowie den Einfluß der Planeten.

$\gamma$) *Die Nutationsellipse* (Fig. 55). Integriert man die mit $\cos \tau$ und $\sin \tau$ behafteten Glieder von $\dot{\xi}$ und $\dot{\eta}$ in (46.2), so ergibt sich

$$\xi = \frac{3}{4} \frac{M_\mathbb{C}}{M_\mathbb{C} + M_\oplus} \frac{\Omega_\mathbb{C}^2}{\omega} \frac{C-A}{C} \cos 2\varepsilon_\odot \sin 2i \frac{T_{\odot \mathbb{C}}}{2\pi} \sin \frac{2\pi}{T_{\odot \mathbb{C}}} t + \cdots,$$

$$\eta = -\frac{3}{4} \frac{M_\mathbb{C}}{M_\mathbb{C} + M_\oplus} \frac{\Omega_\mathbb{C}^2}{\omega} \frac{C-A}{C} \cos^2 \varepsilon_\odot \sin 2i \cdot \frac{T_{\odot \mathbb{C}}}{2\pi} \cos \frac{2\pi}{T_{\odot \mathbb{C}}} t + \cdots \quad (46.4)$$

---

[1] [22], S. 49. Ein Umlauf dauert 25730 Jahre.

mit $T_{\odot\mathbb{C}} = 18{,}6$ Jahre, $t$ in Jahren, die Gleichung eines elliptischen Umlaufs, der in 18,6 Jahren vor sich geht und sich dem gleichmäßigen Teil der luni-solaren Präzession überlagert. Sie stellt das *Haupt-Nutationsglied* dar. Der Umlauf ist rückläufig, die Halbachsen der Ellipse betragen 6,87″ in der West-Ost-Richtung und 9,21″ in der Nord-Süd-Richtung.

**47. Die Bewegung des Mondes und die Abplattung der Erde.** Erde, Mond und Sonne[1] bilden ein abgeschlossenes System. Sein Gesamtdrehimpuls ist konstant. Der Drehimpuls der präzessierenden Erde ist nicht konstant, daher gibt es Änderungen der Drehimpulse der Sonne und des Mondes. Ihre Ursache sind die Drehmomente der vom Äquatorwulst der Erde ausgeübten Anziehungen.

Die Wirkung auf die Sonne ist unmerklich klein. Der Mond aber wird in meßbarer Weise beeinflußt, und die Schwankungen der Mondbewegung können zur Bestimmung der Erdabplattung herangezogen werden[2].

# D. Neuzeitliche Probleme und Entwicklungen.

**48. Geodätische Fragen des Schwerefeldes.** Das größte Hindernis für die Ermittlung der Figur der Erde ist die ungleichmäßige und lückenhafte Verteilung der Schweremessungen. Sie muß vor allem behoben werden. Einige Tausend neue Schweremessungen, die man so verteilt, daß auf je 10 Quadratgrad der Nordhalbkugel und je 25 Quadratgrad der Südhalbkugel ein repräsentativer Schwerewert fällt, können weitgehend Abhilfe schaffen. Um jeden Lotabweichungspunkt sind zusätzlich lokale Vermessungen auszuführen, wofür einige Hundert Schwerestationen genügen[3]. Lokale Vermessungen dürften auch nötig sein, wenn es fraglich ist, ob eine Station des regionalen Netzes als repräsentativ angesehen werden darf.

Für die geodätische Auswertung braucht man die Schwereunterschiede nicht mit höchster Genauigkeit zu messen, es genügt, wenn ihre Ungenauigkeit einige Zehntelmilligal nicht übersteigt. Wesentlich ist, daß die Messungsergebnisse keine regionalen systematischen Fehler enthalten. Daher ist die Verbindung der Anschlußstationen mit Sorgfalt auszuführen und oft zu kontrollieren.

Bei der Entwicklung nach Kugelfunktionen und in den STOKESschen Formeln tritt die absolute Schwere nur als Faktor vor der Summe und den Integralen auf. Es genügt bei weitem, wenn sie mit einer relativen Genauigkeit von $10^{-4}$ angegeben werden kann. Die Widersprüche zwischen den absoluten Schwerebestimmungen stören hier nicht, da sie kleiner sind. Auch die Gezeitenschwankungen des Schwerefeldes haben keinen störenden Einfluß.

Wesentliche Schwierigkeiten potentialtheoretischer Art könnten umgangen werden, wenn man die Ermittlung des Außenraumpotentials auf die Randwerte der Erdoberfläche, also unmittelbar auf die gemessenen Schwerewerte, gründen könnte. Dann fiele die Verlagerung der topographischen Massen mit den Problemen des indirekten Effektes weg. Unerläßlich wäre aber eine genauere Bestimmung der vertikalen Schwereänderung, die man — da sie nicht genau genug gemessen werden kann — mit Integration über die Erdoberfläche berechnen muß. Im Gebirge ist zu beachten, daß sich die Geoidundulation mit der Höhe ändert[4].

Mit der weltweiten Ermittlung der Geoidgestalt löst sich die Frage des dreiachsigen Erdellipsoids. Auch wird die Abplattung zuverlässiger bestimmt.

Die Potentialtheorie des hydrostatisch aufgebauten Erdkörpers führt auf Grenzen der Abplattung. Der aus den Beobachtungen abgeleitete Abplattungswert 1/297 liegt oft am Rand des zugestandenen Bereichs, bisweilen sogar außerhalb [*19*, S. 123/124]. Es ist noch nicht geklärt, ob solche Widersprüche auf der Unvollständigkeit des Beobachtungsmaterials beruhen oder die Voraussetzungen der Theorie nicht zutreffen. Man neigt im allgemeinen zu der zweiten Erklärung.

---

[1] Hinzu kommen noch die Planeten, soweit sie auf die Präzession der Erde Einfluß haben.

[2] [*5*, II], S. 460f. — H. JEFFREYS. The figures of the earth and moon. Month. Not. Roy. Astronom. Soc., Geophys. Suppl. **5**, 219 (1948).

[3] W. A. HEISKANEN: On the World Geodetic System. Veröff. finn. Geodät. Inst. **1951**, Nr. 39. — R. A. HIRVONEN: Obtainable accuracy in the gravimetrical determination of geoid. Bull. Géod. **1955**, Nr. 36, 61. Vgl. auch das letzte Zitat auf S. 593.

[4] A. H. COOK: Approximations in the calculation of the form of the geoid from gravity anomalies. Month. Not. Roy Astronom. Soc., Geophys. Suppl. **6**, 442 (1953).

**49. Geologisch-geodätische Beziehungen.** Falsche Annahmen über die Gesteinsdichte führen zu schwer abschätzbaren Ungenauigkeiten bei der rechnerischen Beseitigung der topographischen Massen. Dichtebestimmungen, die eine geologisch-physikalische Aufgabe sind, sollten so oft wie möglich vorgenommen werden.

Weniger bedeutend für die geodätische, aber ausschlaggebend für die geologische Beurteilung der Schwereanomalien ist die Frage, wie man die Lage der Kompensationsmassen anzunehmen hat. Ihre Beantwortung aus der Schwereverteilung allein ist mit schwer kontrollierbaren Hypothesen behaftet und unsicher. Man wird die Ergebnisse geologischer Forschungen, vor allem aber auch die experimentelle Seismik zu Rate ziehen müssen. Es wird sich nicht vermeiden lassen, daß man die noch gebräuchlichen, sehr schematischen Ansätze aufgibt und durch Annahmen ersetzt, die der komplizierten Vielseitigkeit des Erdkrustenaufbaues besser entsprechen. Erhöhte Rechenarbeit kann wohl mit neuzeitlichen Rechenmaschinen bewältigt werden[1].

Interessante Probleme bietet auch der indirekte Effekt, sowie man die Aufgabe der Geoidbestimmung mit der Ermittlung einer geologischen Normalschwere verbindet. Dann kommt es darauf an, die topographischen Massen so zu verlagern, daß man eine homogene, vom Co-Geoid begrenzte Erdkruste erhält. Wie die Verteilung der verschobenen Massen vorzunehmen ist, hängt davon ab, was man sich über die Massen der wirklichen Erdkruste und die Herausbildung ihrer Anordnung vorstellen kann. Läßt man, wie es meist üblich ist, nur vertikale Verschiebungen starrer Erdkrustenblöcke zu, so treten Verlagerungen der Niveauflächen in allen Tiefen des Erdkörpers auf. Nimmt man an, daß sich das hydrostatische Erdinnere ihnen anpaßt, so treten merkliche indirekte Effekte auf, wie VENING MEINESZ in einer ausführlichen Abhandlung gezeigt hat[2].

Von diesen Schwierigkeiten ist man frei, wenn man bei der Reduktion die topographische Masse innerhalb des Geoids so verteilt, daß sich das Potential im tiefen Erdinnern nicht ändert[3]. Das ist mit reinen Vertikalverschiebungen nicht möglich. Eine solche Reduktion entspricht der Vorstellung, daß sich die wirkliche Erdkruste aus einer konzentrisch geschichteten Kruste durch Massenverschiebungen gebildet hat, die bestimmte horizontale Bewegungskomponenten haben. Solche Strömungen mögen im subkrustalen Material möglich sein.

**50. Astronomisch-geodätische und physikalisch-geodätische Messungen.** Die Entwicklung der astronomischen Beobachtungsverfahren, insbesondere der Photographie und der Zeitbestimmung, hat dazu geführt, daß man astronomische Objekte mit Vorteil in das irdische Vermessungsnetz einbeziehen kann und vielleicht die Möglichkeit erhält, die Ozeane geodätisch zu überbrücken. Als aussichtsreich haben sich erwiesen[4]:

der Vorwärtseinschnitt einer *Rakete* mit photographischem Anschluß an den Sternenhimmel;

Aufnahmen und Zeitbestimmungen bei Anfang und Ende von *Sonnenfinsternissen*;

Zeitbestimmungen bei *Sternbedeckungen* durch den Mond, der Vorwärtseinschnitt des *Mondes* mit photographischem Anschluß an den Sternenhimmel.

Bei der ersten Methode kann die Rakete durch einen künstlichen *Satelliten* ersetzt werden, auch ist an den photographischen Rückwärtseinschnitt vom Satelliten aus zu denken.

Ergänzend tritt die Entfernungsbestimmung mit elektromagnetischen Wellen hinzu. Als erfolgreich werden *Shoran, Hiran, Decca* und *Geoidmeter* genannt. Da es sich um geometrische Verfahren handelt, muß dieser Hinweis genügen.

---

[1] W. HEISKANEN: Ann. Acad. Sci. fenn., Ser. A, III **1953**, 33. — T. J. KUKKAMÄKI: Ann. Acad. Sci. fenn., Ser. A, III **1955**, 42.

[2] F. A. VENING MEINESZ: The indirect isostatic or BOWIE reduction and the equilibrium figure of the earth. Bull. Géod. **1946**, Nr. 1, 33. — F. A. VENING MEINESZ: The isostatic reduction and the indirect BOWIE effect. Bull. Géod. **1949**, Nr. 12, 170.

[3] K. JUNG: Über vollständig isostatische Reduktion. Z. Geophys. **14**, 27 (1938).

[4] W. A. HEISKANEN: New Era of Geodesy. Science, Lancaster, Pa. **121**, 48 (1955). — Symposium: New Era of Geodesy. Bull. Géod. **1955**, Nr. 36,55. — W. D. LAMBERT: Geodetic applications of eclipses and occultations. Bull. Géod. **1949**, Nr. 13, 274

## Anhang I[1].

### 51. Tabelle 4. *Kugelfunktionen erster Art, normiert nach* Ad. Schmidt[2].

| $\vartheta$ | $P_{10}$ | $P_{20}$ | $P_{21}$ | $P_{22}$ | $P_{30}$ | $P_{31}$ | $P_{32}$ | $P_{33}$ | $P_{40}$ | $P_{41}$ | $\vartheta$ |
|---|---|---|---|---|---|---|---|---|---|---|---|
| 0° | 1,0000 | 1,0000 | 0 | 0 | 1,0000 | 0 | 0 | 0 | 1,0000 | 0 | 180° |
| 15° | 0,9659 | 0,8995 | 0,4330 | 0,0580 | 0,8042 | 0,5809 | 0,1253 | 0,0137 | 0,6847 | 0,6979 | 165° |
| 30° | 0,8660 | 0,6250 | 0,7500 | 0,2165 | 0,3248 | 0,8420 | 0,4193 | 0,0988 | 0,0234 | 0,7702 | 150° |
| 45° | 0,7071 | 0,2500 | 0,8660 | 0,4330 | −0,1768 | 0,6495 | 0,6847 | 0,2795 | 0,4062 | 0,1976 | 135° |
| 60° | 0,5000 | −0,1250 | 0,7500 | 0,6495 | −0,4375 | 0,1326 | 0,7262 | 0,5135 | −0,2891 | −0,4279 | 120° |
| 75° | 0,2588 | −0,3995 | 0,4330 | 0,8080 | −0,3449 | −0,3934 | 0,4676 | 0,7125 | 0,1434 | −0,5002 | 105° |
| 90° | 0 | −0,5000 | 0 | 0,8660 | 0 | −0,6124 | 0 | 0,7906 | 0,3750 | 0 | 90° |
| | $-P_{10}$ | $P_{20}$ | $-P_{21}$ | $P_{22}$ | $-P_{30}$ | $P_{31}$ | $-P_{32}$ | $P_{33}$ | $P_{40}$ | $-P_{41}$ | $\vartheta$ |

| $\vartheta$ | $P_{42}$ | $P_{43}$ | $P_{44}$ | $P_{50}$ | $P_{51}$ | $P_{52}$ | $P_{53}$ | $P_{54}$ | $P_{55}$ | $P_{60}$ | $P_{61}$ | $\vartheta$ |
|---|---|---|---|---|---|---|---|---|---|---|---|---|
| 0° | 0 | 0 | 0 | 1,0000 | 0 | 0 | 0 | 0 | 0 | 1,0000 | 0 | 180° |
| 15° | 0,2071 | 0,0350 | 0,0033 | 0,5471 | 0,7792 | 0,2982 | 0,0671 | 0,0096 | 0,0008 | 0,3983 | 0,8215 | 165° |
| 30° | 0,5940 | 0,2264 | 0,0462 | −0,2233 | 0,5598 | 0,6933 | 0,3758 | 0,1201 | 0,0219 | −0,3740 | 0,2635 | 150° |
| 45° | 0,6988 | 0,5229 | 0,1849 | −0,3756 | −0,2567 | 0,4529 | 0,6471 | 0,3922 | 0,1240 | −0,1484 | −0,5012 | 135° |
| 60° | 0,3144 | 0,6793 | 0,4160 | 0,0898 | −0,4979 | −0,2402 | 0,4246 | 0,6240 | 0,3418 | 0,3232 | −0,1085 | 120° |
| 75° | −0,2770 | 0,4879 | 0,6438 | 0,3427 | 0,0731 | −0,4943 | −0,1871 | 0,4998 | 0,5899 | 0,0431 | 0,4494 | 105° |
| 90° | −0,5590 | 0 | 0,7395 | 0 | 0,4841 | 0 | −0,5229 | 0 | 0,7016 | −0,3125 | 0 | 90° |
| | $P_{42}$ | $-P_{43}$ | $P_{44}$ | $-P_{50}$ | $P_{51}$ | $-P_{52}$ | $P_{53}$ | $-P_{54}$ | $P_{55}$ | $P_{60}$ | $-P_{61}$ | $\vartheta$ |

[1] $\bar{5}$ (kleine 5) entstanden durch Aufrundung, $\dot{5}$ (große 5) entstanden durch Abrundung. Ebenso in Anhang II und IV.
[2] Bis zur 6. Ordnung nach Ad. Schmidt [27], die 7. Ordnung neu berechnet. Die 8. Ordnung nach B. Haurwitz und R. A. Craig, Geophysical Research Papers No. 14, Cambridge, Mass. 1952, dort Werte von 5° zu 5°.

*Tabelle 4.* (Fortsetzung.)

| $\vartheta$ | $P_{62}$ | $P_{63}$ | $P_{64}$ | $P_{65}$ | $P_{66}$ | $P_{70}$ | $P_{71}$ | $P_{72}$ | $P_{73}$ | $P_{74}$ | $P_{75}$ | $\vartheta$ |
|---|---|---|---|---|---|---|---|---|---|---|---|---|
| 0° | 0 | 0 | 0 | 0 | 0 | 1,0000 | 0 | 0 | 0 | 0 | 0 | 180° |
| 15° | 0,3923 | 0,1102 | 0,0206 | 0,0026 | 0,0002 | 0,2456 | 0,8236 | 0,4829 | 0,1636 | 0,0376 | 0,0062 | 165° |
| 30° | 0,6864 | 0,5147 | 0,2248 | 0,0630 | 0,0105 | −0,4102 | −0,0497 | 0,5673 | 0,6075 | 0,3471 | 0,1299 | 150° |
| 45° | 0,0566 | 0,5661 | 0,5581 | 0,2908 | 0,0840 | 0,1271 | −0,4458 | −0,3043 | 0,2911 | 0,5877 | 0,4618 | 135° |
| 60° | −0,4882 | −0,0735 | 0,4883 | 0,5667 | 0,2834 | 0,2231 | 0,3233 | −0,2706 | −0,4244 | 0,0668 | 0,5206 | 120° |
| 75° | −0,0244 | −0,4781 | −0,1136 | 0,5064 | 0,5456 | −0,2731 | 0,1558 | 0,4046 | −0,1006 | −0,4557 | −0,0516 | 105° |
| 90° | 0,4529 | 0 | −0,4961 | 0 | 0,6717 | 0 | −0,4134 | 0 | 0,4296 | 0 | −0,4750 | 90° |
| | $-P_{62}$ | $-P_{63}$ | $P_{64}$ | $-P_{65}$ | $P_{66}$ | $-P_{70}$ | $P_{71}$ | $-P_{72}$ | $P_{73}$ | $-P_{74}$ | $P_{75}$ | |

| $\vartheta$ | $P_{76}$ | $P_{77}$ | $P_{80}$ | $P_{81}$ | $P_{82}$ | $P_{83}$ | $P_{84}$ | $P_{85}$ | $P_{86}$ | $P_{87}$ | $P_{88}$ | $\vartheta$ |
|---|---|---|---|---|---|---|---|---|---|---|---|---|
| 0° | 0 | 0 | 1,0000 | 0 | 0 | 0 | 0 | 0 | 0 | 0 | 0 | 180° |
| 15° | 0,0007 | 0,0001 | 0,0962 | 0,7863 | 0,5635 | 0,2257 | 0,0615 | 0,0122 | 0,0018 | 0,0002 | 0,0000 | 165° |
| 30° | 0,0328 | 0,0051 | −0,3388 | −0,3114 | 0,3570 | 0,6251 | 0,4644 | 0,2208 | 0,0733 | 0,0170 | 0,0024 | 150° |
| 45° | 0,2141 | 0,0572 | 0,2983 | −0,1582 | −0,4657 | −0,0664 | 0,4370 | 0,5562 | 0,3719 | 0,1567 | 0,0392 | 135° |
| 60° | 0,5109 | 0,2365 | −0,0736 | 0,4002 | 0,1609 | −0,3665 | −0,3326 | 0,1806 | 0,5310 | 0,4579 | 0,1983 | 120° |
| 75° | 0,5090 | 0,5078 | −0,1702 | −0,3161 | 0,2239 | 0,3551 | −0,1611 | −0,4293 | 0,0018 | 0,5090 | 0,4749 | 105° |
| 90° | 0 | 0,6472 | 0,2734 | 0,0000 | −0,3922 | 0,0000 | 0,4113 | 0,0000 | −0,4577 | 0,0000 | 0,6267 | 90° |
| | $-P_{76}$ | $P_{77}$ | $P_{80}$ | $-P_{81}$ | $P_{82}$ | $-P_{83}$ | $P_{84}$ | $-P_{85}$ | $P_{86}$ | $-P_{87}$ | $P_{88}$ | |

# Anhang II.

## 52. Koeffizienten des Reliefs der Erde nach A. Prey [28] *.

*Tabelle 5.* Vgl. S. 582.

| $n$ | $\nu$ | $\sqrt{\dfrac{1}{x_\nu}\dfrac{(n+\nu)!}{(n-\nu)!}}$ | $\sqrt{2n+1}$ | Lithosphäre ($L$) | | Hydrosphäre ($M$) | | $\left(L-\dfrac{1,03}{2,7}M\right)^2$ | |
|---|---|---|---|---|---|---|---|---|---|
| | | | | $_cL_{n\nu}$ km | $_sL_{n\nu}$ km | $_cM_{n\nu}$ km | $_sM_{n\nu}$ km | (c) km² | (s) km² |
| 0 | 0 | 1,00000 | 1,000 | − 2,456 | * | − 2,681 | * | + 4,365 | * |
| 1 | 0 | 1,00000 | 1,732 | + 1,264 | * | + 1,112 | * | − 1,443 | * |
| | 1 | 1,00000 | 1,732 | + 1,130 | + 0,664 | + 1,006 | + 0,564 | − 2,032 | − 0,774 |
| 2 | 0 | 1,00000 | 2,236 | + 1,134 | * | + 1,037 | * | − 1,804 | * |
| | 1 | 1,7320$\overline{5}$ | 2,236 | + 0,783 | + 0,78$\overline{5}$ | + 0,758 | + 0,613 | − 1,587 | − 0.610 |
| | 2 | 3,46410 | 2,236 | − 0,906 | − 0,222 | − 0,74$\overline{5}$ | − 0,272 | + 1,576 | + 0,769 |
| 3 | 0 | 1,00000 | 2,646 | − 0,123 | * | + 0,054 | * | − 0,627 | * |
| | 1 | 2,44949 | 2,646 | − 0,473 | + 0,183 | − 0,390 | + 0,098 | + 0,693 | + 0,066 |
| | 2 | 7,74597 | 2,646 | − 1,172 | + 1,410 | − 0,912 | + 1,22$\overline{5}$ | + 1,273 | − 2,156 |
| | 3 | 18,9737 | 2,646 | + 0,362 | + 1,514 | + 0,378 | + 1,377 | − 0,944 | − 3,076 |
| 4 | 0 | 1,00000 | 3.000 | + 0,930 | * | + 0,88$\overline{5}$ | * | − 2,017 | * |
| | 1 | 3,16228 | 3,000 | − 0,531 | − 0,539 | − 0,522 | − 0,349 | + 1,103 | + 0,300 |
| | 2 | 13,4164 | 3,000 | − 1,138 | + 0,242 | − 0,867 | + 0,199 | + 0,801 | − 0,203 |
| | 3 | 50,1996 | 3,000 | + 1,153 | − 0,430 | + 1,028 | − 0,270 | − 2,093 | + 0,457 |
| | 4 | 141,986 | 3,000 | − 0,193 | + 1,472 | − 0,224 | + 1,272 | + 0,511 | − 1,788 |
| 5 | 0 | 1,00000 | 3,317 | − 1,314 | * | − 1,047 | * | + 1,614 | * |
| | 1 | 3,87298 | 3,317 | − 0,171 | − 0,493 | − 0,182 | − 0,340 | + 0,581 | + 0,294 |
| | 2 | 20,4939 | 3,317 | − 0,129 | − 0,431 | − 0,068 | − 0,367 | − 0,652 | + 0,600 |
| | 3 | 100,399 | 3,317 | + 0,367 | + 0,207 | + 0,279 | + 0,369 | − 0,611 | − 1,04$\overline{5}$ |
| | 4 | 425,958 | 3,317 | + 1,827 | − 0,230 | + 1,576 | − 0,049 | − 2,504 | − 0,277 |
| | 5 | 1347,00 | 3,317 | − 0,179 | + 0,633 | − 0,148 | + 0,486 | + 0,210 | − 0,302 |
| 6 | 0 | 1,00000 | 3,606 | + 0,864 | * | + 0,558 | * | − 0,192 | * |
| | 1 | 4,58258 | 3,606 | + 0,161 | − 0,396 | + 0,126 | − 0 189 | − 0 192 | − 0,21$\overline{5}$ |
| | 2 | 28,9828 | 3,606 | − 0,076 | − 0,398 | − 0,219 | − 0,31$\overline{5}$ | + 0,727 | + 0,867 |
| | 3 | 173,897 | 3,606 | + 0,281 | + 0,683 | + 0,221 | + 0,704 | − 0,501 | − 1,422 |
| | 4 | 952,470 | 3,606 | + 0,642 | − 0,583 | + 0,447 | − 0,458 | − 0,294 | + 0,882 |
| | 5 | 4467,48 | 3,606 | − 0,477 | − 0,847 | − 0,321 | − 0,845 | + 0,214 | + 2,368 |
| | 6 | 15475,8 | 3,606 | + 0,159 | + 0,022 | + 0,254 | + 0,056 | − 0,836 | + 0,046 |
| 7 | 0 | 1,00000 | 3,873 | − 0,752 | * | − 0,758 | * | + 1,598 | * |
| | 1 | 5,29150 | 3,873 | − 0,138 | + 0,522 | − 0,251 | + 0,467 | + 0,778 | − 0,529 |
| | 2 | 38,8844 | 3,873 | + 0,573 | − 0,044 | + 0,342 | + 0,009 | + 0,027 | + 0,039 |
| | 3 | 274,95$\overline{5}$ | 3,873 | − 0,194 | + 0,074 | − 0,287 | + 0,032 | + 0,753 | + 0,547 |
| | 4 | 1823,84 | 3,873 | − 0,941 | + 0,012 | − 0,999 | + 0,041 | + 2,687 | − 0,197 |
| | 5 | 10943,1 | 3,873 | + 0,046 | + 0,001 | + 0,089 | − 0,139 | − 0,031 | + 0,538 |
| | 6 | 55798,8 | 3,873 | − 0,209 | − 0,526 | − 0,122 | − 0,513 | − 0,050 | + 0,608 |
| | 7 | 208780 | 3,873 | − 0,109 | − 0,378 | − 0,061 | − 0,232 | + 0,228 | − 0,328 |
| 8 | 0 | 1,00000 | 4,123 | + 0,098 | * | − 0,078 | * | + 0,599 | * |
| | 1 | 6,00000 | 4,123 | + 0,040 | − 0,160 | − 0,06$\overline{5}$ | − 0,152 | + 0,462 | + 0,354 |
| | 2 | 50,1996 | 4,123 | + 0,543 | + 0,092 | + 0,471 | + 0,224 | − 0,878 | − 0,994 |
| | 3 | 407,823 | 4,123 | + 0,192 | + 0,113 | + 0,151 | + 0,039 | − 0,041 | + 0,55$\overline{5}$ |
| | 4 | 3158,99 | 4,123 | − 0,006 | + 0,370 | + 0,030 | + 0,409 | − 0,051 | − 0,673 |
| | 5 | 22779,8 | 4,123 | − 0,19$\overline{5}$ | − 0,106 | − 0,136 | − 0,263 | − 0,050 | + 0,911 |
| | 6 | 147630 | 4,123 | + 0,292 | + 0,344 | + 0,304 | + 0,272 | − 0,98$\overline{5}$ | − 0,278 |
| | 7 | 808602 | 4,123 | + 0,688 | + 0,206 | + 0,476 | + 0,233 | − 0,437 | − 0,251 |
| | 8 | 3234410 | 4,123 | − 0,289 | − 0,499 | − 0,227 | − 0,477 | + 0,327 | − 0,770 |

* Durch Multiplikation mit $\sqrt{\dfrac{1}{x_\nu}\dfrac{(n+\nu)!}{(n-\nu)!}}$ für normierte Kugelfunktionen umgerechnet.

Man erfaßt die Bedeutung der einzelnen Glieder, wenn man die Koeffizienten durch $\sqrt{2n+1}$ dividiert.

# Anhang III. Formeln.

**53. Rotationssymmetrische Ellipsoid-Koordinaten** (Fig. 11, S. 544; Fig. 56, Fig. 57)

$$x = \sqrt{\varrho^2 + e^2}\,\sin\bar\vartheta\,\cos\lambda, \qquad y = \sqrt{\varrho^2 + e^2}\,\sin\bar\vartheta\,\sin\lambda, \qquad z = \varrho\cos\bar\vartheta.$$

$$l = \sqrt{x^2 + y^2 + z^2} = \sqrt{\varrho^2 + e^2\sin^2\bar\vartheta},$$

$\varrho$ = kleine Achse des Koordinaten-Ellipsoids,

$\sqrt{\varrho^2 + e^2}$ = große Achse des Koordinaten-Ellipsoids,

$e$ = halber Brennpunktsabstand der Meridian-Ellipsen,

$\bar\vartheta$ = reduzierter Polabstand,     $\lambda$ = Länge,     $l$ = Radiusvektor,

$$dn^2 = (dx^2 + dy^2 + dz^2)_{\substack{\bar\vartheta=\text{const}\\ \lambda=\text{const}}} = \frac{\varrho^2 + e^2\cos^2\bar\vartheta}{\varrho^2 + e^2}\,d\varrho^2,$$

$$ds^2 = (dx^2 + dy^2 + dz^2)_{\substack{\varrho=\text{const}\\ \lambda=\text{const}}} = (\varrho^2 + e^2\cos^2\bar\vartheta)\,d\bar\vartheta^2,$$

$$do^2 = (dx^2 + dy^2 + dz^2)_{\substack{\varrho=\text{const}\\ \vartheta=\text{const}}} = (\varrho^2 + e^2)\sin^2\bar\vartheta\,d\lambda^2.$$

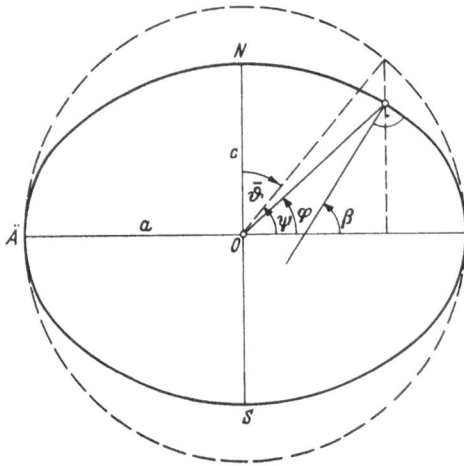

Fig. 56. $a$ Äquatorradius, $c$ Polradius, $\beta$ geographische Breite, $\varphi$ geozentrische Breite, $\psi$ reduzierte Breite, $\bar\vartheta$ reduzierter Polabstand.

Flächenelement:

$$df = ds\,do$$
$$= \sqrt{\varrho^2 + e^2\cos^2\bar\vartheta}\cdot\sqrt{\varrho^2 + e^2}\,\sin\bar\vartheta\,d\bar\vartheta\,d\lambda.$$

Volumenelement:

$$dv = dn\,ds\,do$$
$$= (\varrho^2 + e^2\cos^2\bar\vartheta)\sin\bar\vartheta\,d\varrho\,d\bar\vartheta\,d\lambda.$$

**54. Koordinaten-Transformationen** (Fig. 13, S. 549)

$\bar\vartheta$ = reduzierter Polabstand,

$\vartheta$ = geozentrischer Polabstand,

$\varphi$ = geozentrische Breite,

$\beta$ = geographische Breite,

$\varrho$ = kleine Achse des Koordinaten-Ellipsoids,

$l$ = Radiusvektor,

$$\tan\bar\vartheta = \frac{\varrho}{\sqrt{\varrho^2 + e^2}}\tan\vartheta = \frac{\varrho}{\sqrt{\varrho^2 + e^2}}\cot\varphi, \qquad \tan\bar\vartheta = \frac{\sqrt{\varrho^2 + e^2}}{\varrho}\cot\beta,$$

$$\tan\beta = \frac{\sqrt{\varrho^2 + e^2}}{\varrho}\cot\bar\vartheta, \qquad \tan\beta = \frac{\varrho^2 + e^2}{\varrho^2}\cot\vartheta = \frac{\varrho^2 + e^2}{\varrho^2}\tan\varphi,$$

$$l^2 = \varrho^2 + e^2\sin^2\bar\vartheta.$$

**55. Kugelfunktionen** (normiert nach Ad. Schmidt)

$$P_n(\xi) = P_{n0}(\xi) = \frac{1}{2^n n!}\frac{d^n(\xi^2 - 1)^n}{d\xi^n},$$

$$P_{n\nu}(\xi) = \sqrt{\frac{\varkappa_\nu(n-\nu)!}{(n+\nu)!}}\,\sqrt{1 - \xi^2}^{\,\nu}\,\frac{d^\nu P_n(\xi)}{d\xi^\nu},$$

$$Q_n(\xi) = Q_{n0}(\xi) = \frac{1}{2} P_n(\xi) \cdot \ln \frac{\xi+1}{\xi-1} -$$

$$- \left( \frac{2n-1}{1 \cdot n} P_{n-1}(\xi) + \frac{2n-5}{3(n-1)} P_{n-3}(\xi) + \frac{2n-9}{5(n-2)} P_{n-5}(\xi) + \cdots \right).$$

(Nach dem letzten positiven Koffizienten ist in der Klammer abzubrechen.)

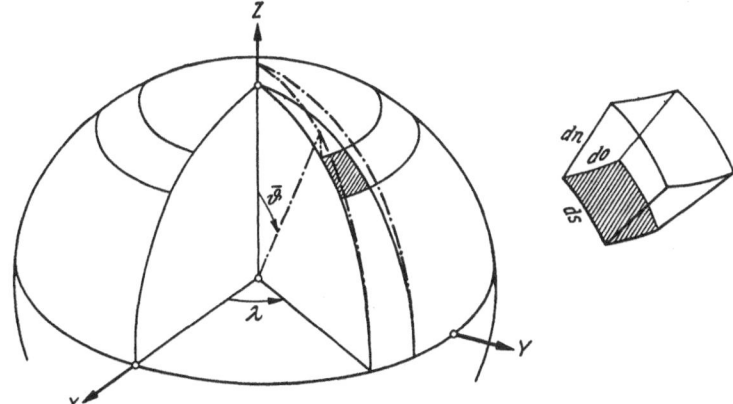

Fig. 57. Flächen- und Volumenelement in rotationssymmetrischen Ellipsoidkoordinaten.

$$Q_{n\nu}(\xi) = \sqrt{\frac{\varkappa_\nu (n-\nu)!}{(n+\nu)!}} \sqrt{1-\xi^2}^{\,\nu} \frac{d^\nu Q_n(\xi)}{d\xi^\nu},$$

$$\varkappa_\nu = \begin{cases} 1, & \text{wenn} \quad \nu = 0 \\ 2, & \text{wenn} \quad \nu \neq 0 \end{cases}$$

$n, \nu$ ganze positive Zahlen (einschließlich 0)

$$Q'_{n\nu}(\xi) = \frac{d}{d\xi} Q_{n\nu}(\xi).$$

$$P_{00} = 1,$$

$$Q_{00}(\xi) = \frac{1}{2} \ln \frac{\xi+1}{\xi-1},$$

$$P_{10}(\xi) = \xi,$$

$$Q_{10}(\xi) = \frac{1}{2} \xi \ln \frac{\xi+1}{\xi-1} - 1,$$

$$P_{11}(\xi) = \sqrt{1-\xi^2},$$

$$Q_{11}(\xi) = \frac{1}{2} \sqrt{1-\xi^2} \ln \frac{\xi+1}{\xi-1} + \frac{\xi}{\sqrt{1-\xi^2}},$$

$$P_{20}(\xi) = \frac{3}{2}\left(\xi^2 - \frac{1}{3}\right),$$

$$Q_{20}(\xi) = \frac{3}{4}\left(\xi^2 - \frac{1}{3}\right) \ln \frac{\xi+1}{\xi-1},$$

$$P_{22}(\xi) = \frac{\sqrt{3}}{2}(1-\xi^2),$$

$$Q_{22}(\xi) = \frac{\sqrt{3}}{4}(1-\xi^2) \ln \frac{\xi+1}{\xi-1} -$$

$$- \frac{\sqrt{3}}{2}\xi + \sqrt{\frac{1}{12}} \frac{\xi}{\xi^2-1},$$

$$P_{30}(\xi) = \frac{5}{2}\left(\xi^3 - \frac{3}{5}\xi\right),$$

$$Q_{30}(\xi) = \frac{5}{4}\left(\xi^3 - \frac{3}{5}\xi\right) \ln \frac{\xi+1}{\xi-1} -$$

$$- \frac{5}{2}\left(\xi^2 - \frac{1}{3}\right) - \frac{1}{6},$$

$$P_{40}(\xi) = \frac{35}{8}\left(\xi^4 - \frac{6}{7}\xi^2 + \frac{3}{35}\right),$$

$$Q_{40}(\xi) = \frac{35}{16}\left(\xi^4 - \frac{6}{7}\xi^2 + \frac{3}{35}\right) \ln \frac{\xi+1}{\xi-1} -$$

$$- \frac{35}{8}\left(\xi^3 - \frac{3}{5}\xi\right) - \frac{1}{3}\xi,$$

$$\xi = P_{10}(\xi), \qquad \xi^2 = \frac{2}{3} P_{20}(\xi) + \frac{1}{3},$$

$$\xi^3 = \frac{2}{5} P_{30}(\xi) + \frac{3}{5} P_{10}(\xi), \qquad \xi^4 = \frac{8}{35} P_{40}(\xi) + \frac{4}{7} P_{20}(\xi) + \frac{1}{5},$$

$$Q_{00}(i\xi) = - i \arctan \frac{1}{\xi},$$

$$Q_{10}(i\xi) = \xi \arctan \frac{1}{\xi} - 1,$$

$$Q_{11}(i\xi) = - i \sqrt{1 + \xi^2} \arctan \frac{1}{\xi} + \frac{i\xi}{\sqrt{1 + \xi^2}},$$

$$Q_{20}(i\xi) = i \frac{3}{2} \left(\xi^2 + \frac{1}{3}\right) \arctan \frac{1}{\xi} - i \frac{3}{2} \xi,$$

$$Q_{22}(i\xi) = - i \frac{\sqrt{3}}{2} (1 + \xi^2) \arctan \frac{1}{\xi} - i \frac{\sqrt{3}}{2} \xi + i \sqrt{\frac{1}{12}} \frac{\xi}{1 + \xi^2},$$

$$Q_{30}(i\xi) = - \frac{5}{2} \left(\xi^3 + \frac{3}{5} \xi\right) \arctan \frac{1}{\xi} + \frac{5}{2} \left(\xi^2 + \frac{1}{3}\right) - \frac{1}{6},$$

$$Q_{40}(i\xi) = i \frac{35}{8} \left(\xi^4 + \frac{6}{7} \xi^2 + \frac{3}{35}\right) \arctan \frac{1}{\xi} + i \frac{35}{8} \left(\xi^3 + \frac{3}{5} \xi\right) - i \frac{1}{3} \xi.$$

In den Entwicklungen treten die imaginären Kugelfunktionen mit imaginären Koeffizienten auf. Die Produkte der Kugelfunktionen mit ihren Koeffizienten sind reell.

**56. Reihenentwicklungen.** α) *Ausgangsentwicklungen.*

*Binomische Reihe:*

$$(1 + \xi)^n = 1 + n\xi + \frac{n(n-1)}{1 \cdot 2} \xi^2 + \frac{n(n-1)(n-2)}{1 \cdot 2 \cdot 3} \xi^3 + \cdots \qquad (|\xi| < 1).$$

*Reihe des natürlichen Logarithmus:*

$$\frac{1}{2} \ln \frac{1 + \xi}{1 - \xi} = \xi + \frac{1}{3} \xi^3 + \frac{1}{5} \xi^5 + \cdots \qquad (|\xi| < 1).$$

*Reihe für arc tan ξ:*

$$\arctan \xi = \xi - \frac{1}{3} \xi^3 + \frac{1}{5} \xi^5 \mp \cdots \qquad (|\xi| \leq 1).$$

*Reihe für Ar Sin ξ:*

$$\text{Ar Sin } \xi = \xi - \frac{1}{2} \cdot \frac{1}{3} \xi^3 + \frac{1 \cdot 3}{2 \cdot 4} \cdot \frac{1}{5} \xi \frac{5}{+} \cdots \qquad (|\xi| \leq 1)$$

β) *Kugelfunktionen zweiter Art.*

$$Q_{20}(i/\varepsilon) \approx i \frac{2}{15} \varepsilon^3 \left(1 - \frac{6}{7} \varepsilon^2 + \frac{5}{7} \varepsilon^4\right),$$

$$Q_{22}(i/\varepsilon) \approx - i \sqrt{\frac{1}{3}} \cdot \frac{4}{5} \varepsilon^3 \left(1 - \frac{8}{7} \varepsilon^2\right),$$

$$Q_{40}(i/\varepsilon) \approx i \frac{8}{815} \varepsilon^5 \left(1 - \frac{15}{11} \varepsilon^2\right),$$

$$Q'_{20}(i/\varepsilon) \approx - \frac{2}{5} \varepsilon^4 \left(1 - \frac{10}{7} \varepsilon^2 + \frac{5}{3} \varepsilon^4\right),$$

$$\frac{Q'_{20}(i/\varepsilon)}{Q_{20}(i/\varepsilon)} \approx i \, 3\varepsilon \left(1 - \frac{4}{7} \varepsilon^2 + \frac{68}{147} \varepsilon^4\right),$$

$$Q'_{40}(i/\varepsilon) \approx \frac{8}{63} \varepsilon^6 \left(1 - \frac{21}{11} \varepsilon^2\right).$$

Solange es — wie üblich — nur auf die Größenordnungen bis herab zu $\alpha^2$ ankommt, werden in den Klammern die letzten Glieder weggelassen. Sie sind hier nur zur Beurteilung der Vernachlässigungen beigefügt.

Bei der Erde ist $\varepsilon^2 \approx \dfrac{1}{150}$, $\alpha \approx \dfrac{1}{300}$.

$\gamma$) *Reihenentwicklungen bei Koordinaten-Umformungen.* Hier werden alle Glieder gebraucht, wenn es noch auf die Größenanordnung $\alpha^2$ ankommt.

*Abkürzungen*[1]    $\varepsilon = \dfrac{e}{c}$,    $(\varepsilon) = \dfrac{e}{\varrho}$,    $\eta = \dfrac{e}{l}$.

$$c \approx a\left(1 - \frac{1}{2}\varepsilon^2 + \frac{3}{8}\varepsilon^4\right), \quad c^2 \approx a^2(1 - \varepsilon^2 + \varepsilon^4), \quad \frac{1}{c} \approx \frac{1}{a}\left(1 + \frac{1}{2}\varepsilon^2 - \frac{1}{8}\varepsilon^4\right)$$

$$\frac{1}{c^2} = \frac{1}{a^2}(1 + \varepsilon^2), \quad \frac{1}{c^3} \approx \frac{1}{a^3}\left(1 + \frac{3}{2}\varepsilon^2\right),$$

$$\alpha = \frac{a-c}{a} \approx \frac{1}{2}\varepsilon^2 - \frac{3}{8}\varepsilon^4, \quad \alpha^2 \approx \frac{1}{4}\varepsilon^4,$$

$$\varepsilon^2 \approx 2\alpha + 3\alpha^2, \quad \varepsilon^2 \approx 4\alpha^2;$$

$$\arctan(\varepsilon) \approx \eta\left(1 + \frac{1}{6}\eta^2 - \frac{1}{2}\eta^2\cos^2\vartheta + \frac{3}{40}\eta^4 - \frac{3}{4}\eta^4\cos^2\vartheta + \frac{7}{8}\eta^4\cos^4\vartheta\right),$$

$$Q_{20}\left(\frac{i}{(\varepsilon)}\right) \approx i\,\frac{2}{15}\eta^3\left(1 + \frac{9}{14}\eta^2 - \frac{3}{2}\eta^2\cos^2\vartheta\right), \quad Q_{40}\left(\frac{i}{(\varepsilon)}\right) \approx -i\,\frac{8}{315}\eta^5,$$

$$Q'_{20}\left(\frac{i}{(\varepsilon)}\right) \approx -\frac{2}{5}\eta^4\left(1 + \frac{4}{7}\eta^2 - 2\eta^2\cos^2\vartheta\right), \quad Q'_{40}\left(\frac{i}{(\varepsilon)}\right) \approx \frac{8}{63}\eta^6,$$

$$\cos^2\overline{\vartheta} \approx \cos^2\vartheta \cdot (1 + (\varepsilon)^2 - (\varepsilon)^2\cos^2\vartheta - (\varepsilon)^4\cos^2\vartheta + (\varepsilon)^4\cos^4\vartheta),$$

$$\cos^2\overline{\vartheta} \approx \cos^2\vartheta \cdot (1 + \eta^2 - \eta^2\cos^2\vartheta),$$

$$\sin^2\beta \approx \cos^2\overline{\vartheta} \cdot (1 + (\varepsilon)^2 - (\varepsilon)^2\cos^2\overline{\vartheta} - (\varepsilon)^4\cos^2\overline{\vartheta} + (\varepsilon)^4\cos^4\overline{\vartheta}$$

$$\sin^2\beta \approx \cos^2\vartheta \cdot (1 + 2(\varepsilon)^2 - 2(\varepsilon)^2\cos^2\vartheta + (\varepsilon)^4 - 5(\varepsilon)^4\cos^2\vartheta + 4(\varepsilon)^4\cos^4\vartheta)$$

$$P_{20}(\cos\overline{\vartheta}) \approx \frac{3}{2}\left(\cos^2\vartheta - \frac{1}{3}\right) + \frac{3}{2}\eta^2\cos^2\vartheta - \frac{3}{2}\eta^2\cos^4\vartheta$$
$$= P_{20}(\cos\vartheta) + \frac{1}{5}\eta^2 + \frac{4}{7}\eta^2 P_{20}(\cos\vartheta) - \frac{12}{35}\eta^2 P_{40}(\cos\vartheta),$$

$$P_{40}(\cos\overline{\vartheta}) \approx P_{40}(\cos\vartheta) \approx P_{40}(\sin\beta),$$

$$P_{20}(\sin\beta) \approx \frac{3}{2}\left(\cos^2\overline{\vartheta} - \frac{1}{3}\right) + \frac{3}{2}(\varepsilon)^2\cos^2\overline{\vartheta} - \frac{3}{2}(\varepsilon)^2\cos^4\overline{\vartheta}$$
$$= P_{20}(\cos\overline{\vartheta}) + \frac{1}{5}(\varepsilon)^2 + \frac{4}{7}(\varepsilon)^2 P_{20}(\cos\overline{\vartheta}) - \frac{12}{35}(\varepsilon)^2 P_{40}(\cos\overline{\vartheta}),$$

$$P_{20}(\sin\beta) \approx \frac{3}{2}\left(\cos^2\vartheta - \frac{1}{3}\right) + 3(\varepsilon)^2\cos^2\vartheta - 3(\varepsilon)^2\cos^4\vartheta$$
$$= P_{20}(\cos\vartheta) + \frac{2}{5}(\varepsilon)^2 + \frac{8}{7}(\varepsilon)^2 P_{20}(\cos\vartheta) - \frac{24}{35}(\varepsilon)^2 P_{40}(\cos\vartheta),$$

$$Q_{20}\left(\frac{i}{(\varepsilon)}\right) \cdot P_{20}(\cos\overline{\vartheta}) \approx i\,\frac{2}{15}\eta^3\left(P_{20}(\cos\vartheta) - \frac{6}{7}\eta^2 P_{40}(\cos\vartheta)\right),$$

$$Q_{40}\left(\frac{i}{(\varepsilon)}\right) \cdot P_{40}(\cos\overline{\vartheta}) \approx i\,\frac{8}{315}\eta^5 P_{40}(\cos\vartheta).$$

$$(\varepsilon) \approx \eta\left(1 + \frac{1}{2}\eta^2 - \frac{1}{2}\eta^2\cos^2\vartheta + \frac{3}{8}\eta^4 - \frac{3}{4}\eta^4\cos^2\vartheta + \frac{7}{8}\eta^4\cos^4\vartheta\right),$$

$$(\varepsilon)^2 \approx \eta^2(1 + \eta^2 - \eta^2\cos^2\vartheta).$$

$$l \approx \varrho\left(1 + \frac{1}{8}(\varepsilon)^4 - \frac{1}{4}(\varepsilon)^2 P_{20}(\cos\overline{\vartheta}) - \frac{1}{15}(\varepsilon)^4 + \frac{2}{21}(\varepsilon)^4 P_{20}(\cos\overline{\vartheta}) - \frac{1}{35}(\varepsilon)^4 P_{40}(\cos\overline{\vartheta})\right),$$

$$_E l \approx a\left(1 - \frac{1}{2}\varepsilon^2\cos^2\vartheta + \frac{3}{8}\varepsilon^4\cos^4\vartheta\right),$$

$$_E l \approx a\left(1 - \frac{1}{2}\eta^2\cos^2\vartheta - \frac{1}{2}\eta^4\cos^2\vartheta + \frac{7}{8}\eta^4\cos^4\vartheta\right),$$

$$_E l \approx a\left(1 - \left(\alpha + \frac{3}{2}\alpha^2\right)\cos^2\vartheta + \frac{3}{2}\alpha^2\cos^4\vartheta\right),$$

Der Index $_E$ vor einem Symbol zeigt an, daß es sich auf das Ellipsoid $\varrho = c$ bezieht.

---

[1] $\varepsilon$ ist konstant. $(\varepsilon)$ kann vom Polabstand abhängig sein, $\eta$ ist vom Polabstand abhängig. In vielen Fällen kann $\varepsilon$ für $(\varepsilon)$ gesetzt werden.

$$\frac{1}{\varrho^2+e^2}\sqrt{\frac{\varrho^2+e^2}{\varrho^2+e^2\cos^2\overline{\vartheta}}}\approx\frac{1}{l^2}\left(1+\frac{1}{2}\eta^2-\frac{3}{2}\eta^2\cos^2\vartheta+\frac{3}{8}\eta^4-\frac{13}{4}\eta^4\cos^2\vartheta+\frac{31}{8}\eta^4\cos^4\vartheta\right),$$

$$\varrho\sqrt{\frac{\varrho^2+e^2}{\varrho^2+e^2\cos^2\overline{\vartheta}}}\approx l,$$

$$\frac{1}{e^4}Q_{20}'\left(\frac{i\varrho}{e}\right)\cdot P_{20}(\cos\overline{\vartheta})\cdot\sqrt{\frac{\varrho^2+e^2}{\varrho^2+e^2\cos^2\overline{\vartheta}}}$$

$$\approx\frac{1}{5}\frac{1}{l^4}\cdot\left(1-3\cos^2\vartheta+\frac{15}{14}\eta^2-\frac{61}{7}\eta^2\cos^2\vartheta+\frac{21}{2}\eta^2\cos^4\vartheta\right)$$

$$\varrho\,P_{20}(\cos\overline{\vartheta})\cdot\sqrt{\frac{\varrho^2+e^2}{\varrho^2+e^2\cos^2\overline{\vartheta}}}\approx\frac{1}{2}l\left((3\cos^2\vartheta-1)+3\eta^2\cos^2\vartheta-3\eta^2\cos^4\vartheta\right).$$

Mit $r(\overline{\vartheta})=c+h(\overline{\vartheta})$ gilt:

$$l\approx c\left(1+\frac{1}{3}\varepsilon^2-\frac{1}{3}\varepsilon^2P_{20}(\cos\overline{\vartheta})-\frac{1}{15}\varepsilon^4+\frac{2}{21}\varepsilon^4P_{20}(\cos\overline{\vartheta})-\frac{1}{35}P_{40}(\cos\overline{\vartheta})+\frac{h}{c}\right)$$

$$fM\frac{1}{e}\arctan\frac{e}{r}\approx\frac{fM}{c}\left(1-\frac{1}{3}\varepsilon^2+\frac{1}{5}\varepsilon^4-\frac{h}{c}\right),$$

$$\frac{fM}{r^2+e^2}\sqrt{\frac{r^2+e^2}{r^2+e^2\cos^2\overline{\vartheta}}}$$

$$\approx\frac{fM}{c^2}\left\{1-\frac{2}{3}\varepsilon^2-\frac{1}{3}\varepsilon^2P_{20}(\cos\overline{\vartheta})+\frac{8}{15}\varepsilon^4+\frac{8}{21}\varepsilon^4P_{20}(\cos\overline{\vartheta})+\frac{3}{35}\varepsilon^4P_{40}(\cos\overline{\vartheta})-2\frac{h}{c}\right\}.$$

$$\frac{2}{3}\omega^2c\sqrt{\frac{c^2+e^2}{c^2+e^2\cos^2\overline{\vartheta}}}\approx\frac{2}{3}\omega^2c\left(1+\frac{1}{3}\varepsilon^2-\frac{1}{3}\varepsilon^2P_{20}(\cos\overline{\vartheta})\right),$$

$$P_{20}(\cos\overline{\vartheta})\cdot\sqrt{\frac{c^2+e^2}{c^2+e^2\cos^2\overline{\vartheta}}}\approx P_{20}(\cos\overline{\vartheta})-\frac{1}{15}\varepsilon^2+\frac{5}{21}\varepsilon^2P_{20}(\cos\overline{\vartheta})-\frac{6}{35}\varepsilon^2P_{40}(\cos\overline{\vartheta}).$$

**57. Die Kugelfunktionen und ihre Differentialquotienten.** Der Ausdruck

$$P_{n\nu}(\xi)=\sqrt{\frac{\varkappa_\nu(n-\nu)!}{(n+\nu)!}}\frac{1\cdot3\ldots(2n-1)}{(n-\nu)!}\sqrt{1-\xi^2}^\nu(\xi^{n-\nu}-p_2\xi^{n-\nu-2}+p_4\xi^{n-\nu-4}\mp\cdots),$$

$$\varkappa_0=1,\quad\varkappa_1=\varkappa_2=\varkappa_3=\cdots=\varkappa_n=2,\quad n=0,1,2,\ldots,\infty;\quad\nu=0,1,2,\ldots,n,$$

$$p_2=\frac{(n-\nu)(n-\nu-1)}{2\cdot(2n-1)},$$

$$\frac{p_{2l}}{p_{2l-2}}=\begin{cases}\dfrac{(n-\nu-(2l-2))(n-\nu-(2l-1))}{2l\cdot(2n-(2l-1))}, & \text{wenn}\quad n-\nu>2l-1,\\[2mm]0, & \text{wenn}\quad n-\nu\le2l-1\end{cases}$$

gilt für jedes $\xi$. Die Reihe

$$Q_{nr}(\xi)=\sqrt{\frac{\varkappa_\nu(n-\nu)!}{(n+\nu)!}}\frac{(n+\nu)!}{1\cdot3\ldots(2n+1)}\sqrt{1-\xi^2}^\nu\left(\frac{1}{\xi^{n+\nu+1}}+\frac{q_2}{\xi^{n+\nu+3}}+\frac{q_4}{\xi^{n+\nu+5}}+\cdots\right),$$

$$q_2=\frac{(n+\nu+1)(n+\nu+2)}{2\cdot(2n+3)},$$

$$\frac{q_{2l}}{q_{2l-2}}=\frac{(n+\nu+(2l-1))(n+\nu+2l)}{2l\cdot(2n+(2l+1))}$$

konvergiert absolut, wenn $|\xi|>1$[1]. Differenziert man nach $\xi$ und faßt die Glieder in geeigneter Weise zusammen, so erhält man ohne Vernachlässigung:

$$P_{\nu n}'(\xi)=\frac{n}{\xi}P_{n\nu}(\xi)\cdot\left\{1+\frac{1}{n}\left[-\frac{\nu}{1-\xi^2}+\frac{2p_2}{\xi^2}\left(1-\frac{p_2}{\xi^2}+\frac{p_4}{\xi^4}\mp\cdots\right)^{-1}\times\right.\right.$$

$$\left.\left.\times\left(1-\frac{2p_4}{p_2\xi^2}+\frac{3p_6}{p_2\xi^4}\mp\cdots\right)\right]\right\},$$

---

[1] [*26*, II], S. 54 und 55. Es handelt sich um hypergeometrische Reihen [*24*, I], [*25*].

$$Q'_{n\nu}(\xi) = -\frac{n+1}{\xi}\, Q_{n\nu}(\xi) \cdot \left\{ 1 + \frac{1}{n+1}\left[\frac{\nu}{1-\xi^2} + \frac{2q_2}{\xi^2}\left(1 + \frac{q_2}{\xi^2} + \frac{q_4}{\xi^4} + \cdots\right)^{-1} \times \right.\right.$$
$$\left.\left. \times \left(1 + \frac{2q_4}{q_2\xi^2} + \frac{3q_6}{q_2\xi^4} + \cdots\right)\right]\right\}.$$

Mit $\xi = i/\varepsilon$ ergibt sich

$$P_{n\nu}\left(\frac{i}{\varepsilon}\right) = i^{n-\nu}\sqrt{\frac{\varkappa_\nu(n-\nu)!}{(n+\nu)!}}\,\frac{1\cdot 3\cdots(2n-1)}{(n-\nu)!}\,\frac{\sqrt{1+\varepsilon^2}^\nu}{\varepsilon^n}(1 + p_2\varepsilon^2 + p_4\varepsilon^4 + \cdots),$$

$$Q_{n\nu}\left(\frac{i}{\varepsilon}\right) = i^{-(n+\nu+1)}\sqrt{\frac{\varkappa_\nu(n-\nu)!}{(n+\nu)!}}\,\frac{(n+\nu)!}{1\cdot 3\ldots(2n+1)}\,\sqrt{1+\varepsilon^2}^\nu\,\varepsilon^{n+1}(1 - q_2\varepsilon^2 + q_4\varepsilon^4 \mp \cdots),$$

$$P'_{n\nu}\left(\frac{i}{\varepsilon}\right) = -i\,n\,\varepsilon\, P_{n\nu}\left(\frac{i}{\varepsilon}\right)\cdot\left(1 - \frac{\nu}{n}\frac{\varepsilon^2}{1+\varepsilon^2} - \frac{2p_2\varepsilon^2}{n}\frac{{}_pZ}{{}_pN}\right),$$

$$Q'_{n\nu}\left(\frac{i}{\varepsilon}\right) = i\,(n+1)\,\varepsilon\, Q_{n\nu}\left(\frac{i}{\varepsilon}\right)\cdot\left(1 + \frac{\nu}{n+1}\frac{\varepsilon^2}{1+\varepsilon^2} - \frac{2q_2\varepsilon^2}{n+1}\frac{{}_qZ}{{}_pN}\right),$$

wobei

$${}_pZ = 1 + \frac{2p_4}{p_2}\varepsilon^2 + \frac{3p_6}{p_2}\varepsilon^4 + \cdots, \qquad {}_pN = 1 + p_2\varepsilon^2 + p_4\varepsilon^4 + \cdots,$$

$${}_qZ = 1 - \frac{2q_4}{q_2}\varepsilon^2 + \frac{3q_6}{q_2}\varepsilon^4 \mp \cdots, \qquad {}_qN = 1 - q_2\varepsilon^2 + q_4\varepsilon^4 \mp \cdots.$$

Zur Abschätzung von ${}_pZ$ hat man

$$\frac{p_{2l}}{p_{2l-2}}\begin{cases} < \dfrac{1}{2l}(n-\nu-(2l-2))\dfrac{2n-\nu-(2l-1)}{2n-(2l-1)} < \dfrac{n-\nu}{2l}, & \text{wenn } n-\nu > 2l-1, \\[2mm] = 0, & \text{wenn } n-\nu \leq 2l-1, \end{cases}$$

und findet hiermit

$$1 \leq {}_pZ = \sum_{l=1}^{\infty}\frac{l\,p_{2l}}{p_2}\varepsilon^{2l-2} < \sum_{l=1}^{\infty}\frac{\left(\dfrac{n-\nu}{2}\varepsilon^2\right)^{l-1}}{(l-1)!} = e^{\frac{n-\nu}{2}\varepsilon^2}.$$

Mit ${}_pN \geq 1$ gilt sodann

$$1 \leq \frac{{}_pZ}{{}_pN} < e^{\frac{n-\nu}{2}\varepsilon^2}.$$

Eine ähnliche Abschätzung für ${}_qZ$ führt auf eine $e$-Funktion mit $n+\nu$ im Exponent. Für die alternierende Reihe ist diese Abschätzung zu grob. Man wird daher ${}_qZ$ und ${}_qN$ nach den angegebenen Formeln berechnen. Die Zahlenergebnisse der Abschätzungen sind in Tabelle 6 eingetragen. Sie zeigen unmittelbar, daß man nur relative Vernachlässigungen von der Größenordnung $\varepsilon^2$ einführt, wenn

$$P'_{n\nu}\left(\frac{i}{\varepsilon}\right) = -i\,n\,\varepsilon\, P_{n\nu}\left(\frac{i}{\varepsilon}\right), \qquad Q'_{n\nu}\left(\frac{i}{\varepsilon}\right) = i\,(n+1)\,\varepsilon\, Q_{n\nu}\left(\frac{i}{\varepsilon}\right)$$

gesetzt wird. Mit asymptotischen Ausdrücken[1] weist man nach, daß diese Beziehungen auch bei sehr großem $n$ gültig sind, wenn $\varepsilon^2$ sehr klein ist.

**58. Das Additionstheorem der normierten Kugelfunktionen erster Art.** Es sei

$$\cos\gamma = \cos\vartheta_1\cdot\cos\vartheta_2 + \sin\vartheta_1\cdot\sin\vartheta_2\cdot\cos(\lambda_2 - \lambda_1),$$

dann ist

$$P_n(\cos\gamma) = \sum_{\nu=0}^{n}\cos\nu(\lambda_2 - \lambda_1)\cdot P_{n\nu}(\cos\vartheta_1)\cdot P_{n\nu}(\cos\vartheta_2).$$

---

[1] Zum Beispiel [29], S. 117.

Tabelle 6. *Zur Abschätzung von* $_pZ/_pN$ *und* $_qZ/_qN$.

| $n$ | $\nu$ | $e^{\frac{n-\nu}{2}\varepsilon^2}$ | $-\frac{\nu}{n}-\frac{2p_1}{n}e^{\frac{n-\nu}{2}\varepsilon^2}$ | $\frac{_qZ}{_qN}$ | $\frac{\nu}{n+1}-\frac{2q_1}{n+1}\frac{_qZ}{_qN}$ |
|---|---|---|---|---|---|
| 2 | 0 | 1,007 | $-0,336$ | 0,9946 | 0,568 |
| 2 | 2 | 1 | $-1$ | 0,9935 | $-0,753$ |
| 16 | 0 | 1,056 | $-0,511$ | 0,9949 | $-0,512$ |
| 16 | 16 | 1 | $-1$ | 0,9933 | $-0,932$ |
| 100 | 0 | 1,403 | $-0,698$ | 0,9949 | $-0,500$ |
| 100 | 100 | 1 | $-1$ | 0,9933 | $-0,977$ |
| 400 | 0 | 3,871 | $-1,933$ | 0,9950 | $-0,498$ |
| 400 | 400 | 1 | $-1$ | 0,9933 | $-0,984$ |

$$(\varepsilon^2 = 0,006\,768\,2)$$

**59. Der Integralsatz der normierten Kugelfunktionen I. Art.** Es sei

$$X(\vartheta, \lambda) = \sum_n X_n(\vartheta, \lambda)$$

$$Y(\vartheta, \lambda) = \sum_n Y_n(\vartheta, \lambda)$$

$$X_n = \sum_{\nu=0}^{n} (_cX_{n\nu} \cos \nu \lambda + _sX_{n\nu} \sin \nu \lambda) \, P_{n\nu}(\cos \vartheta)$$

$$Y_n = \sum_{\nu=0}^{n} (_cY_{n\nu} \cos \nu \lambda + _sY_{n\nu} \sin \nu \lambda) \, P_{n\nu}(\cos \vartheta),$$

dann gilt

$$\int_{\vartheta'=0}^{\pi} \int_{\lambda'=0}^{2\pi} X_n(\vartheta', \lambda') \cdot Y_n(\vartheta', \lambda') \sin \vartheta' \, d\vartheta' = \frac{4\pi}{2n+1} \sum_{\nu=0}^{n} (_cX_{n\nu} \cdot _cY_{n\nu} + _sX_{n\nu} \cdot _sY_{n\nu}),$$

speziell

$$\int_{\vartheta'=0}^{\pi} \int_{\lambda'=0}^{2\pi} X_n(\vartheta', \lambda') \begin{Bmatrix} \cos \nu \lambda' \\ \sin \nu \lambda' \end{Bmatrix} P_{n\nu}(\cos \vartheta') \sin \vartheta' \, d\vartheta' = \frac{4\pi}{2n+1} \cdot \begin{Bmatrix} _cX_{n\nu} \\ _sX_{n\nu} \end{Bmatrix},$$

und es ist

$$\int_{\vartheta'=0}^{\pi} \int_{\lambda'=0}^{2\pi} X_{n\nu}(\vartheta', \lambda') \cdot Y_{m\mu}(\vartheta', \lambda') \sin \vartheta' \, d\vartheta' = 0, \qquad \text{wenn } m \neq n \text{ oder } \mu \neq \nu.$$

# Anhang IV. Zahlenwerte.

**60. Bewegung der Erde [22].**

1 Sterntag $= 86164,0906$ sec mittlere Sonnenzeit.

Rotationsperiode der Erde: 1 Sterntag $+ 0,0091$ sec.

Winkelgeschwindigkeit der Erdrotation: $\omega = 7,292\,115\,08 \cdot 10^{-5}$ sec$^{-1}$.

Entfernung des Sonnenmittelpunkts vom Erdmittelpunkt:

> Mittel:             149504000 km
> Maximum (Aphel):   152006000 km
> Minimum (Perihel): 147002000 km $(\pm 1^0/_{00})$.

Entfernung des Mondmittelpunkts vom Erdmittelpunkt:

> Mittel:   384400 km
> Maximum: 406740 km
> Minimum: 356410 km $(\pm 0,1^0/_{00})$.

Masse der Sonne:   332290 Erdmassen $= 1,9848 \cdot 10^{33}$ g $(\pm 2^0/_{00})$.

Masse des Mondes:   $^1/_{81,53}$ Erdmassen $= 7,326 \cdot 10^{25}$ g $(\pm 2^0/_{00})$.

Schiefe der Ekliptik: $\varepsilon = 23° 27' 8,26'' - 0,4684'' (t-1900)$   $t =$ Jahreszahl.

Neigung der Mondbahn gegen die Ekliptik: $5° 8' 40''$.
Periode des Mondknotenumlaufs 18,60 Jahre.

**61. Die Erd-Dimensionen nach Gradmessungen[1] [22].**

*Tabelle 7.*

| | $a$<br>m | $c$<br>m | $1/a$ |
|---|---|---|---|
| Bessel[2] 1841 . . . . . | 6 377 397,2 | 6 356 079,0 | 299,2 |
| Clarke[3] 1866 . . . . . | 6 378 206,4 | 6 356 583,0 | 295,0 |
| Clarke[3] 1880 . . . . | 6 378 249,1 | 6 356 515,0 | 293,5 |
| Bonsdorff 1888 . . . . | 6 378 444 | | 298,6 |
| Hayford 1906 . . . . . | 6 378 283 | | 297,8 |
| Helmert[2] 1907 . . . . | 6 378 200 | | 298,3 |
| Hayford 1909 . . . . . | 6 378 388,4 | 6 356 908,8 | 297,0 |
| Heiskanen 1926 . . . . | 6 378 397 | | 297<br>[angenommen] |
| Krassowski 1938 . . . | 6 378 245 | | 298,3 |
| Jeffreys[4] 1949 . . . . | 6 378 099<br>± 116 | | 297,10<br>± 36 |
| Ledersteger 1951 . . . | 6 378 298<br>± 34 | | 297<br>[angenommen] |
| Internationales Ellipsoid<br>1924 . . . . . . . | 6 378 388<br>[genau] | 6 356 911,95 | 297<br>[genau] |

**62. Internationales Ellipsoid [22].**

Definition: Äquatorradius $a = 6 378 388$ m (genau).

$$\text{Abplattung: } a = \frac{a-c}{a} = \frac{1}{297} \text{ (genau).}$$

Polradius: $c = 6 356 911,946 128$ m,

$a - c = 21 476,053 872$ m,

$$\varepsilon^2 = \frac{a^2 - c^2}{c^2} = 0,006 768 170 197 224$$

Äquatorquadrant: 　　10 019 148,441 m,

Meridianquadrant: 　　10 002 288,299 m,

Äquatorgrad: 　　　　111 323,872 m,

mittl. Meridiangrad: 　111 136,537 m

$(= \frac{1}{90}$ Meridianquadrant$)$.

1 Grad auf der volumengleichen Kugel: 　111 198,789 m,

mittl. Radius: 　$\frac{a+a+c}{3} = 6 371 229,315$ m.

Radius der oberflächengleichen Kugel: 6 371 227,709 m,

Radius der volumengleichen Kugel: $R = 6 371 221,266$ m,

mittlere Breite: (Tabelle 8)

1. $P_{20}(\sin \varphi) = 0$:　$\varphi = 35° 15' 51,8''$,

2. Radiusvektor $= R$ für geogr. $\beta = 35° 24' 4,0''$,

geozentr. $\varphi = 35° 13' 7,8''$,

reduziert: $\frac{\pi}{2} - \bar{\vartheta} = 35° 18' 35,7''$.

---

[1] W. Heiskanen: Veröff. finn. Geodät. Inst. 1951, Nr. 39.

[2] „legale" Meter (1,00001329 intern. m).

[3] 1 m = 1,00001444 intern. m.

[4] Beruht auf geodätischen und astronomischen Messungen.

Oberfläche: 510100933,5 km²,

Volumen: 1083319780000 km³.

Tabelle 8. *Mittlere Breite und mittlere Schwere auf dem internationalen Ellipsoid.*

| Definition | Geographische Breite $\beta$ | Geozentrische Breite $\varphi = 90° - \vartheta$ | Reduzierte Breite $\psi = 90° - \bar{\vartheta}$ | Schwere $\gamma$ Gal |
|---|---|---|---|---|
| $\gamma_m = \dfrac{1}{O} \displaystyle\int\!\!\int \gamma \, df$ | $\beta_m = 35°21'41''$ | $\varphi_m = 35°10'45''$ | $\psi_m = 35°16'13''$ | $\gamma_m = 979,776_2$ |
| $P_{20}(\cos\vartheta) = 0$ | $\beta_1 = 35°26'48''$ | $\varphi_1 = 35°15'52''$ | $\psi_1 = 35°21'20''$ | $\gamma_1 = 979,783_5$ |
| $P_{20}(\cos\bar{\vartheta}) = 0$ | $\beta_2 = 35°21'20''$ | $\varphi_2 = 35°10'24''$ | $\psi_2 = 35°15'52''$ | $\gamma_2 = 979,775_7$ |
| $P_{20}(\sin\beta) = 0$ | $\beta_3 = 35°15'52''$ | $\varphi_3 = 35°04'57''$ | $\psi_3 = 35°10'24''$ | $\gamma_3 = 979,768_0$ |
| $l = R$ | $\beta_m^{(R)} = 35°24'04''$ | $\varphi_m^{(R)} = 35°13'08''$ | $\psi_m^{(R)} = 35°18'36''$ | $\gamma_m^{(R)} = 979,779_8$ |

Tabelle 9. *Die Äquatorellipse des dreiachsigen Erdellipsoids.*

$c_{22} =$ Koeffizient von $\sin^2\vartheta \cos 2(\lambda - \lambda_{22})$; $\lambda_{22} =$ Meridian der großen Äquatorachse; $a, b =$ großer und kleiner Äquatorradius; $A, B =$ Hauptträgheitsmomente der Erde, bezogen auf die Äquatorachsen $[B > A]$.

| | $c_{22}$ mgal | $\lambda_{22}$ | $a - b$ m | $B - A$ gcm² |
|---|---|---|---|---|
| HELMERT (1915) . . . . . . . | 18 | 17° W | 230 | $5,8 \cdot 10^{40}$ |
| BERROTH (1916). . . . . . . . | 16 | 10° W | 200 | 5,2 |
| HEISKANEN (1924) . . . . . . | 26 | 18° E | 340 | 8,8 |
| HEISKANEN (1928) . . . . . | 19 | 0° | 240 | 6,1 |
| HEISKANEN (1938) . . . . . | 27 | 25° W | 350 | 8,9 |
| JEFFREYS (1942) . . . . . . . | 12 | $\approx 0°$ | 160 | 4,0 |
| K. JUNG (1943) aus Material von ZHURAVLEV (1940) . . . | 28 | 25° W | 370 | 9,3 |
| K. JUNG (1943) aus Material von LUOMA (1941) . . . . . | 15 | 10° W | 200 | 5,0 |
| NISKANEN (1945) . . . . . . . | 22 | 4° W | 290 | 7,4 |
| K. JUNG (1955) aus Material von TANNI (1948) . . . . . | 10 | 10° W | 130 | 3,2 |

Tabelle 10. *Niveauflächen[1] und volumengleiche Koordinaten-Ellipsoide im Erdinnern.*

$a_N, c_N$ Äquator- und Polradien der Niveauflächen; $\alpha_N$ Abplattungen der Niveauflächen; $a, c$ Äquator- und Polradien der Koordinaten-Ellipsoide; $\alpha$ Abplattungen der Koordinaten-Ellipsoide; $R$ Radien der volumengleichen Kugeln. $T$ („Tiefe") $= 6371$ km $- R$.

| $T$ km | $R$ km | $R - c_N$ km | $a_N - R$ km | $1/\alpha_N$ | $R - c$ km | $a - R$ km | $1/\alpha$ |
|---|---|---|---|---|---|---|---|
| 0 | 6371 | 14,29 | 7,16 | 297 | 14,29 | 7,16 | 297 |
| 33 | 6338 | 14,18 | 7,10 | 298 | 14,37 | 7,20 | 294 |
| 80 | 6291 | 14,02 | 7,03 | 299 | 14,48 | 7,25 | 290 |
| 200 | 6171 | 13,61 | 6,82 | 302 | 14,76 | 7,39 | 279 |
| 600 | 5771 | 12,27 | 6,14 | 314 | 15,78 | 7,91 | 244 |
| 1000 | 5371 | 10,96 | 5,49 | 327 | 16,96 | 8,50 | 211 |
| 2000 | 4371 | 7,97 | 3,99 | 366 | 20,83 | 10,46 | 140 |
| 3000 | 3371 | 5,63 | 2,82 | 399 | 27,01 | 13,59 | 83,4 |
| 4000 | 2371 | 3,78 | 1,89 | 419 | 38,41 | 19,44 | 41,3 |
| 4800 | 1571 | 2,32 | 1,16 | 451 | 57,94 | 29,80 | 18,2 |
| 5400 | 971 | 1,38 | 0,69 | 468 | 93,47 | 50,40 | 7,10 |
| 6000 | 371 | 0,52 | 0,26 | 476 | 201,82 | 178,39 | 1,44 |
| 6200 | 171 | 0,24 | 0,12 | 477 | 152,72 | 351,01 | 1,04 |
| 6371 | 0 | 0 | 0 | 477 | 0 | 522,69 | 1 |

[1] Entspricht der Dichteverteilung nach BULLEN (strichpunktierte Kurve in Fig. 39). Nach W. D. LAMBERT und F. W. DARLING: Deutsche Übersetzung aus "Internal constitution of the earth". Deutsche Geodätische Kommission, Reihe A, Nr. 8 (1953), Tabelle 6.

## 63. Die wichtigsten Schwereformeln und ihre Niveau-Sphäroide.

Tabelle 11. $a\omega^2 = 3.391704$ Gal.

| Namen | Reduktion | $\gamma_a$ Gal | b | $b' = \frac{1}{4}\,b_4$ | Herkunft von b | c | $\frac{1}{a}$ | $h_0$ m |
|---|---|---|---|---|---|---|---|---|
| Helmert 1884 | Kondensation in der Tiefe $Ra = 21$ km | 978,00 | 0,005310 | — | — | 0,0034680 | 299,0 | +12,2 |
| Helmert 1901 | Freiluftreduktion[1] | 978,030 ±11 | 0,005302 ±12 | 0,000007 | Wiechert-Darwin | 0,0034679 | 298,1 | − 2,4 |
| Helmert 1901 | Freiluftreduktion[1] | 978,028 ±11 | 0,005300 ±13 | 0,000002 ±13 | Ausgleichung | 0,0034679 | 298,1 | + 8,2 |
| Helmert 1901 | Freiluftreduktion[1] | 978,028 | 0,005300 | 0 | — | 0,0034679 | 298,2 | +12,5 |
| Bowie 1917 | isostatisch (Pratt-Hayford) | 978,039 ±4 | 0,005294 ±12 | 0,000007 | Wiechert-Darwin | 0,0034679 | 297,4 | − 2,3 |
| Heiskanen 1924 | isostatisch (Airy-Heiskanen) | 978,048 ±3 | 0,005293 ±6 | 0,000007 | Wiechert-Darwin | 0,0034678 | 297,4 | − 2,3 |
| Heiskanen 1928 | isostatisch | 978,049 ±3 | 0,005293 ±4 | 0,000007 | Wiechert-Darwin | 0,0034678 | 297,4 | − 2,3 |
| Heiskanen 1928 | isostatisch | 978,049 ±1 | 0,005289 [aus $\alpha = 1/297$ berechnet] | 0,000007 | Wiechert-Darwin | 0,0034678 | (297,0) | − 2,3 |
| Heiskanen 1938 | istostatisch | 978,0451 | 0,0053027 | 0,0000059 | internationales Ellipsoid | 0,0034678 | 298,2 | − 0,1 |
| Niskanen 1945 | isostatisch | 978,0468 ±12 | 0,0052978 ±27 | 0,0000059 | internationales Ellipsoid | 0,0034678 | 297,8 | 0 |
| Jeffreys 1949 | Freiluftreduktion[2] | 978,0544[2] ±2 | 0,0052790[2] | 0,0000059[2] | internationales Ellipsoid | 0,0034678 | 296,2 | 0 |
| Schütte (Zhuravlev) 1950 | Freiluftreduktion | 978,0698[3] ±57 | 0,0052639[3] ±113 | 0,0000059 | internationales Ellipsoid | 0,0034677 | 294,9 | − 0,1 |
| Schütte (Zhuravlev) 1950 | Freiluftreduktion | 978,0520[4] ±33 | 0,0052827[4] ±60 | 0,0000059 | internationales Ellipsoid | 0,0034678 | 296,5 | − 0,6 |
| Darwin 1900 | — | — | 0,0052910 [berechnet aus Massenverteilung] | 0,0000067 [berechnet aus Massenverteilung] | — | 0,0034678 [angenommen] | — | − 2,9 |
| Darwin 1900 | — | — | 0,0052910 [berechnet aus Massenverteilung] | −0,0000017 [verb. von Ansel] | — | 0,0034678 [angenommen] | — | +16,2 |

[1] = Kondensation ins Meeresniveau.   [2] Berechnet aus $B_{00} = 979{,}7725$; $B_{20} = 3{,}4399$; $B_{40} = 0{,}0053$.   [3] „ohne Gewichte": die zwischen gleichabständigen Breitenkreisen gelegenen Zonen werden mit gleichem Gewicht in die Ausgleichung eingeführt.   [4] „mit Gewichten": die Breitenzonen erhalten Gewichte, die der jeweiligen Anzahl von Schwerestationen proportional sind.

**64. Internationale Schwereformel (Gal) [22].**

$$\gamma = 978{,}049000 \,(1 + 0{,}005\,288\,384 \, \sin^2 \beta$$
$$- 0{,}000\,005\,869 \, \sin^2 2\beta$$
$$- 0{,}000\,000\,032 \, \sin^2 \beta \, \sin^2 2\beta) \; [1]$$

$$= 980{,}632\,272 - 2{,}586\,146 \cos 2\beta + 0{,}002\,878 \cos 4\beta - 0{,}000\,004 \cos 6\beta$$

$$= 978{,}049000 \,(1 + 0{,}005\,264\,908 \, \sin^2 \beta$$
$$+ 0{,}000\,023\,348 \, \sin^4 \beta$$
$$+ 0{,}000\,000\,128 \, \sin^6 \beta)$$

$$= 979{,}770\,031 + 3{,}446\,000 \, P_{20}(\sin \beta)$$
$$+ 0{,}005\,259 \, P_{40}(\sin \beta)$$
$$+ 0{,}000\,009 \, P_{60}(\sin \beta).$$

Normale Schwere in 45° geogr. Breite: 980,629 39 Gal,

Arithm. Mittel von Äquator- und Polschwere: 980,635 85 Gal,

Mittl. Schwere (Integral über die Erdoberfläche dividiert durch die Erdoberfläche) 979,776 23 Gal [2],

Gravitation an der Oberfläche der nichtrotierenden, volumen- und massengleichen Kugel 982,036 82 Gal.

Tabelle 12.

*Die Normalschwere nach der internationalen Schwereformel und der normale Schweregradient.*

$$\gamma_{90°} - \gamma_{0°} = 5{,}1723 \text{ Gal.}$$

Anteil der Zentrifugalbeschleunigung an diesem Schwereunterschied: 3,3918 Gal.

| $\beta$ | $\gamma_{Gal}$ | $\varDelta$ | $\varDelta\varDelta$ | Gradient mgal/km | $\varDelta$ | $\varDelta\varDelta$ |
|---|---|---|---|---|---|---|
| 0° | 978,0490 | | 782 | 0 | | 0 |
| 5° | 978,0881 | 391 | 771 | 0,141 | 0,141 | − 0,004 |
| 10° | 978,2043 | 1162 | 735 | 0,278 | 0,137 | − 0,009 |
| 15° | 978,3940 | 1897 | 680 | 0,406 | 0,128 | − 0,012 |
| 20° | 978,6517 | 2577 | 600 | 0,522 | 0,116 | − 0,016 |
| 25° | 978,9694 | 3177 | 507 | 0,622 | 0,100 | − 0,018 |
| 30° | 979,3378 | 3684 | 394 | 0,704 | 0,082 | − 0,022 |
| 35° | 979,7456 | 4078 | 271 | 0,764 | 0,060 | − 0,023 |
| 40° | 980,1805 | 4349 | 140 | 0,801 | 0,037 | − 0,025 |
| 45° | 980,6294 | 4489 | 3 | 0,813 | 0,012 | − 0,025 |
| 50° | 981,0786 | 4492 | − 132 | 0,801 | − 0,012 | − 0,025 |
| 55° | 981,5146 | 4360 | − 267 | 0,764 | − 0,037 | − 0,023 |
| 60° | 981,9239 | 4093 | − 391 | 0,704 | − 0,060 | − 0,022 |
| 65° | 982,2941 | 3702 | − 504 | 0,622 | − 0,082 | − 0,018 |
| 70° | 982,6139 | 3198 | − 603 | 0,522 | − 0,100 | − 0,016 |
| 75° | 982,8734 | 2595 | − 682 | 0,406 | − 0,116 | − 0,012 |
| 80° | 983,0647 | 1913 | − 742 | 0,278 | − 0,128 | − 0,009 |
| 85° | 983,1818 | 1171 | − 776 | 0,141 | − 0,137 | − 0,004 |
| 90° | 983,2213 | 395 | − 790 | 0 | − 0,141 | − 0 |

**65. Masse und Trägheitsmomente der Erde [3] [22].**

$$f \cdot M = 3{,}986\,329 \cdot 10^{20} \text{ cm}^3 \text{ sec}^{-2} \qquad \text{rel. mittl. Fehler } 8 \cdot 10^{-5}.$$

---

[1] In [22] Druckfehler: im letzten Glied fehlt $\sin^2\beta$.

[2] Der abweichende Wert in [22] entstand durch Integration über die Erdkugel, während hier über das Erdellipsoid integriert wurde.

[3] Beste, den Beobachtungen angepaßte Werte A. BERROTH: Z. Geophys. **18**, 106 (1943).

Mit $f = 6{,}670 \cdot 10^{-8}\,\mathrm{cm^3\,g^{-1}\,sec^{-2}}$ ist

$$M = 5{,}977 \cdot 10^{27}\,\mathrm{g} \qquad \text{(Masse der Erde)},$$
$$\sigma_m = 5{,}517\,\mathrm{g \cdot cm^{-3}} \qquad \text{(mittl. Dichte der Erde)}.$$

*Hauptträgheitsmomente* $(B = A)$:

$$\frac{C - A}{M a^2} = \begin{cases} 0{,}001\,106 \pm 0{,}000\,01 & \text{allein aus astronomischen Beobachtungen} \\ 0{,}001\,092\,1 & \text{aus dem internationalen Ellipsoid berechnet,} \end{cases}$$

$$\frac{C}{C - A} = 305{,}8 \pm 0{,}8,$$

$$\frac{A}{C - A} = 304{,}8 \pm 0{,}8,$$

$$\frac{C}{M a^2} = \begin{cases} 0{,}3381 \pm 0{,}0037 & \text{allein aus astronomischen Beobachtungen} \\ 0{,}3339 & \text{aus } (C - A)/M a^2 = 0{,}001\,092\,1 \text{ berechnet,} \end{cases}$$

$$\frac{A}{M a^2} = \begin{cases} 0{,}3370 \pm 0{,}0037 & \text{allein aus astronomischen Beobachtungen} \\ 0{,}3328 & \text{aus } (C - A)/M a^2 = 0{,}001\,092\,1 \text{ berechnet.} \end{cases}$$

Für eine homogene Erde (McLAURINsches Ellipsoid) wäre $\dfrac{C}{M a^2} = \dfrac{2}{5}$.

$$\left.\begin{array}{l} C = 8{,}118 \cdot 10^{44}\,\mathrm{g\,cm^2} \\ A = 8{,}092 \cdot 10^{44}\,\mathrm{g\,cm^2} \\ C - A = 2{,}656 \cdot 10^{42}\,\mathrm{g\,cm^2} \end{array}\right\} \begin{array}{l} \text{dem internationalen Ellipsoid} \\ \text{angepaßte Werte.} \end{array}$$

# Anhang V.

## Verzeichnis häufig gebrauchter Bezeichnungen.

$W$　Potential der Schwere
$U$　Potential der Normalschwere
$V$　Gravitations-Potential
$Z$　Potential der Zentrifugalbeschleunigung
$T$　Potential der Schwereanomalie

$g$　Schwere-Vektor
$\gamma$　Vektor der Normalschwere
$\gamma_{\ddot{a}}$　Äquatorschwere $\left.\right\}$ $\mathfrak{b} = \dfrac{\gamma_p - \gamma_{\ddot{a}}}{\gamma_{\ddot{a}}}$
$\gamma_p$　Polschwere

$\varepsilon$　Lotabweichung
$\varepsilon_{(x)}, \varepsilon_{(y)}$　Komponenten der Lotabweichung
$\zeta$　Geoidundulation

$\omega$　Winkelbeschleunigung der Erdrotation

$\lambda$　Länge
$\beta$　geographische $\left.\right\}$
$\varphi$　geozentrische $\left.\right\}$ Breite
$\psi$　reduzierte
$\vartheta$　geozentrischer $\left.\right\}$ Polabstand (Komple-
$\bar{\vartheta}$　reduzierter $\left.\right\}$ ment der Breite)
$H$　Höhenunterschied
$h$　entsprechender Unterschied der Polradien der Koordinatenellipsoide
$a$　Äquatorradius der Erde

$c$　Polradius der Erde
$e$　halber Brennpunktsabstand der Meridianellipsen
$\alpha = \dfrac{a - c}{a}$　Abplattung der Erde
$\varepsilon = e/c$
$\varrho$　Polradius der Koordinatenellipsoide
$(\varepsilon) = e/\varrho$
$R$　Radius der Erdkugel
$l$　Abstand vom Erdmittelpunkt
$\eta = e/l$
$O$　Erdoberfläche
$df$　Flächenelement
$dv$　Volumenelement
$dm$　Massenelement
$\sigma$　Gesteinsdichte
$\sigma_m$　mittlere Dichte der Erde
$f$　Gravitationskonstante
$A \leqq B < C$　(Haupt)Trägheitsmomente der Erde
$M, M_{\delta}$　Masse der Erde
$M_{\odot}, M_{\mathbb{C}}$　Masse von Sonne und Mond
$\mathbf{M}$　Drehmoment der äußeren Kräfte
$\mathbf{S}$　Drehimpuls
$\Psi$　Präzessionswinkel,
$\dot{\Psi} = d\Psi/dt$　$(t = \text{Zeit})$

$\Omega, \Omega_\odot$ mittlere Winkelgeschwindigkeit der
    Erde in ihrer Bahn um die Sonne
$\Omega_{\mathbb{C}}$ mittlere Winkelgeschwindigkeit des
    Mondes in seiner Bahn um die Erde
$\varepsilon$   Schiefe der Ekliptik
$i$   Winkel zwischen Mondbahn und Ekliptik
$i = \sqrt{-1}$

$P_{n\nu}(\xi)$ LEGENDREsche Kugelfunktion erster
    Art
$Q_{n\nu}(\xi)$ LEGENDREsche Kugelfunktion zweiter
    Art

$$P'_{n\nu}(\xi) = \frac{d}{d\xi} P_{n\nu}(\xi)$$

$$Q'_{n\nu}(\xi) = \frac{d}{d\xi} Q_{n\nu}(\xi)$$

Vorübergehend werden einige der hier angeführten Symbole auch zur Bezeichnung anderer Größen gebraucht.

$\mathfrak{s}$ (große 5) durch *Ab*rundung entstanden; $\mathfrak{z}$ (kleine 5) durch *Auf*rundung entstanden.

## Literatur.

### Grundlegende und historisch wichtige Schriften.

[1] CLAIRAUT: Théorie de la figure de la terre, tirée des principes de l'hydrostatique. Paris 1743. Unveränderter Abdruck 1808. Deutsche Übersetzung: Ostwald's Klassiker Nr.189. Leipzig 1913.

[2] BOUGUER, P.: La figure de la terre déterminée par les observations de M. M. BOUGUER et DE LA CONDAMINE etc. Paris 1749.

[3] STOKES, C. G.: On the variation of gravity and the surface of the earth. Cambridge Phil. Trans. 8, 672 (1849).

[4] BRUNS, H.: Die Figur der Erde. Berlin 1878.

### Zusammenfassende Darstellungen.

[5] HELMERT, F. R.: Die mathematischen und physikalischen Theorien der Höheren Geodäsie. I. Teil: Die mathematischen Theorien. Leipzig 1880. II. Teil: Die physikalischen Theorien. Leipzig 1884. (Selbst nach über 70 Jahren nicht veraltet, soweit es nicht die Beobachtungsergebnisse betrifft. Eine unübertroffen gründliche Darstellung der mathematischen und potentialtheoretischen Beziehungen.)

[6] TISSERAND, F.: Traité de mécanique céleste. Tome II: Théorie de la figure des corps célestes et de leur mouvement de rotation. Paris 1891. (Gehört wie [5] auch heute noch zu den wichtigsten Darstellungen der Grundlagen).

[7] HELMERT, F. R.: Die Schwerkraft und die Massenverteilung der Erde. Enzykl. math. Wiss. 6, 1 B, 85—177 (1910).

[8] BERROTH, A.: Schweremessungen. In Handbuch der Physik, Bd. II, S. 416—486. 1926.

[9] SCHMEHL, H., u. K. JUNG: Figur, Schwere und Massenverteilung der Erde. In Handbuch der Experimentalphysik, Bd. XXV, Teil 2, S. 141—357. 1931.

[10] HOPFNER, F.: Figur der Erde, Dichte und Druck im Erdinnern. In Handbuch der Geophysik, Bd. I, S. 139—308. Berlin: Gebr. Borntraeger 1936.

[11] ANSEL, E. A.: Zur Theorie des irdischen Schwerefeldes. In Handbuch der Geophysik, Bd. I, S. 536—730. 1936.

[12] HEISKANEN, W.: Die Lotabweichungen. In Handbuch der Geophysik, Bd. I, S. 842 bis 877. 1936.

[13] JORDAN-EGGERT: Handbuch der Vermessungskunde, 9. Aufl. Bd. 3, 1. u. 2. Halbband. Stuttgart 1948.

[14] HOPFNER, F.: Physikalische Geodäsie. Leipzig 1933. (Zur Einführung in die potentialtheoretischen Beziehungen sehr geeignet, wenn man berücksichtigt, daß die Behauptung des Verfassers, die im Außenraum gültigen Reihenentwicklungen gälten auch in Innern der Massen, nicht zu recht besteht.)

[15] HOPFNER, F.: Grundlagen der Höheren Geodäsie. Wien 1949.

[16] BAESCHLIN, C. F.: Lehrbuch der Geodäsie. Zürich 1948. (Behandelt die Höhere Geodäsie in einer zur Einführung besonders geeigneten, ausführlichen Weise.)

[17] PERRIER, G.: Petite histoire de la géodésie. Comment l'homme a mesuré et pése la terre. Paris 1939. Deutsche Übersetzung von E. GIGAS: ,,Wie der Mensch die Erde gemessen und gewogen hat. Kurze Geschichte der Geodäsie". Bamberg 1950.

[18] PREY, A.: Anwendung der Methoden der Erdmessung auf geophysische Probleme. In A. PREY, C. MAINKA, E. TAMS, Einführung in die Geophysik, Bd. I. Berlin: Springer 1922.

[19] WAVRE, R.: Figures planétaires et géodésie. Paris 1932.

[20] LEJAY, P.: Développements modernes de la gravimétrie (Kap. V, VI, VII). Paris 1947.

[21] JEFFREYS, H.: The Earth (Kap. IV, V). Third edition. Cambridge 1952.

*Tabellenwerk.*

[22] LANDOLT-BÖRNSTEIN. Zahlenwerte und Funktionen aus Physik, Chemie, Astronomie, Geophysik, Technik, 6. Aufl. Bd. III: Astronomie und Geophysik. Berlin-Göttingen-Heidelberg 1952. (Insbesondere S. 45—52 und 256—270.)

*Zur Einführung in die mathematischen Entwicklungen* (neben allgemeinen Lehrbüchern der Mathematik).

[23] KELLOG, O. D.: Foundations of Potential Theory. Berlin 1929.
[24] HEINE, E.: Handbuch der Kugelfunktionen. 2 Bde. 2. Aufl. Berlin 1878 u. 1881.
[25] LENSE, J.: Kugelfunktionen. Leipzig 1954.
[26] WANGERIN, A.: Theorie des Potentials und der Kugelfunktionen. 2 Bde. Sammlung Schubert LVII u. LIX. Berlin u. Leipzig 1922 u. 1921.
[27] SCHMIDT, AD.: Tafeln der normierten Kugelfunktionen. Gotha 1935.
[28] PREY, A.: Darstellung der Höhen- und Tiefenverhältnisse der Erde durch eine Entwicklung nach Kugelfunktionen bis zur 16. Ordnung. Abh. Ges. Wiss. Göttingen, Math.-phys. Kl., N. F. 11, 1 (1922).
[29] JAHNKE, E., u. F. EMDE: Funktionentafeln mit Formeln und Kurven, 3. Aufl. Leipzig u. Berlin 1938.

# Sachverzeichnis.

## (Deutsch-Englisch.)

Bei gleicher Schreibweise in beiden Sprachen sind die Stichwörter nur einmal aufgeführt.

Ablenkung mit Kappe, *deflection, capped* 259.
Abplattung, *ellipticity* 93, 553, 560, 635.
—, Änderung mit der Höhe, *change with height* 560.
Actinium 300.
Airborne magnetometer 500.
AIRY 232.
Algonkian 292.
Alpen, *alps* 239, 259 , 268.
Alpha-Aktivität, *alpha activity* 310.
Alter der Erde, *age of the earth* 294, 333.
— —, alte physikalische Argumente, *early physical arguments* 291.
— —, Annahme bei der Berechnung, *assumptions made in calculating* 334.
— —, aus der Entwicklungsgeschichte, *from rate of evolution* 291.
— —, aus Sedimentation und Verwitterung, *from sedimentation and weathering* 291.
— —, Lord KELVIN, *Lord* KELVIN 291.
— der Kruste, *of the crust* 335.
— des Mantels, *of the mantle* 335.
— der Minerale, *of minerals* 288.
— der Welt, *of the universe* 294.
Altersbestimmung, radioaktive, *age determination, radioactive* 288.
— Zerfall von Uran und Thorium zu Helium und Blei, *decay of uranium and thorium to helium and lead* 319.
Amplituden-Regelung, automatische, *gain-control, automatic* 155.
ANDERSON, Verwerfungstheorie, ANDERSON, *faulting theory* 264.
Anomalien, isostatische, *anomalies, isostatic* 233, 578.
Anti-Ferromagnetismus, *anti-ferromagnetism* 472.
Appalachians 268.
Äquipotentialfläche, Krümmung, *equi-potential surface, curvature* 217.
Archeozoic 292.
Argon-40 297, 336.
Asthenosphäre, *asthenosphere* 246.
Astronomisches Jahrbuch 1.
Atmosphäre, Drehmoment, *atmosphere, angular momentum* 20.
Atom-Bomben-Explosion, *atom-bomb-explosion* 82.
AUGER-Elektronen, AUGER-*electrons* 307.
Ausdehnungskoeffizient, *thermal expansion, coefficient* 386.
Ausgleichstiefe, *compensation, depth of* 238.
Auslaugung, *leaching* 323.

BABCOCK, H. W. 523.
Band-Paß-Filter, *band-pass-filter* 155.
Basalt 475.
basisches Gestein, *basic rocks* 313.
Baugrund, *building sites* 178.
BENIOFF 63.
Bergketten, unterseeische, *mountain chains, submarine* 253.
Berg-Züge, *mountain ranges* 259.
Beryllium-7 301.
Biblische Berichte, *biblical accounts* 289.
Biege-Schwingungen, *flexural waves* 184.
Biegewellen, *flexural waves* 139.
BLACKETT, P. M. S. 523.
Blattverschiebung, *fault, transcurrent* 258, 264.
Blei-210, *lead-210* 325.
Blei, anomales, *lead, anomalous* 329, 332.
—, gewöhnliches, *common* 319.
—, Isotopen-Analyse, *isotopic analysis* 320.
—, meteorisches, *meteoric* 327.
—, radiogenes, *radiogenic* 319.
—, ursprüngliches, *primeval* 326.
Blei-Datierungs-Methoden, *lead dating methods* 328.
Bleigehalt, ursprünglicher, *lead abundance, primeval* 330.
Bleiglanz, *galena* 326, 350.
Blei-Verhältnis, *lead-ratio* 324.
B M Z 499.
Boden-Rollen, *ground roll* 138.
Bogen, umgekehrter, *arc, reversed* 259.
Bohrloch-Schießen, *well-shooting* 154, 162.
Bohrloch-Untersuchungs-Methoden, *bore hole logging methods* 312.
BOLTWOOD 292.
BOUGUER-Anomalie, BOUGUER-*anomaly* 223, 254.
BOUGUERsche Reduktion, BOUGUER's *reduction* 579.
Bowie-Effekt, BOWIE *effect* 577.
Brandung, *surf* 151.
Brechungs-Gesetz, SNELL's *law* 90, 164.
Breite, Änderung mit der Höhe, *latitude, change with height* 560.
—, geographische, *geographic* 93.
—, geozentrische, *geocentric* 93, 549.
—, mittlere, *mean* 562, 634.
—, reduzierte, *reduced* 544.
Breitendienst, internationaler, *latitude service, international* 17, 606.
Breitenschwankung, *latitude changes* 606.
BRIDGMAN 172.

# Subject Index.

## (English-German.)

Where English and German spelling of a word is identical the German version is omitted.

# Table des matières

pour les contributions écrites en français:

J. COULOMB: Séismometrie,

J. COULOMB: Agitation microséismique,

L. CAGNIARD: Electricité tellurique.